SPRINGER SERIES ON ENVIRONMENTAL MANAGEMENT

DAVID E. ALEXANDER

Series Editor

Springer

New York
Berlin
Heidelberg
Barcelona
Budapest
Hong Kong
London
Milan
Paris
Santa Clara
Singapore
Tokyo

Springer Series on Environmental Management
David E. Alexander, Series Editor

Gradient Modeling: Resources and Fire Management (1979) S.R. Kessell

Disaster Planning: The Preservation of Life and Property (1980) H.D. Foster

Air Pollution and Forests: Interactions between Air Contaminants and Forest Ecosystems (1981) W.H. Smith

Natural Hazard Risk Assessment and Public Policy: Anticipating the Unexpected (1982) W.J. Petak and A.A. Atkisson

Environmental Effects of Off-Road Vehicles: Impacts and Management in Arid Regions (1983) R.H. Webb and H.G. Wilshire (eds.)

Global Fisheries: Perspectives for the '80s (1983) B.J. Rosthschild (ed.)

Heavy Metals in Natural Waters: Applied Monitoring and Impact Assessment (1984) J.W. Moore and S. Ramamoorthy

Landscape Ecology: Theory and Applications (1984) Z. Naveh and A.S. Lieberman

Organic Chemicals in Natural Waters: Applied Monitoring and Impact Assessment (1984) J.W. Moore and S. Ramamoorthy

The Hudson River Ecosystem (1986) K.E. Limburg, M.A. Moran, and W.H. McDowell

Human System Responses to Disaster: An Inventory of Sociological Findings (1986) T.E. Drabek

The Changing Environment (1986) J.W. Moore

Balancing the Needs of Water Use (1988) J.W. Moore

The Professional Practice of Environmental Management (1989) R.S. Dorney and L. Dorney (eds.)

Chemicals in the Aquatic Environment: Advanced Hazard Assessment (1989) L. Landner (ed.)

Inorganic Contaminants of Surface Water: Research and Monitoring Priorities (1991) J.W. Moore

Chernobyl: A Policy Response Study (1991) B. Segerståhl (ed.)

Long-Term Consequences of Disasters: The Reconstruction of Friuli, Italy, in its International Context, 1976-1988 (1991) R. Geipel

Food Web Management: A Case Study of Lake Mendota (1992) J.F. Kitchell (ed.)

Restoration and Recovery of an Industrial Region: Progress in Restoring the Smelter-Damaged Landscape near Sudbury, Canada (1995) J.M. Gunn (ed.)

Limnological and Engineering Analysis of a Polluted Urban Lake: Prelude to Environmental Management of Onondaga Lake, New York (1996) S.W. Effler (ed.)

Steven W. Effler
Editor

Limnological and Engineering Analysis of a Polluted Urban Lake

Prelude to Environmental Management of Onondaga Lake, New York

With 557 Illustrations

 Springer

Steven W. Effler
Upstate Freshwater Institute, Inc.
Syracuse, NY 13214, USA

Series Editor:
David E. Alexander
University of Massachusetts
Department of Geology and
  Geography
Amherst, MA 01003, USA

Cover Art: Satellite photograph of Onondaga Lake, NY, and its urban setting (April 14, 1986).

Contribution No. 150 of the Upstate Freshwater Institute, Box 506, Syracuse, NY 13214, USA

Library of Congress Cataloging-in-Publication Data
Limnological and engineering analysis of a polluted urban lake: prelude
  to environmental management of Onondaga Lake, New York / Steven W.
  Effler, editor.
      p.   cm.—(Springer series on environmental management)
    Includes bibliographical references and index.
    ISBN 0-387-94383-8
    1. Water quality—New York (State)—Onondaga Lake.   2. Sewage
disposal in rivers, lakes, etc.—New York (State)—Onondaga Lake.
3. Soda industry—Waste disposal—Environmental aspects—New York
(State)—Onondaga Lake.   4. Limnology—New York (State)—Onondaga
Lake.   5. Solvay Process Company.   I. Effler, Steven W.
II. Series.
TD224.N7L55   1995
628.1′6837—dc20

                                                                95-5933
                                                                  CIP
                                                                  REV

Printed on acid-free paper.

Acquiring Editor: Robert C. Garber.
Production coordinated by Chernow Editorial Services, Inc. and managed by Bill Imbornoni; manufacturing supervised by Joe Quatela.
Typeset by Best-set Typesetter Ltd., Hong Kong.
Printed and bound by Braun-Brumfield, Inc. Ann Arbor, MI.
Printed in the United States of America.

9 8 7 6 5 4 3 2 1

ISBN 0-387-94383-8 Springer-Verlag New York Berlin Heidelberg SPIN 10480456

Kathy, Carol, Herm, and Notre Dame

Series Preface

This series is concerned with humanity's stewardship of the environment, our use of natural resources, and the ways in which we can mitigate environmental hazards and reduce risks. Thus it is concerned with applied ecology in the widest sense of the term, in theory and in practice, and above all in the marriage of sound principles with pragmatic innovation. It focuses on the definition and monitoring of environmental problems and the search for solutions to them at scales that vary from the global to the local according to the scope of analysis. No particular academic discipline dominates the series, for environmental problems are interdisciplinary almost by definition. Hence a wide variety of specialties are represented, from oceanography to economics, sociology to silviculture, toxicology to policy studies.

In the modern world, increasing rates of resource use, population growth and armed conflict have tended to magnify and complicate environmental problems that were already difficult to solve a century ago. Moreover, attempts to modify nature for the benefit of humankind have often had unintended consequences, especially in the disruption of natural equilibria. Yet at the same time human ingenuity has been brought to bear in developing a new range of sophisticated and powerful techniques for solving environmental problems – for example, pollution monitoring, restoration ecology, landscape planning, risk management, and impact assessment. Books in this series will shed light on the problems of the modern environment and contribute to the further development of the solutions. They will contribute to the immense effort by scholars and professionals of all persuasions and in all countries to nurture an environment that is both stable and productive.

David E. Alexander
Amherst, Massachusetts

Preface

This book is a case study of the scientific and engineering analysis of a single ecosystem, Onondaga Lake, located in metropolitan Syracuse, New York. The single-system focus of this book allows greater attention to details and more effective integration of findings from various disciplines into a holistic comparative studies perspective, as compared to texts that draw on observations from a number of examples. The contributors to this volume believe that the reader will find this case study rich in lessons for a range of disciplines involved in water quality issues.

Onondaga Lake is a particularly appropriate subject for a detailed case study. The lake is replete with environmental problems as a result of the input of large quantities of municipal and industrial waste for more than a century. Many readers will be surprised, if not shocked, that there is a lake in the United States that remains as profoundly degraded as Onondaga Lake. Despite mandated reductions in pollutant loading and reductions brought about by the closure of a major industrial polluter, Onondaga Lake is arguably the most polluted lake in the United States. For example, effluent from the adjoining municipal wastewater treatment plant (METRO) represents nearly 20% of the total annual inflow to the lake. The myriad of environmental problems that prevail and the strong signatures of degradation observed in Onondaga Lake offer a unique opportunity for research and education and many challenges for scientists, engineers, and lake managers.

This analysis builds from basic disciplinary studies of Onondaga Lake in the early chapters (Chapters 2–8) to an interdisciplinary synthesis in the last two chapters (Chapters 9 and 10). The book includes historical (Chapter 1) and hydrogeologic (Chapter 2) descriptions of the area, an analysis of hydrologic and pollutant loadings to the lake (Chapter 3) and treatments of hydrodynamics (Chapter 4), chemistry (Chapter 5), biology (Chapter 6), optics (Chapter 7), and sediments (Chapter 8) of the lake. Chapter 9 presents the development, testing, and application of mechanistic water quality models for the system. The models are in fact a synthesis of our scientific and engineering understanding of the lake. Chapter 10 summarizes the foregoing material and places it in a management perspective.

Research continues on this fascinating lake. Two recent findings, not treated in the text, are noteworthy from a management perspective. First, during the late summer and early fall of 1995, Onondaga Lake suffered a severe lakewide oxygen depletion event in the upper waters. The New York State minimum oxygen concentration standard ($4mg \cdot L^{-1}$) was violated for more than a month. The oxidation of large quantities of ammonia (nitrification) received from METRO contributed greatly to this event. Second, recent research from the laboratory of Martin T. Auer at Michigan Technological University has established that such violations of the oxygen standard would be eliminated by the diversion of the METRO effluent around the lake.

I began research on Onondaga Lake in the mid-1970s as part of my doctoral studies. My research of the lake through the mid-1980s was largely unfunded. Since 1981, Onondaga Lake research has been conducted under the auspices of the Upstate Freshwater Institute, a not-for-profit corporation. Only with time did I understand the connection between the lack of funding and the extreme degree of the lake's polluted state. Partly in response to political and regulatory pressure, research funding became available between 1987 and 1994. This brought an extremely talented array of investigators into the Onondaga Lake research arena; they greatly accelerated and broadened the investigation of the lake and contributed greatly to this book. Unfortunately, there has been an unmistakable resistance to the technical findings and management implications that emerged from this research; advocacy consultants and lawyers now abound. Funding for independent research on the lake dried up in 1995.

The Onondaga Lake story, however, is far from over. Regulators and polluters are negotiating, a ruling from a federal judge is forthcoming, and management decisions may be made in the near future. The lake has many more lessons to give. The Institute, and the contributors to this book, will continue to study this unique system.

Several of the water quality models developed and tested for the lake in Chapter 9 are being used by environmental managers to evaluate a range of remediation alternatives related to the METRO discharge. A number of these alternatives are quite expensive. It is necessary for me to state, on behalf of all the authors of the book, that we are not as a group, nor individually, advocates for a specific management plan for the lake. We are, however, strong advocates for these decisions being made within the context of all the available technical information. The Clean Water Act represents another important constraint on these management deliberations.

This book would not have been possible without the input of many people. I thank the many students who gave their time and effort to study Onondaga Lake, often without pay. I thank the board of directors of the Upstate Freshwater Institute, who have continued to support work on the lake. Several of the contributing authors of this book are members of the board. I thank all of the authors, not only for the quality of their work, but for their patience during this long process. The authors will receive no royalties from the sale of this book, instead all royalties will be donated to the Water Foundation of Central New York, Inc. (a not-for-profit organization). Four research scientists have made very special contributions to the work of the Upstate Freshwater Institute and this book: Mary Gail Perkins, Carol Brooks, Susan Doerr, and Bruce Wagner. We thank Elizabeth Miller, who typed the original manuscript and the endless revisions.

Three people have been particularly instrumental in guiding me professionally, for which I am grateful; Myrton C. Rand, J. Charles Jennett, and Robert D. Hennigan. I am especially grateful to two collaborating researchers on the Onondaga Lake work, Stephen D. Field and Martin T. Auer. Steve helped start the research program and usually found ways to keep it going in the 1970s. He made lasting contributions to the spirit of this on-going work. Marty, more than any other individual, is responsible for raising the level of research at the Institute. I look forward to our continuing collaboration for the next twenty years of the Onondaga Lake story.

Steven W. Effler

Contents

Series Preface vii
Preface ix
Contributors xix

1. **Background** 1
   STEVEN W. EFFLER AND GENA HARNETT

   1.1 Location and Morphometry 1
   1.2 Tributaries and Subbasins 3
   1.3 Climate 5
   1.4 Historic Account 5
   1.5 Specific Waste Sources 10
   1.6 Technical Studies of Onondaga Lake and Its
       Tributaries 18
   1.7 Demography and Land Use 21
   1.8 Government Involvement and a Community's
       Vision 22
   1.9 Summary 26
       References 29

2. **Hydrogeologic Setting** 32
   MARY GAIL PERKINS AND EDWIN A. ROMANOWICZ

   2.1 Background Geology 32
   2.2 Geology and Hydrogeology of the Major
       Tributaries to Onondaga Lake 41
   2.3 Special Topics Related to the Hydrogeology of
       Onondaga Lake 58
   2.4 Onondaga Lake 79
   2.5 Basic Concepts in Hydrogeology 83
   2.6 Summary 89
       References 93

3. Tributaries and Discharges 97
STEVEN W. EFFLER AND KEITH A. WHITEHEAD

   3.1   Hydrology of Onondaga Lake 97
   3.2   Material Loading 110
   3.3   Summary 189
         References 196

4. Hydrodynamics and Transport 200
EMMET M. OWENS AND STEVEN W. EFFLER

   4.1   Introduction 200
   4.2   Lake Inflows and Outflow 201
   4.3   Lake Temperature, Salinity, and Density
         Stratification 213
   4.4   Lake Circulation 227
   4.5   Modeling Stratification and Vertical Transport .. 238
   4.6   Horizontal Mass Transport Model 252
   4.7   Summary 257
         References 261

5. Chemistry 263

   5.1   Salinity.................................. 263
         STEVEN W. EFFLER
   5.2   Dissolved Oxygen 272
         STEVEN W. EFFLER
   5.3   Inorganic Carbon, Ca^{2+}, $CaCO_{3(s)}$, and pH 283
         CHARLES T. DRISCOLL, STEVEN W. EFFLER,
         AND SUSAN M. DOERR
   5.4   Nitrogen Species 294
         CAROL M. BROOKS AND STEVEN W. EFFLER
   5.5   Phosphorus 307
         STEVEN W. EFFLER, CHARLES T. DRISCOLL,
         SUSAN M. DOERR, MARTIN T. AUER, AND
         BRUCE A. WAGNER
   5.6   Anoxic Organic Carbon Decomposition and the
         Distribution of Related Chemical Species 324
         CHARLES T. DRISCOLL, STEVEN W. EFFLER,
         SUSAN M. DOERR, JEFFREY ADDESS, AND
         CAROL M. BROOKS
   5.7   Mercury.................................. 352
         CHARLES T. DRISCOLL AND WEI WANG
   5.8   Particle Chemistry 359
         DAVID L. JOHNSON, J. JIAO, SAUL G. DOS SANTOS,
         AND STEVEN W. EFFLER
   5.9   Summary 368
         References 374

6. Biology 384

6.1 Phytoplankton 384
MARTIN T. AUER, STEVEN W. EFFLER,
MICHELLE L. STOREY, SUSAN D. CONNORS,
AND PHILIP SZE

6.2 Zooplankton 421
CLIFFORD A. SIEGFRIED, NANCY A. AUER,
AND STEVEN W. EFFLER

6.3 Aquatic Macrophytes 436
JOHN D. MADSEN, R. MICHAEL SMART,
LAWRENCE W. EICHLER, CHARLES W. BOYLEN,
JEFFREY W. SUTHERLAND, AND JAY A. BLOOMFIELD

6.4 Benthic Macroinvertebrates 446
BRUCE A. WAGNER, ROBERT DANEHY,
NEIL A. RINGLER, AND STEVEN W. EFFLER

6.5 Fish Communities and Habitats in Onondaga
Lake, Adjoining Portions of the Seneca River,
and Lake Tributaries........................ 453
NEIL A. RINGLER, CHRISTOPHER GANDINO,
PRADEEP HIRETHOTA, ROBERT DANEHY, PETER TANGO,
CHARLES MORGAN, CHRISTOPHER MILLARD,
MARGARET MURPHY, MARK A. ARRIGO,
RONALD J. SLOAN, AND STEVEN W. EFFLER

6.6 Indicator Bacteria 494
MARTIN T. AUER, STEVEN W. EFFLER,
STEPHEN L. NIEHAUS, AND KEITH A. WHITEHEAD

6.7 Summary................................. 515
References 522

7. Optics 535
STEVEN W. EFFLER AND MARY GAIL PERKINS

7.1 Introduction 535
7.2 Optical Measurements 536
7.3 Optical Properties of Water 539
7.4 Historic Changes in Apparent Optical Properties 542
7.5 Components of Attenuation/Evaluation of
Empirical Relationships 549
7.6 Regional Comparison 556
7.7 Angular Distribution of Underwater Irradiance .. 562
7.8 Checks on Optical Measurements and Estimates 564
7.9 Estimates of a and b 565
7.10 Spectroradiometer Measurements............ 569
7.11 Partitioning a............................. 574
7.12 Partitioning b............................. 579
7.13 Components of Attenuation, Models, and
Analysis of Scenarios 584
7.14 Summary................................. 592
References 597

8. Sediments 600

8.1 Deposition 600
STEVEN W. EFFLER

8.2 Surficial Sediments 611
MARTIN T. AUER, NED JOHNSON, MICHAEL PENN,
AND STEVEN W. EFFLER

8.3 Sediment Stratigraphy 622
H. CHANDLER ROWELL AND STEVEN W. EFFLER

8.4 Summary 655
References 659

**9. Mechanistic Modeling of Water Quality in
Onondaga Lake** 667

9.1 Background and Evolution of Model
Frameworks 667
STEVEN W. EFFLER

9.2 Chloride Model 672
SUSAN M. DOERR, STEVEN W. EFFLER,
AND MARTIN T. AUER

9.3 Total Phosphorus Model 679
SUSAN M. DOERR, RAYMOND P. CANALE,
MARTIN T. AUER, AND STEVEN W. EFFLER

9.4 Nitrogen Model 690
RAYMOND P. CANALE, RAKESH K. GELDA,
AND STEVEN W. EFFLER

9.5 Dissolved Oxygen Model 702
RAKESH K. GELDA, MARTIN T. AUER,
RAYMOND P. CANALE, AND STEVEN W. EFFLER

9.6 Fecal Coliform Bacteria Model 714
RAYMOND P. CANALE, EMMET M. OWENS,
MARTIN T. AUER, THOMAS M. HEIDTKE,
AND STEVEN W. EFFLER

9.7 Water Quality Model for the Seneca and
Oswego Rivers 723
RAYMOND P. CANALE, EMMET M. OWENS,
MARTIN T. AUER, AND STEVEN W. EFFLER

9.8 Application of Models 743
STEVEN W. EFFLER AND SUSAN M. DOERR

9.9 Summary 776
References 782

10. Synthesis and Perspectives 789
STEVEN W. EFFLER

10.1 Impact of the Soda Ash/Chlor-alkali Facility
on Onondaga Lake and Adjoining Systems:
Update 789

10.2 The Polluted State of Onondaga Lake:
How Bad Is It? 798

10.3   But Has Not the Quality of Onondaga Lake
       Improved?   800
10.4   The METRO Discharge Is Too Much for a
       Small Lake   803
10.5   Diversion of METRO   805
10.6   Where Do We Go from Here?   808
       References   809

Index   813

Contributors

JEFFREY ADDESS
General Chemical Corporation, Syracuse, NY 13202, USA

MARK A. ARRIGO
College of Environmental Science and Forestry, State University of
New York, Syracuse, NY 13210, USA

MARTIN T. AUER
Michigan Technological University, Houghton, MI 49931, USA

NANCY A. AUER
Michigan Technological University, Houghton, MI 49931, USA

JAY A. BLOOMFIELD
New York State Department of Environmental Conservation,
Albany, NY 12233-0001, USA

CHARLES W. BOYLEN
Rensselear Polytechnic Institute, Albany, NY 12180, USA

CAROL M. BROOKS
Upstate Freshwater Institute, Inc., Syracuse, NY 13214, USA

RAYMOND P. CANALE
University of Michigan, Ann Arbor, MI 48109, USA

SUSAN D. CONNORS
CH2M Hill, Reston, VA 22090-1483, USA

ROBERT DANEHY
College of Environmental Science and Forestry, State University of
New York, Syracuse, NY 13210, USA

SUSAN M. DOERR
Upstate Freshwater Institute, Inc., Syracuse, NY 13214, USA

SAUL G. DOS SANTOS
College of Environmental Science and Forestry, State University of
New York, Syracuse, NY 13210, USA

CHARLES T. DRISCOLL
Syracuse University, Syracuse, NY 13210, USA

STEVEN W. EFFLER
Upstate Freshwater Institute, Inc., Syracuse, NY 13214, USA

LAWRENCE W. EICHLER
Rensselear Polytechnic Institute, Albany, NY 12180, USA

CHRISTOPHER GANDINO
College of Environmental Science and Forestry, State University of
New York, Syracuse, NY 13210, USA

RAKESH K. GELDA
Michigan Technological University, Houghton, MI 49931, USA

GENA HARNETT
The Write Design Co., Albany, NY 12205, USA

THOMAS M. HEIDTKE
Department of Civil Engineering, Wayne State University, Detroit,
MI 48202, USA

PRADEEP HIRETHOTA
College of Environmental Science and Forestry, State University of
New York, Syracuse, NY 13210, USA

J. JIAO
College of Environmental Science and Forestry, State University of
New York, Syracuse, NY 13210, USA

DAVID L. JOHNSON
College of Environmental Science and Forestry, State University of
New York, Syracuse, NY 13210, USA

NED JOHNSON
CH2M Hill, Reston, VA 22090, USA

JOHN D. MADSEN
U.S. Army Corps of Engineers, Louisville, TX 75056, USA

CHRISTOPHER MILLARD
College of Environmental Science and Forestry, State University of
New York, Syracuse, NY 13210, USA

CHARLES MORGAN
College of Environmental Science and Forestry, State University of New York, Syracuse, NY 13210, USA

MARGARET MURPHY
College of Environmental Science and Forestry, State University of New York, Syracuse, NY 13210, USA

STEPHEN L. NIEHAUS
Gosling Czubak Associates, P.C., Traverse City, MI 49684, USA

EMMET M. OWENS
Syracuse University, Syracuse, NY 13210, USA

MICHAEL PENN
Michigan Technological University, Houghton, MI 49931, USA

MARY GAIL PERKINS
Upstate Freshwater Institute, Inc., Syracuse, NY 13214, USA

NEIL A. RINGLER
College of Environmental Science and Forestry, State University of New York, Syracuse, NY 13210, USA

EDWIN A. ROMANOWICZ
Upstate Freshwater Institute, Inc., Syracuse, NY 13214, USA

H. CHANDLER ROWELL
Onondaga Lake Management Conference, Syracuse, NY 13261, USA

CLIFFORD A. SIEGFRIED
Biological Survey, New York State Museum, Albany, NY 12230, USA

R. MICHAEL SMART
U.S. Army Corps of Engineers, Louisville, TX 75056, USA

RONALD J. SLOAN
New York State Department of Environmental Conservation, Albany, NY 12233, USA

MICHELLE L. STOREY
Kieser and Associates, Kalamazoo, MI 49007, USA

JEFFREY W. SUTHERLAND
New York State Department of Environmental Conservation, Albany, NY 12233, USA

PHILIP SZE
Georgetown University, Washington, DC 20057, USA

PETER TANGO
College of Environmental Science and Forestry, State University of
New York, Syracuse, NY 13210, USA

BRUCE A. WAGNER
Upstate Freshwater Institute, Inc., Syracuse, NY 13214, USA

WEI WANG
Syracuse University, Syracuse, NY 13210, USA

KEITH A. WHITEHEAD
Upstate Freshwater Institute, Inc., Syracuse, NY 13214, USA

1
Background

Steven W. Effler and Gena Harnett

1.1 Location and Morphometry

Onondaga Lake is located (lat. 43°06′54″; long. 76°14′34″) immediately north of the City of Syracuse, in Onondaga County, in the middle of the most urbanized area of Central New York (Figure 1.1). Onondaga Lake is in the Oswego River drainage basin (see Figure 1.1, inset). The outflow from the lake exits through a single outlet at its northern end and enters the Seneca River. The Seneca River combines with the Oneida River to form the Oswego River, which flows north, entering Lake Ontario at the City of Oswego (Figure 1.1).

A bathymetric map of Onondaga Lake, published in 1971 (Onondaga County 1971), was based on a survey conducted in 1968. Morphometric information for the lake was updated with a survey conducted in 1987. The bathymetric map (Owens 1987) of Onondaga Lake, based on this recent survey, is presented in Figure 1.2. The details of the development of this map from the survey data were described by Owens (1987). Eight hundred and sixty-three measurements were made with electronic depth finders (recorded to the nearest foot) to support the development of this map. Locations (latitude/longitude) were established with Loran C. The shoreline was defined by the most recent U.S. Geological Survey (USGS) quadrangle map. The average lake elevation during the survey was 362.75 ft.

The lake is oriented along a northwest–southeast axis (Figure 1.2). It has a length along this axis of 7.6 km and a maximum width of 2 km. It lacks dendritic irregularities. The lake is commonly described as having two basins, the south and north, that are separated by a modest "saddle" region that is located approximately 3.6 km from the outlet. Other morphometric characteristics of the lake are presented in Table 1.1, along with values for several other lakes in the region. The morphometry of Onondaga Lake is similar to that of Otisco Lake (Table 1.1). Onondaga Lake is substantially smaller and shallower than Owasco and Skaneateles Lakes, but deeper and larger than Cross and Cazenovia Lakes.

Hypsographic information (depth distribution of contour area and volume) for Onondaga Lake, based on the bathymetric map in Figure 1.2, is presented in Figure 1.3. Such detailed morphometric information is essential to estimate the total content of various materials in the lake from measured vertical profiles.

The most recent bathymetric analysis yields a somewhat smaller volume for the lake than that associated with the bathymetric map presented in 1971 (Onondaga County 1971). The volume obtained from this older map, according to the same general procedure (Owens 1987), and corrected for the small difference in reference surface elevation (363.0 ft for the 1968 survey) (Onondaga County 1971), was $143 \times 10^6 \, m^3$. Approximately 40% of the apparent reduction in volume (Table 1.1) is associated with depths below 18 m (Owens 1987). The reduction in volume probably largely reflects the effects of procedural differences in the development of the bathymetric maps, although deposition also contributed. The maximum depth reported in 1971 (Onondaga County 1971), corrected for the slightly dif-

ferent lake surface elevation, was 20.4 m. Even greater maximum depths have been reported for the lake elsewhere. A maximum depth of 73 ft (22.3 m) appeared on a map of the lake in the New York State Barge Canal Atlas published in 1977, a maximum of 71 ft (21.6 m) appears on a New York State Department of Transportation map published in 1968. This disparity in maximum depths remains unexplained.

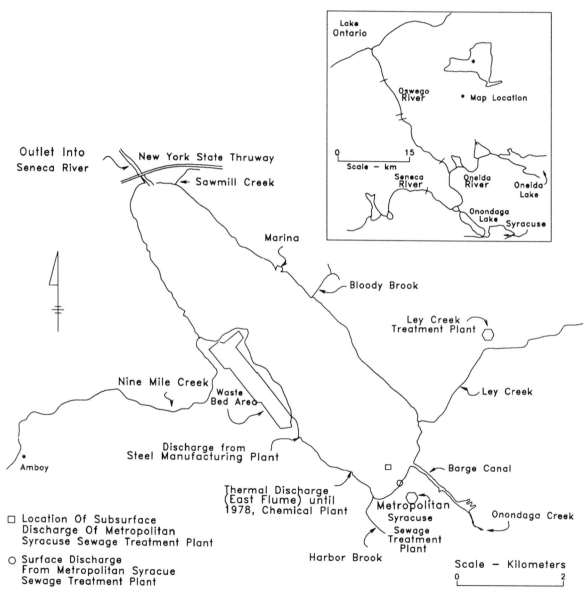

FIGURE 1.1. Location map for Onondaga Lake, with important features of its setting.

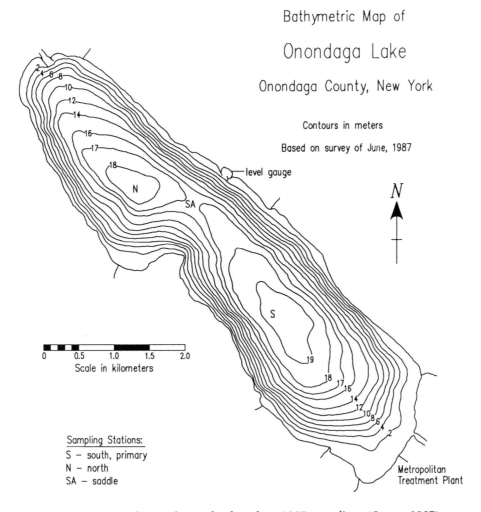

Bathymetric Map of

Onondaga Lake

Onondaga County, New York

Contours in meters

Based on survey of June, 1987

FIGURE 1.2. Bathymetric map of Onondaga Lake, based on 1987 soundings (Owens 1987).

1.2 Tributaries and Subbasins

The major hydrologic inputs to Onondaga Lake are Ninemile Creek, Onondaga Creek, the Metropolitan Syracuse Sewage Treatment Plant (METRO), Ley Creek, Bloody Brook, and Harbor Brook. Other minor tributaries include Sawmill Creek, Tributary 5A, and the East Flume. The locations of the mouths of the tributaries and the outlet are shown in Figure 1.1. A more complete map of the lake's trib-

utaries, including the major watershed subbasins, appears in Figure 1.4. The Onondaga Lake watershed contains approximately 642 km² and lies entirely within Onondaga County, with the exception of a small portion (approximately 2 km²) that lies in north central Cortland County.

Ninemile Creek empties into Onondaga Lake on its western shore; its watershed drains approximately 298 km² both south and west of Onondaga Lake. The Ninemile Creek watershed originates at the outlet of Otisco Lake. The total length of the Ninemile Creek main-stem is 55.2 km. From its source at the Otisco

TABLE 1.1. Selected morphometric characteristics of Onondaga Lake and other Central New York lakes.

Lake	Surface area ($\times 10^6 \, m^2$)	Volume ($\times 10^6 \, m^3$)	Mean depth (m)	Maximum depth (m)
Onondaga	12.0	131	10.9	19.5
Otisco	7.6	79	10.2	20.1
Owasco	26.7	781	29.3	54.0
Skaneateles	35.9	1563	43.5	90.5
Cross	9.0	51	5.5	17.0
Cazenovia	4.4	35	7.9	14.7

Lake outlet, the creek flows in a northerly direction through the village of Marcellus, then in a more easterly direction through the village of Camillus, and eventually through the Lakeland area to Onondaga Lake. The change in configuration of the lower reaches of Ninemile Creek has been associated with the disposal of waste by a chemical manufacturer. Waste beds associated with soda ash production adjoin the lower 3 km of the stream.

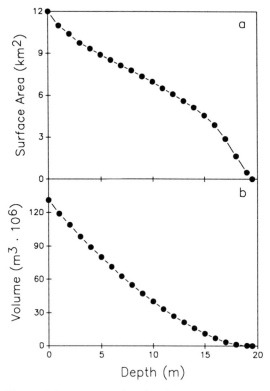

FIGURE 1.3. Hypsographic data for Onondaga Lake, from bathymetric map and analysis of Owens (1987; see Figure 1.2) (a) depth distribution of surface area, and (b) depth distribution of volume.

Onondaga Creek drains an area of approximately 298 km^2 and has a mainstem length of approximately 44.2 km. The Tully mud boils, a major source of sediment to the creek and lake (Effler et al. 1992), are located about 33 km above the creek mouth. Onondaga Creek empties into the southern end of Onondaga Lake. The lower reaches of the creek drain a significant portion of the City of Syracuse. The watershed extends southward until just north of the Tully Lakes area and westward to the drainage divide with the Ninemile Creek watershed. The western branch of Onondaga Creek runs in a generally easterly direction; it joins the mainstem in the Onondaga Indian Reservation. The configuration of lower Onondaga Creek, particularly near its mouth, has been changed substantially, in part to minimize flooding.

Ley Creek's watershed is approximately 77.4 km^2; it extends eastward from the southern end of Onondaga Lake. This lake plain region is residential and industrial in character, with the exception of the headwaters that are located primarily in wetlands. This watershed has become increasingly developed in recent decades.

Harbor Brook extends to the southwest from its mouth on the southernmost end of Onondaga Lake. It has a long and narrow watershed of approximately 29.3 km^2. The lower reach drains a portion of the City of Syracuse while the headwaters drain a mixture of residential, agricultural, and pasture lands. This area has become more residential during the last decade.

Bloody Brook enters Onondaga Lake at roughly the midpoint of its eastern shore in Onondaga Lake Park. The watershed of ap-

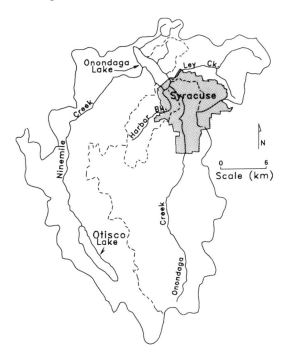

FIGURE 1.4. Watersheds in Onondaga County, including drainage basin and subbasins for Onondaga Lake.

North American continent (Ruffner and Blair 1987). The area's climate is strongly influenced by geographic proximity to Lake Ontario, as well as physiographic differences between the Ontario Lowlands of the northern portion of the drainage basin and the Appalachian Upland to the south.

Lake Ontario influences the local weather by moderating air temperatures and reducing both hot and cold temperature extremes. The average daily high temperature for Syracuse in July is 27.7°C, with an average January low temperature of −9.4°C. (Ruffner and Blair 1987). The proximity of Lake Ontario also results in considerable cloudiness, especially during the winter months. Cold air from the west and northwest often moves across the large, unfrozen surface of Lake Ontario acquiring moisture. This creates "lake-effect" snow squalls and storms during the mid-October to mid-March period (Winkley 1989). This causes rather localized differences in annual snowfall, with the highest average annual totals of 254 to 305 cm occurring in the northern and eastern parts of Onondaga County (Hutton and Rice 1977).

The upland areas, primarily in the southern portion of the drainage basin, tend to be both cooler and wetter than the Ontario Lowlands. Precipitation in the Onondaga Lake drainage area is generally evenly distributed, with the summer months being the driest on average. However, substantial year-to-year variations in rainfall are common to this region, as illustrated by the annual precipitation data presented for the 1961 to 1990 interval in Figure 1.5. Detailed meteorologic data are utilized in subsequent chapters to specify forcing conditions for various lake processes.

proximately 29.3 km² extends to the northeast draining the lake plain area, which is heavily residential.

The setting of the lake within the Three Rivers system is illustrated as an inset in Figure 1.1. The lake's only outlet, located at its northwesterly extreme, discharges into the Seneca River. The point of confluence of the Seneca and Oneida Rivers' is known as the Three Rivers junction. The overall river system is often referred to as the Three Rivers System. The Oswego River is the largest fluvial discharge to Lake Ontario that originates within the Lake Ontario drainage basin.

1.3 Climate

The climate of the Onondaga Lake drainage basin can best be described as continental and somewhat humid. Rapidly changing weather conditions and most of the precipitation are the result of cyclonic storms passing through the Great Lakes basin from the interior of the

1.4 Historic Account

1.4.1 Cultural and Industrial Development

The cultural and industrial history of the Onondaga Lake watershed is summarized in the time line of Figure 1.6. The first appearance of humans in Onondaga County followed the retreat of the last Pleistocene glacier (ap-

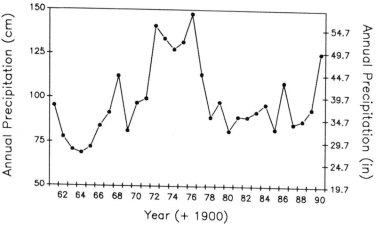

FIGURE 1.5. Annual rainfall at Hancock Airport (approximately 9 km from the City of Syracuse) for the period 1961 to 1990.

proximately 8000 BC). From around 1300 BC to 1000 BC until the arrival of white people, Indian cultures flourished and the development of agriculture and village life advanced. The Onondaga Nation of the Iroquois Confederacy, living south of what is now Syracuse, was the best known of these. Onondaga Lake was the Council Fire site for the Iroquois Nation.

The first recorded European visit to the area was in 1615 by Samuel Champlain, who led his Huron allies on a raid against their Iroquois enemies. In 1654, during one of the occasional truces with the Iroquois, French Jesuit missionaries under Father Simon LeMoyne established a settlement named Ste. Marie de Gannentaha near Onondaga Lake. Father LeMoyne and the missionaries are credited with the discovery of the salt springs near the mouth of Onondaga Creek.

The first European settler in the area is thought to have been Ephraim Webster, who set up a trading post in 1786 near the mouth of Onondaga Creek (Figure 1.1). The major impetus to development around Onondaga Lake was the salt industry. By 1794, James Geddes had set up the first large-scale salt manufacturing operation by Harbor Brook.

The creation of the canal transportation system resulted in the first major alteration of Onondaga Lake's morphology. In 1822, canal commissioners had a channel cut to permit the lake's surface elevation to drop to that of the Seneca River. As a result, an estimated 20% decrease in the surface area and a decrease in depth of 0.6 m occurred, draining wetlands at the southern end of the lake (Clark 1849; Sweet 1874). This action apparently also helped the prevailing malaria problem. By 1828 the Erie and Oswego Canals had linked the Hudson and Mohawk Rivers with Lake Erie and Lake Ontario. The Oswego Canal ran along the east shore of Onondaga Lake and became the prime supply and shipping route for the salt industry that adjoined Onondaga Lake. Eventually, much of the southern half of the Onondaga Lake shoreline was involved with salt manufacturing. The industry reached its peak in 1862.

The railroads reached Syracuse in 1838, which provided an additional impetus for the development of industry. The City of Syracuse was incorporated in 1848. The population, along with industry, grew significantly in Syracuse following the Civil War years. Local salt and limestone deposits provided the basis for the development of important industries. Most notable was the establishment of soda ash production on the western shore of the lake in 1884, which initially utilized salt (NaCl) wells adjoining the lake, and limestone quarries near Dewitt. As the nearby brine and lime-

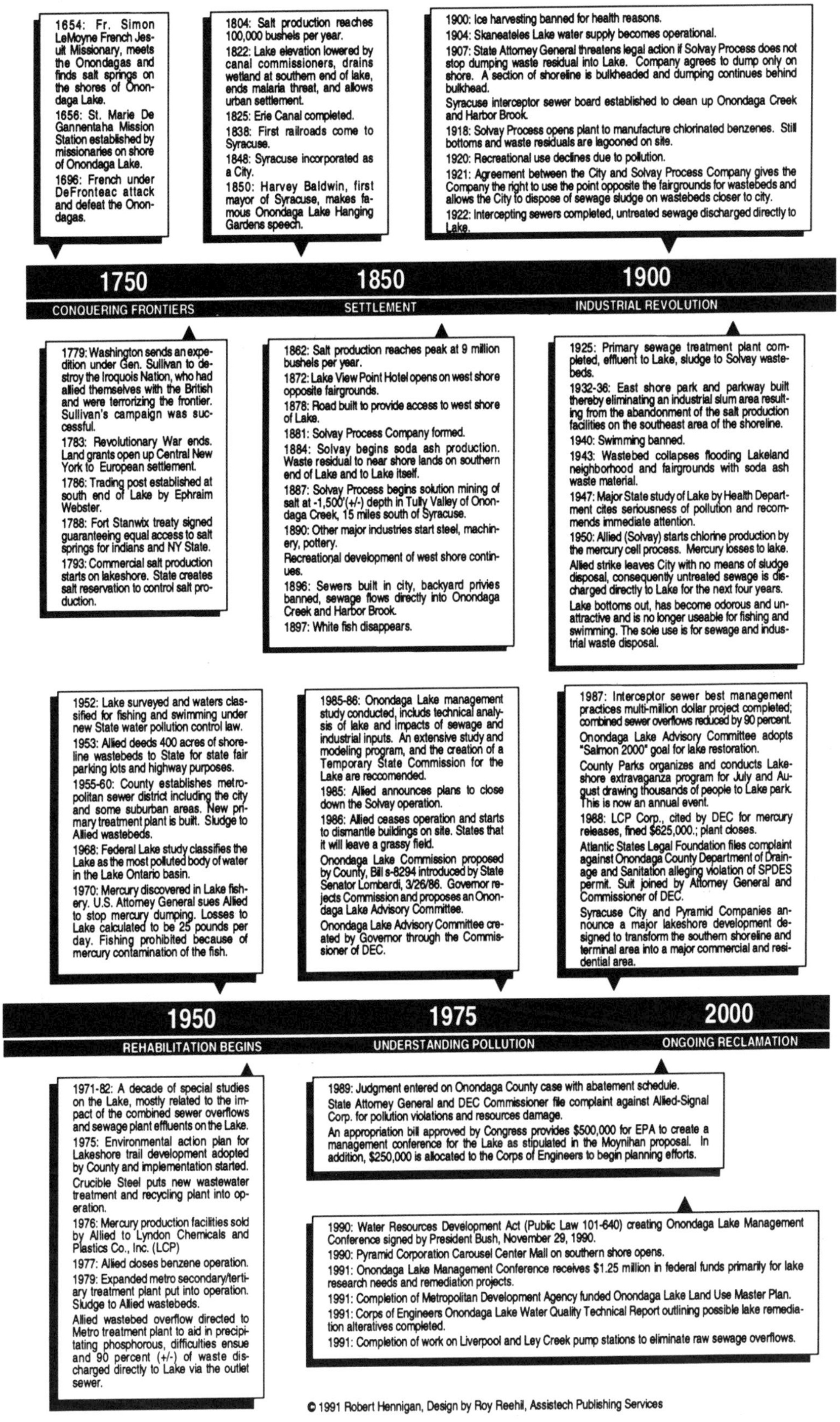

1654: Fr. Simon LeMoyne French Jesuit Missionary, meets the Onondagas and finds salt springs on the shores of Onondaga Lake.
1656: St. Marie De Gannentaha Mission Station established by missionaries on shore of Onondaga Lake.
1696: French under DeFronteac attack and defeat the Onondagas.

1804: Salt production reaches 100,000 bushels per year.
1822: Lake elevation lowered by canal commissioners, drains wetland at southern end of lake, ends malaria threat, and allows urban settlement.
1825: Erie Canal completed.
1838: First railroads come to Syracuse.
1848: Syracuse incorporated as a City.
1850: Harvey Baldwin, first mayor of Syracuse, makes famous Onondaga Lake Hanging Gardens speech.

1900: Ice harvesting banned for health reasons.
1904: Skaneateles Lake water supply becomes operational.
1907: State Attorney General threatens legal action if Solvay Process does not stop dumping waste residual into Lake. Company agrees to dump only on shore. A section of shoreline is bulkheaded and dumping continues behind bulkhead.
Syracuse interceptor sewer board established to clean up Onondaga Creek and Harbor Brook.
1918: Solvay Process opens plant to manufacture chlorinated benzenes. Still bottoms and waste residuals are lagooned on site.
1920: Recreational use declines due to pollution.
1921: Agreement between the City and Solvay Process Company gives the Company the right to use the point opposite the fairgrounds for wastebeds and allows the City to dispose of sewage sludge on wastebeds closer to city.
1922: Intercepting sewers completed, untreated sewage discharged directly to Lake.

1750
CONQUERING FRONTIERS

1850
SETTLEMENT

1900
INDUSTRIAL REVOLUTION

1779: Washington sends an expedition under Gen. Sullivan to destroy the Iroquois Nation, who had allied themselves with the British and were terrorizing the frontier. Sullivan's campaign was successful.
1783: Revolutionary War ends. Land grants open up Central New York to European settlement.
1786: Trading post established at south end of Lake by Ephraim Webster.
1788: Fort Stanwix treaty signed guaranteeing equal access to salt springs for Indians and NY State.
1793: Commercial salt production starts on lakeshore. State creates salt reservation to control salt production.

1862: Salt production reaches peak at 9 million bushels per year.
1872: Lake View Point Hotel opens on west shore opposite fairgrounds.
1878: Road built to provide access to west shore of Lake.
1881: Solvay Process Company formed.
1884: Solvay begins soda ash production. Waste residual to near shore lands on southern end of Lake and to Lake itself.
1887: Solvay Process begins solution mining of salt at -1,500'(+/-) depth in Tully Valley of Onondaga Creek, 15 miles south of Syracuse.
1890: Other major industries start steel, machinery, pottery.
Recreational development of west shore continues.
1896: Sewers built in city, backyard privies banned, sewage flows directly into Onondaga Creek and Harbor Brook.
1897: White fish disappears.

1925: Primary sewage treatment plant completed, effluent to Lake, sludge to Solvay wastebeds.
1932-36: East shore park and parkway built thereby eliminating an industrial slum area resulting from the abandonment of the salt production facilities on the southeast area of the shoreline.
1940: Swimming banned.
1943: Wastebed collapses flooding Lakeland neighborhood and fairgrounds with soda ash waste material.
1947: Major State study of Lake by Health Department cites seriousness of pollution and recommends immediate attention.
1950: Allied (Solvay) starts chlorine production by the mercury cell process. Mercury losses to lake.
Allied strike leaves City with no means of sludge disposal, consequently untreated sewage is discharged directly to Lake for the next four years.
Lake bottoms out, has become odorous and unattractive and is no longer useable for fishing and swimming. The sole use is for sewage and industrial waste disposal.

1952: Lake surveyed and waters classified for fishing and swimming under new State water pollution control law.
1953: Allied deeds 400 acres of shoreline wastebeds to State for state fair parking lots and highway purposes.
1955-60: County establishes metropolitan sewer district including the city and some suburban areas. New primary treatment plant is built. Sludge to Allied wastebeds.
1968: Federal Lake study classifies the Lake as the most polluted body of water in the Lake Ontario basin.
1970: Mercury discovered in Lake fishery. U.S. Attorney General sues Allied to stop mercury dumping. Losses to Lake calculated to be 25 pounds per day. Fishing prohibited because of mercury contamination of the fish.

1985-86: Onondaga Lake management study conducted, includs technical analysis of lake and impacts of sewage and industrial inputs. An extensive study and modeling program, and the creation of a Temporary State Commission for the Lake are recommended.
1985: Allied announces plans to close down the Solvay operation.
1986: Allied ceases operation and starts to dismantle buildings on site. States that it will leave a grassy field.
Onondaga Lake Commission proposed by County, Bill s-8294 introduced by State Senator Lombardi, 3/26/86. Governor rejects Commission and proposes an Onondaga Lake Advisory Committee.
Onondaga Lake Advisory Committee created by Governor through the Commissioner of DEC.

1987: Interceptor sewer best management practices multi-million dollar project completed; combined sewer overflows reduced by 90 percent.
Onondaga Lake Advisory Committee adopts "Salmon 2000" goal for lake restoration.
County Parks organizes and conducts Lakeshore extravaganza program for July and August drawing thousands of people to Lake park. This is now an annual event.
1988: LCP Corp., cited by DEC for mercury releases, fined $625,000.; plant closes.
Atlantic States Legal Foundation files complaint against Onondaga County Department of Drainage and Sanitation alleging violation of SPDES permit. Suit joined by Attorney General and Commissioner of DEC.
Syracuse City and Pyramid Companies announce a major lakeshore development designed to transform the southern shoreline and terminal area into a major commercial and residential area.

1950
REHABILITATION BEGINS

1975
UNDERSTANDING POLLUTION

2000
ONGOING RECLAMATION

1971-82: A decade of special studies on the Lake, mostly related to the impact of the combined sewer overflows and sewage plant effluents on the Lake.
1975: Environmental action plan for Lakeshore trail development adopted by County and implementation started.
Crucible Steel puts new wastewater treatment and recycling plant into operation.
1976: Mercury production facilities sold by Allied to Lyndon Chemicals and Plastics Co., Inc. (LCP)
1977: Allied closes benzene operation.
1979: Expanded metro secondary/tertiary treatment plant put into operation. Sludge to Allied wastebeds.
Allied wastebed overflow directed to Metro treatment plant to aid in precipitating phosphorous, difficulties ensue and 90 percent (+/-) of waste discharged directly to Lake via the outlet sewer.

1989: Judgment entered on Onondaga County case with abatement schedule.
State Attorney General and DEC Commissioner file complaint against Allied-Signal Corp. for pollution violations and resources damage.
An appropriation bill approved by Congress provides $500,000 for EPA to create a management conference for the Lake as stipulated in the Moynihan proposal. In addition, $250,000 is allocated to the Corps of Engineers to begin planning efforts.

1990: Water Resources Development Act (Public Law 101-640) creating Onondaga Lake Management Conference signed by President Bush, November 29, 1990.
1990: Pyramid Corporation Carousel Center Mall on southern shore opens.
1991: Onondaga Lake Management Conference receives $1.25 million in federal funds primarily for lake research needs and remediation projects.
1991: Completion of Metropolitan Development Agency funded Onondaga Lake Land Use Master Plan.
1991: Corps of Engineers Onondaga Lake Water Quality Technical Report outlining possible lake remediation alteratives completed.
1991: Completion of work on Liverpool and Ley Creek pump stations to eliminate raw sewage overflows.

© 1991 Robert Hennigan, Design by Roy Reehil, Assistech Publishing Services

FIGURE 1.6. Historic time line for Onondaga Lake (from Hennigan 1991).

stone deposits were exhausted, salt production wells were developed in the Tully Valley in the 1880s and limestone quarrying activities were moved to Jamesville in 1912. A more detailed description of the soda ash process and other activities at this chemical production facility are provided later in this chapter.

The extent to which Onondaga Lake was enriched with salt before the opening of the Solvay process plant has been open to substantial debate. Recent generations of central New Yorkers have been lead to believe that as much as 50% of the highly elevated salinities that prevailed in the lake before closure of the soda ash facility were a result of natural inputs. In subsequent chapters scientific evidence is presented based on tributary loading, lake water column chemistry, and the sedimentary record, concerning this contention. Here we will briefly review earlier historic accounts concerning the salt content of the lake. This anecdotal information is far from scientifically satisfying. However, it serves to establish that, despite the well-documented salt springs and related salt mining activities that lined the southern shore, there was no historic unanimity concerning the salt content of the lake proper. It is likely that the lake was somewhat enriched in certain ions compared to most other local hard water lakes, such as the eastern Finger Lakes (see Chapters 2 and 3). It should be noted that there are large differences in threshold ionic concentrations that can impart tastes and the much higher concentrations (e.g., Effler and Driscoll 1985) that prevailed during soda ash manufacture. Findings of the historic review are presented in Table 1.2.

TABLE 1.2. Selected historic accounts of salt springs adjoining and salt content of Onondaga Lake.

No.	Comment	Interpretation	Source
1.	Initial description of salt springs along lake, by Father Jerome Lallemont, in *Jesuit Publications* between 1645 and 1646		Higgins 1955
2. a.	Numerous salt springs around southern end of lake, from Conrad Weiser, in April 1737		Chase 1924
b.	Springs on banks of lake from Liverpool south around to Ninemile Creek outlet		Clark 1849
3.	Most salt springs in marshy, swamp area in southeast corner		DeWitt 1801
4.	Some springs identified as bubbling up from near shore sediments (i.e., under water)	Shallow water	DeWitt 1801
5.	Shallow wells varied in salt concentration with time and position		Gordon 1836
6.	Brines began to fail in quantity and quality by 1820s		Higgins 1955
7.	Communication existed between salt production areas and the lake	Natural salinity of lake probably raised by runoff of pumped brines	DeWitt 1801
8.	Observations after 1822 indicate freshness* (low salinity) of lake waters (probably surface)	Lowering of lake in 1822 may have reduced inputs from springs and brining operations	Beck 1826 Clark 1849 Bruce 1896
9.	Contradictory reports on occurrence of chemical stratification		
a.	Pro		DeWitt 1801 Beck 1826 Gordon 1836
b.	Con		Schultz 1810 (as described by Murphy 1978)

* Statements concerning freshness and salt may refer to taste; calcium and chloride concentrations of 250 mg · L^{-1} can impart an unacceptable taste to water.

The growing industrialization provided more leisure time for the populace. In the 1870s and 1880s, a number of resorts were built along the undeveloped northwest shoreline. The "resort era," however, was short lived, as it reached its peak around the turn of the century. The demise of the resorts after World War I was due to the mobility afforded by the automobile in making more distant locations accessible, and to the increasing pollution of Onondaga Lake, which made the lake undesirable for recreational activity. During the decades from 1900 until World War II, a number of manufacturing facilities developed within the watershed. These included major steel, pottery, pharmaceutical, air conditioning, general appliance, and electrical manufacturing facilities. A second era of recreational (noncontact) use began in the depths of the Great Depression in the 1930s. The eastern shore was redeveloped along the abandoned Oswego Canal through a series of public works projects, which included the construction of the Salt Museum, the Ste. Marie de Gannentha mission, and Onondaga Lake Park.

1.4.2 Lake Degradation and Reductions in Pollutant Loading

Events highlighting the degradation of Onondaga Lake as well as steps taken to reduce pollutant loadings are shown in Figure 1.6. By the turn of the century, sewage and industrial pollution had already had a profound impact on Onondaga Lake. By the late 1890s, for example, the lake had lost its coldwater fishery. Particularly noteworthy was the disappearance of the whitefish, a commercially important cold water species (Lipe et al. 1983). Murphy (1978) referred to a memorandum from the soda ash manufacturer that reports hydrogen sulfide was present (i.e., oxygen absent) in "mill" water from the lake (presumably from deep cold lake layers) as early as August 22, 1910. Such chemical conditions would be consistent with the absence of a coldwater fishery. In 1900, ice harvesting was banned for health reasons, further attesting to the lake's increasing degradation. Despite the installation of interceptor sewers and the construction and operation of a primary treatment

plant to treat most of the sewage of Syracuse by the mid 1920s, the lake was increasingly recognized as degraded. Swimming was banned in 1940 for public health reasons. "Lake reclamation" efforts prior to the 1950s focused largely on improving shoreline use for recreational purposes and were not directed at the degraded water quality or ecological conditions.

The interceptor sewers have hydraulic relief structures known as combined sewer overflows (CSOs), through which dilute raw sewage is discharged during major runoff events. Sixty-six CSOs are presently located along Onondaga Creek, Ley Creek, and Harbor Brook; most of these are located along Onondaga Creek. It was estimated that wet weather overflows occurred on the average of nine times per month for a total duration of approximately 24 hr per month in the early 1970s (O'Brien & Gere Engineers, Inc. 1973), and that several billion gallons were discharged through the CSOs annually to the receiving streams and subsequently Onondaga Lake (O'Brien & Gere Engineers, Inc. 1987). This form of sewage pollution is a ubiquitous problem for the older cities of the northeastern United States.

Efforts to reduce pollutant loading to the lake became more pronounced in the late 1950s when Onondaga County established the Metropolitan Sewer District, which encompassed the City of Syracuse and some surrounding suburban areas. The district constructed a large primary sewage treatment plant that was completed in 1960 (METRO). Nevertheless, the nadir of Onondaga Lake's water quality status was reached in the 1950s and 1960s during the advent of federal legislation directed at water pollution control and rehabilitation. The era of concern for nonconventional pollutants began in the lake with the discovery of high levels of mercury in fish flesh and the establishment of a fishing ban in 1970 (Kilborne 1970).

Efforts to reduce pollutant loading increased during the 1970s and 1980s. Onondaga County implemented a ban on high phosphorus detergents in 1972. METRO was upgraded to secondary treatment in 1978, and tertiary treatment in 1981. These actions resulted in major reductions in BOD, suspended solids, and phosphorus loading from the facility. The implementation by Onondaga County of a "Best Management Practices" (BMP) program on the CSO (1983 to 1985) system reduced loading of dilute sewage by reducing overflows significantly (O'Brien & Gere Engineers, Inc. 1987). In 1982, Onondaga County implemented the industrial waste pretreatment program, which reduced heavy metal loadings, particularly lead, chromium, and nickel, to the lake from METRO. A decrease in the use of chemical-intensive industrial processes has also reduced these loadings. In 1991, Onondaga County completed improvements to the Ley Creek and Liverpool Pump Stations (to METRO) to eliminate raw sewage discharges during storm events.

A number of industry abatement programs have taken place as well. Under the first "National Pollution Discharge Elimination System" (NPDES) permit issued to Allied Chemical Corporation, mercury discharges to the lake decreased significantly. These inputs were reduced further in 1977 with the implementation of "Best Practical Technology" (BPT) by Allied. In 1974 a wastewater treatment plant was built for the adjoining steel manufacturing facility. This reduced iron and chromium loading to the lake. This wastewater treatment plant was upgraded in 1981 and again in 1986 to meet more stringent discharge limitations. In 1986, the soda ash/chlor-alkali manufacturer ceased operations. This was followed by the closure of the Linden Chemical and Plastics (LCP) operations in 1988 for illegal discharges of mercury in Onondaga Lake.

Despite added regulatory pressure and a substantial expenditure by the private and public sectors aimed at remediation ($280 million by 1983 [Lipe et al. 1983]), the lake has recently been described as "America's Dirtiest Lake" (Hennigan 1991).

1.5 Specific Waste Sources

Onondaga Lake has been the principal receptacle for domestic waste and much of the industrial wastes from the metropolitan area from the early development of the region to the present. The most clearly manifested and pervasive industrial impacts have been associated with a chemical manufacturing facility (soda ash and other products) on the west shore (Effler 1987). Here a brief review of some of the basic features of the operation of this facility is presented to facilitate an understanding of the character of its impact. Further, an expanded description of the recent history of domestic waste inputs to the lake is presented.

1.5.1 Allied Signal

The chemical plant on the western shore of the lake, originally named the Solvay Process Co. (later part of Allied Chemical Co., and finally part of Allied Signal Co.), has had a

TABLE 1.3. Chemicals produced at the Allied Chemical Co. facility at Syracuse, New York.

Chemical		Period of production
Common name	Formula	
Soda ash	Na_2CO_3	1884–1986
Coke		1892–1924
Producer gas	mixture of CO, H_2, CH_4, N_2, CO_2, and O_2	1892–1924
Caustic soda	NaOH	1896–1968
Baking soda	$NaHCO_3$	1896–1985
Sodium sesqui carbonate		1896–1985
Calcium chloride	$CaCl_2$	1896–1986
Chlorine	Cl_2	1918–1977
Chlorinated benzenes	C_6H_5Cl, $C_6H_4Cl_2$	1918–1977
Muriatic acid	HCl	1918–1977
Caustic potash	KOH	1918–1977
Ammonium chloride	NH_4Cl	1914–1985
Ammonium bicarbonate	NH_4HCO_3	1914–1985
Potassium carbonate	K_2CO_3	1924–1977
Potassium bicarbonate	$KHCO_3$	1963–1977
Sodium nitrite	$NaNO_2$	1920– present
Phenol	C_6HOH	1942–1946
Picric acid	—	1908–1915
Nitric acid	HNO_3	1908–1915
Salicylic acid	—	1915–1918
Methy salicylate	—	1915–1918
Benzyl chloride	—	1917–1920
Benzaldehyde	—	1917–1924
Benzoic acid	—	1917–1924
Phthalic anhydride	—	1917–1924

major environmental impact on the Syracuse area. The plant was originally built to produce sodium carbonate (Na_2CO_3), commonly referred to as soda ash. It was an ideal location because of the local abundance of the necessary raw materials and ample opportunity to dispose of associated wastes. The availability of these raw materials and by-products of the soda ash reactions lead to an impressive diversification in chemical manufacturing at this facility. More than 30 chemicals were manufactured at the plant over its 102 yr tenure. A listing of the products of this facility, and the period of their manufacture is presented in Table 1.3. The chemistry, waste generation, and other selected features of two of the most important processes that have impacted the lake, the production of soda ash, and Cl_2 gas and NaOH (chlor-alkali process), are described here. Henceforth, the chemical production plant is referred to simply as the soda ash/chlor-alkali facility.

1.5.1.1 Soda Ash: Chemistry and Waste Production

Soda ash is an extremely valuable material. It is used in softening water and in the manufacture of glass, soap, and paper. In 1865 Ernest Solvay developed the Solvay process to produce Na_2CO_3 from calcium carbonate ($CaCO_3$) and sodium chloride (NaCl). The simple overall reaction is

$$CaCO_3 + 2NaCl \rightarrow Na_2CO_3 + CaCl_2 \quad (1.1)$$

The abundance of the reactants in the Syracuse area in the form of limestone, and NaCl brines and deposits, and the proximity of the lake for disposal of wastes and as a source of cooling water, made the shores of the lake an ideal location for the production of soda ash. The Solvay Process Company began production on the western shore of Onondaga Lake in 1884.

Calcium carbonate is not used directly; it is heated to release carbon dioxide and leave solid calcium oxide (CaO, lime).

$$CaCO_3(s) \xrightarrow{heat} CaO(s) + CO_2(g) \quad (1.2)$$

The CO_2 and ammonia gas (NH_3) are bubbled through brine solution, where the following reaction occurs:

$$NH_3(g) + CO_2(g) + NaCl(aq) + H_2O(l) \rightarrow \\ NaHCO_3(s) + NH_4Cl(aq) \quad (1.3)$$

This results in the production of sodium bicarbonate ($NaHCO_3(s)$) and ammonium chloride ($NH_4Cl(aq)$). Sodium bicarbonate precipitates from solution, separating as a solid; $NH_4Cl(aq)$ is left behind in solution. Solid $NaHCO_3$ is removed by filtration and washed. This material has product value (e.g., baking soda); however, most of it was converted to Na_2CO_3 by heating ($\sim$300°C):

$$2NaHCO_3 \xrightarrow{heat} Na_2CO_3(s) + H_2O(g) + CO_2(g) \quad (1.4)$$

Ammonia gas used in the Solvay process was recycled by reacting two of the intermediates, NH_4Cl and CaO, according to

$$2NH_4Cl(aq) + CaO(s) \rightarrow \\ CaCl_2(aq) + 2NH_3(g) + H_2O(l) \quad (1.5)$$

Calcium chloride, one of the overall products of the Solvay process (Eq. [1]), also has value (e.g., as a drying agent in concrete mixtures), but not the demand that soda ash has; 25% $\leq$ of the $CaCl_2$ was recovered and sold. The remainder was disposed of as waste. In 1971 there were eleven Solvay process soda ash production facilities in the United States. The Syracuse plant was the last operating facility. Reduction in the demand for soda ash and cheaper alternative sources (e.g., trona deposits [mineral, $Na_2CO_3 \cdot NaHCO_3 \cdot 2H_2O$] in western states) contributed to the closing of the facility in 1986.

The production of large quantities of waste accompanied the soda ash manufacturing process. A waste slurry (5 to 10% suspended solids [Blasland & Bouck Engineers 1989]), containing $CaCl_2$, excess CaO, unreacted $CaCO_3$ and NaCl, $CaSO_4$, and lime impurities, was pumped to waste beds where the soluble fraction (waste bed overflow) drained off and entered the lake (mostly via Ninemile Creek

TABLE 1.4A. Composition of waste-bed overflow, reported from different sources (see footnotes).

Constituent	Concentrations reported in references (mg·L⁻¹)		
	(1)	(2)	(3)
Cl	62,980	53,700	55,388
Na	11,795	11,000	12,720
Ca	25,971	20,000	20,761
SO₄	918	—	1,009
CO₃	60	—	318
OH	276	—	54
Mg	—	—	220

TABLE 1.4B. Ion ratios (mg·L⁻¹) for waste-bed overflow, reported from different sources (see footnotes).

Ion ratio	Ratio (mg·L⁻¹) Values reported in references		
	(1)	(2)	(3)
Na/Ca	0.45	0.55	0.61
Cl/Na	5.34	4.88	4.35
Cl/Ca	2.43	2.69	2.67

(1) O'Brien & Gere 1969.
(2) USEPA 1974.
(3) Blasland & Bouck Engineers 1989.

since the early 1940s). The waste bed overflow was enriched in Cl^-, Na^+, and Ca^{2+}. The composition of the overflow reported from several sources is presented in Table 1.4A. Ratios of the principal ions form a valuable signature to evaluate the origins of these ions in Ninemile Creek and the lake. Values for the ratios Na/Ca, Cl/Na, and Cl/Ca are presented for the various data sources in Table 1.4B. The production of ionic waste has been described as a nearly stoichiometric relationship with soda ash production (USEPA 1974); for each kg of soda ash produced, approximately 0.5 kg of NaCl and 1.0 kg of $CaCl_2$ is released. Monitoring data for the interval 1970–1980 (Onondaga County 1971–1981) indicate the average daily load of the ionic waste to the lake for the period was approximately $2.5 \times 10^6 \, kg \cdot d^{-1}$ (Effler 1987).

1.5.1.2 Solvay Waste

The solid phase left behind after drainage of the waste-bed overflow is described as Solvay waste. The areal distribution of this material is shown in Figure 1.7. Deposits of Solvay waste surround about 30% of the lake; the most recent waste beds are located along Ninemile Creek. More than 2000 acres (8.1 km²) are

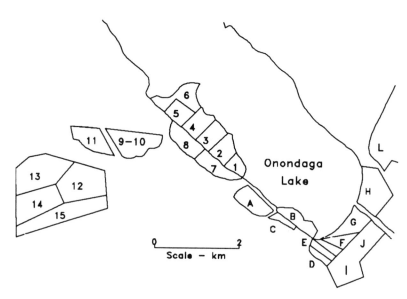

FIGURE 1.7. Areal distribution of Solvay waste deposits (modified from Blasland & Bouck 1989).

covered with Solvay waste. The depths of these deposits vary greatly. The filling/draining cycle continued for the more recent Solvay waste beds until the height reached about 21 m (beds 9, 10, 11 in Figure 1.7). Earlier deposits are as shallow as 2 m (bed G, Figure 1.7). No artificial impermeable material was used to line the waste beds. The mineral composition of the Solvay waste obtained by x-ray diffraction has been reported (Kulhawy et al. 1977) to be 20% $CaCO_3$, 17% Ca silicate, 10% MgOH, 8% $CaO \cdot CaCl_2$, 7% SiO_2, 6% NaCl, 6% $CaCl_2$, 6% alk and Fe oxide, 4% $Ca(OH)_2$, 4% $CaSO_4$, and 12% H_2O of hydration. Other materials deposited along with Solvay waste included fly ash and clinker from the facility's coal burning plant, as well as small quantities of insoluble mercury, lead, and asbestos from the Allied sodium hydroxide and chlorine production facilities (Blasland & Bouck Engineers 1989).

The earliest disposal of Solvay waste along the Onondaga Lake shoreline was in waste beds A to E. Bed A was used for Solvay waste disposal prior to 1920. From the beginning of the Solvay Process Company operations in 1884 until around 1916, the waste was pumped as a slurry into the surrounding marshlands. Disposal was not done in any planned or coordinated manner, but was carried out until a low, swampy area had been filled. In 1893, a road, approximately 10 km in length, was built along the lake's west shore from Syracuse to the Long Branch resort. The road ran immediately adjacent to the lake and its construction required shoreline filling with Solvay waste. Lake "marl" dredged from the lake was also used.

Use of Bed B began around the turn of the century (between 1898 and 1908) and was partially or wholly inactive by 1926. Sewage sludge was disposed of in this area by the City of Syracuse from the late 1950s until about 1966. Bed C was in use for Solvay waste by 1908 and was probably not used after 1926. Between 1959 and 1966, buildings and storage tanks were constructed on Bed C. Although it is not known when beds D and E were put into use, filling was completed by 1926. In the 1950s, interstate highway 690 and an accom-

panying road interchange were constructed on the northeast portion of Bed E. Several buildings were constructed on a portion of Bed D between 1959 and 1966. Varying amounts of waste were disposed of in areas F and G, in the vicinity of METRO.

Beds 1 to 6 were in use by 1926, while beds 7 and 8, on the southwest side of 1 to 6, were not used until 1939. In 1944, around the time of a dike failure that led to flooding along a portion of State Fair Boulevard, deposition of Solvay waste material in beds 1–8 was terminated. Bed 5 has been used for landfilling by the local steel manufacturer, parking lots for the New York State Fair, and sewage sludge disposal by Onondaga County.

Beds 9 to 15 cover approximately 662.6 acres (2.7 km^2) of land. Collectively, Solvay waste was deposited in beds 9–15 from approximately 1944 to 1985. Sequence of use progressed numerically (9 first, 15 last). Bed 15 began accepting demolition debris from the Allied facility beginning in the summer of 1987. Beds 9 to 15 are currently owned by Allied-Signal Corporation.

Less extensive alterations have occurred in the area northwest of the mouth of Ninemile Creek, where the resort beaches of the late nineteenth century were located. In the 1960s, the lake bottom off of the mouth of Ninemile Creek was dredged; spoils were eventually used by 1973 to create the western shore portion of the Onondaga Lake Park (USGS 1978; Blasland & Bouck 1989).

1.5.1.3 Cooling Water

Cooling water for the soda ash process was taken from the lake. In recent years, there have been three different intakes. One is in shallow water, at a depth of approximately 3 m below the normal lake level. The other two are in deeper water at a nearly equal, but uncertain, depth (10.7 to 12.2 m; personal communication, plant personnel). With-drawal from the deeper intakes was preferred because of the colder temperature. However, the upper intake was used more in late summer for a number of years to avoid the corrosive high concentrations of hydrogen sulfide (H_2S) that developed in the lower waters. Later the facility added chlorine to avoid this problem (oxidize the H_2S).

Heated water was discharged directly back to the lake via the East Flume (Figure 1.1) and to Ninemile Creek, upstream of Lakeland, via the West Flume. The East Flume shoreline discharge of spent cooling water was discontinued in 1978, in favor of a multiport diffuser discharge, 900 m offshore, at a depth of 4.6 m. The West Flume discharge was discontinued in early 1980. The West Flume discharge represented about 35% of the total flow of heated water from the facility over the 1976–1979 period (Onondaga County 1977–1980). The estimated average total discharge rate of spent cooling water from these two routes over the same four-year period was about 3.3 $m^3 \cdot s^{-1}$, similar to the discharge rate from METRO (Chapter 3).

1.5.1.4 The Chlor-Alkali Process

The availability of salt also promoted the establishment of a chlor-alkali process at the Allied facility. The products of the process are elementary chlorine and an alkali-sodium hydroxide (NaOH). The process involves the electrolysis of a water solution of NaCl; a direct current is passed through the solution. The overall reaction is

$$2NaCl(aq) + 2H_2O(l) \rightarrow$$
$$2NaOH(aq) + Cl_2(g) + H_2(g) \quad (1.6)$$

The oxidation reaction at the positively charged graphite electrode is

$$2Cl^-(aq) \rightarrow Cl_2(g) + 2e^- \quad (1.7)$$

The elemental chlorine gas was collected above the chamber. Mercury was used as the cathode, where reduction took place, according to

$$2Na^+(aq) + 2e^- \rightarrow 2Na(Hg) \quad (1.8)$$

As the sodium metal forms, it dissolves in Hg, creating an amalgam (Na(Hg); Hg as the solvent; 0.2–0.4% as Na). The amalgam is contacted with water, yielding $H_2(g)$, and a 50% aqueous solution of NaOH:

$$2Na(Hg) + 2H_2O(l) \rightarrow 2NaOH(aq) + H_2(g) \quad (1.9)$$

The chlor-alkali process is continuous; NaCl solution flows in one end of the cell, the NaOH solution flows out the other, and $Cl_2(g)$ and $H_2(g)$ are collected in between.

Mercury is recirculated in the process. However there were losses due to leakage and dumping, as the cells were cleaned or replaced. Mercury waste was released from the chlor-alkali facility at the Allied Chemical plant to Onondaga Lake from 1946 to 1986. Linden Chemical Products Co. operated the facility from 1986 to 1988. One mercury-electrode chlorine production cell was started in 1946, a second was added in 1953 (USEPA 1973). The daily load of Hg to the lake was estimated to be approximately 10 $kg \cdot d^{-1}$ (USEPA 1973), when the U.S. Department of Justice took legal action against the facility in the summer of 1970. The load was reduced to less than 0.5 $kg \cdot d^{-1}$ soon after through modifications in the process, and to less than 0.2 $kg \cdot d^{-1}$ in 1977. It has been estimated that approximately 75,000 kg of Hg were discharged to Onondaga Lake by the chlor-alkali facility over the period 1946 to 1970 (USEPA 1973).

1.5.2 Domestic Waste

1.5.2.1 Before METRO

Much of this text is concerned with the impact and legacy of historic, as well as ongoing, discharges of treated and untreated domestic waste to Onondaga Lake. The lake received increasing amounts of untreated domestic waste via the tributaries of Onondaga Creek, Harbor Brook, and Ley Creek from the time of the early settlement of the surrounding area by Europeans through the early twentieth century. The first strides to treat sewage were taken in 1907 with the creation of the Syracuse Intercepting Board, which constructed two trunk sewers paralleling Onondaga Creek and Harbor Brook. The "interceptor sewer system" was completed in 1922. The sewage was discharged to the lake following screening and disinfection. A primary sewage treatment facility was completed in 1925 adjoining the southern shore to serve the interceptor system, which carried 90% of the City's sewage. By 1928 the facility was overloaded because of the rapid growth of the City. Interceptor sewers of this period were the combined type, intended to carry both sanitary and stormwater flows (combined sewage); thus, the evolution of the 66 CSO system that exists today.

The pollution of Ley Creek was also recognized by the Syracuse Intercepting Board as a serious problem during the late 1920s. In 1933, at the request of Onondaga County, and with the formation of a sanitary district in mind, the State legislature created the Onondaga County Sanitary Sewer and Public Works Commission. The Commission was granted authority to plan and construct sewage treatment facilities, subject only to the approval of the County Board of Supervisors. The Commission extended the trunk sewers. The construction of lateral sewers was left to local governments. The Commission also built the Ley Creek Sewage Treatment Plant in 1934. This secondary treatment (activated sludge) facility was expanded to an average capacity of 9 million gallons per day (MGD). An additional 11 MGD primary treatment capacity was added in 1950, particularly in response to the increasing industrial load from Bristol Laboratories (manufacturer of penicillin and other pharmaceutical products). However, by 1974 the plant was described as both hydraulically and organically overloaded (USEPA 1974). Sewage collection and treatment facilities outside of the City and areas served by the Ley Creek plant in the 1930s, 1940s, and 1950s were very poor.

After World War II, the Onondaga County Public Works Commission recommended a metropolitan-wide course of action for the problem of sewage treatment, because growing suburban areas outside of the City would soon need sewage collection and treatment facilities. As such, it was recommended that the City not enlarge its treatment facilities independent of the County because the problem of pollution extended beyond City limits (New York State Departments of Health and Conservation 1947). It was believed that even if the City undertook corrective action alone, the continued pollutant loading to the lake's tributaries from outlying areas would still leave the lake in poor condition (New York State Departments of Health and Conservation 1947). However, cooperative action was delayed until the mid-1950s. Apparently no single community or the Public Works Commission was willing to assume responsibility for the treatment of sewage and pollution of the lake. Finally the Metropolitan Sewer District was formed by Onondaga County; it encompassed the City as well as certain of the surrounding suburbs.

1.5.2.2 METRO and CSOs

In 1960 Onondaga County completed construction of a primary treatment plant (METRO) on the southeastern shore of the lake to serve this area. This action was spurred by the four-year shutdown of the City's facility in the early 1950s, which was associated with the lack of sludge disposal, during which raw sewage was discharged to the lake. According to the original plans for the facility, the METRO effluent was to be pumped around the lake, combined with the Ley Creek plant effluent, and discharged to the Seneca River (see Figure 1.1). Later a lake discharge was selected for METRO instead, apparently as a cost-saving measure. Diversion of wastewater from the METRO site for discharge to the Seneca River remains a prominent feature of certain management alternatives presently being considered for the lake.

METRO was designed to treat 50 MGD of sewage. It could accommodate a peak flow of 170 MGD. Flow in excess of this quantity was bypassed directly to the lake following screening. By the early 1970s the facility was reported to be hydraulically overloaded, serving a population of about 261,000. In 1972 the average flow to the plant was about 35% above the design flow (USEPA 1974). A portion of the overloading was effluent received from the Ley Creek plant. The overloading resulted in decreased treatment efficiencies; e.g., in 1972 the suspended solids and BOD (5-day) removals were 51 and 26, respectively (USEPA 1974). Efficiently operated primary sedimentation tanks are expected to remove 50–65% of suspended solids and 25 to 40% of BOD (5-day) (Metcalf & Eddy, Inc. 1972). The average loading rate from the facility to the lake in 1972 was about $18,000 \, kg \cdot d^{-1}$ of suspended solids and $29,000 \, kg \cdot d^{-1}$ BOD (5-day). METRO had two 150 cm diameter outfalls in Onondaga Lake—one extending 520 m offshore that discharged at a depth of about 5 m, the other was a shoreline surface discharge. Usually the deeper outfall was used for treated effluent, and the shoreline outfall was only used to discharge bypassing flows (>170 MGD) during runoff events.

Major upgrades to METRO were made in the late 1970s and early 1980s. The facility was upgraded to secondary treatment in 1979 and "advanced" or tertiary treatment (phosphorus removal) in 1981. The facility was designed to treat an average flow of 80 MGD; flows of up to 120 MGD receive full treatment. Peak (e.g., storm) flows of 223 MGD can be accommodated. Flows in excess of 120 MGD receive incomplete treatment (primary treatment and chlorination). Secondary treatment is the contact stabilization modification of activated sludge. By design, chemical precipitation was to be affected by addition (6.5 MGD) of the calcium-rich waste from the settling lagoon overflow of the Solvay waste to the secondary effluent, followed by settling (using the clarifiers from the old facility). The layout of the upgraded METRO facility is presented in Figure 1.8. The effluent standard for phosphorus to be met, established for facilities of this size in the Great Lakes watershed, is $1.0 \, mg \cdot L^{-1}$. Pollutant loadings to the lake anticipated for the plant were 4580, 9810, and $330 \, kg \cdot d^{-1}$ of BOD (5-day), suspended solids, and phosphorus, respectively (USEPA 1974). The estimated phosphorus load from the facility before the upgrade was $1000 \, kg \cdot d^{-1}$ (USEPA 1974). Discharge of the effluent to the Seneca River was dismissed because the river's assimilative capacity was judged to be inadequate (USEPA 1974). It was concluded that discharge to Lake Ontario would not significantly impair the lake, but it was considered too expensive (USEPA 1974). The effluent is discharged to the southern end of Onondaga Lake via a 240-cm diameter shoreline outfall (to discourage "plunging" associated with the elevated density of the waste stream caused by the addition of ionic waste from the soda ash/chlor-alkali facility; see Chapter 4). Flows in excess of 120 MGD are discharged via the old 150-cm diameter deep-water discharge. Occurrences of ammonia concentrations in Onondaga Lake in excess of those known to be toxic to certain aquatic organisms were considered a distinct possibility following the METRO upgrade (USEPA 1974), as substantial nitrification was not anticipated.

Substantial consolidation of domestic waste treatment within the County has been achieved at the upgraded METRO plant. Eight municipal wastewater treatment plants existed in the Onondaga Lake Drainage Basin in the early 1970s. Four of these plants have ceased operation, including the Ley Creek facility in 1980, in favor of treatment at METRO.

The early performance of the innovative method of phosphorus removal was very good. However, the soda ash waste caused operational problems at METRO that resulted in interruptions in phosphorus treatment. These problems were resolved, but the closure of the soda ash/chlor-alkali facility in 1986 required the development of an alternative chemical treatment methodology. Phosphorus removal is now achieved by chemical precipitation with ferrous sulfate (within secondary clarifiers). Average concentrations of phosphorus in the METRO effluent in the early 1990s have been about $0.6 \, mg \cdot L^{-1}$, substantially below the permit requirement of $1.0 \, mg \cdot L^{-1}$. A citizen's lawsuit was brought against METRO in 1987 because of failures to meet State permit requirements for certain other constituents. A summary of annual average influent and effluent loadings for METRO for the 1986–1989 interval is presented in Table 1.5. Effluent loadings from this facility, and the implications for lake quality, are considered in detail in subsequent chapters. Annual average flows for METRO for the four years were 78.7, 74.0, 74.6,

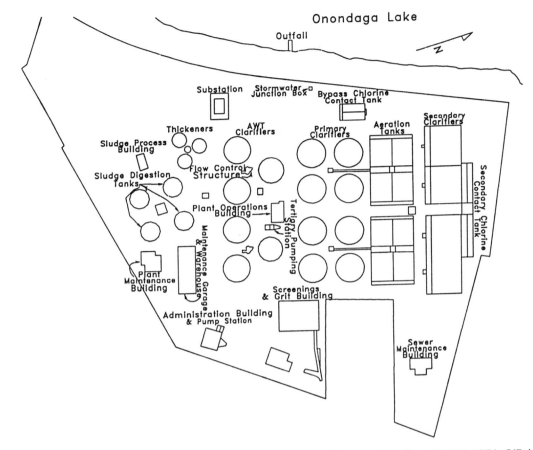

FIGURE 1.8. Metropolitan Syracuse wastewater treatment plant layout, 1981 (from USEPA 1974; O'Brien & Gere Engineers, Inc. 1973).

and 76.1 MGD (Stearns & Wheler Engineers 1991). Both phosphorus and suspended solids loadings to the lake have decreased since the discontinuation of the use of ionic waste from the soda ash/chlor-alkali manufacturer for phosphorus removal in 1986 (Table 1.5). Note that the lack of substantial ammonia nitrogen removal is consistent with the METRO design (e.g., USEPA 1974). Minor treatment process modifications were made at METRO during the 1992–1994 interval.

The combined sewer system tributary to METRO encompasses an area of 6,812 acres, or approximately 26 km²; it is located totally within the City of Syracuse. In 1991, 45 CSOs discharged to Onondaga Creek, 19 to Harbor

TABLE 1.5. METRO annual average influent and effluent loadings, 1986–1989 (from Stearns & Wheler Engineers 1991).

Constituent	Influent (kg·d⁻¹ × 10³)				Effluent (kg·d⁻¹ × 10³)				% Removal			
	1986	1987	1988	1989	1986	1987	1988	1989	1986	1987	1988	1989
BOD₅	64.0	40.4	49.0	59.0	5.9	4.9	5.0	4.5	91	88	90	92
phosphorus	1.35	1.05	1.15	1.20	0.29	0.27	0.21	0.16	79	74	82	87
TKN*	10.0	8.6	9.3	9.9	4.7	4.9	4.7	4.3	53	43	49	56
ammonia	4.3	4.7	5.6	5.3	4.0	3.8	3.9	3.4	7	19	30	36
suspended solids	64.5	39.5	46.3	63.1	5.9	2.5	2.9	3.3	91	94	94	95

*Total Kjeldahl nitrogen.

Brook, and 2 to Ley Creek (Moffa and Associates and Blasland & Bouck Engineers 1991). Presently they discharge combined sewage (dilute raw sewage) to these lake tributaries about 50 times a year. Swirl concentrator pilot studies have been conducted at two locations to evaluate this technology for remediation of the CSO problem. Upgrades of the Liverpool and Ley Creek Pump Stations in the 1990–1992 interval reduced related discharges of raw sewage.

1.6 Technical Studies of Onondaga Lake and Its Tributaries

1.6.1 Background

Measurements of certain water quality parameters are necessary to protect the public for desired resource uses. Further, it is important to protect the aquatic biota from the potential impacts of pollution. Systematic programs of measurements become "studies." The public can become frustrated by environmental studies. Continuing studies are often interpreted as wrongly replacing remediation. If insightfully designed, they can play a critical role in guiding remediation efforts.

The scope of studies should reflect the number and magnitude of pollution problems and the need to adequately support related remediation efforts. Monitoring programs of adequate scope can assess the status of a lake with respect to water quality standards and criteria set by regulators. However, in cases where significant problems (e.g., contravention of standards) prevail, monitoring studies can rarely provide the understanding necessary to support effective remediation. Research studies that focus on important processes and establish cause-and-effect relationships are usually required. Ultimately these efforts are synthesized in the development and testing of quantitative (empirical or deterministic) frameworks, called models, that are utilized to support management decisions regarding remediation.

1.6.2 History and Categories

Onondaga County (Department of Drainage and Sanitation) maintains a valuable compilation of the work conducted on Onondaga Lake, its tributaries, and the Seneca and Oswego Rivers ("Bibliography of Technical Material Pertaining to Onondaga Lake New York"). As of April 1992, a total of 228 reports and manuscripts were listed. Sixty-nine of these had appeared in peer-reviewed journals, and thus had been subjected to independent anonymous multiple peer review. The remainder represents a heterogeneous group, including engineering reports for adjoining facilities that discharge wastes to the lake, graduate theses and related academic studies, and analyses by environmental regulators. A substantial fraction of the contributions (particularly the journal articles before 1985) were done without funding. More than half (120) of the reports and manuscripts have been completed since 1980. The recent proliferation of studies undoubtedly reflects both the increased interest in remediating the lake's problems and the continued emergence of new pollution problems as studies proceeded. The study program for the lake has become more coordinated and funding has improved since the mid-1980s in an effort to support the development and testing of water quality models that would support reliable management decisions for remediation. Domestic waste-related problems have been pursued under funding from local, state, and federal governments. Potential problems associated with the discharges from the recently closed adjoining soda ash/chlor-alkali facility have heretofore been studied largely without funding. Presently a number of problems related to this facility are being studied by consultants under contract to the chemical manufacturer, as part of a negotiated agreement with the New York State Attorney General.

One of the earliest water quality studies of Onondaga Lake was conducted by the Syracuse University Research Corporation (SURC 1966; now Syracuse Research Corporation) for Onondaga County. The scope of the study was limited compared to later efforts, focusing mostly on salinity and phytoplankton growth. The report apparently represented a scientific basis to reject an earlier recommendation (1952) by a local consulting firm to divert sewage effluent to the Seneca River, concluding "little beneficial effect on the quality and general composition of the south end of Onondaga Lake would be achieved by the diversion," and that "the concentration of inorganic salts" (mostly from soda ash production) "would undoubtedly rise if the MSTP flow were diverted" (cover letter, SURC 1966). The second conclusion had an interesting premise. Apparently, the discharge of treated sewage to the lake was viewed as having the benefit of diluting the industrial waste input from the soda ash/chlor-alkali facility.

The first comprehensive limnological and water quality study of the lake, *The Onondaga Lake Study* (Onondaga County 1971), was not conducted until the late 1960s. Physical, chemical, and biological aspects of the lake were investigated by a team of researchers. Comprehensive monitoring of the principal tributaries to the lake was also included to support the estimation of pollutant loads. The lake's degraded state was coupled to elevated external loads of pollutants. Subsequent remediation efforts have focused on reduction of these loads (e.g., Chapter 3; Canale and Effler 1989, Devan and Effler 1984). The lake's ionically enriched condition and various manifestations of its hypereutrophic state were clearly depicted (Onondaga County 1971). Important, and in several cases unusual, conditions and processes were identified as part of this study (e.g., gas ebullition, chemical contribution to density stratification, continuous oversaturation with respect to the solubility of $CaCO_3$, hypolimnetic accumulation of H_2S), which formed the basis for a number of more detailed studies of related phenomena and forcing conditions in later years. The findings of this study continue to serve as an invaluable baseline in documenting changes in a number of features of the water quality of the lake. This baseline study also formed the basis for the design of Onondaga County's "Onondaga Lake Monitoring Program."

The "Onondaga Lake Monitoring Program," administered by Onondaga County's Department of Drainage and Sanitation until 1993, has been conducted annually since 1970. The program was funded solely by Onondaga County through 1990. Since 1991 federal funds, administered by the Onondaga Lake Management Conference, have contributed to the support of this on-going effort. An annual report has been prepared that summarizes observations, and more recently has identified trends and evaluated the status of the lake with respect to certain criteria and standards (Onondaga County 1971–1991). Other monitoring programs, of more limited scope and somewhat different foci have been conducted over portions of this period by the New York State Department of Environmental Conservation, in cooperation with the Onondaga County Health Department, and by the Upstate Freshwater Institute as part of research studies of lake processes. Specific features of these monitoring programs are manifested in various portions of this volume and described in detail by Walker (1991). These databases have been invaluable in supporting trend analyses and evaluations of water quality presented in various of the following chapters. Process and modeling studies for Onondaga Lake have largely been undertaken since the mid-to-late 1980s.

A selective listing of the timing of important findings in the study of Onondaga Lake, and related references and chapters of this volume, is presented in Table 1.6. It serves to illustrate the general progression of understanding for certain of the lake's problems, and how recently some of this work has been completed. A number of studies still need to be conducted or completed. The fish flesh of the lake was not recognized to be contaminated with mercury until 1970, yet it is probable that this condition prevailed soon after the chlor-alkali process was started at the adjoining chemical plant in 1946. The routine contraventions of the New York State minimum oxygen standard during fall mixing and the USEPA free ammonia criteria during stratification were not fully identified and documented until the late 1980s and early 1990s (Table

TABLE 1.6. Timing of important findings and developments in the study of Onondaga Lake and chapters of this volume where more information can be found.

Item	Subtopic/date/status	References	Chapter no.
Baseline study	Field program in 1968/1969; broad-based limnological and tributary loading study	1. Onondaga County 1971	—
Mercury contamination	1. Fish flesh—1970; lake closed to fishing 2. Sediments—1973 3. Water—1990	1. Kilborne 1970 2. USEPA 1973 3. Bloom & Effler 1990	6 7 5
Dissolved oxygen	1. Hypolimnetic anoxia 2. Lake-wide depletion in fall; contravention of standards 3. Recurrence of conditions	1. Onondaga County 1971 2. Effler et al. 1988 3. This volume	5 5 5
Free ammonia	1. Routine contravention of toxicity criteria for fish 2. Recurrence 3. Ammonia management model	1. Effler et al. 1990 2. This volume 3. Canale et al. 1993b	5 5 9
Nitrite	1. Routine contravention of toxicity standard for fish 2. Recurrence	1. Brooks & Effler 1990 2. This volume	5 5
Bacteria (indicator)	1. Contravention of standards lakewide after rainfall events 2. Contravention of standards in south basin after execution of "best management practices" program 3. Fecal coliform management model	1. Stearns & Wheler Engineers 1979 2. Canale et al. 1993a 3. Canale et al. 1993a	6 6 — 9
Clarity	1. Routinely low values 2. Improvements in late 1980s 3. Partitioning of contributions of various materials to clarity problem	1. Onondaga County 1971 2. Auer et al. 1990 3. This volume	8 6, 8 8
Sediment interactions	1. Phosphorus release 2. Oxygen demand	1. Auer et al. 1993 2. This volume	5 5
Salinity/chloride	1. Documentation of high concentrations and qualitative contribution to stratification 2. Quantification of contribution to stratification 3. Origins of lake concentrations	1. Onondaga County 1971 2. Effler et al. 1986 3. Doerr et al. 1994	5 4 3, 9
Waste-bed loadings	Chloride and ammonia	1. Effler et al. 1991	2, 3
"Mud boils"	Impact on suspended solids loading and clarity in Onondaga Creek	2. Effler et al. 1992	2, 3

1.6). The prevailing phosphorus release rate from the sediments and the oxygen demand of the sediments were determined in the late 1980s to early 1990s. But process and modeling studies necessary to predict changes in these important sediment fluxes, as external loads are reduced, have yet to be conducted. Deterministic models for the various problems associated with the discharge of domestic waste to the lake (e.g., fecal bacteria, phosphorus, oxygen and ammonia), became available during the 1989–1993 interval. The dominant role played by the ionic waste discharge from the soda ash/chlor-alkali facility in regulating the lake's salinity and stratification regime was not quantified until the mid-1980s (Table 1.6). Characterization of pollutant loadings and related impacts from the "mud boils" (in the upper reaches of Onondaga Creek) and the waste-bed area of Ninemile Creek have only recently been initiated (Table 1.6). Sedimentary deposits of synthetic organics were discovered along the western shore of the lake in 1991. Those deposits have yet to be fully chemically characterized. Further, the extent and potential impact of the deposits and remediation alternatives need to be assessed.

The severity and number of pollution problems of Onondaga Lake indicated here (Table 1.6) and described in the remainder of this volume have necessitated numerous studies to document, understand, and model these problems. These studies, particularly related to domestic waste impacts, have come very late for this lake. In general these issues were addressed in other lakes in the 1970s. As documented in the following chapters, the data, interpretations, process information, and quantitative frameworks (models) are now available to support management decisions for remediation of a number of the important water quality problems related to continuing domestic waste inputs. Completion of sediment feedback studies and related submodels will enhance these management tools. A monitoring program of related parameters will need to continue, through the response period of the lake to remediation actions, and perhaps indefinitely. Additional related process and modeling studies may be desirable in the future, particularly as new study techniques emerge, to enhance model credibility and capabilities.

The status of studies to evaluate the impacts of industrial waste on the lake and its tributaries is less complete, and the probability for the emergence of additional water quality problems related to these discharges is distinctly greater. A number of additional studies are necessary to fully understand and support effective remediation of these problems. Allied Chemical Co. initiated a number of related studies in 1992. A model for mercury cycling in the lake, intended to support remediation of the fish contamination problem, is not anticipated until 1997. Serious study of organic contamination of the lake is only just beginning. Further, the potential interplay between the industrial and domestic waste problems needs to be resolved.

Water quality managers need to fairly represent the role of studies of Onondaga Lake to the public and the need for continuing studies to avoid making costly errors in remediation and unrealistic promises concerning the future quality of the lake. While certain problems can now be effectively remediated, others will require further study to support continuing management efforts. Further, on a more global basis, Onondaga Lake, because of the degree and diversity of its pollution problems, deserves special consideration at state and federal levels for continued study as a national prototype, and education opportunity, for remediation of degraded aquatic ecosystems.

1.7 Demography and Land Use

Onondaga County's population grew until 1970 as a direct result of the post World War II economic boom. In the past twenty years there has been a "no-growth" pattern. A significant redistribution of population has occurred within Onondaga County as the population of the City of Syracuse has declined significantly while increasing in the suburbs. This is reflected in the 1990 census data showing that Onondaga County's population increased 5,053 (1.09%) from 463,920 in 1980 to 468,973 in the 1990 census, with the City of Syracuse experiencing a 3.67% decrease and several towns showing increases. A small but steady rise in the County's population is anticipated over the next two decades (Onondaga County 1990).

Employment by industry sector during the past decade is shown in Table 1.7. The service sector employed the most workers by 1987, displacing manufacturing. Of 114 occupations most in demand in the Syracuse area, over 60% are from the service sector. The second largest employment industry in 1987—manufacturing—has declined from its high

TABLE 1.7. Employment by industry in Onondaga County*.

Type	1980	1986	1987	Rank**
Durables	36,825	32,858	31,414	8
Nondurables	12,189	11,424	10,956	9
Total manufacturing	49,014	44,282	42,370	9
Transportation and public utilities	11,968	13,252	14,527	8
Construction	7,102	10,497	11,373	10
Agriculture	456	885	927	10
Wholesale trade	15,706	16,872	16,145	8
Retail trade	32,093	40,083	41,363	9
Service	39,093	53,229	55,765	10
Finance, insurance, and real estate	14,877	17,645	17,963	9
Government	34,673	33,700	34,912	9

*Source: NYS Department of Economic Development, *NYS 1989 County Profiles*.
** Rank among NYS Counties in 1987; Onondaga County ranks 10th in terms of population.

point in 1966. Retail trade, the County's third largest employer after manufacturing, increased during the last decade and is expected to surpass manufacturing in the 1990s due to the construction of several new shopping malls, notably the Carousel Center adjacent to Onondaga Lake, the expansion of existing malls, as well as due to the creation of small new retail centers (Onondaga County 1990).

Land use development has followed a pattern of decentralization for the past several decades, leading to expansion in the suburban towns and a mixed pattern of stability, decline, and redevelopment in Syracuse. Nearly 80% of all parcels in the City of Syracuse are residential; land outside of the County Sanitary District, or rural land use, is dominated by residential parcels and agricultural acreage. Land use in the Onondaga Lake watershed proper is shown in Table 1.8. The Cortland County portion of the watershed is classified equally between woodland and agricultural use (N. Jarvis personal communication, 1991).

Recently, a land use inventory history and informational database for the Oil City area (Figure 1.1) have been completed, and a "geographic information system" (GIS) has been developed (Smardon et al. 1989). A pilot project to examine the use of digitized land use information in a GIS format and the feasibility to refine the predictive capabilities of nonpoint source water quality models was completed for selected portions of the Onondaga Creek drainage basin.

1.8 Government Involvement and a Community's Vision

Onondaga Lake is almost entirely surrounded by developed areas and lies within the jurisdictional boundaries of the City of Syracuse, the towns of Salina and Geddes, and the villages of Solvay and Liverpool. More than 75% of the lake shoreline is classified as parkland owned by Onondaga County. The County's lake park system is visited each year by, among others, joggers, picnickers, and bicyclists. However, with the exception of the developed parkland, marina, and bicycle path on the northern side of the lake, the majority of the lakefront remains underdeveloped. After more than a century of damaging industrial and domestic waste-disposal practices, the lakefront is now experiencing the beneficial effects of community-wide interest in restoration. With passage of the Clean Water Act in

TABLE 1.8. Land use in the Onondaga Lake watershed.

Land use	Hectares	Total acreage (%)
Cropland	24,610	33
Urban development	20,273	28
Woodland	16,516	22
Idle	5,828	8
Indian reservation	2,420	3
Water	2,308	3
Orchard	741	1
Other	656	1
Pasture	372	1
Total	73,723	100

Source: Onondaga County Soil and Water Conservation District.

1972, direct intervention of federal and state governments into local water-pollution problems increased. The passage of RCRA in 1978, CERCLA in 1980, and comparable state laws greatly enhanced the enforcement capabilities of state and federal authorities. Today administrative orders, court orders, and pending lawsuits are in place to address some of the lake's most important pollution problems. One of the two most important legal initiatives is concerned with the effects of METRO and the CSOs on the lake; the other addresses the impact of the waste discharges and deposits of the soda ash/chlor-alkali facility.

Congressional approval of the Great Lakes Critical Programs Act of 1990 established the Onondaga Lake Management Conference (OLMC), which provides a framework for federal, state and local governments to cooperate in the cleanup of the lake and for the revitalization of the Onondaga Lake waterfront. The conference funds research intended to support effective management decisions for the lake. In 1993 the OLMC (OLMC 1993a) published a pamphlet entitled *The State of Onondaga Lake*, which presented a simplified and selective summary of the lake's water quality problems and causes of pollution. The material was drawn from the detailed analyses presented subsequently in this text. The pamphlet is intended to communicate salient technical information and issues to the community. The Conference in the same year released its *Management Plan* for the lake (OLMC 1993b). It includes 53 specific recommendations intended to lead to the reclamation of many of the resources of the lake.

In the wake of pollution control efforts, local governments, businesses, and developers are responding to the potential for investment in this urban waterfront. There has been considerable development of compatible land use in the neighborhoods surrounding the lake in recent years (Figure 1.9). The shores of Onondaga Lake are being converted to land that supports commercial development and outdoor recreation. The regulation of this transition is almost entirely the responsibility of local government. In recognition of the scope of present and future development opportunities adjoining Onondaga Lake, the Metropolitan Development Association (MDA), the City of Syracuse, Onondaga County, and the New York State Urban Development Corporation jointly sponsored the preparation of a Land Use Master Plan (Reimann Buechner Partnership 1992) that is to be used by the community as a tool to establish policies for the coordinated development of the lakeshore and adjacent properties. The Land Use Master Plan has incorporated preliminary plans for many area projects. These projects are illustrated as short-term and long-term plans in Figures 1.9 and 1.10, respectively.

The City of Syracuse has also developed an Action Plan for Lakefront Development that documents the City's plans for the Onondaga Lake Inner Harbor Area (Figure 1.9). The City of Syracuse proposed a two-phase $126 million redevelopment of the New York State Thruway Authority Barge Canal facility located near the mouth of Onondaga Creek. The first phase of the Syracuse Inner Harbor project calls for the development of a pleasure boat marina, restaurants and retail establishments, and a water-oriented recreation and cultural park. There is also a long-term component of the plan that envisions the construction of waterfront housing, corporate offices, a hotel, and expansion of the marina. The project promises to produce direct annual economic benefits of $6.75 million, 2,000 construction jobs, and 535 permanent employment positions. It is expected the project will increase the City's tax base, create and retain jobs for area residents, provide recreational and cultural uses along the lake, and open the lakefront to the public (Syracuse Office of Lakefront Development 1991).

Plans for the construction of a freshwater education, research, and vistor's center have been underway since 1992. The multifunctional facility will be located in the Inner Harbor area now under development by the City of Syracuse (Figure 1.9). The project would provide a facility and associated re-

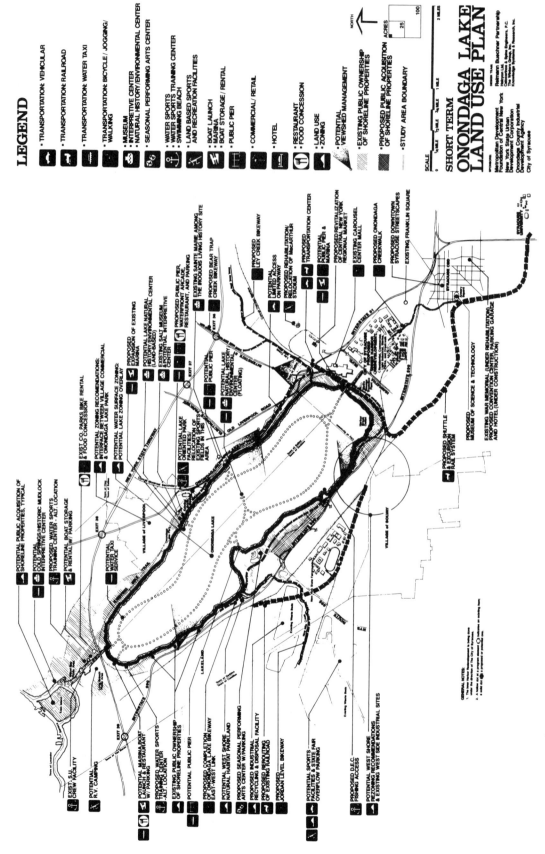

FIGURE 1.9. Short-term Onondaga Lake land use plan (from Reimann Buechner Partnership 1992).

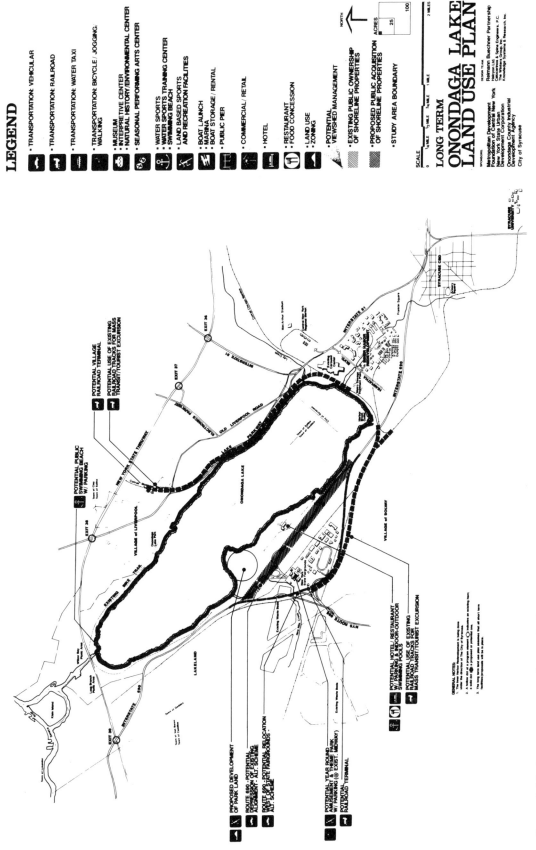

FIGURE 1.10. Long-term Onondaga Lake land use plan (from Reimann Buechner Partnership 1992).

sources to perpetuate scientific and engineering research on freshwater systems, establish a vehicle to promote public education related to freshwater ecosystems, and fulfill a need for a focused research center in a region rich in freshwater resources.

In an effort to gauge the public's perception of the reclamation of the lake and readaption of the Onondaga Lake waterfront for recreation, the OLMC conducted several polls and surveys from 1992 to 1993. Results show that local citizens are eager to use Onondaga Lake for recreation purposes, but are concerned that its present condition may pose an unacceptable risk to human health and aquatic life. In 1992, the OLMC circulated a nonscientific questionnaire to the citizens of Onondaga County. Out of the 1161 questionnaires collected, 1045 said that they would spend more time at Onondaga Lake if it was made safe for swimming. Eighty-five percent of the respondents said they would swim at a public beach on the lake if it were made safe for swimming (OLMC 1992).

The statute that created the OLMC established a Citizens Advisory Committee (CAC) to represent a cross-section of community interests. The committee included representatives from business, labor, and environmental groups. The CAC developed five objectives that reflect a desire by the community to reclaim the resources of Onondaga Lake.

1. The aesthetic qualities of the surface water and shoreline of Onondaga Lake shall be enhanced and improved.
2. Onondaga Lake shall be made fit for contact recreation from the mouth of Onondaga Creek to the Seneca River outlet.
3. The wildlife habitat of Onondaga Lake shall be restored and enhanced to sustain a thriving ecosystem in the lower tributaries of Onondaga Lake and the lake proper.
4. Any water quality remediation of Onondaga Lake shall not result in an adverse impact to the Seneca and Oswego River system.
5. The remediation of Onondaga Lake shall ultimately allow consumption of fish from the lake.

1.9 Summary

1.9.1 Location and Morphometry

Onondaga Lake is located in metropolitan Syracuse, New York—the most urbanized area of central New York. The lake is oriented along a northwest–southeast axis. The length along this axis is 7.6 km; it has a maximum width of 2 km. The lake has two basins. The surface area of the lake is $12.0 \times 10^6 \, m^2$, the volume is $131 \times 10^6 \, m^3$, the mean depth is 10.9 m, and the maximum depth is 19.5 m. The watershed area of the lake is 642 km^2; it lies almost entirely within Onondaga County. The climate within the watershed can be described as continental and somewhat humid. Substantial year-to-year variations in rainfall are experienced in the watershed.

The lake discharges to the Seneca River through a single outlet. The Seneca River combines with the Oneida River at the Three Rivers junction to form the Oswego River. The Oswego River flows into Lake Ontario at the City of Oswego.

1.9.2 Historic Setting, Degradation, and Efforts to Reduce Pollutant Loading

The early development of the NaCl industry adjoining Onondaga Lake was largely responsible for the growth of Syracuse in the late 1700s and the 1800s. The NaCl was taken from wells that adjoined the lake. Much of the southern half of the lake's shoreline was at one time involved with salt manufacture. Peak salt production was reached in 1862. The extent to which the lake was enriched with salt in this era is open to debate.

The availability of transportation, first through the canal system and later through the railroads, supported industrial and population growth. Local salt and limestone deposits and the availability of Onondaga Lake for the disposal of related wastes fostered the establishment of the soda ash industry. Production by the Solvay Process began on the western shore of the lake in 1884. Later this facility expanded into a larger and more diverse chemical manufacturing operation, described here as the soda ash/chlor-alkali facility. During the 1900 to 1940 period, a number of other industries were established in the region, including steel, pottery, pharmaceutical, air conditioning, appliance, and electrical manufacturing facilities.

A number of resorts were built along the northwest shoreline of the lake in the 1870s and 1880s, but the resort era was short lived. The demise of the resorts occurred soon after World War I, in part because of the increasing pollution of the lake.

Onondaga Lake has received the municipal effluent and a large portion of the industrial waste of the region for more than a century. By the turn of the century, pollution already had a profound impact on the lake. By the late 1890s the lake lost its commercially valuable cold-water fishery. Ice harvesting was banned for health reasons in 1900. Untreated sewage entered the lake via its tributaries until after 1900. Swimming in the lake was banned in 1940. Fishing was banned in the lake in 1970, after discovery of high concentrations of mercury in fish flesh.

A combined type sewer system conveyed both sanitary and stormwater flows to treatment facilities on the southern shore of the lake in the early 1900s. Domestic waste treatment in the surrounding metropolitan area has been consolidated and increasing levels of treatment have been applied over the years, although the effluent continues to be discharged to the southern end of the lake. Dilute raw sewage continues to enter tributaries to the lake during runoff events via 66 hydraulic relief structures in the sewer system known as combined sewer overflows (CSOs). Additional reductions in pollutant loadings have been achieved through treatment of industrial wastes, a legislative ban on high phosphorus detergents in 1972, execution of a best management practices program to reduce loadings from CSOs, and closure of the soda ash/chlor-alkali facility in 1986.

1.9.3 Waste Sources

1.9.3.1 Soda Ash/Chlor-Alkali Plant

More than 30 chemicals were manufactured at the chemical manufacturing facility on the west shore of the lake over the 102-year tenure of the plant. This facility was the last Solvay Process soda ash ($NaCO_3$) plant in the United States. Large quantities of solid and liquid waste were produced from this manufacturing process. Waste slurry was pumped to waste beds. The soluble fraction (waste-bed overflow), enriched in three ionic constituents, Cl^-, Na^+, and Ca^{2+}, drained off and entered the lake. For each kg of soda ash produced, approximately 0.5 kg of NaCl and 1.0 kg of $CaCl_2$ of ionic waste was released. The average daily loading of this ionic waste to the lake during the 1970 to 1980 period was about $2.5 \times 10^6 \, kg \cdot d^{-1}$. The solid phase waste, manifested as the Solvay waste beds, presently cover more than 2,000 acres adjoining the lake. Approximately 30% of the lake is bounded by these waste deposits. Earlier beds were only about 2 m high; more recent beds exceed 20 m in height. Cooling water for the soda ash process was taken from the lake and returned at an elevated temperature.

Mercury was used as the cathode for the chlor-alkali process. However, recovery of Hg was incomplete. Mercury waste was released from the Allied Chemical facility from 1946 to 1986. The daily load of Hg to the lake was estimated to be approximately $10 \, kg \cdot d^{-1}$ when the U.S. Department of Justice took legal action against the facility in the summer of 1970. The load was reduced by 95% soon after. Approximately 75,000 kg of Hg were discharged to the lake by the soda ash/chlor-alkali facility during the 1946–1970 period.

1.9.3.2 Domestic Waste

A selective history of the handling and treatment of domestic waste in metropolitan Syracuse is presented in Table 1.9. Before the construction of the sewer system, domestic wastes entered tributaries to Onondaga Lake untreated. Earlier treatment facilities became overloaded rapidly due to the growth of in-

TABLE 1.9. Selective history of domestic waste handling and treatment.

Date(s)	Feature
1907	Creation of Syracuse Intercepting Board
1922	Interceptor sewage system completed
1925	Completion of primary treatment facility for City of Syracuse
1933	Creation of Onondaga County Sanitary Sewer and Public Works Commission
1934	Completion of Ley Creek Plant
1950	Expansion of Ley Creek Plant, particularly to accommodate Bristol Laboratories load.
1950–1953	Discharge of untreated sewage from the Syracuse Treatment Plant because of lack of sludge disposal
1960	Completion of expanded Onondaga County primary 50 MGD treatment facility (METRO); replaces City facility
1979	Upgrade of METRO to secondary treatment (80 MGD)
1980	Closure of Ley Creek plant; influent diverted to METRO
1981	Upgrade of METRO to tertiary treatment; use of ionic waste from soda ash/chlor-alkali facility for phosphorus removal
1983–1985	"Best Management Practices" program on CSO system to reduce occurrence and flows
1986–1988	Phosphorus removal by iron salts following closure of soda ash/chlor-alkali facility; ferrous sulfate, starting in 1988
1990–1992	Upgrading of the Liverpool and Ley Creek pump stations

dustry and the population. Originally (early 1950s) the plan was to discharge the METRO effluent to the Seneca River. A lake discharge was selected instead as a cost-saving measure. Diversion of wastewater from the METRO site for discharge to the Seneca River remains a prominent feature of certain management alternatives presently being considered for the lake.

1.9.4 Technical Studies

A number of studies of Onondaga Lake and its tributaries have been conducted to assess its status with respect to water quality standards, to characterize the lake's biota, to identify and quantify important processes, to establish cause-and-effect relationships, and, most recently, to support the development of mechanistic water quality models. Execution of these studies has lagged behind efforts on other highly impacted systems. The first comprehensive limnological and water quality study of the lake was not conducted until the late 1960s. This effort provided the first documentation of the degraded condition of the lake, related its condition to loadings of both domestic and industrial wastes, and served as a prototype for the continuing Onondaga Lake Monitoring Program.

Speciality studies of the lake proliferated in the 1980s. In the early 1990s, enough information was available to support the development of credible water models for issues related to domestic waste discharges. The database for industrial waste problems is less strong. The emergence of additional water quality problems associated with these discharges is a distinct possibility. Continued studies of the lake and tributaries will be necessary through the period of response to remediation actions and to fill remaining gaps in the understanding of the system (e.g., sediment feedback). Onondaga Lake, because of the degree and diversity of its pollution problems, deserves special consideration by state and federal levels for continued study as a national prototype and education opportunity for remediation of degraded aquatic ecosystems.

1.9.5 Government Involvement and a Community's Vision

Lawsuits and a court order are presently in place to address the impact and related remediation of pollution from the soda ash/chlor-alkali facility and METRO. The Onondaga Lake Management Conference (OLMC) was established in 1990 by Congress to provide a framework for federal, state, and local governments to cooperate in the cleanup of the lake. The OLMC has funded research, published findings, conducted public opinion surveys, and developed a management plan to guide reclamation of the lake. Survey results indicate local citizens are eager to use Onondaga Lake for recreation purposes, but are concerned that its present condition may pose an unacceptable risk to human health and aquatic life. The Citizen's Advisory Committee of the OLMC recommended reclamation of the lake, without causing adverse impact to the river system.

More than 75% of the lake's shoreline is classified as parkland owned by Onondaga County. However, the majority of the lakefront remains underdeveloped. Redevelopment of the Inner Harbor, bounding the mouth of Onondaga Creek, and adjoining areas has already begun. Local government and community leaders have developed a "land use master plan" and a "lakefront development plan" to guide these on-going, as well as future, activities. Plans for the construction of a freshwater education and research center, to be located in the Inner Harbor, are also underway.

References

Auer, M.T., Johnson, N., Penn, M.R., and Effler, S.W. 1993. Measurement and verification of rates of sediment phosphorus release for a hypereutrophic urban lake. *Hydrobiologia* 253: 301–309.

Auer, M.T., Storey, M.L., Effler, S.W., Auer, N.A., and Sze, P. 1990. Zooplankton impacts on chlorophyll and transparency in Onondaga Lake, N.Y., U.S.A. *Hydrobiologia* 200/201: 603–617.

Beck, L.C. 1826. *An Account of the Salt Springs at Salina, in Onondaga County New York with a Chemical Examination of the Water and Several Varieties of Salt Manufactured at Salina and Syracuse.* J. Seymour, New York, NY.

Blasland & Bouck Engineers. 1989. Hydrogeological Assessment of the Allied Waste Beds in the Syracuse Area: Volume 1. Submitted to Allied-Signal, Inc. Solvay, NY.

Bloom, N.S., and Effler, S.W. 1990. Seasonal variability in mercury speciation of Onondaga Lake (New York). *Water Air Soil Pollut.* 53: 251–265.

Brooks, C.M., and Effler, S.W. 1990. The distribution of nitrogen species in polluted Onondaga Lake, N.Y., U.S.A. *Water Air Soil Pollut.* 52: 247–262.

Bruce, D.H. 1896. *Onondaga's Centennial: Gleaning of a Century; Volume 1.* Boston History Company, Boston, MA.

Canale, R.P., Auer, M.T., Owens, E.M., Heidtke, T.M., and Effler, S.W. 1993a. Modeling fecal coliform bacteria-II. Model development and application. *Water Res* 27: 703–714.

Canale, R.P., and Effler, S.W. 1989. Stochastic phosphorus model for Onondaga Lake. *Water Res.* 23: 1009–1016.

Canale, R.P., Gelda, R.K., and Effler, S.W. 1993b. A Nitrogen Model for Onondaga Lake. Submitted to Central New York Regional Planning and Development Board, Syracuse, NY.

Chase, F.H. 1924. *Syracuse and its Environs: A History: Volume 1.* Lewis Historical Publishing Company, Inc.

Clark, J.V.H. 1849. *Onondaga; or Reminiscences of Earlier and Later Times; and Oswego.* Stoddard and Babcock, Syracuse, NY.

Devan, S.P., and Effler, S.W. 1984. The recent history of phosphorus loading to Onondaga Lake. *J Environ Engr.* ASCE 110: 93–109.

DeWitt. 1801. Memoir on Onondaga salt springs and salt manufacturers in the state. *Transactions of the Society for the Promotion of Agriculture, Arts, and Manufactures in the State of New York*, Volume 1.

Doerr, S.M., Effler, S.W., Whitehead, K.A., Auer, M.T., Perkins, M.G., and Heidtke, T.M. 1994. Chloride model for polluted Onondaga Lake. *Water Res.* 28: 849–861.

Effler, S.W. 1987. The impact of a chlor-alkali plant on Onondaga Lake and adjoining systems. *Water Air Soil Pollut.* 33: 85–115.

Effler, S.W., Brooks, C.M., Addess, J.M., Doerr, S.M., Storey, M.L., and Wagner, B.A. 1991. Pollutant loadings from Solvay waste beds to lower Ninemile Creek, New York. *Water Air Soil Pollut.* 55: 427–444.

Effler, S.W., Brooks, C.M., Auer, M.T., and Doerr, S.M. 1990. Free ammonia and toxicity criteria in a polluted urban lake. *Res J Water Pollut Contr Fed.* 62: 771–779.

Effler, S.W., and Driscoll, C.T. 1985. A chloride budget for Onondaga Lake, New York, U.S.A. *Water Air Soil Pollut.* 27: 29–44.

Effler, S.W., Hassett, J.P., Auer, M.T., and Johnson, N. 1988. Depletion of epilimnetic oxygen and accumulation of hydrogen sulfide in the hypolimnion of Onondaga Lake, NY, U.S.A. *Water Air Soil Pollut.* 39: 59–74.

Effler, S.W., Johnson, D.L., Jianfu, F.J., and Perkins, M.G. 1992. Optical impacts and sources of tripton in Onondaga Creek, U.S.A. *Water Res Bull.* 28: 251–262.

Effler, S.W., Owens, E.M., and Schimel, K.A. 1986. Density stratification in ionically enriched

Onondaga Lake, U.S.A. *Water Air Soil Pollut.* 27: 247–258.

Gordon, T.F. 1836. *Gazetteer of the State of New York.* T.K. and P.G. Collins, Philadelphia, PA.

Hennigan, R.D. 1991. American's dirtiest lake. *Clearwaters* 19: 8–13.

Higgins, G.L. 1955. Saline Ground Water at Syracuse, New York. Masters Thesis. Syracuse University, Syracuse, NY.

Hutton, F.Z., and Rice, E.C. 1977. Soil Survey of Onondaga County, New York. U.S. Department of Agriculture, Soil Conservation Service, 235p.

Kilborne, R.S. 1970. Order: Prohibiting Fishing in Onondaga Lake. State of New York Conservation Department. Albany, NY.

Kulhawy, F.H., Sangrey, D.A., and Grove, C.P. 1977. Geotechnical behavior of Solvay Process wastes. Proceedings, ASCE Conference on Geotechnical Practice for Disposal of Solid Waste Materials, pp. 118–135. Ann Arbor, MI.

Lipe, W.C., Haley, J.J., and Ripberger, R.R. 1983. Report of the Onondaga Lake Subcommittee to the Onondaga County Legislature Committee on Public Works, Onondaga County, NY.

Metcalf & Eddy, Inc. 1972. *Wastewater Engineering: Collection, Treatment, Disposal.* McGraw-Hill, New York.

Moffa and Associates and Blasland & Bouck Engineers. 1991. Combined Sewer Overflow Facility Plan. Prepared for Onondaga County, Submitted to the New York State Department of Environmental Conservation, Syracuse, NY.

Murphy, C.B. 1978. Onondaga Lake, *In*: J.A. Bloomfield (Ed.), *Lakes of New York State, Vol. II,* John Wiley & Sons, New York, pp. 223–363.

New York State Departments of Health and Conservation. 1947. Onondaga Lake drainage basin report, Oswego River drainage basin survey series, Report No. 1. Albany, NY.

O'Brien & Gere Engineers, Inc. 1969. Report on Pilot Plant Study. O'Brien & Gere Engineers, Inc. Syracuse, NY.

O'Brien & Gere Engineers, Inc. 1973. Environmental Assessment Statement. Syracuse Metropolitan Sewage Treatment Plant and the West Side Pump Station and Force Main. Report to Onondaga County, Syracuse, NY.

O'Brien & Gere Engineers, Inc. 1987. Combined Sewer Overflow Abatement Program Post BMP Assessment. Report to Onondaga County, Syracuse, NY.

Onondaga County. 1971. Onondaga Lake Study. Project No. 11060, FAE 4/71. Water Quality Office, United States Environmental Protection Agency, Water Pollution Control Research Series.

Onondaga County. 1990. The Onondaga County Plan: A Development Guide for 2010, Report 1; Syracuse-Onondaga County Planning Agency, Syracuse, NY.

Onondaga County. 1971–1991. Onondaga Lake Monitoring Program Annual Report. Onondaga County, Syracuse, NY.

Onondaga Lake Management Conference. 1992. Onondaga Lake Management Conference Questionnaire, June 10, 1992. Final Report. Syracuse, NY.

Onondaga Lake Management Conference. 1993a. *The State of Onondaga Lake* (pamphlet). Syracuse, NY. 20p.

Onondaga Lake Management Conference. 1993b. Onondaga Lake, A Plan For Action. Onondaga Lake Management Plan, Draft for Public Comment, Syracuse, NY. 41p.

Owens, E.M. 1987. Bathymetric Survey and Mapping of Onondaga Lake, New York. Submitted to Onondaga County by the Upstate Freshwater Institute, Syracuse, NY.

Reimann Buechner Partnership. 1992. 1991 Onondaga Lake Development Plan. Metropolitan Development Foundation of CNY, NYS Urban Development Corp., Onondaga County Industrial Development Agency, City of Syracuse, Syracuse, NY, 109p.

Ruffner, J.A., and Blair, F.E. (editors). 1987. *The Weather Almanac.* Gale Research, Detroit, MI.

Schultz, O. (as described by Murphy (1978)). 1810. *Travels on an Inland Voyage Through the States of New York, Pennsylvania, Virginia, Kentucky, and Tennessee, and Through the Territories of Indiana, Louisiana, Mississippi and New Orleans—performed in the years 1807 and 1808; Including a Tour of Nearly Six Thousand Miles,* Vol. I, pp. 29–34. Isaac Riley, New York.

Smardon, R., Drake, B.J., and Kalinoski, R. 1989. Report to the City of Syracuse on the Oil City Remediation Workshop, Vol. 1, IEPP Pub. No. 89-001.

Stearns & Wheler Engineers. 1979. Onondaga Lake Storms Impact Study. Stearns and Wheler, Civil and Sanitary Engineers, Cazenovia, NY.

Stearns & Wheler Engineers. 1991. Projected Flows and Loadings METRO Engineering Alternatives. Onondaga County. Syracuse, NY.

Sweet. 1874. Maps of the Towns of Geddes, Salina and City of Syracuse. Scale, 150 m Rods to the inch. Part of the Township of Manlius and Onondaga Salt Spring Reservation.

Syracuse Office of Lakefront Development. 1991. Action Plan for the Syracuse Lakefront. Mayor's Lakefront Advisory Committee Inner Harbor Task Force Strategic Planning Team, Syracuse, NY. 96p.

Syracuse University Research Corporation. 1966. Onondaga Lake Survey, 1964–1965. Final Report. Prepared for Onondaga County Department of Public Works. Syracuse, NY.

USEPA (United States Environmental Protection Agency). 1973. Report of Mercury Source Investigation Onondaga Lake, New York and Allied Chemical Corporation, Solvay, New York. National Field Investigations Center, Cincinnati, OH and Region II, New York, NY.

USEPA (United States Environmental Protection Agency). 1974. Environmental Impact Statement on Wastewater Treatment Facilities Construction Grants for the Onondaga Lake Drainage Basin, Region II, New York, NY.

USGS (United States Geological Survey). 1978. Syracuse West Quadrangle, New York. 7.5 Minute Series (topographic), scale of 1:24,000 (1973, revised in 1978).

Walker, W.W. 1991. Compilation and Review of Onondaga Lake Water Quality Data. Onondaga County Department of Drainage and Sanitation, Syracuse, NY.

Winkley, S.J. 1989. The Hydrogeology of Onondaga County, New York. Master's Thesis, Syracuse University, Syracuse, NY.

2
Hydrogeologic Setting

Mary Gail Perkins and Edwin A. Romanowicz

This chapter provides a review of available information pertaining to the geology and hydrogeology of Onondaga Lake and its tributaries. New information related to the hydrogeochemistry of both Ninemile Creek and Onondaga Creek, along with an overall interpretation of the general flow system surrounding Onondaga Lake, are also presented. The chemistry of tributary inflows into Onondaga Lake, and the chemistry of Onondaga Lake itself, is directly affected by the geologic setting. Concentrated urban and industrial development in areas which may be in hydrologic connection to the lake and its main tributaries suggests that groundwater in selected areas requires careful evaluation as a potential medium of contaminant transport.

2.1 Background Geology

2.1.1 Bedrock Geology

The bedrock in Onondaga County is comprised of Paleozoic sedimentary rocks (refer to Section 2.5; Figure 2.41 for geologic time scale). A generalized listing of bedrock formations in Onondaga County is presented in Figure 2.1. Knowledge of both bedrock and unconsolidated surficial stratigraphy is important because water chemistry is affected as ground water flows through different bedrock

and surficial units before discharging to a surface stream or lake. The correlation between the regional "water bearing" units for the entire Eastern Oswego River Basin (Kantrowitz 1970) and for the more localized "hydrostratigraphic" units for Onondaga County (Winkley 1989) is presented in Figure 2.2. Hydrostratigraphic units are groupings of rock units with significant lateral extent that exhibit similar distinct hydrologic properties (Maxey 1964). Bedrock ages in the area range from Silurian to Devonian, becoming progressively younger southward due to the regional southerly dip (3–5°) (Figure 2.3). Because the bedrock is nearly horizontal, bedrock outcrops are topographically, rather than structurally, controlled. This means that along any east–west transect, bedrock contacts will occur at specific elevations. The Appalachian Upland (Figure 2.4), located in the southern half of Onondaga county, is capped by the Devonian Hamilton Shales and Onondaga Limestone. The highly resistive Onondaga Limestone, Oriskany Sandstone, and limestones and dolostones of the Helderberg Group form the Helderberg Escarpment. This escarpment defines the northern edge of the Appalachian Upland. The Salina Group outcrops along the base of the escarpment. The Salina Group is comprised of dolostones, shales and the evaporites of the Syracuse Formation. Onondaga Lake is situated on the Vernon Shale, the

PERIOD	GROUP	FORMATION	THICKNESS	LITHOLOGY
Upper Devonian	Genesee	Sherburne	200–220'	shale & siltstone
		Genesee	70–80'	black shale
	–	Tully	30'	limestone
Middle Devonian	Hamilton	Moscow	190'	shale & siltstone
		Ludlowville	230–270'	shale & siltstone
		Marcellus	290–360'	black shale
	–	Onondaga	63–88'	limestone
	–	Oriskany	0–20'	sandstone
Lower Devonian	Helder-berg	Manlius	47–129'	limestone
		Roundout	30–40'	dolostone
Upper Silurian	–	Cobleskill	15–25'	dolostone
	Salina	Bertie	64–97'	dolostone & evaporites
		Camillus	160–190'	shale
		Syracuse	150–220'	dolostone, shale & evaporites
		Vernon	500–600'	red and green shale
Middle Silurian	Lockport	Sconondoa	120–140'	dolostone & limestone
	Clinton	Rochester	120–150'	shale
		Irondaquoit	0–25'	limestone
		Williamson	40–50'	green shale
		Sasquoit	30–70'	shale

FIGURE 2.1. Generalized stratigraphic bedrock cross section for Onondaga County (modified from Winkley 1989).

lowest member of the Salina Group (Rickard and Fisher 1970; Rickard and Fisher 1975).

The Syracuse Formation is an evaporitic member of the Salina Group; it has played a very important role in the history of Syracuse and Onondaga Lake (refer to Chapter 1). The brines found around Onondaga Lake originate in the Syracuse Formation (Higgins 1955; Kantrowitz 1970). Halite (NaCl, rock salt) and gypsum ($CaSO_4 \cdot 2H_2O$), commonly found in the Syracuse Formation, are precipitates which remain after saline water, such as sea water, evaporates. Anhydrite ($CaSO_4$) is also found in the Syracuse Formation. Anhydrite may precipitate from highly saline waters or it may be a diagenetic mineral produced by the dehydration of gyspum (Deer et al. 1980). Evaporites are easily dissolved minerals that, once dissolved, leave behind void spaces. The void spaces are known as salt hoppers or gypsum molds depending on the original evaporite present. There are many of these voids found in the Syracuse Formation (Rickard and Fisher 1975), offering a conduit

Hydro-Strat		FORMATION	THICKNESS	LITHOLOGY		Water-Bearing
Devonian Shale Unit		Sherburne	200–220'	shale & siltstone		Upper Shale Unit
		Genesee	70–80'	black shale		
		Tully	30'	limestone		
		Moscow	190'	shale & siltstone		
		Ludlowville	230–270'	shale & siltstone		
		Marcellus	290–360'	black shale		
Silurian Devonian Carbonate Unit		Onondaga	63–88'	limestone		Upper Carbonate Unit
		Oriskany	0–20'	sandstone		
		Manlius	47–129'	limestone		
		Roundout	30–40'	dolostone		
		Cobleskill	15–25'	dolostone		
Post Vernon Evaporitic Unit		Bertie	64–97'	dolostone & evaporites		Middle Shale Unit
		Camillus	160–190'	shale		
		Syracuse	150–220'	dolostone, shale & evaporites		
Vernon Shale Unit		Vernon	500–600'	red and green shale		
Lockport Carbate Unit		Sconondoa	120–140'	dolostone & limestone		Lower Carbonate Unit
Clinton Shale Unit		Rochester	120–150'	shale		Lower Shale Unit
		Irondaquoit	0–25'	limestone		
		Williamson	40–50'	green shale		
		Sasquoit	30–70'	shale		

FIGURE 2.2. Correlation between the regional "water bearing" units for the entire Eastern Oswego River Basin and the more localized "hydrostratigraphic" units for Onondaga County (modified from Winkley 1989).

for the rapid transportation of ground water (Phillips 1955).

The Vernon Shale outcrops in the flat lake plains in the northern part of the county (Figures 2.3 and 2.4). The Vernon Shale underlying Onondaga Lake is made up of gray, red, and green mudstones (Rickard and Fisher 1970) containing discontinuous gypsum seams (Mozola 1938) that historically may have contained halite (Apfel and Associates, Inc. 1968). Dissolution of these evaporates in the Vernon shale resulted in the formation of secondary porosity (Winkley 1989). The low relief and low permeability of the Vernon

Shale produces very small hydraulic gradients (Kantrowitz 1970). The low permeability and limited vertical flow in the Vernon shale impedes the percolation of meteoric (atmospheric) water deep into the shale, thereby reducing dilution effects on the brines. Natural brines are located at shallow depths (near the surface) in the Vernon Shale because of the low permeability and small hydraulic gradient. It is likely that there is some hydrologic communication between the Vernon Shale and the overlying glacial and lake sediments. Although the extent to which the natural brines from the Vernon shale may be con-

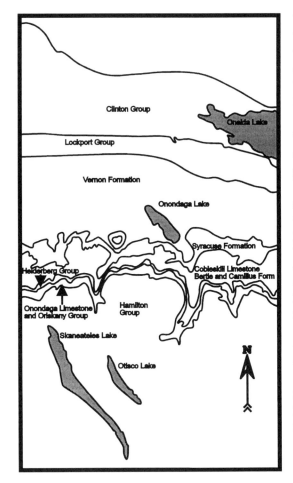

FIGURE 2.3. Generalized bedrock map for the area around Onondaga Lake (modified from Rickard and Fisher 1970).

Winkley also states that the Vernon shale has "excessively hard water" due to high concentrations of gypsum and calcite. Gypsum is also contained in the bedding planes and joints of the Camillus shale which outcrops approximately 2 km (1.2 mi) south of the lake.

2.1.2 Recent Glacial History

Onondaga County is located at the convergence of two distinct physiographic provinces (Figure 2.4). The northern portion of the county is covered by the lake plains of the Ontario Lowlands. These plains, remnants of post glacial Lake Iroquois, extend northward to Lake Ontario. The Appalachian Upland extends from Pennsylvania into southern Onondaga County. This part of the county has high topographic relief, more than a thousand feet (>305 m) within a few miles (Muller 1964). Much of the surficial expression in Onondaga County is a result of the Wisconsin Glaciation (12,000 to 14,500 years ago) (Winkley 1989). Moraines, meltwater channels, drumlins, and U-shaped valleys can be found throughout the county. A complete review of the glacial history of Onondaga

tributing to the salinity of the lake itself (either historically or presently) is unknown, a chloride model (Doerr et al. 1994) for Onondaga Lake indicates the contribution is relatively minor. This issue is discussed in more detail in Section 2.4.2 of this chapter and in Chapter 9.

Onondaga Lake has unusually high concentrations of sulfate ($\sim$145 mg $\cdot$ L^{-1}; UFI 1988, unpublished data). A potential source is gypsum dissolution by ground water, which subsequently flows into Onondaga Lake and its tributaries. The Vernon shale, underlying Onondaga Lake, no longer contains halite deposits but does contain quantities of discontinuous gypsum masses (Winkley 1989).

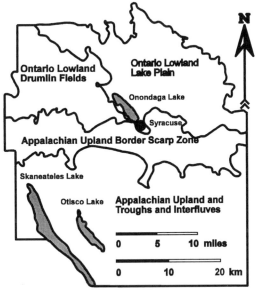

FIGURE 2.4. Locations and boundaries of the different physiographic provinces of Onondaga County (modified from Winkley 1989).

County can be found in "The Hydrogeology of Onondaga County" by Winkley (1989). Below is a brief summary of that information which pertains directly to Onondaga Lake.

Hyper-Lake Iroquois was a large glacial lake located in front of the ice margin; it drained eastward to an outlet at Rome. Main stage Lake Iroquois developed nearly 12,500 years ago, producing well-defined shoreline features with a low-relief lake bottom located at elevations below 410 ft (125 m; USGS datum). Lake Iroquois' extent largely coincides with the distribution of a considerable accumulation of glaciolacustrine sediments across the Ontario Lowland. Lake Iroquois deposited silt and clay over much of the previously deposited delta sand and gravels (Cosner 1984). Approximately 12,000 years ago, the level of Lake Iroquois dropped and a number of post-Iroquois lakes remained within depressions across the Ontario Lowland. As the glacial ice retreated, lakes located within the depressions shallowed as postglacial rebound continued. These isolated lakes became quite eutrophic, eventually filling in with marl and peat sediments. The Onondaga Lake basin is relatively deep; in this particular basin, clay and marl have accumulated as bottom sediments.

2.1.3 Geology of Onondaga Lake

The tributaries to Onondaga Lake all occupy glacial meltwater channels or troughs (Figure 2.5a). Onondaga Lake itself occupies a depression located in the northern terminous of the Onondaga Trough. Most of these meltwater drainage channels and glacial troughs are filled with unconsolidated interbedded sands and gravels as shown in Figure 2.5b (Winkley 1989), forming some of the most important water bearing geologic deposits. The geology of the individual major tributary valleys is discussed later in this chapter (Section 2.2). Knowledge of the geology of Onondaga Lake comes mainly from historic and recent well logs. Most of the well logs are located around the southwestern and southern portion of the lake due to the extensive urban and industrial development in those areas.

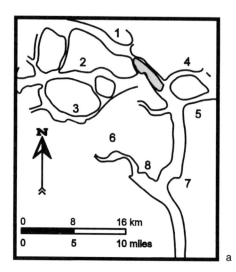

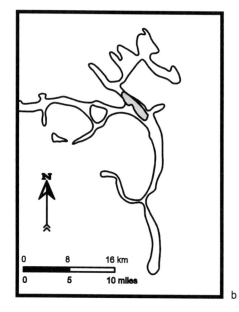

FIGURE 2.5. Selected glacial features and deposits in Onondaga County. (**a**) Location of glacial meltwater troughs and channels. (**b**) Areas where sand and gravel may underlie finer grained sediments (modified from Winkley 1989). Note the similarities in location of glacial troughs and channels and of the sand and gravel deposits. Meltwater channels: (1) Baldwinsville, (2) Jordan-Warners, (3) Elbridge-Camillus, (4) Ley Creek, (5) Erie Canal, (6) Pumpkin Hollow, Glacial Troughs (northward draining), (7) Onondaga, (8) Cedarvale.

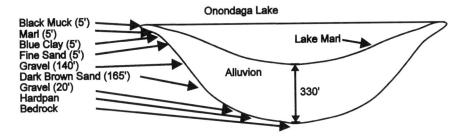

FIGURE 2.6. Drawing not to scale. Cross section of sediments underlying the south-eastern end of Onondaga Lake (modified from Clark 1849).

2.1.3.1 General Description of Unconsolidated Deposits Underlying Onondage Lake

An annotated historical account of the geology underneath Onondaga Lake is found in Clark (1849) indicating a red sandstone bedrock base overlain by a "tenacious" clay. The clay is overlain by alluvium composed of gravel, sand, chocolate-colored clay, marl, and a black swamp muck. Clark (1849) provides a cross section of the lake and the underlying deposits (Figure 2.6). A second historic account (Clayton 1878) includes a table (presented here as Table 2.1) of sediments encountered in a 414 ft (126 m) deep well drilled near the head of the lake in the area of the Barge Canal, most likely near Bear Street. Clayton (1878) states "... every boring in this basin gives this result, the only variations being in the thickness of the several strata, not their character."

TABLE 2.1. Listing of sediments and their thicknesses from a well drilled near "the head of the lake . . . intended to be in the middle of the valley."

Sediment type	Thickness (ft)	(m)
White and beach sand	34	10.4
Blue clay	100	30.5
Light-colored clay	48	14.6
Sand, coarse enough for mortar	209	63.7
Clear gravel	6	1.8
Quick sand	11	3.4
Cemented gravel	2	0.6
Red clay	3	0.9
Red clay (hard)	1	0.3

Modified from Clayton 1878.

An investigation by Thomsen Associates (1982) provides a plan view of the bedrock surface at the mouth of Ninemile Creek (Figure 2.7). This is the remnant of the glacial meltwater channel that fed into the Onondaga Trough. Two geologic cross sections through the surficial material located near the mouth of Ninemile Creek are shown in Figure 2.8.

A number of well logs and their locations are shown in Figure 2.9 (Winkley, unpublished compilation). Many areas are covered with varying depths of artificial fill, interbedded layers of marl, sands, gravels, blue, brown and reddish brown clay, and silt. It is also apparent from Figure 9 that from 6 to 20 ft (1.8 to

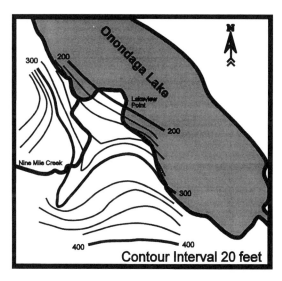

FIGURE 2.7. Elevation of bedrock surface at the mouth of Ninemile Creek (contours in feet) (modified from Thomsen Associates 1982).

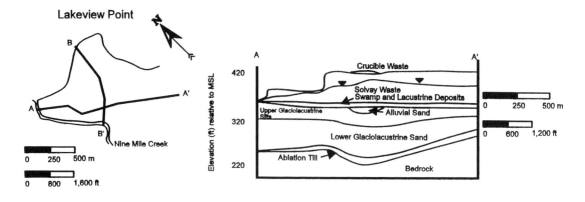

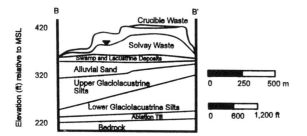

FIGURE 2.8. Two cross sections and their locations through waste beds 1 to 8 (modified from Thomsen Associates 1982).

6.1 m) of marl can be found in areas adjacent to the lake. Marl forms in open waters which are supersaturated with calcite; significant accumulations of marl have been deposited in many of the lakes in Onondaga County (Winkley 1989).

The saturated stratified glacial drift overlying the bedrock varies extensively in thickness. Near the mouth of Onondaga Creek the saturated drift is over 300 ft (91.4 m) thick. The saturated drift rapidly thins and narrows approximately one-third of the way into the lake, with the majority of the lake being underlain by approximately 100 ft (30.5 m) of saturated stratified drift (Winkley 1989). The surficial sediments surrounding the lake consist of peat and/or marl, lodgement till, and glaciolacustrine silt and clay (Winkley 1989). The peat and/or marl consists of peat, clay, and marl formed in the postglacial wetlands. This material is found along the lake perimeter from an area near Ley Creek southward past

Onondaga Creek, continuing to the outlet and then to an area near Sawmill Creek. The glaciolacustrine silt and clay is found from Sawmill Creek down to Liverpool and is frequently laminated, containing some very fine sand with traces of gravel. These deposits are formed in the open water areas of glacial lakes. The lodgement till is found near Liverpool and extends down to Ley Creek. The till is a dense mixture of boulders, cobbles, sand, and gravel-sized particles in a silt and clay matrix. These sediments are believed to have formed under the glacial ice.

2.1.3.2 General Description of Glacial Outwash Deposits Surrounding Onondaga Lake

The location and areal extent of extensive glacial outwash deposits surrounding Onondaga Lake are shown in Figure 2.10. The outwash deposits, consisting mostly of sands

FIGURE 2.9. Selected well logs and their location around Onondaga Lake (from Winkley, personal communication). NOTE: All well log depths are ft.

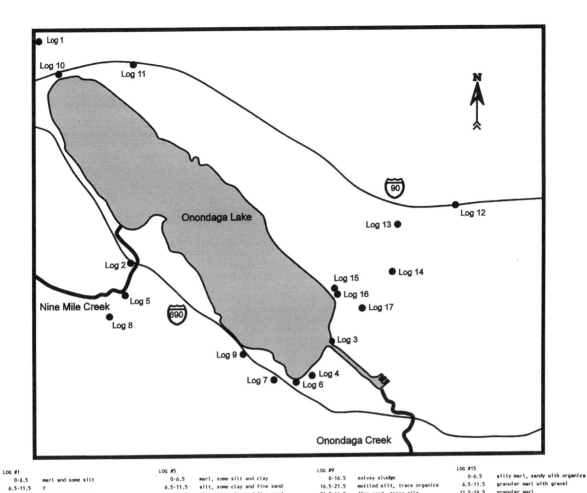

Onondaga Lake

Log 1
Log 10
Log 11
Log 12
Log 13
Log 2
Log 5
Log 8
Log 15
Log 16
Log 14
Log 17
Log 3
Log 9
Log 4
Log 7
Log 6

Nine Mile Creek
Onondaga Creek

LOG #1

0-6.5	marl and some silt
6.5-11.5	?
11.5-21.5	silt and fine sand
11.5-21.5	silt and fine sand
21.5-26.5	fine sand and some silt
26.5-31.5	fine sand and some silt
31.5-36.5	sand, trace fine gravel and silt
36.5-41.5	sand and trace silt
56.5-61.5	sand and silt, trace fine gravel
61.5-65.5	sand, some fine gravel and silt
65.5-71.5	sand, some fine gravel and silt
71.5-76.5	sand and gravel, some silt
76.5-81.5	silt, sand, gravel
81.5-100.4	gravelly glacial till
100.4-104	red shale

LOG #2

0-9	fill
9-15	silt, some clay
15-20	layered muck, silt and marl
20-23	silt and fine sand, trace gravel
23-28	fine gravel, some sand, trace gravel
28-49	fine sand, trace silt
49-51.5	silt, trace fine sand
51.5-59	layered silt, trace fine sand
59-74	fine sand, some silt
74-83	fine gravel, some sand, trace silt
83-103.5	dense gravel and silt, some sand, trace clay

LOG #3

0-6	fill
6-14	solvay refuse
14-20	fine gray sand
20-46	sandy marl and peat
46-57	light brown clay
57-102	dark blue clay
102-119	brown clay
119-123	fine sand and clay
123-153	fine sand

LOG #4

0-4	fill
4-19	solvay refuse
19-57	solvay refuse, shells, fine sand
57-89	blue clay
89-130	brown clay
130-160	clay, fine sand

LOG #5

0-6.5	marl, some silt and clay
6.5-11.5	silt, some clay and fine sand
11.5-16.5	silt, some clay and fine sand trace muck and gravel
16.5-40.5	fine sand, trace to some silt
40.5-45.5	medium sand, trace silt
45.5-55.5	fine sand, trace silt
55.5-60.5	fine to medium sand, trace silt
60.5-65.5	coarse sand, some gravel and silt
65.5-73	coarse to medium sand, trace gravel and silt
73-76	shale and rock boulder

LOG #6

0-9	fill
9-32	solvay refuse
32-45	fine gray sand
45-68	soft blue clay
68-98	reddish brown clay
98-121	red clay and boulders

LOG #7

0-6	fill
6-11	gray marl
11-16	mottled silt and clay
16-21	silt and clay with thin sand lenses
21-26	silt, some fine sand, trace clay
26-28	cemented sand, trace gravel and silt
28-34	weathered shale
34-44	shale

LOG #8

0-6.5	silt, clay, organics
6.5-11.5	silt and clay
11.5-16.5	silt, clay, trace fine sand
16.5-41.5	silt and fine sand
41.5-51.5	till
51.5-60	weathered shale
60-70	shale

LOG #9

0-16.5	solvay sludge
16.5-21.5	mottled silt, trace organics
21.5-46.5	fine sand, trace silt
46.5-60.5	fine sand, trace silt
60.5-70	sand, some silt, trace clay and gravel
70-73.5	gravel, some sand and silt
73.5-81.5	gravel, sand, silt
81.5-86.5	weathered shale
86.5-91.5	shale and sandstone

LOG #10

0-2	fill
2-30	marl
30-43	silt, some clay
43-52	clay and silt
52-73	silt, some fine sand
73-75	gravel, some sand and silt

LOG #11

0-5	brown silt
5-10	gray silt
10-16	light red silt
16-32	soft weathered gray shale
32-46	hard gray shale

LOG #12

0-6	silt, some sand and gravel
6-35	soft shale and silt, some clay
35-55	soft red shale
55-60	green hard shale with gypsum

LOG #13

0-7.5	black muck
7.5-13	gray brown clay
13-34	yellow gray clay
34-42	yellow gray clay
42-108	gray silty clay
108-125	gray sand

LOG #14

0-26	fill
26-33.5	black muck, blue clay
33.5-77	clay and fine sand
77-90.5	gray fine sand
90.5	hardpan

LOG #15

0-6.5	silty marl, sandy with organics
6.5-11.5	granular marl with gravel
11.5-16.5	granular marl
16.5-21.5	clayey silt with wood pieces
21.5-26.5	clayey silt
26.5-31.5	silty clay
31.5-36.5	gray silt
36.5-46.5	fine sandy silt
46.5-56.5	gravelly sand, silty
56.5-65.5	green decomposed shale
65.5-80	red and green shale

LOG #16

	very similar to #15

LOG #17

0-7.5	fill
7.5-18.5	shells, solvay refuse, fine sand
18.5-75	fine sand
75-103	coarse sand and gravel
103-112	bedrock (refusal)

KEY:

"some" = 20-35% by weight
"little" = 10-20% by weight
"trace" = 1-10% by weight

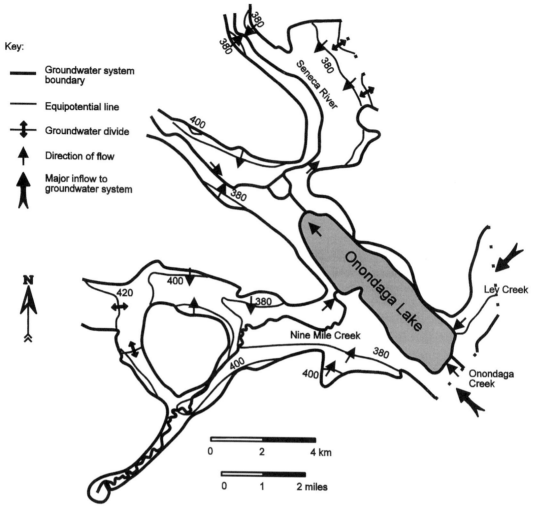

FIGURE 2.10. Potentiometric surface and direction of flow of ground water in the Baldwinsville aquifer (modified from Cosner 1984).

and gravels (Figure 2.5b), underlie the major tributaries to Onondaga Lake and Onondaga Lake itself. A definitive statement of how easily the more permeable sediments may be contaminated, or the rate of potential contaminant transport, cannot be made due to a lack of information regarding the extent to which the patterns of permeable deposits are interconnected and integrated into an aquifer system as a whole. The description that follows is a general characterization based on the limited amount of available information.

Large areas of artificial fill, made land, and urban land are adjacent to most of Onondaga

Lake. Generally, the thickest water-bearing deposits occur in places along the Seneca River and in the meltwater channels leading to Onondaga Lake (Cosner 1984; Figure 2.5). The groundwater gradient is very flat in most of the areas near the lake as shown in Figure 2.10 (Cosner 1984; Pagano et al. 1986; Winkley 1989); however, this map was based on widely scattered well logs and is incapable of showing the specific details of groundwater flow around the lake. In general, water flows directly into the outwash deposits from the uplands by both surface and groundwater flow; in addition, direct rainfall and stream bank storage during high stream flow can also recharge the groundwater system. Groundwater leaves the system mainly by discharge into the overlying surface waters.

2.2 Geology and Hydrogeology of the Major Tributaries to Onondaga Lake

This section presents a summary review of the geology and hydrogeology of the three major surface water tributaries to Onondaga Lake: Onondaga Creek, Ley Creek, and Ninemile Creek. Onondaga Lake is the major discharge point for the entire watershed. As such, the hydrology and chemistry of the lake basin is greatly influenced by its tributary contact points and water inputs. Some information on specific hydrogeologic aspects of both Onondaga Creek and Ninemile Creek exists in such quantities that it will be covered briefly here, and subsequently discussed in detail in Section 2.3.

2.2.1 Onondaga Creek

2.2.1.1 Geology of Selected Areas Along Onondaga Creek

The Onondaga glacial trough (Figure 2.5), the deepest trough in Onondaga County (Winkley 1989), is a narrow, steep-sided bedrock valley (Tanner 1991). The Onondaga Trough is composed of two valleys, the Tully Valley and the Onondaga Valley, and extends from the Valley Heads Moraine northward to Syracuse, with Onondaga Lake occupying its northern terminus. The Late Wisconsin glaciation period is believed to be largely responsible for the U-shaped trough and most of the unconsolidated sediments in the valley (von Engeln 1961; Getchell 1983). Steep valley walls with gradients of 5 to 18° are characteristic of the Tully Valley's catenary transverse profile; the sides steepen with depth before leveling out at the bedrock bottom (Getchell 1983). The unconsolidated valley sediments are underlain

by Upper and Middle Devonian bedrock in the Tully Valley, and Upper Silurian sedimentary beds in the Onondaga Valley (Kantrowitz 1970). A stratigraphic section of the Onondaga Creek watershed (Johnston et al. 1975) is shown in Figure 2.11 and a three-dimensional cross section of the Onondaga Trough (Tanner 1991) is presented in Figure 2.12. The valley floor is a relatively flat, proglacial lake plain which slopes gently northward (Getchell 1983). Bedrock sedimentary structures are relatively flat lying with a strike of N65°W and dip of 1°SW (Getchell 1983).

The bedrock surface of the Onondaga Trough is covered with an assemblage of boulders, cobbles, gravel, and sand (Higgins 1955; Tanner 1991). This layer is overlain by a thick sequence of glacial drift, transported from northern sources (Winkley 1989), that can reach thicknesses of 122 to 153 m (400 to 502 ft) (Getchell 1983). The base of the unconsolidated deposits consists of the previously mentioned mixed drift with an approximate thickness of 72 m (236 ft), overlain by stratified fluvio-glacial sands and gravels that grade upwards into sand beds. The fluvio-glacial deposits range in thickness from 6 to 17 m (20 to 56 ft). Glaciolacustrine laminated clays and silts ranging in thickness from 25 to 31 m (82 to 102 ft) overlie the fluvio-glacial deposits. The uppermost material is the recent alluvium consisting mainly of fluvially reworked glacial material from the valley walls and floor with sediments ranging in size from clays to boulders. The thickness of the alluvium can vary from a thin veneer of a few centimeters to accumulations >30 m (>98 ft). Lastly, a highly compacted till with a high clay content is the major deposit found along the valley walls. The till can vary in thickness from <1 to >30 m (Higgins 1955).

2.2.1.2 Hydrogeology of Selected Areas Along Onondaga Creek

Onondaga Creek and many smaller tributaries drain the Valley Heads moraine and valley sides and flow northward, following the hydraulic gradient, eventually discharging to

PERIOD	GROUP	FORMATION	THICKNESS (feet)
Upper Devonian	Genesee	Genesee Shales	75
	–	Tully Limestone	23
Middle Devonian	Hamilton	Moscow Shales	180
		Ludlowville Shale	350
		Skaneateles Shale	335
		Marcellus Shale	290
	–	Onondaga Limestone	65-70
Lower Devonian	Helderberg	Oriskany Sandstone	12
		Coeymans Limestone	40-50
		Manlius Limestone	12
		Roundout waterlime	30-40
Upper Silurian	Salina	Cobleskill Dolostone	6
		Bartie waterlime	6-10
		Camillus Shale	160-190
		Syracuse Formation	200
		Vernon Shale	500-600

FIGURE 2.11. A stratigraphic cross section of the Onondaga Creek watershed (modified from Johnson et al. 1975).

Onondaga Lake. In the Tully Valley, where the slope of the valley sides is much greater than the valley floor longitudinal slope, lateral contributions from the local flow system take on greater importance. The beds of the streams draining the valley sides consist either of bedrock or a thin vaneer of till overlying bedrock, whereas the beds of all the streams occupying the valley floor consist of uncon-solidated material (Getchell 1983). Onondaga Creek, downstream of the Cedarvale Valley confluence, receives influent flow primarily from shallow unconsolidated aquifers which are affected by the local flow system and high rates of runoff (Kantrowitz 1964; Higgins 1955). In the Tully Valley, Onondaga Creek is underlain by relatively impermeable, finely bedded pink lacustrine clays and is unlikely

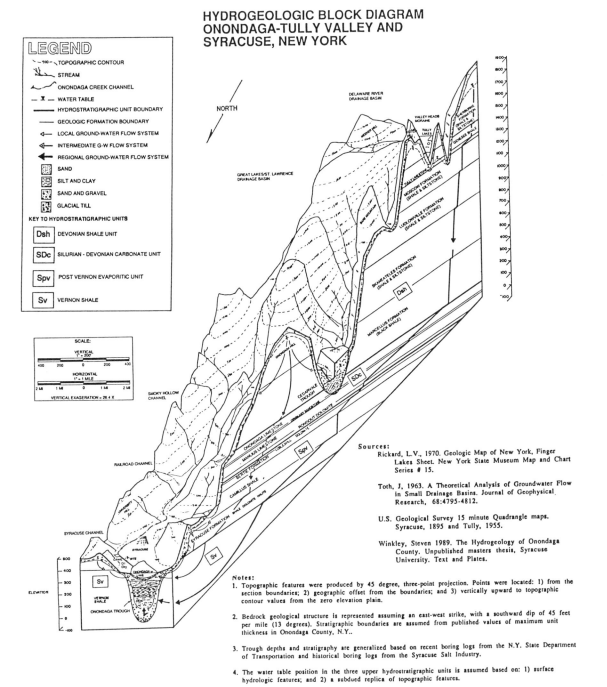

FIGURE 2.12. A three-dimensional view of the stratigraphy of the Onondaga Creek watershed from Tully northward to Onondaga Lake (from Tanner 1991).

to have good hydrologic communication with the groundwater system (NYSDEC 1991). The surface water chemistry of the tributaries to Onondaga Creek generally reflects the mineralogy of the stream beds and locally influent shallow aquifers (Getchell 1983).

Saline waters have been found in the past in the sand and gravel-bearing formations within the Onondaga Valley deposits in the vicinity of Syracuse (Higgins 1955). Tanner (1991) recently found saline water in the Onondaga Valley deposits at depths <30 ft (<9 m) within 3000 ft (914 m) of the southern shore of Onondaga Lake. Salt beds of commercial size (Salina Group) are located at depths averaging 1200 ft (366 m) south of the natural solutioning front, which is located south of Route 20 (Kantrowitz 1970). The saline character of the groundwater in the Syracuse area is due to dissolution of the Salina Group bedrock salt deposits during the ground waters migration northward from the higher elevations of the Tully Valley. Based on the relatively rapid flushing rate of valley sand and gravel deposits and the low probability of the presence of evaporite deposits in the valley fill, Noble (1990) concluded that the occurrence of saline groundwaters in the unconsolidated valley sediments is more dependent on hydrologic communication between the bedrock and the valley fill deposits (flow regime) rather than on the valley fill (aquifer) mineralogy.

The water table in the Tully Valley is a subdued replica of the topography (Koch 1932; Getchell 1983). Along the valley floor, in a north–south direction, the water table slopes less than the surface gradient, whereas along lines parallel to the slopes of the valley sides, the water table has a steeper gradient (Higgins 1955). A piezometric surface resulting from high-pressure head conditions within confined aquifers of the unconsolidated and bedrock aquifer systems appears to exist throughout the valley (Getchell 1983). The piezometric surface, at some locations, extends above the ground surface and indicates flowing artesian conditions in the Tully Valley (Getchell 1983). Kantrowitz (1970) and Phillips (1955) both refer to the saline nature

of some portions of the Onondaga Creek drainage basin sand and gravel aquifer as evidence for artesian conditions existing in the limestone and middle shale bedrock units (Figure 2.2). This groundwater flow system requires water movement updip of the southwest dipping bedrock strata. This updip movement is supported by Herak and String-field's (1972) observations which indicate that under artesian conditions, groundwater movement in jointed and fractured rock is in accordance with the hydraulic gradient between recharge and discharge areas of the regional flow system, independent of dip.

Prior to the 1940s, Onondaga Creek was insulated from the water table by low-permeability lacustrine clays. The construction of the Onondaga Creek Flood Control Dam (located on the Onondaga Indian Reservation) in the late 1940s required the deepening, widening, and rerouting of the Creek, resulting in the breaking of this clay insulation. The net result was additional amounts of freshwater leaking from the creek, into the subsurface, and subsequently being added to the water table. This resulted in a dilution of the saline water reservoir (Higgins 1955). The saline brine in the Onondaga Trough presently has a total dissolved solids (TDS) concentration of greater than $10,000 \, mg \cdot L^{-1}$ (Winkley 1989). The movement of the saline water is toward the valley center and then northward. The valley floor, with its deep deposits of lacustrine and fluvial sediments acts as a channel for groundwater flow (Higgins 1955). In the Syracuse area, channel deposits act as a reservoir for the northward moving groundwater. The reservoir occupies an area underlying most of the city and extending northward under Onondaga Lake. Higgins (1955) described the saline content of Onondaga Lake as less than would be expected if it were in direct contact with the underlying saline reservoir, and attributed this to the presence of an insulating marl bed under the lake.

2.2.1.2.1 Source of Sediment Loading to Onondaga Creek

Slightly more than 50% of the annual external sediment loading to Onondaga Lake has been

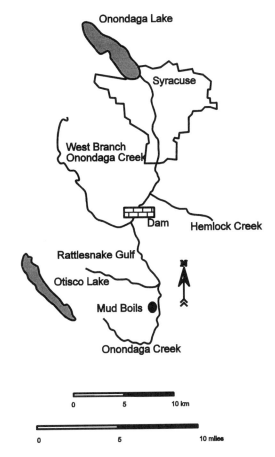

FIGURE 2.13. Location of the Tully Valley mud boil "field" in relation to Onondaga Lake (modified from Effler et al. 1992).

attributed to inputs from Onondaga Creek (Yin and Johnson 1984). In 1987, measurements of suspended solids in Onondaga Creek after several storm events document that greater than 50% of the total suspended solids in Onondaga Creek can originate from the Tully Valley (Figure 2.13) from "mud boils" (Effler et al. 1992). The occurrence and mechanics of mud boil formation are discussed in detail in Section 2.3.3. The Tully Valley mud boils produce a mauve, tannish-colored turbid streamflow that subsequently enters Onondaga Creek and completely mixes with the Onondaga Creek streamflow within about 1.2 km. The color is the result of the high concentrations of Otisco Valley clay found in suspension (Getchell 1983). The mud boil turbidity can be clearly observed in streamflow

for several kilometers downstream under low-flow conditions, and down to the mouth of Onondaga Creek during runoff events. Loading and concentration profiles from above and below the mud boils, collected over a range of flow conditions, indicate that most of the sediment load received at the Creek mouth during runoff events is resuspended sediment and eroded bank material originating from the Tully Valley mud boils (Effler et al. 1992). The impacts of the sediment load in the creek are covered in detail in Chapter 3.

2.2.1.3 Oil City

"Oil City" is an industrial and commercial complex located at the southeastern end of Onondaga Lake around the mouth of Onondaga Creek. Oil City has been used as a bulk storage and transfer facility for numerous industries since 1926. Industrial compounds utilized and stored in the area included the bulk storage of fuel related hydrocarbons, limited location and storage of synthetic organic chemicals, and polychlorinated biphenyls (PCBs) (NYSDEC 1989). Recent efforts to reclaim and revitalize the Oil City area have resulted in several voluntary groundwater investigations by companies located in Oil City. Investigations concentrate on defining the degree of existing contamination by the various storage facilities. The following is a review of that information.

2.2.1.3.1 Geology of the Oil City Area

The sediments underlying the Oil City area fit into six general categories: 10 to 25 ft (33 to 82 m) of fill (usually a sand/gravel mixture with varying amounts of silt, brick, cinder, wood pieces, and plastic); 8 to 20 ft (2.4 to 6.1 m) of Solvay waste; 6 to 35 ft (1.8 to 11 m) of gray sand; 0 to 25 ft (0 to 7.6 m) of peat and marl with varying amounts of sand, silt, and shells; 65 to 73 ft (20 to 22 m) of blue, red, brown, or black clay; and 0 to 25 ft (0 to 8 m) of coarse sands and gravels (Dunn Geoscience Corp. 1988a,b,c; Empire Soils Investigations, Inc. 1990; Blasland and Bouck Engineers 1990; Tanner 1991). Although a very generalized stratigraphic section for the entire area

TABLE 2.2. Summary of information relating to industrial groundwater contamination around the southern portion of Onondaga Lake.

Name/location	# Monitoring/ test borings/ piezometers	Type of conductivity test	Range in values for conductivity	Major contaminant/ $\bar{x}$ conc.	Primary discharge point	Source of information
[†]McKesson Corp. Bear street	22/36/13	Slug	$10^{-5}–10^{-4}$ cm·s^{-1} 0.03–0.3 ft·d^{-1}	hydrocarbons, methanol methylene chloride n,n-dimethylaniline	Barge Canal Onondaga Lake	Blasland and Bouck Engineers 1990
[†]Mobil Oil Bear/Solar Street	17/12/#	Slug	$10^{-5}–10^{-3}$ cm·s^{-1} 0.03–3 ft·d^{-1}	fuel oil, gasoline kerosene	Barge Canal	Empire Soils Investigations, Inc. 1990
[†]Marley Property W. Hiawatha Boulevard	25/#/#	Slug	$10^{-4}–10^{-2}$ cm·s^{-1} 0.3–108 ft·d^{-1}	fuel related compounds TPH* = 3–30 mg·L^{-1}	Barge Canal	Dunn Geoscience Corp. 1988a
[†]Clark Property W. Hiawatha Boulevard	10/13/#	Slug	$10^{-4}–10^{-2}$ cm·s^{-1} 0.3–15 ft·d^{-1}	lubricating oil TPH = 0–15 mg·L^{-1} >200 ppm chlorinated/ halogenated hydrocarbons	Barge Canal	Dunn Geoscience Corp. 1988b
[†]Buckeye Property W. Hiawatha Boulevard	7/#/#	Slug	10^{-2} cm·s^{-1} 15 ft·d^{-1}	lubricating oil TPH = 0–4.4 mg·L^{-1}	Barge Canal	Dunn Geoscience Corp. 1988c
[††]Regional Market Park Street	7/24/#			refuse, CH$_4$, H$_2$S, TPH = ~1 mg·L^{-1} PCBs = <3 µg·L^{-1}	Onondaga Lake Ley Creek (?)	Calocerinos & Spina Engineers 1988
[††]Crouse-Hinds Landfill 7th North Street	11/#/# (North landfill)		$10^{-6}–10^{-3}$ cm·s^{-1} 0.02–7.7 ft·d^{-1}	phenol, benzene, toluene xylene, cyanides	Ley Creek	Engineering Science, Inc. 1983
[††]GM-Fisher Guide Factory Avenue	14/23/#		$10^{-6}–10^{-3}$ cm·s^{-1} 0.003–3 ft·d^{-1}	PCBs = 0–180 ppm	Ley Creek Load ~0.15 g·d^{-1}	O'Brien & Gere 1989a
[†††]Allied Chemical Willis Avenue	26/#/#	Pump (shallow aquifer)	$10^{-6}–10^{-4}$ cm·s^{-1} 0.003–0.3 ft·d^{-1}	halogenated hydrocarbons chlorinated hydrocarbons	Onondaga Lake Load .46 kg·d^{-1}	Groundwater Technology 1984b

* TPH = Total Petroleum Hydrocarbons.
[†] Oil City.
[††] Ley Creek.
[†††] Allied.

can be drawn (as the preceding description indicates), not all areas contain the same layer sequence or the same thickness of layers. At the present time the complex spatial and vertical distribution of the sediments for the area cannot be given in detail.

2.2.1.3.2 Hydrogeology of the Oil City Area

Because of the long history and complex nature of the industrial uses in Oil City, along with the historical anthropogenic changes and effects on the surficial sediment stratigraphy, an extremely complex hydrogeologic and geochemical situation has evolved. Strong spatial heterogeneities in contaminant concentrations (Blasland and Bouck Engineers 1990; Empire Soils Investigations, Inc. 1990; Dunn Geoscience Corp. 1988a,b,c) are probably the result of various industries utilizing a number of locations for the storage/transfer of different chemical products. The assessment

of potential contaminant transport pathways is a difficult problem in this area due to the highly variable surficial geology. Contaminant transport can occur through permeable sediments in the old Onondaga Creek streambed channels (the mouth of the creek has been moved several times (Holmes 1927; Higgins 1955)), along buried sewer and utility pipelines (Blasland and Bouck Engineers 1990), and through abandoned and buried drainage channels from the salt production industry.

A generalized summary of available information on contaminated areas around the southern end of Onondaga Lake is presented in Table 2.2, along with information regarding the measured horizontal hydraulic conductivities and primary discharge points for groundwater. In the area of Oil City groundwater moves at a rate of 10^{-5} to $10^2 \, \mathrm{cm \cdot s^{-1}}$ (0.003 to $>100 \, \mathrm{ft \cdot d^{-1}}$) and discharges primarily to the Barge Canal and Onondaga Lake (Figure 2.14). Table 2.3 provides a partial listing of

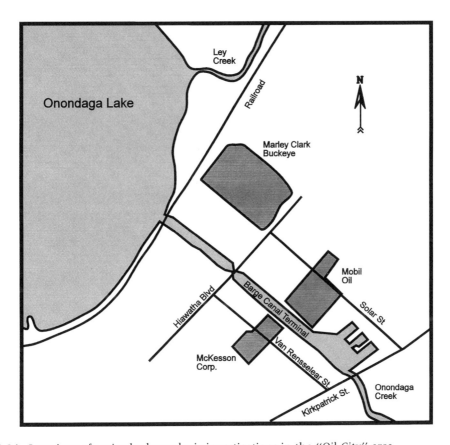

FIGURE 2.14. Locations of major hydrogeologic investigations in the "Oil City" area.

TABLE 2.3. Partial listing of known organic pollutants in the Oil City area and their potential mobility class.

Compound	Mobility class*
Acetone	very high
Aniline	very high
Chloromethane	very high
Methylene chloride	very high
Phenol	very high
1,1 Dichlorothane	very high
Trans-1,2 Dichloroethene	very high
Benzene	high
Benzoic acid	high
Chorobenzene	moderate
Toluene	moderate
Trichloroethene	moderate
1,1,1 Trichloroethane	moderate
1,2 Dichlorobenzene	moderate
2-Butanone	moderate
Ethylbenzene	low
Naphthalene	low
2-Methylnaphthalene	slight

* Mobility rates are based on a low permeability glacial till with an organic carbon content of approximately 0.5% (modified from Fetter 1988).

organic pollutants found in groundwater samples taken from Oil City, along with an assigned potential mobility class based on flow through a low permeable glacial till. It should be kept in mind that the types and concentrations of contaminants are often site specific rather than dispersed throughout the entire area. The potential mobility rates given in Table 2.3 are examples of the potential mobility rates of contaminants under the general geologic composition of glacial till; the actual mobility rates of the contaminants will depend on the specific chemical, concentration, the sediment composition of the pathway, and whether the pathway includes sewer and utility pipelines.

Combined, Tables 2.2 and 2.3 indicate the potential for moderate-to-rapid transport of fuel-related compounds and synthetic organic chemicals to adjacent surface waters. At one site, initial chemical analysis confirmed the migration of hydrocarbon contamination toward the Barge Canal (Empire Soils Investigations, Inc. 1990). Additional investigation resulted in observations of a "sheen" on the

water seeping into five test pits dug near the Barge Canal. Analytical results indicated groundwater samples collected within 60 ft of the Barge Canal had total petroleum hydrocarbon (TPH) concentrations of $>200{,}000\,\mu g \cdot L^{-1}$ (one sample had a TPH concentration of $2{,}300{,}000\,\mu g \cdot L^{-1}$).

Based on the information available, it is reasonable to conclude that the Oil City area is a potential source of chemical contamination to Onondaga Lake via groundwater transport. Although it presently cannot be determined what the exact pathways of transport are, whether patterns of contaminants exist, and at what rates the contaminants move, there is enough evidence to warrant a more intensive compilation of the data and to continue further investigations of this area.

2.2.2 Ley Creek

Located within Ley Creek's drainage basin are a number of large industrial and commercial complexes and several landfills. Available information includes descriptions of geologic conditions, types of contaminants, and directions of groundwater flow. Hydrogeologic assessments of several locations indicate a range of contamination that includes PCBs, benzene, toluene, xylenes, phenols, municipal refuse, and gas (CH_4 and H_2S) accumulations in and around landfill sites (Engineering-Science, Inc. 1983; Calocerinos and Spina Engineers 1988; O'Brien and Gere Engineers 1989a). The following is a summary of several hydrologic investigations undertaken along Ley Creek.

2.2.2.1 Geology of Selected Areas Along Ley Creek

Subsurface investigations at several sites along Ley Creek (Figure 2.15a) reveal a varied subsurface stratigraphy (O'Brien and Gere Engineers 1989a; Calocerinos and Spina Engineers 1988; Engineering Science, Inc. 1983). Approximately 3.8 mi (6 km) upstream from Onondaga Lake, driller's well logs show up to 35 ft (11 m) of dredged fill material (fine grain fluvial and lacustrine deposits, silts and

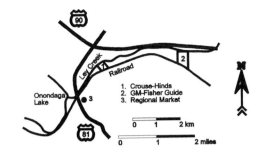

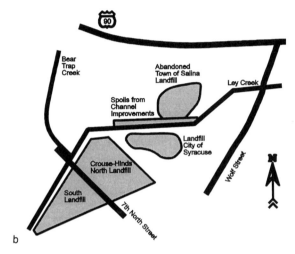

FIGURE 2.15. (a) Locations of major hydrogeologic investigations in the Ley Creek watershed. (b) Locations of known landfills in the Ley Creek watershed that are near Onondaga Lake (modified from Engineering Science, Inc. 1983).

rubble), underlain by 12 ft (4 m) of a gray, wet, soft marl with some sand and peat. At a depth of approximately 20 ft (6 m), 5 ft (1.5 m) of marl grading to silt occurrs and is underlain by over 60 ft (19 m) of silt grading to sand and sand grading to fine gravel (Calorcerinos and Spina Engineers 1988). It is probable that the bedrock underlying the sand and gravel is Vernon shale; it likely occurs at a depth of approximately 100 ft (30 m).

2.2.2.2 Hydrogeology of Selected Areas Along Ley Creek

Much of the land adjacent to Ley Creek from the south branch headwaters downstream to its outlet was originally marsh that was subsequently reclaimed for landfill areas and industrial parks. Reclamation was usually accomplished by filling in the wetland areas with municipal and industrial solid wastes, along with common construction debris. Table 2.2 (refer to Section 2.2.1.3.2) includes information regarding hydraulic conductivities and major contaminants cited from the various hydrogeologic assessments along Ley Creek. The following material is a summary of the available information regarding the locations of contaminated landfills, and one area known to be contaminated with PCBs.

General Motors-Inland Fisher Guide Division is located along Ley Creek and Factory Ave (Figure 2.15a). Soils on the site, along with portions of Ley Creek, are known to be contaminated with PCBs as a result of the disposal and dredging of materials used in hydraulic die casting operations at the Inland Fisher Guide Facility (IFG) (O'Brien and Gere Engineers 1989a). Table 2.4 summarizes the degree of PCB contamination at the site. The mean concentration of PCB's in the ground water at the IFG site is $3.62\,\mu g \cdot L^{-1}$. Mass transport of PCB's from the ground water into Ley Creek is estimated to be $0.15\,gm \cdot d^{-1}$.

Landfills located near the outlet of Ley Creek are listed in Table 2.5 and their locations are shown in Figure 2.15b. The most complete set of information relates to the

clays) underlain by a dense red-brown clay containing silt, sand and embedded gravel fragments (O'Brien and Gere Engineers 1989a). Approximately 1.1 mi (1.8 km) upstream of Onondaga Lake, the stratigraphic column shows 15 ft (5 m) of fill material (foundry sands, municipal, wastes, and wood) underlain by 5 ft (1.5 m) of a high-conductivity peat (2.7×10^{-3} cm $\cdot$ s^{-1}; 7.7 ft $\cdot$ d^{-1}) and 30 ft (9 m) of sand and silt. At the base of the sand and silt unit is approximately 50 ft (15 m) of sand and gravel underlain by bedrock (Vernon shale) at an approximate depth of 100 ft (31 m) (Engineering Science, Inc. 1983). Approximately 2,000 ft (0.6 km) northeast of Onondaga Lake, the surficial materials consist of 8 ft (2 m) of fill (cinders, silt, fine-to-coarse sand, fine-to-medium gravel, wood, and

TABLE 2.4. Summary of suspected PCB contamination at the IFG site.

Sample type	Concentration of PCBs
On-site soils	ND*-180 ppm
Ley Creek sediments	ND-8.3 mg·kg^{-1}
Groundwater samples[†]	ND-21 µg·L^{-1}
Stream water samples	ND-1.4 µg·L^{-1}
Air samples	<0.001 mg·m^{-3}

Modified from O'Brien and Gere Engineers 1989a.
* ND indicates that a value was below the detection limit.
[†] NYS Class GA groundwater standard for PCBs = 0.01 µg·L^{-1}.

Crouse-Hinds Landfill. Two aquifers are located beneath the Crouse-Hinds Landfill; a shallow aquifer composed of peat and located within and directly beneath the fill material at depths of 4 to 8 ft (1.2 to 2.4 m), and a lower aquifer composed of deep sands and gravels (Engineering Science, Inc. 1983; Thomsen Associates 1983). The two aquifers are separated by 12 to 32 ft (3.7 to 9.8 m) of silt and clay (Thomsen Associates 1983). The water table is 4.5 to 6 ft (1.4 to 1.8 m) below the land surface and the surface of the shallow aquifer is at the same depth (Engineering-Science, Inc. 1983). Groundwater flows toward Ley Creek, located 581 ft (170 m) west of the site, at an average rate of 6 ft · yr^{-1}.

A summary of the groundwater chemistry from the Crouse-Hinds north landfill is given in Table 2.6. High TDS concentrations in the leachate (due primarily to calcium, sodium, and potassium) have been attributed to municipal wastes (Engineering Science, Inc. 1983). Due to the shallow screening of the monitoring wells, much of the shallow aquifer was not sampled and groundwater contamination may be worse than initially suspected (Table 2.6; Engineering Science, Inc. 1983). It is also unknown whether the lower sand and gravel aquifer is hydrologically connected to Ley Creek and whether this lower aquifer has been contaminated (Engineering Science, Inc. 1983).

2.2.3 Ninemile Creek

Originating from Otisco Lake, Ninemile Creek flows north over the Appalachian Uplands and down the Helderberg Escarpment near Marcellus. It then flows onto the lake plain, finally entering Onondaga Lake near Lakeland. Ninemile Creek presently contributes over 50% of the surface water chloride load to Onondaga Lake (Auer et al. 1991). Surface and groundwaters north of Warners-Amboy have high ionic concentrations (Kantrowitz 1970; Winkley 1989; Noble 1990). An im-

TABLE 2.5. Landfills located near the outlet of Ley Creek.

Landfill/location	Size	Dates of operation	Type of contaminant	Source
Crouse-Hinds Landfill 7th North Street	North: 22 acres	1950–?	Primarily inorganic materials Also fly-ash, zinc hydroxide, and foundry sands, phenols, benzenes and toluene	1
	South: 21 acres	1960–1969	Unknown quantities of inorganic and organic materials	
Regional Market Park Street		1946–present	Up to 16 ft of miscellaneous fill, trash, CH_4, H_2S PCBs: <3µg·L^{-1}	2
Salina Landfill Wolf Street/Thruway	100 acres		Municipal waste with some PCBs: concentrations range from nondetectable to 270 ppm Also semivolatile and volatile organics and heavy metals	3
Syracuse Landfill Wolf Street		1960–1964	Unknown quantities of municipal wastes construction demolition materials and garbage	1

1. Engineering Science, Inc. 1983; 2. Calocerinos & Spina Engineers 1988; 3. NYSDEC, personal communication.

TABLE 2.6. Summary of ground water chemical analyses from the north Crouse-Hinds landfill.

Constituent	x̄	Range	$n^\dagger$	Units
Conductance	3220	520–8100	22	μmhos·cm^{-1}
Phenol	0.058	<0.01–0.262	22	mg·L^{-1}
Iron	3.7	<0.01–27	22	mg·L^{-1}
Manganese	0.11	<0.01–0.38	22	mg·L^{-1}
Cyanide	—	<0.004	11	mg·L^{-1}
Oil and Grease	5.7	1.5–21.9	7	mg·L^{-1}
Benzene	29	<1.0–220	18	μg·L^{-1}
Toluene	8.1	<1.0–33	18	μg·L^{-1}
Xylene	36	<1.0–270	18	μg·L^{-1}
Total BTX*	70	<1.0–282	18	μg·L^{-1}

Modified from Engineering Science, Inc. 1983.
$^\dagger n$ = number of samples.
* Sum of Benzene, Toluene, Xylene components.

portant question that needs to be addressed in the Ninemile Creek valley is whether the source of the high ionic concentrations found in the ground- and surfacewaters originate from naturally occurring brines or leachate from the extensive (8.2 km^2) waste beds of the chlor-alkali facility located at the mouth of the Ninemile Creek Valley.

Detailed hydrogeologic studies of the Ninemile Creek Valley are limited. The most recent and extensive study conducted by Blasland and Bouck Engineers (1989) was designed to determine possible impacts of the waste beds on Onondaga Lake and propose remediation strategies. Local groundwater flow, hydraulic conductivities, hydraulic head gradients, infiltration rates, surface and groundwater chemistry were determined. Blasland and Bouck Engineers (1989) conclude there is minimal hydraulic communication between groundwater in the valley fill sediments and infiltrated water in the waste beds. They assumed most water infiltrating through the waste beds flows out at seepage faces along the base of the waste beds and becomes surface runoff, ultimately flowing into Ninemile Creek, and a negligible amount of water flows through the waste beds to the surficial deposits. The study further concludes that 83% of the total chloride loading to Ninemile Creek in 1987 originated from saline groundwater discharges; of this, greater than

50% of the chloride load is due to Allieds past operation. Chloride partitioning in Ninemile Creek was determined using hydrologic and chloride budgets.

Geologic and water chemistry data is presented in this section and Section 2.3.4 to demonstrate that at least 90% of the ion load carried by Ninemile Creek has a chemical signature that suggests it is attributable to leachate from the wast beds rather than natural sources. Furthermore, unlike the Onondaga Valley, the Ninemile Creek Valley is not in direct contact with the salt bearing Syracuse Formation until north of Warners-Amboy and therefore should receive a minimal amount of chloride loading from natural sources.

2.2.3.1 Geology of Selected Areas of Ninemile Creek

The following descriptive information about the geology of the Ninemile Creek valley was taken from Muller and Cadwell (1986) and Rickard and Fisher (1970). The upper reaches of Ninemile Creek cut through calcarious lacustrine (lake deposits) clay and silt. Ninemile Creek passes through outwash sands and gravels and alluvial fan deposits south of Marcellus. Alluvial fan deposits are sands, gravels, clay, and silt that are deposited by rivers in channels or flood plains. In the vicinity of Marcellus, Ninemile Creek crosses the fall line, dropping off the Appalachian Plateau and into the Ontario Lowlands. A fall line is an imaginary line that denotes a sudden drop in elevation. At the fall line, Ninemile Creek cuts a 0.4 km (0.25 mi)-wide valley (Blasland and Bouck Engineers 1989) through exposed bedrock (Onondaga Limestone and Helderberg group) on the Helderberg Escarpment. On the Ontario Lowlands, Ninemile Creek cuts through sands and gravels, and finally through lacustrine clays and silts. The valley eventually widens to 0.8 km (0.5 mi), about 1.2 km (4,000 ft) downstream from Camillus (Blasland and Bouck Engineers 1989). The bedrock (Camillus and Syracuse Formation) near Camillus is about 15 m (50 ft) below valley fill (Blasland and Bouck Engineers 1989). Near Onondaga Lake, the Ninemile Creek valley is nearly 0.6 km (1 mi) wide and depth to bedrock is about 38 m (125 ft) (Blasland and

Bouck Engineers 1989). The bedrock is a light-to-dark gray-fractured shale (Syracuse Formation), possibly becoming more fractured with depth (Blasland and Bouck Engineers 1989). Joints in the bedrock contain gypsum. At some sites the shale is overlain with a fractured limestone (Cobleskill Limestone). Fractures in the limestone are filled with calcite ($CaCO_3$) (Blasland and Bouck Engineers 1989). The bedrock is covered by a discontinuous layer of till. The till thins out and disappears completely along the flanks of bedrock slopes (Kantrowitz 1970; Blasland and Bouck Engineers 1989) (Figure 2.16). Along the axis of the valley and in bedrock depressions the till thickens. The thicker till retards groundwater flow from bedrock aquifers to the surficial deposits. Well logs (Blasland and Bouck Engineers 1989) in the vicinity of the waste beds show that surficial deposits consist of interbedded silt, clays, sand, and gravels with very little lateral continuity

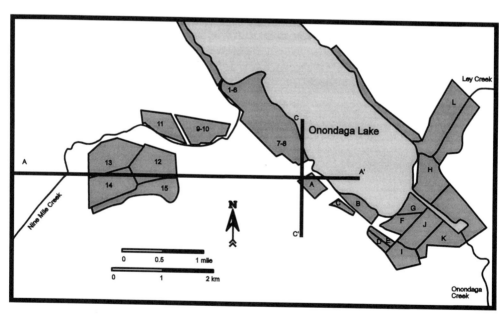

a

FIGURE 2.16. (a) Map of the Allied waste beds area showing the locations of cross sections A and B. (b) Cross section A, perpendicular to the axis of the Ninemile Creek Valley. Shows the thickness of the till and bedrock elevations. Data inferred from well logs. (c) Cross section B, parallel to the axis of the Ninemile Creek Valley. Shows the thinning of the till on the bedrock flanks going into the Onondaga Trough. Data inferred from well logs (modified from Blasland and Bouck Engineers 1989).

Ninemile Creek has been redirected and channelized as it flows through the Allied Waste beds near Onondaga Lake (Figure 2.17). The waste beds (refer to Chapter 1 for a detailed discussion of the waste beds) occupy $8.2 \times 10^6 \, m^2$ ($8.8 \times 10^7 \, ft^2$) along the southwest and southeast shores of Onondaga Lake (Figure 2.17). Thickness ranges from 1.8 m (6 ft) to 21.3 m (70 ft). The total volume is $6.5 \times 10^7 \, m^3$ ($2.3 \times 10^9 \, ft^3$) (Blasland and Bouck Engineers 1989). The waste beds were used largely as a depository for wastes produced from Allied's Solvay process since the

1880s. The waste beds were built directly on the surficial sediments. These sediments range from silty sands to gravels depending on the location. Seepage faces may occur along the base of the waste beds. Many of the waste beds are presently vegetated.

Geologic evidence based on inferred and measured elevations of top of bedrock in the Ninemile Creek valley indicate that natural sources of brines are limited to valley reaches north of Warners-Amboy. The southern reaches of the valley have sufficiently high elevations that the valley does not intersect the salt bearing Syracuse Formation. North of Warners-Amboy the elevation of the valley is

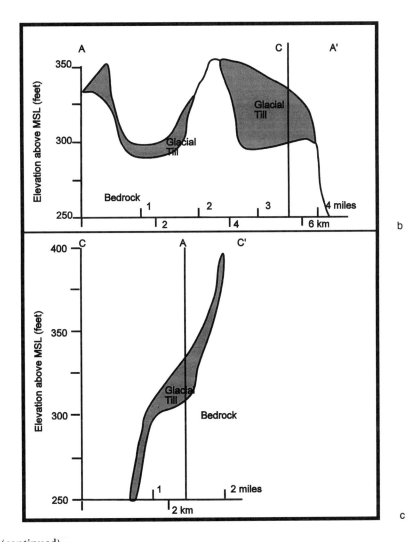

FIGURE 2.16 (continued).

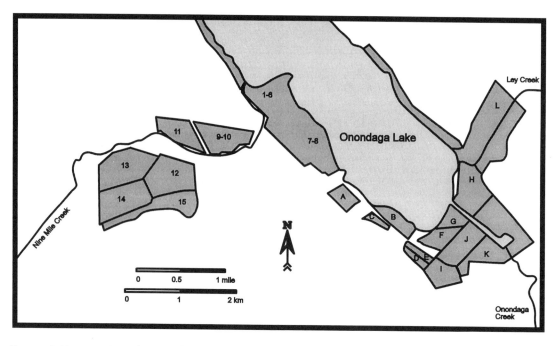

FIGURE 2.17. Location of Ninemile Creek and the Allied waste beds around Onondaga Lake (modified from Blasland and Bouck Engineers 1989).

low enough to intersect the Syracuse Formation. Few wells penetrate to the bedrock, so much of the information concerning depth to bedrock in the Ninemile Creek Valley is inferred (Kantrowitz 1970; Blasland and Bouck Engineers 1989). The elevation of the top of the Syracuse Formation near Onondaga Lake is about 137 m (450 ft). The elevation of the surface of Onondaga Lake is 110 m (361 ft). Because Syracuse Formation dips to the south, its elevation decreases southward. Kantrowitz (1970) infers that the depth to the top of bedrock (Skaneateles Formation of the Hamilton Group) in the Ninemile Creek valley between Marcellus and Otisco Lake is about 45 m (150 ft) with an elevation of 183 m (600 ft). At Warners-Amboy the inferred elevation of the top of bedrock is 69 m (225 ft) above sea level (53 m or 175 ft below the surface) (Kantrowitz 1970). It is not until Warners-Amboy that the elevation of the valley floor (130 m or 425 ft) approaches the elevation of the Syracuse Formation. Consequently, Ninemile Creek valley does not intersect the Syracuse Formation until north of Warners-Amboy.

2.2.3.2 Hydrogeology of Selected Areas of the Ninemile Creek Valley

The hydrogeology of Ninemile Creek Valley is probably very similar to other valleys in the area. Groundwater flow rates in the surficial valley fill vary greatly, depending on the media through which the water flows, as well as adjacent units. Groundwater is recharged along the walls of the valley with flow downward toward the center of the valley. Deeper flow systems may be discharging from the bedrock into overlying surficial valley fill deposits in the lower reaches of the valley, near Onondaga Lake (Blasland and Bouck Engineers 1989). The hydraulic head gradient follows the topography, so flow is northwards from the Appalachian Upland toward Onondaga Lake.

It is uncertain if Ninemile Creek loses to (losing stream) or gains water from (gaining stream) the groundwater in the upper reaches. If the water table elevation is higher than the stream, then the stream gains water. A lower water table elevation in relation to the stream level indicates a losing stream. In the upper reaches, Ninemile Creek may vary from a gaining stream to a losing stream, depending on fluctuations in the water table elevation due to seasonal variation of precipitation (Kantrowitz 1970). Blasland and Bouck Engineers (1989) report that near the waste beds, Ninemile Creek becomes a gaining stream. Groundwater discharge to the stream probably originates from regional groundwater discharge and groundwater mounding under the adjacent waste beds (Figure 2.18). The height of the waste beds range from 1.8 m (6 ft) to 21.3 m (70 ft).

2.2.3.3 Hydrogeology of the Waste Beds

Using 12 wells screened at different elevations, Blasland and Bouck Engineers (1989) identified two different flow systems in the surficial valley fill near the waste beds: shallow and deep. The flow systems have distinguishing hydraulic gradients. Near Amboy the shallow flow system has a hydraulic gradient of $0.0011\,\text{ft}\cdot\text{ft}^{-1}$ changing to $0.0004\,\text{ft}\cdot\text{ft}^{-1}$ near the waste beds. The hydraulic gradient of the deeper flow system changes from 0.0016

$\text{ft}\cdot\text{ft}^{-1}$ near Amboy to $0.0004\,\text{ft}\cdot\text{ft}^{-1}$ near the waste beds. The small difference in hydraulic gradients between the two flow systems may indicate density differences in the groundwater due to different water chemistries.

Blasland and Bouck Engineers (1989) report groundwater mounding occurring under some of the waste beds (Figure 2.18). Mounding is more pronounced in the shallow flow system than in the deeper flow system. Horizontal and vertical hydraulic head gradients of the groundwater mounds beneath some of the waste beds were determined. Horizontal hydraulic gradients range from 0.01 to $0.06\,\text{ft}\cdot\text{ft}^{-1}$. Vertical gradients are downward, ranging from 0.12 to $0.97\,\text{ft}\cdot\text{ft}^{-1}$ (averaging $0.255\,\text{ft}\cdot\text{ft}^{-1}$) indicating groundwater recharge. Blasland and Bouck Engineers (1989) report that the average vertical hydraulic gradient from the waste beds to the natural ground is $0.8\,\text{ft}\cdot\text{ft}^{-1}$.

In situ slug tests were conducted to measure horizontal hydraulic conductivity in the vicinity of the waste beds (Blasland and Bouck Engineers 1989). Hydraulic conductivity ranges from $1.1\times10^{-4}\,\text{cm}\cdot\text{sec}^{-1}$ to $4.2\times$

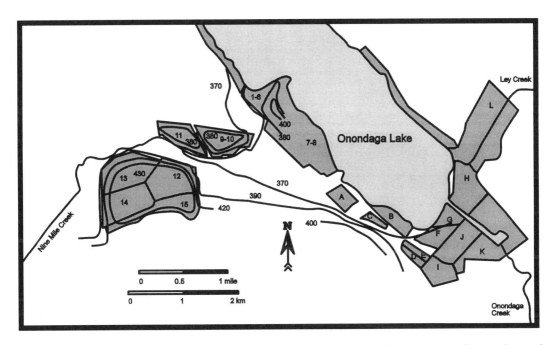

FIGURE 2.18. Potentiometric map of shallow groundwater flow system showing mounding underneath waste beds (modified from Blasland and Bouck Engineers 1989).

10^{-3} cm $\cdot$ sec^{-1}. Field measurements of hydraulic conductivity may vary as much as a factor of 100 from the actual hydraulic conductivity (Dagan 1986). Hydraulic conductivities determined from multiwell pump tests may more accurately reflect inhomogeneities in the surficial deposits (see Section 2.5.2.1). Horizontal hydraulic conductivity measured in the waste beds range from 10^{-4} to 3×10^{-6} cm $\cdot$ sec^{-1}.

Vertical conductivities were not measured for either the waste beds or surficial valley fill. However, vertical hydraulic conductivity is usually a fraction of the horizontal conductivity because of depositional orientation of sediments comprising either bedrock or unconsolidated surficial aquifers.

Using Darcy's Equation (See Section 2.5.1.1), flow rates can be estimated using observed aquifer geometry, hydraulic head gradients, and hydraulic conductivities. Blasland and Bouck Engineers (1989) determined a flow rate of 0.000074 m$^3 \cdot$ s^{-1} (0.026 ft$^3 \cdot$ s^{-1}) through a 5390 m^2 (58,000 ft^2) saturated cross section of Ninemile Creek valley. The flow rate appears unusually small for this geologic setting. Considering potential inaccuracies with in situ hydraulic conductivity measurements (Dagan 1986) the actual flow through the above mentioned cross section may be much greater.

Blasland and Bouck Engineers (1989) estimated the flux of water from the waste beds to the groundwater in the surficial sediments. Flux calculations were based on water budgets and the HELP infiltration model. The estimated flux rate of water discharging from waste beds 11 to 15 to the groundwater is 0.003 m$^3 \cdot$ sec^{-1} (0.11 ft$^3 \cdot$ sec^{-1}), assuming the infiltration rate of precipitation is 0.048 m$^3 \cdot$ sec^{-1} (1.71 ft$^3 \cdot$ sec^{-1}) (Blasland and Bouck Engineers 1989).

The rate of flow from the waste beds to the groundwater can be calculated using a simple Darcy's Equation (Section 2.5.1.1) using hydraulic conductivity and vertical head gradients instead of a flux rate based on numerical models that require many hard to quantify parameters. Table 2.7 shows calculated flux rates using Darcy's Equation from the waste beds to the surficial deposits, using hydraulic conductivities, the average vertical head gradient between the waste beds and surficial deposits and bed areas reported by Blasland and Bouck Engineers (1989). The distribution of wells was not sufficient to determine hydraulic conductivities in each waste bed, so an average hydraulic conductivity (4×10^{-5} cm $\cdot$ sec^{-1}) was used for the calculations. The total flux to the groundwater from the waste bed varies depending on the anisotropy of hydraulic conductivity (the ratio K_h/K_v) of the horizontal and vertical hydraulic conductivities. As K_h/K_v decreases, the flux will increase. The total flux from waste beds 1 to 10 ranges from 0.49 to 0.005 m$^3 \cdot$ sec^{-1} (17.3 to 0.17 ft$^3 \cdot$ sec^{-1}), for waste beds 11 to 15 the total flux ranges from 0.8 to 0.008 m$^3 \cdot$ sec^{-1} (28.3 to 0.3 ft$^3 \cdot$ sec^{-1}).

TABLE 2.7. Water flux from waste beds into natural ground.

Waste bed	Area (m^2)	Flux (m$^3 \cdot$ sec^{-1}) $K_h/K_v = 1.0$	Total flux (m$^3 \cdot$ yr^{-1})		
			$K_h/K_v = 1.0$	$K_h/K_v = 0.01$	$K_h/K_v = 0.001$
1–8	1.24E6	0.40	1.3E7	1.3E5	1.3E4
9–10	2.97E6	0.09	3.0E6	3.0E4	3.0E3
11	2.14E5	0.07	2.2E6	2.2E4	2.2E3
12	5.20E5	0.17	5.3E6	5.3E4	5.3E3
13	6.60E5	0.21	6.7E6	6.7E4	6.7E3
14	5.40E5	0.17	5.4E6	5.4E4	6.4E3
15	4.51E5	0.14	4.5E4	4.5E2	4.5E1
Total	6.60E6	1.25	4.0E7	4.0E5	4.0E4

The flux of water from the waste beds to the surficial deposits could be as large as 25% of the total waste bed infiltration. The remaining 75% is discharged from seepage faces in the waste beds. For instance Blasland and Bouck Engineers (1989) estimate that the total rate of infiltration for waste beds 1 to 15 is 0.04 $m^3 \cdot sec^{-1}$ (1.59 $ft^3 \cdot sec^{-1}$). Assuming that K_h/K_v is 0.01, the calculated flux from Darcy's Equation for waste beds 1 to 15 is 0.013 $m^3 \cdot sec^{-1}$ (4 × 10^5 $m^3 \cdot yr^{-1}$) or about 25% of the infiltration. Thompson and Associates (1982) report that the hydraulic heads in the waste beds can vary from 1.5 to 7 m (5 to 23 ft) during the year. Fluctuation in the hydraulic heads could change vertical hydraulic head gradients resulting in a change in the rate of flux of water from the waste beds to the surficial deposits. Much of the waste bed leachate that makes its way to the surficial deposits ultimately discharges into Ninemile Creek.

Without consideration of sources, Kantrowitz (1970) reports that salty groundwater is present only downstream of Warners-Amboy. Noble (1990) reports a chloride concentration of 0.39 $meq \cdot L^{-1}$ (13.8 $mg \cdot L^{-1}$) and a sodium concentration of 0.27 $meq \cdot L^{-1}$ (6.2 $mg \cdot L^{-1}$) in groundwater samples from glacial till midway between Otisco Lake and Marcellus. Surface and groundwater quality data in the Ninemile Creek valley demonstrates it is unlikely that significant natural chloride loading to the valley occurs south of Warners-Amboy. Minor loading of natural salt to the groundwater in Ninemile Creek valley may occur from groundwater discharging from bedrock aquifers to the surficial deposits. However, discharging groundwater from bedrock aquifers is limited to bedrock areas which have little or no overlying till (Figure 2.16).

Blasland and Bouck Engineers (1989) used chloride and hydrologic budgets to partition ionic rich waters in the lower reaches of Ninemile Creek Valley. They attribute nearly 83% of the chloride load to Onondaga Lake from Ninemile Creek in 1987 to chloride originating from saline groundwater (3.35 ×

10^5 $kg \cdot day^{-1}$ of total chloride load of 4.03 × 10^5 $kg \cdot day^{-1}$). Blasland and Bouck Engineers (1989) assumed that all water infiltrating through the waste beds is discharged through seepage faces as surface runoff. They further assumed that groundwater is discharged to Ninemile Creek at the rate of 0.5 $m^3 \cdot sec^{-1}$ (16.4 $ft^3 \cdot sec^{-1}$). The groundwater contribution to Ninemile Creek was based on excess flow measured in Ninemile Creek at the U.S. Geologic Survey stream gauging station at State Fair Boulevard. However, this gauging station has a poor rating curve because of back-flushing of lake water in the Ninemile Creek channel (USGS, personal communications). Errors in flow measurements at this station are ± 20% of the mean annual flow. In 1987 the mean annual flow in Ninemile Creek was 3.52 $m^3 \cdot sec^{-1}$ (124.25 $ft^3 \cdot sec^{-1}$) (Blasland and Bouck Engineers 1989), resulting in an error of ±0.7 $m^3 \cdot sec^{-1}$ (24.9 $ft^3 \cdot sec^{-1}$) as compared to 0.5 $m^3 \cdot sec^{-1}$ (16.4 $ft^3 \cdot sec^{-1}$) excess reported by Blasland and Bouck Engineers (1989). Because of the poor rating curve at the State Fair Boulevard gauging station, hydrologic budgets using the gauging station at State Fair Boulevard will have insufficient resolution to determine the flux rate of groundwater to Ninemile Creek. Errors in stream-flow measurements at the State Fair Boulevard gauging station will also result in changes in calculated ion loading to the lake. The relative contribution of ion loading from the surface flows and groundwater flow to Ninemile Creek will change as a result of flow measurement errors as well.

Subsequently (Section 2.3.4) it is demonstrated that the chloride, calcium, and sodium ratios in the ground and surface waters in the Ninemile Creek valley near the waste beds are essentially the same as the ratios calculated from historic data related to the chemical composition of the waste bed overflow. Based on mixing models, it is highly improbable that natural loading contributes more than 10% of the chloride load carried by Ninemile Creek; the remainder originates (ultimately) from the waste beds.

2.3 Special Topics Related to the Hydrogeology of Onondaga Lake

2.3.1 The Tar Beds

2.3.1.1 Background

Another source of industrial contamination along the southwestern shore of Onondaga Lake is located near Willis Avenue. Several reports have been prepared over the last ten years that identify and describe groundwater contamination around the Willis Avenue chlorobenzene production plant and the Semet Residue disposal ponds (Geraghty and Miller 1982; Groundwater Technology 1984a,b; O'Brien and Gere Engineers 1989b, 1990). The site contains a number of potential groundwater contamination zones, highlighted in Figure 2.19. Chlorobenzene isomers apparently entered the soil and groundwater as a result of leakage between the different production buildings and the main plant grounds (O'Brien and Gere Engineers 1990). The Benzol Plant supplied benzene to the Willis Avenue plant as a raw material. The Benzol plant operated from 1915 to 1970 and produced benzene, toluene, xylene, and naphthalene (O'Brien and Gere Engineers 1990). When the Benzol plant was demolished in 1973, five petroleum storage tanks containing #2 fuel oil were left standing (Petroleum Storage Facility; Figure 2.19) (O'Brien and Gere Engineers 1990). The storage tanks were dismantled during the final facility closure. The following discussion is concerned mainly with summarizing available information on groundwater contamination beneath the production facility and the Semet Residue Ponds.

From 1910 to 1915 chlor-alkali waste, consisting mostly of $CaCO_3$ and miscellaneous fill material, was disposed of in an area historically designated as "Bed A" (Figure 2.17) (O'Brien and Gere Engineers 1989b). Bed A is located 400 ft (122 m) south of Onondaga Lake on the southwest shore. From 1917 to 1970, organic-based residues generated from the acid washing of coke light oil, prior to fractionation, were deposited in "ponds". The ponds were created by partially excavating the chlor-alkali waste to an unknown depth

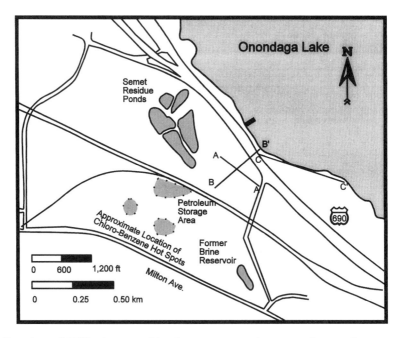

FIGURE 2.19. Location of Willis Avenue chlorobenzene manufacturing plant and Semet Residue Ponds (modified from O'Brien and Gere Engineers 1989b).

TABLE 2.8. Characteristics of Bed A and the Semet residue ponds.

span of use for Bed A	1910–1970
surface area of chlor-alkali waste	38.3 acres ($154{,}998\,m^2$)
mean thickness of chlor-alkali waste	30 ft (9.14 m)
volume of chlor-alkali waste	$50.1 \times 10^6\,ft^3$ ($1.4 \times 10^6\,m^3$)
number of semete residue ponds	5 large, several small
total area of Semet residue ponds	22 acres ($89{,}033\,m^2$)
mean thickness of Semet residue	20 ft (6.10 m)
minimum thickness of Semet residue	10 ft (3.05 m)
maximum thickness of Semet residue	40 ft (12.2 m)
volume of Semet residue	$\approx 73 \times 10^6\,gal$ ($2.8 \times 10^5\,m^3$)

and building dikes from excavated material and other fill material that included concrete rubble, ashes, cinders, soil, bricks, old electrolytic cell parts, etc. (O'Brien and Gere Engineers 1989b). As the level of organic residues increased, the dikes were raised to contain the material (Blasland and Bouck Engineers 1989). The ponds are known as the "Semet Residue Ponds" (O'Brien and Gere Engineers 1989b) or as the "Willis Ave Tar Beds" (Geraghty and Miller 1982; Groundwater Technology 1984). The organic residues contain chlorobenzenes (mono-, di-, and tri-), benzene, toluene, xylene, and naphthalene (Geraghty and Miller 1982). The present consistency of the residues in the ponds ranges from a granular powder to a viscous material resembling tar sludge at the bottom (O'Brien and Gere Engineers 1989). Table 2.8 contains specific details of Bed A and the residue ponds.

2.3.1.2 Geology of the Tar Bed Area

A review of the geologic conditions underlying the Willis Avenue chlorobenzene plant (Figure 2.17) is given by O'Brien and Gere Engineers (1990). The plant is located on top of an artificially elevated fill surface. Approximately 60 to 100 ft (16 to 30 m) of unconsolidated sediments underlie the site and are summarized in Table 2.9. The bedrock surface (Vernon Formations) is at an elevation of 300 to 350 ft (91 to 107 m).

Geologic cross sections through the disposal area are shown in Figure 2.20. There is no clear definition of the variable nature or thickness of the organic residues in the ponds (O'Brien and Gere Engineers 1989b).

Groundwater in both the bedrock and the unconsolidated deposits underlying the Semet Residue Ponds flows in a northerly direction with Onondaga Lake acting as the primary area of groundwater discharge (O'Brien and Gere Engineers 1989b). The water table elevation is 5 to 10 ft (1.5 to 3 m) below the land surface in areas adjacent to the ponds (O'Brien and Gere Engineers 1990) and is located within the artificial fill material at the pond site (O'Brien and Gere Engineers 1989b). There is also evidence of groundwater mounding and radial flow associated with the topographically

TABLE 2.9. Summary of unconsolidated sediments located beneath the Willis Avenue chlorobenzene plant.

	Geologic material	Thickness	
		ft	m
Ground surface	fill material	<1–8	<0.3–2.4
↑	Solvay waste, cinders and ash	≤30	≤9.1
	peat	1–8	0.3–2.4
	marl and clay	1–15	0.3–4.6
	gravel and fine grain sand	1–10	0.3–3.0
↓	silt with clay lenses	10–60	3.0–12.3
bedrock	glacial till	25	7.6

Modified from O'Brien and Gere Engineers 1990.

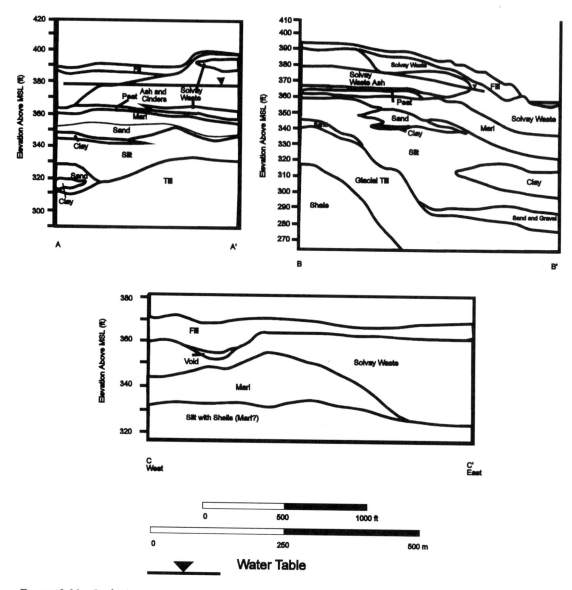

FIGURE 2.20. Geologic cross sections for areas under the Willis Avenue chlorobenzene manufacturing plant (modified from O'Brien and Gere Engineers 1989b).

elevated ponds (O'Brien and Gere Engineers 1989b). Horizontal surface drainage from the ponds is minimized by berms that surround the ponds (O'Brien and Gere Engineers 1989b). Several pump tests were performed on the sediments underlying the chlorobenzene plant and hydrocarbon loading to Onondaga Lake was estimated using the pump test data. The results, summarized in Table 2.10 indicate an estimated total of 146 kg · day^{-1} (53 mT · yr^{-1}) of total dissolved hydrocarbons and 0.195 kg · day^{-1} (0.071 mT · yr^{-1}) of chlorinated hydrocarbons enter Onondaga Lake via groundwater transport. Although the lake sediments do show evidence of contamination by chlorobenzenes, benzenes, toluene, and naphthalene; lake water sample results indicated the total volatile organic compounds had concentrations less than 10 μg · l^{-1} (Geraghty and Miller 1982).

TABLE 2.10. Summary of pump test data and contaminant loads to Onondaga Lake.

Aquifer	Location of pump test	Hydraulic conductivity (gpd·ft²)	Transmissivity (gpd·ft)	Load	Contaminant	Primary discharge
shallow[1]				146 kg·d⁻¹	total dissolved hydrocarbons	Onondaga Lake
	Cinder & Ash	1,500	15,000			
	Marl (basal sediment unit)	2	45		chlori- ated hydro- carbons	
deep[2]				195 gr·d⁻¹		Onondaga Lake
	Willis Avenue	61	2,750			
	Lakeshore	185	7,400			

[1] Groundwater Technology 1984a.
[2] Groundwater Technology 1984b.

2.3.2 History of the Salt Mining Process in the Tully Valley

Salt beds of commercial size (Syracuse Formation of the Salina Group) underlie about 8500 mi² (22,015 km²) of New York State, with thickness varying from 1 to >500 ft (0.3 to >152 m). The Solvay Process Company began withdrawing salt brines in Syracuse near Onondaga Lake when it began the commercial production of soda ash in 1884 (Tully 1985). In the late 1880s, salt beds were traced 36 km southward to the Tully Valley and found at depths from 1100 to 1400 ft (335 to 427 m) below the valley floor and in layers up to 100 ft (31 m) thick (Tully 1985). The Solvay Process Company was utilizing its first brine production well in 1889 and by 1899 the company had 41 wells reaching an average depth of 1200 ft (366 m). Salt brines from the Syracuse Formation (Salina Group) were continuously solution mined from the Tully Valley by the Solvay Process Company (later known as the Allied Chemical Company) from 1889 to 1986 when Allied closed down its operations. From 1986 through 1988, LCP Chemical Company continued the operation, although at a much reduced rate. All brining operations in the Tully Valley ceased in July 1988 (NYSDEC 1991).

The solution mining or "brining operation" utilized five different Brine Field Areas in the 3000 acre (12.14 km²) Tully Brine Field (Tully 1985). The five areas are known as the East-side, the Vesper, the West-side north, the West-side south and the Central Valley. There were also seven core holes drilled, four in the Central Valley, two on the West-side flank, and one on the moraine. Four exploratory salt wells were also drilled in the Tully Valley. Information on the exploratory wells is provided by Fitzpatrick and Callaway (1991). The first exploratory well was drilled in the middle of the Tully Valley near Cardiff, sometime between 1882 and 1888. The well reached a depth of 425 ft (129 m) and stopped due to collapsed pipe. It is rumored among long-time valley residents that the sinkhole at the natural salt springs near Route 20 may be the site of this well. In 1888 a second exploratory well was drilled on the moraine at the south end of the valley, penetrating through 400 ft (122 m) of gravel and quicksand without reaching bedrock. The third exploratory well was drilled just south of Cardiff in 1889, and is known as the "Cardiff Well." This well is notable because it reached a depth of 844 ft (257 m) without finding salt at the expected depth yet discharged a large volume of brine (approximately 50 gal·min⁻¹ (0.19 m³·min⁻¹) (Callaway 1991, personal communication)) until it was plugged in 1968. The Cardiff Well is believed to have intersected the natural solutioning front. The fourth exploratory well, known as the "Benjamin Farm Well," was drilled in 1934 with the specific intention of locating natural brine pools. The brine found in this well was only 60% saturated and of no use to the brining operation. The Benjamin Farm Well is located on the eastern edge of the

TABLE 2.11. Summary information regarding brinefield operation.

Category	# of wells	Date of operation	Duration of operation (yrs)
Brine field areas:			
East-side	65	1889–1957	69
Vesper	27	1895–1944	50
West-side (north)	26	1961–1988	28
West-side (south)	37	1927–1986	60
Central Valley	3	1956–1985	30
Core Holes	7		
Exploratory Wells	4		
Disposal Well	1		
Total number of wells	170		

Modified from Tully 1985.

valley, about a mile north of the brine field and less than 1800 ft (550 m) due east of main mud boil subsidence area (mudboils are discussed in more detail in Section 2.3.3.2). Lastly, one well was also drilled in the West-side south area that was to be used as a disposal well, but construction problems precluded the well from being used. None of the abandoned wells were properly plugged by Allied after the plant shut down operations (Callaway 1991, personal communication). Wells are currently under a NYSDEC program of plugging and abandonment. Table 2.11 provides details on the number of wells in each area, while Figure 2.21 shows the approximate location of each of these areas. Brining operations progressed down valley, toward the northern boundary of Allied Chemical's property (Tully 1985). During 97 years of brine field operation, approximately 31,500 acre · ft $(1.37 \times 10^9 \, \text{ft}^3 \, [3.89 \times 10^7 \, \text{m}^3])$ of rock salt was removed from the Syracuse Formation of the Upper Silurian Salina Group (Tully 1985). According to Tully (1985), nearly 19,500 acre · ft $(8.49 \times 10^8 \, \text{ft}^3 \, [2.41 \times 10^7 \, \text{m}^3])$ of salt was removed from the East-side and the Vesper groups by 1959; and since 1959, approximately 12,500 acre · ft $(5.45 \times 10^8 \, \text{ft}^3 \, [1.54 \times 10^7 \, \text{m}^3])$ of salt was removed from the West-side well group. Tully (1985) states that approximately 1 billion gallons or 3,220 acre · ft $(1.40 \times 10^8 \, \text{ft}^3 \, [3.97 \times 10^6 \, \text{m}^3])$ of nearly saturated brine was produced and con-

veyed to Solvay on a annual basis in recent years. Tully (1985) estimates this process dissolved about 464 acre · ft $(2.02 \times 10^7 \, \text{ft}^3 \, [5.72 \times 10^5 \, \text{m}^3])$ of salt from beds with an estimated average soluble thickness of 210 ft (64 m). This corresponds to an average rate of areal expansion of 2.2 acres · yr^{-1} (96,247 $\text{ft}^2 \cdot \text{yr}^{-1}$ [8,900 $\text{m}^2 \cdot \text{yr}^{-1}$]) for the West-side brine field for the time period 1959 to 1983.

As previously stated, brining operations were confined to five Areas; East-side, Vesper, West-side (north and south) and Central Valley (Figure 2.21). According to Tully (1985), the Vesper and East-side well groups used water imported from a company-owned pond located south of the brine field. Water was pumped down a well that penetrated the salt formation and dissolved the salt to form a brine solution in large cavitites created by prior brining activity. As new wells were developed to join exisiting cavities, the saturated brine was pumped out to a collection area on the surface.

Rock formations overlying the Tully Brine Field Areas were not completely watertight as previously implied by Tully (1983), as water was often encountered in the upper limestones in all areas of the brine field during drilling and probably entered into the brine caverns through leaks in uncemented or broken well casings (NYSDEC 1991). Shaffer (1984) also mentions that in the late 1940s fissures began to open up in the rock along the

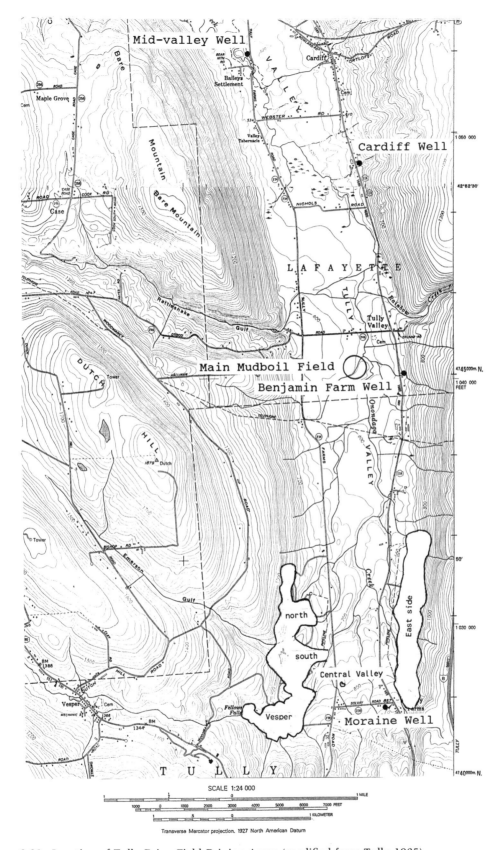

FIGURE 2.21. Location of Tully Brine Field Brining Areas (modified from Tully 1985).

East-side wells, with large cracks extending to the surface. He also notes that a number of sinkholes began developing suddenly beginning in 1949. As of 1984, surveys indicated that subsidence in the East-side well groups had almost stabilized (Shaffer 1984). With regard to the source of water being supplied to the brine cavities, Tully later (1985) states,

Prior to the late 1950s all brine wells were furnished with imported water from the company owned Gatehouse Pond.... During the ten year period centering on about 1959, brining at some West-side south wells was furnished by groundwater in the developing brine cavity while at other wells imported water was still used. Along with subsidence, the West-side south brine cavity soon developed a pattern of fracture zones and interconnections which allowed increasingly greater quantities of surface waters and aquifer groundwaters to flow into the brine cavity.

Tully (1983) previously implied the increased groundwater entry in the West-side brine field signified different geologic conditions. NYSDEC (1991) believes the same rock formations overlie the West-side brine field as overlie the other areas of the Tully brine field; and the solution mining and resultant subsidence resulted in increased fracturing. The increased fracturing in the West-side field is most probably due to the close spacing (150 ft, 46 m) of the "M" well group (NYSDEC 1991). These wells, drilled in the 1920s were spaced unusually close so that they would quickly dissolve a connecting tunnel in the bottom of the salt bed. However, frequent caving took place because of the inability to control the salt dissolution at the top of the cavity and the group did not work out as planned (Shaffer 1984). The resultant subsidence and fracturing most likely led to increased groundwater entry, in addition to leakage along broken well casings, which eventually allowed the importation of surface water to stop (NYSDEC 1991). Tully (1985) remarks: "Groundwater recharge through major fractures now provides abundant inflow to the main West-side cavity.... The one billion gallons of water needed (annually) to supply the brining operations comes directly from these recharges."

Tully (1985) States that virtually all groundwater recharge to the system flows northward through bedrock or the overlying unconsolidated sediments. Once brining operations stopped in 1988, the one billion gallons of groundwater that were previously withdrawn annually then initially became integrated into the existing groundwater storage and flow regime. There is no evidence to support the idea that the groundwater flow regime will revert back to the prebrining state due to the development of fissures after 1959 that presently allow surface water to flow directly into the groundwater system. Tully (1985) estimates the additional water would cause a small rise in the hydrostatic pressure in the West-side brine field; although over a period of time (such as several years), it seems reasonable that the overall hydrostatic head in the entire brine field would rise. Tully (1985) believes that the impact of the hydrostatic rise over the West-field would be to allow a greater opportunity for interaction of diffuse weak brine with the overlying bedrock aquifers and the deep sand and gravel aquifer. In addition, even a modest rise in hydrostatic pressure would transmit increases in localized pore pressure over short distances, thus potentially increasing the hydraulic head in all the overlying aquifers. The increased communication between aquifer systems results from the formation of sinkhole fractures that provide "about 40 times the opportunity for communication between aquifers" (Tully 1985).

2.3.3 Occurrence of Mud Volcanoes and Mud Boils

2.3.3.1 Background

Approximately 52% of the annual external sediment loading to Onondaga Lake has been attributed to inputs from Onondaga Creek (Yin and Johnson 1984; refer to discussion in Chapter 3). During high-flow runoff events a substantial fraction (>50%) of the suspended solids in Onondaga Creek has been found to originate in the Tully Valley (Figure 2.13) from "mud volcanoes" and "mud boils"

(Effler et al. 1992). Mud and sand volcanoes have been given a large variety of names depending on their size, shape, composition, and origin. They have been generically classified according to size: microvolcanoes include those with a diameter <1 m, mesovolcanoes range in diameter from 1–10 m, and megavolcanoes have diameters >10 m (Dionne 1973). Mud and sand volcanoes have been found worldwide and documentation regarding them dates back to at least 1878 (Dionne 1973). Sediment volcanoes are cone-shaped structures formed by the effusion of soft sediments originating in subsurface deposits (Washburn 1956; Washburn 1973; Dionne 1973). Volcanoes are composed of varying percentages of sand, silt, and clay (Dionne 1973; Shilts 1978; Tuttle et al. 1990; Wills and Manson 1990). The lack of sediment accumulation around the point of effusion (vent) makes the structure a boil rather than a volcano (Quirke 1930; Shilts 1978). Shilts (1978) states that many lakes in the Arctic mainland become turbid during the thaw season due to large quantities of sediment washing downslope from active volcanoes and boils.

Mud and sand volcanoes and boils are formed by increased groundwater pore pressures resulting in liquefaction of sediments. The subsequent surficial discharge of sediments as a result of pressure release forms a mud volcano or mud boil (Washburn 1956; Washburn 1973; Lundqvist 1962; Dionne 1973; Shilts 1978; Tuttle et al. 1990; Wills and Manson 1990). Sources of increased pore pressure vary: heavy rain storms, onset of spring runoff, melting of ice lenses, subsidence, and earthquake-induced horizontal accelerations of the ground (Dionne 1973; Shilts 1978; Getchell 1983; Wills and Manson 1990). Liquefaction is the transformation of a granular material from a solid state into a liquified state as a consequence of increased pressure exerted by ground water within the voids among mineral grains (Youd 1973). The subsequent eruption of sediment slurry to the surface results from weaknesses in the impermeable material overlying the unconsolidated deposits (Washburn 1956; Dionne 1973; Shilts 1978).

The discharge is a slurry of sediment and water which may accumulate to form a cone as the material selectively settles out of suspension.

Results of grain size analysis on mud and sand volcanoes of different origins vary. Wills and Manson (1990) found 87% of the sediment had a grain size <0.074 mm diameter and was composed of a sandy silt, while Tuttle et al. (1990) found a mean grain size of 0.10 to 0.04 mm (very fine sand to coarse silt) and a clay content of <5%. Shilts (1978) characterized the Arctic volcanoes and boils as being poorly sorted sediments (mud) with a significant silt and clay fraction. The formation of mud and sand volcanoes and boils tends to occur along fissures and slope breaks in topography (Shilts 1978; Wills and Manson 1990; Tuttle et al. 1990). Shilts (1978) also observed that volcanoes pass through identifiable stages ranging from "young" active formations, which produce larger quantities of sediment, to "old" inactive formations. Most volcanoes stabilize and are reactivated only under exceptional conditions at progressively longer intervals (Shilts 1978). This is not always the case with the Tully Valley mud boils and mud volcanoes.

2.3.3.2 Mud Volcanoes and Mud Boils in the Tully Valley

2.3.3.2.1 Description of Tully Valley Mud Boils

In referring to the sediment volcanoes and boils in the Tully Valley, Getchell (1983) has adopted Shilts (1978) and Allen's (1982) nomenclature of "mud volcano" and "mud boil"; the same nomenclature will be used here. The mud volcanoes in the Tully Valley (Figure 2.21) range in diameter from <1 m to >4 m. The vent is usually <0.3 m in diameter but has been observed to be >1 m. The angle of sediment accumulation varies from 0 to 12°. The mud boils are apparently precursors to volcanoes, although not all boils develop into volcanoes. Mud boils in the Tully valley are often <1 m in diameter, and frequently <0.5 m in diameter. Figure 2.22 shows a

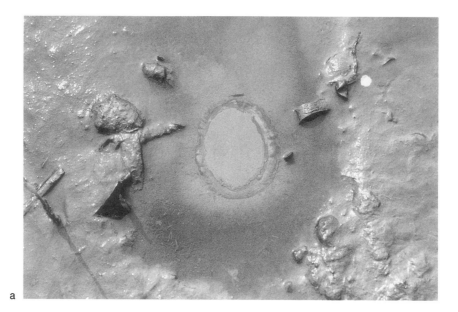

FIGURE **2.22.** Photograph of (**a**) Tully mud boil and (**b**) Tully mud volcano.

typical mud boil (Figure 2.22a) and a typical mud volcano (Figure 2.22b).

Historical evidence (Post Standard Newspaper 1899) documents the presence of a mud volcano, located immediately south of Otisco Valley Road and west of Onondaga Creek during October of 1899. This particular mud volcano destroyed the highway bed and observations indicated the nearby bluff was sub-siding. Other sources (U.S. Army Corps 1991; NYSDOH 1951) document the presence of a few mud boils, located south of Otisco Valley Road and west of Onondaga Creek, before 1913 and persisting through the late 1930s and 1940s. Mud boil activity was also documented during the mid-late 1970s in this same area (Waller 1983). The mud volcanoes and mud boils are presently found concentrated in an area from Otisco Valley Road extending southward approximately 3 km and west of Onondaga Creek towards Tully Farms Road (Figure 2.21). Table 2.12 presents a brief summary of information regarding the location, size, and dates of activity for the more established mud volcanoes and mud boils in the valley near Otisco Road. Small (<0.5 m in diameter) mud boils have also been observed at one location in Rattlesnake Gulf (Perkins and Romanowicz 1990, personal observation). It is not evident whether these mudboils are related in any way to the mudboils located in the Tully Valley. Getchell (1983) notes, as did Shilts (1978), that the activity of the mud boils and volcanoes is dependent on the artesian pressure head; thus they are most active following spring thaw or heavy rainstorms. Personal observations by Waller (1991, personal communication) during the period 1974 through 1986 and Perkins and Romanowicz in 1990 and 1991 also indicate a high level of activity in terms of the occurrence of new mud boil formations and increased flow in the early spring, and a subsequent reduction in activity as summer progresses. Although mud boil activity in the past has been greatest during spring thaw, observations made by the New York State Department of Environmental Conservation indicate the degree of seasonality has decreased over the last three years (Callaway 1991, personal communication). Observations by Romanowicz and Perkins during 1992 also indicate mud boil activity continuing at higher levels throughout the summer months, indicating a reduction in the level of seasonality associated with mud boil formation.

If more than one mud volcano or mud boil is located within an area, individual discharges have been observed to subsequently combine to form small "tributaries" that flow into Onondaga Creek. These inputs can have a significant impact on the water quality of Onondaga Creek. In one study (Effler et al. 1992), suspended solids were measured in an established mud boil tributary to Onondaga Creek, and above and below the mud boil tributary input. Suspended solids concentra-

tions ranged from 760 to 871 mg $\cdot$ L^{-1} in the mud boil tributary. Concentrations of suspended solids in Onondaga Creek just above the mud boil input ranged from 15.9 to 18.5 mg $\cdot$ L^{-1} and below the mud boil input suspended solids ranged in concentration from 102 to 114 mg $\cdot$ L^{-1}.

The Tully Valley mud volcanoes and mud boils produce a mauve, tannish-colored, turbid streamflow that subsequently enters Onondaga Creek and completely mixes with the Onondaga Creek streamflow within about 1.2 km. The color is the result of the high concentrations of Otisco Valley clay found in suspension (Getchell 1983). The mudboil turbidity can be clearly observed in streamflow for several kilometers downstream under low flow conditions, and down to the mouth of Onondaga Creek during runoff events. Loading and concentration profiles from above and below the mudboils, collected over a range of flow conditions, indicate that most of the sediment load received at the creek mouth during runoff events is resuspended sediment and eroded bed material originating from the mud boils (Effler et al. 1992).

2.3.3.2.2 Origin and Continuing Impacts of Mud Boils and Mud Volcanoes

Mud boils and mud volcanoes in the Tully Valley conform to the same basic formation mechanisms described in Section 2.3.3.1. The localized subsidence associated with the mud boil areas is most likely due to the volume of sediment discharged by the mud boils themselves (NYSDEC 1991). Whether the mud boils and mud volcanoes have a natural (Tully 1983) or anthropogenic (Getchell 1983; Kosinski 1985) origin is a continuing point of contention. Additional debate is also focused on whether the integrity of the geologic framework in the southern portion of the Tully Valley has been significantly affected by the long term brining operations that took place, and whether or not the mud boils are currently interconnected to or influenced by the same hydrologic flow regime that is linked to the brine cavities and overlying bedrock aquifers.

TABLE 2.12. Anecdotal information regarding the history of selected mud boils along the western side of Onondaga Creek and south of Otisco Road.

Location	# mud boils/volcanoes	Approximate size (m)	Approximate time of activity	Comments
1	1 volcano	large	1899	Located near Rattlesnake Gulf and the Tully Valley Grist Mill on the hard highway bed. Documented in the Post Standard (article dated Oct. 20, 1899).
2	"field"	large	1970s	Located adjacent to Otisco Road. Noted to be active during the 1970s (NYSDEC 1991), although by 1987 it was apparently inactive and overgrown to such a degree that its location was not readily evident to the casual viewer.
3	"field"	~750 × 300	1970–present	Located near the Sun Pipeline right of way, south of Otisco Road. The field has experienced continued growth and is actively discharging water via a small tributary at the rate of ~0.11 $m^3 \cdot s^{-1}$ (Effler et al. 1992). Sometime after 1987, a second field developed adjacent to the first. Rapid growth has caused the two fields to merge.
4	2 volcanoes	~6 m dia. ~3 m dia.	?–present	Located ~30 m south of Otisco Road, on the west side of Onondaga Creek. Initially appeared (1990) to be mature, relatively inactive, and somewhat overgrown with vegetation. In the spring of 1991 both became very active and were observed discharging water and sediment in a volume large enough to form a small tributary to Onondaga Creek.
5	1 volcano 3 boils	~6 m dia. all <0.5 m dia.	?–present	Located in a depression south of Otisco Road and on the west side of Onondaga Creek. The volcano appeared relatively inactive although it was not vegetated. The mud boils were located approximately 2 m to the SE and were discharging variable quantities of water and sediment. Drainage to Onondaga Creek was not apparent.
6	1 volcano	~6 m dia.	late summer 1991–present	Located in Onondaga Creek, 28 m south of Otisco Road. Experienced rapid development and growth (over the period of <1 week). Flow of the creek is currently around the mud volcano. Formation of volcano may be related to same subsidence event that resulted in the collapse of the Otisco Road bridge over Onondaga Creek.

2.3.4 Distinctive Chemical Constituents of Ground- and Surface Waters

Hydrostratigraphic units are rock or surficial units of significant lateral continuity, exhibiting distinct hydrologic properties. The major hydrostratigraphic units identified in Onondaga County (Winkley 1989) are listed in Table 2.13. Groundwaters from different sources often have characteristic chemical signatures. These chemical signatures are valuable in determining the different sources for a particular mixture of water, and in predicting the characteristics of a mixture of different waters. Noble (1990) analyzed the groundwater chemistry for each of the hydrostratigraphic units in Onondaga County and assigned each unit a characteristic "dominant water type" (e.g., Table 2.13). Dominant water types reflect those cations and anions that contribute at least 50% of the concentration (meq $\cdot$ L^{-1}) of either the cations or anions (Morgan and Winner 1962; Back 1966). Dominant water types can easily be found on Piper plots (refer to Section 2.6.3.2 for a discussion of Piper plots). A Piper plot of the different water types found in Onondaga County is shown in Figure 2.23.

The origins of the high ionic concentrations found in waters entering Onondaga Lake need to be identified and partitioned. Two waters in Onondaga County have sufficient ionic concentrations to be considered as chemical signature endmembers; the Tully Brines and the waste-bed overflow. The Tully Brines originate in the Syracuse Formation near Tully, New York (Blasland and Bouck Engineers 1989). Halite in the Syracuse Formation is currently dissolving along a solution front. This solution front has been moving down dip in the Syracuse Formation. It is presently located along US Highway 20, in the southern part of Onondaga County. Halite is absent from the Syracuse Formation north of the solution front; however, halite is still in the formation south of the solution front. Ionic concentrations of a Tully Brine sample are reported in Figure 2.24 and Table 2.14. It is assumed that the Tully Brines are typical of naturally occurring brines in Onondaga County (Blasland and Bouck Engineers 1989).

The term "waste-bed overflow" is used to mean overflow occurring at the Allied waste beds during use. A range of values for the waste-bed overflow chemistry have been reported (O'Brien and Gere Engineers 1969; USEPA Region II 1974; Blasland and Bouck Engineers 1989) and are consequently shown as a band in the figures presented in this section. The Lockport Formation that underlies the Vernon Shales may potentially contribute groundwater to the Onondaga Lake area. Studies in the Niagara Falls area of New York and Ontario where the Lockport outcrops show that fresh water and saline water is mixing within the Lockport (Noll 1989). It appears that the saline end member's chemistry is similar to the Tully Brines and is distinct from that of the waste-bed overflow. Although other natural groundwaters are likely mixing, they lack sufficient ionic concentrations to significantly affect water chemistry. When mixed with the waste-bed overflow, the waste-bed overflow would become diluted, preserving relative concentrations of ions.

Calcium, chloride, and sodium have been identified as the most significant elements for use in the separation of the Tully Brines from brines originating from the Allied waste beds. Three different ratios (Ca/Cl; Na/Cl; and Na/Ca) for each of the groundwater types

TABLE 2.13. Major hydrostratigraphic units in Onondaga County and the most dominant water types associated with each unit.

Hydrostratigraphic units	Dominant water type
Sands and Gravel	calcium-bicarbonate
Glacial Till	calcium-bicarbonate
Devonian Shales	calcium-bicarbonate
Silurian-Devonian Carbonates	calcium-bicarbonate
Post-Vernon Evaporites[1]	calcium-sulfate
Post-Vernon Evaporites[2]*	sodium-chloride
Vernon Shale	calcium-sulfate
Allied Waste Beds*	calcium-bicarbonate

From Winkley (1989); Noble (1990).
[1] north of the solution front (north of US Highway 20).
[2] south of the solution front (Tully Brines).
* not included in Winkley (1989) and Noble (1990).

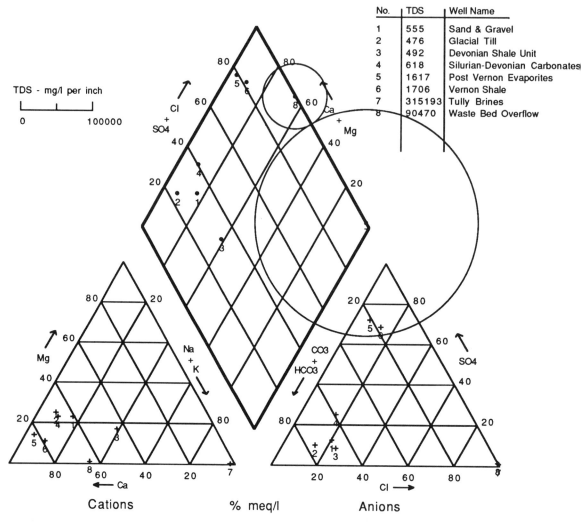

No.	TDS	Well Name
1	555	Sand & Gravel
2	476	Glacial Till
3	492	Devonian Shale Unit
4	618	Silurian-Devonian Carbonates
5	1617	Post Vernon Evaporites
6	1706	Vernon Shale
7	315193	Tully Brines
8	90470	Waste Bed Overflow

FIGURE 2.23. The Piper plot shows the different natural water types from different hydrostratigraphic units in Onondaga County (data from Winkley 1989; Noble 1990; Blasland and Bouck Engineers 1989).

from Table 2.13, including the waste-bed overflow and the Tully Brines (Blasland and Bouck Engineers 1989), are shown in Figure 2.24. The geometric mean and ionic concentration ratios for each of these three constituents for each of the water types is listed in Table 2.14. It is more useful to express concentrations as ratios rather than as absolute concentrations, as ratios accommodate simple dilution effects. Absolute concentrations will decrease with dilution, whereas ratios will remain constant with dilution. Graphically, the ratios are the slopes of lines through data

points as one element is plotted against another (Figure 2.24). From Figure 2.24 and Table 2.14, it is clear that with the exception of the Tully Brines, the natural waters have low concentrations of calcium, chloride, and sodium when compared to the waste-bed overflow.

During the summer of 1991 it was discovered that Niagara Mohawk (an office complex located in downtown Syracuse) was discharging ion-rich water to Onondaga Creek. This water was pumped from the ground and used as a coolant for the complex's air conditioning

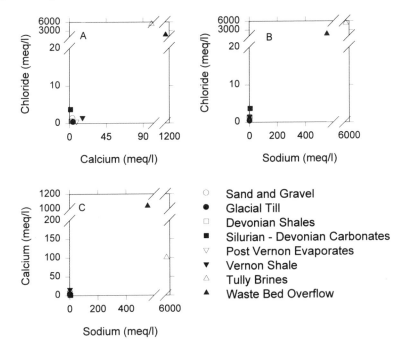

FIGURE 2.24. Plots of Ca, Cl, and Na concentration relationships in the different water types in Onondaga County, including the Tully Brines and the waste-bed overflow (data from Noble 1990, Blasland and Bouck Engineers 1989).

systems. Results of the chemical analyses of this water from different buildings in the complex are presented in Table 2.15. The sodium-chloride ratios are very similar to the sodium-chloride ratios of surface water samples from Onondaga Creek (Table 2.16) and the Tully Brines (Table 2.14). The similarity between the saline groundwater in the Onondaga Creek Valley and the Tully Brines is consistent with the regional ground-

TABLE 2.14. Geometric means of the concentrations and ratios of Ca, Cl, and Na in each of the hydrostratigraphic units in Onondaga County, including the Tully Brines and the Allied Waste-Bed Overflow.

Hydrostratigraphic units	Concentration (meq·l^{-1})			Ratios		
	Ca	Cl	Na	Ca/Cl	Na/Cl	Na/Ca
Sands and Gravel[a]	3.5	1.5	1.6	2.3	1.1	0.5
Glacial Till[a]	4.3	0.5	0.3	8.6	0.6	0.07
Devonian Shales[a]	2.1	0.6	1.1	3.5	1.8	0.5
Silurian-Devonian Carbonates[a]	1.1	3.7	5.0	0.3	1.4	4.5
Post Vernon Evaporites[a]	8.5	0.6	0.6	14.2	1.0	0.07
Vernon Shale[a]	16.1	1.6	1.2	10.1	0.8	0.07
Tully Brines[b]	101	5318	5265	0.02	1.0	52.4
Allied Waste Beds[c]	1295	1774	513	0.7	0.3	0.4
Allied Waste Beds[d]	998	1513	478	0.7	0.3	0.5
Allied Waste Beds[b]	1036	1562	553	0.7	0.4	0.5

[a] Noble (1990).
[b] Blasland and Bouck Engineers (1989) (only one sample).
[c] O'Brien and Gere Engineers (1969) (only one sample).
[d] EPA (1974) (only one sample).

TABLE 2.15. Concentrations of Ca, Na, and Cl and ratios from the Niagara Mohawk Complex in downtown Syracuse (July 16, 1991).

Building	Ca (meq·L^{-1})	Na (meq·L^{-1})	Cl (meq·L^{-1})	Ca/Cl ($\times 10^{-4}$)	Na/Cl	Na/Ca
A	0.19	513.27	502.12	3.78	1.02	2664.7
B	0.17	482.82	544.43	3.12	0.89	2888.3
C	0.27	398.43	321.58	8.40	1.24	1484.1

water flow in the valley. Groundwater flow is from the south, where salt in the Syracuse Formation is currently being dissolved into the groundwater solution front near Route 20. The calcium-chloride and sodium-calcium ratios at the Niagara Mohawk complex and those found in the Tully Brines and Onondaga Creek are not similar.

Similar plots of chloride, calcium, and sodium for surface waters in Ninemile Creek (adjacent to the Allied waste beds), Onondaga Creek at Spencer Street, and Onondaga Lake (6 and 18 m depth) are presented in Figures 2.25 to 2.29. The acceptability criterion for data sets used in this analysis was a charge balance within 10%. The Tully Brines and the waste-bed overflow are the most likely sources of calcium, chloride, and sodium in the above mentioned surface waters (Blasland and Bouck Engineers 1989).

The Ninemile Creek surface water ion ratios (Figure 2.25) before and after closure of the soda ash/chlor-alkali plant equals that of the waste-bed overflow. Concentrations of calcium, chloride, and sodium in the Ninemile Creek surface waters are less than concentrations in the waste-bed overflow, indicating that the waters are diluted with less ionically enriched waters. Further, substantial decreases in the concentrations of these ions occurred following closure of the facility (Chapter 1). Despite the lower concentrations of calcium, chloride, and sodium found in Ninemile Creek and the decreased loadings to the surface waters of Ninemile Creek since closure of the chlor-alkali facility, the unchanged ionic ratios establish that the continuing ionic loading originates from the waste-bed overflow. A Piper plot (Figure 2.30) indicates that the composition of the Ninemile Creek surface

TABLE 2.16. Annual average concentrations and ratios of Ca, Cl, and Na reported in Ninemile Creek, Onondaga Creek, and Onondaga Lake (6 and 18 m depth) for the years 1981–1989 (to 1990 for Onondaga Lake) (UFI 1981–1990).

Ninemile Creek

	Concentration (meq·1^{-1})			Ratios		
	Ca	Cl	Na	Ca/Cl	Na/Cl	Na/Ca
1981	70.7	103.2	37.6	0.69	0.36	0.53
1982	64.2	100.5	40.5	0.64	0.40	0.63
1983	104.6	155.4	57.8	0.67	0.37	0.55
1984	31.5	41.6	17.1	0.76	0.41	0.54
1985	41.6	63.4	27.2	0.66	0.43	0.65
1986	26.8	38.0	15.4	0.71	0.41	0.57
1987	22.5	38.7	18.7	0.58	0.48	0.83
1988	26.8	37.2	16.1	0.72	0.43	0.60
1989	20.1	26.1	10.7	0.77	0.41	0.53
81–85	70.4	105.1	40.4	0.68	0.38	0.57
87–89	23.7	33.6	14.7	0.71	0.44	0.62

TABLE 2.16. *Continued*

Onondaga Creek

	Concentration (meq · l^{-1})			Ratios		
	Ca	Cl	Na	Ca/Cl	Na/Cl	Na/Ca
1981	6.0	7.7	6.8	0.78	0.88	1.13
1982	6.7	9.5	8.5	0.71	0.89	1.27
1983	6.1	11.1	10.1	0.55	0.91	1.66
1984	5.9	8.7	8.1	0.68	0.93	1.37
1985	6.7	15.2	13.7	0.44	0.90	2.04
1986	5.8	11.2	10.0	0.52	0.89	1.72
1987	6.4	15.1	14.0	0.42	0.93	2.19
1988	6.2	12.4	11.6	0.50	0.94	1.87
1989	5.8	10.4	9.6	0.56	0.92	1.66
81–85	6.3	10.9	9.9	0.58	0.91	1.57
87–89	6.1	12.6	11.7	0.48	0.93	1.92

Onondaga Lake (6 m)

	Concentration (meq · l^{-1})			Ratios		
	Ca	Cl	Na	Ca/Cl	Na/Cl	Na/Ca
1981	29.0	46.7	23.1	0.62	0.49	0.80
1982	23.6	38.1	19.5	0.62	0.51	0.83
1983	22.6	39.6	19.6	0.57	0.49	0.87
1984	20.4	36.8	17.5	0.55	0.48	0.86
1985	24.5	45.7	23.1	0.54	0.51	0.94
1986	12.6	23.1	13.0	0.55	0.56	1.03
1987	9.7	17.4	10.9	0.56	0.63	1.12
1988	9.3	16.7	11.5	0.56	0.69	1.24
1989	8.0	13.0	8.6	0.62	0.66	1.08
1990	7.8	11.5	7.8	0.68	0.68	1.00
81–85	24.3	41.8	20.8	0.58	0.50	0.86
87–90	8.8	15.0	9.9	0.59	0.66	1.13

Onondaga Lake (18 m)

	Concentration (meq · l^{-1})			Ratios		
	Ca	Cl	Na	Ca/Cl	Na/Cl	Na/Ca
1981	35.5	57.6	26.8	0.62	0.47	0.75
1982	26.5	43.8	21.2	0.61	0.48	0.80
1983	31.3	51.4	24.9	0.61	0.48	0.80
1984	28.0	47.5	21.4	0.59	0.45	0.76
1985	28.5	49.4	23.7	0.58	0.48	0.83
1986	21.8	38.6	19.9	0.56	0.52	0.91
1987	11.6	18.7	10.5	0.62	0.56	0.91
1988	10.6	17.7	12.1	0.60	0.68	1.14
1989	8.8	14.9	9.6	0.59	0.64	1.09
1990	7.9	11.4	7.4	0.69	0.65	0.94
81–85	30.6	50.8	24.1	0.60	0.47	0.79
87–90	9.6	15.7	10.1	0.61	0.64	1.05

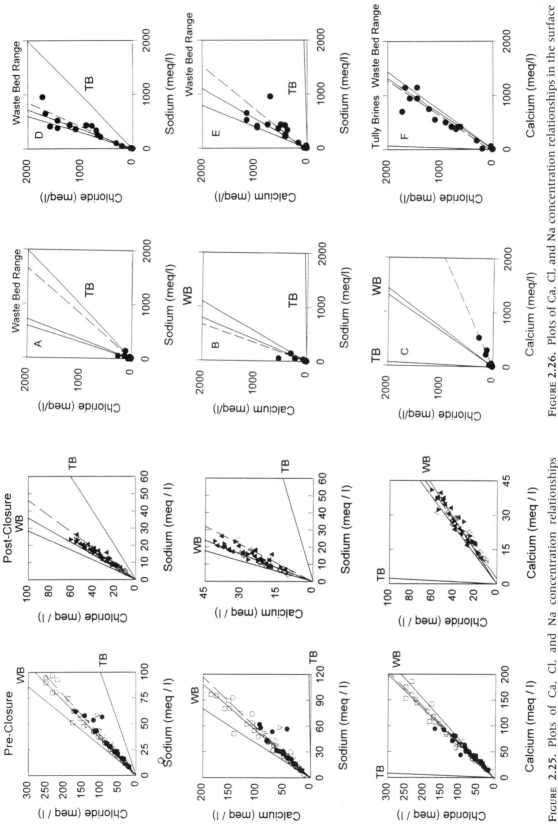

FIGURE 2.25. Plots of Ca, Cl, and Na concentration relationships in the surface water in Ninemile Creek (1980–1989) (data from Onondaga County 1980–1989).

FIGURE 2.26. Plots of Ca, Cl, and Na concentration relationships in the surface water (a–c) and groundwater (d–f) near the Allied waste beds (data from Blasland and Bouck Engineers 1989).

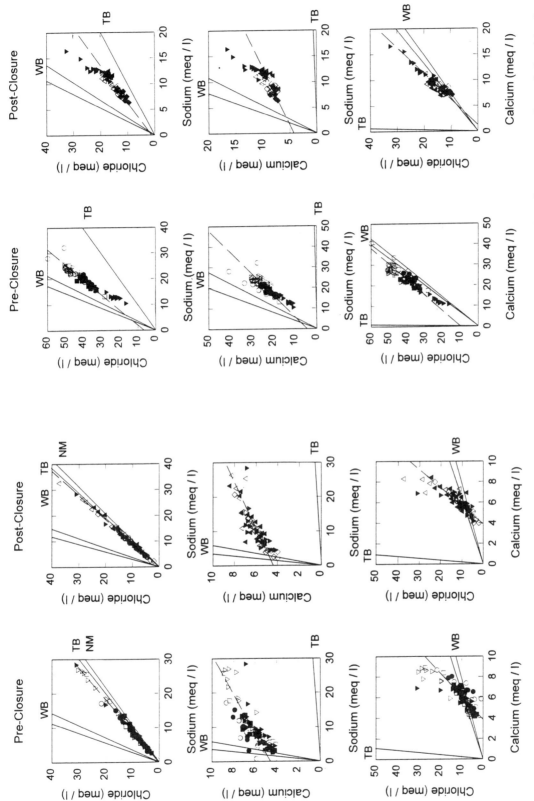

FIGURE 2.28. Plot of Ca, Cl, and Na concentration relationships in Onondaga Lake at a depth of 6 m (1980–1990) (data from Onondaga County 1980–1990).

FIGURE 2.27. Plots of Ca, Cl, and Na concentration relationships in the surface water in Onondaga Creek (1980–1989) (data from Onondaga County 1980–1989).

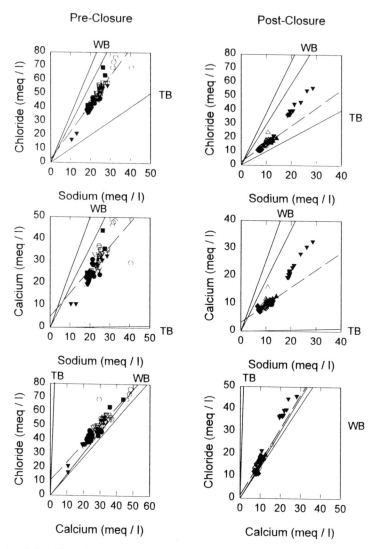

FIGURE 2.29. Plots of Ca, Cl, and Na concentration relationships in Onondaga Lake at a depth of 18 m (1980–1990) (data from Onondaga County 1980–1990).

waters and waste-bed overflow are essentially the same; both are composed of the same unique combination of ions. Finally, the surface and groundwater chemistry in the vicinity of the Allied waste beds, as reported by Blasland and Bouck Engineers (1989), is very similar to the waste-bed overflow (Figure 2.26).

Similar plots (Figure 2.27) show that ion ratios in Onondaga Creek are very similar to the Tully Brines. This is expected since the brines found in the salt springs located adjacent to Onondaga Lake originated in the Tully Valley and flowed to the lake through the Onondaga Trough. A Piper plot of Onondaga Creek surface water (Figure 2.31) indicates these waters are sodium-chloride type, like the Tully Brines. The elevated calcium content of the creek most likely results from the mixing of water from the Devonian Shale Units (Figures 2.23 and 2.31).

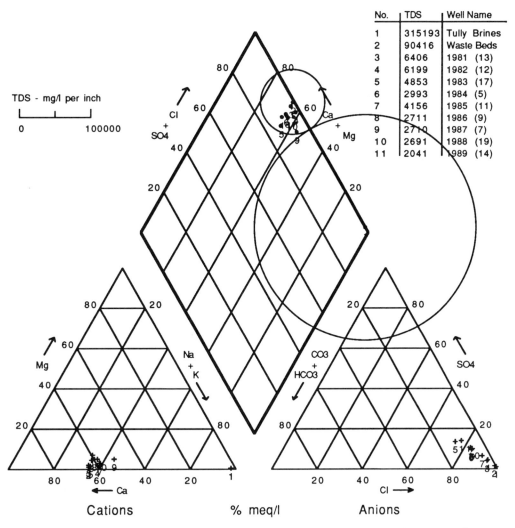

No.	TDS	Well Name
1	315193	Tully Brines
2	90416	Waste Beds
3	6406	1981 (13)
4	6199	1982 (12)
5	4853	1983 (17)
6	2993	1984 (5)
7	4156	1985 (11)
8	2711	1986 (9)
9	2710	1987 (7)
10	2691	1988 (19)
11	2041	1989 (14)

FIGURE 2.30. Piper plot of surface water in Ninemile Creek (1980–1989) (data from Onondaga County 1980–1989).

The plots (Figures 2.28 and 2.29) for Onondaga Lake do not show significant differences between the 6 and 18 m depths. These waters were dominated by calcium-chloride before closure of the Allied soda ash/chlor-alkali facility (Figures 2.32 and 2.33). The similarity to the waste-bed overflow ratios indicates the dominant role the waste beds and Ninemile Creek played in the ionic composition of the lake in that period. In more recent years the water chemistry has shifted due to the greater relative contribution of the Tully Brines (Figures 2.28 and 2.29).

This is particularly noticeable in the sodium-chloride plots. Mean average concentrations and ratios of Ca, Cl, and Na from Ninemile Creek, Onondaga Creek, and Onondaga Lake are reported in Table 2.16.

It is particularly appropriate here to analyze mixing trends of the Tully Brines and waste-bed overflow, as the ions are largely conservative. Mixing analyses are a useful and a widely used method to determine the percentage of each end-member water that has mixed with other water, forming the final water mixture (Freeze and Cherry 1979).

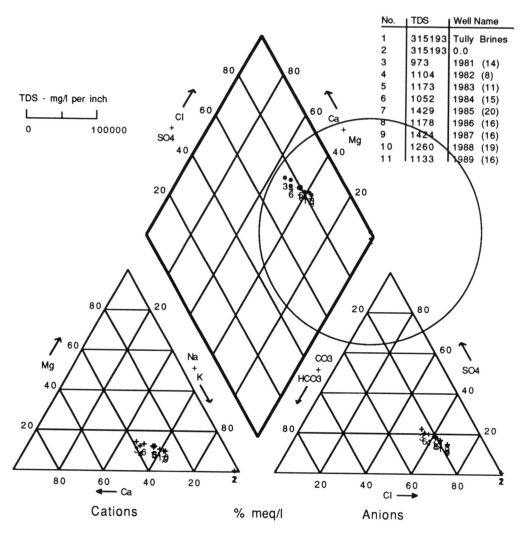

No.	TDS	Well Name
1	315193	Tully Brines
2	315193	0.0
3	973	1981 (14)
4	1104	1982 (8)
5	1173	1983 (11)
6	1052	1984 (15)
7	1429	1985 (20)
8	1178	1986 (16)
9	1424	1987 (16)
10	1260	1988 (19)
11	1133	1989 (16)

FIGURE 2.31. Piper plot of surface water in Onondaga Creek (1980–1989) (data from Onondaga County 1980–1989).

Concentrations resulting from the mixing of two end-member waters can be calculated from Equation 2.1:

$$C_c = (C_aV_a + C_bV_b) \cdot (V_a + V_b)^{-1} \quad (2.1)$$

in which $C_{a,b}$ = concentration of a particular ion in waters a and b respectively, and $V_{a,b}$ equals the volume of waters a and b respectively resulting in a total volume (c) of the final mixture (C).

Mixing trends for two endmembers, the Tully Brines and the waste-bed overflow are presented in Figure 2.34. A mixture of at least a 10% contribution from the Tully Brines and a 90% contribution from the waste-bed overflow would be required to deviate visibly ion ratios from the waste-bed overflow. In Onondaga Lake, where both the Tully Brines and the waste-bed overflow mix, lake chemistry ion ratios are most similar to the waste-bed overflow ratio. Data plotted on the mixing curve falls through the waste-bed overflow (Figure 2.34). Based on the mixing trend, it appears that very small volumes of the Tully Brines are entering the lake (less than 10% of the total volume) compared to

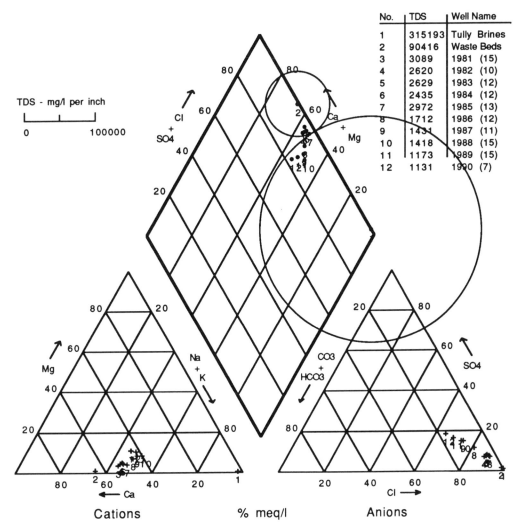

FIGURE 2.32. Piper plot of water in Onondaga Lake at a depth of 6 m (1980–1990) (data from Onondaga County 1980–1990).

the volume of waste-bed overflow entering the lake (around 90%). This data also suggests that the salinity levels in the lake are still due primarily to inputs originating from the waste-bed overflow.

2.4 Onondaga Lake

2.4.1 Groundwater–Lake Water Interactions

Lakes are in hydrologic communication with the groundwater flow system. The function of lakes in the groundwater system can vary. There are groundwater recharge lakes, groundwater discharge lakes, and through-flow lakes. Studies show that the lake–groundwater interaction is complicated and depends on many variables: magnitude of local flow systems, lake geometry, anisotropies in the media, depth to surface water (Winter 1976), and changes in the water table configuration (Winter 1983). The recharge–discharge function of a lake is determined by the local flow system in which the lake is situated. If the lake is encompassed entirely by

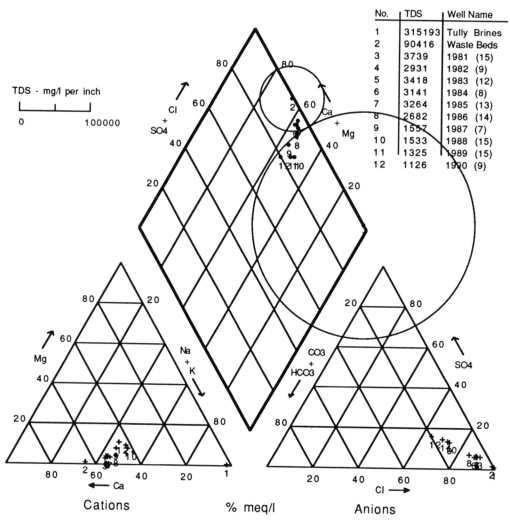

No.	TDS	Well Name
1	315193	Tully Brines
2	90416	Waste Beds
3	3739	1981 (15)
4	2931	1982 (9)
5	3418	1983 (12)
6	3141	1984 (8)
7	3264	1985 (13)
8	2682	1986 (14)
9	1557	1987 (7)
10	1533	1988 (15)
11	1325	1989 (15)
12	1126	1990 (9)

FIGURE 2.33. Piper plot of water in Onondaga Lake at a depth of 18 m (1980–1990) (data from Onondaga County 1980–1990).

a local flow system, then depending on the hydraulic heads of the system the lake is recharging or discharging. Lakes that are not entirely surrounded by a single local-flow system could be either recharge, discharge, or through-flow lakes. Local flow systems around a lake are important in identifying the hydrologic setting of the lake. Local groundwater flow is a small identifiable flow system that includes recharge and discharge areas. Local flow systems are separated from more regional flow systems by flow lines. An understanding of the hydrogeology around

lakes may be important in determining an accurate water budget for a lake. Typically, the groundwater contribution to the lake water budget is assumed to be the residual after the surface and atmospheric inputs and outputs of the lake budget have been quantified. Therefore,

$$S_i + P - ET - S_o = G_i - G_o \quad (2.2)$$

in which S_i and S_o equal the surface inflow and outflows respectively, G_i and G_o equal the groundwater inflows and outflows, ET equals evapotranspiration, and P equals direct

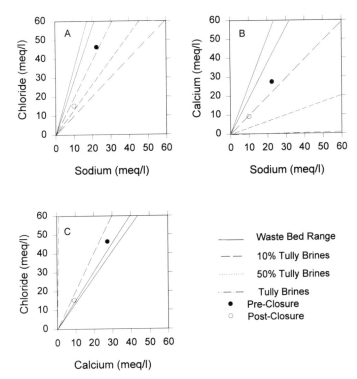

FIGURE 2.34. Plots of Ca, Cl, and Na ratios resulting from 10 and 50% mixes of the Tully Brines with the waste-bed overflow. A circle indicates the average lake ratio before and after closure of the chlor-alkali manufacturer.

precipitation. This expression assumes there is no change in the lake storage. The contribution of groundwater to the hydrologic budget of lakes has rarely been determined directly; the contribution in glaciated terrains has been found to be substantial in several cases (e.g., 13–25%; Cook et al. 1973; Winter and Woo, 1990).

2.4.2 Groundwater Interactions With Onondaga Lake

Previous studies assumed groundwater contributions to Onondaga Lake to be negligible (e.g., Effler and Driscoll 1986; Doerr et al. 1994). A chloride budget for Onondaga Lake (Doerr et al. 1994) suggests that there are insufficient known chloride loadings from surface water and sediment pore waters (Effler et al. 1990) to account for the present chloride concentration in Onondaga Lake. This chlo-

ride deficiency, although minor ($\approx 5\%$), could be made up from chloride loading from groundwater.

Groundwater models (Winter 1981) simulating lakes situated mid-slope of a regional hydraulic gradient (an occurrence common in glaciated terrains) predict groundwater flow into the lake from the down gradient direction because of groundwater mounding at the down gradient end of the lake (Figure 2.35). In glaciated terrains the geometry of the lake is important. Lakes that are situated on highly permeable surficial fill, which is not as deep as the width of the lake, often receive groundwater seepage in the near shore areas (McBride and Pfannkuch 1975). Actual groundwater fluxes can be determined with seepage meters installed near the shore (Lee 1977).

To directly calculate the potential groundwater flow to and from a lake, the hydrologic

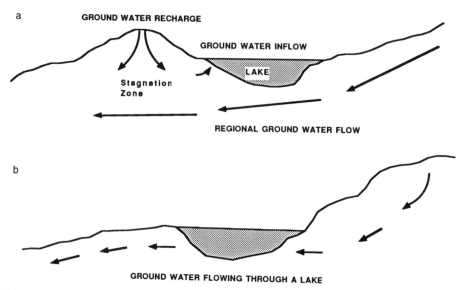

FIGURE 2.35. (a) This lake is located on the mid-slope of the hydraulic gradient. Groundwater recharge is occurring from the down-gradient end of the lake because of the groundwater mound. (b) This lake is similarly located mid-slope of the hydraulic gradient, but no groundwater recharge is occurring from the down-gradient end of the lake because there is no groundwater mound.

setting of a lake must be studied in detail. Regional and local flow systems must be identified, including areas of recharge and discharge. Hydraulic head measurements around the lake are necessary to detect minor groundwater mounding which may recharge a lake.

The regional groundwater flow systems in Onondaga County are driven by the high hydraulic heads on the Appalachian Plateau; regional groundwater flows northward from the south. There are many localized flow systems as well. These flow systems are driven by small recharge mounds and discharge to near-by areas. Local flow systems around Onondaga Lake probably recharge in the higher elevations to the south of the lake and discharge near Onondaga Lake and north of Syracuse. Most hydrogeologic studies near Onondaga Lake have been site specific. There have been no extensive studies of the hydrogeologic setting of the lake. Regional studies (Pagano et al. 1986; Winkley 1989) are too large scaled to provide the detail needed to identify the hydrologic setting of Onondaga Lake.

Potentiometric maps of the area around Onondaga Lake (Pagano et al. 1986; Winkley 1989) indicate that regional groundwater flow follows the Onondaga Trough from the south. The potentiometric surface near Onondaga Lake's outlet shows that groundwater flow continues northward. A minor component of groundwater discharging to the surficial material from the bedrock may exist, as in other areas of the Onondaga Trough.

Groundwater seepage into Onondaga Lake probably occurs near shore adjacent to Onondaga Creek and Ninemile Creek, where groundwater flowing north in the troughs intersect the lake. Regional groundwater flow in bedrock may discharge to the lake along the flanks of the bedrock troughs, where the till thins. Historically it is known that there were naturally occurring brine springs along the southeast shore of the lake (Chapter 1). These springs may be structurally controlled by the Vernon Formation. The contrast in hydraulic conductivity between the Syracuse Formation and the Vernon Shale often results in groundwater seepage along the bedrock contact (Winkley 1989). Onondaga Lake is situated just north of this contact (Figure 2.3).

One critical area for study is the outlet of Onondaga Lake. Regional groundwater flow in the Onondaga Trough continues northward from the outlet. It is possible that Onondaga Lake may actually be a through-flow lake, receiving groundwater from near the southeast shores and releasing groundwater from the northwest shores, near the outlet. The lack of significant topographic relief near the outlet precludes the possibility of any significant groundwater mounding. Without groundwater mounding the mechanism for groundwater flow into the lake in that area (Winter and Woo 1990) does not exist. The residual from a water budget based on surface exchanges will not reflect the groundwater flowing into Onondaga Lake, but the difference between groundwater inflow and outflow. However, if Onondaga Lake is a through-flow lake, it is unlikely that the water and chloride budgets would be significantly affected since a current chloride model, assuming no groundwater contribution (Doerr et al. 1994) predicts within 5% present chloride concentrations in Onondaga Lake. Groundwater input and outputs to Onondaga are probably a minor component (<5%) of the water budget.

2.4.3 Direct Waste-Bed Erosion to Onondaga Lake

The older waste beds (1 to 6, B, F, G, H, and L) define part of the Onondaga Lake shoreline (Figure 2.17). Waste beds 1 to 6 make up the promontory extending from the south west shoreline into Onondaga Lake. Based on aerial photographs, Blasland and Bouck Engineers (1989) determined that as much as 40 ft (12.2 m) of the waste beds have eroded into the lake along a 440 ft (134 m) shoreline. The detritus has likely been transported southeast by longshore currents forming a spit extending south east into the lake (Blasland and Bouck Engineers 1989). The erosional faces along the waste-beds can be easily observed looking southwest across the lake from the marina and Salt Museum. Due to the age of waste beds 1 to 6, chloride loading to Onondaga Lake from waste-bed erosion is

probably minimal since the chloride concentrations of the waste material decrease with age (Blasland and Bouck Engineers 1989). Furthermore, Doerr et al. (1994) have a calibrated chloride budget (within 5%) for Onondaga Lake that does not take into consideration chloride from waste-bed erosion.

2.5 Basic Concepts in Hydrogeology

2.5.1 Background

2.5.1.1 Concepts of Fluid Flow

The analysis of fluid flow requires the recognition of a potential energy gradient that drives fluid flow from a high potential to a lower potential (Hubbert 1940). Concepts of groundwater flow have their basis in fluid mechanics and the Bernoulli Equation.

Groundwater motion can be described by Laplace's Equation,

$$\frac{\delta^2 h}{\delta x^2} \frac{\delta^2 h}{\delta y^2} \frac{\delta^2 h}{\delta z^2} = 0 \qquad (2.3)$$

where h equals hydraulic head, and x, y, and z are the three axes of a Cartesian coordinate system. This is fundamentally the same equation that is used to describe heat flow. A simpler analytical equation to describe groundwater flow is Darcy's Law.

$$Q = -KIA \qquad (2.4)$$

The variables in Darcy's Law are the rate of flow (Q), hydraulic conductivity (K), hydraulic gradient I (dh/dx), and cross-sectional area normal to the flow of water (A).

Hydraulic conductivity describes the rate at which water passes through a porous medium. It is dependent on porosity (the ratio of void spaces in a rock or sediment to the total volume of the rock or sediment) and permeability (the quantity of interconnected void spaces).

The hydraulic gradient is the change in elevation of the hydraulic head over an horizontal distance. The hydraulic head is the

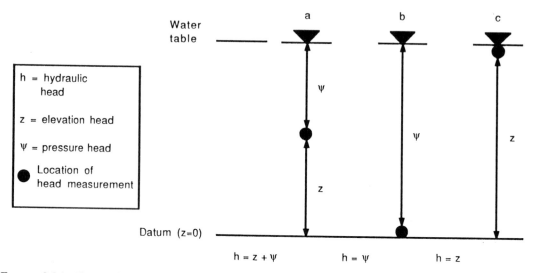

FIGURE 2.36. Figure showing the difference between pressure head and elevation head. In an unconfined aquifer with lateral flow or no flow the hydraulic head will be equal to the water table elevation at any depth. The elevation and pressure contributions to the hydraulic head will vary depending on the depth of piezometer. (a) Combination elevation and pressure head. (b) Pressure head only, due to the weight of overlying water. (c) Elevation head only, no pressure head.

height water will rise in a piezometer (hollow tube). There are two components to hydraulic head: the elevation head and the pressure head (Figure 2.36). The hydraulic head is the energy grade line. Groundwater always flows in the direction of high hydraulic head to low hydraulic head. The hydraulic head is measured by inserting a piezometer into the ground to the desired depth. The height the water rises in the piezometer is measured. This height can easily be corrected for any datum. If the elevation of the piezometer is known, hydraulic head can be measured relative to sea level.

2.5.1.2 Identification of Common Hydrogeologic Terms

These terms are diagrammed in Figure 2.37.

Groundwater is water below the unsaturated zone and water table. Groundwater may be stored for a long period of time, transported from area to area, or released either to the atmosphere through evapotranspiration or to surface water through discharge. Water can be added to the ground in areas of recharge.

Consequently, groundwater should be considered in any surface-water budget.

Aquifer is a saturated permeable geologic unit that can transmit and store significant quantities of water under ordinary hydraulic gradients.

Confined aquifer is an aquifer that is overlain by a geologic unit with a low permeability.

Recharge area is an area in which there are downward components of flow into the aquifer. Percolating water moves downward into the deeper parts of an aquifer in a recharge area.

Water Table is the level at which water lies in an unconfined aquifer. The water table's hydraulic head is all elevation head because the pore water pressure at the water table is atmospheric.

Potentiometric head or surface is the surface that represents the level to which water will rise in tightly cased wells. If the head varies significantly with depth then there may be components of vertical flow in the groundwater flow system. The water table is a particular potentiometric surface for an unconfined aquifer. Figure 2.38 shows different

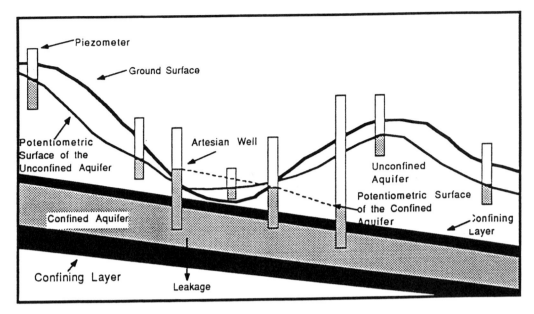

FIGURE 2.37. Common hydrologic terms.

vertical hydraulic heads and how they determine the direction of fluid flow.

Artesian flow occurs when the potentiometric head is higher than the ground elevation.

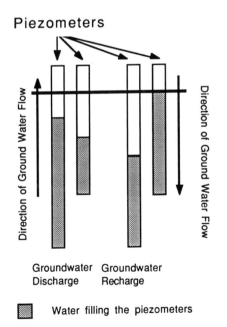

FIGURE 2.38. Vertical hydraulic heads and their relationship to groundwater recharge and discharge.

2.5.2 Methods for Measuring Hydrogeologic Parameters and the Concept of Representative Elemental Volumes

2.5.2.1 Methods for Measuring Hydrogeologic Parameters

The hydrogeologic parameters used in Darcy's Law are conceptually simple. However, field methods for quantifying these parameters are difficult and inaccurate (Driscoll 1986). The field measurements can be very costly, requiring the installation of several wells. An inherent problem with field measurements is the need to quantify inhomogeneity in hydrogeologic parameters over a large area using observations made in a small area. It is necessary determine what minimum size area is necessary to best approximate the overall hydrologic conditions.

There are many ways to estimate hydraulic conductivity using laboratory tests and field tests. Field tests include single well slug test and multiwell pump tests. In a slug test, water is either added or removed from a well. As water in the well recovers, water height in the well is plotted as a function of time to give a recovery curve. This recovery curve is then used to estimate the hydraulic conductivity. In a pump test, one well is pumped to remove water. As water is removed a drawdown cone forms about the well. By observing the changes in water depth from another observation well at a distance, the hydraulic conductivity can be calculated. Despite these simple tests, hydraulic conductivity is very difficult to measure accurately. Errors in hydraulic conductivity of a couple orders of magnitude are not only common, but are probably the rule.

2.5.2.2 Representative Elemental Volume

It is difficult or impossible to measure hydrogeologic parameters over an entire study area.

So, it is necessary to select a smaller area for investigation that is representative of the overall study area. For instance, the hydraulic conductivity measured using multiwell pump test is actually the overall conductivity of the area inscribed by the wells. To apply this conductivity to the entire area, it is necessary to be certain that the area inscribed by the wells is representative of the whole study area. A minimum area (or volume, when considering aquifer thickness) that best replicates the overall conditions of the study area is called the "representative elemental volume" or "REV."

A poor selection of the REV can easily lead to misleading results. Determining the hydraulic conductivity of a highly fractured

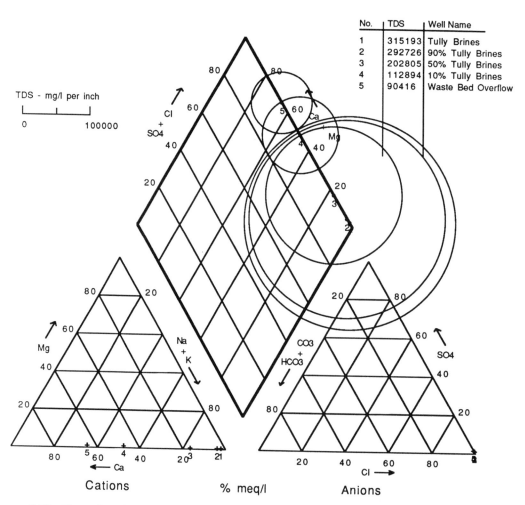

FIGURE 2.39. Piper plot of Tully Brines and waste-bed overflow mixing.

shale using a multiwell pump test removed from any fractures would indicate conductivities significantly lower than the large-scale conductivity of the area, as the fracture provide easy flow paths for water. Similarly, a hydraulic conductivity measurement made in a highly fractured area, in an otherwise unfractured area, would have conductivities higher than the effective conductivity.

2.5.3 Water Quality

2.5.3.1 Milliequivalents

Concentrations are often expressed in milliequivalents/liter $(meq \cdot l^{-1})$. The mass of the compounds are normalized by their respective atomic weights. Milliequivalent concentrations are not artificially elevated or lowered because of differences in the atomic weight of the compounds. So, comparisons between different compounds are meaningful. Equivalents are useful for determining if reactions are taking place involving any of the compounds in question. Milliequivalent concentrations can be calculated by Eq. 2.5.

$$\frac{mg \cdot l^{-1}}{(GMW/charge)} = meq \cdot l^{-1} \qquad (2.5)$$

where GMW equals the gram molecular formula weight and charge equals atomic charge.

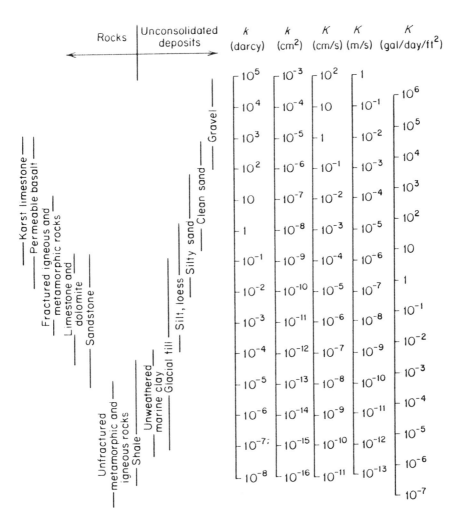

FIGURE 2.40. Range of values of permeability and hydraulic conductivity (from Freeze and Cherry 1979).

2.5.3.2 Piper Plots

An effective method, to show how ground-water is mixing, is to use changes in water chemistry. To do this one must identify and isolate "end-member" waters. The "end-member" waters are the unmixed "source" waters. Once the "end-member" waters have been identified, it is possible to demonstrate how waters are mixing by studying changes in water chemistry.

The most useful method for showing mixing trends are Piper plots (Piper 1944). Piper plots allow us to graphically show water types and how waters are mixing (Figure 2.39). The major anion and cations are separated on two different trilinear plots. Percentage of total concentrations (meq $\cdot$ l^{-1}) of the anions Cl, SO$_4$, CH$_3$, and HCO$_3$, and Mg, Na, K, and Ca cations are plotted. Each of these trilinear plots are separated into major water types based on a minimum 50% con-

centration of an anion and cation. For instance a calcium-sulfate water has at least 50% cation concentration of calcium and at least a 50% anion concentration of sulfate. The anions and cations together are plotted on a tetragon. As two different waters mix, changes in the water chemistry plot along a line connecting the two end-member water types (Figure 2.39). Consequently, once end-members are identified, mixes involving these two end-members can be recognized easily.

2.5.3.3 Values for Permeability and Hydraulic Conductivity

Figure 2.40 presents ranges of values for both permeability and hydraulic conductivity as a function of rock/unconsolidated deposit type.

2.5.3.4 Geologic Time Scale

See Figure 2.41.

Eon	Era	Period	Epoch	Time (mil. of years)
Phanerozoic	Cenozoic	Quaternary	Holocene	0.01
			Pleistocene	1.6
		Tertiary — Neogene	Pliocene	5.3
			Miocene	
			Oligocene	23.7
		Tertiary — Palaeo-gene	Eocene	36.6
				57.8
			Palaeocene	66.4
	Mesozoic	Cretaceous	late Cretaceous	97.5
			early Cretaceous	144
		Jurassic	late Jurassic	163
			middle Jurassic	187
			early Jurassic	208
		Triassic		245
	Palaeozoic	Permian		286
		Carbon-iferous — Pennsylvanian		320
		Carbon-iferous — Mississippian		360
		Devonian		408
		Silurian		438
		Ordovician		505
		Cambrian		570
Proterozoic		Precambrian		
Archaean				4,600

FIGURE 2.41. Geologic time scale (from Driscoll 1986).

2.6 Summary

2.6.1 Geology and Hydrogeology of Onondaga Lake Drainage Basin

2.6.1.1 Bedrock Geology

The bedrock in Onondaga County is comprised of Paleozoic sedimentary rocks. Bedrock ages in the area range from Silurian to Devonian, becoming progressively younger southward due to the regional southerly 3 to 5° dip. The Appalachian Upland, located in the southern half of Onondaga County, is capped by the highly resistive Helderberg Group. The Helderberg Group forms an escarpment that defines the northern edge of the Appalachian Upland. The Salina Group, outcropping along the base of the Helderberg Escarpment, contains the evaporitic (halite and gypsum) bearing Syracuse Formation. Onondaga Lake is situated on the lowest member of the Salina Group, the Vernon Shale, which outcrops in the flat Ontario Lowlands lake plain in the northern half of the county. The Vernon Shale contains high concentrations of sulfate and calcite, which may contribute to the high sulfate and calcium concentrations in surface waters. Although halite, a natural source of sodium and chloride, is present south of Route 20, at depths greater than 1,000 ft (305 m), it is not present in the rock stratigraphy underlying Onondaga Lake.

2.6.1.2 Surficial Geology

Much of the surficial geologic expression in Onondaga County is a result of the Wisconsin Glaciation (12,000 to 14,500 years ago). Approximately 12,000 years ago the level of glacial Lake Iroquois dropped and a number of post-Iroquois lakes remained within depressions across the Ontario Lowlands. Onondaga Lake is located within one of the deeper remnant lake basins at the northern terminus of the Onondaga Trough. Most of the tributaries to Onondaga Lake lie in glacial meltwater drainage channels and glacial troughs that are filled with unconsolidated interbedded sands and gravels. Strata underlying Onondaga Lake consist of red sandstone overlain by a "tenacious" clay and alluvium (composed of gravel, sand, chocolate-colored clay, marl, and black swamp muck). Different well logs also note the presence of red clay, quicksand, sand, light-colored clay, and blue clay. The depth of the unconsolidated deposits has been measured to at least 414 ft (126 m). Surficial sediments surrounding Onondaga Lake consist of artificial fill, peat, marl, lodgement till, and glaciolacustrine silt and clay.

The thickest water bearing deposits adjacent to Onondaga Lake usually occur in places along the Seneca River and in the meltwater channels leading to Onondaga Lake. The water table gradient is very shallow in most areas near the lake. Water generally flows directly into the outwash deposits from the uplands by both surface and groundwater flow. Groundwater leaves the system mainly by discharge into the overlying surface water tributaries. Onondaga Lake is the ultimate discharge point for the surrounding watersheds.

2.6.2 Hydrogeology of Onondaga Lake Tributaries

2.6.2.1 Hydrogeology of Onondaga Creek

Salt beds of commercial size (Salina Group) are located at depths averaging 1,200 ft (366 m) in the Tully Valley. A natural solution front is located south of Route 20; bedrock salt deposits have been dissolved north of the solution front (i.e., no bedrock salt exists north of Route 20). The movement of the saline water is toward the valley center and then northward. The valley floor, with deep deposits of lacustrine and fluvial sediments, acts as a channel for groundwater flow. In the Tully Valley, Onondaga Creek is underlain by relatively impermeable, fine-grained pink lacustrine clays and is unlikely to have good hyrologic communication with the groundwater system.

Downstream of the Cedervale Valley confluence, Onondaga Creek receives influent primary from the shallow unconsolidated aquifer which is affected by local flow systems and high rates of runoff. The saline character of the groundwater in the Syracuse area is due to dissolution of the Salina Group bedrock salt deposits during the groundwaters natural migration northward from the higher elevations in the Tully Valley. In the Syracuse area, channel deposits act as a reservoir for the northward moving groundwater. The reservoir occupies an area underlying most of the city and extending northward under Onondaga Lake. The occurrence of saline groundwater in the unconsolidated valley sediments is more dependent on hydrologic communication between the bedrock and valley fill deposits (flow regime) rather than on the valley fill (aquifer) mineralogy. The saline content of Onondaga Lake has been described as less than would be expected if it were in direct contact with the underlying saline reservoir. This has been attributed to the presence of an insulating marl bed under the lake.

"Oil City" is an industrial and commercial complex located at the southeastern end of Onondaga Lake around the mouth of Onondaga Creek. Oil City has been used as a bulk storage and transfer facility for numerous industries since 1926. Industrial compounds utilized and stored in that area included fuel related hydrocarbons, synthetic organic chemicals, and PCBs. The assessment of potential contaminant transport pathways is difficult due to the long history and complex nature of the industrial uses in Oil City, multiple historic anthropogenic changes, and strong spatial heterogeneities in contaminant concentrations. Contaminant transport can occur through permeable sediments in the old Onondaga Creek streambed channels, along buried sewer and utility pipelines, and through abandoned and buried drainage channels from the salt production industry. In the area of Oil City, groundwater moves at a rate of 10^{-5} to 10^2 cm $\cdot$ s^{-1} and discharges primarily to the Barge Canal and Onondaga Lake. Organic contaminants found in the groundwater underlying Oil City include fuel oil, gasoline, kerosene, BTEXs, naphthalene, methylene chloride, acetone, and phenols. The types and concentrations are often site specific and their mobility rate will depend on the specific chemical, concentration, sediment composition in the contaminant pathway, and whether the pathway includes buried sewer

and utility pipelines. Based on available information, it is reasonable to conclude that the Oil City area is a potential source of chemical contamination to Onondaga Lake via groundwater transport.

2.6.2.2 Hydrogeology of Ley Creek

Located within the Ley Creek drainage basin are a number of large industrial and commercial complexes and several landfills. Available information includes descriptions of geologic conditions, types of contaminants, and directions of groundwater flow. Sites reviewed include Inland Fisher Guide (a known source of PCB contamination to Ley Creek), Crouse Hinds Landfill (contains inorganic materials, foundry sands, phenols, benzenes, and toluene), the Regional Market (miscellaneous fill materials, trash, methane and sulfide gases, and traces of PCBs), Salina Landfill (municipal wastes, some PCBs, heavy metals, semivolatile and volatile organics), and the Syracuse Landfill (unknown quantities of municipal wastes).

2.6.2.3 Hydrogeology of Ninemile Creek

Ninemile Creek originates at the mouth of Otisco Lake and flows north over the Appalachian Upland and down the Helderberg Escarpment. It then flows along the Ontario Lowlands lake plain through the Allied wastebed disposal area, and enters Onondaga Lake near Lakeland. Ninemile Creek presently contributes about 50% of the total chloride load to Onondaga Lake. Geologic evidence based on inferred and measured elevations of top of bedrock in the Ninemile Creek Valley indicate sources of salt are limited to valley reaches north of Warners-Amboy. Both regional groundwater discharge and groundwater mounding under the adjacent waste beds contributes to the groundwater discharge to Ninemile Creek north of Warners-Amboy. The waste beds were built directly on the surficial sediments (silty sands to gravels). Using Darcy's Equation, flux rates between the waste beds and the surficial deposits were calculated. The flux of water from waste beds to the surficial deposits could be as large as

25% of the total waste-bed precipitation infiltration. The remaining 75% is discharged from seepage faces in the waste beds. Much of the waste-bed leachate that enters the surficial deposits ultimately discharges to Ninemile Creek.

A chemical mixing model was used to determine whether the origin of the sources of salt to Ninemile Creek were naturally occuring brines or anthropogenic inputs. The mixing model used ionic ratios found in groundwater and surface-water samples upstream and downstream of the Warners-Amboy boundary and compared them to ionic ratios of waste-bed leachate and Tully brines. Based on the mixing model, it is highly probable that natural brine loading contributes less than 10% to the prevailing chloride load carried by Ninemile Creek. The remaining $\approx 90\%$ ultimately originates from waste-bed leachate.

2.6.3 Special Topics

2.6.3.1 The Tar Beds

Several studies have been prepared over the last ten years that identify and describe the degree of groundwater contamination around the Willis Avenue chlorobenzene production plant and the Semet Residue disposal ponds (tar beds) located along the southwestern shore of Onondaga Lake. From 1917 to 1970, organic based residues generated from the acid washing of coke light oils, prior to fractionation, were deposited in "ponds" that had been excavated in an area known as "Bed A" (a preexisting artificial fill area). The organic residues contain chlorobenzene, benzene, toluene, xylene, and naphthalene. Chlorobenzene isomers apparently entered the soil and groundwater system as a result of leakage between the different production buildings and the main plant grounds. The present consistency of the residues in the ponds ranges from a granular powder to a viscous material resembling tar sludge. The ponds cover an area of $89 \times 10^3 \, \text{m}^2$ (22 acres) with an average thickness of approximately 6 m (20 ft). The ponds contain approximately $2.8 \times 10^5 \, \text{m}^3$ of waste product.

Groundwater underlying the ponds flows in a northerly direction with Onondaga Lake acting as the primary discharge area. There is evidence of groundwater mounding and radial flow associated with the topographically elevated ponds. Pump tests were performed on the sediments underlying the chlorobenzene plant and hydrocarbon loading to Onondaga Lake was estimated. A total of $146 \, \text{kg} \cdot \text{d}^{-1}$ ($53 \, \text{mT} \cdot \text{yr}^{-1}$) of total dissolved hydrocarbons and $0.195 \, \text{kg} \cdot \text{d}^{-1}$ ($0.07 \, \text{mT} \cdot \text{yr}^{-1}$) of chlorinated hydrocarbons have been estimated to enter Onondaga Lake via groundwater transport. Although lake sediments do show evidence of contamination, lake water samples have had concentrations of total volatile organic compounds of less than $10 \, \mu\text{g} \cdot \text{L}^{-1}$.

2.6.3.2 History of the Salt Mining Process in the Tully Valley

Salt beds of commercial size underlie about $22,015 \, \text{km}^2$ of New York State, with thicknesses varying from 0.3 to $>152 \, \text{m}$. The Solvay Process Company began withdrawing salt brines in Syracuse near Onondaga Lake when it began the commercial production of soda ash in 1884. In the late 1880s salt beds were traced 36 km southward to the Tully Valley and found at depths of 335 to 427 m below the valley floor and in layers up to 31 m thick. The Solvay Process Company (later known as the Allied Chemical Company) was utilizing its first brine production well in 1889. By 1899 the company had 41 wells reaching an average depth of 366 m. The solution mining or "brining operations" utilized five different Brine Field Areas in the $12.4 \, \text{km}^2$ Tully Valley. The brining operation ran until 1985 and utilized a total of 170 wells (in addition to 7 core holes, 4 exploratory wells, and 1 disposal well). Approximately $3.89 \times 10^7 \, \text{m}^3$ of rock salt was removed from the Syracuse Formation. In recent years, approximately one billion gallons of nearly saturated brine was produced and conveyed to Solvay on an annual basis. From 1986 through 1988, LCP Chemical Company continued the operation, although at a much reduced rate. All brining operations in the Tully Valley ceased in July 1988. The extensive removal of brine, coupled with close spacing of wells, is believed to have caused localized subsidence and increased fracturing of the overlying rock formations.

2.6.3.3 Occurrence of Mud Volcanoes and Mud Boils in the Tully Valley

Mud and sand volcanoes and boils are formed by increased groundwater pore pressures resulting in liquefaction of sediments. Sources of increased pore pressures typically include heavy rainstorms, the onset of spring runoff, and subsidence of overlying sediments. The subsequent surfical discharge of sediments as a result of the pressure release forms a mud/sand volcano or boil (characterized by the absence of a sediment cone). In the Tully Valley, mud volcanoes range in diameter from <1 to >4 m. The angle of sediment accumulation varies from 0 to 12°. Mud boils are often <1 m in diameter and frequently <0.5 m in diameter. The mud boils and volcanoes produce a mauve, tannish colored turbid streamflow as a result of the high concentration of Otisco Valley clay found in suspension. The turbid streamflow enters and mixes with Onondaga Creek. In 1987 measurements of suspended solids in Onondaga Creek after storm events document that greater than 50% of the total suspended solid load in Onondaga Creek originated from the Tulley Valley.

Historical evidence first documents the presence of a mud volcano in the Tully Valley in October 1899. Other sources document mud boil and volcano activity before 1913 and persisting through the late 1930s and 1940s. Activity was again documented in the mid-to-late 1970s and again in the 1980s and 1990s. Whether the mud boils/volcanoes have a natural or anthropogenic origin is a continuing point of contention. Additional debate is also focused on whether the integrity of the geologic framework in the southern portion of the Tully Valley has been significantly affected by the long-term brining operations that took place, and whether the mud boils are

currently innerconnected with or influenced by the same hydrologic flow regime that is linked to the brine cavities and overlying bedrock aquifers.

2.6.4 Groundwater–Lake Water Interactions in Onondaga Lake

Lakes are in hydrologic communication with the underlying groundwater flow system. The role of lakes within the groundwater system can vary widely. Previous studies have assumed groundwater contributions to Onondaga Lake are negligible. A chloride budget for Onondaga Lake demonstrated a minor deficiency in chloride; although minor (<5%), the difference could be made up from chloride loading from groundwater. In order to directly calculate the potential groundwater flow to and from a lake, the hydrogeologic setting of the lake must be studied in detail. Regional and local flow systems must be identified along with areas of recharge and discharge. Hydraulic head measurements around the lake are necessary to detect groundwater mounding. Most hydrogeologic studies near Onondaga Lake have been site specific and have not addressed Onondaga Lake as a whole. Regional studies which include Onondaga Lake lack the detail needed to determine the hydrogeologic setting of Onondaga Lake. Study of regional maps have led to speculation that groundwater seepage into Onondaga Lake probably occurs in nearshore areas adjacent to Onondaga Creek and Ninemile Creek, where groundwater flowing northward in the glacial troughs intersects the lake. Regional groundwater flow in bedrock may discharge to Onondaga Lake along the flanks of the bedrock troughs, where till thins. One critical area for study is the outlet of Onondaga Lake. Regional groundwater flow in the Onondaga trough continues northward from the outlet. This would suggest that Onondaga Lake may actually be a flow through lake, receiving groundwater from the southern shores of the lake and releasing water from the northwestern shore near the outlet. The lack of significant topographic relief near the outlet precludes the possibility of any significant groundwater mounding. Without groundwater mounding, the mechanism for groundwater flow into the lake in that area does not exist.

References

Allen, R.L. 1982. Developments in Sedimentology. Elsevier Scientific, NY.

Apfel and Associates, Inc. 1968. Groundwater Geology of Onondaga County *In*: O'Brien and Gere Engineers, Onondaga County Comprehensive Public Water Supply Study, CPWS-21, Appendices, p. F1–F38.

Auer, M.T., Effler, S.W., Heidtke, T.M., and Doerr, S.M. 1991. Hydrologic Budget Considerations and Mass Balance Modeling of Onondaga Lake. Submitted to the Onondaga Lake Management Conference. Syracuse, NY.

Back, W. 1966. Hydrogeochemical Facies and Ground-water flow patterns in Northern Part of Atlantic Coastal Plain. U.S. Geological Survey Professional Paper 498-A. 42p.

Blasland and Bouck Engineers. 1990. Final Remediation Investigation Report: McKesson Corp., Bear Street Facility. Submitted to New York State Department of Environmental Conservation, Region 7, Syracuse, NY.

Blasland and Bouck Engineers. 1989. Hydrogeologic Assessment of the Allied Waste Beds in the Syracuse Area: Volume 1. Submitted to Allied-Signal, Inc. Solvay, NY.

Calocerinos and Spina Engineers. 1988. Preliminary Report on Subsurface Investigation Program: Regional Market, Syracuse, New York. Syracuse, NY.

Clark, J.V.H. 1849. Onondaga; or Reminiscences of Earlier and Later Times; and Oswego. Stoddard and Babcock, Syracuse, NY.

Clayton, W.W. 1878. History of Onondaga County, New York. D. Mason and Co. Publishers, Syracuse, NY.

Cooke, G.G., McComas, M., Bhargava, T.N., and Heath, R. 1973. Monitoring and Nutrient Inactivation on Two Glacial Lakes Before and After Nutrient Diversion. Kent State University, Center for Urban Regionalism Interim Research Report.

Cosner, O.J. 1984. Atlas of Four Selected Aquifers in New York. United States Environmental Protection Agency, Water Management Division. Contract Number 68-01-6389.

Dagan, G. 1986. Statistical theory of ground water flow and transport: pore to laboratory, laboratory to formation and formation to regional scale. *Water Resources Res.* 22: 120S–134S.

Deer, W.A., Howie, R.A., and Zussman, J. 1980. *An Introduction to the Rock Forming Minerals.* Longman, London.

Dionne, J.C. 1973. Monroes: a type of so-called mud volcanoes in tidal flats. *J Sediment Petrol.* 43: 848–856.

Doerr, S.M., Effler, S.W., Whitehead, K.A., Auer, M.T., Perkins, M.G., and Heidtke, T.M. 1994. Chloride model for polluted Onondaga Lake. *Water Res.* 28: 849–861.

Driscoll, F.G. 1986. *Groundwater and Wells.* Johnson Filtration Systems, St. Paul, MN,

Dunn Geoscience Corp. 1988a. Hydrogeologic Conditions at the Clark Property. Submitted to New York State Department of Environmental Conservation, Region 7, Syracuse, NY.

Dunn Geoscience Corp. 1988b. Hydrogeologic Conditions at the Marley Property: Proposed Carousel Center Mall, Syracuse, New York. Submitted to New York State Department of Environmental Conservation, Region 7, Syracuse, NY.

Dunn Geoscience Corp. 1988c. Hydrogeologic Conditions at the Buckeye Property, Syracuse, New York. Submitted to New York State Department of Environmental Conservation, Region 7, Syracuse, NY.

Effler, S.W., Doerr, S.M., Brooks, C.M., and Rowell, H.C. 1990. Chloride in the pore water and water column of Onondaga Lake, NY, U.S.A. *Water Air Soil Pollut.* 51: 316–326.

Effler, S.W., and Driscoll, C.T. 1986. A chloride budget for Onondaga Lake, New York, USA. *Water Air Soil Pollut.* 27: 29–44.

Effler, S.W., Johnson, D.L., Jiao, J.F., and Perkins, M.G. 1992. Optical Impacts and Sources of suspended solids in Onondaga Creek, U.S.A. *Water Res Bull.* 28: 251–262.

Empire Soils Investigations, Inc. 1990. Hydrogeologic Assessment: Mobil Oil Terminal, Bear and Solar Streets, Syracuse, New York. Submitted to New York State Department of Environmental Conservation, Region 7, Syracuse, NY.

Engineering Science, Inc. 1983. Engineering Investigations and Evaluations at Inactive Hazardous Waste Disposal Sites—Crouse Hinds, Onondaga County, New York. Submitted in association with Dames and Moore to New York

State Department of Environmental Conservation, Syracuse, NY.

Fetter, C.W., Jr. 1988. *Applied Hydrogeology.* Merrill, Columbus, OH.

Fitzpatrick, K. and Callaway, T.C. 1991. Memo to M.G. Perkins titled, "Tully Valley Exploratory Wells." Dated 12/9/91.

Freeze, R.A. and Cherry, J.A. 1979. *Groundwater,* Prentice-Hall, Englewood Cliffs, NJ.

Geraghty and Miller, Inc. 1982. Ground Water Quality Conditions of the Former Willis Avenue Plant, Allied Chemical Corp, Solvay, NY.

Getchell, F.A. 1983. Subsidence in the Tully Valley, New York. Master's Thesis, Syracuse University, Syracuse, NY.

Groundwater Technology. 1984a. Hydrogeologic Investigation: Allied Chemical-Willis Avenue and Lakeshore Area, Solvay, New York. Submitted to Allied-Chemical Corp. Solvay, NY.

Groundwater Technology. 1984b. Shallow Aquifer Investigation: Allied Chemical-Willis Avenue and Lakeshore Area, Solvay, New York. Submitted to Allied-Chemical Corp. Solvay, NY.

Herak, M., and Stringfield, V.T. 1972. *Karst, important karst regions of the Northern Hemisphere.* Elsevier, Amsterdam.

Higgins, G.L. 1955. Saline Ground Water at Syracuse, New York. Master's Thesis, Syracuse Unversity, Syracuse, NY.

Holmes, G.D. 1927. Report of Investigation by the Syracuse Intercepting Sewer Board: Onondaga Creek Flood Prevention, Syracuse. NY. Submitted to the Common Council of the City of Syracuse, Syracuse, NY.

Hubbert, M.K. 1940. The theory of groundwater motion. *J Geol.* 48: 785–944.

Johnston, L.R., Stutzbach, S.J., Weaver, P.L., and Zebuhr, R.H. 1975. Watershed Analysis for Onondaga Creek. SUNY College of Environmental Science and Forestry, Syracuse, NY.

Kantrowitz, I.H. 1964. Ground-Water Resources of the Syracuse Area: NYSGA Guidebook, 36th Annual Meeting, pp. 36–38.

Kantrowitz, I.H. 1970. Ground-Water Resources in the Eastern Oswego River Basin, New York. State of New York Conservation Department Water Resources Commission, Basin Planning Report ORB-2, Albany, NY. 129p.

Koch, G.H. 1932. The Hydrology of the Onondaga Drainage Basin. Master's Thesis, Syracuse University, Syracuse, NY.

Kosinski, A.J. 1985. "Analytical Critique" of Summary Consultants Report: Relationship of Brining Operations in the Tully Valley to

the Behavior of Groundwater and Geological Resources. Albany, NY.

Lee, D.R. 1977. A Device for Measuring Seepage Flux in Lakes and Estuaries. *Limnol Oceanog.* 22: 140–147.

Lundqvist, J. 1962. Patterned Ground and Related Frost Phenomena in Sweden. *Sviriges Geologiska Undersokning Arsbok* 55(7): 101.

Maxey, G.B. 1964. Hydrostratigraphic Units. *J Hydrology.* 4: 38–62.

McBride, M.S., and Pfannkuch, H.O. 1975. The distribution of seepage within lake beds. *U.S. Geol. Survey, J Res.* 3: 505–512.

Morgan, C.O., and Winner, M.D., Jr. 1962. Hydrogeochemical Facies in the 400 foot and 600 foot sands of the Baton Rouge Area, Lousiana. U.S. Geological Survey, Professional Paper 450-B.

Mozola, A.J. 1938. Contributions on the Origin of the Vernon Shale. Master's Thesis, Syracuse University, Syracuse, NY.

Muller, E.H. 1964. Surficial Geology of the Syracuse Field Area, Guidebook, New York State Geological Association, 36th Annual Meeting, pp. 25–35.

Muller, E.H., and D.H. Cadwell, 1986. Surficial Geologic Map of New York: Finger Lakes Sheet, 1:250,000, New York State Museum-Geological Survey, Map and Chart Series #40.

NYSDEC. 1989. Volume 1: Engineering Investigations at Inactive Hazardous Waste Sites: Phase II Investigation-Mercury Sediments, Onondaga Lake. Albany, NY.

NYSDEC. 1991. Memo to Onondaga Lake Management Conference: comments on Draft State of the Lake Chapter: Hydrogeologic Setting of Onondaga Lake. Dated 6/13/91.

New York State Department of Health. 1951. The Onondaga Lake Drainage Basin. Recommended Classifications and Assignments of Standards of Quality and Purity for Designated Waters of New York State. Oswego River Basin Survey Series Report No. 1. Albany, NY.

Noble, J.M. 1990. A Reconnaissance of the Natural Groundwater Geochemistry of Onondaga County, New York. Master's Thesis, Syracuse University, Syracuse, NY.

Noll, R.S. 1989. Geochemistry and Hydrology of Groundwater Flow Systems in the Lockport Dolomite, near Niagara Falls, New York. Master's Thesis, Syracuse University, Syracuse, NY.

O'Brien and Gere Engineers. 1969. Metropolitan Syracuse Sewage Treatment Plant Wastewater Treatment Facilities. Syracuse, NY.

O'Brien and Gere Engineers. 1989a. Ley Creek Dredged Material Area Field Investigation: General Motors Corp.-Fisher Guide Division, Syracuse, NY.

O'Brien and Gere Engineers. 1989b. History of the Semet Residue Ponds; Geddes, New York. Submitted to Allied-Signal Inc., Solvay, NY.

O'Brien and Gere Engineers. 1990. History of the Willis Avenue Plant, Petroleum Storage Facility, and Associated "Hot Spots"; Geddes, New York. Submitted to Allied-Signal Inc., Solvay, NY.

Onondaga County. 1971–1991. Annual Monitoring Report. Syracuse, NY.

Pagano, T.S., Terry, D.B., and Ingram, A.W. 1986. Geohydrology of the Glacial-Outwash Aquifer in the Baldwinsville Area, Seneca River, Onondaga County, New York. United States Geological Survey Water Resources Investigation Open File Report 85-4094.

Phillips, J.S. 1955. Origin and Significance of Subsidence Structures in Carbonate Rocks Overlying Silurian Evaporites in Onondaga County, Central New York. Master's Thesis, Syracuse University, Syracuse, NY.

Piper, A.M. 1944. A graphic procedure in the geochemical interpretation of water analyses. *Trans Amer Geophys Union.* 25: 914–923.

Post Standard Newspaper, October 20, 1899. Syracuse, NY.

Quirke, T.T. 1930. Spring Pits: Sedimentation Phenomena. *J Geol.* 38: 88–91.

Rickard, L.V., and Fisher, D.W. 1970. Geologic Map of New York, Finger Lakes Sheet (1:250,000). New York State Museum and Science Service Map and Chart Series Number 15.

Rickard, L.V., and Fisher, D.W. 1975. Correlation of the Silurian and Devonian Rocks in New York State. New York State Museum and Science Service Map and Chart Series Number 24. 16p.

Shaffer, A.L. 1984. Tully Brine Field History. New York State Department of Environmental Conservation, Albany, NY.

Shilts, W.W. 1978. Genesis of Mudboils. *Can J Earth Sci.* 15: 1053–1068.

Tanner, P. 1991. Inorganic Ground-Water Chemistry at an Industrial Site Near Oil City, and Onondaga Lake, Syracuse, New York. Master's Thesis, State University of New York College of Environmental Science and Forestry, Syracuse, NY.

Thomsen Associates. 1982. Phase II Geotechnical Investigations: Crucible, Inc. Solid Waste Management Facility. Submitted to New York State Department of Environmental Conservation, Region 7, Syracuse, NY.

Thomsen Associates. 1983. Memo to Crouse-Hinds Corporation regarding Preliminary Water Quality Interpretation of Monitoring Results from Landfill. Dated 4/26/83.

Tully, W.P. 1983. Summary Consultants Report. Relationship of Brining Operations in the Tully Valley to the Behavior of Groundwater and Geologic Resources. Submitted to Allied Chemical Corporation, Solvay, NY.

Tully, W.P. 1985. Draft Environmental Impact Statement for the Permitting of Drilling Two Wells for the Purpose of Solution Mining of Salt at the Tully Brine Fields in the Town of Tully, New York. Prepared for Allied Chemical Corporation, Solvay, NY.

Tuttle, M., Law, K.T., Seeber, L., and Jacob, K. 1990. Liquefaction and Ground Failure Induced by the 1988 Saguenay, Quebec, Earthquake. *Geotech J.* 27: 580–589.

United States Army Corp. 1991. Onondaga Lake, New York: Reconnaissance Report. Appendix A: Water Quality Technical Report. Submitted to the Onondaga Lake Management Conference, Syracuse, NY.

United States Environmental Protection Agency, Region II. 1974. Environmental Impact Statement on the Wastewater Treatment Facilities Construction Grants for the Onondaga Lake Drainage Basin.

Upstate Freshwater Institute, Inc. 1981–1990. Unpublished data.

von Engeln, O.D. 1961. *The Finger Lakes Region: Its Origin and Nature.* Cornell University Press, Ithca, NY.

Waller, R.M. 1983. Memo to NYSDEC Region 7, Syracuse, NY.

Washburn, A.L. 1956. Classification of Patterned Ground. *GSA Bull.* 67: 823–865.

Washburn, A.L. 1973. *Periglacial Processes and Environments.* Edward Arnold, London.

Wills, C.J., and Manson, M.W. 1990. Liquefaction at Soda Lake: Effects of the Chettenden Earthquake Swarm of April 18, 1990. *California Geology.* 88: 225–232.

Winkley, S.J. 1989. The Hydrogeolgy of Onondaga County, New York. Master's Thesis, Syracuse University, Syracuse, NY.

Winter, T.C. 1976. Numerical Simulation Analysis of the Interaction of Lakes and Ground Water. United States Geological Survey Professional Paper 1001.

Winter, T.C. 1981. Effects of Water-table Configuration on Seepage Through Lakebeds. *Limnol Oceanogr.* 26: 925–934.

Winter, T.C. 1983. The Interaction of Lakes With Variably Saturated Porous Media. *Water Resources Res.* 19: 1203–1218.

Winter, T.C., and Woo, M.K. 1990. Hydrology of Lakes and Wetlands. *In*: *Surface Water Hydrology*, M.G. Wolman and H.C. Riggs (eds.), Volume O-1 of The Geology of North America, Geological Society of America, pp. 159–187. Boulder, CO.

Yin, C., and Johnson, D.L. 1984. An Individual Particle Analysis and Budget Study of Onondaga Lake Sediments. *Limnol Oceanogr.* 29: 1193–1201.

Youd, T.L. 1973. Liquefaction, Flow and Associated Ground Failure. United States Geological Survey Circular 688, 12p.

3
Tributaries and Discharges

Steven W. Effler and Keith A. Whitehead

3.1 Hydrology of Onondaga Lake

3.1.1 Discharges: Measurements and Estimates

3.1.1.1 Gauged Inflows

Discharge to Onondaga Lake has been well monitored since the late 1960s/early 1970s, as the principal surface inflows have been nearly continuously gauged since that time. The United States Geological Survey (USGS) presently maintains nine continuous gauging stations in the watershed: three on Ninemile Creek, two each on Onondaga Creek and Harbor Brook, one on Ley Creek, and lake level is monitored at the marina on the east shore (Figure 3.1). A station is located proximate to the mouth of each of these four tributaries (Figure 3.1). The upstream station on Onondaga Creek (located at the city limit of Syracuse, above the combined sewer overflows) has played a particularly important role in the assessment of the relative contributions of the urban area in the loading of certain key pollutants from Onondaga Creek. The discharge from the metropolitan sewage treatment plant (METRO) has also been monitored continuously.

Important features of the flow monitoring program for the gauges proximate to the tributary mouths and the upstream Onondaga Creek station, including tenure, quality of measurements, and predictive expressions for missing data, are presented in Table 3.1. The upstream station on Onondaga Creek has been monitored the longest—since 1951. The station on Ley Creek was the last one established (1973). Considerable disparity in the quality of flow measurements prevails for the various stations (Table 3.1). The highest quality stream flow measurements are made on Onondaga Creek; measurements at intervals as short as 15 min are available and have been used in certain loading analyses (e.g., Heidtke 1989). In contrast, the accuracy of measurements at Ninemile Creek, Lakeland, is poor, largely as a result of backflow effects from the lake during low flow periods. Staff from USGS recommend that data from this station not be used at time scales of less than one day (e.g., daily average flows).

3.1.1.2 Ungauged Inflows

Ungauged inflows to Onondaga Lake include Bloody Brook, Sawmill Creek, Tributary 5A, East Flume, and the Seneca River (Figure 3.1). Instantaneous measurements in the outlet and aerial photography indicate the irregular occurrence of inflow from the Seneca River, into the northern end of Onondaga Lake, a situation that is promoted by the nearly equal

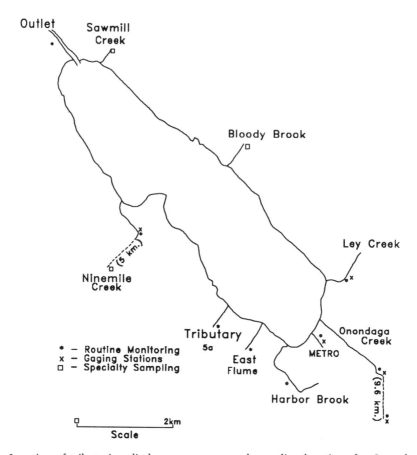

FIGURE 3.1. Location of tributaries, discharges, gauges, and sampling locations for Onondaga Lake, NY.

TABLE 3.1. Gauging of hydrologic inputs to Onondaga Lake.

Tributary	Descriptor	Gauge		Predictive expressions**
		Date established	Rating*	
Metropolitan sewage treatment plant	METRO			
Onondaga Creek, downstream	ONOND	9/70	good	
Onondage Creek, upstream	ONONU	9/51	good	
Ninemile Creek, downstream	NINE	12/70	poor	NINE = 0.853 · ONONU + 1.091 R^2 = 0.756
Ley Creek	LEY	12/72	good	LEY = 0.294 · ONONU + 0.632 R^2 = 0.632
Harbor Brook, downstream	HAR	10/70	fair	

* Categories of USGS, according to 1990 rating
   good: 95% of daily discharge values are within 10% of true value
   fair: 95% of daily discharge values are within 15% of true value
   poor: less than fair accuracy
** Expressions are for daily average flow in $m^3 \cdot s^{-1}$; expression for NINE based on regression of paired observations with ONONU for period July 1, 1975 to September 30, 1981; expression for LEY based on regression of paired observations with ONONU for period October 1, 1973 to September 30, 1981

surface elevations of the river and the lake. This phenomenon is described and analyzed in more detail in Chapter 4. Greater river inflow is expected in dry periods (e.g., summer) when the lake's tributary flows are reduced. Owens (personal communication) has estimated, based on a chloride mass balance conducted around the lake outlet, that about 29% of the surface inflow to the lake over the June–September interval of 1991 was from the Seneca River. This contribution is probably higher than for most years, as tributary inflow to the lake for that interval is ranked sixth driest out of a 19 year period. Sawmill Creek has a staff gauge near its mouth. Heidtke (1989) investigated flow conditions in Bloody Brook, Sawmill Creek, and Tributary 5A as part of a detailed analysis of phosphorus and fecal coliform loading to the lake conducted over the summer and fall period of 1987. Sixteen instantaneous flow measurements were made for each of these three tributaries. Correlations between these inflows and gauged discharges were evaluated through linear least squares regression; the resulting relationships are presented in Table 3.2.

The expressions of Table 3.2 are valuable in estimating flows from these ungauged tributaries in the absence of direct measurements. All of the populations on which these expressions are based are far from normally distributed (e.g., one or two high observations from a total of 16; Heidtke 1989), thus the high R^2 values are somewhat misleading. The flow from the East Flume has been reported as a constant in most years since closure of the

soda ash/chlor-alkali plant; a value of approximately $0.09 \, m^3 \cdot s^{-1}$ was reported for 1990.

3.1.2 Evaporation and Precipitation

The evaporative flux of water across a lake surface (J_v, $kg \cdot m^{-2} \cdot s^{-1}$) can be estimated according to the following expression (Harbeck 1958)

$$J_v = \frac{k_a}{R \cdot T_a} \cdot (e_s - e_a) \quad (3.1)$$

in which k_a equals mass transfer coefficient ($m \cdot s^{-1}$), R equals gas constant (4.62 millibars $\cdot m^3 \cdot kg^{-1} \cdot K^{-1}$), T_a equals absolute temperature of air (K), e_s equals saturation vapor pressure at the water surface temperature (millibars), and e_a equals actual air vapor pressure (millibars). Various empirical expressions to estimate k_a for lakes exist (Bowie et al. 1985). These are reviewed, and a function specific to Onondaga Lake is developed, in Chapter 9. However, to support the development of J_v, we have here adopted the predictive expression of Harbeck (1958),

$$k_a = 0.0021 \cdot W \quad (3.2)$$

in which W equals windspeed ($m \cdot s^{-1}$). The values of e_a and e_s can be estimated from dewpoint temperature (T_d, °C) and the water surface temperature (T_s, °C), utilizing the following expressions developed by Environmental Laboratory (1986)

$$e_a = 2.171 \cdot 10^8 \cdot e[-4157/(T_d + 239)] \quad (3.3)$$
$$e_s = 2.171 \cdot 10^8 \cdot e[-4157/(T_s + 239)] \quad (3.4)$$

The evaporative loss from the lake (E, $m^3 \cdot s^{-1}$) is calculated from J_v according to

$$E = \frac{J_v \cdot A_s}{Rho} \quad (3.5)$$

in which A_s equals surface area of the lake (m^2) and Rho equals density of water (= $1000 \, kg \cdot m^{-3}$).

The value of E was calculated daily for the April to October interval of 1987, 1988, and

TABLE 3.2. Predictive expressions for ungauged discharges to Onondaga Lake, based on gauged inflows.

Tributary	Descriptor	Predictive expression*	R^2
Bloody Brook	BLOOD	BLOOD = 2.86(HAR) − 0.21	0.79
Sawmill Creek	SAW	SAW = 0.175(LEY) + 0.17	0.72
Tributary 5A	T5A	T5A = 0.341(HAR) + 0.08	0.80

*In $m^3 \cdot s^{-1}$. See Table 3.1 for symbols of gauged tributaries incorporated in expressions

1989, based on daily meteorological data collected for T_a, W, and T_d at the National Weather Service station at Hancock Airport, and weekly (interpolated to daily) measurements of T_s. A time plot of daily precipitation (directly onto the lake; represented as an average daily flow rate, $m^3 \cdot s^{-1}$), measured at Hancock Airport, minus E for the April to October interval of 1987 is presented in Figure 3.2. Period average conditions for 1987, 1988, and 1989 are presented in Table 3.3. Despite substantial imbalances at shorter time scales, the sink of water from the lake associated with evaporation is approximately balanced with precipitation inputs over this period (e.g., Figure 3.2, Table 3.3). On average, the net deficit of precipitation and evaporation for 1987, a very dry year, was $0.07 \, m^3 \cdot s^{-1}$ (Table 3.3). These processes were nearly in balance in 1988, and in 1989, a wet year, the average net input (precipitation minus evaporation) was $0.09 \, m^3 \cdot s^{-1}$. The daily flow reported for the same period of 1987 at Harbor Brook, the smallest gauged tributary, appears in Figure 3.2 for comparison; the average value for the April–October interval of 1987, 1988, and 1989 appears in Table 3.3. The net inflow/loss associated with the precipitation/evaporation processes is small for Onondaga Lake, as it is substantially less than a minor tributary (see subsequent section for estimate of Harbor Brook's contribution to total lake inflow).

TABLE 3.3. Average conditions for selected hydrologic sources and evaporation, for April–October interval of three years.

Year	Precipitation ($m^3 \cdot s^{-1}$)	Evaporation ($m^3 \cdot s^{-1}$)	Harbor Brook discharge ($m^3 \cdot s^{-1}$)
1987	0.38	0.45	0.22
1988	0.38	0.37	0.18
1989	0.48	0.39	0.27

3.1.3 Hydrologic Loads to Onondaga Lake

3.1.3.1 Long-Term Average and Interannual Variations

Annual and long-term average hydrologic loading estimates are presented here for the 1971–1989 interval. Hydrologic loadings to the lake are estimated by summing the gauged and ungauged inputs. The discharge of spent cooling water from the West Flume upstream of the Lakeland gauge on Ninemile Creek (Chapter 1) until the spring of 1980 requires appropriate corrections for flow and material loading over the preceding period of the monitoring database. Net loading to the lake is determined by subtracting the West Flume loads from the gross loads determined at the Lakeland gauge. Constituent concentrations in the West Flume are assumed to equal lake concentrations. The temporal distributions of monthly average flows in Ninemile Creek at Lakeland and the West Flume over the 1976–1980 interval are presented in Figure 3.3a; the West Flume's contribution to the stream's flow over the same period appears in Figure 3.3b. West Flume discharges tended to be somewhat greater in summer. However, this flow was more seasonally uniform than the stream flow. Thus the cooling water discharge represented much more of the stream flow during the summer low-flow period; it was 40% of the stream flow in some summer months (Figure 3.3b). The average contribution to total stream flow over the 1976–1979 period was 16.5%, the range in annual contributions was 13.1–20.1%.

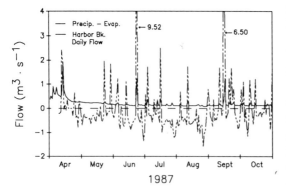

FIGURE 3.2. Measured daily precipitation (Hancock Airport), daily flow near the mouth of Harbor Brook, and estimated daily evaporation for Onondaga Lake, April–October 1987.

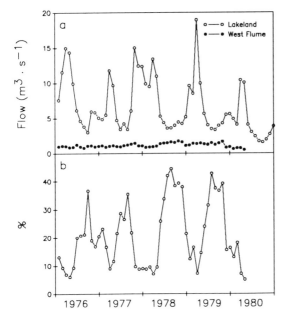

FIGURE 3.3. West Flume monthly flows, 1976–1980: (**a**) comparison to flows in Ninemile Creek at Lakeland, and (**b**) contribution to Lakeland flow.

The hydrologic budget presented for the lake here expands the period of analysis over that covered by Devan and Effler (1984). Further, they failed to accommodate the ungauged tributaries in their analysis. Inflow from the river is not accommodated in the budget presented here because of the lack of detailed estimates for extended periods of time. Thus hydrologic loading and flushing are underestimated, particularly during low tributary flow periods. Evaporation and precipitation are ignored here because the net effect of these processes is small compared to the magnitude of surface inflows (Figure 3.2). For example, this source/sink term is well within the level of uncertainty in measured flows from the gauged tributaries. Direct groundwater exchange with the lake basin is not considered (i.e., it is assumed to be zero). There is presently no evidence that such exchanges (e.g., springs) contribute significantly to the hydrologic budget of the lake. To the contrary, the success of a mass-balance chloride model for the lake that ignored groundwater exchange (Chapter 9) makes it

unlikely that direct exchange with the groundwater system is significant compared to the level of surface discharge received. However, groundwater inputs of low flow, but enriched concentrations, can significantly affect the material budgets of certain constituents (see Chapter 2). An example of this is the saline seeps that discharge to lower Onondaga Creek, which contribute little hydrologically, but are significant to the chloride budget of Onondaga Lake (Doerr et al. 1994; Chapter 9).

Missing daily flows from the gauged tributaries were estimated using the predictive expressions of Table 3.1. Daily flows for Bloody Brook, Sawmill Creek, and Tributary 5A were determined for the entire 1971–1989 interval from daily flows reported for Harbor Brook and Ley Creek according to the predictive equations of Table 3.2 (based on instantaneous measurements). Uncertainty in some of the estimates for the ungauged flows is compounded by the error propagated from expressions used to calculate daily flow during periods of missing data for presently gauged tributaries. A constant *net* flow ($0.09 \, \mathrm{m^3 \cdot s^{-1}}$) from the East Flume was maintained for the calculations.

The estimated long-term hydrologic loading conditions for Onondaga Lake are presented in Table 3.4. The average annual flows for the 1971–1989 period are presented for each gauged inflow and the sum of the ungauged inflows. The average total inflow to the lake on an annual basis has been $16.48 \, \mathrm{m^3 \cdot s^{-1}}$. The variability in annual inflow from each source is represented by the standard deviation and the relative standard deviation. The estimated percent contribution to flow (see pie chart of Figure 3.4) is based on the estimated total inflow received by the lake over the 19 yr period; a range of annual contributions is also presented (Table 3.4). The largest sources of water annually to the lake by a wide margin are Ninemile Creek and Onondaga Creek; together they represent nearly 62% of the surface flow received over the 1971–1989 interval. The domestic waste effluent from METRO represented about 19% of the inflow over this period. METRO's annual contribution ranged from 12 to 28%.

TABLE 3.4. Annual flow conditions for surface inflows to Onondaga Lake, and contributions to total inflow, for the period 1971–1989.

Tributary	Average	Annual flow ($m^3 \cdot s^{-1}$)		Watershed yield ($m \cdot yr^{-1}$)	% Contribution to total inflow	
		Standard deviation	Standard deviation/ average		Average	Range
Ninemile Creek	5.05	1.72	0.25	0.59	30.4	23.7–34.1
Onondaga Creek	5.22	1.31	0.25	0.55	31.4	27.6–34.1
METRO	2.99*	0.33	0.11	—	18.9	11.7–28.3
Ley Creek	1.28	0.33	0.26	0.52	7.7	5.9–9.5
Harbor Brook	0.38	0.15	0.38	0.41	2.2	1.6–3.6
Others**	1.56	0.51	0.32	—	9.3	7.3–13.4

*Does not include bypass discharges that occur during certain rainfall events; average value of 0.059 $m^3 \cdot s^{-1}$ over the period 1986–1990
**Sum of bloody Brook, Sawmill Creek, Tributary 5A, and the East Flume

The ungauged tributaries are estimated to have contributed more than 9% of the total inflow, nearly equal to the sum of the gauged flows from Ley Creek and Harbor Brook. Note that tributary flows have varied substantially over the 1971–1989 (e.g., relative standard deviations ≥0.25, Table 3.4), as a result of the variability in precipitation common to this region. By comparison, the discharge rate from METRO has remained rather uniform.

The average watershed yield (Table 3.4) was calculated by dividing the long-term average flow from the four gauged tributaries by the respective watershed areas. Harbor Brook had the lowest yield; Ninemile Creek had the highest. Differences in groundwater

inputs may explain the differences in yield reported. Most of the water from METRO originates from outside of the Onondaga Lake watershed.

The time course of total annual inflow and the annual inflows from the major sources is presented in Figure 3.5. Note that the Harbor Brook and Ley Creek inputs have been combined with the ungauged flow in this plot. The highest total inflow occurred in 1972 (annual average >24 $m^3 \cdot s^{-1}$), the year of Hurricane Agnes; the lowest in 1988 (~11 $m^3 \cdot s^{-1}$). Year-to-year variations were generally reflected in all of the gauged tributaries, as indicated in the correlation matrix of Table 3.5. These interannual differences were a

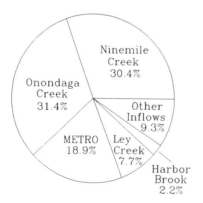

FIGURE 3.4. Long-term (1971–1989) annual contributions to total surface inflow to Onondaga Lake. Other inflows include the East Flume, Tributary 5A, Bloody Brook, and Sawmill Creek.

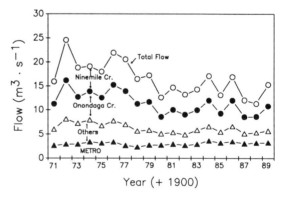

FIGURE 3.5. Annual surface flows to Onondaga Lake: total and major components for the period 1971–1989.

TABLE 3.5. Correlation matrix for annual flows and rainfall (at Hancock Airport).

No. Variable (flows and rainfall)	Correlation coefficients					
1. Total inflow	1.00					
2. Ninemile Creek flow	0.98	1.00				
3. Onondaga Creek flow	0.98	0.96	1.00			
4. Ley Creek flow	0.98	0.73	0.83	1.00		
5. Harbor Brook flow	0.90	0.98	0.82	0.66	1.00	
6. Rainfall	0.88	0.81	0.83	0.81	0.83	1.00
Variable no.	1.	2.	3.	4.	5.	6.

result of variations in precipitation, as shown by the high correlations between annual flow and annual precipitation measured at Hancock Airport. Inflow to the lake was lower in the 1980s than the 1970s because of reduced precipitation in the region.

3.1.3.2 Seasonal Hydrologic Loading

Strong seasonal variations in stream flow are commonly observed in temperate climates. The seasonality in hydrologic loading to Onondaga Lake is depicted in Figure 3.6a; the average monthly inflow from all tributaries (METRO *not* included) over the period 1971–1989 is plotted, with ±1 standard deviation bars. The monthly inflow from METRO is also included (Figure 3.6a). Standard deviations are not evident for this source because of the relatively uniform volume of waste discharged (e.g., ±1 standard deviation bars are small by comparison to dimensions of symbols used in Figure 3.6a) from this facility. Strong seasonal variations in hydrologic loading from the tributaries of the lake occur. The highest rates of inflow occur in March and April. Inflow decreases through late spring and summer. The minimum usually occurs in July, August, or September. Inflow increases to an intermediate level in late fall and winter. On average, for the 1971–1989 period, the tributary inflow received in April is nearly 4 times greater than that received in August. The details of the seasonality represented by the monthly means clearly have varied year to year, as reflected by the magnitudes of the

standard deviations. Relative standard deviations for the various months exceed 0.5.

The relatively uniform discharge from METRO, in combination with the strong seasonality in tributary hydrologic loading (Figure 3.6a), results in strong seasonality in

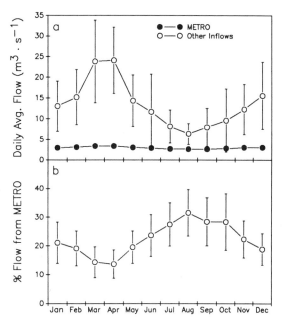

FIGURE 3.6. Seasonality in surface flows to Onondaga Lake: (**a**) average monthly total tributary and METRO inflows for the period 1971–1989, with ±1 standard deviation bars (modified from Effler and Hennigan 1995), and (**b**) average monthly percentage contribution of METRO inflow to total inflow for the period 1917–1989, with ±1 standard deviation bars (modified from Effler et al. 1995a).

the contribution of METRO to the total surface inflow (Figure 3.6b). METRO's contribution is the smallest in March and April, and the largest in July, August, and September, when primary productivity (Wetzel 1983) and recreational demands are greatest. The METRO discharge has contributed 28%, on average, of the surface inflow to the lake for the June–September interval. This discharge has been the single largest source of water to the lake for the month of August in 14 years over the 1971–1989 interval. This seasonality in the contribution of METRO, in combination with the lake's rapid flushing rate (see subsequent treatment), has important water quality implications for the lake.

3.1.3.3 Hydrologic Events

The short-term hydrologic characteristics of individual tributaries to Onondaga Lake can play an important role in determining the magnitude and temporal distribution of constituent loadings and associated water quality impacts. For any precipitation event, the characteristic shape of the response hydrograph at a particular location is strongly influenced not only by the character of the event and antecedent rainfall/runoff conditions, but by specific features of the contributing drainage area, including land use, topography, and soil type (Viessman et al. 1977). Thus tributaries that differ in these respects can be expected to exhibit different hydrologic responses to a similar precipitation input. The accuracy of tributary loading estimates, calculated as the product of flow and instream concentration, can be affected by the application of short-versus long-term hydrologic data. The significance of this effect will vary with the constituent of interest and the time scale established for the water quality assessment.

Stream-flow response at the two gauged stations on Onondaga Creek, Dorwin Avenue and Spencer Street, are presented, along with rainfall hyetographs based on Hancock Airport data (located 8.5 km from the creek mouth), for three rainfall events in the summer of 1990 (Figure 3.7). This treatment represents a

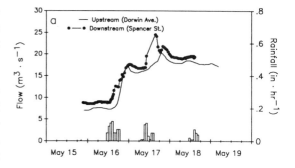

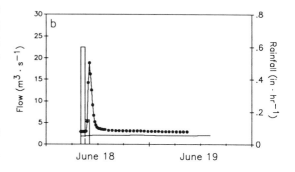

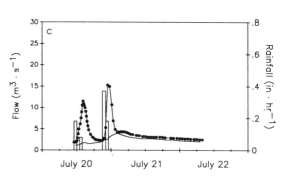

FIGURE 3.7. Hyetographs (at Hancock Airport) and hydrographs for Dorwin Avenue and Spencer Street for three events in 1990: (**a**) May 16–18, (**b**) June 18, and (**c**) July 20.

range of runoff events for two locations with rather disparate hydrologic settings. The contributing area upstream of Dorwin Avenue is mainly rural land and the impervious area is relatively small. Spencer Street is located just above the mouth of Onondaga Creek. The watershed of Onondaga Creek between these two stations is largely urban with a high fraction of impervious surface, in sharp contrast to the conditions above Dorwin Avenue. Additionally, a number of CSOs are located along the creek between these two stations. These differences in the character of the contributing watershed are usually clearly manifested in the hydrographs for the two locations. The time of concentration for the urban portion of the watershed is relatively short compared to that for the rural portion of the watershed. The hydrograph is generally more abrupt at Spencer Street, showing rapidly rising and receding limbs and a much higher instantaneous peak than observed at Dorwin Avenue (Figure 3.7).

A total rainfall of 1.19 in. was recorded at Hancock Airport for the event of May 16–18, 1990. The rainfall was distributed over three 6 to 8 hr intervals over a 56 hr period (Figure 3.7a). The urban site responded faster to the initial rainfall and more abruptly to the second increment of rain (Figure 3.7a). However, for the most part, the responses at these two locations were similar. The similarity of the responses for this event was probably largely a result of a preceding rainfall event of 1.22 in., within a 20 hr period on May 12–13. This earlier rainfall probably resulted in relatively high soil moisture and low infiltration rates within the normally pervious areas upstream of the Dorwin Avenue site, thereby causing an unusually abrupt stream-flow response similar to that of the downstream urban location. The stream response at these locations diverged greatly for the other events considered in Figure 3.7b and c.

The second rainfall event evaluated here was short, but intense: 0.75 in. of rain fell within 2 hr on June 18, 1990 (Figure 3.7b). Negligible rain was recorded for the previous nine days; thus, antecedent soil moisture was probably low and infiltration rates high within the pervious portions of the basin. Very disparate hydrographs were observed for the two stations for this event (Figure 3.7b). A minimal response occurred at the Dorwin Avenue site; stream flow remained virtually unchanged $(2.0–2.1 \, m^3 \cdot s^{-1})$. In sharp contrast, the hydrograph at Spencer Street demonstrated a rapid and abrupt increase in flow, from 2.8 to $18.9 \, m^3 \cdot s^{-1}$ within 2 to 3 hr of the initial precipitation. The rate of recession was equally rapid; the stream returned to a base flow of

about $3.5 \, m^3 \cdot s^{-1}$ within 2–3 hr of the peak flow. This pattern reflects a short period of intense surface runoff from the impervious areas below Dorwin Avenue in response to the correspondingly short, intense rainfall. Rainfall intensity varied substantially across the basin for this event (Smithgall et al. 1991).

Two distinct intervals of precipitation occurred during the third rainfall event of July 20: 0.42 in. of rain were measured over a 3 hr interval at the beginning of the event, and 0.55 in. were recorded, beginning 7 hr later, within a 2 hr period (Figure 3.7c). The antecedent soil moisture content was probably low and infiltration rates high within the pervious areas of the basin for this event, as there had been an absence of rainfall over an extended period of time. A gradual, comparatively sluggish flow response, with a single peak, was observed at Dorwin Avenue. Flow increased from $1.1 \, m^3 \cdot s^{-1}$ to a peak of $3.9 \, m^3 \cdot s^{-1}$, approximately 15 hr after the onset of precipitation (Figure 3.7c) and receded slowly thereafter. In contrast, the hydrograph at Spencer Street closely tracked the specific rainfall pattern for this event; two abrupt increases and subsequent decreases in flow are clearly resolved in the hydrograph (Figure 3.7c). The response was very rapid (2 to 3 hr) at this station for both periods of rainfall. The first interval of rainfall produced an increase in flow from 1.8 to a peak of $11.3 \, m^3 \cdot s^{-1}$; the second rainfall resulted in an increase from 2.3 to $15.3 \, m^3 \cdot s^{-1}$. Surface runoff contributions appeared to terminate at this urban location within 3 to 4 hr after the rainfall ceased (Figure 3.7c).

The flow responses presented for these two stations illustrate basic characteristics observed for other tributaries in the watershed. Specifically, Harbor Brook and Ley Creek (Figure 3.1) demonstrate flow responses similar to those observed on Onondaga Creek at Spencer Street because of their largely urban settings. However, the flow response to rainfall events at the Lakeland Station on Ninemile Creek tends to be more like that observed at Dorwin Avenue because much of the upstream area is nonurban and pervious.

3.1.4 Lake Level

The elevation of the surface of Onondaga Lake has been monitored inside the small harbor located on the east shore of the lake (Figure 3.1) since 1970. The harbor provides good isolation from the influence of wind-generated waves. The distribution of the yearly average lake elevation for the 1971–1989 period is presented in Figure 3.8. The range of these annual averages is only about 0.33 m. The maximum was in 1972, the year of Hurricane Agnes, and the minimum was in 1988. This yearly distribution is strongly correlated (R = 0.90) with the annual total tributary inflow to the lake (Figure 3.5). The linear least-squares regression expression that describes the relationship is

$$El = 0.023 \cdot Q_y + 110.37 \qquad (3.6)$$

in which El equals the elevation of surface water (m) and Q_y equals the total annual inflow rate ($m^3 \cdot s^{-1}$). This interplay largely reflects the regulation of the lake level by the control of the Seneca River level, associated with dam operation on the river system. In other words, increased flows result in increased water storage in the lake. However, only a very small percentage of increased flows are retained in the lake basin. For example, less than 1% of the increment in annual surface inflow received in 1972 compared to 1988 was reflected in increased in-lake storage in 1972.

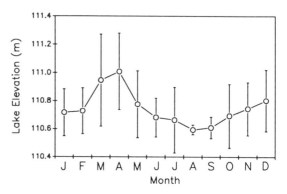

FIGURE 3.9. Seasonality in surface elevation of Onondaga Lake as the monthly average for the 1971–1989 interval, with ±1 standard deviation bars.

There is also a tendency for seasonal variations in lake elevation, associated with the seasonal hydrologic loading from surface inputs. The average lake surface elevation observed monthly over the 1971–1989 interval is presented in Figure 3.9. The bars about the average values represent limits of 1 standard deviation. The seasonality of average surface elevation has tracked rather closely (R = 0.86) the seasonal dynamics of total inflow (Figure 3.6). The highest elevations occur most often in April and March; the minimum usually in August and September. On average, this seasonal difference (at the resolution of one month) has been about 0.4 m. The seasonal storage of water (Figure 3.9) is also very small compared to the magnitude of the seasonal variation in total inflow (e.g., storage accounts for <3% of the difference in inflow, on average, between April and August (Figure 3.6)). Substantial year-to-year variability in elevation has been common for most of the months, except in late summer (Figure 3.9).

The daily distribution of the elevation of the surface of Onondaga Lake is presented for 1987 and 1989 in Figure 3.10. The year 1987 was a low inflow year, the second lowest over the 1971 to 1989 interval; 1989 approached average conditions for that period. Major differences in short-term structure are evident for these two years. It is probable that the distribution for each year is unique at this time scale. Note, for instance, the abrupt rise

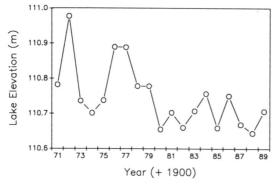

FIGURE 3.8. Yearly average surface elevation of Onondaga Lake for the 1971–1989 period.

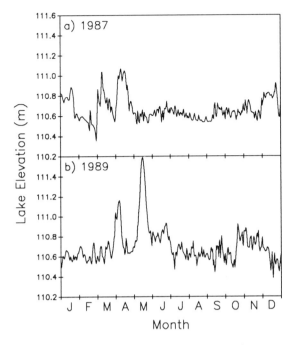

FIGURE 3.10. Daily distribution of surface elevation of Onondaga Lake.

in surface elevation of nearly 1.0 m in May 1989 that occurred in response to a major runoff event (Figure 3.10b). However, certain features of the long-term seasonality are manifested. The highest elevations occurred in March and April in 1987 (Figure 3.10a), and in April and May in 1989 (Figure 3.10b). Low values were observed in August and May of 1987, but also note the major, and largely progressive, decrease (>0.4 m) during the period of ice cover, from late January through early March. Lake surface elevation was low during the ice-cover period in 1989 and again in late summer and December. The major spring peaks are in response to runoff events. We attribute most of the smaller scale variations to the operation of the Seneca River.

The rather substantial variations in surface elevation experienced by Onondaga Lake may have significant ecological implications. Some of these will be addressed in subsequent chapters. For example, the changes in elevation represent changes in hydrostatic pressure. These pressure changes may contribute importantly to the dynamics in gas ebullition

(gases released from the sediment that pass through the surface) observed for the system (Chapter 5). Further, the changes in elevation impact the flora and fauna of the near shore area. For example, periods of drawdown during ice cover may damage or destroy the root systems of exposed aquatic plants (Chapter 6).

3.1.5 Lake Flushing

The most detailed analysis of lake flushing has been that of Devan and Effler (1984); they presented annual complete-mixed flushing rate estimates for the 1970–1981 period. The analysis presented here supersedes that effort, as it: (1) includes surface inflows not previously considered, (2) covers the interval 1971–1989, (3) considers seasonal flushing on a complete-mixed basis, (4) considers flushing of the epilimnion, and (5) evaluates the influence of the METRO discharge on flushing.

3.1.5.1 Complete-Mixed Flushing

The yearly complete-mixed flushing rates for the 1971–1989 period are presented in Figure 3.11. These values were determined from the annual total surface inflows of Figure 3.6 and the lake volume ($131 \times 10^6\,\text{m}^3$) reported by Owens (1987), corresponding to a lake surface elevation of 110.55 m. These did not differ significantly from estimates that utilized lake volumes that had been adjusted for interannual (Figure 3.8) and seasonal (Figure 3.9)

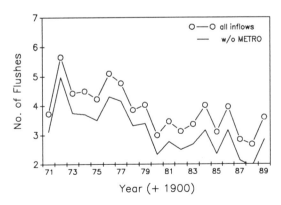

FIGURE 3.11. Annual complete-mixed flushing rates for Onondaga Lake for the 1971–1989 period.

differences in lake surface elevation. The flushing rate estimates for the 1971–1981 period are about 10% higher than previously reported (Devan and Effler 1984) due to the inclusion of inflows previously not considered ("other" inflows of Figure 3.4). Onondaga Lake can be described as a high flushing rate lake. For example, it greatly exceeds the rates of flushing of all the Finger Lakes (Schaffner and Olgesby 1987). The average flushing rate for the 1971–1989 period was 3.9 flushes · yr^{-1}; the range was 2.7 to 5.7 flushes · yr^{-1}. Note the lower flushing rates that prevailed in the 1980s compared to the 1970s due to reduced rainfall in the later period (Table 3.5). The high flushing rate of the lake has important implications for remediation efforts, as the response time (the time it takes to reach a new steady state) is short. Treating the lake as a completely-mixed system, and for the range of flushing rates presented in Figure 3.11, it would take only 0.5 to 1.1 yr to reach 95% of a new steady-state concentration following an abrupt change in the input of a conservative substance (e.g., chloride). The response time is even shorter for reactive materials (Thomann and Mueller 1987).

An analysis of seasonal flushing in Onondaga Lake for the complete-mixed assumption for the 1971–1989 interval is presented in Figure 3.12; the seasonality tracks that presented for seasonal inflows in Figure 3.6a. The highest rates of flushing generally

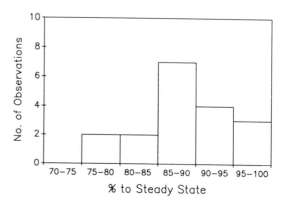

FIGURE 3.13. Distribution of percentage to steady-state conditions for conservative substances in Onondaga Lake over the November 1 to April 28 interval for the 1971–1989 period.

occur in March and April (~0.52 flushes per month, on average); the lowest in July, August, and September (~0.2 flushes per month on average). The high rate of flushing of the lake raises the question "to what extent do conditions in fall, at the onset of fall turn-over (entrainment of hypolimnetic accumulations of various substances), influence those in spring (e.g., onset of stratification) of the following year." To address this issue the complete-mixed flushes were calculated for the November 1–April 28 (reasonable boundaries for the onset of complete fall turn-over and summer stratification since 1987 (Chapter 4)) interval (n = 18) over the 1971–1989 period. The extent to which steady state is approached for a conservative substance (e.g., chloride), for the complete-mixed assumption can be calculated according to

$$\text{percent of steady state} = (1 - e^{-F}) \cdot 100 \tag{3.7}$$

in which F equals number of flushes. The median value of F for the November 1–April 28 interval, over the 1971–1989 period (n = 18), was 2.3, the range was 1.55 to 3.44; the corresponding median and range of percentages to steady state are 90% and 79 to 97%. The distribution of percentage to steady-state conditions for the hydrologic conditions of 1971–1989 is presented in Figure 3.13. In 14 of the 18 intervals, the percentage to steady

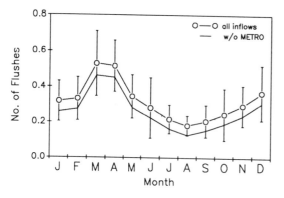

FIGURE 3.12. Seasonality in complete-mixed flushing rates for Onondaga Lake for the 1971–1989 period, with ±1 standard deviation bars.

state was greater than 85%. This analysis indicates that only a small faction of the conservative substances present in the lake at the onset of fall turnover would be present by the onset of stratification the following spring. The analysis is conservative (i.e., effective removal would be greater) for reactive materials. However, this tends to be compensated for by the incomplete mixing that undoubtedly prevails during periods of ice cover.

3.1.5.2 Epilimnetic Flushing

Surface discharges to stratifying lakes are not completely mixed with lake waters during stratification periods, but rather mix with, and are largely incorporated within, the warm less-dense waters of the epilimnion. This seasonal feature is not of great significance to many water quality characteristics of lakes with long detention times (i.e., low flushing rates). However, in high flushing rate systems; such as Onondaga Lake, substantial fractions of various materials received early in the stratification period may pass through the lake before the onset of fall turnover, without being incorporated into the total lake volume. This perspective was complicated for Onondaga Lake by the plunging inflow phenomenon during the operation of the soda ash/chlor-alkali manufacturer (Owens and Effler 1989), but this complication has been significantly ameliorated by the closure of this facility (Chapter 4).

An analysis of the flushing of the epilimnion of Onondaga Lake has been conducted for the May to September interval of three years (Figure 3.14), based on the temporal distribution of the lower boundary of the epilimnion (e.g., see Chapter 4) and the coupled volume of this layer specified by the hypsographic data of Owens (1987), and the daily summed surface inflow rate. The year-to-year differences are associated almost entirely with differences in flow, as the variation of the volume of the epilimnion (attributable to the stochastic character of forcing meteorological conditions (see Chapter 4)) is much smaller by comparison. The epilimnion of the lake had

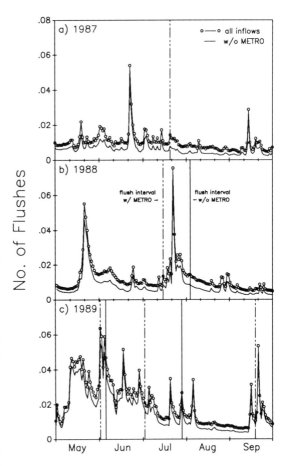

FIGURE 3.14. Analysis of the flushing of the epilimnion of Onondaga Lake over the May–September interval.

been flushed through by about mid-July of the low inflow (Figure 3.4) years of 1987 (Figure 3.14a) and 1988 (Figure 3.14b). The epilimnion had nearly flushed through a second time in 1988 by the end of September. In the more typical inflow year of 1989, the epilimnion had already flushed twice by early July, and flushed a third time by about mid-September (Figure 3.14c).

3.1.5.3 Influence of METRO on Flushing

Annual and seasonal flushing rates determined for the case of the absence of the METRO discharge, for the hydrologic record of 1971–1989, are included in Figures 3.11 and 3.12. The fraction of the existing inflow from METRO that would continue to enter the lake in the absence of the facility is unknown. In these analyses total elimination has been assumed. The average annual flushing rate for this case is 3.2 flushes/yr; the range is 1.9 to 5.0 flushes/yr. Suppression of the METRO inflow results in a nearly uniform incremental seasonal reduction in flushing rate. Without the METRO inflow the epilimnion would not have flushed completely in the May to September interval of 1987. The first flush would have been delayed until after the runoff event of early August in 1988, and only two complete flushes of the epilimnion would have occurred over that interval in 1989.

3.2 Material Loading

3.2.1 Introduction

Lake water quality is largely a manifestation of material loading rates. There are both external and internal (e.g., via sediment release processes) sources of materials to the water columns of lakes. However, internal inputs for the most part reflect earlier external inputs. Internal loading of various key materials in Onondaga Lake is described and quantified in subsequent chapters. Entry routes for external loads include inflows from tributaries and discharges, atmospheric deposition, and direct groundwater inputs to the lake basin. Our focus here is on the specification of the concentrations of selected materials of water quality concern in the tributaries and inflows to Onondaga Lake, and the development of accurate mass loading rates for these materials. Atmospheric and direct groundwater inputs for the most part are so relatively small for materials of present water quality concern that they can be ignored. External loading estimates are invaluable in establishing cause-and-effect relationships for various lake water quality problems, and essential to drive related mass-balance water quality models.

Inputs are usually divided into two broad categories: point sources and nonpoint sources. Point sources are those inputs that have a well-defined point of discharge. The

METRO discharge, which enters the lake directly (Figure 3.1), is an example of this type of discharge. The overflow(s) from the Solvay waste beds during their active filling phase (most recently on Ninemile Creek), an industrial discharge, was also a point source input. The "combined sewer overflows" (CSOs) represent a set of dispersed point sources along Onondaga Creek, Ley Creek, and Harbor Brook. Nonpoint sources include: (1) agriculture, (2) silviculture, (3) urban and suburban runoff, (4) groundwater, and (5) atmospheric deposition (Thomann and Mueller 1987). In contrast to point sources, nonpoint sources are diffuse in nature, though the space scale can vary greatly. For example, the space scale of rural runoff is quite large compared to that associated with leaching from the Solvay waste beds. A number of nonpoint inputs are often transient, associated with runoff events. The Onondaga Lake surface inputs represent a particularly interesting suite of sources for a detailed loading analysis because of the diversity of input types and materials of water quality concern, and the major changes that have occurred in recent history (e.g., since the late 1960s) in response to reclamation efforts and closure of the soda ash/chlor-alkali facility.

The temporal character of loading inputs and related lake impacts strongly influences the sampling needs of tributary monitoring programs intended to establish cause-and-effect relationships between inputs and receiving water quality. The vehicle to evaluate cause-and-effect relationships is usually a model, often in the form of a mechanistic mass-balance model. Several of these models have been developed for the lake. Water quality samples should be collected over the range of flow and time critical to the identified water quality issue. For a given number of concentration samples, loading estimates will usually be of greater precision if the sampling schedule is weighted toward high-flow seasons and storm events, as these intervals often account for a high percentage of the annual seasonal loading (Walker 1987). Sampling programs to support the evaluation of short-term impacts must be time-intensive enough to fully resolve lake impacts and accurately specify the contaminant load responsible for the impact. For example, high-frequency sampling (e.g., hourly) is required to accurately specify indicator bacteria loads received from CSOs during major runoff events that cause contravention of public health standards in Onondaga Lake (Canale et al. 1993). However, other water quality issues with longer time scales (nutrients) generally require less detailed tributary sampling, although the program must always be scrutinized to assure that representative estimates of material loading emerge.

3.2.2 Description of Monitoring Programs

The tributaries of the lake and METRO have been monitored by various parties over the period 1968–1990, mostly with the intention of supporting the development of loading estimates for various materials of water quality interest. These monitoring programs have differed greatly with respect to tenure, inflows sampled, parameters measured, and sampling frequency. Selected features of some of these programs are summarized in Table 3.6; sampling stations are located on Figure 3.1. All of the details of the monitoring programs cannot be included within the constraints of Table 3.6. Further, less extensive programs have not been included. For example, the New York State Department of Environmental Conservation (NYSDEC) has recently monitored Ninemile Creek at Amboy and Lakeland to assess inputs related to the waste beds (five times per year, 1987–1990). The NYSDEC has also conducted sampling profiles along Onondaga Creek (four profiles over the period summer 1989 to spring 1991) to assess impacts associated with the mud boils. Onondaga and Ninemile Creeks have also been included in the Rotating Intensive Basin Studies (RIBS) monitoring program conducted by the NYSDEC. The program incorporates water column, bottom sediment, and macroinvertebrate and fish monitoring in an effort to produce an integrated assessment of ambient water quality. Water samples were collected

up to ten times per year. Metals, organics, and conventional pollutants are included in the program.

The long-term program of Onondaga County, initiated in 1968 (Table 3.6), is unique with respect to its tenure and number of parameters monitored (Onondaga County 1988). A number of major changes in pollutant loading, associated with reclamation measures and changes in industrial activity, can be documented from this database (see subsequent treatment). Further, results from this program provided direction for subsequent speciality loading studies conducted over the period 1987–1990. These additional speciality studies focused on fewer water quality issues. They were conducted to support the development of mechanistic water quality models and to isolate the location of entry of selected pollutants along the length of the two major natural tributaries to the lake. Monitoring programs have been conducted at a particularly high frequency to quantify fecal coliform (FC) and total phosphorus (TP) inputs associated with major runoff events. The METRO effluent has been monitored intensively, outside of the facility's permit requirements, for nitrogen and phosphorus in recent years because of the major role this source plays in overall loading of these important materials to the lake (see subsequent treatment in this chapter). Flow- weighted, 24-hr, composite samples have been collected at METRO for the recent specialty loading studies. All other samples have been grab type. Parameters measured routinely by the Department of Drainage and Sanitation as part of its permit requirements are listed in Table 3.7. Most of these data are available since 1981. In addition to the parameters listed in Table 3.7, the plant is also required to monitor certain heavy metals, organics, and toxics in the effluent, as well as to test the effluent's toxicity.

3.2.3 Load Calculation Techniques

Loading estimates are often highly sensitive to the specific calculation procedure. Selection of

TABLE 3.6. Description of tributary monitoring programs.

Program no.	Description	Tenure	Parameters	Frequency	Responsible party
1	Long-term, all significant tributaries	Since 1968	Nutrients, major ions, metals, indicator bacteria, pH, and more*	$2 \cdot mo^{-1}$	D&S**
2	Specialty loading studies	Summer 1987– spring 1990			
a.	Bacteria (FC) loading, all major sources	Summer 1987	Fecal coliforms (FC)	$1-10\,d^{-1}$	D&S/UFI***
b.	Phosphorus loading				
(1)	All significant tributaries	Summer 1987	Total phosphorus (TP)	$1-10\,d^{-1}$	UFI/D&S
(2)	METRO effluent	1988 1989 1990	TP	$1\,wk^{-1}$ $5\,wk^{-1}$ $5\,wk^{-1}$	UFI
(3)	Onondaga Creek upstream (Dorwin) versus downstream (Spencer)				
	(a) Background	1988–1889	TP	$3\,wk^{-1}$-$2\,d^{-1}$	UFI
	(b) Runoff events	1989–1990	TP	$1-2\,hr^{-1}$	NYSDEC****
(4)	Ninemile Creek upstream (Amboy) versus downstream (Lakeland)	Spring 1989– Spring 1990	TP, soluble reactive phosphorus (SRP)	$3-5\,wk^{-1}$	UFI
c.	Nitrogen loading				UFI
(1)	METRO effluent	Mid-summer 1988–1990	NO_3, NO_2, NH_3, less TKN	$5\,wk^{-1}$	
(2)	Ley and Onondaga Creeks	Summer–fall 1988	NO_3, NO_2, NH_3, less TKN	$5\,wk^{-1}$	
(3)	Ninemile Creek upstream (Amboy) downstream (Lakeland)	Spring–fall 1989 Summer–fall 1988 Spring–fall 1989	NO_3, NO_2, NH_3, less TKN	$5\,wk^{-1}$	
d.	Chloride loading		Chloride (Cl)		UFI
(1)	Ninemile Creek				
(a)	Lakeland	Summers of 1980 and 1981		$4\,d^{-1}$	
		Spring 1989– fall 1990		$5\,wk^{-1}$	
(b)	Amboy	Spring 1989– spring 1990		$5\,wk^{-1}$	
(c)	Other inflows	Spring 1990– fall 1990		$1-5\,wk^{-1}$	
e.	Sulfate loading		Sulfate (SO_4)		UFI
(1)	Ninemile (Lakeland and Amboy)	Spring–fall 1989		$2-5\,wk^{-1}$	
(2)	Other inflows	Spring–fall 1989		$1\,wk^{-1}$	

* See recent lake monitoring reports (e.g., Onondaga County 1988)
** Department of Drainage and Sanitation, Onondaga County
*** Upstate Freshwater Institute
**** New York State Department of Environmental Conservation

TABLE 3.7. Monitoring of METRO effluent required by permit.

Parameter	Frequency	Sample type
BOD_5	$1 \cdot day^{-1}$	24-hr composite
Suspended solids	$1 \cdot day^{-1}$	24-hr composite
Fecal coliform	$1 \cdot day^{-1}$	grab
Total Kjeldahl nitrogen	$2 \cdot wk^{-1}$	24-hr composite
Ammonia	$2 \cdot wk^{-1}$ before March 1989, $1 \cdot d^{-1}$ thereafter	24-hr composite
pH	$6 \cdot d^{-1}$	grab
Settleable solids	$6 \cdot d^{-1}$	grab
Residual chlorine	$6 \cdot d^{-1}$	grab
Total phosphorus	$1 \cdot d^{-1}$	24-hr composite
Total dissolved solids	$2 \cdot wk^{-1}$	24-hr composite

a specific loading calculation method should depend on the temporal goals of the estimates (e.g., daily, seasonal, or annual), and the relationships between constituent concentration and flow and between concentration and time. Desired properties of the loading estimates include minimum bias and minimum variance. A biased procedure yields an incorrect answer, regardless of the number of observations, whereas variance can be reduced usually by increasing the number of random samples. Bias in loading estimates has two sources: unrepresentative sampling and the use of an inappropriate calculation method (Walker 1987). Commonly, unrepresentative sampling results from differences in the distributions of flow during the times of sampling and the entire loading calculation period. However, this can be accommodated usually by the selection of an appropriate calculation procedure.

Walker (1987) has developed interactive, user-friendly software, entitled FLUX, that provides several procedures to estimate loadings. The program is intended to estimate loading from intermittent grab or event sampling over a period for which complete flow data (e.g., daily average or detailed instantaneous flow data from a USGS monitoring station) exists. Seven alternative calculation methods are provided. The first six provide a range of options to weigh limited observations of concentration with an essentially continuous flow record. FLUX provides

an option to stratify the samples into groups based on flow or time for the first six methods. Stratifying the sample often increases accuracy and reduces potential biases in loading estimates. The variances of the estimated mean loadings are calculated for the first six methods to provide relative indications of error. FLUX also includes a variety of graphic and statistical diagnostics to assist the user in evaluating data adequacy and in selecting the most appropriate calculation method and stratification scheme for loading estimates.

The first method, calculation of the average loading as the simple mean of individual determinations (e.g., product of concentration from grab sample and daily flow), is the simplest of the calculation schemes. This method has been adopted by the County in the annual monitoring program (Onondaga County 1971–1991) to calculate annual loads. This completely ignores the unsampled flow record and often has higher variance than the other methods (Walker 1987). The method gives unbiased results only if the samples are taken randomly with respect to the flow regime. Simulation analyses (Walker 1987) indicate that this method is most appropriate for situations in which concentration tends to be inversely related to flow (i.e., loading does not vary greatly with flow). The five other (more sophisticated) concentration-flow algorithms (e.g., Bodo and Unny 1983; Mosteller and Tukey 1978; Walker 1987) included in FLUX are described by Walker

(1987). The last calculation procedure option provided by FLUX, interpolation, differs fundamentally from the other six methods. In this procedure, the sample data are treated as a time series; concentrations on unsampled dates are estimated by interpolation between samples dates. This approach is appropriate in situations in which there is a significant trend or seasonal component of the concentration variance that is independent of flow or when samples are collected frequently. This approach has been used here to develop high-frequency loading estimates for a number of parameters that drive related dynamics water quality models for Onondaga Lake.

A fundamental first step in developing loading estimates is to investigate the relationship between constituent concentration and flow. A strong relationship becomes the basis for predicting concentrations for periods of no samples from flow data, thereby generating a continuous distribution of concentration to match flow to support loading calculations. A number of functionalities are commonly investigated that may have a strictly empirical basis (e.g., Walker 1987) or a more mechanistic basis (e.g., Johnson 1979; Manczak and Florczyk 1971). Particular emphasis has been placed on evaluation and development of functionalities for various constituents for the two largest tributaries to Onondaga Lake—Onondaga Creek and Ninemile Creek. Empirical relationships (segmented or unsegmented) that emerge from plots of log concentration versus log flow are perhaps the most widely used to support loading estimates (Thomann and Mueller 1987). FLUX supports log–log relationships in loading estimates. A recurring relationship observed for several constituents (particularly the major ions) in the Onondaga Lake tributaries is decreasing concentration with increasing flow, manifested as a hyperbola, with an equation $y = a/x + b + d$ where y equals concentration and x equals flow; Manczak and Florczyk 1971). The relationship is linear when concentration is plotted against the inverse of flow (subsequently referred to as an inverse flow relationship). The relation-ship usually reflects dilution (often referred to as a dilution model) of localized inputs such as wastewater discharges and enriched groundwater seeps. The dilution model is not presently supported by FLUX.

3.2.4 Suspended Solids

3.2.4.1 Background

Sediment received in tributary (i.e., alloch-thonous) inputs can be important with respect to sedimentation and clarity in receiving lakes. Further, these particles provide surfaces for potentially important processes involving the cycling of nutrients (Kramer et al. 1972) and organic pollutants (Loder and Liss 1985), and may function as nuclei for in-lake precipitation of minerals (Stumm and Morgan 1981). Only recently (1988) have suspended solids (SS) measurements been incorporated into the routine monitoring program for Onondaga Lake. Certain earlier specialty studies included SS measurements. Yin and Johnson (1984) developed a SS budget for Onondaga Lake for 1981 and partitioned the particles into 19 chemistry-based classes. They used scanning electron microscopy with automated image analysis and x-ray energy spectroscopy (SAX), a technique of individual particle character-ization (see Chapter 5 for application of this technique to characterization of the particle population of the lake). Suspended solids concentrations were also monitored in the principal tributaries of the lake during several storm events during the summer of 1976.

A SS loading budget is presented here based on the most recent tributary monitoring data. This budget is compared to the one prepared by Yin and Johnson (1984) for 1981. The chemical partitioning of particle types in the 1981 allochthonous sediment load established by Yin and Johnson (1984) is reviewed. This is updated for Onondaga Creek by more recent measurements. Finally, we document the impact of SS loading on the optical aesthetics at the mouth of Onondaga Creek and investi-gate the upstream origins of this material using both aggregate measures and SAX.

3.2.4.2 *Suspended Solids Loading Budgets*

Grab samples have been collected for SS analysis approximately twice per month from positions near the mouth (Figure 3.1) of the major hydrologic inputs since 1988. Statistics describing the concentration of SS in these tributaries over this period are presented in Table 3.8. The highest concentrations are observed in Onondaga Creek; the lowest in Harbor Brook. The relationship between the instantaneous concentrations and daily average flow is evaluated for Onondaga Creek and Ninemile Creek in a log–log format in Figure 3.15. A distinct positive relationship is apparent for Onondaga Creek, but the relationship between concentration and flow is not strong for Ninemile Creek. Measurements made in 1976 have been added to the Onondaga Creek plot for comparison. Averages of suspended solids concentrations have been used for the 1976 data set when more than one measurement was made in a day. These earlier observations fall within the range of the recent data, though they tend to be slightly higher. The relationship between SS concentration and flow for Onondaga Creek over the 1988–1990 period is described by

$$[SS] = 14.76\,Q^{0.558} \qquad (3.8)$$

The correlation coefficient (R) for the log-transformed version of this expression is 0.504.

FLUX was used to develop loading estimates for the inputs of Table 3.8 for the 1988–1990 period, based on the discrete measurements of SS and daily average flow rates. Estimates of

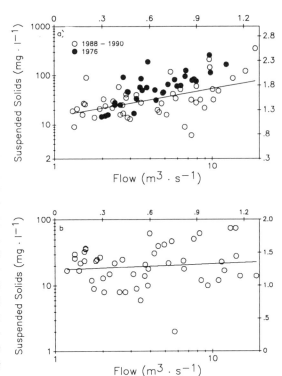

FIGURE 3.15. Evaluation of relationship between concentration of suspended solids and flow, for the 1988–1990 period: (**a**) Onondaga Creek, and (**b**) Ninemile Creek.

total suspended solids loading (for Onondaga Creek and Ninemile Creek) obtained with the various computational techniques offered in FLUX are presented in Table 3.9, along with the associated values of the measure of performance, the coefficient of variation. The best estimates of loading were obtained with the "jackknife" computation method (method 6 REG-3, Table 3.9). Note that use of the

TABLE 3.8. Suspended solids concentrations in surface discharges to Onondaga Lake, 1988–1990.

Source	Suspended solids concentrations (mg · L⁻¹)				
	Average	Median	Maximum	Minimum	n
Onondaga Creek	47.9	29	356	6	48
Ninemile Creek	24.7	22	74	2	49
METRO	13.8	12	86	5	49
Ley Creek	25.1	20	103	3	49
Harbor Brook	20.1	9	157	1	49

TABLE **3.9.** Suspended solids loadings for Onondaga and Ninemile Creeks, 1988–1990, from FLUX.

Method	Onondaga Creek		Ninemile Creek	
	Total load (kg)	CV	Total load (kg)	CV
1 AVLOAD	28,186,850	0.362	9,890,458	0.208
2 QWTDC	27,005,130	0.306	9,365,701	0.161
3 IJC	27,458,950	0.312	9,385,334	0.163
4 REG-1	26,367,220	0.286	9,320,055	0.158
5 REG-2	26,856,460	0.269	9,363,564	0.164
6 REG-3	20,171,300	0.189	8,968,005	0.146

simplest loading estimate technique ("average load") resulted in substantial overestimation of the loads.

The contributions of the five sources to the total load over the 1988–1990 interval are represented in the pie chart of Figure 3.16. Onondaga Creek makes a disproportionately high contribution to the SS load to Onondaga Lake relative to its contribution to total inflow. Onondaga Creek contributed about 57% of the total load. Ninemile Creek contributed about 25%. A comparison of these loading estimates to those made by Yin and Johnson (1984) for 1981 conditions is presented in Table 3.10. The percentage contribution results for the two largest sources—Onondaga Creek and Ninemile Creek—were remarkably similar for these two different efforts. The higher loading from METRO in 1981 may have been due in part to the release of $CaCO_3$ precipitate, which formed as a result of the use of soda ash

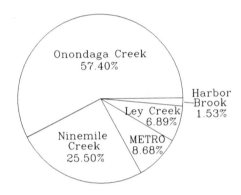

FIGURE **3.16.** Contributions to total (minor tributaries of Figure 3.3 not considered) suspended solids loading to Onondaga Lake for the period 1988–1990.

TABLE **3.10.** Estimates of suspended solids loading to Onondaga Lake.

Source	1988–1990*			1981**	
	Average SS Load, $\times 10^3$ kg · d^{-1}	CV %	% Contribution	Average SS Load, $\times 10^3$ kg · d^{-1}	% Contribution
Onondaga Creek	25.7	0.189	57.4	31.2	55.8
Ninemile Creek	11.4	0.146	25.5	14.8[†]	26.5
METRO	3.9	0.094	8.7	6.8	12.2
Ley Creek	3.1	0.136	6.9	2.8	5.0
Harbor Brook	0.7	0.243	1.5	0.3	0.5
Sum of inputs	44.8			55.9[††]	

* January 1988–September 1990
** Obtained from Yin and Johnson's (1984) annual estimate
[†] Determined as sum of Ninemile Creek (above waste-bed area) and "industrial discharge" (discharge from waste beds to Ninemile Creek) values reported by Yin and Johnson (1984)
[††] Yin and Johnson (1984) included three additional small sources that would make the total 60.0 × 10³ kg · d⁻¹

waste for tertiary treatment at that time. The summed loading rate from these sources was about 25% greater in the earlier effort. This difference is not explained by the respective tributary flow conditions, as the average total inflow rate was $13.68 \, m^3 \cdot s^{-1}$ in 1981 and $14.11 \, m^3 \cdot s^{-1}$ in the 1988–1990 period. Factors that may have contributed to the differences in SS loading estimates (Table 3.10) include the closure of the soda ash/chlor-alkali facility (e.g., reduction in calcium carbonate loading), differences in the sizes of the databases on which estimates were based (64 observations by Yin and Johnson (1984) over approximately 12 months versus 49 observations in 27 months for the 1988–1990 period), and differences in computation procedures used in developing loading estimates.

Time plots of the monthly loads from Onondaga Creek, the sum of the other four monitored inputs, and the sum of all five inputs are presented in Figure 3.17. The monthly distribution of the summed flow from these sources is presented in Figure 3.17d. Almost all of the SS inputs (e.g., >95%) to the lake are included in the "total" loading estimates of Figure 3.17c, as the other minor tributaries are not enriched in SS. Strong seasonal and year-to-year variations in the loads are observed (Figure 3.17) that are largely driven by natural variations in runoff. Variations in the summed monthly inflow explain 99% of the variations observed in monthly SS loads by linear least-squares regression. Thus the increasing trend in loading over the three-year period (Figure 3.17) is not interpreted as reflecting a worsening in related pollution problems, but rather as a manifestation of increases in runoff over the same period. The Onondaga Creek load was more than the sum received from the other four sources through most of the 1988–1990 interval (particularly during high-flow periods).

The relationship between loading and flow was evaluated for Onondaga Creek near its mouth by Yin and Johnson (1984). The expression developed from linear least-squares regression of log (SS load) versus log (flow rate) data (14 paired observations) was

$$SS_L = 528 \, Q_{ON}^{2.08} \qquad (3.9)$$

in which SS_L equals suspended solids load

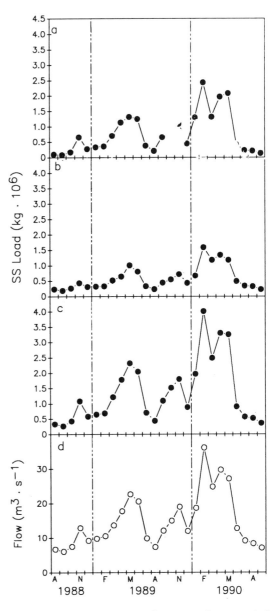

FIGURE 3.17. Time plots of estimated suspended solids loads to Onondaga Lake for the 1988–1990 period: (a) Onondaga Creek, (b) sum of Ninemile Creek, Ley Creek, METRO, and Harbor Brook (minus Onondaga Creek), and (c) sum of (a) and (b) (Total), and (d) summed inflow from these sources.

from Onondaga Creek $(kg \cdot ^{-1})$ and Q_{ON} equals average daily flow for Onondaga Creek $(m^3 \cdot s^{-1})$. The value of SS_L is the product of the instantaneous concentration and the daily average flow. A correlation coefficient of 0.88 was reported for the log transformed form of this relationship. However, flows less than $3 \, m^3 \cdot s^{-1}$ were under-represented in the population. A plot of log (SS_L) versus log (Q_{ON}) is presented for the 1988–1990 database in Figure 3.18; flows less than $3 \, m^3 \cdot s^{-1}$ are better represented in this larger database. The relationship between SS_L and Q_{ON} based on this population, derived from linear least-squares regression, is

$$SS_L = 1288 \, Q_{ON}{}^{1.56} \qquad (3.10)$$

The log transformed form of this expression has a correlation coefficient of 0.85. The relationship of Yin and Johnson (1984) has been added to Figure 3.18 for comparison. The relationships are similar over the range of the most frequently observed conditions. This similarity, and the fact that concentrations measured in 1976 are largely consistent with the prevailing concentration–flow relationship for Onondaga Creek (Figure 3.18), indicates that the potential for SS loading from this primary source has probably remained essentially unchanged over the last 10 to 15 years.

3.2.4.3 Composition of Inflowing Particles

The classification scheme used by Yin and Johnson (1984) to chemically characterize the particles entering Onondaga Lake in 1981

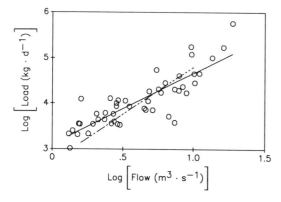

FIGURE 3.18. Evaluation of relationship between suspended solids load and daily average flow in Onondaga Creek, as a log–log plot, for two data sets.

appears as the left-hand column of Table 3.11. The no/low x-rays class contains mostly organic particles. The Ca-only class contains mostly calcium carbonate particles. The Ca-rich class contains a mixture of calcium carbonate and other noncarbonate minerals. The Fe-Mn-rich class contains $Fe(OH)_3$, Fe_2O_3, $FeOOH$, MnO_2, and other Fe and Mn compounds. The Si-only class contains quartz, diatom particles, and other amorphous silica. The Si-rich class is a mixture of quartz or clay with other minor minerals. Clays 1, 2, 3, 4, and 5 have chemical compositions close to kaolinite, illite, montmorillonite, biotite, and muscovite. The Miscellaneous Al-Si class probably contains mostly clays.

More than 28,000 particles from 53 inflow samples were analyzed by the SAX technique; paired measurements of suspended solids were made to support loading estimates of the different particle classes from the SAX composition data. The specifics of sample preparation, and the instrumentation and related operating conditions, were documented by Yin and Johnson (1984).

Particle composition results for the major inflows, according to these particle classifications, are presented in Table 3.11. Clays were the dominant component in Onondaga Creek and Ley Creek and upstream in Ninemile Creek above the waste beds. Si-only particles were the second most important class from these sources. The particles in the discharge from the waste beds were almost entirely calcium carbonate. Organic particles were dominant in the discharge from METRO. The second highest component from this source was calcium carbonate, probably formed in tertiary treatment as a result of the use of the calcium enriched wastewater from the soda ash/chlor-alkali manufacturer.

The estimated loadings of the various particle types (metric tons per year, $MT \cdot y^{-1}$) to Onondaga Lake according to source (and the summation) are presented in Table 3.12. This summary has been simplified by the omission of some minor sources; only sources included in the SS loading budget presented previously for the 1988–1990 interval are considered. Note that the loading at the mouth of Ninemile

TABLE 3.11. Weight percent composition of Onondaga Lake input sediments.

Particle type	Onondaga Creek	Ninemile Creek	Industrial** discharge	METRO	Ley Creek	Harbor Brook
No/low	0.6	1.3	*	80.0	0.3	0.4
Ca-P	0.2	0.2	*	0.9	*	1.8
HM-bearing	0.1	*	*	0.3	0.3	2.7
Dolomitelike	3.1	1.7	*	0.2	4.9	5.5
Pyritelike	0.1	*	*	*	0.1	*
Ca only	3.0	5.4	>99.0	10.1	6.0	28.3
Ca-rich	4.1	6.1	*	1.8	4.9	19.1
Clay 1	6.1	2.8	*	0.2	4.2	3.5
Clay 2	38.3	32.9	*	0.4	40.4	13.1
Clay 3	0.8	8.1	*	0.4	3.5	1.9
Clay 4	3.8	0.1	*	*	1.4	0.1
Clay 5	5.3	4.9	*	0.3	3.5	1.6
Misc. Al-Si	9.6	11.2	*	0.3	6.9	5.9
Fe-Mn-rich	2.8	2.1	*	0.1	2.1	1.8
Ti-rich	0.2	0.1	*	*	0.1	0.7
Si only	15.0	14.1	*	2.4	13.9	6.2
Si-rich	6.0	6.5	*	1.0	5.0	3.7
Al-rich	*	0.1	*	0.1	0.1	0.1
Misc. macro.	0.8	2.5	*	1.6	2.2	3.7
Total	100	100	100	100	100	100

Modified from Yin and Johnson 1984
* Means <0.1%
** Discharge from waste beds

TABLE 3.12. Total sediment load into the Onondaga Lake (tonnes·yr^{-1}).

Particle type	Onondaga Creek	Ninemile Creek	Industrial discharge	METRO	Ley Creek	Harbor Brook	Total
No/low	70	31	*	1992[†]	3	*	2,094
Ca-P	18	5	*	22	*	2	47
HM-bearing	16	1	*	7	3	3	34
Dolomitelike	357	52	*	6	49	6	468
Pyritelike	13	1	*	*	1	*	14
Ca only	343	163	2350	252	61	29	3,197
Ca-rich	463	185	1	44	50	20	757
Clay 1	698	84	*	4	43	4	833
Clay 2	4,350	1000	1	9	407	14	5,781
Clay 3	90	246	*	9	35	2	384
Clay 4	430	2	*	*	14	*	449
Clay 5	600	149	*	6	36	2	789
Misc. Al-Si	1,090	339	*	9	70	6	1,512
Fe-Mn-rich	315	63	*	2	22	2	405
Ti-rich	25	3	*	1	1	*	33
Si only	1,700	427	1	60	140	6	2,339
Si-rich	685	199	*	25	50	4	960
Al-rich	*	2	*	2	1	*	8
Misc. macro	87	76	*	39	23	4	233
Total	11,400	3040	2360	2490	1010	100	20,300

Modified from Yin and Johnson 1984
* Means <1 tonne
[†] Have been corrected

Creek was approximated as the summation of the Ninemile Creek and waste-bed discharge columns of Table 3.12. Accordingly, calcium carbonate was probably the dominant particle class at the mouth of Ninemile Creek. Nearly 50% of the allochthonous sediment load to the lake was estimated to be clays and more than 70% of this clay load is received from Onondaga Creek. The combination of the Si-only and Si-rich classes represented more than 15% of the total load. Calcium carbonate particles represented nearly 20% of the allochthonous load of sediment; about 55% of this load was estimated to have been contributed by the waste-bed discharge. Organic sediment represented only 10% of the external load; this originated almost entirely from METRO.

The gross composition of the particle assemblage at the the mouth of Onondaga Creek was updated by Johnson through SAX analyses performed on four samples collected in 1987 and 1988. The comparison is presented in Table 3.13. A much broader classification scheme (5 classes) has been used here (Table 3.13). The Ca-particle type includes the Ca-only and Ca-rich classes of Tables 3.11 and 3.12. The Si particle type includes the Si-only and Si-rich groups. Clay is made up of all 5 clays of Tables 3.11 and 3.12, plus the Misc. Al-Si group. All other particle types are included in "Other." The particle population from this source appears to have remained unchanged over the 1981–1988 interval. The largest difference is in the contribution of organic particles; although still a relatively

small component, the relative increase in this particle type appears to be substantial. This may be an artifact of the SAX technique, as improvements have been made in detection and quantification of organic particles.

3.2.4.4 Optical Aesthetics in the Mouth of Onondaga Creek

The water quality of Onondaga Creek has historically been of concern largely because of its potential effects on Onondaga Lake. It is not only the major source of allochthonous sediments, but an important source of other pollutants to the lake (Canale et al. 1993; Canale and Effler 1989). Thus the creek is one of the foci of lake remediation efforts. However, recently developed plans for the creek, which include the reestablishment of an Atlantic Salmon run up the creek and redevelopment of recreational and commercial activities in the vicinity of the creek mouth, have brought attention to the prevailing habitat and aesthetic conditions of the stream. The optical aesthetics of the creek mouth are routinely deteriorated after runoff events; the water has a turbid brown "muddy" appearance.

Secchi disc transparency (SD; m) and turbidity (T_n; NTU) measurements were made in the mouth of Onondaga Creek at a frequency of approximately once per day for a three month period in the fall of 1988 to assess the dynamics of these optical characteristics. Turbidity is widely used as a surrogate measure of the scattering coefficient (m^{-1}) (DiToro 1978; Effler 1988; Kirk 1981; see expanded optical treatment in Chapter 7).

Paired time plots of rainfall (regional National Weather Service Station at Hancock Airport, 8.5 km from creek mouth), daily average flow at Spencer Street, and T_n and SD at the creek mouth, are presented in Figure 3.19. The major increases in stream discharge in late October and early November (Figure 3.19b) corresponded to rainfalls of 3.8 and 5.7 cm (Figure 3.19a), respectively. The deleterious effects runoff events have on the turbidity and clarity characteristics in the mouth of Onondaga Creek are illustrated in

TABLE 3.13. Comparison of the composition of sediments in the mouth of Onondaga Creek, 1981 versus 1987 and 1988, as weight percentage.

Particle type	1981 (%)*	1987–1988 (%)**
No/Low	0.6	4.7
Ca	10.2	13.5
Si	21.0	16.1
Clay	63.9	60.4
Other	4.3	5.3

* See Table 3.11; based on an analysis of 11 samples
** Average of samples collected on June 24 and September 19, 1987 and on October 23 and November 3, 1988

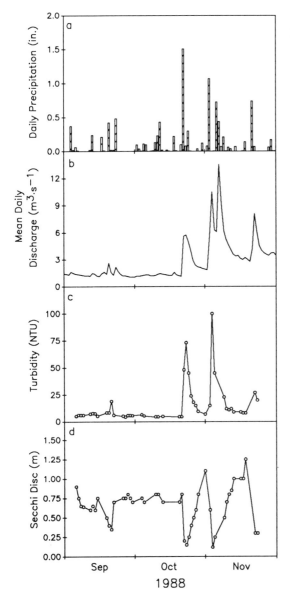

FIGURE 3.19. Time plots of rainfall, flow and selected optical conditions at the mouth of Onondaga Creek in the fall of 1988: (a) daily precipitation, (b) mean daily discharge, (c) turbidity and (d) Secchi disc transparency.

Figure 3.19c and d. In all cases, the runoff events (Figure 3.19a) were accompanied by abrupt major increases in turbidity (Figure 3.19c) and reductions in clarity (Figure 3.19d). The effects were particularly strong for the largest runoff events of late October and early November. However, the negative impacts of even the smaller events of late September and late November were clearly manifested (Figure 3.19). These deleterious effects are observed rather frequently during the year. For example the peak daily average flow (at Spencer Street) observed during the late October runoff event $(5.2\,\mathrm{m^{-3} \cdot s^{-1}})$ is exceeded approximately 35% of the time (USGS, 1972–1989). The period of degradation was approximately 5 days for both of the two major runoff events (Figure 3.19). The maximum values of turbidity during the late October and early November events were approximately 75 and 100 NTUs, respectively; the corresponding minima in Secchi depth were 0.15 and 0.1 m. Quite uniform conditions prevailed in the stream mouth during the low-flow period from late September to late October (Figure 3.19); turbidity ranged from 4.5 to 6.7 NTUs, and Secchi disc transparency ranged from 0.7 to 0.8 m.

The changes in Secchi disc transparency were largely driven by changes in the intensity of light scattering, associated with the changes in the concentrations of particles. This is reflected by the strong inverse relationship observed between Secchi depth and turbidity (Effler 1988). Variations in turbidity (Figure 3.19c) explained 94% of the observed variations in Secchi depth (Figure 3.19d), according to the relationship, $SD = N''/T_n$ (N'' is a coefficient with units of $m \cdot NTU$), developed from optical theory (Effler 1988; see Chapter 7). The value of N'' was 8.7, within the range expected from theory (Effler 1988).

A second undesirable feature of the influx of high concentrations of particles to the mouth of Onondaga Creek is the brown "muddy" appearance imparted to the water. This is widely observed where terrigenous materials represent the major component of the particle assemblage. Kirk (1985) has attributed this largely to elevated concentrations of particles to which humic substances are adsorbed. The water appears brown, as mediated by the spectral distribution of emergent radiation (passing upward through the water surface), because of the strong selective absorption of blue light by the particles (Kirk 1983, 1985).

3.2.4.5 Upstream Origins of Suspended Solids in Onondaga Creek

3.2.4.5.1 Upstream Surveys of Flow and Suspended Solids

Three synoptic surveys of Onondaga Creek, from the "mud boils" to Spencer Street, just above the creek mouth, were conducted in the fall of 1988 to identify sources of suspended solids and characterize the transport of this material downstream. The three surveys were conducted on days of average (November 15), moderately high (upper 35% of mean daily flows at Spencer Street; October 23), and very high (upper 10%; November 3) flow. Stations included in those surveys are shown in Figure 3.20. The array of stations included in the three surveys differ somewhat for the main stem of Onondaga Creek. However, the mouths of the three principal tributaries to the creek, Rattlesnake Gulf (R), West Branch Onondaga Creek (WB), and Hemlock Creek

(H), were included in each case (Figure 3.20). At certain stations samples were collected for suspended solids, without making flow measurements. Flows monitored continuously at Dorwin Avenue and Spencer Street were augmented by instantaneous determinations at other stations based on velocity meter measurements.

The "mud boils" are unusual point source inputs of suspended solids to the creek located approximately 33 km upstream of Onondaga Lake (Figure 3.20). The term "mud boil" denotes the effusion of soft sediments brought to the surface by artesian discharge (Shilts 1978). The hydrogeological setting of the "mud boils" is discussed in Chapter 2. Detailed seasonal loading studies of these inputs were not undertaken until 1992. Earlier qualitative observations indicate inputs are greater in spring when ground water flow is elevated. The water in the creek just below the "mud boils" (e.g., site OR, Figure 3.20) routinely has low clarity and a turbid brown "muddy" appearance. Numerous observations indicate this condition is attenuated downstream (e.g., Dorwin Avenue, site D; Figure 3.20) to a more clear and less "muddy" appearance under low-flow conditions, but it persists along the entire main stem of the stream during elevated runoff periods.

Flow data for the three surveys appear in Table 3.14. Flow increased by a factor of approximately six from site OR to site S on November 15. The relative increases were greater on the higher flow days; the factors were approximately 11 on October 23 and 3 on November 3 (Table 14). The three major tributaries together contributed approximately 60, 45, and 35% of the flow recorded at Site S on October 23, November 3, and November 15, respectively (Table 3.14).

Profiles of the concentration of suspended solids and the estimated load of suspended solids for Onondaga Creek for the three surveys are presented in Figure 3.21. The corresponding concentrations for the three principle tributaries, proximate to their point of entry to Onondaga Creek, are presented in Table 3.15. The concentrations upstream of the "mud boils" in Onondaga Creek (Figure

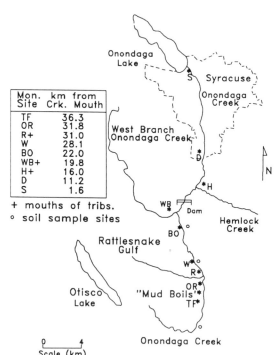

Mon. Site	km from Crk. Mouth
TF	36.3
OR	31.8
R+	31.0
W	28.1
BO	22.0
WB+	19.8
H+	16.0
D	11.2
S	1.6

+ mouths of tribs.
° soil sample sites

FIGURE 3.20. Onondaga Creek, with tributaries, sample site key for stream profiles, and the location of the "mud boils" (from Effler et al. 1992).

TABLE 3.14. Flow in Onondaga Creek and its tributaries on three days in 1988.

Station*	Stream** distance (km)	FLOW (m³·s⁻¹)		
		October 23	November 3	November 15
TF	36.3		0.36	0.27
OR	31.8	0.47	0.82	0.49
[R]†	31.0	[0.17]	[0.52]	[0.23]
W	28.1	0.93		
BO	22.0		2.69	0.96
[WB]	19.8	[2.40]	[3.82]	[0.54]
[H]	16.0	[0.69]	[0.81]	[0.25]
D	11.2	4.81	9.97	2.52
S	1.6	5.24	10.76	2.89

* See Figure 3.20
** Distance from Onondaga Lake (km)
† [ ]s indicate tributaries to Onondaga Creek

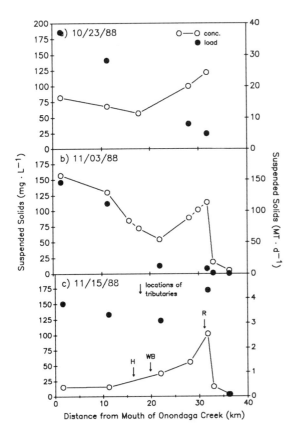

FIGURE 3.21. Profiles of instantaneous concentrations and loads of suspended solids in Onondaga Creek for three days in 1988 (from Effler et al. 1992).

3.21) and in the tributaries (Table 3.15) were low compared to concentrations observed immediately below the "mud boils" on all occasions, and compared to concentrations observed for the entire reach below the "mud boils" under the high-flow conditions of October 23 and November 3 (Figure 3.21a and b). The three principal tributaries for flow apparently contribute very little to the total sediment load of Onondaga Creek in the range of flows encountered in the three surveys. The total contribution of the tributaries (Table 3.15) to the load at Spencer Street (Figure 3.21) was less than 10% for each of the three surveys.

The concentration of suspended solids just below the "mud boils" (site OR, Figure 3.20) was between 100 and 125 mg · L⁻¹ for all three surveys. Concentrations of from 760 to 860 mg · L⁻¹ were measured in small tributaries (located 33 km upstream of the lake) that carried portions of the "mud boil" discharge to Onondaga Creek during the surveys.

Widely different profiles for suspended solids concentration and loading were manifested in the creek (Figure 3.21) under the wide range of flows encountered in the three surveys. Note that the scaling used for loads for the three days in Figure 3.21 differs greatly. The concentration of suspended solids decreased

TABLE **3.15.** Suspended solids concentrations and loads for tributaries to Onondaga Creek.

Tributary	Site	Suspended solids					
		Concentration (mg·L^{-1})			Load (MT·d^{-1})		
		October 23	November 3	November 15	October 23	November 3	November 15
Rattlesnake Gulf	R	32	17	5	0.46	0.78	0.10
West Branch Onondaga Creek	WB	10	7	2	2.06	2.40	0.10
Hemlock Creek	H	20	18	12	1.18	1.24	0.24

progressively from below the "mud boils" to the mouth of the creek under the relatively low-flow conditions of November 15 (Figure 3.21c). The load of suspended solids remained relatively uniform (3.0 to 4.0 MT · d^{-1}) in the lower reaches of the creek after an initial modest decrease below the "mud boils", on this date (Figure 3.21c). The decrease is more likely significant at lower stream flows and, perhaps, when "mud boil" inputs are greater. In strong contrast, the concentration of suspended solids increased downstream, after an initial decrease below the "mud boils," on the two days of higher-flows (Figure 3.21a and b). The downstream increase was particularly great on the highest flow day of November 3 (Figure 3.21b). Further, progressive increases in the load of suspended solids occurred below the "mud boils" on the two high-flow days. The increases were particularly great in the lower 20 km of the creek on these two occasions (Figure 3.21a and b). The estimated instantaneous loads just above the creek mouth were approximately 150 MT · d^{-1} on November 3 and 37 MT · d^{-1} on October 23—many times greater than the load measured under the relatively low-flow conditions on November 15. This supports earlier analyses, which indicated that much of the sediment load received at the creek mouth is delivered during high flow events. Apparently, the urban area contributes little to the suspended solids load at the creek mouth (e.g., absence of major increases in load between Dorwin Avenue and Spencer Street). The apparent load from the "mud boils" for the three

surveys (4–8 MT · d^{-1}) should not be considered representative of the entire range of loads that occurs, as the loading is known to vary with changes in groundwater level (Getchell 1983), and variations in this level are known to be often strongly out of phase with changes in surface water flow.

Most of the suspended solids received at the creek mouth during runoff events are apparently from along the creek, as the downstream load greatly exceeds the "mud boil" and tributary inputs. The suspended solids profiles indicate these inputs are distributed along the creek. It's probable that bank erosion and resuspension of deposited sediments are the primary sources of this distributed load during high-flow (i.e., elevated water velocities) periods. Copious quantities of sediment cover the stream bottom and the banks of the creek downstream of the "mud boils." Further, the appearance and consistency of the deposits throughout much of the length of the creek below the "mud boils" is quite similar to that of the deposits immediately below the "mud boils." The water has the same "muddy" brown appearance from just below the "mud boils" (site OR) to the mouth of the creek during high runoff events. We hypothesize that most of the recent deposits in the creek are composed of material discharged to the stream from the "mud boils." This sediment is probably deposited during periods of low-to-moderate flow. Further, we hypothesize that much of the suspended solids load received at the mouth of Onondaga Lake during runoff events is sediment that has been resuspended from the stream bed or eroded from its banks and ultimately originates from the "mud boils." The SAX results reviewed in the subsequent section represent an initial testing of these hypotheses.

3.2.4.5.2 Individual Particle Analysis

The use of aggregate measures of particle content, such as the concentration of suspended solids and turbidity, can be valuable in tracking the origins of particles within a watershed. However, these aggregate measures are limited in their ability to resolve the relative importance of different sources when the potential for multiple sources exists. Individual particle analysis (IPA), by scanning election microscopy equipped with automated image analysis and x-ray energy spectroscopy (SAX), can provide a more powerful basis to evaluate the importance of individual sources by providing detailed morphometric and chemical characterization of large numbers of particles. Resolution of the contribution of specific sources is facilitated if substantial heterogeneity in particle composition prevails between the sources. In recent years SAX has been used increasingly to gain insights into particle dynamics in lakes (e.g., Effler and Johnson 1987; Johnson et al. 1991; Yin and Johnson 1984) and the ocean (Bernard et al. 1986; Carder et al. 1986). This represents perhaps the first application to stream systems.

Particulate material samples were obtained for SAX analysis for the stream profiles of October 23, November 3, and November 15 of 1988. These were supplemented by creek specimens on May 14 and September 19, 1987, creek mouth samples for a storm event in late June of 1987, watershed soils (Figure 3.20), and collections of samples from the "mud boils." Forty-nine samples from the Onondaga Creek watershed were analyzed. More than 500 particles were characterized from each sample. Nineteen particle-type classes, similar to these adopted by Yin and Johnson (1984; Table 3.11), were used in this analysis. Results are presented as the percentage of the total particle cross sectional area present in the sample.

Unfortunately, strongly heterogeneous particle populations, particularly for the "mud boils," do not emerge from the particulate materials of Onondaga Creek. An ordination was performed to facilitate identification of temporal or spatial inhomogeneity in the particle populations. This form of analysis frequently facilitates interpretation by reducing dimensionality and by identifying gradational relationships among samples that may be obscured by such classification techniques as cluster analysis. The samples were arranged in a two-dimensional continuum (Parks 1968), using principal component analysis, so that the proximity of any two samples is proportional to their similarity. Thus, similar samples will be close to one another. The two components ("x" and "y" of Figure 3.22) are

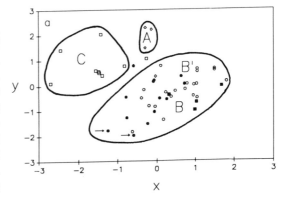

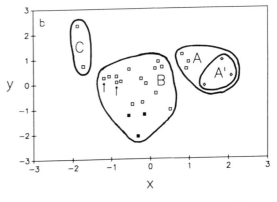

FIGURE 3.22. (a) Ordination plots of principal components, with sample groups identified for all IPA results, and (b) modified sample population and principal component analysis (from Effler et al. 1992).

linear combinations of the original 19 observation variables (particle classes).

The ordination plot for all the samples, plus the derived (Effler et al. 1992) urban and stream background signatures, is presented in Figure 3.22a. Group A consists of two "mud boil" sediment samples and an OR (Figure 3.20) sample collected under the high-flow conditions of November 3, 1988. The distinctive nature of these samples is evident in the plot (Figure 3.22a). Most of the remaining samples are contained within group B. The derived stream background sample is the value group B. The signatures for the two urban samples are farthest from group A (Figure 3.22a). The five samples in subgroup B', with "y" ordination values closest to those of group A were all taken at site S during the high-flow event of June 22–24, 1987, a period during which the transport of "mud boil" materials downstream to the creek mouth would be expected to be highest. Their position in the ordination indicates substantial similarity to the "mud boil" sediments, implying significant contributions of "mud boil" particles to the particle population at the creek mouth during runoff events.

An additional ordination analysis, from which selected samples (25) were eliminated (for details, see Effler et al. 1992), has more fully delineated the characteristic nature of "mud boil" sediments (Figure 3.22b). Three groups of samples are identified in this ordination. Group A, with the highest "x" ordination values, contains two "mud boil" samples and four samples from site OR, immediately downstream of the "mud boil" site (Figure 3.20). Subgroup A' (Figure 3.22b) contains the same three samples included in group A of the first ordination (Figure 3.22a). Group C of Figure 3.22b is comprised of samples from Spencer Street collected on two low-flow days in 1987. Thus it is expected that they are least like, or have the smallest content of, the "mud boil" sediments. Group B (Figure 3.22b), between these extremes, contains the remaining watershed samples.

Two different transport models were developed and applied to the time series SAX data from site S collected during the storm runoff period of June 22–24, 1987 to evaluate the contribution "mud boil" sediment made to the particle population at that site. The first model (tracer method) is based on the dilution of characteristic tracer particles present as a component of the "mud boil" sediments. The second model calculates the fraction of "mud boil" sediment which when mixed with the average stream background particle assemblage gives the greatest Chi-square homogeneity value in a comparison of predicted and observed particle-type distributions (minimum marginal homogeneity). Each of these models used the three "mud boil" source samples defined in subgroup A' of Figure 3.22b. The sample ordination results (Figure 3.22b) and the particle-type distribution summary, presented by Effler et al. (1992), show that dolomite particles are the most distinctive "tracer" particles for the "mud boil" sediments. Although not unique, they are enriched in the "mud boil" materials by a factor of 2.5 to 9 compared with the other watershed samples (Effler et al. 1992); this is substantially higher than for any other particle-type. Assuming the "mud boil" sediments to be the only source of tracer particles, the fraction of "mud boil" materials present in samples from Spencer Street (F_{mb}) is simply the ratio of dolomite particles measured at Spencer Street to that present in the "mud boil" or source material. This fraction was then applied to each of the other particle type distribution values for the "mud boil" source material to estimate the total contribution of particle projected area present in Spencer Street samples from the "mud boil" sediments.

To avoid overestimation of the "mud boil" contribution, associated with variable composition (Figure 3.22b) and contributions of dolomite from other sources, the calculated particle type contributions were compared with those observed for the storm runoff event samples on a class-by-class basis. The fraction of the "mud boil" contribution was reduced until the maximum overestimate (if any) for any particle type was less than 1.5% (actual). This constraint resulted in a conservative estimate of "mud boil" contribution to the suspended sediments at Spencer Street (Effler et al. 1992).

The development and application of the Chi-square homogeneity model to estimate the contribution of "mud boil" sediments to the particle assemblage at Spencer Street is described elsewhere (Effler et al. 1992). This approach probably yields estimates of the "mud boil" contribution that approach a reasonable upper bound.

Estimates of the percent contribution of "mud boil" particles at Spencer Street for the late June event of 1987, according to both models, are presented in Figure 3.23a; the hydrograph for the event appears in Figure 3.23b. The range of values for each sample represents the values computed by the two different models. The first and last values shown in Figure 3.23a are for samples taken during low-flow periods on May 14 and

September 19, 1987, respectively. They contain minimal amounts of "mud boil" sediment materials (Figure 3.22b). They probably represent reasonable estimates of transport under low-to-median stream-flow conditions; the contribution of "mud boils" is less than 10% under these conditions. However, during the middle of the storm runoff event, near the peak of the secondary maximum in the stream hydrograph (Figure 3.23b), the model calculations show that more than 50% of the suspended sediments at Spencer Street is of "mud boil" origin. Samples from Spencer Street during this event were in subgroup B' in the ordination plot of Figure 3.22a. Application of these simple estimating techniques to the SAX database for the Onondaga Creek watershed support the position that during high stream-flow periods, the resuspension and transport of "mud boil" sediment materials contributes importantly to the degradation of aesthetics and the elevated sediment loading at the mouth of Onondaga Creek.

3.2.5 Nitrogen

3.2.5.1 Background

The external inputs of nitrogen have great potential importance because various of the species demand oxygen (Bowie et al. 1985), have toxic effects on fish at elevated concentrations (Russo et al. 1974; USEPA 1985), and are critical plant nutrients (Harris 1986). Further, recent studies have documented that certain of these problems prevail in Onondaga Lake. For example, free ammonia (NH_3) and nitrite (NO_2^-) toxicity criteria are routinely exceeded (Brooks and Effler 1990; Effler et al. 1990), and hypolimnetic accumulation of ammonia (total; $T-NH_3$) contributes to the severe lakewide oxygen depletion observed annually (Brooks and Effler 1990) during the fall mixing period (Chapter 5). Thus, it is important to establish the loading of the various nitrogen species and determine the contributions of the different sources.

Concentration/flow and concentration/time relationships are evaluated here for the im-

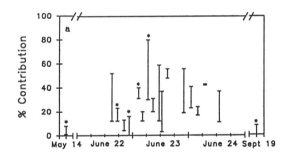

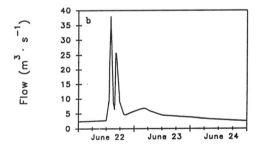

FIGURE 3.23. Temporal distributions at the mouth of Onondaga Creek in response to a rainfall event in June 1987: (**a**) percentage contribution of mud boil particles to total particle projected area, by the models (cases where tracer method estimate exceeded value obtained from Chi-square homogeneity approach are indicated), and (**b**) hydrograph (from Effler et al. 1992).

portant sources. Historic changes in certain of these relationships are identified. Estimates of loading for two recent years, 1989 and 1990, are presented, which will support the testing of a seasonal mechanistic model for T-NH$_3$ and other nitrogen species (Chapter 9). Total annual nitrogen loading is partitioned according to nitrogen specie and source for 1989. Changes in annual loadings from certain sources are identified, and the contribution of T-NH$_3$ to Ninemile Creek from the waste-bed area is quantified for the March–October interval of 1989.

3.2.5.2 Tributary Concentrations: Time and Flow Relationships

Mean and median concentrations of T-NH$_3$, organic-N, and T-NO$_x$ for 1989 are presented for METRO and the four major tributaries in Table 3.16. The minor tributaries are not considered in this nitrogen loading analysis because of their small contribution. The East Flume has high concentrations of T-NH$_3$ and T-NO$_x$ (Onondaga County 1991), but its low flow makes it presently unimportant in terms of contribution to total loading. The highest concentrations of T-NH$_3$, by a wide margin, are routinely observed in the METRO effluent. The concentrations in Ley Creek and Ninemile Creek are higher than in Onondaga Creek and Harbor Brook. METRO also has the highest concentrations of organic-N and T-NO$_x$ by a large margin (Table 3.16). Note that the mean and median values for T-NO$_x$ in the METRO effluent differ greatly, indicating that these

concentrations are not normally distributed. Onondaga and Ley Creeks have organic-N concentrations somewhat higher than Ninemile Creek and Harbor Brook. Harbor Brook and Onondaga Creek have higher T-NO$_x$ concentrations than Ninemile and Ley Creeks.

3.2.5.2.1 Metro

Significant concentration/flow relationships are not observed for nitrogen species (or other constituents considered in this chapter) in the METRO effluent, because of the rather uniform flow maintained for this discharge and the rather wide variations in concentrations encountered. Major recurring variations are observed in the concentrations of the nitrogen species in the METRO effluent as a result of the seasonal operation of nitrification. The temporal distributions of weekly average concentrations of organic-N, the sum of organic-N plus T-NO$_x$, and total N (sum of T-NH$_3$, T-NO$_x$, and organic-N) in the METRO effluent in 1989 are presented in Figure 3.24a (also see Effler et al. 1995a). Weekly averages are presented to more clearly resolve the seasonality in the distributions. Each value of T-NH$_3$ and T-NO$_x$ used in the preparation of the figure is the average of five (out of seven) 24-hr flow-weighted composite samples collected each week; the organic-N values are the average of measurements on two composites from each week. The total nitrogen concentration remains relatively uniform; the average value in 1989 was 17.9 mgN · L^{-1} and the relative standard deviation was 0.13. The distribution

TABLE 3.16. Nitrogen species concentrations in surface discharges to Onondaga Lake, 1989.

Source	T-NH$_3$			Organic-N			T-NO$_x$		
	$\overline{X}$	median	n	$\overline{X}$	median	n	$\overline{X}$	median	n
MERTO	9.71	10.27	230	3.45	3.45	108	5.06	1.55	218
Ninemile Creek	0.44	0.41	145	0.45	0.41	22	0.65	0.63	144
Onondaga Creek	0.17	0.10	22	0.57	0.43	22	0.98	0.99	22
Ley Creek	0.62	0.61	22	0.75	0.65	22	0.55	0.54	22
Harbor Brook	0.22	0.22	22	0.42	0.35	22	1.41	1.48	22

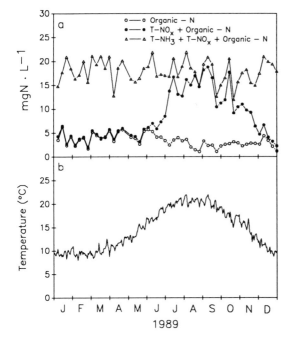

FIGURE 3.24. Temporal distributions in the METRO effluent in 1989: (**a**) concentrations of nitrogen species, and (**b**) temperature (modified from Effler et al. 1995a).

among the species remained relatively uniform from January through May of 1989 and T-NH₃ was the dominant species; the average concentrations of T-NH₃, organic-N and T-NO$_x$ for the period were 13.6, 4.1, and 0.34 mgN · L^{-1}, respectively. Thereafter strong redistribution in the species (particularly T-NH₃ and T-NO$_x$) occurred (Figure 3.24a), as a result of the nitrification process. Largely progressive increases in T-NO$_x$ and decreases in T-NH₃ were observed from June through August. Nearly complete nitrification was achieved from mid-August through mid-September; note that reductions in organic-N were also observed in this period (Figure 3.24a). Generally progressive decreases in T-NO$_x$ and increases in T-NH₃ occurred from September to early December of 1989.

The distinct seasonality in nitrification observed in 1989 (Figure 3.24a) reflects to a large extent the influence of temperature (Figure 3.24b) on the kinetics of the nitrification process (Bowie et al. 1985) that is widely observed in contact stabilization facilities in north temperate climates (USEPA 1975). Note that the greatest nitrification was achieved during the period of peak effluent temperature (Figure 3.24). Though the general character of the seasonality in the distribution of T-NH₃ and T-NO$_x$ in the METRO effluent has been observed annually in recent years, the details of timing and extent of nitrification differ year to year. For example, contrast the distributions of these species in the METRO effluent in 1988 (Figure 3.25) to those presented for the same period in 1989 (Figure 3.24). Despite very similar temperature conditions, the extent of nitrification was less complete in the summer of 1988 than in 1989 (i.e., higher T-NH₃ concentrations were maintained in summer of 1988), and the onset of significant nitrification started later and ended sooner. Differences in plant operation probably contribute to these year-to-year variations.

The distribution of nitrogen species in the METRO effluent has changed significantly in

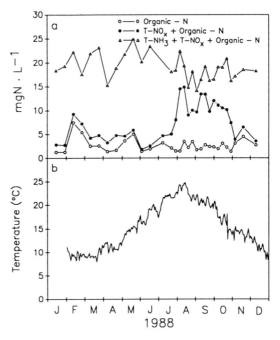

FIGURE 3.25. Temporal distributions in the METRO effluent for a portion of 1988: (**a**) concentrations of nitrogen species and (**b**) temperature (modified from Effler et al. 1995a).

recent years, as exemplified by 1979 and 1985 in Figure 3.26. No significant nitrification was affected in these earlier years; more than 90% of the nitrogen discharged was as organic-N or T-NH$_3$. Substantial redistribution between these two forms of nitrogen occurred from the late 1970s to mid-1980s. For example, more than 50% of the nitrogen was as organic-N in 1979 (Figure 3.26a; except in late March), however, T-NH$_3$ was the dominant species in 1985 (Figure 3.26b). The greater contribution of T-NH$_3$ to TKN in the 1980s in the METRO effluent reflects increased biological treatment.

3.2.5.2.2 Ninemile Creek

The most intensive monitoring of nitrogen species (T-NH$_3$ and T-NO$_x$) on this tributary was conducted over the March to October period of 1989 as part of a larger effort (more constituents) to assess the impact of the waste beds on material loading from Ninemile Creek to Onondaga Lake (Effler et al. 1991). The positions of two stations with respect to the adjoining waste beds, the creek, and the lake,

FIGURE 3.27. Sampling locations above and below the waste beds along Ninemile Creek monitored for the May–October interval of 1989 (modified from Effler et al. 1991).

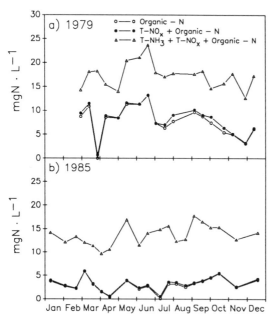

FIGURE 3.26. Temporal distributions in the METRO effluent of concentrations of nitrogen species (modified from Effler et al. 1995a).

are presented in Figure 3.27. The two stations, located 5 km apart, bracket the most recently active waste beds from soda ash production. The downstream station is the same location included in the long-term monitoring effort (Onondaga County 1971–1991). One hundred and forty-five paired grab samples were collected from these two stations during the study period. Time plots of the concentrations of T-NH$_3$ and T-NO$_x$ for the two sites are presented in Figure 3.28, along with daily average flow measured at the Lakeland station. Strikingly higher (average factor of 7.6 for the study period) concentrations of T-NH$_3$ were observed at the downstream station throughout the study period (Figure 3.28a). The increase clearly indicates the influx of large quantities of T-NH$_3$ over the study reach. Further, the temporal distributions of T-NH$_3$ and Cl were highly correlated at the Lakeland station, indicating these materials have the same origins (Effler et al. 1991). In sharp contrast, the distributions of T-NO$_x$ were similar for the two stations (Figure 3.28b), indicating the concentration of this constituent

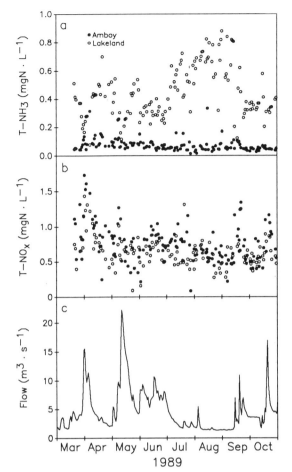

FIGURE **3.28.** Paired distributions of concentrations for two stations on Ninemile Creek: (**a**) T-NO$_x$ and (**b**) T-NH$_3$ (**c**) distribution of daily flow at Lakeland station (modified from Effler et al. 1991).

at the Lakeland station is largely regulated by inputs received upstream of Amboy.

Concentrations of free ammonia (NH$_3$) were calculated according to methods described in Chapter 5 (Effler et al. 1990; Messer et al. 1984). The water quality status of Ninemile Creek at Lakeland with respect to NH$_3$ was evaluated by comparing the calculated concentrations of NH$_3$ to USEPA toxicity criteria (non-salmonid and salmonid), which are functions of pH and temperature (USEPA 1985). pH measurements were made on all samples (90% of the Lakeland observations were within the range 7.8–8.2); temperature data were obtained from Onondaga County

(1991). The NH$_3$ concentration at Lakeland was calculated to be greater than 0.025 mgN · L^{-1} on several occasions during low flow periods (Effler et al. 1991), but remained below the criteria to protect fish against the toxic effects of this specie. A more complete treatment of NH$_3$ toxicity criteria is presented subsequently for the lake (Chapter 5).

Relationships between instantaneous concentrations and daily average flows are evaluated as log–log plots in Figure 3.29. A weak positive relationship between the concentration of T-NO$_x$ and flow at Lakeland prevailed (Figure 3.29a); the relationship at Amboy (not shown) was similar. The relationships between the concentration of T-NH$_3$ and flow at both Amboy and Lakeland are evaluated in Figure 3.29b. The relationship between T-NH$_3$ concentration and flow at Lakeland sharply contrasted the other cases, not only because it was distinctly stronger, but also because of the indicated decreases in concentration with increased flow. The relationship between the concentration of T- NH$_3$ and flow at Lakeland in 1989, according to Figure 3.29b, is described by

$$[\text{T-NH}_3] = 0.813 \, Q^{-0.526} \quad (3.11)$$

in which [T-NH$_3$] equals concentration of T-NH$_3$ (mgN · L^{-1}), and Q equals flow (m^3 · s^{-1}).

The T-NH$_3$ concentrations and daily average flows at Lakeland form a hyperbola on a linear axis (Figure 3.30a), widely identified (e.g., Johnson 1979; Manczak and Florczyk 1971) as a dilution curve. These data are replotted in the concentration versus inverse flow plot of Figure 3.30b to facilitate quantification of the relationship. The expression determined from the 1989 data set from linear least-squares regression is (Figure 3.30b)

$$[\text{T-NH}_3] = 0.73 \, Q^{-1} + 0.20 \quad (3.12)$$

This relationship explained 71% of the variability observed in the concentration of T-NH$_3$ over the monitored period of 1989. The performance of this expression in supporting the estimation of loading at the Lakeland site over the mid-March to November interval of 1989 was essentially equivalent to that of the

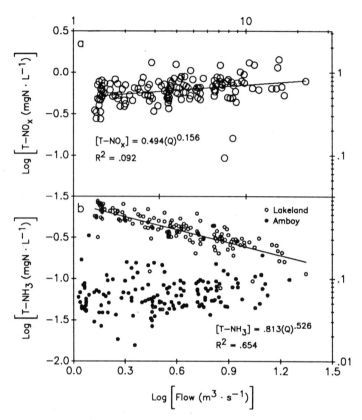

FIGURE 3.29. Evaluation of flow concentration relationships for Ninemile Creek in 1989: (**a**) T-NO$_x$ at Lakeland, and (**b**) T-NH$_3$ at Lakeland and Amboy.

log–log expression (Equation (3.11)). Total loading estimates for the interval were within 3% of each other based on these two expressions (Equations (3.11) and (3.12)). However, the dilution curve (i.e., inverse flow) expression (Equation (3.12)) has been adopted here, as it is considered a more fundamental (and mechanistic) representation of the relationship between concentration and flow. This relationship supersedes the expression presented earlier by Effler et al. (1991) for this data set (Equation (3.11)), that was developed from a log–log evaluation. Further, it represents an effective basis to stay apprised of the capacity for discharge of T-NH$_3$ from the region bracketed by these sites. A number of other constituents in Ninemile Creek and Onondaga Creek are subsequently demonstrated to reflect dilution curve flow relationships in this chapter, as manifested by the linearity of concentration versus inverse flow plots.

The relationship determined for 1989 is compared (as a line; Equation (3.12)) to observations made in three other periods in Figure 3.31. The 1988 data set, collected to support model testing for the spring to fall period, had higher concentrations at low flows than observed in 1989. Elevated flows were not well represented in these data. Higher concentrations clearly prevailed before closure of the soda ash/chlor-alkali facility (Figure 3.31), undoubtedly as a result of the greater discharge of T-NH$_3$ from the waste-bed area during the operation of the facility. The average and median T-NH$_3$ concentrations observed for the intensive data sets of 1988 and 1989 were 0.60 and 0.57 mgN · L^{-1}, compared to values of 2.0 and 1.4 mgN · L^{-1} for the 1979–1985 period. The greater variability

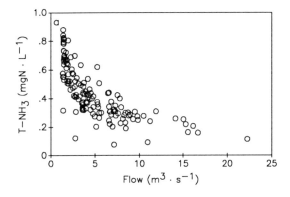

in the concentration/flow relationship during that period (Figure 3.31 versus Figure 3.30b) implies the release of this material was irregular compared to conditions that have prevailed since closure (Figure 3.30b). Particularly high T-NH$_3$ concentrations (e.g., >4 mgN · L$^-$) were observed on several occasions during low-flow conditions. The present NH$_3$ criteria (USEPA 1985) would have been exceeded frequently at Lakeland during low flow periods before the closure of the soda ash/chlor-alkali facility.

3.2.5.2.3 Onondaga Creek

Time plots of concentrations of T-NH$_3$ and T-NO$_x$ (based on ~100 grab samples), and daily average flow, measured near the mouth of Onondaga Creek (Spencer Street) for the May–November period of 1988 are presented in Figure 3.32. This is the most intensive data set for these species for this tributary. Highly variable concentrations prevailed during this period. Very high concentrations of NH$_3$ (e.g., >0.5 mgN · L^{-1}) were observed on several occasions; the possibility that the USEPA criteria for NH$_3$ are exceeded occasionally should not be discounted. The relationships between concentration of these species and

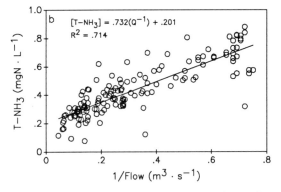

FIGURE 3.30. Evaluation of the relationship between T-NH$_3$ concentration and flow in Ninemile Creek at Lakeland in 1989: (a) linear format, and (b) inverse flow format.

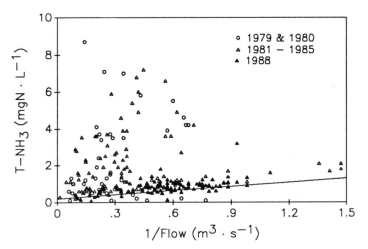

FIGURE 3.31. Comparison of concentration/flow relationship for Ninemile Creek developed from the 1989 data set (represented by the line; see Figure 3.27) to data collected in 1988 and before diversion of the soda ash waste to METRO.

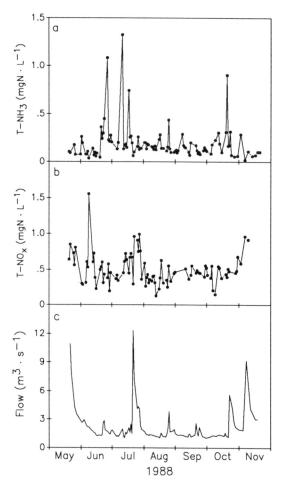

FIGURE 3.32. Temporal distributions for the May–November period of 1988 at Spencer Street on Onondaga Creek: (**a**) concentration of T-NH₃, (**b**) concentration of T-NO$_x$, and (**c**) flow.

mechanistic model for nitrogen species in the lake (Canale et al. 1995, Chapter 9). Three data sets of concentration for the various species were available for this analysis: Onondaga County's tributary monitoring data (approximately biweekly monitoring of METRO and major tributaries), METRO's Discharge Monitoring Reports (daily 24-hr composite sampling of T-NH₃ and twice per week 24-hr composite sampling of TKN at METRO only), and intensive independent monitoring of various species for specific periods of 1989 at METRO and Ninemile Creek. All of these monitoring programs have been discussed earlier in this chapter. The periods over which these various data sets were used to calculate loadings are specified in Table 3.17. Except for T-NH₃ for Ninemile Creek, the interpolation option of FLUX was used to generate concentrations on days that were not sampled. Loading calculations used flows available from METRO and the gauged tributaries.

daily average flow are evaluated in Figure 3.33. Both relationships were weak (e.g., compared to T-NH₃ at Lakeland). A negative relationship was observed for T-NH₃ (Figure 3.33a); a somewhat stronger positive relationship was observed for T-NO$_x$ (Figure 3.33b).

3.2.5.3 Nitrogen Loading

3.2.5.3.1 Estimates to Support Lake Nitrogen Model

Daily external loadings of T-NH₃, organic-N, and T-NO$_x$ to Onondaga Lake in 1989 and 1990 were developed to support testing of a

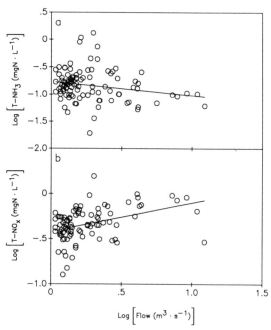

FIGURE 3.33. Evaluation of flow concentration relationships for Onondaga Creek (Spencer Street): (**a**) T-NH₃, and (**b**) T-NO$_x$.

TABLE 3.17. Concentration data sources for nitrogen loads: 1989 and 1990.

	METRO	Ninemile	Onondaga	Ley	Harbor
NH₃					
UFI	1/1/89–12/27/89	3/21/89–10/31/89			
D&S Biweekly		1/01/89–3/20/89 11/01/89–12/31/89	1/01/89–12/31/90	1/01/89–12/31/90	1/01/90–12/31/90
D&S Daily	12/28/89–12/31/90				
TKN					
UFI	1/01/89–12/31/89				
D&S Biweekly		1/01/89–12/31/90	1/01/89–12/31/90	1/01/89–12/31/90	1/01/89–12/31/90
D&S Daily	1/01/90–12/31/90				
NO₂ + NO₃					
UFI	1/01/89–12/24/89	3/21/89–10/31/89			
D&S Biweekly	12/25/89–12/31/90	1/01/89–3/20/89 11/01/89–12/31/90	1/01/89–12/31/90	1/01/89–12/31/90	

The daily loading estimates for these species are presented for 1989 in Figure 3.34 for METRO and the sum of four tributaries; the sum of these loads is the total load. METRO clearly plays a dominant role in the magnitude and seasonality of the T-NH₃ and T-NO$_x$ loadings (compare Figure 3.34 and Figure 3.24). Total loading of T-NH₃ was high and total T-NO$_x$ loading was low from January through mid-June. Total loadings of T-NH₃ and T-NO$_x$ were dramatically lower and higher, respectively, from July through September, as a result of the seasonal operation of nitrification at METRO. The other four tributaries made a significant contribution to organic-N loading throughout the May–December interval.

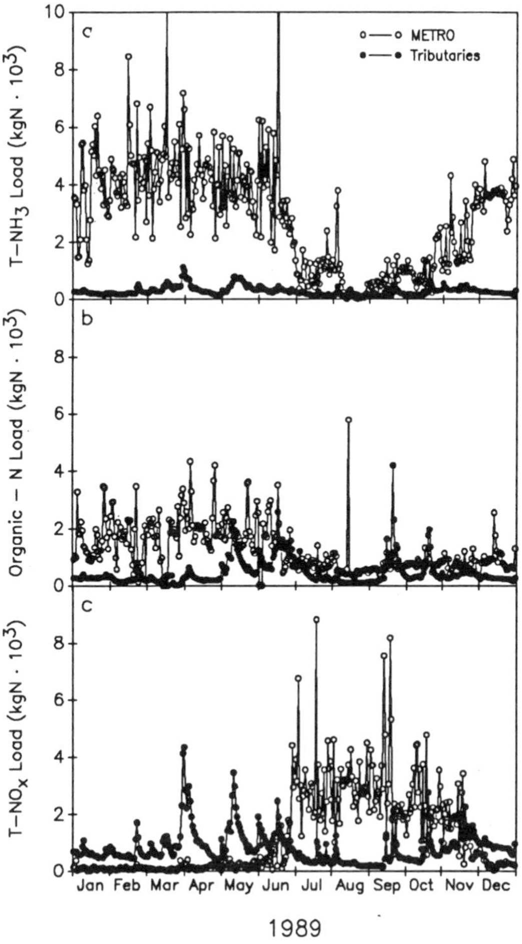

FIGURE 3.34. Daily estimates of loading to Onondaga Lake from METRO and the sum of four tributaries for 1989: (**a**) T-NH₃, (**b**) organic-N, and (**c**) T-NO$_x$ (modified from Canale et al. 1995).

1989

These four tributaries were the major source of T-NO$_x$ to the lake through early June. Peaks in T-NO$_x$ (Figure 3.34c) and organic-N (Figure 3.34b) tributary loading in early April and mid-May were associated with major runoff events. The dramatic short-term (e.g., day-to-day) variations in T-NH$_3$ and T-NO$_x$ loading are largely manifestations of the major short-term variations measured for these constituents in the METRO effluent.

The total external loads for 1989 and 1990 for the three nitrogen species are compared in Figure 3.35. There was substantial disparity in loading of these species in the two years,

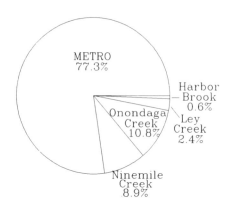

FIGURE 3.36. Contributions of four fluvial discharges and METRO to total nitrogen loading in 1987.

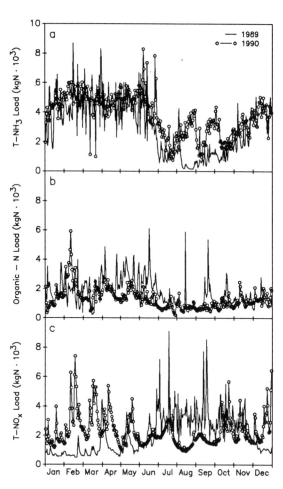

FIGURE 3.35. Daily estimates of total loading to Onondaga Lake for 1989 and 1990: (a) T-NH$_3$, (b) organic-N, and (c) T-NO$_x$ (modified from Canale et al. 1995).

largely as result of the differences in METRO effluent concentrations. The T-NH$_3$ and organic-N loads remained higher and the T-NO$_x$ load lower in the summer and fall of 1990, apparently as a result of less complete nitrification at METRO. In addition, higher loadings of T-NO$_x$ were experienced in the spring of 1990, perhaps associated with the higher tributary flows compared to 1989. The impact these substantial differences in seasonal loading had on the distributions of these species in the lake is evaluated within the framework of a mechanistic model in Chapter 9.

3.2.5.3.2 Annual Nitrogen Loading Estimates

3.2.5.3.2.1 External Loading Budget, 1989. The contributions of the four fluvial discharges and METRO to total nitrogen loading (sum of T-NH$_3$, organic-N, and T-NO$_x$) in 1989 are presented in the pie chart of Figure 3.36 (also see Effler et al. 1995a). The analysis for this year is supported by the most complete data set for METRO and Ninemile Creek, and it is considered representative of recent loading conditions for Onondaga Lake. METRO contributes about 75% of the nitrogen that the lake receives. Onondaga Creek and Ninemile Creek are the second and third contributors; together these tributaries represent about 20% of the nitrogen load. The ungauged inputs probably contribute less than the sum of the Harbor Brook and Ley Creek inputs.

The nitrogen species are partitioned according to T-NH$_3$, T-NO$_x$, and organic-N, and the same five sources, in Figure 3.37. The areas of the pie charts (3) have been proportioned according to the contributions of the three species to total nitrogen loading; the corresponding breakdown in loads appears in Table 3.18. Onondaga Lake received about 2.5 × 10^6 kg of nitrogen from these five sources in 1989. This corresponds to an extremely high areal loading rate of ~200 gN · m^{-2} · y^{-1}. For example, this loading exceeds by a factor of 1.5 the highest value included in the compilation of Brezonik (1972), which addressed the implications of nitrogen loading on trophic state. The contributions of T-NH$_3$, T-NO$_x$, and organic nitrogen to total nitrogen loading in 1989 were 45, 30, and 25%, respectively. METRO contributed more than 90% of the T-NH$_3$ load, more than 50% of the T-NO$_x$ load, and more than 70% of the organic-N load.

The second largest source of T-NH$_3$ was Ninemile Creek. Onondaga Creek was the second largest source of T-NO$_x$ and organic-N.

3.2.5.3.2.2 Historic Changes. Nitrogen species concentration data are available for the inputs to the lake since 1979. A recurring issue in the development of loading estimates for nitrogen species, as well as other pollutants received by the lake, is the stability of loading conditions. This dictates the time interval(s) over which certain concentration/flow relationships are evaluated and applied. Some of the implications of the loading/stability issue are illustrated in the comparison of annual loading estimates for Ninemile Creek for the 1979–1989 interval in Figure 3.38a. Three different sets of loading estimates are presented: those developed by Onondaga County (Method 1 of FLUX) annually, those based on a single concentration/flow model for the entire 1979–1989 interval ("grouped model," Figure 3.38a), and estimates based on individual year

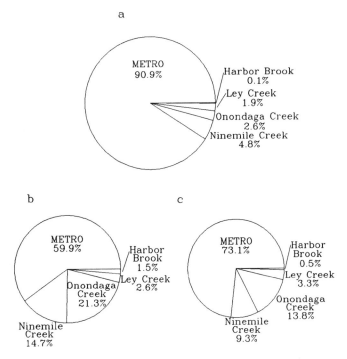

FIGURE 3.37. Contributions of four fluvial discharges and METRO to loading of three nitrogen species in 1989: (**a**) T-NH$_3$, (**b**) T-NO$_x$, and (**c**) organic-N. Relative areas of "pies" are proportional to contribution of the three species to total nitrogen loading.

TABLE 3.18. Contributions to total nitrogen loading to Onondaga lake in 1989, according to nitrogen species and source.

Source	Species							
	T-NH$_3$ kg ($\times 10^3$)	%	T-NO$_x$ kg ($\times 10^3$)	%	Organic-N kg ($\times 10^3$)	%	Total kg ($\times 10^3$)	%
METRO	1037.2	90.9	447.8	59.9	457.9	73.1	1942.9	77.3
Ninemile Creek	55.2	4.8	110.0	14.7	58.3	9.3	223.5	8.9
Onondaga Creek	26.4	2.6	159.0	21.3	86.0	13.8	271.5	10.8
Ley Creek	21.1	1.9	19.2	2.6	29.5	3.3	60.8	2.4
Harbor Brook	1.6	0.1	11.3	1.5	2.9	0.5	15.8	0.6
Total	1141.5		747.3		625.6		2514.4	
Total %	45.4		29.7		24.9			

concentration/flow models for the 1979–1986 interval, combined with a single concentration/flow model for 1987–1989 ("yearly model," Figure 3.38a). The latter model is considered the primary (i.e., most representative) estimate. Note that the annual loads using the overall concentration/flow model differ greatly in a number of years, as a result of invoking stable loading conditions over the 10 yr interval. This approach systematically underestimates the T-NH$_3$ loading during the 1979–1983 interval and overestimates loading for the 1984–1989 period. The estimates from Onondaga County track much more closely the dynamics depicted by the yearly concentration/flow model analysis, particularly over the 1983–1989 interval. Differences of approximately 10% were observed in 1981 and 1982; a maximum difference of about 25% occurred in 1980. This similarly of performance was observed for other inputs and nitrogen species; the paired estimates were particularly close for METRO (Figure 3.38b). Based on this, the summary annual loads for nitrogen species available from Onondaga County (1971–1991) have been used to represent earlier loading conditions over a period (1971–1979) for which the raw input data are not available.

The history of annual loadings reported by the Onondaga County monitoring program for T-NO$_x$, organic-N, and T-NH$_3$ for the major sources is presented in Figure 3.39. The organic-N load is the difference between the T-NO$_x$ load and the sum of the organic-N and T-NO$_x$ loads. Similarly, the T-NH$_3$ load is the difference between the sum of the T-NH$_3$, organic-N, and T-NO$_x$ loads, and the sum of the T-NO$_x$ and organic-N loads. Note that different "scaling" has been used for the various inputs. Rather substantial changes in loading from the various sources have occurred over the twenty-year period, though total loading has not changed dramatically (Figure 3.39g). In general, there appears to have been a shift in loading in recent years from the

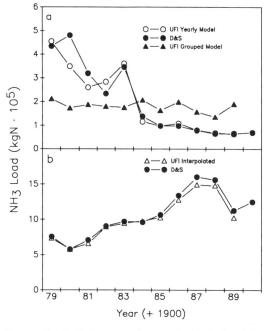

FIGURE 3.38. Estimates of annual T-NH$_3$ load for inputs to Onondaga Lake for the 1979–1990 interval, with different calculation methods: (a) Ninemile Creek, and (b) METRO.

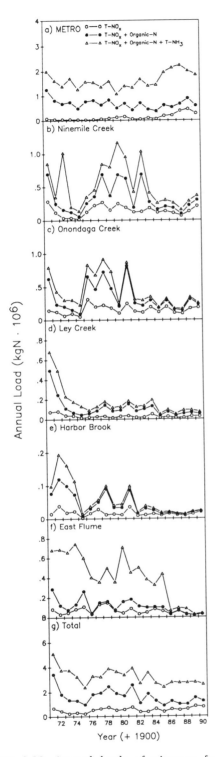

Figure 3.39. Annual loads of nitrogen for the 1971–1990 interval, total and for six different inputs.

tributaries to METRO. The input from the East Flume is not included in the total load, as it represented an artificial form of recycle without significant net loading to the lake. During much of the summer period, when cooling water was drawn from the deeper depth, it represented cycling from the enriched hypolimnion to the epilimnion. In periods when the shallow intake was used, it represented cycling of epilimnetic water. Note the abrupt reduction of this input in 1986 with the closure of the soda ash/chlor-alkali facility. The overall nitrogen load from METRO increased by about a third since the early 1980s (Figure 3.39a). This may in part reflect the retention of more sewage in the sewer system associated with the BMP program, though coincident reductions in nitrogen loads carried by Onondaga Creek, Ley Creek, and Harbor Brook appear to be inadequate to explain the METRO increases. Major redistribution among the nitrogen species has been noted at METRO, which reflects an increased level of treatment. The first signs of nitrification were apparent after the introduction of secondary treatment in 1978, although nitrification has only become significant (i.e., $T-NO_x$ a significant component of the nitrogen load) in the late 1980s. Organic-N was 50% or more of the METRO load until after introduction of secondary treatment in 1978. It decreased to less than 15% of the METRO load in the late 1980s. This redistribution has been largely reflected in the same shifts in the summed loading from all external sources (Figure 3.39g), because of the dominant role METRO (e.g., Figure 3.37) plays in overall loading.

The major changes in nitrogen loading from the tributaries indicated in Figure 3.39 cannot be explained by natural variations in fluvial discharge. More reasonably they reflect changes in pollutant inputs to these streams (e.g., $T-NH_3$ loading from Ninemile Creek, 1979–1990 (Figure 3.38)). The major decrease in loading from Ley Creek in the early 1970s (Figure 3.39d) probably reflects stabilization and scouring losses of organic deposits known to exist below the old Ley Creek sewage treatment plant. However, the dynamics observed in Ninemile Creek, Onondaga Creek, and

Harbor Brook through the early 1980s remain largely unexplained.

3.2.5.3.3 T-NH₃ Load from Waste-Bed Area on Ninemile Creek

The $T-NH_3$ load received in the stream reach bordering the waste beds on Ninemile Creek was estimated for the March–October interval of 1989 based on the paired upstream and downstream measurements (n = 145) of $T-NH_3$ concentration (Figure 3.28b; Effler et al. 1991). Loads were calculated as the product of the daily average flow and measured instantaneous concentrations. The estimates for Lakeland were based on the flows (Q_L) reported by the USGS. Corresponding daily average flows at Amboy were estimated according to the simple linear expression

$$Q_A = 0.80 \times Q_L \qquad (3.12)$$

The value 0.80 equals the ratio of upstream watershed areas of the creek for Amboy and Lakeland locations. Ninemile Creek is accepted to be a "gaining" stream in its lower reaches (Chapter 2). The incremental input received from the waste-bed area was calculated by first subtracting the paired daily loads (n = 145) and then taking the average of the differences. The average load at Lakeland over the study period was $145 \, kg \cdot d^{-1}$; the average at Amboy was $20 \, kg \cdot d^{-1}$. The estimated input to the reach bordering the waste beds was therefore $125 \, kg \cdot d^{-1}$. Extending this estimate to an entire year corresponds to an annual load of $43.8 \times 10^3 \, kg$, or about 85% of the annual load reported for Ninemile Creek (Lakeland) in 1989 (Table 3.18). The input of $T-NH_3$ (and Cl) in this reach has been attributed to leachate from the waste beds, as $T-NH_3$ (and Cl) is common to Solvay waste (Effler et al. 1991). The ultimate waste origin of this material does not appear to be in doubt (Chapter 2). However, the "immediate" origins of this material may be better described as a combination of leachate and contaminated groundwater (Chapter 2). The waste-bed area is presently the second largest source of $T-NH_3$ to Onondaga Lake (Table 3.18).

3.2.6 Phosphorus

3.2.6.1 Background

Phosphorus (P) loading is the principal regulator of primary productivity in most lakes (Dillon 1975; Vollenweider 1968). Several of Onondaga Lake's prevailing water quality problems are largely manifestations of its hypereutrophic state, and have been attributed to excessive phosphorus loading, including: (1) high standing crops of phytoplankton (Auer et al. 1990; Chapter 6), (2) poor clarity (Auer et al. 1990; Chapter 7), (3) rapid depletion of hypolimnetic oxygen (Effler et al. 1986; Chapter 5), and (4) lakewide depletion of oxygen during fall turnover (Effler et al. 1988; Chapter 5). Several measures have been taken (e.g., Canale and Effler 1989; Devan and Effler 1984; Chapter 1), at great expense, to remediate these problems through reduction of phosphorus loading.

Prevailing and historic external P loadings to Onondaga Lake are documented here, as well as the unusual internal loading that was associated with the use of the lake as a source of cooling water by the soda ash/chlor-alkali manufacturer (Effler and Owens 1987). Reductions in total phosphorus (TP) concentrations in METRO's effluent, and therefore loadings from this facility, are documented. Relationships between P concentration and flow are evaluated for the two largest fluvial inputs, Ninemile Creek and Onondaga Creek. Detailed daily total external loading estimates are developed for the spring 1987 through 1990 interval, which are subsequently used to support the testing of a seasonal mechanistic model for TP. Preliminary partitioning of loads along the lengths of Onondaga Creek and Ninemile Creek is also documented.

3.2.6.2 Input Concentrations

3.2.6.2.1 Total Phosphorus Versus Total Inorganic Phosphorus

A parameter described as total inorganic phosphorus (TIP) was used in the County monitoring program as a surrogate measure of TP over the period 1970–1989. The TIP method, described as "Total Acid Hydrolyzable

Phosphorus" in the most recent Standard Methods (Method 4500-P B2; APHA 1989), uses a hot sulfuric acid digestion procedure. It includes reactive phosphorus as well as acid hydrolyzable forms of phosphorus (condensed phosphates and higher molecular weight inorganic species). Some natural waters contain organic phosphate compounds that are hydrolyzed to orthophosphate under these digestion conditions. The TIP method is not intended to include particulate organic forms of phosphorus, in particular microbial biomass. The most widely used analytical method for TP in limnological and water quality work utilizes persulfate digestion (e.g., Method 4500-P B5; APHA 1989), a more severe digestion, which includes microbial biomass. The TP measurement has recently (since July 1989) been incorporated in the County program. The residual between the TP and TIP determinations is a measurement of organic P. It is important to evaluate the relationship between TIP and TP for the major inputs to the lake (as well as the lake itself; see Chapter 5) to support historic analyses, as TP loading has been assessed for many lakes and is widely used in empirical loading/trophic state relationships (e.g., Vollenweider 1968, 1975; OECD 1982), as well as to drive mechanistic eutrophication models (e.g., Chapra and Reckhow 1983; Thomann and Mueller 1987).

The relationship between TP and TIP is evaluated for four locations, representing the three largest external sources of P to the lake, in Figure 3.40. Equivalency (diagonal) lines are included in each of the plots for reference. TIP appears to approach TP to varying extents and with varying degrees of precision for these inputs. Note that in no case was TIP found to exceed TP, consistent with the characteristics of the analyses (APHA 1989). The most variable relationship was manifested for the METRO effluent (Figure 3.40a). The majority of the measurements for this source approach equivalence. However, infrequently organic P represents a significant fraction of the TP discharge (probably in the form of biomass from secondary treatment) from the plant, as indicated by the substantial differences between TP and TIP. Organic P is irregularly also a significant component of TP in Ninemile Creek (Figure 3.40b) and Onondaga Creek at Spencer Street (Figure 3.40c). In contrast, organic P apparently only rarely contributed significantly to TP in Onondaga Creek at Dorwin Avenue (Figure 3.40d).

The relationships determined from the recent paired TP and TIP analyses (Figure 3.40a–c) represent a basis to recreate the historic TP loading picture for the lake over the period of TIP monitoring. However, it is important to identify the inherent limitations of such a rectified record. First, the reliability of estimates of TP loading based on these expressions is limited at least by the prevailing uncertainties in the relationships between TIP and TP (Figure 3.40a–d). The related uncertainties in loading estimates could be exacerbated if differences in TIP and TP are autocorrelated with flow, though no such relationships emerge from the limited data base. Additionally, and perhaps more importantly, historic variations in the mechanistic bases of these relationships (e.g., changes in the composition of P) could compromise their applicability. One can only speculate about the character and degree of variability of the chemical associations of P in the major inputs over the last 20 years. The possibility that significant changes in the composition of P entering the lake from METRO (e.g., changes in the level of treatment and chemicals used in tertiary treatment) and the major tributaries occurred cannot be discounted.

3.2.6.2.2 Metro

A special emphasis has been placed on establishing the concentration of P in the METRO discharge because this input has been recognized as the principal source of P to Onondaga Lake (Canale and Effler 1989, Devan and Effler 1984). Samples have been analyzed for TP (persulfate method) daily since 1980 by the Department of Drainage and Sanitation of Onondaga County. It should be noted that P loads from the plant reported by the County (1970–1990) have not been based on these daily measurements, but have instead been estimated (as TIP loads) based on the biweekly sampling program. A separate

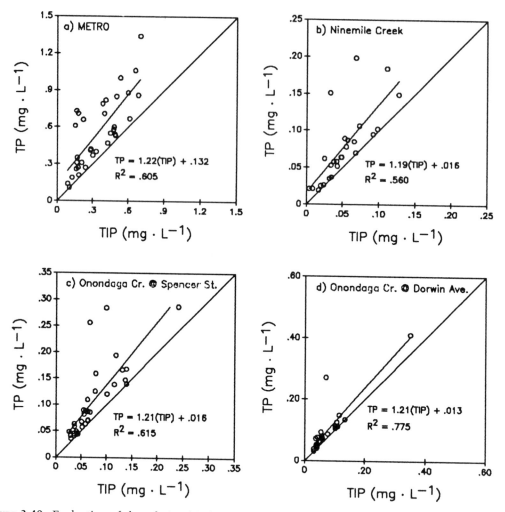

FIGURE 3.40. Evaluation of the relationship between TP and TIP for Onondaga Lake inputs.

laboratory analyzed 202 of the daily samples in 1989 and 251 in 1990 for the concentration of TP. The results for the two different laboratories were quite similar (Figure 3.41). The differences are not considered significant; for example replacement of the County monitoring data with the alternate laboratory measurements in 1989 and 1990 resulted in differences in annual loading from the facility of only 6 and 4%, respectively.

Following the discontinuation of the routine diversion of ionic waste from the Solvay waste beds to METRO, $FeSO_4$ was initially used to remove P at the sewage treatment plant. As part of an effort to optimize the removal of P from the facility's effluent, an evaluation of

chemical treatment schemes was conducted over the period July 1988–March 1991. Various chemicals ($FeSO_4$, $FeCl_3$, alum, and polymers) and points of addition within the facility were evaluated (Stearns and Wheler Engineers 1991). Ferrous sulfate treatment has been adopted at the facility, based on the outcome of the study (Stearns and Wheler Engineers 1991).

The results of detailed monitoring of the TP concentration of the METRO effluent over the 1980–1990 period are presented in Figure 3.42. The distributions throughout the entire period are characterized by relatively large short-term (e.g., daily) variations. Substantial variations over longer time scales have also

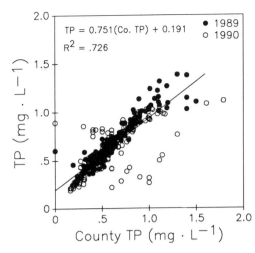

FIGURE 3.41. Comparison of TP concentrations determined by two laboratories on paired samples of METRO effluent, 1989 and 1990. Linear regression expression is included.

occurred (Figure 3.42). Tertiary treatment was initiated in mid-May of 1981. The concentration of TP was rarely below $1.0 \, mg \cdot L^{-1}$ in 1980 and early 1981. Abrupt major decreases in TP concentration (as low as $0.2 \, mg \cdot L^{-1}$) were achieved initially with the beginning of tertiary treatment (Figure 3.42). However, the TP concentration exceeded $1 \, mg \cdot L^{-1}$ for major portions of 1981, 1982, and 1983, mostly as a result of protracted interruptions in the reception of ionic waste from the soda ash/chlor-alkali facility (usually as a result of operational problems at METRO associated with the reception of this waste). Plant performance was decidedly improved in 1984 and 1985, when the ionic waste was accommodated more continuously. Some deterioration in this feature of the quality of the effluent occurred in 1986 and 1987 (Figure 3.42) with the discontinuation of the use of ionic waste for tertiary treatment, and the inception of the use of commercially available iron salts as a replacement. Major reductions in the effluent TP concentration were achieved over the mid-July 1988 through 1990 period (Figure 3.42), associated with the effort to optimize the removal of phosphorus at METRO. Even through the recent and most successful years, short-term increases (e.g., TP

concentration $>1.0 \, mg \cdot L^{-1}$) have been observed. These are likely attributable to irregular discharges of high concentrations of solids enriched in phosphorus. Notable periods of elevated concentrations in recent years have been June of 1988, late October through November in 1989, August and September of 1990, and late January and early February of 1991. Particularly low TP concentrations were achieved in October of 1990 and May and July 1991.

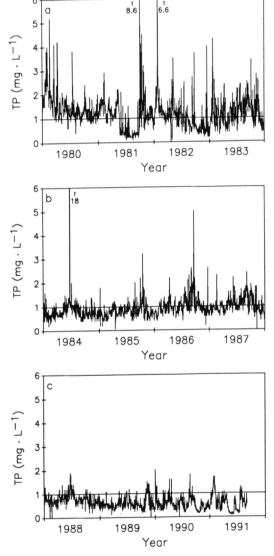

FIGURE 3.42. Daily TP concentration in METRO effluent since 1980.

The estimated annual average TP concentrations in the METRO effluent are presented for the 1970–1990 period in Figure 3.43a (also see Effler et al. 1995a); the corresponding annual average flows from the facility over the same interval appear in Figure 3.43b. Substantial disparity exists within this record with respect to the reliability of the estimates, associated with analytical techniques (e.g., TIP versus TP) and the sampling regime. The values over the period 1969–1979 are based on the biweekly measurements of TIP in grab samples reported by Onondaga County (1971–1980; i.e., have been rectified based on the TIP–TP relationship established since 1989; Figure 3.40a). The estimates for 1980–1988 (mid-October) are based on daily TP measurements on normal composite samples. Values presented for 1989 and 1990 are based on daily TP measurements of flow-weighted composite samples. The y-axis has been segmented with time, because of the major reduction in P concentration achieved over the period of record, to facilitate resolution of recent decreases. Further, the timing

of certain notable management actions is included for reference. Reductions in concentrations have occurred in response to the detergent ban, addition of secondary treatment, addition of tertiary treatment and, most recently, improvements in tertiary treatment (1988–1990). Nearly a twenty-fold reduction in the annual average TP concentration in the facility's effluent has been achieved since 1970.

Average concentrations for the April–August interval, particularly critical to summer phytoplankton growth in this high flushing rate system (e.g., Auer et al. 1990), have been included in Figure 3.43 for the 1980–1990 interval. Concentrations over this portion of the year were within 10% of the annual average, except in 1981, the first year of operation of tertiary treatment. The substantially lower concentrations in the critical growing seasonal of this year reflect the limited operation of the process in that first year. Those concentrations of TP were not achieved in the plant's effluent again until 1989 (Figure 3.43).

The flow from METRO (Figure 3.43b) has remained rather uniform by comparison. Flows were somewhat higher in the mid-1970s, a period when runoff was the highest reported over the 21 year period. The increasing trend in flow since the late 1970s (despite reduced runoff in the region) largely reflects modest increases in the population served and significant improvement in the efficiency of sewage collection (e.g., improvement in CSO efficiency). Thus loading reductions from the facility are somewhat less than the decreases in TP concentration (see subsequent treatment).

3.2.6.2.3 Ninemile Creek

This tributary has been monitored intensively for P over two intervals in recent years; during the June–August interval of 1987, to support validation of a TP model for the lake (Chapter 9; Auer et al. 1994), and from March 1989 to February 1990, as part of a larger effort to assess the impact of the waste beds on material loading (Effler et al. 1991). Total phosphorus was measured in grab samples

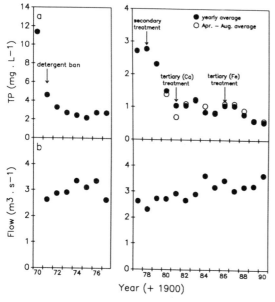

FIGURE 3.43. Annual average conditions in METRO's effluent over the period 1970–1990: (a) TP concentration (modified from Effler et al. 1995a), and (b) flow.

collected at the Lakeland station in both cases. Additionally, in the later effort paired samples were collected at Amboy (Figure 3.27), and soluble reactive phosphorus (SRP) measurements were made on all samples as well. Paired samples were collected at Amboy and Lakeland on approximately 100 days in the 1989/1990 effort. The data base for 1987 includes 144 measurements; however, this population has been reduced to 95 daily values by averaging multiple observations within a day (Heidtke 1989a) because of the limitations in flow measurements for this site. No strong relationship between either P species and flow emerged. This is exemplified by the log (TP)−log (flow) plot for Lakeland in Figure 3.44.

The waste beds apparently have no major effect on the concentration of TP. On average, the concentration of TP was about 10% lower at Lakeland than Amboy (Effler et al. 1991). The paired observations at these two sites were highly correlated ($TP_{Amboy} = 0.837$ $TP_{Lakeland} + 0.015$; $R^2 = 0.77$). However, the stream deposits adjoining and downstream of the waste beds, which are enriched in $CaCO_3$, may reduce the SRP concentration and load carried by the creek. Paired distributions of SRP measured at the two stations appear in Figure 3.45; the daily average flow at Lakeland is presented in Figure 3.45b. On average, the SRP concentration decreased by a factor of 2.0 over the 5 km stream reach bracketed by the sampling sites. The disparity

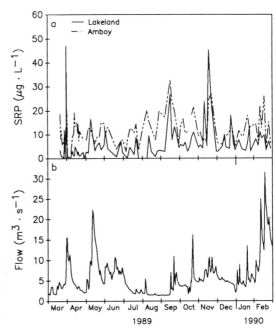

FIGURE 3.45. Distributions for Ninemile Creek for the March 1989−February 1990 period: (**a**) paired concentrations of SRP at Amboy and Lakeland, and (**b**) daily average flow at Lakeland.

was particularly great during the summer low-flow period (Figure 3.45). The reduction has little consequence from a lake management perspective as its magnitude is small compared to TP loadings received from this and other inputs (Effler et al. 1991; see subsequently presented material). However the phenomenon is interesting. Effler et al. (1991) speculated that the reduction in SRP is due to adsorption to, or coprecipitation with, the $CaCO_3$ stream deposits that blanket much of this stream reach. This form of interaction is known to occur in certain hardwater lakes (Murphy et al. 1983; Otsuki and Wetzel 1972). The greater reductions in SRP observed over the study reach during low flows (Figure 3.45) may reflect the effect of increased contact time between the water and the deposits.

3.2.6.2.4 Onondaga Creek

3.2.6.2.4.1 Background. Onondaga Creek has drawn special attention with regards to TP concentrations and loading for two reasons:

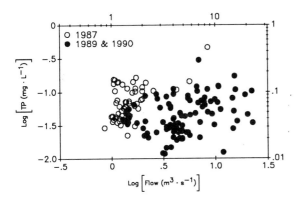

FIGURE 3.44. Evaluation of the relationship between the concentration of TP and daily average flow for Ninemile Creek.

(1) it has been demonstrated to be the second largest source of external TP loading (Heidtke and Auer 1992), and (2) the strong differences in land uses and TP loading (e.g., on a "unit area load" (UAL) basis) within the watershed of the creek. Recall that though most of the watershed is rural (~80%), the stream passes through the City of Syracuse just before entering Onondaga Lake. The rural portion of the creek receives runoff from agricultural and wooded lands and a high sediment load from the "mud boils" (Effler et al. 1992). Recall further that large quantities of "mud boil" deposits are scoured from portions of the creek below these discharges during high-flow events (see earlier sections of this chapter; Effler et al. 1992). The urban portion of the stream receives surface runoff, combined sewer overflows and probably inputs from leaking sewers.

Selected features of various monitoring programs conducted on the creek are presented in Table 3.19. The creek was monitored approximately daily at Spencer Street for TP during dry weather periods in the June–early September interval of 1987 to support testing of a mechanistic TP model for the lake. More frequent monitoring was conducted during two intense storms. Subsequent efforts have focused on partitioning the rural and urban components of the creek load and on more fully resolving wet weather contributions. The second program is limited in the frequency of monitoring. However, its intended continuity will make it valuable in identifying major changes. The creek was monitored regularly

TABLE 3.20. Description of event sampling.

Event	Number of observations		Sampling interval
	Dorwin	Spencer	
1	80	78	May 10–12, 1990
2	123	126	May 15–19, 1990
3	101	99	June 8–11, 1990
4	39	47	June 18–19, 1990
5	69	67	July 20–22, 1990
6	21	15	Sept 14–15, 1989

at both Spencer Street and Dorwin Avenue, usually at a frequency of twice per day, over the period April 1988–October 1989 (program 3, Table 3.19). Samples were collected with an automatic sampler, except during the winter, when samples were collected manually. The concentration of TP was measured on 692 samples from the Dorwin Avenue site over this interval; 674 samples were analyzed from Spencer Street. Additionally, six storm events were monitored in detail (≥1 sample/hr) at the two sites in the spring/summer interval of 1989 and 1990, using automatic sampling equipment. Features of this monitoring program are described in Table 3.20. A single event received modest coverage in 1989. Approximately 430 samples were collected and analyzed for TP from the two stations for these events. More exact start and stop times for the events are presented elsewhere (Heidtke 1991).

3.2.6.2.4.2 Intensive 1988/1989 and Long-Term Programs. The sharply contrasting time series

TABLE 3.19. Onondaga Creek monitoring programs for TP.

Program	Period	Site(s)	Frequency	Description/source
1	Summer 1987	Spencer street	$\geq 1 \cdot day^{-1}$	Support of TP model development (Heidtke 1989a)
2	Since 1990	Spencer Street Dorwin Avenue	biweekly	Long-term Onondaga County program (Onondaga County 1990, 1991)
3	April 1988– October 1989	Spencer Street Dorwin Avenue	$1-2 \cdot day^{-1}$	Support for partitioning of rural/urban contributions (Heidtke and Auer 1992)
4	Summer 1990	Spencer Street/ Dorwin Avenue	$\geq 24 \cdot day^{-1}$, irregularly	Rural/urban contributions for storm events (Heidtke 1991; Smithgall et al. 1991)
5	1990 and 1991	Spencer Street	11 times in 1990	NYSDEC (Rotating Intensive Basin Survey: RIBS)

of TP concentrations observed for Dorwin Avenue and Spencer Street during the intensive April 1988–October 1989 program are presented in Figure 3.46 (note the different scaling used to resolve the distributions). The concentrations at Spencer Street greatly exceeded those at Dorwin Avenue for most of the study period; the differences were particularly great in August and September of 1988. Daily average flows at the two sites over the same period are presented in Figure 3.46c. Relatively large inputs are implied for the urban portion of the watershed by the mag-

nitude of the concentration differences, as the differences in flow for the two sites were small by comparison. Smithgall (1989) indicated that a major break in the sewer system was at least partially responsible for the extremely high TP concentrations observed at Spencer Street in August and September of 1988. The relationships between TP concentration and flow at the two sites are evaluated in a log–log format in Figure 3.47. Data available for the summer of 1987 (program 1; Table 3.19) have also been included for Spencer Street (Figure 3.47b), as they extend the observations for low-flow conditions. No strong functionality between TP concentration and flow emerged for either site. Higher variability in TP concentrations with flow was indicated for Spencer Street than Dorwin Avenue. Note, however, that most of the highest concentrations observed at Dorwin Avenue occurred at some of the highest flows (Figure 3.47a).

The distributions of TP concentrations measured at the two Onondaga Creek stations during the April 1988–October 1989 (1988/1989) program are presented in an alternate format in Figure 3.48. The distributions for these stations were strikingly different. None of the observations at Dorwin Avenue exceeded $0.8\,mg \cdot L^{-1}$; 98% of the concentrations were less than $0.2\,mg \cdot L^{-1}$, and 90% were less than $0.1\,mg \cdot L^{-1}$ (Figure 3.48a). The distribution of TP at Spencer Street is illustrated for the entire 1988/1989 data set (Figure 3.48b) and the last 12 month interval (October 1, 1988 through September 30, 1989; Figure 3.48c). This time segmentation isolates the period believed to be influenced by a break in the sewer system (Smithgall 1989). The distributions of concentrations at Spencer Street for both the entire population and the 12 month segment were skewed to substantially higher values; e.g., more than 50% of the observations were greater than $0.1\,mg \cdot L^{-1}$ (Figure 3.48b and c). Six percent of the observations for the entire Spencer Street population were greater than $1\,mg \cdot L^{-1}$, and thus fall outside of the bounds of the figure. The distributions of the subsets within the most common concentration interval for

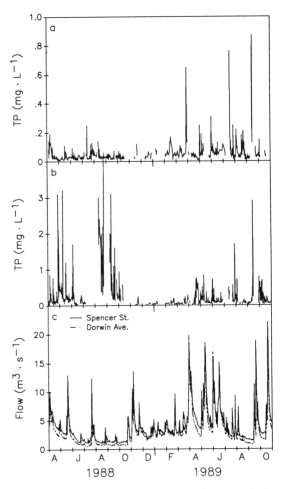

FIGURE 3.46. Time series for Onondaga Creek for the April 1988–October 1989: (**a**) instantaneous concentrations of TP at Dorwin Avenue, (**b**) instantaneous concentrations of TP at Spencer Street, and (**c**) daily average flow at Dorwin Avenue and Spencer Street.

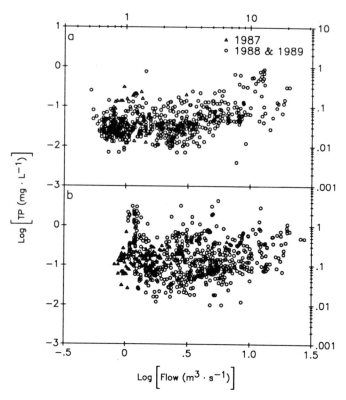

FIGURE 3.47. Evaluation of the relationships between TP concentration and flow in Onondaga Creek: (a) Dorwin Avenue, and (b) Spencer Street.

both sites, 0.0–0.1 mg · L^{-1}, have been further resolved as insets of Figure 3.48a–c. Twenty-five percent of all the observations at Dorwin Avenue fell within the narrow range of 0.02–0.03 mg · L^{-1}, and more than 70% were within the range 0.00–0.05 mg · L^{-1} (Figure 3.48a, inset). Only about 15% of all the observations at Spencer Street were within the 0 to 0.05 mg · L^{-1} interval (Figure 3.48b, inset). Elimination of the first six months of TP measurements at Spencer Street caused the distribution of concentrations to be shifted to somewhat lower concentrations (Figure 3.48c); for example, 45% of the observations exceeded 0.1 mg · L^{-1} and less than 2% exceeded 1 mg · L^{-1}. The mean concentrations of TP over the entire program were 0.05 mg · L^{-1} at Dorwin Avenue and 0.28 mg · L^{-1} at Spencer Street. The mean concentration at Spencer Street for the 12 month segment was 0.15 mg · L^{-1}.

The flow conditions under which the grab samples were collected for this program were generally representative of flow conditions over the monitoring period. The mean instantaneous flow rate corresponding to concentration measurements at Dorwin Avenue was 2.83 m^3 · s^{-1}; individual values ranged from 0.51 to 20.70 m^3 · s^{-1}. A mean annual flow of 2.55 m^3 · s^{-1} was reported by USGS for 1988; a mean flow of 3.74 m^3 · s^{-1} was reported for the 1989 water year. The daily minimum and maximum reported for the 1989 water year were 0.57 and 19.82 m^3 · s^{-1}. The mean instantaneous flow rate corresponding to measured concentrations at Spencer Street was 4.28 m^3 · s^{-1}, with a range of 1.05 to 20.10 m^3 · s^{-1}. These values closely resemble published USGS flow values for the 1989 water year mean (4.50 m^3 · s^{-1}), daily minimum (1.08 m^3 · s^{-1}), and daily maximum (18.83 m^3 · s^{-1}) flows. The average flows en-

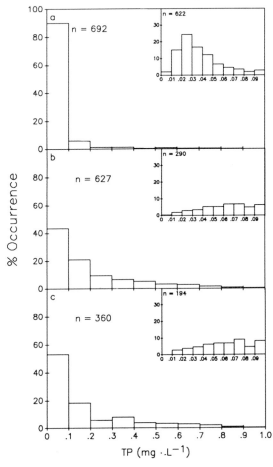

FIGURE **3.48.** Distributions of TP concentrations in Onondaga Creek: (**a**) Dorwin Avenue, over the interval April 1988–October 1989, (**b**) Spencer Street, over the interval April 1988–October 1989, and (**c**) Spencer Street, over the interval October 1988–September 1989.

TABLE **3.21.** Comparison of flow(s) during the 1988/1989 program (No. 3, Table 3.19) to documented long-term conditions.

Statistics	Dorwin Avenue	Spencer Street
Average flow ($m^3 \cdot s^{-1}$)		
*1988/1989	3.22	4.03
**1968(71)–1989	3.65	4.93
Rank		
Entire program	8th of 21	5th of 18
12 month segment	9th of 23	7th of 20

* April 1988–October 1989
** Since 1968 for Dorwin Avenue; since 1971 for Spencer Street

A replicate sampling and TP analysis program was conducted over the April–December interval of 1989 at both stations, as part of the intensive 1988/1989 effort (program 3, Table 3.19), to evaluate the representativeness of the results of single-grab samples. On approximately 40 occasions three separate grab samples were collected at each site. The three samples were collected as three separate "grabs" from different portions of the stream cross section, within about 5 min of each other. The precision of the TP results was high for both sites. Less than 25% of the triplicates from Dorwin Avenue had a relative error greater than 10%; only 3% had a relative error greater than 15%. Less than 25% of the triplicates from Spencer Street had a relative error greater than 10%; only 10% had a relative error greater than 15%. The relative error was observed to exceed 5% at Dorwin Avenue only when the TP concentration was greater than $0.05 \, mg \cdot L^{-1}$. In contrast, relative errors in excess of 10% were observed at Spencer Street only at concentrations less than $0.05 \, mg \cdot L^{-1}$. The representativeness of grab samples for TP at the Onondaga Creek monitoring sites has been further tested by the NYDEC through their RIBS monitoring program, which utilizes a USGS method to obtain a flow-composited sample across a stream cross section, simultaneously with the collection of grab samples. Results from grab samples were supported as representative, based on the nearly equivalent values obtained with the two sampling methods on five

countered during the April 1988 to October 1989 interval, and for the 12 month segment, are compared to conditions documented for the corresponding intervals included in the longer-term flow monitoring data base in Table 3.21. Average flows at both stations during the 1988/1989 program were less than the longer-term means. Flows over the entire intensive study period rank eighth highest of the last twenty-one intervals for Dorwin Avenue, and fifth of the last eighteen for Spencer Street.

different occasions in 1990 at Spencer Street (over a range of concentrations of 0.05 to 0.43 mg · L⁻¹), and on three occasions at both sites in 1991.

The distributions of the most recent measurements (1990–1991) of TP at Dorwin Avenue and Spencer Street (Figure 3.49), despite the rather small population sizes, support the observations for the intensive 1988/1989 program at Dorwin Avenue, but indicate a potential shift to lower concentrations at Spencer Street from those that prevailed during the last 12 months of the intensive program. Subsequently presented analyses indicate that more intensive sampling would be necessary to substantiate this position. Perhaps further decreases at Spencer Street have occurred as a result of additional reductions in sewage inputs.

3.2.6.2.4.3 Events. Time series plots of instantaneous flows, measured concentrations and calculated loads are presented for Dorwin Avenue and Spencer Street for three of the six

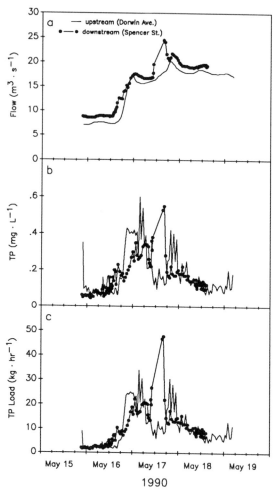

FIGURE 3.50. Time series of instantaneous loading conditions for Onondaga Creek for the May 15–May 19 interval for Dorwin Avenue and Spencer Street: (a) flow, (b) TP concentration, and (c) TP loads.

monitored events (Nos. 2, 4, and 5; Table 3.20) in Figures 3.50, 3.51 and 3.52. The hydrographs have appeared earlier in this chapter. Plots for all six events appear elsewhere (Heidtke 1991). The mid-May event (No. 2; Figure 3.50) was the largest monitored in 1990. Substantial increases in TP concentration were observed at both stations with the increase in flow. Concentrations at Dorwin Avenue were greater than at Spencer Street for much of the event. However, at mid-day on May 17, concentrations at Spencer Street were greater, probably as a result of a CSO

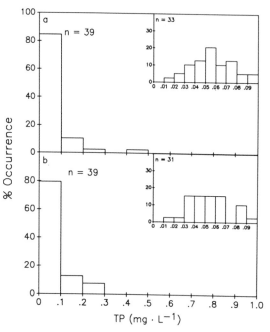

FIGURE 3.49. Distributions of TP concentrations in Onondaga Creek over the 1990–1991 interval: (a) Dorwin Avenue, and (b) Spencer Street.

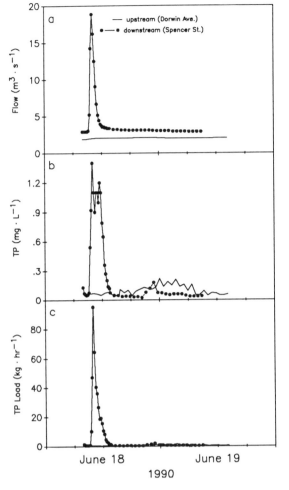

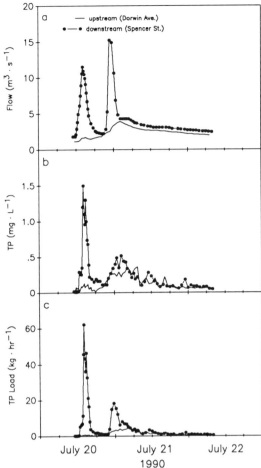

FIGURE **3.51.** Time series of instantaneous loading conditions for Onondaga Creek for the June 18–June 19 interval for Dorwin Avenue and Spencer Street: (**a**) flow, (**b**) TP concentration, and (**c**) TP load.

FIGURE **3.52.** Time series of instantaneous loading conditions for Onondaga Creek for the July 20–July 22 interval for Dorwin Avenue and Spencer Street: (**a**) flow, (**b**) TP concentration, and (**c**) TP loads.

event (Smithgall et al. 1991). Note that concentrations generally decreased through May 18, despite the continued elevated flows. Recall that the mid-June (No. 4) event was modest in comparison, with rainfall intensity that varied substantially across the basin (Smithgall et al. 1991). The time series of TP concentrations at Spencer Street tracked very closely the hydrograph for that site (Figure 3.51a and b). A modest delayed increase in TP was observed at Dorwin Avenue. The mid-July event included two rainfall events, that are particularly well resolved in the hydro-

graph for Spencer Street (Figure 3.52a). The time series of TP concentrations at both sites (Figure 3.52b) again tracked the respective hydrographs closely.

A mean concentration was determined for each event by integrating the "loadograph" (the time series of instantaneous loads; e.g., Figures 3.50c, 3.51c, and 3.52c), and dividing the resulting cumulative mass by the associated water volume (by integrating the event hydrograph) for the event. Analogously, the mean event loading rate was calculated as the area under the loadograph divided by the event duration. This approach for assessing event-based loading is described as the "sequential discrete procedure". Results of this analysis are presented in Table 3.22. The mean concentration of TP at Dorwin Avenue for the events ranged from 0.05 (a minor event) to 0.52 mg · L^{-1} (the largest event). This represents as much as a ten-fold increase from the average (0.05 mg · L^{-1}) reported for the intensive 1988/1989 program. Events 2 and 6 are henceforth described as the "major events"; note the much higher loads and higher concentrations that prevailed for these two cases. These two events had significantly higher average and instantaneous peak flows than the others; the peak instantaneous flows were approximately 24.6 and 27.8 m^3 · s^{-1} for events 2 and 6, respectively. None of the other events exceeded 20 m^3 · s^{-1}, and only one exceeded 15.5 m^3 · s^{-1}. The mean flow for these two events ranged from more than 2 to 6 times that of the other events (Table 3.22).

3.2.6.3 Loadings

3.2.6.3.1 Ninemile Creek

Time series of estimated daily loads of TP and SRP for the March 1989–February 1990 interval, and TP loads during the summer of 1987, are presented for the Lakeland site on Ninemile Creek in Figure 3.53. The estimates were made with the interpolation option of FLUX. The peaks in loading were essentially

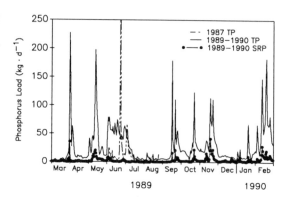

FIGURE 3.53. Time series of estimated daily loads of TP and SRP in Ninemile Creek at Lakeland for the March 1989–February 1990 interval, and for TP loads for the June–August interval of 1987.

manifestations of peaks in flow. The daily TP load exceeded 100 kg · d^{-1} on sixteen occasions over the 1989/1990 interval, and once during the summer of 1987. The estimated total TP load for the 1 yr period was 10,180 kg; which corresponds to an average daily load of 27.9 kg · d^{-1}. This loading rate may be higher than average conditions, for the activities presently prevailing in this watershed, as the total flow for this 1 yr period ranked fourth highest over the last nineteen intervals (1971–1990). Note the distinctly different structure in TP loading encountered in the summer of 1987. The mean daily load for this period was estimated to be 15.2 kg · d^{-1}, substantially lower than deter-

TABLE 3.22. Comparison of TP loading data at upstream and downstream sites on Onondaga Creek for events[a].

Event	Dorwin Avenue			Spencer Street		
	$\bar{Q}$* (m^3·s^{-1})	$\bar{C}$** (mg·L^{-1})	$\overline{W}$*** (kgP·d^{-1})	$\bar{Q}$* (m^3·s^{-1})	$\bar{C}$** (mg·L^{-1})	$\overline{W}$*** (kgP·d^{-1})
1	5.49	0.09	42	6.42	0.28	158
2	14.83	0.19	245	15.90	0.22	304
3	3.14	0.05	12	4.39	0.09	32
4	2.04	0.10	17	3.76	0.34	111
5	2.43	0.15	32	4.05	0.31	107
6	10.78	0.52	483	15.42	0.75	995

[a] Modified from Heidtke 1991
* Calculated as the area under the event hydrograph divided by event duration
** Calculated as the area under the event loadograph divided by the event hydrograph
*** Calculated as the area under the event loadograph divided by event duration

mined for the same period in 1989 ($27.5\,kg \cdot d^{-1}$). The interannual differences in flow for this period explain much of this disparity. The average flow over this period in 1987 was $1.53\,m^3 \cdot s^{-1}$, whereas in 1989 it was $4.67\,m^3 \cdot s^{-1}$.

To estimate the union area loads (UAL) for the Ninemile Creek watershed, the export from Otisco Lake is treated as an upstream load. The discharge from Otisco Lake was monitored for the concentration of TP over the period April 1988–March 1989; 101 grab samples were collected over this 1 yr period. Flow is monitored continuously just below the discharge from the lake. The estimated export of TP from Otisco Lake to Ninemile Creek over the 1 yr period, calculated with the interpolation option of FLUX, was 388 kg. Modest variation in the export of TP can be expected as a result of year-to-year differences in runoff. Based on the estimate of export for this single year, the contribution of TP loading from the Otisco Lake discharge to the Ninemile Creek load at Onondaga Lake is minor (e.g., ~5%). The calculated UAL for the Ninemile Creek watershed downstream of Otisco Lake ($298\,km^2$), based on the 12 month loading estimate for March 1989–February 1990 (10,180 kg), corrected for the Otisco Lake discharge, is $33\,kg \cdot km^{-2} \cdot yr^{-1}$. This value is typical of UAL's reported for rural areas in the literature (Beaulac and Reckhow 1982; Clesceri et al. 1986; Hartigan et al. 1983; Johnson et al. 1978).

The distribution of SRP loading tracked that of TP, though the SRP load remained substantially lower throughout the one year period (Figure 3.53). Several of the daily peaks in SRP loading exceeded $20\,kg \cdot d^{-1}$. The loading rates for SRP were particularly low during low flow periods in summer, as a result of the lower concentrations of SRP reported in those intervals. The estimated total load of SRP over the one year period was 1195 kg, approximately 12% of the TP load.

3.2.6.3.2 Onondaga Creek

3.2.6.3.2.1 In-stream. The time series of instantaneous TP loads, calculated as the product of the instantaneous concentrations and instantaneous flows, determined for the Dorwin Avenue and Spencer Street sites are presented in Figure 3.54. The peaks in loads at Dorwin Avenue coincided with hydrologic peaks (Figure 3.54a). However, Spencer Street had a number of additional peaks (Figure 3.54b), associated with elevated concentrations of TP (e.g., late summer of 1988 (Figure 3.46b)). Long-term (e.g., annual) loading rates are presented here (Table 3.23) for Dorwin Avenue, Spencer Street, and the urban portion of the creek (by difference; Spencer Street minus Dorwin Avenue), for three different periods, April 1988–October 1989, October 1988–October 1989, and mid-February–December 1990. The second period is a subset of the first period, that eliminates a large number of very high concentrations from the population that may be attributable to breaks in the sewer system (Smithgall 1989). The inclusion of the third period is important because of the implied shift to lower concentrations at Spencer Street

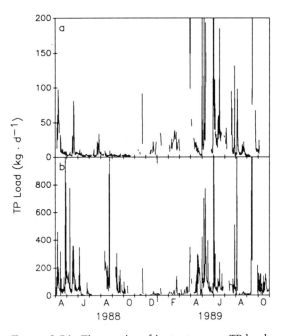

FIGURE 3.54. Time series of instantaneous TP loads for Onondaga Creek for the April 1988–October 1989 interval: (**a**) Dorwin Avenue, and (**b**) Spencer Street.

TABLE **3.23.** Daily average loads for Onondaga Creek for three time intervals by three calculation procedures.

Interval	Loading rates (kg·d⁻¹)								
	Dorwin Avenue			Spencer Street			Urban Load*		
	W_1	W_2	W_3	W_1	W_2	W_3	W_1	W_2	W_3
April 1988– October 1989	17.2	18.2	19.0	91.9	78.1	78.7	74.7	59.9	59.7
October 1988– September 1989	27.2	26.9	25.2	89.8	70.3	64.1	62.6	43.4	38.9
1990**	—	31.6	37.2	—	55.4	66.6	—	23.8	29.4

*Urban load = Spencer Street load−Dorwin Avenue load
**Mid-February–December 1990

(Figure 3.49b), though the small number of observations weakens the associated loading estimates (see subsequent analysis of impact of sampling frequency on loading estimates for Spencer Street). Three different methods were used to estimate the average loading rates: (1) the average of all instantaneous loading rates (W_1; Heidtke 1991), (2) the average of all daily loads, in which the daily load is calculated as the product of the daily flow and the daily average concentration (W_2), and (3) the average of all daily loads as determined with the interpolation option of FLUX (W_3). Only the last two methods were utilized for the 1990 data set.

The loading rate at Dorwin Avenue was higher for the second interval than the first interval (Table 3.23). However, the loading rate at Spencer Street was reduced for the second interval (Table 3.23). The combined effect of these differences was a substantial reduction in the urban contribution to total loading. The estimates based on instantaneous loads (W_1) were higher at Spencer Street for the first two time intervals (Table 3.23). These may be overestimates, associated with more frequent measurements being made during periods of particularly high concentrations. The average daily load calculated from daily flow and daily average concentration (W_2) has been adopted here as the basis of partitioning TP loading for the creek. These estimates were about 15% less than the FLUX based estimates (W_3) for the 1990 period. The increase in loading rate at Dorwin Avenue,

from about $18 \text{ kg} \cdot \text{d}^{-1}$ for the first interval to $27 \text{ kg} \cdot \text{d}^{-1}$ for the October 1988–September 1989 subset, and to $32 \text{ kg} \cdot \text{d}^{-1}$ for 1990, largely tracked increases in flow (average flows for the respective intervals were 4.0, 4.5, and $6.5 \text{ m}^3 \cdot \text{s}^{-1}$). The decreasing trend in loading rate for Spencer Street (and therefore the urban load) demonstrated over the same intervals (Table 3.23) probably reflects reductions in waste inputs (e.g., broken sewers) within the City of Syracuse.

The results of an analysis of the impact of sampling frequency on the total TP load for the April 1988–October 1989 data set for Spencer Street and Dorwin Avenue are presented in Figure 3.55. The loads were calculated using the interpolation method supported by FLUX (see previous text). To

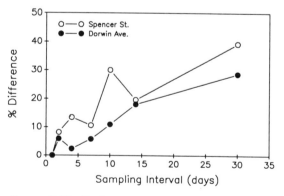

FIGURE **3.55.** Evaluation of the dependence of the level of uncertainty of TP loading estimates for Onondaga Creek on sampling interval.

facilitate the analysis, the following modifications in the 1988/1989 data set were made: (1) single concentrations were established for days in which multiple (usually two) measurements were made by averaging, and (2) single daily concentrations were determined by linear interpolation and inserted for days on which sampling was not conducted. The dependence of loading estimates on sampling frequency for Spencer Street was more erratic than for Dorwin Avenue (Figure 3.55; note the variations found between 1 and 2 weeks), probably because of the more variable TP concentrations encountered at the downstream site (Figure 3.49a and b). The loading estimate deviated from the value based on daily observations by more than 10% at Spencer Street for a sampling interval of four days; it exceeded 10% at Dorwin Avenue at a sampling interval of 10 days. A two week sampling interval would have resulted in a difference of approximately 20% for both sites. A weekly sampling interval is probably adequate to support reasonable estimates of prevailing TP loading at Dorwin Avenue, although subsequently presented analyses indicate additional sampling is warranted for major runoff events. Daily sampling for an extended period (e.g., >6 mos) is recommended for updates of long-term, as well as short-term (e.g., to support dynamic seasonal models), TP loads at Spencer Street. Therefore, until the distribution of TP concentrations at Spencer Street is established to be substantially more uniform, loading estimates at Spencer Street based on biweekly sampling (e.g., 1990, Table 3.23) must be considered substantially more uncertain than those presented for the earlier two intervals.

Water quality impacts of P loading to lakes are primarily determined by long-term average rates. Although short-term, event-based variations in loading can contribute significantly to long-term rates and influence lake quality over brief periods, the magnitude of seasonal or annual contributions is most important to lake response over the same time scales. For this reason it is useful to compare the magnitude of rural and urban loading contributions for events (such as those monitored in 1990) to that demonstrated over longer (e.g., annual) periods. The data collected over the April 1988–October 1989 interval reflect a broad range and distribution of flows, concentrations, and loads, and thus can be considered representative of long-term conditions over the monitored period. The less comprehensive 1990 TP data base continues to be included to reflect possible recent changes in these loading conditions. Comparisons of the long-term (W_2) and event-loading rates and loads for the two sites (i.e., and respective portions of the watershed) are presented in Table 3.24. UAL comparisons are presented only for the long-term loading information. The UAL associated with the TP load at Dorwin Avenue is based on the upstream drainage area (229 km^2). The UAL presented for Spencer Street (Table 3.24) reflects loading from the urban area (difference between Spencer Street and Dorwin Avenue loads), distributed over the incremental area of the watershed (53 km^2).

Despite the fact that the area bounded by Dorwin Avenue and Spencer Street only represents about 20% of the total watershed, this region contributes a major portion of the TP load received by the lake from the creek. This urban area contributed nearly 80% of the creek's load over the 19 month interval of 1988 and 1989 (Table 3.24) and more than 60% over the last 12 months of that interval. The limited data for 1990 indicate the contribution from this area may have decreased further. However, prevailing conditions need to be more fully resolved by an intensive monitoring program, particularly at Spencer Street. Further, given the implied improvement is real, some consideration should be given to the recurrence frequency and magnitude of sewer breaks. The intensity of loading from this region, represented in terms of UALs, ranged from 165 to 410 kg·km^{-2}·yr^{-1}. These values are not unusual for urban land areas (Johnson et al. 1978; Marselak 1978), but they contrast strongly with the estimates of UAL presented for the rural portion of the watershed (upstream of Dorwin Avenue; Table 3.24). The

TABLE 3.24. Long-term and event TP loading rates, loads, and UALs for Onondaga Creek.

Program/event	Rural			Urban			Urban/total (ratio)
	Loading rate (kg·d^{-1})	Load (kg)	UAL (kg·km^{-2}·yr^{-1})	Loading rate (kg·d^{-1})	Load (kg)	UAL** (kg·km^{-2}·yr^{-1})	
Long term							
1988–1989 (1)	18.2	6,645*	29	78.1	28,500*	410	0.77
1988–1989 (2)	26.9	9,820*	43	70.3	25,650*	300	0.62
1990	31.6	11,534	50	55.4	20,220*	165	0.43
Events							
1	42	93	—	158	349	—	0.73
2 (major)	245	735	—	304	912	—	0.19
3	12	33	—	32	88	—	0.63
4	17	19	—	111	125	—	0.85
5	32	59	—	107	196	—	0.70
6 (major)	483	322	—	995	663	—	0.51

(1) April 1988–October 1989
(2) October 1988–October 1989
* Annual load
** UAL for fraction of watershed bounded by Dorwin Avenue and Spencer Street

estimated range of UAL values for the rural area was about $30-50\,kg\cdot km^{-2}\cdot yr^{-1}$, consistent with literature values reported for rural land (Beaulac and Reckhow 1982; Clesceri et al. 1986; Hartigan et al. 1983; Johnson et al. 1978).

The great disparity in loads received during the six events at the two sites on the creek is evident in Table 3.24. The total loads delivered during these events were about 1260 kg at Dorwin Avenue and 2330 kg at Spencer Street. These represented from 11 to 19% of the annual loads presented for Dorwin Avenue, and 8 to 12% of those presented for Spencer Street (Table 3.23). Rural contributions, as measured at Dorwin Avenue, accounted for approximately 25% of the cumulative TP mass measured at Spencer Street for the four minor events (204 out of 758 kg); that is, urban sources between the two sites accounted for about 75% of the creek's load. However, a dramatic shift in the contributions occurred for the two major events. The urban contributions for these events were 19 and 51%; overall, the urban contribution for the major events was 33% (518 out of 1575 kg). The cause for the major shift is presently unknown. The coincident transport of mud boil deposits from upstream of Dorwin to the lake during major events

(Effler et al. 1992; also see earlier sections of the chapter) warrants investigation of the possible influence this phenomenon has on the observed shift in the partitioning of TP loading along the creek for major runoff events. The rural contribution to loading for all six events was 54%. The apparent shift in urban/rural contributions associated with major runoff events may influence the partitioning observed in dry versus wet years. However, the effect is not expected to be great in light of the modest contribution events apparently make to long-term TP loading (Table 3.24). The urban component of the Onondaga Creek load appears to be the appropriate focus for reduction of inputs from this tributary to the lake. Execution of Best Management Practices (BMP) programs for rural areas may be expected to achieve a maximum reduction of 10 to 25% (PLUARG 1978), depending on the fraction of land involved in various uses.

3.2.6.3.2.2 Combined Sewer Overflows, Event Loads. Estimates of event-specific CSO TP loadings have been reported for events 2 through 5 (Smithgall et al. 1991). Estimates were based on monitoring of three of the major CSO discharges on Onondaga Creek. Volume discharges for unmonitored CSOs in the drainage basin were estimated using the Storm Water Management Model (SWMM). The monitored CSO discharges represent 542 hectares (1340 acres), which is slightly greater than 25% of the Onondaga Creek CSO area. Moreover, the monitored overflows were determined to be qualitatively representative of typical overflows in the Onondaga Creek basin (Smithgall et al. 1991).

The estimated total CSO TP loads for events 2 through 5 are compared to the urban loads determined from coincident monitoring at Dorwin Avenue and Spencer Street (by difference) in Table 3.25. Note that Heidtke (1991; see Table 3.24) and Smithgall et al. (1991) obtained nearly equivalent loads from the analysis of the same data sets for three of the four events (Table 3.25). The urban load is expected to be somewhat greater than the CSO load because of contributions from other than CSOs. Projections for events 2, 3, and 4 support this expectation; the estimates were 71, 51, and 95% of the observed urban loads (based on the Smithgall et al. (1991) estimates) for these events. A significant overprediction of the CSO load was made for event 5, which had two distinct peaks in discharge (Figure 3.52a). This is almost certainly due to the use of TP concentrations measured during the first peak ("first flush effect") for the second discharge peak as well, when concentrations were probably lower. Based on the comparisons of Table 3.25, it is apparent that the majority of the TP load received in the urban reach of Onondaga Creek during events is supplied via the CSOs. It should be noted that CSOs occur relatively infrequently. Simulations of the occurrences of CSO events with SWMM, based on long-term rainfall data for the region, indicate these discharges occur on an average of $225\,hr \cdot yr^{-1}$ (Smithgall et al. 1991). The average event lasts for 4 hr and occurs 4.7 times mo^{-1}; slightly greater frequencies prevail during the summer months (Smithgall et al. 1991).

These CSO loads and attendant rainfall data, along with long-term rainfall data for the region, can support approximate estimates of annual TP loads from CSOs. Events 2 through 5 resulted from rainfall that totaled 8.64 cm (Smithgall et al. 1991). The average annual rainfall is 97.8 cm. Thus the annual TP load delivered to Onondaga Creek via CSO's is about 6,650 kg (= (97.8/8.64) × 587). Other independent estimates of Onondaga Creek CSO TP annual loads have been presented by Smithgall et al. (1991). These estimates were based on SWMM simulations of CSO volume discharge responses to 38 yr of rainfall data (from Hancock Airport), and the 1990 CSO event TP concentration data base (Smithgall et al. 1991). The average load from this analysis, 9,330 kg, is considered an upper bound, as the second of back-to-back events (e.g., event 5; Figure 3.51a) almost certainly has lower TP concentrations (recall the "first flush" phenomenon) than the average values invoked in the analysis. Most likely the annual average load from the CSOs on the creek is between 7,000 and 8,000 kg; a reasonable range is from 4,500 (dry year) to 10,000 kg (wet year). Estimates of TP load reductions associated with various CSO abatement alternatives have been presented by Smithgall et al. (1991).

TABLE 3.25. Comparison of TP loads from CSOs to Onondaga Creek during runoff events in 1990 with urban loads to the creek determined from in-stream measurements.

Event	Urban load* (kg)		Estimated CSO load (kg)
	Heidtke (1991)	Smithgall et al. (1991)	
2	177	116.2	82.5
3	55	51.6	26.5
4	106	92.6	87.7
5	137	145.4	390.0
Total		405.80	586.7

Modified from Smithgall et al. 1991
* Urban load = Spencer Street load minus Dorwin Avenue load

3.2.6.3.3 METRO By-Pass

The wastewater that is by-passed to the stormwater outfall of METRO irregularly, associated with rainfall/runoff events, represents a small fraction (~2%, for 1986–1990 interval) of the annual discharge from the facility (e.g., Stearns and Wheler Engineers 1991). However, this fraction is of interest because of the high TP concentrations that prevail in this partially treated (primary treatment and chlorination) wastewater, and the timing of these events, particularly with respect to the productive summer months. Substantial year-to-year variations in the annual by-pass volume and the timing and magnitude of the events is to be expected, as a result of natural variations in rainfall.

Data documenting the volume of flow by-passed to the stormwater outfall annually (and during the summer), average TP concentrations in the by-passed flow, and related estimates of TP loading, are presented in Table 3.26. The annual and summertime by-passed flows varied by about a factor of two over the 1986–1990 interval. The average TP concentration of this discharge decreased progressively over this period. The annual load varied from 2.9×10^3 kg, in 1987, to 7.4×10^3 kg, in 1986. The summer load varied from 1.3×10^3 kg, in 1987 and 1988, to 2.6×10^3 kg in 1986. These estimates probably bound loads to be expected from the stormwater outfall of METRO in most years. However, yet higher (perhaps 1.5 times) loads could occur under the higher runoff conditions observed for the region in the mid-1970s (see Figure 3.5). The average annual TP load from this source for the four years (4.7×10^3 kg) was less than 8% of the annual METRO load in 1989 and 1990 (see subsequent treatment).

3.2.6.3.4 Internal Loading: Spent Cooling Water

The use of the hypolimnion as a source of cooling water by the soda ash/chlor-alkali facility resulted in the net transfer of water and various of its constituents, including P, from the lower layers(s) to the productive epilimnion because of reductions in density associated with increases in temperature (Effler and Owens 1987; Chapter 4). This artificial cycling of water was of particular concern for Onondaga Lake because of the elevated P concentrations that develop in the hypolimnion (see Chapter 5; Driscoll et al. 1992; Effler et al. 1985). Recall (Chapter 1) that there are three intakes for withdrawal of lake water—one in shallow water, and two in deeper water at a nearly equal but uncertain depth (10.7–12.2 m; personal communication, plant personnel). The cooling water circulation did not stop until late summer of 1986, despite closure of the facility in February, probably in an effort to facilitate fall turnover.

Effler and Owens (1987) developed estimates of the associated loading of P to the lake's epilimnion over the April–October interval of 1980 and 1981 from cooling water intake flow rate data (measured at the facility) and lake P data. Water was withdrawn almost entirely from the deep intakes in this interval in both years, at a rate of $2.3–3.6 \, \text{m}^3 \cdot \text{s}^{-1}$. The uncertainty in the depth of the intakes

TABLE 3.26. METRO by-pass data for the 1986–1990 period: Flow, total phosphorus concentration, and estimated loads.

Year	By-pass flow ($m^3 \times 10^6$)		Average (±std. dev.) TP concentration ($mg \cdot L^{-1}$)	TP load (kg $\times 10^3$)	
	Annual	Summer*		Annual	Summer
1986	2.47	0.87	3.0 (±2.5)	7.41	2.61
1987	1.12	0.51	2.6 (±0.6)	2.91	1.33
1988	1.30	0.53	2.5 (±0.45)	3.79	2.10
1990	2.54	0.84	1.8 (±0.71)	4.67	1.54

From Stearns and Wheler Engineers, 1991
* May–September

causes significant uncertainty in associated phosphorus loading estimates because of the strong vertical gradients in P (mostly as SRP) concentration that develop annually in the hypolimnion (Chapter 5; Wodka et al. 1983). The estimated TP loads to the epilimnion over the April–October interval of 1980 and 1981 associated with the cooling water operation were 17×10^3 and 12.5×10^3 kg, which represented about 20% of the total external loading over that period in both years. Circulated cooling water was estimated to be the single largest source of P to the epilimnion of the lake in the June to August period of 1981 (during the initial operation of tertiary treatment at METRO; Effler and Owens 1987). This artificially enhanced "feedback" of P from the enriched hypolimnion may have contributed to maintaining phytoplankton growth following the initial spring bloom (see Chapter 6).

3.2.6.3.5 Minor Tributaries and Comparisons

The most intensive monitoring of the minor tributaries (i.e., other than Onondaga and Ninemile Creeks) for TP was conducted in the summer (mid-June to mid-August) of 1987 (Heidtke 1989a). Average loading rates and related support data determined for that period for all the tributaries, as well as values obtained for the April–September intervals of 1989 and 1990 for selected tributaries, are presented in Table 3.27. The loading rates presented for 1987 are those reported by (Heidtke 1989a). The loading rates for 1989 and 1990 were based on estimates using the interpolation option of FLUX. Flow-weighted (total load divided by total flow) concentrations are also presented in Table 3.27.

Bloody Brook, Sawmill Creek, and Tributary 5A together only represented about 6% of the tributary load in 1987, and only about 1% of the total external load during that period (Heidtke 1989a). The load from these tributaries and the East Flume ($\sim$2.4 kg $\cdot$ d^{-1}, Heidtke 1989a) was found to be less than the contribution from the METRO overflow during the June–August period of 1987 (Heidtke 1989). The lowest average concentration of the four principal (e.g., routinely monitored) tributaries was observed in Harbor Brook; the smallest contributor ($\sim$2%) of these inputs (Table 3.27). The uniformity of the flow-weighted concentrations for the two study periods indicates year-to-year variations in loading from this source can be expected to track changes in runoff. In contrast, the volume weighted concentrations of TP in Ley Creek decreased substantially with the increased flow of 1990. The high TP concentrations observed for Ley Creek were matched only by the levels reported for Onondaga Creek in 1987 and 1989. The origins of the high TP concentrations in Ley Creek deserve investigation. Based on the estimated annual loads of TP for Ley Creek and Harbor Brook in

TABLE 3.27. Average loading rates of Onondaga Lake tributaries for portions (April–September) of 1987, 1989, and 1990.

Tributary	No observation			Average flow (m$^3 \cdot$ s^{-1})			Average load (kg $\cdot$ d^{-1})			Concentration TP* (mg $\cdot$ L^{-1})		
	1987	1989	1990	1987	1989	1990	1987	1989	1990	1987	1989	1990
Onondaga Creek**	272	136	12	1.62	5.39	5.32	23.7	98.6	49.6	0.169	0.212	0.108
Ninemile Creek	193	63	12	1.47	4.90	5.19	14.9	30.8	45.6	0.117	0.072	0.102
Ley Creek	78	—	12	0.52	—	1.27	10.0	—	16.3	0.221	—	0.148
Harbor Brook	81	—	12	0.16	—	0.37	1.3	—	2.5	0.095	—	0.077
Bloody Brook	77	—	—	—	—	—	1.6	—	—	—	—	—
Sawmill Creek	77	—	—	—	—	—	1.3	—	—	—	—	—
Tributary 5A	79	—	—	—	—	—	0.5	—	—	—	—	—

*Flow weighted; calculated by dividing total load by total volume
**At Spencer Street

1990 of 5,690 and 946 kg (Onondaga Country 1991), the associated estimates of UALs are 74 and 32 kg · km^{-2} · yr^{-1}. The value for Harbor Brook is comparatively low for an urbanized area (e.g., Johnson et al. 1978; Marselak 1978), and is more typical of values reported for rural areas with limited agricultural activity (Beaulac and Reckhow 1982; Clesceri et al. 1986; Hartigan et al. 1983; Johnson et al. 1978). However, the Ley Creek value reflects additional inputs (perhaps agricultural or urban).

The data of Table 3.27 further support the position that year-to-year differences in loading from Ninemile Creek are associated largely with natural variations in flow, as the volume-weighted concentration remained rather uniform for the different flow conditions encountered in the three years. The decrease from 1989 to 1990 in Onondaga Creek has been considered previously. Note the highest volume-weighted concentration of the three years occurred in 1989, despite the relatively elevated flow. The substantially lower load determined for 1987 is partly attributable to the reduced flows experienced in that period. Concentrations and loadings were substantially greater in Onondaga Creek than Ninemile Creek during the April–September periods of 1987 and 1989, however, these differences appear to have been largely eliminated by 1990 (Table 3.25), apparently due to reductions in urban inputs.

3.2.6.3.6 Time Series of Loads, 1987–1990

Time series of estimated total external loading of TP to Onondaga Lake were developed for

four years (1987–1990) to support testing of a mechanistic model for TP for the lake. There were three major sources for the concentration data used to support this analysis, all of which have been discussed earlier in this text: METRO's discharge monitoring data (for METRO effluent only), Onondaga County's tributary monitoring data (biweekly), and intensive independent monitoring data on Onondaga and Ninemile Creeks. The periods over which these data were employed in the loading analyses are summarized in Table 3.28.

In addition, loadings were developed for the minor tributaries (Bloody Brook, Sawmill Creek, Tributary 5A, METRO by-pass and the East Flume) for the mid-May to mid-October period of 1987 (Heidtke 1989a). These inputs, with the exception of the METRO by-pass (considered insignificant to the total load), were accommodated for other portions of the modeled period by uniformly applying the average minor tributary load of 5.65 kg · d^{-1}.

With the exception of the mid-May to mid-October interval of 1987, all loads used to support this modeling effort were developed using the interpolation option of FLUX. The loads for that 1987 period were those reported by Heidtke (1989a); these loads were also based on an interpolation technique. These loads were left intact to preserve consistency with an earlier modeling effort.

The calculated time series of daily external TP loading to Onondaga Lake for the spring 1987 through 1990 interval are presented in Figure 3.56; the total and METRO loads are shown (i.e., tributary load = total load − METRO load). Rather substantial variations in

TABLE 3.28. Concentration data sources for phosphorus model loads.

Sources	METRO	Ninemile	Onondaga	Ley	Harbor
UFI	5/13/87–10/19/87	5/13/87–10/19/87	5/13/87–10/19/87	5/13/87–10/19/87	5/13/87–10/19/87
UFI and D&S Combined		3/08/89–3/07/90	4/1/88–12/28/88 1/02/89–11/15/89		
D&S biweekly		1/01/87–5/12/87 10/20/87–3/08/89 3/08/90–12/31/90	1/01/87–5/12/87 10/20/87–3/31/88 12/29/88–1/01/89 11/16/89–12/31/90	1/01/87–5/12/87 10/20/87–12/31/90	1/01/87–5/12/87 10/20/87–12/31/90
METRO DMR	1/01/87–5/12/87 10/20/87–12/31/90				

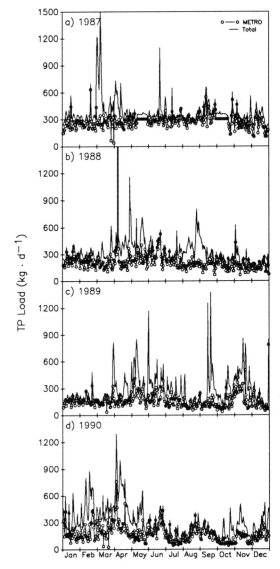

FIGURE 3.56. Time series of estimated daily total and METRO TP loading to Onondaga Lake for four years (modified from Doerr et al. 1995).

The annual versus April–September contributions from METRO were similar except in 1988 (Figure 3.57b), in part because of the very high contributions reported for Onondaga Creek in late summer of that year (Figure 3.56b; also note the great disparity in contributions from Onondaga Creek on an annual versus seasonal basis in the same year). The tributary load was the lowest in 1987, a low runoff year. However, this low rate is in part an artifact of the omission of April and May, a period characterized by elevated tributary loading (Figure 3.56). Note that reductions in METRO loading from 1987 to 1988 were nearly compensated for by increases in loading from

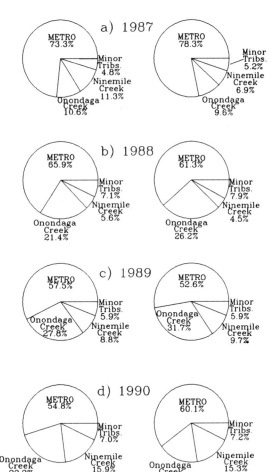

FIGURE 3.57. Contributions of METRO and four tributaries of TP loading, annually (left) and for the April–September (right) interval, for four years.

total loading, the time course of METRO and tributary loadings, and the contribution of METRO versus tributaries were experienced in these four years. A breakdown of the contributions to total loading is presented on an annual and spring to early fall basis for the four years in Figure 3.57. Average loading and flow rate conditions for the four years are presented in Table 3.29.

TABLE **3.29.** Average total phosphorus loading and flow rates to Onondaga Lake for the April–September interval, 1987–1990.

Year	Flow (m³·s⁻¹)		Total phosphorus loading rates (kg·d⁻¹)			
	Total	Tributary	Total	METRO	Onondaga Creek	Ninemile Creek
1987*	11.6	8.6	375	275	40	42
1988	11.0	7.8	331	218	71	19
1989	14.8	11.6	286	164	80	25
1990	20.6	16.9	316	173	71	50

*June–December

Onondaga Creek (Table 3.29); the added inputs from this tributary were particularly striking in late summer (Figure 3.56b). The years 1989 and 1990 represent a striking contrast in material and hydrologic loading to the preceding two years (Table 3.29), and therefore an opportunity for expanded model testing (e.g., verification). Substantial loading reductions were achieved, mostly as a result of decreases in METRO's inputs, and increased flushing was provided by increased tributary flows (Table 3.29).

3.2.6.3.7 History of Total Phosphorus Loading

The history of annual TP loading to Onondaga Lake for the 1970 to 1990 interval is presented in Figure 3.58. Note that the x-axis has been split to accommodate the major reductions in loading achieved over the period and more

clearly resolve the improvements in recent years. The total load presented is exclusive of the minor tributaries (Bloody Brook, Sawmill Creek, Tributary 5A, METRO by-pass, and East Flume). The loading summary of Figure 3.58 represents an update and improvement over earlier summaries (Devan and Effler 1984; Canale and Effler 1989). Earlier analyses underestimated TP loads because of the use of TIP concentration data (e.g., see Figure 3.40a–c). Recall the inherent limitations, identified previously, in recreating historic TP loadings from TIP data (1970–1979), based on recently collected paired data (Figure 3.40). Further, the confidence in the loads since 1980 is greater because of the daily measurement of TP in the METRO effluent, the dominant source.

The changes in total loading over the period of record have been regulated largely by changes in the METRO discharge (Figure

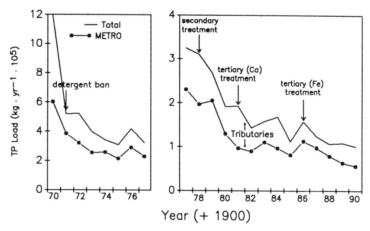

FIGURE **3.58.** History of annual external TP loading to Onondaga Lake over the 1970–1990 interval; total and METRO loads. Timing of major management actions is included.

3.58); variations in the METRO load explain 95% of the variations observed in total loading. Only before the detergent ban and since 1989 was METRO estimated to represent less than 60% of the total load. The reductions in loading from METRO largely track the reductions in TP concentrations described earlier (Figure 3.43) for the facility. The differences in relative reductions between concentration (factor of 20) and load (factor of 12) reflect increases in flow from the facility. A two-fold reduction in total loading was achieved within a year of the establishment of the high-P detergent ban. Total loading was further reduced by a third by the addition of secondary treatment, and by a similar amount (though variable) by the addition of tertiary treatment with the ionic waste from the soda ash/chlor-alkali manufacturer. Yet further reductions were achieved since 1987 through efforts to optimize P removal at METRO (Stearns and Wheler Engineers 1991). Note that the total and METRO loads have decreased by more than a factor of 10 over the period of record.

Reductions in tributary TP loading (total load minus METRO load in Figure 3.58) are also indicated, particularly since the mid-1970s. However, these reductions may be largely a manifestation of lower rainfall and runoff in the 1980s. Estimates of annual TP loading from the tributaries and annual tributary flow are positively correlated (Figure 3.59). Two sets of loading estimates are presented: (1) those generated from Onondaga County's annual TIP loads, according to the TP/TIP corrections developed previously (Figure 3.40), and (2) those obtained from the detailed daily loads developed for 1987–1990 (Figure 3.56). Note that the annual tributary loads obtained in these two different ways (four years, 1987–1990) were in general similar (Figure 3.59). The overall data (1972–1990) set supports a rather strong positive relationship between tributary flow and load. However, analysis of the last four years, for which there is the highest confidence in loading estimates, implies a much different relationship; for example, loads nearly independent of flow. The load for 1990 is particularly pivotal to this analysis, as it deviates the most from previous estimates. The success of the lake TP model in simulating lake concentrations in 1990 (Chapter 9) supports the veracity of the estimate(s) for that year.

A breakdown of the percent contributions of the various sources to TP loading is presented for four selected years in Figure 3.60. The reduced contributions of METRO in 1989 is noteworthy, as it reflects the further reductions in TP in METRO's effluent. Some

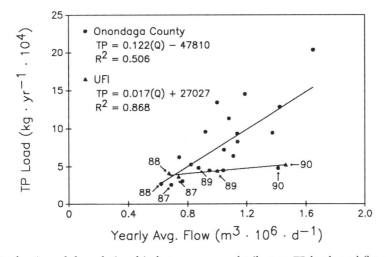

FIGURE 3.59. Evaluation of the relationship between annual tributary TP loads and flows.

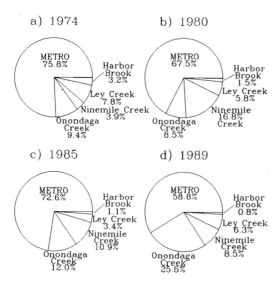

FIGURE 3.60. Contributions to annual TP loads for four years.

reordering of the contributions of the various tributaries is indicated, for which detailed explanations are not available. For example, rather substantial year-to-year variability in the relative roles of the two largest natural inflows, Onondaga and Ninemile Creeks, is indicated over the period of record, which does not appear to have occurred since 1987 (Table 3.29). The most reliable breakdown is probably given for 1989, based on the higher number of observations made in that year. However, as demonstrated earlier, the Onondaga Creek load was unusually high in that year. Some reduction in the Ninemile Creek load since the early 1980s has probably occurred as a result of the elimination of upstream wastewater discharges.

3.2.7 Major Ions

3.2.7.1 Background

External loading of the major ionic constituents of Onondaga Lake—Cl^-, SO_4^{2-}, HCO_3^- (as alkalinity), Ca^{2+}, and Na^+—is addressed in this section. Ionic loading is an extremely important issue for this system, because of the water quality implications the elevated concentrations have had on the lake. For example chemical stratification developed annually

before the closure of the soda ash/chlor-alkali facility as a result of the particularly high concentrations of Cl^-, Ca^{2+}, and Na^+ that prevailed in certain discharge(s) to the lake (Effler and Owens 1986; Effler et al. 1986; Owens and Effler 1989; see Chapter 4). This altered the stratification regime of the lake and thereby exacerbated the system's problems of limited oxygen resources (Effler 1987; Owens and Effler 1989). High rates of $CaCO_3$ precipitation and deposition have prevailed in this alkaline system (Effler and Driscoll 1985a; also see Chapter 8).

Calcium carbonate dominates the recent deposits of the lake (Chapter 8). The net sedimentation rate is unusually high (Chapter 8), largely because of the high concentrations of Ca^{2+} that are maintained as a result of industrial discharges. The high SO_4^{2-} concentrations that prevail exacerbate the oxygen resource problem of the lake (Effler et al. 1988), and may influence the cycling of sedimentary P (Driscoll et al. 1993; Chapter 5). Further, the position that Onondaga Lake was, and is, a naturally saline system has recently been questioned (Effler and Driscoll 1985b, Effler et al. 1990), despite earlier determinations (O'Brien and Gere Engineers Inc. 1973; Rooney 1973) that approximately half of the salinity in the lake (before closure of the soda ash/chlor-alkali facility) had natural origins.

Here we review the concentrations of the major ions that occur in the major inflows to the lake, and identify important relationships between the ionic species concentrations and flow. Historic changes in relationships are evaluated. These analyses, along with tributary monitoring data, are used to support estimation of daily loads of Cl^- for the 1973– 1991 interval, to support testing of a mass balance Cl^- model. The history of annual loading of the various ionic species is reviewed on both a total budget and individual source basis. Particular emphasis is given to Cl^- loading conditions because of the confidence in the data base for this analyte and the reliability of this species as a surrogate measure of salinity in the lake (Chapter 5). Finally, the role localized inputs of certain species have played in contributing to the loads emanating from Onondaga and Ninemile Creeks is documented.

3.2.7.2 Tributary Concentrations

Average annual concentrations of the major ions in the three principal surface inflows to the lake are presented for three different years in Table 3.30; conditions for two minor tributaries, Ley Creek and Harbor Brook, are also included for 1989. All of these inputs are enriched in the listed ionic species compared to average concentrations reported for river waters from around the world (Wetzel 1983). The lake receives alkaline hardwater inflows, enriched in SO_4^{2-}, Na^+, and Cl^-. Ionic waste contributions are manifested in the particularly high concentrations of Cl^-, Na^+, and Ca^{2+} in Ninemile Creek, and irregularly found in METRO's effluent (e.g., 1981), and perhaps in the high concentrations of Cl^- and Na^+ in Onondaga Creek. The minor tributaries may also be influenced by waste inputs.

The three years represent different cases with respect to the input of the Cl^-, Na^+, and Ca^{2+} enriched wastewater from the Solvay process. The differences between 1989 and the other two years are largely a result of the closure of the soda ash/chlor-alkali manufacturer. The differences between 1980 and 1981, manifested in Ninemile Creek and METRO, reflect the partial diversion of ionic waste from the Solvay waste beds to METRO in 1981 to affect P removal. The dynamics of the ionic waste flow to METRO during the period of diversion are presented, at a monthly

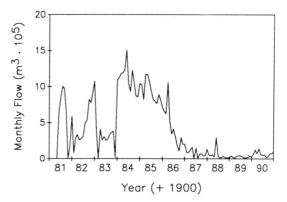

FIGURE 3.61. Monthly discharge of ionic waste from the soda ash/chlor-alkali facility to METRO, 1981–1990 (modified from Effler et al. 1995b).

resolution, in Figure 3.61. The implications of this diversion and its temporal structure for the time course of tributary concentrations and loading of the major ions are described in detail in the following sections. Note the substantial variability in the diversion flow over the 1981–1985 interval, and the continued discharge following closure of the soda ash/chlor-alkali facility in early 1986.

3.2.7.2.1 Ninemile Creek

The most temporally detailed monitoring of this inflow was conducted during the mid-April through mid-November 1980 interval. Samples were collected every 6 hr at the Lakeland station with an ISCO automatic sampler; the only analyte was Cl^-. Time plots of the daily average concentration (average of four measurements) and flow are presented for this period in Figure 3.62. Variations in Cl^-

TABLE 3.30. Annual average major ion concentrations ($mg \cdot L^{-1}$) in surface discharges to Onondaga Lake, 1980, 1981, and 1989 (Onondaga County 1981, 1982, and 1990).

Species	Ninemile			Onondaga			METRO			Ley	Harbor
	80	81	89	80	81	89	80	81	89	89	89
Cl^-	6913	3448	768	414	344	318	344	2501	234	251	195
Na^+	—	797	210	—	167	188	—	692	132	135	93
Ca^{2+}	—	1208	321	—	124	109	—	948	104	107	217
*Alk	215	186	188	225	210	217	187	165	182	183	233
**SO_4^{2-}	—	—	227	—	—	158	—	—	179	143	423

* Alk = alkalinity, as $mg \cdot L^{-1}$ $CaCO_3$
** SO_4^{2-} = measured in separate program (e.g., samples collected on different days)

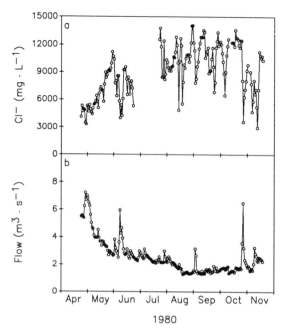

FIGURE 3.62. Temporal distributions for Ninemile Creek at Lakeland 1980: (**a**) daily average Cl⁻ concentration, and (**b**) daily average flow.

3.62a) in the absence of significant changes in flow (Figure 3.62b).

The relationship between Cl⁻ concentration and flow for this station, before diversion of the ionic waste to METRO, is further evaluated in Figure 3.63b, based on biweekly sampling conducted by the County over the period 1973–1980. Note that this data set includes much higher flows, but has a smaller representation of low flows compared to the more intensive 1980 effort. Recall that higher flows prevailed in the mid-1970s, associated with higher rainfall, whereas 1980 was one of the lowest flow years in the 1971–1989 period (Figure 3.5). The mean and median concentrations for this population are 4468 and 3950 mg · L⁻¹, respectively. Substantial variability in the Cl⁻ concentration–flow relationship is again indicated.

A program of the same intensity (4 samples per day) as the 1980 effort, but shorter duration (April to mid-July), was conducted in 1981, to document the impact of the diversion

concentration within a day were generally low. The daily average concentrations ranged from 2980 to 14,060 mg · L⁻¹; the median and mean values were 9180 and 9040 mg · L⁻¹. The highest concentrations were observed during the summer low-flow period; the lowest were observed during peak-flow periods. Various concentration-flow relationships were evaluated. The best performance was obtained by plotting concentration versus inverse flow (Figure 3.63a), consistent with the dilution model (Manczak and Florczyk 1971). The best fit linear regression expression for these data is

$$[Cl^-] = 10{,}839\,Q^{-1} + 3969 \quad (3.13)$$

in which $[Cl^-]$ equals the concentration of Cl⁻ (mg · L⁻¹) and Q equals the daily average flow (m³ · s⁻¹). The relationship is not particularly strong ($R^2 = 0.52$). Much of the variability is probably associated with variations in the production of the waste and operation of the waste beds. Interruptions in the discharge of ionic waste are apparent as abrupt decreases in Cl⁻ concentration (Figure

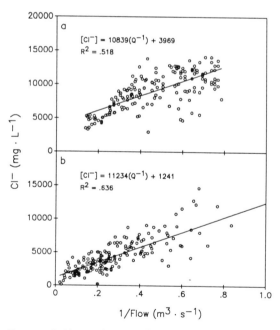

FIGURE 3.63. Evaluation of the relationship between Cl⁻ concentration and flow in Ninemile Creek: (**a**) April–November 1980, and (**b**) 1973–1980 (biweekly sampling) (modified from Effler et al. 1995b).

of ionic waste to METRO on the Cl^- concentration and load of Ninemile Creek. Time plots of the Cl^- concentration at Lakeland and the daily flow of the ionic waste from the Solvay waste beds to METRO for that period of 1981 are presented in Figure 3.64. The decrease in the Cl^- concentration in the creek, starting in mid-May, closely tracked the diversion of the ionic waste, further establishing the extent to which the operation of the waste beds regulated concentrations of the ions common to the Solvay Process waste.

Temporal distributions of Cl^- concentration, and the attendant flow, are presented for portions of 1989 and 1990 in Figure 3.65, to illustrate conditions following closure of the soda ash/chlor-alkali facility. Major reductions in Cl^- concentration have occurred since the closure of this facility. The range of concentrations for this population is 140 to $1750\,mg \cdot L^{-1}$; the median and mean values are 903 and $927\,mg \cdot L^{-1}$. The relationship between concentration and flow continues

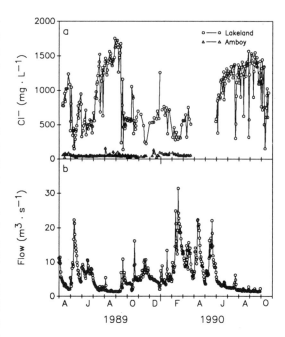

FIGURE 3.65. Temporal distributions for Ninemile Creek for portions of 1989 and 1990: (**a**) Cl^- concentrations for portions of 1989 and 1990 at Lakeland and Amboy, and (**b**) daily average flow.

to be best described by the dilution model (Figure 3.66). The best fit linear regression expression for the combined 1989 and 1990 data sets is

$$[Cl^-[= 1756\,Q^{-1} + 288 \qquad (3.14)$$

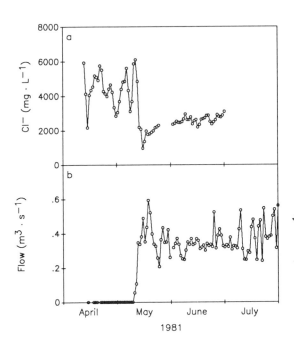

FIGURE 3.64. Temporal distributions for the April–July interval of 1981: (**a**) daily average Cl^- concentration in Ninemile Creek at Lakeland, and (**b**) daily average flow of ionic waste from soda ash/chlor-alkali facility to METRO.

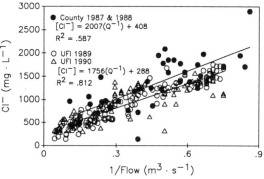

FIGURE 3.66. Evaluation of the relationship between Cl^- concentration and flow in Ninemile Creek at Lakeland, after closure of the soda ash chlor-alkali facility.

Earlier, Effler et al. (1991) utilized a log–log format to describe the relationship between Cl^- concentration and flow at Lakeland for the 1989 data set. The resulting expression ($[Cl^-] = 1972\,Q^{-0.746}$) produces essentially equivalent seasonal and annual loading estimates. However, the inverse relationship is preferred, as in the case of $T\text{-}NH_3$ (Figure 3.30b), because it is considered a more fundamental representation of the character of the input. While irregularities still exist for this recent data set (e.g., particularly the abrupt decreases in 1990 in the absence of major variations in flow), this relationship is markedly stronger ($R^2 = 0.81$) than that observed during active operation of the waste beds (Figure 3.63), implying the continuing source is more constant in character now than before closure of the soda ash/chlor-alkali facility.

Less frequent observations made in 1987 and 1988 have also been included on Figure 3.66. There is some indication of a modest reduction in input of Cl^- to the creek over the brief interval since closure of the soda ash/chlor-alkali facility. Further, it is important to note that the temporal distributions of Cl^- and $T\text{-}NH_3$ at Lakeland were found to be highly correlated ($R = 0.90$) in 1989, yet distinctly different from the distributions of constituents not related to soda ash manufacture (e.g., NO_3^-, SO_4^{2-}), indicating Cl^- and $T\text{-}NH_3$ have the same origins within the creek's watershed (Effler et al. 1991).

The time course of Cl^- concentrations in Ninemile Creek at Lakeland is depicted for the period of record (Onondaga County database) as the annual mean and median concentrations in Figure 3.67. Variations through 1980 reflect year-to-year differences in dilution (stream flow) and probably variations in waste production. The lower concentrations in 1981 and 1982 reflect partial diversion of the ionic waste to METRO (Figure 3.61). Diversion was reduced in 1983, but more complete in 1984 and 1985. The lowest concentrations have been observed following closure of the soda ash/chlor-alkali facility.

3.2.7.2.1.1 Other Ions and Upstream Conditions. The concentrations of Na^+, Ca^{2+}, and SO_4^{2-}

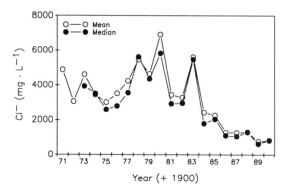

FIGURE **3.67.** Temporal distribution of the annual mean and median concentrations of Cl^- in Ninemile Creek at Lakeland.

in Ninemile Creek at Lakeland have been, and continue to be, subject to substantial variations. The concentrations of Na^+ and Ca^{2+} have also systematically decreased with the closure of the soda ash/chlor-alkali facility (Table 3.30). However, the concentrations of these species and SO_4^{2-} demonstrate substantial temporal variations as a result of the wide variations in stream flow that are common in this stream.

A constant relationship between Cl^-, Na^+, and Ca^{2+} has been, and continues to be, maintained in Ninemile Creek at Lakeland, which approaches that of the Solvay waste-bed overflow (Chapter 2). This is demonstrated for the 1981–1989 interval in Figure 3.68; note the inclusion of lines that bracket reported relationships for the waste. The stoichiometry of NaCl brine has been added to Figure 3.68a for comparison. Recall that the 1981 to 1985 interval represents a period of active production of the waste, though a substantial fraction of the ionic waste was being diverted to METRO. The 1986–1989 period follows the discontinuation of soda ash production. Only data sets which met the QA/QC criterion of charge balance within 10% were used in the analysis; 41% of the observations were rejected based on this criterion. The relationships between the three ions were all strong (Figure 3.68). The maintenance of the ion ratios of the Solvay waste-bed overflow for these constituents, despite the reductions in absolute concentrations, establishes their

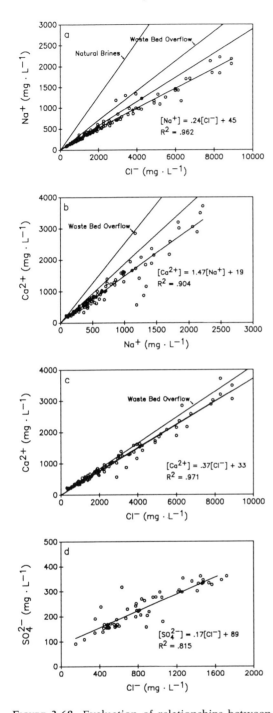

ultimate wastewater origins (Chapter 2). Leachate from the waste beds and discharges of groundwater contaminated with the ionic constituents of the Solvay waste are probable routes by which this material continues to enter the stream. The uniform character of the concentration-flow relationship (e.g., demonstrated for Cl^- in Figure 3.66) implies an important role for groundwater inputs.

Sulfate has not been included in the long-term monitoring program. However, the results of paired SO_4^{2-} and Cl^- measurements made in 1989 (Figure 3.68d) indicate that the distribution of SO_4^{2-} at Lakeland following closure of the soda ash/chlor-alkali facility is highly correlated to the distribution of Cl^-, Na^+, and Ca^{2+}. This relationship almost certainly differed during the operation of that facility as the origins of the SO_4^{2-} are largely upstream of the Solvay waste beds (see subsequently presented material).

The distribution of Cl^- at an upstream (above the waste beds) station in Amboy over the April 1989 to March 1990 interval is included in Figure 3.65a, to document the strong enrichment of this species in Ninemile Creek in the waste-bed area (see Figure 3.27). The average Cl^- concentration at Amboy for the 162 observations was $54\,mg \cdot L^{-1}$ (Effler et al. 1991); less than 7% of the average observed at Lakeland ($821\,mg \cdot L^{-1}$) over the same period. Further, the Cl^- concentration at Amboy was much more independent of flow (Figure 3.69) than observed at Lakeland. A less intensive monitoring effort conducted by the NYSDEC at these two stations over the 1987–1990 period documented the enrichment of this stream reach with Na^+ and Ca^{2+}, as well as Cl^- (Table 3.31). The average Na^+ concentration at the upstream station ($20\,mg \cdot L^{-1}$) was less than 6% of the average value measured at Lakeland ($359\,mg \cdot L^{-1}$; Table 3.31). The average Ca^{2+} concentration at the upstream station ($151\,mg \cdot L^{-1}$) was about 30% of the downstream station concentration ($518\,mg \cdot L^{-1}$). The relationships between these ions also appear relatively uniform at Amboy. The average of the Na^+/Cl^- ratio at this site was 0.47 (on a $mg \cdot L^{-1}$ basis); the relative standard deviation was 0.10. The average of the Ca^{2+}/Cl^- ratio at this

FIGURE 3.68. Evaluation of relationships between concentrations of major ions in Ninemile Creek: (**a**) Na^+ versus Cl^-, (**b**) Ca^{2+} versus Cl^-, (**c**) Ca^{2+} versus Cl^-, and (**d**) SO_4^{2-} versus Cl^-. Plots (**a**), (**b**), and (**c**) based on biweekly data over 1981–1989 interval (modified from Effler et al. 1995b); plot (**d**) based on paired measurements made in 1989.

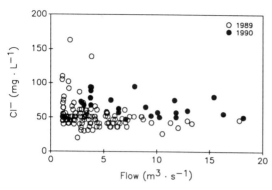

FIGURE 3.69. Evaluation of the relationship between Cl⁻ concentration and daily average flow in Ninemile Creek at Amboy. Flow at Amboy is assumed to be 80% of gauged flow at Lakeland (Effler et al. 1991).

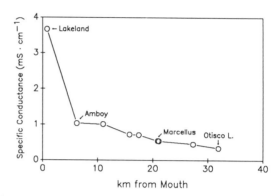

FIGURE 3.70. Specific conductance profile of Ninemile Creek, October, 18, 1991.

site was 3.79; the relative standard deviation was 0.16. Profiles of specific conductance made by the NYSDEC since closure of the soda ash/chlor-alkali facility from above the waste beds to the creek mouth indicate the ionic content remains largely unchanged from Lakeland to the lake, though measurements made close to the lake under low-flow conditions may be confounded by backflow from the lake.

A preliminary characterization of the location of ionic inputs to Ninemile Creek is presented as a profile of specific conductance in Figure 3.70. Progressive increases in ionic content are apparent from the Otisco Lake outlet to a position ~5 km upstream of Amboy. No significant ionic waste discharges are known in this reach. The zones of inflection presumably reflect regions in which saline groundwater enters the stream. Note that by

far the greatest inflection in ionic content for the entire profile occurs in the reach bounding the most recently active waste beds, between Amboy and Lakeland.

Sulfate concentrations were observed to decrease with increasing flows in Ninemile Creek in 1989, in a manner similar to that observed for Cl⁻, Na⁺, Ca²⁺, and T-NH₃. The relationship between SO_4^{2-} concentration and flow at Lakeland is evaluated according to the dilution model in Figure 3.71. The best fit linear regression expression is

$$[SO_4^{2-}] = 309\,Q^{-1} + 128 \qquad (3.15)$$

in which $[SO_4^{2-}]$ equals the concentration of SO_4^{2-} (mg · L⁻¹). However, unlike the findings for Cl⁻, Na⁺, (most of) Ca²⁺, and T-NH₃, the

TABLE 3.31. Paired concentrations of Cl⁻, Na⁺ and Ca²⁺ in Ninemile Creek at Amboy and Lakeland (1987–1989), as monitored by the NYSDEC.

Parameter	Amboy		Lakeland		n*
	$\bar{x}$	range	$\bar{x}$	range	
Cl⁻	40	32–49	1017	408–1750	15
Na⁺	20	14–23	359	110–590	13
Ca²⁺	151	108–174	518	202–700	12

*n—No. of paired observations

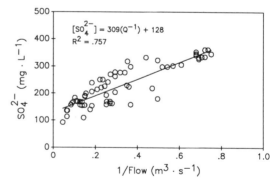

FIGURE 3.71. Evaluation of the relationship between SO_4^{2-} concentration and flow in Ninemile Creek at Lakeland.

principal origins of SO_4^{2-} are apparently upstream of the waste-bed area, as detailed monitoring in 1989 (65 paired samples over the April–October period) documented the near-equivalent concentrations for this constituent at the Amboy and Lakeland sites. The average ratio of the Amboy to the Lakeland value was 0.93; 91% of the observations fell within the range of ratio values of 0.8 to 1.1. Inputs of SO_4^{2-} over this stream reach apparently compensate approximately for the hydrologic gain (approximately 20%, see Effler et al. 1991).

Assuming the SO_4^{2-} concentrations measured at Amboy reflect the dissolution of gypsum, and the daily average flow at Amboy is 80% of that reported at Lakeland (Effler et al. 1991), the SO_4^{2-} concentration–flow relationship for Amboy (Figure 3.72a) can be used to estimate the dynamics for the associated fraction of Ca^{2+}. The predicted concentration–flow relation for this Ca^{2+} fraction appears in Figure 3.72b, along with the limited number of measurements of Ca^{2+} for that location.

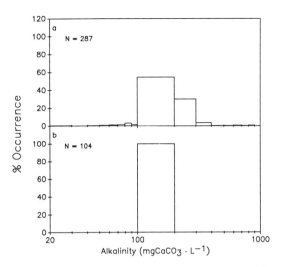

FIGURE 3.73. Distribution of alkalinity in Ninemile Creek at Lakeland for two intervals, based on biweekly sampling: (a) 1973–February 1986, and (b) March 1986–1990.

The difference between the prevailing concentrations and these predictions represents the Ca^{2+} fraction not associated with gypsum; the average value of this residual for flows between 1.4 and 8.0 $m^3 \cdot s^{-1}$ (n = 7) was 60 mg $\cdot L^{-1}$.

Distributions of alkalinity at Lakeland are presented for two intervals in Figure 3.73a and b, a period before closure of the soda ash/chlor-alkali plant (1973–February 1986), and following closure of the facility. The log scale accommodates the very broad distribution observed during the operation of the facility (e.g., minimum of 25 mg $CaCO_3 \cdot L^{-1}$, maximum of 4064 mg $CaCO_3 \cdot L^{-1}$), compared to the narrower distribution that has prevailed since. More than 35% of the observations for the earlier interval were greater than 200 mg $CaCO_3 \cdot L^{-1}$, and a number of observations were less than 100 mg $CaCO_3 \cdot L^{-1}$. The higher alkalinities encountered before closure reflect the irregular release of high-alkalinity (e.g., caustic) waste to the stream. The deviation from postclosure alkalinity levels was particularly striking in the late summer and early fall of 1980 (Figure 3.74a). The pH levels were also elevated over much of this period (Figure 3.74b). These are clear

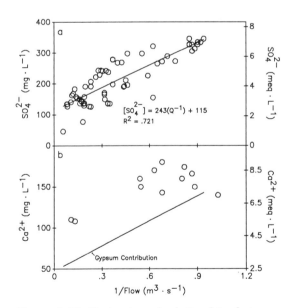

FIGURE 3.72. Evaluation of relationships between concentration and daily average flow in Ninemile Creek at Amboy: (a) SO_4^{2+}, 1989, and (b) Ca^{2+}, observed, and predicted for fraction associated with gypsum dissolution. Based on assumption, Amboy flow = 0.8X Lakeland flow.

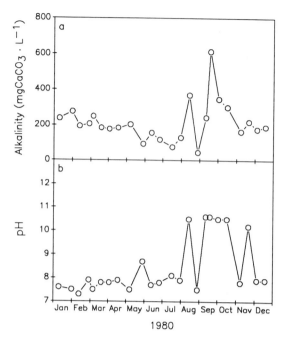

FIGURE 3.74. Distributions in Ninemile Creek at Lakeland in 1980: (**a**) alkalinity, and (**b**) pH.

indications of the discharge of caustic waste. However, the coincidence of elevated alkalinities and pH at the Lakeland station is not recurring throughout the database for the preclosure period.

The much higher and lower concentrations observed irregularly before closure of the facility (Figure 3.73a) indicates the operation of source and sink processes (respectively) at levels that exceed those that have prevailed since. The calcium carbonate deposits that blanket nearly the entire reach downstream of the waste beds, and along the near-shore area of the lake adjoining the creek mouth, are clear evidence of the operation of a sink process for alkalinity, in the form of calcium carbonate precipitation. Reductions in Ca^{2+} concentrations and elimination of particularly high in-stream pH values since closure of the soda ash/chlor-alkali facility have probably resulted in reductions in $CaCO_3$ precipitation.

The concentration of alkalinity in Ninemile Creek since closure of the soda ash/chlor-alkali facility remains relatively uniform over a wide range of flow, as illustrated by the log concentration–log flow plot for the intensive

1989 monitoring data in Figure 3.75. More than 85% of the alkalinity values from this data set were within the rather narrow range of 165 to 210 mg $CaCO_3 \cdot L^{-1}$. The waste-bed area presently does not greatly influence alkalinity concentrations in Ninemile Creek, as paired (n = 120) measurements in 1989 at Amboy and Lakeland were very similar. This is reflected in the following linear regression expression ($R^2 = 0.80$)

$$[A]_{Amboy} = 1.12[A]_{Lakeland} - 22.2 \quad (3.16)$$

in which: $[A]_{Amboy}$ and $[A]_{Lakeland}$ equals the alkalinity concentrations (mg $CaCO_3 \cdot L^{-1}$) at Amboy and Lakeland, respectively.

3.2.7.2.2 Onondaga Creek

3.2.7.2.2.1 Chloride. The most intensive monitoring of Cl^- concentration in Onondaga Creek was conducted over the 1989–1990 period at Spencer Street; grab samples were collected on 49 different days in 1989 and on 101 days in 1990. Monitoring was most intense in the late spring to fall interval. The temporal distribution of Cl^- concentration over the 1989–1990 period is presented in Figure 3.76a; the distribution of daily average flow over the same period is presented in Figure 3.76b. Concentrations of Cl^- decreased with increasing flow. A plot of concentration versus inverse daily average flow (Figure 3.77) supports a dilution model (Manczak

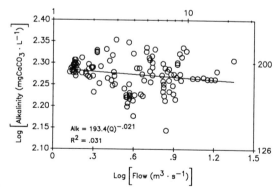

FIGURE 3.75. Evaluation of the relationship between alkalinity and daily average flow in Ninemile Creek at Lakeland in 1989.

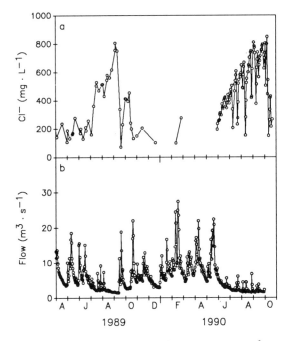

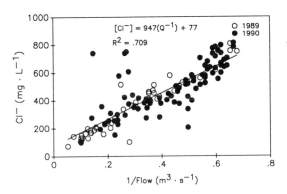

FIGURE 3.76. Temporal distributions in Onondaga Creek at Spencer Street for portions of 1989 and 1990: (**a**) Cl⁻ concentration and (**b**) daily average flow.

and Florczyk 1971) for this constituent, as observed for Cl⁻ and several other materials in Ninemile Creek. A single expression appears appropriate to describe the relationship for both years. Changes in flow (as inverse flow) explained about 70% of the variability observed in Cl⁻ concentration (Figure 3.77),

according to the linear least squares regression expression

$$[Cl^-] = 947\,Q^{-1} + 77 \qquad (3.17)$$

Apparently, an inverse flow relationship is generally recurring, as illustrated by the presentation of the Onondaga County data base for the 1973–1989 period in Figure 3.78. The differences between the relationships for these two databases are not considered great. The degree of scatter in this long-term data base, and the differences from the more recent intense data set (Figure 3.77), indicate real variability in the concentration-flow relationship has occurred over the period of monitoring.

The dilution character of the Cl⁻ concentration-flow relationship in Onondaga Creek, implies a localized input (i.e., point source(s)) such as waste discharges or groundwater discharges of Cl⁻ to the stream. This prompted a preliminary investigation of the origin(s) of ·the input(s). Substantially higher Cl⁻ concentrations have been observed at Spencer Street than at Dorwin Avenue. Statistics summarizing the 17 paired measurements (on grab samples collected within 0.5 hr of each other) made over the early May 1989 to mid-July 1991 period appear in Table 3.32. The average concentration at Spencer Street was more than four times that reported for Dorwin Avenue.

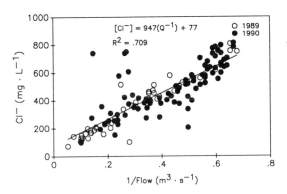

FIGURE 3.77. Evaluation of the relationship between Cl⁻ concentration and flow in Onondaga Creek at Spencer, for 1989 and 1990 conditions.

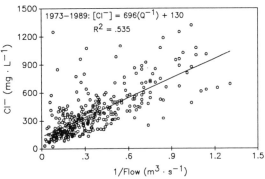

FIGURE 3.78. Evaluation of the relationship between Cl⁻ concentration and flow in Onondaga Creek at Spencer Street, over the period 1973–1989 (based on biweekly sampling).

TABLE 3.32. Summary of paired measurements (n = 17) of Cl^- made at Dorwin Avenue and Spencer Street over the May 1989–mid-July 1991 period.

Statistic	Dorwin Avenue	Spencer Street
n	17	17
Range (mg·L^{-1})	51–184	100–1340
Mean (mg·L^{-1})	97	427
Standard deviation	39.5	317
Median	94	329

Specific conductivity measurements were made in the creek at bridges on July 12, 1991 from Tallman Street (3.8 km above the creek mouth), working downstream, to Spencer Street. The resulting stream profile appears in Figure 3.79. The profile was flat until West Genesee Street (2.2 km above the creek mouth), where a nearly four-fold increase in specific conductance occurred, that remained unchanged to the routine monitoring location (Spencer Street) just above the creek mouth. On July 14, 1991 the Cl^- concentration was observed to increase from 220 to 1250 mg·L^{-1} over a distance of 0.3 km, from an upstream location to the West Genesee Street crossing. The saline input was tracked to the adjoining Niagara Mohawk Power Company facility, where saline groundwater was being discharged to Onondaga Creek. The discharge has been described as seasonal by the power company, as the groundwater was being used

to support cooling needs of the corporate offices (i.e., particularly during the summer months). However, stratification of the Cl^- concentration and flow populations at Spencer Street on a summer (May–September)/winter basis, does not reflect a significant seasonal difference in Cl^- loading from this facility (Figure 3.80). Ground water was used for cooling over the period 1960–summer of 1991. Documentation provided by the power company reflected substantial year-to-year variations in the pumping of the groundwater, which undoubtedly contributed to the scatter observed in the Cl^- concentration–flow relationship (Figures 3.77, 3.78, and 3.80). Three different point source (pipes) inputs are located in close proximity to the facility. On July 14, 1991, one of the outfalls had a Cl^- concentration of 17,920 mg·L^{-1}. The concentrations of major ions in the various wells used by the facility, as determined from samples collected on a single day, were reported in Chapter 2. The major constituents were Cl^- and Na^+. However, no results were reported for two potentially important constituents, SO_4^{2-} and HCO_3^-. The power company discontinued the practice of using ground-water for cooling in August 1991. Although this has eliminated the point discharge of Cl^- and Na^+ enriched brine to the creek and lake, it will be important to assess the extent to which this material may continue to reach the creek.

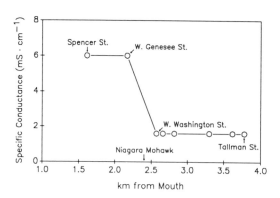

FIGURE 3.79. Profile of specific conductance for Onondaga Creek, within the City of Syracuse, July 12, 1991.

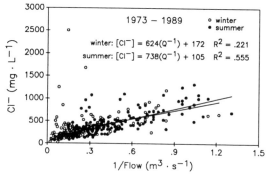

FIGURE 3.80. Evaluation of the relationship between Cl^- concentration and flow in Onondaga Creek at Spencer Street for two time strata, summer and winter, over the 1973–1989 period.

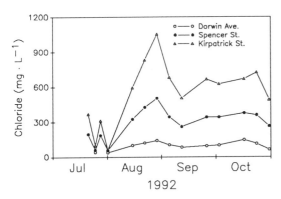

FIGURE 3.81. Comparison of time distributions of Cl⁻ concentration in Onondaga Creek for three sites, July–October 1992.

Additional significant loading of Cl⁻ to the creek further downstream (between Spencer Street and Kirpatrick Street; Kirpatrick Street. borders the N.Y.S. Department of Transportation barge canal terminal) was documented in 1992. The enrichment of the creek from Dorwin Avenue to Spencer Street, and from Spencer Street to Kirpatrick Street for the mid-July to October 1992 interval is depicted in Figure 3.81. The Cl⁻ concentration was on average 1.8 times higher (n = 14) at Kirpatrick Street than at Spencer Street. This relationship remained rather uniform over that interval; the relative standard deviation of the ratio was only 0.14. The small-scale character of the enrichment was delineated by specific conductance profiling in both lateral and longitudinal directions downstream of Spencer Street. A three-dimensional "surface" depicting the results is presented in Figure 3.82. A particularly sharp increase occurred about 60 m downstream. However, some of the inputs appear to be diffuse, as manifested by the "lobes" bordering both sides of the stream. Visual evidence (e.g., bubbles) of the entry of ground-water was encountered at several locations. This reach is within an area previously identified as having salt springs, before extensive urban development (Chap-

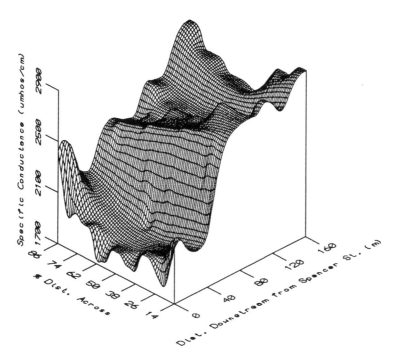

FIGURE 3.82. Specific conductance surface downstream of Spencer Street on August 21, 1992, depicting spatial character of salt inputs.

ters 1 and 2). The extent to which the recent discontinuation of the cooling water operation of Niagara Mohawk Power Company has interacted with this flux is unknown.

3.2.7.2.2.2 Other Major Ions. The concentrations of Cl^-, SO_4^{2-}, Na^+, and Ca^{2+} in Onondaga Creek have all demonstrated substantial variability, regulated primarily by stream flow. Recall that variations in the power company discharge may also have contributed. The distributions of these constituents are all correlated, either because of their identical origins or similarities in loading characteristics of the sources. Thus the demonstrated appropriateness of the dilution model for the Cl^- concentration–flow relationship (e.g., Figure 3.77) should be expected to apply to these other constituents as well in Onondaga Creek. The relationships between these ionic species at Spencer Street are depicted in Figure 3.83 (also see treatment for Cl^-, Na^+, and Ca^{2+} in Chapter 2). The first three relationships evaluated (Figure 3.83a–c) are based on monitoring (paired measurements) conducted over the period 1981–1989 by Onondaga County. Observations (26% of the total) that did not meet the criterion of charge balance within 10% were not included in the analysis. The evaluation of the relationship between Cl^- and SO_4^{2-} (Figure 3.83d) is based on paired measurements made in a separate program in 1989. The relationships between Na^+ and Cl^- and SO_4^{2-} and Cl^- were stronger than those that included Ca^{2+}. The relationship between Na^+ and Cl^- (Figure 3.83a) approaches the stoichiometry of NaCl (pure NaCl would have a slope of 0.65 in the format of Figure 3.83a), associated mostly with ground-water utilized and discharged by Niagara Mohawk Power Company to the creek.

The relationship between Ca^{2+} and SO_4^{2-} is evaluated indirectly through their dependencies on flow, as paired observations are not available. Plots of concentrations versus inverse flow at Spencer Street are presented in Figure 3.84. The relationship for SO_4^{2-} is much stronger. Concentrations appear also in units of $meq \cdot L^{-1}$ for this analysis. If it is assumed that SO_4^{2-} reflects dissolution of gypsum

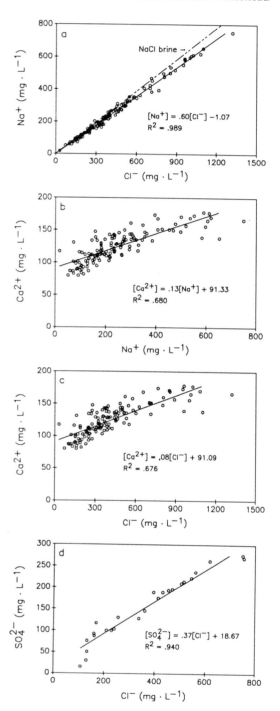

FIGURE 3.83. Evaluation of relationships between concentrations of major ions in Onondaga Creek: (**a**) Na^+ versus Cl^-, (**b**) Ca^{2+} versus Na^+, (**c**) Ca^{2+} versus Cl^-, and (**d**) SO_4^{2-} versus Cl^-. Plots (**a**), (**b**), and (**c**) based on biweekly data over 1981–1989 interval, (**d**) based on paired measurements made in 1989.

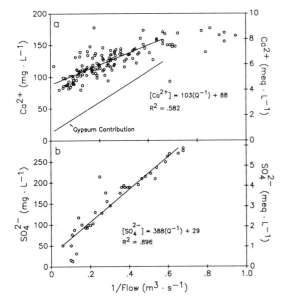

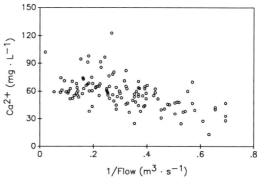

FIGURE 3.85. Evaluation of the relationship between the concentration of the nongypsum fraction of Ca^{2+} and flow in Onondaga Creek at Spencer Street.

FIGURE 3.84. Evaluation of relationships between concentration and daily average flow in Onondaga Creek at Spencer Street: (**a**) Ca^{2+}, observed (1981–1989), and predicted for fraction associated with gypsum dissolution, and (**b**) SO_4^{2-} 1989.

($CaCO_4$), the prevailing concentrations of Ca^{2+} can more than accommodate the observed SO_4^{2-} concentrations. A line has been added to the Ca^{2+} plot to reflect the distribution that would be observed if all the Ca^{2+} was derived from dissolution of gypsum. The concentration–flow relationship of the Ca^{2+} fraction not associated with gypsum dissolution, generated by difference, appears in Figure 3.85. This fraction is more invariant; somewhat lower concentrations of this fraction are indicated at lower flows. Most of the nongypsum fraction of Ca^{2+} is probably associated with HCO_3^-.

The dilution model was not supported for alkalinity in Onondaga Creek. The relationship between alkalinity and flow is evaluated in a log–log format in Figure 3.86, based on biweekly monitoring data collected over the period 1973–1989. Alkalinity is insensitive to flow at lower levels of flow, and appears to decrease slightly at higher discharge levels. Both overall and flow-stratified relationships are presented in Figure 3.86.

3.2.7.2.3 METRO

Major changes in the concentrations of Cl^-, Na^+, and Ca^{2+} have occurred in the METRO effluent from the early 1970s to the present, and major short-term variations continue to occur, due to inputs from the soda ash/chlor-alkali manufacturer's waste-bed lagoon. This is an outcome of the implementation of the cotreatment (domestic and industrial waste; USEPA 1974) scheme at METRO. While phosphorus removal was affected (but not without operational problems at METRO; see Chapter 1), the reduction in ionic waste loading (almost entirely as Ca^{2+}) was minimal. Just as much reduction in Ca^{2+} loading may

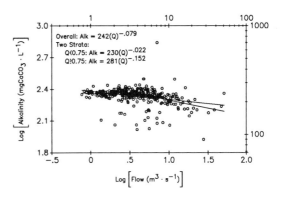

FIGURE 3.86. Evaluation of relationship between alkalinity and flow in Onondaga Creek, based on biweekly data over the period 1973–1989.

have occurred before diversion of the ionic waste to METRO, by precipitation and deposition in Ninemile Creek below the waste beds.

Changes in the distribution of Cl⁻ concentration in the METRO effluent related to the reception of this industrial waste, during the operation of soda ash facility, and following its closure, are illustrated in the plots of Figure 3.87. A log concentration scale has been utilized to facilitate resolution of variations within the broad range of concentrations. The last interval includes frequent observations (n = 99) made during the summer of 1990.

Broad distributions were apparent for all three time periods. The most commonly observed Cl⁻ concentrations before reception of the ionic waste were in the range 500–600 mg · L⁻¹ (Figure 3.87a). The distribution was shifted to much higher concentrations during the interval of diversion of the waste (and continued manufacture of soda ash). The

most commonly observed Cl⁻ concentrations during this interval were in the range 3000 to 4000 mg · L⁻¹ (Figure 3.87b). However, lower concentrations, corresponding approximately to values observed in earlier years, were also observed occasionally, during periods when the diversion was temporarily discontinued (usually operational problems at METRO associated with the reception of this waste). The dynamics in ionic waste flow to METRO responsible for this distribution were presented earlier in Figure 3.61. Protracted periods of discontinued ionic waste input to METRO occurred in 1981, 1982, and 1983. A number of short-term (e.g., day-to-day) interruptions in the flow also occurred that are not resolved in Figure 3.61. The distribution since closure of the facility has shifted strongly to lower concentrations; concentrations are most often between 100 and 200 mg · L⁻¹. The infrequent occurrence of high concentrations reflects the continuing, but irregular, discharge of ionic waste (Figure 3.87c) from a lagoon that adjoins the most recently active Solvay waste beds. An unexplained feature of the distributions is the higher concentrations that prevailed before reception of the ionic waste (Figure 3.87a) compared to after closure of the soda ash/chlor-alkali facility (Figure 3.87c). This apparent anomaly may reflect shortcomings in the earlier sampling program (e.g., inclusion of lake water in samples collected at METRO).

The continuing irregular discharge of ionic waste to METRO poses difficulties in accurately specifying the related loading to the lake. The problem is illustrated in Figure 3.88a, where the temporal distribution of Cl⁻ concentration in the METRO effluent (flow-weighted composite samples) is compared for two different sampling frequencies; daily (weekdays) for the late June–October period of 1990, and biweekly over the same interval. The pumping schedule from the waste-bed lagoon for that period is presented in Figure 3.88. The major short-term peaks in Cl⁻ concentration are clearly a result of the irregular inputs of ionic waste from the soda ash/chloralkali facility, as all the perturbations in METRO concentrations coincide with intervals

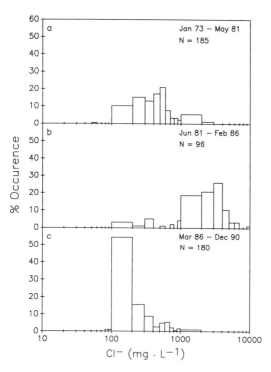

FIGURE 3.87. Distribution of Cl⁻ concentration in the METRO effluent: (a) January 1973–May 1981, (b) June 1981–February 1986, and (c) March 1986–December 1990.

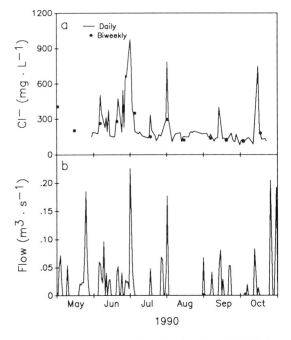

FIGURE 3.88. Temporal distribution for METRO for the May–October period of 1990: (**a**) Cl^- concentrations in the effluent, and (**b**) discharge of ionic waste to METRO (modified from Effler et al. 1995b).

of pumping of the ionic waste from the lagoon. Loading estimates of the ionic waste constituents are presently highly uncertain, because of the event-like character of the discharge to METRO. For example, these events may not be captured in a biweekly sampling program (e.g., Figure 3.88a), resulting in underestimates of related loads. Alternately, inclusion of one or more of these peaks in such a low-frequency monitoring program can result in systematic overestimation of the loads (e.g., using interpolation methods, see fecal coliform bacteria loading section for more details). The average Cl^- concentration measured on days when the ionic waste was not discharged to METRO was $167\,mg \cdot L^{-1}$; substantial variability was noted in these background conditions (relative standard deviation = 0.45).

The characteristics described here for Cl^{-1} concentration for the routine diversion and post-soda ash production intervals are representative of conditions observed for Na^+ and Ca^{2+}, as well, largely as a result of coupled distributions of these species in the ionic waste. Statistics describing the distributions of these three species and alkalinity in the METRO effluent are presented for three intervals in Table 3.33.

The average and median values of alkalinity have remained largely unchanged over the three intervals, although conditions were more uniform after closure of the soda ash/ chlor-alkali plant (Table 3.33). The absence of much lower concentrations in the mid-May 1981 to mid-February 1986 interval is surprising and unexplained, in light of the significant $CaCO_3$ precipitation that was known to occur in the facility during that period.

TABLE 3.33. Statistics describing the distributions of concentrations $(mg \cdot L^-)$ of Cl^-, Na^+, Ca^{2+}, and alkalinity in the METRO effluent for three intervals (biweekly sampling program).

Statistic	Cl^-			Na^+			Ca^{2+}			Alkalinity*		
	1	2	3	1	2	3	1	2	3	1	2	3
Average	502	2744	336	—	788	166	—	980	135	210	203	205
Standard deviation	314	1641	233	—	450	94	—	589	53	49	53	39
Median	464	2900	240	—	820	132	—	985	113	213	209	218
Minimum	56	135	93	—	84	32	—	85	64	53	69	91
Maximum	2357	9360	1200	—	2648	540	—	3475	338	650	456	267
n	185	96	104	—	95	102	—	96	102	191	96	104

1 1973–mid-May 1981
2 mid-May 1981–mid-February 1986
3 mid-February 1986–1990
* As $CaCO_3$

3.2.7.2.4 Major Ion Loading

3.2.7.2.4.1 Loading to Support Models and Mass Balance Analyses. Time series of daily external Cl^- loads were developed for the 1973–1991 interval, using FLUX, to support testing of a mass balance Cl^- model for the lake (Chapter 9). The biweekly monitoring data of Onondaga County were used solely for calculations for Ninemile Creek, Onondaga Creek, METRO, Ley Creek, and Harbor Brook. The ungauged tributary loading rate estimated by Auer et al. (1992) was applied uniformly throughout the interval. This load is relatively small (5%) for the period following closure of the soda ash/chlor-alkali facility. The interpolation option of FLUX was used to estimate loadings from METRO, Harbor Brook, Ley Creek, and Onondaga Creek. Ninemile Creek loads were calculated according to year-specific concentration/flow (log–log) relationships (e.g., Figure 3.29b; also see Effler et al. 1991), except for three years (1981, 1872, and 1986) in which abrupt changes in Cl^- concentration, unrelated to flow, occurred. The Ninemile Creek loads were corrected for the West Flume (i.e., recycle of lake water; see Chapter 1) contribution for the 1973–early 1980 interval (Doerr et al. 1994). The calculated time series of total external Cl^- load for the 1973–1991 period is presented in Figure 3.89. The major reduction in external loading that accompanied closure of the soda ash/chlor-alkali facility in 1986 is clearly resolved. The short-term peaks are largely associated with runoff events.

Time series of daily external alkalinity loads were developed for the spring–fall interval of two years, bracketing the closure of the facility, 1985 and 1989. The calculated distributions are presented in Figure 3.90. Both the seasonal and short-term variations in loading were driven by variations in hydrologic loading.

3.2.7.2.4.2 History of Annual Loading: Cl^-, Na^+, Ca^{2+}, and Alkalinity. The annual loads of Cl^-, Na^+, Ca^{2+}, and alkalinity from surface inflows to the lake have been determined for the period of record (Effer et al. 1995b). Most of the estimates are based on interpolation using FLUX. Ninemile Creek loads for the 1971–

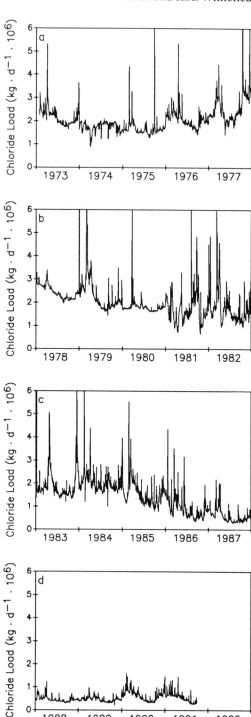

Figure 3.89. Estimated time series of daily external Cl^- loads to Onondaga Lake for the period 1973–1991.

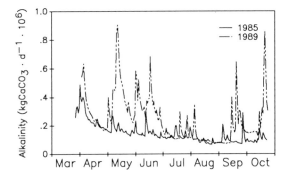

FIGURE 3.90. Estimated time series of daily alkalinity loads to Onondaga Lake for the spring–fall interval of 1985 and 1989.

1980 interval were corrected for cooling water recycle inputs from the West Flume. Uniform concentrations within years (the annual average value reported by Onondaga County (1971–1990)), were invoked in calculating the recycle loads to the creek. This simplification has a very minor effect on the subsequently presented annual loads. Annual loading estimates reported by Onondaga County have been used to fill gaps where the original data are no longer available. These do not differ substantially from values obtained using some of the successful concentration/inverse flow (dilution model) relationships developed for these constituents for Ninemile and Onondaga Creeks in earlier intervals. Inputs not considered in these analyses for Na^+, Ca^{2+}, and alkalinity include Trib 5A, East Flume, Sawmill Creek, and Bloody Brook. These sources make only minor (e.g., <10%) contributions to the total external loading of these materials. The estimates of annual loads are presented in Figure 3.91; the contributions of the major sources are presented in Figure 3.92. Reliable estimates of Na^+ and Ca^{2+} loading were not available for certain years. Note that loading estimates for Ca^{2+} (Figure 3.91c) from Onondaga County have been included for the entire period of the analysis. The similarity of the two different estimates in recent years supports the pattern for earlier years based only on Onondaga County estimates.

The calculated distributions for Cl^-, Na^+, and Ca^{2+} reflect the dominant role the dis-

charge from the most recently formed Solvay waste beds (along Ninemile Creek) has played in the loading of these materials to the lake. Variations in the loading of these three constituents have tracked each other over the period of record. Abrupt decreases in loadings occurred in 1986 with the closure of the soda

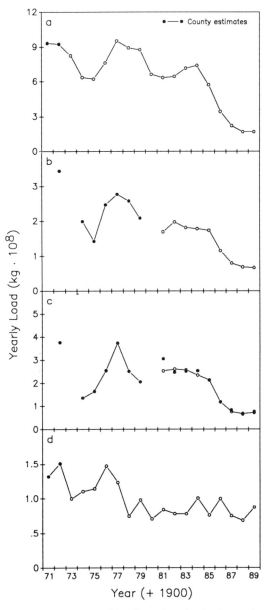

FIGURE 3.91. Annual loading of major ionic species to Onondaga Lake over the period 1971–1989; (a) Cl^-, (b) Na^+, (c) Ca^{2+}, and (d) alkalinity (as $CaCO_3$) (modified from Effler et al. 1995b).

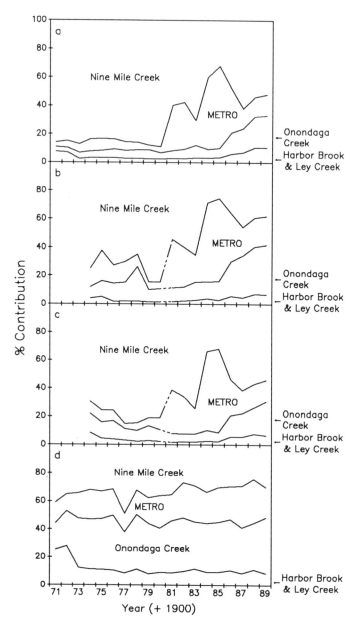

FIGURE 3.92. Contributions to annual loadings, according to surface inputs, for the period 1971–1989: (a) Cl^-, (b) Na^+, (c) Ca^{2+}, and (d) alkalinity (modified from Effler et al. 1995b).

ash/chlor-alkali facility (Figure 3.91a–c). The variations in loading before closure were probably mostly due to variations in the production of the industrial waste (i.e., the production of soda ash). The range of the summed annual loading of Cl^-, Na^+, and Ca^{2+} over the interval 1974–1985 was 9.5 – 16.4 × 10^8 kg. Over the 1974–1980 period, Ninemile Creek contributed approximately 86, 71, and 78% of the annual loads of Cl^-, Na^+, and Ca^{2+} to the lake, respectively. The diversion of the discharge (1981–1986) from the waste beds to METRO caused a major shift from Ninemile Creek to METRO in the contribution of these two inflows to the loading of these constituents (Figure 3.92a−c). However, the diversion did not result in a significant reduction in the overall loading of these materials to the lake. The substantial interannual variations in the relative contributions of these ions from Ninemile Creek and METRO are largely due to year-to-year differences in the quantity of ionic waste diverted to METRO (see Figure 3.61). Variations in the annual volume of waste diverted to METRO accounted for more than 90% of the variations observed in the annual Cl^- load of the facility over the 1981−1986 period. The annual average load of ionic waste from soda ash production over the 1974−1985 period was about 1.1 million metric tons.

Since closure of the soda ash/chlor-alkali facility, there has been a major reduction in total loading of these constituents and redistribution of the relative contributions of the surface inflows to the loading of these materials. By 1989 the annual loading of Cl^-, Na^+, and Ca^{2+} had decreased by 79, 67, and 70% from the average documented for the 1974−1985 period, as a result of the closure of the soda ash/chlor-alkali plant. The residual annual load of ionic waste from soda ash production by 1989 was about 0.14 million metric tons, or about 13% of the preclosure load.

Contributions to annual material loading by the five gauged surface inflows are presented for the major ion species for 1989 in Figure 3.93a−e; the hydrologic contributions of these inflows are presented in Figure 3.93f for comparison. Ninemile Creek made disproportionate (with respect to flow) contributions to the overall loads of Cl^-, Ca^{2+}, and SO_4^{2-} (Figure 3.93a, c, and e). Ninemile Creek and Onondaga Creek contributed nearly equally to the Na^+ loading (Figure 3.93b). The greatest alkalinity load emanated from Onondaga Creek (Figure 3.93d). Onondaga Creek and METRO contributed nearly equally to SO_4^{2-} loading (Figure 3.93e); METRO was the third contributor for the other four constituents. Substantial decreases in the contribution of Onondaga Creek to Cl^- and Na^+ loading may occur as result of the discontinuation of the

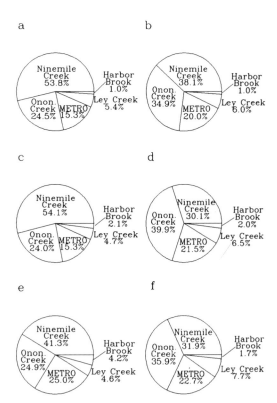

FIGURE 3.93. Contributions to material and hydrologic loading for Onondaga Lake inflows in 1989: (a) Cl^- load (load = 1.6×10^8 kg), (b) Na^+ load (load = 6.7×10^7 kg), (c) Ca^{2+} load (load = 7.2×10^7 kg), (d) alkalinity load (load = 8.9×10^7 kg), (e) SO_4^{2+} load (load = 6.8×10^7 kg), and (f) hydrologic load (load = 4.5×10^8 m^3).

use of ionically enriched ground-water by Niagara Mohawk Power Company for cooling.

3.2.7.2.4.3 Continuing Anthropogenic Contributions of Cl^-, Na^+, and Ca^{2+}. There have been easily identified anthropogenic inputs of Cl^-, Na^+, and Ca^{2+} that continued to reach Onondaga Lake via surface inflows following closure of the soda ash/chlor-alkali facility, including: (1) inputs from the waste-bed area to Ninemile Creek, (2) inputs of Cl^- and Na^+ from wells under the Niagara Mohawk Power Company facility to Onondaga Creek, and (3) the discharge of the sewage treatment plant. The high concentrations of Cl^-, Na^+, and Ca^{2+} at the mouths of Ley Creek and Harbor Brook indicate a portion of these smaller loads may also have an anthropogenic basis. However, here we focus on the three primary surface inflows to the lake—Ninemile Creek, Onondaga Creek, and METRO. Estimates of anthrogenic contributions to these inputs are provided. These estimates are conservative to the extent that the smaller tributaries and potential direct inputs from the adjoining waste beds (e.g., those that border the lake) are not considered.

Estimates of the continuing anthropogenic contributions to the surface loadings to Onondaga Lake of Cl^-, Na^+, and Ca^{2+} are presented in Table 3.34. The partitioning of total loading for the three tributaries presented for the ions for 1989 in Figure 3.93 was a principal basis for the analysis. The analysis was largely based on special loading studies conducted on these inputs in 1989 and 1990, discussed previously in this section. The estimates for Cl^- are the most reliable, as they are supported by the largest databases. The total contributions presented (Table 3.34) are the sum of the Ninemile Creek, Onondaga Creek, and METRO estimates. The scenario of treating the loads of Cl^- and Na^+ at Dorwin Avenue on Onondaga Creek as an antropogenic contribution is included, as these loads essentially reflect inputs from the "mud boils" (which have yet to be established as an anthropogenic impact). The contributions given for this scenario (Table 3.34) probably represent upper bounds. The contribution from the soda ash/chlor-alkali facility was calculated as the sum of the METRO component estimated to be received from that facility and the anthropogenic (waste-bed area) Ninemile Creek contribution.

The most definitive information is available for the waste-bed area of Ninemile Creek, based on intensive monitoring of Cl^- above and below the most recently active waste beds (Figure 3.65), less intensive monitoring of Na^+ and Ca^{2+} for the same two sites (Table 3.31), and the well-established ratios between these species that are maintained at Lakeland (Figure 3.68). The Cl^- load emanating from the waste-bed area is quantified as the differences in the upstream load and the downstream load. The interpolation option of FLUX was used to calculate loads over the intensive monitoring period of April 1989 through February 1990. Daily average flows measured at Lakeland ($Q_{Lakeland}$) were used to support the calculations; the average daily flow at Amboy (Q_{Amboy}) was estimated according to the simple linear expression (Effler et al. 1991),

$$Q_{Amboy} = 0.80 \times Q_{Lakeland} \qquad (3.18)$$

The value 0.80 equals the ratio of the upstream watershed areas of the creek for the two stations. The estimated load at Lakeland for that period was 9.98×10^7 kg, the estimate for Amboy was 0.82×10^7 kg. Thus, the waste-bed area represented 92% of the total Ninemile Creek Cl^- load. The associated loads of Ca^{2+} and Na^+ at Lakeland were estimated according to the Na^+/Cl^-, Ca^{2+}/Cl^- relationships established in Figure 3.68. The upstream loads were determined based on the average concentrations reported for these species in Table 3.31.

METRO's contribution to the loading of these materials in 1989 appeared in Figure 3.93; recall these estimates were only ap-

TABLE 3.34. Anthropogenic contributions to major ion loadings.

Ion	Anthropogenic contributions to total annual loads (%)				
	Total	Ninemile Creek	Onondaga Creek	METRO	Soda ash facility
Cl^-	81 (89)*	50	17 (24)*	15 (5)**	55 (63)*
Na^+	79 (89)*	35	24 (35)*	20 (7)**	42 (52)*
Ca^{2+}	40	25	0	15 (5)**	30

( )* Contribution including load at Dorwin Avenue on Onondaga Creek
( )** Component of METRO contribution received from inactive soda ash/chlor-alkali facility

proximate because of the limitations of the sampling program (e.g., Figure 3.88). The fraction of these inputs attributable to the soda ash/chlor-alkali facility was estimated from the intensive Cl^- data set collected on METRO's effluent over the June–October interval of 1990 (Figure 3.88). The total load from METRO determined for that interval, with the interpolation option of FLUX, was 8.5×10^6 kg. The load that would have occurred in the absence of the saline input from the soda ash/chlor-alkali was estimated as the summation of the products of the average concentration, for days on which no input was received (167 mg · L^{-1}), and daily flow. This estimated load was 5.75×10^6 kg; that is, the saline input from the soda ash/chlor-alkali facility represented 32% of the Cl^- load discharged from METRO over that period. This value is expected to be a reasonable estimate of the contribution that continues to be made (e.g., on an annual basis) from this source to the METRO discharge. This represents about 5% of the total annual Cl^- load to the lake (Figure 3.93 and Table 3.34). The contributions of Na^+ and Ca^{2+} were estimated according to the Na^+/Cl^- and Ca^{2+}/Cl^- ratios established for Ninemile Creek at Lakeland (Figure 3.68).

The contribution of the saline groundwater discharge from Ninemile Mohawk Power Company to Onondaga Creek was based on the difference in Cl^- loads at Spencer Street and Dorwin Avenue. This represents a reasonable approximation based on the "flatness" of the specific conductance profile upstream of the discharge (e.g., Figure 3.79). The load at Dorwin Avenue was determined as the product of the average Cl^- concentration (97 mg · L^{-1}) observed over the May 1989–July 1991 interval (Table 3.32) times the daily average flows measured at that location in 1989. Approximately 30% of the load at Spencer Street emanated from above Dorwin Avenue (mostly from the mud boils). Thus about 70% of the Onondaga Creek Cl^- load and 17% of the total load to the lake (Table 3.34) was attributable to the power company discharge. The Na^+ load at Spencer Street was based on the established (Figure 3.83a) $Na^+/$

Cl^- ratio. The Na^+ load at Dorwin Avenue was calculated from the Cl^- load, assuming the stoichiometry of NaCl. Based on the recently documented influx of Cl^- further downstream at Kirpatrick Street (Figures 3.81 and 3.82), the possibility that a significant fraction of this cooling water discharge would have reached the creek via some other route cannot be discounted. Thus the cooling water discharge load may not have been a net anthropogenic input.

At least 55% of the Cl^-, 42% of the Na^+, and 30% of the Ca^{2+}, entering the lake (i.e., from soda ash production) via surface inputs in 1989 were derived from anthropogenic sources (Table 3.34). Assuming the cooling water discharge is a net anthropogenic input, but the mud boil inputs are natural, approximately 80% of the Cl^- and Na^+ and 40% of the Ca^{2+}, entering the lake via surface inputs in 1989 are estimated to be pollution inputs (Table 3.34). However, the anthropogenic contributions could approach 90% for Cl^- and Na^+ if the mud boils are found to be a result of solution mining (Table 3.34). Most of this ionic loading is associated with the soda ash/chlor-alkali facility, most of which enters the lake via Ninemile Creek (Table 3.34). If the mud boils are found to be a result of solution mining, the contributions of Cl^- and Na^+ from the activities of that facility increase significantly (Table 3.34). The groundwater discharge from the power company represented about 17% of the total Cl^- load in 1989, and approximately 24% of the Na^+ load. The partitioning presented in Table 3.34 is probably representative of conditions that prevailed since closure of soda ash/chlor-alkali facility until the summer of 1991, when the groundwater discharge from the power company was discontinued. The relative contribution of the chemical facility to the anthropogenic loading of Cl^- and Na^+ may increase with the elimination of the power company discharge. Minor year-to-year variability in the presented partitioning is to be expected as a result of natural variations in runoff. More systematic changes could occur in the future as pollutant reservoirs are depleted or remediated (e.g., waste beds).

3.2.8 Fecal Coliform Bacteria

3.2.8.1 Background

External inputs of fecal coliform bacteria result in frequent contravention of public health standards for bathing beaches in the southernmost portion of the lake even during dry weather periods (Canale et al. 1993). The area of impact expands to include much of the lake following major rainfall events (Chapter 6), as a result of episodic loading of fecal coliform bacteria associated mostly with discharges of raw sewage from combined sewer overflows. It is important to quantify the loading of fecal coliforms to the lake and partition it according to the sources. This will support related management efforts to remediate the lake's public health bacteria problems within the framework of a mechanistic model for fecal coliform bacteria (Canale et al. 1993; Chapter 9). The effort to develop fecal coliform loading information adequate to support model development differs fundamentally from that required for other water quality constituents considered heretofore with respect to time scale. The time scale of wet-weather loading of fecal coliform bacteria is relatively short (usually hours), and related impacts on the lake are transient, lasting only several days (Canale et al. 1993).

Here we document fecal coliform concentrations that prevailed in the various inputs to the lake during dry and wet weather intervals of the intensive study period of the summer of 1987. These concentrations and available flow data are used to estimate loads and partition overall loading among the sources for a dry weather period and two wet weather periods. Finally, historic data are analyzed to identify changes in indicator concentrations in Onondaga Creek that may be attributable to related management efforts.

3.2.8.2 Review of 1987 Monitoring Program

Ten tributaries and discharges to Onondaga Lake (Figure 3.1) were intensively monitored for fecal coliform bacteria over the period June 1–August 27, 1987, to support the development of fecal coliform loads to the lake, and thereby support the testing of a mechanistic model for fecal coliform bacteria for the lake. Fecal coliform concentrations were measured by the membrane filter technique (APHA 1985); units are "number of colony forming units per 100 ml" ($cfu \cdot 100\,ml^{-1}$). Dry weather samples were collected daily (weekdays). During two storm events, the inputs were sampled at 1 to 2 hr intervals the first day, 3 hr intervals the second day, 6 hr intervals the third day, and 12 hr intervals thereafter, until background flows were reached.

Two substantial wet weather events were encountered during the monitored period of 1987. The first event, hereafter designated "storm 1," occurred on June 22; 2.66 inches of rain (as measured at Hancock Airport) fell over a period of 8 hr, 0.95 inches of rain fell between 1200 and 1300 hr. The storm had a return frequency of 7 yr based on an average precipitation of 0.38 in./hr over a 7 hr period, however, the rainfall proximate to the lake probably had a return frequency of 3 yr (Moffa and Associates, personal communication). The second event, hereafter designated "storm 2," occurred on July 14; 0.95 in. of rain (as measured at Hancock Airport) fell over a 5 hr period, the maximum rainfall was 0.72 in. between 1100 and 1200 hr. The storm had a return frequency of 1 yr based on an average precipitation of 0.72 in./hr for 1 hr period (Moffa and Associates, personal communication).

Significant wet weather days during the 1987 study period were identified (Heidtke 1989) based on regional precipitation data. The identified days were June–24, July 2–4, July 12–14 August 9–11, and August 23–24. The tributary concentration data were stratified into wet and dry weather periods based on these designated "wet" days (Heidtke 1989b).

3.2.8.3 Concentrations

The breakdown on average and maximum concentrations of fecal coliform bacteria in the tributaries during a dry weather monitoring period of July 29–August 7 is presented in

Table 3.35. This period represents one of three loading intervals for which the fecal coliform model was tested (Chapter 9). Fecal coliform concentrations in the METRO effluent are very low during dry and wet weather as a result of disinfection. The METRO overflow does not discharge in dry weather. The lowest concentrations observed during dry weather were in Ninemile Creek and Sawmill Creek. The low concentrations in Ninemile Creek may reflect the influence of elevated chloride concentrations (Mancini 1978). The low levels in Sawmill Creek are probably indicative of uncontaminated conditions. The concentrations were somewhat higher during dry weather in Ley Creek, Bloody Brook, Tributary 5A, and the East Flume, indicating the presence of fecal contamination. Onondaga Creek and Harbor Brook have the highest dry weather fecal coliform bacteria concentrations of the inputs (Table 3.35). These concentrations indicate the presence of substantial fecal contamination during dry weather periods.

The maximum and average concentrations of fecal coliform bacteria observed in the tributaries during the two wet weather events of 1987 are presented in Table 3.36. Analysis of data from storm 1 is complicated by uncertainty in certain measurements; failure to provide adequate dilution in some measurements caused some values to be reported as

TABLE 3.36. Average and maximum concentrations of fecal coliforms bacteria in tributaries and discharges to Onondaga Lake during two wet weather events in the summer of 1987[a].

Tributary/ discharge	Fecal coliform concentrations cfu·100 ml⁻¹			
	Average		Maximum	
	Storm 1	Storm 2	Storm 1	Storm 2
Onondaga Creek	45,600	244,046	100,000	1,250,000
Harbor Brook	59,800	101,112	200,000	1,010,000
Ley Creek	24,583	73,867	47,000	165,000
Ninemile Creek	17,536	16,411	51,000	124,000
Bloody Brook	8,720	10,139	10,000	21,300
Sawmill Creek	9,180	14,220	10,000	77,000
Tributary 5A	7,715	11,281	10,000	60,000
East Flume	17,582	42,128	39,100	105,000
METRO Effluent	80	106	80	465
METRO Overflow	ND	8,080	ND	32,100

[a] Modified from Canale et al. (1993)

"greater than" (>) a particular concentration (e.g., some plates had colonies "too numerous too count" (TNTC)). Substantially higher concentrations were reported for the discharges during storms 1 and 2 (Table 3.36) than observed during dry weather (Table 3.35). Particularly dramatic increases were observed for storm 1 in Sawmill and Ninemile Creeks, indicating the presence of important storm-related inputs and, in Ninemile Creek, a diminution of bactercidal effects of chloride. No strong relationships between fecal coliform concentrations and flow were observed for any of the inputs. A detailed time plot of fecal coliform concentrations measured in Onondaga Creek during the monitoring period of 1987 is presented in Figure 3.94. The storm related peaks in this tributary (and Harbor Brook and Tributary 5A) were short lived, with a duration of about one day (Figure 3.94). The time for return to baseline levels in Ninemile Creek, Sawmill Creek, Bloody Brook, and Ley Creek was about two days. Note the very substantial variations in Onondaga Creek even during dry weather periods (Figure 3.94).

The "best management practices" (BMP) program for the combined sewer system has reduced inputs from the combined sewer overflows to the tributaries (O'Brien and Gere Engineers Inc. 1987). The data base from

TABLE 3.35. Average and maximum concentrations of fecal coliform bacteria in tributaries and discharges to Onondaga Lake during a dry weather period (July 29–August 7, 1987).

Tributary/discharge	Fecal coliform concentration (cfu·100 ml⁻¹)	
	Average	Maximum
Onondaga Creek	25,525	56,400
Harbor Brook	14,344	35,400
Ley Creek	2,595	17,000
Ninemile Creek	108	436
Bloody Brook	3,747	21,800
Sawmill Creek	554	1,440
Tributary 5A	2,448	11,800
East Flume	4,737	8,180
METRO Effluent	14	45
METRO overflow	0	0

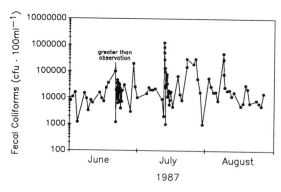

FIGURE 3.94. Temporal distribution of the concentration of fecal coliform bacteria in Onondaga Creek (at Spencer Street) in the summer of 1987.

Onondaga County's long-term monitoring program (biweekly sampling) is an appropriate vehicle to demonstrate the impact of this management program. The distribution of fecal coliform bacteria concentrations in Onondaga Creek before the BMP program is compared to the distribution after completion of the BMP program in Figure 3.95. Despite the fact that the concentrations remain high in the creek, the levels have decreased sub-

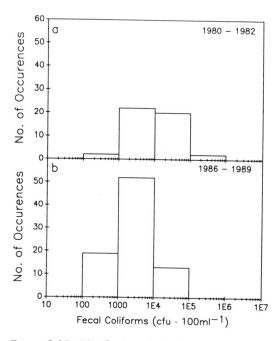

FIGURE 3.95. Distribution of fecal coliform bacteria concentrations in Onondaga Creek (at Spencer Street) for two periods: (a) 1979–1982, and (b) 1986–1989.

stantially since before the BMP program. The median concentration over the 1979–1982 interval was 18,000 cfu $\cdot$ 100 ml^{-1}; the median for the 1986–1989 interval was 2640 cfu $\cdot$ 100 ml^{-1}.

3.2.8.4 Loads

Fecal coliform bacteria loads were estimated for all (10) inputs monitored for the one dry period and the two wet weather intervals of 1987. A time series of hourly loads was calculated by interpolation of paired measurements of flow (where appropriate) and concentration for each input. Adjustments were made to the raw database to minimize inaccuracies resulting from insufficient sampling immediately prior to, during, or following a wet weather event. These adjustments were necessary to improve the accuracy of loading estimates, as the episodic nature of the wet weather events caused systematic problems for estimates based on interpolation. Overestimates in loading that result from failure to make measurements just before and just after the events are illustrated in Figure 3.96a. Absence of measured instantaneous loads specified at time t_2 (data point B) and t_4 (point D) in a given time series results in systematic overestimates of loads (compare the area under the curve formed by A–C–E to that under the curve formed by B–C–D). To correct for this problem, each tributary database was adjusted by inserting a set of pre- and poststorm values determined from analysis of measured bacteria concentrations during dry and wet weather (i.e., two strata). This was done by inserting the average dry weather concentration at the low flows bounding the storm hydrograph. Days included in the two strata have been identified previously; fifty or more observations were included in both strata for each of the important sources (Heidtke 1989b). Similarly, the absence of a concentration measurement at or near the instantaneous loading peak may cause the interpolation-based load calculation method to underestimate the load over a specified time interval (Figure 3.96b). In this case, the average wet weather concentration is inserted at the peak instantaneous flow. "Daily adjust-

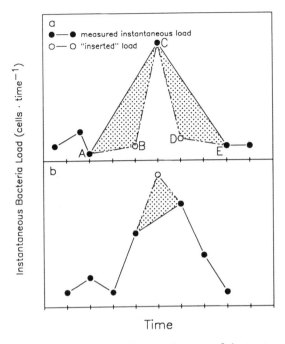

FIGURE 3.96. Cases of misestimates of instantaneous bacteria loads: (**a**) conditions leading to overestimates and (**b**) conditions leading to underestimates.

ments" were also made for days without measurements (dry weather), by inserting the product of daily average flow and the mean dry weather concentration.

A summary of fecal coliform loading results for the two storms and the overall monitoring period is presented in Table 3.37. Note that the storm periods, as specified in Table 3.37, are extended periods that include dry weather periods immediately prior to and following the runoff event. The "effective contributing period" (ECP), the time during which the stream pollutographs are significantly affected by the rainfall event, was in each case much shorter (June 22–24 and July 14–15). Two different loading estimates are provided for storm 1, because of the occurrence of TNTC observations in some of the tributary samples that are believed to bound the conditions that prevailed during that event. Onondaga Creek is seen to be the dominant source of fecal coliform bacteria to Onondaga Lake. This source represented 60 to 85% of the load for storm 1, more than 70% of the load for storm 2, and from 75 to 85% of the total load for

the 85 day monitoring period. Ley Creek, Harbor Brook, and Ninemile Creek together accounted for 15 to 20% of the load. Approximately 50% of the total load for the study period was received during these two storms (about 5% of the total 85 day monitoring period). The ECP accounted for approximately 85 to 95% of the overall storm contributions for Onondaga Creek, Ninemile Creek, and Sawmill Creek. The ECP represented from 60 to 85% of the total fecal coliform bacteria load from all sources in storm 1; the ECP accounted for 75% of the load from all sources in storm 2.

Fecal coliform bacteria loads have been calculated previously for several of the inputs to Onondaga Lake (O'Brien and Gere Engineers Inc. 1987, Stearns and Wheler Engineers 1979). These programs differed significantly in their design. Provisional USGS flow data were utilized to estimate loadings for five different events in 1976 (Stearns and Wheler Engineers 1979). Interpolation was used in these efforts to estimate concentrations for periods when samples were not collected. O'Brien and Gere Engineers Inc. (1987) estimated loads for Onondaga Creek, Ley Creek, and Harbor Brook for wet weather by collecting samples upstream and downstream of the CSO service area on each stream on six wet weather and two dry weather dates, after the implementation of the BMP program. Methods used in calculating the loads were not described. Loads determined for Onondaga Creek in the three different programs are compared in Figure 3.97. Despite problems in comparing these results, it appears loads are somewhat lower than in 1976. This may reflect improvements associated with implementation of the BMP program on the sewer system.

3.3 Summary

3.3.1 Hydrology of Onondaga Lake

A hydrologic analysis of Onondaga Lake was conducted based on a comprehensive twenty year record of continuous gauging of the principal inflows and lake level. The

TABLE 3.37. Summary of fecal coliform bacteria loading to Onondaga Lake, over the period June 1–August 27, 1987.

(a) Storm 1: June 19–July 2

Tributary	Average load (cfu·d^{-1})	Total load (cfu)	Contribution (%)	
			Lower bound	Upper bound
Onondaga Creek (lower)	9.9E + 13	1.4E + 15	61.8	—
Onondaga Creek (upper)	3.5E + 14	4.9E + 15	—	85.2
Ninemile Creek	1.6E + 13	2.2E + 14	10.0	3.9
Harbor Brook	1.3E + 13	1.8E + 14	7.9	3.1
Ley Creek	2.3E + 13	3.2E + 14	14.3	5.5
Others	0.6E + 13	0.8E + 14	6.0	2.3
total (lower bound)		2.2E + 15		
total (upper bound)		5.7E + 15		

(b) Storm 2: July 9–July 22

Tributary	Average load (cfu·d^{-1})	Total load (cfu)	Contribution (%)
Onondaga Creek	8.7E + 13	1.2E + 15	72.8
Ninemile Creek	2.8E + 12	3.9E + 13	2.5
Harbor Brook	1.0E + 13	1.4E + 14	9.1
Ley Creek	1.1E + 13	1.6E + 14	10.1
Others	0.4E + 13	0.6E + 14	5.5
total		1.6E + 15	

(c) Overall: June 1–August 27

Tributary	Average load (cfu·d^{-1})	Total load (cfu)	Contribution (%)	
			Lower bound	Upper bound
Onondaga Creek (lower)	6.3E + 13	5.4E + 15	75.1	—
Onondaga Creek (upper)	1.0E + 14	8.9E + 15	—	83.3
Ninemile Creek	3.3E + 12	2.8E + 14	4.0	2.7
Harbor Brook	6.1E + 12	5.2E + 14	7.3	4.9
Ley Creek	6.8E + 12	5.8E + 14	8.1	5.4
Others	4.9E + 12	4.2E + 14	5.5	3.7
total (lower bound)		7.2E + 15		
total (upper bound)		1.1E + 16		

average surface inflow for the 1971–1989 period was 16.5 m³·s^{-1}. Less than 10% of this inflow was not continuously gauged. The net inflow/loss associated with precipitation/evaporation processes is very small by comparison, and there is no evidence that exchange with the ground-water system is a significant component of the lake's hydrologic budget. The largest sources of water annually to the lake are Ninemile Creek and Onondaga Creek; together they contributed nearly 65% of the surface flow received over the 1971–1989 interval. The third largest source of water

(19%) has been the METRO effluent; much of this water comes from outside of the Onondaga Lake watershed. Substantial inter-annual variations in inflow are observed due to year-to-year differences in runoff.

Strong seasonal variations in hydrologic loading from the fluvial inputs are observed. However, the METRO discharge remains uniform by comparison. The highest rates of tributary inflow occur in March and April. The minimum usually occurs over the July–September interval. Thus the METRO discharge contributes relatively more to total

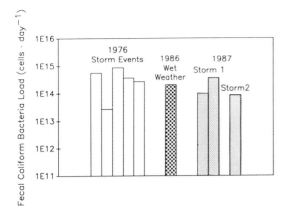

FIGURE 3.97. Comparison of fecal coliform loads for Onondaga Creek for different years by different programs.

3.3.2 Constituent Concentrations in Onondaga Lake Inflows

The tributaries of the lake are naturally enriched in alkalinity, calcium, and sulfate. However, Ninemile Creek has had high ionic concentrations as a result of anthropagenic discharges. The METRO discharge is enriched in domestic waste constituents. Fecal contamination enters primarily from Onondaga Creek, particularly after runoff events. This tributary irregularly also has high concentrations of other domestic waste constituents. A number of monitoring programs have been, and continue to be, conducted to support estimates of seasonal and annual loading of pollutants to Onondaga Lake, that in turn support several related mass-balance water quality models and analyses of long-term trends in loading. Grab samples have been collected over a wide range of frequencies (e.g., $24\,d^{-1}$ to biweekly). Analyses of composite samples from the METRO effluent have been conducted for a number of parameters since the early 1980s as part of the facility's permit requirements. This analysis of material concentrations and loading has focused on the following parameters:

1. Suspended solids (SS).
2. Nitrogen species, including nitrate and nitrite (T-NO$_x$), total ammonia (T-NH$_3$), and total Kjehdahl nitrogen (TKN).
3. Phosphorus species, primarily total phosphorus (TP).
4. Major ions, including chloride (Cl$^-$), sulfate (SO$_4^{2-}$), alkalinity (Alk), calcium (Ca^{2+}), and sodium (Na$^+$).
5. Fecal coliform bacteria ((FC) a widely used indicator of fecal contamination).

3.3.2.1 Concentration/Flow Relationships

A fundamental first step in the material loading analysis was the evaluation of the concentration/flow and concentration/time relationships for the above constituents. Expressions developed from the available databases describing the relationships between constituent concentrations and flow,

inflow during the critical water quality period of summer (25% on average over the June–September interval). The METRO discharge has been the single largest source of water to the lake for the month of August in 7 yr over the 1971–1989 period.

Lake level, on both an annual and seasonal basis, is strongly positively correlated to total inflow. This is largely a manifestation of the regulation of the lake level by dam control of the level of the Seneca River. The range in annual average lake level over the 1971–1989 period was only about 0.34 m. However, short-term variations of 1 m, in response to major runoff events, have not been unusual.

Onondaga Lake is a high flushing-rate lake; the average flushing rate for the 1971–1989 period (assuming a completely mixed system) was 3.9 flushes · yr^{-1}. This greatly exceeds rates of flushing reported for the Finger Lakes. The average annual flushing rate of the lake would be about 3.2 flushes · yr^{-1} without the METRO discharge. The high flushing rate has important implications for remediation efforts, as it causes the response time (the time it takes to reach a new steady state) to be short (<1 year). On average, the lake flushes through more than once during the March–April period. During the summer stratification period, the epilimnion is flushed through about three times under average flow conditions.

and among constituents, in Ninemile Creek and Onondaga Creek are presented in Table 3.38. The two stations are close to the stream mouths. Such relationships do not emerge for METRO as its discharge flow is relatively uniform. The relationships of Table 3.38 are valuable in estimating concentrations, and thereby loads, over periods in which samples have not been collected. With the exception of SS in Onondaga Creek, the constituent concentration expressions of Table 3.38 are linearly proportional to the inverse of flow (i.e., concentrations increase as flow decreases); hyperbolae are formed on linear concentration versus flow plots. These relationships are described as dilution curves. They characteristically are manifested downstream of localized continuous inputs, such as wastewater discharges and groundwater seeps. Localized inputs of T-NH$_3$, Cl$^-$, Na$^+$, and Ca^{2+}—materials common to Solvay process waste—have been, and continue to be, received by Ninemile Creek in the area of the waste beds. The relationships between the concentrations of Cl$^-$, Na$^+$, and Ca^{2+} have remained unchanged at Lakeland since before closure of the soda ash/chlor-alkali, despite reductions in concentrations. These relationships (Table 3.38) are very similar to those documented for the Solvay waste-bed overflow during the facility's operation, thereby

establishing the wastewater origins of this material. Note that the concentration/flow relationship for Cl$^-$ (and thereby Na$^+$ and Ca^{2+}) has changed dramatically since the Solvay waste-bed overflow was received solely by Ninemile Creek (Table 3.38). Sulfate appears to have natural groundwater origins (e.g., gypsum dissolution) in both the Ninemile Creek and Onondaga Creek watersheds.

The dilution curve characteristics observed for Cl$^-$ (and Na$^+$) in Onondaga Creek (Table 3.38) were largely a manifestation of the discharge of spent cooling water, enriched in NaCl brine, from the Niagara Power Company facility. The relationships between Cl$^-$ and Ca^{2+}, and Cl$^-$ and SO$_4^{2-}$, are probably largely casual. The discontinuation of the use of saline ground-water for cooling (in 1991) may result in changes in a number of the expressions for Onondaga Creek (Table 3.38). Recently (1992) additional groundwater inputs of Cl$^-$ have been found just above the creek mouth. The expression for SS in Onondaga Creek reflects increases in concentrations encountered during high flows, associated largely with the "scouring" of upstream deposits, much of which originate from sediment discharges from the mud boils in the Tully Valley. The other constituents considered have not demonstrated strong concentration/flow relationships.

TABLE 3.38. Concentration ([ ], mg·L^{-1})/flow (Q, m^3·s^{-1}) relationships and relationships among constituents for Ninemile Creek and Onondaga Creek.

Tributary	Constituent(s)	Period	Relationship	Database
Ninemile Creek at Lakeland*	T-NH$_3$	1989	[T-NH$_3$] = 0.73Q^{-1} + 0.20	Intensive
	Cl$^-$	1973–1980	[Cl$^-$] = 11234Q^{-1} + 1241	Biweekly
		1989–1990	[Cl$^-$] = 1,756Q^{-1} + 288	Intensive
	SO$_4^{2-}$	1989	[SO$_4^{2-}$] = 309Q^{-1} + 128	Intensive
	Na$^+$/Cl$^-$	1981–1989	[Na$^+$] = 0.24 [Cl$^-$] + 45	Biweekly
	Ca^{2+}/Cl$^-$	1981–1989	[Ca^{2+}] = 0.37 [Cl$^-$] + 33	Biweekly
	SO$_4^{2-}$/Cl$^-$	1989	[SO$_4^{2-}$] = 0.17 [Cl$^-$] + 89	Intensive
Onondaga Creek at Spencer Street	Cl$^-$	1989–1990	[Cl$^-$] = 947Q^{-1} + 77	Intensive
		1973–1989	[Cl$^-$] = 696Q^{-1} + 130	Biweekly
	SO$_4^{2-}$	1989	[SO$_4^{2-}$] = 388Q^{-1} + 29	Intensive
	Na$^+$/Cl$^-$	1973–1989	[Na$^+$] = 0.60 [Cl$^-$] − 1.1	Biweekly
	Ca^{2+}/Cl$^-$	1973–1989	[Ca^{2+}] = 0.08 [Cl$^-$] + 91	Biweekly
	SO$_4^{2-}$/Cl$^-$	1989	[SO$_4^{2-}$] = 0.37 [Cl$^-$] + 19	Intensive
	SS	1988–1990	SS = 14.76Q$^{0.558}$	Biweekly

*Downstream of the most recently active waste beds

3.3.2.2 Concentration/Time Relationships

Important changes in pollutant concentrations have occurred, and will continue to occur, over a wide range of time scales. Several of these are permanent in nature, in response to the implementation of management actions, and the closure of the adjoining soda ash/chlor-alkali plant. Some of the more important features of the temporal changes are reviewed in Table 3.39.

Extremely high concentrations of FCs are observed in Onondaga Creek, Ley Creek, and Harbor Brook following runoff events, mostly as a result of inputs from CSOs. Frequent sampling through and after the events is necessary to support estimates of these indicator bacteria. Total phosphorus concentrations also increase in Onondaga Creek during these events in response to upstream and CSO inputs. Irregular occurrences of high concentrations of Cl^- in the METRO effluent

TABLE 3.39. Temporal changes in concentrations of pollutants in Onondaga Lake inflows.

Time scale	Parameter(s)	Inflow(s)	Description
Irregular; hrs/days	FC, TP	Onondaga Creek (Harbor Brook, Ley Creek)	Major increases in urban areas in response to runoff-induced CSO events; also increases in TP in upstream area of Onondaga Creek for major events
	Cl^- (Na^+ and Ca^{2+})	METRO, since early 1986	Coupled to irregular discharge from waste-bed lagoon
Seasonal	$T-NH_3/T-NO_x$	METRO	Decrease in $T-NH_3$ and increase in $T-NO_x$ during warmer months due to operation of nitrification
Irregular; within a year	TP (and other domestic waste constituents)	Onondaga Creek (Harbor Brook, Ley Creek)	Increases in TP (and other domestic waste constituents) as a result of sewer line breaks; e.g., Onondaga Creek in April–September 1988
Irregular; year to year	Cl^-, Na^+, Ca^2	Ninemile Creek	Associated with differences in industrial discharge, before 1981
		Ninemile Creek/METRO	Irregular diversion to MERO from waste beds over summer 1981–late winter 1986 period
Long-term	TP	METRO	Reduction in annual averages concentration in effluent by factor of 20 over 1970–1990 interval
	Shift in TKN components	METRO	Decrease in organic-N, increase in $T-NH_3$, with addition of secondary treatment
	FC	Onondaga Creek	Decrease in FC concentrations since early 1980s due to BMP program
	$T-NH_3$	Ninemile Creek	Decrease following diversion of waste-bed overflow to METRO, and again following closure of soda ash/chlor-alkali facility
	Cl^-, Na^+, Ca^{2+}	Ninemile Creek/METRO	Decreases in both inputs following closure of soda ash/chlor-alkali facility

* CSO: combined sewer overflow

are coupled to irregular discharges from the soda ash/chlor-alkali manufacturer's lagoon. As a result of these perturbations the Cl^- concentration in daily composite effluent samples can increase from about $165\,mg \cdot L^{-1}$ to nearly $1000\,mg \cdot L^{-1}$. The seasonal operation of nitrification at METRO since the late 1980s has caused effluent $T-NH_3$ concentrations to decrease from 15 to $20\,mg \cdot L^{-1}$ in winter and early spring to less than $5\,mgN \cdot L^{-1}$ in summer.

Total phosphorus concentrations tend to increase in Onondaga Creek, as it passes from a rural area through the downstream urban area. The increases can be dramatic under low-flow conditions if adjoining sewer lines are broken; for example, TP concentrations in excess of $2\,mg \cdot L^{-1}$ were observed during low-flow periods of the May–September interval of 1988, apparently as a result of the sewer line break. Abrupt changes in Cl^- (and Na^+ and Ca^{2+}) concentration occurred in Ninemile Creek before diversion of the ionic waste from the waste beds to METRO in 1981, presumably as a result of interruptions in the discharge of the wastewater. Further, the irregular occurence of very high concentrations of alkalinity (e.g., $>300\,mgCaCO_3 \cdot L^{-1}$) and high pH in the creek over the same period indicates the discharge of caustic waste. Increases in concentrations of Cl^- (Na^+ and Ca^{2+}) observed in the METRO effluent and decreases in Ninemile Creek over the mid-summer 1981–latewinter 1986 period were essentially coupled, reflecting the diversion of the waste to METRO to affect precipitation of phosphorus. Substantial interannual, and even greater day-to-day, differences in concentrations resulted from interruptions in the diversion.

Reductions in TP concentrations in METRO's effluent have occurred in response to the detergent ban, addition of secondary treatment, addition of tertiary treatment and, most recently (1988–1990), improvements in tertiary treatment. A twenty-fold reduction has been achieved over the 1970–1990 interval. The annual average TP concentration in the effluent in 1990 was approximately $0.6\,mg \cdot L^{-1}$, well below the existing permit

requirement of $1.0\,mg \cdot L^{-1}$. The concentration of FC bacteria in Onondaga Creek at Spencer Street has decreased (in dry and wet weather) apparently as a result of the execution of the BMP program on the sewer system. The closure of the soda ash/chlor-alkali facility (February 1986) resulted in dramatic reductions in the concentrations of Cl^-, Na^+, and Ca^{2+} in Ninemile Creek and METRO, and in $T-NH_3$ concentrations in Ninemile Creek. For example, in 1981 the mean concentrations of Cl^- in Ninemile Creek and METRO's effluent were approximately 3450 and $2500\,mg \cdot L^{-1}$, respectively, whereas the mean concentrations in 1989 were about 770 and $235\,mg \cdot L^{-1}$.

3.3.3 Constituent Loading

Daily time series of loads have been developed from both special intensive and routine (e.g., biweekly) monitoring programs, to support year-round mass-balance analyses and predictive water quality models, for the following materials: (1) nitrogen species, including $T-NH_3$, organic-N, and $T-NO_x$ (1989 and 1990), (2) TP (1987–1990), (3) Cl^- (1973–1991), and (4) alkalinity (1985 and 1989). Estimates of FC loads for two wet weather events, one dry weather period, and the entire June 1–August 27 interval of the summer of 1987, were reviewed. The histories of annual loading of most of these constituents, as well as for SS, Na^+, and Ca^{2+}, have been presented. A summary of loading rates developed mostly for 1989 is presented in Table 3.40, along with a breakdown of the contributions from the various inputs. Other minor tributaries not included in this analysis contribute less than 10% to the total loadings of the constituents considered.

Onondaga Creek makes a disproportionately high contribution to the SS load (57%) to the lake, relative to its contribution to inflow (Table 3.40), associated mostly with the discharge of sediment to the stream from the mud boils, located more than 32 km upstream of the lake. There is no evidence that the SS load from the creek has changed significantly since the early 1980s. Substantial year-to-year

TABLE 3.40. Summary of loading rates for 1989 (and other specified periods) and contributions of sources.

Constituent	Period	Loading rate (kg·d^{-1})				Contribution (%)			
		METRO	Onondaga Creek	Ninemile Creek	Other**	METRO	Onondaga Creek	Ninemile Creek	Other**
SS	1988–1990	3900	25,700	11,400	3,800	9	57	25	9
T-NH$_3$	1989	2780	78	142	58	90	3	5	2
T-NO$_3$	1989	967	203	277	107	63	13	18	6
organic-N	1989	907	416	269	82	74	12	10	4
TP	1989	169	73	24	20	59	26	9	6
Cl$^-$	1989	57×10^3	107×10^3	236×10^3	28×10^3	15	25	54	6
Na$^+$	1989	37×10^3	64×10^3	70×10^3	13×10^3	20	35	38	7
Ca^{2+}	1989	30×10^3	47×10^3	107×10^3	13×10^3	15	24	54	7
Alk*	1989	52×10^3	97×10^3	73×10^3	21×10^3	21	40	30	9
SO$_4^{2-}$	1989	47×10^3	46×10^3	77×10^3	16×10^3	25	25	41	9
FC***	June 1–August 27, 1987	—	—	—	—	0	83	3	14

* As CaCO$_3$
** Includes Ley Creek and Harbor Brook
*** Based on upper loading bound estimate for the period

differences in loading from this source will occur associated with natural variations in rainfall/runoff common to the region. METRO is the major source of all of the nitrogen species (Table 3.40). Nearly 75% of the total nitrogen (sum of organic-N, T-NH$_3$, and T-NO$_x$) load, and 90% of the T-NH$_3$ load, received by the lake is from METRO. The second largest source of T-NH$_3$ is the waste-bed area of Ninemile Creek.

The annual TP load to the lake has been reduced by about a factor of 12 over the 1970–1990 interval, largely as a result of reductions achieved at METRO. However, this facility remains the dominant source (nearly 60% of the load in 1989; Table 3.40) and thereby the appropriate focus for continued reductions in loading of this critical nutrient. Loadings from the tributaries are subject to substantial year-to-year variability, associated with natural variations in rainfall/runoff. The second largest source of phosphorus to the lake is Onondaga Creek, mostly because of the substantial inputs received by the urban portion (City of Syracuse) of the stream. The urban area of the watershed of Onondaga Creek (about 20% of the total watershed) is estimated to contribute about 60% of the TP load carried by the stream. However, the partitioning of TP loading between the

upstream rural area and the downstream urban area is subject to substantial variation during runoff events, and infrequently for longer periods (e.g., sewer line breaks). The intensities of loading from the rural and urban areas, represented as UALs of 30 to 50 kg·km^{-2}·yr^{-1} and about 300 kg·km^{-2}·yr^{-1} respectively, are consistent with values reported elsewhere for similar land uses. Note that the estimated UAL for the entire Ninemile Creek watershed, of about 35 kg·km^{-2}·yr^{-1} is similar to that reported for the rural portion of the Onondaga Creek watershed.

The discharge from the Solvay waste beds has played a dominant role in the loadings of Cl$^-$, Na$^+$, and Ca^{2+}. Substantial variations in the loading of these materials occurred before the closure of the soda ash/chlor-alkali facility, apparently due to variation in the production of the industrial waste. The annual average load of ionic waste from soda ash production over the 1974–1985 period was about 1.1 million metric tons. Diversion of the ionic waste to METRO changed its point of entry, but did not significantly reduce the associated loading to the lake. Major reductions in the loadings of these materials occurred following the closure of the soda ash/chlor-alkali facility. The residual annual load of ionic waste from soda ash production by 1989 was about 0.14 million metric tons, or about 13% of the preclosure load. At least 55% of the Cl^-, 42% of the Na^+, and 30% of the Ca^{2+} entering the lake by 1989 was derived from soda ash production.

References

APHA. 1985, 1989. *Standard Methods for the Examination of Water and Wastewater, 16th and 17th Editions*. American Public Health Association, Washington, D.C.

Auer, M.T., Doerr, S.M., and Effler, S.W. 1994. A zero degree of freedom total phosphorus model. *J Environ Engr ASCE* (in review).

Auer, M.T., Effler, S.W., Heidtke, T.M., and Doerr, S.M. 1992. Hydrologic budget considerations and mass balance modeling of Onondaga Lake. Submitted to the Onondaga Lake Management Conference, Syracuse, NY.

Auer, M.T., Storey, M.L., Effler, S.W., Auer, N.A., and Sze, P. 1990. Zooplankton impacts on chlorphyll and transparency in Onondaga Lake, New York, U.S.A. *Hydrobiology*. 200/201: 603–617.

Beaulac, M.N., and Reckhow, K.H. 1982. An examination of land use-nutrient export relationships. *Water Resources Bull.* 18: 1013–1024.

Bernard, P.C., Van Greiken, R.E., and Eisma, D. 1986. Classification of estuarine particles using automated electron microprobe analysis and multivariate techniques. *Environ Sci Technol.* 20: 267–273.

Bodo, B., and Unny, T.B. 1983. Sampling strategies for mass-discharge estimation. *J Environ Engr ASCE.* 198: 812–829.

Bowie, G.L., Mills, W.B., Porcella, D.B., Campbell, C.L., Pagenkopf, J.C., Rupp, G.L., Johnson, K.M., Chan, P.W.H., Gherini, S.A., and Chamberlain, C. 1985. Rates, Constants, and Kinetics Formulation in Surface Water Quality Modeling, 2d ed. EPA/6090/3-85/040. U.S. Environmental Protection Agency, Athens, Georgia, 455p.

Brezonik, P.L. 1972. Nitrogen: sources and transformations in natural waters, *In*: H.E. Allen and T.R. Kramer (eds.), *Nutrients in Natural Waters*. John Wiley & Sons, New York, pp 1–50.

Brooks, C.M., and Effler, S.W. 1990. The distribution of nitrogen species in polluted Onondaga Lake, N.Y., U.S.A. *Water Air and Soil Poll.* 52: 247–262.

Canale, R.P., Auer, M.T., Owens, E.M., Heidtke, T.M., and Effler, S.W. 1993. Modeling fecal coliform bacteria: II Model development and application. *Water Res.* 27: 703–714.

Canale, R.P., and Effler, S.W. 1989. Stochastic phosphorus model for Onondaga Lake. *Water Res.* 23: 1009–1016.

Canale, R.P., Gelda, R., and Effler, S.W. 1995. Development and testing of a nitrogen model for Onondage Lake, NY. *Lake Reservoir Manag.* (in review).

Carder, K.L., Steward, R.G., Johnson, D.L., and Prospero, J.M. 1986. Dynamics and composition of particles from anaeolian input event to the Sargasso Sea. *J Geophys Res.* 91(D1): 1055–1066.

Chapra, S.C., and Reckhow, K.H. 1983. *Engineering Approaches for Lake Management. Volume 1: Data Analysis and Empirical Modeling*. Butterworth, Boston, MA.

Clesceri, N.L., Curan, S.J., and Sedlak, R.I. 1986. Nutrient loads to Wisconsin lakes: Part I. Nitrogen and phosphorus export coefficients. *Water Resources Bull.* 22: 983–990.

Devan, S.P., and Effler, S.W. 1984. History of phosphorus loading to Onondaga Lake. *J Environ Engr ASCE* 110: 93–109.

Dillon, P.J. 1975. The phosphorus budget of Cameron Lake, Ontario: The importance of flushing rate to the degree of eutrophy of Lakes. *Limnol Oceanogr.* 20: 28–45.

DiToro, D.M. 1978. Optics of turbid estuarine water: approximations and applications. *Water Res.* 12: 1059–1068.

Driscoll, C.T., Effler, S.W., Auer, M.T., Doerr, S.M., and Penn, M.R. 1993. Supply of phosphorus to the water column of a productive hardwater lake: controlling mechanisms and management considerations. *Hydrobiology* 253: 61–72.

Doerr, S.M., Canale, R.P., and Effler, S.W. 1995. Development and testing of a total phosphorus model for Onondage Lake, N.Y. *Lake Reservoir Manag.* (in review).

Doerr, S.M., Effler, S.W., Whitehead, K.A., Auer, M.T., Perkins, M.G., and Heidtke, T.M. 1994. Chloride model for polluted Onondaga Lake. *Water Res.* 28: 849–861.

Effler, S.W. 1988. Secchi disc transparency and turbidity. *J Environ Engr ASCE* 114: 1336–1447.

Effler, S.W., Brooks, C.M., Addess, J.M., Doerr, S.M., Storey, M.L., and Wagner, B.A. 1991. Pollutant loadings from Solvay waste beds to lower Ninemile Creek, New York. *Water Air Soil Pollut.* 55: 427–444.

Effler, S.W., Brooks, C.B., Auer, M.T., and Doerr, S.M. 1990. Free Ammonia and toxicity criteria in a polluted urban lake. *Res J Water Pollut Contr Fed.* 62: 771–779.

Effler, S.W., Brooks, C.M., and Whitehead, K.A. 1995a. Domestic waste inputs of nitrogen and phosphorus to Onondaga Lake, and water quality impheations. *Lake Reservoir Manag.* (in review).

Effler, S.W., and Driscoll, C.T. 1985a. Calcium chemistry and deposition in ionically polluted Onondaga Lake, NY. *Environ Sci Technol.* 19: 716–720.

Effler, S.W., and Driscoll, C.T. 1985b. A chloride budget for Onondaga Lake, NY, U.S.A. *Water Air Soil Pollut.* 27: 29–44.

Effler, S.W., Driscoll, C.T., Wodka, M.C., Honstein, R., Devan, S.P., Jaran, P., and Edwards, T. 1985. Phosphorus cycling in ionically polluted Onondaga Lake, New York. *Water Air Soil Pollut.* 24: 121–130.

Effler, S.W., Hassett, J.P., Auer, M.T., and Johnson, N. 1988. Depletion of epilimnetic oxygen and accummulation of hydrogen sulfide in the hypolimnion of Onondaga Lake, NY, U.S.A. *Water Air Soil Pollut.* 39: 59–74.

Effler, S.W., and Hennigan, R. 1995. Onondaga Lake: Legacy of pollution. *Lake Reservoir Manag.* (accepted).

Effler, S.W., and Johnson, D.L. 1987. Calcium carbonate precipitation and turbidity measurements in Otisco Lake, N.Y. *Water Resources Bull.* 23: 73–77.

Effler, S.W., Johnson, D.L., Jiao, J.F., and Perkins, M.G. 1992. Optical impacts and sources of suspended solids in Onondaga Creek, U.S.A. *Water Resources Bull.* 28: 251–262.

Effler, S.W., and Owens, E.M. 1986. The density of inflows to Onondaga Lake, U.S.A., 1980 and 1981. *Water Air Soil Pollut.* 28: 105–115.·

Effler, S.W., and Owens, E.M. 1987. Modification in phosphorus loading to Onondaga Lake, U.S.A., associated with alkali manufacturing. *Water Air Soil Pollut.* 32: 177–182.

Effler, S.W., Perkins, M.G., and Brooks, C. 1986. The oxygen resources of the hypolimnion of ionically enriched Onondaga Lake, NY, U.S.A. *Water Air Soil Pollut.* 29: 93–108.

Effler, S.W., Perkins, M.G., Whitehead, K.A., and Romanowicz, E.A. 1995b. Ionic Inputs to Onondaga Lake: origins, character, changes, and selected implications. *Lake Reservoir Manag.* (in review).

Environmental Laboratory. 1986. CE-QUAL-R1: A numeric One-dimensional Model of Reservoir Water Quality: User's Manual, Instruction Report E-82-1, U.S. Army Engineer Waterways Experiment Station, Vicksburg, MS.

Getchell, F.A. 1983. Subsidence in the Tully Valley, New York, Masters Thesis, Syracuse University, Syracuse, NY.

Harbeck, G.E. 1958. Water-loss investigations: Lake Mead Studies. *U.S. Geol. Survey Prof. Pap.* 298: 29–37.

Harris, G.P. 1986. *Phytoplankton Ecology, Structure, Function, and Fluctuation.* Chapman and Hall, New York.

Hartigan, J.P., Quasebarth, T.F., and Southerland, E. 1983. Calibration of NPS model loading factors. *J Environ Engr ASCE* 109: 1259–1272.

Heidtke, T.M. 1989a. Onondaga Lake Loading Analysis: Total Phosphorus. Submitted to Central New York Regional Planning and Development Board, Syracuse, New York.

Heidtke, T.M. 1989b. Onondaga Lake Loading Analysis: Fecal Coliforms. Submitted to Central New York Regional Planning and Development Agency, Syracuse, NY.

Heidtke, T.M. 1991. Partitioning Total Phosphorus Loadings from Onondaga Creek: An Assessment of Rural and Urban Contributions. Submitted to the Onondaga Lake Management Conference, Syracuse, NY.

Heidtke, T.M., and Auer, M.T. 1992. Partitioning phosphorus loads: implications for lake restoration. *J Water Resour Plan Manag. (ASCE)* 118: 562–579.

Johnson, A.H. 1979. Estimating solute transport in streams from grab samples. *Water Resources Res.* 15: 1224–1228.

Johnson, D.L., Jiao, J.F., Dos Santos, S.G., and Effler, S.W. 1991. Individual particle analysis of suspended materials in Onondaga Lake. *Environ Sci Technol.* 25: 736–744.

Kirk, J.T.O. 1981. Estimation of the scattering coefficient of natural waters using underwater irradiance measurements. *Austral J Mar Freshwater Res.* 32: 533–539.

Kirk, J.T.O. 1983. *Light and Photosynthesis in Aquatic Ecosystems.* Cambridge University Press, London.

Kirk, J.T.O. 1985. Effects of suspensoids (turbidity) on penetration of solar radiation in aquatic ecosystems. *Hydrobiologia* 125: 195–208.

Kramer, J.R., Heibes, S.E., and Allen, H.E. 1972. Phosphorus: analysis of water, biomass, and sediment. In: *Nutrients in Natural Waters*, Wiley Interscience, New York, pp. 51–100.

Loder, T.C., and Liss, P.S. 1985. Control by organic coatings of the surface charge of estuarine suspended particles. *Limnol Oceanogr.* 30: 418–421.

Mancini, J.L. 1978. Numerical estimates of coliform mortality rates under various conditions. *J Wat Pollut Contr Fed.* 50: 2477–2484.

Manczak, H., and Florczyk, L. 1971. Interpretation of results from the studies of pollution of surface flowing waters. *Water Res.* 5: 575–584.

Marselak, J. 1978. Pollution Due to Urban Runoff: Unit Loads and Abatement Measures. Report to the International Joint Commision. Windsor, Ontario.

Messer, J.J., Ho, J., and Grenney, W.J. 1984. Ionic strength correction for extent of ammonia ionization in freshwater. *J Can Fish Aquat Sci.* 41: 811–815.

Mosteller, F., and Tukey, J.W. 1978. *Data Analysis and Regression: A Second Course in Statistics.* Addison-Wesley, Reading, MA.

Murphy, T.P., Hall, K.J., and Yesaki, I. 1983. Coprecipitation of phosphate with calcite in a naturally eutrophic lake. *Limnol Oceanogr.* 28: 58–69.

O'Brien and Gere Engineers Inc. 1973. Environmental Assessment Statement, Syracuse Metropohtan Sewage Treatment Plant and the West Side Pump Station and Force Main. Submitted to Onondaga County, Syracuse, NY.

O'Brien and Gere Engineers Inc. 1987. Combined Sewer Overthow Abatcment Program Post-BMP Assessment. Submitted to the Department of Drainage and Sanitation of Onondage County, Syracuse, NY.

Organization for Economic Co-operation and Development (OECD). 1982. Eutrophication of Waters: Monitoring, Assessment and Control. Director of Information, OECD, Paris, 154p.

Onondaga County. 1971–1991. Onondaga Lake Monitoring Program Annual Reports. Onondaga County Department of Drainage and Sanitation, Syracuse, NY.

Onondaga County. 1988. Onondaga Lake Monitoring Program, Annual Report 1987. Department of Drainage and Sanitation, Onondaga County, Syracuse, NY.

Otsuki, A., and Wetzel, R.G. 1972. Coprecipitation of phospate with carbonates in a marl lake. *Limnol Oceanogr.* 17: 763–767.

Owens, E.M. 1987. Bathymetric Survey and Mapping of Onondaga Lake, New York. Submitted to Onondaga County by the Upstate Freshwater Institute, Syracuse, NY.

Owens, E.M., and Effler, S.W. 1989. Changes in stratification in Onondaga Lake, New York. *Water Resources Bull.* 25: 587–597.

Parks, R.A. 1968. Paleoecology of *Venericardia sensu lato* (Pelecypoda) in the Atlantic and Gulf Coastal Province: an application of paleosynecologic methods. *J Paleo.* 42: 955–989.

PLUARG, (Johnson, M.G., Corneau, J.C., Heidtke, T.M., Sonzogni, W., and Stahlbaum, B.) 1978. Management Information Base and Overview Modeling. PLUARG Special Report to the International Joint Commision. Windsor, Ontario.

Rooney, J. 1973. Memorandum (cited by USEPA, 1974, Environmental Impart Statement on Wastewater Treatment Facilities Construction Grants for the Onondage Lake Dramage Basin).

Russo, R.C., Smith, C.E., and Thurston, R.V. 1974. Acute toxicity of nitrite to rainbow trout (*Salmo gairdneri*). *J Fish Res Bd Can.* 31: 1653–1655.

Schaffner, W.C., and Olgesby, R.T. 1987. Phosphorus loadings to lakes and some of their responses, Part 1. A new calculation of phosphorus loading and its application to 13 New York lakes. *Limnol Oceanogr.* 23: 120–134.

Shilts, W.W. 1978. Genesis of mud boils. *Can J Earth Sci.* 15: 1053–1068.

Smithgall, C. 1989 (April). Analysis of NYSDEC 1988. Tributary Phosphorus Data (memo). Moffa and Associates Inc. Syracuse, NY.

Smithgall, C., Moffa, P.E., and Klosowski, R. 1991 (March). 1990 Onondaga Creek Phosphorus Loading Evaluations (memo). Moffa and Associates Inc. Syracuse, NY.

Stearns and Wheler Engineers. 1979. Onondaga Lake Storms Impact Study. Submitted to Onondaga County, Department of Drainage and Sanitation. Syracuse, NY.

Stearns and Wheler Engineers. 1991. Projected Flows and Loadings METRO Engineering Alternatives. Submitted to Onondaga County Department of Drainage and Sanitation. Syracuse, NY.

Stumm, W., and Morgan, J.J. 1981. *Aquatic Chemistry.* Wiley Interscience, New York.

Thomann, R.V., and Mueller, J.A. 1987. *Principles of Surface Water Quality Modeling and Control.* Harper and Row, New York.

United States Environmental Protection Agency (USEPA). 1985. *Ambient water quality criteria for ammonia—1984.* Office of Water Regulations and Standards Criteria and Standards Division, Washington, D.C.

United States Environmental Protection Agency. 1975. Process Design Manual for Nitrogen Control. Technology Transfer Document. Washington, D.C.

United States Geological Survey, 1972–1989. Water Resources Data for New York, Water Years 1971–1988, Western New York. Albany, New York.

Viessmann, W., Knapp, J.W., Lewis, G.L., and Harbaugh, T.E. 1977. *Introduction to Hydrology*, 2d ed. Harper and Row. New York.

Vollenweider, R.A. 1968. Scientific Fundamentals of the Eutrophication of Lakes and Flowing Waters with Particular Reference to Nitrogen and Phosphorus as Factors in Eutrophication. Technical Report DAS/C81/68, Organization for Economic Cooperation and Development, Paris.

Vollenweider, R.A. 1975. Input–output models with special reference to the phosphorus loading concept in limnology. *Schwerz Z Hydrol.* 37: 53–83.

Walker, W.W. 1987. Empirical Methods for Predicting Eutrophication in Impoundments. Report 4: Phase III: Applications Manual. Technical Report E-81-9. US Army Engineer Waterways Experimental Station, Vicksburg, MS.

Wetzel, R.G. 1983. Limnology. Saunders, Philadelphia.

Wodka, M.C., Effler, S.W., Driscoll, C.T., Fields, S.D., and Devan, S.P. 1983. Diffusivity-based flux of phosphorus in Onondaga Lake. *J Environ Engr ASCE.* 109: 1403–1415.

Yin, C.Q., and Johnson, D.L. 1984. Sedimentation and particle class balances in Onondaga Lake, N.Y. *Limnol Oceanogr.* 29: 1193–1201.

4
Hydrodynamics and Transport

Emmet M. Owens and Steven W. Effler

4.1 Introduction

This chapter deals with physical processes occurring within the basin of Onondaga Lake and the lake outlet. These processes are important in mediating the manifestations of the lake's water quality problems. Further, certain pollutant inputs have had rather distinct impacts on the hydrodynamics regime of the lake, which are documented here.

Important morphometric features of the basin (see Chapter 1) with respect to hydrodynamics and transport are as follows. First, as is the case with most lakes, the basin of Onondaga Lake is much larger in the horizontal direction than in the vertical. The length:depth and width:depth ratios for the lake basin are roughly 500:1 and 100:1, respectively. It is then expected that water motion in the lake is largely horizontal, with vertical velocity sufficiently small so that a hydrostatic pressure distribution is maintained under all conditions. The lake is also sufficiently deep so that thermal stratification would be expected to form in the water column in response to the meteorological variations that typically prevail in the north temperate climate of central New York. Data are presented later in this chapter which show that Onondaga Lake in recent years has demonstrated a typical dimictic stratification regime (experiences two periods of mixing, one in

spring and one in fall), though this was not always the case due to anthropogenic effects.

Detention time (or flushing rate) is another characteristic which can affect physical processes in a lake basin. The detention time of Onondaga Lake, under average streamflow conditions, is approximately 90 days (see Chapter 3). Orlob (1983) has presented a criterion to evaluate the relative importance of stratification and lake inflow/outflow in controlling the distribution of mass constituents in a lake. The densimetric Froude number F_D is defined by

$$F_D = (L/h_m)(Q/V_T)\frac{\rho h_m}{\Delta \rho g} \qquad (4.1)$$

in which L = lake length; h_m = lake mean depth; Q = average lake inflow/outflow; V_T = lake volume; g = acceleration of gravity; ρ = density of water; and $\Delta\rho$ = surface-to-bottom density difference. Recall for Onondaga Lake, $L/h_m \approx 500$ and $V_T/Q \approx 90$ days. Under average summer conditions, $\Delta\rho/\rho \approx 4 \times 10^{-4}$, resulting in $F_D \approx 2 \times 10^{-3}$. Orlob (1983) indicated that for $F_D \ll 1$, horizontal distributions of dissolved and suspended mass tend to be rather uniform, while significant vertical variations exist.

Onondaga Lake, with horizontal dimensions of roughly 10 km and 2 km along its long and short axes, is classified as a lake of medium size (Spigel and Imberger 1980). With such fetch and the fact that the topography surrounding the lake basin is relatively low, it is expected that wind stress on the water surface has a significant effect on water motion and heat and mass transport in the lake basin. The length of the detention time of the lake basin is an indication that lake inflows and the outlet have a relatively small effect on water motion and transport in the lake. The region of the lake basin in which lake inflows or the outlet have a significant impact on motion and transport is known as the "near-field" of such inflows. Outside of the near-field, the motion and transport is largely determined by wind speed and direction. In lakes with the characteristics described above, it would normally be expected that the near-field associated with each inflow to the lake, and that associated with the lake outlet, would be small relative to the overall size of the lake. In fact, this is the case with many of the inflows to the lake. However, due to the existence of significant density differences between the lake surface waters and that of certain lake inflows, and between lake surface waters and that of the Seneca River, inflows have had a significant impact on water motion and heat and mass transport in the lake. These effects are described in the following section.

4.2 Lake Inflows and Outflow

The extent to which inflow and outflow affect these characteristics is dependent on (Ford and Johnson 1983): (1) the relative magnitude of the inflow/outflow rate and the lake volume, as quantified by the detention time; and (2) the density difference between inflows and the lake waters. Water density is related to its temperature and dissolved solids concentration (see Chapter 5). Density differences between inflows and lake surface waters commonly exist due to differences in the relative rates at which temperature of streams and lake surface waters change in response to changes in meteorological conditions over time scales ranging from diurnal to seasonal. These sea-

sonal differences are usually sufficiently small so that, after mixing in the vicinity of the mouth of the inflow (so-called "near-field" mixing), the density difference is not significant. However, the presence of high concentrations of dissolved salts in some lake inflows (see Chapter 3) results in relative differences after near-field mixing that are substantially greater (Effler and Owens 1986). These density differences are sufficient to cause motion of the inflow outside the region that would normally be considered to be the near field; this motion is driven by the remaining relative density of the inflow after inflow mixing. If the inflow is more dense than the lake surface waters (as has been the case with Ninemile Creek and METRO), the inflow forms a plunging inflow or "underflow", as shown in Figure 4.1. The dense inflow flows down the sloping bottom of the lake until it reaches a depth in the lake where the underflow density equals the lake density at that depth, or it reaches the lake bottom. If the inflow is less dense (such as the East Flume discharge containing waste heat), an "overflow" forms (Figure 4.1).

The existence of either underflows or overflows does not have a significant impact on horizontal transport, largely because the velocity of the density-driven motion is relatively small and does not influence the wind-induced water motion in the lake. However, vertical transport in the lake may be strongly influenced by such density currents. In the case of an underflow, mass constituents are transported directly (by advection) to a greater depth in the lake than would otherwise occur. In the case of an overflow, the stability of the

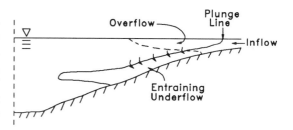

FIGURE 4.1. Basic inflow patterns for entraining underflows and overflows.

water column is increased by the buoyant surface layer, making it more resistant to wind mixing.

The density differences that prevailed between the major inflows and the lake before closure of the soda ash/chlor-alkali facility have been described (Effler and Owens 1986). These findings are summarized here, along with an updated analysis, to establish changes in these relationships since closure of this facility. Density was calculated using measurements of the temperature and chloride concentration of lake and tributary waters, utilizing an equation of state (see Chapter 5). Linear interpolations of density with time were made (sampling frequencies varied from three times per week to twice per month) for the lake and inflows to yield estimates of density differences between inflows and the lake. The results of these calculations are shown in Figure 4.2; positive density difference indicates inflow density is greater than lake density.

4.2.1 Cooling Water Intakes and Discharge

The soda ash/chlor-alkali facility removed water from the lake through three intakes, used the water primarily for cooling and released the water to the lake through a shoreline discharge, called the East Flume, and to Ninemile Creek via the West Flume. In 1978, a submerged diffuser was installed for the spent cooling water discharge 900 m offshore. The estimated density difference between the lake and the East Flume discharge for a year before installation of the diffuser is shown in Figure 4.2a; the lower density of the discharge is a result of its elevated temperature.

Thermal imagery clearly depicts the entry and surface pooling of warm inflows in the early winter of 1970, in Figure 4.3 (from Stewart 1978). The thermal plumes of the cooling water discharge and the submerged METRO discharge are clearly evident. The heated discharge tended to stabilize the water column by placing warm, buoyant water at the surface. This type of discharge can be expected to cause decreased mixing in the surface layer (epilimnion), probably resulting

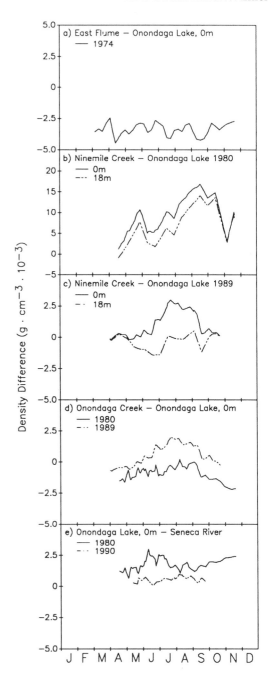

FIGURE 4.2. Density differences between inflows and the surface waters of Onondaga Lake.

in a somewhat shallower layer than would otherwise occur. Buoyancy effects of the soda ash/chlor-alkali plant's cooling water discharge were mitigated by the installation of the diffuser. Analysis of the performance of the

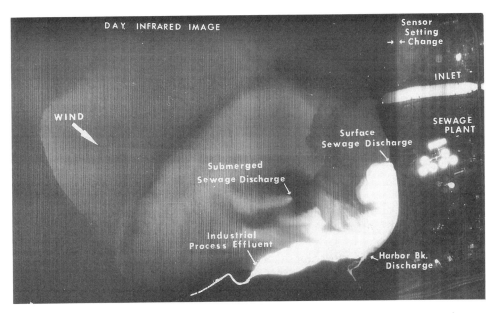

FIGURE 4.3. Thermal imagery at 10:30 hr on December 11, 1970 showing distribution of heated discharge from METRO and the soda ash/chlor-alkali facility. Dark is cold and light is hot (from Stewart 1978).

diffuser on a single day indicated that the discharge was incorporated into the upper mixed layer (NALCO Environmental Sciences Co. 1978). Much of the south basin failed to establish ice-cover during the winter before the diffuser was installed (e.g., Stewart 1978); however, ice-cover was substantially more complete after installation of the diffuser.

The soda ash/chlor-alkali facility withdrew lake water from three intakes for cooling. One intake was located in shallow water (3 m), and two were positioned in deeper water of uncertain depth, estimated to be at a depth of 10.7 to 12.2 m by plant personnel. When the deep intakes were used the cooling water operation resulted in net cycling of water (and associated materials) from stratified layers to the upper layers of the lake (Effler and Owens 1986), regardless of the operation of the diffuser. Uncertainty in the depth of the deep intakes confounds accurate estimates of the coupled artificial internal material loading associated with this practice, because of the strong vertical gradients that have prevailed

for a number of important water quality constituents in the hypolimnion (see Chapter 5).

4.2.2 Plunging Inflows: Ninemile Creek and METRO

The density differences between Ninemile Creek and the lake are presented from the perspective of lake conditions before and after closure of the soda ash/chlor-alkali facility in Figure 4.2b and 4.2c, respectively. The magnitude of density stratification in the lake is represented by density difference between the creek and the surface and bottom waters of the lake. Ninemile Creek was substantially more dense than the waters of the lake before closure of the facility due to its elevated salinity, which was often 10 ppt (parts per thousand) greater than lake surface waters. Note the different scaling used in Figure 4.2b to accommodate the great density difference that prevailed before closure of the facility. The specific case of Figure 4.2b (1980) may approach a worst case, as the mitigating effect of somewhat higher temperatures, associated with the West Flume discharge (see Chapter 1), was eliminated in early 1980. The impact of this dense inflow on the stratification regime of the lake is documented subsequently. The difference in density between Ninemile Creek and the lake has decreased greatly since closure of the facility; however, its density remains greater than the surface waters of the lake (Figures 4.2b and c). The fluctuations in the creek−lake density difference, particularly before the facility closure, are associated with the diluting effects of runoff events.

An aerial photograph of the inflow of Ninemile Creek to the lake in early summer during operation of the soda ash/chlor-alkali facility (June 3, 1969) is presented in Figure 4.4. The region where the light-colored inflow is visible at the surface is the near-field, where dilution of the inflow with lake surface water is occurring. The sharpness of the boundary of the near field cloud reflects the plunging (underflow of Figure 4.1) phenomenon. This photograph may be considered typical of summertime conditions that prevailed for the Ninemile Creek inflow during the period of

operation of the waste beds along the creek (mid-1940s−1981).

METRO discharged treated sewage primarily from a submerged outfall (6.1 m deep about 520 m from shore) until 1980. Wastewater not receiving full treatment during runoff events was discharged from a shoreline location. Thereafter fully treated wastewater was discharged from a shoreline position, and portions not receiving full treatment during runoff events were released via the deep discharge. Beginning in 1981, a portion of the overflow from the waste beds operated by the soda ash/chlor-alkali facility was diverted from Ninemile Creek to METRO for use in phosphorus removal. From 1981 until the closure of the facility in 1986, except for brief periods of discontinued diversion, the METRO effluent was also more dense than the surface waters of the lake due to its elevated salinity. Thus, from 1981 to 1986, there were two negatively buoyant (plunging) inflows into the lake. Of course the density of Ninemile Creek was reduced during this period as a result of diversion. Following closure of the soda ash/chloralkali facility, the METRO dis-

FIGURE 4.4. Aerial photograph of plunging inflow to Onondaga Lake, June 3, 1969, (from Effler and Owens 1986; original picture from K.M. Stewart).

charge has effectively acted as a neutrally buoyant inflow as described in the following section.

This depth of entry of an underflow, or the depth of neutral buoyancy, depends on the near-field mixing and the attendant density stratification in the lake. The near-field mixing depends on characteristics of the inflow and the receiving lake, including the flow rate of the inflow, the density difference between the inflow and the upper waters of the lake, and the geometry of the inflow zone. As described subsequently, the depth of neutral buoyancy was in the area of the thermocline before closure of the soda ash/chlor-alkali facility, except when thermal stratification was very weak or nonexistent. The plunging inflow(s) effectively flowed directly to the bottom of the lake, when the thermal stratification was weak.

4.2.3 Other Inflows

Onondaga Creek is also ionically enriched (see Chapter 3), though the extent to which this is anthropogenically based is uncertain. This inflow was less dense than the lake for most of the year before closure of the soda ash/chloralkali facility, however, it is often more dense than the lake since closure (Figure 4.2d) because of the associated reduction in the salinity of the lake (Doerr et al. 1994). The salinity, and therefore density, of Onondaga Creek remains lower than that of Ninemile Creek (Figure 4.2). Further, the harbor and channel at the mouth of Onondaga Creek probably serves to mitigate the underflow tendencies of this inflow by facilitating active mixing with lake waters before entry into the lake basin.

The remaining minor inflows to the lake, Ley Creek, Harbor Brook, Sawmill Creek, Bloody Brook, and Tributary 5A, were decidedly less dense than the lake (e.g., overflows) before closure, but are more nearly neutrally buoyant since closure. The effects of these nearly neutrally buoyant inflows on water motion is restricted to near-field regions that are quite small (probably on the order of 100 to 200 m for the largest tributaries). Within these near-field regions, the water motion

associated with the inflow is dissipated to the open waters of the lake and internal and bottom friction. Depending on the streamflow, the mixing within the near-field is relatively vigorous. For most mass concentrations of interest (an exception being fecal coliform bacteria, see Canale et al. 1993 and Chapter 9), any differences in concentration between the inflow and the lake are largely eliminated at the outer limits of the near-field.

4.2.4 Lake Outlet Flow Conditions

Neglecting evaporation and possible groundwater losses, all water entering Onondaga Lake flows out of the basin through the outlet channel at the northwest corner of the lake. The configuration of the 1.9 km outlet channel that connects the lake to the Seneca River is shown in Figure 4.5. Klein Island is an artificial feature formed by the excavation of the channel on its northwest side as a part of the Barge Canal construction in the early twentieth century. Based on the assumption that water flowing downstream seeks the shortest path, the channel south of the island, referred to herein as the "South Channel," most likely carries little flow. Thus the confluence of the river and outlet channels is effectively at the northern tip of the island.

Unusual flow conditions are common to the outlet, due not only to the elevated density of the lake from industrial discharges, but also because of the elimination of the natural gradient along the outlet from the lake to the river. The river has been channelized and its elevation/flow is regulated for hydropower and navigation purposes. Further, the lake level has been lowered, in part to make Onondaga Lake part of the same navigation system, so that the lake and river elevations are now nearly equal. The density difference between the lake and Seneca River water is shown in Figure 4.2e for a year before (1980) and after (1990) closure of the soda ash/chloralkali facility. The reduction in density difference after closure is due to the reduced salinity of lake waters.

The dynamics of the density differences between the lake and river water in the lake outlet are clearly manifested in data from a biweekly monitoring program (see Chapter 3), for water temperature (T) and chloride (Cl^-) concentration measured at depths of 0.6 and 3.7 m at the Long Branch Road bridge over the outlet channel (Figure 4.5). The results of this monitoring for 1986 (the last year of operation of the chlor-alkali manufacturer) and 1989 are shown in Figures 4.6 and 4.7, respectively. Chloride concentrations at the two depths in the outlet channel, and the lake surface concentrations, are shown in Figures 4.6a and 4.7a. With the exception of one measurement in 1986, the chloride concentration at the 3.7 m depth in the outlet channel (outlet bottom) was approximately equal to the lake surface concentration. However, at various times during these two years, the surface concentration in the outlet channel was substantially less than lake surface concentrations. Note that Cl^- concentrations in the lake increased with depth before closure of the facility (see subsequent sections of this chapter). Thus, it is clear that, at certain times, water from some source other than the lake enters the outlet channel. This source is the

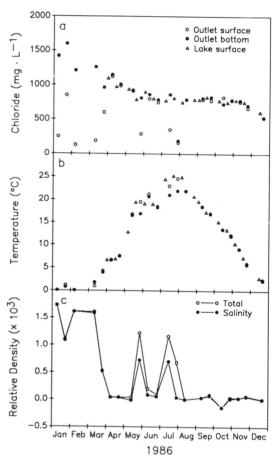

FIGURE 4.6. Onondaga Lake outlet 1986; inflow/outflow dynamics as indicated by differences between outlet surface and outlet bottom conditions: (**a**) chloride concentration, (**b**) temperature, and (**c**) relative density.

Seneca River. Apparently on rare occasions only river water is found in the outlet; e.g., on July 29, 1986 (Figure 4.6a) the Cl^- concentrations at the surface and bottom of the outlet channel were substantially less than lake concentrations. Generally the water temperatures are nearly equal with depth, with the exception that when Cl^- stratification is present in the channel, the temperatures at the bottom of the outlet channel tend to be slightly lower (Figures 4.6b and 4.7b).

Combining the outlet temperature and Cl^- measurements, the water density at the two depths in the outlet channel may be determined, using the equation of state developed

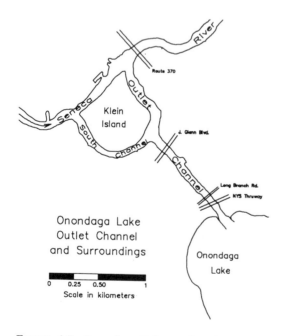

FIGURE 4.5. Onondaga Lake outlet channel and surroundings.

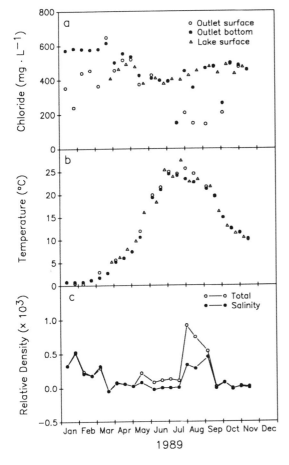

FIGURE 4.7. Onondaga Lake outlet 1989; inflow/outflow dynamics as indicated by differences between outlet surface and outlet bottom conditions: (a) chloride concentration, (b) temperature, and (c) relative density.

the relative density due to Cl^-, designated ε_C, is defined as

$$\varepsilon_C = \frac{\rho(Cl_1, T_1) - \rho(Cl_2, T_1)}{\rho(Cl_1, T_1)} \qquad (4.3)$$

The variation of ε and ε_C is shown in Figures 4.6c and 4.7c. Most, but not all, of the density difference that occurred in the outlet channel during these two years was attributable to differences in salinity (as measured by Cl^-; see Chapter 5).

Daily (weekdays) samples were collected for Cl^- analysis at the southeast end of the lake outlet channel at depths of 0 and 3 m during the summer months of 1990. The results are shown Figure 4.8. The measurements show the dynamic nature of conditions in the outlet channel, particularly at the channel surface. The Cl^- concentration at the surface shows significant fluctuation between values in the range of 150 to 250 $mg \cdot L^{-1}$ (representing largely Seneca River water), and values in the range of 350 to 500 $mg \cdot L^{-1}$, which is largely lake water. On several occasions, the bottom waters in the outlet also are in the river concentration range. These data (Figure 4.8) support the earlier observations (Figures 4.6 and 4.7) of stratified flow, but more completely document the level of temporal variability that prevails.

Further evidence of unusual flow conditions in the outlet channel is provided by velocity

in Chapter 5. If $\rho(Cl, T)$ represents the functional relationship between the water density ρ and the water T and Cl^- concentration expressed by the equation of state, the relative density ε is defined by

$$\varepsilon = \frac{\rho(Cl_1, T_1) - \rho(Cl_2, T_2)}{\rho(Cl_1, T_1)} = \frac{\rho_1 - \rho_2}{\rho_1} \qquad (4.2)$$

in which the subscript "2" refers to the surface water and "1" to the bottom water in the outlet channel. The relative density ε is then a dimensionless measure of the density difference between the surface and bottom waters. In order to isolate the effect of Cl^- on density,

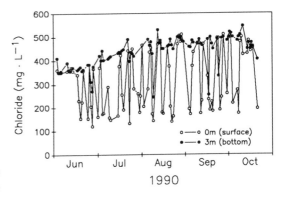

FIGURE 4.8. Onondaga Lake outlet 1990; inflow/outflow dynamics as indicated by differences between outlet surface and outlet bottom chloride concentrations.

measurements made during the period 1975 through 1985 (Onondaga County 1976–1986). Measurements of current speed and direction were made at 0.6 m depth intervals at verticals located 7.6 m apart along the length of the Long Branch Road bridge (Figure 4.5). These data indicate that three flow conditions occur at this location: (1) flow out of the lake (to the northwest) over the entire cross section; (2) flow out of the lake in the lower (deeper) portion together with flow in the opposite direction (toward the lake) in the upper (surface) portion of the cross section; and (3) flow toward the lake over the entire cross section. These conditions are consistent with the Cl$^-$ and temperature monitoring data collected for the outlet (Figures 4.6 to 4.8). These data indicate that the exiting water flow (to the northwest) is lake water, while that flowing towards the lake is water from the Seneca River upstream of the confluence with the outlet channel. Water movement from the Seneca River into Onondaga Lake is apparent in aerial images. Infrared aerial imagery (Figure 4.9) and aerial photography show that

water from the river has moved into the surface waters of the lake.

As a part of a hydrodynamic study of the Seneca River and its interaction with Onondaga Lake, additional temperature and salinity measurements were made in the outlet channel (Owens 1991, unpublished data). These measurements, together with the earlier observations, indicate that the flow conditions in the lake outlet channel may be phenomenologically described as follows. When the flow in the Seneca River and the flow out of the lake are sufficiently large, turbulent mixing in the Seneca River channel downstream of Klein Island homogenizes the lake and river flows. Within the outlet channel, lake water flows out of the lake throughout the channel; this is designated as outlet flow regime A, as shown in Figure 4.10. As the river and lake outlet flows decrease, this mixing decreases. At some critical condition, the mixing cannot overcome the density difference between the lake and river waters, and a stratified flow regime is formed, with less dense river water flowing over the

FIGURE 4.9. Documentation of inflow to Onondaga Lake from the Seneca River. Thermal imagery (8–14 μm) at 21:20 hrs on July 8, 1970 in which a cool reverse flow is indicated at the surface of the lake outlet. Dark is cold and light is hot (from Stewart 1978).

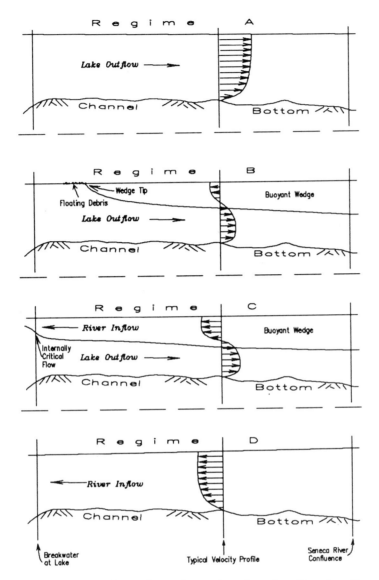

FIGURE 4.10. Flow regimes in Onondaga Lake outlet: regime A, regime B, regime C, and regime D.

top of water flowing out of the lake. Measurement of stratification conditions in the Seneca River in 1990 indicates that salinity stratification forms in the river downstream of the confluence with the outlet channel when the upstream river flow falls below about $85 \, m^3 \cdot sec^{-1}$ ($3000 \, ft^3 \cdot sec^{-1}$), which is about 6 times the MA7CD10 flow of the river.

As the lake outlet flow decreases further, a "buoyant wedge" of river water begins to intrude up into the outlet channel; this condi-

tion is designated as regime B (Figure 4.10). Salinity profiles taken along the outlet channel have documented the existence of the wedge. The tip of the wedge can often be seen as a line of floating debris which has been carried out of the lake by the outlet flow, but is trapped at the tip of the wedge by the weak upstream currents within the wedge.

As the lake outflow continues to decrease, another critical condition is reached when the tip of the buoyant wedge reaches the threshold between the outlet channel and the lake, located at the end of the breakwaters at the southeast end of the outlet channel. Further decreases in lake outflow beyond this critical value result in inflow from the river into the lake, a condition designated as regime C (Figure 4.10). Regime C is similar to the classical "lock exchange" hydraulics problem (Schijf and Schönfeld 1953) which occurs when a barrier between fresh and salt water is opened. In fact, regimes B and C can only occur if a sufficiently large density difference exists between lake and river water, differences primarily due to differences in salinity.

Based on monitoring data and hydrodynamic theory which is introduced later, regimes A, B, or C may exist for extended periods of time. An additional flow condition designated regime D exists when the flow is to the southeast (from the river to the lake) throughout the entire channel. Aside from unlikely dynamic effects, regime D can only occur if the water surface elevation at the river-outlet channel confluence is greater than that in the lake at its northwest corner. Tributary and discharge flows into the lake basin, particularly the relatively stable flow from METRO, prevents the existence of regime D for an extended period of time. It is most likely that this regime occurs during periods of increasing water surface elevation in the river, perhaps aided by short-term decreases in the lake elevation such as that caused by a seiche or surface wind stress on the lake outlet channel. Inflow into the lake over the entire outlet cross section, together with other lake inflows, can only cause the water surface elevation of the lake to increase, which eventually must cause water to flow out of the lake.

Efforts to establish a water balance for the lake based on limited (e.g., biweekly) instantaneous velocity measurements have not been successful. The inability to establish such a balance is largely due to the highly dynamic nature of outlet channel flow conditions. Using the same outlet velocity measurement procedures outlined above, Seger (1980) measured outlet flows at the Long Branch Road bridge more intensively over time over a one-month period in mid-summer of 1978.

The results are shown in Table 4.1. It may be seen that the flow regime, and the magnitude of flows in either direction in the outlet channel, was quite variable over this period. In particular, the variation of flows over the 8-hr interval on August 16 is striking. With this type of variability, it can be expected that nearly continuous flow measurements would be required in order to define the net flow in the outlet channel based on measurements alone.

It should be recognized that direct measurements of water motion to the southeast at the Long Branch Road bridge in the outlet channel do not necessarily indicate that water from the Seneca River is mixing with the main body of water in the lake. In other words, water from the river was not necessarily mixing with water in the lake at the rates shown in Table 4.1 on the indicated dates. This is due to two factors. First, the occurrence of stratified flow at some point along the outlet channel cannot discriminate between flow regime 2 or 3; only measurements at the end of the channel can directly show this. In addition,

TABLE 4.1. Measured flows in lake outlet channel, 1978.

Date	Time	River inflow $(m^3 \cdot sec^{-1})$	Lake outflow $(m^3 \cdot sec^{-1})$
July 17	15.00	5.7	12.8
July 18	13.00	2.9	13.9
July 19	13.00	3.5	11.0
July 20	14.00	2.5	10.5
Jyly 23	13.00	8.4	9.0
Joly 23	16.00	8.8	8.1
July 23	19.00	7.3	3.1
July 25	18.00	0.5	17.6
Jyly 26	9.00	6.3	11.8
Jyly 28	11.00	0	27.6
August 1	15.00	6.8	11.6
August 2	11.00	6.4	11.8
August 4	20.00	1.7	17.0
August 6	20.00	5.6	13.1
August 8	11.00	6.1	13.0
August 9	13.00	5.1	13.9
August 9	14.00	5.0	14.7
August 16	13.00	11.7	2.1
August 16	15.00	7.5	8.6
August 16	18.00	0	15.9
August 16	21.00	0	17.1

From Seger 1980

even if water from the river actually enters the lake at the end of the outlet channel, flow and mixing in the lake close to the end of the channel ("near-field" mixing) may cause a portion of the water from the river to flow back out of the lake.

For the purpose of quantitative analysis, the flow in the lake outlet channel may be idealized as being made up of two distinct layers. Herein, h_2 and h_1 define the depths of flow of the upper (representing river water of density ρ_2) and lower (water exiting the lake of density ρ_1) layers in the outlet channel of the lake. The average velocity of flow in the upper and lower layers is V_2 and V_1, respectively, and the discharge or flow rate in each layer is Q_2 and Q_1. These quantities are generally functions of longitudinal channel distance x (positive to the northwest in the outlet channel), and time t, with the velocities and flow rates having positive values when flowing from the lake to the river (to the northwest). In the analysis described here, it is assumed that conditions are at a steady state; such an assumption certainly conflicts with the highly dynamic conditions that are sometimes observed, but is consistent with the preliminary nature of this analysis. Utilizing the fact that the channel width is much greater than the depth, the variation of the layer depths with distance along the length of the outlet channel can be described by steady-state equations which balance the acceleration of water associated with changing water depth with forces resulting from shear stresses at the upper and lower boundary of the two layers and gravity. These equations are

$$\frac{dh_2}{dx} = \frac{(S + S_S - S_{I2})(1 - F_1^2) - S - S_{I1} + S_B}{\varepsilon - F_1^2 - F_2^2 + F_1^2 F_2^2} \quad (4.4)$$

$$\frac{dh_1}{dx} = \frac{(S + S_{I1} - S_B)(1 - F_2^2) - S - S_S + S_{I2}}{\varepsilon - F_1^2 - F_2^2 + F_1^2 F_2^2} \quad (4.5)$$

where ε is the relative density defined above; $F_1 = V_1/(gh_1)^{0.5}$ and $F_2 = V_2/(gh_2)^{0.5}$ are the Froude numbers for each layer; S = bottom slope of the outlet channel; $S_S = \tau_S/\rho g h_2$; $S_{I1} = \tau_I/\rho g h_1$; $S_{I2} = \tau_I/\rho g h_2$; $S_B = \tau_B/\rho g h_1$; $\tau_S =$ surface shear stress due to wind; $\tau_I =$ shear stress at the interface between the two layers;

and $\tau_B =$ shear stress at the channel bottom. Schijf and Schönfeld (1953) were among the first to present equations (4.4) and (4.5); these relationships have been used by many others subsequently in the analysis of stratified flow problems, particularly in estuaries. The stresses are computed using

$$\tau_S = C_D \rho_a |W_c| W_c \quad (4.6)$$

$$\tau_I = C_I \rho |(V_1 - V_2)|(V_1 - V_2) \quad (4.7)$$

$$\tau_B = C_B \rho |V_1| V_1 \quad (4.8)$$

where $\rho_a =$ density of air; C_D, C_I, and C_B are dimensionless friction coefficients; and $W_c =$ component of wind speed along the centerline of the channel.

When the river flow falls below about $85 \, m^3 \cdot sec^{-1}$, the stratified channel flow conditions in the outlet and river channels may be described by these equations. In the Seneca River channel downstream of the confluence, the flow in the upper layer (Q_2) is equal to the river flow from upstream, and the lower layer flow (Q_1) is the outflow from the lake. The channel conditions (cross sectional shape, bottom roughness, and slope) control the values of h_1 and h_2 at the channel confluence. These values of h_1 and h_2 may then be used as boundary conditions for the solution of equations (4) and (5) moving upstream in the outlet channel.

Field studies were undertaken during 1991 in order to determine applicable values of the friction coefficients C_D, C_B, and C_I. In addition, data on the channel shape and slope were compiled in order to solve these equations for local conditions. Example solutions of these equations are shown here for the purpose of demonstrating the conditions which control the flow regime in the lake outlet channel; research on this issue continues. These example calculations are based on the following assumptions. The channels are assumed to be of rectangular shape, with the Seneca River channel in the vicinity of the confluence being 5.5 m deep and 75 m wide; the outlet channel is 3.7 m deep and 45 m wide. The flow in the Seneca River upstream of the lake was assumed to be $28 \, m^3 \cdot sec^{-1}$, a typical summer low flow, with h_1 and h_2 determined by assuming the stratified channel flow to be uniform, so that dh_1/dx and dh_2/dx are zero. The friction coefficient values used were $C_B = 0.01$, $C_I = 0.005$, and $C_D = 0.001$. Based on the observed salinities of lake and river water since closure of the soda ash/chlor-alkali facility, the relative density $\varepsilon = 0.0001$. All conditions are assumed to be at steady state.

Simulations of the effect of the rate of outflow from the lake are presented in Figure 4.11, which shows the position of the tip of the buoyant wedge in the outlet channel as a function of the lake outflow, with no wind. The wedge tip position is expressed as the distance upstream from the confluence. A decrease in the flow from 7.1 to $4.3\,m^3 \cdot sec^{-1}$ is sufficient to move the tip of the wedge from the vicinity of the John Glenn Blvd. bridge to the end of the channel at the lake (Figure 4.5). Under all these conditions, the difference in water surface elevation from one end of the outlet channel to the other is less than $0.003\,m$, a result of the very low flow velocities under the conditions of stratified flow.

In these example calculations, as lake outflow decreases below $4.3\,m^3 \cdot sec^{-1}$, the flow

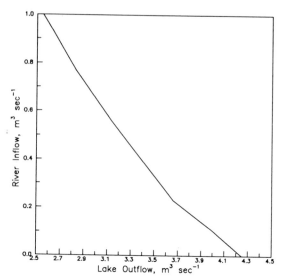

FIGURE 4.12. Steady-state reduction of relationship between outflow from Onondaga Lake and inflow to the lake from the Seneca River.

changes from regime 2 to regime 3 (Figure 4.10), and inflow from the river to the lake commences. Due to the abrupt change in the width of the channel at the entrance to the outlet channel (essentially from the width of the outlet channel to the width of the lake), an internally critical flow condition is reached at the transition. The critical condition occurs when the denominator of equations (4.4) and (4.5) are zero, or when

$$\varepsilon - F_1^2 - F_2^2 + F_1^2 F_2^2 = 0 \qquad (4.9)$$

It is this critical condition which controls the rate of flow from the river into the lake under the so-called "bidirectional" flow regime. Figure 4.12 shows the rate of inflow from the outlet channel into the lake (Q_2) as the lake outflow Q_1 decreases from 4.3 to $2.6\,m^3 \cdot sec^{-1}$. At a lake outflow of $2.6\,m^3 \cdot sec^{-1}$, the computed river inflow is $1\,m^3 \cdot sec^{-1}$, or about 40% of the lake inflow. Wind conditions can easily effect the outlet channel flow, because the water surface slope of the outlet channel is very small. The predicted effect of wind along the length of the outlet channel in either direction for a lake outflow of $2.6\,m^3 \cdot sec^{-1}$ is presented in Figure 4.13. Winds of increasing speed from the southeast tend to decrease the

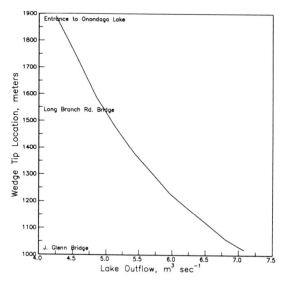

FIGURE 4.11. Steady-state prediction of relationship between outflow from Onondaga Lake and the location of the tip of the buoyant (Seneca River water) wedge in the outlet channel (condition of no wind).

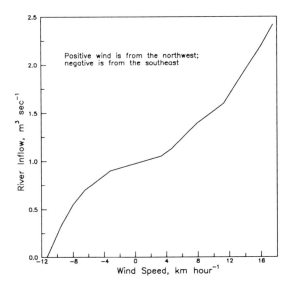

FIGURE 4.13. Steady-state prediction of relationship between wind speed and inflow to the lake from the Seneca River.

river inflow to the lake, with a wind of $11.3\,\mathrm{km \cdot hr^{-1}}$ decreasing the inflow to zero, the transition between regimes 2 and 3. Conversely, winds from the northwest increase the rate of river inflow to the lake; a wind of $12.9\,\mathrm{km \cdot hr^{-1}}$ doubles the inflow from 1.0 to $2.0\,\mathrm{m^3 \cdot sec^{-1}}$.

Despite its preliminary nature, this analysis leads to several important conclusions. First, the general outlet flow conditions can be predicted with the hydrodynamic theory presented above. Such calculations have not been previously attempted. Further, the analysis supports the position that the lake outlet conditions are quite sensitive to a number of environmental conditions that naturally vary over time. Thus, the outflow conditions should be expected to be dynamic, particularly during the summer when stream flows are low. Of course, the rate of tributary inflow to the lake basin is variable in response to rainfall in the watersheds of these tributaries. Winds are also variable in speed and direction. Under stratified flow conditions in the outlet the water surface in the channel is quite flat; under unsteady conditions, fluctuations in the water surface elevation at either end of the outlet channel can be expected to have a strong impact on

the flow regime and rate of river inflow to the lake. Fluctuations in water surface elevation in the river may occur as a result of rainfall, gate operations at the NYSDOT dams of the Three Rivers system or hydropower operations on the Seneca River. Fluctuations in water surface elevations in the lake can be caused by lake inflow, or by winds over the lake and resulting seiche activity. The reduction in the salinity of lake caused by the closure of the soda ash/chlor-alkali facility is expected to have decreased the frequency of occurrence of bi-directional flow in the outlet channel (flow regimes 2 or 3) and the rate of river inflow under those flow regimes. It would not be expected that changes in lake salinity would effect the occurrence of regime 4.

The simple steady-state model described above is sufficient to demonstrate the various outlet flow regimes and the sensitivity of flows to various environmental conditions. However, a more useful predictive model would account for time variations in environmental and flow conditions, as steady state conditions are rarely encountered. In addition, the actual shape and variation of the cross section of the outlet channel must be incorporated into such a model.

4.3 Lake Temperature, Salinity, and Density Stratification

4.3.1 Introduction

Density stratification occurs in deep lakes in temperate climates. This stratification is regulated almost entirely by vertical temperature differences in the vast majority of lakes. Dimictic stratification regimes are observed in lakes with the dimensions of Onondaga Lake in this region; i.e., characterized by strong stratification in the summer, weak inverse stratification under the ice in winter, and turnover (vertically isothermal conditions) in spring and fall. Stratification has been described as the most fundamental characteristic regulating the overall metabolism of lakes, as it;

1. affects the magnitude of vertical mixing, and thereby the cycling of key materials;

2. establishes the dimensions and duration of stratification, and thereby the oxygen resources of the hypolimnion and the upper level of phytoplankton biomass in the epilimnion; and

3. establishes the temperature of the layers, and thereby the rates of biochemical reactions.

The dependence of lake water quality on stratification has been described by Stauffer and Lee (1973), Stefan et al. (1976), Orlob (1983), and others. The features of stratification are dependent on several uncontrollable characteristics, such as basin morphometry and setting, hydrology, and meteorological conditions (Harleman 1982). The stratification regime is sensitive to meteorological conditions; substantial year-to-year variations in the regime are to be expected in a temperate climate as a result of natural variations in meteorological conditions (Effler et al. 1986a; Owens and Effler 1989). Further, the stratification regime may also be influenced by pollution inputs such as those that cause changes in light penetration (e.g., Effler and Owens 1985), or result in buoyant overflows or dense underflows (Owens and Effler 1989), as described earlier. The effects of meteorological variability make it difficult to determine the impact of modifications to lake inflows or outflow based only on lake data taken before and after such modifications. The need to determine the effect of such changes is an important motivation for the development of stratification and vertical transport models of the lake, as described in this chapter. This section is intended to:

1. establish the homogeneity that prevails between the two major basins of the lake;

2. document the salinity stratification that prevailed during much of the period of the operation of the soda ash/chlor-alkali plant and partition the contributions temperature and salinity stratification made to overall density stratification;

3. document changes in stratification that have occurred since closure of the soda ash/chlor-alkali plant; and

4. evaluate the regularity of spring turnover during the operation of the facility.

4.3.2 Extent of Spatial Homogeneity

Longitudinal uniformity in mass constituents and heat are expected in the horizontal directions in Onondaga Lake based on an order-of-magnitude force balance presented earlier in this chapter. Some of the most detailed measurements of both horizontal and vertical variations in physical and related chemical properties in Onondaga Lake were made by Stewart (1971, 1978). The appearance of an unusual lens of warm and cold water in thermal profiles collected in 1968 prompted Stewart (1971, 1978) to further investigate spatial heterogeneity in stratification through a series of 17 thermal transects made along the major axis of the lake in 1968 and 1969. Some of the observations from this work include (Stewart 1971):

1. modest tilting of isotherms in response to wind;

2. localization of warm water in the southern end of the lake as a result of warm inputs in that region, particularly from the input of spent cooling water from the East Flume; and

3. lens of water in the lake, that, on a strictly temperature basis, appear to be unstable.

The results of one of Stewart's thermal transects (3 December 1968) is presented in Figure 4.14 (Onondaga County 1971). Apparent unstable conditions (temperature increasing with depth) are evident throughout much of the water column, as well as the localized impact of the cooling water discharge in the southeastern portion of the lake (Figure 14). It was recognized (Stewart 1971; Rand 1971) that these temperature inversions were a manifestation of substantial salinity stratification, which in fact resulted in a stable water column.

The vertical variation of temperature and mass concentrations of lakes the size and shape of Onondaga Lake can usually be assumed to

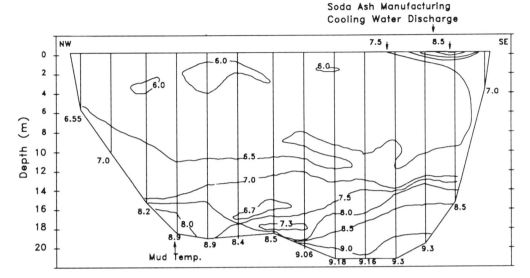

FIGURE 4.14. Thermal transect of Onondaga Lake on December 3, 1968 (by Stewart; modified from Figure 4.3, p. 143, Onondaga County 1971).

be essentially uniform throughout the basin; e.g., these lakes are often assumed to be "one dimensional." Temporary tilting of the stratified layers due to internal seiche activity occurs in all lakes following wind events. During a seiche some horizontal heterogeneity exists due to vertical displacement of portions of the profile, but the duration of this tilting is relatively brief and the average displacement of the profiles is zero.

The presence of buoyant and dense inflows is not expected to substantially affect the appropriateness of the one-dimensional assumption for Onondaga Lake. Stewart's transects in 1968 and 1969 demonstrated some localized heterogeneity in stratification along the main axis of the lake (e.g., Figures 4.3 and 4.14; see 16 other transects by (Stewart 1971, pp. 140–148)). However, the vast majority of the lake demonstrated largely uniform conditions despite the various intrusions. Thus, stratification data collected routinely at a single deep water location, such as the routine monitoring site located in the middle of the south basin, can be expected to accurately reflect the seasonal dynamics of this regime for essentially the entire lake. The lake can be described as a one-dimensional system for the

purpose of describing distributions of heat and of most substances.

The one-dimensional assumption was subsequently tested by 73 measuring Cl^- and thermal stratification at both the north and south deep stations (Field 1980). The assumption was supported by the near-equivalence observed in 50 paired observations of Cl^- concentrations in the near-surface waters (0 to 5 m intervals; Figure 4.15a) and at a depth of 15 m (Figure 4.15b) for the two stations, over the May through October interval of 1978. The near equivalence of the thermal component of stratification over this period is illustrated in three selected sets of paired temperature profiles for the two stations in Figure 4.16. More recent comparisons of paired measurements made for the two stations in later years (e.g., following closure of the soda ash/chlor-alkali facility) continue to support the one dimensional assumption for the stratification regime of the lake. Thus stratification data presented for the south deep station are representative of lake-wide conditions.

The only constituent of water quality interest for which significant horizontal variation within the lake basin has been demonstrated to date is fecal coliform bacteria. Lake moni-

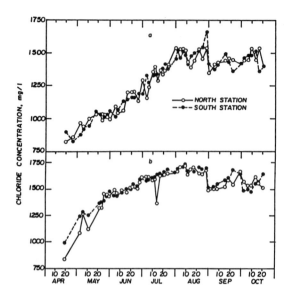

FIGURE 4.15. Comparisons of the vertical distributions of Cl at the south and north deep stations of Onondaga Lake, 1978: (**a**) 0–5 m average and (**b**) 15 m (from Effler and Driscoll 1985).

toring following combined sewer overflow events (see Chapter 1) indicates a significant gradient in fecal coliform concentrations prevails from the south end of the lake (proximate to most sources) along the major axis of the lake (see Chapter 6). The presence of such horizontal variations is due to the relatively brief duration of the overflow events and the extremely large difference in concen-

tration of fecal coliform in the inflows relative to lake concentrations (inflow concentrations during storms are commonly a factor of 10^4 to 10^5 greater than lake concentrations; see Chapters 3 and 6).

4.3.3 Salinity and Thermal Stratification

The vertical distributions of Cl^- concentration at the south deep station are presented for the April through November interval of 1980 in Figure 4.17 as isopleths; this regime is also represented as 19 selected individual profiles in Figure 4.18. Chloride concentration is an excellent surrogate measure of salinity in Onondaga Lake (Effler et al. 1986c; see Chapter 5). This is the most temporally and vertically intensive specification of the distribution of salinity available for the lake during the period of the operation of the soda ash/chlor-alkali facility; 54 profiles, at 1 m intervals, were collected. Strong salinity stratification was established at the start of the measurements in mid-April (vertical salinity difference of 2.6 ppt). The higher concentrations at the bottom reflect the effect of the plunging underflow from Ninemile Creek. Salinity levels remained relatively uniform in the lower waters until nearly the end of the monitoring period of 1980. However, salinity levels increased progressively in the upper

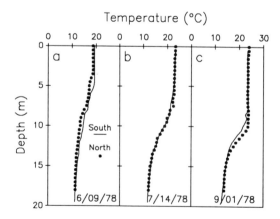

FIGURE 4.16. Comparison of temperature profiles at the south and north stations of Onondaga Lake, 1978: (**a**) June 9, (**b**) July 14, and (**c**) September 1.

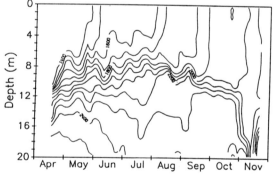

FIGURE 4.17. Isopleths of Cl at the south deep station of Onondaga Lake for the April–November interval of 1980.

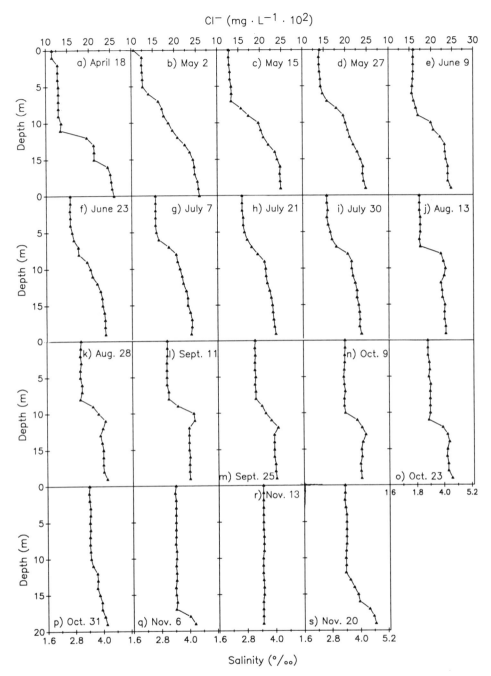

FIGURE 4.18. Profiles (19) of Cl at the south deep station of Onondaga Lake for the April–November interval of 1980.

waters (Figures 4.17 and 4.18) as a result of the entry of the salinity-enriched underflow into the upper waters during this period. This occurred because, following near-field mixing of the dense inflow with lake surface waters, the plunging inflow could not penetrate the well developed density stratification of the lake (see subsequent description of overall density stratification). Reformation of salinity stratification after the onset of fall turnover in 1980 was observed (Figures 4.17 and 4.18). A limited number of Cl⁻ measurements lead one researcher (Jackson 1968) to the incorrect conclusion that Onondaga Lake was meromictic.

The sharply contrasting distribution of salinity in Onondaga Lake following the closure of the soda ash/chlor-alkali plant is exemplified by the Cl⁻ isopleths of 1989 presented in Figure 4.19. Salinity was reduced by about 65% compared to 1980 and salinity stratification was almost absent. These are manifestations of the reduction of salinity loading to the lake (see Chapter 3). The reduction in the density difference between Ninemile Creek and the lake following closure of the soda ash/chlor-alkali facility (Figure 4.2b and c) apparently has greatly reduced the extent of plunging of this inflow. However, there is evidence that the plunging phenomenon for Ninemile Creek has not been totally eliminated by the reduction in density of the creek inflow. For example, plunging to mid-depths in late summer is indicated in Figure 4.20a, and plunging to the bottom is indicated after the onset of complete fall turnover and under the ice in Figure 4.20b and c, respectively. The continuation of this phenomenon is attributable to the continued entry of ionic waste from soda ash manufacturing from the region of the waste beds on Ninemile Creek (Effler et al. 1991; see Chapter 3).

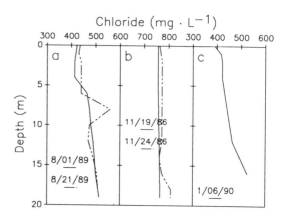

FIGURE 4.20. Profiles of Cl at the south deep station of Onondaga Lake indicating continued plunging of Ninemile Creek: (**a**) August 1 and 29, 1989, (**b**) November 19 and 24, 1986, and (**c**) January 6, 1990.

The features of the thermal stratification regime for 1980 are presented in the isotherms of Figure 4.21. Certain of these characteristics are typical of stratifying north temperature lakes (e.g., Wetzel 1983). Peak heat content developed in July and August. The lower layers warmed slowly during the stratification period, indicating limited vertical mixing. Progressive deepening of the epilimnion occurred from late August until the onset of complete turnover in November, associated with elevated kinetic energy inputs from increased wind and reduced air temperatures. Spring turnover was either extremely brief, or did not occur (Effler and Perkins 1987), as the

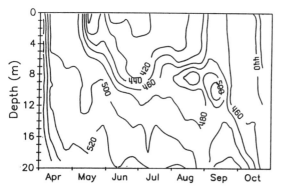

FIGURE 4.19. Isopleths of Cl at the south deep station of Onondaga Lake for the April–October interval of 1989.

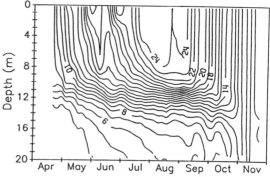

FIGURE 4.21. Isotherms at the south deep station of Onondaga Lake for the April–November interval of 1980.

lower waters were quite cold (~4°C) through April. Further, modest temperature inversions were observed at mid-depths and near the bottom (not resolved in isotherms of Figure 4.21, but see subsequent example profile); i.e., unstable if temperature is considered the sole regulator of density. This condition is of course unusual for stratifying lakes, but was not unusual during the operation of the soda ash/chlor-alkali facility, as documented earlier by Stewart (1971, 1978).

4.3.4 Density Stratification and Its Components

4.3.4.1 Detailed Analysis of Profiles

Density stratification in Onondaga Lake can be analyzed using measurements of T and Cl^- concentration over depth and by determination of water density using the equation of state (see Chapter 5). Individual estimates of density ρ (g · cm^{-3}) are expected to be accurate within 0.005 g · cm^{-3}, based on realistic ranges of sampling and measurement error (e.g., Cl^- concentration within 20 mg · L^{-1} and T within 0.1°C). Density gradients (g · cm^{-3} · m^{-1}), for the entire watercolumn have been calculated over the period of record at the vertical resolution of measurements (usually 1 m). The contribution of the salinity and T components of stratification were evaluated by calculating the total density gradient at a depth z_i (m) as

$$\left(\frac{d\rho}{dz}\right)_{i-1/2} \approx \frac{\rho(Cl_i,T_i) - \rho(Cl_{i-1},T_{i-1})}{z_i - z_{i-1}} \quad (4.10)$$

in which z_i = depth of measurement of Cl_i and T_i. The density gradient due to temperature alone was calculated by suppressing the observed gradient in Cl^-. The $i - \frac{1}{2}$ subscript indicates that the density gradient is estimated at the midpoint between the measurements i and $i - 1$. The following expression represents this contribution

$$\left(\frac{d\rho}{dz}\right)_{i-/2,T} \approx \frac{\rho(Cl_i,T_i) - \rho(Cl_i,T_{i-1})}{z_i - z_{i-1}} \quad (4.11)$$

Similarly the expression to represent the contribution of salinity gradients to the density gradient is

$$\left(\frac{d\rho}{dz}\right)_{i-/2,Cl} \approx \frac{\rho(Cl_i,T_i) - \rho(Cl_{i-1},T_i)}{z_i - z_{i-1}} \quad (4.12)$$

Detailed resolutions of the components of density stratification are presented for six selected days in 1980 in Figures 4.22 and 4.23. Part (a) of each figure presents the measured stratification of the density components, temperature, and salinity (as measured by Cl^-). Part (b) shows the corresponding calculated density profile. Part (c) shows the depth distribution of the total density gradient, and that associated with the attendant thermal stratification; the difference is due to salinity stratification. Paired measurements of T and Cl^- were made at 1 m intervals in 1980 (Effler et al. 1986b).

Density stratification was well established on April 18 and was almost entirely due to salinity stratification (Figure 4.22), caused by the plunging of ionically enriched Ninemile Creek. Density stratification increased by June 23, because of the increasing contribution of thermal stratification; chemical stratification decreased from the earlier observation (Figure 4.22). However, salinity stratification still represented nearly one third of the overall density stratification. Note the thermal stratification that was localized in the near surface waters on this day, commonly described as secondary stratification. This temporary condition develops in lakes during calm hot periods. Overall density stratification continued to increase through July 30 due to increasing thermal stratification, that more than compensated for the further decrease in salinity stratification (Figure 4.22); salinity stratification represented less than 30% of the density stratification on this date. The vertical density differences were largely located between 6 and 14 m. The peak salinity gradient was shallower than the peak thermal gradient (Figure 4.22). The deepening of the epilimnion that accompanies the seasonal increase in kinetic energy input and surface heat loss is apparent in the profiles of September 11 (Figure 4.23). The peak in salinity from 10 to

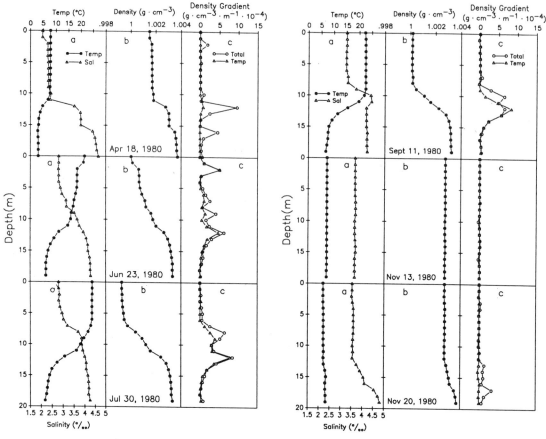

FIGURE 4.22. Resolution of density stratification components in Onondaga Lake in 1980, for April 18, June 23, and July 30 (modefied from Effler and Oweds 1995), respectively: (a) profiles of density components, (b) overall density profile, and (c) depth distribution of total and thermal density gradients.

FIGURE 4.23. Resolution of density stratification components in Onondaga Lake in 1980, for September 11, November 13, and November 20, respectively: (a) profiles of density components, (b) overall density profile, and (c) depth distribution of total and thermal density gradients.

11m, located above the maximum thermal gradient (Figure 4.23), is a manifestation of the plunging underflows from Ninemile Creek (e.g., unable to penetrate the thermal-based density gradient). Such maxima were commonly observed in late summer and early fall during the operation of the soda ash/chlor-alkali facility. Vertical uniformity in temperature, salinity, and therefore density, was observed on November 13 (Figure 4.23). Active vertical circulation (fall turnover) is expected under these conditions. However, these conditions also made the lake particularly susceptible to reformation of salinity based density

stratification as a result of the plunging saline underflow. By November 20 salinity-based density stratification was reestablished (Figure 4.23). Note also that temperatures were slightly higher in the lowermost waters, but that this apparent instability was more than compensated for by the elevated salinity. This reformation of stratification (i.e., interrupted or abbreviated periods of turnover) had negative implications for the oxygen resources of the lowermost layers, as the attendant reduced vertical mixing allowed localized depletion of oxygen associated with decomposition processes concentrated within the sediments.

Temporal structure and frequency of occurrence of the reformation of stratification during typical turnover periods while the soda ash/chlor-alkali plant was operating were affected greatly by ambient turbulence. For example, the reformation of stratification between November 13 and 20 (Figure 4.23) occurred during a rather calm period.

4.3.4.2 Seasonal Transformations

A representation of the partitioning of density stratification between thermal and salinity stratification for the entire spring through late fall monitoring interval (54 paired profiles) of 1980 appears in Figure 4.24. Both the total and salinity component plots represent the total watercolumn (i.e., top-to-bottom) differences. The difference between the two is the thermal component. The seasonality in the two components is clearly depicted. The magnitude of salinity-based density stratification and its relative contribution to overall density stratification was a maximum in spring. This component of density stratification decreased in a nearly linear manner until the onset of fall turnover in mid-November (Figure 4.24). The salinity component represented nearly 40% of the density stratification for the study period. The temporal structure of overall density stratification was regulated largely by the dynamics in the thermal component. The fine scale (abrupt) variations in total density

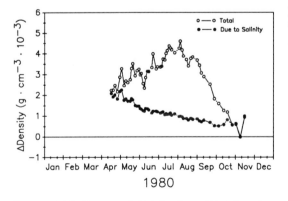

FIGURE 4.24. Temporal distribution of total water-column differences in density, and the salinity component of density, for the watercolumn of Onondaga Lake in 1980.

differences (Figure 4.24) reflect the development and subsequent breaking-up of secondary thermal stratification (cycling between calm and windy periods) and the occurrence of major mixing (entrainment) events associated with the passage of weather fronts. The seasonal structure in overall density stratification and the thermal component reflects the annual heating and cooling cycle common to deep lakes in this region.

The partitioning of density stratification is expanded to include the entire data base since 1968 in Figure 4.25 (also see Effler and Owens 1995). Certain years (1971, 1978, 1979, and 1982) have been omitted from the analysis due to shortcomings in the Cl^- data (also see Doerr et al. 1994). Clearly there was a significant salinity component of overall density stratification annually over the monitored portion of the period of soda ash manufacturing (Figure 4.25). These is no reason to assume this was not a recurring impact through the entire tenure of this operation. There was no significant salinity component of density stratification over the monitored intervals of 1987 through 1990 (Figure 4.25p–s); i.e., after the shut down of the soda ash manufacturing process. However, recall that salinity-based density stratification continues to develop under the ice. Rather strong differences in the magnitude of salinity-based density stratification, and its seasonal dynamics were observed during the period of operation of the facility (Figure 4.25a–o). The salinity component represented from 13% (1975, Figure 4.25g) to 50% (1972, Figure 4.25d) of the overall density stratification over the May through September interval for the period 1968 through 1986; the average was 30%. The year-to-year differences are largely associated with the magnitude of the initial salinity stratification established in spring of each year. This can be expected to have a stochastic character because of its dependence on various meteorologically related factors, such as ambient turbulence, flow rate of the plunging inflow, and density (temperature and salinity) difference between the dense inflow and the lake.

Progressive decreases in the salinity components of stratification were observed from

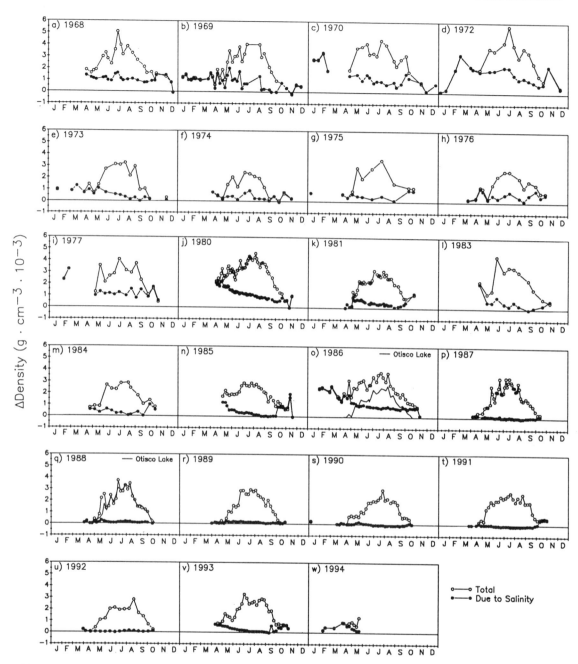

FIGURE 4.25(a–w). Temporal distributions of total watercolumn differences in density, and the salinity component of density, for the watercolumn of Onondaga Lake, for most years in the period 1968–1994. Distribution for Otisco Lake in 1986 (o) and 1988 (q) (modified from Effler and Owens 1995).

spring through fall of most years during the operation of the soda ash/chlor-alkali facility (Figure 4.25). Reformation of salinity-based stratification was noted in the late fall of several years (e.g., 1969, 1970, 1972, 1980, 1981, and 1985), reflecting the increased plunging of the dense saline inflow during that period of reduced watercolumn stability. The period of stratification has been about 5 months over the first six years since closure (1987–1992, Figure 4.25p–u). This represents a substantial reduction from the conditions that prevailed during the operation of the facility. Quantification of the duration of stratification before closure is difficult because of the reformation of stratification and the fact that the monitoring period was shorter than the duration of stratification in several instances. With the exception of 1974 (Figure 4.25f), the period of stratification was more than 6 months, usually more than 7 months, and in some cases spring turnover did not occur or was extremely brief (e.g., Figure 4.25d and o; also see subsequent section on failure of spring turnover).

Clearer signatures of the continued entry of saline plunging underflows have been manifested in monitoring conducted since 1991. Reformation of salinity-based stratification was documented in the late fall of 1991 (Figure 4.25t) and 1993 (Figure 4.25v). The lake failed to completely turnover in the spring of 1993 (Figure 4.25v) and 1994 (Figure 4.25w) because the S-based stratification established during the ice cover was not broken up by kinetic energy inputs in spring (see subsequent section). These lingering effects of the residual ionic waste discharge (see Chapter 3) are not attributable to increases in ionic polluted loading. Rather, they apparently are manifested only under certain meteorological conditions. For example, ice cover persisted into the spring of 1993 and 1994 later than usual (K. Stewart, personal communication), and unusually high runoff was experienced in the spring of 1993. The impacts of the ionic waste discharge from the soda ash/chlor-alkali manufacturer on the lake's stratification regime have been substantially ameliorated by the closure of the facility (e.g., Figure 4.25). However, the residual waste loading (see Chapter 3) continues to irregularly impact density stratification in the lake, which in turn has negative implications for its oxygen resources (see Chapter 5).

The total watercolumn density difference in mid-summer (an indicator of watercolumn stability) has decreased since the loss of salinity stratification that accompanied closure of the chemical facility. The average for July for the 1968 through 1985 period, for the years documented in Figure 4.25, was $3.1 \times 10^{-3} g \cdot cm^{-3}$, the range was 1.3 to $5.0 \times 10^{-3} g \cdot m^{-3}$. The average for the 1987 through 1990 period was $2.7 \times 10^{-3} g \cdot cm^{-3}$, the range was 2.3 to $2.9 \times 10^{-3} g \cdot cm^{-3}$. The stratification regime of Onondaga Lake (e.g., total density difference and timing of stratification) over the spring to late fall interval has recently been observed to be quite similar to that of Otisco Lake (Effler et al. 1989; Figure 4.25q). However, note the strong divergence between the lakes in 1986 (Figure 4.25o), when salinity stratification contributed significantly to density stratification in Onondaga Lake. More limited comparisons in the spring of 1993 and 1994 documented strong differences between these two systems, associated with the failure of turnover in Onondaga Lake (Effler and Owens, 1995).

4.3.4.3 Layer Dimensions: Chemocline and Thermocline

The dynamics of the plunging of the dense saline industrial discharge affected not only the duration and magnitude of overall density stratification, but the dimensions of the vertical layers. The density gradient barrier established by the thermal component of stratification (thermocline) in spring promoted the development of a chemocline (salinity gradient) above it; e.g., accumulation of plunging salinity. For example, note the chemocline was located above the thermocline on July 30, 1980 (Figure 4.22) and on September 11, 1980 (Figure 4.25). This was observed throughout the May–October period of 1980 (Figure 4.26), and apparently was a recurring feature of the stratification regime during the tenure of the soda ash/chlor-alkali plant. The disparity in vertical positions of the chemocline and thermocline became great during calm periods. Occurrences of near-coincidence of these depths reflect antecedent periods of high kinetic energy inputs (Figure 4.26). This displacement

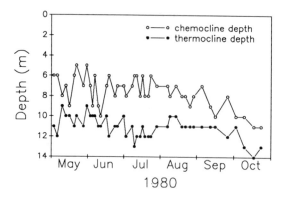

FIGURE 4.26. Distributions of the positions of the thermocline and chemocline in Onondaga Lake for the May–October period of 1980.

of the chemocline above the thermocline encouraged shallow mixed-layer depths, which tends to increase the maximum standing crop of phytoplankton in nutrient-saturated systems (Stefan et al. 1976).

4.3.5 Failure of Spring Turnover

4.3.5.1 Background and 1986 Observations

Turnover is an extremely important aspect of the annual stratification cycle, particularly with respect to the oxygen conditions in the lower layers of the lake. During this time the oxygen resources of the lower layers of productive lakes, particularly those with small-to-medium size hypolimnia, are replenished. The unusually elevated stability that developed under the ice in Onondaga Lake before closure (Figure 4.27), and that continues to a lesser extent since closure, required greater kinetic energy inputs to achieve turnover, and thereby replenish the dissolved oxygen of the lower layers. Strong salinity stratification, often combined with an isolated setting that shelters a lake from wind-driven kinetic energy inputs, is the basis for the essentially permanent stratification observed in meromictic lakes (Hutchinson 1957; Wetzel 1983). The salinity-enriched lowermost waters of meromictic lakes (the monimolimnion) are permanently devoid of oxygen, and usually enriched in oxygen-demanding reduced substances (e.g.,

H_2S and CH_4; Hutchinson 1957; Wetzel 1983). The closure of the soda ash/chlor-alkali facility in 1986 prompted the investigation of the impact this would have on the occurrence of spring (vernal) turnover, and an analysis of related historic data.

Stratification conditions were again quantified by paired profiles of Cl^- and T using the equation of state of Effler et al (1986c). The failure of the lake to turnover in the spring of 1986 is clearly documented by the stratification information presented in Figure 4.28a–c. Thus the lake was mono-mictic (instead of dimictic) in this year as a result of the ionic pollution. The salinity stratification that developed under the ice as a result of the plunging of the industrial ionic discharge (e.g., Effler et al. 1986b) failed to break-up following the loss of ice from the lake ("ice-out"; Figure 4.28a). The salinity stratification was forced deeper in the lake, vertical gradients became stronger, and the concentration of Cl^- in the lower-most layers decreased following "ice-out" (Figure 4.28a), as a result of wind-driven vertical mixing. Vertical irregularities in April and early May in the deeper layers (Figure 4.28a) may be manifestations of the time structure of wind (mixing) events.

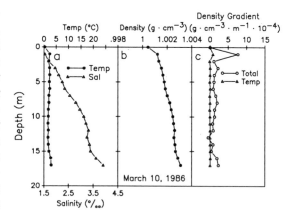

FIGURE 4.27. Resolution of density stratification components in Onondaga Lake for March 10, 1986 (ice-cover): (a) profiles of density components, (b) overall density profile, and (c) depth distribution of total and thermal density gradients.

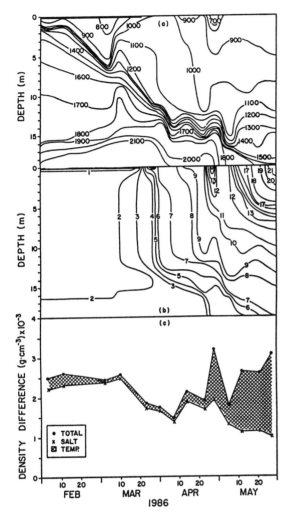

FIGURE 4.28. Stratification in Onondaga Lake for the February–May interval of 1986: (a) Cl isopleths in g · m^{-3}, (b) isotherms in °C, and (c) total density difference, and temperature and salinity components (from Effler and Perkins 1987).

The absence of approximately 4°C temperatures in the deeper layers under the ice (Figure 4.28b) was due in part to the depression of the temperature of maximum density (by approximately 0.8°C) associated with the ionic enrichment of the lake (see Chapter 5). A manifestation of the failure of spring turnover to occur was the low temperatures that persisted in the lower waters into late spring. Other lakes in the region of similar morphometry are typically 3–5°C warmer at the bottom by early May (e.g., Effler et al. 1989)

as a result of heating of the entire water column during spring turnover. For example, Otisco Lake, located 22 km to the south, was approximately 3.4°C warmer than Onondaga Lake at the bottom on May 12, 1986 (Effler et al. 1987). Note the loss of thermal stratification within the upper layers of the lake that occurred in early May as a result of high winds. Thermal stratification was subsequently reformed in mid-May (Figure 4.28b).

The salinity component of stratification was dominant under the ice, and remained a major component throughout the spring monitoring period (Figure 4.28c). The dynamics of density stratification observed following "ice-out" (Figure 4.28c) reflect mixing, lake flushing, and heating processes. Short-term variations in the magnitude of salinity stratification reflect imbalances in wind-driven vertical mixing (which tended to decrease salinity stratification) and flushing of the upper layers with more dilute water (because of closure of the soda ash/chlor-alkali facility, which tended to increase salinity stratification). The magnitude of thermal stratification reflected typical seasonal heating and mixing. Note the substantial thermal component of density stratification in late April (Figure 4.28c), localized in the upper layers (Figure 4.28b), that acted to preserve the salinity component of the deeper layers (Figure 4.28c) through the late-April/early-May period of high kinetic energy input. The failure of spring turnover to occur had negative implications for the oxygen resources of the lower layers. Oxygen was depleted from these layers under the ice and anoxia developed before ice-out. These layers are not replenished with oxygen in the spring when turnover does not occur. The anoxia has value as an indicator for other years in which spring turnover did not occur. Other impacts of the altered stratification regime on the oxygen resource of the lake are identified in Chapter 5.

4.3.5.2 Failure of, or Abbreviated, Spring Turnover: History

The coincidence of low temperature (~4°C) and anoxia in the lower waters in late April reflects a negative impact of the plunging inflow, associated either with the failure of spring turnover to occur (e.g., Figure 4.28b and c) or a very abbreviated spring turnover. Here we use these coincident conditions as a basis to identify other years in which this impact was manifested, from a data/base which has less temporally detailed monitoring data. The magnitude of salinity stratification in late April is not a good indicator of the occurrence of turnover, as salinity stratification redeveloped rapidly following turnover during the operation of the soda ash/chlor-alkali facility.

Paired observations of the temperature and dissolved oxygen concentration at 18 m in late April for the period of record are presented in Figure 4.29a and b. The occurrences of anoxia and lower temperatures have clearly been coupled in Onondaga Lake (Figure 4.29a and b). The demarcating temperature appears to be approximately 4.5°C (e.g., anoxia observed at this time if temperature ⩽4.5°C). Such an empirical approach is of course imperfect, as year-to-year differences in the date of "ice-out," wind-driven mixing, and water/column heating can be expected to influence these conditions. Despite these limitations, it appears that the lake either failed to turnover in the spring of a number of years, or ex-

perienced a very brief turnover period, during the operation of the soda ash/chlor-alkali facility, and that this had a negative effect on the oxygen resources of the lake. The coincident occurrence of anoxia and temperatures less than 4.5°C was observed at 18 m in late April in 7 of the 19 yr of record during the chemical plant's operation (1968–1986; Figure 4.29a and b). Note that the conditions reported for 1974 did not fit the empirical relationship (Figure 4.29a and b). Spring turnover appeared to have occurred more regularly after the early 1970s (though substantial year-to-year differences in the oxygen resources of the lower layers are evident). Perhaps antecedent (under the ice) salinity stratification was stronger in the earlier years. There was a greater propensity for the failure of spring turnover to occur in 1986 due to the shut down of the soda ash/chlor-alkali facility, associated with the flushing of the upper layers with more dilute inputs (Effler and Perkins, 1987).

The effect was not observed over the 1987–1992 interval (Figure 4.29a and b) following closure of the soda ash/chlor-alkali facility, indicating the occurrence of spring turnover. This undoubtedly was a result of the reduction of salinity-based density stratification under the ice associated with closure. However, the signature of anoxia and low temperature for the bottom waters in spring reemerged for the failures of spring turnover in 1993 and 1994 (Effler and Owens 1995).

4.3.6 Summer Heat Income

The summer heat income Θ_s(cal · cm^{-2}) is the amount of heat necessary to raise the temperature of the lake from an isothermal condition at 4°C to the maximum observed heat content, normalized by the surface area of the lake (Wetzel 1983). Heat content is of importance to a lake because of the strong dependence of biological activity on temperature. The summer heat income at any given time is determined by

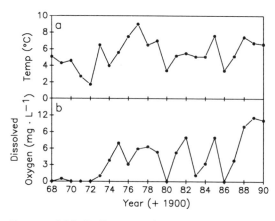

FIGURE 4.29. Indicators of spring turnover in Onondaga Lake, 1968–1990, measured in late April at a depth of 18 m: (**a**) temperature and (**b**) DO.

$$\Theta_3 = \frac{\rho c}{A_s} \left[\int_o^{z_m} TA dz - T_m \int_o^{z_m} A dz \right] \quad (4.13)$$

in which c = specific heat of water; A_s = surface area of the lake; z_m = the maximum depth of the lake (m); A = plan area of the lake; z = vertical distance above the lake bottom; and T_m = the temperature of maximum density (~4°C). Both T and A are functions of z; Equation (4.13) is commonly evaluated using discrete measurements of T over depth. The value of Θ_s is dependent on lake morphometry, setting, and climate (Wetzel 1983).

The values of Θ_s calculated for Onondaga Lake for 13 years are shown in Table 4.2. These are the years for which the most comprehensive (vertically and temporally) temperature profiles are available. The estimated Θ_s value for Lake Mendota, Wisconsin (18250 cal · cm^{-2}; Wetzel 1983) falls in the range shown for Onondaga lake. The value of Θ_s for the years before closure of the soda ash/chloralkali facility (1978–1985) was distinctly greater (8.8% higher on average) than observed since closure (1987–1990). Note there is only one (1983) overlap in these two populations; they represent statistically significant conditions (P < 0.0001, student T-test, null hypothesis of the means being equal). The decrease in Θ_s is attributable to the discontinuation of the discharge of cooling water by the soda ash/chlor-alkali facility.

TABLE 4.2. Summer heat income Θ_s for Onondaga Lake for selected years.

Year	Θ_s cal · cm^{-2}
1978	19,700
1979	18,300
1980	19,100
1981	19,400
1982	18,500
1983	17,000
1984	18,500
1985	19,400
1986	16,700
1987	17,400
1988	17,900
1989	17,300
1990	17,400

4.4 Lake Circulation

In order to quantitatively describe the dynamics of temperature and concentrations of various pollutants in the lake, mathematical models of these properties have been developed (see Chapter 9). An important component of the fecal coliform model is transport, or the manner in which heat and mass move within the lake. A prerequisite for the application of a heat or mass transport model is a model of water motion in the lake. This section describes a circulation model (Owens 1989) which was developed in order to quantify water motion and horizontal dispersive transport in the surface layer (epilimnion) of the lake. Dispersion in the surface layer was computed from the horizontal flow field. This model has been used as the basis for a mass transport model of fecal coliform bacteria in the lake, as subsequently described in detail.

A schematic side view of a typical temperature profile in a lake during summertime, and an associated typical velocity profile due to wind-induced flow, are presented in Figure 4.30. The base of the surface layer is located at the thermocline. The wind-induced currents are generally in the direction of the wind near the water surface, and are in the opposite direction near the bottom of the layer. The reversal in current direction in part allows for conservation of water volume in the enclosed lake basin. The goal of a circulation model is the prediction of this water motion as a function of the wind speed and direction over the lake surface.

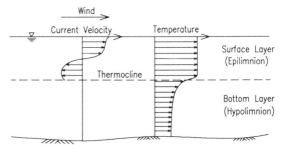

FIGURE 4.30. Typical vertical temperature and velocity profiles for a stratified lake.

4.4.1 Overview of Existing Predictive Models

Cheng et al. (1976) have described the various types of models that may be used to predict circulation in a lake. Two characteristics define the major differences and capabilities of the models. The first characteristic is that of single versus multilayer models. In stratified lakes, simulation of circulation over the entire lake depth requires the use of a multilayer model in order to describe the effect of stratification on the flow. Multilayer models are fully three dimensional and are thus complex in construction and expensive to operate. Single layer models have generally been applied to unstratified lakes, but may also be used in describing only the epilimnion of a stratified lake (Walker et al. 1986). With the primary concern being transport and water quality conditions in the surface waters, a single layer model of the epilimnion (surface mixed layer) of Onondaga Lake was developed. This model allows the realistic, physically based prediction of transport in the epilimnion of the lake as driven by wind, shoreline inflows, and outflow.

The second characteristic is related to the manner in which time variation of wind conditions is considered. The more general time-variable models predict the response of the lake over time to forcing conditions which are also a function of time. Steady-state models predict constant circulation caused by a single set of forcing conditions. The time-variable models are generally more time consuming and expensive to operate. Cheng (1977) has presented physical and mathematical arguments indicating that in many situations the response of lake circulation to time-variable winds may be simulated using a steady-state model. In this approach, transient circulation is simulated by a series of steady states. Each steady-state period in the series is a period of time over which the wind speed and direction are relatively constant. This approach was adopted in this study. Use of the series-of-steady-states solution combines the simplicity of the steady-state solution with the ability to simulate

wind and streamflow conditions which can generally be expected to vary.

4.4.2 Description of the Steady-State Model

The governing equations for this model were first proposed by Ekman (1905) in his classical analysis of ocean currents; a model based on these equations has been described as an "Ekman-type" model (Cheng et al. 1976). As in any hydrodynamic model, the governing equations are those of motion and continuity for water, which for this model are

$$-fv = -\frac{1\partial p}{\rho \partial x} + \frac{\partial}{\partial z} K_m \frac{\partial u}{\partial z} \qquad (4.14)$$

$$fu = -\frac{1\partial p}{\rho \partial y} + \frac{\partial}{\partial z} K_m \frac{\partial v}{\partial z} \qquad (4.15)$$

$$g = -\frac{1\partial p}{\rho \partial z} \qquad (4.16)$$

$$\frac{\partial u}{\partial x} + \frac{\partial v}{\partial y} + \frac{\partial w}{\partial z} = 0 \qquad (4.17)$$

in which u, v, and w correspond to the x, y, and z components of water velocity, respectively $(m \cdot hr^{-1})$, with x positive to the southeast along the long axis of the lake, y positive to the northeast normal to x, and z positive upward and zero at the water surface (m); f = Coriolis parameter which accommodates the effect of the Earth's rotation on the flow (hr^{-1}); p = local pressure $(N \cdot m^{-2})$; K_m = vertical kinematic eddy viscosity $(m^2 \cdot hr^{-1})$; and g = acceleration of gravity $(m \cdot hr^{-2})$. Equations (4.14) through (4.16) are the x, y and z components of the equations of conservation of momentum, and Equation (4.17) is the flow continuity equation. The momentum equations have been linearized by assuming that the nonlinear acceleration terms are small compared to the Coriolis acceleration, which is generally a good assumption for a lake the size of Onondaga Lake (Liggett and Hadjitheodorou 1969). In addition, the lake is considered to be shallow so that the dominant frictional forces are the vertical diffusion of momentum due to wind shear at the surface and bottom friction; the horizontal diffusion of momentum is neglected. Also, vertical accelerations are neglected and the pressure distribution is hydrostatic (Equation (4.16)). These assumptions are accurate for lakes the size of Onondaga Lake.

Boundary conditions at the water surface ($z = 0$) and at the layer bottom ($z = -h$) are required for their solution. At the water surface,

$$\rho K_m \frac{\partial u}{\partial z} = \tau_{sx} \qquad (4.18)$$

$$\rho K_m \frac{\partial v}{\partial z} = \tau_{sy} \qquad (4.19)$$

in which τ_{sx} and τ_{sy} = x and y-components of the wind-induced shear stress at the water surface ($N \cdot m^{-2}$). The components of the wind stress were computed by

$$\tau_{sx} = -C_D \rho_a W^2 \sin\theta \qquad (4.20)$$

$$\tau_{sy} = -C_D \rho_a W^2 \cos\theta \qquad (4.21)$$

in which C_D = dimensionless drag coefficient; W = wind speed ($m \cdot hr^{-1}$), and θ is the wind direction. Equations 4.18 and 4.19 represent the most important forcing term for this lake circulation model, that due to wind.

At the bottom of the layer, the following boundary conditions were used:

$$K_m \frac{\partial u}{\partial z} = Ku \qquad (4.22)$$

$$K_m \frac{\partial v}{\partial z} = Kv \qquad (4.23)$$

in which K = empirical bottom friction coefficient with units of velocity ($m \cdot hr^{-1}$). An equation of this form has been used to describe the shear stress at an internal density interface in various stratified flow models, such as Schijf and Schönfeld (1953) and Arita and Jirka (1987). The empirical coefficient K was first applied to the lake circulation problem by Cheng (1977). This representation of bottom friction allows the upper (no slip condition, $K \gg 1$) and lower (no stress condition, $K \ll 1$) bounds to be covered.

A "local" solution to Equations 4.14 through 4.17 can be computed. This solution is applicable to water bodies of large horizontal extent, where the flow does not "feel" the effect of the shoreline, and where the water depth and wind stress are uniform. Ekman's (1905) analytical solution for large water depth is applicable to the open ocean; Welander's (1957) analytical solution is for limited depth. These solutions are termed local because they do not consider any variability in the x and y directions and they are based on the assumption that the depth-averaged flow is zero (no advection). Despite the possible shortcomings in a lake basin of medium size, the local solution is of interest because: (1) the local solution for circulation and transport may, under some circumstances, be a good approximation of that determined by a complete solution of the equations for a lake basin; and (2) the local solution for these boundary conditions is an analytical one, and is relatively easy to compute. This solution is given by Cheng (1977).

In the more general case where u, v, p are functions of x, y, and z and τ_{sx}, τ_{sy}, K, and h are functions of x and y, the equations presented above are three dimensional. Models based on these same equations have been proposed and used by Liggett and Hadjitheodorou (1969), Cheng (1977), Shanahan and Harleman (1982), and others. Many of these workers have transformed the governing equations into a two-dimensional form by vertical integration over the depth of flow. An important simplifying assumption that is required for this integration is that the eddy viscosity is constant with depth. The eddy viscosity describes the vertical diffusion of momentum, or the extent to which water at the surface, when stressed by the wind, tends to drag along water beneath it. Due to the very large Reynold's number of lake circulation flows, the flow is fully turbulent. In such a turbulent boundary layer, the eddy viscosity is generally expected to be a function of depth and of the applied wind stress. The assumption of a spatially constant eddy viscosity, which was adopted in this study, is one of the available techniques described by

Rodi (1987) and ASCE Task Committee (1988) to describe the effect of turbulence in a circulation model. The other available options are unnecessarily complicated for this application, in large part because they require the use of a fully three dimensional numerical solution of the governing equations, and they are more difficult to verify. In addition, Svensson (1979) has shown that a properly chosen constant viscosity can give nearly the same results as a more sophisticated turbulence model which allows a spatially variable viscosity.

Thus, a depth-independent relationship for the eddy viscosity must be determined. Basic theories of turbulent flow have lead to the development of predictive formulae for diffusion as the product of a characteristic turbulent velocity scale and a characteristic turbulent length scale. Csanady (1973), Fisher et al. (1979), and Svensson (1979) have investigated such relationships in surface water flows. For this problem, the appropriate velocity scale is the wind-induced shear velocity u_*, where

$$u_*^2 \equiv \frac{1}{\rho} \sqrt{\tau_x^2 + \tau_y^2} \qquad (4.24)$$

The appropriate length scale is h, the depth of flow. Thus, for Onondaga Lake, the predictive expression for the eddy viscosity is

$$K_m = C_V u_* h \qquad (4.25)$$

where C_V is a dimensionless coefficient to be determined. Expressions of the same form have been used to predict diffusion in a variety of laboratory and field-scale flows (Fischer et al. 1979).

The integrated equations of motion are most conveniently expressed in terms of a stream function ψ for the depth-integrated flow, where ψ ($m^3 \cdot hr^{-1}$) is defined by

$$\bar{u}h \equiv \int_{-h}^{o} u dz \equiv \frac{\partial \psi}{\partial y} \qquad (4.26)$$

$$\bar{v}h \equiv \int_{-h}^{o} v dz \equiv -\frac{\partial \psi}{\partial x} \qquad (4.27)$$

in which $\bar{u}$ and $\bar{v}$ = depth-averaged velocities ($m \cdot hr^{-1}$). The integrated equations of motion can then be written in terms of ψ as

$$\frac{\partial^2 \psi}{\partial x^2} + \frac{\partial^2 \psi}{\partial y^2} + A\frac{\partial \psi}{\partial x} + B\frac{\partial \psi}{\partial y} + C = 0 \qquad (4.28)$$

The details of this integration process and the development of Equation (4.28) are given by Liggett and Hadjitheodorou (1969). In Equation (4.28), the quantities A, B, and C are functions of the layer depth h, viscosity K_m, and bottom friction coefficient K, and are thus functions of position in the lake (x and y). The quantity C is also related to the wind stress conditions. In the single layer approach used here, the layer depth h is taken to be the local water depth if the local water depth is less than the depth of the epilimnion. If the local water depth is greater than that of the epilimnion, the layer depth h is equal to the depth of the epilimnion.

From Equations (4.26) and (4.27) it can be shown that the difference in the values of the stream function between any two points in the lake is equal to the net flow between these points. Thus, the solution of Equation (4.28) directly yields the advective component of transport for a depth-averaged model. Solution of Equation (4.28) requires specification of the values of ψ along the shoreline of the lake, which defines the region of (x, y) over which this equation is solved. Thus, shoreline inflows and outflows are introduced into the problem by specifying that ψ changes when moving along the shore only at points where inflows and outflows are located, and the magnitude of the change is equal to the flow rate at that point.

Exact solutions to Equation (4.28) can only be achieved for the aforementioned "local" conditions. Solutions for natural, irregular basins must use approximate, numerical methods. In this study, the finite element method was used because it has been found to be the most accurate and flexible in simulating such basins (Cheng et al. 1976). The quadratic quadrilateral element, as applied by Gallagher and Chan (1973), was used here. In this method, the water surface of the lake is divided into a number of eight-sided polygons. Within each of these polygons, the stream function ψ is assumed to vary quadratically in space. Application of the method of weighted residuals to Equation (4.28) yields a system of eight linear equations for each element, with the eight unknowns in these equations being the values of the stream function at the element nodes, which are located at the junction of the element sides. When done for all elements covering the lake surface, and assigning values of the stream function on the lake shore based on the known inflows and outflow, a system of linear equations for the stream function at all nodes inside of the lakeshore is obtained. This system of linear equations is solved by standard techniques, yielding values of ψ at the nodes of the finite element mesh. By reversing the integration process used to obtain Equation (4.28), the local pressure and velocity components u and v can be obtained at any point in the epilimnion of the lake.

A computer program based on this mathematical model was written in mixed-language code using the Fortran and Basic computer languages. The program was designed to run on an IBM-PC compatible microcomputer. The program was applied to two test cases to verify the accuracy of the computer code in solving the governing equations. The first test was that of the circulation in a shallow basin with constant depth. As discussed above, this is the "local" solution for which an analytical solution is available, that of the Ekman spiral in finite depth (Welander 1957; Cheng 1977). The numerical solution computed by the model was compared to the analytical solution; the two were identical. A typical "Ekman spiral" computed in such a constant depth basin is shown in Figure 4.31. The current speed is highest at the surface and reverses with depth. The direction of the current is turned clockwise from the wind direction at the surface, and rotates in a clockwise spiral with increasing depth due to the influence of the Coriolis acceleration. For the case of sufficiently large K (Equations (4.22) and (4.23)), the velocity approaches zero at the bottom of the layer.

The second test case is one that has been used in the development of several other lake circulation models that have appeared in the literature. This test case, which was first used

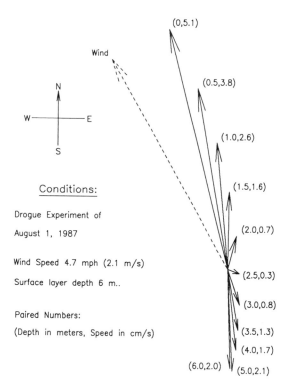

FIGURE 4.31. Ekman spiral of current speeds with depth in a constant depth basin.

by Liggett and Hadjitheodorou (1969), is a rectangular basin roughly the shape of Lake Michigan. It is 500 km × 125 km, with a maximum depth of 80 m. The bottom slopes uniformly from the shore to the middle of the lake from the sides and the end. Liggett and Hadjitheodorou (1969) approached this problem using the above model equation solved by the finite difference method, while Gallagher and Chan (1973) used quadratic quadrilateral finite elements in their solution. The model developed in this study was also applied to this same test case. In Figure 4.32a, the computed stream function at two lake cross sections using 40 elements and 149 unknown nodes is compared to that obtained by Liggett and Hadjitheodorou (1969) using a very dense grid of 1700 node points. Despite the significantly lower number of unknowns, the finite element model solution is very close to the finite difference solution. Figure 4.32b compares computed surface velocities at two cross sections for the two models. Again, a

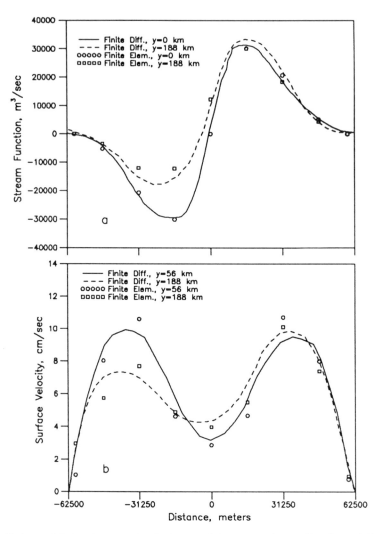

FIGURE 4.32. Predictions of finite element circulation model compared to earlier finite difference solutions for test problem of Liggett and Hadjitheodorou (1969): (**a**) stream function and (**b**) surface velocity.

good comparison is shown. Based on these results, it was concluded that the model was constructed properly, and that the solution techniques and computer code used to solve the model equations were operating correctly.

4.4.3 Application to Onondaga Lake

4.4.3.1 Finite Element Grid

The first step in the process of applying the circulation model to Onondaga Lake was to

develop a grid of finite elements covering the surface of the lake. The selection of a grid was governed not only by the needs of the circulation model, but was also affected by the structure of the fecal coliform bacteria model with which it was to be interfaced (see Canale et al. 1993; see Chapter 9). Computationally, it is most convenient if the boundaries separating segments in the water quality model coincide with the edges of the finite elements used in the circulation model.

The number of elements used affects the accuracy of the solution of the circulation model. The numerical solution converges toward the exact solution of the model equations as the number of discrete solution nodes increases. However, the computer memory and computation time needed to compute a solution also increases with this same change. Thus, a good finite element grid is one which achieves an accurate solution without making the calculations unnecessarily lengthy.

The finite element grid for the Onondaga Lake circulation model is presented in Figure 4.33. It is made up of 93 elements (eight-sided polygons). The nodes are located at the intersections of the sides. The entire lake is covered by the configuration of interconnected elements. This configuration was found to provide accurate definition of the shape of the lake for the circulation model.

4.4.3.2 Model Calibration to Drogue Observations

The credibility of the circulation model was established by successful simulation of water motion, as directly measured through the tracking of deployed drogues, over a range of wind conditions. Drogues were released, and their positions tracked over an interval of several hours, on eight days during the late summer/early fall period of 1987. A summary of wind conditions and epilimnion depths on the days of the deployments is presented in Table 4.3. The raw data on position of the drogues over time, along with plots of the wind conditions during the experiments, are presented by Owens (1989). Wind direction is reported here as the compass direction in degrees from which the wind is blowing with

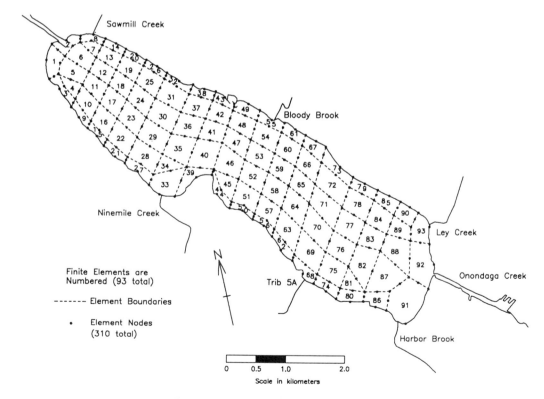

FIGURE 4.33. Finite element configuration for Onondaga Lake circulation model.

TABLE **4.3.** Summary of ambient conditions for 1987 drogue experiments.

Date	Wind speed (m·sec⁻¹)	Wind direction (degrees)	Epilimnion (Depth, m)	Number of drogues
August 1	2.1	150	6.0	2
August 28	3.1	60	9.0	2
September 12	2.3	135	9.0	2
September 18	5.9	70	10.5	2
September 23	4.3	250	10.5	3
September 25	6.2	315	10.5	2
October 1	5.0	300	13.0	2
October 8	7.7	300	13.5	2

0 degrees indicating from the north, 90 indicating from the east, and so on. As seen in Table 4.3, the drogue experiments covered a fairly wide range of environmental conditions in terms of wind speed, direction, and epilimnion depth.

Data from Hancock Airport were used to specify wind conditions during the first drogue deployment. Wind speed and direction at 3-hr intervals were recorded by the National Weather Service at the airport. Shortly afterward, a recording wind gauge was installed, 9.5 m above the ground, at a location adjoining the New York State Barge Canal Terminal, about 1.5 km southeast of the mouth of the Onondaga Creek. Wind speed and direction were recorded at 10-min intervals at this wind gauge. The drogues were tracked for a period of time averaging 3 hr, with range of duration from 1.5 to 5 hr. At the time of release, and at a number of intervals over the duration of the experiment, the location of the drogue was determined using Loran position equipment. Based on these measurements, the average speed and direction of the drogue over the duration of the experiment were determined. These results are given in Table 4.4.

The values of three model coefficients were determined by model calibration; i.e., ap-

TABLE **4.4.** Predicted and measured drogue speed and direction for 1987 experiments.

Date	Element (number)	Depth (m)	Speed (cm·s⁻¹) Predicted	Speed (cm·s⁻¹) Measured	Direction (predicted)	Degrees (measured)
August 1	65	0	5.9	5.1	145	165
	72	2	1.9	0.7	200	195
August 28	70	0	10.3	7.6	80	76
	71	3	4.5	1.1	70	106
September 12	65	0	9.5	5.4	150	155
	65	3	4.2	0.8	150	194
September 18	23	0	8.0	15.0	90	81
	23	3	7.6	3.9	100	94
September 23	29	0	9.6	10.9	280	264
	29	3	3.7	2.7	280	281
	64	0	8.4	10.9	310	264
September 25	52	0	15.4	16.3	300	325
	58	3	11.8	4.1	300	337
October 1	40	0	13.3	12.6	305	315
	40	3	12.1	4.7	305	328
October 8	64	0	9.9	20.0	300	311
	58	4	7.7	4.3	300	325

plying the model to conditions which occurred during the drogue experiments, and adjusting these coefficients to minimize the difference between predicted and measured drogue speed and direction. The coefficients adjusted were the water surface drag coefficient, C_D (Equations (4.20) and (4.21)), the bottom friction factor K (Equations (4.22) and (4.23)), and the constant C_V in the expression for eddy viscosity (Equation (4.25)). Svensson (1979) used the relationship of Equation (4.25) in a study of circulation in the coastal ocean, and determined a value of C_V of 0.026 based on numerical experiments with a sophisticated turbulence model. There is very little other guidance in the literature as to the range of C_V. The value of K depends on the nature of the bottom conditions. Most of the "bottom" of the layer to which the circulation model was applied is actually the interface between the epilimnion and hypolimnion. A number of studies have investigated shear at such interfaces, many in the context of flow in stratified estuaries. A number of studies, mostly in ocean applications, have been done to determine drag coefficients as a function of wind speed (Wu 1969). The values of C_D determined in these studies fall in a fairly narrow range of 0.001 to 0.003. However, all such studies utilized wind data that were measured directly over the water surface. In this study, wind data were collected over land. In addition, the limited fetch of Onondaga Lake may cause some variation from ocean values of C_D.

Over the duration of a drogue experiment, a single drogue typically traversed across one or more of the finite elements covering the surface of the lake (Figure 4.33). The single element which contained the largest portion of the path of the drogue was identified, and the predicted speed and direction at the center of that element was compared to the measured speed and direction of the drogue.

A single set of model coefficient values, C_V = 0.026, C_D = 0.004, and K = 10^{-4} m · hr^{-1}, determined through the calibration process, was found to produce reasonably accurate predictions of the water motion measured in all eight drogue deployments. Predictions were most sensitive to the value of C_V, less sensitive to C_D, and only moderately sensitive to K. The spiral velocity profile shown in Figure 4.31 is a local solution using these coefficient values with a layer depth of 6 m. The accuracy of the predictions of drogue movement was generally good (Table 4.4). The results show that, on several dates, the predicted currents at depths below the water surface tend to be rotated more from the wind direction and to be lower in magnitude than observed in the drogue tracks. This result is likely to be due to windage (drogue motion caused directly by wind, not water, movement). Some of this difference may also be due to the assumed constant eddy viscosity over depth. Nonetheless, the agreement between predictions and measurements is good, and the circulation model, with the coefficient values listed above, is judged to be accurate for use in predicting water motion in the upper mixed layer of Onondaga Lake.

4.4.3.3 Circulation Predictions for Various Conditions

Predictions of the circulation model are displayed in the form of arrows, which indicate the speed and direction of the predicted motion at a point located at the base of each arrow. The length of the arrow scales to the speed of the predicted current, using the scale shown on the plot. The predicted circulation for the wind conditions that occurred on August 1, 1987, the date of the first drogue experiment, is shown in Figure 4.34. The wind was from the south-southeast (direction 150 degrees) at 2.1 m · sec^{-1} during this period, and streamflows were relatively low. Circulation at the surface, 3 m, 5 m, and the depth-averaged circulation is shown in Figure 4.34a–d, respectively. The currents are generally in the direction of the wind, with some offset in the nearshore areas, at the water surface (Figure 4.34a). Very close to the shore, the velocity turns parallel to the shore so that no flow passes through the shoreline (except at points of inflow or outflow). The current turns in the opposite direction of the wind at depth (Figure 4.34b and c). The

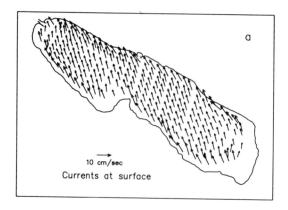

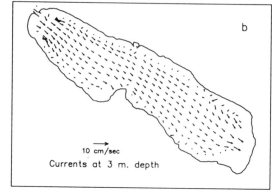

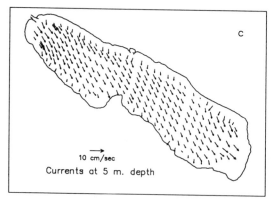

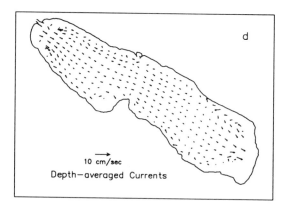

FIGURE 4.34. Predictions of circulation model for Onondaga Lake for wind/stratification conditions of August 1, 1987: (**a**) of lake surface, (**b**) at depth of 3 m, (**c**) at depth of 5 m, and (**d**) depth-averaged within the epilimnion (depth of 6 m).

depth-averaged circulation, shown in Figure 4.34d, indicates the circulation pattern is such that the depth-average flow is in the direction of the wind in the shallower shoreline areas, and is opposite to the wind direction in the middle of the lake.

The series of plots in Figure 4.35 show predictions for the same wind speed depicted in Figure 4.34, but with the wind from the southeast (direction 225°). The same general pattern of flow variation with depth is seen here (Figure 4.35a–c) as for the south-southeast wind. The influence of the shape of the lake basin is seen more strongly in the plot of depth-averaged circulation shown in Figure 4.35d. Gyres are seen to form at the two ends of the lake, and around the point on the southwest shore of the lake. Comparison of Figures 4.34d and 4.35d show that the depth averaged circulation is substantially greater

through most of the lake when winds are nearly along the major axis of the lake basin (northwest-southeast) compared to wind across the lake. Sensitivity analyses indicate that the velocity of water motion increases somewhat as the surface layer (epilimnetic) depth increases. This is expected due to the effects of internal and bottom friction.

Transport in Onondaga Lake is dominated by wind-induced motion, except under calm conditions. For example, peak streamflows (the one-in-seven year storm in June 1987; Heidtke 1989) had an imperceptible effect on the predicted circulation for the wind conditions of August 1, 1987 (see Figure 5 in Owens 1989). The influence of streamflow alone on circulation is simulated using these streamflows in combination with no wind; the depth-averaged predictions are presented in Figure 4.36. Note that the scale for the

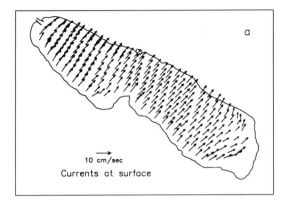

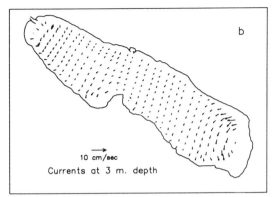

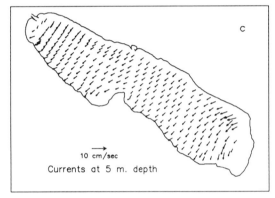

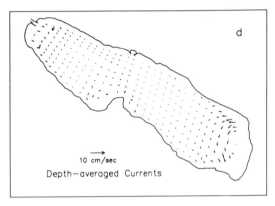

FIGURE 4.35. Predictions of circulation model for Onondaga Lake for wind (except direction of 225 degrees)/stratification conditions of August 1, 1987: (**a**) at lake surface, (**b**) at depth of 3 m, (**c**) at depth of 5 m, and (**d**) depth-averaged within the epilimnion (depth of 6 m).

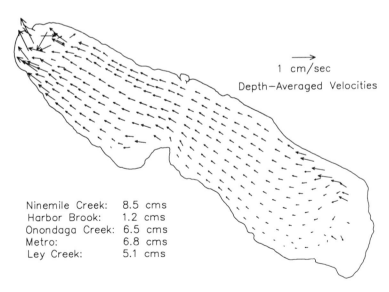

Ninemile Creek:	8.5 cms
Harbor Brook:	1.2 cms
Onondaga Creek:	6.5 cms
Metro:	6.8 cms
Ley Creek:	5.1 cms

FIGURE 4.36. Predictions of circulation model for Onondaga Lake for no wind but high runoff, epilimnion depth of 6 m; depth-averaged velocities.

velocity vectors in Figure 4.36 is one-tenth that used in Figures 4.34 and 4.35. A somewhat uniform pattern of flow up the lake, toward the outlet, is predicted under these conditions (Figure 4.36).

4.5 Modeling Stratification and Vertical Transport

4.5.1 Introduction

This section describes two models that can be used to analyze stratification and vertical transport of heat and mass in the lake. The first is a diagnostic model which can be used to determine the time and depth variation of the turbulent diffusion coefficient in the stratified layers of the lake. The term "diagnostic" is used to differentiate it from predictive models, in that it is used to compute values of the turbulent diffusion coefficient from existing in-lake measurements. In addition, a predictive stratification and vertical transport model has also been developed. This model is predictive in nature, and allows for calculation of vertical profiles of temperature, Cl^- concentration, and vertical diffusion coefficients over time for any (historical or hypothetical) set of meteorological and inflow/ outflow conditions.

Both models are based on the equations which describe the conservation of heat and mass in the lake. Based on the small densimetric Froude number computed earlier in the chapter, these equations assume that conditions are uniform in the horizontal plane, and account for only vertical and time variations. This assumption is commonly made for lakes and reservoirs (Orlob 1983; Harleman 1982). Monitoring of temperature and mass concentrations indicate that it is a good engineering assumption for Onondaga Lake (e.g., Figures 4.15 and 4.16). The heat conservation equation used in both models is

$$\frac{\partial T}{\partial t} + w\frac{\partial T}{\partial z} = \frac{1}{A}\frac{\partial}{\partial z}\left[A(D_T + K_s)\frac{\partial T}{\partial z}\right]$$

$$\frac{1}{\rho CA}\frac{\partial}{\partial z}(A\phi_s) + \frac{q_I}{A}(T_I - T) \qquad (4.29)$$

in which w = vertical velocity; z = elevation (positive upward); D_T and K_s = molecular and turbulent diffusion coefficients for heat; ϕ_s = flux of solar radiation in the water column; q_I = rate of inflow into the water column of the lake per unit vertical distance; and T_I = temperature of inflow. This equation has been widely used as the basis for thermal stratification models (Harleman 1982). Likewise, the conservation of Cl^- ion in the water column is described by

$$\frac{\partial C}{\partial t} + w\frac{\partial C}{\partial z} = \frac{1}{A}\frac{\partial}{\partial z}\left[A(D_C + K_s)\frac{\partial C}{\partial z}\right]$$

$$+ \frac{q_I}{A}(C_I - C) \qquad (4.30)$$

in which C = Cl^- concentration; D_C = molecular diffusion coefficient for Cl^-; and C_I = Cl^- concentration of inflows. The vertical velocity w(z,t) is an areally-averaged value, and thus represents the net effects of inflow and outflow at depth. It is defined by

$$w = \frac{1}{A}\int_o^z (q_I - q_o)dz \qquad (4.31)$$

where q_o = lake outflow per unit vertical distance. In order to use these equations, the turbulent diffusion coefficient must be determined. This is a common element of the following models.

4.5.2 Diagnostic Vertical Transport Model

Based on the work of Jassby and Powell (1975) and others, Equations (4.29) and (4.30) may be modified in order to allow direct calculation of the turbulent diffusion coefficient K_s. This technique has previously been applied to Onondaga Lake by Wodka et al. (1983). The analysis presented here is a revised calculation using new and different data than those used by Wodka et al. (1983), and is based on both T and Cl^- measurements.

If Equations (4.29) and (4.30) are multiplied by the plan area A and integrated over elevation from the deepest part of the lake basin at z = 0 to a general elevation z, the following two relationships for K_s result

$$K_s = \frac{1}{\dfrac{\partial T}{\partial z}}\left[\frac{1}{A}\frac{d}{dt}\left(\int_o^z A T dz\right) + \frac{\phi_s}{\rho c}\right] - D_T$$

(4.32)

$$K_s = \frac{1}{\dfrac{\partial C}{\partial z}}\left[\frac{1}{A}\frac{d}{dt}\left(\int_o^z A C dz\right)\right] - D_C \quad (4.33)$$

where it is assumed that q_I and q_o, and thus w, are zero over the entire vertical distance from the lake bottom to the position z at which these equations are applied. The quantity in the square brackets of each of these equations is the flux of temperature or Cl^- due to diffusion in the water column; dividing this flux by the local gradient yields the diffusion coefficient. Thus, this technique is known as the "flux-gradient" method. It is normally applied to estimate the turbulent diffusion coefficient at depths below the thermocline in lakes; at such depths in Onondaga Lake, light attenuation is such that ϕ_s in Equation (4.32), is negligible. In the calculations described below, the values of the molecular diffusion coefficients used are $D_T = 5.0 \times 10^{-4}\,m^2 \cdot hr^{-1}$ and $D_C = 4.5 \times 10^{-6}\,m^2 \cdot hr^{-1}$.

The values of the heat or Cl^- flux and vertical gradient may be estimated over depth and time using T and Cl^- profiles from the lake through the stratification season. The heat content, H(z,t), and mass content, M(z,t), are defined as the following integrals

$$H \equiv \int_o^z A T dz \tag{4.34}$$

$$M \equiv \int_o^z A C dz \tag{4.35}$$

Using discrete measurements of T and Cl^- concentration from profiles and bathymetric data for the lake basin, the integrals may be approximated as

$$H(z_{i+1/2}, t_k) \approx \sum_{j=1}^{j=i} T_j V_j \tag{4.36}$$

$$M(z_{i+1/2}, t_k) \approx \sum_{j=1}^{j=i} C_j V_j \tag{4.37}$$

where T_j, C_j, and V_j = temperature, Cl^- concentration, and volume of the jth layer,

respectively, where j = 1 is the bottom-most layer, and $M(z_{i+1/2}, t_k)$ refers to a calculation at the interface between the ith and (i + 1)th layers at time t_k. At a fixed depth, a linear, least-squares regression of H or M versus time t can be computed using at least two computed values of H or M. The slope obtained from this regression is an estimate of dH/dt or dM/dt; when divided by the plan area $A(z_{i+1/2})$, estimates of the temperature or mass flux (the quantities in square brackets in Equations (4.32) and (4.33)) result. This regression includes all values of H and M which are within the time interval $(t_k - \Delta t/2)$ $\leq t \leq (t_k + \Delta t/2)$. If heat transfer is occurring only by the process of diffusion, it is expected that H increases continuously with time which is associated with the warming of water below the thermocline through the stratification season. The seasonal variation of M is dependent on the sources of salinity to the lake and is not as well defined as that for temperature (H). Due to noise in the profile data, which is likely to be primarily associated with seiche activity, the computed time series of H and M in a particular year may exhibit variation which is not due to vertical diffusion of heat or mass. Consequently, a relatively long time interval Δt may be necessary to smooth out such noise. Here a time interval $\Delta t = 30$ days was used.

Estimates of the vertical gradient in T, $\partial T/\partial z$, or Cl^- concentration, $\partial C/\partial z$, can be similarly computed. Again to smooth out data noise, the temperature measurements $T(z_{i-1}, t_k)$, $T(z_i, t_k)$, $T(z_{i+1}, t_k)$, and $T(z_{i+2}, t_k)$ are regressed against depth z; the slope of the resulting straight line is then an estimate of the vertical gradient in T at position $z_{i+1/2}$ and time t_k; Cl^- gradients were estimated similarly. The values of gradients were then averaged over the same time interval Δt used in determining the average flux.

With the fluxes and vertical gradient so computed, a value of K_s ($z_{i+1/2}$, t_k) may be computed separately from T and Cl- measurements using Equations (4.32) and (4.33). These calculations have been performed for various years of historical data. However, the assumption that q_I and q_O are zero in the hypolimnion of Onondaga Lake is likely to be valid only for the period following the closure of the soda ash/chlor-alkali plant. During its operation, the intake of cooling water from the hypolimnion of the lake, and its subsequent discharge through a diffuser in the epilimnion, resulted in a downward transport of heat and mass (e.g., Cl^-) by advection. In addition, plunging inflows from either Ninemile Creek or METRO at times resulted in a direct inflow at depth in the water column. Thus certain portions of the data sets before 1987 may have systematic limitations that violate the constructs of the flux gradient approach. However, since the closure of the soda ash/chlor-alkali facility, the vertical gradients in Cl^- concentration have become sufficiently weak that calculation of K_s using Equation (5.33) is difficult. Calculation of K_s based on Cl^- profiles was attempted for several years prior to closure of the soda ash/chlor-alkali facility, but the results must be viewed with some skepticism. During all periods, individual profiles of temperature or Cl^- were removed from the calculation if that profile was not consistent with the general trend of transport. Again, such inconsistencies are likely due to vertical displacement of the profiles which occurs during an internal seiche.

Figure 4.37 shows the variation of K_s at three elevations (10, 14, 18 m) in the hypolimnion of the lake based on T measurements, for five years since closure of the soda ash/chlor-alkali manufacturer (1987–1991). The computed diffusivities are shown as the ratio K_s/D_T, or the ratio of the turbulent diffusivity to the molecular diffusity. Several general observations can be made. A consistent seasonal variation in K_s is seen, with relatively high values (10 to 100 times D_T) occurring in late April through early June, and values generally less than 10 times D_T in late June through August. Diffusivities are then generally on the increase moving into early autumn. This seasonal variation has been widely observed (Jassby and Powell 1975), and generally indicates that K_s is inversely proportional to the magnitude of stratification (surface to bottom density difference) at any

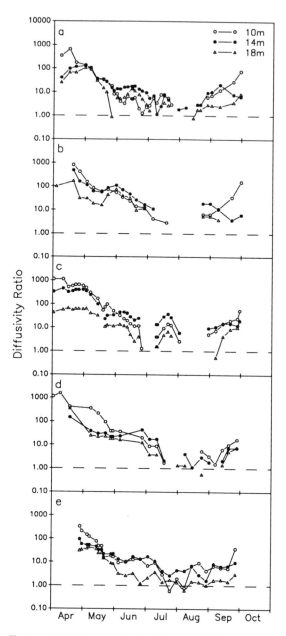

FIGURE 4.37. Temporal distributions of the vertical diffusity ratio (turbulent/molecular diffusivity), based on temperature profiles, in Onondaga Lake at three depths: (**a**) 1987, (**b**) 1988, (**c**) 1989, (**d**) 1990, and (**e**) 1991.

given time. Also, K_s is generally lowest near the lake bottom (18 m depth). As the T gradient appears in the denominator of Equation (4.32), K_s can only be computed when significant thermal stratification has been established, and when the thermocline was located above the point at which K_s is calculated. These restrictions define the portion of the year for which calculations have been performed.

For all years except 1991, diffusivity is not shown for periods in July and August. During this part of the year vertical heat (and mass) flux in the hypolimnion is relatively small because stratification is at a maximum. Thus, the slow, long-term increase in H associated with diffusion of heat is easily corrupted by the short-term variation in H associated with seiche motion. During such periods, computed values of K_s are not shown. This problem occurred in all years with the exception of 1991 (Figure 4.37e). The seasonal variation in the turbulent diffusion coefficient is due primarily to the attendant variation in the magnitude of stratification in the water column of the lake; the seasonal variation of winds has a secondary role.

Figure 4.38 shows the variation of K_s at 14 m for the years 1987, 1989, and 1991, respectively; these three years were chosen to show the year-to-year variation in K_s. Noting that the diffusivity ratio is plotted on a logarithmic axis, moderate variation in K_s is seen. The year-to-year variation in the vertical diffusion coefficient shown in Figure 4.38 can be attributed to variation in meteorological conditions.

These calculations, which are based on regular measured T profiles, indicate that turbulent mixing and associated vertical transport occurs in the hypolimnion of Onondaga Lake during the summer stratification season. Due to the expected uncertainties in the calculations associated with the relatively infrequent measurements of T profiles, accurate determination of the magnitude of the turbulent diffusion coefficient during midsummer is often not possible using the flux-gradient technique. Noting that the internal seiche period in Onondaga Lake during midsummer is roughly 18 hours, the frequency of profile measurements necessary to identify and remove the effect of the internal seiche would have to be greatly increased so that a number of profiles are measured over the period of the seiche.

Values of the ratio K_s/D_C computed from Cl^- profiles (Equation (4.33)) for depths of 14, 16, and 18 m, for the years 1980, 1981, 1985, and 1986 are presented in Figure 4.39. These four years were selected as relatively frequent Cl^- profiles were measured, with concentration measured at depth intervals of at least 1 m. Values of K_s were estimated to be relatively constant with time during 1980 at roughly 1000 times the molecular diffusity D_C (Figure 4.39a). Recall that 1980 was the last year in which most of the ionic waste input to the lake from the soda ash/chlor-alkali facility entered through Ninemile Creek. During August the computed fluxes were sufficiently low so that meaningful values of K_s could not be determined. During 1980, the computed flux of Cl^- at these depths was positive (upward), indicating that Cl^- which had accumulated in the hypolimnion during winter and spring was diffusing upward during the summer.

Calculations of K_s for 1981 and 1985, shown in Figure 4.39b and c, did not yield reasonable results. For, example, the calculations for 1981 show the highest values in July and August, when the minimum values are expected. The results are likely a manifestation of the presence of vertical advection or plunging inflows in the water column. The calculations for 1986, shown in Figure 4.39d,

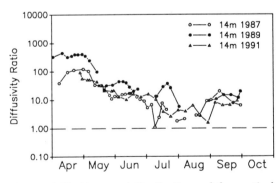

FIGURE 4.38. Temporal distributions of the vertical diffusivity ratio (turbulent/molecular diffusivity), based on temperature profiles, in Onondaga Lake at a depth of 14 m, for three years.

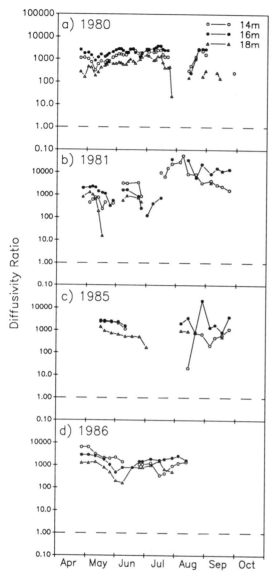

FIGURE 4.39. Temporal distributions of the vertical diffusity ratio (turbulent/molecular diffusivity), based on chloride profiles, in Onondaga Lake at three depths.

and 18 m, for the years 1980, 1981, 1985, and 1986, are presented in Figure 4.40. In this figure, the value of K_s, rather than the ratio to the molecular diffusivity, is plotted over time. If transport is dominated by the process of turbulent diffusion, then the values of K_s should be essentially the same regardless of whether T or Cl^- profiles are used. Some differences in the value of K_s computed by the two equations is to be expected, associated with noise in the data. The similarity in the distributions observed in 1980 and 1986 (Figure 4.40) support the position that vertical transport at these depths was regulated largely by turbulent diffusion (i.e., not molecular diffusion). The poor agreement in 1981 (Figure 4.40) and 1985 (Figure 4.40) probably reflects the failure of the system to meet the constraints of the calculations in those years.

4.5.3 Predictive Vertical Transport Model: Stratification Model UFILS4

Predictive stratification and vertical transport models provide a method for prediction of K_s that is not dependent on measurements of T or C, and that solve Equations (4.29) and (4.30) for T and C as a function of elevation z and time t. The mechanistic mathematical model of density stratification UFILS4 (Upstate Freshwater Institute Lake Stratification Model No. 4), was developed and calibrated for Onondaga Lake (Owens and Effler 1985) to quantify the impact of ionic waste input and the cooling water operation of the soda ash/chlor-alkali facility on the lake's stratification regime. The model is a one-dimensional integral model which is one of several types of models which describe vertical transport and stratification in lakes and reservoirs (Harleman 1982). UFILS4 evolved from CE-THERM-RI (Environmental Laboratory 1982), which was developed from other earlier (e.g., Ford and Stefan 1980, Harleman 1982) vertical transport models.

Integral models, such as UFILS4, are the most commonly used model type for studying stratification and vertical transport in lakes

are reasonable and show a value of K_s of roughly 1000 times D_C, which decreases slowly during May through mid-August. The closure of the soda ash/chlor-alkali facility earlier in the year probably eliminated the plunging phenomenon during the period of calculations in 1986.

Comparisons of values of K_s computed based on T and Cl^- profiles at depths of 14, 16,

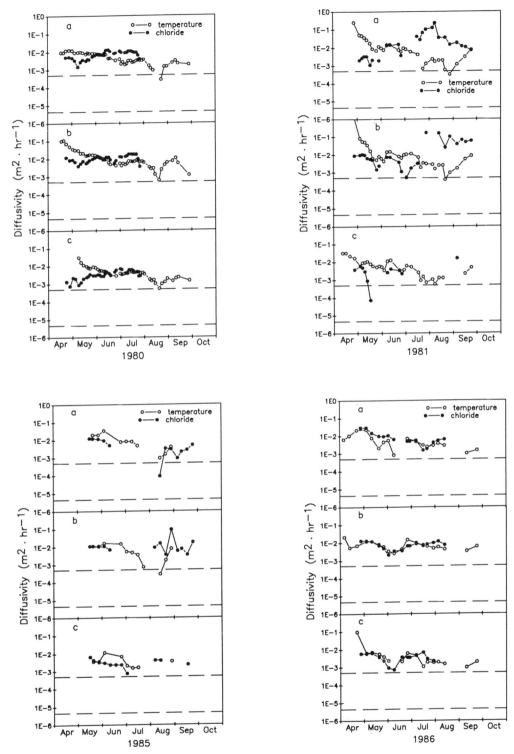

FIGURE 4.40. Four comparisons of temporal distributions of the vertical turbulent diffusion coefficient for Onondaga Lake for 1980, 1981, 1985, and 1986, based on temperature versus chloride profiles: (a) 14 m, (b) 16 m, and (c) 18 m.

and reservoirs which exhibit horizontal homogeneity; these are also used in water quality studies (Harleman 1982, Rodi 1987). The distinguishing characteristic of integral models is that it is assumed that a well-mixed surface layer of some depth exists; within this layer, temperature and all mass concentrations are assumed to be uniformly mixed. UFILS4 is different from other integral models in that the well-mixed condition is obtained in solutions by assigning a sufficiently large value of the turbulent diffusion coefficient K_s within the "mixed-layer", so that solutions of the conservation equations produce uniform T and concentration. The additional requirement of integral models is that a method for determining the depth of this well-mixed layer be specified. UFILS4 has several components or "submodels" which are described below.

4.5.3.1 Water Surface Mass and Heat Transfer

A fundamental transfer which occurs at the water surface is loss of water volume from the lake due to evaporation. In lakes, evaporation is related to the moisture content of the air, the wind conditions, and the stability of the atmosphere. Average wind conditions in central New York are sufficiently high that "force convection" due to wind dominates the "free convection" effects of a potentially unstable atmosphere. Thus, the evaporation from Onondaga Lake is controlled by winds and atmosphere moisture. The following relationship is used in the stratification model to quantify f, the rate of loss water in the lake per unit surface area $(m \cdot hr^{-1})$

$$f = \frac{k_a}{\rho_a R T_a}(e_s - e_a) \qquad (4.38)$$

in which k_a = a transfer coefficient $(m \cdot hr^{-1})$; R = universal gas constant; T_a = absolute temperature (K); e_s = saturated partial pressure of water vapor at the temperature of the water surface; and e_a = partial pressure of water vapor in the air. The transfer coefficient k_a was assumed to be a linear function of wind speed, as described by Environmental

Laboratory (1982). In Onondaga Lake, evaporation and direct precipitation make up a very small part of the water budget. Thus evaporation is only considered in the model in terms of the associated heat transfer.

The heat flux at the surface of a water body has five components, which are the net (incident minus reflected) solar (short-wave) radiation ϕ_{SO}, the net (incident minus reflected) atmosphere (long-wave) radiation ϕ_A, the (long-wave) back radiation from the water surface ϕ_B, the heat loss associated with evaporation ϕ_E, and the sensible or conductive flux ϕ_C. With the convention that upward fluxes are positive, ϕ_{SO} and ϕ_A are always negative, ϕ_B and ϕ_E are always positive, and ϕ_C may have either sign.

For use in simulating historical periods, incident solar radiation is best measured directly. Alternatively, average daily values can be estimated based on latitude, longitude, julian day, and cloud over (Environmental Laboratory 1982). For the historical simulations of Onondaga Lake, direct measurement are available for most periods. Incident atmospheric radiation is similar, but direct measurements are not available, and a predictive formula was used. The formula used (Environmental Laboratory 1982) relates incident atmospheric radiation to air temperature and moisture (or cloud cover). Back radiation from the water surface can be predicted accurately based on the assumption that a water surface radiates as a near-perfect blackbody; thus back radiation is a function of only the absolute temperature of the water surface.

The heat flux associated with evaporation is directly related to the evaporation rate by ϕ_E = $f/\rho L_V$, in which L_V = latent heat of vaporization $(kcal \cdot kg^{-1})$. The conductive heat flux ϕ_C is assumed to be driven by the same mechanism as evaporation, so that

$$\phi_C = \rho_a c_a k_a (T_S - T_A) \qquad (4.39)$$

in which c_a = specific heat of air $(kcal \cdot kg^{-1} \cdot °C^{-1})$, and T_A = air temperature (°C).

Combining the five components, ϕ_N, the net heat flux at the water surface, is defined as

$$\phi_N = \phi_{SO} + \phi_A + \phi_B + \phi_E + \phi_C \quad (4.40)$$

A portion of the solar radiation penetrates the water surface and is absorbed with depth, as represented by the solar radiation term in Equation (4.29). If a fraction β_S of ϕ_S is absorbed at the water surface, the absorption with depth can be described by

$$\phi_S = \phi_{SO}(1 - \beta_S)e^{-k_d(z_s-z)} \quad (4.41)$$

in which ϕ_S = solar radiation at elevation z (below water surface); k_d = light extinction coefficient (m^{-1}); and z_s = elevation of the water surface. The method of determination of k_d and values obtained for Onondaga Lake are presented in Chapter 7.

For given weather conditions, the heat flux at the water surface is then a function only of the water surface temperature. Thus, the boundary condition at the water surface (z = z_s) used in the solution of Equation (4.29) is

$$(D_T + K_S)\frac{\partial T}{\partial z} = -(\phi_N - \beta_S\phi_{SO}) \quad (4.42)$$

It is assumed that any heat flux at the lake bottom is quite small compared to the water surface heat flux, and can then be neglected. Thus the other boundary condition needed for the solution of Equation (4.29) is

$$(D_T + K_S)\frac{\partial T}{\partial z} = 0 \quad (4.43)$$

The weather data which are used to determine evaporation and heat flux in these models are incident solar radiation, cloud cover fraction, dry-bulb and dew point air temperatures, and wind speed. All data, with the exception of solar radiation, are routinely measured by the National Weather Service at Hancock Airport, located 8.5 km northeast of the lake. Solar radiation has been measured by UFI at various locations in close proximity to the lake.

4.5.3.2 Near-Field Mixing: Inflow and Outflow

Inflow and outflow are considered in UFILS4 in the quantities q_I, T_I, C_I, and q_o (Equations (4.29) through (4.31)). For the case of neutrally-buoyant inflows, the water from tributaries enters and mixes with epilimnetic water. The quantities T_I and C_I are set to the temperature and Cl^- concentration of an inflow as it enters the lake, and q_I is computed by distributing the total flow evenly over the depth of the epilimnion (Environmental Laboratory 1982). For a negatively buoyant (plunging) inflow, substantial near-field mixing must be considered in a one-dimensional model, which modifies the value of q_I, T_I, and C_I. UFILS4 contains an inflow mixing algorithm that accommodates plunging inflows, which is based on theory described by Hebbert et al. (1979) and Ford and Johnson (1983).

When an inflow which is more dense than the surface waters of the lake enters at the shoreline, the inflow is assumed to plunge downward along the sloping bottom of the lake. As it flows downward, it entrains water from the water column of the lake; this entrainment has two effects. First the density of the inflow is reduced by mixing with the relatively buoyant lake water. As the inflow often has a temperature that is relatively close to that of the lake surface waters, this reduction in density occurs due to entrainment of lake water which is lower in salinity. Second, the effective flow rate of the inflow increases. As the inflow continues to plunge, it either reaches an elevation where the inflow density equals that of the water column, or it flows to the bottom of the lake. In either case, it is the character of the inflow at this point, as it effectively enters the water column of the lake, where the quantities q_I, T_I and C_I are computed.

Monitoring data for the lake and its outlet, described previously in this chapter, support the assumptions made in the model that flow out of the lake occurs from the epilimnion. UFILS4 is based on the assumption that the rates of inflow and outflow are equal, so that the lake level does not change with time (see Chapter 3). The value of q_o in Equation (4.31) is computed using the selective withdrawal algorithm developed by the U.S. Army Corps of Engineers (Environmental Laboratory 1982).

4.5.3.3 Mixing in the Epilimnion

All integral models base the determination of the mixed-layer depth on an equation which describes a turbulent kinetic energy budget for the layer. Components of this budget include production of turbulent kinetic energy by wind stress and associated velocity shear in the layer, production or dissipation of turbulent kinetic energy by unstable or stable stratification in the mixed layer, viscous dissipation and other losses of turbulent kinetic energy, and conversion of turbulent kinetic energy to potential energy through growth of the layer. UFILS4 uses the particular budget expression proposed by Bloss and Harleman (1979), which can be expressed as

$$C_T \frac{\sigma^2}{h} \frac{dh}{dt} = C_F \frac{\sigma^3}{h} - \frac{\Delta\rho}{\rho} g \frac{dh}{dt} - C_D \sigma^2 \sqrt{\frac{\Delta\rho g}{\rho h}}$$

(4.44)

in which σ = a velocity scale which accounts for the net input of turbulent kinetic energy due to wind stress and stratification; h = mixed layer depth; $\Delta\rho$ = density difference which exists at the base of the mixed layer; and C_T, C_F, and C_D are dimensionless coefficients whose values were determined from various experiments by Bloss and Harleman (1979). Each term of this equation has dimensions of rate of energy input/dissipation per unit mass. This equation can be best visualized for the case of a growing mixed layer. Under such conditions, the term on the left side represents the "spin-up" of water that is entrained into the mixed layer to the level of turbulence that exists in the layer; the terms on the right side represent the net input of turbulent kinetic energy from wind and stratification, the conversion of turbulent kinetic energy to potential energy of the growing layer, and a dissipation/loss term. Rearranging this equation into a dimensionless form yields

$$\frac{1}{\sigma} \frac{dh}{dt} = \frac{C_F - C_D \sqrt{Ri}}{C_T + Ri}$$

(4.45)

in which

$$Ri \equiv \frac{\Delta\rho g h}{\rho \sigma^2}$$

(4.46)

In conjunction with the heat and Cl^- conservation equations, this equation can be used to solve for the layer depth h as a function of time.

4.5.3.4 Mixing in the Stratified Layers

Turbulent diffusion below the mixed layer is also included in UFILS4. The following simple algebraic expression for the turbulent diffusion coefficient is used

$$K_S = C_H \frac{u_*^3}{h_m} \frac{1}{N^2 + N_o^2}$$

(4.47)

in which C_H is an empirical coefficient; and N is the Brunt-Vaisala frequency, defined by

$$N^2 \equiv -\frac{g}{\rho} \frac{\partial\rho}{\partial z}$$

(4.48)

and N_o^2 = the value of N^2 associated with the compression of water with increasing depth in the water column. This is a largely empirical equation which includes all of the important processes which are known to effect turbulent diffusion in lakes, including wind stress, as quantified by u_*, and the local stratification (N^2) in the water column. In UFILS4, K_s is a function of both depth and time.

4.5.3.5 Calibration/Verification

Selected features of the simulated stratification regime of Onondaga Lake are compared to observations for the spring to fall interval of 1980 and 1981 in Figure 4.41a–h. The model simulated with good accuracy the features of the complex stratification of the lake for the stratification periods of 1980 and 1981 (Figure 4.41; Owens and Effler 1989). The substantial differences in the stratification regime of the lake observed for these two years largely reflect differences in meteorological conditions between the two years (Owens and Effler 1985). The model performed well with regard to the simulation of: (1) the dimensions of the stratified layers (Figure 4.41a and e), (2) the thermal component of stratification (Figure 4.41b and f), (3) the chemical component of stratification

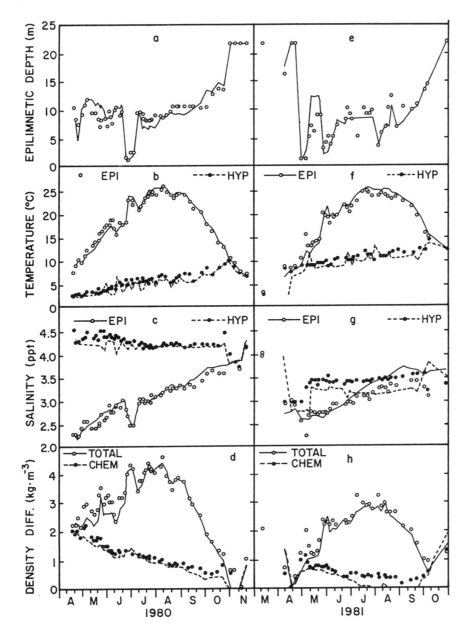

FIGURE 4.41. Model performance of selected features of stratification in Onondaga Lake: (**a**) epilimnetic depth, 1980, (**b**) epilimnetic and hypolimnetic temperature, 1980, (**c**) eplimnetic and hypolimnetic salinity, 1980, (**d**) total and chemical density difference, 1980, (**e**) eplimnetic depth, 1981, (**f**) epilimnetic and hypolimnetic temperature, 1981, (**g**) eplimnetic and hypolimnetic salinity, 1981, and (**h**) total and chemical density difference, 1981. Simulations as solid and dashed lines (from Owens and Effler 1989).

(Figure 4.41c and g), (4) overall density stratification (Figure 4.41d and h), and (5) the contribution of the chemical component to overall density stratification. The model also performed well in simulating the onset of stratification in spring (Figure 4.41h), the onset of complete fall turnover (Figure 4.41a), and the redevelopment of stratification as a result of the plunging of ionic waste (Figure 4.41d). The better performance of the model for the thermal component of stratification than the chemical component (Figure 4.41) is to expected, based on the various shortcomings in existing theory to describe plunging inflows (French 1985). The model performed somewhat better in 1980 than 1981 with regard to the chemical component of stratification (Figure 4.41c and g).

4.5.3.6 Meteorological Data

Meteorological data available from the Syracuse, N.Y., NWS station (Hancock Airport), located 8.5 km northeast of Onondaga Lake, are appropriate for specification of meteorological forcing conditions for the lake (Owens and Effler 1985). Thirty years of data, from the period 1945 through 1983, were available to define the substantial meteorological variability experienced at the lake.

The mean and the mean ±1 standard deviation of the 30 values of air temperatures and wind speed on each day throughout the year (presented in Figure 4.42a and b, respectively) are typical of north temperature climate (Effler et al. 1986a; Stauffer 1980). Maximum air temperature and minimum wind speeds occur, on average, in mid-to-late summer. The variability of wind speed and air temperature is greater in spring and fall (Figure 4.41a and b). Heat budget and mixing calculations in the model are particularly sensitive to these two parameters (Effler et al. 1986a).

4.5.3.7 Specification of Initial and Forcing Conditions

Three different sets of model runs were made to evaluate the impact that closure of the soda/chlor-alkali facility would have on the stratification regime of the lake. The simulations were designed to define the regime during most of the recent period of soda ash manufacture (entry of ionic waste via Ninemile Creek, designated Case 1), during the transient period following closure of the facility in 1986 (Case 2), and following the flushing of the high ionic concentrations from the lake (Case 3). The combinations of

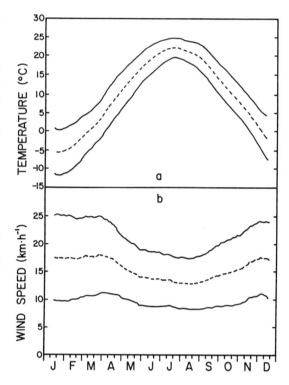

FIGURE **4.42.** Temporal distributions of mean daily meteorological conditions ±1 standard deviation: (**a**) air temperature and (**b**) wind speed (from Owens and Effler 1989).

initial stratification and inflow conditions utilized in the simulations are indicated in Table 4.5. All cases were evaluated for the entire 30-yr meteorological database to place the projected impacts in the perspective of natural meteorological variations (e.g., Effler and Owens 1989). Saline inflow from Ninemile Creek was included in the "alkali manufacturing" case (Case 1), according to the conditions documented in 1980 (Effler and Owens 1986).

Simulations were conducted for open water periods only. Simulations were initiated on the day ice-cover was lost ("ice-out") from the lake and terminated on the day of ice formation the following winter. The dates of "ice-out" for Cases 1 and 3 (Table 4.1) were estimated for each of the 30 years of weather data using empirical relationships between antecedent air temperature and date of ice-out, developed in a manner described by

TABLE **4.5.** Conditions specified for stratification model simulations.

Condition	Case 1 during alkali manufacturing	Case 2 postmanufacturing (short-term)	Case 3 postmanufacturing (long-term)
Dense inflow	Yes	No	No
Initial conditions	Estimated chemical stratification	Measured chemical stratification	No stratification
Inflows and outflows	Monthly average of historical data	Same as Case 2, but salinity at background level; cooling water flow at 80% of Case 1	Same as Case 1, but no cooling flow

Modified from Owens and Effler 1989

Ragotzkie (1978). This calculation assumes that when the average air temperature over some number of days exceeds a prescribed value, the lake will experience loss of ice at the end of that period. Based on six observations of the date of "ice-out" for Onondaga Lake (K. Stewart, personal communication), it was estimated that ice-out occurred when the air temperature for 10 consecutive days exceeded 3°C during the period of soda ash manufacturing. The average error in the predicted date of ice-out using these values was less than 3 days. For Case 3, ice-out was predicted to occur when the average air temperature for 40 consecutive days exceeds 4°C, based on data for Cazenovia Lake (30 observations; D. Hart, personal communication), a lake of similar size and morphometry that is located 29 km southeast of Onondaga Lake. The average error in the predicted date of ice-out for Cazenovia Lake using these values was approximately 6 days. The earlier date of ice-out experienced by Onondaga Lake during soda ash/chlor-alkali production was due largely to the cooling water (East Flume) discharge. The thermal discharge was assumed to operate at 1980 conditions for Case 1, at a rate reduced by 80% for Case 2, and was not in operation for Case 3.

Initial stratification conditions (just before ice-out) were dissimilar for the three sets of model runs because of the differences in the occurrence and character of chemical stratification (Figure 4.43). Initial chemical stratification for the soda ash manufacturing era (Figure 4.43, Case 1) represents average conditions (five different years) observed in late February to mid-March beneath ice cover (Onondaga County 1973, 1977, 1978, 1981, 1982). The stratification conditions used in Case 2 (Figure 4.43) were those measured on March 10, 1986, 10 days before ice-out. The slightly greater chemical stratification observed on this date probably reflects the dilution of the upper waters that occurred in the month that followed the shutdown of soda ash manufacturing on February 6, 1986. The lack of chemical stratification for long-term projections (Figure 4.43, Case 3) is con-

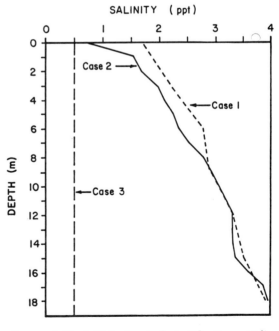

FIGURE **4.43.** Initial chemical stratification conditions for three sets (cases) of model runs (from Owens and Effler 1989).

sistent with the absence of plunging saline inflows. Conditions documented in 1990 under the ice approached Case 3, as the total salinity difference for the entire watercolumn was less than 0.2 ppt. Larger differences may develop in years with longer periods of ice-cover. The differences in initial thermal stratification conditions for the three cases were minor.

The densities of the inflows and various lake layers were determined with the equation of state developed specifically for the lake (see Chapter 5). The densities of the nonsaline inflows were specified from available monitoring data (Onondaga County 1971–1982). Other conditions specified in the model simulations of stratification include: light attenuation coefficient of $2.0\,m^{-1}$ (e.g., Effler et al. 1984), model solution and meteorological update intervals of 1 day, and a vertical simulation interval of 1 m.

4.5.3.8 Model Simulations

Simulations with the calibrated/verified stratification model for the 30-yr weather database indicate that the closure of the soda ash/chlor-alkali facility would have profound effects on the regularity of turnover and the duration of stratification in Onondaga Lake. It was predicted that the lake will turn over regularly in spring and fall following the cessation of plunging (dense ionic) inflows (Case 3); i.e., as a result of the discontinuation of the discharge of ionic waste from soda ash manufacture. A major feature of the transient case is the likelihood that spring turnover would occur in only 16 of the 30 years. The failure of Onondaga Lake to turn over in the spring of 1986 (i.e., following ice-out) has been documented (Effler and Perkins 1987). Fall turnover appeared likely for the transient case as the occurrence of fall turnover was predicted for 28 of the 30 years. The probability that turnover would not occur in the fall of 1986 was approximately 14% (2/14 × 100), as the model predicted that fall turnover failed to occur only in years when spring turnover did not occur (2 out of 14). Turnover occurred in Onondaga Lake in the fall of 1986, and in

the spring and fall of four successive years (1987–1990) since closure.

The simulations indicate both spring and fall mixing have not been regular occurrences in Onondaga Lake during the period of ionic waste discharge by the soda ash/chlor-alkali plant. The model predicted that spring turnover failed to occur in 6 years. Analysis of monitoring data (Onondaga County 1971–1982) support the position that spring turnover either did not occur regularly in the lake or was brief in a number of years. The reduced propensity for the occurrence of spring turnover for the transient case was a manifestation of the flushing of the upper layers with more dilute inputs in the period following closure. However, the likelihood of fall turnover was lower during the operation of the alkali facility because the propensity of the ion-enriched inflow to dive deep in the lake increased as density stratification was reduced in fall (e.g., Effler et al. 1986b).

The differences in the duration of stratification for the three cases are depicted in a probabilistic format, based on model simulations for 30 years of meteorological data, in Figure 4.44. The duration of stratification was predicted to decrease greatly after closure of the soda ash/chlor-alkali facility; on the average it is predicted that it would be approximately 45% shorter. The average duration of stratification was predicted to be approximately seven months before the closure of the facility; this is consistent with observations from that period. The average duration of stratification in the future was predicted to be about four months. Only rarely (approximately 10% probability) were meteorological conditions such that the predicted duration of stratification was less than five months during the operation of the soda ash/chlor-alkali facility. The duration of stratification following closure was predicted to be less than five months in more than 80% of the years (Figure 4.45). The average duration of stratification observed for the lake since closure (1987–1990) has been about five months. A major reduction in the period of stratification since closure of the facility has been noted. However, the degree of reduction appears to

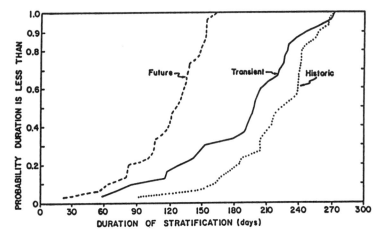

FIGURE 4.44. Probability that the duration of stratification is less than a specified number of days, for three cases (from Owens and Effler 1989).

have been somewhat overpredicted (four versus five months).

The distribution of the mean values of selected characteristics of stratification throughout the open water period of each of the three cases is shown in Figure 4.45. On any given day the plotted values are the arithmetic mean of that characteristic from the 30 simulations and may include periods of turnover or stratification. The closure of the

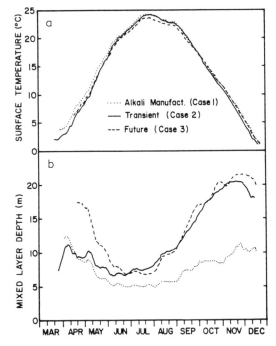

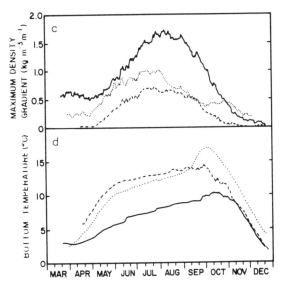

FIGURE 4.45. Model simulations of the distribution of the mean values of selected characteristics of stratification for each of three cases: (**a**) surface temperature, (**b**) mixed layer depth, (**c**) maximum density gradients, and (**d**) bottom temperature (from Owens and Effler 1989).

chemical facility was predicted not to signi-ficantly affect the T of the epilimnion of the lake (Figure 4.45a). The depth of the upper mixed layer was predicted to be less during soda ash manufacturing (Figure 4.45b) due to (1) the plunging of saline inflows to the lake bottom in spring and fall, which tended to prevent turnover, and (2) the plunging and entry of saline inflows above the thermocline in summer, which tended to act as additional barrier to the deepening of the mixed layer.

The magnitude of the maximum density gradient in the water column was predicted to be substantially greater during the transient period than that which prevailed during soda ash manufacturing (Figure 4.45c), as a result of the flushing of the upper layers with more dilute inflows. In general, the magnitude of the maximum density gradient is predicted to be lower following closure of the facility (Figure 4.45c). The approximately equal magnitudes of maximum density gradients in the historic and future cases during late summer are likely due to increased mixing induced by saline inflows entering the metal-imnion during the operation of the soda ash/chlor-alkali facility.

The low values of lake bottom water T in the transient case (Figure 4.45d) are due to failures of spring turnover and the limited vertical transport of heat brought about by large density gradients. The lower waters were rapidly heated in early fall during the period of soda ash production (Figure 4.45d) as a result of the plunging of ion enriched Ninemile Creek (warm by comparison).

The long-term impacts of the closure on two fundamental features of stratification, mixed layer depth (Figure 4.46a) and the maximum density gradient (Figure 4.46b), are presented from the perspective of the vari-ability associated with meteorological vari-ability in Figure 4.46a and b. The envelopes in Figure 4.46 represent the bounds of the mean + 1 std. dev. The broad "bands" shown reflect the important influence of meteorological variability (also see Owens and Effler 1989). They should not be confused with bands formed from more traditional sensitivity analysis of model coefficients, which are designed to reflect the level of uncertainty in model simulations associated with uncertainty in the values of the coefficients; these bands are very narrow compared to the bands of Figure 4.46. Substantial overlap is predicted for these features of stratification (Figure 4.46) because of the important influence of weather, i.e., normal annual variations in weather can cause variations in the stratifica-tion regime that at times will exceed the systematic changes caused by the closure of the soda ash/chlor-alkali plant. Thus, com-paring individual years from during and after operation of the facility could be misleading. However, on the average, the mixed layer depth will be greater (Figures 4.45 and 4.46a). The greatest similarity with historic conditions for mixed layer depth is predicted for early summer; however, the magnitude and vari-ability (Figure 4.46a) is expected to be some-what greater during this period. The maximum density gradient conditions were predicted to be similar in magnitude and variability in late summer to those experienced during soda ash manufacturing (Figure 4.46b).

The continuing residual loading of ionic waste from soda ash manufacturing (~10% of preclosure levels; Effler and Whitehead 1995), not anticipated in the analysis of Owens and Effler (1989), is causing continuing impact on the lake's stratification regime (Effler and Owens 1995). Though these impacts are in general diminished compared to preclosure conditions; e.g., clearly mani-fested under only certain meteorological conditions (Effler and Owens 1995).

4.6 Horizontal Mass Transport Model

The fecal coliform bacteria contamination pro-blem of the lake that prevails for several days following combined sewer overflow events (e.g., Canale et al. 1993) poses special chal-lenges for the supporting transport modeling, because of the shorter time scale (days versus seasonal) and the need for longitudinal re-solution. Unlike other water quality param-

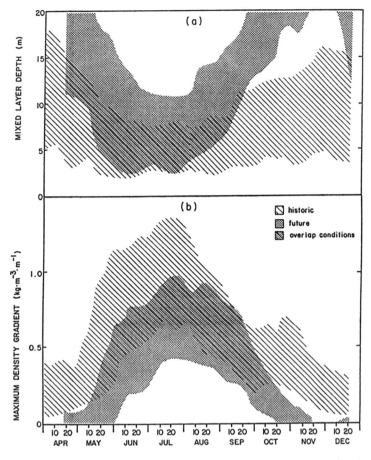

FIGURE 4.46. Model simulation of comparisons of historic and future distributions, as ±1 standard deviation envelopes, of: (a) upper mixed layer depth and (b) maximum density gradient (from Owens and Effler 1989).

eters of interest, such as phosphorus and ammonia, distinct south to north basin gradients in fecal coliform concentrations develop after the events. Two characteristics of the combined overflow events are responsible for the disparity. First, concentrations of fecal coliforms in the lake's tributaries during the events are commonly 3 to 4 orders of magnitude larger than lake concentrations prior to the event (see Chapter 3). Further, the magnitude of the contaminated inflow is relatively high, but of limited duration. By comparison, the differences in tributary and lake concentrations for other water quality constituents are small. Thus, in order to simulate the temporal and spatial distributions of fecal

coliform bacteria concentrations in response to event inputs and ambient environmental conditions (e.g., turbulent mixing and transport), a horizontal mass transport model was necessary. The circulation model described earlier in this chapter was developed for the purpose of supporting the development of such a model.

The most commonly used form of transport model is the multiple-box (or tanks-in-series) type model. The lake is spatially segmented into boxes in one, two, or three dimensions. Spatial averaging of transport characteristics in these boxes leads to the apparent transport known as dispersion. Dispersion arises form spatial variations in velocity and concentrations over the dimensions of the boxes; the magnitude of the dispersion coefficient is a function of the scale of segmentation (box size). In order to apply a multiple box transport model to a particular lake, the advective, and dispersive transport components must be specified over each of the interfaces or boundaries between adjacent boxes. The advective component is represented by a flow rate or velocity averaged over each interface. The dispersive component is determined by specifying a dispersion coefficient, or equal and opposite exchange flow, across each interface (Shanahan and Harleman 1982). A statement of mass conservation for a multiple box model is given by

$$\frac{dC_i}{dt} = \frac{1}{V_i}\left[\sum_{in}Q_{ij}C_j - \sum_{out}Q_{ij}C_i\right]$$
$$+ \sum\left[\frac{E_{ij}A_{ij}}{L_{ij}}(C_j - C_i)\right] - kC_i \quad (4.49)$$

where C_i = concentration in box i; V_i = volume of box i; Q_{ij} = flow from box i to box j, averaged over the area A_{ij}, the area of the interface between the two boxes; E_{ij} = dispersion coefficient over the interface; and L_{ij} = distance between centers of boxes; and k = a first order decay coefficient which represents settling and death of bacteria cells. The first summation on the right side of Equation (4.49) is for interfaces where the average flow direction is from adjoining box j into box i, while the second is used when the flow is out of box i into adjoining box j. In this model, the transport components that must be determined to simulate water quality are Q_{ij}, which represents advection, and the dispersion coefficient E_{ij}.

It is desirable to establish the credibility of the model independent of the water quality model, and to relate the exchanges between the boxes to quantifiable features of the hydrodynamics of the lake. The approach used here generally follows the approach of Shanahan and Harleman (1982). The flow field of Onondaga Lake was simulated with the three-dimensional circulation model described earlier; the circulation model output was then used to rigorously determine advec-

tion and dispersion for the multiple-box transport model.

4.6.1 Segmentation and Specification of Advection

The spatial segmentation of the transport model is illustrated in Figures 4.47 and 4.48. The segmentation is two dimensional; there are two layers in the vertical to reflect the epilimnion and the hypolimnion. The boxes are aligned along the major axis of the lake; 8 boxes in series of approximately equal length in the epilimnion, and 3 boxes in series of about equal length in the hypolimnion. Thus the model has no lateral resolution. This configuration is consistent with the focus of water quality concerns for the lake (e.g., Canale et al. 1993) and the most important features of horizontal differences in fecal coliform concentrations. The primary concerns are with conditions in the epilimnion.

The plan view presentation of Figure 4.48 is most valuable to identify the various (4) transport processes operating in the Onondaga Lake transport model. Net flow of water across the box interfaces associated with advection is accommodated in the surface layer (epilimnion, component A of Figure 4.48), as is longitudinal dispersion (exchange flow) in the epilimnion (component B) and the hypolimnion (component D), and vertical diffusion between the epilimnion and hypolimnion (component C).

The average flow of water at the interfaces between surface boxes (component A in Figure 4.48) can be determined directly from the inflow values. Only the major contributors of inflow volume were considered in the circulation model; these are Ninemile Creek, Harbor Brook, METRO, Onondaga Creek, and Ley Creek. Table 4.6 defines the calculation of Q_{ij}, the advective flow of water from box i to box j in the fecal coliform model (see also Canale et al. 1993).

4.6.2 Calculation of Dispersion

Dispersion results from transport being averaged over a finite area within the frame-

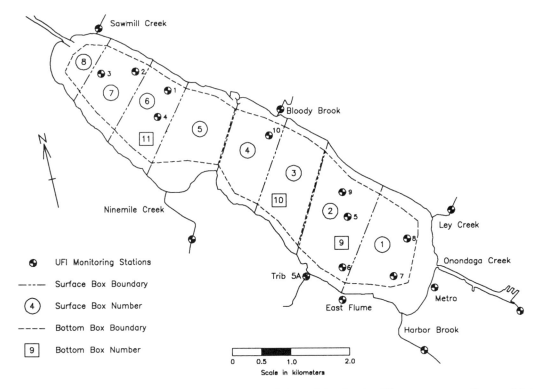

FIGURE 4.47. Spatial segmentation of transport model used for fecal coliform water quality model for Onondaga Lake Monitoring stations for water quality model shown.

work of the model (the interfacial areas between adjoining boxes), where the velocity distribution is nonuniform. Using techniques described by Csanady (1973), Fischer et al. (1979), and Shanahan and Harleman (1982), the value of the dispersion coefficient over finite areas can be computed if the velocity distribution and diffusion coefficient are known. The circulation model predicts these quantities, and thus the dispersion coefficient

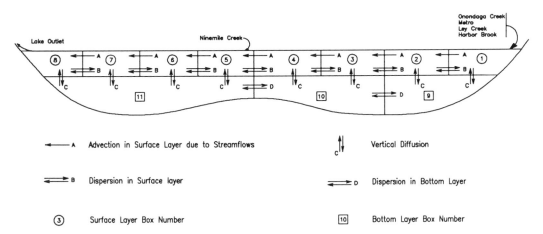

FIGURE 4.48. Plan view of spatial segmentation of transport model used for fecal coliform water quality model for Onondaga Lake. Transport processes shown.

TABLE 4.6. Equations used in calculations of advective transport.

From box	To box	Variable name	Formula for calculation
8	7	Q_{78}	$(Q_{harbor} + Q_{metro} + Q_{onondaga} + Q_{ley})$
7	6	Q_{67}	Q_{78}
6	5	Q_{56}	Q_{67}
5	4	Q_{45}	Q_{56}
4	3	Q_{34}	$(Q_{45} + Q_{ninemile})$
3	2	Q_{23}	Q_{34}
2	1	Q_{12}	Q_{23}

can be computed directly from the circulation predictions.

The circulation model predicts the horizontal velocities u and v continuously over depth at each node of the finite element mesh; based on the quadratic basis functions, the continuous velocity distribution can also be computed at all points along the sides of an element. Note that each boundary of the multiple box fecal coliform model is coincident with one or more sides of the finite element mesh. The calculation of dispersion coefficients is a two-step process, where the effects of first vertical and then horizontal averaging are computed. Based on the computed flow distribution along the sides of elements which make up the boundary between adjacent boxes, the dispersion coefficients in the x and y directions, associated with vertical (depth) averaging, denoted E_{xx} and E_{yy}, may be computed as

$$E_{xx} = -\frac{1}{h}\int_{-h}^{0} \dot{u}(z)\left[\int_{-h}^{z} \frac{1}{K_s(\xi)}\left[\int_{-h}^{\xi} \dot{u}(\varsigma)d\varsigma\right]d\xi\right]dz \quad (4.50)$$

$$E_{yy} = -\frac{1}{h}\int_{-h}^{0} \dot{v}(z)\left[\int_{-h}^{z} \frac{1}{K_s(\xi)}\left[\int_{-h}^{\xi} \dot{v}(\varsigma)d\varsigma\right]d\xi\right]dz \quad (4.51)$$

where $K_s(z)$ = the vertical turbulent diffusivity; ξ and ς are dummy variables for integration over depth; and $\dot{u}$ and $\dot{v}$ are the deviations from the depth-average velocities defined by

$$\dot{u}(z) \equiv \bar{u} - u(z) \quad (4.52)$$

$$\dot{v}(z) \equiv \bar{v} - v(z) \quad (4.53)$$

where $\bar{u}$ and $\bar{v}$ are the depth-averaged velocities defined in Equations (4.26) and (4.27). Consistent with the assumption of constant diffusion of momentum (the turbulent viscosity K_m) over depth, K_s is also assumed to not be a function of z. By definition, the diffusivity K_s and viscosity K_m are related by

$$\frac{K_m}{K_s} \equiv Sc \quad (4.54)$$

where Sc = the turbulent Schmidt number, which generally has a value slightly less than unity in turbulent flows. Here it is assumed that Sc = 1, so that K_s can be computed from Equation (4.25). Inspection of Equations (4.50) and (4.51) indicates that it is shear in the vertical direction (values of $\dot{u}$ and $\dot{v}$) and the vertical diffusion coefficient K_s which control the magnitude of dispersion. Due to the complex nature of the integrals in Equations (4.50) and (4.51), they were evaluated numerically using the trapezoidal rule.

The dispersion coefficients E_{zz} and E_{yy} account only for averaging in the vertical direction; in the multiple box model, averaging over the width of the interface between boxes is also used. Fischer et al. (1979) discuss the problem of averaging over a horizontal direction when considering longitudinal dispersion in a river. The same approach has been used in lake modeling by Shanahan and Harleman (1982). In the present case, the x-direction is along the long axis of the lake, so horizontal averaging is being performed in the y direction. The width-averaged, longitudinal dispersion coefficient is then given by

$$E_L = \frac{1}{A}\int_{y_w}^{y_e} E_{xx}dy - \frac{1}{A}\int_{y_w}^{y_e} \dot{U}(y)\left[\int_{y_w}^{y} \frac{1}{E_{yy}h}\left[\int_{y_w}^{\xi} \dot{U}(\varsigma)d\varsigma\right]d\xi\right]dy \quad (4.55)$$

where A = the area of the interface over which averaging is being computed; y_w and y_e = values of y which define the west and east shoreline of the lake; and $\dot{U}(y)$ = local deviation of the depth-integrated velocity U from the average over the area A. This deviation is defined as

$$\dot{U}(y) = \bar{U} - U(y) \qquad (4.56)$$

where $U \equiv \bar{u}h$ and

$$\bar{U} \equiv \int_{y_w}^{y_e} \bar{u}dy \qquad (4.57)$$

Again, Equation (4.55) was solved numerically over each interface between boxes in the epilimnion of the lake (Figure 4.48).

For any given wind conditions, the equations above can be used to compute a dispersion coefficient E_L for each interface. This coefficient may then be used in a solution of Equation (4.49) to compute the variation of fecal coliform concentration in each box over time. In order to simplify the calculation of dispersion coefficients, the following simple relationship for E_L may be used

$$E_L = C_L u_* h \qquad (4.58)$$

where C_L is a dimensionless coefficient. This same relationship is commonly used to describe longitudinal dispersion in rivers (Fischer et al. 1979). Using numerous solutions of the circulation model and calculation of the dispersion coefficient (Equation 4.55), it was determined that a value of $C_L = 5000$ was reasonably accurate in approximating the values of E_L. Use of the simple expression of Equation (4.58) in place of the more complicated relationship (Equation (4.55)) also effectively "decoupled" the circulation and fecal coliform models, which was desirable for practical purposes. The determination of the coefficient C_L is described in more detail by Owens (1989).

Horizontal transport between adjacent boxes in the hypolimnion was not computed by any rigorous method; the horizontal dispersion coefficient in the hypolimnion was simply set to one-tenth the value computed from Equation (4.58). The predicted fecal coliform concentration in the epilimnion, the only lake layer of significant interest, was rather insensitive to the value of horizontal dispersion in the hypolimnion. Vertical transport was computed using flux-gradient calculations of K_s, as described earlier. Results of simulations of the fecal coliform water quality model are presented in Chapter 9.

4.7 Summary

4.7.1 Configuration: Lake Inflows and Outflow

The lake is open to wind-generated turbulence that discourages development of longitudinal heterogeneity for most water quality constituents. The momentum of inflows is dissipated within relatively small "mixing zones" in the vicinity of the tributary mouths or outfall, because the magnitude of inflow to the lake is small relative to the volume of the lake. However, density differences between inflows and the lake caused by anthropogenic influences have resulted in significant lake-wide effects on vertical transport of heat and mass.

Ionic waste discharge generated from soda ash production has caused inflows carrying this waste to be more dense than the lake, and enter the lake as plunging inflows or "underflows". Underflows flow down the sloping bottom of the basin, mixing with ambient lake waters, and intrude into the water column at a depth where the density of the inflow equals that of the lake layer. The plunging inflow phenomenon has allowed the direct input of inflowing water and associated materials to the bottom of the lake when stratification is weak or nonexistent. Upon establishment of stratification, the underflow(s) entered to depths in the vicinity of the thermocline. As the principle pathway for the ionic waste discharge since the mid-1940s, Ninemile Creek has entered the lake routinely as an underflow. Before 1981, the salinity difference between the creek and the lake was typically 10 ppt. The associated underflow effect was ameliorated during periods of high runoff. Over the period 1981 to early 1986, both the METRO discharge and the creek entered as underflows, because a fraction of the ionic waste was diverted to the wastewater facility to affect tertiary treatment. The closure of the soda ash/chlor-alkali facility greatly ameliorated, but has not eliminated, the plunging inflow phenomenon in the lake, as ionic waste-based density differences between Ninemile Creek and the lake persist.

The discharge of warm cooling water (taken from the lake) from the soda ash/chlor-alkali facility, entered the lake as a buoyant overflow (entered across surface of more dense lake) until a submerged diffuser was installed in 1978. Ice-cover was substantially more complete in the south basin of the lake following the installation. When the deeper intakes were used, the cooling water operation resulted in net transport of water and associated materials from stratified layers to the upper waters of the lake. The summer heat income of the lake has decreased on average by about 9% since the discontinuation of the discharge of cooling water that accompanied closure of the facility.

Unusual flow conditions prevail in the lake's outlet to the Seneca River as a result of two anthropogenic effects: (1) the elevated density of the lake caused by ionic waste discharges, and (2) the nearly equal surface elevations of the lake and river associated with the use of the system for navigation and hydropower. Four flow regimes occur in the outlet: (1) flow from the lake to the river over the entire outlet channel, (2) stratified conditions in the outlet channel, where a "buoyant wedge" of river water has intruded into the outlet channel but is not flowing into the lake, (3) similar to regime 2, but the buoyant wedge extends beyond the breakwaters at the boundary to the lake, so that river water is flowing into the lake, and (4) water from the river is flowing toward the lake over the entire depth of the outlet channel. Regimes 2 and 3 are particularly prevalent during summer low flow periods. Flow conditions are highly dynamic in the outlet, often changing several times within a day.

A simple predictive two-layer, steady-state model of the flow regime in the outlet channel has been developed. The model predicts the inflow of the buoyant wedge of river water into the lake as lake outflow (i.e., lake tribu-tary flow) decreases below some critical value ($\sim 4.3\,\mathrm{m}^3 \cdot \mathrm{sec}^{-1}$). Further decrease in lake outflow is predicted to result in increased inflow from the river to the lake. The model also shows that winds from the northwest tend to increase river inflow to the lake.

4.7.2 Density Stratification: Components

Detailed transects of thermal profiles in the late 1960s, and temperature and salinity profiles from the centers of the north and south basins since the late 1960s, indicate longitudinal uniformity, supporting the "one-dimensional" (i.e., significant variations in density components only in the vertical) description for Onondaga Lake. The ionic waste discharge and the resulting plunging inflow phenomenon caused the natural thermally-based stratification regime to be modified by contributions from salinity stratification. Rather strong salinity-based stratification was initially established in early spring when the underflow(s) entered at the bottom of the lake. Late in spring, when stratification was well established, the underflow entered above the thermocline, as it could not penetrate the density gradient. As a result, the chemocline (depth of maximum density gradient due to salinity) in summer was at a depth equal to or less than the depth of the thermocline (depth of maximum density gradient due to temperature). Thus the ionic waste discharge had the effect of reducing the depth of the upper mixed layer of the lake during summer.

During the operation of the soda ash/chlor-alkali facility, the contribution of the salinity component to overall density stratification decreased progressively from spring until late summer, as a result of the seasonal progression of the thermal component of stratification and the entry of the ionic discharge into the upper layers. Rather strong differences in the magnitude of salinity-based density stratification and its seasonal dynamics were observed during the period of operation of the facility. The salinity component represented from 13 to 50% of the overall density stratification over the May through September interval for the 1968–1986 interval; the average was 30%. Salinity-based restratification developed in periods of low ambient turbulence during fall turnover as a result of the underflow phenomenon. Salinity-based stratification developed progressively under the ice. The unusually high stability of the water column at the time of ice-out required greater kinetic (e.g., wind-based) energy inputs to achieve spring turnover. As a result, the lake either failed to turnover in the spring, or experienced a very brief turnover period, in 7 of the last 19 years of the operation of the soda ash/chloralkali plant.

Monitoring conducted over the 1987–1990 interval indicates the impact of the ionic waste discharge on the stratification regime of the lake has been greatly ameliorated by the loading reductions (see Chapter 3) that accompanied closure of the soda ash/chloralkali facility. Spring turnover was observed over the 1987–1990 interval, following closure of the facility. Further, there was no significant chemical component of stratification over this interval. Thus there was a reduction in water column stability during summer. The duration of stratification was reduced to about 5 months, similar to Otisco Lake, a nearby lake of similar morphometry. Restratification during fall turnover, resulting from the continuing underflow phenomenon, is still observed, but is less strongly manifested. More recent observations in 1993 and 1994 depict lingering impacts on the lake's stratification regime from the residual ionic waste loading from soda ash manufacturing.

4.7.3 Lake Circulation and Transport Models

Predictive models of water motion and mass and heat transport in the lake have been developed. A steady-state model of water motion predicts circulation in the upper mixed layer of the lake. The model is an "Ekman-type" model, and is based on the equations of motion which neglect the nonlinear acceleration and horizontal diffusion of momentum

terms. Water motion in the model is forced by wind stress at the water surface and is modulated by Coriolis acceleration, linearized bottom friction, and internal friction characterized by a constant eddy viscosity. Based on these assumptions, the equations of motion are recast in stream function form, and were solved using the finite element method, using 8-node quadratic elements. A finite element grid, made up of 93 elements and 310 nodes, was developed for the lake. The circulation model was applied to this grid and used to simulate water motion in the lake measured according to drogue releases on 8 different days in the summer and fall of 1987. Values of three dimensionless model coefficients were determined by model calibration: (1) a drag coefficient relating measured wind speed to computed surface shear stress, (2) a coefficient relating shear stress at the lake bottom or at the thermocline to water motion at that depth, and (3) a coefficient relating the constant eddy viscosity to wind stress and surface layer depth. The model reproduced the measurements with a reasonable degree of accuracy using a single set of model coefficients. The circulation model generally predicts water motion in the direction of the wind at the surface, and opposite the wind at depth. This pattern is modified by the sloping lake bottom and the shoreline. Application of the model demonstrated transport in the lake is dominated by wind-induced motion. Tributary inflows, even at high (e.g., storm) flow levels, have little impact on predicted velocities at significant distances from the mouths of the tributaries.

Prior to closure of the soda ash/chlor-alkali facility, the operation of the deep cooling water intake and discharge resulted in a significant vertical advection downward. Since closure, turbulent diffusion is the only process by which significant vertical transport occurs in the lake. Two vertical heat and mass transport models have been developed for the lake. Both models are based on the one-dimensional mass and heat conservation equations. The diagnostic vertical transport model, based on the "flux-gradient" procedure, supports estimation of the turbulent diffusion coefficient as a function of depth and time, given measured profiles of temperature and chloride concentration over time. Temporal distributions of diffusion coefficients computed from temperature profiles measured over the spring to early fall interval of 1987–1990 showed seasonality typical of lakes in this region; the magnitude of the coefficient was generally inversely proportional to the magnitude of density stratification. Year-to-year differences in the distributions were probably driven by meteorological variability. Diffusion coefficients could be computed based on temperature and chloride profiles for some years before closure of the soda ash/chlor-alkali facility. Coefficient values obtained from the two different sets of profiles agreed well in some years, while the comparison was not good in other years. Effects of the cooling water operation may be responsible for errors in the calculations for years prior to closure.

A predictive integral, or "mixed layer," vertical transport (stratification) model has been calibrated and verified for the lake. The model considers numerous processes, including water surface heat transfer (including radiation, evaporative, and conductive components), water surface evaporation, and inflow mixing. The depth of the upper mixed layer is computed using an entrainment law based on a turbulent kinetic energy budget, and turbulent diffusion in the stratified layers is related to wind stress and the magnitude of local density stratification. The model was calibrated for the stratification regime of 1980, and verified for the conditions of 1981; two years for which substantial differences meteorological and stratification driven were observed. Site-specific coefficients incorporated in expressions relating wind speed to shear stress, and the turbulent diffusion coefficient to wind stress and local stratification, were determined in the calibration process. The model performed well in simulating the features of the complex stratification regime of the lake, including: (1) the dimensions of the stratified layers, (2) the thermal and salinity components of stratification, (3) overall density stratification, and (4) the timing of stratification, including the redevelopment of chemical stratification in fall.

The calibrated and verified model was used to evaluate the impact of the soda ash/chlor-alkali facility on stratification in the lake. The analysis accommodated the effect of variation in weather conditions, as quantified by meteorological data collected at Hancock Airport in Syracuse for 30 years during the period 1945 through 1983. Simulations were made for a period of operation of the facility (ionic waste discharge to Ninemile Creek), the transient period of 1986, and after closure of the facility, following the flushing of elevated salinity from the lake. Simulations indicated the closure would result in the more regular occurrence of spring turnover and reduction of the average duration of stratification from 7 to 4 months. Other impacts of the facility identified in the model analysis were reduced depth of the upper mixed layer, increased overall density stratification, and cooler bottom water temperatures.

A horizontal mass transport model was also developed for Onondaga Lake for the purpose of simulating the concentration of fecal coliform bacteria in the lake resulting from combined sewer overflow events. The mass transport model was based on the multiple box approach, wherein the total volume of the lake is discretized into a series of adjoining volumes. The fecal coliform model used an array of eight boxes oriented along the long axis of the lake in the epilimnion, and three similar boxes in the hypolimnion. The steady-state circulation model described in this chapter was used as the basis for computing dispersion coefficients at the interfaces between adjoining boxes in this model. The interfaces in the mass transport model were coincident with element boundaries in the circulation model. Dispersion coefficients were computed using classical Taylor–Fischer theory for dispersion, and considered the averaging of transport over the depth and width of the interface between adjoining boxes. After calculation of dispersion coefficients in the surface layer of the lake for a variety of wind speeds and directions, a simple equation for estimating dispersion was developed. This equation related the longitudinal dispersion coefficient to the shear stress at the water surface and the surface layer

4. Hydrodynamics and Transport

depth. The value of the dimensionless coefficient in this relationship was determined from the computed values of the dispersion coefficient based on circulation predictions. This simple equation for dispersion was used in the fecal coliform model described in Chapter 9.

References

Arita, M., and Jirka, G.H. 1987. Two layer model of saline wedge. I. Entrainment and interfacial friction. *J Hydr Engr ASCE* 113: 1229–1248.

ASCE Task Committee. 1988. Turbulence modeling of surface water flow and transport: Parts I through V. *J Hydr Engr ASCE* 114: 970–1073.

Bloss, S., and Harleman, D.R.F. 1979. Effect of wind mixing on thermocline formation in lakes and reservoirs. *Tech Rep.* 249, R.M. Parsons Lab, Massachusetts Institute of Technology, Cambridge, MA.

Canale, R.P., Auer, M.T., Owens, E.M., Heidtke, T.M., and Effler, S.W. 1993. Modeling fecal coliform bacteria-II. Model development and application. *Water Res.* 27: 703–714.

Cheng, R.T. 1977. Transient, three-dimensional circulation of lakes. *J Engr Mech ASCE* 103: 17–34.

Cheng, R.T., Powell, T.M., and Dillon, T.M. 1976. Numerical models of wind-driven circulation in lakes. *Appl Math Modeling.* 1: 141–159.

Csanady, G.T. 1973. Turbulent Diffusion in the Environment, D Reidel Publishers, New York.

Doerr, S.M., Effler, S.W., Whitehead, K.A., Auer, M.T., Perkins, M.G., and Heidtke, T.M. 1994. Chloride model for polluted Onondaga Lake. *Wat Res.* 28: 849–861.

Effler, S.W., Brooks, C.M., Addess, J.M., Doerr, S.M., Storey, M.L., and Wagner, B.A. 1991. Pollutants loadings from the Solvay waste beds to lower Ninemile Creek, New York. *Water Air Soil Pollut.* 55: 427–444.

Effler, S.W., and Driscoll, C.T. 1985. A chloride budget for Onondaga Lake, NY, U.S.A. *Water Soct Pollut.* 27: 29–44.

Effler, S.W., and Owens, E.M. 1986. The density of inflows to Onondaga Lake, USA, 1980 and 1981. *Water Air Soil Pollut.* 28: 105–115.

Effler, S.W., and Owens, E.M. 1995. Density stratification in Onondaga Lake: 1968–1994. *Lake Reservoir Manag.* (in review).

Effler, S.W., Owens, E.M., and Schimel, K.A. 1986b. Density stratification in ionically enriched Onondaga Lake, USA. *Water Air Soil Pollut.* 27: 247–258.

Effler, S.W., Owens, E.M., Schimel, K.A., and Dobi, J. 1986a. Weather-based variations in thermal stratification. *J Hydr Engr ASCE* 112: 159–165.

Effler, S.W., and Perkins, M.G. 1987. Failure of spring turnover in Onondaga Lake, NY (USA). *Water Air Soil Pollut.* 34: 285–291.

Effler, S.W., Perkins, M.G., Garofalo, J.E., and Roop, R. 1987. Limnological Analysis of Otisco Lake for 1986. Submitted to Onondaga County, Syracuse, NY.

Effler, S.W., Perkins, M.G., Wagner, B., and Green, H. 1989. Limnological Analysis of Otisco Lake, 1988. Submitted to the Onondaga County Water Quality Management Agency, Syracuse, NY.

Effler, S.W., Schimel, K., and Millero, F.J. 1986c. Salinity, chloride, and density relationships in ion-enriched Onondaga Lake, NY. *Water Air Soil Pollut.* 27: 169–180.

Effler, S.W., Wodka, M.C., and Field S.D. 1984. Scattering and absorption of light in Onondaga Lake. *J Environ Engr ASCE* 110: 1134–1145.

Ekman, V.W. 1905. On the influence of the Earth's rotation on ocean currents. *Arkiv Met Astr Fusik.* 2: 1–53.

Environmental Laboratory 1982. CE-QUAL-R1: A Numerical One-Dimensional Model of Reservoir Water Quality: User's Manual, *Instruction Report* E-82-1, US Army Engineer Waterways Experiment Station, Vicksburg, MS.

Field, S.D. 1980. *Nutrient-saturated growth in hypereutrophic Onondaga Lake, Syracuse, NY.* Ph.D. Thesis presented to Departmented of Civil Engineering, Syracuse University, Syracuse, NY.

Fischer, H.G., List, E.J., Koh, R.C.Y., Imberger, J., and Brooks, N.H. 1979. *Mixing in Inland and Coastal Waters.* Academic Press, New York.

Ford, D.E., and Johnson, M.C. 1983. Assessment of Reservoir Density Currents and Inflow Processes. *Technical Report E-83-7.* Environmental Laboratory. Waterway Experimental Station, Vicksburg, MS.

Ford, D.E., and Stefan, H.G. 1980. Thermal predictions using integral energy model. *J Hydr Engr ASCE* 106: 39–55.

French, R.K. 1985. Discussion of plunging flow into a reservoir. *J Hydr Engr ASCE* 111: 175–176.

Gallagher, R.H., and Chan, S.T.K. 1973. Higher-order element analysis of lake circulation. *Computers & Fluids.* 1: 119–132.

Harleman, D.R.F. 1982. Hydrothermal analysis of lakes and reservoirs. *J Hydr Engr ASCE* 108: 302–325.

Hebbert, R., Imberger, J., Loh, I., and Patterson, J. 1979. Collie River underflow into Wellington Reservoir. *J Hydr Engr ASCE* 105: 533–546.

Heidtke, T. 1989. *Onondaga Lake Loading Analysis: Fecal Coliforms.* Upstate Freshwater Institute, Syracuse, NY.

Hutchinson, G.E. 1957. *A Treatise on Limnology. Vol. I. Geography, Physics and Chemistry.* John Wiley and Sons, New York.

Imberger, J., Patterson, J., Hebbert, B., and Loh, I. 1978. Dynamics of a reservoir of medium size. *J Hydr Engr ASCE* 104: 725–743.

Jackson, D.F. 1968. Onondaga Lake, New York— An unusual algal environment, in *Algae, Man, and the Environment.* Syracuse University Press, Syracuse, NY.

Jassby, A., and Powell, T.M. 1975. Vertical patterns of eddy diffusion during stratification in Castle Lake, California. *Limnol Oceanogr.* 20: 530–543.

Liggett, J.A., and Hadjitheodorou, C. 1969. Circulation in shallow homogeneous lakes. *J Hydr Engr ASCE* 95: 609–620.

NALCO Environmental Sciences Co. 1978. *Thermal Plume in Onondaga Lake Near Allied Chemical Plant Solvay, New York.* Report to Allied Chemical Co., Solvay, NY.

Onondaga County 1971. Onondaga Lake Study. Project No. 11060, FAE 4/71, Water Quality Office, United States Environmental Protection Agency, Water Pollution Control Research Series. Circinate OH.

Onondaga County 1971–1991. *Onondaga Lake Monitoring Program.* Department of Drainage and Sanitation, Onondaga County, NY.

Orlob, G.T. 1983. One-dimensional models for simulation of water quality in lakes and reservoirs. Chap. 7 of *Mathematical Modeling of Water Quality: Streams, Lakes, and Reservoirs* G.T. Orlob, ed. John Wiley and Sons, New York.

Owens, E.M. 1989. *A Model of Transport Model of Fecal Coliform Bacteria in Onondaga Lake, New York.* Report submitted to Upstate Freshwater Institute, Syracuse, NY.

Owens, E.M., and Effler, S.W. 1985. *A Stratification model for Onondaga Lake, NY.* Central New York Regional Planning and Development Board, Syracuse, NY.

Owens, E.M., and Effler, S.W. 1989. Changes in stratification in Onondaga Lake, New York. *Water Res Bull.* 25: 587–597.

Rand, M.C. 1971. Chemical considerations, in *Onondaga Lake Study* Project No. 11060, Water Quality Office, U.S. Environmental Projection Agency, Washington, DC.

Ragotzkie, R.A. 1978. Heat budget for lakes. In: *Lakes: Chemistry, Geology, and Physics*, A. Lerman, ed. Spring-Verlag, New York, pp. 1–19.

Rodi, W. 1987. Examples of calculation methods for flow and mixing in stratified fluids. *J Geophys Res.* 92: 5305–5328.

Schijf, J.B., and Schönfeld, J.C. 1953. Theoretical considerations on the motion of salt and fresh water. *Proc IAHR Hydraulics Convention*, Minneapolis, MN.

Seger, E.S. 1980. *The fate of heavy metals in Onondaga Lake.* M.S. Thesis, Department of Civil Engineering, Clarkson College, Potsdam, NY.

Shanahan, P., and Harleman, D.R.F. 1982. Transport in lake water quality modeling. *J Environ Engr ASCE* 110: 42–57.

Spigel, R.H., and Imberger, J. 1980. The classification of mixed layer dynamics in lakes of small to medium size. *J Phys Oceanogr.* 10: 1104–1121.

Stauffer, R.E. 1980. Windpower time series above a temperate lake. *Limnol Oceanogr.* 25: 513–528.

Stauffer, R.E., and Lee, G.F. 1973. The role of thermocline migration in regulating algal blooms. In: *Modeling the Eutrophication Process*, E.J. Middlebrooks, D.H. Falkenborg, and T.E. Maloney, eds. Ann Arbor Science Co., Ann Arbor, MI, pp. 73–82.

Stefan, H., Skoglund, T., and Megard, R.O. 1976. Wind control of algae growth in eutrophic lakes. *J Envir Engr ASCE* 102: 1210–1213.

Stewart, K.M. 1971. Other Physical Considerations. In: *Onondaga Lake Study*, Water Quality Office. Environmental Protection Agency, Project No. 11060 FAE 4/71, 0-439-910. U.S. Government Printing Office, Washington DC.

Stewart, K.M. 1978. Infrared imagery and thermal transects of Onondaga Lake, NY. *Verh Internat Verin Limnol.* 20: 496–501.

Svensson, U. 1979. The structure of the turbulent Ekman layer. *Tellus* 31: 340–350.

Walker, W.W., Laible, J.P. Owens, E.M., and Effler, S.W., 1986. *Impact of an Offshore Hatchery Discharge on Phosphorus Concentrations in and around Hawkins Bay, Lake Champlain.* Prepared for the Department of Fish and Wildlife, State of Vermont.

Welander, P. 1957. Wind action on a shallow sea: Some generalizations of Ekman's theory. *Tellus* 9: 45–52.

Wetzel, R.G. 1983. *Limnology*, 2d ed. Saunders College Publishing, Philadelphia, PA.

Wodka, M.C., Effler, S.W., Driscoll, C.T., Fields. S.D, and Devan, S.P. 1983. Diffusivity-based flux of phosphorus in Onondaga Lake. *J Environ Engr ASCE* 109: 1403–1415.

Wu, J. 1969. Wind stress and surface roughness at the air-sea interface. *J Geophys Res.* 74: 444.

5
Chemistry

Steven W. Effler, Charles T. Driscoll, Susan M. Doerr, Carol M. Brooks,
Martin T. Auer, Bruce A. Wagner, Jeffrey Addess, Wei Wang,
David L. Johnson, J. Jiao, and Saul G. Dos Santos

5.1 Salinity

5.1.1 Composition

The concentrations of four major cations, Ca^{2+}, Mg^{2+}, Na^+, K^+, and three major anions, HCO_3^-, SO_4^{2-}, and Cl^-, essentially constitute the total ionic salinity of most fresh waters, as other ions make very minor contributions (Wetzel 1983). Undissociated materials, such as silicic acid, make insignificant contributions to the total dissolved solids in moderate to higher ionic strength waters. Four of the major ions, Mg^{2+}, Na^+, K^+ and Cl^-, are largely conservative (i.e., unreactive) in lake ecosystems. The salinity of most lakes is governed by the composition of inflowing waters (Wetzel 1983).

The composition of salinity in Onondaga Lake for late July of four different years is shown on a weight basis in Figure 5.1, and on a milliequivalent basis in Figure 5.2. These are volume-weighted water column concentrations based on profiles that included 7 depths (0, 3, 6, 9, 12, 15, and 18 m) and hypsographic data for the lake (Owens 1987). The relative contributions of the major solutes, Ca^{2+}, Na^+, and Cl^-, remain largely unchanged with depth. Further, only modest changes in composition occur seasonally. Composition for average world river water (Livingstone 1963; as reported by Wetzel 1983) and nearby

Otisco Lake, an alkaline hardwater system (Schaffner and Oglesby 1979), are included for further comparison. Otisco, Owasco, and Skaneateles Lakes have very similar ionic compositions (Schaffner and Oglesby 1979). Water chemistry conditions in Onondaga Lake in 1972, 1981, and 1985 (Figures 5.1a–c and 5.2a–c) were quite similar, and are generally representative of the period before closure of the soda ash/chlor-alkali facility. Sodium represents a much greater fraction of the salinity in Onondaga Lake than for the average world river water and local Otisco Lake (Figures 5.1 and 5.2). Magnesium and K^+ contributed very little to salinity in Onondaga Lake prior to closure of the chemical processing facility. Chloride represented the vast majority of anionic salinity in the lake, whereas it is a minor component in average world river water and local Otisco Lake (Figures 5.1 and 5.2). The dominance of Ca^{2+}, Na^+, and Cl^- reflects the very high loads of these materials received as waste from the production of soda ash (see Chapter 3).

The composition "pie" graphs for 1989 (Figures 5.1d and 5.2d) are generally representative of conditions since closure of the soda ash/chlor-alkali facility. The decreases in the relative contributions of Ca^{2+}, Na^+, and Cl^- by 1989 reflect the major reductions in external loading of these materials associated with the closure of the soda ash/chlor-alkali

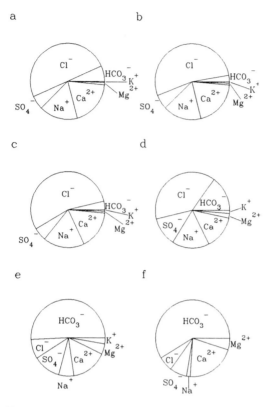

FIGURE 5.1. Composition of salinity in Onondaga Lake, for different years, and other systems: (**a**) Onondaga Lake, July 20, 1972, (**b**) Onondaga Lake, July 22, 1981, (**c**) Onondaga Lake, July 31, 1985, (**d**) Onondaga Lake, July 26, 1989, (**e**) average world river, and (**f**) Otisco Lake.

facility (see Chapter 3). For example, Cl^- decreased from representing nearly 55% of the salinity, on a weight basis, to less than 40% (Figure 5.1). The summer contributions of Ca^{2+}, Na^+, and Cl^- to salinity on a weight basis decreased from more than 85% to less than 70%.

Typical volume-weighted concentrations of the major ions in Onondaga Lake before and after closure of the soda ash/chlor-alkali plant are compared to concentrations for average world river water and Otisco Lake in Table 5.1. Note that the salinity (S, ‰) of Onondaga Lake before closure of the soda ash/chlor-alkali facility was about 13 times greater than that of Otisco Lake; it remains about 5 times greater after closure of the facility. Clearly

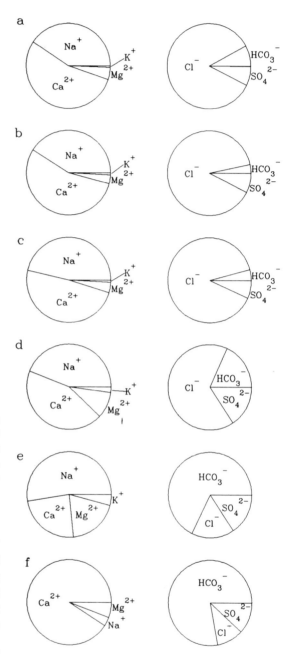

FIGURE 5.2. Equivalent fraction of cations (meg-left) and anions (meg-right) in Onondaga Lake, for different years, and other systems: (**a**) Onondaga Lake, July 20, 1972, (**b**) Onondaga Lake, July 22, 1981, (**c**) Onondaga Lake, July 31, 1985, (**d**) Onondaga Lake, July 26, 1989, (**e**) average world river, and (**f**) Otisco Lake.

TABLE 5.1. Typical concentrations for major ions in Onondaga Lake compared to concentrations in average world river water and Otisco Lake.

System	Concentrations (meq·L^{-1})							
	Ca^{2+}	Mg^{2+}	Na$^+$	K$^+$	HCO$_3^-$(CO$_3^{2-}$)	SO$_4^{2-}$	Cl$^-$	Salinity (‰)*
Onondaga Lake								
July 22, 1981	32.1	2.0	24.0	0.4	1.85	4.12	48.7	3.25
July 26, 1989	8.7	2.0	8.8	0.4	3.7	3.2	13.5	1.23
Everage world river**	0.75	0.34	0.27	0.06	0.96	0.23	0.22	0.11
Otisco Lake	2.4	0.9	0.15	—	2.65	0.35	0.4	0.25

* ‰ = ppt (parts per thousand)
** Livingstone (1963), as reported by Wetzel (1983)

Onondaga Lake was, and remains, highly enriched with Ca^{2+}, Na$^+$, and Cl$^-$; the major dissolved species of Solvay waste. The earlier Ca^{2+} concentration of the lake exceeded the average world river concentration by more than a factor of 30, and that of Otisco Lake by a factor of 10. The relative enrichment of the lake with Na$^+$ and Cl$^-$ was even greater (e.g., concentrations exceeded those of Otisco Lake by approximately a factor of 100). In contrast, the values of alkalinities are similar for Onondaga and Otisco Lakes. Major reductions in Ca^{2+}, Na$^+$, and Cl$^-$ occurred with the closure of the chemical facility (Table 5.1; discussed in more detail subsequently). The lower concentrations of Cl$^-$ observed recently in Onondaga Lake still greatly exceed the notably (Schaffner and Oglesby 1979) high level (105 mg·L^{-1}) reported for Cayuga Lake for the 1965–1970 period (Effler et al. 1989a). Approximately 70% of the elevated concentrations in Cayuga Lake was attributed to waste input from a NaCl mining facility located on the southeastern shore of the lake (Effler et al. 1989a).

The high SO$_4^{2-}$ concentrations of Onondaga Lake are largely a result of its hydrogeologic setting (see Chapters 2 and 3). These elevated concentrations allow for sulfate reduction as an important anaerobic metabolic pathway in the hypolimnion of the lake following the onset of anoxia (see later sections in this chapter). Bicarbonate/CO$_3^{2-}$ are also highly reactive species in Onondaga Lake; the cycling of these species will also be considered subsequently in this chapter.

The high concentrations of Ca^{2+}, Na$^+$, and Cl$^-$ that have prevailed in Onondaga Lake have had important implications (Effler 1987a). This enrichment has undoubtedly altered the biological diversity in the lake. Species diversity has been reported to be strongly reduced at the salinity levels that prevailed in Onondaga Lake through 1985 (Remane and Schleiper 1971). Increases in the diversity of certain populations since 1986 are documented subsequently (see Chapter 6), that are at least in part due to the reductions in salinity. Chloride is an often-used indicator of pollution that is generally considered symptomatic of a variety of cultural activities. For example increases in the Laurentian Great Lakes have been noted, largely as a result of inputs from domestic wastes, industries, and de-icing activities (Sonzogni et al. 1983). The phytoplankton assemblage is known to be influenced by Cl$^-$ concentration (Stoermer 1978). Calcium is particularly important in hard water lakes, as a significant fraction is often observed to precipitate from the upper waters as CaCO$_3$ during the productive summer months (Effler 1987b; Wetzel 1983). This process influences optics and a number of element cycles (Effler 1987b). The elevated concentrations of Ca^{2+}, Na$^+$, and Cl$^-$ in the lake and enriched tributaries have increased water density (see subsequent treatment in this section), and thereby altered the hydrodynamics of the lake (see Chapter 4). The discharge of the ion-enriched waters of Onondaga Lake has contributed importantly to related material budgets for downstream systems. For example, the mean annual release of Cl^- from the lake (7.65×10^5 metric tons) over the period 1970–1981 represented approximately 12% of the total load to Lake Ontario over that period (Effler et al. 1985a).

5.1.2 Chloride Relationships

The distributions of all three of the major ions, and S are highly correlated in Onondaga Lake (Effler et al. 1986d). It is desirable to establish Cl^- concentration as a reliable surrogate of S in the lake because of the ease of analysis and large data base for this constituent. The validation of a dynamic mass balance model for Cl^- for the lake for the 1973–1991 period is documented in Chapter 9. This model is a quantitative framework to evaluate the origins of Cl^- in the lake. The relationship between S (summation of seven ions) and Cl^- was evaluated for three years before closure of the soda ash/chlor-alkali facility (together) in Figure 5.3a, and for three years after closure in Figure 5.3b. The rather large range in ionic concentrations before closure of the facility is a manifestation of chemical stratification that commonly prevailed (see Chapter 4). A strong relationship between Cl^- and S was evident though the relationship shifted somewhat following closure as a result of changes in the contribution of Cl^- to S (Figure 5.1d). Much of the limited variability in the linear relationships (Figure 5.3a and b) can be attributed to analytical uncertainty. The relationships for the three earlier years (1972, 1981, and 1985) were essentially equivalent. Measurements for intervening years also essentially fit the relationship of Figure 5.3a,

$$S = 0.00152Cl + 0.59 \qquad (5.1)$$

in which $Cl = Cl^-$ ($mg \cdot L^{-1}$). Thus this expression is considered representative of the monitored interval before closure of the facility (1968–1985); it supplants an earlier relationship presented by Effler et al. (1986d). The two relationships (Figure 5.3a and b) are particularly valuable in describing the influence of salinity on density and density

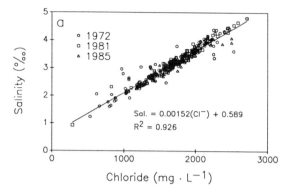

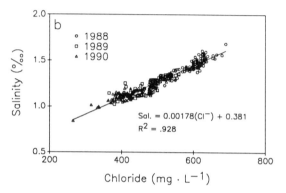

FIGURE 5.3. Evaluation of the relationship between Cl^- concentration and salinity (S) in Onondaga Lake: (**a**) before closure of soda ash/chlor-alkali facility, for three years, 1972, 1981 and 1985, and (**b**) after closure of the facility, for three years, 1988, 1989 and 1990.

stratification (Effler et al. 1986a; Owens and Effler 1989).

Ionic strength, I ($mg \cdot L^{-1}$), is a measure of the interionic effect resulting primarily from electrical attractions and repulsions between the various ions (Stumm and Morgan 1981). It is defined by

$$I = 0.5 \sum_i C_i Z_i^2 \qquad (5.2)$$

in which C_i = concentration of ionic species (anions and cations) i ($mol \cdot L^{-1}$), and Z_i = unit charge of species i. The value of I is valuable in estimating activity coefficients for ionic species (Stumm and Morgan 1981) and various thermodynamic properties of solutions (Millero 1974). The strong relationships demonstrated between Cl^- and salinity (Figure

5.3) indicate that Cl^- is probably also a good predictor of I in Onondaga Lake. The relationships between I and Cl^-, for the same ionic data used to investigate relationships between S and Cl^-, are evaluated in Figure 5.4a and b. The relationship before closure of the soda ash/chlor-alkali facility was

$$I = 3.20 \times 10^{-5}Cl + 0.0168 \quad (5.3)$$

The relationship after closure is

$$I = 3.78 \times 10^{-5}Cl + 0.0087 \quad (5.4)$$

Note Cl has units of $mg \cdot L^{-1}$. These equations represent an expedient method to estimate I in Onondaga Lake that requires only Cl^- data. Equation (5.3) supplants an expression developed earlier (Effler et al. 1986d).

5.1.3 Density Effects

Dissolved substances affect the thermodynamic properties of lake water. Even in rather dilute waters the shift in density (Chen and Millero 1977a) and related properties (Chen and Millero 1977b) from dilute conditions can be important. The following equation of state, that incorporates the pure water density–temperature relationship presented by Millero et al. (1976a) and the S functionality developed by Chen and Millero (1978), can be used to estimate the density of many fresh waters based on S and temperature data.

$$\begin{aligned}
d = &(0.99983952 + 1.6945176 \times 10^{-2}T \\
&- 7.9870401 \times 10^{-6}T^2 - 4.6170461 \\
&\times 10^{-8}T^3 + 1.0556302 \times 10^{-10}T^4 \\
&- 2.8054253 \times 10^{-13}T^5)(1 + 1.6879850 \\
&\times 10^{-2}T)^{-1} + (8.25917 \times 10^{-4} \\
&- 4.4490 \times 10^{-6}T + 1.0485 \times 10^{-7}T^2 \\
&- 1.2580 \times 10^{-9}T^3 + 3.315 \\
&\times 10^{-12}T^4)S + (-6.33761 \times 10^{-6} \\
&+ 2.8441 \times 10^{-7}T - 1.6871 \\
&\times 10^{-8}T^2 + 2.83258 \times 10^{-10}T^3)S^{3/2} \\
&+ (5.4705 \times 10^{-7} - 1.97975 \times 10^{-8}T \\
&+ 1.6641 \times 10^{-9}T^2 - 3.1203 \\
&\times 10^{-11}T^3)S^2 \quad (5.5)
\end{aligned}$$

in which T = temperature (°C).

Despite substantial differences in total mass fractions of ions between most fresh waters and the sea, the above equation has been successfully applied to rivers (Millero et al. 1976b), estuaries (Millero et al. 1976b), lakes (Millero 1975), and hypersaline lagoons (Fernandez et al. 1982). The anionic composition of Onondaga Lake is similar to that of seawater, with a Cl^- fraction much greater than typical fresh waters (Figure 5.2). However, the lake's cation composition differs greatly from both seawater and typical freshwater, with the Na^+ fraction much greater than fresh waters, but less than seawater, the Mg^{2+} fraction much less than found in either system, and the Ca^{2+} fraction much greater than found in seawater (Effler et al. 1986d).

The density of water of known ionic composition can be estimated by using the general

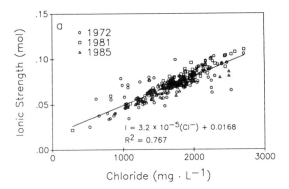

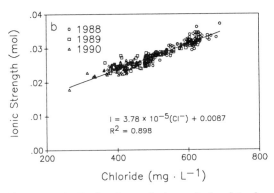

FIGURE 5.4. Evaluation of the relationship between Cl^- concentration and ionic strength (I) in Onondaga Lake: (a) before closure of soda ash/chlor-alkali facility, for three years, 1972, 1981 and 1985, and (b) after closure of the facility, for three years, 1988, 1989 and 1990.

additivity rule for apparent molal properties (Young's rule; Young 1951; Millero 1975)

$$\phi_v = \sum_{MX} E_M E_X \phi_{MX} \qquad (5.6)$$

in which ϕ_v = apparent equivalent volume of salts ($cm^3 \cdot eq^{-1}$); E_M and E_X = equivalent fractions of cation M and anion X; and ϕ_{MX} = the apparent equivalent volume of MX at the ionic strength of the mixture ($cm^3 \cdot eq^{-1}$). Binary solution data for the major sea salts are used in the calculations. Density can be calculated at 25°C using the approach of Wirth (1940) and Millero et al. (1976a), according to the equation

$$d - d_o = 10^{-3}(M - d_o\phi_v)N \qquad (5.7)$$

in which d_o = density of pure water at 25°C ($=0.9979449\,g \cdot cm^{-3}$ (Kelly 1967)); M = mean equivalent weight of lake ions = $\sum E_i M_i$ and $N = 1/2\Sigma e_i$ (eq $\cdot L^{-1}$; e_i = equivalents of ion i). The ϕ_v of the lake can be estimated by the Redlich equation (Millero 1973; Millero et al. 1976a), which accounts for ion–ion interactions

$$\phi_v = \phi_v^O + S_v I^{0.5} + b_v I \qquad (5.8)$$

in which ϕ_v^O = apparent equivalent volume at infinite dilution = $\sum E_i \phi_i^O$ (ϕ_i^O = apparent equivalent volume of ion i($cm^3 \cdot eq^{-1}$)); S_v = theoretical Debye–Hückel limiting law slope = $\Sigma E_i S_{vi}$ (S_{vi} = theoretical Debye–Hückel limiting law slope of ion i), 2.45 for Onondaga Lake; and b_v = an empirical parameter related to deviations from the limiting law = $\Sigma E_i b_{vi}$ (b_{vi} = parameter value for ion i).

Typical major ionic compositions of an epilimetic depth and a hypolimnetic depth observed in mid-summer before closure of the soda ash/chlor-alkali facility are presented in Table 5.2, along with associated calculated values of S, N, and I and measured Ts. Calculated values of ϕ_v^O, b_v, and M, for 25°C, are presented for these two compositions in Table 5.3. Values of ϕ_v of 12.419 and 12.711 were calculated for the epilimnetic and hypolimnetic depths (Table 5.2). The densities obtained from these values, according to Equation (5.7), were compared to those estimated with Equation (5.5) (i.e., equation for

TABLE 5.2. Typical major ion composition of Onondaga Lake (measured June 18, 1980).

(a) Epilimnion (5 m depth)

Ion	mg$\cdot L^{-1}$	mmol$\cdot L^{-1}$	meq$\cdot L^{-1}$
Cl^-	1660	46.9	46.9
HCO_3^-	127	2.08	2.08
SO_4^{2+}	132	1.37	2.75
Na^+	593	21.9	21.9
Ca^{2+}	564	14.1	28.2
Mg^{2+}	20	0.82	1.64

Parameter	Value
Temperature	17.4°C
S	3.01‰
N	$51.74 \times 10^{-3}\,eq \cdot L^{-1}$
I_v	$68.0 \times 10^{-3}\,mol \cdot L^{-1}$

(b) Hypolimnion (15 m depth)

Ion	mg$\cdot L^{-1}$	mmol$\cdot L^{-1}$	meq$\cdot L^{-1}$
Cl^-	2360	66.7	66.7
HCO_3^-	205	3.36	3.36
SO_4^{2+}	152	1.58	3.16
Na^+	714	31.0	31.0
Ca^{2+}	809	20.2	40.4
Mg^{2+}	20	0.82	1.64

Parameter	Value
Temperature	6.8°C
S	4.26‰
N	$73.13 \times 10^{-3}\,eq \cdot L^{-1}$
I_v	$95.7 \times 10^{-3}\,mol \cdot L^{-1}$

From Effler et al. 1986d

seawater, with dilution to a salinity equivalent to that of Onondaga Lake levels). Density determined in these two ways differed by about 110 ppm, at 25°C, largely due to the high equivalent fraction of Ca^{2+} in the lake compared to seawater (Effler et al. 1986d). A more exact equation of state, specific to the ion distribution of Onondaga Lake is developed below. This relationship has been used in the evaluation of components of density stratification in the lake (Effler et al. 1986a), and in estimating the density of ion-enriched inflows to the lake. Despite the modest shifts in ionic equivalent fractions from the lake to Ninemile Creek, and in the lake since closure of the soda ash/chlor-alkali plant (e.g., Figure 5.1), the site-specific relationship remains more accurate than the generalized expression

TABLE 5.3. Calculation of M, ϕ_v^0, b_v for Onondaga Lake at 25°C (June 18, 1980).

Ion	E_i	M_i	E_iM_i	ϕ_i	$E_i\phi_i$	b_{vi}	E_ib_{vi}
(a) Epilimnion (5 m depth)							
Ca^{2+}	0.545	20.04	10.922	−8.93	−4.867	0.242	0.132
Mg^{2+}	0.032	12.15	0.389	−10.59	−0.339	−0.197	−0.006
Na^+	0.423	22.99	9.724	−1.21	−0.512	1.078	0.456
HCO_3^-	0.040	61.02	2.441	24.29	0.972	2.122	0.085
SO_4^{2-}	0.053	48.03	2.546	6.99	0.370	0.134	0.007
Cl^-	0.907	35.45	32.153	17.83	16.172	−1.030	−0.934
			58.175		11.796		−0.260

Ion	E_i	M_i	E_iM_i	ϕ_i	$E_i\phi_i$		E_ib_{vi}
(b) Hypolimnion (15 m depth)							
Ca^{2+}	0.553		11.082		−4.938		0.134
Mg^{2+}	0.022		0.267		−0.232		−0.004
Na^+	0.424		9.748		−0.513		0.457
HCO_3^-	0.046		2.807		1.117		0.098
SO_4^{2-}	0.043		2.065		0.301		0.006
Cl^-	0.911		32.298		16.243		−0.938
			58.267		11.978		−0.247
Mean			58.223		11.887		−0.254
			±0.048		±0.09		±0.006
Error in 10^6 d							

From Effler et al. 1986d

(Equation (5.5)). Note that Equation (5.5) is adequate to address most related water quality issues in Onondaga Lake, as estimates of density differences, either with depth or between inflows and the lake, are nearly equivalent whether a generalized (Equation (5.5)) or the subsequently developed system-specific relationship is used.

The system-specific relationship for Onondaga Lake allows for the prediction of density as a function of Cl^- concentration (expressed as $g \cdot Kg^{-1}$ (instead of $mg \cdot L^{-1}$), $=Cl^-$) and T. It was developed by using the Masson Equation to estimate ϕ_{v_o} ($\phi_v = \phi_v^0 + S_vI_v^{0.5}$) and the temperature distributions of

ϕ_{vi} and S_{vi} reported by Millero (1974). The equation of state is of the form

$$d = d_o + A_vCl' + B_vCl'^{3/2} \quad (5.9)$$

in which $A_v = 10^{-3}(M − \phi_v^0d_o(N/Cl'))$; and $B_v = −10^{-3}S_vd_o(N/Cl')(I/Cl')^{0.5}$. Based on the composition data (Tables 2 and 3), $N/Cl' = 0.0318$, $I/Cl' = 0.0407$, and $M = 58.223$. The parameter A_v is related to ion-water interactions and B_v is related to ion-ion interactions of the major components (Millero 1974). The temperature distributions of d_o, ϕ_v^0, S_v, 10^3A_v, and $10^{-3}B_v$ are presented in Table 5.4. Reliable Masson values of ϕ_{vi}^0 and S_{vi} for the major ionic components are only

TABLE 5.4. Temperature distributions of D_o, ϕ_v^0, S_v, 10^3A_v, and -10^3Bv for Onondaga Lake.

Temperature (°C)	d_o ($g \cdot cm^{-3}$)	ϕ_v^0	S_v	10^3A_v	-10^3B_v
0	0.999840	8.563	3.009	15.435	0.5970
25	0.997048	12.018	1.976	14.372	0.3909
50	0.988038	12.792	1.794	14.168	0.3517

From Effler et al. 1986a

available for 0, 25, and 50°C (Millero 1974). By least squares fit

$$10^3 A_v = 1.5435 - 5.970 \times 10^{-3}t$$
$$+ 6.872 \times 10^{-5}t^2 \qquad (5.10a)$$
$$-10^3 B_v = 0.0189 - 3.662 \times 10^{-4}t$$
$$+ 4.216 \times 10^{-6}t^2 \qquad (5.10b)$$

The performance of this expression (Equation (5.9)) is substantially better than the seawater relationship for Onondaga Lake; density estimates agree well ($\Delta d = 10\,ppm$) with the more exact (Equation (5.7); but experimentally more demanding and costly) estimates. The relationship presented here supplants the one presented earlier by Effler et al. (1986d).

A fundamental characteristic of water is the temperature of maximum density (T_{MD}, °C); for pure water the value is 3.98°C. Increases in salinity and pressure cause decreases in T_{MD}, according to the expression (Chen and Millero 1977b)

$$T_{MD} = 3.9839 - 1.9911 \times 10^{-2}p - 5.822$$
$$\times 10^{-6}p^2 - (0.2230 + 1.111$$
$$\times 10^{-4}p)S \qquad (5.11)$$

in which p = pressure in bars. The value of T_{MD} is of particular interest at the end of fall turnover, before the onset of ice-cover. A typical value of S during this period, before closure of the soda ash/chlor-alkali plant, was about 3.6‰. The T_{MD} at this salinity (and at a pressure of 1 atm) is estimated to be 3.18°C, according to Equation (5.11). This value represents a substantial depression in T_{MD} of 0.80°C, which far exceeds ($\times 12$) the maximum effect of pressure (at a maximum lake depth of 19.5 m) in Onondaga Lake. The S during late fall after closure of the facility decreased to 1.3‰. This value of S, typical of present conditions, coincides with a T_{MD} depression of about 0.30°C (at a pressure of 1 atm), which is still substantially greater than the maximum effect of pressure.

5.1.4 Distributions of Salinity

5.1.4.1 Seasonal

Isopleths of S, based on detailed measurements of Cl⁻ (see Chapter 4), are presented for 1980

in Figure 5.5. The strong salinity stratification depicted was a recurring manifestation of the plunging of the dense saline waste in the lake before closure of the soda ash/chlor-alkali plant. The hydrodynamics that regulated the seasonal dynamics of this chemical stratification, interannual variations in its character, and the impact it had on the stratification regime and related features of the water quality of the lake are addressed in Chapter 4 and elsewhere in this chapter.

Seasonal variations in S in the water column (as volume-weighted value) are presented for most years in the 1968–1990 interval in Figure 5.6. Results for 1978, 1979, and 1982 are not presented because of apparent errors in the Cl⁻ data (e.g., unreasonable temporal variations in Cl⁻). Most of the abrupt changes apparent in the earlier years (e.g., Figure 5.6a–c) are probably artifacts associated with analytical limitations. Seasonal variations in S were at least in part coupled to the generally recurring seasonality in hydrologic loading (see Chapter 3), and thereby lake flushing. Reductions in S were apparent in late winter to spring for several years (e.g., Figure 5.6d, e, g, j, n), coincident with elevated tributary flows. Increases in S occurred during the summer and early fall of most years due to reduced tributary inflows during that period. The progressive decreases in S in 1986 reflect the flushing of the lake following the closure of the soda ash/chlor-alkali facility in February of that year. Progressive increases in S resumed

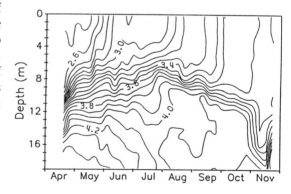

FIGURE 5.5. Isopleths of salinity (‰) in Onondaga Lake for 1980; from Cl⁻ concentration data.

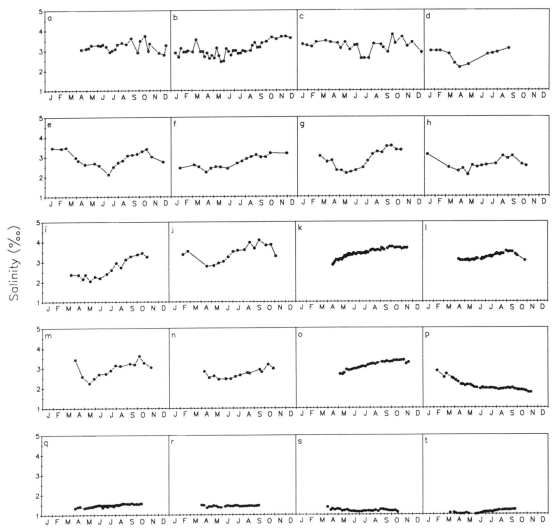

FIGURE 5.6. Seasonal variations in salinity (S; volume weighted, from profiles of Cl⁻) in Onondaga Lake for 20 years of the 1968–1990 interval (a) 1968, (b) 1969, (c) 1970, (d) 1971, (e) 1972, (f) 1973, (g) 1974, (h) 1975, (i) 1976, (j) 1977, (k) 1980, (l) 1981, (m) 1983, (n) 1984, (o) 1985, (p) 1986, (q) 1987, (r) 1988, (s) 1989, and (t) 1990.

during the summer following closure of the plant (e.g., Figure 5.6q, r, and t), as before, though these are not as clearly manifested because of the scaling limitations of Figure 5.6.

5.1.4.2 Long Term

Longer-term variations in S are manifested in the time plot of Figure 5.7; the S values pre-sented are the volume-weighted averages for the May through September interval. Chloride data in 1979 were inadequate to support the estimate of S. The average S for the monitored period before closure of the plant (1968–1985) was about 3.0‰; after closure (1987–1990) the average has decreased to 1.3‰, a reduction of more than 55%. The average S for 1990 was about 1.2‰. Rather substantial variations in S occurred during the operation of the

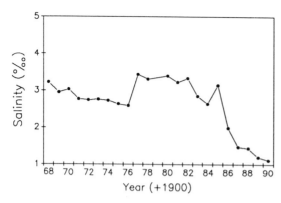

FIGURE 5.7. Long-term variations in salinity (S; volume weighted averages for the May—September interval of each year) in Onondaga Lake over the period 1968–1990.

facility. The peak value of 3.4‰, observed in 1977 and 1980, was 1.3 times greater than the minimum (2.6‰) of 1976. At least some of these variations were associated with natural variations in hydrologic loading (see Chapter 3) common to this region (also see Cl⁻ model in Chapter 9).

5.2 Dissolved Oxygen

5.2.1 Background

The concentration of dissolved oxygen (DO) is perhaps the single most important feature of water quality; it is a critical parameter which dictates the resource potential of surface waters, and it is an important regulator of chemical processes and biological activity. Oxygen is required for most forms of aquatic life. For example, certain combinations of low temperature and high DO concentrations are required for the maintenance of a cold water sport fishery. Oxygen is produced through plant photosynthesis within the depth interval of adequate light (photic zone). Oxygen is consumed mostly by microbial respiratory and organic stabilization processes. Oxygen is moderately soluble in water. The equilibrium between the liquid (freshwater) and gas phases is regulated largely by temperature, as specified by the following expression (APHA 1985).

$$\ln C_{sf} = -139.34411 + \frac{1.575701 \times 10^5}{T_K}$$
$$- \frac{6.642308 \times 10^7}{T_K^2}$$
$$+ \frac{2.243800 \times 10^{10}}{T_K^3}$$
$$- \frac{8.621949 \times 10^{11}}{T_k^4} \qquad (5.12)$$

in which C_{sf} = freshwater DO equilibrium (saturation) concentration in $mg \cdot L^{-1}$ at 1 atm, and T_k = temperature (°K; $T_k = T(°C) + 273.150$). Exchange across the air–water interface occurs when disequilibrium prevails in the surface waters (e.g., oxygen inputs by reaeration under undersaturated conditions).

Upon the onset of thermal stratification the hypolimnion becomes largely isolated from sources of DO. Oxygen is generally progressively depleted from hypolimnia through the summer stratification period because the demand for DO associated with oxidation processes exceeds the sources of DO (Wetzel 1983). In the case of severe DO depletion, anaerobic conditions can develop in the hypolimnion. These conditions encourage sediment releases and subsequent hypolimnetic accumulations of nutrients (Freedman and Canale 1977, Mortimer 1941) and reduced chemical species (e.g., Fe^{2+}, H_2S, CH_4; Mortimer 1941, 1971). The exertion of oxygen demand by the reduced species following entrainment during the fall mixing period can cause substantial oxygen depletion throughout the deepening epilimnion (e.g., Effler et al. 1988).

Oxygen demand in the hypolimnion is usually localized at the sediment water interface (parametrized as sediment oxygen demand (SOD; $g \cdot m^{-2} \cdot d^{-1}$)). The magnitude of SOD is largely a function of the deposition of organic material from the epilimnion (DiToro et al. 1990; Snodgrass 1987) and thereby the productivity of the trophogenic zone. Factors affecting the O_2 resources of the hypolimnion in addition to SOD, include the volume of the hypolimnion and the extent of vertical mixing between the epilimnion and the hypolimnion. The depletion of hypolimnetic DO is most often parametrized as the "areal hypolimnetic oxygen deficit" (AHOD; $g \cdot m^{-2} \cdot d^{-1}$). Areal hypolimnetic oxygen deficit has been used as an indicator of trophic state (Hutchinson 1938; Mortimer 1941). Mortimer (1941) proposed limits of $0.25 \, g \cdot m^{-2} \cdot d^{-1}$ for the upper limit of oligotrophy and $0.55 \, g \cdot m^{-2} \cdot d^{-1}$ for the lower limit of eutrophy.

5.2.2 Recent Seasonal and Vertical Distributions

The vertical and temporal distributions of DO in 1988 at the south deep station of Onondaga Lake are presented as isopleths in Figure 5.8a; just below it (Figure 5.8b) the attendant thermal stratification regime is described by isotherms. The stratification regime is presented because of its importance in regulating the distribution of oxygen, particularly in the lower layers. These distributions are generally representative of conditions since closure of the soda ash/chlor-alkali facility. Recall that

since 1987 density stratification in the lake has been regulated almost solely by thermal stratification (see Chapter 4). The limited vertical transport from the epilimnion that occurs across the density gradients of the metalimnion is the principal source of O_2 to the hypolimnion of the lake. Photosynthetic inputs to the hypolimnion are limited to the periods when light penetration is sufficient (see Chapter 7). Features of the stratification regime that influence the distribution of oxygen in the hypolimnion include (Owens and Effler 1989; DiToro and Connolly 1980): the timing and duration of stratification, the dimensions of the layers, the magnitude of the density gradient across the metalimnion, and the temperature of the hypolimnion. Thus the features of the oxygen resources of hypolimnia, particularly of eutrophic lakes, can be expected to vary substantially year-to-year due to natural variations in the stratification regime associated with meteorological variability (Effler et al. 1986b; Owens and Effler 1989).

Strong oxygen depletion commenced in the lowermost waters of the lake in early May of 1988 (Figure 5.8a) with the establishment of only a minor degree of stratification (Figure 5.8b). Hypolimnetic DO depletion proceeded rapidly; anoxia was first observed just above the sediments in mid-May. Oxygen concentrations in the hypolimnion always decreased with depth (until depth of anoxia was reached). This vertical distribution is widely observed in eutrophic lakes (Wetzel 1983); it reflects the localization of oxygen demanding stabilization processes at the sediment-water interface and limited mixing within the hypolimnion (Hutchinson 1957; Wetzel 1983). The depth interval of anoxia expanded in Onondaga Lake through early August to within 6 m of the surface (i.e., bottom 13 m of the watercolumn were anoxic). The oxia/anoxia boundary was within the metalimnion for most of the summer stratification period (Figure 5.8a). The upper boundary of anoxia deepened as the upper mixed layer deepened (Figure 5.8b) during the fall mixing period. The rate of deepening became very rapid just before the onset of complete fall turnover.

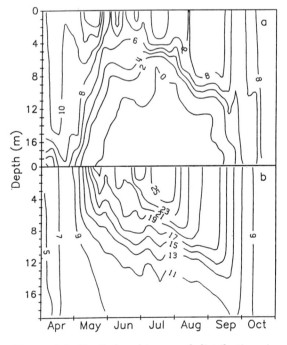

FIGURE 5.8. Vertical and temporal distributions in Onondaga Lake in 1988: (a) DO ($mg \cdot L^{-1}$), as isopleths, and (b) temperature (°C), as isotherms.

Strong temporal variations in DO concentrations occurred in the near surface waters of the lake through mid-summer (Figure 5.8a). The fine details of the DO distributions cannot be accurately resolved in the isopleths. However, the maximum concentrations (often greatly exceeding saturation) were always observed in the surface waters. These dynamics were regulated largely by phytoplankton activity. For example, the highest concentrations of DO were observed on calm days during phytoplankton blooms. Progressive depletion in DO concentrations in the upper waters (Figure 5.8a) occurred with the approach to fall turnover (Figure 5.9), because of the entrainment of hypolimnetic layers enriched in oxygen-demanding reduced chemical species (Effler et al. 1988). The DO concentration of the entire watercolumn was approximately $4\,mg \cdot L^{-1}$ at the onset of fall turnover. Concentrations subsequently increased throughout the entire water column as fall turnover proceeded, largely as a result of atmospheric reaeration (see Chapter 9). A more detailed treatment of the features of the DO distributions of 1988 and other years is described subsequently.

5.2.3 Disequilibrium in Surface Waters

The percent saturation of DO concentrations with respect to equilibrium with atmospheric

O_2 in the surface waters at the south deep station of Onondaga Lake is represented for the April through November interval of two selected years in Figure 5.9. Oxygen measurements were made at about 1000 hr in both years. Conditions observed in Otisco Lake in 1988 are presented for comparison. Clearly disequilibrium with respect to atmospheric O_2 has been common in the surface waters of Onondaga Lake (Figure 5.9). Dissolved oxygen concentrations were within 10% of equilibrium on only 10 of 59 occasions in 1980, and 9 of 36 occasions in 1988. Saturation conditions were more closely maintained in Otisco Lake; 21 of 30 observations were within 10% of atmospheric equilibrium. The strongly undersaturated conditions that developed in the fall of each year in Onondaga Lake (Figure 5.9) during the period of rapid deepening of the upper mixed layer is recurring (Effler et al. 1988), as described below. The minimum percent saturation reached in Otisco Lake during fall turnover was approximately 75%. Oxygen concentrations in Onondaga Lake tracked saturation the most closely in September of both years (Figure 5.9). Very abrupt changes in the degree of disequilibrium occurred in the April through August interval of each year, associated with abrupt changes in primary production. These can occur as a result of changes in phytoplankton biomass, and incident light. The surface waters of the lake approached or exceeded 200% saturation on numerous occasions in 1980, and on several occasions in 1988.

The distributions of equilibrium conditions with respect to DO concentrations in the surface waters of the lake are described for the April through September interval for two different periods in Figure 5.10a and b. The time segmentation about 1986/1987 is recurring in a number of data analyses presented in this work. It represents a substantial demarcation with respect to the salinity of the lake, and related features of stratification. This analysis of a larger data set (e.g., compared to Figure 5.9) supports the position that disequilibrium with respect to the concentration of DO was (Figure 5.10a), and continues to be (Figure 5.10b), the prevailing condition in the surface

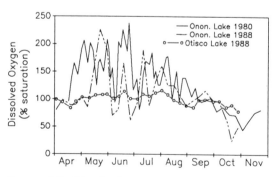

FIGURE 5.9. Dissolved oxygen equilibrium conditions in the surface waters; Onondaga Lake for two years (1980 and 1988), and Otisco Lake for 1988.

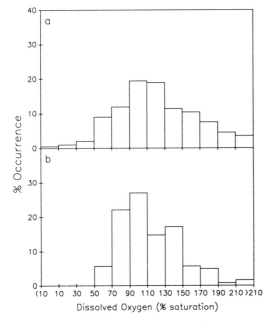

FIGURE 5.10. Distributions of DO equilibrium conditions in the surface waters of Onondaga Lake for two periods: (**a**) 1978–1981, 1985, 1986, and (**b**) 1987–1990.

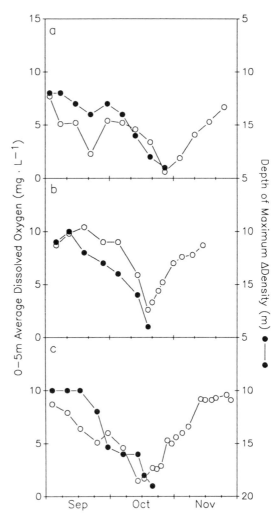

FIGURE 5.11. Depletion of DO in the upper waters of Onondaga Lake with the approach to complete fall turnover, and subsequent recovery, for three years: (**a**) 1985, (**b**) 1988, and (**c**) 1990.

waters of Onondaga Lake. Only 20% of the observations were within 10% of saturation in the five years included in the earlier interval (Figure 5.10a). An increased tendency towards equilibrium conditions is apparent since 1987. Approximately 28% of the observations in this recent interval were with 10% of saturation; 40% of the observations were more than 110% saturation in this latter interval, compared to 56% in the earlier interval. The shift in the distribution of saturation conditions (Figure 5.10a and b) is consistent with the reductions of phytoplankton standing crop since 1987 (see Chapter 6).

The features of the recurring DO depletion in the upper waters (average DO concentration in the 0 to 5 m depth interval), that accompanies the deepening of the epilimnion (represented as the depth of maximum density gradient) with the approach to fall turnover, are presented in more detail for three selected years in Figure 5.11a–c. The minima in DO concentrations occur at, or slightly before, the onset of complete turnover (Figure 5.11).

However, year-to-year differences in the details of the "oxygen sag" experienced in the upper waters (e.g., timing, rates of depletion and recovery, minimum DO concentration) are undoubtedly strongly influenced by the time course of wind-based mixing and antecedent stratification, and probably also affected by the detailed time structure of primary productivity (see Chapter 9). Thus the details of the "oxygen sag" should be expected to vary substantially year-to-year due to natural meteorological variations. For example, note the interannual differences in the time course of the approach to fall turnover for the selected 3 years (Figure 5.11). Thus great care should be taken to avoid placing too much emphasis on year-to-year differences in the fall DO minimum, as they may only reflect natural meteorological variability.

The most severe DO depletion of the three years occurred in 1985 (Figure 5.11a). The minimum concentration approached $0\,mg \cdot L^{-1}$ in late October. An earlier minimum occurred in September of that year, associated with entrainment of approximately 3 m of the hypolimnion. The subsequent increase in DO concentration in early October reflects input(s) from reaeration and probably photosynthesis. The minimum (approximately $2.5\,mg \cdot L^{-1}$) developed slightly earlier in 1988 (Figure 5.11b), but it was not as severe as in 1985 (Figure 5.11a). The minimum (approximately $1\,mg \cdot L^{-1}$) in 1990 occurred even earlier (Figure 5.11c). Oxygen recovery was rapid after turnover, particularly in 1988 and 1990.

The recurrence of the fall DO depletion in the upper waters of Onondaga Lake is presented in Figure 5.12, as a histogram of the minima observed in the 1973–1992 interval. There is some disparity in the reliability of the yearly minimum statistic in this population because of year-to-year differences in the frequency of monitoring. In 13 of the years since 1978 the frequency was at least weekly through the fall minimum period. However,

TABLE 5.5. Dissolved oxygen minima in the upper waters of Onondaga Lake during the fall mixing period, 1987–1992.

Year	Minimum DO ($mg \cdot L^{-1}$)
1987	2.7
1988	2.6
1989	3.2
1990	1.6
1991	1.9
1992	4.9

the minima obtained in most of the other years are less reliable as the monitoring frequency was twice per month or less. Regardless of the limitations of the data, clearly severe DO depletion in the upper waters of Onondaga Lake during fall turnover is a recurring problem (Figure 5.12). In 13 of the 18 years for which appropriate monitoring data exist the DO concentration decreased to concentrations less than $4\,mg \cdot L^{-1}$. A potential change in this feature of the oxygen resources of the lake since 1986 cannot be definitively resolved at this time because of the limitations of the data base (e.g., monitoring frequency) and the effects of meteorological variability. The year-to-year variability in this severe manifestation of the degraded oxygen resources of the lake is illustrated in Table 5.5 for the 1987–1992 period, over which conditions approaching phosphorus—saturated phytoplankton growth have been maintained (see Chapter 6).

5.2.4 Depletion of Oxygen in the Lower Waters

5.2.4.1 Impacts Associated with Salinity Stratification

Oxygen is depleted from the lower waters under the ice, primarily associated with the oxidation of organic matter within the sediments, as illustrated by the isopleths presented for the winter-spring interval of 1986 in Figure 5.13. Note these data support previously presented density stratification data (see Chapter 4) that indicated the lake did not turn over in the spring of this year. This failure

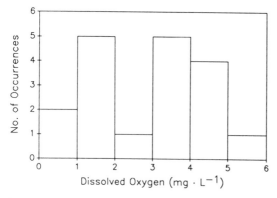

FIGURE 5.12. Distribution of annual minima DO in the epilimnion of Onondaga Lake for the period 1973–1990.

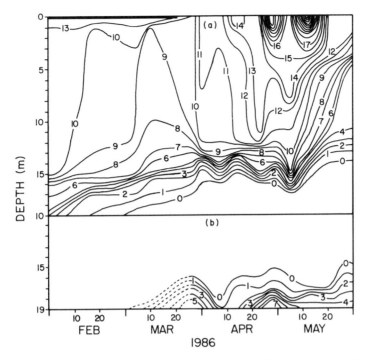

FIGURE 5.13. Isopleths for Onondaga Lake for the winter through spring interval of 1986: (a) DO, mg · L⁻¹, and H₂S, mg · L⁻¹ (from Effler and Perkins 1987).

to turn over was a result of the elevated salinity-based density stratification that developed under the ice from the operation of the soda ash/chlor-alkali facility. This was not an uncommon phenomenon during the operation of the facility (see Chapter 4). The failure of the lake to turn over in spring did not allow the lower layers to become replenished with DO. The bottom 2 to 4 m of the lake were anoxic from late March through mid-May. The short-term variations in the vertical limits of the oxic/anoxic boundary following ice-out through early May (Figure 5.13) reflect the effects of vertical mixing that causes turnover in other local lakes with similar dimensions. The failure to replenish the oxygen resources of the lower layers in spring had negative implications later in the year as the extension of the period of anoxia promoted the accumulation of more oxygen-demanding by-products of anaerobic metabolism, and thus greater demand on the oxygen resources of the upper waters during fall turnover (see subsequent sections of this chapter). This effect

may have been offset by lower temperature in the bottom waters (Owens and Effler 1989), which slows reaction rates (Bowie et al. 1985).

The importance of meteorological variability in regulating year-to-year differences in the oxygen resources of Onondaga Lake during the operation of the soda ash/chlor-alkali facility is illustrated by the disparity in density stratification and oxygen concentration in the hypolimnion of the lake for two successive years with distinctly different wind power conditions in the spring (Figures 5.14 and 5.15). Windpower (a function of wind speed (Wu 1969)) was substantially higher in April of 1981 (Figure 5.14). Turnover was observed in mid-April of this year, as evidenced in the vertically uniform profiles of both density (Figure 5.15a) and DO (Figure 5.15b). In strong contrast, substantial density (salinity-based) stratification and DO depletion within the lower stratified layers prevailed at the same time in 1980 (Figure 5.15a and b). The very low DO concentration(s) and temperature of the lower layer in 1980 indicate either spring

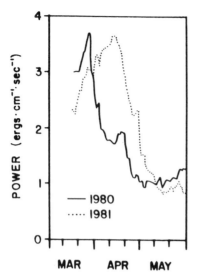

FIGURE 5.14. Comparison of estimated wind power at the surface of Onondaga Lake during the spring of 1980 and 1981 (from Effler et al. 1986c).

turnover did not occur, or was very brief (see Chapter 4).

The rapid reformation of salinity-based density stratification that occurred during the operation of the soda ash/chlor-alkali facility had negative impact on the oxygen resources of the lower layers that were subject to re-stratification. Oxygen was rapidly depleted in the relatively small restratified layer, associated with the high oxygen demand exerted at the sediment-water interface. Two examples are presented for mid-November, when lakes of this size in this region are turning over, in Figure 5.16. The first example (Figure 5.16a) is during the operation of the soda ash/chlor-alkali facility, the second is after closure of the plant (Figure 5.16b). In both cases the lake had restratified within only several days and DO was strongly depleted in the lower stratified layer. The propensity for restratification, and

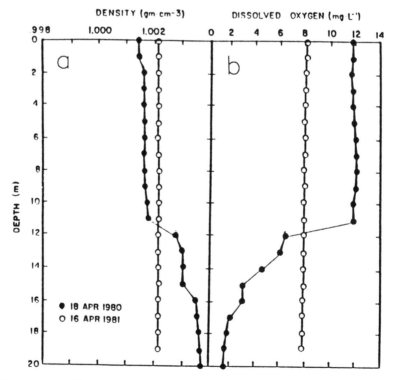

FIGURE 5.15. Comparison of density and DO stratification in Onondaga Lake in mid-April: (a) 1980, and (b) 1981 (from Effler et al. 1986c).

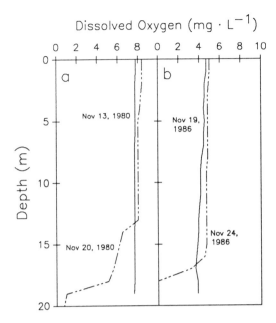

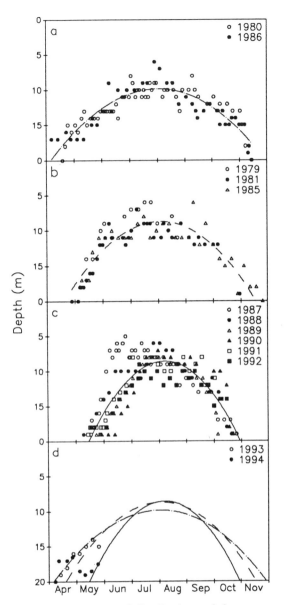

FIGURE 5.16. Examples of rapid DO depletion in the bottom waters of Onondaga Lake during periods of salinity-based reformation of density stratification: (**a**) before closure of the soda ash/chlor-alkali facility, November, 1980, and (**b**) after closure of the facility, November 1986.

its impact on lake oxygen resources, has decreased but not been eliminated (e.g., Figure 5.16b), since the closure of the facility.

The great sensitivity of the oxygen resources of the lower waters of Onondaga Lake to attendant stratification is further manifested in changes in the duration of anoxia associated with changes in the duration of stratification. Before closure, Effler (1987a) speculated that the prolonged period of stratification, associated with the saline discharge from soda ash production, exacerbated the lake's problem of limited oxygen resources by extending the period of anoxia. The comparison of the temporal distribution of the upper bound of anoxia in the lake from before (and the year of) closure to postclosure conditions presented in Figure 5.17 supports this position. The preclosure data are partitioned into the "failure of spring turnover" case (Figure 5.17a) and years in which spring turnover occurred (Figure 5.17b). Anoxia extended earlier into spring in the "failure of spring turnover" case.

FIGURE 5.17. Temporal distributions of the upper depth boundary of anoxia in Onondaga Lake: (**a**) before closure of soda ash facility, case of "failure of spring turnover", (**b**) before closure, case of occurrence of spring turnover, (**c**) after closure, 1987–1992, and (**d**) comparison of polynomial curve fits of distributions **a**–**c** and April and May observations of 1993 and 1994.

The substantial breadth of the envelopes formed by these distributions is probably mostly attributable to variations in the stratification regime driven by natural meteorological variability. The distribution for postclosure conditions for the 1987–1993 interval (Figure 5.17c) is clearly narrower than the preclosure distribution(s) (Figure 5.17a and b); polynomial best fit lines based on least squares regression of the distributions are compared in Figure 5.17d. Thus the lower layers generally remained oxygenated longer in spring and early summer and were replenished earlier in fall in the 1987–1992 interval (Figure 5.17d), consistent with the shorter period of stratification observed in those years (see Chapter 4). The possibility that reductions in oxygen demand coincident with the closure may have contributed to the observed changes in oxygen resources can not be eliminated. However, the abruptness of the change seems inconsistent with the time scale of sediment processes (e.g., Chapra and Canale 1991). Further, the reversion in the onset of anoxia in spring of 1993 and 1994, two years in which spring turnover was again not observed, to preclosure conditions (Figure 5.17d) does not support such an explanation. Observations in April of both years and May of 1993 track very closely the preclosure distributions, and fall outside the envelope of 1987–1992 observations (Figure 5.17c), because of the failure of spring turnover. The conditions documented for late May of 1994 approached the postclosure case of Figure 5.17c, because of high levels of vertical exchange over that period. The evidence is compelling that the ionic waste from soda ash production exacerbated the lake's problem of limited oxygen resources, and that lingering impacts continue because of the continued loading, albeit reduced, of this waste.

5.2.4.2 Areal Hypolimnetic Oxygen Deficit

The depletion of dissolved oxygen from the hypolimnion of Onondaga Lake is depicted for six different years in Figure 5.18a–f, as the time plot of volume-weighted DO concentrations below a depth of 10 m. Each data point is based on an oxygen profile collected at the south deep station at 1 m intervals (e.g., Effler et al. 1986c), and the hypsographic data

of Owens (1987). The AHOD was calculated (Wetzel and Likens 1991) as the product of the slope of each depletion plot ($g \cdot m^{-3} \cdot d^{-1}$) and the mean depth of the hypolimnion ($Z_h = V_h/A_h$; in which V_h = volume of the hypolimnion, and A_h = surface area of plane delimiting the upper boundary of the hypolimnion).

The AHOD could not be estimated in years in which the lake failed to turnover in spring or the turnover was brief (i.e., significant number of years during the operation of the soda ash/chlor-alkali facility), due to the occurrence of DO deficits in the spring. The high rate of DO depletion experienced in the hypolimnion of the lake (Figure 5.18a–f) confounds the reliable determination of AHOD. First, frequent measurements are required to faithfully resolve the time course of the depletion because the hypolimnion becomes anoxic so rapidly. Secondly, the depletion occurs in the early part of the stratification period when the watercolumn stability is still relatively low. During this period the lake is susceptible to interruptions in the depletion due to influxes of oxygen that can accompany mixing events. Cornett and Rigler (1980) identified the omission of system specific vertical mixing conditions as a potentially significant shortcoming of empirical AHOD models. In an earlier analysis of the hypolimnetic oxygen resources of Onondaga Lake, Effler et al. (1986c) observed that year-to-year differences in the AHOD of the lake were in part due to differences in the magnitude of vertical mixing.

Results are presented here only for years in which spring turnover was captured in the monitoring program. Depletion was progressive for the most part in the six years shown in Figure 5.18a–f. Anomolous increases in DO occurred on a single occasion in 1981 (Figure 5.18a) and 1990 (Figure 5.18d), and on two occasions in 1987 (Figure 5.19b). The increases in 1987 probably reflect the dynamics of attendant vertical mixing. The values of AHOD were determined by linear regression over the portion of the depletion curves indicated in Figure 5.18. Volume-weighted DO concentrations less than

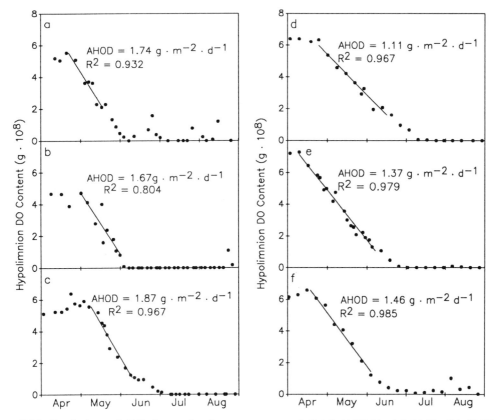

FIGURE 5.18. Depletion of DO from the hypolimnion (10 m to bottom) of Onondaga Lake for 6 years, with determinations of "areal hypolimnetic oxygen deficit" (AHOD): (a) 1981, (b) 1987, (c) 1989, (d) 1990, (e) 1991, and (f) 1992.

$2 \, mg \cdot L^{-1}$ were not included in the regressions, as there were indications that the depletion rate decreased at these lower concentrations (e.g., Figure 5.18a and c), consistent with related kinetic treatments (Bowie et al. 1985; see Chapter 9). The best estimates of AHOD for the six years appear on Figure 5.18. The average for these years is $1.54 \, g \cdot m^{-2} \cdot d^{-1}$, the range is 1.1 to $1.87 \, g \cdot m^{-2} \cdot d^{-1}$. These values are indicative of a highly eutrophic system (e.g., Mortimer 1941); they approach the highest values incorporated in data bases used to develop predictive empirical expressions for AHOD (Cornett and Rigler 1980; Walker 1979; Welch and Perkins 1979). The value determined for 1990 is distinctly lower than the other years (Figure 5.18d). As is the case for the DO minimum during the fall mixing period, no clear trend emerges from

this population of AHOD values, similarly, the year-to-year variability in this feature of the oxygen resources of the lake is attributed largely to meteorological variability.

Comparisons of the estimates of AHOD for Onondaga Lake are made to estimates available for Otisco Lake and Owasco Lake in Table 5.6. The values for Onondaga Lake are, with the exception of the 1990 estimate, substantially greater than those obtained for the other two lakes. These differences are consistent with the established differences in trophic state of the lakes (Table 5.6), though the AHOD value of Owasco Lake is greater than expected for its trophic state. The differences reported for Otisco Lake are also considered indicative of natural variations, associated with the influence of meteorological variability on the stratification regime. Note, based on the limited

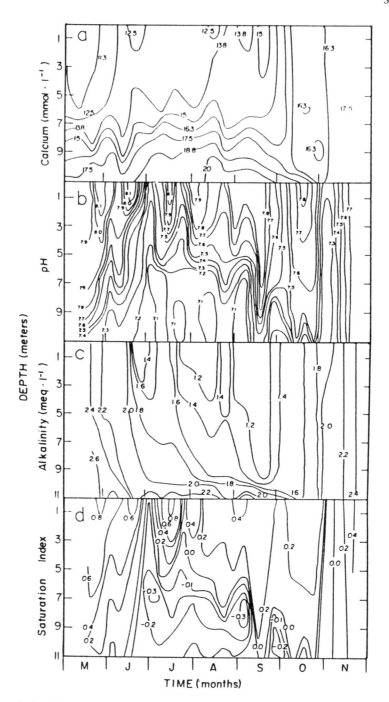

FIGURE 5.19. Isopleths of: (**a**) calcium, (**b**) pH, (**c**) alkalinity, and (**d**) calcite saturation index values, for Onondaga lake during a portion of 1980 from (Effler and Driscoll 1985).

TABLE 5.6. Comparison of AHOD values: Onondaga Lake, Otisco Lake, and Owasco Lake.

Lake	Year	AHOD $(g \cdot m^{-2} \cdot d^{-1})$	Source
Onondaga (hypereutrophic)*	1981	1.7	Figure 5.18
	1987	1.7	Figure 5.18
	1989	1.9	Figure 5.18
	1990	1.1	Figure 5.18
	1991	1.4	Figure 5.18
	1992	1.5	Figure 5.18
Otisco (mesotrophic)	1983	1.0	Effler et al. 1985c
	1986	0.8	Effler et al. 1987a
	1988	0.8	Effler et al. 1989d
Owasco (oligomesotrophic)	1987	0.62	Effler et al. 1987b

*Trophic state

populations of Table 5.6, that the relative variability reported for Otisco Lake is substantially less than observed in Onondaga Lake. Greater variability for Onondaga Lake is probably explained by the depletion of O_2 in the earlier, less stable, portion of the stratification period.

5.3 Inorganic Carbon, Ca^{2+}, $CaCO_{3(s)}$, and pH

5.3.1 Background

Inorganic carbon constitutes the major pH buffering system in fresh waters (Table 5.7). pH is an important regulator of chemical reactions and influence on aquatic biota (Stumm and Morgan 1981). The temporal and vertical distributions of pH in productive systems are mediated through the dynamics of photo synthetic consumption and respiratory/decomposition production of CO_2 (Wetzel 1983). The kinetics of the aqueous reactions are rapid (Table 5.7); i.e., equilibrium conditions among these constituents are maintained. However, the two-phase reactions are frequently observed to be in disequilibrium. The epilimnia of many hard-water lakes become oversaturated with respect to the solubility of $CaCO_3$ (calcite) during the productive summer period, as a result of increased temperature (Brunskill 1969; Strong and Eadie 1978) and the photosynthetic process (Effler 1984; Effler and Driscoll 1985; Effler et al. 1981, 1982; Otsuki and Wetzel 1974).

The precipitation and deposition of $CaCO_3$ (usually as calcite) is a widely observed phenomenon in oversaturated hard-water lakes (Effler 1987b; Kelts and Hsù 1978; Jones and Bowser 1978). This process results in the removal of aqueous Ca^{2+} and dissolved inorganic carbon (DIC = HCO_3^- + CO_3^{2-} + $H_2CO_3^*$), and often the accumulation of $CaCO_3$ in underlying sediments (Jones and Bowser 1978). Calcium carbonate has been reported to be a major component of the sediments of many hardwater lakes (e.g., Jones and Bowser 1978). Enhanced $CaCO_3$ deposition can be caused by cultural eutrophication. For example, the annual rate of epilimnetic decalcification in Gull Lake (Michigan) increased as the rate of productivity increased (Moss et al. 1983). The $CaCO_3$ precipitation phenomenon has broader importance than its effect on deposition rate, as it may influence the cycling of other constituents such as P (Otsuki and Wetzel 1972; Murphy et al. 1983; Wodka et al. 1985; see subsequent section of this chapter), DOC (Otsuki and Wetzel 1973), and other particles that serve as nuclei for precipitation (Johnson et al. 1991; see subsequent section of this chapter). Further, water clarity may decrease greatly during periods of $CaCO_3$

TABLE 5.7. Aqueous reactions involving inorganic carbon, and associated thermodynamic equilibrium constants[+].

Phase	Equilibria	Thermodynamic constants	Equation numbers
Gas−liquid	$CO_2(g) \rightleftharpoons CO_2(aq)$	$K_H = \dfrac{[CO_2(aq)]}{P_{CO_2}}$	(5.13)
Liquid	$CO_2(aq) + H_2O \rightleftharpoons H_2CO_3$	$K = \dfrac{[CO_2(aq)]}{[H_2CO_3]}$	(5.14)
Liquid	$H_2CO_3 \rightleftharpoons H^+ + HCO_3^-$	$K_{H_2CO_3} = \dfrac{[H^+][HCO_3^-]}{[H_2CO_3]}$	(5.15)
		$K_1 = \dfrac{[H^+][HCO_3^-]}{[H_2CO_3*]}$	(5.16)
Liquid	$HCO_3^- \rightleftharpoons H^+ + CO_3^{2-}$	$K_2 = \dfrac{[H^+][CO_3^{2-}]}{[HCO_3^-]}$	(5.17)
Liquid	$H_2O \rightleftharpoons H^+ + OH^-$	$K_W = [H^+][OH^-]$	(5.18)
Liquid-solid	$Ca^{2+} + CO_3^{2-} \rightleftharpoons CaCO_3$	$K_S = [Ca^{2+}][CO_3^{2-}]$	(5.19)

[+] Species $CO_2(aq)$, H_2CO_3, HCO_3^-, CO_3^{2-}, H^+, OH^-, Ca^{2+}
$[H_2CO_3^*] = [CO_2(aq)] + [H_2CO_3]$

precipitation due to increases in light scattering (Effler et al. 1987c; Effler et al. 1991b; Weidemann et al. 1985; see Chapter 7).

Here we describe findings of an on-going analysis of the distribution of pH, inorganic carbon, Ca^{2+}, and $CaCO_3$ equilibrium conditions, in the lake that spans the 1980–1990 interval. The interplay between these distributions and limnological processes is described. Changes, particularly in the epilimnion, are documented that resulted from reductions in ionic waste loading associated with the closure of the soda ash/chlor-alkali facility. Sediment trap fluxes and sedimentary chemistry data, developed in more detail in Chapter 8, are used to corroborate water column data indicating the occurrence of $CaCO_3$ precipitation/deposition and reductions in these processes since the closure.

Chemical equilibrium calculations were conducted with the model MINEQL (Westall et al. 1976). Thermodynamic data used in the analysis were obtained from the chemical equilibrium model WATEQ (Ball et al. 1980). Chemical speciation was calculated for fixed (ambient) pH conditions in which chemical precipitation was not allowed to occur. Dissolved inorganic carbon concentrations and values of the partial pressure of CO_2 (P_{CO_2})

were calculated from measured values of alkalinity and pH, with correction for temperature and ionic strength. Saturation index values (SI) were determined from these calculations, according to

$$SI = \log(Q_p/K_p) \qquad (5.20)$$

where SI = saturation index with respect to calcite solubility, Q_p = ion activity product ($\{Ca^{2+}\}\{CO_3^{2-}\}$) of the solution, and K_p = thermodynamic solubility constant for calcite ($pK_{so} = 8.475$ (Ball et al. 1980)). Positive, negative and zero SI values suggest that a solution is oversaturated, undersaturated, or in equilibrium, respectively, with respect to calcite solubility.

The principal watercolumn analytes for this study were pH, alkalinity, and Ca^{2+}. Other constituents measured that were utilized to support equilibrium model calculations included Cl^-, Na^+, Mg^{2+}, and SO_4^{2-}. Features of the supporting surficial sediment characterization and sediment trap programs are described in Chapter 8. pH profiles before 1991 were measured in the laboratory on samples collected in opaque bottles. Bottles were completely filled. Measurements commenced immediately upon return to the laboratory, generally within 1 to 2 hours of collection. Since 1991, pH profiles have also been measured *in situ* with a HYDROLAB (SURVEYOR). The accuracy of the laboratory pH probe(s) was 0.05 to 0.10 pH units; the advertised accuracy of the HYDROLAB probe was 0.2 pH units. The potential implications of systematic differences in these two types of pH measurements on calculated distributions of inorganic carbon and free ammonia in the lake are evaluated subsequently in this chapter.

5.3.2 1980

Pronounced temporal and vertical variations in Ca^{2+}, pH, alkalinity, and calcite equilibrium conditions were observed in the upper waters (11 m) of Onondaga Lake over the spring–fall interval of 1980 (Figure 5.19). The very high Ca^{2+} concentrations (Figure 5.19a) compared to other local hard-water lakes (e.g., Table 5.1; also see Effler et al. 1981, 1982; Effler 1987a) was largely a manifestation of industrial waste loading from the soda ash/chlor-alkali facility (see Chapter 3). Calcium (as well as Cl^- (Chapters 4 and 9) and Na^+ (Owens and Effler 1989)) concentrations increased significantly during the stratification period. The progressive enrichment with Ca^{2+} reflects the hydrodynamic implications (Effler et al. 1986a; Owens and Effler 1989; also see Chapter 4), as well as approximate temporal uniformity, of ionic waste loading to the lake, combined with the reduced flushing rate during the summer. Recall inputs of the high density saline waste (Effler and Owens 1986) plunged to the lower waters during periods of (vertical) isodensity, but low ambient turbulence (Owens and Effler 1989). Thereafter, the waste entered the upper layers (Owens and Effler 1989), resulting in the progressive enrichment of Ca^{2+}, thereby masking potential decreases in concentrations of Ca^{2+} associated with $CaCO_3$ precipitation.

The upper waters of the lake are alkaline (Figure 5.19b), not unlike other hard-water alkaline lakes of the central New York region (Effler et al. 1981, 1982; Effler 1987b). Solution pH values were generally highest at the lake surface and declined with increasing depth. The pH increased to maximum values

(>8.1) in late spring/early summer. Subsequently the pH declined to <7.4 in the surface waters in late June/early July and then increased again to higher values in mid-July (Figure 5.19b). The temporal and vertical features of the pH distribution track the seasonal dynamics and vertical extent of phytoplankton activity; the peak pH values coincided with phytoplankton blooms, and the minimum of late June and early July occurred during a minimum in phytoplankton biomass (Effler and Driscoll 1985). Increases in pH are the result of photosynthetic uptake of CO_2 exceeding input of CO_2, thereby causing decreases in the concentration of $H_2CO_3^*$ (Table 5.7).

A two-fold reduction in alkalinity occurred in the epilimnion during the summer period (Figure 5.19c). This major depletion of alkalinity is clear evidence (e.g., Effler et al. 1981, 1982) of the major decalcification of the lake. Similar (e.g., nearly stoichiometric) depletions of Ca^{2+} are observed in hard-water lakes where seasonal $CaCO_3$ precipitation occurs (e.g., Effler et al. 1981, 1982). However, this was not observed in Onondaga Lake because of the high waste loading of Ca^{2+}, relative to precipitation/deposition losses, that masked such a signature.

The upper waters of the lake were oversaturated with respect to the solubility of calcite (SI > 0) throughout the study period of 1980 (Figure 5.19d). The temporal and vertical patterns of SI largely tracked the pH distribution (Figure 5.19b and d), and the dynamics of phytoplankton biomass (Effler and Driscoll 1985). This coupling has been reported for other productive systems (Effler 1984; Effler et al. 1981, 1982; Kelts and Hsù 1978) and suggests regulation of the dynamics by photosynthetic (phytoplankton) activity.

Oversaturation with respect to calcite solubility is common for hard-water lakes. The maximum degree of disequilibrium with respect to the solubility of calcite in Onondaga Lake is compared to conditions documented for other hard-water systems of various trophic states in Table 5.8. Note that the maximum SI values for eutrophic/hypereutrophic systems are not much greater than some of those

TABLE 5.8. Maximum calcite saturation index (SI) values for selected lakes.

Lake	Maximum SI	Trophic state	Reference
Cazenovia, N.Y.	0.78	mesotrophic	Effler et al. 1982
Constance (Germany)	0.95	mesotrophic	Stabel 1986
Erie	0.74	mesotrophic	Strong and Eadie 1978
Green			
Fayetteville	0.95	oligotrophic	Brunskill 1969
Jamesville	0.82	mesotrophic	Effler et al. 1981
Green Bay (L. Michigan)	1.29	eutrophic	Unpublished, Effler and Auer
Michigan	0.60	oligo-mesotrophic	Strong and Eadie 1978
Ontario	0.78	mesotrophic	Strong and Eadie 1978
Onondaga (1980)	1.10	hypereutropic	(Figure 5.1d; Effler and Driscoll 1985)
Owasco, N.Y.	1.05	oligo-mesotrophic	Unpublished
Saginaw Bay	1.30	eutrophic	Effler 1984
Zurich, Switzerland	1.04	mesotrophic	Kelts and Hsù 1978

Modified from Effler and Driscoll 1985

observed for less productive systems. Further, despite the elevated concentrations of Ca^{2+}, the extent of oversaturation with respect to calcite solubility in Onondaga Lake was not high by comparison. Differences in SI values among the systems (Table 5.8) may reflect ecosystem-specific differences in the rates of calcite formation that are affected by a number of processes. Factors such as the concentration and nature of dissolved organic matter, the availability of nucleation sites, and biological activity may affect the degree of disequilibrium and rates of $CaCO_{3(s)}$ formation (Strong and Eadie 1978).

5.3.3 Analysis of 1980–1990 Interval

5.3.3.1 Changes in Concentrations and $CaCO_3$ Deposition

The marked reduction in ionic loading to the lake that occurred as a result of the closure of

the soda ash/chlor-alkali facility is depicted in the form of three year averages before (1983–1985) and after (1987–1989) closure for the three major ionic constituents in Table 5.9 (also see Chapter 3). The closure has presented an opportunity to examine the impact of the waste discharge on the in-lake concentrations of Ca^{2+} and inorganic carbon species, and on the precipitation and deposition of $CaCO_3$. Additionally, through these changes it is possible to gain insights into factors regulating DIC and Ca^{2+} chemistry, and $CaCO_3$ precipitation in hard-water lakes.

Temporal patterns in the Ca^{2+}, alkalinity, and DIC concentrations, and the ratio Ca^{2+}: DIC, for the surface waters of the lake over the 1980–1990 interval are presented in Figure 5.20a–d. Concentrations of Ca^{2+} were high ($9-17\,mmol \cdot L^{-1}$; $360-680\,mg \cdot L^{-1}$) before closure of the facility. The progressive increase in concentration during summer (Figure 5.19a) is shown to be a recurring trend before

TABLE 5.9. Average annual loads to Onondaga Lake for Ca^{2+}, Na^+, and Cl^-, for two three-year intervals: 1983–1985 and 1987–1989.

Constituent	Loading				% Reduction
	1983–1985		1987–1989		
	($\times 10^9\,mol \cdot yr^{-1}$)	($\times 10^7\,kg \cdot yr^{-1}$)	($\times 10^9\,mol \cdot yr^{-1}$)	($\times 10^7\,kg \cdot yr^{-1}$)	
Ca^{2+}	5.9	2.4	1.8	0.7	70
Na^+	7.7	1.8	3.1	0.7	59
Cl^{-1}	18.9	6.7	4.6	1.6	76

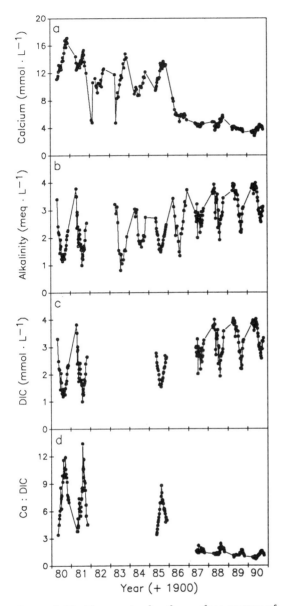

facility, as demonstrated earlier for S (Figure 5.7). The rapid response is a manifestation of the high flushing rate of the lake (Devan and Effler 1984). This has been quantified within the framework of a mass balance Cl^- model in Chapter 9 (also see Doerr et al. 1994). The average Ca^{2+} concentrations over the 1983–1985 and 1987–1989 periods for the surface waters were 11.3 and 4.3 $mmol \cdot L^{-1}$, respectively. Thus the reduction in Ca^{2+} concentrations (62%) was similar to that reported for loading over the same interval (Table 5.9)

Decalcification of the upper waters of the lake has been clearly manifested annually over the entire 1980–1990 period by the marked depletion of alkalinity (Figure 5.20b) and DIC (Figure 5.20c) during summer stratification. The extent of DIC and alkalinity depletion has decreased since closure of the soda ash/chlor-alkali facility, indicating a reduction in $CaCO_3$ precipitation in the lake since the closure. Vertical structure in these characteristics before and after closure is compared by summertime profiles for 1980 and 1989 in Figure 5.21. Greater DIC (Figure 5.21a–c) and alkalinity (Figure 5.21d–f) depletion throughout the epilimnion is depicted for 1980 than for 1989 in these profiles. Since closure, decalcification has not been clearly manifested until late summer. Note the sharp decline in decalcification coincided with the onset of reduced Ca^{2+} concentrations and Ca^{2+}: DIC ratios in the upper waters (Figure 5.20).

Average deposition rates for the May through September interval of 1985 and 1989 are presented for dry weight and particulate inorganic carbon (PIC) in Table 5.10 (detailed time distributions are presented in Chapter 8). These downward fluxes are high compared to values reported in the literature (see Chapter 8). The decrease in decalcification following closure is corroborated by reductions in downward flux of PIC determined from analysis of sediment trap collections (Table 5.10). The ionic waste discharge from the soda ash/chlor-alkali facility clearly promoted precipitation and deposition of $CaCO_3$ in the lake. The PIC deposition rate decreased by a factor of about 3 from 1985 to 1989. Calcium carbonate deposition played a

FIGURE 5.20. Time series for the surface waters of Onondaga Lake: (a) Ca^{2+}, (b) alkalinity, (c) dissolved inorganic carbon (DIC), and (d) molar ratio, Ca: DIC (modified from Driscoll et al. 1994a).

closure (Figure 5.20a). The three conspicuously lower observations in 1982 and 1983 reflect short-term runoff (dilution) events (Figure 5.20a). The major feature of the time series is the large and abrupt decrease in concentration that immediately followed the closure of the

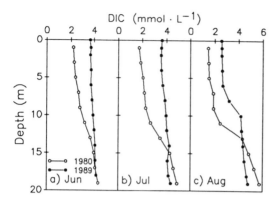

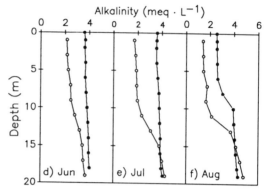

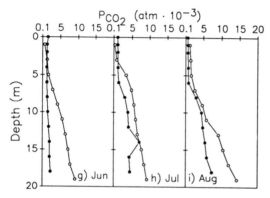

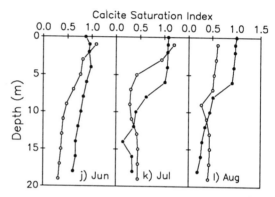

FIGURE 5.21. Selected profiles of 1980 and 1989 conditions in Onondaga Lake for the months of June, July, and August: (a–c) DIC, (d–f) alkalinity, (g–i) P_{CO_2}, (j–l) calcite saturation index (SI) (modified from Driscoll et al. 1994a).

dominant role in overall deposition in the lake in both years; it represented approximately 60% of the total weight deposition for both study periods (Table 5.10). Calcium carbonate also represents about 60% of the surficial sediments of the deposition basin of the lake; also quite high by comparison to concentrations reported in the literature (Chapter 8). The decrease in $CaCO_3$ deposition from 1985 to 1989 (64%, Table 5.4) matches the relative decrease in external loading (70%) and epilimnetic Ca^{2+} concentrations (62%) observed since closure of the soda ash/chlor-alkali facility. The reduction in Ca^{2+} loading associated with the closure is likely the primary cause of the observed (Table 5.10, Figure 5.20b) decrease in $CaCO_3$ (and dry weight) deposition in the lake. Yet further decreases in

deposition are to be expected if effective remediation of the continuing input of ionic waste (mostly via Ninemile Creek) is achieved, as this source is estimated to represent about 35% of the present Ca^{2+} load (Chapter 3, also see Effler et al. 1991a).

The possibility that a decrease in primary production coincident with the closure of the

TABLE 5.10. Average downward fluxes in Onondaga Lake determined from sediment trap collections over the May–September interval, 1985 and 1989.

Year	Suspended solids $(mg \cdot m^{-2} \cdot d^{-1})$	PIC $(mmol\, C \cdot m^{-2} \cdot d^{-1})$
1985	39,400	215
1989	13,230	78

soda ash/chlor-alkali facility contributed to the reduction in CaCO₃ deposition cannot be discounted. A significant decrease in particulate organic carbon (POC) deposition, an indicator of primary productivity (Kelly and Chynoweth 1981), was observed over the same interval (Chapter 8). This pattern is somewhat problematic. The reduction in POC flux seems inconsistent with the observation that the phytoplankton of the lake appear to have remained essentially nutrient-saturated since the time of closure (see Chapter 6). Further, the only significant change in nutrient loading keyed to the closure was the discontinued artificial internal load associated with the use of hypolimnetic water for cooling by the soda ash/chlor-alkali facility (Effler and Owens 1987). There are at least two factors, largely unrelated to primary productivity, that could explain the coincident reduction in POC deposition. Increases in zooplankton grazing since 1986 (Chapter 6) would be accompanied by increases in organic carbon losses to support grazing activities. Further, the higher POC deposition rate before closure may in part be a manifestation of the elevated CaCO₃ precipitation, as organic particles are known to serve as nuclei for formation of this solid phase in Onondaga Lake (see subsequent section of this chapter, Johnson et al. 1991).

Calculations of P_{CO_2} indicate near equilibrium with atmospheric CO_2 in the upper waters and major increases with depth within the hypolimnion (Figure 5.21g–i). The water column of the lake was oversaturated with respect to atmospheric CO_2 for nearly all dates sampled (86 out of 88) in 1980, 1981, 1985, and 1989. The accumulation of P_{CO_2} in the hypolimnion declined markedly from 1980 to 1989 (Figure 5.21g–i), presumably due to declines in the deposition of POC (Chapter 8).

5.3.3.2 CaCO₃ Solubility

Chemical equilibrium model calculations suggest the water column of Onondaga Lake has been oversaturated with respect to calcite solubility prior to and following closure of the soda ash/chlor-alkali facility. Results are presented for the upper waters for 4 years as

distributions in Figure 5.22 (a total of 90 monitoring days), and as vertical profiles for the summer months of 1980 and 1989 in Figure 5.21j–l. Average and median values of SI did not differ greatly in the upper waters (particularly in light of the great dependency on the details of the dynamics of phytoplankton activity; e.g., Figure 5.19) despite the much lower Ca^{2+} concentrations in 1989. This lack of substantial change in SI implies a compensating shift to higher pH (Table 5.7).

Oversaturation with respect to calcite solubility throughout the water column demonstrated for Onondaga Lake (Figure 5.21j–l) is unusual. Moreover, water column

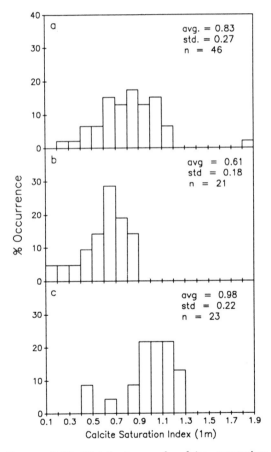

FIGURE 5.22. Distributions of calcite saturation index (SI) values for surface waters of Ononodaga Lake for the May–September interval: (a) 1980 and 1981, (b) 1985, and (c) 1989 (from Driscoll et al. 1994a).

profiles of Ca^{2+} do not depict any increase in lower water concentrations during summer stratification. Thus, deposited $CaCO_3$ does not appear to be remobilized through dissolution reactions in the hypolimnion. This pattern is atypical. Generally, in hard-water lakes, dissolution of at least of portion of the deposited $CaCO_3$ occurs by late summer due to production of CO_2 and decrease in pH in the lower waters (e.g., Effler et al. 1981, 1982). This process is usually manifested as decreases in SI in the hypolimnion with time to negative values, and coupled increases in Ca^{2+} concentration.

The kinetics of calcite precipitation have been described as being controlled by surface reactions (Morel 1983). Most surface controlled reactions can be described by an empirical rate law of the form

$$\frac{dC}{dt} = K(C - C_s)^\alpha \tag{5.21a}$$

where C = bulk solution concentration of solute, C_s = concentration of solute in equilibrium with solid, K = rate constant, and α = order of reaction. If changes in concentrations of two reactants have to be considered, another form of the above expression is

$$\frac{dC}{dt} = - K \, (10^{SI} - 1)^\alpha \tag{5.21b}$$

Morel (1983) indicates that the precipitation of calcite is usually considered to obey second order kinetics (i.e., $\alpha = 2$).

The lack of significant change in SI (Figure 5.22) despite the major reduction in epilimnetic decalcification (Figure 5.20b and c) and PIC deposition (Table 5.10) does not support predictive precipitation rate expressions based on SI (e.g., Equation (21b)). In other words, SI is not a good indicator of the rate of precipitation of $CaCO_3$ in Onondaga Lake. This is illustrated in the plot of PIC downward flux versus SI for weekly deployment intervals in 1985 and 1989 in Figure 5.23. This lack of relationship may extend to other hardwater lakes that experience $CaCO_3$ precipitation. For example, Stabel (1986) reported that seasonal $CaCO_3$ SI maxima did not coincide with maxima in downward fluxes of $CaCO_3$ in

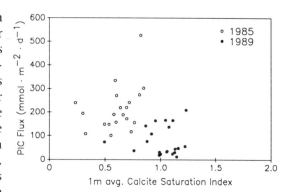

FIGURE 5.23. Evaluation of the relaitonship between the deposition rate of particulate inorganic carbon (PIC) from the upper waters and the calcite saturation index (SI) value in the surface waters for Onondaga Lake for 1985 and 1989.

Lake Constance. Effler and Driscoll (1985) demonstrated that seasonal variation in $CaCO_3$ deposition and calcite SI were only weakly correlated in Onondaga Lake in 1980.

5.3.3.3 Shifts in pH Since Closure of the Soda Ash/Chlor-Alkali Facility

One of the most intriguing effects of changes in Ca^{2+} loading and $CaCO_3$ deposition in Onondaga Lake has been the shift in acid/base chemistry. Plant metabolism is an important regulator of the dynamics of pH in the epilimnia of productive lakes (Effler 1984; Effler et al. 1981, 1982). Thus, the temporal patterns of pH in productive epilimnia are recurring only in so far as the temporal structures of plant photosynthesis and respiration are recurring. For example, in most years Onondaga Lakes experiences a spring phytoplankton bloom (e.g., Auer et al. 1990; see Chapter 6) during which some of the highest pH values occur (Effler and Driscoll 1985). However, other temporal features are not recurring, largely because of the stochastic nature of important regulating environmental forcing conditions.

Two notable changes in the distribution of epilimnetic pH have occurred over the 1980–1989 period (Figure 5.24) that are coupled to the closure of the soda ash/chlor-alkali facility. First, pH had a broader distribution before

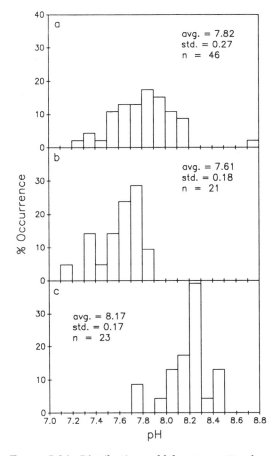

FIGURE 5.24. Distributions of laboratory pH values for the surface waters of Onondaga Lake for the May–September interval: (**a**) 1980 and 1981, (**b**) 1985, and (**c**) 1989 (from Driscoll et al. 1994a).

Thus, the decrease in $CaCO_3$ precipitation following closure (Table 5.10) would be accompanied by reduced H^+ production, and thereby a tendency for increased pH.

This mechanism is demonstrated with a theoretical "titration" for Onondaga Lake made using MINEQL. In this calculation, Ca^{2+}, added as a Cl^- salt, was varied from 1.25 to 22.4 $mmol \cdot L^{-1}$ to assess changes in the watercolumn chemistry that occurred due to changes in the industrial Ca^{2+} loading to the system. Calculations were made using water chemistry components (other than Ca^{2+} and Cl^-) that are average values for 1989 (Driscoll et al. 1994a) and assuming the system was in equilibrium with atmospheric CO_2 ($P_{CO_2} = 10^{-3.5}$ atm). The results of the titration are presented in Figure 5.25. Added $CaCl_2$ resulted in increases in Ca^{2+} concentration (Figure 5.25a) and the precipitation of $CaCO_3$, as the solution was saturated with

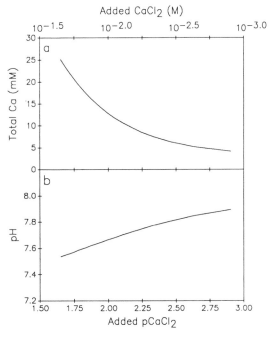

FIGURE 5.25. Model predictions for the upper waters of Onondaga lake for theoretical titration for additions of $CaCl_2$: (**a**) concentration of Ca^{2+} and (**b**) pH. The system was assumed to be in equilibrium with atmospheric CO_2 (i.e., $P_{CO_2} = 10^{-3.5}$ atm; from Driscoll et al. 1994a).

closure than following closure. Second, pH values have increased following closure. The first change reflects the lower buffering capacity before closure that resulted from the lower alkalinity and DIC caused by the higher deposition of $CaCO_3$. The median pH value for the 1980 and 1981 observations was 7.82, the value for 1989 and 1990 (not shown) was 8.22. The shift is attributed to the reduction in the rate of precipitation of $CaCO_3$. Inputs of Ca^{2+} as a neutral salt (e.g., $CaCl_2$) and related $CaCO_3$ precipitation results in the production of H^+, according to

$$Ca^{2+} + Cl^- + HCO_3^- \rightarrow CaCO_3 \downarrow + H^+ + Cl^- \tag{5.22}$$

respect to the solubility of calcite throughout the titration. Model predictions support the above mechanism (Equation (5.22)) for the observed increase in pH, as decreases in pH are predicted for increased additions of $CaCl_2$ (Figure 5.25b). Future shifts to higher pH that would be expected to accompany additional reductions in ionic loading (Chapter 3), as a result of reduced $CaCO_3$ precipitation, would exacerbate the free ammonia toxicity problem (Effler et al. 1990; later in this chapter) in the lake.

5.3.4 *In Situ* pH Measurements, Implications for Equilibrium Calculations

The temporal distributions of pH based on laboratory and field measurements are compared for the study period of 1992 for lake depths of 1 and 18 m in Figure 5.26a and b. *In situ* pH measurements are lower than laboratory values. Differences were greater in

the bottom waters than in the near-surface layer (Figure 5.26). Bottom water laboratory values were often 0.4 pH units higher than *in situ* measurements; differences of 0.1 to 0.2 pH units were more common in the upper waters.

The character of these distributions indicates the systematically higher laboratory pH values are a result of degasing of CO_2 from lake samples. Recall that calculations presented earlier for P_{CO_2} (Figure 5.22g–i) reflected highly oversaturated conditions in the hypolimnion; presumably a manifestation of decomposition. Thus, the tendency for degasing is greater for hypolimnion water samples. Sources and sinks of CO_2 are expected to be more in balance in the upper waters, as the sinks of photosynthetic uptake and loss across the air-water interface are operative, and less decomposition probably occurs compared to the hypolimnion. Periods of convergence in the upper waters reflect short-term increases in the sink processes (e.g., phytoplankton blooms, wind events). Differences of ≤ 0.1 pH units between the two forms of measurement in the upper waters should not be considered significant in light of the accuracy limitations of the pH probes (≤ 0.1 pH units for laboratory, 0.2 pH units for field).

The effect of the lower *in situ* pH measurements on calculations of DIC was insignificant for the upper waters. Estimates of DIC based on laboratory measurements were about 10% lower than those determined from *in situ* measurements. The lower *in situ* pH values lead to higher values of P_{CO_2}; MINEQL calculations based on laboratory and *in situ* measurements are compared for the 1 and 18 m depths in Figure 5.27a and b. Laboratory measurement of pH apparently leads to major underestimation of P_{CO_2} (factor of 3–4) in the bottom waters, particularly in late summer (Figure 5.27b). Values of P_{CO_2} based on field pH measurements exceeded laboratory-based estimates by much smaller margins in the upper waters from May through mid-August; differences were greater during the fall mixing period (Figure 5.27a).

Implications of the disparity in pH measurements on calculations of equilibrium conditions (MINEQL) with respect to the solubility

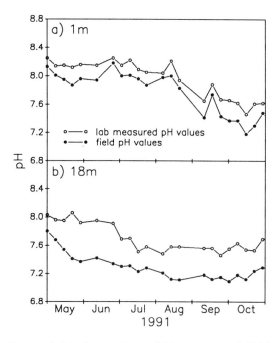

FIGURE 5.26. Comparison of laboratory and field pH measurements for the May–October interval of 1991: (**a**) 1 m, and (**b**) 18 m.

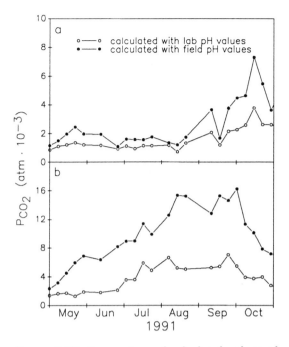

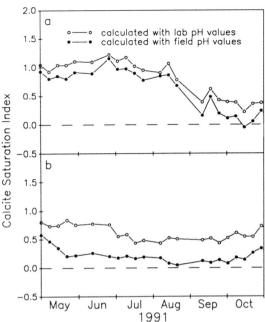

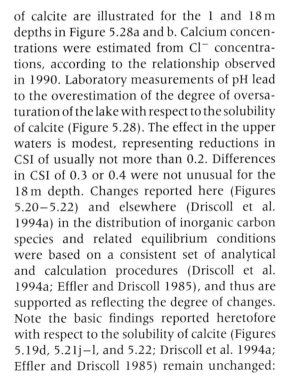

FIGURE **5.27.** Comparison of calculated values of P_{CO_2} from laboratory and field pH measurements for the May–October interval 1991: (**a**) 1 m, and (**b**) 18 m.

FIGURE **5.28.** Comparison of calculated calcite saturation index values from laboratory and field pH measurements for the May–October interval of 1991: (**a**) 1 m, and (**b**) 18 m.

of calcite are illustrated for the 1 and 18 m depths in Figure 5.28a and b. Calcium concentrations were estimated from Cl^- concentrations, according to the relationship observed in 1990. Laboratory measurements of pH lead to the overestimation of the degree of oversaturation of the lake with respect to the solubility of calcite (Figure 5.28). The effect in the upper waters is modest, representing reductions in CSI of usually not more than 0.2. Differences in CSI of 0.3 or 0.4 were not unusual for the 18 m depth. Changes reported here (Figures 5.20–5.22) and elsewhere (Driscoll et al. 1994a) in the distribution of inorganic carbon species and related equilibrium conditions were based on a consistent set of analytical and calculation procedures (Driscoll et al. 1994a; Effler and Driscoll 1985), and thus are supported as reflecting the degree of changes. Note the basic findings reported heretofore with respect to the solubility of calcite (Figures 5.19d, 5.21j–l, and 5.22; Driscoll et al. 1994a; Effler and Driscoll 1985) remain unchanged:

(1) the lake is oversaturated in the upper and lower waters, and (2) the degree of oversaturation is greater in the epilimnion than the hypolimnion (Figure 5.28).

5.3.5 Formation of Alewife Concretions

Dead alewives (*Alosa pseudoharengus*) were observed to wash ashore in large numbers in Onondaga Lake as combustible, chalk-like, concretions in the 1950s, 1960s, and 1970s (Dence 1956; Sondheimer et al. 1966; Wilcox and Effler 1981). The concretions retained much of the fish's original structure, though they lacked heads, tails, and vertebral columns. The ventral scutes, ribs, and pelvic fins remained. The muscles had been replaced by a chalky substance, pale yellow in color, which Dence (1956) described as resembling candle wax in texture. Similar concretions have been found on rare occasions on ocean shores (e.g., Wells and Erickson 1933). The alewife is a lipid-rich marine fish that became established in a number of upstate New York lakes following invasion of the Great Lakes. The alewife often manifested its presence by high mortality; dead carcasses were often observed to wash ashore in great numbers (particularly in the Great Lakes). Alewives are no longer a significant component of the fish population of Onondaga Lake (see Chapter 6), and concretions are no longer found.

Sondheimer et al. (1966) published comparative chemical analyses of Onondaga Lake concretions and dead carcasses collected from the shore of Lake Ontario; certain of these findings are presented in Table 5.11. The concretions had lost most of the proteinaceous material and were enriched in Ca and C (Table 5.11), particularly as fatty acids (Sondheimer et al. 1966). Sondheimer et al. (1966) estimated that the fatty acid content, mostly as palmitic acid, was more than 30 times that of the carcasses. The proposed mechanism for the formation of calcium-fatty acid concretions (Berner 1968), includes the processes of bacterial decay, precipitation of $CaCO_3$, and reaction of fatty acids with $CaCO_3$ to form calcium salts of fatty acids. The proposed overall stoichiometry of the reaction is

$$2RCOOH + CaCO_3 \rightarrow$$
$$Ca(RCOO)_2 + CO_2 + H_2O \quad (5.23)$$

Anaerobic conditions apparently are also required for structure retaining concretions (Wilcox and Effler 1981). The mode of $CaCO_3$ formation has been considered problematic (Berner 1968), as environments that are anaerobic are rarely oversaturated with respect to the solubility of $CaCO_3$ minerals. However, these conditions have prevailed in the hypolimnion of Onondaga Lake for a number of years (Driscoll et al. 1994a; Effler et al. 1988; also see other sections of this chapter).

The common occurrence of alewife concretions in Onondaga Lake in the 1950s, 1960s, and 1970s apparently was a manifestation of the uniquely polluted (e.g., Ca^{2+} waste from soda ash manufacture, anaerobic conditions from domestic waste discharge) state of the ecosystem, combined with the invasion of the lipid-rich alewife. The absence of these concretions in recent years is probably primarily a result of the major reduction in the alewife population in the lake (Chapter 6). The basic environmental requirements, of anaerobiosis and oversaturation with respect to the solubility of calcite, continue to prevail in the lake's hypolimnion in summer. However, conditions may be less conducive now than during the period of common observation of the concretions because of the following changes in related environmental conditions in the lake's hypolimnion (see other sections of this chapter): (1) reduction in the period of anaerobiosis, (2) increases in redox potential (e.g., somewhat less "reducing" conditions), and (3) reductions in Ca^{2+} concentrations.

5.4 Nitrogen Species

5.4.1 Background

Nitrogen exists in a number of different forms in aqueous environments. Various of these forms have important water quality implications. For example, NH_4^+ and NO_3^- are the principal forms of N used for plant nutrition (Harris 1986; Wetzel 1983), NH_4^+ and its organic precursors can represent an important sink for O_2 (Bowie et al. 1985; Harris 1986), unionized ammonia (here designated NH_3;

TABLE 5.11. Elemental and gross fatty acid composition of alewife concretions from Onondaga Lake and carcasses from Lake Ontario.

Alewife form	C	H	N	P	Ca	Mg	K	U	O	Unsaturated[†] fatty acid	Saturated[†] fatty acid
Concretion	6.63	10.5	0.3	0.1	6.0	0.07	0.003	0	15.9	7.7	74.2
Carcass	42.5	6.7	11.3	7.8	2.0	0.27	0.30	0	0	2.4	5.1

[†] Modified from Sondheimer et al. 1966

USEPA 1985) and NO_2^- (Lewis and Morris 1986) are toxic to fish at rather low concentrations. Ammonia is a terminal product in the decomposition of organic material (Kelly et al. 1988). The distribution among the different forms of N is mediated largely by a number of biochemical processes (e.g., plant assimilation, nitrification, denitrification, and ammonification) that operate in the N cycle, as well as the input of these forms of N from the watershed and discharges. The N cycle is considered to be complex because of the large number of chemical species of N (Harris 1986) and biochemical processes (Sprent 1987) involved in the cycle, and the great sensitivity of these processes to ambient environmental conditions (Hutchinson 1957; Sprent 1987; Wetzel 1983).

Determination of the distribution among the species of N and its dynamics is valuable in identifying the operation of various processes and their relationship to ambient conditions, as well as in assessing the related features of water quality. The extremely high external loading of nitrogen that has prevailed for Onondaga Lake (see Chapter 3) makes it essential to assess the distribution of N species in the lake, from a water quality perspective. Further, recent alterations in the distribution of the loading amongst the forms offers a special opportunity to evaluate the sensitivity of in-lake conditions to such changes. The distributions of five forms of N are described here: NO_3^-, NO_2^-, T-NH_3 (sum of NH_4^+ and NH_3), NH_3, and organic N. Emphasis is placed on the most recent conditions (1988–1992), however, some longer term characteristics are considered. The distributions are analyzed to: (1) evaluate the extent to which they reflect attendant limnological conditions and the levels of external loading, (2) to identify the operation of certain processes in the N cycle, and (3) identify water quality problems of the lake associated with N species.

5.4.2 Partitioning T-NH_3: NH_4^+, NH_3

In aqueous solutions, NH_3 exists in equilibrium with NH_4^+ and OH^-. The following equation expresses this equilibrium:

$$NH_3(g) + nH_2O(l) \leftrightarrow NH_3 \cdot nH_2O(aq) \leftrightarrow$$
$$NH_4^+ + OH^- + (n-1) H_2O(l) \quad (5.24)$$

Dissolved NH_3 exists in hydrated form (Equation (5.24)). Commonly used analytical methods determine total ammonia (represented here as T-NH_3). The distribution between the concentration of T-NH_3 and NH_3 is well described by Equation (5.25) of Table 5.12. The procedure used to calculate NH_3 in Onondaga Lake is presented in Table 5.12; the specified protocol is appropriate for other fresh waters, except the relationship used to estimate ionic strength (I; Equation (5.30)) is specific to Onondaga Lake. The concentration of NH_3 is calculated from a suite of measurements that includes T-NH_3, pH, temperature, and the concentration of Cl^- (as an estimator of ionic strength).

The concentration of NH_3, associated with a specified concentration of T-NH_3, increases with increases in pH and temperature, and reduction in ionic strength (Table 5.12). Under constant pH, temperature and ionic strength conditions (uniform f (Table 5.12)), the concentration of NH_3 increases as the concentration of T-NH_3 increases. The calculation is particularly sensitive to pH and least sensitive to ionic strength changes observed in the lake (Brooks and Effler 1990).

5.4.3 Temporal and Vertical Distributions of Species

The temporal and vertical distributions of the N species in Onondaga Lake have certain recurring patterns. However, the details of these distributions are extremely sensitive to a number of features of the attendant limnological conditions (Brooks and Effler 1990), that are subject to substantial meteorologically based year-to-year variations (e.g., Owens and Effler 1989). For that reason, stratification, phytoplankton biomass, dissolved oxygen, and pH conditions are presented for the spring to fall period of 1989 (Figure 5.29a–d) to support the presentation of the distribution of N species for this period. Particular attention should be paid to the features of stratification with respect to timing and layer dimensions

TABLE 5.12. Calculation procedure for the concentration of free ammonia in Onondaga Lake*.

Equation number	Relationship	Definition of symbols	Reference
5.25	$f = 1/(10^{pka-pH-s} + 1)$	f = fraction of T-NH_3 as NH_3 pk_a = negative logarithm of dissociation constant of Equation 5.24 s = salinity correction term	Messer et al. 1984
5.26	$pk_a = 0.09018 + 2729.92/T$	T = temperature (°K)	Emerson et al. 1975
5.27	$s = \dfrac{-A'I^{0.5}}{I^{0.5} + 1}$	I = ionic strength A' = coefficient	Messer et al. 1984
5.28	$A' = 1.824483 \times 10^6 [E(T_c + 273.160)]^{-1.5}$	E = dietectric constant T_c = temperature (°C)	Messer et al. 1984
5.29	$E = 87.74 - 0.4008T_c + 9.398 \times 10^{-4}T_c^2 - 1.41 \times 10^{-6}T_c^3$		Truesdall and Jones et al. 1974
5.30	$I = 1.456[Cl] - 0.0390 \times 10^{-3}$	$[Cl]$ = molar concentration of chloride	Effler et al. 1986d
5.31	$(NH_3) = f(T\text{-}NH_3)$		Messer et al. 1984

* Modified from Effler et al. 1990

Appropriate for most fresh waters, with the exception of Equation 5.30, which is specific to Onondaga Lake

(Figure 5.29a), as they largely regulate the timing and dimensions of anoxia (Figure 5.29b), and the development of vertical differences in pH (Figure 5.29c). The importance of phytoplankton activity (e.g., biomass in Figure 5.29d) in regulating the temporal structure of pH and DO in the near surface waters of the lake has been noted earlier.

The measured temporal and vertical patterns in the concentrations of NO_3^-, NO_2^-, and T-NH_3 (as mgN · L^{-1}) for the April through October period of 1989 are presented as isopleths in Figure 5.30a–c. Distributions for these species for the same period of 1988 were presented by Brooks and Effler (1990). The temporal distributions of organic N for the 1989 period are presented for depths of 1 and 16 m in Figure 5.30d. Concentrations of all the measured forms of N were uniform with depth during spring and fall turnover (Figure 5.30). However, pronounced temporal and vertical variations in all of these forms are observed during the period of stratification. The very sharp vertical gradients observed in late summer and early fall are a manifestation of entrainment-based mixing that prevails in this period annually (e.g., Brooks and Effler 1990).

Both NO_3^- and NO_2^- were depleted rapidly from the hypolimnion (Figure 5.30a and b) following the onset of anoxia in 1989 (Figure 5.29b). Formation of these species was eliminated with the onset of anoxia. The subsequent depletion of these species is usually attributed to denitrification (e.g., Wetzel 1983), though the possible contribution of bacterial assimilation of NO_3^- to this depletion cannot be discounted. The hypolimnion was essentially devoid of NO_3^- and NO_2^- by late June in 1989 (Figure 5.30a and b). The depths over which these species were detectable tracked approximately the upper boundary of anoxia (Figure 5.29b) for most of the stratification period. Depletion of NO_3^- and NO_2^- progressed from the lowermost layers upward within the hypolimnion, indicating the localization of denitrification at the sediment–water interface.

Only minor changes in the concentration of NO_3^- occurred in the epilimnion until late June in 1989 (Figure 5.30a). Subsequently NO_3^- increased rapidly from less than 1.5 to 3.0 mgN · L^{-1} in mid-August. The concentration decreased with the approach to turnover as a result of dilution with the totally

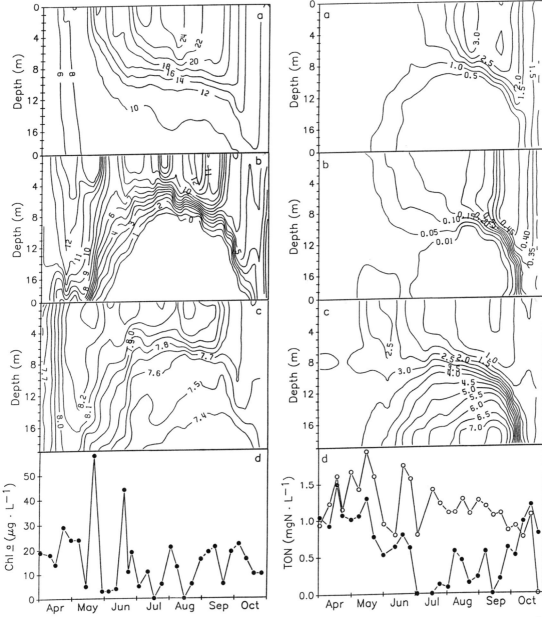

FIGURE 5.29. Selected limnological characteristics of Onondaga Lake during the 1989 study period: (a) isotherms (°C), (b) isopleths of dissolved oxygen (mg · L⁻¹), (c) isopleths of pH, and (d) concentration of chlorophyll at a depth of 1 m.

FIGURE 5.30. Distributions of N species (mgN · L⁻¹) in Onondaga Lake during the 1989 study period: (a) isopleths of NO_3^-, (b) isopleths of NO_2^-, (c) isopleths of T-NH_3, and (d) concentration of organic N at depths of 1 and 16 m.

depleted lower waters. Lake-wide increases were observed following the onset of complete fall turnover in mid-October. Nitrite concentrations increased progressively in the epilim-

nion with the onset of stratification from about 0.05 to 0.25 mgN · L⁻¹ (Figure 5.30b). Nitrate and NO_2^- maxima at the interface of the oxic and anoxic layers, observed in a number of productive lakes with anoxic hypolimnia (Hutchinson 1957; Wetzel 1983), were not found in Onondaga Lake.

The concentrations of NO_2^- that prevailed in the epilimnion are extremely high compared to those reported for other lakes; usually concentrations fall in the range of 0 to $0.01 \, \text{mgN} \cdot L^{-1}$ (Hutchinson 1957; Wetzel 1983). Hutchinson (1957) indicates the occurrence of appreciable NO_2^- levels in surface water is a manifestation of sewage contamination. Nirite is formed by NO_3^- reduction (Sprent 1987; Hutchinson 1957) and as an intermediate product in the nitrification process (Keeney 1973; Sprent 1987; Wetzel 1983). It is considered to be a labile material in oxic water because of its propensity to be oxidized to NO_3^-. The increase in epilimnetic NO_2^- reflects a condition of sources exceeding sinks; the sources include external inputs (particularly METRO) and perhaps internal production. The high concentrations of NH_3 that prevailed in the epilimnion during the period (see subsequently presented material) may have inhibited the second stage of the nitrification process, as NH_3 has been described as toxic to *Nitrobacter* (responsible for second stage of nitrification; Anthonisen et al. 1976; Sprent 1987).

The concentration of T-NH_3 declined in a mostly progressive manner in the epilimnion in 1989, from a maximum of about 3.0 mg-N $\cdot$ L^{-1} in May to a minimum of approximately $0.5 \, \text{mgN} \cdot L^{-1}$ in mid-September (Figure 5.30c). The lake-wide concentration during fall turnover ($\sim 1.5 \, \text{mgN} \cdot L^{-1}$) was about one-half the concentration observed during spring turnover. The concentrations of T-NH_3 and NO_3^- that were maintained in the epilimnion were extremely high, indicating internal losses, particularly associated with primary production, were compensated by continuing inputs. Of these two forms, T-NH_3 is preferred by phytoplankton for energetic reasons (Harris 1986; Sprent 1987; Wetzel 1983). The sum of the concentrations of T-NH_3 and NO_3^- in the upper productive waters of Onondaga Lake ranged from about 2.5 to 4.9 mgN $\cdot$ L^{-1} during the monitored period of 1989; substantially above concentrations considered limiting to

algae growth (see Chapter 6). These saturated conditions with respect to N species available to support phytoplankton growth are recurring in the lake. Typically, both T-NH_3 and NO_3^- become greatly depleted by phytoplankton uptake in the epilimnia of eutrophic lakes during summer stratification (Wetzel 1983).

Concentrations of T-NH_3 in the hypolimnion increased progressively with the establishment of stratification in May (Figure 5.30c). Concentrations of T-NH_3 increased with depth in the hypolimnion, indicating the localization of ammonification at the sediment–water interface and within the sediments. Concentrations of more than $8.0 \, \text{mgN} \cdot L^{-1}$ were measured in the lowermost waters before the onset of complete fall turnover; a concentration that approaches the highest hypolimnetic values reported in the literature (e.g., Hutchinson 1957; Wetzel 1983).

Substantial variability in the concentration of organic N was observed at depths of 1 and 16 m during 1989 (Figure 5.30d). The near zero values obtained for the hypolimnion depth on several occasions from mid-to late summer may represent analytical shortcomings associated with the very large contributions of T-NH_3 to TKN during this period. The variations at 1 m are largely associated with changes in phytoplankton biomass (Figure 5.29d). The average concentration at 1 m was $1.17 \, \text{mgN} \cdot L^{-1}$; the average for 16 m was $0.63 \, \text{mgN} \cdot L^{-1}$.

5.4.4 Interannual Differences in N-Species Distributions

Interannual differences in the temporal distributions of NO_3^-, NO_2^-, and T-NH_3 in the upper waters (1 m) of the lake are reflected by the differences shown for the spring-fall interval of 1988, 1989, and 1990 in Figure 5.31a–c; the distributions for the sum of these three species is presented in Figure 5.31d. Average and range statistics for these N species for the three years are presented in Table 5.13. The striking differences in the distributions, particularly for NO_3^- and T-NH_3 (Figure 5.31a and c), for the three years represent in part

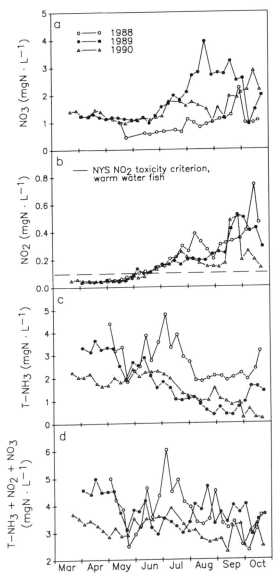

FIGURE 5.31. Temporal distributions of N species ($mgN \cdot L^{-1}$) at a depth of 1 m in Onondaga Lake for 1988, 1989 and 1990: (a) NO_3^-, (b) NO_2^- (c) T-NH_3, and (d) summation of NO_3^-, NO_2^- and T-NH_3.

summer and early fall of 1989, the period in which the lowest T-NH_3 concentrations were observed. Differences between the years were not always compensated for by complimentary shifts in other species (e.g., Figure 5.31d). For example, NO_3^- concentrations were very similar in the upper waters through mid-July of 1989 and 1990, but substantial differences in T-NH_3 concentrations prevailed over much of the same period of these two years. The distributions of NO_2^- concentration for the three years were uniform by comparison, particularly through July (Figure 5.31b). Abrupt changes in the temporal distributions of NO_2^- were observed in the fall of all three years. A unifying characteristic of the distributions, despite the documented interannual differences (Figure 5.31), is the exceedingly high concentrations (e.g., Hutchinson 1957; Wetzel 1983) of NO_3^-, NO_2^- and T-NH_3 that prevail in the upper waters of Onondaga Lake (e.g., Table 5.13).

The temporal distributions of T-NH_3 at a depth of 16 m are presented for the same three years in Figure 5.32. Progressive increases in T-NH_3 were observed during the period of stratification in all three years. The rate of increase was the greatest in 1988 and the lowest in 1990. The roles the different loading conditions, tributary flows, (see Chapter 3) and in-lake processes (e.g., nitrification) played in regulating the observed year-to-year differences (e.g., Figures 5.31 and 5.32) are evaluated within the quantitative framework of a mechanistic model in Chapter 9.

5.4.5 NH_3

The temporal and vertical patterns in the concentration of NH_3 are dependent on ambient conditions, as well as on the concentrations of T-NH_3. The influence of the ambient conditions is quantified by the value of f (Equations (5.25), Table 5.12), the fraction of T-NH_3 that exists as NH_3. The temporal and vertical distributions of f in Onondaga Lake for 1989 are presented in Figure 5.33a as isopleths. The earlier presentation of the distribution of f in the lake for 1988 (Brooks and Effler 1990) is the only other detailed documentation for this

shifts in the contribution of the individual species to the summed total (Figure 5.31d). For example, the lowest concentrations of NO_3^- were observed in mid-summer of 1988, the same period in which the highest T-NH_3 concentrations were observed. The highest NO_3^- concentrations were observed in late

TABLE 5.13. Statistics of N species distribution in the upper waters (1 m) of Onondaga Lake for 1988, 1989, and 1990.

Year	N species (mgN · L^{-1})					
	NO$_3^-$		NO$_2^-$		T-NH$_3$	
	$\bar{X}$	range	$\bar{X}$	range	$\bar{X}$	range
1988	0.93	0.44–2.23	0.24	0.05–0.74	2.75	1.82–4.78
1989	1.82	0.91–3.93	0.19	0.04–0.52	1.78	0.37–3.65
1990	1.48	0.96–2.86	0.15	0.04–0.51	1.51	0.20–2.37

parameter we are aware of for any lake. The general seasonal and vertical patterns and the magnitudes of f documented here (Figure 5.33a) are probably representative of conditions in many productive dimictic hardwater alkaline lakes, as the pH and temperature conditions measured for Onondaga Lake (Figure 5.27a and c) are typical of such systems (Hutchinson 1957; Wetzel 1983).

Strong vertical differences in the value of f prevailed throughout the stratification period. The average value of f for the upper 5 m of the water column during the stratification period was more than seven times greater than the average value for the lower 5 m. Further, the value of f varied greatly with time in the upper productive layers (Figure 5.33a). The principal factor regulating the vertical and temporal character of these trends was pH (note the exponential dependence of f on pH (equation (5.25)); Table 5.12), although temperature also contributed significantly. The abrupt

increases in f that occurred in the upper waters were the result of abrupt increases in pH (Figure 5.29c) caused by increased phytoplankton activity (e.g., Figure 5.29c and d, Effler et al. 1990).

Substantial variations in the value of f, and therefore in the concentrations of NH$_3$ (Equation (5.31), Table 5.12), can occur within a day in productive waters as a result of pH changes that are caused by diurnal variations in photosynthetic activity (Crumpton and

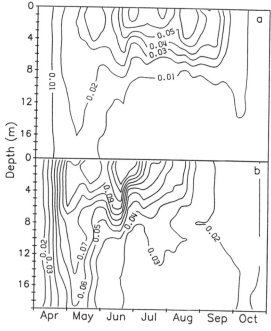

FIGURE 5.33. Temporal and vertical distributions in Onondaga Lake in 1989 of: (**a**) f (fraction of T-NH$_3$ as NH$_3$), as isopleths, and (**b**) NH$_3$ (mgN · L^{-1}), as isopleths.

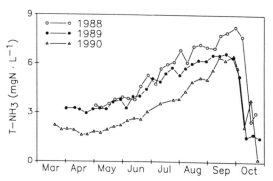

FIGURE 5.32. Temporal distributions of T-NH$_3$ in Onondaga Lake at a depth of 16 m for 1988, 1989, and 1990.

Isenhart 1988; Talling 1974; Verduin 1960). For example, a diurnal change of ± 0.2 pH units from the early summer (temperature of 18.7°C, chloride of 12.9 mmoles $\cdot$ L^{-1}) pH value of 8.5 (f = 0.118), results in f varying from 0.078 to 0.176 This variation in pH is expected to be of a magnitude commonly experienced by the lake. The great sensitivity of f to pH has important implications with respect to the need for accurate pH measurements to support the calculation of f and the concentration of NH$_3$, and with respect to the time of day of sampling. Subsequently, the implications of *in situ* versus laboratory pH measurements on the calculated concentrations of NH$_3$, and the status of the lake with respect to related criteria, are demonstrated. The magnitude of the diurnal variations in pH and in the concentration of NH$_3$ are expected to be the greatest on days of peak photosynthetic rate and minimum surface turbulence (e.g., a clear, calm day during a phytoplankton bloom). The usual time of lake monitoring (~1000 hours) may cause some slight negative bias in reflecting daily average pH (e.g., Talling 1974).

The temporal and vertical patterns of the calculated concentrations of NH$_3$ for the study period of 1989 are presented in Figure 5.33b. These distributions represent the product of the distributions (Equation (5.31), Table 5.12) of the T-NH$_3$ concentration (Figure 5.30c) and f (Figure 5.33a). The temporal and vertical patterns in the concentration of NH$_3$ tracked most closely the structure of f because of the much stronger vertical gradients and greater temporal variations within the epilimnion of f. As a result, the distributions of T-NH$_3$ and NH$_3$ concentrations contrasted each other strongly. For example the upper waters had much higher concentrations of NH$_3$ than the lower waters throughout the stratification period. Note the much more abrupt changes in the concentration NH$_3$ that occurred in the upper waters as a result of photosynthetically induced changes in pH. The range in the concentration of NH$_3$ in the epilimnion during the stratification period was from 0.007 to 0.156 mgN $\cdot$ L^{-1}. Despite the substantial increase in the hypolimnetic concentration of T-NH$_3$ during the

summer, only a modest increase in the concentration NH$_3$ occurred, in part because a decrease in pH (Figure 5.29c), and thereby f (Figure 5.33a), occurred over the same period.

The general features of the distributions of f and NH$_3$ in the lake are recurring (e.g., compare Figure 5.33 to distributions presented for 1988 by Brooks and Effler (1990)). The temporal distributions of f and NH$_3$ in the upper waters (1 m) of the lake are compared for 1988, 1989, and 1990 in Figure 5.34a and b. Differences in NH$_3$ concentrations in the upper waters among these three years are due to differences in both f (Figure 5.34a) and T-NH$_3$ concentrations (Figure 5.31c). Summary statistics intended to elucidate the basis for the observed differences in NH$_3$ concentrations are presented in Table 5.14; comparisons (for the May–September interval) of T-NH$_3$ concentrations, f, and the most important environmental parameters regulating f, are included. Note that the NH$_3$ concentrations in 1988 were on average about twice the levels observed in 1989 and about 2.7 times greater than the concentrations found in 1990. The principal

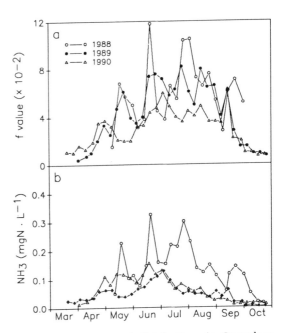

FIGURE 5.34. Temporal distributions in Onondaga Lake at a depth of 1 m for 1988, 1989, and 1990: (**a**) f, and (**b**) NH$_3$.

TABLE 5.14. Comparison of NH_3 concentrations and regulating parameters, in the upper waters (1 m) of Onondaga Lake for the May–September interval of 1988, and 1989, 1990.

Year	Temperature (°C)		
	$\bar{x}$	range	med.
1988 n = 21	21.2	8.5–28.0	21.9
1989 n = 22	19.9	9.3–25.3	20.0
1990 e n = 22	19.8	12.5–265	21.8

Year	pH		
	$\bar{x}$	range	med.
1988 n = 21	8.16	7.79–8.52	8.17
1989 n = 22	8.16	7.74–8.44	8.21
1990 e n = 22	8.01	7.72–8.24	8.01

Year	f		
	$\bar{x}$	range	med.
1988 n = 21	0.061	0.015–0.118	0.057
1989 n = 22	0.053	0.015–0.082	0.058
1990 e n = 22	0.037	0.020–0.063	0.036

Year	T-NH_3 (mgN·L^{-1})		
	$\bar{x}$	range	med.
1988 n = 21	2.71	1.82–4.78	2.50
1989 n = 22	1.57	0.37–3.33	1.13
1990 e n = 22	1.61	0.81–237	1.65

Year	NH_3 (mgN·L^{-1})		
	$\bar{x}$	range	med.
1988 n = 21	0.16	0.05–0.33	0.14
1989 n = 22	0.08	0.01–0.16	0.08
1990 e n = 22	0.06	0.02–0.13	0.06

reason for the higher NH_3 concentrations in 1988 compared to 1989 is the elevated concentrations of T-NH_3 in 1988, though the higher f value in 1988, associated mostly with the higher temperature, also contributed (Table 5.14). The lower NH_3 concentrations in 1990 are mostly associated with the modest reductions in pH values (Table 5.14).

5.4.6 Status with Respect to Toxicity Criteria

5.4.6.1 NH_3

Free ammonia is toxic to aquatic life, particularly fish (USEPA 1985). Free ammonia has been described as one of the "two [the other being chloride] greatest challenges to aquatic life in the immediate vicinity of wastewater outfalls" (Lewis 1988). High concentrations of NH_3 are undesirable and inconsistent with the maintenance or development of a healthy fishery. The U.S. Environmental Protection Agency (EPA) has recently promulgated new criteria for NH_3 concentrations to protect aquatic life against its toxic effects (USEPA 1985), which have been adopted as water quality standards by the State of New York. The equations used to calculate USEPA's final chronic value (FCV) and final acute value (FAV) of NH_3, to protect against chronic and acute toxicity, are presented in Table 5.15. Note that in all cases the salmonid criteria are more stringent than the nonsalmonid criteria. The criteria require that four-day average concentrations of NH_3 not exceed, more often on average than once every three years, the FCV, and that one-hour average concentrations not exceed, more often on average than once every three years, one-half of the FAV (0.5 × FAV).

The new criteria have been described as "more realistic and better justified from a conceptual viewpoint" (Lewis 1988) than the earlier, simpler criteria because of the incorporation of important environmental factors that influence toxicity (e.g., pH and temperature; see Table 5.15). Although the new criteria are functions of pH and temperature, they are less sensitive than the value of f (and therefore

TABLE 5.15. Calculation of EPA (1985) final chronic value (FCV) and final acute value (FAV) for (NH_3).

Relationship	Definition of symbols	Equation number
$FCV = \dfrac{0.05}{FPH \cdot FT}$; for pH $\geq$ 7.7	FPH = pH adjustment factor FT = temperature adjustment factor	(5.32)
$FCV = \dfrac{0.042}{FT \cdot 10^{7.7 \cdot pH}}$; for pH $<$ 7.7		
where: FPH = 1.0; for 8.0 $<$ pH $\leq$ 9.0 $FPH = \dfrac{1.0 + 10^{7.4-pH}}{1.25}$; for 6.5 $\leq$ pH $\leq$ 8.0		(5.33)

$FT = 10^{0.03(20-TCAP)}$; for TCAP $\leq T_c \leq 30°$
$FT = 10^{0.03(20-Tc)}$; for 0 $\leq T_c \leq$ TCAP
	Acute
	    TCAP = 20°C for salmonids
	    TCAP = 25°C for nonsalmonids
	Chronic
	    TCAP = 15°C for salmonids
	    TCAP = 20°C for nonsalmonids
		(5.34)

$FAV = \dfrac{0.52}{FPH \cdot FT}$		(5.35)

the concentration of NH_3; Table 5.12) to changes in these parameters. In other words, the offsets for pH and temperature included in the criteria (Table 5.15) do not match the increase in NH_3 concentration that results from increases in these parameters (Table 5.12). The new criteria vary in time and space as pH and temperature change, accommodating the widely observed (see review of USEPA (1985)) reduction in toxicity (increase in LC_{50}) with increases in pH and temperature.

The status of the upper waters (1 m) of Onondaga Lake with respect to the nonsalmonid chronic and acute standards for NH_3 toxicity for 1988, 1989, and 1990 is presented in Figure 5.35a–c (also see Effler et al. 1996). Vertical profiles describing the status of the lake with respect to these criteria are presented for dates in mid-July of these three years in Figure 5.36a–c. Note the strong seasonal (Figure 5.35) and vertical (Figure 5.36) variations in the criteria associated with seasonal and vertical differences in pH and temperature (Figure 5.29a and c). The differences in NH_3

concentrations in the upper waters in the three years resulted in marked differences in the status of the lake with respect to the state standards. The FCVs were exceeded throughout the lake on all monitoring occasions in 1988, with the exception of the last sampling date during fall turnover (e.g., Figures 5.33a and 5.34a; Effler et al. 1990). This pattern indicates that the state chronic standards for NH_3 for salmonids and nonsalmonids were violated continuously throughout the lake from spring to fall turnover. The most severe contravention occurs in the upper waters (Figure 5.36 a–c), because of the higher pH values and temperatures (Figure 5.29a and c). Contravention of the state acute toxicity standards for NH_3 for nonsalmonids was indicated on three occasions 1988 in the upper waters (Figure 5.35a) during the phytoplankton blooms of May, June, and July (Effler et al. 1990).

The concentration of NH_3 in the uppermost waters equaled or exceeded the acute nonsalmonid standard on two occasions in the

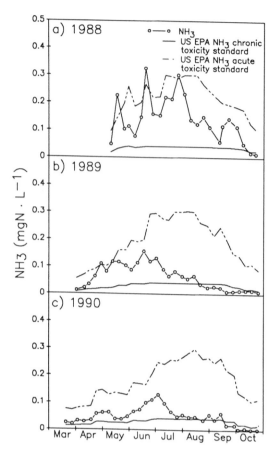

FIGURE 5.35. Status of the upper waters (1 m) of Onondaga Lake with respect to the nonsalmonid chronic and acute criteria for NH₃ toxicity: (**a**) 1988, (**b**) 1989, and (**c**) 1990 (modified from Effler et al. 1996).

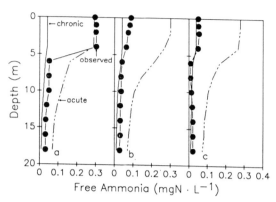

FIGURE 5.36. Watercolumn status of Onondaga Lake in mid-July with respect to the nonsalmonid chronic and acute criteria for NH₃ toxicity, as profiles: (**a**) July 25, 1988, (**b**) July 24, 1989, and (**c**) July 25, 1990.

indicated in Figure 5.35 as a result of photosynthetically induced diurnal increases in pH. Conditions documented for the upper waters subsequently (e.g., ≥1991) do not support the superficial improving trend indicated here for the 1988–1990 period.

The implications of the lower pH values determined from *in situ* versus laboratory measurements (Figure 5.26) on the calculated distributions of NH₃, and status with respect to the chronic criterion for nonsalmonids, at depths of 1 and 18 m in 1991 are illustrated in Figure 5.37. Both NH₃ concentrations and the nonsalmonid standards were lower when calculated (Tables 5.12 and 5.15) from lower field pH values. However, the effect on the determined status with respect to the standard was modest because of the compensating effects of pH on these calculations (Table 5.12 and 5.15). The temporal structure of the violations in the upper waters, including the duration of violation, was similar for the two pH data sets (Figure 5.37a and b). Whereas the margin of violation was relatively greater later in the summer in the bottom waters based on the *in situ* pH data (Figure 5.37c and d). The average margins of violation (free ammonia concentration divided by standard) for the May–September interval were shifted somewhat lower for both the near surface and bottom waters based on the *in situ* pH data

spring of 1989 (Figure 5.35b). The FCV was exceeded from late March through mid-August; thereafter the NH₃ concentration was less than the FCV. The reduced severity of the NH₃ problem in 1989 compared to 1988 was largely attributable to the lower T-NH₃ concentrations in the epilimnion in 1989 (Table 5.14, Figure 5.31c). Further improvements with regard to violation of the standards occurred in 1990; violation of the acute standard was not observed and the margin of exceedance of the FCV was generally less than in 1989. However, the FCV value was equaled or exceeded on three occasions in September of 1990. The acute standard was probably violated more often, or more nearly approached than

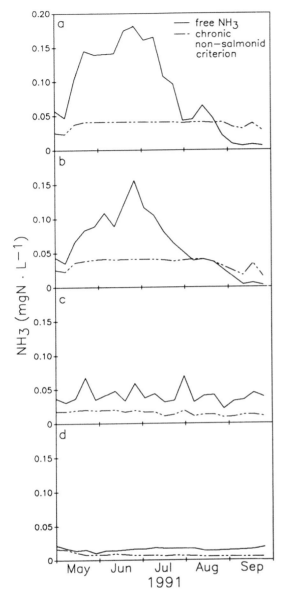

FIGURE 5.37. Comparison of status of Onondaga Lake with respect to the nonsalmonid chronic criterion for NH_3 toxicity for laboratory and field pH measurements: (a) Lab pH, 1 m, (b) *in situ* pH, 1 m, (c) lab pH, 18 m, and (d) *in situ* pH, 18 m.

base. The average margin of violation at a depth of 1 m over the May–September interval of 1991 based on laboratory pH measurements was a factor of 2.2; the average for the *in situ* measurements was 1.7. For the bottom waters the average margin was a factor of 2.7 based

on laboratory pH; the average for the *in situ* measurements was 2.3. This analysis indicates that the NH_3 problem of Onondaga Lake is only slightly less severe if calculations are based on *in situ* measurements. Based on these results and the generally lower performance (accuracy) characteristics of the *in situ* measurements, there appears to be little to choose between laboratory and *in situ* measurements with regard to the NH_3 problem of the lake.

The prevalence and margin of violation of the chronic toxicity criteria for NH_3 and the more infrequent occurrence of violation of the acute standard for nonsalmonids in the upper waters of Onondaga Lake are unusual, even for a hypereutrophic lake (Effler et al. 1990). The extent to which the prevailing high NH_3 concentrations impact the diversity and abundance of aquatic life (particularly fish) in the lake is presently unknown. Evaluation of this problem is complicated by the myriad of other water quality problems of Onondaga Lake (Effler 1987a; Effler et al. 1986a; Effler et al. 1988). Violations in the anoxic hypolimnia of eutrophic lakes should probably not be considered resource limiting; the fundamental limiting condition is anoxia. Reclamation of the oxygen resources of these hypolimnia would in most cases also eliminate the NH_3 problem of the lower layer (e.g., reestablishing nitrification and reducing ammonification).

5.4.6.2 NO_2^-

Nitrite can be toxic to fish at low concentrations (Lewis and Morris 1986). Salmonids, channel catfish, and fathead minnows have been observed to be particularly susceptible (Lewis and Morris 1986). As is the case for NH_3, there is little scientific information on the chronic effects of NO_2^-. The water quality standards for class A, B and C surface waters in New York State (NYSDEC 1984) are 100 and $20\,\mu gN \cdot L^{-1}$ for warmwater and coldwater fisheries, respectively. The coldwater standard was derived by applying a factor of 0.05 to the value of $0.4\,mgN \cdot L^{-1}$ as, a concentration within the range of acute values reported for rainbow trout and chinook salmon (Russo et al. 1974; Smith and Williams 1974).

The warmwater standard was established by applying a factor of 0.05 to the value of $2 \, \text{mgN} \cdot \text{L}^{-1}$, described as (NYSDEC 1984) within the range of acute values reported for a number of warmwater fish (McCoy 1972; Wallen et al. 1957).

The warmwater standard for NO_2^- has been added to Figure 5.31b. This standard was exceeded in Onondaga lake throughout the mid-summer to fall interval of 1988, 1989, and 1990 at a depth of 1 m (Figure 5.31b), and throughout the oxygenated layers (Figure 5.30b). The margin of violation increases progressively from mid-through late summer, associated with the progressive increases in NO_2^- concentrations over that period. The maximum margins of violation of the warmwater standard observed in 1988, 1989, and 1990 at 1 m (Figure 5.31b) were factors of 7.4, 5.2, and 5.1 respectively.

A number of materials have been identified as mitigating against the toxic effects of NO_2^- (Lewis and Morris 1986). The strongest documented interaction is with Cl^- (Lewis and Morris 1986; Russo and Thurston 1977). The strength of this external environmental offset it greatest for the least sensitive species and smallest for the most sensitive species (Lewis and Morris 1986). "The effect of chloride on the toxicity of NO_2^- is now known to be so great that experiments in which Cl^- concentrations are not documented are of very little value because they cannot be meaningfully compared with the results of other studies" (Lewis and Morris 1986, p. 187). Crawford and Allen (1977) demonstrated that the mortality of small chinook salmon in seawater occurred at NO_2^- concentrations 50 to 100 times higher than in fresh water. Based on regression analyses of various data sets, Lewis and Morris (1996) found that the addition of $1 \, \text{mg} \cdot \text{L}^{-1} \, Cl^-$ increases the 96-h LC50 of NO_2^- by 0.088 to 0.61 $\text{mgN} \cdot \text{L}^{-1}$. Utilizing this relationship and a Cl^- concentration of $400 \, \text{mg} \cdot \text{L}^{-1}$ (representative of present prevailing conditions), the 96−h LC50 for NO_2^- would range from 35 to 244 $\text{mg} \cdot \text{L}^{-1}$; the values for $100 \, \text{mg} \cdot \text{L}^{-1} \, Cl^-$ would range from 8.8 to 61 $\text{mg} \cdot \text{L}^{-1}$. If this offset were applied to

existing Onondaga Lake conditions (or even at a Cl^- concentration of $100 \, \text{mg} \cdot \text{L}^{-1}$), including the factor utilized by the NYSDEC (0.05) in the derivation of the standards, the prevailing NO_2^- concentrations would not be considered toxic. The New York State standards for NO_2^- presently do not consider the effects of Cl^- (NYSDEC 1984). Other ions mitigate against the toxic effects of NO_2^-, in particular HCO_3^- and Ca^{2+}. However, they are much less effective than Cl^- (Lewis and Morris 1986).

5.4.7 Longer Term Conditions

Summertime conditions that prevailed in earlier years with respect to potentially toxic species, NH_3 and NO_2^-, are in part manifested in data collected as part of Onondaga County's lake monitoring program (Onondaga County 1971–1991; Walker 1991). Median values of summer (May–September) measurements of T-NH_3 and NO_2^- from the lake's upper waters are presented for the 1968–1992 period in Figure 5.38a and b. Onondaga County data are from a 3 m sampling depth. Upstate Freshwater Institute (UFI) data (1988–1992) are from 2 m. The median statistic is used here because of its "robustness"; that is, the influence of analytical outliers is reduced. The yearly population sizes varied greatly over the period of record; measurements have been made most frequently during the 1988–1992 interval.

No effort is made here to explain the details of the highly variable (often year-to-year) distributions observed during the period of record, nor can the possibility of the inclusion of faulty observations within earlier portions of the data set be eliminated. The most important point is that NH_3 and NO_2^- problems have prevailed in the upper waters of the lake for many years. While modest year-to-year variability (e.g., Table 5.14) and systematic shifts (Figure 5.24) in pH values have occurred in the upper waters over the period of record, clearly the summertime T-NH_3 concentrations have remained high enough since at least the late 1960s (Figure 5.38a) to represent routine

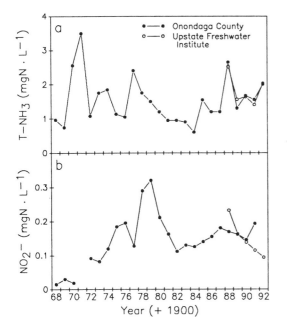

FIGURE 5.38. Long-term trends in N specie concentrations in the upper waters (Onondaga County-3 m, UFI-2 m) of Onondaga Lake, as median values of summer observations: (a) T-NH₃, and (b) NO₂⁻.

violations of the recently established New York State standards to protect against the toxic effects of NH₃. Violations of the warm-water fish NO₂⁻ standard have routinely occurred in the summer in Onondaga Lake since at least the early 1970s (Figure 5.38b). The concentrations reported for the 1968–1970 interval are much lower than those measured subsequently. This may reflect early analytical problems. Factors contributing to the dynamics (Figure 5.38a and b) include variations in external loading, tributary flow (Chapter 3), and in-lake biochemical processes.

5.5 Phosphorus

5.5.1 General

Phosphorus (P) has probably received more attention than any other element in freshwater studies because of its critical role in plant

metabolism and the major degradations in water quality that occur in response to anthropogenic inputs (Wetzel 1983). Phosphorus has long been recognized as the most critical nutrient controlling phytoplankton growth in most lakes (Carlson 1977; Dillon 1975; Hutchinson 1973; Vollenweider 1968). Elevated inputs of P to lake ecosystems, associated with man's activities (e.g., disposal of sewage, agricultural wastes, etc.), enhances phytoplankton growth (see Chapter 6), and associated deleterious consequences, such as reduced clarity (see Chapter 7) and increased hypolimnetic oxygen depletion (see earlier section of this chapter). This accelerated "aging" of lakes associated with elevated loadings of P has been described as cultural eutrophication.

At least three different forms of P are routinely measured in limnological and water quality studies. Total phosphorus (TP) is a widely used measure of trophic state (e.g., Carlson 1977; Vollenweider 1975, 1982). Total dissolved phosphorus (TDP) is measured on filtered (0.45 μm) samples. Most TDP is assumed to be ultimately available to support phytoplankton growth. A third commonly measured form of P is soluble reactive phosphorus (SRP); this form is a component of TDP that is usually assumed to be immediately available for phytoplankton. Particulate phosphorus (PP), a measure of phytoplankton biomass in many productive systems, is calculated as the difference of paired TP and TDP analyses (PP = TP − TDP).

Here the temporal and vertical distributions of several important forms of P are documented for Onondaga Lake. Recent changes are identified, and longer-term databases (e.g., since the late 1960s) are reviewed. Several features of the P cycle in the lake are depicted. Particular emphasis is given to quantifying sediment P release and identifying regulating mechanisms. Sedimentary P data are presented in Chapter 8. External loadings (see Chapter 3), in-lake concentrations, and key cycling processes, have been synthesized within the framework of a mechanistic mass balance model for P for the lake, that is developed and tested in Chapter 9.

5.5.2 Description of Recent Distributions

5.5.2.1 Typical Profiles

Temporal and vertical features of the distributions of TP and SRP that have prevailed in Onondaga Lake over the spring to fall interval in recent years are exemplified by the profiles (12) of Figure 5.39. Temperature profiles are included for certain of the dates to illustrate the important role the stratification regime plays in regulating the vertical distribution of the P species. A number of the features of the distributions for 1989 are recurring for the lake and are generally characteristic of eutrophic and hypereutrophic lakes (e.g., Wetzel 1983). Phosphorus, particularly SRP, is depleted from the epilimnion from spring through summer (Figure 5.39 a–f), associated with phytoplankton uptake and deposition. The interplay between the dynamics of phytoplankton growth and the various forms of P, and the status of this community with respect to the availability of P to support of growth, are described in detail in Chapter 6. During algae blooms, a metalimnetic minimum in TP sometimes is observed (Figure 5.39d). Phosphorus concentrations increase progressively in the hypolimnion soon after the onset of stratification, mostly in the form of SRP. Increasing concentrations of SRP occur with depth in the hypolimnion. This skewed vertical distribution reflects sediment release of P, combined with limited vertical mixing within the hypolimnion. Concentrations of SRP $>$ 1000 $\mu g \cdot L^{-1}$ have been observed annually in the lowermost layers in late summer and early fall since at least 1987. Substantial quantities of P are entrained into the epilimnion as it deepens with the approach to fall turnover (Figure 5.39).

The relationship between PP and chlorophyll (the most widely used measure of phytoplankton biomass) is shown in Figure 5.40, based on four years (1987–1990) of paired TP, TDP, and chlorophyll measurements made ·on samples from 1 m depth. The populations of these measurements are not evenly distributed throughout the range of observations.

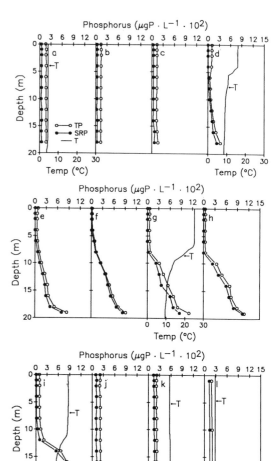

FIGURE 5.39. Characteristic vertical distributions of TP and SRP in Onondaga Lake, measured at the south deep station in 1989, with temperature profiles on six selected dates: (**a**) April 3, (**b**) April 24, (**c**) May 8, (**d**) May 22, (**e**) June 26, (**f**) July 17, (**g**) August 7, (**h**) August 29, (**i**) September 26, (**j**) October 10, (**k**) October 24, and (**l**) November 27.

Concentrations of chlorophyll explained 70% of the variability observed in PP, by linear least squares regression analysis (Figure 5.40). Variations in the cellular content of P and chlorophyll in phytoplankton, as well as temporally irregular contributions of abiotic forms of PP (see Chapter 8), are probably largely responsible for the variability in the relationship between PP and chlorophyll. Despite these

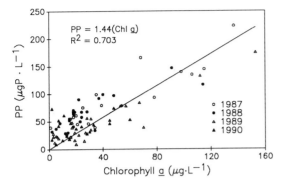

FIGURE 5.40. Evaluation of the relationship between PP and chlorophyll *a* in the upper waters of Onondaga Lake, for the period 1987–1990.

5.5.2.2 Distributions in the Upper Layer

The temporal distributions of TP, TDP, and PP in the upper waters (1 m) over the April–October interval of 1987, 1988, 1989, and 1990 are presented in Figure 5.42a–d. Means of triplicate analyses are presented with ±1 deviation bars; bars often do not appear because

limitations it is clear that the dynamics of PP in the upper waters of the lake largely track the temporal variations of phytoplankton biomass. The very high concentrations of PP in the upper waters in mid-June of 1987 (Figure 5.41a) reflect a major phytoplankton bloom. Lower concentrations were observed during a minimum in phytoplankton biomass (see Chapter 6) in mid-July (Figure 5.41b), and intermediate concentrations of PP were observed over the deeper mixed layer of late August (Figure 5.41c) during a secondary phytoplankton bloom. Particulate P concentrations in the hypolimnion largely reflect depositing material.

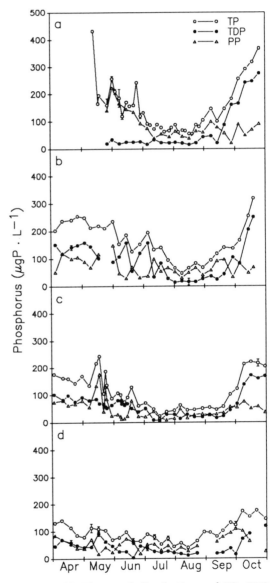

FIGURE 5.42. Temporal distributions of TP, TDP, and PP in the upper waters (1 m) of Onondaga Lake (south station): (**a**) 1987, (**b**) 1988, (**c**) 1989, and (**d**) 1990.

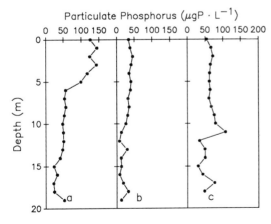

FIGURE 5.41. Profiles of PP in Onondaga Lake in 1987: (**a**) June 15, (**b**) July 13, and (**c**) August 24.

of the high precision of the analyses. Imbalance in the sources and sinks of TP in the upper waters over the spring to fall interval has been manifested by the rather strong seasonality in TP concentrations observed in recent years (Figure 5.42). The highest concentrations were observed near periods of turnover, the lowest in July and August (Figure 5.42 a–d). The abrupt increases observed in late spring and early summer in some years (e.g., Figure 5.42a and c) may reflect inputs from storm events (e.g., late June 1987), as well as temporary decreases in phytoplankton uptake and deposition. Dissolved P represents a substantial portion of TP in the spring to fall interval (38, 50, 55, and 42% of the TP on average in 1987, 1988, 1989, and 1990, respectively). This form of P is the dominant fraction at fall turnover and irregularly during stratification (Figure 5.42). A substantial fraction of TDP is probably organic, reflecting mineralization of algal biomass and zooplankton excretion. Large decreases in TP concentrations were achieved in the upper waters from 1988 to 1990, particularly in spring and early summer, as a result of reductions in P loading from METRO (see Chapters 3 and 9). Concentrations of PP show marked variation over the summer season, as well as considerable year-to-year variability (Figure 5.42), generally tracking the dynamics of phytoplankton biomass (see Chapter 6). The major peaks in PP have occurred either in the late spring to early summer or late summer to early fall intervals, periods when phytoplankton blooms commonly occur in the lake (see Chapter 6).

TABLE 5.16. Total phosphorus (TP) concentrations for boundary values for trophic categories suggested in the literature.

Trophic state	TP ($\mu g \cdot L^{-1}$)			
	(1)	(2)	(3)	(4)
Oligotrophic	4.3–11.5	10<	10<	9<
Mesotrophic	11.5–37.5	10–21.7	10–20	9–35
Eutrophic	>37.5	>21.7	>20	>35

(1) Auer et al. 1986
(2) Chapra and Dobson 1981
(3) Vollenweider 1975
(4) Vollenweider 1982

Concentrations of TP corresponding to trophic state boundaries have been developed by several researchers (Table 5.16). The most significant differences in these proposed delineations are associated with the boundary between mesotrophy and eutrophy. The most stringent limit is the earlier value ($20 \mu g \cdot L^{-1}$) of Vollenweider (1975). The New York State Department of Environmental Conservation (1993) recently established a "guidance value" for TP of $20 \mu g \cdot L^{-1}$ (epilimnetic summer mean). Average TP concentrations in the upper waters (1 m) of Onondaga Lake for the May–September interval of 1987, 1988, 1989 and 1990 are presented in Table 5.17, along with values obtained recently for other Central New York systems. Reductions in TP loading from METRO and repair of sewer breaks (see Chapter 3) caused the approximately 40% reduction in the epilimnetic summer average TP concentrations in Onondaga Lake from 1987 and 1988 to 1989 and 1990 (Table 5.17;

TABLE 5.17. Summer average TP concentrations in the upper waters of Onondaga Lake and other lakes in Central New York.

Lake	Year	TP ($\mu g \cdot L^{-1}$)	Source
Onondaga Lake	1987	131	—
Onondaga Lake	1988	142	—
Onondaga Lake	1989	93	—
Onondaga Lake	1990	81	—
Owasco Lake	1986	4.5	Effler et al. 1987
Otisco Lake	1988	17	Effler et al. 1989d
Little Sodus Bay	1988	42	Effler et al. 1989c
Cross Lake	1988	66	Effler et al. 1989b

see P model in Chapter 9). However, even at the 1990 TP concentrations, the lake exceeds the most lenient mesotrophy boundary by a factor of more than two, and the most stringent boundary and the New York State guidance value by a factor of four. The 1989 and 1990 concentrations continued to exceed those measured in other local lakes.

Healey and Hendzel (1980) proposed using seston composition ratios as an indicator of nutrient deficiency for phytoplankton growth (Table 5.18). Composition ratios of N:P and C:P are indicative of P deficiencies. Comparison of these composition ratios with values for seston in the watercolumn and in sediment traps for 1987, 1988, and 1989 demonstrate very low composition ratios for Onondaga Lake. Seston in Onondaga Lake are highly enriched in P. Carbon:P ratios for Onondaga Lake are much lower than other values reported in the literature (Hecky et al. 1993). This analysis is consistent with water column data of elevated TP concentrations and suggests that phytoplankton growth is not deficient of P in Onondaga Lake.

5.5.2.3 Water Column Content

The total water column content of TP and SRP are compared for the April–October interval of 1988 and 1990 in Figure 5.43 (a and b). The SRP content of the 10–19 m depth interval is included to illustrate the contribution of hypolimnetic concentrations. Other components of dissolved P make a very small contribution to the TDP of the lake's hypolimnion

(e.g., SRP represented 93% (n = 33) of TDP at a depth of 16 m in the summers of 1988 and 1989). The much lower TP content of the lake during spring turnover in 1990 was manifested mostly in the reduction in SRP content throughout the watercolumn. Most of the SRP was lost from the upper waters by mid-May of 1988, presumably due to phytoplankton uptake and deposition to the sediments. Soluble reactive P accumulated progressively in the hypolimnion in both years; the SRP in the hypolimnion in 1990 approached concentrations observed in 1988 by mid-summer. Most of the TP in the lake in the summer of both years was hypolimnetic SRP. Note that the initially lower TP content of the spring of 1990, brought about by reductions in external loading (see Chapters 3 and 9), was masked by late summer and early fall as a result of the high contribution of the hypolimnetic SRP pool to total lake content (Figure 5.43). Further, SRP from the 10–19 m depth interval made a smaller contribution in late October following the entrainment that accompanied fall turnover (Figure 5.43; also see Figure 5.39).

There has been disagreement concerning the extent to which the SRP that accumulates in anoxic hypolimnetic waters is available to support plant growth upon reaching the upper productive layer (via vertical mixing). Hutchinson (1957) indicates that much of the P released from the sediments of most lakes is reprecipitated with the attendant Fe^{2+} upon reaching oxygenated layers as a particulate $Fe(III)-PO_4$ complex and is largely unavailable for phytoplantkon. Some researchers (e.g.,

TABLE 5.18. Composition ratios used as indicators for nutrient deficiency for phytoplankton growth and water column and sediment trap values for Onondaga Lake composition ratios of C:N, N:P, C:P $(mol \cdot mol^{-1})$ and C:Chl$a$ $(mol \cdot g^{-1})$.

Ratio	Deficiency	Degree of nutrient deficiency			Onondaga Lake	
		None	Moderate	Severe	Water column	Sediment traps
C:N	N	<8.3	8.3–14	>14.3	4.4	5.0
N:P	P	<22	—	>22	8.4	7.9
C:P	P	<129	129–258	>258	37	39
C:CHLa	General	<4.2	4.2–8.3	>83	3.3*	3.5*

* After Healey and Hendzel 1980
ratio based on total chlorophyll, not chlorophyll a, no deficiency would still be indicated if adjustment were made (see Chapter 6)

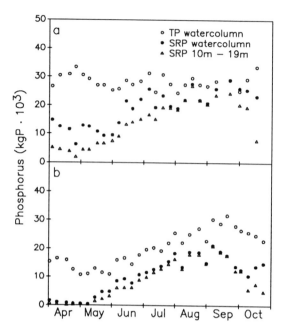

FIGURE 5.43. Temporal distributions of the total water column content of TP and SRP, and the content of SRP in the depth interval 10–19 m: (a) 1988, and (b) 1990.

Hutchinson 1957) have indicated the complex may slowly hydrolyze and restore some available P to the upper waters. Others (Kortmann et al. 1987; Stauffer and Lee 1974) have argued that this hypolimnetic P is incorporated into phytoplankton (i.e., totally available) upon entering the epilimnion. Nürnberg (1985) conducted an intensive P mass balance around the fall mixing period in Lake Magog and found that, despite high hypolimnetic Fe concentrations, no more than 30% of the hypolimnetic P formed $Fe(III)$-PO_4 (i.e., unavailable), while 30% was incorporated into plankton and 30% remained as SRP.

Review of the water column content of TP and SRP from late summer stratification through the onset of complete fall turnover in October in recent years (Figure 5.44a–d) is valuable in assessing the extent to which the hypolimnetic accumulations in Onondaga Lake may be precipitated upon reaching oxygenated layers. More than 50% of the P in the lake in August occurs as SRP, almost entirely in the hypolimnion (e.g., 10–19 m interval).

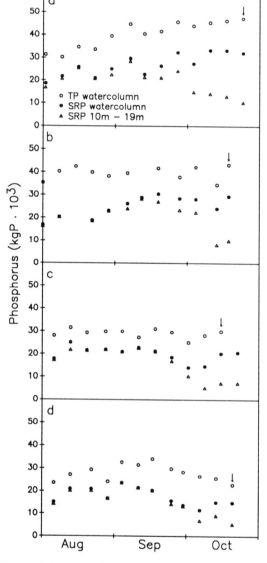

FIGURE 5.44. Transformations in the water column content of TP and SRP in Onondaga Lake from late summer through the onset of complete fall turnover (vertical arrow): (a) 1987, (b) 1988, (c) 1989, and (d) 1990.

Entrainment of SRP is apparent by its vertical redistribution with the approach to turnover; manifested by a smaller contribution from the lower layers (10–19 m) to the SRP content of the lake (Figure 5.44). The water column does not exhibit an abrupt decrease in the P content that would be expected if a substantial fraction of the entrained SRP was rapidly precipitated and deposited to the sediments. Decreases in TP coupled to entrainment events have been minor and not recurring (Figure 5.44). Although we cannot eliminate the possibility of precipitation of entrained SRP with Fe during the fall mixing period, such a process apparently is not a major sink for SRP during this interval. Most of the SRP that accumulates in the hypolimnion and subsequently is transported to the upper layers, appears to be available to support phytoplankton growth. This pattern is probably a manifestation of the high H_2S concentrations that develop in the hypolimnion (and ultimately the high SO_4^{2-} concentrations common to the lake, and the hypereutrophic state of the lake; see subsequent section of this chapter) which apparently limit the release of

Fe^{2+} to the water column. Hutchinson (1957) and Wetzel (1983) report that SO_4^{2-} has been added to certain small lakes to enhance primary productivity; presumably SO_4^{2-} is reduced to H_2S which binds with Fe^{2+}, thereby enhancing release of P from the sediments to the water column.

5.5.3 Recent History of Lake Concentrations: Upper Water

Historic P concentration data for the upper waters of Onondaga Lake are represented as summertime (May–September) median values for the 1968–1993 periods in Figure 5.45a. Annual TP loads estimated for the 1970–1990 period are presented for reference in Figure 5.45b (modified from Figure 3.58). Three different organizations have monitored

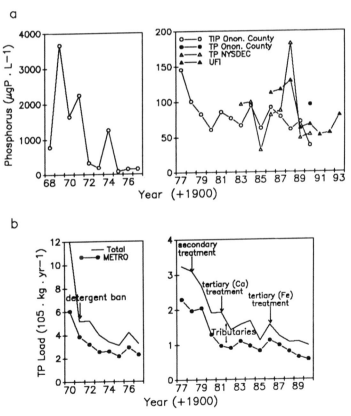

FIGURE 5.45. (a) Long-term trends in lake P concentrations in the upper waters (2 or 3 m) of Onondaga Lake as median values of summer concentrations, and (b) coupled temporal distribution of annual TP loading from METRO (see Chapter 3, modified from Effler et al. 1996).

P concentrations in the lake. Onondaga County data (Figure 5.45a) are measurements made on samples from a depth of 3 m, NYSDEC data are based on surface collections, and Upstate Freshwater Institute (UFI) values are based on samples collected from a depth of 2 m. These different monitoring depths do no significantly affect long-term trends (e.g., see Figure 5.39) among the databases. The median concentrations are used in the analysis of long-term trends (as previously for selected N species, Figure 5.38) because of the "robustness" of this statistic. The frequency of monitoring (i.e., population sizes) differed greatly among the three programs; Onondaga County sampled 7 to 8 times in most summers, NYSDEC only 3 to 4 times, and UFI at least 15 times. In light of the strong dynamics in P concentrations that are observed (e.g., Figure 5.42), these differences in temporal coverage could have significant effects on median concentrations determined for the summer period.

Two different forms of P are represented in the long-term data; TP and total inorganic P (TIP; Figure 5.45a). Most monitoring programs have adopted TP as the primary measure of P related to trophic state. Features of the TIP measure were reviewed in Chapter 3. The value of TIP is not a good measure of TP as it systematically excludes most forms of organic P. This exclusion is particularly troublesome for eutrophic and hypereutrophic lakes, because much of the P in the upper productive layers is often associated with algal biomass. Further, the fraction not detected in TIP measurements is subject to major variations because of seasonality in phytoplankton biomass. The analysis of paired TP and TIP data collected by Onondaga County from the upper waters (0, 3, and 6 m depths) of the lake in 1990 (Figure 5.46) illustrates the limitations of the TIP parameter. Total inorganic P was in general substantially lower than TP and it represented a variable fraction of TP (Figure 5.46). Further, the relationship between TIP and TP (Figure 5.46) may have limited utility in generating estimates of TP concentrations for earlier (e.g., before 1986) years, as there appears to be little reason to assume it would remain uniform over the decreases in P loading (Figure 5.45b)

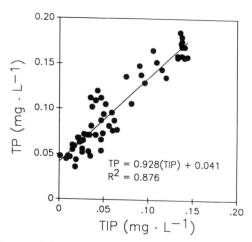

FIGURE 5.46. Evaluation of the relationship between TIP (total inorganic phosphorus) and TP in the upper waters (0, 3, and 6 m) of Onondaga Lake in 1990.

and in-lake concentrations (Figure 5.45a) that have occurred. In contrast, TIP is a good estimator of TP in the hypolimnion of Onondaga Lake, because most of the P in this layer is in the form of SRP.

Note the x-axis has been split in Figure 5.45a (as in Figure 5.45b) to accommodate the apparent major reductions in lake P concentrations over the period of record and to more clearly resolve annual variations in recent years. A number of inconsistencies emerge from a review of these data. The early lake TIP data (left panel of Figure 5.45a) are suspect because the abrupt changes observed are unrealistically large (e.g., increase from about 750 to 3600 $\mu g \cdot L^{-1}$ from 1968 to 1969 is not plausible, even for a grossly polluted urban lake before the detergent ban) and they are not generally correlated to annual loading estimates (Figure 5.45b). Undoubtedly a substantial decrease in TP occurred over this interval. The decreases in reported TIP concentrations in the late 1970s appear generally consistent with the reductions in TP loading (Figure 5.45b) over the same period. The lack of a consistent pattern between the TIP and NYSDEC TP data for the various years (Figure 5.45a) could reflect the differences in sampling frequency and/or analytical shortcomings. This problem is manifested for only 1 of 5 years (1989) when TIP and UFI TP concentrations are compared over the 1986–1990 period. Unfortunately, the overall P data base for the upper waters of the lake does not yield a clear and temporally detailed picture of the changes that have doubtless occurred in response to the generally progressive decreases in external loading (Figure 5.45). The annual dynamics manifested in the UFI lake TP data base over the 1987–1990 interval are supported by the success of a mechanistic TP model in simulating in-lake concentrations over that period (see Chapter 9). The reductions in lake concentrations starting in 1989 have previously been attributed in part to further reductions in loading from METRO.

5.5.4 Sediment Release of P

5.5.4.1 Background

Sediment release of P represents an internal load, in contrast to external loads that enter a lake via tributaries or discharges. Reclamation efforts for culturally eutrophic lakes have focused on the reductions of external loads of P (Uttormark 1978). Response to these efforts has been highly variable. Lake Washington responded rapidly and very positively to sewage diversion (Edmondson and Lehman 1981). However, other systems have responded more slowly and less positively (e.g., Ahlgren 1977; Larsen et al. 1981; Welch et al. 1986) to reductions in external loading, apparently in part due to elevated levels of internal loading from sediments (sediment feedback). Earlier, Devan and Effler (1984) documented the failure of Onondaga Lake's clarity to improve over the 1970–1981 period despite a ten-fold reduction in external P loading. Thus, for very practical, management-based reasons, lake scientists have given great attention to resolving, on a system-specific basis, the role sediment exchange plays in P cycling, to effectively anticipate response to reclamation efforts (Auer et al. 1993; Driscoll et al. 1993; Golterman 1977; Psenner and Gunatilaka 1988; Sly 1986).

Earlier analyses (Devan and Effler 1984; Effler et al. 1985b; Honstein 1981; Murphy 1978) of the sediment P release issue for Onondaga Lake implied an unusually low efflux from the sediments prevailed, because of regulation by $Ca-PO_4$ minerals, associated with the lake's very high Ca^{2+} concentrations. However, detailed lake data collected subsequently (e.g., Figures 5.39 and 5.43) offer clear evidence to the contrary, showing accumulation of SRP in the hypolimnion is a prominent feature of Onondaga Lake during summer stratification. Research in the late 1980s and early 1990s has continued to assess the magnitude of sediment P release in Onondaga Lake and to evaluate regulatory processes and mechanisms. Findings from three different components of this work are reviewed here: (1) analysis of water column data, (2) theoretical considerations, including chemical equilibrium modeling, and (3) microcosm experiments conducted to directly quantify release rates.

5.5.4.2 Review of Potential Regulating Processes

It is essential to understand and quantify the processes regulating sediment release, in order to effectively support (usually costly) management decisions to control P loading to lake systems. A number of mechanisms have been identified as potentially regulating the supply of P from sediments, including: redox (oxidation-reduction) reactions, adsorption, mineral phase solubility, mineralization of organic matter, and turbulence (e.g., Böstrom et al. 1988; Caraco et al. 1991; Wetzel 1983). Related factors that have been proposed as controlling sediment P release include: O_2, (e.g, Einsele 1936; Mortimer 1941; 1942), SO_4^{2-} concentration (e.g., Caraco et al. 1989, 1991; Hawke et al. 1989), pH (e.g., Anderson 1975; Stauffer 1985), and sediment Fe content (e.g., Baccini 1985; Lean et al. 1986).

Redox reactions, keyed to anoxia, have been found to regulate sediment release of phosphorus in most lakes (Mortimer 1971; Wetzel 1983). Much greater release of P from bottom sediments is generally observed in lakes which experience hypolimnetic anoxia than in lakes that remain oxygenated throughout the hypolimnion (Mortimer 1941, 1942, 1971; Nürnberg 1984). Early work by Einsele (1936) and Mortimer (1941, 1942) remains the most accepted statement of the processes regulating the release of P from lake sediments in which redox conditions are the primary factor (described as the Einsele/Mortimer model). An oxidized microzone (usually <50 nm thick) effectively limits the release of reduced materials (e.g., Fe^{2+}) and associated forms of dissolved P in lakes where oxygenated conditions prevail at the sediment-water interface. "The oxidized layer forms an efficient trap for iron and manganese, as well as phosphate, thereby greatly reducing transport of materials into the water..." (Wetzel 1983, p. 261). Phosphorus is believed to be mostly sorbed to oxidized forms of Fe (e.g., Fe oxyhydroxides) in this layer. However, in systems with more limited oxygen resources, abrupt releases of Fe^{2+} and dissolved P are observed following the loss of the microzone (reduction/dissolution of Fe oxides, hydroxides, and complexes) that accompanies the onset of watercolumn anoxia. The release of Fe^{2+} generally commences slightly before that of SRP (Mortimer 1941, 1942). Release generally continues during the remainder of the stratification period, manifested as progressively increasing quantities of SRP and Fe^{2+} in the anoxic layers. The molar quantity of Fe^{2+} released is characteristically much greater than the quantity of SRP (Larsen et al. 1981; Mortimer 1941, 1942).

The timing of Fe^{2+} and SRP releases subsequent to the development of anoxia, has been generally accepted (e.g., Wetzel 1983) as supporting the Einsele/Mortimer model (i.e., Fe activity) as regulatory with respect to sedimentary release of P. Time series plots of DO, Fe^{2+}, pH and SRP from just above the sediments of Otisco Lake in 1986 (Figure 5.47) depict an example of a lake in Central New York in which mobilization of sedimentary P is consistent with Mortimer's (1971) conceptualization of the microzone. Several weeks after the onset of anoxia (Figure 5.47a), the release of Fe^{2+} (Figure 5.47b) and subsequently SRP (Figure 5.47d) commenced. Note that the molar concentration of Fe^{2+} was substantially

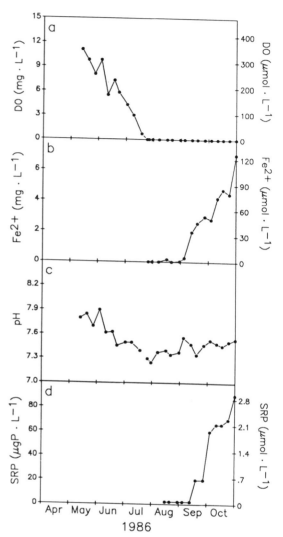

FIGURE 5.47. Temporal distributions of selected parameters in the bottom waters (18 m) of Otisco Lake in 1986: (**a**) DO, (**b**) Fe^{2+}, (**c**) pH, and (**d**) SRP.

greater than that of SRP. Phosphorus release was not coupled to changes in pH (Figure 5.47c) in this system.

According to the Einsele/Mortimer model, maintenance of oxygenated conditions at the sediment−water interface is critical to minimize or eliminate P release. Thus artificial oxygenation of anoxic hypolimnia (as a lake management option) should decrease the P loading by reducing the supply of P from sediments to the water column (Cooke et al. 1986; Gächter 1987; Taggart 1984).

The Einsele/Mortimer model has recently been criticized. Caraco et al. (1989, 1991) found that SO_4^{2-} concentrations explained more of the variability in sediment release of P across 23 study lakes than O_2 concentrations. Elevated rates of P supply have been observed in some lakes that remain aerobic (Böstrom et al. 1988). Böstrom et al. (1988) emphasized the importance of considering other chemical processes, as well as physical and biological processes when assessing sediment P release.

There is evidence that P cycling in calcareous lakes may differ from noncalcareous lakes (Böstrom et al. 1988, Stauffer 1985). For example, P availability in the sediments and overlying water of calcareous Lake Kinneret has been found to be regulated by Ca-PO$_4$ minerals (Staudinger et al. 1990). For systems exhibiting Ca-PO$_4$ mineral control of solution P concentrations, pH is an important water chemistry parameter. Gunatilaka (1982) reported that the adsorption of total PO_4^{3-} to the calcareous sediments of Neusiedlersee (Austria, Hungry) was strongly influenced by pH. Stumm and Leckie (1971) also showed that for calcareous systems total PO_4^{3-} was released (desorbed) from calcite with decreasing pH.

5.5.4.3 Evaluations of Regulatory Processes Based on Water-Column Data

Temporal trends in the concentrations of a number of the chemical parameters considered in the evaluation of P cycling in Onondaga Lake at a depth just above the sediment–water interface (19 m) are presented for the April–October interval of 1990 in Figure 5.48. Oxygen was depleted rapidly from the bottom waters following the onset of stratification in early May; anoxia developed within 5 to 6 weeks (Figure 5.48a). Within about 1 week of the onset of anoxia total Fe^{2+} was evident (Figure 5.48b), and soon after total H_2S release was observed (Figure 5.48c). The short time interval between the initial detection of the by-products of Fe^{3+} and SO_4^{2-} reduction, despite the rather significant difference in the

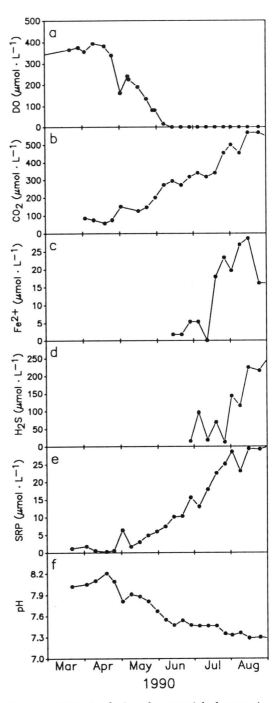

FIGURE 5.48. Analysis of potential factors influencing the release of P from the sediments of Onondaga Lake; bottom water (19 m) conditions from late March–August of 1990: (**a**) DO, (**b**) CO$_2$, (**c**) Fe^{2+}, (**d**) H$_2$S, (**e**) SRP, and (**f**) pH (from Driscoll et al. 1993).

redox potentials of these processes (Stumm and Morgan 1981; Böstrom et al. 1988), reflects the very large demand for electron acceptors in the lower waters due to the high level of productivity and subsequent deposition of organic carbon (see subsequent sections of this chapter).

The concentrations of total Fe^{2+} were considerably lower than values typically reported during anoxia for systems following the Einsele/Mortimer model (e.g., Mortimer 1941, 1942; Larsen et al. 1981). Further, these low concentrations appear to be inconsistent with the rather high SRP concentrations that develop in the bottom waters; the molar ratio of total Fe^{2+} to SRP was less than 1 for most of the stratification period (mean ratio ± std dev. = 0.52 ± 0.31). The relatively low total Fe^{2+} concentrations that prevailed were almost certainly a result of control by the very high total H_2S concentrations. Molar ratios of total Fe^{2+} to total H_2S were less than 1, and for most observations were less than 0.1 (mean ratio ± std. dev = 0.096 ± 0.1). Calculations with MINEQL (Driscoll et al. 1993) indicate that the bottom waters were oversaturated with respect to the solubility of FeS during most of the period of anoxia (mean SI ± std dev. = 1.4 ± 0.41). Individual particle analysis techniques have documented the presence of FeS precipitate in the bottom waters and lake sediments (Yin and Johnson 1984). These patterns suggest that elevated total H_2S concentrations restrict total Fe^{2+} availability in the lower waters during anoxic conditions.

The most conspicuous inconsistency in the Onondaga Lake conditions with respect to the Einsele/Mortimer model is the timing of P release (Figure 5.48d). The onset of the release of P coincided with the onset of thermal stratification and the initial increases in CO_2 (Figure 5.48b) and decreases in pH (Figure 5.48e). This occurred more than 5 weeks before the development of anoxia and more than 6 weeks before the appearance of measurable quantities of total Fe^{2+}. Clearly there are features of the sediment release of P in Onondaga Lake that do not fit the Einsele/Mortimer model; other processes apparently are important in regulating the mobilization of P from

the sediments of Onondaga Lake. Following turnover the pH of the lower waters was near 8.0. During stratification the increase in CO2 resulted in pH decreases to about 7.3. The coincidence of the decrease in pH and release of P is consistent with both the influence of pH on the solubility of Ca-P minerals (Stumm and Leckie 1971; Staudinger et al. 1990) and the progression of decomposition processes.

5.5.4.4 Evaluating the Role of Mineralization

Decomposition processes result in the mobilization of P incorporated in deposited organic material. Caraco et al. (1989, 1991) developed an approach to evaluate the potential importance of this recycle pathway. Depositing organic particles in most lakes have organic C:P ratios between 106 and 333 mol · mol^{-1}P (Caraco et al. 1991). The C:P value of 106 mol · mol^{-1}P corresponds to the stoichiometry expected for algal biomass growing under optimum nutrient conditions (Redfield ratio; Redfield 1958). Values of C:P in algal biomass greater than 100 are expected in lakes which are deficient in P (Table 5.18; Hecky et al. 1993). Recall that the ratio for particles in Onondaga Lake is substantially lower (C:P = 37, Table 5.18), indicating they are highly enriched in P. When the ratio of DIC + CH_4 release to SRP release in the lower waters falls near the C:P ratio of seston, mineralization reactions alone can explain P release from sediments. Release ratios (DIC + CH_4: SRP) greater than the C:P ratio of seston suggest that mineralized P is immobilized within sediments, whereas C:P release ratios less than the C:P ratio of seston indicate the release of P from sediments that was previously immobilized (Caraco et al. 1989, e.g., adsorption, mineral formation).

The relationship between DIC + CH_4 release and SRP release in Onondaga Lake, determined by monitoring the overlying water for the April–September interval of 1989, 1990, and 1991, is shown in Figure 5.49a–c. Values of DIC were calculated using thermodynamic relationships from measured values of alkalinity and pH, while correcting for temperature

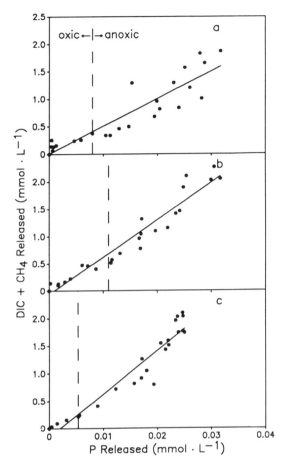

FIGURE 5.49. Concentrations of released carbon (DIC + CH$_4$) as a function of released phosphorus in Onondaga Lake at a depth of 18 m over the spring to late summer interval: (**a**) 1989, (**b**) 1990, and (**c**) 1991.

and ionic strength (Stumm and Morgan 1981) as described previously (e.g., Figure 5.20c). Demarcation of observations made under oxic and anoxic conditions is shown; note that considerable P release occurred before the onset of anoxia during the study years. The mean release of DIC + CH$_4$ relative to P release over the stratification period for the three years was 66 mol C · mol^{-1}P. Adjusting for ebullitive loss of CH$_4$ (33% of dissolved CH$_4$ release; see subsequent section), the sediment release ratio of DIC + CH$_4$ to P for 1989, 1990, and 1991 increases somewhat to 56, 74, and 87 mol C · mol^{-1}P, respectively (mean value of 72 mol C · mol^{-1}P for the three study years).

While the DIC + CH$_4$:SRP ratios from the lower waters are comparable or even less than values expected from mineralization of algal biomass with element stoichiometry corresponding to optium nutrition (106 mol C · mol^{-1}P), they are considerably greater than the C:P ratios of seston in Onondaga Lake (water column C:P 37 mol C · mol^{-1}P; sediment trap C:P 38 mol C · mol^{-1}P), indicating immobilization of P compared to organic C. This is consistent with sedimentary P retention implied from whole-lake mass balance analyses (Devan and Effler 1984; also see TP model of Chapter 9). This analysis of C:P ratios suggests that in Onondaga Lake P is supplied from sediments to the water column largely by mineralization of seston detritus. The release of P and CO$_2$ (DIC) from sediments to the water column under oxic conditions likely reflects the by-products of microbial mineralization of organic detritus. Evidently, the rate of P release from mineralization exceeds the rate of PO$_4$ retention by abiotic processes (e.g., adsorption, precipitation) resulting in the mobilization of SRP (a fraction of the deposited P) to the water column even under oxic conditions. The most likely processes (abioitic) responsible for immobilization of P in the lake sediments (the fraction of the deposited P not released) are either precipitation of Ca-PO$_4$ minerals or absorption of PO$_4$ to sediment surfaces.

Shifts in the composition of seston and related P cycling could accompany future reductions in external loading. As limiting concentrations of P for phytoplankton growth are approached (e.g., Chapter 6), the cellular content of P could decrease (i.e., C:P ratio increases), thereby decreasing the deposition of P from the upper waters (see Chapter 9) and perhaps decreasing sediment feedback.

5.5.4.5 Chemical Equilibrium Calculations: Ca-PO$_4$ Mineral Control

There are many Ca-PO$_4$ minerals that can potentially form in calcareous environments (e.g., monetite, brushite, octacalcium phosphate, whitlockite, hydroxyapatite, fluorapatite). In addition, total PO$_4^{3-}$ may be regulated by adsorption on calcite particles (Stumm and Leckie 1971). There is no direct evidence of specific Ca-PO$_4$ mineral formation in Onondaga Lake. However, using individual particle analysis, Honstein (1981) found that P in Onondaga Lake sediments was largely associated with Ca rich particles. Chemical equilibrium calculations with MINEQL showed conditions of oversaturation with respect to a number of relatively insoluble Ca-PO$_4$ minerals (e.g., hydroxyapatite, fluorapatite). The lower waters of Onondaga Lake are generally slightly oversaturated with respect to the solubility of whitlockite (Figure 5.50; mean SI ± std. dev. for the period from early May−early October 1989 = 0.48 ± 0.20). Conditions of oversaturation were evident in recent years as well as prior to the closure of the soda ash/chlor-alkali facility, when Ca^{2+} concentrations were higher but pH values were lower (1981 shown as an example year). As a result, whitlockite is used here as a reference mineral.

A chemical equilibrium "titration" was conducted with MINEQL to evaluate the potential response of Onondaga Lake to changes in water chemistry (Driscoll et al. 1993). This titration involved addition of CaCl$_2$ to simulate the response of Onondaga Lake to variations in Ca^{2+} loading due to changes in industrial discharge (Figure 5.51). The Ca^{2+} concentrations observed in the chemical equilibrium titration spanned from lower concentrations

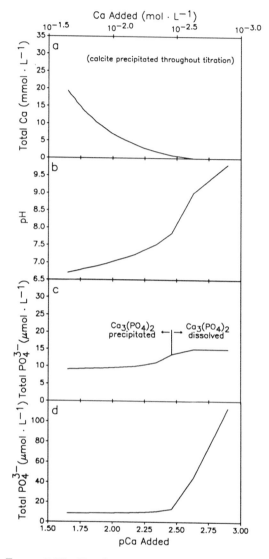

FIGURE 5.51. Simulations of the chemical titration of Onondaga Lake waters with incremental addition of CaCl$_2$ using the chemical equilibrium model MINEQL: (a) total calcium, (b) pH, (c) total PO$_4^{3-}$, whitlockite as a dissolved solid, and (d) total PO$_4^{3-}$, whitlockite as a fixed solid. (from Driscoll et al. 1993).

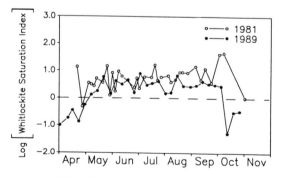

FIGURE 5.50. Values of saturation indexes (SI) with respect to the solubility of whitlockite (Ca$_3$(PO$_4$)$_2$) for the lower waters (19 m) of Onondaga Lake during the summer stratification period, 1981 and 1989 (from Driscoll et al. 1993).

observed in recent years following the closure of the industrial facility (e.g., 1989) to very high concentrations typical of lake conditions when the industrial facility was in full operation (e.g., 1980).

Calcium concentrations increased with addition of CaCl$_2$. Due to saturation of the system with respect to CaCO$_3$(s), most of the added Ca^{2+} precipitated as $CaCO_3(s)$ (Figure 5.51a). The increase in Ca^{2+} concentrations is coincident with decreases in pH and increased solubility of $CaCO_3(s)$. Acid production occurs when inputs of $CaCl_2$ are lost from solution by $CaCO_3(s)$ precipitation (see Equation (22)). The predicted response of total PO_4^{3-} to changes in added Ca^{2+} is interesting and may provide insight into the patterns in SRP in the lower waters of Onondaga Lake (Figure 5.51c and d). Total PO_4^{3-} showed little change in concentration over the high range of the $CaCl_2$ titration (added Ca from 4 to 22.4 mmol/L), despite a wide variation in Ca^{2+} and pH. This response is due to conditions of oversaturation and precipitation of $Ca_3(PO_4)_2(s)$. This pattern is consistent with observations that SRP concentrations released from Onondaga Lake sediments today are comparable to values in the early 1980s (e.g., 1980); although P inputs to the lake, have decreased in recent years. The similarity in the concentrations of SRP released from sediments under highly varying loading of P (Chapter 3) is consistent with Ca-PO_4 mineral control (though the time frame of equilibration of the sediments to changing deposition must also be considered), but also with the continued nutrient saturated status of the lake's phytoplankton (Chapter 6). At lower inputs of $CaCl_2$ (somewhat lower than currently experienced by Onondaga Lake) the solution becomes undersaturated with respect to the solubility of $Ca_3(PO_4)_2(s)$ and concentrations of total PO_4^{3-} increase. The extent of this increase depends on the pool of $Ca_3(PO_4)_2(s)$ available for dissolution. If the pool is limited, release of total PO_4^{3-} will be limited by the quantity available (e.g., dissolved $Ca_3(PO_4)_2(s)$ titration; Figure 5.51c). If the pool is large (e.g., fixed $Ca_3(PO_4)_2(s)$ titration) then a large increase in total PO_4^{3-} will occur (Figure 5.51c).

5.5.4.6 Estimation of Sediment P Release Rate(s)

Nürnberg (1987) has identified three methods for estimating internal P loads:

1. calculation of the change in the mass of P in the hypolimnion during anoxia from vertical concentration profiles,

2. by direct laboratory measurements using intact sediment cores, and

3. calculation by difference, using the annual P budget, with all other sources and sinks being determined independently.

Each of these methods has shortcomings. The third method is least desirable, because as a residual its accuracy depends on the accuracy of all the other inputs to the mass balance analysis. The other two approaches are more commonly used to support the development of mass balance P models. The laboratory measurement approach is perhaps most attractive to support modeling efforts as it is independent of the data against which predictions are to be compared. However, there are procedural concerns for the laboratory approach, including obtaining an undisturbed core with an intact sediment−water interface, and achieving experimental conditions in the laboratory that adequately reflect the ambient environment of the lake. Further, the appropriate application of representative laboratory results (typical units, $mg \cdot m^{-2} \cdot d^{-1}$) to lake-wide conditions must accurately accommodate areal extent and duration of release. The hypolimnetic accumulation approach has the disadvantage of failing to isolate the effect of sediment release, as dissolved P is transported upward from the hypolimnion via the processes of diffusion (e.g., Wodka et al. 1983) and entrainment (Effler et al. 1986e) and some P is undoubtedly released to the hypolimnion by decaying settling material. These two shortcomings are to some extent compensating (one is a sink; the other a source).

5.5.4.6.1 Hypolimnetic Accumulation of SRP

The progressive accumulation of SRP in the hypolimnion of the lake was previously illustrated (Figures 5.39 and 5.43). Accumulation is depicted for four consecutive years (1987−1990) in Figure 5.52(a−d), for which comprehensive SRP data are available. The demarcation depth for the upper bound of the hypolimnion specified here was 8 m, consistent with the observed vertical distribution of SRP (e.g., Figure 5.39 f−h) and the vertical

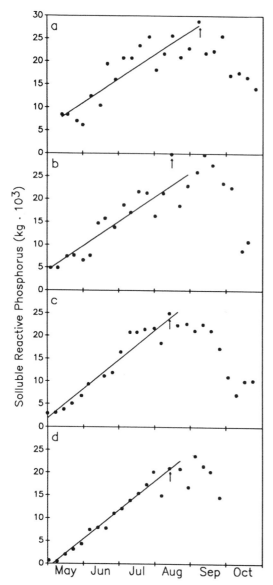

FIGURE 5.52. Net accumulation of SRP in the hypolimnion (8 m to bottom) of Onondaga Lake for four years: (**a**) 1987, (**b**) 1988, (**c**) 1989, and (**d**) 1990. Yearly maxima for estimates of release rate, according to Nürnberg (1987), identified with arrows.

(Owens 1987) were used, along with the SRP profiles, to estimate the total SRP content below 8 m.

The decreases observed annually in September reflect losses to the overlying waters associated with the deepening of the epilimnion with the approach to fall turnover. Nürnberg (1987, p. 1160) recommended that the internal load be "determined as the difference between maximum hypolimnetic P mass and "background" P mass just before the onset of anoxia". The maximum values invoked in our application of Nürnberg's approach are identified by arrows in Figure 5.52. The internal P loads calculated in this way are presented in Table 5.19 in units of $kg \cdot yr^{-1}$ (or summer $^{-1}$), and in units of $kg \cdot d^{-1}$ over the period of progressive. accumulation. The latter representation appears appropriate for Onondaga Lake, based on the nearly linear time course of accumulation observed annually. The net accumulation rate of SRP, determined as the slope of the linear least squares regression of the hypolimnetic total SRP content plots (Figure 5.52 a−d) for the June−August intervals of each year, is also presented in Table 5.19, along with the standard error. These two approaches yield similar internal loading rates for these data sets. We prefer the linear regression approach, as it utilizes much of the annual data base and reduces the possibility of an unrealistic value based on a single outlier. An apparent areal release rate ($mgP \cdot m^{-2} \cdot d^{-1}$) is presented in Table 5.19, that is based on the accumulation rate and utilizes the contour area corresponding to a depth of 8 m. The range of estimates for the four years obtained in this way is $22-27 \, mgP \cdot m^{-2} \cdot d^{-1}$ (Table 5.19).

Differences in the calculated release rate of P among the four years by hypolimnetic accumulation may reflect analytical and sampling limitations, and the influence of vertical mixing events (e.g., irregular occurrence of entrainment based mixing events). The linearity of the accumulation was particularly strong in 1989 and 1990 (Figure 5.52c and d). The relative variability among years is small (Table 5.19) compared to values reported by Nürnberg (1987). These results

segmentation adopted in the TP model for the lake (see Chapter 9). Measurements of SRP were made at 1 m depth intervals in 1987 and 1988, and at 2 m depth intervals in 1989 and 1990 (also at 19 m on a number of occasions). The recent hypsographic data for the lake

TABLE **5.19.** Sediment phosphorus release rates for Onondaga Lake, determined from hypolimnetic accumulations in two ways, for four years.

Year	P-Release rate			
	Nürnberg (1987) kg · summer^{-1}	Technique kg · d^{-1}	Linear regression kg · d^{-1} (std. error)	Areal rate* mgP · m^{-2} · d^{-1}
1987	21,800	210	171	22	22
1988	25,000	213	174	21	22
1989	22,200	178	209	15	27
1990	20,000	200	206	7	26

*Calculated from linear regression flux (kg · d^{-1}) and the sediment contour area for a depth of 8 m.

reflect essentially equivalent sediment release rates for the summer stratification intervals of the four years. Assessment of P deposition over the same period, through analyses of archived sediment trap collections, together with the apparent release rates, will indicate the extent to which the P cycle of the hypolimnion has responded on the shortterm to changes in external loading over the same period (see Chapter 3).

5.5.4.6.2 Laboratory Determinations from Core Samples

Phosphorus release experiments have been conducted on intact (effective recovery of sediment–water interface) core samples by Auer et al. (1993) and Penn et al. (1993). Procedural details are presented in each of these sources. Gases were bubbled into the water overlying the sediments during the experiment to maintain the desired oxygen (oxic/anoxic) and pH conditions, and to enhance mixing (without disturbing the sediment–water interface). The earlier experiments of Auer et al. (1993) were conducted in 1987, on core samples collected from three deep water locations (south deep, mid-lake, and north deep), in late July and early September. Release rates were determined for 17 core samples. The fluxes (of SRP) ranged from 9.0 to 21.3 mgP · m^{-2} · d^{-1}; the average was 13.4 mgP · m^{-2} · d^{-1} (Auer et al. 1993). The reproducibility of the experiments compared favorably with results published for other systems (Nürnberg 1987). No statistically

significant differences in release rates were observed among the three coring locations, or between the July and September collections.

The results of Auer et al. (1993) are compared to release rates reported for lakes of different trophic states in the review of Nürnberg (1988) in Figure 5.53. The findings for Onondaga Lake fall in a range expected for eutrophic to hypereutrophic lakes (Figure 5.53). Note however, that these fluxes are less than those estimated from hypolimnion accumulations (Table 5.19). Auer et al. (1993) calculated that this internal load (based on a mean rate of 13 mgP · m^{-2} · d^{-1}) was the second largest source of P to the lake over the mid-May–mid-October interval of 1987 after METRO, representing nearly 25% of the total

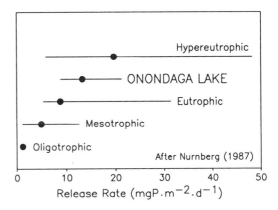

FIGURE **5.53.** Laboratory sediment release results for Onondaga Lake (1987) from a perspective of rates reported for lakes of different trophic states (after Nürnberg 1987).

TABLE 5.20. Lower water (19 m) chemistry and stratification conditions at time of core sample collections in 1992.

Sampling date	Stratification status	SRP ($\mu g \cdot L^{-1}$)	Oxygen ($mg \cdot L^{-1}$)	pH
May 1*	Prestratification	50	10	8.0
June 23	Stratified	700	0	7.7
Aug. 6	Stratified	900	0	7.5
Nov. 9**	Post-turnover	500	7	7.9

From Penn et al. 1993
* Cores had ~1 mm thick brown (e.g., "ferric") layer at interface
** Cores had no "ferric" layer, but apparently a microbial microaerophilic mat

input. However, only a fraction of the P released from sediments makes its way to the upper productive waters during summer stratification (mediated by limited vertical mixing; see Chapter 9).

Additional sediment core experiments were conducted in 1992. In these experiments the seasonality of the P release rate was evaluated by conducting experiments on cores collected over a broader time interval. The timing of core collections represented a wide range of lower water conditions that may influence P release from sediment; salient water chemistry and stratification characteristics of the overlying water at the time of sampling are summarized in Table 5.20.

Strong seasonality of sediment P release is implied by the results (Figure 5.54). A continuous distribution, which should be considered speculative, has been fit to the limited temporal data (Figure 5.54). It is likely that the sediment area (i.e., range of lake depths) over which the fluxes of Figure 5.54 apply varies seasonally; e.g., as the depth boundary of anoxia changes. A peak in release (>30 $mgP \cdot m^{-2} \cdot d^{-1}$), of uncertain duration (e.g., days to weeks), accompanies the onset of anoxia in spring. Rates then decrease to a level of 9–15 $mgP \cdot m^{-2} \cdot d^{-1}$, that appears to prevail for the anoxic summer stratification period (Figure 5.54). These later fluxes are consistent with those reported by Auer et al. (1993) for the summer of 1987. The average of the 6 measurements (Figure 5.54) is 15.1 $mg \cdot m^{-2} \cdot d^{-1}$; this may be a representative average for the spring through fall interval. Conditions in the late fall to spring interval are highly uncertain (dashed line, Figure 5.54). There are not recent detailed oxygen data from the lake during ice-cover to establish the persistence of oxygen at the lake bottom through this interval. The processes responsible for the apparent seasonality in the core release experiments have yet to be resolved.

5.6 Anoxic Organic Carbon Decomposition and the Distribution of Related Chemical Species

5.6.1 Introduction

The decomposition of organic carbon is a fundamental process in aquatic ecosystems, as it results in the release of inorganic carbon

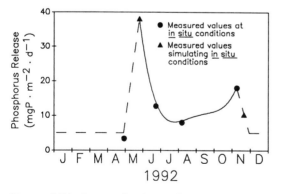

FIGURE 5.54. Seasonality in P release from Onondaga Lake sediments, according to laboratory experiments on sediment core samples, 1992.

and nutrients and provides electrons to electron acceptors such as oxygen and oxidized forms of sulfur, nitrogen, iron and manganese (Kelly et al. 1988). Decomposition processes include reduction of O_2 (aerobic decomposition), Mn^{4+} (Mn reduction), NO_3^- (denitrification), Fe^{3+} (Fe reduction), and SO_4^{2-} (SO_4^{2-} reduction), and CH_4 production (methanogenesis). The thermodynamic propensity for the various reactions is important in regulating the order in which processes proceed. A simple plausible model of organic material decomposition, that has been in part supported by field studies (e.g., Froelich et al. 1979; Kelly et al. 1988), is based strictly on thermodynamic considerations. In other words, certain electron acceptors are energetically favored and therefore are preferentially reduced until they are consumed. Then the next most energetically favorable oxidant is utilized. In nature the different reduction processes have been observed to proceed simultaneously in certain systems (e.g., Schafran and Driscoll 1987). The predicted sequence of reactions is presented in Table 5.21 (most thermodynamically favored at top of table) for organic matter of the Redfield composition (Redfield 1958; $(CH_2O)_{106}(NH_3)_{16}$ (H_3PO_4); the oxidation state of organic C is zero). The decomposition/ reduction processes are largely localized in the sediments or at the sediment-water interface (e.g., Kelly et al. 1988; Schafran and Driscoll 1987).

Oxygen is the most electronegative acceptor, and is the most important electron acceptor in oxic aquatic environments, despite the presence of other electron acceptors (Kelly et al. 1988). Once O_2 is consumed in the hypolimnion (see earlier sections of this chapter), anoxic decomposition processes proceed, in a sequence that tends to track the free energy considerations presented in Table 5.21 (see subsequent discussion). Note, there is little difference in the free energy associated with SO_4^{2-} reduction and methanogenesis. Other factors, besides free energy, can affect the occurrence and relative importance of the various decomposition processes, including concentrations of organic carbon (the major electron donor) and the various electron acceptors, the availability and sites of reduction of the electron acceptors (Kelly et al. 1988), and solution composition, particularly pH. Note that the electron acceptors of Table 5.21 fall into two groups: (1) those that are relatively soluble in the hypolimnion and diffuse into the sediments (O_2, SO_4^{2-}, NO_3^-), and (2) those that are highly insoluble (Fe^{3+} and Mn^{4+}) and are reduced within the sediments or at the sediment–water interface (Kelly et al. 1988). Moreover, the by-products of electron acceptors can be subdivided into

TABLE 5.21. Oxidative reactions for organic material of the Redfield (1958) composition.

	Equation number
$(CH_2O)_{106}(NH_3)_{16}(H_3PO_4) + 138O_2 \rightarrow 106CO_2 + 16HNO_3 + H_3PO_4 + 122H_2O$ $\quad G° = -3190 \, kJ \cdot mol^{-1}$ (glucose) $\qquad CO_2:O_2 = 0.77$	(5.36)
$(CH_2O)_{106}(NH_3)_{16}(H_3PO_4) + 236MnO_2 + 472H^+ \rightarrow 236Mn^{+2} + 106CO_2 + 8N_2 + H_3PO_4 + 366H_2O$ $\quad G° = -3090 \, kJ \cdot mol^{-1}$ (Birnessite) $\qquad CO_2:Mn^{2+} = 0.45$	(5.37)
$(CH_2O)_{106}(NH_3)_{16}(H_3PO_4) + 84.8HNO_3 \rightarrow 42.4N_2 + 106CO_2 + 16NH_3 + H_3PO_4 + 148.4H_2O$ $\quad G° = -2750 \, kJ \cdot mol^{-1}$ $\qquad CO_2:NO_3^- = 1.25$	(5.38)
$(CH_2O)_{106}(NH_3)_{16}(H_3PO_4) + 424FeOOH + 848 H^+ \rightarrow 424Fe^{2+} + 106CO_2 + 16NH_3 + H_3PO_4 + 742H_2O$ $\quad G° = -1330 \, kJ \cdot mol^{-1}$ (FeOOH) $\qquad CO_2:Fe^{2+} = 0.25$	(5.39)
$(CH_2O)_{106}(NH_3)_{16}(H_3PO_4) + 53SO_4^{2-} + 106H^+ \rightarrow 53H_2S + 106CO_2 + 16NH_3 + H_3PO_4 + 106H_2O$ $\quad G° = -380 \, kJ \cdot mol^{-1}$ $\qquad CO_2:SO_4^{2-} = 2$	(5.40)
$(CH_2O)_{106}(NH_3)_{16}(H_3PO_4) \rightarrow 53CH_4 + 53CO_2 + 16NH_3 + H_3PO_4$ $\quad G° = -350 \, kJ \cdot mol^{-1}$ $\qquad CO_2:CH_4 = 1$	(5.41)

Equations from Froelich et al. 1979

two groups: (1) those materials which can be released to the water column and subsequently reoxidized (e.g., Fe^{2+}, H_2S, dissolved CH_4), and (2) those materials which are permanently lost (i.e., N_2 formation by denitrification, ebullition of CH_4, precipitation of FeS).

Heretofore, only Kelly et al. (1988), Schafran and Driscoll (1987), and Mattson and Likens (1993) have presented quantitative data demonstrating the relative importance of all the electron acceptors (Table 5.21) to overall anoxic decomposition for lakes. The three test systems of Kelley et al. (1988) were soft water Canadian Shield Lakes: two were artificially eutrophied and one was artificially acidified. Rates of reduction of individual anoxic terminal electron acceptors and the rate of accumulation of CH_4 in the hypolimnia were measured and corrected for inputs and losses via vertical mixing (Kelly et al. 1988). Methanogenesis accounted for most (72–82%) of the anoxic decomposition; sulfate reduction was the next most important process (16–20%; Kelly et al. 1988). In Mirror Lake, New Hampshire, Mattson and Likens (1993) observed an average decomposition rate of $5.3\,\text{mmol} \cdot m^{-2} \cdot d^{-1}$. Of this, 43% occurred through aerobic decomposition, while alternate electron acceptors (SO_4^{2-}, NO_3^-, Fe^{3+}, methanogenesis) together accounted for 20%. In this investigation 37% of the observed decomposition could not be accounted for by depletion of electron acceptors. Other investigators have measured the relative importance to organic carbon decomposition of selected pairs of electron acceptors (Cappenberg et al. 1984; Ingvorsen and Brock 1982; Jones and Simon 1980; Martens and Klump 1980). In this section we:

1. review the nature of the various anoxic decomposition processes,
2. describe the temporal and vertical patters of anoxic terminal electron acceptors and coupled reduced species in the hypolimnion of Onondaga Lake (see a previous section for description of the loss of O_2), and thereby the sequence of the decomposition processes, and

3. quantify the relative contribution of each electron acceptor and related metabolic processes to decomposition of organic C and development of hypolimnion oxygen demand in Onondaga Lake.

5.6.2 Review of Anoxic Decomposition Processes
5.6.2.1 Denitrification

Denitrification is conducted by many heterotrophic facultative anaerobic bacteria under very low DO ($\sim0.2\,\text{mg} \cdot L^{-1}$) or anaerobic conditions (Seitzinger 1988). These bacteria utilize NO_2^- or NO_3^- as the terminal electron acceptor during the oxidation of organic matter and produce N_2, NO, or N_2O (Payne 1973):

$$NO_3^- \rightarrow NO_2^- \rightarrow NO \rightarrow N_2O \rightarrow N_2 \quad (5.42)$$

Nitrogen gas (N_2) is the dominant end product, which is largely lost from the system via ebullition. The amount of organic carbon oxidized by denitrification can be estimated according to equation (38) of Table 5.21; approximately 1.25 g-atoms of organic carbon are stabilized per mole of NO_3^- reduced.

There are two sources of NO_3^- for sediment denitrification: NO_3^- diffusing into the sediments from the water column, and NO_3^- produced in the sediments via nitrification of NH_4^+ (Seitzinger 1988). Groundwater inputs can represent a third source in certain systems. Methods used to assess denitrification rates have been reviewed by Knowles (1982) and Seitzinger (1988). Estimates based on the loss of NO_3^- and NO_2^- from anoxic hypolimnia may be false low in many cases, as the production of NO_3^- within the oxygenated layers of the sediments is not accommodated (Seitzinger 1988).

Denitrification occurs in sediments and in hypolimnia with low or no oxygen (Brezonik and Lee 1968; Keeney 1973; Seitzinger 1988). However, the rates of denitrification in sediments is generally much higher than found in the water column (Seitzinger 1988; Wetzel 1983). Seitzinger (1988) reports a range of denitrification for lake sediments of 2 to

$171\,\mu mol\,N \cdot m^{-2} \cdot hr^{-1}$. The range for anoxic hypolimnetic water reported is 0.2 to 1.9 mol $N \cdot L^{-1} \cdot d^{-1}$, not including the unusually high value reported for Lake Kinneret (Seitzinger 1988). Note the units reported for sediment de-nitrification is consistent with a zero order reaction, whereas the units for the watercolumn rate reflect first order kinetics. Seitzinger (1988) observed no obvious relationship between the rate of denitrification and trophic state.

5.6.2.2 Sulfate Reduction

Sulfate reduction is conducted by heterotrophic anaerobic bacteria. The best known SO_4^{2-} reducing bacteria belong to the genus *Desulfovibrio* (Hutchinson 1957). These bacteria use SO_4^{2-} as the terminal electron acceptor during the oxidation of organic matter and produce H_2S, as represented by (Westrich and Berner 1984)

$$2CH_2O + SO_4^{2-} \rightarrow H_2S + 2HCO_3^- \quad (5.43)$$

where CH_2O represents the solid-phase organic detritus (Ramm and Bella 1974; Westrich and Berner 1974). Note one mole of H_2S is produced per mole of SO_4^{2-} reduced (Equation (5.43)). Two g-atoms of organic carbon are stabilized per mole of SO_4^{2-} reduced (Equations (5.40) and (5.43)). Sulfate reduction is the major anaerobic metabolic pathway in marine systems. However, it has been found to be a minor to secondary pathway compared to methanogenesis in lakes that have been studied to date (e.g., Ingvorsen and Brock 1982; Kelley et al. 1988). This limited contribution to anaerobic decomposition in lakes has largely been attributed to low inputs of SO_4^{2-} in most freshwater systems (Ingvorsen et al. 1981; Molongoski and Klug 1980; Smith and Klug 1981) and the fact that SO_4^{2-} is a weak oxidant relative to other electron acceptors. A SO_4^{2-} concentration of 0.2 mmol $\cdot L^{-1}$ (higher than found in most lakes; Wetzel 1983) has been reported to be non-limiting to SO_4^{2-} reduction and inhibiting to methanogenesis (Winfrey and Zeikus 1977). Other investigators have presented evidence that the two anaerobic processes proceed in parallel in certain lakes (e.g., Ingvorsen and Brock 1982; Kelly et al. 1988).

Hydrogen sulfide can also be released as a by-product of the degradation of protein-bound S; however, the principal source of H_2S production is the SO_4^{2-} reduction process (Effler et al. 1988; Ingvorsen and Brock 1982; Ramm and Bella 1974). Hydrogen sulfide released early after the onset of anoxia commonly forms insoluble precipitates with metals, particularly FeS (Hutchinson 1957). Hydrogen sulfide released in excess of metal concentrations accumulates during the stratification period. Sulfate reduction has been documented in lake sediments (Ingvorsen et al. 1981; Molongoski and Klug 1980) and in anoxic hypolimnia (e.g., Ingvorsen and Brock 1982). The vast majority of SO_4^{2-} reduction occurs in the sediments if SO_4^{2-} is available. For example, only 15% of the depletion in SO_4^{2-} in the hypolimnion of Lake Mendota could be accounted for by SO_4^{2-} reduction in the water (Ingvorsen and Brock 1982). Factors controlling the rate of SO_4^{2-} reduction, include the concentration and metabolizability of the organic substrates (Westrich and Berner 1984), and temperature (Aller and Yingst 1980; Ingvorsen et al. 1981; Jorgensen 1977). Sulfate reduction increases exponentially with temperature; Q_{10} values obtained for different systems and a variety of experimental techniques fall within the rather narrow range of 2.9 to 3.5 (Abdollahi and Nedwell 1979; Aller and Yinst 1980; Ingvorsen et al. 1981; Jorgensen 1977).

5.6.2.3 Methanogenesis

Methanogenesis is conducted by heterotrophic anaerobic bacteria. There are four major genera and eight taxonomically described species (Rudd and Taylor 1980). The stoichiometry of the methanogenesis process is described by Equation (5.41) of Table 5.21. Note that two moles of organic carbon are stabilized for every mole of CH_4 produced. The final step in CH_4 formation takes place by one of two processes, lithotrophic reduction of

CO_2 by enzymatic addition of hydrogen derived from organic acids

$$CO_2 + 8H \rightarrow CH_4 + 2H_2O \quad (5.44)$$

or acetoclastic conversion of organic acids to CH_4 and CO_2

$$*CH_3COOH \rightarrow *CH_4 + CO_2 \quad (5.45)$$

The second process is usually considered to be dominant (Rudd and Taylor 1980).

Methanogenesis is an ubiquitous process in anaerobic freshwater and marine sediments (Rudd and Taylor 1980). Methane production is not quantitatively important within the water columns of anoxic hypolimnia (Rudd and Hamilton 1978; Zeikus and Winfrey 1976). In eutrophic lakes with anoxic hypolimnia, CH_4 diffusing upward from the sediments accumulates in the stratified layers. However, unlike H_2S, CH_4 has low solubility. The solubility of CH_4 with respect to temperature and pressure (depth) has been described by DiToro et al. (1990) using the following expression, based on the solubility data of Yamamoto et al. (1976),

$$C_{s\text{-}CH4} = 25\left[1 + \frac{H_o}{10}\right] \cdot 1.024^{(20\text{-}T)} \quad (5.46)$$

in which $C_{s\text{-}CH4}$ = saturation concentration of CH_4 (mg $CH_4 \cdot L^{-1}$), H_o = depth of the water column (m), and T = temperature (°C).

When the production of CH_4 within the sediments exceeds the rate of its diffusive removal, the total gas pressure within the pore water begins to exceed ambient pressure (the sum of hydrostatic and atmospheric pressures), a critical concentration for bubble formation is reached (Klots 1961) and ebullition (escape for gas bubbles) can occur. It is because of this phenomenon that interstitial waters of lake sediments are rarely oversaturated with respect to the solubility of CH_4 (Rudd and Taylor 1980). The flux of CH_4 from sediments as ebullitive gas represents a loss of carbon from the system, because most of the CH_4 released as bubbles escapes to the atmosphere. During vertical transport a fraction (usually minor) of the gas phase redissolves into the water column. This contributes to the accumulation of CH_4 in anoxic hypo-

limnia, while the fraction that dissolves in the epilimnion is either aerobically oxidized or evades to the atmosphere (Rudd and Taylor 1980). Other gases are routinely found in ebullitive gases, because they are stripped out of interstitial, and to a lesser extent overlying, water. Usually N_2 is present at concentrations greater than all other gases except CH_4. Most of the N_2 lost via ebullition is produced through denitrification.

Methane production has been shown to be a function of the deposition of organic carbon (Robertson 1979), a flux that is directly related to trophic state. Based on a study of five eutrophic lakes, Kelly and Chynoweth (1981) concluded that CH_4 production is linearly related to organic deposition. Methane production is also a function of temperature (Zeikus and Winfrey 1976). However, this factor is probably of secondary importance in most deep north temperate lakes, as the temperatures of overlying hypolimnetic waters are low and tend to be similar from lake to lake and year to year. Thus, substantial changes in CH_4 release for a lake over a period of years is probably a good indicator of a change in trophic state (e.g., Kelly and Chynoweth 1981). The contribution of ebullition to overall CH_4 release is expected to be greater as the level of primary productivity increases.

5.6.2.4 Iron and Manganese Reduction

These reduction processes are also conducted by heterotrophic bacteria. The stoichiometry of these oxidation reactions (redox) is represented in Table 5.21 (Equations (5.37) and (5.39)). The contribution of these processes to overall stabilization of deposited organics appears to be impeded by the concentrations of oxidized Mn and Fe in the near-surface sediments and the relatively low availability (due to low solubility) of these materials (Kelly et al. 1988). Kelly et al. (1988) observed Fe^{2+} to be mobilized at a higher rate than Mn^{2+}, indicating slower effective kinetics for the Mn reduction process. Some Fe reduction has been observed to occur in an anoxic hypolimnion (e.g., depositing material; Davidson et al. 1981). Kelly et al. (1988) found that Fe reduction accounted for from 1.3 to 4.5% of the anoxic decomposition in three productive Canadian Shield lakes, Mn reduction accounted for 0.2 to 0.6% of the anoxic decomposition.

5.6.3 Identification of Decomposition Processes in Onondaga Lake

Profiles of temperature and selected chemical species are presented for four dates covering the mid-May–late August period of 1989 in Figure 5.55, to illustrate the extent and progression of the various decomposition processes that occur annually in Onondaga Lake. Temperature profiles are based on measurements made at 0.5 m intervals; DO, Fe^{2+}, H_2S and CH_4 measurements were made at 1.0 m intervals, and the other parameters (with the exception of Mn; see Figures 5.55b–2 and 5.55d–2) were obtained at 2.0 m intervals. More systematic descriptions of the distributions of these chemical species are presented elsewhere in this chapter. Year-to-year differences in the timing of the distributions occur as a result of natural variations in the stratification regime (Owens and Effler 1989) associated with meteorological variability. The selected dates are representative of conditions observed in late spring (just after the onset of stratification; Figure 5.55a), early summer (Figure 5.55b), mid-summer (Figure 5.55c), and late summer (Figure 5.55d). Note that depletion of oxygen and NO_3^- plus NO_2^- and accumulations of the reduced substances and decomposition end-products are generally skewed towards the sediments, indicating the localization of the related decomposition processes at the sediment–water interface or within the sediments.

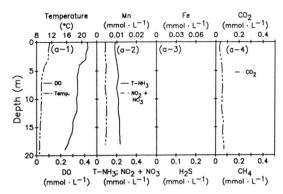

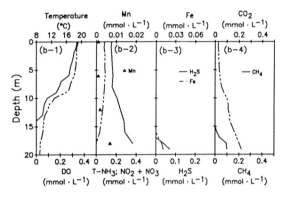

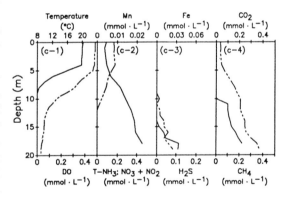

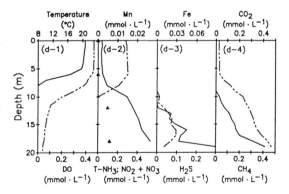

FIGURE **5.55.** Profiles of temperature and selected chemical species (DO, Mn, NO_3^- plus NO_2^- , Fe^{2+}, H_2S, CH_4 and CO_2) in Onondaga Lake for four dates in 1989: (**a**) May 15, (**b**) June 19, (**c**) July 17, and (**d**) August 21, 1989.

The time progression of species concentrations represented by the profiles of Figure 5.55a–d is consistent with the sequence of decomposition processes according to thermodynamic considerations presented in Table 5.21. However the temporal separation of the processes is somewhat confounded by: (1) the high rates of decomposition in the lake, (2) the occurrence of a range of redox conditions in sediments which allows for the energetically different processes, to proceed simultaneously, and (3) the long time interval (monthly) between measurements utilized in Figure 5.55. Some oxygen depletion was apparent in the hypolimnion soon after the onset of stratification (Figure 5.55a–1), which was accompanied by some accumulation of CO_2 (Figure 5.55a–4) associated with aerobic decomposition. No manifestations of denitrification (Figure 5.55a–2), or less energetically favored decomposition processes (Figure 5.55a–3, a–4), were evident in the hypolimnion under aerobic conditions.

The lower 6 m of the lake were anoxic by mid-June (Figure 5.55b–1). Sedimentary release of Mn (presumably in the form of Mn^{2+}; Figure 5.55b–2), depletion of NO_3^- plus NO_2^- (Figure 5.55b–2), release of Fe^{2+}, H_2S (Figure 5.55b–3), and CH_4 (Figure 5.55b–4), and the further accumulation of CO_2, was evident on this date. These transformations indicate the occurrence of all of the decomposition processes (6) listed in Table 5.21 in Onondaga Lake. Note that the concentrations of Mn and Fe^{2+} remained much lower than the concentrations of T-NH_3, H_2S, CH_4 and CO_2. The depth interval of NO_3^- plus NO_2^- depletion was somewhat smaller than that of O_2 throughout the summer, associated with the lower electronegativity of the denitrification process. Accumulation of ferrous iron, H_2S and CH_4 was limited to the anoxic layers. The depth interval of anoxia expanded during the summer, as did the intervals of NO_3^- plus NO_2^- depletion, and reduced substances accumulation (Figure 5.55c and d). The progressive accumulation of the various reduced species in the hypolimnion through the summer reflects the continuing decom-

position of organic material by the various processes identified in Table 5.21.

The major decomposition processes after the onset of anoxia and the depletion of NO_3^- and NO_2^-, were SO_4^{2-} reduction and methanogenesis. The much smaller molar concentrations of Mn and Fe^{2+} were probably a manifestation of the limited availability of oxidized forms of these materials in the lake's sediments. The availability of Fe may be further diminished by the high ambient H_2S concentrations, as FeS is highly insoluble (Stumm and Morgan 1981). Precipitation of metal sulfides may affect the vertical concentration profiles of these constituents, and cause underestimation of the role Mn and Fe reduction play in decomposition (e.g., Kelly et al. 1988) based on the accumulation of the reduced metal species in the watercolumn. For example, in the mid-to-late summer interval, Fe^{2+} maxima at intermediate hypolimnetic depths have been observed (e.g., Figure 5.55d–3). Recall the molar ratio of total Fe to total H_2S in the lowermost anoxic layers is usually less than 0.1 (Driscoll et al. 1993). The decrease in Fe^{2+} in the bottom waters likely reflects the formation of iron sulfide precipitates. Ferrous iron is also evident at shallower depths in the hypolimnion than H_2S. This feature may also reflect the higher redox potential of Fe reduction than SO_4^{2-} reduction.

5.6.4 Distribution of Decomposition By-Products

5.6.4.1 Loss of NO_3^- and Accumulation of T-NH_3

The temporal distributions of the total mass of NO_3^- in the bottom waters (below 8 m) of Onondaga Lake for the May–August interval of 1989, 1990, and 1991 are presented in Figure 5.56. Thermal stratification was established by mid-May of 1989; anoxia was evident in the lowermost layers two weeks thereafter. Stratification was established by early May in 1990, but the onset of anoxia was delayed to early June. The decreases in

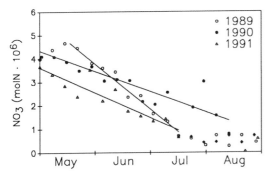

FIGURE 5.56. Temporal distributions of the total mass of NO_3^- nitrogen below 8 m in Onondaga Lake for the May–August interval of 1989, 1990, and 1991.

the stock of NO_3^- before the onset of stratification were observed uniformly throughout the watercolumn. This pattern may reflect the extent of the denitrification process, as the entire water column comes into contact with the deep water sediments periodically during spring turnover, though the nitrification process almost certainly operates simultaneously in the upper sediments during oxic periods. Dilution from tributary inflows may also contribute to the depletion observed before stratification.

The continued depletion of NO_3^- from the lower waters after the onset of stratification and anoxia (Figure 5.56) is a clear manifestation of denitrification. Nitrite was also depleted, but it is not included in Figure 5.56 because this pool is small compared to NO_3^-. The rate of NO_3^- loss increased after the onset of anoxia (i.e., elimination of the nitrification and O_2 reduction processes), particularly in 1989. The apparent rate of denitrification during the anoxic period of 1989, determined from the slope of the linear least squares regression of Figure 5.56, was 6.3×10^4 mol $\cdot$ d^{-1}.

Ammonia is formed as a result of the stabilization process of ammonification. Since much of the decomposition is localized at the sediment–water interface and within the sediments of the lake, T-NH$_3$ is effectively released from the sediments. The progressive accumulation of T-NH$_3$ in the hypolimnion

(depths >8 m) of Onondaga Lake is depicted for three different years in Figure 5.57. Near-linear increases in the mass of T-NH$_3$ occur (Figure 5.57). The estimated rate of accumulation in 1989 was 6.0×10^4 mol $\cdot$ d^{-1}.

Composition ratios of C:N have been used to examine N deficiency for phytoplankton growth (Table 5.18). Carbon:N composition ratios of seston in the water column (4.4) and in sediment trap collections (5.0) from the lake are very low, indicating that phytoplankton growth is not deficient in N in Onondaga Lake (also see Chapter 6). Ratios of C:N for Onondaga Lake are considerably lower than any values reported by Hecky et al. (1993) in a review of the stoichiometry of seston in lakes. These observations suggest that phytoplankton in Onondaga Lake are highly enriched in N. This is probably a manifestation of the unusually high concentrations of T-NH$_3$ and NO_3^-, the primary forms of N used by algae, that prevail in the productive layers of the lake (Brooks and Effler 1990; also see Figures 5.30 and 5.31).

To investigate the mineralization of deposited seston as a mechanism supplying T-NH$_3$ from sediments to the lower waters, paired measurements of T-NH$_3$ and DIC+CH$_4$ concentrations were examined for the 18 m depth during the summer stratification period for 1989, 1990 and 1991 (Figure 5.58a–c). Concentrations of DIC were calculated from

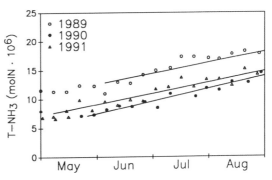

FIGURE 5.57. Accumulation of T-NH$_3$ in the hypolimnion (depths $\geq$ 10 m) of Onondaga Lake for the May–August interval of 1989, 1990, and 1991.

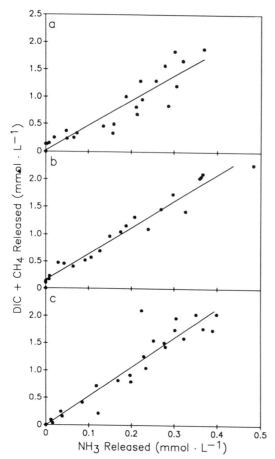

FIGURE 5.58. Evaluation of the relationship between 18 m concentrations of carbon (DIC plus CH_4) and T-NH_3 to support investigation of the mineralization process in Onondaga Lake for three years: (**a**) 1989, (**b**) 1990, and (**c**) 1991. Adjustment for CH_4 ebullitive loses are not included.

These C:N values are very low and consistent with the low C:N values of seston in Onondaga Lake. This analysis suggests that release of T-NH_3 from sediments to the water column is due to mineralization of seston deposited from the watercolumn. These calculations indicate a slight increase in the C:N ratio occurs from the watercolumn (4.4) and sediment trap (5.0) seston to values released during sediment decomposition. This may be due to a modest level of retention (e.g., 10−15%) of N in sediment (e.g., associated with production of microbial biomass) and/or errors in the calculation of DIC+CH_4 release (e.g., the assumption of 33% of CH_4 loss due to ebullition). Changes in the N content of seston could have important implications for N cycling. For example, if the C:N ratio of seston decreases, say in response to a decrease in water column concentrations of T-NH_3 and NO_3^-, the deposition of N from the upper waters (see Chapter 9), as well as feedback from the sediments, could decrease.

5.6.4.2 Accumulation of H_2S and Depletion of SO_4^{2-}

The temporal and vertical patterns of H_2S and SO_4^{2-} concentrations in Onondaga Lake for the spring to fall interval of 1980 are presented as isopleths in Figure 5.59. Concentrations of H_2S increased progressively in the hypolimnion from late May−late September. Recall that the period of stratification was longer then because of the chemical stratification that developed annually before closure of the soda ash/chlor-alkali facility (see Chapter 4). Further, spring turnover either did not occur or was extremely brief in 1980 (cf. Effler and Perkins 1987; Chapter 4). Hydrogen sulfide concentrations were always highest just above the sediments, and decreased progressively with decreasing depth in the hypolimnion (Figure 5.59a). Concomitant decreases in SO_4^{2-} were observed (Figure 5.59b) in the hypolimnion as the concentration of H_2S increased. The pronounced decreases in SO_4^{2-} can only be explained by SO_4^{2-} reduction, as S is only required in small amounts by algae and the

chemical equilibrium relationships as before. A linear relationship is demonstrated between the releases of T-NH_3 and DIC+CH_4 (Figure 5.58a−c); the slope is indicative of the stoichiometry of the mineralization process. The values of the C:N ratio obtained from this analysis were 4.6, 4.8 and 5.3 for the years 1989, 1990, and 1991, respectively. When these values are corrected for ebullition of CH_4 by assuming that 33% of the CH_4 released from sediments is lost via ebullition (see subsequent section), the C:N ratios increase to 5.1, 5.2, and 5.9 for 1989, 1990, and 1991, respectively.

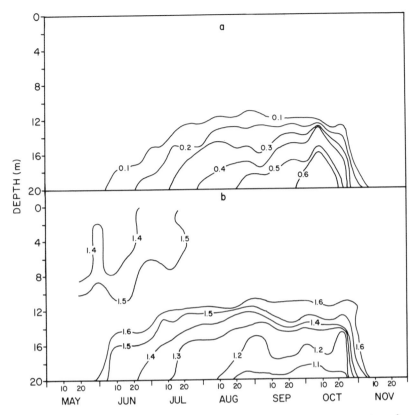

FIGURE 5.59. Isopleths of sulfur species in Onondaga Lake in 1980: (a) H$_2$S (mmol · L^{-1}), and (b) SO$_4^{2-}$ (mmol · L^{-1}) (from Effler et al. 1988).

water column of Onondaga Lake is undersaturated with respect to the solubility of SO$_4^{2-}$ minerals. The greatest depletions of SO$_4^{2-}$ were observed in the lowermost waters. The vertical profiles of H$_2$S and SO$_4^{2-}$ in the hypolimnion demonstrate that most of the SO$_4^{2-}$ reduction occurs within the sediments or at the sediment–water interface. Despite the progressive depletion of SO$_4^{2-}$ in the hypolimnion, concentrations throughout this layer in 1980, and all other years of record, remained above the value at which SO$_4^{2-}$ reduction is limited (0.2 mmol · L^{-1}; Ingvorsen et al. 1981). The depths at which H$_2$S was evident decreased rapidly in late October, and this specie was eliminated by early November, as a result of entrainment and oxidation during the fall mixing period.

The temporal and vertical features of the concentrations of H$_2$S and O$_2$ in the lake for the mid-June to mid-October interval of 1990 are represented by the profiles of Figure 5.60. Hydrogen sulfide was not detected (e.g., H$_2$S > 1.5 × 10^{-2} mmol · L^{-1}) in oxygenated layers over the period of record. Detectable concentrations of H$_2$S were usually vertically separated from the anoxic/oxic boundary by at least several meters, except when the epilimnion deepened rapidly during late summer and early fall (e.g., Figure 5.60k and l). Other reduced species (e.g., Fe^{2+} and CH$_4$) were found closer to the oxic/anoxic interface in early summer. This phenomenon may be influenced by the precipitation of H$_2$S with Fe^{2+}. Certain features of the concentration profiles of H$_2$S were similar in 1990 to patterns in 1980 (Figure 5.59). However, there were important differences between the two years. The onset of H$_2$S accumulation was significantly later and the concentrations

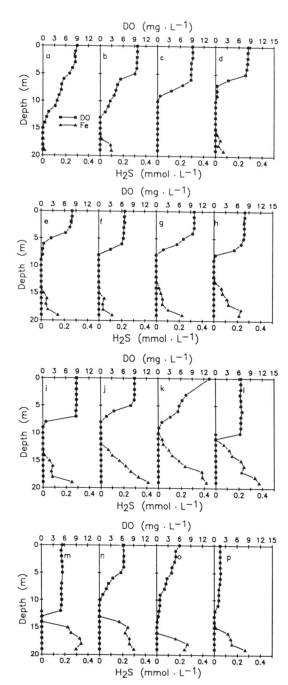

FIGURE 5.60. Profiles of H_2S and O_2 in Onondaga Lake for the mid-June–mid-October interval of 1990: (**a**) June 27, (**b**) July 4, (**c**) July 11, (**d**) July 18, (**e**) July 31, (**f**) August 7, (**g**) August 14, (**h**) August 23, (**i**) August 30, (**j**) September 5, (**k**) September 12, (**l**) September 18, (**m**) September 26, (**n**) October 1, (**o**) October 8, and (**p**) October 15.

remained lower in 1990 (Figure 5.60 versus Figure 5.59). These differences were at least in part due to the establishment of anoxic conditions later in 1990, due largely to the development of a more normal stratification regime in the interim (see Chapter 4).

Hydrogen sulfide profiles from mid-September for eight years of the 1980–1990 interval are presented in Figure 5.61a–h. The highest concentrations were observed in 1985 (Figure 5.61c). Concentrations were decidedly lower in both 1989 and 1990 than in the early and mid-1980s (Figure 5.61g and h). The highest concentration of H_2S observed over the eight-year period of record was 1.55 mmol $\cdot L^{-1}$ (instead of the 1.65 mmol $\cdot L^{-1}$ reported by Effler et al. (1988)), in early fall of 1985. The lowest maximum value was 0.4 mmol $\cdot L^{-1}$ in 1990. High concentrations of H_2S occur widely in the monimolimnia of meromictic lakes (Hutchinson 1957; Wetzel 1983). The development of high concentrations of H_2S in holomictic lakes is rare, being limited to hypereutrophic systems with adequate

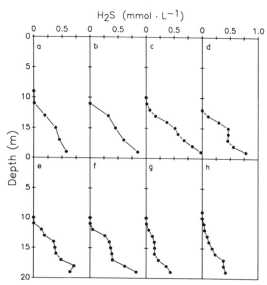

FIGURE 5.61. Profiles of H_2S in mid-September in Onondaga Lake for eight years: (**a**) September 13, 1980, (**b**) September 15, 1981, (**c**) September 23, 1985, (**d**) September 15, 1986, (**e**) September 14, 1987, (**f**) September 19, 1988, (**g**) September 12, 1989, and (**h**) September 12, 1990.

concentrations of SO_4^{2-} (Hutchinson 1957). For example, progressive depletion of SO_4^{2-} and accumulation of H_2S were observed annually in the hypolimnia of hypereutrophic Lake Mendota (Ingvorsen et al. 1981) and Wintergreen Lake (Molongoski and Klug 1980). However, even the lower maxima observed in Onondaga Lake in 1989 and 1990 exceed those of Lake Mendota (2.5×) and Wintergreen Lake (1.3×). The higher level of H_2S production in Onondaga Lake compared to other hypereutrophic lakes is likely attributable to the higher SO_4^{2-} concentrations rather than to differences in trophic state.

The progressive accumulation of H_2S in the hypolimnion of the lake is illustrated for three years in Figure 5.62. The mass of H_2S increased in a nearly linear manner through late summer. The subsequent leveling off and decrease reflected first an approximate balance between continued H_2S production and transport losses to (and subsequent oxidation) overlying layers and later, decreases in H_2S stocks due to mixing of oxygenated water exceeding rates of H_2S production. The accumulation rates determined for the eight years of record, as the slope of the linear least squares regression of time series plots of the hypolimnetic pool of H_2S (e.g., Figure 5.62), are presented in Table 5.22, along with correlation coefficients. The maximum annual stock of H_2S is also presented, as well as the average temperature at 19m for the June–September interval. Temperatures were lower in 1980 and 1986 than other years because spring turnover failed to occur (Effler and Perkins 1987). Large variations in the rate and

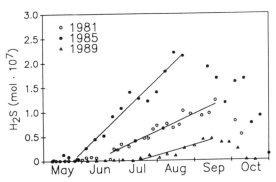

FIGURE 5.62. Accumulation of H_2S in the hypolimnion of Onondaga Lake for three years, 1981, 1985, and 1989. Linear least squares regression lines included to describe rate of accumulation.

maximum pool of H_2S accumulation have occurred over the 11 year interval. The highest rates of accumulation and maxiumum pool occurred in 1985 (Figure 5.62) and 1987 (Table 5.22). Note that though the rate of accumulation was higher in 1987, a lower maximum pool occurred as nearly linear increases proceeded over a longer period in 1985. The lowest rate of accumulation (Figure 5.62) and pool of H_2S occurred in 1989 (Table 5.22). The onset of H_2S accumulation was greatly delayed in 1989 (Figure 5.62) and 1990 compared to the other years because anoxic conditions developed later. This was largely responsible for the lower values of maximum H_2S pool observed in these two years (Table 5.22).

Several factors may have contributed to the differences in H_2S accumulation, including changes in the stratification regime (e.g.,

TABLE 5.22. Hypolimnetic accumulation rates of H_2S in Onondaga Lake.

Year	Accumulation rate (mol $\times$ $10^4 \cdot d^{-1}$)	R^2	Maximum pool (mol $\times$ 10^6)	Temperature (°C)
1980	9.0	0.945	13.5	5.8
1981	10.7	0.945	12.0	10.7
1985	20.6	0.895	21.4	10.7
1986	7.4	0.771	11.0	7.8
1987	22.5	0.832	15.2	9.1
1988	10.9	0.943	10.8	9.8
1989	6.1	0.869	4.3	9.4
1990	9.9	0.821	5.2	10.6

duration of stratification and temperature at the sediment–water interface), and changes in trophic state. The causes for the differences have yet to be clearly established. The features of accumulation were not strikingly different for the two years in which the lowermost waters remained colder (Table 5.22). The lower annual maxima in H_2S content in 1989 and 1990 appear to be consistent with documented reductions in the deposition of organic carbon (Chapter 8). However, these changes may also in part reflect longer term diagenetic (sedimentary) processes.

5.6.4.3 Accumulation of Iron and Manganese

Detailed Fe^{2+} data from the water column of Onondaga Lake (e.g., for 1990 in Figure 5.63), and more limited data for Mn (Figure 5.55), depict the mobilization of Fe^{2+} and Mn (presumably as Mn^{2+}) from the sediments of the lake during summer anoxia, presumably as a result of Fe and Mn reduction (Table 5.21). Concentrations of Fe were low in oxygenated portions of the water column; no Fe^{2+} was detected within the oxic layers of the lake. Two features of the hypolimnetic Fe^{2+} distributions differ from those of the other accumulating reduced species. First, the molar concentrations of Fe^{2+} in the hypolimnion are considerably lower than concentrations observed for the other reduced species (Figure 5.55). Second, unlike the other species, maxima of Fe^{2+} are manifested distinctly above the sediment-water interface by late summer. For example, the Fe^{2+} maxima on August 30 (Figure 5.63i) and September 12 (Figure 5.63k), of 1990 occurred at depths of 17 and 14 m, respectively. Both of these disparate features of the hypolimnetic distributions of Fe^{2+} are likely a manifestation of interaction with the high concentrations of H_2S that develop.

Progressive accumulation of Fe^{2+} occurred throughout the anoxic hypolimnion until mid-August of 1990 (Figure 5.63a–g). The maximum concentration was about 0.03 $mmol \cdot L^{-1}$. This is about one-quarter of the

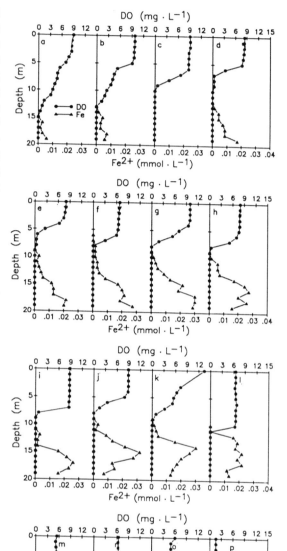

FIGURE 5.63. Profiles of Fe^{2+} and O_2 in Onondaga Lake for the late June–mid-October interval of 1990: (**a**) June 27, (**b**) July 4, (**c**) July 11, (**d**) July 18, (**e**) July 31, (**f**) August 7, (**g**) August 14, (**h**) August 23, (**i**) August 30, (**j**) September 5, (**k**) September 12, (**l**) September 18, (**m**) September 26, (**n**) October 1, (**o**) October 8, and (**p**) October 15.

maximum observed in late summer in the anoxic layer (just above the sediments) of nearby Otisco Lake (see Figure 5.47). Over the next month concentrations of Fe^{2+} decreased in the lowermost layers, creating a maximum that was displaced from the sediment water interface (Figure 5.63h–l). Concentrations of H_2S continued to increase strongly in the lowermost layers over the same interval (Figure 5.60h–l). Molar concentrations of H_2S (maximum $>0.4\,mmol \cdot L^{-1}$) were much greater at these depths over this period than Fe^{2+} concentrations; the mean molar ratio of Fe: H_2S in the bottom waters over the anoxic interval was about 0.1. Ratio values $\geqslant 0.2$ were generally observed at the depths of the Fe^{2+} maxima. This pattern is consistent with the observed oversaturated conditions with respect to the solubility of FeS (Driscoll et al. 1993) and the presence of FeS particles (Yin and Johnson 1984) in the bottom anoxic waters of the lake. Evidently the decrease in Fe^{2+} concentrations at the deepest anoxic depths in late summer reflects the immobilization of Fe^{2+} by H_2S, to form precipitated FeS.

The accumulation of the hypolimnetic Fe^{2+} pool is depicted for two years in Figure 5.64, 1985 and 1989. The distributions of the pools for the two years differ in three important ways. First, accumulation started earlier in 1985, associated with the earlier onset of anoxia. Second, the initial accumulation rate was higher in 1985, indicating a higher rate of

Fe reduction; recall this was also observed for the H_2S pool in 1985 (Figure 5.62). Finally, there was a clear reduction in the Fe^{2+} pool after July in 1985 before the onset of significant entrainment and oxidation losses observed in early fall (e.g., late September in 1989). This is in contrast to the continued increase in the H_2S pool in 1985 through August. The decrease in the Fe^{2+} pool in August in 1985 probably reflects higher FeS precipitation rates associated with the much higher H_2S concentrations that developed in 1985 (Figure 5.62). Reductions in the Fe^{2+} pool occurred in early fall of 1989 as Fe^{2+} was entrained into the epilimnion and oxygen transport into the hypolimnion increased (Figure 5.64). During this period most of the decrease in Fe^{2+} was probably due to the oxidation of soluble Fe^{2+} to highly insoluble Fe(III). Ferric iron precipitates from the water column as a Fe oxyhydroxide mineral.

Manganese data for the water column of Onondaga Lake are much more limited. Hypolimnetic concentrations of Mn (Onondaga County 1990) during anoxia, presumably as Mn^{2+}, are elevated (maximum of about $0.008\,mmol \cdot L^{-1}$) compared to levels in the epilimnion, but are about an order of magnitude lower than Fe^{2+} concentrations (Figure 5.55). Thus Mn^{2+} is an insignificant electron donor in this system (Figure 5.64). Apparently Fe^{2+} is mobilized at a much higher rate than Mn^{2+} in Onondaga Lake, consistent with the observations of Kelly et al. (1988) for three other study systems.

5.6.4.4 CH_4

5.6.4.4.1 Accumulation of Dissolved CH_4

The temporal and vertical distributions of dissolved CH_4 and O_2 in Onondaga Lake for the mid-June–late October interval of 1989 are presented in the profiles of Figure 5.65a–r. This is the first year that systematic measurements of this specie were made (Addess 1990; Addess and Effler 1996). The character of the distribution of CH_4 was similar to that reported previously for H_2S (Figure 5.60). Dissolved CH_4 was not detected in oxygenated layers

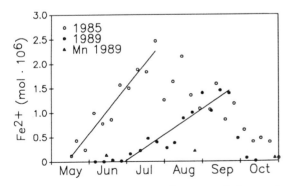

FIGURE 5.64. Temporal distributions of the pool of Fe^{2+} in the hypolimnion (depths $\geqslant 10\,m$) of Onondaga Lake for two years, 1985 and 1989. Estimate of Mn pool for 3 days in 1989 included.

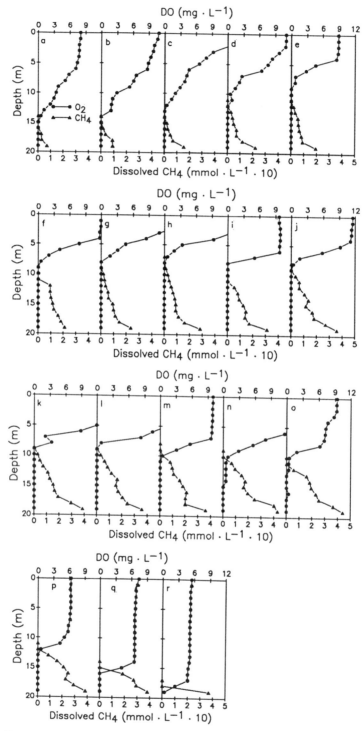

FIGURE 5.65. Profiles of dissolved CH_4 and O_2 in Onondaga Lake for the mid-June–late October period of 1989: (**a**) June 15, (**b**) June 19, (**c**) June 26, (**d**) July 3, (**e**) July 11, (**f**) July 17, (**g**) July 24, (**h**) August 1, (**i**) August 7, (**j**) August 14, (**k**) August 21, (**l**) August 29, (**m**) September 5, (**n**) September 12, (**o**) September 19, (**p**) September 26, (**q**) October 3, and (**r**) October 10 (modified from Addess and Effler 1996).

during the study period (Figure 5.65a–r). Dissolved CH_4 was first detected on June 15, in the lowermost layers, approximately one week after the onset of anoxia. Concentrations of CH_4 increased progressively in the hypolimnion from mid-June–late September, and reached a maximum of $0.47 \, mmol \cdot L^{-1}$ ($7.53 \, mg \cdot L^{-1}$) just above the sediment. The vertical extent of CH_4 often included the entire anoxic interval (up to a depth of 9 m on August 1; Figure 5.65h). The depth interval of detectable CH_4 narrowed as the epilimnion deepened during the fall mixing period. Methane concentrations were always highest near the sediment and decreased progressively within the hypolimnion with the approach to the boundary of oxia and anoxia. This vertical profile is consistent with a sedimentary source of CH_4 and incomplete vertical mixing in the hypolimnion. Similar temporal and vertical distributions of dissolved CH_4 were documented in 1990, indicating the profiles presented in Figure 5.65 are recurring for the lake.

The dissolved CH_4 concentrations reported here are not unusually high compared to values reported in the literature for the hypolimnia of other eutrophic lakes. Ingvorsen and Brock (1982) reported a maximum concentration of $0.40 \, mmol \cdot L^{-1}$ for Lake Mendota. Rudd and Hamilton (1975) reported a maximum concentration of $1.28 \, mmol \cdot L^{-1}$ in eutrophic Lake 227. Strayer and Tiedje (1978) reported maximum concentrations of 2.5 and $5.0 \, mmol \cdot L^{-1}$ for two consecutive years in hypereutrophic Wintergreen Lake. The maximum concentration measured in Onondaga Lake in 1990 was $0.45 \, mmol \cdot L^{-1}$, very similar to the 1989 maximum.

The accumulations of CH_4 stocks in the hypolimnion in 1989 and 1990 are presented in Figure 5.66. These calculations are based on CH_4 profiles (e.g., Figure 5.65) and hypsographic data for the lake (Owens 1987). Methane accumulated at nearly linear, and approximately equal, rates in the two years ($6.63 \times 10^4 \, mol \cdot d^{-1}$ in 1989 and $6.74 \times 10^4 \, mol \cdot d^{-1}$ in 1990). The coincident accumulations of CH_4 and H_2S in the hypolimnion (Figures 5.62 and 5.66) indicates that

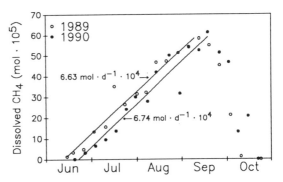

FIGURE 5.66. Accumulation of CH_4 in the hypolimnion of Onondaga Lake in two years, 1989 and 1990. Linear least squares regression lines included to describe accumulation rates.

the processes of SO_4^{2-} reduction and methanogenesis proceed in parallel in the sediments of Onondaga Lake. The estimated maxima for the CH_4 pool for the two years were $5.9 \times 10^6 \, mol$ in 1989 and $6.1 \times 10^6 \, mol$ in 1990. The modest differences in the timing of these distributions for the two years (e.g., start of accumulation and date of peak content) were largely attributable to natural year-to-year differences in the onset of stratification. These accumulations are subsequently incorporated into a watercolumn budget for CH_4 for the lake.

5.6.4.4.2 Ebullition of CH_4

Ebullitive gas was collected with inverted cone devices (Addess 1990) located approximately 1.5 m above the sediments, at 2 to 5 day intervals from April–October of 1989 and 1990. Collections were made routinely at the south deep station, and occasionally at other sites and further up in the water column. The dynamics of total ebullitive flux and the CH_4 fraction observed in 1989 at the south station are presented in Figure 5.67a (Addess and Effler 1996). Standard deviation bars are presented for those periods when gas collections were made in triplicate. Variability represented by these bars probably reflects small-scale heterogeneity in ebullitive flux. This variability is minor compared to the temporal variability that prevailed for most of the study period (Figure 5.67a). Variability at this time scale has not been addressed previously for a stratifying lake. Total gas flux for the study period averaged 7.3 mmol $\cdot$ m^{-2} $\cdot$ d^{-1}; the maximum was 34.2 mmol $\cdot$ m^{-2} $\cdot$ d^{-1}. Ebullitive CH$_4$ flux averaged 6.0 mmol $\cdot$ m^{-2} $\cdot$ d^{-1}; the maximum was 30.9 mmol $\cdot$ m^{-2} $\cdot$ d^{-1}.

It has been hypothesized that changes in pressure, associated mostly with changes in lake elevation (hydrostatic pressure, Figure 5.67b), contributed to this strong temporal variability. Changes in atmospheric pressure were small compared to those for hydro-static pressure (Addess 1990). The events of greatest ebullitive flux (late May, early and late July, and mid-August) occurred over periods of substantial decreases in hydrostatic pressure, or when hydrostatic pressure was at a minimum. Lower fluxes were generally observed during periods when the lake level was increasing. However, the relationship between ebullitive flux and hydrostatic pres-sure is either not simple or other factors contribute importantly to these dynamics, as changes in flux over the entire study period were not strongly correlated to hydrostatic pressure. Changes in hydrostatic pressure have been shown to exert a substantial effect on ebullitive rates in shallow systems. Chanton et al. (1989) found that a reduc-tion in the hydrostatic pressure of 5 to 7% brought about by tides was enough to cause distinct diurnal increases in ebullition. The phenomenon was also observed by Martens and Klump (1980). The changes experienced in Onondaga Lake were less abrupt; e.g., a 4% decrease in hydrostatic pressure occurred over the period of May 15–19, a 2% decrease occurred over the period June 20–July 4, and 1% decreases occurred later in the study (Figure 5.67b).

The frequency distribution of % CH$_4$ in the ebullitive gas for the 1989 study period is presented in Figure 5.68 for measurements made at all sites (Addess and Effler 1996).

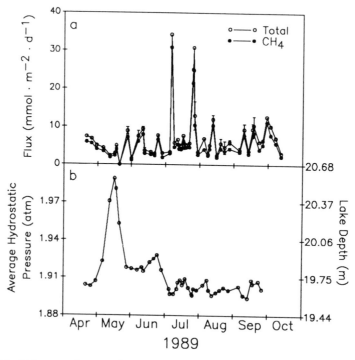

FIGURE 5.67. (a) Ebullitive gas flux at the south deep station 1989. Bars represent ±1 standard deviation limits. (b) Average hydrostatic pressure for the intervals that flux measurements were taken in 1989 (modified from Addess and Effler 1996).

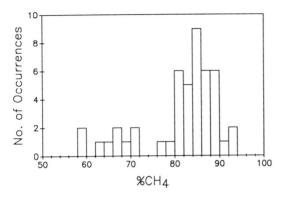

FIGURE 5.68. Frequency distribution for ebullitive gas composition, from all sites, in 1989 collections (modified from Addess and Effler 1996).

A bimodal distribution is indicated; the dominant mode was centered around a composition of 86% CH_4, the secondary mode was centered around a composition of 67%. The balance of the ebullitive gas was mostly N_2; CO_2 was detected ($\leq 1\%$) on a few of occasions. Variations in composition of the ebullitive gas apparently occurred lake-wide, as the dynamics of the south and north deep stations tracked each other (Addess 1990).

The strong temporal variations observed in ebullitive flux (Figure 5.67a) also apparently occurred lake-wide, as the same temporal structure was observed for the south and north deep stations (Addess 1990). There were some indications that the flux at the south station was somewhat higher than in the north basin, though the level of precision of the flux measurements (e.g., Figure 5.67a) confounded quantification of these differences (Addess 1990). Strong variations in ebullitive flux were also observed in 1990; several of the higher releases again coincided with decreases in hydrostatic pressure.

5.6.4.4.3 CH_4 Budget and Comparison to Other Lakes

The CH_4 budget for the lake for the period June 15–September 12, 1989 is presented in Figure 5.69 (Addess and Effler 1996). The salient features of the budget are summarized in Table 5.23. The demarcation depth between the epilimnion and hypolimnion was assumed to be 10 m. Methane released as ebullitive gas was estimated to average 5.3 $mmol \cdot m^{-2} \cdot d^{-1}$; this represented a total sedimentary release of 3.34×10^6 mol CH_4 over that period. Redissolution of ebullitive CH_4 into the hypolimnion, estimated from paired deployment of the gas collection devices at just above the sediments and at a depth of 10 m, represented approximately 10% (3.27×10^5 mol) of the overall ebullitive release. This estimate should be considered approximate because of the limited level of precision in the collection of the gas (Addess 1990). However, the estimate is similar to that obtained by Robertson (1979). Applying the same vertical loss rate for CH_4 gas in the epilimnion, an estimated 5.3×10^5 mol of CH_4 dissolved into the epilimnion, where it was oxidized to CO_2. The atmospheric loss of CH_4 is estimated to be 2.5×10^6 mol; about 74% of the ebullitive release. This represents a loss of organic carbon from the system (Figure 5.69).

The total diffusive release ($CH_{4(d)}$) of dissolved CH_4 from the sediments was estimated according to the following expression

$$CH_{4(s)} = CH_{4(ha)} + CH_{4(e)} - CH_{4(eb)} \quad (5.47)$$

in which $CH_{4(s)}$ = dissolved CH_4 released from sediments (mol), $CH_{4(ha)}$ = net hypolimnetic accumulation of CH_4 (mol), $CH_{4(e)}$ = CH_4 lost from hypolimnion by diffusion into epilimnion (mol), and $CH_{4(eb)}$ = CH_4 redissolved into hypolimnion from escaping ebullitive gas (mol). The daily rate of mass transfer of dissolved CH_4 ($W_{v\text{-}CH4}$; $mol \cdot d^{-1}$) into the overlying epilimnion, mediated by vertical diffusivity, was determined according to (Chapra and Reckhow 1983)

$$W_{v\text{-}CH_4} = v_t \cdot A_t \cdot (C_{h\text{-}CH_4} - C_{e\text{-}CH_4}) \quad (5.48)$$

in which v_t = vertical heat exchange coefficient ($m \cdot d^{-1}$), A_t = surface area at the thermocline, and $C_{h\text{-}CH_4}$ and $C_{e\text{-}CH_4}$ = concentrations of CH_4 (volume-weighted) in hypolimnion and epilimnion (= 0; $mol \cdot m^{-3}$). The value of $CH_{4(e)}$ is estimated by summing $W_{v\text{-}CH_4}$ values over intervals included in the June–September period. The estimation of v_t

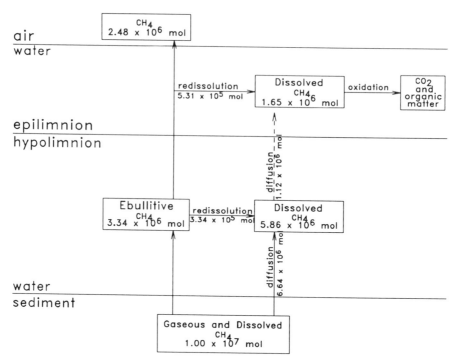

FIGURE 5.69. Methane (CH_4) budget for Onondaga Lake, mid-June–mid-September interval of 1989 (modified from Addess and Effler 1996).

has been described in detail in Chapters 4 and 9. The total upward loss of CH_4 from the hypolimnion over the budget calculation period was estimated to be 1.12×10^6 mol CH_4, or 17% of the net accumulation. Rudd and Hamilton (1978) estimated that only 10% of the diffusive release from the sediments of Lake 227 during summer stratification diffused across the thermocline. In contrast, Fallon et al. (1980) estimated approximately 45% of the CH_4 release from hypolimnetic sediments of Lake Mendota diffused into the epilimnion. The total diffusive release of dissolved CH_4 from the sediments (from Equation (5.47)) was estimated to be 6.65×10^6 mol; approximately 88% of this accumulated in the hypolimnion (Figure 5.69). The total release of CH_4 from the hypolimnetic sediments (equal to the sum of the diffusive and ebullitive releases) is estimated to be 1.0×10^7 mol

TABLE 5.23. Features of a CH_4 budget for Onondaga Lake (June 15–September 12, 1989).

Component	Mass (mol $\times 10^6$)	Comments
a. Ebullition		
1. release from sediments	3.34	From gas trap field collections
2. dissolved into hypolimnion	0.33	By flux differences measured *in situ* at 18 m and 10 m
3. dissolved/oxidized into epilimnion	0.53	Assumes same rate of loss as in 2.
4. release to atmosphere	2.48	Calculated as 1. minus the sum of 2. and 3.
b. Diffusive Inputs/Dissolved CH_4		
1. diffused/oxidized into epilimnion	1.12	Calculated from v_t and CH_4 vertical gradients
2. accumulation (net) in hypolimnion	5.86	Calculated from depth profiles of CH_4
3. released from sediments	6.64	Calculated from Equation (5.47)

(16 mmol $\cdot$ m^{-2} $\cdot$ d^{-1}; i.e., diffusion represents two-thirds of the total).

A comparison of the CH$_4$ flux determined for Onondaga Lake to values reported for other systems, and the fraction of the organic deposition released as CH$_4$, is presented in Table 5.24. All of the other systems included in the comparison have been described as eutrophic by the respective authors. However, they apparently represent a range of primary productivity, as differences in organic deposition of more than an order of magnitude are included (e.g., Wintergreen Lake is hypereutrophic). The flux of CH$_4$ in Onondaga Lake is in the middle of the range of observations reported to date (Table 5.24). The CH$_4$ release from the sediments of the lake during the June–September interval of 1989 was 30% of the organic carbon deposited over the same period. This value is somewhat lower than others reported previously (Table 5.24). While the reason for this discrepancy is not clear, the presence of elevated concentrations of SO$_4^{2-}$ and NO$_3^-$ may contribute. These solutes are important electron acceptors and oxidize a significant amount of organic matter deposited in the sediments of Onondaga Lake. These processes are probably not so prominent in the other lakes studied (Table 5.24). As a result, methanogenesis is a less significant pathway for organic matter decomposition in Onondaga Lake, in comparison with these other lakes.

5.6.5 Summary of Organic Matter Oxidation Processes

In an effort to synthesize and integrate the information on decomposition and redox processes in the hypolimnion of Onondaga Lake, organic carbon and electron budgets were developed for the summer stratification period. The production or loss rates of solutes that are products or reactants in decomposition/redox reactions (Table 5.21) were determined from observed accumulation (e.g., Figures 5.57, 5.62, 5.65, and 5.67) and depletion (e.g., Figure 5.56) rates, adjusted for vertical losses (via vertical mixing) to the epilimnion (as described for CH$_4$ in Equations (5.47) and (5.48)). The analysis was conducted for the summer stratification interval of 1981, 1989, 1990, and 1991. Rates are presented on an areal basis, stipulating the contour area for the 8.5 m depth of the lake. This vertical segmentation is consistent with the commonly observed position of the thermocline (Chapter 4), and it was also adopted for water quality models for the lake (Chapter 9). The rates for the various constituents, for the respective periods of production/depletion, are presented in Table 5.25. The determination of the rates of production of reduced species and depletion of electron acceptors, adjusted for vertical exchange, is illustrated for 1989 in Figure 5.70.

TABLE 5.24. Comparison of methane flux and percentage of organic carbon deposition recycled via this flux.

Lake	Organic deposition (mmol C $\cdot$ m^{-2} $\cdot$ d^{-1})	Methane flux (mmol C $\cdot$ m^{-2} $\cdot$ d^{-1})	Release %	Reference
Onondaga Lake	54*	16	30	Addess 1990
Third Sister (1976)	15	5.4	36	Kelly and Chynoweth (1981)
Third Sister (1977)	9.2	3.5	38	Kelly and Chynoweth (1981)
227 (Ontario) (1974)	19	10.8	57	Rudd and Hamilton (1978)
Frain's (1977)	33	19	58	Kelly and Chynoweth (1981)
Frain's (1977)	25	12	48	Kelly and Chynoweth (1981)
Mendota (1977)	67.5	35.8	53	Fallon et al. (1980)
Wintergreen (1976)	127	49	39	Kelly and Chynoweth (1981)

Modified from Addess and Effler 1996
* See POC fluxes from sediment traps in Chapter 8

TABLE **5.25.** Rates of solute accumulation or loss (−) in the hypolimnion of Onondaga Lake.

Constituent	Rate (mmol·m^{-2}·d^{-1}) Year			
	1981	1989	1990	1991
DIC	35	32	49	42
T-NH$_3$	5.3	11.8	10.9	10.0
O$_2$	−62	−63	−41	−48
NO$_3^-$	—	−8.8	−4.7	−5.4
Fe^{2+}	0.9	1.6	1.1	1.6
SO$_4^{2-}$	—	−11.2	−11.0	−7.0
H$_2$S	16.5	9.9	14.7	13.5
CH$_4^*$	—	15.7 (10.5)	14.5 (9.7)	18.1 (12.1)

*Includes ebullitive loss, assumed to be 33% of total (Figure 5.70); soluble component in parentheses

Substantial year-to-year variability in the rates is apparent. However, there are no indications of major systematic changes (e.g., decreases) from 1981 to the latter years (Table 5.25), that would be expected if major reductions in primary productivity and the attendant downward flux of organic carbon had occurred. This pattern is consistent with recent C:P ratio data presented for seston earlier in the' chapter (Table 5.18) and an analysis presented in Chapter 6 that concludes the phytoplankton community of the lake remains essentially P saturated despite the major reductions in external P loading achieved since 1981 (Figure 5.45b). Note the rates in 1981 are near or within the range observed for the later three years.

Rates of H$_2$S accumulation (mean value for 1989–1991 interval, 12.7 mol · m^{-2} · d^{-1}) are generally comparable to rates of SO$_4^{2-}$ loss (mean value 1989–1991, 9.7 mol · m^{-2} · d^{-1}). Note it is difficult to accurately quantify rates of SO$_4^{2-}$ loss in the watercolumn of Onondaga Lake because background concentrations of SO$_4^{2-}$ are high. Moreover, analytical determinations of SO$_4^{2-}$ were made by the turbidimeter procedure which is not a very precise analytical method, particularly in the concentration range common to Onondaga Lake. Nevertheless, the general agreement between values of SO$_4^{2-}$ loss and H$_2$S accumulation suggests that most of the SO$_4^{2-}$ reduced in hypolimnetic sediments is quantitatively released to the watercolumn. Apparently

relatively little S is retained in sediments as organic S or FeS. Only limited accumulation of S as FeS is expected despite the fact that the lower waters are oversaturated with respect to the solubility of FeS (Driscoll et al. 1993), because of the limited availability of Fe^{2+}. Recall that ratios of total Fe^{2+} to total H$_2$S remain generally low in the anoxic hypolimnetic waters of the lake (Figure 5.55; Driscoll et al. 1993).

5.6.5.1 Organic C Budget for Hypolimnion

The downward flux of organic C into the hypolimnion of the lake over the May to mid-September interval of 1989 has been determined through the analysis (POC) of sediment trap collections made below the epilimnion (see Chapter 8 for detailed presentation of results, as well as procedural features of the sediment trap program). Loss of deposited organic C through decomposition processes over that same period is manifested via oxidation to DIC, and production of CH$_4$ (diffusing upward from the sediments, or lost via ebullition).

The mean rate of deposition of organic C was 54 mmol C · m^{-2} · d^{-1} over the May–mid-September interval of 1989. This value exceeds the sum of the production rates of DIC (32 mmol · m^{-2} · d^{-1}) and CH$_4$ (10.9 mmol · m^{-2} · d^{-1}; obtained by adjusting the methanogenesis rate of Table 5.25 to represent

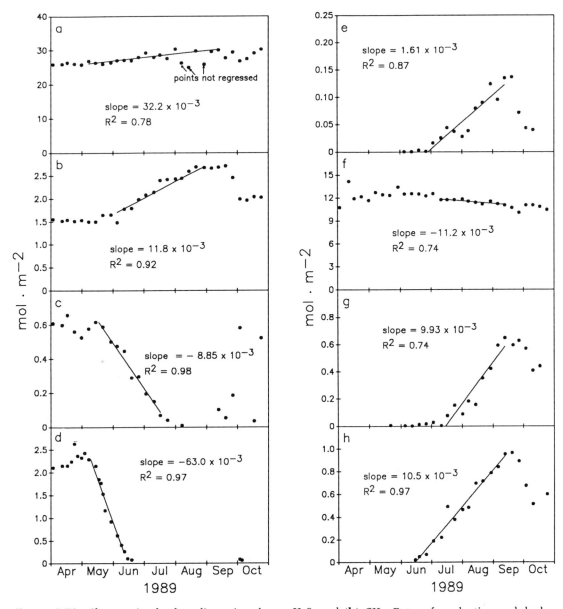

FIGURE 5.70. Changes in the hypolimnetic solute pools of Onondaga Lake in 1989: (a) DIC, (b) T-NH$_3$, (c) NO$_3^-$, (d) DO, (e) Fe^{2+}, (f) SO$_4^{2-}$, (g) H$_2$S, and (h) CH$_4$. Rates of production and depletion as slopes from regression analyses.

the entire stratification period). According to these fluxes, 59% of the organic C input to the hypolimnion was oxidized to DIC and 20% was decomposed by methanogenesis. The discrepancy in this budget calculation, represents an estimate of the rate of organic C accumulation in the sediments (i.e., not decomposed) over the interval of the analysis (11.1 mmol $\cdot$ m^{-2} $\cdot$ d^{-1}; 21%). The general magnitude of this estimate of seasonal organic C accumulation rate is supported by its similarity to the estimate of accumulation of organic C in recent lake sediments (20.3 mmol $\cdot$ m^{-2} $\cdot$ d^{-1}), albeit crude, based on the organic C content and sedimentation rate determined through analyses of core samples from the lake (see Chapter 8).

5.6.5.2 Hypolimnetic Electron Budget

The cumulative consumption of individual electron acceptors for a hypolimnetic electron budget was determined using the rates of solute production/loss calculated for 1989, 1990, and 1991 (Table 5.25) and the stoichiometry of electron transfer reactions (Table 5.21). This budget can be expressed on an electron equivalence (eeq) basis or on a DIC equivalence basis using the stoichiometry of Table 5.21. A complete budget can only be presented for 1989 because that is the only year organic C inputs to the hypolimnion are presently available (analysis of sediment trap collections).

Molecular oxygen was initially used exclusively as the electron acceptor in hypolmnetic decomposition reactions from the period when thermal stratification was established in early May of 1989 until mid-June when the pool was depleted (Figure 5.70). This resulted in 8.8 eeq · m^{-2} (2.2 mol DIC · m^{-2}) of electron donor oxidized (Figure 5.71). Denitrification commenced in mid-May, prior to the complete removal of DO. The hypolimnetic pool of NO$_3^-$ was exhausted by mid-July, consuming 2.8 eeq · m^{-2} (0.7 mol DIC · m^{-2}) of electron donor. Methanogenesis was evident by mid-June and proceeded through

the stratification period, representing 5.6 eeq · m^{-2} (1.4 mol DIC · m^{-2}). Sulfate was the last electron acceptor utilized. Sulfate reduction commenced in mid-July and continued until a turnover, resulting in 5.0 eeq · m^{-2} (1.25 mol DIC · m^{-2}) of electron donor oxidized.

The apparent order of electron acceptor utilized in the hypolimnion of Onondaga Lake occurred in a somewhat different pattern than anticipated from thermodynamic considerations (Figure 5.71; Table 5.21). Molecular oxygen and NO$_3$ were consumed first, as expected. However, the use of these oxidants was followed sequentially by the emergence of Fe^{2+}, CH$_4$, and H$_2$S. Based on the energetics of electron transfer reactions, Fe reduction is expected to occur well before methanogenesis and SO$_4^{2-}$ reduction (at a higher redox potential). Moreover, it is anticipated that methanogenesis and SO$_4^{2-}$ reduction would occur at about the same redox potential. The timing of the reduced specie signatures of these anaerobic processes may be somewhat misleading with respect to the sequence of the processes. What appears to be thermodynamically anomalous, is probably an artifact of the spatial separation of the processes within the sediments. Methanogenesis occurs continuously in the subsurface sediments; CH$_4$ is manifested in the overlying anoxic water-column with the loss of oxygen from the surface layer. Thus CH$_4$ can start to accumulate before the redox potential of the overlying water reaches the value corresponding to the process. Conversely, SO$_4^{2-}$ reduction is supported in the upper sediments by downward diffusion of SO$_4^{2-}$ from the watercolumn. Thus H$_2$S should not be expected to be manifested until these sediment depths reach the appropriate redox potential. The situation for Fe reduction is similar, as Fe in the deeper sediments is probably not available for reduction/solubilization (e.g., as FeS). Newly deposited Fe is mobilized soon after the onset of anoxia (but at a higher potential than SO$_4^{2-}$ reduction, i.e., appears sooner).

If the deposited organic C is assumed to have an oxidation state of zero (Table 5.21), the average rate of the electron equivalence of electron donor entering the hypolimnion

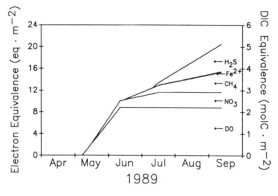

FIGURE 5.71. Hypolimnion accumulation of total electron acceptors, for Onondaga Lake 1989.

during summer stratification in 1989 was $216 \, meeq \cdot m^{-2} \cdot d^{-1}$. Much of this organic matter was, in turn, oxidized by a suite of electron acceptors. The sum of these individual reduction reactions over the stratification period, resulted in a total rate of electron acceptor consumption of $174 \, meeq \cdot m^{-2} \cdot d^{-1}$. The discrepancy between inputs of the electron donor (organic C) to the hypolimnion and the electron acceptor reactions represents the net transfer of electrons to the sediment ($42 \, meeq \cdot m^{-2} \cdot d^{-1}$).

Oxygen was the most important electron acceptor in the hypolimnion during summer stratification, representing 32% of the total electron equivalence of electron donor for 1989 (Table 5.26). After O_2 reduction, denitrification (10%), methanogenesis (20%), and SO_4^{2-} reduction (18%) were all quantitatively important reduction reactions in the electron balance of the hypolimnion (Table 5.26). The DIC equivalence calculated from the sum of the electron acceptor fluxes from sediments ($44 \, mmol \, DIC \cdot m^2 \cdot d^{-1}$) exceeded the measured rate of DIC release from sediments ($32 \, mmol \cdot m^{-2} \cdot d^{-1}$) by about 38%.

Although information on inputs of organic C to the hypolmnion was only available for 1989, rates of measured and calculated DIC accumulation and fluxes of electron acceptors were determined for 1990 and 1991, as well (Table 5.26). This analysis showed some year-to-year variation in the rates of DIC accumulation ($32-49 \, mmol \cdot m^{-2} \cdot d^{-1}$, 40%) and in the total fluxes of electron acceptors utilized ($161-198 \, meeq \cdot m^{-2} \cdot d^{-1}$, 21%). Year-to-year differences in the contribution of individual electron acceptors utilized were also evident. For example, in 1989 organic matter oxidized by denitrification was $22 \, meeq \cdot m^{-2} \cdot d^{-1}$, representing 10% of the total flux of electron acceptors. This rate was much lower in 1991 ($13 \, meeq \cdot m^{-2} \cdot d^{-1}$), contributing only 6% of the total flux of electron acceptors. In contrast the flux of organic matter oxidized by SO_4^{2-} reduction was low in 1989 ($39 \, meeq \cdot m^{-2} \cdot d^{-1}$; 18% of total). This flux was much larger in 1991 ($65 \, meeq \cdot m^{-2} \cdot d^{-1}$) representing 32% of the total flux of electron acceptors.

During all three study years a discrepancy was evident between measured DIC production and values calculated based on the stoichiometry of electron transfer reactions (Table 5.26). If calculated DIC production exceeds measured rates of accumulation, this pattern suggests a partial oxidation of organic detritus during sediment burial. However, if measured rates of hypolimnetic DIC accumulation exceed calculated values, this discrepancy is indicative of a reduction of sediment organic matter. In 1989 and 1991 the measured rate of DIC accumulation in the hypolimnion was less than calculated values. However, this pattern was reversed in 1990.

It is interesting to compare the electron budget of Onondaga Lake with other investigations (Table 5.26). There have been only a few electron budgets published in the literature and they demonstrate considerable variability in the use of electron acceptors for the oxidation of organic matter that occurs in lake ecosystems. Of the sites evaluated, Onondaga Lake exhibited the highest organic C deposition and the highest DIC accumulation in the hypolimnion. Like Onondaga Lake, none of the other studies show an important role of Fe^{2+} reduction in electron transfer reactions. Several of the studies show an imbalance between measured DIC accumulation and calculated values expected from the consumption of electron acceptors. Lake 226N in the Experimental Lake area of Ontario exhibited calculated DIC production in excess of measured values, while in Mirror Lake, NH, measured DIC production was greatly in excess of calculated values. Other lakes (Dart's and L227) show good agreement. It is not clear if the discrepancy between measured and calculated rates of hypolimnetic DIC accumulation reflect any real biogeochemical process(es) or merely errors associated with electron budget calculations.

5.6.6 Oxidation of Reduced Species

The reduced species T-NH₃, CH_4, H_2S, Fe^{2+}, and Mn^{2+}, which accumulate in the hypolimnion of Onondaga Lake during the anoxic

TABLE 5.26. Comparison of hypolimnetic electron budgets, including organic C sedimentation rates during summer stratification, for several lakes. Fluxes of electron acceptors and percentage of total electron acceptors utilized (in parentheses). Measured DIC production and calculated values of DIC based on stoichiometry of electron transfer reactions.

Lake	Organic C Input (mmol·m⁻²·d⁻¹)	DIC Accumulation (mmol·m⁻²·d⁻¹)		Fluxes Electron Acceptors (meq·m⁻²·d⁻¹)						Reference
		Measured	Calculated	O_2	NO_3^-	Fe^{2+}	SO_4^{2-}	CH_4	Total	
Blelham Tarn	38.2	10.7	9.2	18 (49)	7.3 (20)	ND	0.85 (2)	10.7 (29)	36.8	Jones and Simon 1980
Dart's	4.7	4.7	4.7	15.2 (81)	2.8 (14)	0.4 (2)	0.7 (4)	—	18.8	Schafran and Driscoll 1987
Mirror	10	5.33	3.4	9.2 (69)	0 (0)	0.43 (3)	1.5 (20)	2.3 (17)	13.4	Mattson and Likens 1993
L226	ND	8.1	13	15.2 (30)	4.2 (8)	0.97 (2)	8.1 (16)	23 (45)	51.8	Kelly et al. 1988
L227	21	7.38	7.8	1.6 (5)	0.3 (1)	0.6 (2)	8.8 (28)	20.1 (64)	31.3	Kelly et al. 1988
L223	ND	7.62	6.7	2.7 (10)	0.3 (1)	1.8 (7)	9.1 (34)	12.8 (48)	26.8	Kelly et al. 1988
Onondaga Lake										
1989	54	32	44	69 (39)	22 (13)	0.98 (1)	39 (22)	44 (25)	174	This study
1990	—	49	40	63 (39)	18 (11)	0.53 (0.3)	44 (27)	36 (22)	161	
1991	—	42	50	75 (38)	13 (7)	0.85 (0.4)	65 (33)	44 (22)	198	

summer period, are all subject to oxidation when they are entrained into the overlying oxygenated waters during the fall mixing period. The accumulation of large quantities of T-NH$_3$ (Figure 5.57), CH$_4$ (Figure 5.66), and H$_2$S (Figure 5.62) in the hypolimnion of Onondaga Lake during the summer represents a substantial potential oxygen demand for the overlying epilimnetic waters. Some demand is exerted during the stratification period by diffusion of these species to the upper waters. However, most of the demand is exerted during a rather brief period (e.g., from mid-September to mid-October) with the approach to fall turnover, as the enriched lower layers are progressively entrained into the deepening upper mixed layer. Here we review oxidation processes for CH$_4$, H$_2$S, Fe^{2+}, and T-NH$_3$, and describe the role these processes play with respect to the O$_2$ resources of Onondaga Lake.

5.6.6.1 Description of Oxidation Processes

Aerobic CH$_4$ oxidation is a bacterially mediated process. Apparently a large number of bacteria are capable of aerobic CH$_4$ oxidation; over 100 strains have been isolated (Rudd and Taylor 1980). During stratification, CH$_4$ oxidizing bacterial activity has been found to be localized in the metalimnia of some lakes, where the dissolved inorganic nitrogen (DIN) concentrations are typically low in the overlying epilimnion due to depletion by phytoplankton (e.g., Rudd et al. 1976). More vertically uniform CH$_4$ oxidation is observed in these systems during turnover due to elevated DIN concentrations throughout the water column (Rudd et al. 1976). Methane oxidation should be expected to be more evenly distributed in the epilimnion of Onondaga Lake because of the unusually high DIN concentrations that prevail in the epilimnion (Figure 5.31). The overall oxidation process can be represented by the following stoichiometric equation (DiToro et al. 1990; Sweerts et al. 1991).

$$CH_4 + 2O_2 \rightarrow CO_2 + 2H_2O \quad (5.49)$$

Methane is oxidized in the presence of O$_2$ at a rapid rate. Jannasch (1975) estimated the rate of CH$_4$ oxidation in an interval above the oxic/anoxic interface of Lake Kivu to be 0.48 mmol $\cdot$ m^{-3} $\cdot$ d^{-1}. Rudd et al. (1976) measured a metaliminetic peak rate of about 1.2 mmol $\cdot$ m^{-3} $\cdot$ d^{-1} during summer in eutrophic Lake 227, and a range of 3.6 to 12 mmol $\cdot$ m^{-3} $\cdot$ d^{-1} throughout the water column after the onset of fall turnover. The failure to detect CH$_4$ in the oxygenated layers of Onondaga Lake reflects the rapid rate of CH$_4$ oxidation in this system, and indicates that evasion is not a significant loss pathway.

Hydrogen sulfide is also extremely labile in the presence of O$_2$. Hydrogen sulfide is oxidized by abiotic and microbially mediated processes (Jorgensen et al. 1979). Jorgensen et al. (1979) determined the half-life of H$_2$S just above the oxic/anoxic boundary of saline Solar Lake to be approximately 5–10 min. The estimated sulfide oxidation rate was 13–5 mmol $\cdot$ m^{-2} $\cdot$ d^{-1} (Jorgenson et al. 1979). The failure to detect H$_2$S in the oxygenated layers of Onondaga Lake reflects rapid rates of oxidation. The overall stoichiometric reaction for oxidation of H$_2$S is represented by

$$H_2S + 2O_2 \rightarrow SO_4^{2-} + 2H^+ \quad (5.50)$$

Two moles of O$_2$ are required to oxidize one mole of H$_2$S compared to 1.6 moles of O$_2$ per mole of CH$_4$ (Equation (5.49)). The unusually high contribution of SO$_4^{2-}$ reduction to the anaerobic metabolism in Onondaga Lake, associated with its usually high SO$_4^{2-}$ concentration, may have deleterious implications for the O$_2$ resources of the lake because H$_2$S is much more soluble than CH$_4$. Recall a significant fraction of the CH$_4$ released from the sediments of the lake is lost from the system through ebullition (Figure (5.69).

Nitrification is a two-stage process. The first stage is the oxidation of ammonia to nitrite by the bacteria *Nitrosomonas*.

$$NH_4^+ + 1.5O_2 \rightarrow NO_2^- + H_2O + 2H^+ \quad (5.51)$$

During the second stage of nitrification *Nitrobacter* oxidizes NO$_2^-$ to NO$_3^-$.

$$NO_2^+ + \tfrac{1}{2}O_2 \rightarrow NO_3^- \quad (5.52)$$

The complete oxidation of ammonia (combining Equations (5.51) and (5.52) can be represented by:

$$NH_4^+ + 2O_2 \rightarrow NO_3^- + H_2O + 2H^+ \quad (5.53)$$

Thus 4.57 g of O_2 are required for complete oxidation of 1 g of NH_4^+. Several researchers have reported nitrification to be localized at the sediment-water interface (Cavari 1977; Curtis et al. 1975; Hall 1986). Results of laboratory microcosm experiments indicate no significant nitrification occurs within the watercolumn of the lake. System-specific kinetic results (see Chapter 9), time series T-NH_3 data for the lake in certain years after the onset of fall turnover, and the literature (Bowie et al. 1985) indicate the kinetics of nitrification are slow compared to the oxidation of CH_4 and H_2S.

The iron oxidation pathway has been assumed to be (Adams et al. 1982; Sweerts et al. 1991).

$$4Fe^{2+} + O_2 + 6H_2O \rightarrow 4FeOOH + 8H^+$$
$$(5.54)$$

The following pathway has been proposed for manganese oxidation (Adams et al. 1982)

$$2Mn^{2+} + O_2 + 2H_2O \rightarrow 2MnO_2 + 4H^+$$
$$(5.55)$$

Our analysis, and those of others (Table 5.26), indicates that Fe and Mn reduction are minor electron transfer pathways in lake systems. The potential oxygen demand associated with the hypolimnetic accumulations observed for these two species in Onondaga Lake is minor compared to the other reduced species.

5.6.6.2 Interplay With Lake-Wide Fall DO Depletion

The large pool of T-NH_3 that accumulates in the hypolimnion of Onondaga Lake represents a large "potential" oxygen sink that is not realized during the fall mixing period. To some extent this is due to the relatively slow kinetics of the nitrification process. More importantly, in certain years T-NH_3 appears to be largely depleted by an alternate sink

process—phytoplankton uptake. Evidence for this emerges from the strong variations in fall (from mid-September to late October) depletion of watercolumn T-NH_3 that have been observed to be highly correlated (R = 0.94) to the average fall concentration of chlorophyll (Figure 5.72). Thus the magnitude of the phytoplankton sink for T-NH_3 during the fall mixing period depends on the intensity of the fall bloom.

The lake-wide DO depletion observed during the fall mixing period, parametrized as the mass of DO by which the lake is undersaturated at the onset of complete fall turnover, can be explained by the nearly immediate DO demand (as oxygen equivalents) exerted with the entrainment of CH_4 and H_2S through the mixing period (mid-September to late October; Figure 5.73). In all four years over the 1989–1992 interval, the oxygen demand exerted by these two reduced species was nearly equal to the mixing period DO deficit. The mean fall O_2 demand associated with CH_4 and H_2S for the four years was about 7.6 × 10^8 g, while the coincident DO deficit was 8.5 × 10^8 g. This crude mass balance analysis (Figure 5.73; e.g., omission of other time variable O_2 sinks and sources) serves to demonstrate the central role these by-products of anaerobic metabolism play in the severe lake-wide DO depletion observed annually in Onondaga Lake (Figure 5.12). This issue is

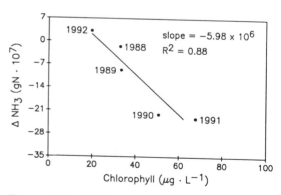

FIGURE 5.72. Evaluation of the relationship between the fall (mid-September–late October) depletion of the T-NH_3 pool and the average fall concentration of chlorophyll (0–8 m, volume-weighted average) in Onondaga Lake.

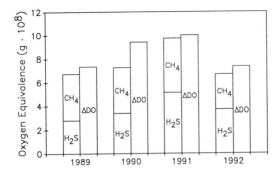

FIGURE 5.73. Oxygen demand (summation of CH$_4$ and H$_2$S pools) for fall mixing period and deficit of the O$_2$ pool at the onset of complete fall turnover, for Onondaga Lake, for four years.

addressed in a more comprehensive and quantitative fashion within the framework of a dynamic mechanistic model for DO in Chapter 9.

The H$_2$S pool represented on average 50% of the nearly immediate fall O$_2$ demand estimated for the four years (Figure 5.73). The major role this constituent plays is a manifestation of the unusually high (but natural; see Chapter 3) concentrations of SO$_4^{2-}$ that prevail in the lake. Its interesting to speculate on what conditions would occur if the SO$_4^{2-}$ concentrations were much lower, so that SO$_4^{2-}$ reduction was limited, as is the case in many lakes. In this case, most of the corresponding decomposition would likely be diverted to methanogenesis. Much of the additional CH$_4$ production would escape the system through ebullition, and thus would not exert the associated O$_2$ demand during the critical fall mixing period. Accordingly, the lake-wide DO depletion would be less severe. However, for much lower SO$_4^{2-}$ concentrations, greater concentrations, and associated oxygen demand, of Fe^{2+} could occur.

5.6.6.3 Hypolimnetic Oxygen Demand

The electron budget conducted for the lake's hypolimnion and the assumed oxidation pathways of the reduced species can be used to estimate the associated oxygen demand, and partition the demand according to its components. Oxygen demanding processes

considered here include aerobic oxidation, oxidation of Fe^{2+}, CH$_4$ and H$_2$S, and nitrification. Note that reductive pathways which involve ebullitive loss (i.e., denitrification, the fraction of CH$_4$ lost via ebullition) are not included. The total hypolimnetic oxygen demand (HOD) was determined as a rate (g $\cdot$ m^{-2} $\cdot$ d^{-1}) based on the sum of electrons transferred via the various metabolic processes at the end of summer stratification, according to:

$$HOD = a_1 \cdot J_{O_2} + a_2 J_{Fe^{2+}} + a_3 J_{CH_4} + a_4 J_{H_2S} + a_5 J_{T\text{-}NH_3} \quad (5.56)$$

in which a_1, a_2, a_3, a_4, and a_5 = stoichiometric coefficients to convert concentrations to oxygen demand equivalents (1, 0.25, 2, 2, and 2, respectively), and J_x = flux of component $\times$ for the hypolimnion, adjusted for vertical losses and inputs to the overlying epilimnion during the stratification period. Note that the value of J_{O_2} (2.02 g $\cdot$ m^{-2} $\cdot$ d^{-1} in 1989) is slightly greater than AHOD (1.9 g $\cdot$ m^{-2} $\cdot$ d^{-1} in 1989; Table 5.6), because estimated inputs of DO from the epilimnion over the period of depletion have been included in J_{O2}.

Estimates of HOD and the contributions of the components are presented for 1989, 1990 and 1991 in Table 5.27, along with estimates of AHOD, and the mean of sediment oxygen demand (SOD) measurements (see Chapter 9 for experimental details). Values of HOD were similar for the four years; the mean was 2.1 g $\cdot$ m^{-2} $\cdot$ d^{-1}. The carbonaceous component (sum of first four terms of equation (55) comprised about 66% of the total, the remainder is associated with T-NH$_3$ (the nitrogenous component). About 40% of the carbonaceous component is aerobic oxidation; H$_2$S (27%) and CH$_4$ (31%) oxidation also make significant contributions to the carbonaceous component of the oxygen demand (Table 5.27). The general magnitude of the estimated HOD, and the partitioning of components, is supported by the similarity to the determinations of AHOD and the measurements of SOD (Table 5.27). The HOD is expected to be somewhat greater than AHOD because of the inclusion of vertical exchange between the epilimnion and hypolimnion.

TABLE 5.27. Hypolimnetic oxygen demand, and partitioning of components, for Onondaga Lake for three years, and comparison to other measures of oxygen demand.

Component/measure	HOD ($g \cdot m^{-2} \cdot d^{-1}$)			
	1989	1990	1991	Mean
Aerobic oxidation	0.55 (26%)	0.50 (26%)	0.60 (27%)	0.55 (26%)
Fe^{2+}	0.008 (0.4%)	0.004 (0.2%)	0.007 (0.3%)	0.006 (0.3%)
CH_4	0.46 (22%)	0.38 (20%)	0.46 (21%)	0.43 (20%)
H_2S	0.31 (15%)	0.35 (18%)	0.52 (23%)	0.40 (19%)
Carbonaceous*	1.32 (63%)	1.23 (64%)	1.59 (72%)	1.38 (66%)
Nitrogenous (T-NH_3)	0.75 (36%)	0.70 (36%)	0.64 (29%)	0.70 (33%)
Total**	2.08	1.93	2.22	2.10
AHOD	1.9	1.1	1.4	1.47
SOD				1.68

* Sum of the preceding four components
** Sum of carbonaceous and nitrogenous

The value of HOD is also expected to be greater than SOD. Although SOD is the dominant O_2 sink in the hypolimnion of the lake, some demand is also exerted within the water column of the hypolimnion (see DO model in Chapter 9).

5.7 Mercury

Onondaga Lake contains elevated concentrations of mercury (Hg) in its water column (Bloom and Effler 1990), sediment (NYSDEC 1990; see Chapter 8) and fish tissue (see Chapter 6). It is estimated that approximately 76,000 kg of Hg was discharged into the lake by the adjoining soda ash/chlor-alkali facility from 1946 to 1970 (USEPA 1973; see Chapter 1). Although the discharge of Hg was reduced by more than 95% soon after the U.S. Department of Justice took action against the facility in 1970, and was further reduced with closure of the chlor-alkali operation in 1988, concentrations of Hg in a large fraction of the legal-size fish in the lake continue to exceed the U.S. Food and Drug Administration (FDA) action level (i.e., 1.0 ppm; NYSDEC 1987; see Chapter 6).

The Hg cycle in aquatic ecosystems is complicated due to the myriad of species and pathways (Figure 5.74). Our understanding of the biogeochemistry of Hg has increased markedly over the last ten years with development of clean protocols for the sampling and analysis of Hg (Gill and Fitzgerald 1985, 1987). There is little confidence in measurements of aqueous Hg prior to 1985, due to the likelihood of sample contamination (Fitzgerald and Watras 1989). Atmospheric deposition of Hg to lakes largely occurs as inorganic Hg (Winfrey and Rudd 1990), although inputs of methyl Hg occur (Bloom and Watras 1989). Within oxygenated waters, Hg(II) will complex with inorganic ligands (e.g., Cl^-, OH^-), bind with dissolved organic carbon (DOC) or sorb to particulate matter.

Mercuric ion can be reduced microbially to form elemental Hg (Hg^o). Most waters are oversaturated with respect to the solubility of atmospheric Hg^o, and Hg^o is volatilized to the atmosphere (Vandal et al. 1991). Within anoxic zones, Hg forms strong aqueous complexes with sulfide and precipitates as HgS. Within anaerobic environments or within anoxic microzones in aerobic environments, Hg(II) can be converted to methyl Hg. Sulfate reducing bacteria appear to be important in the methylation of Hg (Compeau and Bartha 1985; Gilmour et al. 1992). Methyl Hg may bind to DOC or be demethylated by microbial processes. Methyl Hg is generally thought to be the form of Hg which bioconcentrates in

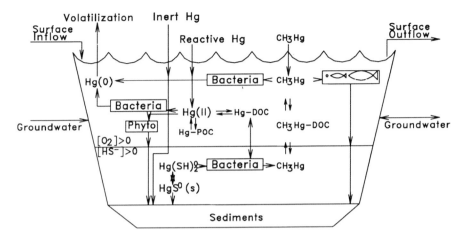

FIGURE 5.74. Schematic diagram summarizing the Hg cycle in lake ecosystems.

fish. Gill and Bruland (1990) reported a strong relationship between concentrations of organo Hg in lakes and concentrations of Hg in fish.

This section on the chemistry of Hg includes: (1) theoretical chemical equilibrium calculations on the speciation of Hg that are relevant to Onondaga Lake, (2) a discussion of investigations by Bloom and Effler (1990) and Wang (1993) which include information on the concentration and speciation of Hg, and (3) a mass balance of Hg for Onondaga Lake.

5.7.1 Theoretical Chemical Equilibrium Calculations

In order to gain insight on Hg chemistry in Onondaga Lake, theoretical calculations were conducted with the chemical equilibrium model MINEQL (Schecher and McAvoy 1992). Three chemical equilibrium calculations were conducted for this analysis. A Hg pe-Ph (pe = −log electron activity) predominance area diagram was constructed for conditions that are applicable to Onondaga Lake ($Hg_T = 10^{-10}$ mol $\cdot$ L^{-1}, $Cl_T = 10^{-2}$ mol $\cdot$ L^{-1}, $S_T = 2 \times 10^{-3}$ mol $\cdot$ L^{-1}). In addition, a chemical equilibrium "titration" was conducted, in which fixed pe conditions were varied from 14 to −4. In the titration pH was fixed at 7.5, the temperature was assumed to be 8°C and total Hg (Hg_T) is 10^{-10} mol $\cdot$ L^{-1}.

According to the Hg pe-pH predominance diagram, the stable species of Hg in the oxidizing upper waters of Onondaga Lake (Eh > 0.4 V; pe > 9), is $HgCl_2$ at neutral and lower pH values, and $Hg(OH)_2$ at higher pH conditions (Figure 5.75). With decreases in pe the thermodynamically stable form of Hg shifts to elemental Hg (Hg°). With further decreases in pe associated with conditions experienced in the hypolimnion during summer stratification

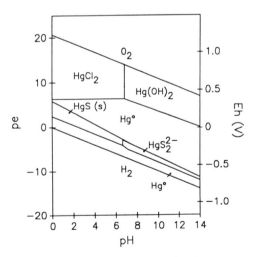

FIGURE 5.75. Mercury pe-pH predominance area diagram for conditions that are typical for the lower waters of Onondaga Lake ($Hg_T = 10^{-10}$ mol $\cdot$ L^{-1}, $Cl_T^- = 10^{-2}$ (mol$\cdot$L^{-1}), $S_T = 2 \times 10^{-3}$ mol $\cdot$ L^{-1}.

(Eh = −0.23 V; pe = −4.1), and pH = 7.3, the predominant form of Hg is a Hg polysulfide complex (HgS_2^{2-}).

With a pe-pH diagram, the stable form of Hg under specific pH and redox conditions (pe) is defined. Unfortunately, this approach does not permit a detailed examination of the composition of Hg in a system nor an evaluation of changes in the speciation of Hg under varying chemical conditions. For this purpose, the chemical equilibrium "titration", under varying conditions of pe was conducted. Like the predominance area diagram, the results of the pe "titration" indicate shifts in the speciation of Hg with changing environmental conditions that occur in Onondaga Lake. Mercuric chloride ($HgCl_x$) complexes and mercuric hydroxide ($Hg(OH)_x$) complexes predominate under high pe conditions that are anticipated in the epilimnion and in the mixed waters during spring and fall (Figure 5.76). Aqueous Hg predominantly occurs as soluble complexes in Onondaga Lake water; concentrations of aquo Hg (Hg^{2+}) are calculated to be several orders of magnitude lower than the concentration of the dominant species. As pe decreases, the composition of Hg species does not change markedly until pe is equal to 2.5. At this redox potential, concentrations of Hg-Cl and Hg-OH complexes decrease moderately until the pe declines to −2.5 (Figure 5.76). Elemental mercury (Hg°) predominates in the system when pe is in the range of 2.5 to −2.5 (Figure 5.75). When pe declines below −2.5, the concentration of SO_4^{2-} decreases dramatically due to SO_4^{2-} reduction and concentrations of H_2S and HS^- increase markedly. Under these conditions the speciation of Hg shifts so that soluble Hg sulfide complexes (HgS_2^{2-} and $Hg(HS)_2$) become significant, resulting in marked decreases in Hg^{2+}, $HgCl_x$ and $Hg(OH)_x$.

5.7.2 Measurements of Mercury in Onondaga Lake

Because the Hg content in fish is directly affected by the concentration and speciation of Hg in the watercolumn, a preliminary study of the chemistry of Hg was conducted in Onondaga Lake. In this investigation four collections were made at three depths in the watercolumn from the spring to fall 1989 (Bloom and Effler 1990). Clean techniques were used during all phases of sample collection and handling. The species of Hg investigated in the study included total Hg (Hg_T), reactive Hg (Hg_R), monomethyl Hg (CH_3Hg), elemental Hg (Hg°), and dimethly Hg (($CH_3)_2Hg$) (Table 5.28). These were analyzed by using the Hg fractionation procedure developed by Bloom (1989).

The results of this investigation revealed seasonal changes in the concentration and vertical patterns of Hg species in the lake. Concentrations of Hg_T were very high in Onondaga Lake, ranging from 7 to 26 ng·L^{-1}. Concentrations of Hg_T were highest throughout the watercolumn in the spring. From spring to summer concentrations of Hg_T increased with increasing depth in the lake. This pattern was not due to increases in concentra-

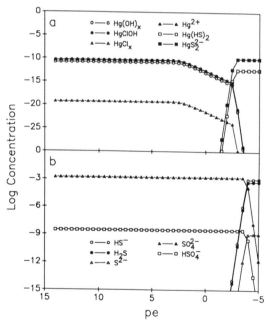

FIGURE 5.76. Changes in the speciation of Hg(a) and S(b) with variations in pe for conditions that are typical for the lower waters of Onondaga Lake ($Hg_T = 10^{-10}$ mol·L^{-1}, $Cl_T = 10^{-2}$ mol·L^{-1}, $S_T = 2 \times 10^{-3}$ mol·L^{-1}, temperature 8°C).

TABLE 5.28. Summary of mean concentrations of total Hg and fractions of Hg for 0, 10, and 18 m depths of Onondaga Lake for 1989 (in $ng\,Hg \cdot L^{-1}$; Bloom and Effler 1990).

Mercury fracation	Depth (m)		
	0	10	18
Reactive (Hg_R)	3.3 (29%)	2.3 (21%)	3.9 (18%)
Monomethyl (CH_3Hg)	1.2 (11%)	2.2 (20%)	4.5 (21%)
Dimethyl (($CH_3)_2Hg$)	0.003 (<1%)	0.003 (<1%)	0.007 (<1%)
Elemental ($Hg°$)	0.11 (1%)	0.072 (41%)	0.040 (4%)
Total (Hg_T)	11.2 (100%)	11.1 (100%)	21.0 (100%)

tions of Hg_T in the lower waters, but rather decreases in Hg_T in the epilimnion. During fall turnover concentrations were uniform in the watercolumn.

The concentrations of Hg_R, an analog for weakly complexed ionic species, increased progressively from April to August. Concentrations were highest throughout the lake in August. Generally, Hg_R comprised a larger fraction of Hg_T in the upper waters than the lower waters (Table 5.28; Bloom and Effler 1990). Elevated concentrations of total CH_3Hg were also evident in Onondaga Lake (0.34–9.7 $ng \cdot L^{-1}$). Like Hg_T, concentrations of CH_3Hg increased with increasing depth. The highest concentrations of CH_3Hg were observed in the lower waters during summer stratification. The fraction of Hg_T occurring as CH_3Hg increased from about 11% in the oxic upper waters to about 20% in the anoxic lower waters.

The watercolumn of Onondaga Lake contained low concentrations of the volatile Hg species, $Hg°$ and $(CH_3)_2Hg$. The mean concentration of $Hg°$ was 0.07 $ng\,Hg \cdot L^{-1}$ and concentrations of $(CH_3)_2Hg$ were near the detection limit. Assuming the atmospheric concentration of $Hg°$ of 2.3 $ng \cdot m^{-3}$ (Fitzgerald 1986) and the Henry's Law constant of 0.3 (Lindqvist et al. 1984), the saturation index (SI = log(observed $Hg°$ concentration/saturated $Hg°$ concentration)) of $Hg°$ for the upper water of the lake was 1.04 to 1.22. This pattern indicates that the concentration of $Hg°$ in the lake was oversaturated with respect to the solubility of atmospheric $Hg°$. Therefore, the lake releases $Hg°$ to the atmosphere by evasion.

Wang (1993) conducted a detailed investigation of temporal and spatial patterns of Hg_T. In this study, samples were collected for Hg_T and other water chemistry parameters at the inlet tributaries and in the water column of Onondaga Lake during 1992. Inlet streams and the lake outlet were monitored by grab samples collected on ten dates from February to August 1992. These sites included Otisco Lake outlet, Ninemile Creek at Amboy, Ninemile Creek at Lakeland, Onondaga Creek at South Onondaga, Onondaga Creek at Spencer Street, upstream Ley Creek at Schuler Road (10 km from the lake inlet), Ley Creek at Park Street (inlet of Onondaga Lake), and METRO effluent. In addition, watercolumn samples were collected on eight dates from April to November 1992 at the southern basin site of the lake. These samples were taken from three depths, the surface (0 m), 10 m, which approximates the thermocline, and the lower most waters (19 m). The depth samples were collected with a submersible pump and hose system. Samples were collected and processed using clean techniques.

Stream samples showed highly variable concentrations of Hg_T (Table 5.29). Total Hg concentrations were lowest near headwater sources and increased with distance downstream. For example, in Onondaga Creek mean concentrations of Hg_T were 1.4 $ng\,Hg \cdot L^{-1}$ at South Onondaga and increased to 3.6 $ng\,Hg \cdot L^{-1}$ at Spencer Street, just upstream from Onondaga Lake. In Ninemile Creek, mean concentrations of Hg_T increased markedly from 1.7 $ng\,Hg \cdot L^{-1}$ at the Otisco Lake outlet to 4.0 $ng\,Hg \cdot L^{-1}$ at Amboy to 13 $ng\,Hg \cdot L^{-1}$ at Lakeland prior to discharge in the

TABLE 5.29. Summary of the concentrations of Hg_T in inlet and outlet waters of Onondaga Lake (in $ng\,Hg\cdot L^{-1}$) for 1992. These values are compared with concentrations of Hg species measured by Bloom (1990) on November 10, 1989. DL indicates concentrations occurring below the detection limit.

| Site | 1992 | | | | 1989 | | |
	Mean	Std. Deviation	Minimum	Maximum	Hg_T	Hg_R	CH_3Hg
Ninemile—Otisco	1.7	1.0	0.6	4.1	—	—	—
Ninemile—Amboy	4.0	3.2	DL	12.8	5.1	0.29	0.06
Ninemile—Lakeland	13	7.1	DL	21	18	0.99	0.10
Onondaga Creek—South Onondaga	1.4	1.2	0.086	4.0	—	—	—
Onondaga Creek—Spencer Street	3.6	2.6	DL	8.0	3.6	0.67	0.29
Ley Creek—Schuyler Road	2.9	2.0	0.19	7.1	—	—	—
Ley Creek—Park Street	4.3	2.7	1.2	9.7	—	—	—
METRO	15	10	5.0	35	32	16	0.38
Outlet	2.3	1.1	0.92	3.8	7.3	1.8	1.05

lake. The mean values of Hg_T measured in the tributaries in 1992 were similar to the values analyzed by Bloom (1989) for the same sites on a single date in 1989 (Table 5.29).

Many of the stream sites exhibited patterns of increasing concentrations of Hg_T with increasing concentrations of dissolved organic carbon (DOC; e.g., Ninemile Creek at Amboy $R^2 = 0.34$; Ninemile Creek at Lakeland $R^2 = 0.30$; upstream Ley Creek $R^2 = 0.42$; Onondaga Creek at Spencer Street $R^2 = 0.44$). Several investigators have reported empirical relationships between concentrations of Hg and DOC in surface waters (Lindqvist 1991; Mierle and Ingram 1991; Driscoll et al. 1994b). Mercury strongly associates with naturally occurring organic matter (Andersson 1979; Benes and Havlik 1979). It seems likely that stream transport of Hg_T is closely linked with DOC.

Water column concentrations of Hg_T in the upper waters of Onondaga Lake were highly variable in 1992 (Figure 5.77). Values ranged from 2.6 to $19.2\,ng\,Hg\cdot L^{-1}$, with a mean concentration of $9.5\,ng\,Hg\cdot L^{-1}$. Temporal patterns in Hg_T concentrations were not strongly manifested. The highest concentrations of Hg_T occurred in the upper waters in the fall (2 September, $22.5\,ng\,Hg\cdot L^{-1}$; 4 November, $12.2\,ng\,Hg\cdot L^{-1}$; 18 November, $19.2\,ng\,Hg\cdot L^{-1}$). This fall increase in Hg may be associated with increases in concentrations of suspended solids that occur in the lake at this time (Wang 1993) or the mixing of lower

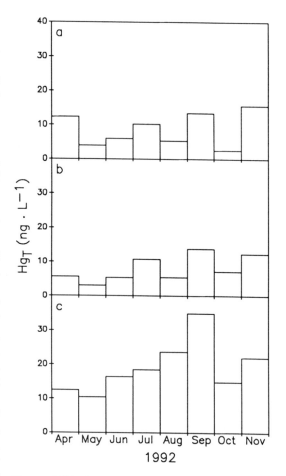

FIGURE 5.77. Concentrations of Hg_T in the (**a**) 0 m, (**b**) 1 m and (**c**) 19 m waters of Onondaga Lake, 1992.

waters with elevated concentrations of Hg_T during fall turnover. Like the upper water values, concentrations of Hg_T at the 10 m depth of the lake were variable. The mean Hg_T concentration at 10 m was 9.1 ng Hg $\cdot$ L^{-1}, similar to values at the lake surface.

The highest Hg_T concentrations in Onondaga Lake occurred in the lower waters, with a mean concentration of 21.6 ng Hg $\cdot$ L^{-1} (Figure 5.77). During spring turnover, Hg_T concentrations were relatively low in the hypolimnion (12.6 ng Hg $\cdot$ L^{-1}). As summer stratification proceeded, concentrations of Hg_T increased, reaching a maximum value of 35 ng Hg $\cdot$ L^{-1} in mid September. Concentrations of Hg_T decreased somewhat during fall turnover.

The increase of Hg_T in the lower waters of the lake closely coincided with increases in H_2S concentration (Figure 5.78). As shown in the theoretical chemical equilibrium calculations (Figures 5.75 and 5.76), Hg forms strong complexes in the presence of H_2S. HgS_2^{2-} is generally the predominant form of Hg when molar concentrations of H_2S exceed concentrations of Hg_T. The coincidence of increases of Hg_T and H_2S in the lower waters suggests that H_2S production and mobilization to the water column facilitates upward transport of Hg_T from sediments through the formation of soluble complexes.

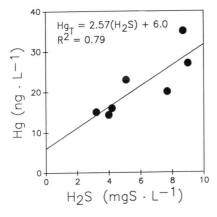

FIGURE 5.78. Relationship between concentrations of Hg_T and concentrations of H_2S_T in the lower waters of Onondaga Lake.

Studies of the chemistry of Hg by Bloom and Effler (1990) and Wang (1993) showed similar results. Concentrations of Hg_T measured in both studies were similar. Moreover, these investigations demonstrated a pattern of increasing concentrations of Hg_T with increasing depth in the lake. The more temporally intensive study of Wang (1993) revealed the accumulation of very high concentrations of Hg_T in the lower waters during summer stratification that were not evident in the few collections made by Bloom and Effler (1990).

5.7.3 Mercury Budget for Onondaga Lake

Using information on Hg_T in inflowing streams and hydrologic data, a Hg_T budget was developed for Onondaga Lake for 1992. During 1992 the discharge at Ninemile Creek was 456,000 m$^3 \cdot$ d^{-1}, Onondaga Creek 476,000 m$^3 \cdot$ d^{-1}, Ley Creek 123,000 m$^3 \cdot$ d^{-1}, and METRO was 276,000 m$^3 \cdot$ d^{-1}. The annual discharge of Ninemile Creek at Amboy was assumed to be 80% of the measured value at Lakeland based on the percentage of the total watershed area upstream of Amboy (Effler et al. 1991a). The precipitation for the study year was 111 cm.

Values of mass flux of Hg_T were determined by summing the stream flow for the period before and after a given sample collection. The cumulative flow of water for a tributary was multiplied by the sample concentration determined for that collection. The fluxes for the year 1992 were summed to determine an annual stream input. Unfortunately, values of Hg_T were not available for the fall 1992. As a result, Hg_T concentrations obtained for late August were assumed to be representative of the fall period and used in mass balance calculations. For inflows that weren't monitored for Hg_T (e.g., Harbor Brook, Bloody Brook), Hg_T concentrations were assumed to be similar to values measured at Ley Creek. Lake outlet samples contained relatively low concentrations of Hg_T (Table 5.29) and are not considered to be representative of lake outflow because of the bidirectional flow regime in the lake

outlet (near surface samples more representative of Seneca River (see Chapter 4)). As a
result, Hg_T concentrations obtained from the
lake surface were assumed to be representative
of the lake outflow. Outlet discharge was
assumed to be equal to the sum of the discharge of all the lake inlet tributaries.

Precipitation concentrations of Hg_T were
assumed to be $10\,ng\,Hg \cdot L^{-1}$ (Fogg and Fitzgerald 1979; Fitzgerald 1986; Bloom and
Watras 1989). Evasion of Hg° was calculated
by Bloom (1990) based on measured values
of Hg° in 1989 and assuming a mass transfer
coefficient (Fitzgerald 1986). A value of 0.3
$kg \cdot yr^{-1}$ was obtained from this analysis.
Water column pools for the lake were determined for three depth intervals (surface
$0-7\,m$, $76 \times 10^6\,m^3$; middle $7-15\,m$, $48 \times
10^6\,m^3$; lower $15-22\,m$, $7 \times 10^6\,m^3$). Cal-
culations of Hg_T pools in the lower waters
showed a systematic increase over the summer stratification period. A rate of sediment
release of Hg_T of $280\,ng\,Hg \cdot m^{-2} \cdot d^{-1}$ was
calculated based on the rate of accumulation
of the lower water Hg_T pool (Figure 5.79).
This flux was not corrected for vertical mixing. The sediment pool for $1\,cm$ depth was
calculated from a mean surface sediment
concentrations of $12\,\mu g \cdot g^{-1}$ obtained from a
sediment contour map (Effler 1987). The Hg_T
budget shows stream inputs contribute the

predominant flux of Hg_T to Onondaga Lake
($5.45\,kg\,Hg \cdot yr^{-1}$). Atmospheric deposition to
the lake surface appears to be a minor input
of Hg_T to the lake ($0.13\,kg\,Hg \cdot yr^{-1}$; Figure
5.79). Although sediments represent a large
pool of Hg in the lake ($14,000\,kg\,Hg$) and
release from sediments seems to result in very
high concentrations of Hg_T in the lower waters
(Figure 5.77), the flux of Hg_T release from
sediments was small ($0.28\,kg\,Hg \cdot yr^{-1}$) relative to fluvial sources. Ninemile Creek was the
major external source of Hg_T ($2.91\,kg\,Hg \cdot
yr^{-1}$) representing 48% of the total input.
Note that the flux of Hg_T in Ninemile Creek at
Amboy ($0.59\,kg\,Hg \cdot yr^{-1}$) was small relative
to the input of Hg_T entering the lake. This
pattern suggests that there is a large input of
Hg_T to Ninemile Creek between Amboy and
Lakeland which represents a large fraction of
the total Hg_T input to the lake (38%). Inputs
of Hg_T associated with the METRO effluent
($1.52\,kg\,Hg \cdot yr^{-1}$) also contributed a significant fraction (25%) of the total Hg_T loading.

The mass balance calculations show that
Onondaga Lake is relatively conservative with
respect to inputs of Hg_T. The total input of Hg_T
to the lake ($5.86\,kg\,Hg \cdot yr^{-1}$) was very close
to the total output ($5.75\,kg\,Hg \cdot yr^{-1}$). The
outflow of Hg_T from the lake largely occurred
via the outlet ($5.6\,kg\,Hg \cdot yr^{-1}$). Evasion losses
of degassing Hg° were a minor outflux (0.15

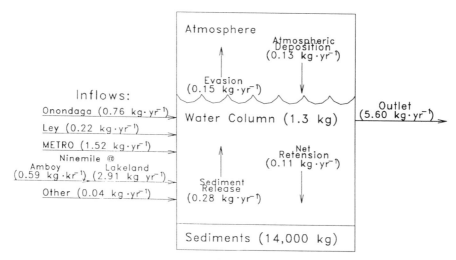

FIGURE 5.79. Mass balance of Hg_T for Onondaga Lake. Pools are shown in kg and fluxes are given in
$kg\,Hg \cdot yr^{-1}$.

TABLE 5.30. Summary of Hg_T concentrations in lakes.

Lake systems	Location	Total Hg	Reference
Remote Lakes			
drainage (14)	Adirondacks	0.8–5.3	Driscoll et al. 1994b
seepage (2)	Adirondacks	0.8	Driscoll et al. 1994b
seepage (4)	Wisconsin	0.9–1.9	Fitzgerald and Watras 1989
drainage	Washington (state)	0.2	Bloom 1989
alpine	California	0.6	Gill and Bruland 1980
drainage (3)	Manitoba	0.2–1.1	Bloom and Effler 1990
Urban Lakes			
Lake Union	Washington (state)	1.7	Bloom 1989
Great Lakes			
Erie	(U.S./Canada)	3.9	Gill and Bruland 1990
Ontario	(U.S./Canada)	0.9	Gill and Bruland 1990
Mining Contaminated Lakes			
Clear Lake	California	3.6–104	Gill and Bruland 1990
Davis Creek Reservoir	California	5.2–6.4	Gill and Bruland 1990
Chlor-alkali Contaminated Lakes			
Clay Lake	Ontario	5–80	Parks et al. 1989
Onondaga Lake	New York	2–35	Bloom and Effler 1990

$kg\,Hg \cdot yr^{-1}$). These calculations suggest that the lake is a slight sink for inputs of Hg_T ($0.11\,kg\,Hg \cdot yr^{-1}$). Note, however, that this discrepancy in the mass balance is well within the errors associated with the calculations.

The watercolumn pool of Hg_T is about $1.3\,kg\,Hg$. This value divided by the total influx of Hg_T, gives a mean residence time of Hg_T in the water column of 0.22 yr. This value is short and comparable to the hydraulic residence time of the lake (see Chapter 3).

It is of interest to compare observations of Hg_T from Onondaga Lake with other systems (Table 5.30). Concentrations of Hg_T in Onondaga Lake are about one order of magnitude greater than those observed in relatively pristine systems. The concentrations and patterns in speciation of Onondaga Lake are similar to those observed in Clay Lake, Ontario (Furutani and Rudd 1980), also a site of past Hg contamination from a chlor-alkali plant.

5.8 Particle Chemistry

5.8.1 Introduction

Particles play an important role in water quality and they may reflect the operation of geochemical processes in the watershed and the lake basin. Particles regulate light scattering and thereby clarity (Kirk 1983). Surface processes, depending on the particle character, can be important in the cycling of nutrients (e.g., Kramer et al. 1972; Murphy et al. 1983) and organics (Loder and Liss 1985). Sedimentation results from the accumulation of deposited particles. The population of particles in the water column at any time is a composite of sediments received as tributary inputs, resuspended lake deposits, and particles produced within the water column. Internally produced particles include phytoplankton and inorganic precipitates (e.g., $CaCO_3$ particles in hard water lakes (Effler and Johnson 1987)). It is valuable to partition the major particle types according to elemental chemistry and origins to evaluate the manageability of related water quality problems and the interplay between the particle assemblage and related system processes.

Individual particle analysis (IPA) with scanning electron microscopy (SEM), interfaced with X-ray energy spectroscopy (XES), provides both a morphological and elemental characterization of particles that has been widely applied to environmental issues concerned with particulates (Yin and Johnson 1984). Computer controlled scanning electron microscopy (CCSEM) interfaced with XES is a time efficient way of providing this type of information. Each particle examined in this fashion contains information on 20 to 30 variables, depending upon the analytical hardware/software configuration. Johnson (1983) has presented a background summary of the technique and its general applicability. Similar methods have been used in the study of particle dynamics in the marine environment (Bernard et al. 1986; Carder et al. 1986), and to document $CaCO_3$ precipitation and the attendant contribution to light scattering (Effler and Johnson 1987; Weidemann et al. 1985) in hard water lakes. In Chapter 3 the results of tributary IPA characterizations made in 1981 (Yin and Johnson 1984), and for Onondaga Creek in 1987 and 1988 (Effler et al. 1992), are reviewed. Here we review CCSEM results for the watercolumn of Onondaga lake based on a monitoring program conducted for the upper waters from mid-May through mid-November of 1987 and limited water column sampling conducted in 1981. Some of the 1987 results are interpreted with respect to monitoring results for transparency, and phytoplankton biomass and composition.

5.8.2 Methods

5.8.2.1 Sampling and Analyses

Details of the sampling, sample handling, CCSEM/XES analysis are described elsewhere (Johnson et al. 1991; Yin and Johnson 1984). Differences in the IPA techniques used on samples collected in 1981 versus 1987 are documented in those references. An abbreviated description is presented here. A total of 28 samples were collected and analyzed from the 10% light level (depth of 1.0 to 4.0 m; information to be subsequently interfaced with optics analysis in Chapter 7) of the south deep station over the six month period in 1987. The CCSEM system consisted of an ETEC Auto-Scan scanning electron microscope interfaced with a KEVEX 7500 X-ray energy spectrometer and a LeMont Scientific DA-10 Image Analysis System. The minimum feature size detectable with the specified operating conditions for the samples collected in 1987 was 0.3 µm (area equivalent diameter).

Analytical errors in 1987 were evaluated with respect to enumeration of particles per unit volume of water (PNV) and also the total particle cross sectional area per unit volume (PAV). The sampling uncertainty for PNV was approximately 10%. The analytical error was evaluated by comparing triplicate analyses of the same field of view in the SEM. A reproducibility of ±0.5% was observed for inorganic particles, and 1.5% for organic particles. The reproducibility of PAV was assessed in a similar manner. Instrumental uncertainty averaged ±1% at the one standard deviation level. At the 95% confidence level, analyses of different fields of view on a given sample filter showed a frame-by-frame-variation of 18%; approximately twice that observed for measures of PNV.

5.8.2.2 Particle Classification

Each particle characterized by CCSEM analysis resulted in digital information for 30 observation variables; including relative net X-ray emission intensity for 25 elements (Na and higher atomic numbers), the gross X-ray count rate, the feature area and area equivalent diameter, as well as feature orientation and perimeter. These variables have been used in a classification scheme for the analysis of particles from several freshwater systems (Effler and Johnson 1987; Weidemann et al. 1985; Yin and Johnson 1984). This classification scheme was slightly modified for studies of Onondaga Creek (Effler et al. 1992); this modified version was adopted for the 1987 lake analyses presented here. Here, 19 classes of particles were defined by using criteria for the ranges of X-ray emission relative intensity. Results were grouped into categories with geochemical significance; six size ranges were accommodated. These classes were aggregated into one of six larger chemical groupings of particles for the 1987 study, as previously described for Onondaga Creek (Chapter 3). These six particle types effectively reflect the major chemical features of the analyses. Specifications of these six major particle categories are presented in Table 5.31.

The NL particle type (Table 5.31) refers to the "No and Low" X-ray particles. These particles are mostly organic, including phytoplankton, detritus, and bacteria. The CA category (Table 5.31) includes mostly $CaCO_3$ particles. The SI particles are those with at least 50% Si X-rays, but with less than 4% Al X-rays ("Si-bearing" class), combined with the "Si-only" particle type which contains over 98% Si X-ray relative emission intensity. This category includes quartz particles (usually of terrigeneous origin), but in Onondaga Lake it is dominated by frustules and fragments of frustules of diatoms. The CL category is made up of five separate alumino-silicate classes associated with "clay" materials found in the Onondaga Lake watershed. The AG category represents $CaCO_3$ precipitate in association with a silica or alumino-silicate matrix; they may also be adventitious aggregates of CaCO3 with Si-rich materials. The OT particle grouping includes all other particle types; it may include Ti-rich particles, particles that are high in Fe, and so on.

The results of the individual particle characterizations for 1987 are summarized both by particle numbers per unit volume (PNV) and according to the two-dimensional projected area of particles per unit volume of water (PAV). These two modes of summary provide different interpretive perspectives. The temporal changes in particles type populations are presented below in terms of PNV as a function of particle size. The size parameter is area

equivalent diameter. The potential environmental implications of suspended particle concentrations, such as light scattering or pollutant adsorption, may be better represented by examining PAV as a function of particle type and size.

5.8.3 Particle Population Dynamics

The total PNV (thousands of particles per ml) is plotted in Figure 5.80 for the monitoring period May 19 through November 15, 1987. Also shown is the arithmetic mean particle size expressed as area equivalent diameter (μm). Total PNV showed a two-fold increase from about 30,000 particles·ml^{-1} to over 60,000 particles·ml^{-1} during the first month of the monitoring period. From mid-June to late June, PNV declined rapidly to values of about 25,000 particles·ml^{-1}. A second peak occurred in late June associated with a major storm runoff event (Effler et al. 1992). Thereafter PNV decreased rapidly to about 10,000 particles·ml^{-1}. Some of the temporal changes in PNV appear to have been influenced by the dynamics of phytoplankton, represented as chlorophyll (method of Parsons et al. 1984; Figure 5.81b) and phytoplankton counts (by major groups, exclusive of diatoms; Figure 5.81c; note phytoplankton count data is less temporally resolved than the chlorophyll data). Secchi disc transparency, however, remained relatively constant at values between 0.5 and 1 m through the end of June (because of the high levels of scattering; Chapter 7).

TABLE 5.31. Generic categories that describe the major types of suspended solids in Onondaga Lake.

Name	CCSEM attributes	Comments
NL	Net X-ray count ≤ 600	"Organic" detritus*
CA	Net Ca X-ray >70% and SI < 5%	$CaCO_3$ precipitate
SI	Net Si X-rays >50% with Al < 3%	Quartz plus diatoms
CL	Presence of Al *plus* Si > 50%	"Clay particles"
AG	Presence of Si *plus* Ca > 80%	Ca ppt. on clay or silica "nucleus"
OT	All other features	Anything else

*This category may also contain low average atomic weight inorganic material as well as true "organic" residues

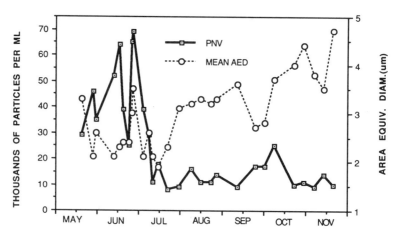

FIGURE 5.80. Temporal trends in PNV and arithmetic mean particle diameter (area equivalent diameter) in the upper waters of Onondaga Lake in 1987 (from Johnson et al. 1991).

The maximum Secchi disc transparency (Figure 5.81a) was coincident with the PNV minimum (Figure 5.80), the chlorophyll minimum (Figure 5.81b) and the lowest phytoplankton population (Figure 5.81c), in early July.

Particle size (as area equivalent diameter) also showed considerable variability during the first two months of the monitoring period (Figure 5.80). The high initial value of about 3.3. µm was associated with a diatom bloom. During the May 19 sampling numerous pennate and centric diatom frustules were present. By the beginning of June, diatoms were rarely encountered in the SEM analyses, and mean size had fallen to about 2.0 µm. During the June 22 and 24 samplings, immediately following the peak storm runoff, particle size again peaked at about 3.5 µm and was coincident with the PNV maximum. The increase in particle numbers and mean particle size associated with the storm runoff period was evidently responsible for the Secchi disc transparency minimum of less than 0.3 m (Figure 5.81a). During July, August, and September, particle numbers in the water column averaged about 12,000 ml^{-1}; this was three to four times lower than the spring concentrations. Except for a brief peak in early October, these low particle concentrations continued through mid-November (Figure 5.80). Over the same time period, there was a progressive increase

in particle size; mean particle size increased about two-fold from 2.2 to 4.5 µm.

The dominant particle categories (cf. Table 5.31) observed in each of the size ranges determined by CCSEM for the 28 samples included in the monitoring program are presented in Table 5.32. Note that the size is not a cumulative value; e.g., entries for the column ≤6.0 µm are for the size range less than 6 µm but greater than 3.0 µm in area equivalent diameter, etc. The smallest size range includes particles >0.3 µm but ≤1.5 µm. Several time periods (A through E) are indicated which reflect significant transformations and show a continuity of particle composition and size. These size/composition regions describe the temporal variability of particle types found in Onondaga Lake.

During the first three weeks (interval A) of the monitoring period, the large particles were dominated by diatom frustules. As the diatoms were replaced by cryptomonads and flagellate greens (Figure 5.81c), the mean particle size decreased and the most common generic particle class became the NL type. But by mid-June, PNV and phytoplankton populations were declining and organic particles were evidently aggregating or growing larger; by June 22 the NL type of particle was the most abundant in all of the size ranges.

Time period B (Table 5.32) was influenced by the storm runoff event. The particulate

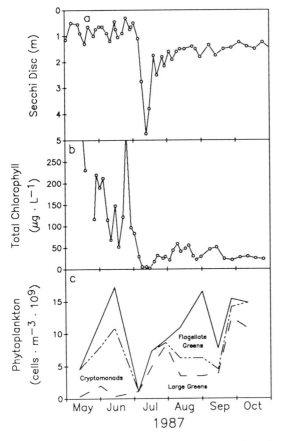

FIGURE 5.81. Temporal trends in selected parameters in Onondaga Lake in 1987: (a) Secchi disc transparency, (b) total chlorophyll at 1 m, and (c) concentration of selected phytoplankton groups in upper waters (from Johnson et al. 1991).

material transported by Onondaga Creek during this period is characterized in Chapter 3 (also see Effler et al. 1992); it consisted mainly of SI, CL, and AG types of particles. The NL and CA particle types were minor components of the tributary sediment load. A greater than two-fold increase in PNV and an increase in mean particle area equivalent diameter occurred during this interval (Figure 5.80). However particles larger than 6 micrometers were dominated by the CA particle family on June 24 and 25 (Table 5.32), immediately following the peak runoff. The smaller size ranges were composed primarily of SI features; these were quartz and not diatom

fragments. In the tributary samples during the storm, 98% of the CA particle type consisted of features smaller than 6 micrometers. Thus, the large CA particles in Onondaga Lake following the storm input of suspended sediment particles apparently were the result of water column precipitation of $CaCO_3$.

The large size of the CA particles indicates that the $CaCO_3$ may have been precipitating on other foreign particles, provided in great abundance (Chapter 3; Effler et al. 1992) during this period. The upper waters of Onondaga Lake are routinely highly over-saturated with respect to calcite (Effler and Driscoll 1985, Driscoll et al. 1994a; also earlier sections of this chapter). In a similar way as a catalyst reduces the activation energy, foreign solids may catalyze the nucleation of over-saturated minerals by reducing the energy barrier (Stumm and Morgan 1981). The key to this process is the degree of match between the foreign surface (heteronucleus (Stumm and Morgan 1981)) and the precipitating crystal; lattice type and atomic distances are more important than chemical similarity. Phase changes in natural waters are almost invariably initiated by heterogeneous solid substrates; e.g., skeletal particles, clays, sands and biocolloids (Stumm and Morgan 1981). The heterogeneous nucleation phenomenon was evaluated in Onondaga Lake samples from June 24 and 25 by examining CA particles after treatment with dilute HNO_3. Hetero-geneous nucleation is demonstrated by the micrographs of Figure 5.82a–d. In this case a SI feature had been coated with $CaCO_3$, as indicated by the micrographs of "before" (Figure 5.82a) and "after" (Figure 5.82b) acid treatment and the associated Ca X-ray maps before (Figure 5.82c) and after (Figure 5.82d) acid treatment. Many such cases of hetero-geneous nucleation were observed in lake samples from these dates, though the majority of heteronuclei were organic, as judged from their lack of backscattered electron image. Sediment trap data (see Chapter 8) indicate $CaCO_3$ precipitates and is deposited from the upper waters of Onondaga Lake throughout the spring–early fall interval. However, the influx of certain particle types (and probably

TABLE 5.32. Particle composition in Onondaga Lake by size (area equivalent diameter) and composition. Particle types refer to those defined in Table 5.31.*

Date	≤1.5 μm*	≤3.0 μm	≤6.0 μm	≤12.0 μm	>12.0 μm	Time period
5/19	NL	NL	AG	SI	SI	
5/27	NL	NL	SI	SI	SI	
5/29	NL	NL	NL	SI	SI	A
6/11	NL	NL	NL	NL	SI	
6/15	NL	NL	NL	NL	OT	
6/18	NL	NL	NL	NL	AG	
6/22	NL	NL	NL	NL	NL	
6/24	NL	NL	NL	NL	CA	
6/25	NL	SI	SI	CA	CA	B
7/2	NL	NL	NL	CA	CA	
7/6	NL	NL	NL	NL	NL	
7/13	NL	NL	NL	NL	NL	
7/19	NL	CA	CA	NL	#	
7/20	NL	CA	CA	NL	OT	C
7/28	NL	NL	NL	NL	NL	
8/6	NL	NL	NL	NL	NL	
8/13	NL	CA	NL	NL	CA	D
8/20	NL	CA	NL	NL	CA	
8/24	NL	CA	NL	NL	CA	
9/18	NL	NL	NL	NL	NL	
9/21	NL	CA	NL	NL	SI	
9/28	NL	CA	CA	NL	SI	
10/5	NL	CA	CA	CA	AG	E
10/19	NL	CA	CA	NL	SI	
10/26	NL	CA	NL	NL	SI	
11/2	NL	CA	NL	NL	SI	
11/9	NL	CA	NL	NL	AG	
11/16	NL	NL	NL	NL	NL	

From Johnson et al. 1991
*Particles greater than 0.3 μm but less than 1.5 μm in area equivalent diameter
No particles in this size range

the internal production of certain types), may promote the irregular occurrence of increased $CaCO_3$ precipitation. Other time-variable phenomena probably contribute to the dynamics observed in $CaCO_3$ precipitation (e.g., Effler 1987b).

During time interval C, the last three weeks of July, PNV remained relatively constant at about 10,000 to 12,000 particles · ml^{-1}. Particle size increased immediately following a second summer storm (return frequency of about 1 yr) in mid-July. The intermediate particle size fractions were dominated by CA particles (Table 5.32) suggesting another, but briefer, increased $CaCO_3$ precipitation event. The smaller size of these CA particles com-

pared to the earlier event (Table 5.32), indicates either small heteronuclei, or reduced growth of the $CaCO_3$ precipitate.

At the beginning of August (interval D, Table 5.32) there was a rapid decline in the number of particles in the smallest size range, while the population of >12.0 μm features remained constant. From mid to late August CA particles were dominant in the 1.5 to 3.0 μm and larger than 12.0 μm size ranges. The disparity of sizes of the $CaCO_3$ precipitation probably reflects differences in the size of heteronuclei. The $CaCO_3$ precipitation may have been promoted by increased production in this period. Note the increase in chlorophyll (Figure 5.81b) and the shift from "large

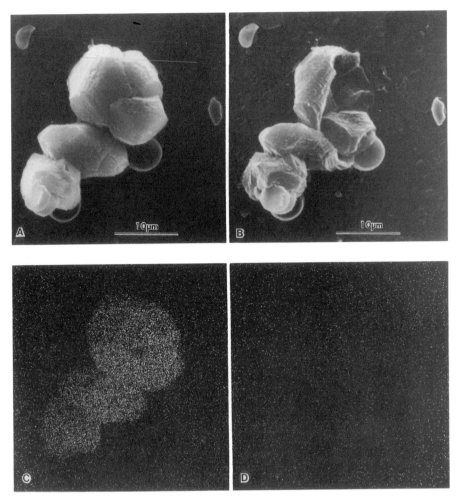

FIGURE **5.82.** Electron micrographs and Ca X-ray maps of a "Ca-rich" feature (CA category) collected on July 25, 1987: (**A**) micrograph before acid treatment, (**B**) micrograph after acid treatment, (**C**) Ca X-ray map before acid treatment, and (**D**) Ca X-ray map after acid treatment (from Johnson et al. 1991).

greens" to "flagellate greens" (Figure 5.81c) in late August.

The most protracted period of $CaCO_3$ precipitation was from mid-September through early November (interval E). The CA particles were the dominant category in the 1.5 to 3.0 μm range throughout this period; CA particles also dominated the 3.0 to 6.0 μm range from late September through mid-October. The large particles during this interval were mostly diatoms, though AG particles were also important. By mid-November the diatoms had largely disappeared, and all particle size ranges were again dominated by the NL particle type.

5.8.4 Particle Projected Area Results

The temporal distributions of total PAV, and PAV for each of five of the particle categories are presented in Figure 5.83. The CL particle type results (not shown in Figure 5.83) demonstrated little variation with time in the 1987 study period, averaging only 3% of the total PAV for the study period. The accuracy of CCSEM imaging of particles is less for organic (NL) materials than for inorganic types of particles (Johnson 1983). However, when measures of particulate organic carbon (POC) were

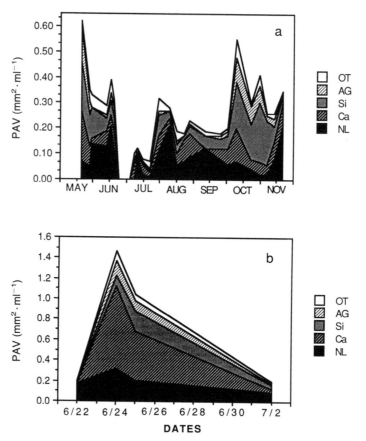

FIGURE 5.83. Temporal distribution of PAV and contributions of five particle categories, in the upper waters of Onondaga Lake in 1987: (**a**) spring through fall interval, exclusive of storm runoff period of late June, and (**b**) storm runoff period of late June (from Johnson et al. 1991).

compared to the PAV fraction contributed by the NL particles, a correlation of r = 0.87 was obtained, indicating a high degree of internal consistency in the PAV measures of the NL particles.

Wide variations in PAV and the relative contribution of the five particle types occurred during the 1987 study period. The value of PAV ranged from a minimum of $0.06\,mm^2 \cdot ml^{-1}$ to over $0.5\,mm^2 \cdot ml^{-1}$ during the spring phytoplankton bloom (Figure 5.83a), and reached more than $1.4\,mm^2 \cdot ml^{-1}$ on June 24—about 24 hours after the peak storm event flow in Onondaga Creek (Effler et al. 1992). More detailed resolution of PAV and the contributing components is presented for the interval around the storm in Figure 5.83b. The apparent effect of nucleation on

$CaCO_3$ precipitation is dramatically portrayed in this latter figure. PAV contributed by CA particle types exceeded 55% of the total on June 24, whereas these particles showed a maximum of 30% from Ninemile Creek and 8% from Onondaga Creek during the storm (see Chapter 3). There were four distinct peaks in the NL contribution to PAV: in mid-June, early August, early September, and mid-November (Figure 5.83a). During these periods more than 50% of the PAV in the upper waters of the lake was associated with organic particles. Temporal variations in phytoplankton (Figure 5.81b and c) contributed to at least some of these dynamics. The AG category made significant contributions to PAV in May and October (Figure 5.83a), when CA particles were also important. The contribution of OT particles to PAV remained small. Many of these particles were composed of Ca, P, and chlorine, but had low gross X-ray count rates indicative of particles of high organic content.

The area-weighted percentage contributions that each of the six particle groups made to PAV in the upper waters of Onondaga Lake in 1987 appear to Table 5.33; temporal variability is reflected by the standard deviation limits. The only category which does not have an in-lake source is CL; its input to Onondaga Lake is strictly from tributary inputs (Yin and Johnson 1984). Thus it can serve as a tracer for the tributary contribution of suspended sediment particles. Comparing the PAV contribution of the CL group from the tributaries to the lake lead Johnson et al. (1991) to estimate that approximately 85% of the lake PAV was associated with in-lake processes in 1987; the remainder was derived from tributary inputs. Assuming that the NL particle group is comprised of organic materials and all other particle categories are inorganic, organic particles contributed nearly 40% of the PAV on average (Table 5.33). A first approximation of the sources of PAV, on a study average basis, is presented in Figure 5.84. Assuming the NL particles have a biological source and that the vast majority of SI particles are diatoms, and all other particles are abiotic, nearly 60% of the PAV in the upper waters in 1987 was associated with biological production. The CA, AG and OT categories include particle types that are believed to be formed in the lake; thus about 25% of the PAV in 1987 was inorganic particles that formed in the water-

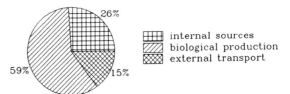

FIGURE 5.84. Relative contributions to PAV in upper waters of Onondaga Lake in 1987, derived from external and internal sources (modified from Johnson et al. 1991).

column of the lake (mostly from precipitation processes). The remaining 15% corresponds to the estimated (Johnson et al. 1991) tributary contribution.

5.8.5 IPA Results for Onondaga Lake, 1981/1982

Yin and Johnson (1984) analyzed several water column samples and six sediment trap collections (see Chapter 8 and Wodka et al. (1985)) for specification of sediment trap techniques) with SAX (SEM with automated image analysis and XED) as part of their individual particle/sediment budget analysis of Onondaga Lake. A total of 1646 particles were analyzed from epilimnetic (2) and hypolimnetic (2) water column samples; 6,334 particles were analyzed from trap samples. The analyzed trap samples covered approximately one-third of the mid-April to mid-September period; samples representing from 3 to 20 d of deployment were included. Nineteen, chemically based, particle classes were used in this work (Yin and Johnson 1984). Particle size distributions of the various classes were not presented, though the common range reported (diameter of 0.2–20 μm) is consistent with the more recent findings. SAX was reported to be inefficient in detecting organic particles in this earlier work (Yin and Johnson 1984).

Calcium carbonate particles were a major component of the particle assemblage of all water column and sediment trap samples. Most of the autochthonous sediment production (83% of the estimated 15,900 MT · yr^{-1})

TABLE 5.33. Mean percentage contribution of each of the six particle types contribution to PAV, upper waters of Onondaga Lake, 1987.

Particle category	Mean % of PAV	Standard deviation (%)
NL	39	22
CA	22	13
SI	23	15
CL	3	2
AG	8	6
OT	5	6

was estimated to be CaCO3 particles. Calcite was the dominant mineral determined in the X-ray diffraction analysis of the flocculent sediments of the lake reported by Yin and Johnson (1984). Direct comparison between the water column results of 1981 (Yin and Johnson 1984) and 1987 (Johnson et al. 1991) is of limited quantitative value; mostly because of the limited number of samples in 1981 and the highly dynamic conditions that are known to prevail (e.g., Figure 5.5), but also due to methodological differences. Yin and Johnson (1984) speculated that a fraction of the Ca particles formed in the epilimnion dissolved once they reached the hypolimnion. However, this is not supported by equilibrium calculations (Effler and Driscoll 1985, Driscoll et al. 1994a) or time series of Ca^{2+} concentrations from the bottom waters of the lake (previously in this chapter). Many particles, other than Ca-only and Ca-rich particles were found to contain some Ca, suggesting that calcite was deposited on the surface of many other particles (Yin and Johnson 1984). This was supported in the later IPA analysis work (e.g., Figure 5.82).

Additionally, FeS particles were found in the hypolimnion of Onondaga Lake in late summer (Yin and Johnson 1984). Ferrous iron and elevated concentrations H_2S were present at that time. Formation of FeS precipitate is consistent with the oversaturated conditions that prevailed with respect to amorphous FeS in the lower anoxic layers during this period (described previously in this chapter).

5.9 Summary

5.9.1 Salinity

The salinity characteristics of Onondaga Lake were reviewed for the late 1960's to 1990 interval. The dominant anion has been Cl^-; the dominant cations have been Ca^{2+} and Na^+. This composition largely reflects the high loads of these materials received as waste from soda ash production (see Chapter 3). The contribution of these three constituents to the

lake's salinity (S (‰)) has decreased since closure of the soda ash/chlor-alkali facility from more than 85% to less than 70%. Chloride represented about 55% of S before closure; now it is about 40%. Before closure of the facility, the S of Onondaga Lake was about 13 times greater than that of nearby hardwater Otisco Lake; it remains about 5 times greater after closure.

The concentration of Cl is demonstrated to be a good surrogate measure of S and ionic strength in Onondaga Lake, though the relationships have shifted since the closure due to the composition changes. A system-specific equation of state, which calculates density as a function of Cl concentration and temperature, is developed. This corrects an earlier presentation of the relationship (Effler et al. 1986d). The contribution of salinity to water density has had important implications for density stratification in the lake and adjoining portions of the Seneca River, and coupled features of water quality (also see Chapters 4 and 9). The high ionic content of the lake shifted the temperature of maximum density about 0.8°C lower during winter months before closure; the depression has been about 0.3°C since closure.

Salinity has varied seasonally and year-to-year in the lake before and after closure, in part because of natural variations in dilution provided by tributary flow. A major systematic reduction in S occurred with the closure of the soda ash/chlor-alkali facility. The average volume-weighted S of the lake for the 1968–1985 interval before closure was about 3‰. By 1990 the S had decreased about 60% to 1.2‰. Remane and Schlieper (1971) have reported biological species diversity to be strongly reduced at S levels that prevailed in the lake before closure of the facility. This is consistent with observations reported in Chapter 6 for Onondaga Lake that bracket the closure.

5.9.2 Dissolved Oxygen

The oxygen resources of Onondaga Lake are extremely limited, indicative of a highly eutrophic system. Dissolved oxygen (DO) is depleted rapidly from the hypolimnion soon after the onset of density stratification. Strong DO stratification is manifested within the hypolimnion, with DO decreasing with increased depth, during the interval of depletion. This vertical distribution within the hypolimnion reflects the localization of demand at the sediment–water interface and limited vertical mixing. Strong vertical gradients are manifested for a number of constituents within the hypolimnion associated with the localization of various source/sink processes at the sediment–water interface. The depth interval of anoxia in the water column expands progressively during summer stratification, reaching as high as 6 m from the surface. The average value of the "areal hypolimnetic oxygen deficit" (AHOD), a widely used parameterization of the rate of depletion of oxygen in the hypolimnion and indicator of trophic state, for six different years over the 1981–1992 interval was $1.54 \, g \cdot m^{-2} \cdot d^{-1}$. This value approaches the highest values incorporated in compilations in the literature, indicating the lake's highly eutrophic character. Substantial year-to-year differences in AHOD are observed, probably as a result of variations in vertical mixing, brought about by meteorological variability.

Disequilibrium, with respect to saturation conditions, is commonly observed in the upper waters of the lake during summer stratification, which is driven largely by phytoplankton activity. Progressive depletion of DO occurs annually in the upper waters of the lake with the approach to fall turnover, associated with the entrainment of hypolimnetic layers enriched in oxygen-demanding reduced substances. The temporal minimum occurs slightly before, or at, the onset of complete fall turnover. Violations of the New York State DO standard are observed nearly lake-wide during the fall mixing period of most years. Year-to-year differences in this degraded feature of the oxygen resources of the lake are probably also largely meteorologically driven.

The ionic waste discharge of the soda ash/chlor-alkali facility has exacerbated the lake's problem of limited oxygen resources, through alteration of the system's density stratification/

mixing regime. The failure of the lake to completely turnover in spring in a number of years eliminated the replenishment of DO in the lowermost layers in those years. The lower layers became anoxic earlier in spring and remained anoxic longer into fall before closure of the facility, because of the longer period of stratification caused by its saline discharge. The occurrence of restratification during periods of typical turnover for other local lakes, caused by plunging of the dense ionic waste before closure, was attended by rapid DO depletion in the bottom stratified layers. Lingering impacts of the continued loading of saline waste on this feature of the lake's oxygen resources have recently been documented.

5.9.3 Inorganic Carbon, Ca^{2+}, $CaCO_{3(s)}$, pH, and Alewife Concretions

An analysis of the distributions of pH (laboratory), inorganic carbon, Ca^{2+}, and calcite equilibrium conditions in Onondaga Lake over the 1980–1990 interval has been presented, with particular emphasis on the changes that have been brought about by the closure of the adjoining soda ash/chlor-alkali facility. Temporal and vertical features of the distribution of pH track seasonal dynamics and the vertical extent of phytoplankton activity. Peak pH values are observed in the near surface waters during phytoplankton blooms. Marked depletions of alkalinity and dissolved inorganic carbon (DIC) are observed annually in the lake's epilimnion. This is clear evidence of substantial decalcification of the upper waters of the lake. The increase in Ca^{2+} concentration observed annually in the summer in the epilimnion before closure of the facility, atypical for lakes experiencing substantial decalcification, reflects continued external waste loading of this constituent and reduced tributary flow during that period. The upper and lower waters of the lake have remained oversaturated with respect to the solubility of calcite. The lack of detectable increases in Ca^{2+} in the lower waters during stratification supports the position that precipitated $CaCO_3$ is not subsequently solubilized at the lake bottom.

The major reduction in external Ca^{2+} loading (~70%) to the lake, associated with the closure of the soda ash/chlor-alkali plant, brought about similar decreases in Ca^{2+} concentration in the upper waters and precipitation of $CaCO_3$ (see Chapter 8), and reduced the extent of alkalinity and DIC depletion in the epilimnion. However, the extent of oversaturation with respect to calcite in the upper waters, as reflected by the saturation index (SI), remained essentially unchanged from before to after closure. Thus, SI is not a good predictor of the rate of precipitation of $CaCO_3$ in Onondaga Lake. The closure of the facility caused two notable shifts in the distribution of pH in the lake's upper waters. First, the distribution is narrower since closure associated with the increased buffering capacity caused by the maintenance of higher DIC concentrations. Second, the pH values are slightly higher because of reduced H^+ generation associated with decreases in the precipitation of $CaCO_3$. Additional shifts of the type described above are to be expected if further reductions in the waste inputs of Ca^{2+} are achieved.

Recent paired laboratory and field pH measurements have demonstrated laboratory values are higher for Onondaga Lake (0.1 to 0.2 units in the upper waters, often 0.4 units in the lower waters), apparently as a result of "degasing" of CO_2 from lake samples. Use of laboratory pH values leads to underestimation of P_{CO_2} and overestimation of the degree of oversaturation with respect to calcite. The effect is modest for the upper waters and more substantial for the bottom waters. However, the basic findings reported for the 1980–1990 interval with respect to the solubility of calcite in the lake remain unchanged: (1) the lake is oversaturated in both the upper and lower waters, and (2) the degree of oversaturation is greater in the epilimnion than the hypolimnion. Furthermore, the overall temporal analysis for the interval is supported because the distributions documented are based on a consistent set of analytical and calculation procedures.

Concretions of the lipid-rich alewife (*Alosa pseudoharengus*) washed ashore in large numbers in Onondaga Lake in the 1950s, 1960s, and 1970s. The concretions were combustible and chalk-like, retaining much of the fish's original structure. The concretions had lost most of the proteinaceous material, and were enriched in Ca and fatty acids. The unusual occurrence of these concretions in the lake has been attributed to the unique combination of anaerobiosis and oversaturation with respect to the solubility of calcite in the lake's hypolimnion, and the invasion of the lipid-rich alewife. The absence of these concretions in recent years is attributed primarily to the major reduction in the lake alewife population.

5.9.4 Nitrogen Species

The temporal and vertical distributions of several important N species, including total ammonia (T-NH_3), free ammonia (NH_3), NO_3^-, and NO_2^-, and the status of the lake with respect to related toxicity standards, were reviewed.

Concentrations of T-NH_3, NH_3, and NO_3^- that are maintained in the epilimnion throughout the productive months, and the concentrations of NO_2^- that develop in the epilimnion, are unusually high compared to concentrations reported for other stratifying lakes in the literature. These high concentrations have important water quality implications. The summed concentrations of T-NH_3 and NO_3^- nitrogen continuously exceed levels associated with the limitation of phytoplankton growth. The concentrations of NH_3 and NO_2^- in the epilimnion routinely exceed standards, often by a wide margin, established to protect nonsalmonid (as well as salmonid) fish from the toxic effects of these species. The appropriate focus for remediation of the severe NH_3 problem of the upper waters of the lake is reduction of the T-NH_3 concentrations. The toxic effects of NO_2^- may be mitigated by the high Cl^- concentrations of the lake. Analysis of long-term data for the lake indicate the NH_3 and NO_2^- problems have prevailed in the upper waters for many years—at least since the late 1960s or early 1970s.

Following the onset of anoxia, NO_3^- and NO_2^- are lost rapidly from the lake's hypolimnion, presumably as a result of denitrification, and T-NH_3 accumulates in this layer (bottom water concentrations in late summer of $\geqslant 8\,mgN \cdot L^{-1}$ are not unusual), apparently as a result of ammonification. The pronounced stratification of these species observed within the hypolimnion indicates the localization of the denitrification and ammonification processes at the sediment–water interface and within the sediments. These transformations are not unusual for productive lakes that experience hypolimnetic anoxia. Despite the higher T-NH_3 concentrations of the hypolimnion, NH_3 concentrations are not as high as in the epilimnion because of the more than compensating effect the higher pH values of the upper waters has on the fraction of T-NH_3 that exists as NH_3. Violations of the NH_3 standards prevail in the hypolimnion, but the margin of violation is lower than in the epilimnion. Further, remediation of the primary limiting resource condition of anoxia in the lower waters may be accompanied by elimination of the violations of the NH_3 standards.

Pronounced seasonal and year-to-year variations in the concentrations of the N species are observed in the upper waters of the lake. Short-term temporal structure is influenced greatly by the dynamics of various limnological processes, but particularly by the time course of phytoplankton activity. Typically the highest concentrations of NH_3 and most severe violations of toxicity standards are observed in early summer, when T-NH_3 concentrations are high (e.g., $2-3\,mgN \cdot L^{-1}$) in the epilimnion and pH is driven high by a phytoplankton bloom. Concentrations of T-NH_3 decline in the epilimnion during summer, and concentrations of NO_3^- and NO_2^- increase. Year-to-year differences in the severity of NH_3 violations have been a result of both differences in T-NH_3 concentrations and pH. The lower pH values determined from *in situ* versus laboratory measurements have only a modest effect on the calculated status of the lake with respect to NH_3 toxicity standards, because of

the somewhat compensating effects of pH on the calculations of NH_3 and the standards. For example the average margin of violation of the nonsalmonid standard in the upper waters over the spring–fall interval of 1991 was reduced by about 15% by use of *in situ*, rather than field, pH data.

5.9.5 Phosphorus

The distributions of four forms of P, total P (TP), total dissolved P (TDP), soluble reactive P (SRP), and particulate P (PP), in the lake are described and evaluated, with particular emphasis on conditions over the 1987–1990 interval. The longer term (back to the late 1960s) P data base for the upper productive layers has limited value because of the failure to measure a form of P that included phytoplankton biomass, and the inclusion of suspect data. Important features of the distributions of the various forms appear to be recurring since the late 1980s, and are generally characteristic of eutrophic and hypereutrophic lakes described in limnology texts.

Concentrations of SRP, a form immediately available to phytoplankton, generally decrease in the epilimnion during the summer. Progressive increases in SRP with time and depth in the hypolimnion clearly depict releases from the sediments. This is the dominant form of P in the hypolimnion; maximum concentrations of $>1\,mg \cdot L^{-1}$ are observed annually in the lowermost layer. By mid-summer the SRP content of the hypolimnion is the dominant fraction of the entire P pool of the lake. The strong dynamics of PP observed in the epilimnion of the lake reflect the temporal pattern of phytoplankton biomass; major year-to-year variations in the fine temporal structure of this form of P occur.

The summer average concentration of TP in the epilimnion decreased from $142\,\mu g \cdot L^{-1}$ in 1988 to $81\,\mu g \cdot L^{-1}$ in 1990, apparently in response to reductions in external loading. The lower 1990 value greatly exceeds the guidance value of $20\,\mu g \cdot L^{-1}$ established by the State of New York, and upper limits established for mesotrophy $(20-37.5\,\mu g \cdot L^{-1})$ by various researchers. The seston of the lake is highly enriched in P, based on $N:P$ and $C:P$ ratios published in the literature, suggesting phytoplankton growth in the lake is not P deficient.

Processes potentially regulating the sediment release of P were reviewed. Redox reactions, keyed to anoxia and the subsequent mobilization of Fe^{2+}, have been found to regulate release of P in most lakes (Einsele/Mortimer model). However, the initiation of substantial releases before the onset of anoxia, and the relatively low Fe^{2+} concentrations compared to SRP in the lowermost anoxic waters, indicate other processes are important in mobilizing P from the sediments of Onondaga Lake. Analysis of $C:P$ ($DIC + CH_4 : SRP$) ratios of the lowermost layer during stratification, and of seston, suggest that in Onondaga Lake P is supplied from the sediments to the watercolumn largely by mineralization of seston. This mechanism is consistent with the observed coincidence of the initiation of P release with an increase in CO_2 and decrease of pH. The $C:P$ ratio analysis indicates a fraction of the deposited P is immobilized. Processes likely responsible for the immobilization include precipitation of $Ca-PO_4$ minerals and adsorption of PO_4^{3-} to sediments. Chemical equilibrium calculations with MINEQ show the lower waters of the lake were (e.g., 1980, when Ca^{2+} concentrations were higher), and continue to be, oversaturated with respect to several $Ca-PO_4$ minerals (e.g., whitlockite).

Progressive accumulations of SRP, of similar magnitude, are observed annually for the hypolimnion of Onondaga Lake. Phosphorus release experiments have been conducted on intact core samples collected from deep-water portions of the lake in the summer of 1987, and over a broader interval of 1992. The average release flux for the 1987 experiments ($n = 17$) was $13.4 \, mg \cdot m^{-2} \cdot d^{-1}$. It was estimated that this internal load represented 25% of the total load received by the lake over the spring–fall interval of 1987. Results of the later experiments imply there is strong seasonality in the magnitude of sediment P release in the lake.

5.9.6 Redox Chemistry

Oxidation-reduction processes are a critical component of the biogeochemistry of lake ecosystems. These reactions influence the cycling of critical elements (e.g., N, S, and P) and may contribute to the depletion of O_2 in the watercolumn. Onondaga Lake is a very productive system exhibiting high deposition of organic matter. Under conditions of summer stratification, inputs of organic matter to the hypolimnion are initially mineralized by aerobic oxidation resulting in rapid depletion of DO and anoxic conditions. When concentrations of DO are low, carbonaceous oxidation occurs via denitrification, Fe^{3+} reduction, SO_4^{2-} reduction, and/or methanogenesis. These processes result in the depletion of water column concentrations of NO_3^- and SO_4^{2-}, and the accumulation of Fe^{2+}, H_2S, and CH_4.

Hydrogen sulfide accumulates to very high concentrations during summer stratification ($1.55 \, mmol \cdot L^{-1}$) due to the high natural inputs of SO_4^{2-} coupled with elevated deposition of organic matter to the hypolimnion. Concentrations of H_2S observed in Onondaga Lake are among the highest values reported in the literature for freshwater systems. Examination of hypolimnetic pools shows that accumulation of H_2S approximately balances losses of SO_4^{2-}, suggesting that only a small fraction of the SO_4^{2-} reduced in the sediments of Onondaga Lake is retained through FeS formation.

In contrast to H_2S, concentrations of Fe^{2+} are relatively low in the lower waters of Onondaga Lake during stratification. Moreover, the pool of Fe^{2+} that accumulates in the water column is low in comparison to other reduced species. These low concentrations of Fe^{2+} are likely due to linkages with H_2S. The $Fe^{2+}:H_2S$ ratio is low (<0.1) in the lowermost anoxic waters. Moreover, these lower waters are highly oversaturated with respect to the solubility of FeS resulting in the retention of Fe in the sediments. Water-column profiles of Fe^{2+} show a maximum above the sediment–water interface, reflecting the precipitation of Fe^{2+} by H_2S in the lowermost waters.

Release of CH_4 from sediments occurs by both diffusive flux and ebullition. Diffusive losses were about 67% of the total flux in 1989. In contrast to H_2S, maximum concentrations of CH_4 in the hypolimnion are not high compared to other values reported for productive lakes. Rates of CH_4 flux from the sediments of Onondaga Lake are near the middle of a wide range of values reported in the literature. However, this flux represents a smaller fraction of the hypolimnitic organic C budget in comparison to values in the literature. This discrepancy may be attributed to elevated concentrations of SO_4^{2-} in Onondaga Lake and the prominent role of SO_4^{2-} reduction in oxidation of organic matter.

In addition to carbonaceous oxidation, mineralization of organic matter results in the release of NH_4^+ from the sediments to the watercolumn. Elevated concentrations of NH_4^+ were evident in the lower waters during summer stratification. Comparison of C:N ratios in seston with T-NH_3:DIC + CH_4 ratios in the lower waters of the lake, shows that the accumulation of T-NH_3 is largely due to mineralization of depositing organic matter. The C:N ratios observed in seston and in the hypolimnion of Onondaga Lake are unusually high and suggest the stoichiometric composition of plankton is highly enriched in N. This pattern is undoubtedly a reflection of the high concentrations of dissolved inorganic N in Onondaga Lake.

In an effort to synthesize the redox chemistry of Onondaga Lake, a hypolimnetic electron budget was developed. Electron inputs in the form of depositing organic matter are in excess of the transfer of electrons to electron acceptors, resulting in the net deposition of organic matter (and electrons) to the sediments. The net deposition of organic C to sediments from the water column was in general agreement with the rate of organic C accumulation in recent sediments. Oxygen was the major electron acceptor used in the oxidation of organic matter. However, significant mineralization of organic matter also occurred by denitrification, SO_4^{2-} reduction, and methanogenesis. Iron reduction is not important in the oxidation of organic

matter in the hypolimnion of Onondaga Lake.

The high rates of organic matter deposition have implications for the O_2 resources of the lake. During summer stratification the inputs of organic matter to the hypolimnion directly consume O_2 by aerobic oxidation and indirectly consume O_2 through the oxidation of reduced materials that are released to the water column as by-products of mineralization reactions (i.e., Fe^{2+}, H_2S, CH_4, and NH_4^+). Of these processes aerobic, H_2S, CH_4, and T-NH_3 oxidation all contribute significantly to the overall hypolimnetic oxygen demand (HOD). The calculated HOD is quantitatively consistent with measured values of AHOD and sediment oxygen demand (SOD) for the lake.

Reduced materials that accumulate in the anoxic hypolimnion during summer stratification are ultimately oxidized during fall turnover. For approximately a month, from mid-September to mid-October, the oxidation of mineralization by-products results in a marked depletion of DO from the watercolumn. A mass balance calculation suggests that this DO depletion is largely attributable to the oxidation of H_2S and CH_4 in the water column. Hypolimnetic accumulations of T-NH_3 do not appear to exert a substantial fraction of the reduced species oxygen demand during the fall mixing period (e.g., nitrification), but instead are in some years depleted by algal assimilation.

5.9.7 Mercury

Concentrations of total Hg (Hg_T) and CH_3Hg in Onondaga Lake are among the highest reported in the literature. These values are indicative of a site experiencing substantial Hg contamination and similar to values reported for lakes with Hg discharge from chlor-alkali facilities or mining activities. Watercolumn observations show a pattern of increasing Hg_T concentrations with increasing lake depth during the summer stratification period. In particular, the lower waters were characterized by marked increases of Hg_T to concentrations of $35 \, ng \cdot Hg \cdot L^{-1}$. This increase in Hg_T coincided with increases in H_2S concentration.

Thermodynamic calculations show that under the conditions that prevail in Onondaga Lake, the stable form of Hg is the soluble complex HgS_2^{2-} when concentrations of H_2S are in excess of Hg_T. It seems likely that the formation of H_2S in sediment and the subsequent mobilization to the watercolumn facilitates the transport of Hg_T through complex formation.

Methyl Hg is a critical species because this fraction of Hg accumulates in fish tissue. In Onondaga Lake CH_3Hg was about 10% of Hg_T in the upper oxic waters, increasing to about 20% in the lower anoxic waters. Concentrations of $(CH_3)_2Hg$ were insignificant. The watercolumn was oversaturated with respect to the solubility of atmospheric Hg^o, indicating evasion of Hg^o and release to the atmosphere.

Tributaries of Onondaga Lake showed highly variable concentrations of Hg_T. Concentrations of Hg_T increased with increasing drainage area. Several of the stream sites exhibited patterns of increasing concentrations of Hg_T with increasing concentrations of DOC, suggesting a linkage between mobilization of naturally occurring organic solutes to drainage water and transport of Hg_T. Concentrations of Hg_T were highest in Ninemile Creek and the METRO effluent.

A Hg budget for Onondaga Lake showed that drainage flows were the major inputs of Hg_T to the system and the principal sink was outflow from the lake. Onondaga Lake appeared to be a slight sink for inputs of Hg_T. Although the lake sediments are highly contaminated with Hg, only a small component of the watercolumn inputs of Hg_T appears to be derived from sediment release. Rather, the major sources of Hg_T to the lake were from Ninemile Creek (48%) and METRO (25%). Inputs of Hg to Ninemile Creek appear to largely occur between Amboy and Lakeland. The results of the Hg budget suggest that the problem of continued contamination of fish flesh in Onondaga Lake may be remediable through control of external loads.

5.9.8 Particle Chemistry

Computer controlled scanning electron microscopy interfaced with X-ray energy spectro-scopy has been used to morphologically and chemically characterize the particle population of the upper waters of Onondaga Lake for the spring to fall interval of 1987. The technique was applied on a more limited basis, but over a greater depth interval and for sediment trap collections, in the early 1980s. These studies have established that the particle population of the lake experiences major variations in numbers, size distribution, and composition in response to the dynamics of autochthonous particle production and external loading. Evidence for the autochthonous production of oversaturated materials, particularly $CaCO_3$, is presented.

The majority of abiotic inorganic particles formed in the epilimnion of the lake in 1987, and undoubtedly in intervening years, were $CaCO_3$. Various inorganic and organic particles function as heteronuclei in the precipitation of $CaCO_3$ in the lake. The sudden influx of particles from tributaries (particularly Onondaga Creek (see Chapter 3)) following major runoff events triggers $CaCO_3$ precipitation events. However, other factors, not yet resolved, also influence the dynamics of concentrations and size of $CaCO_3$ particles in the lake. The presence of FeS particles in the oversaturated anaerobic hypolimnion in late summer has also been documented.

The dynamics of the phytoplankton were an important regulator of the particle population of the upper waters of the lake. The PAV ranged from a minimum of $0.06 \, mm^2 \cdot ml^{-1}$ to more than $0.5 \, mm^2 \cdot ml^{-1}$ over the study period of 1987. Approximately 85% of the PAV of the upper waters of the lake in 1987 was associated with in-lake processes. About 60% is associated with biological (e.g., phytoplankton) production, 25% was attributed to inorganic particle (e.g., $CaCO_3$) formation. The remaining 15% was received from tributary inputs.

References

Abdollahi, H., and Nedwell, D.B. 1979. Seasonal temperature as a factor influencing sulfate reduction in a saltmarsh sediment. *Microb Ecol.* 5: 73–89.

Adams, D.D., Matisoff, G., and Snodgrass, W.J. 1982. Flux of reduced chemical constituents (Fe^{2+}, Mn^{2+}, NH_4^+, and CH_4) and sediment oxygen demand in Lake Erie. *Hydrobiology* 92: 405–414.

Addess, J.M. 1990. Methane cycling in Onondaga Lake, NY. Masters Thesis, College of Environmental Science and Forestry, State University of New York. Syracuse, NY.

Addess, J.M., and Effler S.W. 1996. Summer methane fluxes and fall oxygen resources of Onondaga Lake, New York. *Lake Reservoir Manag.* (in press).

Ahlgren, I. 1977. Role of sediments in the process of recovery of a eutrophic lake. In: H.L. Golterman (ed.), *Interactions between Sediments and Fresh Water*. Junk, Hague, pp. 372–377.

Aller, R.C., and Yingst, J.Y. 1980. Relationships between microbial distribution and the anaerobic decomposition of organic matter in surface sediments of Long Island Sound, U.S.A. *Mar Biol.* 56: 29–42.

Anderson, J.M. 1975. Influence of pH on release of phosphorus from lake sediments. *Arch Hydrobiol.* 76: 411–419.

Andersson, A. 1979. Mercury in soils. In: J.O. Nriagu (ed.), *The Biogeochemistry of Mercury in the Environment*. Elsevier North Holland Biomedical Press, Dordrecht, The Netherlands, pp. 79–106.

Anthonisen, A.C., Loehr, R.C., Prakansam, T.B.S., and Srinath, E.G. 1976. Inhibitions of nitrification by ammonia and nitrous acid. *J Wat Pollut Control Fed.* 48: 835–852.

APHA. 1985. *Standard Methods for the Examination of Water and Wastewater*, 16th ed. American Public Health Association, Washington, DC.

Auer, M.T., Johnson, N., Penn, M.P., and Effler, S.W. 1993. Measurement and verification of rates of sediment phosphorus release for a hypereutrophic urban lake. *Hydrobiology* 253: 301–309.

Auer, M.T., Kieser, M.S., and Canale, R.P. 1986. Identification of critical nutrient levels through field verification of models for phosphorus and phytoplankton growth. *Can J Fish Aquat Sci.* 43: 379–388.

Auer, M.T., Storey, M.L., Effler, S.W., Auer, N.A., and Sze, P. 1990. Zooplankton impacts on chlorophyll and transparency in Onondaga Lake, New York, U.S.A. *Hydrobiology* 200/201: 603–617.

Baccini, P. 1985. Phosphate interactions at the sediment–water interface. In: W. Stumm (ed), *Chemical Processes in Lakes*. Wiley-Interscience, New York, pp. 189–205.

Ball, J.W., Nordstrom, D.K., and Jenne, E.A. 1980. Additional and Revised Thermochemical Data for WATEQZ: a Computerized Model for Trace and Major Element Speciation and Mineral Equilibria of Natural Waters; U.S. Geol. Survey Water Resources Investigations, Report 78–116, Menlo Park, CA.

Benes T., and Havlik, B. 1979. Speciation of mercury in the environment. In: J.O. Nriagu (ed.), *The Biogeochemistry of Mercury in the Environment*. Elsevier North Holland Biomedical Press, Dordrecht, The Netherlands, pp. 175–202.

Bernard, P.C. VanGrieken R.E., and Eisma, D. 1986. *Environ Sci Technol.* 20: 267–273.

Berner, R.A. 1968. Calcium carbonate concretions formed by the decomposition of organic matter. *Science* 159: 195–197.

Bloom, N. 1989. Determination of picogram levels of methylmercury by aqueous phase ethylation, followed by cryogenic gas chromatography with cold vapour atomic fluorescence detection. *Can J Fish Aquat Sci.* 46: 1131–1140.

Bloom, N. 1990. A preliminary mass balance for mercury in Onondaga Lake, New York. In: S.T. Saroff (ed.), *Proceedings of the Onondaga Lake Remediation Conference*. Environmental Protection Bureau, New York State Office of the Attorney General, Albany, NY, pp. 124–128.

Bloom, N., and Effler, S.W. 1990. Seasonal variability in the mercury speciation of Onondaga Lake (New York). *Water Air Soil Pollut.* 53: 251–265.

Bloom, N., and Fitzgerald, W.F. 1988. Determination of volatile mercury species at the picogram level by low-temperature gas chromatography with cold-vapor atomic fluorescence detection. *Anal Chim Acta.* 208: 151–161.

Bloom, N., and Watras, C.J. 1989. Observations of methylmercury in precipitation. *Sci Tot Environ.* 87/88: 199–207.

Böstrom, B., Andersen, J.M., Flerscher, S., and Jansson, M. 1988. Exchange of phosphorus across the sediment–water interface. *Hydrobiology* 170: 229–244.

Bowie, G.L., Milles, W.B., Porcella, D.B., Campbell, C.L., Pagenkopf, J.R., Rupp, G.L., Johnson, K.M., Chan, P.W.H., Gherini, S.A., and Chamberlin, C.E. 1985. Rates, Constants, and Kinetic Formulations in Surface Water Quality Modeling, 2d ed. EPA/600/3-85/040, United States Environmental Protection Agency, Environmental Research Laboratory, Athens, GA 30613.

Brezonik, P.L., and Lee, G.F. 1968. Denitrification as a nitrogen sink in Lake Mendota, Wis. *Environ Sci Technol.* 2: 120–125.

Brooks, C.M., and Effler, S.W. 1990. The distribution of nitrogen species in polluted Onondaga Lake, N.Y., U.S.A. *Water Air Soil Pollut.* 52: 247–262.

Brunskill, G.J. 1969. Fayetterville Green Lake, New York. II. Precipitation and sedimentation of calcite in a meromictic lake with laminated sediments. *Limnol Oceanogr.* 14: 830–847.

Cappenberg, T.E., Hordijk, C.A., and Hagenaars, C.P.M.M. 1984. A comparison of bacterial sulfate reduction and methanogenesis in the anaerobic sediments of a stratified lake ecosystem. *Arch Hydrobiol Beih Ergebn Limnol.* 19: 191–199.

Caraco, N.F., Cole J.J., and Likens, G.E. 1989. Evidence for sulfate-controlled P release from sediments of aquatic systems. *Nature (London)* 341: 316–318.

Caraco, N.F., Cole, J.J., and Likens, G.E. 1991. A cross-system study of phosphorus release from Lake sediments. In: J. Cole, G. Lovett and S. Findlay (eds.), *Comparative Analyses of Ecosystems: Patterns, Mechanisms, and Theories.* Springer-Verlag, New York, pp. 241–258.

Carder, K.L., Steward, R.G., Johnson, D.L., and Prospero. 1986. *Geophys Res.* 91: 1955–1066.

Carlson, R.F. 1977. A trophic status index for lakes. *Limnol Oceanogr.* 22: 361–368.

Cavari, B.Z. 1977. Nitrification potential and factors govering the rate of nitrification in Lake Kinnert. *Oikos* 28: 285–290.

Chanton, J.P., Martens, C.S., and Kelley, C.A. 1989. Gas transport from methane-saturated, tidal freshwater and wetland sediments. *Limnol Oceanogr.* 34: 807–819.

Chapra, S.C., and Canale, R.P. 1991. Long-term phenomenological model of phosphorus and oxygen for stratified lakes. *Water Res.* 25: 707–715.

Chapra, S.C., and Dobson, H.F.H. 1981. Quantification of the lake typologies of Naumann (surface growth) and Thienemann (oxygen) with special reference to the Great Lakes. *J Great Lakes Res.* 7: 182–193.

Chapra, S.C., and Reckhow, K.H. 1983. *Engineering Approaches for Lake Management,* Volume 2: *Mechanistic Modeling.* Butterworth, Boston.

Chen, C.T., and Millero, F.J. 1977a. Precise equation of state for seawater covering only the oceanic range of salinity, temperature and pressure. *Deep Sea Res.* 24: 365–369.

Chen, C.T., and Millero, F.J. 1977b. The use and misuse of pure water PVT properties for lake waters. *Nature* 266: 707–708.

Chen, C.T., and Millero, F.J. 1978. The equation of state of seawater determined from sound speeds. *J Mar Res.* 36: 657–691.

Compeau, G.C., and Bartha, R. 1985. Sulfate-reducing bacteria: principal methylators of mercury in anoxic estuarine sediment. *Appl Environ Microbiol.* 50: 498–502.

Cooke, G.D., Welch, E.B., Perterson, S.A., and Newroth, P.R. 1986. *Lake and Reservoir Restoration.* Butterworth, Boston.

Cornett, R.J., and Rigler, F.H. 1980. The areal hypolimnetic oxygen deficit: an empirical test of the model. *Limnol Oceanogr.* 25: 672–679.

Crawford, R.E., and Allen, G.H. 1977. Seawater inhibition of nitrite toxicity to chinook salmon. *Trans Am Fish Soc.* 106: 105–109.

Crumpton, W.G., and Isenhart, T.M. 1988. Diurnal patterns of ammonia and unionized ammonia in streams receiving secondary treatment effluent. *Bull Environ Contam Toxicol.* 40: 539–544.

Curtis, E.J.C., Durrant, K., and Harman, M.M.I. 1975. Nitrification in rivers in the Trent Basin. *Water Res.* 9: 255–268.

Davidson, W., Heaney, S.I., Talling, J.F., and Rigg, E. 1981. Seasonal transformations and movements of iron in a productive English lake with deep-water anoxia. *Schweiz H Hydrobiol.* 42: 196–224.

Dence, W.A. 1956. Concretions of the alewife, *Pomolobus pseudoharengus* (Wilson) at Onondaga Lake, New York. *Copeia* 3:155–158.

Devan, S.P., and Effler, S.W. 1984. The recent history of phosphorus loading to Onondaga Lake. *J Environ Engr ASCE* 110: 93–109.

Dillon, P.J. 1975. The phosphorus budget of Cameron Lake, Ontario: The importance of flushing rate to the degree of eutrophy of lakes. *Limnol Oceanogr.* 20: 28–39.

DiToro, D.M., and Connolly, J.P. 1980. Mathematical Models of Water Quality in Large Lakes, Part 2: Lake Erie, USEPA, Environmental Research Laboratory, EPA-600/3-80-065, Duluth, MN.

DiToro, D.M., Paquin, R.P., Subburamu, K., and Gruber, D.A. 1990. Sediment oxygen demand: methane and ammonia oxidation. *J Environ. Engr ASCE* 116: 945–986.

Doerr, S.M., Effler, S.W., Whitehead, K.A., Auer, M.T., Perkins, M.G., and Heidtke, T.M. 1994. Chloride model for polluted Onondaga Lake. *Wates Res.* 28: 849–861.

Driscoll, C.T., Effler, S.W., Auer, M.T., Doerr, S.M., and Penn, M.R. 1993. Supply of phosphorus to the water column of a productive hardwater lake: controlling mechanisms and management considerations. *Hydrobiology* 253: 61–72.

Driscoll, C.T., Effler, S.W., and Doerr, S.M. 1994a. Changes in inorganic carbon chemistry and deposition of Onondaga Lake, New York. *Environ Sci Technol.* 28: 1211–1218.

Driscoll, C.T., Yan, C., Schofield, C.L., Munson, R., and Holsapple, J. 1994b. The chemistry and bioavailability of mercury in remote Adirondack lakes. *Environ Sci Technol.* (in press).

Edmondson, W.T., and Lehman, J.T. 1981. The effect of changes in nutrient income on the condition of Lake Washington. *Limnol Oceanogr.* 26: 1–29.

Effler, S.W. 1984. Carbonate equilibrium and the distribution of inorganic carbon in Saginaw Bay. *J Great Lakes Res.* 10: 3–14.

Effler, S.W. 1987a. The impact of a chlor-alkali plant on Onondaga Lake and adjoining systems. *Water Air Soil Pollut.* 33: 85–115.

Effler, S.W. 1987b. The importance of whiting as a component of raw water turbidity. *J Am Water Works Assoc.* 79: 80–82.

Effler, S.W., Auer, M.T., and Johnson, N.A. 1989a. Modeling Cl concentration in Cayuga Lake, U.S.A. *Water Air Soil Pollut.* 44: 347–362.

Effler, S.W., Brooks, C.M., Addess, J.M., Doerr, S.M., Storey, M.L., and Wagner, B.A. 1991a. Pollutant loadings from Solvay waste beds to lower Ninemile Creek, New York. *Water Air Soil Pollut.* 55: 427–444.

Effler, S.W., Brooks, C.M., Auer, M.T., and Doerr, S.M. 1990. Free ammonia in a polluted hypereutrophic, urban lake. *Res J Water Pollut Contr Fed.* 62: 771–779.

Effler, S.W., Brooks, C.M., and Whitehead, K.A. 1996. Domestic waste inputs of nitrogen and phosphorus to Onondaga Lake, and water quality implications. *Lake Reservoir Manag.* (in press).

Effler, S.W., Devan, S.P., and Rodgers, P.W. 1985a. Chloride loading to Lake Ontario from Onondaga Lake, New York. *J Great Lakes Res.* 11: 53–58.

Effler, S.W., and Driscoll, C.T. 1985. Calcium chemistry and deposition in ionically polluted Onondaga Lake, NY. *Environ Sci Technol.* 19:716–720.

Effler, S.W., Driscoll, C.T., Wodka, M.C., Honstein, R., Devan, S.P., Juran, P., and Edwards, T. 1985b. Phosphorus cycling in ionically polluted Onondaga Lake, New York. *Water Air Soil Pollut.* 24: 121–130.

Effler, S.W., Field, S.D., and Quirk, M. 1982. The seasonal cycle of inorganic carbon species in Cazenovia Lake, NY, 1977. *Freshwater Biol.* 12: 139–147.

Effler, S.W., Field, S.D., and Wilcox, D.A. 1981. The carbonate chemistry of Green Lake, Jamesville, NY. *J Freshwater Ecol.* 1: 141–153.

Effler, S.W., Hassett, J.P., Auer, M.T., and Johnson, N. 1988. Depletion of epilimnetic oxygen and accumulation of hydrogen sulfide in the hypolimnion of Onondaga Lake, NY, U.S.A. *Water Air Soil Pollut.* 39: 59–74.

Effler, S.W., and Johnson, D.L. 1987. Calcium carbonate precipitation and turbidity measurements in Otisco Lake, New York. *Water Res Bull.* 23: 73–79.

Effler, S.W., Johnson, D.L., Jiao, J.F., and Perkins, M.G. 1992. Optical impacts and sources of suspended solids in Onondaga Creek, U.S.A. *Water Res Bull.* 28: 251–262.

Effler, S.W., Johnson, D.L., Perkins, M.G., and Brooks, C. 1985c. "A Selective Limnological Analysis of Otisco Lake, NY." Environmental Management Council, Onondaga County, Syracuse, NY.

Effler, S.W., and Owens, E.M. 1986. The density of inflows to Onondaga Lake, U.S.A., 1980 and 1981. *Water Air Soil Pollut.* 28: 105–115.

Effler, S.W., and Owens, E.M. 1987. Modifications in phosphorus loading to Onondaga Lake, U.S.A., associated with alkali manufacturing *Water Air Soil Pollut.* 32: 177–182.

Effler, S.W., Owens, E.M., and Schimel, K.A. 1986a. Density stratification in ionically enriched Onondaga Lake, U.S.A. *Water Air Soil Pollut.* 27: 247–258.

Effler, S.W., Owens, E.M., Schimel, K.A., and Dobi, J. 1986b. Weather-based variations in thermal stratification. *J Hydr Engr ASCE* 112: 159–165.

Effler, S.W., and Perkins, M.G. 1987. Failure of spring turnover in Onondaga Lake, NY, U.S.A. *Water Air Soil Pollut.* 34: 285–291.

Effler, S.W., Perkins, M.G., and Brooks, C.M. 1986c. The oxygen resources of the hypolimnion of ionically enriched Onondaga Lake, NY, U.S.A. *Water Air Soil Pollut.* 29: 93–108.

Effler, S.W., Perkins, M.G., Carter, C., Wagner, B.A. Brooks, C., and Kent, D. 1989b. "Limnology and Water Quality of Cross Lake, 1988". Submitted to Cayuga County Department of Health, Auburn, NY.

Effler, S.W., Perkins, M.G., and Garofalo, J.E. 1987a. "Limnological Analysis of Otisco Lake for 1986". Environmental Management Council, Onondaga County, Syracuse, NY.

Effler, S.W., Perkins, M.G., Garofalo, J.E., Greer, H., Johnson, D.L., and Auer, N. 1987b. "Limnological Analysis of Owasco Lake for 1986." Submitted to Cayuga County Department of Health, Auburn, NY.

Effler, S.W., Perkins, M.G., Greer, H., and Johnson, D.L. 1987c. Effects of whiting on turbidity and optical properties in Owasco Lake, NY. *Water Res Bull.* 23: 189–196.

Effler, S.W., Perkins, M.G., Kent. D., Brooks, C.M., Wagner, B., Storey M.L., and Greer, H. 1989c. "Limnology of Lake Como, Duck Lake, Otter Lake, Parker Pond, and Little Sodus Bay, 1988". Submitted to Cayuga County Department of Health, Auburn, NY.

Effler, S.W., Perkins, M.G., and Wagner, B.A. 1991b. Optics of Little Sodus Bay. *J Great Lakes Res.* 17: 109–119.

Effler, S.W., Perkins, M.G., Wagner, B.A., and Greer, H. 1989d. "Limnological Analysis of Otisco Lake, 1988." Submitted to Water Quality Management Agency of Onondaga County, Syracuse, NY.

Effler, S.W., Schimel, K.A., and Millero, F.J. 1986d. Salinity, chloride, and density relationships in ion enriched Onondaga Lake, NY. *Water Air Soil Pollut.* 27: 169–180.

Effler, S.W., Wodka, M.C., Driscoll, C.T., Brooks, C.M., Perkins, M.G., and Owens, E.M. 1986e. Entrainment-based flux of phosphorus in Onondaga Lake. *J Environ Engr ASCE* 112: 617–622.

Einsele, W. 1936. Über die beziehungen des eisenkreislaufs zum phosphatkreislauf im eutrophen See. *Arch Hydrobiol.* 29: 664–686.

Emerson, K., Russo, R.C., Lund, R.E., and Thurston, R.V. 1975. Aqueous ammonia equilibrium calculations: effect of pH and temperature. *J Fish Res Bd Can.* 32: 2379–2383.

Fallon, R.D., Harrits, S., Hanson, R.S., and Brock, T.D. 1980. The role of methane in internal carbon cycling in Lake Mendota during summer stratification. *Limnol Oceanogr.* 25: 357–360.

Fernandez, H., Vazquez, F., and Millero, F.J. 1982. The density and composition of hypersaline waters of a Mexican lagoon. *Limnol Oceanogr.* 27: 315–321.

Fitzgerald, W.F. 1986. Cycling of mercury between the atmosphere and oceans. In: P. Buat-Menaud (ed.), *The Role of Air-Sea Exchange in Geochemical Cycling.* D. Reidel, Hingham, MA, pp. 363–408.

Fitzgerald, W.F., and Watras, C.J. 1989. Mercury in surficial waters of rural Wisconsin lakes. *Sci Tot Environ.* 87/88: 223–232.

Fogg, T.R., and Fitzgerald, W.F. 1979. Mercury in southern New England coastal rains. *J Geophys Res.* 84: 6987–6989.

Freedman, P.L., and Canale, R.P. 1977. Nutrient release from anaerobic sediments. *J Envir Engr ASCE* 103: 233–244.

Froelich, P.N., Klinkhammer, G.P., Bender, M.L., Luedtke, N.A., Heath, G.R., Cullen, D., Dauphin, P., Hammond, D., Hartman, B., and Maynard, V. 1979. Early oxidation of organic matter in pelagic sediments of the eastern equatorial Atlantic: suboxic diagenesis. *Geochim Cosmochim Acta.* 43: 1075–1090.

Furutani, A., and Rudd, J.W.M. 1980. Measurement of mercury methylation in lake water and sediment samples. *Appl Environ Microbiol.* 40: 770–776.

Gächter, R. 1987. Lake restoration. Why oxygenation and artificial mixing cannot substitute for a decrease in the external phosphorus loading. *Schweiz Z Hydrol.* 49: 170–185.

Gill, G.A., and Bruland, K.W. 1990. Mercury speciation in surface freshwater systems in California and other areas. *Environ Sci Technol.* 24: 1392–1400.

Gill, G.A., and Fitzgerald, W.F. 1985. Mercury sampling of open ocean waters at the picomolar level. *Deep Sea Res.* 32: 287–297.

Gill, G.A., and Fitzgerald, W.F. 1987. Picomolar mercury measurements in seawater and other materials using stannous chloride reduction and two-stage gold amalgamation with gas phase detection. *Mar Chem.* 20: 227–243.

Gilmour, C.C., Henry, E.A., and Mitchell, R. 1992. Sulfate stimulation of mercury methylation in freshwater sediments. *Environ Sci Technol.* 26: 2281–2287.

Golterman, H.L. (ed.). 1977. *Interactions Between Sediments and Freshwater.* Dr. W. Junks, The Hague.

Gunatilaka, A. 1982. Phosphate adsorption kinetics of resuspended sediments in a shallow lake, Neusiedlersee, Austria. *Hydrobiology* 91: 293–298.

Hall, G.H. 1986. Nitrification in lakes. In: J.I. Prosser (ed.), *Nitrification.* IRL Press, Washington, DC, pp. 127–156.

Harris, G.P. 1986. *Phytoplankton Ecology, Structure, Function, and Fluctuation.* Chapman and Hall, New York.

Hawke, D., Carpenter, P.D., and Hunter, K.A. 1989. Competitive adsorption of phosphate on geothite in marine electrolytes. *Environ Sci Technol.* 23: 187–191.

Healey, F.P., and Hendzel, L.L. 1980. Physiological indicators of nutrient deficiency in lake phytoplankton. *Can J Fish Aguat Sci.* 37: 442–453.

Hecky, R.E., Campell, P., and Henzel, L.L. 1993. The stoichiometry of carbon, nitrogen, and phosphorus in particulate matter of lakes and oceans. *Limnol Oceanogr.* 38: 709–724.

Honstein, R.L. 1981. An Assessment of Mechanisms by Which Phosphorus May Be Regulated Within the Sediments of Onondaga Lake, New York. Masters Thesis, Department of Civil Engineering, Syracuse University, Syracuse, NY.

Hutchinson, G.E. 1938. On the relation between oxygen deficit and the productivity and typology of lakes. *Int Rev Gesamten Hydrobiol Hydrogr.* 36: 336–355.

Hutchinson, G.E. 1957. *A Treatise of Limnology.* Volume I: *Geography, Physics and Chemistry.* John Wiley and Sons, New York.

Hutchinson, G.E. 1973. Eutrophication: the scientific background of a contemporary practical problem. *Am Sci.* 61: 269–279.

Ingvorsen, K., and Brock, T.D. 1982. Electron flow via sulfate reduction and methanogensis in the anaerobic hypolimnion of Lake Mendota. *Limnol Oceanogr.* 27: 559–564.

Ingvorsen, K., Zeikus, J.G., and Brock, T.D. 1981. *Appl Environ Microbiol.* 42: 1029–1036.

Jannasch, H.W. 1975. Methane oxidation in Lake Kivu (central Africa). *Limnol Oceanogr.* 20: 860–864.

Johnson, D.L. 1983. Automated scanning microscopic characterization of particle inclusions in biological tissues. *Scan Elect Micros.* 3: 1211–1228.

Johnson, D.L., Jiao, J., DesSantos, S.G., and Effler, S.W. 1991. Individual particle analysis of suspended materials in Onondaga Lake, New York. *Environ Sci Technol.* 25: 736–744.

Jones, B.F., and Bowser, C.J. 1978. The mineralogy and related chemistry of lake sediments. In: A Lerman (ed.), *Lakes: Chemistry, Geology, Physics.* Springer-Verlag, New York, pp. 179–227.

Jones, J.G., and Simon, B.M. 1980. Decomposition processes in the profundal region of Blelham Tarn and the Lund tubes. *J Ecol.* 68: 493–512.

Jorgensen, B.B. 1977. The sulfur cycle of a coastal marine sediment (Limfjorden Denmark). *Limnol Oceanogr.* 22: 814–832.

Jorgensen, B.B., Kuenen, J.G., and Cohen, Y. 1979. Microbial transformations of sulfur compounds in a stratified lake (Solar Lake, Sinai). *Limnol Oceanogr.* 24: 799–822.

Keeney, D.R., 1973. The nitrogen cycle in sediment–water systems. *J Environ Qual.* 2: 15–29.

Kelly, C.A., and Chynoweth, D.P. 1981. The contribution of temperature and of the input of organic matter in controlling rates of sediment methanogenesis. *Limnol Oceanogr.* 26: 891–897.

Kelly, C.A., Rudd, J.W.M., and Schindler, D.W. 1988. Carbon and electron flow via methanogenesis, SO_4^-, NO_3^-, Fe^{3+}, and Mn^{4+} reduction in anoxic hypolimnia of three lakes. *Arch Hydrobiol Beih Ergebn Limnol.* 31: 333–344.

Kelly, G.S. 1967. Precise representation of volume properties of water at one atmosphere. *J Chem Engr Data.* 12: 66–69.

Kelts, K., and Hsù, K.J. 1978. Freshwater carbonate sedimentation. In: A Lerman (ed.), *Lakes: Chemistry, Geology, Physics.* Springer-Verlag, New York, pp. 295–324.

Kirk, J.T.O. 1983. *Light and Photosynthesis in Aquatic Ecosystems.* Cambridge University Press, Cambridge, ENG.

Klots, C.E. 1961. Effect of hydrostatic pressure upon the solubility of gases. *Limnol Oceanogr.* 6: 365–366.

Knowles, R. 1982. Denitrification. *Microbiol Rev.* 46: 43–70.

Kortmann, R.W., Henry, D.D., Kuether, A., and Kaufman, S. 1987. Epilimnetic nutrient loading by metalimnetic erosion and resultant algal responses in Lake Waramaug, Connecticut. *Hydrobiology* 91: 501–510.

Kramer, J.R., Herbes, S.E., and Allen, H.E. 1972. Phosphorus: analysis of water, biomass, and sediment. In: *Nutrients in Natural Waters.* Wiley Interscience, New York, pp. 51–100.

Larsen, D.P., Schultz, D.W., and Malereg, K.W. 1981. Summer internal phosphorus supplies in Shagawa Lake, Minnesota. *Limnol Oceanogr.* 26: 740–753.

Lean, D.R.S., McQueen, D.J., and Story, V.A. 1986. Phosphate transport during hypolimnetic aeration. *Arch Hydrobiol.* 108: 269–280.

Lewis, W.J. 1988. Uncertainty in pH and temperature corrections for ammonia toxicity. *J Water Pollut Contr Fed.* 60: 1922–1929.

Lewis, W.M., and Morris. 1986. Toxicity of nitrite to fish: a review. *Trans Am Fish Soc.* 115: 183–195.

Lindqvist, O. 1991. Mercury in the Swedish environment. Recent research on causes, consequences, and corrective methods. *Water Air Soil Pollut.* 55: 1–2.

Lindqvist, O., Jernelov, A., Johansson, A., and Rodhe, H. 1984. Mercury in the Swedish environment. Global and local sources. Report 1816. National Swedish Environmental Protection Board, Solna, Sweden.

Livingtone, D.A. 1963. Chemical composition of river and lakes. Chapter G. Data of Geochemistry, 6th ed. Prof. Paper. U.S. Geol. Survey 440-G, 64pp.

Loder, T.C., and Liss, P.S. 1985. Control by organic coatings of the surface charge of estuarine suspended particles. *Limnol Oceanogr.* 30: 418–421.

Martens, C.S., and Klump, J.V. 1980. Biogeochemical cycling in an organic-rich coastal marine basin-1. Methane sediment–water exchange processes. *Geochimica et Cosmochimica Acta.* 44: 471–490.

Mattson, M.D., and Likens, G.E. 1993. Redox reactions of organic matter decomposition in a soft water lake. *Biogeochem* 19: 149–172.

McCoy, E.F. 1972. The Role of Bacteria in the Nitrogen Cycle of Lakes. USEPA, Office of Research and Monitoring, Water Pollution Control Research Service 16010 EHR 03/72, Washington, DC.

Messer, J.J., Ho, J., and Grenney, W.J. 1984. Ionic strength correction for extent of ammonia ionization in freshwater. *Can Fish Aquat Sci.* 41: 811–815.

Mierle, G., and Ingram, R. 1991. The role of humic substances in the mobilization of merucry from watersheds. *Water Air Soil Pollut.* 56: 349–357.

Millero, F.J. 1973. Seawater—a test of multicomponent electrolyte solution theories. I. The apparent equivalent volume, expansibility and compressibility, of artificial seawater. *J Solution Chem.* 2: 1–22.

Millero, F.J. 1974. Seawater as a multicomponent electrolyte solution. In: E.D. Goldberg (ed.), *The Sea,* Volume 5. John Wiley and Sons, New York, pp. 3–80.

Millero, F.J. 1975, The physical chemistry of estuaries. In: T.M. Church (ed.), *ACS Symposium Series,* No. 18, American Chemical Society, Washington, DC, pp. 25–55.

Millero, F.J., Gonzalez, A., and Ward, G.K. 1976a. The density of seawater solutions at one atmosphere as a function of temperature and salinity. *J Mar Res.* 34: 61–93.

Millero, F.J., Lawson, D., and Gonzalez, A. 1976b. The density of artificial river and estuarine waters. *J Geophys Res.* 81: 1177–1179.

Molongoski, J.J. and Klug, M.J. 1980. Quantification and characterization of sedimenting particulate organic matter in a shallow hypereutrophic lake. *Freshwater Biol.* 10: 447–506.

Morel, F.M.M. 1983. *Principles of Aquatic Chemistry.* John Wiley and Sons, New York.

Mortimer, C.H. 1941. The exchange of dissolved substances between mud and water (Parts I and II). *J Ecol.* 29: 280–329.

Mortimer, C.H. 1942. The exchange of dissolved substances between mud and water in lakes (Parts III and IV, summary, and references). *J Ecol.* 30: 147–201.

Mortimer, C.H. 1971. Chemical exchanges between the sediments and water in the Great Lakes—speculations on probable regulatory mechanisms. *Limnol Oceanogr.* 16: 387–404.

Moss, B., Wetzel, R.G., and Lauff, G.H. 1983. Annual productivity and phytoplankton changes between 1969 and 1974 in Gull Lake, Michigan. *Freshwater Biol.* 10: 113–121.

Murphy, C.B. Jr. 1978. Onondaga Lake. In: J.A. Bloomfield (ed.), *Lakes of New York State.* Volume II: *Ecology of the Lakes of Western New York.* Academic Press, New York, pp. 224–336.

Murphy, T.P., Hall, K.J., and Yesaki, I. 1983. Coprecipitation of phosphate with calcite in a naturally eutrophic lake. *Limnol Oceanogr.* 28: 58–69.

New York State Department of Environmental Conservation (NYSDEC). 1984. "Fact Sheet." Surface Water Quality Standard Documentation for Nitrite, Albany, NY.

New York State Department on Environmental Conservation (NYSDEC). 1987. An Overview of Mercury Contamination in the Fish of Onondaga Lake. Technical report 87-1 (BEP), Division of Fish and Wildlife, Albany, NY.

New York State Department of Environmental Conservation (NYSDEC). 1990. Mercury Sediments—Onondaga Lake. Engineering Investigations at Inactive Hazardous Waste Sites, Phase II Investigation, New York State Department of Environmental Conservation, Volume 1, Albany, NY.

New York State Department of Environmental Conservation. 1993. New York State Fact Sheet for Phosphorus: Ambient Water Quality Value for Protection of Recreational Uses. Bureau of Technological Services and Research, Albany, NY.

Nürnberg, G.K. 1984. The prediction of internal phosphorus load in lakes with anoxic hypolimnia. *Limnol Oceanogr.* 29: 111–124.

Nürnberg, G.K. 1985. Availability of phosphorus upwelling from iron-rich anoxic hypolimnia. *Arch Hydrobiol.* 104: 459–476.

Nürnberg, G.K. 1987. A comparison of internal phosphorus loads in lake anoxic hypolimnion: Laboratory incubation versus in situ hypolimnetic phosphorus accumulation. *Limnol Oceanogr.* 32: 1160–1164.

Onondaga County. 1971–1991. Onondaga Lake Monitoring Program Annual Reports. Onondaga County Department of Drainage and Sanitation, Syracuse, New York.

Otsuki, A., and Wetzel, R.G. 1972. Coprecipitation of phosphate with carbonates in a marl lake. *Limnol Oceanogr.* 17: 763–767.

Otsuki, A., and Wetzel, R.G. 1973. Interaction of yellow organic acids with calcium carbonate in freshwater. *Limnol Oceanogr.* 18: 490–493.

Otsuki, A., and Wetzel, R.G. 1974. Calcium and total alkalinity budgets and calcium carbonate precipitation of a small hard-water lake. *Arch Hydrobiol.* 73: 14–30.

Owens, E.M. 1987. "Bathymetric Survey and Mapping of Onondaga Lake, New York," submitted to Department of Drainage and Sanitation, Onondaga County, NY.

Owens, E.M., and Effler, S.W. 1989. Changes in stratification in Onondaga Lake, New York. *Water Res Bull.* 25: 587–597.

Parks, J.W., Lutz, A., and Sutton, J.A. 1989. Water column methylmercury in the Wabigoon/English River—Lake system: factors controlling concentrations, speciation and net production. *Can J Fish Aquat Sci.* 46: 2184–2202.

Parsons, T.R., Maita, Y., and Lalli, C.M. 1984. *A Manual of Chemical and Biological Methods for Seawater Analysis.* Pergamon Press, New York.

Payne, W.J. 1973. Reduction of nitrogenous oxides by microorganisms. *Bacteriol Rev.* 37: 409–452.

Penn, M.R., Pauer, J.J., Gelda, R.K., and Auer, M.T. 1993. Laboratory Measurements of Chemical Exchange at the Sediment–Water Interface of Onondaga Lake. Report submitted to Onondaga Lake Management Conference, Syracuse, NY.

Psenner, R., and Gunatilaka, A. (eds.). 1988. Proceedings of the First International Workshop and Sediment Phosphorus. *Arch Hydrobiol.* 30: 1–115.

Ramm, A.E., and Bella, P.A. 1974. Sulfide production in anaerobic microcosms. *Limnol Oceanogr.* 19: 425–441.

Redfield, A.C. 1958. The biological control of chemical factors in the environment. *Am Sci.* 46: 206–226.

Remane, A., and Schlieper, C. 1971. *The Biology of Brackish Water.* John Wiley and Sons, New York.

Robertson, C.K. 1979. Quantitative comparison of the significance of methane in the carbon cycles of two small lakes. *Arch Hydrobiol Beih Ergebn Limnol.* 12: 123–135.

Rudd, J.W.M., Furutani, A., Flett, R.J., and Hamilton, R.D. 1976. Factors controlling methane oxidation in shield lakes: the role of nitrogen fixation and oxygen concentration. *Limnol Oceanogr.* 21: 357–364.

Rudd, J.W.M., and Hamilton, R.D. 1975. Methane oxidation in a eutrophic Canadian Shield lake. *Verh Internat Verein Limnol.* 19: 2669–2673.

Rudd, J.W.M., and Hamilton, R.D. 1978. Methane cycling in a eutrophic shield lake and its effects on whole lake metabolism. *Limnol Oceanogr.* 23: 337–348.

Rudd, J.W.M., and Taylor, C.D. 1980. Methane cycling in aquatic environments. *Adv Aquat Microbiol.* 2: 77–150.

Russo, R.C., Smith, C.E., and Thurston, R.V. 1974. Acute toxicity of nitrite to rainbow trout (*Salmo gairdneri*). *J Fish Res Bd Can.* 31: 1653–1655.

Russo, R.C., and Thurston, R.V. 1977. The acute toxicity of nitrite to fishes. In: R.A. Tubb (ed.) *Recent Advances in Fish Toxicology.* USEPA, Ecological Research Series, EPA-600/3-77-085, Corvalis, OR, pp. 118–131.

Schaffner, W.R., and Oglesby, R.T. 1979. Limnology of Eight Finger Lakes: Hemlock, Canadice, Honeoye, Keuka, Seneca, Owasco, Skaneateles, and Otisco. In: J.A. Bloomfield (ed.), *Lakes of New York State.* Volume 1: *Ecology of the Finger Lakes.* Academic Press, New York, pp. 313–470.

Schafran, G.C., and Driscoll, C.T. 1987. Comparison of terrestrial and hypolimnetic sediment generation neutralizing capacity for an acidic Adirondack lake. *Environ Sci Technol.* 21: 988–993.

Schecher, W.D., and McAvoy, D.C. 1992. MINEQL+: A software environment for chemical equilibrium modeling. *Comput Environ Urban Systems.* 16: 65–76.

Seitzinger, S.P. 1988. Denitrification in freshwater and coastal marine ecosystems: ecological and geochemical significance. *Limnol Oceanogr.* 33: 702–724.

Sly, P.G. (ed.). 1986. *Sediments and Water Interactions.* Springer-Verlag, New York.

Smith, C.E., and Williams, W.G. 1974. Experimental nitrite toxicity in rainbow trout and chinook salmon. *Trans Am Fish Soc.* 103: 389–390.

Smith, R.L., and Klug, M.J. 1981. Reduction of sulfur compounds in the sediments of a eutrophic lake basin. *Appl Environ Microbiol.* 41: 1230–1237.

Snodgrass, W.J. 1987. Analysis of models and measurements for sediment oxygen demand in Lake Erie. *J Great Lakes Res.* 13: 738–756.

Sondheimer, E., Dence, W.A., Mattick, L.R., and Silverman, S.R. 1966. Composition of combustible concretions of the alewife, *Alosa pseudoharengus.* *Science* 152: 221–223.

Sonzogni, W.C., Richardson, W.L., Rodgers, P.W., and Monteith. 1983. Chloride pollution of the Great Lakes: current assessment. *J Wat Pollut Contr Fed.* 55: 513–521.

Sprent, J.I., 1987. *The Ecology of the Nitrogen Cycle.* Cambridge University Press, Cambridge, England.

Stabel, H.H. 1986. Calcite precipitation in Lake Constance: chemical equilibrium, sedimentation, and nucleation by algae. *Limnol Oceanogr.* 31: 1081–1093.

Staudinger, B., Peiffer, S., Avnimelech, Y., and Berman, T. 1990. Phosphorus mobility in interstitial waters of sediments in Lake Kinneret, Israel. *Hydrobiology* 207: 167–177.

Stauffer, R.E. 1985. Relationships between phosphorus loading and trophic state in calcareous lakes of southeast Wisconsin. *Limnol Oceanogr.* 30: 123–145.

Stauffer, R.E., and Lee, G.F. 1974. The role of thermocline migration in regulating algal bloom. In: E.J. Middlebrooks, D.H. Falkenberg, and T.E. Maloney (eds.) *Modeling the Eutrophic Process.* Ann Arbor Science, Ann Arbor, MI, pp. 73–82.

Stoermer, E.F. 1978. Phytoplankton assemblages as indicators of water quality in the Laurentian Great Lakes. *Trans Am Microsci Soc.* 97: 2–16.

Strayer, R.F., and Tiedje, J.M. 1978. In situ methane production in a small, hypereutrophic, hardwater lake: loss of methane from sediments by vertical diffusion and ebullition. *Limnol Oceanogr.* 23: 1201–1206.

Strong, A.E., and Eadie, B.J. 1978. Satellite observations of calcium carbonate precipitations in the Great Lakes. *Limnol Oceanogr.* 23: 877–887.

Stumm, W., and Leckie, J.J. 1971. Phosphate exchange with sediments: its role in the productivity of surface waters. *Adv Wat Pollut Res.* 5; 1970, III-26: 1–16.

Stumm, W., and Morgan, J.J. 1981. *Aquatic Chemistry,* 2d ed. John Wiley and Sons, New York.

Sweerts, J.P.A., Bar-Gilissen, M.J., Corneleses, A.A., and Cappenburg, T.E. 1991. Oxygen consuming processes at the profundal and littoral sediment–water interface of a small meso-eutrophic lake (Lake Vechtan, The Netherlands). *Limnol Oceanogr.* 36: 1124–1133.

Taggart, C.T. 1984. Hypolimnetic aeration and zooplankton distribution: a possible limitation to the restoration of cold-water fish production. *Can J Fish Aquat Sci.* 41: 191–198.

Talling, J.F. 1974. Measurements on non-isolated natural communities: In standing waters. In: R.A. Vollenweider (ed.), *A Manual on Methods for Measuring Primary Production in Aquatic Environments.* IBP Handbook No. 12, Blackwell Scientific, London.

Truesdell, A.H., and Jones, B.F. 1974. WATEQ, a computer program for calculating chemical equilibria in natural waters. *J Res U.S. Geol Surv.* 2: 233–274.

United States Environmental Protedtion Agency (USEPA). 1973. Report of Mercury Source Investigation: Onondaga Lake, New York and Allied Chemical Coroporation, Solvay, New York. National Field Investigation Center, Cincinnati, OH and Region II, New York.

United States Environmental Protection Agency (USEPA). 1985. Ambient Water Quality Criteria for Ammonia—1984. Office of Water Regulations and Standards Criteria and Standards Division, Washington, DC.

Uttormark, F.D. 1978. General Concepts of Lake Degradation and Lake Restoration. USEPA. 740/5-79-001, pp. 65–70.

Vandal, G.M., Mason, R.P., and Fitzgerald, W.F. 1991. Cycling of volatile mercury in temperate lakes. *Water Air Soil Pollut.* 56: 791–803.

Verduin, J. 1960. Phytoplankton Communities in Western Lake Erie and the CO_2 and O_2 changes associated with them. *Limnol Oceanogr.* 5: 372–380.

Vollenweider, R.A. 1968. Scientific Fundamentals of the Eutrophication of Lakes and Flowing Waters with Particular Reference to Nitrogen and Phosphorus as Factors in Eutrophication. Technical Report DAS/C81/68, Organization for Economic Cooperation and Development, Paris, France.

Vollenweider, R.A. 1975. Input–output models with special reference to the phosphorus loading concept in limnology. *Schweiz Z Hydrol.* 33: 53–83.

Vollenweider, R.A. (ed.) 1982. Eutrophication of Waters: monitoring, Assessment and Control. Organization of Economic Cooperation and Development, Paris, France.

Walker, W.W. 1979. Use of hypolimnetic oxygen depletion rate as a trophic state index for lakes. *Water Res Res.* 15: 1463–1470.

Walker, W.W. 1991. Compilation and Review of Onondaga Lake Water Quality Data. Report submitted to Onondaga County Department of Drainage and Sanitation. Syracuse, NY.

Wallen, et al. 1957, Toxicity to *Garnbusia affinis* of certain pure chemicals in turbid waters. *Sewage Industr Wastes.* 29: 695

Wang, W. 1993. Patterns of total mercury concentration in Onondaga Lake. Master's thesis, Syracuse University, Syracuse, NY.

Weidemann, A.D., Bannister, T.T., Effler, S.W., and Johnson, D.L. 1985. Particle and optical properties during $CaCO_3$ precipitation in Otisco Lake. *Limnol Oceanogr.* 30: 1078–1083.

Welch, E.B., and Perkins, M.A. 1979. Oxygen deficit-phosphorus loading relation in lakes. *J Water Pollut Contr Fed.* 51: 2823–2828.

Welch, E.B., Spyridakis, D.E., Shuster, J.I., and Horner, R.R. 1986. Declining lake sediment phosphorus release and oxygen deficit following wastewater diversion. *J Wat Pollut Contr Fed.* 58: 92–96.

Wells, R.D., and Erickson, E.T. 1933. The analysis and composition of fatty material produced by the decomposition of herring in sea water. *J Am Chem Soc.* 55: 338–342.

Westhall, J.C., Zachary, J.C., and Morel, F.M.M. 1976. MINEQL, A Computer Program for the Calculation of Chemical Equilibrium in Aqueous Systems. Ralph M. Parson Laboratory for Water Resources and Environmental Engineering, Technical Note 18. Massachusetts Institute of Technology, Cambridge, MA.

Westrich, J.T., and Berner, R.A. 1984. The role of sedimentary organic matter in bacterial sulfate reduction: the G model tested. *Limnol Oceanogr.* 29: 236–249.

Wetzel, R.G. 1983. *Limnology,* 2d ed. Saunders College, Philadelphia.

Wetzel, R.G., and Likens, G.E. 1991. *Limnological Analyses,* 2d ed. Springer-Verlag, New York.

Wilcox, D.A., and Effler, S.W. 1981. Formation of alewife concretions in polluted Onondaga Lake. *Environ Pollut. (Series B)* 2: 203–215.

Winfrey, M.R., and Rudd, J. 1990. Environmental factors affecting the formation of methylmercury in low pH lakes. *Environ Toxicol Chem.* 9: 853–869.

Winfrey, M.R., and Zeikus, J.G. 1977. Effect of sulfate on carbon and electron flow during microbial methanogensis in freshwater sediments. *Appl Environ Microbiol.* 33: 276–281.

Wirth, H.E. 1940. The problem of the density of seawater. *J Mar Res.* 3: 230–247.

Wodka, M.C., Effler, S.W., and Driscoll, C.T. 1985. Phosphorus deposition from the epilimnion of Onondaga Lake *Limnol Oceanogr.* 30: 833–843.

Wodka, M.C., Effler, S.W., Driscoll, C.T., Field, S.D., and Devan, S.P. 1983. Diffusivity-based flux of phosphorus in Onondaga Lake. *J Environ Engr.* 109: 1403–1445.

Wu, J. 1969. Wind stress and surface roughness at the air-sea interface. *J Geophys Rev.* 74: 444–455.

Yamamoto, S., Alcanskas, J.B., and Crozier, T.E. 1976. Solubility of methane in distilled water and seawater. *J Chem Engr Data.* 21: 78–80.

Yin, C., and Johnson, D.L. 1984. An individual particle analysis and budget study of Onondaga Lake sediments. *Limnol Oceanogr.* 29: 1193–1201.

Young, T.F. 1951. Recent development in the study of interactions between molecules and ions, and the equilibrium in solutions. *Rec Chem Progr.* 12: 81–95.

Zeikus, J.G., and Winfrey, M.R. 1976. Temperature limitations of methanogensis in aquatic sediments. *Appl Environ Microbiol.* 31: 99–107.

6
Biology

Martin T. Auer, Steven W. Effler, Michelle L. Storey, Susan D. Connors,
Philip Sze, Clifford A. Siegfried, Nancy A. Auer, John D. Madsen,
R. Michael Smart, Lawrence W. Eichler, Charles W. Boylen,
Jeffrey W. Sutherland, Jay A. Bloomfield, Bruce A. Wagner,
Robert Danehey, Neil A. Ringler, Christopher Gandino, Pradeep Hirethota,
Peter Tango, Mark A. Arrigo, Charles Morgan, Christopher Millard,
Margaret Murphy, Ronald J. Sloan, Stephen L. Niehaus, and
Keith A. Whitehead

6.1 Phytoplankton

6.1.1 Composition and Abundance

6.1.1.1 Background

The phytoplankton, the algae of the open water, consists of a diverse assemblage of forms, representing all major taxonomic groups. These forms have different physiological requirements, and thereby respond differently to such physical and chemical parameters as light, temperature, and nutrient regime (Wetzel 1983). Furthermore the phytoplankton are often influenced by grazing and other biological interactions. Coexistence of a number of species is generally observed at any given time, although major transformations in the assemblage occur seasonally in temperate lakes as a result of changes in important regulating conditions. This periodicity in the phytoplankton tends to be reasonably constant from year to year for a particular lake unless substantial perturbations are experienced. Such perturbations can be expected to be commonplace in a polluted and culturally eutrophic lake such as Onondaga Lake, where material loadings (Chapter 3) and in-lake concentrations of pollutants continue to change (see Chapter 5).

Much descriptive work has been devoted to establishing the associations of phytoplankton species in lakes of different characteristics, particularly in systems covering a range of trophic states (Wetzel 1983; Reynolds 1984). For example, certain characteristic assemblages of phytoplankton occur repeatedly in lakes of increasing nutrient enrichment. While phytoplankton composition data are limited with respect to providing insight into regulating environmental factors, this information has been valuable in establishing correlations between qualitative and quantitative abundance of the algae, and in establishing the value of the assemblage as a broad indicator of environmental conditions. The seasonality of the phytoplankton of Onondaga Lake, and changes in the assemblage over the 1968–1990 interval, are documented in this section. Speculation on the causes of these changes is also provided.

6.1.1.2 Types of Phytoplankton in Onondaga Lake

Important groups of algae with representatives in the phytoplankton of Onondaga Lake from 1968–1990 have been flagellated green algae, nonflagellated green algae, diatoms, cryptomonads and cyanobacteria. Other groups, such as dinoflagellates, chrysophytes, and euglenoids also have been present, but in lower numbers. A brief introduction to the biology of these groups follows. A summary of characteristics used to distinguish different phyla of planktonic algae is given in Table 6.1. For further information, Sze (1986) and Sandgren (1988) are recommended sources.

The green algae in the Phylum Chlorophyta are characterized by having chlorophylls *a* and *b* as their major photosynthetic pigments, which normally give their cells a grass green color and a starch reserve, which is readily detected by staining cells with iodine. The taxonomy of green algae is undergoing substantial revision based on ultrastructural and molecular studies, and there is disagreement among phycologists whether three, four or five classes should be adopted as representing fundamental divisions within the phylum (Mattox and Stewart 1984). Generally, it is recognized that flagellated forms occur in several different classes, the Prasinophyceae (or Micromonadophyceae), with primitive characteristics, and the more advanced Chlorophyceae. Both prasinophytes and chlorophytes are present in Onondaga Lake. Other planktonic green algae are non-flagellated

cells or colonies belonging, mostly, to the Chlorophyceae.

Among flagellated greens *Heteromastix* and *Tetraselmis* (= *Platymonas*) are unicellular prasinophytes, while *Chlamydomonas* is a unicellular chlorophyte. Because of the large number of species belonging to the genus *Chlamydomonas*, and their resemblance to reproductive stages of other green algae, specific species in Onondaga Lake have not been distinguished. Colonial flagellates are rarely found in the lake. Characteristically, flagellated green algae show a positive phototaxis (swim towards the light) and tend to accumulate near the surface under calm conditions, sometimes producing a foam of light green bubbles when the water is disturbed. Flagellated greens have been an important group in Onondaga Lake in the spring and also have shown brief peaks of abundance during the summer and fall.

The nonflagellated green algae occur as solitary cells, aggregations of cells and well-defined colonies. Members of the Order Chlorococcales (Class Chlorophyceae) are the most common representatives of this type

TABLE 6.1. Characteristics of different phyla of phytoplankton.

Phylum	Major pigments	Carbohydrate reserve	Cell covering	Flagellated stages
Cyanophyta (cyanobacteria)	chl a; phycobilins	Starch (glycogen)	Peptidoglycan wall	Absent
Chlorophyta (green algae)	chl a, b	Starch	Cellulose wall	1–4 smooth flagella
Chrysophyta	chl a, c_1, c_2; fucoxanthin	Chrysolaminarin		
Class Chrysophyceae (golden brown algae)			Scales	Heterokontous (1 smooth flagellum and 1 flagellum with stiff hairs)
Class Bacillariophyceae (diatoms)			Silica frustule	Absent except in gametes
Pyrrophyta (dinoflagellates)	chl a, c_2; peridinin OR chl a, c_1 c_2; fucoxanthin	Starch	Theca (often cellulose plates)	Dinokontous (posteriorly directed flagellum and encircling flagellum)
Cryptophyta (cryptomonads)	chl a, c_2; phycobilins	Starch	Periplast	2 flagella with stiff hairs
Euglenophyta	ch a, b	Paramylon	Pellicle	1–2 smooth flagella

From Sze 1992

in Onondaga Lake. They show various morphologic adaptions to increase the surface area of cells, presumably to increase frictional drag as an aid in flotation. Commonly seen structural modifications include elongate shapes and spines, and arrangement of cells in flattened colonies. Chlorococcalean green algae are abundant in Onondaga Lake during the summer; some species are also common in the spring and fall. Other nonflagellated green algae include desmids, which are only occasionally found (and in low numbers) in the lake, and a filamentous form *Binuclearia* which was very common in the mid-1980s.

The diatoms, belonging to the Class Bacillariophyceae (or Diatomophyceae) in the Phylum Chrysophyta, occur as single cells or small colonies (nonflagellated). Their major pigments are chlorophylls a, c_1 and c_2, and the carotenoid fucoxanthin, which normally gives cells a brown color. The distinguishing feature of diatoms is a silicous wall or frustule, consisting of two large valves and often smaller bands. Because of the nature of the wall, diatoms require an adequate supply of silicon, in the form of silicic acid $Si(OH)_4$, for growth. When the supply is insufficient, diatoms are replaced by algal groups that do not require silicon (Schleske and Stoermer 1971; Kilham et al. 1986). Silicon is normally derived from the natural weathering of rocks rather than from wastewaters. The diatom frustule, especially the valve face, has a sculptured appearance. Various types of markings have been used in the taxonomy of diatoms. Diatoms are divided into two groups, the centric diatoms (Order Centrales) and the pennate diatoms (Order Pennales), based on the symmetry of these markings. The centric diatoms show radial arrangements, while the pennate diatoms show bilateral arrangements about the long axis of a valve. All centric diatoms are planktonic, while the majority of pennate diatoms are benthic. However, a few pennate diatoms are common in the plankton. Both centric and pennate diatoms are annually found in the plankton of Onondaga Lake, usually being most abundant in the spring. Species of *Cyclotella* have been the most common centric diatoms, while common pennate diatoms are *Diatoma*, *Asterionella*, *Synedra*, and *Nitzschia*.

The cryptomonads (Phylum Cryptophyta) are unicellular flagellates that are probably more widespread and common in freshwater environments than generally recognized. However, because of their small size and delicate nature, they are often overlooked. They are characterized by having chlorophylls a and c_2, and either phycoerythrin or phycocyanin as their principal pigments. Although they are often brown, their color may vary greatly depending on which pigments are dominant. Characteristically, their cells are elongated with a asymmetric shape, and flagella arise from slightly below the anterior end. *Cryptomonas* and *Chroomonas* are common in Onondaga Lake throughout the growing season. Their importance was probably underestimated during the earlier years of the study of Onondaga Lake.

A fourth group of importance in Onondaga Lake is the cyanobacteria, formerly called blue–green algae. Unlike the other algal groups, cyanobacteria are prokaryotes, having a simple cell structure similar to eubacteria. Their cells lack a defined nuclear region, chloroplasts, and other complex organelles. Their photosynthetic systems are associated with flattened vesicles in the cells, called thylakoids, and their principal photosynthetic pigments are chlorophyll a and several phycobiliproteins. Often the blue phycobilin phycocyanin predominates. Many, although not all, cyanobacteria are capable of fixing nitrogen (N_2). Nitrogen fixation occurs when other forms of nitrogen, especially nitrate (NO_3^-) and ammonium (NH_4^+) are unavailable. Many filamentous cyanobacteria form special cells, called heterocysts, in which the pathways of N_2 fixation are separated from the normal photosynthetic activity of vegetative cells (N_2 fixation is inhibited by oxygen). Because of the high concentrations of NH_4^+ and NO_3^- that prevail continuously in Onondaga Lake, N_2 fixation is unlikely to be important. Common planktonic cyanobacteria occur in large colonies of cells or in filaments, often surrounded by a sheath of gelatinous mucilage. Their cells contain gas vesicles (collectively the gas vesicles of a cell are its gas vacuole), which are used in buoyancy regulation. Often planktonic cyanobacteria have a positive buoyancy and form a scum on the surface of the water. However, under other circumstances the development of gas vesicles can be adjusted to move cells vertically in the water column. Some strains of cyanobacteria produce toxins (Gorham and Carmichael 1988), but there is no evidence that this occurs in Onondaga Lake. The most common cyanobacterium in Onondaga Lake, and a species that has formed dense summer blooms, is *Aphaizomenon flos-aquae*. The filaments of this species typically occur in bundles, but in preserved samples the bundles often come apart. Other representatives in Onondaga Lake include *Microcystis* (= *Polycystis* = *Anacystis*) and *Anabaena*. Notable has been the late summer abundance of cyanobacteria in 1968–1971, their sporadic occurrence and low numbers in 1972–1989, and a recent reappearance of *Aphanizomenon flos-aquae* in substantial numbers (1990).

Several other groups of flagellated algae are sometimes present in the plankton of Onondaga Lake. However, their abundance is low and rarely have they contributed significantly to blooms. The euglenoids are represented by unicellular green flagellates and the genus *Colacium*, which attaches to the exoskeleton of zooplankton. The most common chrysophyte or golden brown alga is *Synura*, which is a colonial form occurring in the winter and early spring. Several dinoflagellates have been reported in the lake, but only in small numbers. They are unicellular forms with the typical dinokont structure of a posteriorly directed flagellum and a flagellum that encircles the cell in an equatorial groove. Normally dinoflagellates are brown in color.

6.1.1.3 Early Studies: Through 1977

6.1.1.3.1 Prior to 1968

Very little information on the phytoplankton of the lake was available prior to 1968. These data were reviewed by Sze and Kingsbury (1972) and Jackson (1968). Perhaps the earliest published data on algae in Onondaga

Lake is by Hohn (1951), in which he included a list of diatoms found there as part of a broader survey of diatoms in New York State. Summer blooms were first reported in 1962 and consisted of euglenoids and green algae (*Chlamydomonas, Chlorella, Schenedesmus*), with diatoms also common (Compton et al. 1966; Moore 1967; Jackson 1968, 1969). Jackson (1968) commented on the absence of cyanobacteria and suggested that they were inhibited by the salinity of the lake water. However, in 1966 and 1967, cyanobacteria were apparently common in the lake (Moore 1967).

6.1.1.3.2 1968–1971

A regular sampling program for phytoplankton was started in 1968, as part of Onondaga County's monitoring program. A preliminary phase in 1968 preceded a detailed study of conditions in the lake in 1969. This was followed by a less intensive monitoring program (1970 to date). Water samples have been collected, usually biweekly over the spring–early fall interval from the south deep station, at the surface and depths of 3, 6 and 12 m. Higher frequency sampling and more stations were included in the 1969 program. The surface and 3 m samples generally represent activity in the photic zone (Chapter 7). The 6 m and 12 m samples are from the lower part of the epilimnion and below the thermocline for most of the monitoring period (Chapter 4). Samples have been preserved with Lugols iodine and phytoplankton stained with aniline blue, before cells are collected by gentle filtration (see deNoyelles 1968; Sze and Kingsbury 1972; Sze 1975, for more information on procedures). Filters have been mounted on glass slides and cleared for examination under a phase contrast microscope, and the species of phytoplankton identified and counted. Details on phytoplankton abundance during the period from 1968–1971 are given in Sze and Kingsbury (1972); Sze (1972, 1975), the report on the 1969 study (Onondaga County 1971), and annual monitoring reports (Onondaga County 1971, 1972). The summary of phytoplankton occurrence presented for that period in Table 6.2 is derived from these sources.

The two principal sampling locations in 1968 were the south deep and north deep stations. Observations on the phytoplankton began in the spring and were largely qualitative. A seasonal succession was described in which the major groups were diatoms in the spring, green algae in the early summer (June–July), and cyanobacteria in the late summer and fall (August–October). This pattern of seasonal succession is widely observed in eutrophic lakes (Wetzel 1983; Hutchinson 1967). Important taxa included the diatoms *Nitzschia, Diatoma, Cyclotella,* the green algae *Scenedesmus, Chlorella, Ankistrodesmus,* and the cyanobacteria *Microcystis* (reported as *Polycystis*), *Aphanizomenon,* and *Anabaena.* Of significance was an early summer bloom dominated by *Scenedesmus* and *Chlorella,* followed by a late summer bloom in which *Microcystis* was dominant.

In 1969, sampling was weekly throughout the year at both stations, with additional samples taken near the mouths of Ley Creek and Ninemile Creek and at the outlet to the Seneca River. In 1969 diatoms and *Chlamydomonas* dominated in the spring and chlorococcalean green algae in the early summer. A bloom of cyanobacteria persisted from late summer to early fall. Maximum phytoplankton abundance, as estimated by biovolume, occurred in June, July, and August. *Chlamydomonas* spp. showed several peaks during the growing season and was especially abundant in the spring. The diatoms *Cyclotella glomerata* and *Diatoma tenue* were common in the spring. *Chlorella vulgaris, Scenedesmus quadricauda* and *Scenedesmus obliquus** were responsible for the early summer bloom of chlorococccalean green algae. Cyanobac-

---

*There are many described species of *Scenedesmus,* but the work of Trainor (Trainor and Egan 1990) has shown a high degree of polymorphism, bringing into doubt the validity of many of these species. Therefore, in Onondaga Lake samples, the genus *Scenedesmus* has been divided into two types: *S. quadricauda* has ovoid cells, often with spines, and *S. obliquus* has fusiform cells without distinct spines.

teria contributing to the late summer/early fall bloom were *Microcystis aeruginosa* and *Aphanizomenon flos-aquae.* The "filamentous" diatom *Melosira granulata* peaked in September, and *Cryptomonas* was present throughout the growing season and became dominant in October.

The importance of *Chlorella* and *Scenedesmus* in the summer continued a trend that apparently started in the early 1960s. Reports of coccoid cyanobacteria prior to 1968 may refer to *Microcystis* (Moore 1967), which was dominant in 1968 and 1969. After 1969 *Aphanizomenon* was largely responsible for late summer cyanobacterial blooms and *Mirocystis* was of only secondary importance. The abundance of another species of *Melosira* has been linked to periods of intense mixing (Lund 1954, 1971). Although they did not contribute significantly to the overall production in the lake, the occurrence of two algae belonging to primarily marine genera appeared to be related to the high salinity of the lake water (Sze and Kingsbury 1972). The diatom *Chaetoceros* was found in small numbers in the plankton, and the green macroalga *Enteromorpha* in the benthos along the waste beds.

The seasonal patterns of phytoplankton in 1970 and 1971 generally followed the pattern observed in 1969 and provided a measure of the annual variation under the conditions that existed from 1968–1971. The diatom *Nitzschia palea* was common in spring, along with *Cyclotella* and *Diatoma.* In early summer, *Oocystis parva* and *Ankistrodesmus falcatus* were also common but of secondary importance to *Chlorella* and *Scenedesmus.* The cryptomonads *Chroomonas* and *Cryptomonas* were common in the lake during the growing season. *Chlamydomonas* continued to show several peaks during the year. *Aphanizomenon* was the major cyanobacterium in late summer, with *Microcystis* of lesser importance. *Melosira* was not abundant either year.

6.1.1.3.3 1972–1977

Principal references for the phytoplankton for the 1972–1977 period are Sze (1975, 1980).

TABLE **6.2.** Phytoplankton, 1968–1977.

	1968	1969			1970			1971		
	**	M/J	Jy/A	S/O	M/J	Jy/A	S/O	M/J	Jy/A	S/O
FLAGELLATED GREEN ALGAE										
Chlamydomonas spp.	X	XX	XX	XX	XX	XX	X	XX	X	XX
NONFLAGELLATED GREEN ALGAE										
Schroederia setigera			X	X	X	X	X		X	
Pediastrum boryanum	X	X					X		X	X
Pediastrum duplex	X	X	X	X		X	X		X	X
Pediastrum simplex	X		X	X		X	X			
Chlorella vulgaris	X	XX	XX	X	XX	XX	X	XX	XX	XX
Oocystis parva	X	XX	XX	X	X	X	X		XX	X
Ankistrodesmus falcatus	X	X	X	X	X					
Scenedesmus obliquus	X	XX	XX	X	XX	X	X	X	X	X
Scenedesmus quadricauda		X	XX	X	XX	X	X	XX	XX	X
Actinastrum hantzschii										
Closterium gracile				X						
Staurestrum paradoxum				X						
DIATOMS										
Melosira granulate	X	X	X	XX			X		X	X
Cyclotella glomerata		XX	XX	X	XX	X	X	XX	X	X
Cyclotella spp.		X	X	X	X	X	X		X	X
Coscinodiscus subtills	X		X	X			X		X	X
Diatoma tenue		XX	X	X	X			X		
Fragilaria crotonensis			X			X				
Synedra spp.		X			X			X		
Asterionella formosa		X			X					
Entomoneis alata	X	X			X					
Nitzschia palea	X	X			X	X		X	X	X
EUGLENOIDS, CHRYSOPHYTES and DINOFLAGELLATES										
euglenoids										
Syrura uvella		XX								
dinoflagellates	X			X			X			
CRYPTOMONADS										
Chroomonas sp.			X	X		X	XX	X	XX	XX
Cryptomonas spp.	X	X	X	X	XX	X	XX	XX	XX	XX
CYANOBACTERIA										
Microcystis aeruginose	X		X	XX			X			
Merismopedia sp.										
Anabaena circinalis			X				X			
Anabaena flos-aquae	X		X	X			X			
Aphanizomenon flos-aquae	X		X	XX		X	XX		XX	X
NUMBER OF SAMPLE DATES	13	8	9	9	4	2	4	3	4	3

South Station; X = present, XX = common (≥1000 cells/ml)
** Presence only, August–ocober 1968
M/J = May and June, Jy/A = July and August, S/O = September and October

TABLE 6.3. Phytoplankton, 1972–1977.

	1972			1973			1974			1975			1976			1977		
	M/J	Jy/A	S/O	M/J	Jy/A	S/O	M/J	Jy/A	S/O	M/J	Jy/A	S/O	M/J	Jy/A	S/O	M/J	Jy/A	S/O
FLAGELLATED GREEN ALGAE																		
Chlamydomonas spp.	XX	X		X	XX	X	XX	X	XX	XX	XX	XX	XX	XX	X	XX	XX	X
NONFLAGELLATED GREEN ALGAE																		
Schroederia setigera		XX			X	X	X	X							X		X	
Pediastrum boryanum		X														X		
Pediastrum duplex		X		XX	XX	X		X	XX			X			X			
Pediastrum simplex			X						X					X			X	X
Chlorella vulgaris	XX	XX	XX	XX	XX	XX	X	XX	XX	XX	XX	XX	XX	XX	X	XX	XX	XX
Oocystis parva	XX	XX	X	X	XX	X	X	X	XX	XX	XX	X	X	XX	X	X	XX	XX
Ankistrodesmus falcatus	X			X		X	X				X		X	X	X	X	X	
Scenedesmus obliquus	XX	XX	XX	XX	XX	XX	XX	X	XX	XX	XX	XX	XX	XX	X	XX	XX	XX
Scenedesmus quadricauda	X	XX	X	XX	XX	X	X	X	X	XX	X	X	XX	XX	X	XX	XX	XX
Actinestrum hantzschii																		
Closterium gracile		X						X										
Staurastrum paradoxum																		
DIATOMS																		
Melosira granulata								X										
Cyclotella glomerata	XX	X		XX	XX	XX	XX	X	XX	XX	XX	XX	XX	XX	XX	XX	XX	X
Cyclotella spp.	XX	X		X	X	X	X	X	X		X	X	X	X	X	X	X	X
Coscinodiscus subtilis		X		X					X								X	X
Diatome tenue	X			XX			XX		XX	X		X					X	
Fragilaria crotonensis								X	X									
Synedra spp.	X	X		X	X	X	X		X	X	X	X	X	X	X	X	X	
Asterionella formosa	X	X					X		X				X	X		X	X	
Entomoneis alata	X	X				X	X		X	X		X	X			X	X	X
Nitzschia palea	X	X		X		X	X		X		X	X	XX	X	X	X	XX	X

	M/J	Jy/A	S/O	M/J	Jy/A	S/O	M/J	Jy/A	S/O	M/J	Jy/A	S/O	M/J	Jy/A	S/O	
EUGLENOIDS, CHRYSOPHYTES and DINOFLAGELLATES																
euglenoids								XX	X		X			X	X	
Synura uvella																
dinoflagellates			X		X		X	X		X			X	X		
CRYPTOMONADS																
Chroomonas sp.	XX	XX	XX	XX	X	X	X	XX	XX	XX	XX	XX	XX	XX	XX	XX
Cryptomonas spp.	X	X	X	X	XX	XX	XX	XX	X	X	X	X	XX	X	XX	XX
CYANOBACTERIA																
Microcystis aeruginosa									X	XX						
Merismopedia sp.								XX	X	XX			XX	X		
Anabaena circinalis																
Anabaena flos-aquae																
Aphanizomenon flos-aquae		X	X		X											
NUMBER OF SAMPLE DATES	3	5	4	3	4	4	4	5	4	3	4	4	4	4	4	4

South Station; X = present, XX = common (≥1000 cells/ml)

M/J = May and June, Jy/A = July and August, S/O = September and October

In 1972, absence of late summer blooms of cyanobacteria was a major change in the phytoplankton. Abundance of chloroccalean green algae continued throughout the summer and into the fall. Listings of the species identified for the interval appear in Table 6.3. Diatoms and flagellated algae continued to be common in the spring. Important species were *Chlamydomonas* spp., *Cyclotella glomerata*, *Cyclotella meneghiniana*, *Diatoma tenue*, *Nitzschia palea*, *Entomoneis alata* (as *Amphiprora alata*), and *Synedra delicatissima*. Chlorococcalean green algae were the predominant group in the summer and fall. *Chlorella vulgaris* and *Scenedesmus obliquus* were the major species. *Scenedesmus quadricauda* and *Oocystis parva*, the diatoms *Cyclotella glomerata* and *C. meneghiniana*, and *Chroomonas* sp. were also common. During the summers of 1975 and 1976, *Cyclotella* (probably *C. meneghiniana*) was especially abundant. Overall during the period 1972–1977, *Oocystis parva*, *Chroomonas* sp. and *Cyclotella meneghiniana* increased in importance.

The disappearance of the cyanobacteria as a important component of the lake's phytoplankton has been attributed to the major reduction in phosphorus loading (Devan and Effler 1984) associated with the local ban of phosphates in detergents in July 1971 (Murphy 1973; Sze 1975). This elimination was probably caused by an increase in the N/P ratio that resulted from the reduction in phosphorus input. Schindler (1977) and Smith (1983) found cyanobacteria predominate when the N/P ratio is low, and that similar shifts in the phytoplankton assemblage occurred when the ratio was increased. Bioassays with natural assemblages of Onondaga Lake phytoplankton, which measured growth responses to different forms of enrichment, indicated that the chlorococ-

calean green algae were more likely to deplete phosphorus than nitrogen (Sze 1980). Increased grazing pressure on the phytoplankton may have also developed in this period, as the zooplankton biomass in 1978 was found to be much greater than observed in 1969 (Meyer and Effler 1980).

6.1.1.4 1978–1990

Data on the phytoplankton for the 13–year period from 1978–1990 have not been published previously. Thus a more detailed summary is given than for the earlier years. The north deep station was sampled occasionally in this period. In general, the north and south stations were similar in algal composition. Cell counts from samples collected at 0 and 3 m depths have been averaged to give representative values of abundance for the photic zone. The occurrence of phytoplankton species for the 1978–1990 period is summarized in Table 6.4.

6.1.1.4.1 1978–1982

Distributions of the major phytoplankton groups for the 1978–1982 period are summarized in Figure 6.1a–e. The trends observed from 1972–1977 continued during this period. The spring community consisted of flagellated algae (green algae and cryptomonads), diatoms and in some years the colonial green alga *Dictyosphaerium pulchellum* (Chlorococcales). In summer, chlorococcalean green algae bloomed, with dominance shifting from *Scenedesmus obliquus* in early summer to *Oocystis parva* in late summer. Other common green algae were *Chlorella vulgaris* and *Scenedesmus quadricauda*. The cryptomonad *Chroomonas* was also common.

TABLE 6.4. Phytoplankton, 1978–1990 (May–October).

	1978	1979	1980	1981	1982	1983	1984	1985	1986	1987	1988	1989	1990
FLAGELLATED GREEN ALGAE													
Chlamydomonas spp.	XX	XX	XX	XX	XX	XX	XX	XX	XX	XX	XX	XX	X
Heteromastix angulata			XX	X		XX	XX	X	XX	X	X		
Platymones elliptica							X	X	X	X	X	X	XX
Gonium pectorale					X			X				X	

TABLE 6.4. *Continued*

	1978	1979	1980	1981	1982	1983	1984	1985	1986	1987	1988	1989	1990
NONFLAGELLATED GREEN ALGAE													
Binuclearia tectorum					X	XX	XX	XX	XX	X			
Schroederia setigera	X						X		X	X	X	X	X
Dictyosphaerium pulchellum		XX	X	XX	XX	XX	X	XX	X	X	XX	XX	
Pediastrum boryanum									X	X			
Pediastrum duplex	X	X		X	X	X		X	X	XX	XX	X	X
Pediastrum simplex			X		X		X						
Pediastrum tetras					X								
Colestrum microporum										XX	XX	X	X
Chlorella vulgaria	XX	XX	XX	XX	XX	XX	X	X	XX	X	XX	X	
Oocystis parva	XX	XX	XX	XX	XX	XX	XX	XX	X	XX	XX	XX	X
Ankistrodesmus falcatus	XX	XX	XX	XX	XX	X	XX	XX	XX	X	X	X	X
Scenedesmus obliquus	XX	XX	XX	XX	XX	XX	XX	X	XX	X	X	X	X
Scenedesmus quadricauda	XX	XX	XX	XX	XX	XX	XX	X	X	X	X	X	X
Actinastrum hantzschii					X	X	X	XX	XX	X		X	
Kirchneriella elongate			X	X	XX	X	X	X	X	X	X		
Tetraedron sp.	X	X	X	X	X	X	X	X	X				
Quadriqula lacustris										X	X		
Cruciqinia spp.	X		X	X	X	X	X	X		X	X	X	
Selanastrum sp.											X		
Elakatothrix viridis												X	
Cosmarium sp.			X		X	XX	X		X	X	X	X	
Closterium spp.		X	X	X	X		X		X	X	X	X	
Staurastrum sp.										X	X		
DIATOMS													
Melosira granulata			X	X	X					X			X
Coscinodiscus sp.		X	X	X			X			X		X	
Cyclotella meneghiniana	XX	XX	XX	XX	XX	XX	X	XX	XX	X	X	X	X
Other centric diatoms	XX	XX	XX	XX	XX	XX	XX	XX	XX	X	X	X	X
Diatoma tenue	X	X	X	X	XX	X	XX	XX	X	X	X	X	X
Synedra spp.	X	X	X	X	X	X	X	X	X	X	X	X	X
Fragilaria crotonenis	X				X					X	X	XX	X
Asterionella formosa	X			X					X	X	X	X	X
Nitzschia spp.	X	X	X	X	X	X	X	X	X		X	X	X
Entomoneis alata	X		X	X	X		X	X	X				
Other pennate diatoms	X	X		X			X	X	X		X		
EUGLENOIDS, CHRYSOPHYTES and DINOFLAGELLATES													
Euglena sp.	X	X	X	X	X	X	X	X	X	X	X		
Lepocinclis sp.		X	X			X		X	X				
Colacium sp.										X			
Ceratium hirundinella										X	X	X	X
Other dinoflagellates	X	X	X	X	X	X	X	X	X	X	X	X	X
Dinobryon sp.												X	
CRYPTOMONADS													
Chroomonas sp.	XX	XX	XX	XX	XX	XX	XX	XX	XX	XX	XX	XX	XX
Chryptomonas erosa	XX	XX	XX	XX	XX	XX	XX	XX	XX	XX	XX	XX	XX
CYANOBACTERIA													
Merismopedia sp.		X	X	X	X	X	X	X					
Microcystis sp.			X	X						X		X	X
Coelosphaerium sp.					X	X	X	X	X	X	X	X	X
Oscillatoria sp.							X	X	X				
Anabaena ap.					X	X	X	X		X		X	XX
Aphanizomenon flos-aquae									X	X	X	X	XX

South Station; X = present, XX = common (≥1000 cells/ml)

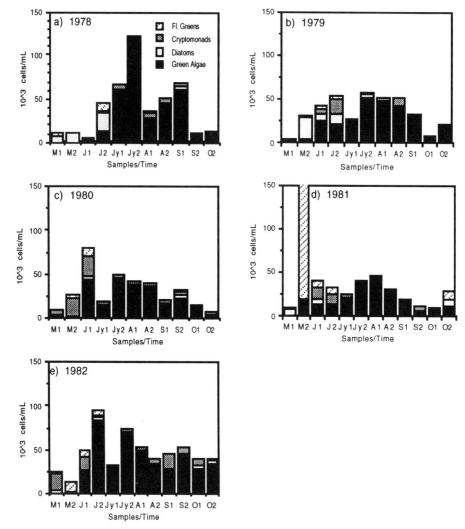

FIGURE 6.1. Seasonality in phytoplankton groups (8), in Onondaga Lake, 1978–1982: (a) 1978: summer peaks of *Scenedesmus obliquus* and *Oocystis*, (b) 1979: summer peak of *S. obliquus* and *Oocysti* common, (c) 1980: *Scenedesmus* spp. and *Oocystic* common, (d) 1981: peak of *Heteromastix* in late May, *Scenedesmus* spp. and *Oocystis* common in summer, and (e) 1982: peak of *Dictyosphaerium* in late June, *Scenedesmus* and *Oocystic* common in summer and fall.

Species showing peaks during the spring period were variable from year to year (some of this variability may be due to missing peaks between sampling dates). A peak of the small flagellated green alga *Heteromastix* occurred in 1981, and *Dictyosphaerium* peaked in 1982. A decline in the importance of centric diatoms as a whole was observed during the growing season from 1978–1982.

6.1.1.4.2 1983–1986

Distributions among the phytoplankton groups for the 1983–1986 period are presented in Figure 6.2a–d. This period was marked by much greater variability in phytoplankton composition during the growing season and an earlier decline in overall phytoplankton abundance in the late summer or fall.

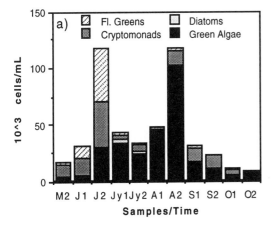

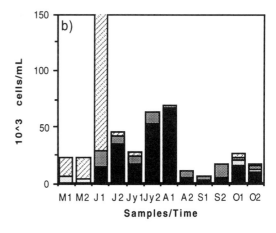

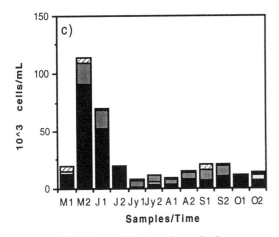

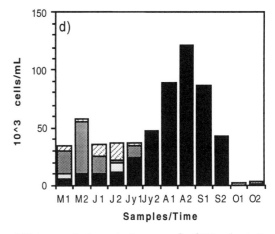

FIGURE **6.2.** Seasonality in phytoplankton groups (4), in Onondaga Lake 1983–1986: (**a**) 1983, peak of *Heteromastix* and cryptomonads in late June, peak of *Binuclearia* in late August, (**b**) 1984, peak of *Heteromastix* in early June, peak of *Binuclearia* in late July–early August, (**c**) 1985 peak of *Dictyosphaerium* in late May–early June, and (**d**) 1986, *Binuclearia* abundant late July–September.

The spring periods (May–June) showed variation from year to year in terms of which species peaked. *Heteromastix* and cryptomonads dominated in this period in 1983, *Heteromastix* in 1984, *Dictyosphaerium* and cryptomonads in 1985, and cryptomonads in 1986.

Either chlorococcalean green algae (primarily Oocystis) or a filamentous green alga *Binuclearia tectorum* was abundant during the summer periods. *Binuclearia* showed distinct peaks in late August 1983 and in late July and early August 1984. In both years phytoplankton abundance was low following the peak of *Binuclearia*, and in 1985 overall abundance was low during the entire summer.

In 1986, there was an intense bloom of *Binuclearia* starting in late July, peaking in August and persisting through September. Notable during this period was the decline in importance of *Chlorella vulgaris*, *Scenedesmus obliquus and S. quadricauda*. *Oocystis parva* continued to be a major contributor to summer blooms in 1983 and 1984, but its abundance was lower in 1985 and 1986.

6.1.1.4.3 1987–1990

Distributions of the phytoplankton groups for the 1987–1990 period are presented in Figure 6.3a–d. This four-year period was marked by

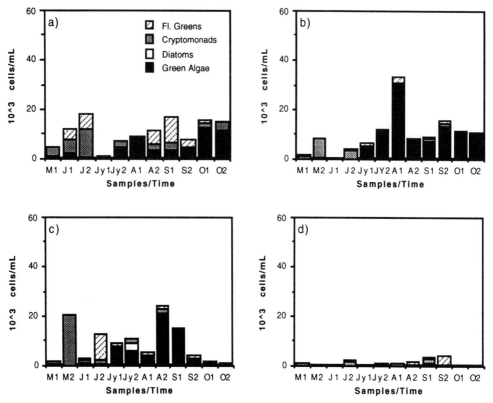

FIGURE 6.3. Seasonality in phytoplankton groups (4), in Onondaga Lake, 1987–1990: (a) 1987, (b) 1988, peak of *Dictyosphaerium* and *Oocystis* in early August, (c) 1989, peak of cryptomonads in late May, peak of *Oocystis* in late August and early September, and (d) 1990.

lower abundance of eukaryotic phytoplankton throughout the growing season and a reappearance of cyanobacteria in substantial numbers in 1990. In the spring period, the important groups were flagellated green algae and cryptomonads. The dominant forms during the summer and fall were chlorococcalean green algae and flagellates. Common summer species were *Chlamydomonas* spp., *Chroomonas* sp., *Oocystis parva* and *Dictyosphaerium pulchellum*. At times, *Oocystis* and *Dictyosphaerium* showed distinct peaks. At the same time, cyanobacteria showed a progressive increase in importance in the summer phytoplankton (Figure 6.4). From 1987–1989, they occurred sporadically in relatively low numbers, but in 1990, *Aphanizomenon flos-aquae* was common during the summer.

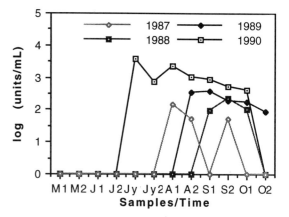

FIGURE 6.4. Cyanobacteria in Onondaga Lake, 1987–1990: primarily *Aphanizomenon flos-aquae*; units counted are colonies or filaments, values of log (units · ml^{-1}) <1.7 are shown as 0.

6.1.1.4.4 Trends in Major Groups (1978–1990)

Trends in abundance of the major groups of phytoplankton for the 1978–1990 period are presented in Figures 6.5 through 6.8. Of particular concern are blooms with peaks exceeding 10^4 cells·ml^{-1}. Flagellated green algae (Figure 6.5) have often been abundant in the spring (May–June) and sporadically common during the summer and fall. Since 1986 their spring peaks have been less frequent and less intense, but since 1983 green flagellates have been more common during the summers. Except for spring peaks of *Heteromastix* (1981, 1983, 1984),

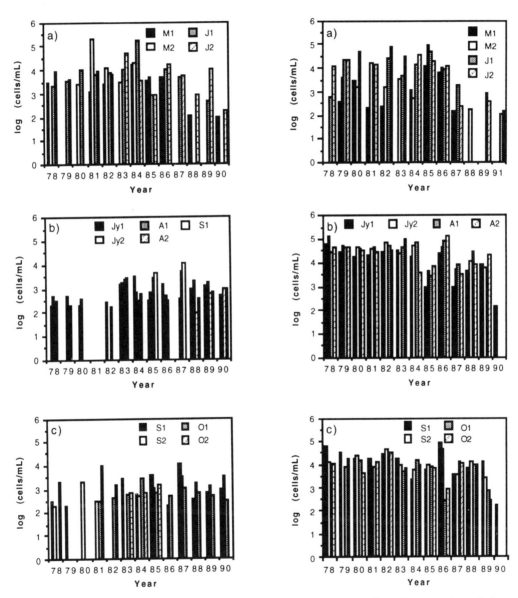

Figure 6.5. Annual and seasonal variations in flagellated green algae in Onondaga Lake, 1978–1990: **(a)** May–June, **(b)** July–August, and **(c)** September–October.

Figure 6.6. Annual and seasonal variations in non-flagellated green algae in Onondaga Lake, 1978–1990: **(a)** May–June, **(b)** July–August, and **(c)** September–October.

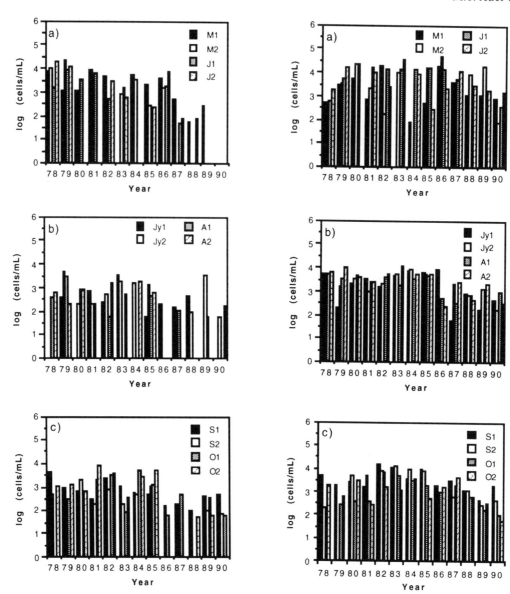

FIGURE 6.7. Annual and seasonal varitions in diatoms in Onondaga Lake, 1978–1990: **(a)** May–June, **(b)** July–August, and **(c)** September–October.

FIGURE 6.8. Annual and seasonal variations in cryptomonads in Onondaga Lake, 1978–1990: **(a)** May–June, **(b)** July–August, and **(c)** September–October.

Chlamydomonas spp. has been the principal green flagellate (Table 6.4).

Nonflagellated green algae have been largely responsible for the noticeable discoloration of the lake water during summers since 1972. Except for *Binuclearia* in 1983, 1984, and 1986, the common species have been members of the Order Chlorococcales.

Since the spring bloom of *Dictyosphaerium* in 1985, there has been a noticeable reduction in abundance of chlorococcalean green algae (the 1986 peak is due to *Binuclearia*). This is especially apparent when the spring and mid-summer periods are compared (Figure 6.6). Very low levels were observed in 1990. Species that showed marked declines since 1985 are

Chlorella vulgaris, Scenedesmus obliquus, and *Scenedesmus quadricauda.* A noticeable decrease in *Oocystis* did not occur until 1990.

There has been a progressive decline in the abundance of diatoms in the spring during the last 13 years and reduced frequency of occurrence during the summer and fall since 1986 (Figure 6.7). The distribution of cryptomonads has remained the most uniform since 1978 (Figure 6.8), with only a slight decrease since 1986.. The principal genera have been *Chroomonas* and *Cryptomonas.*

Overall there appears to be a decline in intensity of blooms by eukaryotic phytoplankton during the last 13 years. However, an increase in abundance of cyanobacteria occurred in 1989 and 1990 (Figure 6.4), reflecting the return of *Aphanizomenon flos-aquae* to importance in the summer. The reemergence of this cyanobacterium to importance does not appear to reflect a change in nutrient conditions; e.g., the N/P ratio remains quite high. Instead the return of importance of *Aphanizomenon flos-aquae* may reflect the operation of a selection process due to the shift to more efficient grazing zooplankton in the late 1980s (see subsequent section), as cyanobacteria may not be a suitable food for zooplankton. The size of the *Aphanizomenon flos-aquae* ''bundles'' may further inhibit grazing.

6.1.2 Biomass and Productivity

6.1.2.1 Chlorophyll

6.1.2.1.1 Background

The concentration of phytoplankton biomass is an important aspect of water quality and is often a major lake management concern, as it affects lake resource utilization. In particular, it is often the principal regulator of water column clarity (e.g., Field and Effler 1983; Megard et al. 1979; see Chapter 7) and its apparent color (Kirk 1985). Further, the deposition and subsequent decomposition of phytoplankton biomass and related detritus is generally primarily responsible for the depletion of oxygen in hypolimnia (Wetzel 1983). However, the proper unit of represen-

tation is indeed problematic. Units of phytoplankton biomass that have been used in limnological studies have included volume (based on cell counts and unit cell volumes), dry weight of carbon, and pigments (particularly chlorophyll *a*). Each has its advantages and disadvantages. Chlorophyll *a* is the most widely used measure of phytoplankton standing crop, although related pigments (e.g., chlorophyll *b* and *c* and phaeopigments) are also often included.

Chlorophyll *a* is the principal photosynthetic pigment and is common to all phytoplankton. The major advantages of chlorophyll *a* as a measure of phytoplankton biomass are: (1) the measurement is relatively simple and direct, (2) it integrates cell types and ages, (3) it accounts, to some extent, for cell viability, and (4) it can be quantitatively coupled to important optical properties (e.g., Weidemann and Bannister 1986). The principal disadvantage is that, because it is a community measure, it does not differentiate phytoplankton type. Further, the cellular content of chlorophyll *a* and other pigments varies with phytoplankton species, and, for a particular species, according to ambient conditions (Kirk 1983).

Phaeopigments are chlorophyll degradation products that are produced primarily by zooplankton grazing and are found mostly in the fecal pellets of zooplankton (Welshmeyer and Lorenzen 1985). Welshmeyer and Lorenzen (1985) indicate that phaeopigments in the watercolumn are almost entirely associated with the very small fecal pellets produced by microzooplankton (e.g., copepod nauplii and rotifers). Phaeopigments found in the fecal pellets of macrozooplankton are rarely encountered in the watercolumn because of the high rate of deposition of the pellets (Welshmeyer and Lorenzen 1985). However, this fraction of the phaeopigments can be efficiently collected in sediment traps deployed below the upper mixed layer of the water column (Welshmeyer and Lorenzen 1985). Welshmeyer and Lorenzen (1985) described, and documented initial applications of a method to assess phytoplankton growth and settling and grazing losses of phytoplankton through a program of monitoring chlorophyll *a* and phaeophytin in productive layers of the watercolumn and in sediment trap collections underlying this layer. The potential power of such an information set, and therefore the effective partitioning of pigments between chlorophyll *a* and phaeophytin, is attractive.

Phaeopigments and other chlorophylls (e.g., chlorophyll *b* and *c*), in addition to chlorophyll *a*, absorb light (Kirk 1983; Tilzer 1983) and therefore contribute to the overall attenuation of light and regulation of clarity. Although the absorption spectra of these pigments differ (Kirk 1983), some summation of these pigments is probably a better representation of the light attenuation implications of phytoplankton pigments than solely chlorophyll *a*.

In the following sections we present a critical review of chlorophyll *a* and related pigment data for Onondaga Lake. The first goal is to identify reliable pigment measures to reflect the impact of phytoplankton standing crop on the lake's quality. An analysis of the available data bases for these parameters is then presented to characterize prevailing conditions and evaluate potential trends.

6.1.2.1.2 Comparison of Results from Different Methods

There are a number of different methods for the measurement of chlorophyll *a*, related pigments, and their degradation products. These methods differ with respect to extraction (e.g., Marker and Jinks 1982; Schanz and Rai 1988), and analytical techniques (spectrophotometry, fluorometry, high-pressure liquid chromatography (HPLC); e.g., Jacobsen 1982; Schanz and Rai 1988). Substantial disparity in analytical results has been reported, for the various techniques, which has been attributed to differences in several features of the overall methodology (e.g., Jacobsen 1982; Schanz and Rai 1988). Because of this and the central role chlorophyll measurements have played in water quality studies, international workshops have been conducted that focused on resolution of these

differences and standardization of methods (Rai 1980; Rai and Marker 1982).

HPLC has generally been accepted as the standard analytical technique for the measurement of chlorophyll *a*, related chlorophylls, and degradation products (e.g., Gieskes and Kraay 1983; Jacobsen 1982) because of its ability to resolve the individual constituents. Jacobsen (1982, p.36) contends that "the presence of all the chlorophylls and their degradation products as found in natural phytoplankton extracts creates a matrix which cannot be resolved by commonly used spectrophotometric and fluormetric analysis." However, the use of HPLC for pigment analysis is relatively limited in water quality monitoring programs, and the method is still considered comparatively complex. Here we compare paired measurements made on 90% acetone extracts using spectrophotometry and HPLC for 1990. Further, we evaluate differences in two spectrophotometric methods; the method of Lorenzen (1967), which claims to separate chlorophyll *a* from its degradation products (phaeopigments), and the method of Parsons et al. (1984), which claims to separate chlorophyll *a* from the auxiliary pigments, chlorophyll *b* (common mostly to Chlorophyta) and chlorophyll *c* (common to Chrysophyta and Pyrrophyta).

The temporal distributions of chlorophyll *a* concentrations determined by HPLC and the two spectrophotometric methods from sample splits for the 1 m depth in 1990 are presented in Figure 6.9a. Chlorophyll *a* measured by the Lorenzen method is designated C_L; this parameter measured by the Parsons et al. technique is designated C_P. The data presented through October for all three cases are means of triplicate analyses. The precision of the Parsons et al. method was substantially better than the Lorenzen results; the precision for the HPLC method was the worst. The average relative standard deviations through October were 0.05, 0.14, and 0.24 for the Parsons et al., Lorenzen, and HPLC methods respectively. The overall precision of the HPLC method for the 1990 study period was influenced strongly by its poor performance in

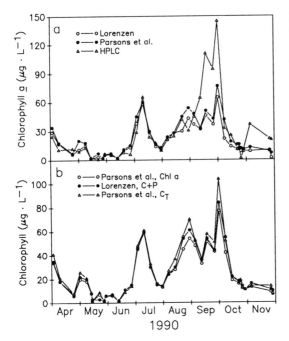

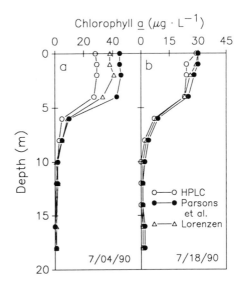

FIGURE 6.9. Temporal distributions of phytoplankton pigments in the upper waters (1 m) of Onondaga Lake for the April–November interval of 1990: (a) chlorophyll *a* according to HPLC (C_{HPLC}) and two different spectrophotometric methods (C_L and C_p), and (b) C_p, C + P, and C_T.

measures of chlorophyll *a* are presented for two different dates in July 1990 in Figure 6.10a and b. Very similar vertical structures were produced by the three different methods on both dates. The results diverged the most on July 4 (Figure 6.10a); the order of measurements in the upper waters was Parsons et al. > Lorenzen > HPLC.

More robust measures of the concentration of phytoplankton pigments, based on spectrophotometric methods, are compared with each other and C_p for the same (1 m) 1990 sample set in Figure 6.9b. The more robust measures are intended to accommodate to some extent the additional light absorption characteristics of pigments other than chlorophyll *a*. Clearly chlorophyll *a* is the principal pigment included in these measurements (Figure 6.9b). One of the robust pigment measures presented is the sum of chlorophyll *a* plus phaeophytin (C + P) determined from the Lorenzen (1967) method, the other is the sum of chlorophyll *a*, *b*, and *c* (C_T) determined according to the method of Parson et al. (1984). Note that these robust measures track very closely the dynamics determined for C_p. The average ratios of C + P : C_p and

late September and October. The results from the three methods tracked each other very closely through mid-August (Figure 6.9a). The HPLC results usually were either essentially equal to, or somewhat less than, the spectrophotometric values over this period. Lower HPLC values for chlorophyll *a* have been reported elsewhere and have been attributed to the inclusion of other related pigments in the spectrophotometric analyses (e.g., Gieskes and Kraay 1983; Jacobsen 1982). The relationships among the three sets of results became more irregular from mid-August through November. The general temporal structures are quite similar, however, the HPLC values were distinctly higher than the spectrophotometric results in the last three weeks of September and in November. These later differences remain unexplained. Comparisons of vertical profiles of these three

FIGURE 6.10. Vertical profiles of C_{HPLC}, C_p, and C_L in Onondaga Lake: (a) July 4, 1990, and (b) July 18, 1990.

$C_T : C_p$ for the data of Figure 6.9b were 1.07 and 1.13, respectively.

The near equivalence of C + P and C_T is supported for the larger four-year database in Figure 6.11a, for the $0–60\,\mu g \cdot L^{-1}$ range (also see Table 6.5). The relationship is stronger ($C_T = 1.016\,(C + P) − 1.2;\ R^2 = 0.95$) over the larger range $0–300\,\mu g \cdot L^{-1}$, but it is unduly biased by the smaller number of observations for the higher concentrations. However, the relationship between the two spectrophotometric measures of chlorophyll a (C_L and C_p) for the larger database (1987–1990) is much worse than observed in 1990 (Figure 6.11; compare to Figure 6.9a). Similarly poor performance is observed between C_L and C + P, and C_L and C_T (Table 6.5). In sharp contrast, strong relationships between C_p and C + P, and C_p and C_T, prevail (Table 6.5). The Lorenzen (1967) spectrophotometric measurement of chlorophyll a appears to be flawed for application to Onondaga Lake, probably as a result of the unreliable partitioning of chlorophyll a and phaeophytin. This method appears to produce false high estimates of phaeophytin and false low estimates of chlorophyll a. HPLC measurements indicated that phaeopigments were essentially absent from the water column of Onondaga Lake in 1990. Average phaeophytin concentrations at a depth of 1 m determined according to Lorenzen (1967) were 33, 24, 6, and $6\,\mu g \cdot L^{-1}$ for 1987, 1988, 1989, and 1990. The longest chlorophyll data base for the lake is based on the Lorenzen (1967) method. Based on the apparent shortcomings in the partitioning of

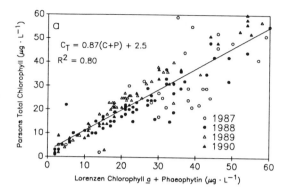

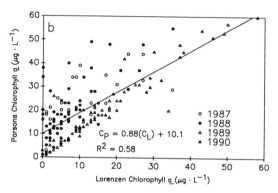

FIGURE 6.11. Evaluation of relationships between different measures of phytoplankton pigments in Onondaga Lake over the 1987–1990 interval: (**a**) C_T versus C + P, and (**b**) C_p and C_L. All values are means of analyses of triplicate samples. Results of linear squares regression analyses included.

TABLE 6.5. Relationships between different pigment measures for Onondaga Lake (1987–1990), according to linear least square regression.

X	Y	Intercept	Slope	R^2	n
C_L	C_p	10.1	0.88	0.58	165
C_L	C_T	1.3	0.99	0.48	160
C_L	C + P	16.2	0.75	0.31	158
C + P	C_T	2.5	0.87	0.80	155
C_p	C_T	1.3	1.17	0.95	161
C_p	C + P	3.2	1.11	0.83	158
C_{HPLC}	C_T	3.6	1.04	0.66	30
C_{HPLC}	C + P	2.7	1.03	0.68	30

chlorophyll a and phaeophytin by this method, and the reliability of the sum C + P (Figure 6.11a), historic trend analysis should be based on C + P. Further, the failure of the Lorenzen (1967) method to reliability partition C and P in the lake eliminates the possibility of application of the Welschmeyer and Lorenzen (1985) method to quantify phytoplankton growth and loss processes.

The parameter C_T has been adopted as the primary measure of phytoplankton biomass and pigment concentration (e.g., see related optical analyses in Chapter 7) in recent research efforts for the lake. Measurements of C_T are strongly correlated to values of chlorophyll a determined by HPLC, the precision of the measurement compares

TABLE 6.6. Precision of pigment analyses of replicate samples for the near surface waters of Onondaga Lake over the 1987–1990 interval.

Pigment neasure	Average relative standard deviation	Number of triplicate samplings
C_L	0.37	141
P	0.36	139
C_P	0.09	138
C + P	0.10	139
C_T	0.11	139

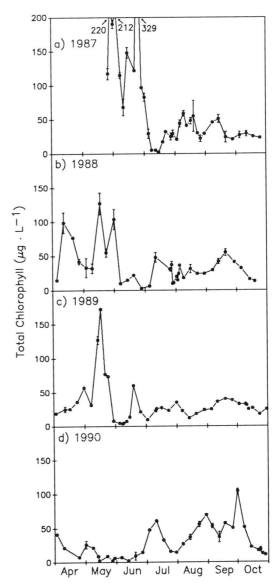

FIGURE 6.12. Temporal distributions of C_T in the upper waters (1 m) of Onondaga Lake for the spring to fall interval.

favorably with other pigment measures (Table 6.6), and it does not suffer from the apparent systematic limitations of the Lorenzen (1967) method.

6.1.2.1.3 Recent Temporal and Vertical Distributions

The temporal distributions of C_T measured on samples collected from 1 m at the south deep station on Onondaga Lake for the spring to fall interval of 1987, 1988, 1989 and 1990 are presented in Figure 6.12a–d. Uncertainty in the measurements is reflected by vertical bars with dimensions equal to ±1 standard deviation, based on analyses of triplicate samples; bars are not apparent for most measurements (Figure 6.12) due to the high precision of the measurements. Biomass levels in this near-surface layer are particularly important with respect to the public's perception of optical aspects of water quality.

Despite the rather striking differences in the fine details of the temporal structures for the four years, there appear to be certain recurring seasonal features for the lake. Strong temporal variations are routinely observed. The highest concentrations of chlorophyll in the upper waters in three of the four years (1987–1989) were observed during the spring phytoplankton bloom (Figure 6.12a–c). The spring bloom is normally attributed to increased light availability and temperature, under relatively nutrient-rich conditions (Wetzel 1983). The year 1990 is conspicuous by the absence of such a clear spring peak (Figure 6.12d). The highest concentrations observed over the four year period (e.g., >200 µg · L^{-1}) were observed

in May and June of 1987. This was followed by a precipitous decrease in chlorophyll in early July. This phenomenon has been termed a "clearing" event in the literature (e.g., Lampert et al. 1986), because of the abrupt increase in clarity that occurs as a result (see Chapter 7 for analysis of optical implications). The clearing event(s) occurred earlier in 1988 and 1989 than in 1987. The clearing event phenomenon is widely observed in productive lakes (Lampert 1978; Lampert et al. 1986; Sommer et al. 1986). A detailed analysis of the factors regulating the dynamics of phytoplankton standing crop, including the clearing phase, in Onondaga Lake has been presented by Auer et al. 1990. Low biomass conditions prevailed from mid-May to mid-June in 1990. Minima in C_T in the upper waters were less than $3 \mu g \cdot L^{-1}$ over the 1987–1990 interval. The highest chlorophyll maxima in 1990 were observed in late summer and early fall (Figure 6.12d). Chlorophyll concentrations were more uniform by comparison in this period in the other three years (Figure 6.12a–c).

The temporal distribution of C + P in the upper waters (0.25 m) of the south basin of the lake in 1978 (Field 1980) is presented in Figure 6.13 as a basis of comparison to the more recent distributions of Figure 6.12. The 1978 distribution is considered to be generally representative of conditions in most years over the interval 1970–1986 (also see historic analysis of transparency in Chapter 7). The highest peaks in biomass in 1978 were also observed in May and June (Figure 6.13). The strongest difference compared to the distributions of 1987–1990 is the absence of a period of very low concentrations of phytoplankton biomass (clearing event); C + P levels remained above $20 \mu g \cdot L^{-1}$ throughout the May–September interval.

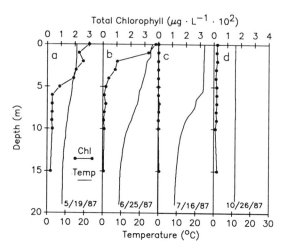

FIGURE 6.14. Vertical distribution of C_T and temperature in Onondaga Lake for four selected dates in 1987: (**a**) May 19, (**b**) June 25, (**c**) July 16, and (**d**) October 26.

The vertical distribution of C_T and temperature is presented for four selected days in 1987, as profiles in Figure 6.14; temporal and vertical distributions of total chlorophyll in 1988 are represented as isopleths in Figure 6.15. Strong vertical gradients in chlorophyll concentrations commonly develop during the late spring/early summer phytoplankton bloom (Figure 6.14a and b, Figure 6.15). The highest concentrations are generally found in the surface waters, where light conditions are most conducive for growth. This vertical structure is a result of high phytoplankton

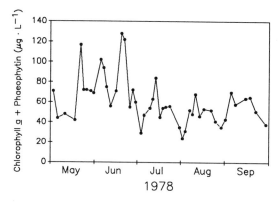

FIGURE 6.13. Temporal distribution of C + P in the upper waters (0.25 m) of the south basin of Onondaga Lake for the May–September interval of 1978.

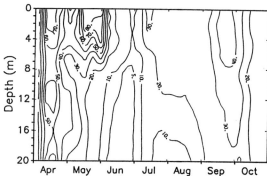

FIGURE 6.15. Isopleths of C_T in Onondaga Lake for the April–October interval of 1988.

growth rates and low light penetration, combined with a shallow upper mixed layer. Conditions of rapid heating of the surface waters and low ambient wind are common during this period. Note the epilimnion is poorly defined in these cases (Figure 6.14a and b). Subsurface (e.g., metalimnetic) maxima have not been observed for the lake, probably because of limited light penetration conditions that prevail for most of the growing season (Chapter 7). Chlorophyll concentrations have been observed to be uniformly low with depth during the "clearing" event period (e.g., Figure 6.14c), and tend to remain more vertically uniform thereafter (Figure 6.14d and Figure 6.15).

The temporal distributions of depth integrated C_T (mg $\cdot$ m^{-2}), for the 0–10 m interval (summed chlorophyll content over that depth interval, per unit area of lake surface), for the spring to fall interval of 1987, 1988, 1989, and 1990, are presented in Figure 6.16a–d. Most of the viable biomass is contained within this depth interval over this period. The disparities between these distributions and those presented earlier for concentrations in the upper waters (Figure 6.12) reflect the influence of the vertical heterogeneity that commonly prevails in the water column, particularly in the early part of the stratification period (Figure 6.15). In particular, the differences in watercolumn biomass between the spring and fall periods are reduced, though the minima in early to mid-summer remain clearly resolved. The differences between 1990 and the other three years are made more striking by this representation. The biomass supported over the July–October interval of 1990 was higher than in previous three years, but total watercolumn biomass levels in May and June were much lower.

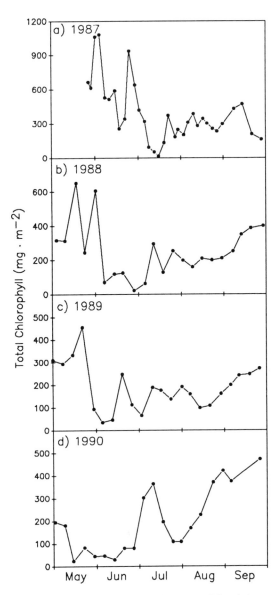

FIGURE 6.16. Temporal distributions of depth integrated (0–10 m) C_T concentrations in Onondaga Lake for the spring to fall interval for four years.

6.1.2.1.4 Spatial Variations

Paired measurements of C + P made on near-surface (0.25 m) samples collected at the north and south deep stations on 46 occasions in 1978 are presented in Figure 6.17a. The pigment concentrations tended to be somewhat higher at the south deep station in this year; 70% of the observations were higher at that site. The median concentration at the south station was 55 μg $\cdot$ L^{-1}, compared to 39 μg $\cdot$ L^{-1} at the north site. The results for fewer paired observations made in the near-surface waters in 1987 (Figure 6.17b) indicate that substantial spatial differences still occasionally occur, but that there is no major

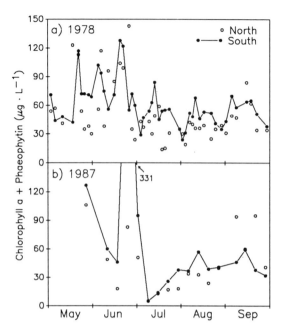

FIGURE 6.17. Comparison of temporal distributions of C+P in the upper waters (0.25 m) of south and north deep stations of Onondaga Lake.

TABLE 6.7. Summary of pigment concentrations in the near-surface waters of Onondaga Lake for the May−September interval.

Year	C + P ($\mu g \cdot L^{-1}$)		C_T ($\mu g \cdot L^{-1}$)	
	$\bar{x}$	Median	$\bar{x}$	Median
1978	70	55		
1985	63	59		
1986	37	20		
1987	73	44	72	46
1988	36	32	33	29
1989	34	24	36	24
1990	26	20	28	21

The median, a more robust measure of the central tendency of such populations (Reckhow and Chapra 1983), may be a better statistic to describe the annual summer biomass conditions under these circumstances. Both mean and median data reflect substantial interannual variations in biomass. Note that C_T and C + P values track each other closely over the 1987−1990 period, supporting the use of the available C + P data to reflect historic conditions. The reduction in C + P observed in 1986 does not appear to be in response to a significant reduction in external phosphorus loading (Chapter 3). Concentrations increased again in 1987. However, concentrations were lower thereafter, particularly in 1990.

Chlorophyll *a* is considered to be the most direct, and therefore probably the best, indicator of trophic state. However, there is not universal agreement on the concentrations of chlorophyll *a* that demarcate trophic states. A summer average chlorophyll *a* value of $2.0 \mu g \cdot L^{-1}$ has been used as the demarcation between oligotrophy and mesotrophy (Dobson et al. 1974; National Academy of Science 1972). There is less agreement for the demarcation between mesotrophy and eutrophy; the bounding summer average value reported from different sources (e.g., Dobson et al. 1974; National Academy of Science 1972; Great Lakes Group 1976) ranges from 8 to $12 \mu g \cdot L^{-1}$. Recall that C_T exceeds the concentration of chlorophyll *a* (e.g., C_p) by about 10%. Clearly Onondaga Lake should be considered highly eutrophic. Even the lowest summer average concentration observed in

prevailing north/south trend. The concentration was strikingly higher at the south site for the measurement made in late June of 1987 (Figure 6.17b). The median concentrations were very similar for the two stations for the monitoring period of 1987; 40 and $37 \mu g \cdot L^{-1}$ at the south and north sites, respectively. The apparent reduction in spatial differences for phytoplankton pigments is consistent with reductions in clarity differences over the same period (Chapter 7).

6.1.2.1.5 Historic Changes and Trophic State

Phytoplankton biomass conditions for the near surface waters of Onondaga Lake for the May−September (summer) interval, for selected years over the 1978−1990 period, are represented as mean and median values in Table 6.7. Note that the mean and median values differ substantially in a number of years, reflecting the nonnormal character of the temporal distributions of phytoplankton biomass that occur (e.g, extreme peaks in the spring/early summer bloom; Figure 6.12a).

1990 exceeds the lower boundary of eutrophy by about a factor of three (Table 6.7). Higher concentrations observed earlier in the lake exceeded this demarcation by as much as a factor of eight (Table 6.7).

6.1.2.2 Productivity Measurements

In situ productivity experiments using the oxygen evolution technique (Vollenweider 1974) were conducted in the lake as part of several studies. Murphy (1978) reports the results of several experiments conducted in 1975. More than 50 experiments were executed at both the north and south deep stations in 1978 (Field and Effler 1983). Additional measurements were made in 1980 and 1981 (Effler and Driscoll 1985). A limited number of measurements were made in 1991 to support testing of a P-I relationship developed in laboratory experiments. Carbon-14 measurements (e.g., Vollenweider 1974) were conducted only during the summer of 1977 (Onondaga County 1978). Numerous laboratory measurements of productivity were made in the 1987–1991 interval to develop kinetic coefficients to quantify phytoplankton growth; these are described subsequently. In this section some of the features of the vertical and temporal features of primary productivity in the lake are described.

6.1.2.2.1. Depth Distribution of Primary Productivity

The depth distribution of primary productivity is strongly influenced by the extent of light penetration and, in oligotrophic systems with a deep photic zone, by features of thermal stratification. Surface photoinhibition is often observed when the penetrating irradiance exceeds the photoinhibition threshold of the productive components of the phytoplankton community (Steele 1962). The depth at which maximum rates of photosyntesis occur varies according to light penetration, and therefore the concentration of light attenuating materials. Thus net phytoplankton growth, manifested as increases in standing crop, feeds back to influence these depth distribution characteristics. Photoinhibition may not be

resolved when extremely high concentrations of attenuating substances prevail, such as during algal blooms. Under higher light penetration conditions a zone of maximum photosynthetic rates, which overlies a zone of near-exponential decline of rates with increasing depth, is often observed. Metalimnetic peaks in photosynthetic rates are observed only in stratified systems with a high degree of light penetration.

Selected vertical profiles of primary productivity are presented, based on both C-14 and the oxygen evolution techniques, in Figure 6.18a–c. Carbon-14 experiments are a mea-

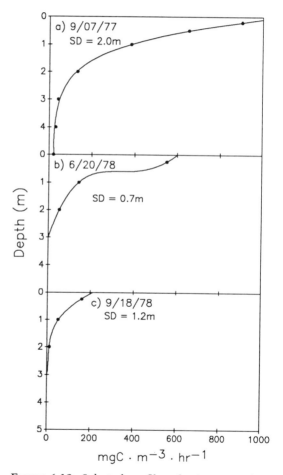

FIGURE 6.18. Selected profiles of primary production in Onondaga Lake: (**a**) net (from C-14, Onondaga County 1978), September 7, 1977, (**b**) gross, June 20, 1978, and (**c**) gross, September 18, 1978. Secchi disc (SD) values shown.

sure of net production; the oxygen evolution technique can be used to estimate both gross and net productivity (Vollenweider 1974). The focus here is on the shape and vertical extent of the photosynthetic rate profiles. Differences in magnitude in the uppermost waters reflect differences in incident light, standing crop, and experimental protocol. Incubations for the C-14 experiments (e.g., Figure 6.18a) were about 1 hour near noon. Incubations for the oxygen evolution measurements (e.g., Figure 6.18b and c) usually extended from about 0800 to 1600. Secchi disc (SD) measurements made during the experiments are included on each plot. Nearly exponentially decreasing rates were observed using both methods under these conditions of very low clarity (Figure 6.18a–c). Photoinhibition at the uppermost incubation depths was observed rarely in the late 1970's (e.g., Field 1980), but was noted in a single experiment conducted in 1990 (Storey et al. 1993). The vertical features of primary productivity encountered most often in Onondaga Lake are characteristic of eutrophic or hypereutrophic systems (Wetzel 1983).

6.1.2.2.2. Temporal Distributions of Areal Productivity

Areal productivity is determined by integration of the volumetric results through depth (e.g.,

euphotic zone). Temporal variations in areal productivity ($mgC \cdot m^{-2} \cdot hr^{-1}$1) determined over the mid-April to mid-October interval of 1978 (Field 1980) are presented in Figure 6.19. Only a coarse level of seasonality is apparent; productivity tended to be somewhat higher in summer than in spring and fall. The most conspicuous feature of the distribution is the frequently observed abrupt short-term changes. This is attributed largely to the highly variable incident light conditions common to this region (Auer and Effler 1989). Field and Effler (1983) described the nutrient conditions (with respect to phosphorous and nitrogen) in the productive layers throughout this period as saturated. The performance of two different light/temperature–productivity models is included in Figure 6.19 (Field and Effler 1983). Both models (invoking nutrient saturation) performed well in simulating the observed variations in areal productivity; they both explained about 75% of the variability in productivity (Field and Effler 1983). Field measurements supporting the testing of the submodel included: chlorophyll a, light attenuation in the watercolumn, water temperature, primary productivity, and total solar radiation. The poorest performances were observed in late May, mid-June, and mid-August, when the models overpredicted productivity (Figure 6.19).

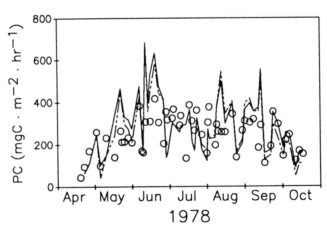

FIGURE 6.19. Temporal variations in areal productivity in Onondaga Lake over the mid-April to mid-October interval of 1978, with simulations from two light/ temperature submodels (from Field and Effler 1983).

6.1.3 Kinetic and Process Studies

6.1.3.1 Background

Phytoplankton grow under appropriate light, temperature, and nutrient availability conditions. Loss processes for phytoplankton operate simultaneously; these include respiration, settling, and grazing. The imbalance between growth and loss processes, net phytoplankton growth, can be described according to

$$\frac{dP}{dt} = (G - D - s - z_g)P \qquad (6.1)$$

in which P = phytoplankton biomass (e.g., μg chlorophyll $\cdot L^{-1}$), G = gross specific growth rate (d^{-1}), D = specific respiration rate (d^{-1}), s = settling rate (d^{-1}), z_g = loss rate due to grazing (d^{-1}), and t = time (d). During periods of increases in phytoplankton biomass dP/dt > 0; phytoplankton biomass declines when dP/dt < 0. In some representations additional (more minor) loss terms are included (e.g., excretion, non-predatory mortality (Bowie et al. 1985; also see Chapter 9)).

Mass balance mathematical models of phytoplankton growth have been developed that simulate seasonal growth and biomass through solution of an expression of the form of Equation (6.1). Phytoplankton growth models have gained wide acceptance as management tools supporting the development of remediation strategies for eutrophic lakes and embayments (Chapra and Reckhow 1983; Thomann and Mueller 1987). All these models share a common need: specification of values of kinetic coefficients that quantify the various source/sink processes for phytoplankton growth (Equation (6.1)). Relationships used to describe various source and sink processes for phytoplankton are reviewed here. Subsequently, the need for site specific measurement of these kinetic coefficients is identified.

6.1.3.1.1 Growth

The rate coefficient for gross growth (G) is mediated by light, temperature, and nutrient (usually phosphorus) supply. This may be modeled using a multiplicative approach (e.g., Chapra and Rechkow 1983; Auer et al. 1986) where the specific gross growth rate coefficient is calculated as a fraction of the maximum rate:

$$G = G_{max,20} \cdot [G(I) \cdot G(T) \cdot G(N)] \quad (6.2)$$

in which $G_{max,20}$ = maximum specific gross growth rate at 20°C (d^{-1}), and G(') = dimensionless functions describing the effects of light (I), temperature (T), and nutrients (N, phosphorus in this case).

Various formulations relating photosynthesis and light (termed P–I curves) have been proposed (cf. Jassby and Platt 1976; Field and Effler 1982). Among the most commonly employed is a Monod-type saturation function (Monod 1942)

$$G(I) = \frac{I}{K_I + I} \qquad (6.3)$$

in which I = light intensity (PAR, $\mu E \cdot m^{-2} \cdot s^{-1}$), and K_I = half-saturation constant for growth; the light intensity where $G = 0.5 \cdot G_{max,20}$. According to this function, growth is a linear function of light at low intensities, then saturates, asymptotically approaching $G_{max,20}$ at higher light intensities.

Temperature effects on growth have been described using linear and exponential functions, and using curves which reflect a growth optimum (cf. Bowie et al. 1985). Exponential functions, in the following form, are quite common

$$G(T) = \Theta_G^{(T-20)} \qquad (6.4)$$

in which Θ_G = a dimensionless constant which describes the variation in the rate constant with temperature. The notation Θ_G is utilized here when it is applied specifically for a temperature correction on the gross specific growth rate.

There are two basic approaches for quantifying the relationship between phytoplankton growth and nutrient availability; both have been widely applied in mathematical models. The first approach, based on Monod (1942) kinetics, relates growth rate to the external (dissolved) nutrient concentration:

$$G(N) = \frac{[P]}{K_P + [P]} \qquad (6.5a)$$

in which $G(N)$ = the fraction of the maximum specific growth rate (range 0 to 1) achieved at phosphorus concentration $[P]$ (mg $\cdot$ L^{-1}) and K_p is the half-saturation constant (mg $\cdot$ L^{-1}), a measure of the organism's efficiency in acquiring phosphorus. Phosphorus available to algae is typically characterized as soluble reactive phosphorus (SRP), however, other dissolved forms (e.g., nonreactive or organic phosphorus) can support growth (Berman 1970; Taft et al. 1977). Soluble reactive P (SRP) is an imperfect representation of the external P pool available for algae uptake. The analyte SRP includes ortho-P and the reactive fraction of the dissolved organic P (DOP) pool. The unreactive fraction of the DOP pool, parametrized as the difference between the concentrations of total dissolved P (TDP) and SRP (TDP-SRP), can be made available to algae through enzymatic hydrolysis (Healey and Hendzel 1979; Gage and Gorham 1985). Enzymes cleave ortho-P groups from large organic P molecules otherwise unavailable to algae (Bentzen et al. 1992). The most well-known of these enzymes are the alkaline phosphatases, which have been the subject of many studies (e.g., Fitzgerald and Nelson 1966; Franco 1983; Hantke and Melzer 1993; Jones 1971) Alkaline phosphatase activity (APA) is thought to play a major role in P regeneration in the water column (Ahn et al. 1993) and may serve to extend the duration of algal blooms in eutrophic lakes (Heath and Cooke 1975). Inverse relationships have been observed between APA and SRP concentration (Hernandez et al. 1992; Pettersson 1980; Reichardt 1971) and between APA and cell quota (Q, see below) (Gage and Gorham 1985; Healy 1973; Wynne 1977), suggesting that APA is an indicator of P-deficient algae (Jansson et al. 1988).

The second approach, for quantifying the interplay between nutrient $\cdot$ availability and growth is based on Droop (1968) kinetics, which relates growth rate to the internal (or stored) nutrient concentration:

$$G(N) = 1 - (Q_o/Q) \qquad (6.5b)$$

in which Q = the cell quota (mass nutrient $\cdot$ unit biomass^{-1}) and Q_o is the minimum cell quota, i.e., the internal phosphorus concentration where growth ceases.

The Monod model offers the advantage of directly relating growth to watercolumn phosphorus levels. The disadvantage of the Monod approach is that it ignores the well-documented phenomenon of luxury uptake, where nutrients including phosphorus are acquired and stored at levels well beyond the immediate demand for growth. By drawing on internal nutrient stores, algae can grow at near maximum rates during periods of water column nutrient depletion. The Droop model overcomes this problem by relating growth to internal nutrient levels, thus accommodating the luxury uptake–storage phenomenon. The Droop approach requires a mass balance on the internal nutrient pool, considering contributions through nutrient uptake from the external environment and losses through the demand for growth.

The additional kinetic complexity required by the Droop model limits its management application. DiToro (1980) recognized this in modeling nutrients and phytoplankton in Lake Erie (see also DiToro and Connolly, 1980). He suggested that since the pool size response to the external environment is rapid compared with the response to demands for growth (nutrient uptake rate $>$ growth rate), algae would achieve a cellular equilibrium when the rate of change in external nutrient levels is slow relative to the rate of growth. Under these conditions, i.e., a relatively stable external nutrient environment, the linkage between growth rate and nutrient resources may be fairly represented by the simpler Monod model.

6.1.3.1.2 Loss Processes

Algal respiration rates have been reported to vary with light (Heichel 1970; Graham et al. 1982; Falkowski et al. 1985) and temperature (Jorgensen 1968; Goldman and Carpenter 1974; Kirk 1983; Harris 1986). Although there is some agreement that respiration rates under conditions of active photosynthesis (e.g., in the light) are higher than rates observed at a basal metabolic state (e.g., in the dark), there is no generally accepted formulation describing this effect. Temperature effects are modeled as for the specific gross growth rate:

$$D = D_{20} \cdot \Theta_D^{(T-20)} \qquad (6.6)$$

in which D_{20} = respiration rate at 20°C (d^{-1}), and Θ_D = dimensionless coefficient which describes the variation in the rate constant with temperature.

Several methods have been used to estimate rates of zooplankton grazing, including direct cell counts (Frost 1972; Poulet 1973), the use of radioactive tracers (Haney 1971), and the application of pigment budget models (Welschmeyer and Lorenzen 1985; Laws et al. 1988). Each has its limitations. The pigment budget approach is attractive as the phytoplankton and zooplankton populations are not manipulated. Though the Welschmeyer and Lorenzen (1985) method cannot be applied because of the shortcomings of the Lorenzen (1967) pigment analysis for Onondaga Lake, pigment analysis of sediment trap collections may support a first approximation of the summed losses associated with both settling and grazing. Grazed chlorophyll is essentially converted to phaeopigments and collected in sediment traps (cylinder deployed vertically below the upper mixed layer, see Chapter 8; Welschmeyer and Lorenzen 1985). Phytoplankton settling losses from the epilimnion can be calculated from the accumulation of sedimented chlorophyll in traps.

The downward flux of C_T ($mg \cdot m^{-2} \cdot d^{-1}$) has been assessed in Onondaga Lake for the spring-fall interval annually over the 1987–1990 period, based on the analysis of sediment trap collections (Chapter 8). This serves as a basis to estimate the sum of the grazing and settling loss rates ($s + z_g$), as both chlorophylls and phaeopigments are included in the C_T measurement. However, the estimate is only approximate, as the lower molecular weight of the phaeopigments (e.g., pheophorbide a) versus chlorophylls is not considered (e.g., phaeopigment concentration has a "chlorophyll equivalent" equal to 1.51 times its

value; Welschmeyer and Lorenzen 1985). Thus changes in the relative contributions of chlorophylls and phaeopigments to the trap collections cannot be accommodated. The settling velocity of C_T is determined from the trap flux according to

$$V_{S/CT} = S_{CT} \div C_{T/AVG} \qquad (6.7)$$

in which $V_{S/CT}$ = settling velocity of C_T ($m \cdot d^{-1}$), S_{CT} = downward flux of C_T ($mg \cdot m^{-2} \cdot d^{-1}$), and $C_{T/AVG}$ = average C_T ($mg \cdot m^{-3}$) over sediment traps during the period of deployment. Then to a first approximation, the summed loss rate coefficient for settling and grazing is calculated according to

$$s + z_g = V_{S/CT} \div Z_e \qquad (6.8)$$

in which Z_e = depth of epilimnion (m).

6.1.3.1.3 On the Value of Site-Specific Kinetic Coefficient Values

Modelers often select phytoplankton kinetic coefficient values for phytoplankton models from literature compilations such as the comprehensive review of Bowie et al. (1985); further refinement is accomplished through model calibration. Alternatively, coefficients may be determined by direct measurement using laboratory cultures or natural phytoplankton assemblages. Given the species-specific response of algae to environmental conditions (Wetzel 1983), it is reasonable to expect that coefficient values will vary widely among lakes, and even seasonally within a lake, due to differences in the composition of the phytoplankton community (Wetzel 1983). Coefficients developed for one site may help to bound conditions in another, but they cannot be confidently transferred among systems. Model coefficients determined through laboratory experimentation (e.g., unialgal cultures) offer a high degree of reliability, but it is not clear that they are representative of any natural phytoplankton assemblage. For these reasons, the evaluation of physiological models and related kinetic coefficients on a site-specific basis, using the natural phytoplankton assemblage, offers an opportunity to enhance model credibility and improve the reliability of model output (Auer and Canale 1986).

Storey et al. (1993) have recently demonstrated with a Monte-Carlo analysis the improved credibility of model predictions of nutrient-saturated areal net production in Green Bay, Lake Michigan, that resulted from the use of system-specific kinetic coefficients determined on natural assemblages (versus the use of literature values). The dramatic improvement is illustrated in Figure 6.20. Variability in three key coefficients was accommodated in the analysis, G_{max} (at 20°C), K_I, and Θ_G (D_{20} was held constant; Storey et al. 1993). The distributions of the experimentally derived coefficients were assumed to be normal, with standard deviations based on experimental results. The distributions for the coefficient populations obtained from the literature were assumed to be uniform over the reported ranges (e.g., Bowie et al. 1985). One thousand calculations of areal production

were performed, with each of the coefficients randomly chosen from the specified distributions for each of the calculations. The substantial differences in the means of the predicted populations (Figure 6.20) indicates the inaccuracy of predictions that results from the use of centrally distributed coefficient values from literature compilations. Further, the much narrower distribution obtained by development of site-specific coefficient values (Figure 6.20a) establishes the greater reliability of these predictions.

6.1.3.2 Determination of Phytoplankton Kinetic Coefficients for Onondaga Lake

6.1.3.2.1 Methods

Features of the laboratory methodology used to develop phytoplankton kinetic coefficients for Onondaga Lake are presented in Table 6.8. Experimental details have been presented by Storey et al. (1993). Rates were fit to functions describing their dependency on light and temperature using least squares nonlinear regression (Sigma Plot 4.0).

6.1.3.2.2 Results and Suitability of Physiological Models

The utility of the Monod model (Equation (6.3)) for describing the light functionality of gross photosynthesis was examined. The fit of measured rates of chlorophyll-specific gross photosynthesis determined for the Onondaga Lake phytoplankton assemblage to the Monod model is presented in Figure 6.21. The mean relative error was 35 and 13% for specific and chlorophyll—specific rates, respectively. Values of $G_{max} = 2.90\,d^{-1}$ (0.708 mgO$_2 \cdot$ μg Chl$^{-1} \cdot d^{-1}$) and $K_I = 28\,\mu E \cdot m^{-2} \cdot s^{-1}$ were determined by least squares nonlinear regression, according to the relationship of Equation (6.3). Coefficients of variation ranged from 8−15% for G_{max} and 27−83% for K_I. The value determined for Onondaga Lake for the carbon-specific G_{max} is well within the rather broad range (0.2−8.0 d^{-1}) reported for this coefficient in other studies (Bowie et al. 1985). This carbon-specific G_{max} value is remarkably similar to the value reported for

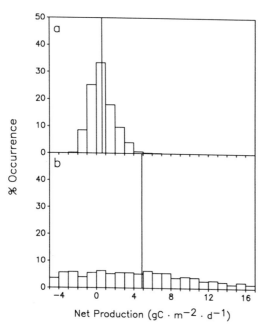

FIGURE 6.20. Monte-Carlo model analysis of Green Bay: simulations of areal net phytoplankton production for (a) sytem-specific kinetic coefficients determined on natural assemblages, and (b) literature values of kinetic coefficients.

TABLE 6.8. Features of methodology for development of phytoplankton kinetic coefficients for Onondaga Lake.

Feature	Description
Experiments	
1. Acclimation of samples	Light (I) and temperature (T), 12 hr
2. Addition of KH_2PO_4	to assure nutrient saturated conditions
3. Measures of biomass	C_T, and particulate organic carbon (POC)
4. Photosynthesis and respiration rates; for matrix of I, T	Time rate of change in DO; net photosynthesis measured over 1–4 hr incubation in light; community respiration measured over 2–4 hr incubation in dark; I = 25, 50, 175, 300, 600, and 1000 $\mu E \cdot m^{-2} \cdot s^{-1}$; T = 10, 15, 20, 25, and 30°C
Calculations	
1. Algal respiration rate	Residual of community respiration rate and rate of DO depletion of filtered (0.45 μ) sample
2. Gross photosynthetic	Sum of net photosynthesis and algal respiration rates
3. Normalization of rates for biomass	
a. Chlorophyll-specific ($mgO_2 \cdot \mu gChl^{-1}d^{-1}$)	Division of volumetric rates by C_T
b. Specific rates (d^{-1})	Multiplication of volumetric rate by photosynthetic quotient (PQ = 1, 12 mgC $\cdot$ 32mgO$_2$$^{-1}$), divided by POC

From Storey et al. 1993

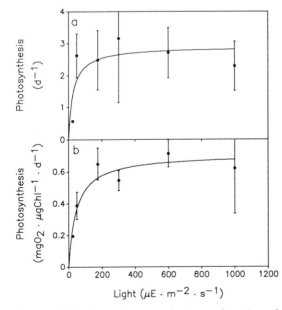

FIGURE 6.21. Gross photosynthesis as a function of light at 20°C for natural phytoplankton assemblages of Onondaga Lake: (**a**) specific photosynthesis (d^{-1}), and (**b**) chlorophyll specific ($mgO_2 \cdot \mu gChl^{-1} \cdot d^{-1}$). Limits of vertical bars equal to ±1 standard deviation.

Green Bay (Storey et al. 1993), that was developed according to essentially the same protocol described by Storey (1989). However, the chlorophyll—specific values for G_{max} were markedly dissimilar for the two systems, associated with their different carbon to chlorophyll ratios. The K_I value falls somewhat below the range (38–115 $\mu E \cdot m^{-2} \cdot s^{-1}$) presented by Bowie et al. (1985), but it seems well justified by the fit of the data (Figure 6.21).

The exponential model presented as Equation (6.4) was examined for its utility as a function relating temperature and rates of gross photosynthesis and algal respiration. The fit of measured rates of carbon-specific gross photosynthesis and algal respiration to this model is presented for the Onondaga Lake phytoplankton assemblage in Figure 6.22. The mean relative error for gross photosynthesis was 26%; the error for respiration was 31%. The fit for the temperature functionality is poorer than for the light (e.g., Figure 6.22 versus Figure 6.21). The results

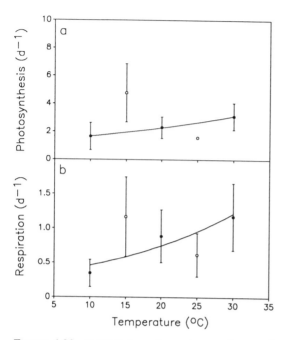

FIGURE 6.22. Temperature dependence of natural phytoplankton assemblages of Onondaga Lake: **(a)** specific photosynthesis (d^{-1}), and **(b)** phytoplankton respiration. Limits of vertical bars equal to ± standard deviation. "Open circle" data not included in regression analysis to determine coefficients.

do, however, suggest increases in rates of gross photosynthesis and algal respiration with increases in temperature. The values determined for Θ_G and Θ_D for the Onondaga Lake assemblage were 1.03 and 1.05, respectively; within the range reported from other studies and summarized by Bowie et al. (1985). The values of D_{20} for Onondaga Lake were determined to be $0.89\,d^{-1}$ and $0.18\,mgO_2 \cdot \mu g\ Chl^{-1} \cdot d^{-1}$ (Storey et al. 1993), slightly higher than the values included in the compilation of Bowie et al. (1985). The site-specific coefficients determined for the phytoplankton assemblage of Onondaga Lake are summarized in Table 6.9.

6.1.3.2.3 Settling

Phytoplankton pigment deposition has been quantified in Onondaga Lake as part of a sediment trap program intended to support the measurement of the downward flux (e.g.,

TABLE 6.9. Site-specific kinetic coefficients for phytoplankton growth in Onondaga Lake.

Coefficient	Units	Value ± 1 S.E.*
$G_{max,20}$	(d^{-1})	2.90 ± 0.45
$G_{max,20}$	$(mgO_2 \cdot \mu g\ Chl^{-1} \cdot d^{-1})$	0.71 ± 0.06
K_I	$(\mu E \cdot m^{-2} \cdot s^{-1})$	28 ± 24
Θ_G	(dimensionless)	1.03 ± 0.00
D_{20}	(d^{-1})	0.89 ± 0.38
D_{20}	$(mgO_2 \cdot \mu g\ Chl^{-1} \cdot d^{-1})$	0.18 ± 0.07
Θ_D	(dimensionless)	1.05 ± 0.02

Modified from Storey et al. 1993
S.E.*—standard error of estimate

$mg \cdot m^{-2} \cdot d^{-1})$ of a number of constituents of water quality interest. Details of this program, and temporal distributions of downward flux of C_T (S_{CT}) in Onondaga Lake over the 1987–1990 period are presented in Chapter 8. The temporal distribution of $V_{S/CT}$ $(m \cdot d^{-1})$ for the epilimnion of Onondaga Lake for the May–October interval of 1987 is presented in Figure 6.23. Summary statistics for $V_{S/CT}$ in the lake for 1987, 1988, and 1989 are presented in Table 6.10.

Recall the deposition of phytoplankton pigments (including phaeopigments) reflects the operation of both phytoplankton settling and zooplankton grazing (Welschmeyer and Lorenzen 1985). Clearly $V_{S/CT}$ varies strongly with time in Onondaga Lake (Figure 6.23). This dynamic character is apparently recurring, as indicated by the large C.V. values observed in each of the three years (Table 6.10). Phyto-

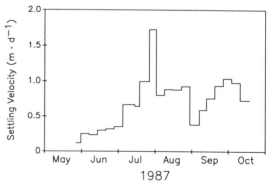

FIGURE 6.23. Temporal distribution of chlorophyll settling velocity in Onondaga Lake for 1987.

TABLE 6.10. Settling velocity of chlorophyll (C_T) from the epilimnion of Onondaga Lake.

Year	$V_{S/CT}$			
	$\bar{x}$ (m·d^{-1})	c.v.*	range (m·d^{-1})	n**
1987	0.69	0.54	0.12–1.73	21
1988	0.78	0.73	0.06–2.42	26
1989	0.73	0.48	0.12–1.44	28

* Coefficient of variation
** Number of sediment trap deployments

plankton settling velocities are influenced by the shape, size, density, and physiological state of phytoplankton (Bowie et al. 1985). Phytoplankton models often invoke a temporally uniform settling rate for phytoplankton (e.g., Bowie et al. 1985). The downward fluxes of C_T assessed with the traps, and therefore the distribution of $V_{S/CT}$, probably reflect significant, and variable contributions of phaeopigments (incorporated in zooplankton fecal pellets). Much of the variation in $V_{S/CT}$ is probably due to the dynamics in the activity of macrozooplankton. The average value of $V_{S/CT}$ remained essentially unchanged over the three year period (Table 6.10), though the average downward flux decreased by more than 20% (Chapter 8).

6.1.3.3 Nutrient Status

6.1.3.3.1 The Limiting Concept and Status with Respect to N

The nutrient present in the shortest supply relative to the needs of the algae is described as the "limiting" nutrient; its availability often is important in controlling growth. The two nutrients which most often are found to limit growth in aquatic systems are nitrogen and phosphorus. In the vast majority of freshwater environments phosphorus is the limiting nutrient (Harris 1986; Wetzel 1983). However, the cases of nitrogen limitation are particularly noteworthy because the proliferation of nuisance filamentous blue–green algae, capable of N_2 fixation, is often observed in the summer in these systems. Phosphorus is usually considered the limiting nutrient when the ratio of fixed inorganic (ammonia plus

nitrate) N to total soluble P is greater than 7–30:1 (7:1 Russell-Hunter 1970; 15:1 Redfield et al. 1963; 30:1 Rhee 1978). Nitrogen to P ratios observed in the euphotic zone of Onondaga Lake (29–290:1, over the 1988–1990 interval) indicate that P is presently the nutrient which is in shortest supply.

In extremely nutrient rich systems all nutrients may be present in concentrations above which limitation of growth occurs. Such a system is described as nutrient saturated. Recall Onondaga Lake was described as a nutrient-saturated system in the late 1970s (e.g., Field and Effler 1983). Under these conditions other environmental conditions, primarily light and temperature, regulate phytoplankton growth.

Nitrogen species that can be utilized to support algal growth (ammonia- and nitrate-N) are abundant in Onondaga Lake. Combined, or fixed inorganic nitrogen (summation of ammonia and nitrate-N) concentrations have remained well above $2\,mgN \cdot L^{-1}$ in the euphotic zone over the entire growing season in recent years (Chapter 5). Utilizing a reasonable half-saturation constant for these species ($K_N = 0.025\,mgN \cdot L^{-1}$; Bowie et al. 1985), these inorganic nitrogen concentrations correspond to essentially saturated (i.e., non-limiting) conditions, as the nitrogen limitation function exceeds a value of 0.99.

6.1.3.3.2 P Pools, APA, and the Extent of P Limitation

In this section the seasonality in the P physiology of the phytoplankton assemblage of the lake is examined, as it is manifested in ambient dissolved and stored P supplies and APA (also see Connors et al. 1996). The extent of P limitation experienced seasonally in the lake is evaluated within the frameworks of the Monod (Equation (6.5a)) and Droop (Equation (6.5b)) models. Kinetic coefficients adopted for the analyses ($K_p = 0.9\,\mu gP \cdot L^{-1}$ and $Q_o = 1.4\,\mu gPP \cdot mgPOC^{-1}$) are those determined by Auer et al. (1986) for Green Bay, Lake Michigan. The value of Q_o has been converted to be consistent with the C_T measure of biomass ($Q_o = $

$0.067\,\mu gPP \cdot \mu gChl^{-1}$; unpublished data S.W. Effler). The actual values for these kinetic coefficients for the phytoplankton assemblage of the lake may differ from those reported for Green Bay and may vary seasonally. However, the analysis serves to provide a comparative evaluation of the two kinetic frameworks, and in characterizing seasonal trends in P limitation. Shortcomings in the use of SRP, within the Monod framework, to quantify the extent of P limitation in the lake are demonstrated. Evidence for the uptake of DOP by the lake's phytoplankton, following the depletion of SRP, and its interplay with APA, are presented. Relationships between APA, SRP, and Q, and the utility of APA as an indicator or P limitation, are established.

Paired seasonal distributions of dissolved P species, Q, and C_T for 1989, 1990, 1991, and 1992 are presented in Figures 6.24a–d. The pattern determined for APA in the summer of 1992 has been included with the plot for Q in Figure 6.24d(2). Despite year-to-year differences in the temporal details of these distributions, associated with interannual variations in the timing and intensity of phytoplankton blooms, several recurring patterns emerge (Figure 6.24). Concentrations of DOP are initially high in spring (e.g., $\sim 50\,\mu gP \cdot L^{-1}$ Figures 6.24a(1), b(1), c(1), and d(1)). SRP concentrations are high at the time of ice-out (usually early to mid-March). Major depletions in SRP associated with the spring phytoplankton bloom (e.g., Figure 6.24a(3), b(3), c(3), and d(3)), were already manifested by the start of monitoring in April 1990 (Figure 6.24b(1)), 1991 (Figure 6.24c(1)), and 1992 (Figure 6.24d(1)). Note that while the timing of the clearing event in 1989 (Figure 6.24a) was apparently well matched by a major depletion of SRP, this was not the case in the other three years. DOP levels remained relatively high in all four years during periods in which clearing events occurred. This supports the view that observed reductions in algal biomass are a manifestation of grazing pressure (see subsequent session), not nutrient limitation (Connors et al. 1996).

Abrupt increases in SRP are quite evident during the clearing events (Figure 6.24a(1), b(1), c(1), and d(1)). These increases presumably reflect a reduction in phytoplankton uptake, with continued external loading, and perhaps an elevated level of recycle within the upper waters (Lehman 1980; Auer et al. 1990). The cellular P content (Q) of the remaining standing crop apparently increases during clearing events (Figure 6.24a(2), b(2), c(2), and d(2)), indicating luxury uptake by the phytoplankton. Though the possibility that this may reflect a larger relative contribution to the PP pool from non-phytoplankton components cannot be discounted. Distinct blooms occurred following clearing events in this high nutrient, high light environment in three of the four years.

Phytoplankton biomass remains relatively high from mid- to late summer (Figure 6.24a(3), b(3), c(3), and d(3)). Decreases are generally not as dramatic or as long as the earlier clearing events. SRP concentrations remain relatively low (mostly $<2\,\mu gP \cdot L^{-1}$ during this period), as do the values of Q (mostly $<2\,\mu gPP \cdot \mu gChl^{-1}$). The DOP pool decreases over this period of depleted SRP, suggesting algal utilization to sustain the elevated levels of phytoplankton. Clear indications of uptake from the DOP pool were also seen during the spring blooms of 1991 (Figure 6.24c(1)) and 1992 (Figure 6.24d(1)). Several researchers have indicated that both SRP and DOP can be utilized by algae (Bentzen et al. 1992; Cotner and Wetzel 1992; Pick 1987). The leveling off or increase in DOP in late summer and early fall may reflect inputs of SRP and DOP originating through entrainment of P-rich hypolimnetic water (e.g., Effler et al. 1986), depletion of the labile portion of the DOP pool, and reduced demand as the environmental conditions supporting phytoplankton growth (e.g., light and temperature) become less favorable.

The APA pattern in 1992 (Figure 6.24d(2)) provides more compelling evidence of the utilization of the DOP pool by the phytoplankton assemblages of Onondaga Lake. The pattern complements the temporal details of the distributions of the dissolved fractions. APA increases in mid-July with the onset of low SRP concentrations, and DOP concen-

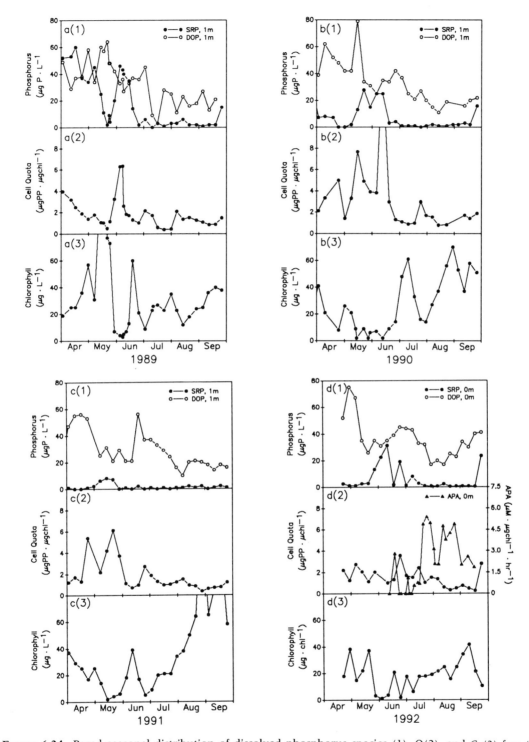

FIGURE 6.24. Pared seasonal distribution of dissolved phosphorus species (1), Q(2), and C$_T$(3) for: (a) 1989, (b) 1990, (c) 1991, and (d) 1992 (modified from Connors et al. 1996).

trations decrease, consistent with increased enzyme activity and subsequent uptake of cleaved ortho-P. The lower Q levels maintained in late summer, when the DOP pool is utilized, suggest that the kinetics of uptake for this pool are less favorable than for SRP. This is consistent with the energetics of the process, as DOP utilization requires the extra step of hydrolysis prior to uptake. The increase in the DOP pool in September parallels the decrease in APA (Figure 6.24d(1) and (2)). Ahn et al. (1993) reported a similar seasonality in APA in Lake Soyang, Korea; activity was low in spring and increased over the summer. High levels of APA were attributed to a shortage of ambient inorganic P and low stored P reserves (Ahn et al. 1993).

Temporal distributions of estimates of G(N) according to both the Monod and Droop expressions are presented for each of the four years in Figure 6.25a–d. The Monod framework depicts a much greater degree of limitation annually than the Droop model (Figure 6.25) and implies a strong seasonality in G(N) which is inconsistent with the dynamics of the phytoplankton standing crop (Figure 6.24a(3), b(3), c(3), and d(3)) and with measures of the flux of algal biomass from the epilimnion (Chapter 8). For example, the Monod model predicts low values of G(N) during extended periods of high C_T. Further, the downward fluxes of C_T and POC (Chapter 8), measures of primary production, remained high through the period of low SRP in the summer of 1989, consistent with conditions of near nutrient-saturated growth. It is noteworthy that the value of K_p adopted here is near the lower end of the range commonly used in phytoplankton models (Bowie et al. 1985). Even greater degrees of limitation would be predicted if values at the upper end of the range were applied.

The inadequacy of the Monod model and SRP in representing the status of P limitation for the phytoplankton of Onondaga Lake as developed above (Figure 6.25) is further supported by the temporal dynamics of the various P pools (Figure 6.24). There is strong evidence that DOP is utilized (SRP is not the sole measure of P availability) and that algal

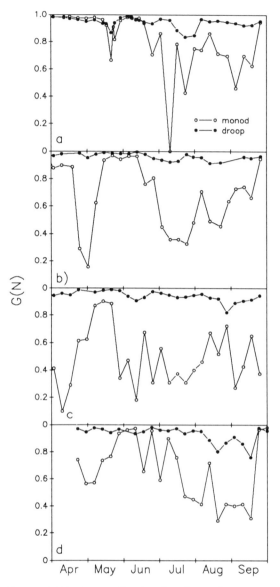

FIGURE 6.25. Seasonality in phosphorus limitation in Onondaga Lake according to Monod and Droop relationships: (a) 1989, (b) 1990, (c) 1991, and (d) 1992 (modified from Connors et al. 1996).

growth is uncoupled from water column SRP (cellular equilibrium is not maintained). Fundamental limitations of the Monod model prevent it from providing a genuine reflection of reality in cases where the assumptions of "cellular equilibrium" (DiToro 1980) are not met. The G(N) distributions which emerge from application of the Droop framework, based instead on the internal nutrient status of the phytoplankton, are by comparison more realistic (Figure 6.25). Internal P concentrations are sufficient to sustain near maximum rates of algal growth over the entire spring to early fall interval, even when SRP becomes depleted (Figure 6.25). Accordingly, the Droop model suggests that P limitation does commonly occur in Onondaga Lake over the spring–fall interval, but that the degree of limitation is quite small.

The response of APA to changes in SRP and Q in Onondaga Lake is examined in Figure 6.26. APA was a strong function of SRP (Figure 6.26a). The shape of the curve suggests initiation of APA at an SRP concentration of about 1 to $2 \mu gP \cdot L^{-1}$, with marked suppression above $2 \mu gP \cdot L^{-1}$. These findings are consistent with those of Pettersson (1980) where

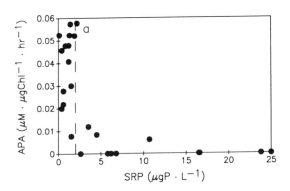

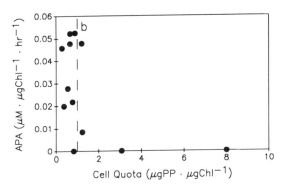

FIGURE 6.26. Relationships for the surface waters of Onondaga Lake between APA and: (**a**) SRP, and (**b**) Q. Dotted lines indicate the apparent threshold for suppression of APA (modified from Connors et al. 1996).

APA was observed to increase at ortho-P concentrations $<1 \mu gP \cdot L^{-1}$.

APA is clearly related to Q as well and is apparently initiated in response to a reduction in the internal P reserves of the algae (Figure 6.26b). The inflection point in Figure 6.26b suggests initiation of APA at a Q value of about $1 \mu g PP \cdot \mu g Chl^{-1}$. Gage and Gorham (1985) and Pettersson (1980) noted a similar inflection in the APA-Q relationship, however, direct comparisons of their results with those presented here are not possible due to the differences in the units of expression of Q. Gage and Gorham (1985) proposed that the inflection point be utilized as a warning level of incipient P starvation.

The relationship between APA and Q should be considered the more fundamental of the two, consistent with concepts of nutrient limitation which propose that the stored nutrient pool size regulates growth (e.g., Auer and Canale 1982; Droop 1968, 1973). Wynne (1981) reported that phosphatases in Lake Kinneret were controlled by changes in stored P and/or by other metabolic processes, rather than by ambient dissolved P concentrations. This was also recognized by Gage and Gorham (1985) for Minnesota lakes, where reductions in APA were used to define the point at which cells began to accumulate surplus cellular P.

For water quality managers, the analysis of APA may offer the best information on conditions of resource limitation. Engendered in response to reduced internal nutrient pool size and attendant reductions in growth rate, the onset of APA signals incipient nutrient stress. The onset of APA (inflection points in Figure 6.26a, $1-2 \mu gSRP \cdot L^{-1}$, and 6.26b, $1 \mu g PP \cdot \mu g Chl^{-1}$), within the context of Monod and Droop kinetics, is portrayed in Figure 6.27a–c. Moving to the left in these figures, away from conditions of nutrient saturation, it can be seen that the onset of APA occurs where: (1) the growth response of the Monod model becomes linear (Figure 6.27a), (2) surplus internal nutrient stores are being accumulated (Figure 6.27b), and (3) the elbow of the growth response curve for the Droop model is approached (Figure 6.27c). In each case, further reductions in nutrient availability

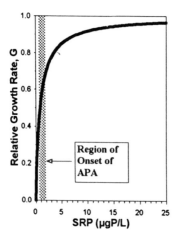

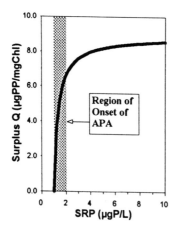

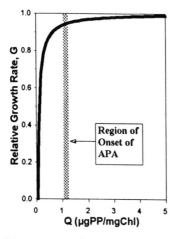

FIGURE 6.27. Onset of APA as related to: (a) growth rate, Monod kinetics, (b) accumulation of surplus nutrient stores, and (c) growth rate, Droop kinetics (from Connors et al. 1996).

may be expected to yield proportional reductions in algal growth. This is a criterion for nutrient limitation which is of value to water quality managers and which may be assayed with some degree of confidence.

Applied here as a basis for assessing nutrient limitation in Onondaga Lake, Figure 6.27 suggests that a modest level of limitation (~75–90% of the maximum specific growth rate) is achieved only in late summer and that the degree of limitation in late summer has increased over the 1989–1992 period (Figure 6.25). Undoubtedly the availability of P to phytoplankton in Onondaga Lake was historically greater, e.g., higher concentrations of dissolved and stored P and more nearly saturated growth at the much greater external loading rates that once prevailed (Chapter 3). Major reductions in loading from METRO since the late 1960s have altered the nutrient status of the system, moving from conditions of saturation and near-saturation toward a modest degree of nutrient limitation. This is consistent with observations that only slight changes in related water quality characteristics (algal standing crop, oxygen resources) have occurred over the period of loading reductions. Apparently further reductions in P loading to Onondaga Lake will be necessary to achieve a substantial degree of nutrient limitation with the related water quality benefits.

Auer et al. (1990) utilized the system-specific kinetic coefficients (Table 9; Storey 1989; Storey et al. 1993), measuring fluxes and related ambient conditions to assess seasonal source and sink processes for phytoplankton for the spring to fall interval of 1987, to identify the important processes regulating the dynamics of phytoplankton standing crop in the lake. The spring phytoplankton bloom was demonstrated to have developed in response to nutrient-saturated conditions, increasing temperature and adequate light. However, abrupt decreases in biomass were attributed to both P limitation and zooplankton grazing, because the Monod relationship was utilized in the analysis. Thus the role of P limitation in regulating the dynamics of phytoplankton biomass was overstated and that of zooplankton grazing understated.

6.2 Zooplankton

6.2.1 Background

Zooplankton are the animal components of the plankton. Two groups of organisms generally dominate the zooplankton of lakes; the Rotifera, and the Crustacea. Of the Crustacea, two types are most common; the Cladocera and Copepoda. Rotifera is a class of the phylum Aschelminthes and includes about 100 completely planktonic species (Ruttner-Kolisko 1974). Planktonic rotifers feed primarily by sedimenting small particles into their mouth orifice by means of their coronal cilia (Edmondson 1959). The constant motion of this wheel of cilia has lead to their common name: "wheel animals." Most rotifers are between 40 and 500 µm in length. The Cladocera is an order of the crustacean subclass Branchipoda. Cladocerans have a distinct head and body covered by a bivalve carapace. Most cladocerans feed by filtering particles from water circulated through the valves by action of their legs. Cladocera generally range in size from ~0.2 to 3.0 mm (Pennak 1978). The Copepoda comprise a subclass of the Crustacea with three suborders of free-living representatives, two of which, Calanoida and Cyclopoida, are planktonic. Copepods are clearly segmented with generally elongate, cylindrical bodies divided into a distinct head, thorax, and abdomen. Calanoids feed primarily by filtration, using their antennae to produce a current past their mouthparts. Calanoid copepods are capable of a great deal of feeding selectivity. Cyclopoid copepods are raptorial feeders, seizing prey with their mouth parts. Copepods are generally from 0.3 to 3.0 mm in length (Pennak 1978).

Zooplankton play an important role in the dynamics of lake systems. The maximum biomass at each trophic level of lake systems is determined by bottom-up factors—ultimately by the supply of critical nutrients (McQueen et al. 1986). Thus, the upper bound for zooplankton biomass is established by phytoplankton biomass. However, the zooplankton biomass actually realized may be further limited, and the composition of the community greatly influenced, by the composition of the phytoplankton community, water quality and top-down (i.e., predation by planktivorous fish) factors. The structure of the herbivorous zooplankton community can also have significant impacts on the composition and biomass of the phytoplankton community (Porter 1977; McCauley and Briand 1979), and thus may affect transparency, an important aspect of the public's perception of water quality.

A number of the water quality problems of Onondaga Lake identified in this volume are known to influence the zooplankton community, including: high concentrations of toxics (e.g., free ammonia and heavy metals), salinity, inorganic particles, and low dissolved oxygen concentrations. The importance of competition and predation in structuring zooplankton communities is well documented (e.g., Hall et al. 1976; Neill 1984; Siegfried 1987; Schneider 1990). Phytoplankton community composition may affect zooplankton grazing and assimilation efficiencies (Gliwicz and Siedlar 1980), and thus affect the outcome of interspecifiic competition (Wetzel 1983). Predation by planktivorous fish can play an important role in limiting the biomass of, and structuring both the size and species composition of, the zooplankton community (Brooks and Dodson 1965; Hutchinson 1971; McQueen et al. 1986; Siegfried 1987; O'Gorman et al. 1991).

Here we document and analyze the major changes in the zooplankton community of Onondaga Lake which have occurred over the period 1969–1989. This analysis includes earlier findings (i.e., Waterman 1971; Meyer and Effler 1980) but emphasizes previously unreported results of monitoring over the 1979–1989 interval. Principal components analysis is used to examine the changes in the structure of the zooplankton community of the lake and relate them to limnological variables. Factors associated with the major changes in the community are identified.

6.2.2 Previous Studies/Methods

The first systematic study of the Onondaga Lake zooplankton community was conducted by Waterman (1971). Waterman reported the results of 12 months of sampling during 1969. Zooplankton were collected by vertical tows with a 23 cm diameter No. 20 mesh (73μm) net. Weekly (summer) to monthly (winter) samples were collected at deep stations in both the south and north basins.

The next systematic study of the zooplankton community of Onondaga Lake took place in 1978 (Meyer and Effler 1980). This study was designed specifically to assess community changes since Waterman's 1969 study. Zooplankton were collected by vertical tows from 10 m to the surface with a 12 cm diameter No. 20 mesh Wisconsin style plankton net. Samples were collected weekly (and sometimes more frequently) from the same sites as in Waterman's (1971) study, from April through October. Zooplankton were enumerated at 30X, using a Bogarov counting trap. At least 200 individuals of each of the dominant species were counted in each sample (Meyer and Effler 1980). Zooplankton population estimates for the north and south basins were not significantly different (Meyer 1979), so the average of the two basins was presented (Meyer and Effler 1980).

The sampling program established in 1978 was continued at the south basin station through 1981, renewed in 1985, and continued through 1989. The frequency of sampling varied somewhat between years but results are generally available for weekly intervals from May through August of each year except for 1985 and 1987, when samples were collected monthly and biweekly, respectively. Samples were field preserved in a 4% formalin solution. Quantitative analysis of samples followed Meyer and Effler (1980). Density estimates were converted to biomass estimates through the use of conversion factors developed from published and empirically determined values (Dumont et al. 1975; Makarewicz and Likens 1979; Yan and Strus 1980; Siegfried unpublished).

Towed nets generally underestimate zooplankton abundance (Rawson 1956; Likens and Gilbert 1970). Net efficiency is assumed to to be consistent from year to year of the Onondaga Lake study, and thus no correction factor is necessary for this comparative analysis.

6.2.3 Zooplankton Species Assemblage

The number of zooplankton species reported from Onondaga Lake has increased significantly over the period of record (Table 6.11). Much of this increase is attributable simply to the increased collection effort. Most of the additions to the species assemblage represent detection of a rare species, however, a number of species new to the lake have become dominant members of the community. *Diaphanosoma leuchtenbergianum* is a good case in point; this cladoceran was not reported by Waterman (1971) or by Meyer and Effler (1980) and did not appear in zooplankton samples until 1986, when it was first detected. It subsequently became a dominant of the fall community in the same year. It was consistently found to be a dominant member of the community in each of the subsequent years of study.

The cladoceran assemblage of Onondaga Lake has changed dramatically over the period of record. *Ceriodaphnia quadrangula* and *Daphnia similis* were the only cladocerans reported as community dominants in Onondaga Lake in 1969 (Waterman 1971). *Ceriodaphnia* continued to be a dominant of the community throughout much of the period of record, but *Daphnia* was not even found in the lake during some study intervals. The reported occurrence of *Daphnia similis*, a cladoceran typical of highly saline environments, may be an artifact. Although reported to occur in Onondaga Lake by Waterman (1971), it has not been reported in subsequent studies (Meyer and Effler 1980; present study). Brooks (1957) considered *D. similis* a western species and did not report it from the New York region. Meyer and Effler (1980) suggested that the earlier report of *D. similis* was most likely due to a taxonomic misidentification of *D. pulex*. However Water-

TABLE 6.11. Zooplankton assemblage of Onondaga Lake, 1969–1989 (r = rare, X = common).

Taxon	1969[1]	1978[2]	1979–1981[3]	1986–1989[3]
CLADOCERA				
Alona affinis		r		
Bosmina longirostris	r	r	r	X
Ceriodaphnia quadrangula	X	X	X	X
Chydorus sphaericus	r	r		
Daphnia galeata				X
Daphnia pulex (pulicaria)	X	X		X
Daphnia similus	X			
Diaphanosoma leuchtenbergianum				X
Eubosmina coregoni				r
Leptodora kindtii				r
COPEPODA				
Diaptomus sicilis	r	r		
Diaptomus siciloides				X
Cyclops bicuspidatus	X	X		r
Cyclops vernalis	X	X	X	X
Mesocyclops edax	r			
ROTIERA				
Ascomorpha sp.			X	
Asplanchna sp.	r	X		
Brachionus angularis				X
Brachionus calyiflorus		X	X	X
Brachionus plicatus		r		
Brachionus variabilis				X
Brachionus sp.	X	X		
Filinia longiseta	r	r		X
Filinia terminalis				X
Kellicottia bostoniensis			r	X
Kellicottia longispina			X	X
Keratella c. cochlearis	r		r	r
Keratella c. robusta			r	X
Keratella c. tecta			r	
Keratella hiemalis	X	X		
Keratella quadrata		X	r	X
Keratella testudo			X	X
Keratella valga		r		
Nothalca squamula				r
Ploesoma truncatum				r
Polyarthra remata			X	X
Polyarthra sp.	X	X		
Synchaeta sp.		X		
Trichocerca multicrinnis				r

1 Waterman 1971
2 Meyer and Effler 1980
3 This analysis

man (1971) did report *D. pulex* from Onondaga Lake as well as *D. similis*, suggesting a recognition of two morphologically distinct species. Since voucher specimens from the 1969 study were not retained, it is not possible to confirm the occurrence of *D. similis* or any other, unidentified species, in Onondaga Lake during the 1969 study.

Daphnia pulex was reported from Onondaga Lake in the 1969 and 1978 studies (Waterman 1971; Meyer and Effler 1980) but was not present in subsequent collections. Recent research has recognized "*D. pulex*" as a complex of at least two species (Brandlova et al. 1972). One, *Daphnia pulex*, is generally an inhabitant of small ponded waters while the other, the morphologically similar *Daphnia pulicaria*, is more typical of large bodies of water (Brandlova et al. 1972; Siegfried and Sutherland 1992). Recent determinations indicate that *D. pulicaria* has been present in Onondaga Lake during the 1986–1989 interval. The earlier reports of *D. pulex* in Onondaga Lake should probably be considered *D. pulicaria*.

Two recent additions to the cladocera of Onondaga Lake, *Eubosmina coregoni* and *Leptodora kindtii*, were found in 1989 samples. Neither species was reported from the lake in earlier studies (Table 6.11).

The copepod assemblage has also changed during the period of record. *Cyclops vernalis* has been the only copepod species present throughout the monitored period. Other cyclopoid species occurred more sporadically. The calanoid copepod, *Diaptomus sicilis*, was reported as a rare species in both 1969 and 1978 (Waterman 1971; Meyer and Effler 1980) but was not observed in subsequent years. No calanoid copepods were recorded from Onondaga Lake during the 1979–1981 study interval. In 1986 *Diaptomus siciloides* was first observed in the lake and by 1987 had become one of the zooplankton community dominants.

The rotifer assemblage reported for Onondaga Lake has expanded from six species in 1969 to the current 24 species. Most of the additions prior to 1986 represent the detection of rare species (Table 6.11). A number of rotifers first observed in the lake during the 1986–1989 study interval have become dominants of the rotifer assemblage. The increase in rotifer species richness may reflect recent improvements in water quality conditions.

In spite of recent increases in species richness the present zooplankton species assemblage of Onondaga Lake is somewhat depa-

uperate compared to other lakes of the region; e.g., Oneida Lake has at least ten copepod species (Mills et al. 1978). Other large lakes of New York have an assemblage of crustaceans similar to that of Oneida Lake, i.e., 6 to 10 copepod and 8 to 12 cladoceran species (Siegfried and Quinn 1987; Siegfried unpublished). Circumneutral lakes of the Adirondack region of New York State averaged 10 crustacean and 16 rotifer species, and as many as 14 crustacean and 21 rotifer species, in single mid-summer collections (Siegfried et al. 1984, 1989; Sutherland et al. 1990). Thus, single visits to many lakes of New York have yielded more zooplankton species than ten years of collections at Onondaga Lake. Even highly acidic lakes of the Adirondack region, when sampled intensively, have yielded more than 20 rotifer species (Siegfried 1991) while large circumneutral lakes of New York may have as many as 50 rotifer species present during an annual cycle (Siegfried and Quinn 1987; Siegfried unpublished).

The low zooplankton species richness of Onondaga Lake is not attributable to the eutrophic condition of the lake. Highly eutrophic lakes often have more than twenty crustacean and more than twenty rotifer species present in an annual cycle of the plankton assemblage (Siegfried and Kopache 1984). Instead, the low species richness of the lake is probably attributable to stresses associated with its polluted state, a frequently reported response of the zooplankton community to various forms of pollution (e.g., Sprules 1975; Siegfried et al. 1989). Potential interactions between various aspects of the lake's water quality and the zooplankton community are addressed subsequently.

6.2.4 Community Dynamics: Seasonal and Annual

6.2.4.1 Total Biomass and Composition

Major changes in the composition and biomass of zooplankton have occurred over the period of record (Figure 6.28). Multiple biomass peaks were evident in each year of study. Biomass maxima were usually recorded in-

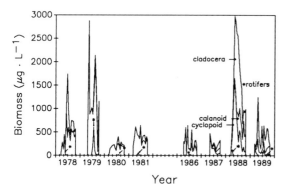

FIGURE 6.28. Distribution of estimated zooplankton biomass in Onondaga Lake over the 1978–1989 interval, partitioned according to zooplankton group.

June or early July. The zooplankton community of Onondaga Lake in 1969 was dominated by rotifers through May (Waterman 1971). Dominance shifted to cyclopoid copepods in early June and then to cladocerans from late June through August. Our estimate of mean June–August 1969 zooplankton biomass was only ~125 µg · L^{-1}; the lowest for the period of record. The maximum biomass (368 µg · L^{-1}) was attained in July, with a secondary peak in late August (265 µg · L^{-1}). Extremely low zooplankton densities were reported for the January–April interval and December 1969, that were attributed to low temperatures and reduced phytoplankton abundance (Waterman 1971).

Zooplankton community composition and biomass levels were dramatically different in 1978 compared to 1969. Although rotifers again dominated the spring community, dominance by cladocera was sharply reduced. The abundance of large daphnids was dramatically reduced in 1978; from a mean of ~10 · L^{-1} in 1969 to less than 1 · L^{-1} in 1978 (Meyer and Effler 1980). Meyer and Effler (1980) also reported a progressive decrease in mean daphnid size over the 1970–1975 interval. Although small cladocera were present in the plankton of Onondaga Lake from June through October 1978, cyclopoid copepods dominated the biomass on most collection dates (Meyer and Effler 1980; Figure 6.28).

The mean summer zooplankton biomass was 625 µg · L^{-1} in 1978, with a late June peak of almost 1800 µg · L^{-1} (Figure 6.28).

Dominance by cyclopoid copepods was even more pronounced from 1979 through 1981 (Figure 6.28). Small cladocera were usually present in July and August but generally did not dominate total community biomass. Rotifers were significant only in July over the 1979–1981 interval. Rotifers are often the first group of zooplankton to increase population levels in the spring. It is possible that a sampling program beginning in May would miss an early spring population increase.

Substantial variation in zooplankton biomass was observed in the 1979–1981 interval (Figure 6.28). The mean biomass for the May–August period of 1979 was 1340 µg · L^{-1}, the second highest over the period of record. Mean biomass was much lower in 1980 and 1981, 220 µg · L^{-1} and 420 µg · L^{-1}, respectively.

Limited samples available for 1985 indicate a zooplankton community similar to that documented for the 1978–1981 interval. Zooplankton community composition during 1986 was also generally similar to that recorded in 1979–1981; dominated by cyclopoid copepods throughout the year but with small cladocerans present at significant levels from mid-July through August and rotifers in July (Figure 6.28). Biomass levels remained low in 1986; the peak was about 650 µg · 1^{-1} and the mean was 280 µg · L^{-1} (Figure 6.28).

Zooplankton community composition was dramatically different in the 1987–1989 period, compared to earlier intervals. Beginning in 1987, and continuing through 1989, cladocera and a calanoid copepod increasingly dominated biomass (Figure 6.28). Cyclopoid copepods were dominant in the spring of 1987 and 1988 but were replaced by the calanoid copepod and cladocera by mid-July in 1987 and by mid-June in 1988 (Figure 6.28). Cladocera and calanoid copepods dominated zooplankton biomass throughout much of 1989. There has been a distinct shift to dominance by large cladocera, i.e. *Daphnia* spp., which represented less than 1% of zooplankton biomass in 1986, but 39%, 43%, and 57% in 1987, 1988, and 1989, respectively.

Rotifers were significant dominants of the spring communities of 1988 and 1989. Rotifers may have also been important in 1987 (sampling did not begin until June).

Zooplankton community biomass remained low through the 1987 monitoring period (mean of 285 μg · L^{-1}). The greatest zooplankton standing crops were observed in 1988; mean of 1612 μg · L^{-1} and maximum of 3070 μg · L^{-1}. Mean biomass decreased in 1989 to 430 μg · L^{-1}. Biomass maxima occurred in late May to early June of 1986–1989.

No trend in summer zooplankton biomass was evident over the 1978–1989 period. Although mean annual zooplankton biomass fluctuated widely from year to year, the mean for the 1978–1981 and 1986–1989 study periods were remarkably similar, i.e., ~650 μg · L^{-1}. This suggests a system of relatively constant productive capacity. Chlorophyll data presented earlier in this chapter support this observation. Chlorophyll concentrations have varied greatly over the spring–fall interval; but the average concentrations for the different years have not varied greatly. The mean chlorophyll concentration was 38 μg · L^{-1} for the 1978 study period, and ranged between 28 and 32 μg · L^{-1} for the 1986–1989 interval.

6.2.4.2 Selected Components

Seasonal and annual dynamics of rotifers in Onondaga Lake have been highly variable (Figure 6.29a). Population maxima occurred in early June to early July from 1978 through 1981 and in 1986. No distinct population peak was evident in 1987 but prominent spring population maxima were evident in 1988 and 1989. *Keratella* spp., primarily *K. testudo*, were generally the most abundant rotifers of the Onondaga Lake zooplankton. *Keratella* dominated the rotifer community on nearly every date from 1979 through 1981, and from 1986 through 1989. *Brachionus* spp. were dominant in May of 1978 and a subdominant in July of 1979. *Brachionus* was not a significant component of the community again until 1989, when *B. calyiflorus* was a codominant of the

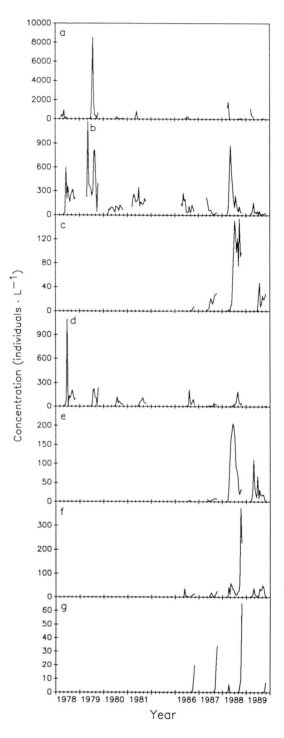

FIGURE 6.29. Distributions of concentrations of selected zooplankton in Onondaga lake over the 1978–1989 interval: (**a**) Rotifera, (**b**) *Cyclops vernalis*, (**c**) *Diaptomus siciloides*, (**d**) *Ceriodaphnia quadrangula*, (**e**) *Daphnia* spp., (**f**) *Bosmina longirostris*, and (**g**) *Diaphanasoma leuchtenbergianum* (modified from Siegfried et al. 1996).

spring rotifer peak. *Polyarthra remata* was consistently represented in the rotifer community through July and August of most years. A small population of *Kellicottia longispina* was present in Onondaga Lake in both 1979 and 1981. In 1986 it occurred as a co-dominant during the June–July rotifer peak. Although *Filinia longiseta* was consistently reported from Onondaga Lake, it was a dominant component of the community only during the 1988 and 1989 spring peaks.

The seasonal dynamics of the cyclopoid copepods were generally similar each year of study, i.e., usually a spring peak in abundance and population levels fluctuating widely thereafter (Figure 6.29b; plotted as the sum of adults and copepodids). *Cyclops vernalis* has been the most abundant cyclopoid copepod throughout the period of record, typically accounting for more than 95% of the cyclopoid population. Late spring or early summer population maxima appear to be typical for *C. vernalis* (Meyer and Effler 1980). The adults of this species are largely carnivorous. The other cyclopoids of Onondaga Lake were rarely encountered.

Calanoid copepods were virtually absent from 1978 through 1981 (Figure 6.28). *Diaptomus siciloides* appeared in samples in late July 1986 and was generally abundant in July and August in the 1987–1989 interval (Figure 6.29c). Two abundance peaks were evident each of these years; one in early July, and another in August (Figure 6.29c).

Ceriodaphnia quadrangula was very abundant in 1978, with peak populations of more than $1000 \cdot L^{-1}$. Concentrations of this species remained lower in other years (Figure 6.29d). Populations of *Ceriodaphnia* were particularly low in 1987 and 1989. Population maxima of about $200 \cdot L^{-1}$ were generally evident in June–July of other years.

Other cladocera were generally abundant only in the 1986–1989 samples. *Daphnia*, the largest of the herbivorous cladocera of Onondaga Lake, although present in 1978–1981 and 1986, and slightly more abundant in 1987, was very abundant only in 1988 and 1989, when it dominated the crustacean biomass (Figure 6.29e). *Bosmina longirostris* (Figure 6.29f) and *Diaphanosoma leuchtenbergianum* (Figure 6.29g) were abundant only

over the 1986–1989 interval. *Diaphanosoma* was not observed in the lake until 1986, and *Bosmina* occurred only sporadically prior to 1986. *Bosmina* populations usually peaked in late spring to early summer, except in 1988 when a large late summer peak was evident. Late summer to fall peaks in *Bosmina* populations are typical in New York lakes (Siegfried and Quinn 1987; Siegfried unpublished). The occurrence of *Diaphanosoma* in Onondaga Lake was limited to late summer to fall population maxima, except in 1988.

6.2.4.3 *Principal Components Analysis*

The structure of the Onondaga Lake zooplankton community was examined by principal components analysis. Relative species biomass values were arcsine transformed and the principal components extracted from the correlation matrix. Patterns in zooplankton community composition were related to selected limnological variables by calculating Pearson correlation coefficients between collection scores on the principal components and limnological variables. The principal components analysis expresses the variation between zooplankton samples by extracting components, or factors, that explain a large part of the underlying variance and covariance of the original samples (Sokol and Rohlf 1981). The analysis of component scores then seeks to identify possible causal factors for the biological variability.

Four components were retained (i.e., eigenvalues >1.0) in the principal component analysis (Table 6.12). These components explained 70% of the variability in zooplankton community composition. Species loadings on the principal components indicate that the most important variation in community structure is between samples dominated by *Daphnia* and *Diaptomus* and those dominated by cyclopoid copepods (Figure 6.30). This component explains 27% of the variation in community composition (Table 6.12). The second most important variation was between samples dominated by the small cladocerans, *Diaphanosoma leuchtenbergianum* and *Ceriodaphnia quadrangula*, and those dominated by rotifers (Figure 6.30). This second principal component explains another 16% of the variation

TABLE 6.12. Structure of Onondaga Lake zooplankton community principal components, relative biomass, summer samples, 1978–1981 and 1986–1989.

Zooplankton species	Pearson correlations with species			
	PC1	PC2	PC3	PC4
Daphnia	0.729	−0.203	−0.451	−0.239
Bosmina	0.282	0.038	0.230	0.738
Diaphanosoma	0.375	0.503	0.261	0.269
Ceriodaphnia	−0.241	0.553	0.364	−0.502
Cyclops vernallis	−0.861	0.047	−0.354	0.285
Diaptomus siciliodes	0.757	0.260	−0.132	−0.005
nauplii	−0.046	0.438	0.199	−0.013
rotifers	0.082	−0.670	0.700	−0.089
Eigenvalue	2.133	1.312	1.128	1.015
Percent	26.66	16.40	14.10	12.69
Cumulative	26.66	43.06	57.16	69.85

ing mean monthly scores for the first two principal components (Figure 6.31). The mean monthly scores for 1978–1981, with the exception of May 1978, are closely clustered, with negative scores on principal component 1 and generally positive scores on the second component (Figure 6.31a). The close clustering of principal component scores reflects a community dominated primarily by cyclopoid copepods throughout each year of study. The one exception to the clustering of scores, the May 1978 score, reflects the complete domin-

in zooplankton community composition. The third component retained in the principal components model contrasts samples dominated by *Daphnia* with those dominated by rotifers, while the fourth component contrasts *Bosmina* with *Ceriodaphnia* dominated samples.

Seasonal and annual variations in community composition were examined by plott-

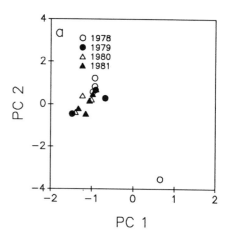

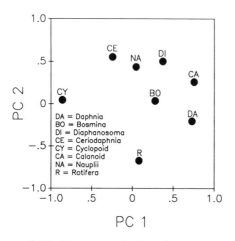

FIGURE 6.30. Structure of Onondaga Lake zooplankton community, first two principal components, relative biomass, summer samples for the 1978–1981 and 1986–1989 intervals (modified from Siegfried et al. 1996).

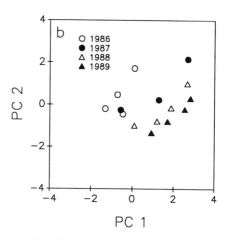

FIGURE 6.31. Plot of mean monthly zooplankton assemblage scores on first two principal components: (**a**) 1978–1981, and (**b**) 1986–1989 (modified from Siegfried et al. 1996).

ance of rotifers and thus strongly negative scores on the second principal component (Figure 6.31a).

Plots of mean monthly scores for collections of 1986–1989 (Figure 6.31b) are quite different from those of 1978–1981 (Figure 6.31a). There are distinct trends in scores from year to year and seasonally within each year. Scores become increasingly positive on both principal components from May through August of each year. This reflects the increased dominance of *Daphnia, Diaptomus,* and *Diaphanosoma* from May to August. The shift from primarily a cyclopoid dominated community in 1986 to progressively greater dominance by cladocerans and calanoid copepods is reflected in the progressive shift toward more positive scores on the first principal component.

Scores on the first principal component were strongly correlated (P < 0.0001) with total turbidity and Secchi depth transparency; negatively with turbidity and positively with transparency (Table 6.13). Significant negative correlations were also evident with chlorophyll concentration and calcite turbidity (Table 6.13). These correlations reflect a decrease in turbidity and accompanying increase in transparency associated with the shift to a zooplankton community dominated by *Daphnia* and calanoid copepods. The negative correlation with chlorophyll may reflect the seasonal shift from cyclopoid dominance in the spring, when chlorophyll levels are high, to periods of cladocera dominance, when lower chlorophyll concentrations prevail. These correlations reflect changes in water quality parameters accompanying, but not necessarily caused by, shifts in zooplankton community composition.

Scores on the second principal component are highly negatively correlated (P < 0.0001) with dissolved oxygen parameters and positively with temperature. This is associated with the seasonal shift from rotifer dominance in the spring, when water temperatures are low and dissolved oxygen concentrations are high, to dominance by small cladocera later in the year when water temperature is higher and dissolved oxygen is lower. The relationship with oxygen is not water quality based in this instance, but rather driven by temperature regulated changes in the solubility of oxygen. Scores on the third principal component reflect a seasonal shift from rotifers to cladocera, as well as an annual shift toward *Daphnia* dominance as transparency improved between 1978 and 1989.

TABLE 6.13. Pearson correlation coefficients for Onondaga Lake zooplankton community principal components scores and limnological parameters (numbers in parenthesis-degrees of freedom).

Parameter	PC1	PC2	PC3	PC4
Total turbidity (49)	−0.62****	−0.04	0.30	0.31
Calcite turbidity (49)	−0.29*	0.11	0.28	0.09
Secchi depth (108)	0.57****	−0.21*	−0.26**	−0.24
DO − 10 m (108)	−0.09	−0.38****	0.13	0.12
DO − 15 m (108)	0.10	−0.44****	0.31**	−0.06
$DO_z = 0\,mg \cdot L^{-1}$ (108)	0.12	−0.49****	0.22*	−0.06
$DO_z < 5\,mg \cdot L^{-1}$ (108)	0.01	−0.37***	0.17	0.15
Stratif.z (107)	−0.14	−0.17	0.42****	−0.05
Chl a − 1 m (66)	−0.35**	−0.31*	0.15	0.12
Chl a − 4 m (58)	−0.26*	0.21	0.15	0.21
Chl a − 10 m (44)	−0.35*	0.15	0.27	0.20
Mean Temp. 0−10 m (107)	0.05	0.53****	−0.06	−0.17

**** P < 0.001
*** P < 0.001
** P < 0.01
* P < 0.5

6.2.5 Trophic Interactions

6.2.5.1 Phytoplankton-Zooplankton Interactions

Some cyanobacteria inhibit feeding of grazing zooplankton and can increase mortality of zooplankton (Lampert 1987; DeBernardi and Giusson 1990). Recent reports indicate that some cyanobacteria are also nutritionally inadequate for crustacean herbivores (Arts et al. 1992). The combination of inhibitory/toxic excretions and general unsuitability as a food source is thought to have contributed to the low zooplankton biomass observed in 1969 (Meyer and Effler 1980). The shift in phytoplankton community composition, from cyanobacteria in 1969 to greens, cryptomonads, and diatoms, would be expected to improve food resources for the herbivorous crustacea. The increase in herbivorous zooplankton biomass since 1969 is consistent with this mechanism.

The return of cyanobacteria to dominance in the 1990 phytoplankton community (see earlier section of this chapter) of Onondaga Lake may affect the composition of the zooplankton community. Daphnid feeding can be inhibited by toxic cyanobacteria, but the feeding of calanoid copepods is generally unaffected (DeMott et al. 1991). This could lead to increased importance of calanoids and a decrease in daphnid abundance. Two species, *Bosmina longirostris* and *Brachionus calyciflorus*, important in the zooplankton in 1989, are known to be resistant to cyanobacteria toxins (Starkweather and Kellar 1983; Fulton and Paerl 1987; Fulton 1988) and may benefit by the recurrence of cyanobacteria in Onondaga Lake.

Zooplankton grazing represents an important "top down" force that influences phytoplankton community structure. Zooplankton grazing can reduce algal biomass and/or, by selective grazing, alter the composition of the phytoplankton community. Grazing was thought to be unimportant in limiting the phytoplankton community in 1969 (Sze and Kingsbury 1972) but by the late 1980s was credited with precipitating "clearing events" (Auer et al. 1990). Analysis of

paired distributions of phytoplankton biomass (as C_T) and herbivorous zooplankton (calculated as total biomass minus adult *Cyclops vernalis* biomass) biomass in later years clearly indicates an effect from grazing. For example, the sharpest decrease in C_T in the summer of 1978 coincided with the greatest increase in herbivorous zooplankton biomass (Figure 6.32a). The decrease in phytoplankton biomass was not great enough to be manifested as a "clearing event" (Figure 6.32b; see related optical analysis in Chapter 8). The effect of grazing on C_T, and thereby clarity, has been greater in recent years (since 1987), as illustrated for 1988 in Figure 6.33a and b. Note that the major (and opposite) inflections in C_T and zooplankton biomass in June (Figure 6.33a) were accompanied by a distinct increase in transparency (Figure 6.33b), and that continued increases in herbivorous zooplankton biomass in July apparently contributed to the further reduction in C_T to its seasonal minimum (much lower than observed in 1978) and the development of the maximum clarity of the "clearing event". The development of a "clearing event" in 1988 (as

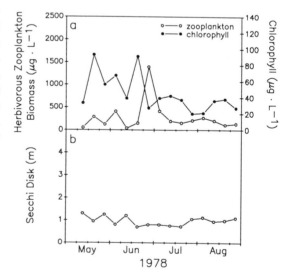

FIGURE 6.32. Temporal distributions for the May–August interval of 1978: (**a**) herbivorous zooplankton biomass and chlorophyll (C_L + P), and (**b**) Secchi disc transparency (modified from Siegfried et al. 1995).

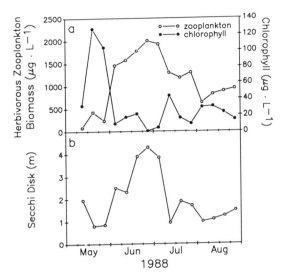

FIGURE 6.33. Temporal distributions for the May–August interval of 1988: (**a**) herbivorous zooplankton biomass and total chlorophyll (C_T), and (**b**) Secchi disc transparency (modified from Siegfried et al. 1996).

opposed to 1978) reflects increased grazing pressure resulting from higher herbivorous zooplankton biomass (compare Figures 32a and 33a) and more efficient grazing, associated with the major shift in the zooplankton assemblage (large daphnids and calanoids, Figures 6.28 and 6.29). The "clearing events" observed annually over the 1987–1989 interval are apparently largely a result of the shift in the zooplankton assemblage to more efficient grazers (Figure 6.29). These events, caused primarily by "top-down" processes, have been perceived by the public as a major improvement in water quality.

Although zooplankton can exert intense grazing pressure on the phytoplankton community, blooms will likely continue to occur (Evans et al. 1992). The composition of the phytoplankton community may, however, be strongly influenced by the selective grazing of zooplankton. Herbivorous zooplankton are capable of considerable grazing selectivity (Svensson and Stenson 1991). Selection, based on algal size, shape, taste, and/or abundance, removes edible phytoplankton species and may benefit competing species.

The recent shift to a late summer phytoplankton community dominated by *Aphanizomenon flos-aquae* is consistent with the increased dominance of the zooplankton community by large daphnids (Figure 6.29e). Large daphnids are extremely efficient grazers on flagellated green algae, cryptomondas, and diatoms (Svensson and Stenson 1991). Removal of these forms from the plankton may facilitate cyanobacteria dominance of the phytoplankton. In lakes where large daphnids dominate the zooplankton, *A. flos-aquae* populations are generally dominated by flake-forming morphs (Porter 1977; Lynch 1980; Andersson and Cronberg 1984). These large "flakes" are, by virtue of their large size, inedible for daphnids and interfere with food collection. The interference of large algae in food collection by *Daphnia* has been detailed by Gliwicz and Siedlar (1980).

Zooplankton may also influence the phytoplankton community through nutrient cycling, by excretion (Lehman 1980, 1984) or through decomposition of fecal pellets (Svensson and Stenson 1991). Zooplankton excretion may shift nutrient limitation between P and N in nutrient enriched systems (Elser et al. 1988; Moegenburg and Vanni 1991). At high N/P ratios zooplankton excretion is likely to reduce P limitation (Moegenburg and Vanni 1991). The fecal pellets of cladocerans are more loose and form less dense packets than the pellets of copepods, and thus are subject to greater decomposition and cycling within the productive upper waters (Svensson and Stenson 1991). Thus increases in herbivorous zooplankton biomass and a shift to increased dominance by cladocera may contribute to dominance by nitrogen fixing cyanobacteria.

6.2.5.2 Fish-Zooplankton Interactions

Planktivorous fish are the principal agents of "top-down" control of zooplankton community structure. The planktivorous fish community of Onondaga Lake has undergone cyclic changes in composition. The alewife, *Alosa pseudoharengus*, was abundant in Onondaga Lake in the late 1950s and into the 1960s

(Dence, 1956). By the late 1960s, during the study of Waterman (1971), alewife were absent from the lake, but had become reestablished by the time of the Meyer and Effler (1980) study of 1978. Alewife are presently virtually absent from the lake (Ringler, personal communication) and are thought to have been absent throughout the 1980s. The planktivorous fish community of Onondaga Lake is presently dominated by gizzard shad, *Dorosoma cepidianum*, and white perch, *Morone americana*.

Models of trophic relationships in pelagic systems conclude that predator mediated interactions are strong at the top of the food web (McQueen et al. 1986). There is a great deal of evidence that pelagic size selective planktivores such as alewife and smelt can have pronounced impacts on the composition of the zooplankton community (Brooks and Dodson 1965; Hutchinson 1971; Siegfried 1987). Feeding by these highly selective predators generally results in a community dominated by small crustaceans.

The pattern of zooplankton community composition changes documented for Onondaga Lake since the late 1960's (Figures 6.28 and 6.29) appears to be related, at least in part, to changes in the abundance of alewife. The dominance of the zooplankton community by *Cyclops vernalis*, known to be moderately successful in avoiding predation (O'Brien 1979), and small cladocera (Figures 6.28 and 6.29) in 1978 is consistent with the observation that alewife were common in that period (Meyer and Effler 1980). Further, the significant large daphid component in 1969 and the late 1980s (Figure 6.29) is consistent with the virtual absence of alewife in those periods (Onondaga County 1971). Unfortunately, detailed fish data are not available for much of the interim period. However, it is probable that alewife, and other efficient pelagic planktivores, were not common in the mid-1980s.

The planktivorous fish that remain in Onondaga Lake are unlikely to impact the zooplankton community in the same manner as the alewife. Although gizzard shad larvae feed on zooplankton, adult gizzard shad are primarily phytophagous (Kutkuhn 1957; Bodola 1966; Tisa and Ney 1991), and would be expected to demonstrate little size selective effects on the zooplankton. The larvae of gizzard shad tend to be concentrated in the littoral zone of lakes (Tisa and Ney 1991) and would not be expected to impact the limnetic zooplankton community. Although large concentrations of gizzard shad might reduce zooplankton biomass, the reduction would occur for all species, not just the large bodied forms. White perch can also be planktivorous, particularly when young. Older perch generally school near shore and would not be expected to have a major impact on the zooplankton of the open waters. The reappearance of daphnids in Onondaga Lake appears to have been facilitated, in part, by the absence of effective planktivorous fish. However the timing of this reappearance indicates it was apparently triggered by other environmental influences.

6.2.6 Water Quality Interactions

6.2.6.1 Ammonia

The free ammonia (NH_3) toxicity data base for zooplankton is much smaller than for fish (USEPA 1985). Chapter 5 should be consulted for the procedures for calculation of the concentration of NH_3 (a fraction of the routinely measured constituent, total ammonia, $T\text{-}NH_3$) and related toxicity criteria for salmonid and non-salmonid fish, and for the evaluation of the status of the lake over the period 1988–1990 with respect to these criteria (also see Effler et al. 1990). The USEPA included the results of six different studies that evaluated the toxicity of NH_3 to cladocerans in their review of NH_3 toxicity to aquatic animals (USEPA 1985). Most of the experiments focused on acute toxicity. The most detailed studies have been conducted on *Daphnia magna*; other test organisms have included *Ceriodaphnia acanthina*, *Daphnis pulicaria*, and *Simocephalus vetulas*.

The limited results indicate that cladocerans are more tolerant of NH_3 than fish (USEPA 1985). Free ammonia was reported to be acutely toxic to cladocerans at concentrations ranging from 0.44 to 2.28 mgN $\cdot$ L^{-1}. Chronic effects for two daphnids were observed at concentrations from 0.25 to 0.99 mgN $\cdot$ L^{-1}. The concentration of NH$_3$ has probably only rarely exceeded 0.25 mgN $\cdot$ L^{-1} in the upper waters of Onondaga Lake over the 1970–1990 period (see Chapter 5); e.g., it exceeded this value on two occasions in mid-summer of 1988 (Chapter 5; Effler et al. 1990). The possibilities that the high NH$_3$ concentrations that have prevailed in the lake have placed selective pressure on the zooplankton assemblage and had chronic effects on the community can not be discounted. Indirect effects of NH$_3$ on the zooplankton community may occur as the more sensitive fish species are affected (USEPA 1985). Consumption of prey by largemouth bass decreases as NH$_3$ concentrations increase (USEPA 1985), suggesting that predator–prey interactions may be affected during periods of high NH$_3$ concentrations. The relatively uniform T-NH$_3$ concentrations that have prevailed since the early 1970s (Chapter 5) indicate that NH$_3$ has not played a major direct role in the strong changes in the zooplankton community observed over the same period. In fact, the highest annual average T-NH$_3$ concentration documented for the upper waters of Onondaga lake was observed in 1988 (Chapter 5), the same year the highest zooplankton biomass was observed (Figure 6.28).

6.2.6.2 Heavy Metals

Heavy metals, in both the cationic and soluble complex form, can be toxic to, or inhibit, zooplankton (Biesinger and Christensen 1972; Andrews et al. 1977). Toxicity data for specific zooplankton species are very limited. High concentrations of heavy metals were reported in the baseline study of Onondaga Lake (Onondaga County 1971); e.g., the epilimnetic mean concentrations of Cu and Cr were 50 and 20 µg $\cdot$ L^{-1}, respectively. Reductions reported by the mid-1970s (e.g., mean epilimnetic concentrations of Cu and Cr of 23 and 5 µg $\cdot$ L^{-1} respectively in 1975 (Onondaga County 1976)), were considered by Meyer

and Effler (1980) in their evaluation of environmental factors contributing to the major increase in zooplankton biomass from 1969 (Waterman 1971) to 1978. The vast majority of heavy metal measurements reported for the lake, including those incorporated in the above results, are based on the analysis of unfiltered samples. This lack of partitioning between total and dissolved fractions (an exception is Seger (1979)) confounds interpretation of the reported reductions with respect to toxicity.

If the sensitivity of Onondaga Lake cladocerans to Cu is similar to that reported for *Daphnia magna* (Biesinger and Cristensen 1972), toxic effects cannot be ruled out. However, Meyer and Effler (1980) indicated that toxic effects were unlikely at the concentrations reported for 1969. Their conclusions were based on the relative success of cladocera in 1969, and the tendency of Cu to become inorganically and organically bound in nontoxic forms (Sylva 1976; Shuman and Woodward 1977). Cyclopoid copepods, including *Cyclops vernalis*, have been found to be substantially more tolerant to copper than cladocera (McIntosh and Kevern 1974).

6.2.6.3 Mineral Particles

Mineral particles in general, and CaCO$_3$ particles specifically, are known to interfere with the feeding of grazing zooplankton (Eadie 1979; Vanderploeg et al. 1987; Hart 1988; Koenings et al. 1990). Mineral (or inorganic) turbidity has been demonstrated to reduce both survival and recruitment of filter feeders because of reduced foraging efficiency. Selective or raptorial feeders, i.e., calanoid copepods and cyclopoid copepods, are generally more efficient feeders in lakes with high mineral turbidity and often dominate the zooplankton communities of these lakes (e.g., Koenings et al. 1990). Many glacial lakes have zooplankton communities consisting of a single macro-zooplanktor, a cyclopoid copepod (Koenings et al. 1990).

The study of Vanderploeg et al. (1987) is particularly noteworthy, as they investigated the effect of calcite on the feeding of *Daphnia pulex* and *Diaptomus sicilis* in Lake Michigan. The net feeding rate of *D. pulex* was found to be reduced by 16% at the concentrations of calcite found in Lake Michigan ($0.46 \, mm^3 \cdot L^{-1}$); at a higher calcite concentration ($0.8 \, mm^3 \cdot L^{-1}$) the feeding rate was reduced 30%, a large enough reduction to inhibit reproduction (Vanderploeg et al. 1987). In addition to lower feeding efficiency, calcite would also slightly increase respiration, owing to increased swimming effort required to compensate for the weight of mineral particles in the gut (Vanderploeg et al. 1987). The coating of cladocerans with $CaCO_3$ reported for the lake before closure (Garofalo and Effler 1987) may also have contributed to increased respiration. In contrast, the feeding of *D. sicilis* was found to be unaffected by the presence of a relatively high concentration of calcite in natural seston (Vanderploeg et al. 1987). The disparate responses of the cladoceran and the copepod to calcite reflects the differences in feeding mechanisms. Cladocerans have limited ability to reject filtered particles, while copepods use long-range olfaction in conjunction with co-ordinated movements of the mouthparts to actively capture food particles (Vanderploeg et al. 1987). Thus the probability of copepods capturing food particles instead of inert particles is greater than for cladocerans. Vanderploeg et al. (1987) speculated that in systems with high calcite concentrations (i.e., higher than Lake Michigan), cladocerans may be at a significant competitive disadvantage relative to copepods.

The limited $CaCO_3$ particle concentration data (as $mm^3 \cdot L^{-1}$) available for the upper waters of Onondaga Lake are presented in Table 6.14 (also see Chapter 5). Only three summer observations are available for 1981,

compared to 20 in 1987 (observations after a major storm that were biased by the $CaCO_3$ coating phenomenon (Johnson et al. 1991; Chapter 5) were not included). Superficially there appears to be little difference between the two years. The small number of 1981 observations and the known temporal variability in the concentrations of $CaCO_3$ (see Chapter 7) eliminate these data as serving as a quantitative basis of comparison before and after closure of the soda ash/chlor-alkali facility. However, it is clear that concentrations that can reduce the feeding rate of *Daphnia pulex* (Vanderploeg et al. 1987), and presumably other large daphnids, occurred in both years (Table 6.14). Further, concentrations occurred in 1987, and probably in 1981, that could inhibit reproduction ($0.8 \, mm^3 \cdot L^{-1}$; Vanderploeg et al. 1987). Recall that a cyclopoid copepod was generally dominant before closure and through mid-summer of 1987 (Figure 6.28), consistent with observations for other lakes enriched in mineral particles (Koenings et al. 1990).

Summer mean and median $CaCO_3$ turbidity (T_c) values for the upper waters for the 1985–1990 interval are presented in Figure 6.34 (see Chapter 7 for detailed temporal distributions within these years). Recall the limitations of turbidity as a measure of particle concentration, because of its dependence on particle size distribution. Conditions remained largely unchanged over the 1985–1987 period (Figure

TABLE 6.14. Comparison of summertime $CaCO_3$ particle concentrations in Onondaga Lake, 1981 and 1987.

Period	Concentration ($mm^3 \cdot L^{-1}$)		n
	Average	Range	
1981	0.40	0.16–0.60	3
1987	0.39	0.02–2.01	20

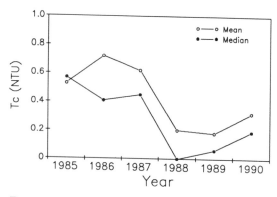

FIGURE 6.34. Temporal distribution of calcite turbidity (T_c) in the upper waters of Onondaga Lake for the spring–fall interval, 1985–1990.

6.34), an interval that bracketed closure of the facility. However, T_c was substantially lower over the 1988–1990 interval (Figure 6.34). The indicated reduction in $CaCO_3$ content is consistent with the major increase in large daphnid biomass and dominance of the zooplankton community observed in 1988 and 1989 (Figures 6.28 and 6.29c). The recurrence of large daphnids in the lake in 1986 and 1987, after a number of years of absence, cannot presently be coupled to changes in $CaCO_3$ particle concentrations. However, the implied reduction in $CaCO_3$ content of the lake's upper waters after 1987 (Figure 6.34) is likely a contributing factor in the dominance of large daphnids in the zooplankton of the lake in 1988 and 1989, and the attendant benefits with respect to clarity.

Daphnia filter feeding is apparently capable of removing considerable amounts of suspended calcite (Vanderploeg et al. 1987), as well as other inorganic particles, from the watercolumns of lakes. The increases in daphnid abundance in the late 1980s may therefore have caused increased losses of abiotic (e.g., mineral) particles from the water column of the lake. Reductions in inorganic forms of turbidity, as well as phytoplankton, have been reported for the clearing event of 1987 (based on IPA techniques, Chapter 7). Thus, zooplankton grazing may contribute to the particularly low turbidities and high Secchi disc transparencies reported for the clearing events in the 1987–1990 period (Chapter 7), by removing minerals and other abiotic particles as well as phytoplankton.

6.2.6.4 Salinity

High salinity (S) is known to exert selective pressure on the zooplankton community (Remane and Schlieper 1971). The majority of cladocera and copepoda are restricted to waters less than 1‰ S, while many of the freshwater rotifers can inhabit waters with S between 0 and 3‰ (Remane and Schlieper 1971). Freshwater cladocera are generally faced with the necessity to hyper-regulate their osmotic balance; salt ingested with food is reabsorbed by the maxillary gland (Aladin

1991). Brackish conditions, such as prevailed in Onondaga Lake prior to the closure of the soda-ash/chlor-alkali plant (Chapter 5), would require cladocera to hyporegulate, i.e., excrete ingested salts, in order to maintain proper osmotic balance. Salinity averaged about 3‰ in Onondaga Lake for the May–September interval before closure of the soda ash/chlor-alkali facility, and has decreased to nearly 1‰ since closure (Chapter 5).

The probable role of the earlier elevated salinity levels in limiting zooplankton diversity in Onondaga Lake was identified by Meyer and Effler (1980). Most of the common forms reported in 1969 and 1978 were identified as unusually tolerant of elevated S (Meyer and Effler 1980). Salinity tolerances of cladocerans and copepods found in Onondaga Lake included in the review of Remane and Schlieper (1971) are presented in Table 6.15. *Cyclops vernalis*, a dominant of the zooplankton of Onondaga Lake before 1988, had not previously been identified with oligohaline systems (Meyer and Effler 1980). In fact, Anderson (1950) reported a toxicity threshold for $CaCl_2$ of $1730 \, mg \cdot L^{-1}$ for *C. vernalis*, substantially less than Onondaga Lake concentra-

TABLE 6.15. Salinity tolerances of the copepods and cladocerans of Onondaga Lake.

Group/species	Salinity tolerance (‰)
Cladocera	
Ceriodaphnia quadrangula	Up to 4–5
Daphnia pulex	Up to 4–5
Alona affinis	Up to 4–5
Chydorus spaericus	Up to 4–5
Bosmina longirostris	Up to 4–5*
Daphnia galeata	No limit found
Diaphanosoma leuchtenbergianum	No limit found
Eubosmina coregoni	No limit found
Leptodora kindtii	Up to 4–5*
Copepoda	
Diaptomus sicilus	Up to 35
Diaptomus siciloides	No limit found
Cyclops bicuspidatus	Up to 10
Cyclops vernalis	No limit found
Mesocyclops edax	No limit found

According to Remane and Schlieper (1971)
*Limit given for genus only

tions before closure of the chemical plant. Three other species for which no salinity limits were found are now common or dominant forms, including *Daphina galeata*, *Diaphanosoma leuchtenbergianum*, and *Diaptomus siciloides*. The absence of these species from the review of Remane and Schlieper (1971; limits for relatively salinity-tolerant forms are emphasized) implies they are probably relatively intolerant to salinity. It is interesting to note that the species *Cyclops bicuspidatus*, which has been described by Remane and Schlieper (1971) as halophilous (e.g., Table 6.15), and was commonly observed in 1969 and 1978, has only rarely been observed since closure of the facility (Table 6.11).

The reductions in S of the lake that resulted from closure of the soda ash/chlor-alkali facility probably contributed to the documented increase in zooplankton species richness in the late 1980s (Table 6.11). Further increases in zooplankton diversity are a reasonable expectation for further reductions in S, as many species are intolerant of the S that continues to prevail (Remane and Schlieper 1971).

6.3 Aquatic Macrophytes

6.3.1 Background

Aquatic macrophytes include plants common to aquatic environments that are visible with the naked eye, these include both vascular plants (flowering and nonflowering plants with water and food-conducting tissues) and nonvascular plants, such as the macroscopic algal charophytes. These species are predominantly rooted or attached to the sediment, although some free-floating forms exist (e.g., duckweed (*Lemna* sp.) and coontail (*Ceratophyllum demersum*)). Aquatic macrophytes are classified as having emergent (i.e., extending out of the water), floating (i.e., floating on the surface of the water), or submersed (i.e., totally under water) leaves. The reader is directed to other sources for treatments of aquatic macrophyte taxonomy (Fassett 1957) and ecology (Wetzel 1983; Hutchinson 1975).

Aquatic macrophytes are an important constituent of lake ecosystems. They produce food for other aquatic organisms, serving as the base of an aquatic food chain; and also provide habitat areas for insects, fish of all age classes, and other resident aquatic and semiaquatic organisms. Littoral zones, and the vegetation that structure them, are a prime area for the spawning of most fish species, including many species important to sport fisheries. Aquatic vegetation also serves to anchor soft sediments, stabilize underwater slopes, and remove suspended particles and nutrients from overlying waters. The absence of littoral vegetation limits diversity as well as production of desirable fish species, and thereby the recreational value of a lake.

6.3.2 Historical Information for Onondaga Lake

Little quantitative or observational data have been recorded for aquatic macrophytes within Onondaga Lake, as noted in the *Proceedings of the Onondaga Lake Remediation Conference* (Saroff 1990), until the work of Madsen et al. (1992). Anecdotal evidence indicated that the lake sustained extensive areas of charophytes, either *Chara* or *Nitella*. This is supported by the observation of charophyte stems serving as the nuclei of many of the of oncolites (algal pisoliths; see Chapter 8) found in the nearshore zones (Dean and Eggleston 1984). These plants are believed to have been lost from the lake soon after the opening of the soda ash/chlor-alkali plant (Dean and Eggleston 1984). Other macroscopic aquatic plants observed prior to 1991 were *Potamogeton pectinatus* (Stone and Pasko 1946), *Myriophyllum spicatum*, and *Ceratophyllum demersum* (J.A. Bloomfield, personal communication). Some additional observations of historical relevance have been assembled by McMullen (1992).

Historical species richness estimates are difficult to determine, and may have fluctuated even before any human influence. Historical references are summarized in Table 6.16. From the list compiled by McMullen (1992), 16 species are listed, although only 13 would definitely be found from the water's edge or deeper. Other species might possibly be added

TABLE **6.16**. Historical references to aquatic macrophyte species in Onondaga Lake, compared to recent finds from 1991 survey.

Species	1991	McMullen	Gilman	Personal
Carex viridual		X		
Ceratophyllum demersum	X			X[2]
Chara sp.				?[4]
Cyperus odoratus		X		
Heteranthera dubia	X	X	X	
Iuncus pelocarpus		X		
Myriophyllum spicatum	X			?[2,5]
Najas flexilis		X		
Najas marina		X		
Nitella sp.				?[4]
Nuphar luteum		X		
Nymphaea odorata		X		
Phragmites communis		X		
Polygonum amphibium			X	
Potamogeton crispus	X			
Potamogeton pectinatus	X	X	X	X[2,3]
Potamogeton robbinsii		X		
Ruppia maritima		X		
Sagittaria rigida		X		
Scirpus americanus		X		
Scirpus validus		X		
Sporobolus cryptandrus				
Typha angustifolia		X	X	
Zannichellia palustris				

[1] Listing of voucher specimens noted by Bruce Gilman (New York State Museum herbarium)
[2] Jay Bloomfield and Jim Sutherland, NYS DEC, Field Notes for May 31, 1990
[3] Stone and Pasko 1946
[4] Dean and Eggleston 1984
[5] Identified at site as *Myriophyllum sibiricum*, later revised to *M. spicatum*

to this list; but the references are unclear, and this is an attempt at a conservative listing. Two species can definitely be added to these 13 from voucher specimens at the State Museum Herbarium, bringing the total number of species to 15. In addition, the record preserved within the oncolites indicates that either or both of the macroalgae genera *Chara* and *Nitella* were common in the lake (Dean and Eggleston 1984). Fifteen species of vascular plants from the lake's edge and deeper is typical, or possibly even low, for a eutrophic lake of this type in New York State (Table 6.17). Of the fifteen species listed in Table 16, eight are submersed species. The average number of submerged species for 31 eutrophic lakes from around the world was 4.4; the average for all 439 lakes examined was 6.6 (Taggett et al. 1990).

6.3.3 Field Survey of 1991

The first systematic study of the macrophyte community of Onondaga Lake was undertaken in 1992 (Madsen et al. 1992). The study included a field survey of presence and abundance of aquatic plants, and laboratory experiments of factors affecting plant growth.

6.3.3.1 Methods

The distribution of submersed littoral vegetation was examined using 40 transects located in a stratified-random manner around Onondaga Lake (Figure 6.35). Transects were placed perpendicular to the shoreline. Each transect was 100 m long, divided into 1 m segments, and extended from the shore to no greater than 5 m deep. At each 1 m interval, a 0.1 m² quadrat was placed. Abundance of each

TABLE 6.17. Species richness of submersed, floating-leaved, emergent, and all components of littoral zone aquatic plant communities of some New York lakes.

Lake	Number of submered species	Number of floating-leaved species	Number of emergent species	Number of aquatic species	Reference
Brant Lake	29	4	5	38	Eichler 1990
Cayuga Lake	10	1	0	11	Miller 1978
Cross Lake	5	3	0	8	Miller 1978
Duck Lake	6	3	1	10	Miller 1978
Eagle Lake (1988)	19	2	1	22	Taggett 1989
Eagle Lake (1989)	23	2	3	28	Eichler and Madsen 1990c
Fourth Lake	15	4	4	23	Eichler and Madsen 1990a
Galway Lake	18	0	1	19	Eichler and Madsen 1990b
Honeoye Lake	16	2	1	19	Gilman 1985
Lake Como	8	3	0	11	Miller 1978
Lake Luzerne	25	2	1	28	Eichler and Madsen 1990a
Little Sodus B	8	9	2	19	Miller 1978
Loon Lake	21	4	6	31	Taggett 1989
Onondaga Lake (historical)	9	3	8	20	This chapter
Onondaga Lake (current)	5	0	0	5	This chapter
Otter Lake	5	3	0	8	Miller 1978
Owasco Lake	14	1	0	15	Miller 1978
Paradox Lake	12	3	4	19	Taggett 1989
Parker Pond	1	3	0	4	Miller 1978
Skaneateles L.	11	0	0	11	Miller 1978
Schroon Lake	18	1	2	21	Taggett 1989
Second Lake	22	4	4	30	Eichler and Madsen 1990a
Third Lake	12	3	3	18	Eichler and Madsen 1990a
New York Average (N = 23)	13.6	2.6	2	18.2	Above refs.
World average (submersed, eutrophic)	4.4 (N = 163)	—	—	—	Taggett et al. 1990
World average (submersed, saline)	3.2 (N = 25)	—	—	—	Taggett et al. 1990
World average (submersed, all lakes)	6.6 (N = 439)	—	—	—	Taggett et al. 1990

species within the quadrats was recorded, based on a Daubenmire scale used by SCUBA divers knowledgeable in aquatic plant identification. This method has been used in a number of other surveys (e.g., Madsen et al. 1989). Transects were examined in late June of 1991. Classification of surficial sediments was made visually for each quadrat. Sediment samples were collected to standardize observations.

6.3.3.2 Species and Depth Distributions

A map of the surficial littoral zone sediments of the lake determined in the survey appears in Figure 6.36. Note that the oncolite zone covers much of the littoral zone, including most of the eastern and northwestern shores. Sandy material is mixed in with oncolites in much of this zone. Silts, soft fine-grained sediments composed mostly of calcium carbonate (Chapter 8), occur primarily on the southwestern shoreline and in the area just north of the waste beds.

Plant leaves and stems were generally covered with carbonate particles. Five species of submersed macrophytes were found in Onondaga Lake (Table 6.18). Interestingly, no species of either floating-leaved or emergent plants were observed. These forms are typical of the transition from the littoral zone to wetlands. The lack of these species may

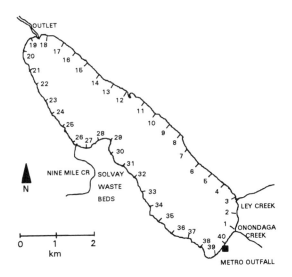

FIGURE 6.35. Map showing the location of vegetation survey transects in Onondaga Lake, New York (modified from Madsen et al. 1996).

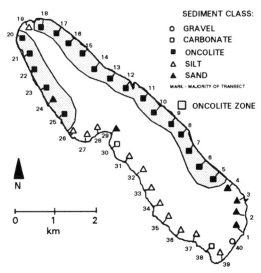

FIGURE 6.36. Map of the surficial zone sediments of Onondaga Lake (modified from Madsen et al. 1996).

be related to water level changes or water chemistry, as well as shoreline changes associated with urbanization and wave scouring of unstable sediments.

Plants were observed in only 13.3% of the quadrats surveyed. *Potamogeton pectinatus* was the most common (11.2% of the quadrats). *Heteranthera dubia* occurred in only 1.8% of the quadrats; the others were rare, occurring in ≤0.3% of the quadrats. All five species were present in depths of 0 to 2 m. In water depths of between 2 and 3 m, only *Heteranthera dubia* and *Ceratophyllum demersum* were observed, and beyond a depth of 3 m, only *Heteranthera dubia* was found.

Species richness of all groups of aquatic plants is very low compared to other New York lakes (Table 6.17). Typically, a New York lake will have an average of 18 species of aquatic plants: 14 submersed species, 2 floating-leaved, and 2 emergent species. With no floating-leaved or emergent species and only 5 submersed species, Onondaga Lake is well below this average. Many factors can contribute to this low species diversity, including degree of eutrophication, salinity, size of lake, inappropriate substrate, and degree of urbanization. Oligo- and mesotrophic lakes tend to have the highest species diversity of submersed plants; eutrophic lakes have sub-

TABLE 6.18. Aquatic macrophyte species list for Onondaga Lake, Ninemile Creek, and Otisco Lake (partial list for Otisco Lake).

Species	Common name	Onondaga Lake	Ninemile Creek	Otisco Lake
Ceratophyllum demersum	Coontail	X		
Heteranthera dubia	Water Stargrass	X		
Myriophyllum spicatum	Eurasian Watermilfoil	X	X	X
Potamogeton crispus	Curly-leaf pondweed	X	X	X
Potamogeton nodosus	American pondweed		X	X
Potamogeton pectinatus	Sago pondweed	X		

stantially lower diversity (Taggett et al. 1990). Saline lakes tend to have even lower diversity of submersed species, since only a few plants tolerate highly saline conditions. Important factors in the decline of species richness for Onondaga Lake were probably the increase in the salinity of the lake, and the attendant increase in precipitation of $CaCO_3$, as a result of industrial discharges, and cultural eutrophication, that doubtless caused decreases in light penetration. Reversals in these conditions should be expected to result in increased species diversity.

Ceratophyllum demersum (coontail) is a non-rooted submersed aquatic plant that was typically found intermingled with other dense growths of other plants, and was not found in great abundance. Most occurrences were in less than 2 m of water, among growths of Sago pondweed.

Heteranthera dubia (water stargrass) had the deepest range, occurring out to depths of 4 m in the northeastern part of the lake. Usually, it was found adjacent to stands of *P. pectinatus* in 1 to 3 m of water.

Myriophyllum spicatum (Eurasian watermilfoil) was observed as small colonies at only two locations, at depths of 0 to 2 m. Eurasian watermilfoil is a nonnative species that creates nuisance problems in many lakes throughout North America, including New York. This species deserves continued monitoring in Onondaga Lake.

Potamogeton crispus (curly-leaf pondweed) is another exotic species that creates nuisance problems in New York. It was found in large quantities at five locations in depths of 0 to 2 m around the lake. This plant may actually be more widespread and more abundant than noted since its growth cycle is such that it often senesces by late June or early July.

Potamogeton pectinatus (Sago pondweed) was the most abundant species in Onondaga Lake, usually occurring in shallow areas from 0 to 2 m, though some plants were found as deep as 3 m. *Potamogeton pectinatus* is widespread in lakes and streams throughout North America, often dominant in eutrophic streams and in alkaline lakes of the prairies. This species has

great value for waterfowl, wildlife and as habitat for fish and benthic organisms.

6.3.3.3 Patterns in Lakewide Distribution

A map of aquatic species distribution in Onondaga Lake documented in the 1991 survey is presented in Figure 6.37. Sites without aquatic macrophytes were largely restricted to those areas with the greatest fetch, or exposure. These sites were predominantly in the southeastern and northwestern areas of the lake. Only once did a site have macrophytes without *P. pectinatus* being present. *Ceratophyllum demersum* was found predominantly along the western shore, with *H. dubia* found predominantly on the eastern shore. *Myriophyllum spicatum* was found at two nearly adjacent sites northwest of the marina on the northeastern shore. *Potamogeton crispus* was distributed sparsely around the entire lake.

Individual sites (transects) exhibited very limited diversity, with an average of slightly more than 1 species per transect (mean = 1.3). The number of species at each site ranged from 0 to 3 (Figure 6.37). *Potamogeton pectinatus* was the most common, occurring at

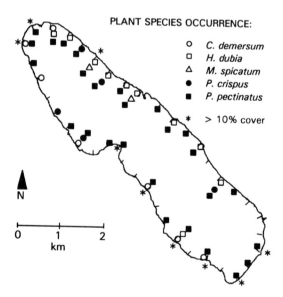

FIGURE 6.37. Distribution of aquatic plant species in Onondaga Lake in late June of 1991 (modified from Madsen et al. 1996).

27 of the 40 transects (68%). The remainder of species were found in the following order: *Heteranthera dubia* (10 of 40 transects, 25%), *Ceratophyllum demersum* (7 transects, 18%), *Potamogeton crispus* (5 transects, 13%) and *Myriophyllum spicatum* (2 transects, 5%).

In most locations, existing *P. pectinatus* populations were very small. Two locations, northwest of the outlet and northwest of Onondaga Creek, had large beds of Sago Pondweed. These two sites had fine sediments (probably enriched in nutrients). At all other sites, beds were small, which seems to indicate that their establishment has been recent. The total areal coverage of *P. pectinatus* was much less than typical lakes, or than depth/light penetration conditions would be expected to support. A large amount of space is available for expansion. The scarcity of other species also supports the position that their establishment has been recent. *Potamogeton pectinatus*, being more tolerant of high salinity than most species, probably recolonized the lake first, or existed in isolated locations, under the earlier more limiting conditions. Reestablishment at other locations has been recent. New species are probably slow to recolonize due to the persistent high salinities and low light penetration. Other factors that may have limited plant distribution include variations in water level, wave action effects of unstable sediments, and the physical character and composition of the near-shore sediments. These factors are discussed below.

6.3.4 Factors Limiting Plant Distribution

6.3.4.1 Light Penetration

Light is a major limiting factor regulating the depth distribution of submersed aquatic plants. This effect is expressed as the maximum depth to which plants grow, as well as different species that may colonize and thrive under differing regimes of light. Submersed aquatic plants can colonize deeper areas as light penetration increases. The influence of light penetration, and recent changes in this

characteristic, on the maximum depth of macrophyte colonization (Z_c, m) in Onondaga Lake is evaluated here utilizing empirical expressions developed in the literature and the optics database for the lake (Chapter 7). The empirical expressions are based on one of two types of light penetration data; the more widely used Secchi disc transparency (Z_{SD}, m), and the light attenuation coefficient (K_d, m^{-1}), a more accurate measurement, determined from measurements with an underwater irradiometer (see Chapter 7).

Canfield et al. (1985) developed the following expression relating Z_c to Z_{SD}, based on data from 100 north temperate lakes.

$$\log(Z_c) = 0.61[\log(Z_{SD})] + 0.26 \quad (6.9)$$

Chambers and Kalff (1985) developed the following relationship, based on observations for 90 lakes. Optical features not

$$Z_c = (1.33[\log(Z_{SD})] + 1.40)^2 \quad (6.10)$$

accommodated in these expressions, such as spectral quality of penetrating light (Chambers and Prepas 1968), may also influence Z_c. The average summer value of Z_{SD} reported for the 1978–1986 period was about 0.90 m (Chapter 7). The associated values of Z_c determined from the above empirical models are 1.7 and 1.8 m, respectively. Clarity has increased in recent years, the average Z_{SD} value for the 1987–1990 was about 1.9 m (Chapter 7). The associated predicted values of Z_c are 2.7 and 3.15 m, respectively. These predictions approach the Z_c values observed, 3 to 4 m, during the 1991 survey. The somewhat deeper observed Z_c values compared to predictions may reflect the influence of periods of particularly deep light penetration ("clearing events") that have been common in the lake in early summer in these recent years (Chapter 7).

Vant et al. (1986) developed the following empirical expression to predict Z_c based on K_d,

$$Z_c = \frac{4.34}{K_d} \quad (6.11)$$

which corresponds to the 1.3% light level (see Chapter 7). The data base supporting this expression is much smaller (15 New Zealand lakes) than those supporting the Z_{SD}-based relationships. The summer average K_d values for Onondaga Lake for 10 years of the 1978–1990 interval are presented in Figure 6.38a, with bars corresponding to the minimum and maximum values (see Chapter 7 for more complete treatment). The increases in light penetration in recent years are clearly depicted by the reductions in K_d. The associated predicted distribution of Z_c values over the same period is presented in Figure 6.38b. The average value of Z_c for 1978–1981, 1985, and 1986 was 2.6 m. The value of Z_c, based on summer average K_d, increased to 4.0 m in the 1988–1990 period, essentially the same value observed during the survey of 1991.

Detailed temporal distributions of K_d are presented for two years in Figure 6.39, 1985 and 1988. The earlier year is representative of conditions documented for the 1978–1986 interval; high K_d values with relatively low

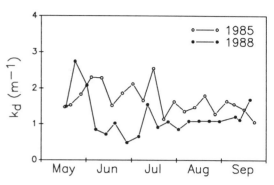

FIGURE 6.39. Temporal distributions of K_d in Onondaga Lake for 1985 and 1988.

temporal variability. The 1988 distribution is similar to those observed over the 1987–1990 interval; the most conspicuous difference is a rather protracted period of low K_d values ("clearing event") in early summer. The "clearing events" represent a window of opportunity which some macrophyte species may be able to exploit. These species would be early season plants, or species that are fully grown and senesce by mid-summer. Examples of species that might do well under these conditions, and are presently found in the lake, are *Potamogeton crispus*, *P. pectinatus*, and possibly *Myriophyllum spicatum*. Conversely, some species with a perennial growth form or ones that dominate later in the growing season would not have an advantage, such as *Elodea canadensis* or *Potamogeton robbinsii*. Other species, such a *Najas* spp., would not be able to complete their life cycle before reduced light penetration would cause parent plant mortality. This "window" was approximately 4 weeks long in 1987, 1988 and 1989, but the "window" in 1990 was almost six weeks long. Undoubtedly these events are significant to aquatic plant growth and can only facilitate revegetation.

6.3.4.2 Salinity

The high levels of salinity that have prevailed in Onondaga Lake (Chapter 5) represent an additional stress to macrophytes. Salinity tolerances of a number of macrophyte species are presented in Table 6.19. A number of species are not tolerant of the salinity that

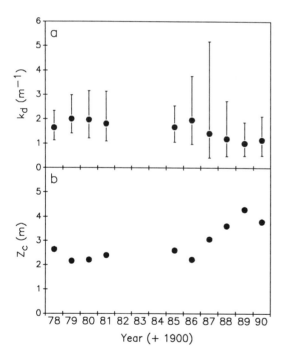

FIGURE 6.38. Summer average conditions in Onondaga Lake: (a) light attenuation coefficient, K_d, with bars corresponding to minimum and maximum values, and (b) predicted maximum depth of macrophyte colonization (Z_c), according to expression of Vant et al. (1986).

TABLE **6.19.** Tolerance range of some tolerant aquatic macrophytes for salinity, Cl^{-1}, and conductivity levels in fresh water.

Species	Salinity $(g \cdot l^{-1})$	Conductivity $(mS \cdot cm^{-1})$	Cl^{-1} $(mg \cdot l^{-1})$	Reference
Brasenia schreberi		<238	<255	Kadono 1982
Ceratophyllum demersum		<16,200	<5790	Kadono 1982
Hydrilla verticillata	<10			Haller et al. 1974
H. verticillata		<10,900	<3910	Kadono 1982
Lemna minor	<16.65			Haller et al. 1974
Lemna minor	<10			Hammer and Heseltine 1988
Myriophyllum spicatum	<16.65			Haller et al. 1974
Myriophyllum spicatum	<10			Hammer and Heseltine 1988
M. spicatum		<15,100	<4830	Kadono 1982
M. verticillatum		<183	<83.3	Kadono 1982
Najas guadalupensis	<13.32			Haller et al. 1974
N. marina		<10,900	<3910	Kadono 1982
Phragmites australis	10–20			Hammer and Heseltine 1988
Potamogeton pectinatus	<58			Hammer and Heseltine 1988
P. richardsonii	<10			Hammer and Heseltine 1988
Ruppia maritima	<70			Hammer and Heseltine 1988
R. occidentalis	<58			Hammer and Heseltine 1988
Trapa natans		<219	<255	Kadono 1982
Typha latifolia	<10			Hammer and Heseltine 1988
Utricularia vulgaris	<10			Hammer and Heseltine 1988
Vallisneria americana	<13.32			Haller et al. 1974
V. asiatica		<9,800	<330	Kadono 1982

occurred in the lake before closure of the soda ash/chlor-alkali plant, or the lower level that has prevailed since closure of the facility (Chapter 5). The water chemistry of Onondaga Lake is more similar to many of the inland saline prairie lakes than lakes in upstate New York. Further, many of the species adapted to saline freshwater environments have no avenue of access to this lake. Past high concentrations of solutes may have been sufficient alone to prohibit many common emergent and floating-leaved species, due to increased stress from evapotranspiration. Several submersed species are notable for being tolerant to high solute concentrations, such as *P. pectinatus*. The dominance of *P. pectinatus* in Onondaga Lake is probably in part due to its tolerance of high solute concentrations. Its current dominance may continue as it has preempted many habitats from other species, although as solute concentrations continue to decrease, other species may increase in abundance. Further decreases in concentrations of solutes may also encourage the growth of emergent and floating-leaved species.

The tolerance data presented in Table 6.19 may be somewhat misleading, because of the different sources of the information. Some of the data are observed ranges at which the species is found. Other data are derived from experimental studies in which competition and other factors were eliminated. Salinities of even one-tenth to one-hundredth the ranges presented in Table 6.19 may be sufficient to reduce the growth and production of aquatic plants. Some floating-leaved plants such as *Brasenia schreberi* and *Trapa natans* (Table 6.19) would not tolerate the Cl^- concentrations that continue to prevail in Onondaga Lake (Chapter 5).

6.3.4.3 Calcium Carbonate Precipitation and Near-Shore Sediments

The extremely high calcium concentrations (a major component of salinity in the lake; Chapter 5) that prevailed during the operation of the soda ash/chlor-alkali facility promoted an extremely high rate of precipitation and deposition of calcium carbonate ($CaCO_3$;

Chapters 5 and 8). Precipitation of $CaCO_3$ onto plant surfaces has been documented in other hard water lakes (Wetzel 1960). This process can be passive, with the plant being a catenation point for precipitation (Smart 1990), or active through bicarbonate use in photosynthesis (Wetzel 1983). If the rate of plant coverage by $CaCO_3$ exceeds the rate of plant growth, plants can become totally coated. It has been speculated that this mechanism may have been responsible for, or contributed to, the disappearance of charophytes from Onondaga Lake (Dean and Eggleston 1984). This mechanism may also be effective against other species, such a *P. pectinatus* and *M. spicatum*. This potential stress on the submersed macrophyte community has diminished following the closure of the soda ash/chlor-alkali facility, as indicated by the major reduction in $CaCO_3$ deposition (Chaptes 5 and 8).

Sediment characteristics play a major role in the distribution of aquatic plants within Onondaga Lake. Gravel sediments supported the smallest populations of aquatic plants, followed closely by oncolite-dominated sediments (Madsen et al. 1992). *Potamogeton pectinatus*, the dominant species, occurred in all sediment types, but occurred significantly less often on oncolites and gravel than other substrates. On oncolite dominated sediments, *Potamogeton pectinatus* and *Heteranthera dubia* were encountered with equal frequency (Madsen et al. 1992). These survey results imply that plants have difficulty in establishing and remaining on oncolite sediments, compared to other sediment types (except gravel). This is probably more related to the physical character of these deposits than their chemical make-up, as the results of greenhouse growth experiments indicate plants can grow to the same abundance on oncolites as on other sediment types of Onondaga Lake (Madsen et al. 1992). Oncolites provide an unstable substrate for plant growth because they are highly susceptible to movement associated with normal wave action. This is a result of their "stone-like," but low density, configuration.

6.3.4.4 Water Level, Invasive Plants and Watershed Effects

Water level data for the lake have been reviewed in Chapter 3. Variations between years and within a given year have usually been less than 1 m. Even this magnitude of variation in water level may have a significant effect on the littoral zone vegetation of the lake, since light penetration restricts the depth of macrophyte growth to such shallow depths (Table 6.17). The impacts of wave action and ice scour interact with those of water level.

The composition of the future macrophyte community is largely dependent on the plant propagules that can reach the lake environment. It is often assumed that waterfowl carrying seeds in their digestive tracts is the single most important mode of transport for aquatic plants propagules, particularly seeds. Although this is a viable method of distribution, the two most important mechanisms for plant dispersal are water movement and man (not necessarily in that order). Human introduction of plants to new environments may be either intentional, such as planting, or inadvertent. The most common mode of aquatic plant dispersal by man is the transport of plant fragments or propagules on boats and boat trailers.

Downstream water movement is the most powerful and consistent carrier of plant propagules. Therefore, we can expect upstream water bodies to have an important impact on the composition of plants in Onondaga Lake. In Table 6.18, some plant species observed in Ninemile Creek and Otisco Lake are recorded (this is by no means an exhaustive list). Two of those species, *M. spicatum* and *P. crispus*, are already found in small numbers in Onondaga Lake. As salinity decreases, the abundance of these two species could possibly increase, eventually leading to a more typical composition for a eutrophic New York lake. Since these species are found in abundance upstream of Onondaga Lake, it is unlikely that their continual introduction can be prevented. The other species sighted in Ninemile Creek, *P.*

nodosus, probably as yet cannot tolerate water conditions in the lake.

6.3.5 Laboratory Growth Bioassays

The limitations of the sediment composition of the near shore zone of Onondaga Lake on macrophyte growth were evaluated in a series of greenhouse bioassays experiments (Madsen et al. 1992). The purposes of the experiments were to evaluate:

1. the potential of Onondaga Lake sediments to grow plants relative to a fertile sediment,
2. the potential of various types of the lake's sediments to support growth,
3. the relationship between the existing plant and sediment distributions, and
4. the potential for revegetation of selected plant species and the development of nuisance (exotic) species populations.

Growth of *Potamogeton pectinatus* on the fertile reference sediment was significantly higher ($\geq$2X) than growth on the Onondaga Lake sediments (Figure 6.40; Madsen et al. 1992). These sediment limiting conditions indicate that remediation efforts directed at improving littorial zone vegetation should consider inclusion of a sediment remediation component. The existing limiting conditions may be related to the very high $CaCO_3$ content of these sediments, that has been attributed in part to an impact of the ionic discharge from the soda ash/chlor-alkali plant.

The various sediment classes distinguished in the plant survey, by visual characterization, were not significantly different with respect to potential for supporting the growth of *P. pectinatus* under greenhouse conditions (Figure 6.40; Madsen et al. 1992). This may be due to similarities in chemical compostion of the classes (all rich in $CaCO_3$), despite the disparities in particle size. Low nutrient content and availability are probably at least part of the sediment limiting conditions.

In another study, sediments were taken from sites with relatively high (mean 32%),

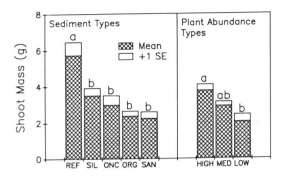

FIGURE **6.40.** Growth of *Potamogeton pectinatus* on a reference (REF) and Onondaga Lake sedments in a greenhouse experiment (from Madsen et al. 1992). Onondaga sediments: SIL (silt), ONC (oncolite), ORG (organic), SAN (sand). *In situ* plant abundance sites: HIGH (high; 32%) plant cover; MED (medium; 9%) plant cover; LOW (low; 0.001%) plant cover. Different letters indicate a significant difference at the p = 0.05 level using a Bonferroni significant difference test, ANOVA (modified from Madsen et al. 1996).

medium (mean 9%), and low (mean 0.001%) percent cover of aquatic plants. The order of growth on these sediments in the greenhouse assays was found to be the same; means of 3.7, 2.9, and 2.1 g were observed for the high, medium, and low percent cover areas, respectively (Figure 6.40; Madsen et al. 1992). This further supports the position that sediment conditions are presently limiting to the macrophyte community in Onondaga Lake.

Eleven different species were tested on Onondaga Lake sediments (Table 6.20). Three exotic species (*M. spicatum, P. crispus,* and *T. natans*) were tested to evaluate the potential for nuisance populations of these species to develop. The other eight species were tested as potential candidates for revegetation efforts. All eleven of the species tested exhibited some growth on Onondaga Lake sediments. However, relative success varied greatly between the species (Table 6.20). The trials of eight native species indicated that three species in particular, *Potamogeton pectinatus, P. nodosus,* and *Sagittaria latifolia,* are the best choices for initial revegetation efforts. These species not only performed well on oncolite sediments, at

TABLE **6.20.** Performance of eleven species on Onondaga Lake sediments.

Species	Oncolite	Organic	Onc + Org
Elodea canadensis	Fair	Poor	Fair
Myriophyllum spicatum	Fair	Fair	Fair
Nymphaea odorata	Poor	**	**
Potamogeton crispus	Fair	Poor	Poor
Potamogeton nodosus	Good	Good	Good
Potamogeton pectinatus	Good	Fair	Good
Sagittaria latifolia	Good	Good	Good
Sagittaria rigida	Poor	**	**
Trapa natans	Poor	Poor	Poor
Typha latifolia	Fair*	Fair*	Fair*
Vallisneria americana	Poor	Fair	Good

* Although *Typha latifolia* had higher biomass than many other species, the plants were chlorotic and the response was poor relative to what should be expected for this species.
** Treatment not performed

current water chemistry conditions, but have propagules that are easily handled for revegetation efforts. Of the three exotic species tested, *T. natans* performed poorly and is unlikely to produce significant nuisance growths under current sediment and water chemistry conditions. *Myriophyllum spicatum* and *P. crispus* exhibited fair growth, but substantial improvement in sediment fertility would need to occur before large nuisance populations would develop.

6.4 Benthic Macroinvertebrates

6.4.1 Background

Benthic macroinvertebrates are animals that inhabit the substratum of lakes and streams. Macroinvertebrates are by definition visible to the unaided eye and are retained on a U.S. Standard No. 30 sieve (0.595 mm openings; APHA 1990). Many biological groups typically contribute to the macroinvertebrate community, including nematodes, aquatic worms (particularly oligochaetes), flat worms, crustacea (e.g., ostracods, isopods, decapods, amphipods), mollusks, and aquatic insects. This community is important in the cycling of organic and inorganic constituents, and it re-

presents an important source of food for many juvenile and adult fish. This community is often studied because of its value as an indicator of ecosystem quality.

The benthic macroinvertebrate community is very sensitive to environmental stress, so that features such as community structure, density, and species richness are useful in assessing the impact of pollution and perturbations in environmental quality (e.g., APHA 1990; Resh and Unzicker 1975; Wiederholm 1984). Assessment of the character of the benthic macroinvertebrate community is particularly valuable because of their limited movements compared to the overlying water mass. The organisms in a particular location are found there for extended periods, often with little change in community structure (Hynes 1970). Because of its relative stability, the benthic community may reflect the worst conditions that prevail at a particular location. However, seasonal dynamics of individual species may result in extreme variation at specific sites within any year.

6.4.2 Onondaga Lake/Near-Shore Zone

6.4.2.1 Previous Study/Methods for 1989 Samples

The benthic macroinvertebrate community of Onondaga Lake has previously received very little attention. In August of 1969 a total of eight Ekman dredge samples were collected; two depths at four different near-shore stations (Onondaga County 1971). No detailed identification or quantification of benthic invertebrates was reported; chironomid larvae and ostracods were the only groups mentioned. Large numbers of benthic organisms were observed at only one of the stations, located near the mouth of Onondaga Creek (Onondaga County 1971).

Presented here are the results of identification and enumeration of benthic macroinvertebrates from four near-shore littoral zone sites (Figure 6.41) collected on June 29, 1989. Two of the sites (W-1, W-2) are located on the western shore on the Allied waste-beds; the

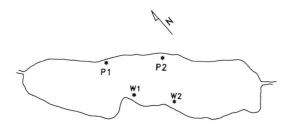

FIGURE 6.41. Benthic macroinverebrate sampling sites for Onondaga Lake, June 1989.

other two sites (P-1, P-2) are located on the east shore in Onondaga Lake Park (Figure 6.41). The physical character of the substrate for the sites was similar. The waste-bed sites consist mainly of small, irregularly shaped pieces composed mostly of calcium carbonate ($CaCO_3$), ranging in size from small pebbles to cobble, on top of a layer of soft sand and silt, also made up mostly of $CaCO_3$. The substrate at these sites appeared to be Allied waste-bed material, based on its similar appearance to the waste bed deposits located along the shore above the waterline and its similar behavior in dilute acid. The Allied waste-beds are known to extend into Onondaga Lake in this region (Blasland and Bouck 1989; Chapter 1). The park sites were dominated by "oncolites" (ovoid $CaCO_3$ enriched structures, see Chapter 8), mostly of pebble size. Gravel-like and china (porcelain) waste materials of similar size were also encountered, as well as some cobble size material. Calcium carbonate enriched sand and silt material occurred around and under these structures. Mats of algae attached to substrate were common at all the sites during the sampling. Six replicate samples were collected randomly from 64 half-meter square quadrants ($16 m^2$ area) from each of the four sites, using a Portable Invertebrate Box Sampler (PIBS; encloses an area of $0.1 m^2$). This sample size has been considered sufficient to characterize benthic communities (Canton and Chadwick 1988). Samples were collected in water depths of 10 to 50 cm. Samples were preserved with 70 to 80% ethanol.

The protocol used in the analysis had some inherent limitations. The sampling method used is not believed to support quantification of mollusks. Worms tended to deteriorate during sample handling and preservation. Quantification of nematodes was made difficult by their small size. Ostracods were not quantified due to their small size, though they were common in all the samples. Bryozoans were also not quantified.

6.4.2.2 Community Structure

The list of benthic macroinvertebrates identified in the near-shore littoral zone of Onondaga Lake in June 1989 is presented in Table 6.21. The level of identification differed substantially for the various groups of orga-

TABLE 6.21. List of near-shore macrobenthic organisms of Onondaga Lake New York, June 1989.

Group/genus/species
Turbellaria spp.
Nematoda spp.
Bryozoa
Plumatella repens
Oligochaeta
Enchytraeidae
Hirudinea
Glossiphoniidae
Gastropoda
Planorbidae *Gyraulus*
Physella spp.
Isopoda
Caecidotea spp.
Amphipoda
Gammarus fasciatus
Hydracarina
Limnesia
Insects
Odonata
Coenagrionidae coenagrion
Hemiptera
Corixidae
Coleoptera
Hydrophilidae spp.
Diptera
Ceratopogonidae
Chironomidae
Cricotopus sylvestris
Chironomus decorus
Glyptotendipes lobiferus
Tanytarsus guerlus
Parachironomus abortivus

nisms. The number of taxa is quite low; a number of common groups are not present. Crayfish, caddisflies and mayflies were absent, and the species richness of chironomids was extremely low (e.g., Simpson and Bode 1980; Winner et al. 1980). Presumably the low number of taxa of benthic macroinvertebrates is a manifestation of the polluted state of Onondaga Lake, including the altered condition of the substrate in the near-shore zone. The zebra mussel (*Dreissena polymorpha*) was found in the north basin of Onondaga Lake in the summer of 1992. This organism has become a nuisance in the Great Lakes and has invaded a number of central New York lakes via the Three Rivers system (also see Chapter 9).

Although the diversity of the population is low, concentrations are rather high. Organism density is presented for the four sites in Table 6.22. The precision of the counts of replicate samples was substantially better for the organisms present in the greatest densities. Note the very few organisms present in quantifiable populations (Table 6.22). The contributions of

the groups of organisms found are compared for the four sites in Figure 6.42a–d. Chironomids were dominant in numbers at all sites; they were overwhelmingly dominant at the waste bed sites. Oligochaetes contributed significantly to the benthic macroinvertebrate population at sites in Onondaga Lake Park but were present in much lower concentrations along the waste beds. The only other significant contributor was the amphipod, *Gammarus fasciatus*. This organism occurs widely in the northeast and is tolerant of high Cl^- concentrations (Bousfield 1973). The amphipod is substantially larger than the oligochaetes and chironomids encountered, therefore its contribution to biomass is greater than indicated by the numeric density data (Table 6.22 and Figure 6.42). Three species dominated the chironomids at all four sites (Table 6.22 and Figure 6.43a–d). All five of the chironomid species are known to be tolerant of polluted conditions (e.g., Simpson and Bode 1980). Substantial small scale (e.g., within $16\,m^2$ quadrants) and larger scale (e.g., distance between sample sites) variations in densities

TABLE 6.22. Density and precision of replicate counts of benthic macroinvertebrates for four stations on Onondaga Lake in June 1989.

| Group/genus/species | \multicolumn{8}{c}{Density (number·m⁻²)/Coefficient of Variation (C.V., fraction)} |
| | P-1 | | P-2 | | W-1 | | W-2 | |
	$\bar{X}$	C.V.	$\bar{X}$	C.V.	$\bar{X}$	C.V.	$\bar{X}$	C.V.
Nematoda	302	1.03	60	0.89	2	2.24	15	0.75
Oligochaeta								
Enchytraeidae	3063	0.46	2770	0.63	83	1.14	288	0.74
Gastropoda								
Planorbidae *Gyraulis*	2	2.24	8	0.82	0	—	60	1.05
Physella spp.	2	2.24	0	—	2	2.24	28	0.87
Amphipoda								
Gammarus fasciatus	715	0.58	377	0.13	130	0.70	360	0.64
Hydrachnidia								
Limnesia	47	0.94	5	1.00	3	2.24	5	1.53
Diptera								
Chironomidae	9280	0.39	4395	0.29	5347	0.48	10118	0.36
Cricotopus sylvestris	4946		1354		2663		3056	
Chironomus decorus	1763		1442		1684		4837	
Glyptotendipes lobiferus	2301		1587		920		1771	
Tanytarsus guerlus	269		13		48		455	
Parachironomus abortivus	204		105		32		0	

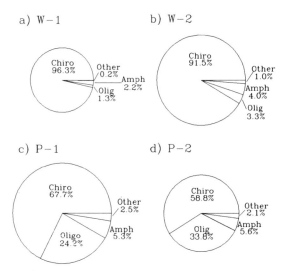

FIGURE 6.42. Contributions of organism groups to total density of benthic macroinvertebrates at four sites in the near-shore zone of Onondaga Lake in June 1989: Areas are proportional to density.

are manifested in the results (Table 6.22). The larger-scale differences are dramatized by the differences in the areas of the pie-charts of Figures 6.42 and 6.43. The results from this limited sampling do not depict a strong differ-

ence in benthic macroinvertebrate density between the shorelines of the park and the waste beds (Table 6.22 and Figure 6.42).

Species (taxon) diversity is represented here by the Shannon-Weiner index; values for the four Onondaga Lake sites and two sites on Cayuga Lake are presented for comparison in Table 6.23. The Cayuga Lake sites were characterized as part of an evaluation of the impact of a small volume NaCl discharge ("outfall," Table 6.23) along the southeastern shore of the lake on the proximate near-shore zone (Ringler and Wagner 1988). Chloride concentrations from 360 to 980 mg $\cdot$ L^{-1} were observed in the immediate vicinity of the outfall in that study (Ringler and Wagner 1988). Ambient lake concentrations (Effler et al. 1989) of $45-50$ mg $\cdot$ L^{-1} prevailed at the reference station in Cayuga Lake. The substrate was mostly cobble at both Cayuga Lake stations; samples were collected from water depths in the ranges $0-25$, $25-50$, and $50-75$ cm. Sampling and handling methods (Ringler and Wagner 1988) were very similar to those used for Onondaga Lake.

The diversity of the near-shore benthic macroinvertebrate population in Onondaga Lake is lower than reported for the reference site of Cayuga Lake (Table 6.23). This is probably a manifestation of the more polluted state of Onondaga Lake. Note that the diversities at the outfall on Cayuga Lake and along Onon-

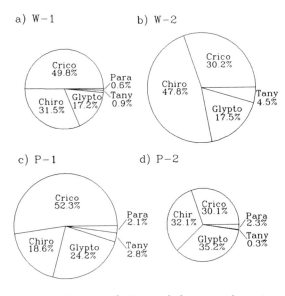

FIGURE 6.43. Contributions of chironomid species to total chironomid population at four sites in the nearshore zone of Onondaga Lake in June 1989: Areas are proportional to density.

TABLE 6.23. Shannon−Weiner Diversity Indices for Onondaga Lake (4) and Cayuga Lake (2) sites.

Lake	Site	Shannon−Weiner Index
Onondaga	W-1	0.18
	W-2	0.35
	P-1	0.84
	P-2	0.90
Cayuga*	Outfall	
	0−25 cm	0.83
	25−50 cm	0.92
	50−75 cm	0.65
	Reference	
	0−25 cm	1.31
	25−50 cm	1.92
	50−75 cm	1.86

* From Ringler and Wagner (1988)

daga Lake Park were similar (Table 6.23). The lower indices at the waste bed sites probably reflect more degraded conditions along that shoreline, perhaps associated with the industrial origins of the substrate in that area.

6.4.3 Onondaga Creek/Above and Below the Mud Boils

6.4.3.1 Previous Studies/Methods for 1991 and 1992 Study

The benthic macroinvertebrates of Onondaga Creek were examined by Simpson (1982) and Bode et al. (1989). The Simpson study was in response to a request by the Onondaga Indian Nation to investigate the effects of increased turbidity in Onondaga Creek below the "mud boils" (see Chapter 2). The survey conducted in 1989 by Bode et al. included all tributaries to Onondaga Lake. Both researchers found that the hyperturbid condition of Onondaga Creek below the "mud boils" had only a slight impact (e.g., "Community slightly but significantly altered from pristine state," Bode et al. 1989).

In 1991 the effects of hyperturbidity on macroinvertebrates were examined as part of a larger study of aquatic resources throughout the watershed of the south branch of Onondaga Creek. Sampling was conducted near the locations of the two previous studies, at approximately the same time of the year as the 1989 survey. The macroinvertebrate communities at two sites on Onondaga Creek in the Tully Valley were compared (Figure 6.44). The Snavlin Farms site is located 1.1 km above the mud boils in Tully NY. The second site was located 2.5 km below the mud boils, 200 m upstream from Webster Road.

Transparency and dominant substrate type at these sites were assessed in 1992. Twenty-two Secchi disk measurements were taken to measure transparency from May to October 1992. Thirty transects were conducted at both sites in 1992 to determine dominant substrate type. The percentage contribution of substrate in the following classes was estimated: boulder >50 cm, rock 15 to 50 cm, rubble 7.5 to 15 cm, gravel 0.6 to 7.5 cm, sand/silt <0.6 cm, and organic muck.

The macroinvertebrate communities were sampled at these sites using a Surber sampler (0.093 m²) on June 3, 1991. Three 2.5 minute samples were taken randomly in one riffle at each site. The samples were fixed with 10%

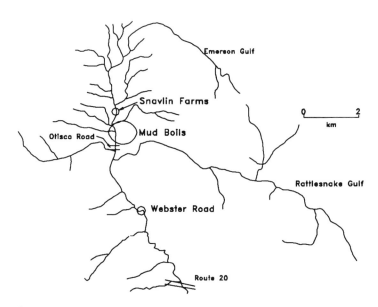

FIGURE 6.44. Benthic macroinvertebrate sampling sites, above and below the "mud boils," on Onondaga Creek, June, 1991.

formalin solution containing rose Bengal dye. Samples were later elutriated, and placed in 70% ethanol, and sorted using a Bausch and Lomb 3×30 stereoscopic dissecting microscope. All invertebrates were identified to family using taxonomic keys by Merritt and Cummins (1984) and Peckarsky (1991).

Macroinvertebrate community structures above and below the mud boils were analyzed using richness, Shannon–Weiner diversity (MacArthur and MacArthur 1961), evenness (Pielou 1966), relative abundance and density. The significance of differences between sites was evaluated using t-tests at the 95% level of confidence (Two Sample Analysis, Statgraphics version 2.6). A Trellis diagram (Wallwork 1970) was constructed to show similarity among samples and dissimilarity between sites.

6.4.3.2 Results and Discussion

Transparency was much lower at the downstream site, but the dominant substrate was quite similar (Table 6.24). There was relatively little variability in the Secchi disk measurements at Webster Road. The maximum measurement of 18 cm occurred during high flow condition, probably the result of dilution from upstream sources. The Secchi disk measurements at the Snavlin Farms site were highly variable with a maximum of greater than 150 cm (disk could be seen on pool substrate) and a minimum of 25 cm. The

minimum measurement occurred after a heavy rain when adjoining fields were freshly plowed. The size composition of the substrate at Snavlin Farms could be visually estimated, whereas the substrate at Webster Road had to be felt by hand to estimate composition due to the high turbidity. Despite the hyperturbid conditions at the Webster Road site, gravel was found frequently (Table 6.24).

There were clear differences in the macrobenthic communities between the two sites (Table 6.25). Sixteen families were found at the Snavlin Farms site and nine at the Webster Road site. The disparity in community indices between sites demonstrated dramatic

TABLE 6.24. Summary of 22 Secchi disk measurements from May to October 1992 and dominant substrate at 30 transects at Webster Road and Snavlin Farms study sites along Onondaga Creek and Tully Valley, New York.

	Webster Road	Snavlin Farms
Secchi Disk		
Mean	12 cm	87 cm
Maximum	18 cm	150 cm
Minimum	8 cm	25 cm
Dominant Substrate (%)		
Rock 6–20″	0	10
Rubble 3–6″	5	21
Gravel 1/4–3″	41	20
Sand/Silt	54	49

TABLE 6.25. List of taxa, their functional group and their abundance found at Snavlin Farms and Webster Road sites.

Taxon	Functional feeding group	Webster Road	Snavlin Farms
Collembola			
Isotomidae	Collector/Gatherer	1	0
Ephemeroptera			
Baetidae	Scraper/Gatherer	2	284
Ephermerellidae	Collect/Gatherer	0	1
Plecoptera			
Leuctridae	Shredder	0	2
Perlidae	Predator	0	3
Trichoptera			
Hydroptilidae	Scraper/Collector	0	13
Hydropsychidae	Collector/Filterer	1	84
Coleoptera			
Elmidae	Collector/Gatherer	1	162
Psephenidae	Scraper	0	1
Diptera			
Tipulidae	Varies	3	40
Ceratopogonidae	Predator/Collector	0	3
Simuliidae	Collector/Filterer	0	10
Chironomidae	Collector/Gatherer	2	94
Athericidae	Predator	0	2
Others			
Gammaridae		7	0
Oligochaeta	Collector/Gatherer	21	59
Platyhelminthes		0	3
Asellidae		1	0
Other		0	7
Total Organisms		39	768
Total Families		9	16

TABLE 6.26. Comparison of average diversity, richness, density, and evenness at Webster Road and Snavlin Farms (Tully Valley, New York, June 1991).

Community index	Site	
	Webster Road	Snavlin Farms
Diversity	0.9032*	1.50*
Richness	5.0*	10.676*
Density	139.78*	2753.68*
Evenness	0.5410	0.552

* Difference is significant at 95% level

changes in community structure above and below the mud boils (Table 6.26). Diversity, richness, and density were significantly different between the two sites. Densities of organisms at Snavlin Farms and Webster Road were 2752 and 139 individuals/m², respectively (Figure 6.45). Webster Road was dominated by Oligochaeta and Snavlin Farms by Ephemeroptera (Figure 6.45). A Trellis diagram further demonstrates the difference between sites above and below the "mudboils" and indicates that samples within sites were more similar than between sites (Figure 6.46).

The life of a lotic organism depends on adaptations to feeding, growing, and maintaining position in flowing water. Clinging and burrowing are primary adaptations for lotic invertebrates. Clingers have certain morphological attributes such as claws and flattened bodies that allow attachment to the substrate.

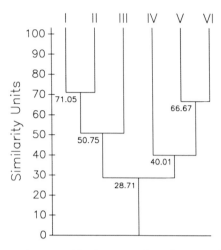

FIGURE 6.46. Trellis diagram of Surber samples at Snavlin Farms (I, II, III) and Webster Road (IV, V, VI), Tully Valley, NY, 1991.

Burrowers avoid the current by living within the hyporheic zone. High levels of accumulated sediment may produce a shifting bottom resulting in conditions that make it difficult for clingers to maintain position in the stream. At Webster Road, 26 of the 39 organisms collected were burrowers. In contrast, at Snavlin Farms clingers (560 individuals) were more abundant than burrowers (206 individuals) (Table 6.25).

The distribution of functional feeding groups (Table 6.25) was influenced by the hyperturbidity originating from the mud boils. Primary producers are an important food source for many aquatic macroinvertebrates such as scrapers and grazers. Very little light penetrates the water at the Webster Road site, so that primary production is reduced. Of the 39 organisms found at Webster Road only a single filterer was collected. Most filterers depend on nets for obtaining food items from the current. The large amounts of sediment would tend to foul these nets, reducing their ability to obtain prey. In contrast two families of filterers, Hydropsychidae and Simuliidae, comprised 13% of the organisms collected at Snavlin Farms. Many invertebrates are visual predators. Consequently, poor visibility due to high levels of suspended sediment may reduce their ability to locate prey. The Snavlin Farms

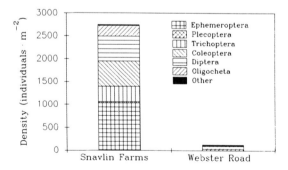

FIGURE 6.45. Density of macrobenthic invertebrate taxa at the Snavlin Farms and Webster Road sites, June 1991.

site had three predator groups, whereas no predators were identified at Webster Road. These differences in functional feeding groups indicate the benthic macroinvertebrate community at Webster Road has been greatly modified as a result of the high levels of suspended sediment discharged upstream from the "mud boils."

The results presented here (Tables 6.25 and 6.26; Figure 6.45) indicate a biological impact of the "mud boils" beyond that reported earlier by Simpson (1982) and Bode et al. (1989). The macrobenthic organisms should be identified to a lower taxonomic level to more fully resolve these differences in findings.

6.5 Fish Communities and Habitats in Onondaga Lake, Adjoining Portions of the Seneca River, and Lake Tributaries

6.5.1 Introduction

The Onondaga Lake system was historically a high quality fishery characterized by large and consistent catches of landlocked Atlantic salmon (*Salmo salar*) and ciscoes (*Coregonus artedii*), known locally as Onondaga Lake whitefish. The presence of ciscoes is indicative of a cold water community. Therefore, dissolved oxygen was present in adequate concentrations (e.g., $>5\,mg \cdot L^{-1}$) in at least portions of the hypolimnion throughout summer. Additional species included American eel (*Anguilla rostrata*) and burbot (*Lota lota*). The Onondaga Lake fauna was included in the New York State Biological Survey of 1927, but the system was already markedly perturbed by that time, with extinctions and establishment of exotic species well underway. Fishery surveys were subsequently carried out in 1946, 1969, 1972, and 1980. Perhaps the general condition of the lake and nearby area diminished interest by fishery scientists intent on studies of more attractive waters. Primarily in response to the challenge of Professor

Robert Hennigan to restore Atlantic salmon to the region by the year 2000, a set of fisheries investigations was initiated in 1987.

This section outlines the historical and current fish community structure of Onondaga Lake, its principal tributaries, and the Seneca River. Movements of individuals within, to and from the lake are considered, basic patterns of age, growth, and survivorship are described, and trophic relationships within the lake and river are considered. Habitat features that appear to be limiting to the development of natural fish communities and a fishery are considered, including low dissolved oxygen levels during fall turnover, and fish flesh contamination by mercury.

6.5.2 Historical Information for Onondaga Lake

The first scientific survey of the fish fauna of the Onondaga Lake system appears to have been carried out by Greeley (1928) as part of a biological survey of the Oswego River watershed (Table 6.27). A variety of written reports document fish and fishing activities in the region nearly three centuries earlier, when a cold water fishery prevailed. Father LeMoyne observed Atlantic salmon in Onondaga Lake in 1654 (Beauchamp 1908). Smith (1892) reported that Jesuits caught adult salmon in the Oswego River in July, 1654; based on fisherman reports he wrote that the salmon entered the Oswego system in June and were present "most of the time." Dewitt Clinton (1815; cited in Webster 1982) noted that the dam built at Baldwinsville did not appear to promote the passage of fish, and a local tavern owner believed that the dam "injured" the salmon fishery. Thurlow Reed, a contemporary of Dewitt Clinton, had observed Onondaga Indians capturing Atlantic salmon at night with clubs and spears at the mouth of Onondaga Creek (Beauchamp 1908).

Atlantic salmon, presumably landlocked populations migratory to Lake Ontario, were a prominent feature of the Onondaga Lake system when the region was first settled. Fox

TABLE 6.27. Fish surveys of Onondaga Lake 1927–1991.

Species	H*	27	46	69	80	89	90	91
					Year			
1. Sea lamprey								x
2. Longnose gar					x	x	x	x
3. Bowfin					x	x	x	x
4. American eel								x
5. Alewife			x		x	x		x
6. Gizzard shad					x	x	x	x
7. Channel catfish			x	x	x	x	x	x
8. Brown bullhead				x	x	x	x	x
9. Yellow bullhead							x	x
10. White sucker		x		x	x	x	x	x
11. Shorthead redhorse		x	x	x	x	x	x	x
11a. redhorse spp.	x	x	x					
12. Carp		x	x	x	x	x	x	x
13. Creek chub								x
14. Spottail shiner								x
15. Spotfin shiner								x
16. Golden shiner		x	x		x	x	x	x
17. Emerald shiner (Buckeye)		x	x	x		x		
18. Bluntnose minnow		x						x
19. Fathead minnow								x
20. Rudd								x
21. Cisco	H							
22. Atlantic salmon	H							
23. Brown trout						x	x	x
24. Rainbow trout								x
25. Tiger trout (hybrid)							x	
26. Brook trout								x
27. Lake trout					x			
28. Rainbow smelt								x
29. Central mudminnow								x
30. Grass pickerel		x						
31. Chain pickerel								x
32. Northern pike			x		x	x	x	x
33. Tiger muskellunge (hybrid)						x	x	x
34. Burbot							x	
35. Banded killifish		x	x			x	x	x
36. Brook silverside						x	x	x
37. Brook stickleback				x		x	x	x
38. White perch				x	x	x	x	x
39. White bass			x			x	x	x
40. Smallmouth bass				x	x	x	x	x
41. Largemouth bass		x			x	x	x	x
42. White crappie					x	x	x	x
43. Black crappie					x	x	x	x
44. Rock bass						x	x	x
45. Bluegill				x	x	x	x	x
46. Green sunfish								x
47. Pumpkinseed		x		x	x	x	x	x
48. Tesselated darter								x
49. Logperch			x				x	x
50. Yellow perch		x	x	x	x	x	x	x
51. Walleye			x	x	x	x	x	x
52. Freshwater drum				x	x	x	x	x

*H = Historical accounts pre-1900

(1930) wrote that salmon were abundant in the Seneca River and Onondaga Lake in 1870s. Webster (1982) considered the Lake Ontario tributaries, which are dominated by the Oswego River system, to have been the most striking example of freshwater colonization by Atlantic salmon anywhere in the world. Onondaga Lake and its tributaries were an essential component of this unique distribution.

Greeley (1928) used seines, gill nets, traps, and set lines to determine the species in the lake and its tributaries. Twelve species were found in the lake (Table 6.27). Lake sturgeon had been recorded in the Seneca and Cayuga Canal near Montezuma and occasionally in the lower Oswego River, but there seem to be no records of its distribution in the Onondaga system proper. Greeley (1928) listed 11 species for Onondaga tributaries. Interestingly, the number of species in the Seneca River at that time (39) was nearly four times as great as in the lake. Diversity measures cannot be applied to the data as reported, but overall the lake community appeared to be relatively depauperate compared to other local lakes of comparable size and volume.

Stone et al. (1946) surveyed Onondaga Lake during June and August 1946. Although their primary emphasis was fishes, they observed a single species of aquatic plant, *Potamogeton pectinatus*, and a sparse benthic community consisting of *Gammarus* sp. and small chironomid larvae, similar to recently reported findings (see earlier sections of this chapter). A total of 413 fish was captured June 18–20, 1946, using multiple mesh (experimental) gill nets at four sites and trap nets at three sites. The 14 species (Table 6.27) exceeded the number reported by Greeley (1928), but more than 93% of the catch was carp (*Cyprinus carpio*). Seining results for August 1946 were largely unquantified; the species identified included logperch (*Percina caprodes*), juvenile sunfishes (*Lepomis* spp.), white bass (*Morone chrysops*), and carp.

Notes entitled "Onondaga Lake trapnet records" attached to the Stone et al. (1946) report provide some data for the 1960s (Table 6.28), a period for which there is little informa-

TABLE 6.28. Trap net catch records for Onondaga Lake, 1963–1964 (NYDEC).

	June 26–28, 1963	June 11–12, 1964
Channel catfish	210	14
Brown bullhead	30	7
White bass	190	0
Smallmouth bass	14	17
Largemouth bass	0	4
Walleye	30	8

tion available. The absence of carp is especially noteworthy, but these data may not represent the total catch from their efforts.

Noble and Forney (1969) surveyed the fish populations of the lake in 1969 using trap and gill nets. Their sample was comprised of 14 species (Table 6.27). They described the Onondaga Lake fishery as a warm water fish community with growth rates comparable to those of other waters in the northeast.

Murphy (1978) concluded, based primarily on Noble and Forney's (1969) analysis, that the only change evident in the Onondaga Lake fish community was the presence of significant numbers of white perch (*Morone americana*) beginning in 1969. Mills and Forney (1988) indicated that white perch reached nearby Oneida Lake from the Mohawk–Hudson River system late in the 1940s, with low numbers present through the late 1960s and early 1970s. With the interconnection of these lakes via the Seneca and Oneida Rivers, similar population levels may have existed in Onondaga Lake at that time. White perch may have been too rare in Onondaga Lake to be caught in the trap nets set in 1963 and 1964 (Table 6.28).

Chiotti (1981) reviewed data based on collections of more than 4800 fish in conjunction with a sampling program to monitor mercury in fish tissue. Trap nets, gill nets and beach seines were employed on July 1, 1980 and June 30, 1981. A total of 22 species was collected in this composite sample (Table 6.27), dominated by white perch (63%) and alewife (*Alosa pseudoharengus*) (14%). The relative contribution of carp to the community declined dramatically sometime between 1946 and 1980. Gizzard shad (*Dorosoma cepedianum*), currently a major component of the ichthyofauna of the lake, represented less than 2% of the sample. The results of seining the most suitable inshore areas led Chiotti (1981) to conclude that significant year classes of young fish were not present in Onondaga Lake in 1980. An examination of size distributions of black crappies (*Pomoxis nigromaculatus*), yellow perch (*Perca flavescens*), and pumpkinseed sunfish (*Lepomis gibbosus*) suggested that they had experienced low recruitment in other years as well. With the exception of white perch, fish reproduction in the early 1980s was characterized as "sporadic" (Chiotti 1980).

6.5.3 Fish Community Structure

6.5.3.1 Onondaga Lake: 1989–1991

Fish were collected at 28 sites along the margin of Onondaga Lake and 12 sites in the limnetic zone (Figure 6.47) in the May–

November interval of each year over the 1989–1991 period. Inshore sampling methods included the use of Indiana, South Dakota, and Oneida trap nets with 23 m leaders and 11.5 m wings with 0.635 cm mesh. The leaders of the trap nets were set perpendicular to shore or other near-shore structures in water 1–3 m in depth; the wings were set at a 45° angle from the leader and trap (Nielson and Johnson 1983). Fish were removed from the traps daily. Traps were checked five days a week and left open on weekends. Inshore sampling was not conducted in the south end of the lake due to unstable substrate conditions.

Seine sweeps were conducted biweekly at four locations around lake margins to examine assemblages of juvenile fish in the littoral zones (Figure 6.47). Two continuous 50 m sweeps were made parallel to the shoreline in water <1 m in depth with a 20 m bag seine constructed of 0.635 nylon mesh.

Experimental gill nets, 50 m in length, with 3.8, 5.1, 6.4, 7.6, 8.9 and 10.2 cm stretch mesh, were set at three week intervals at 5

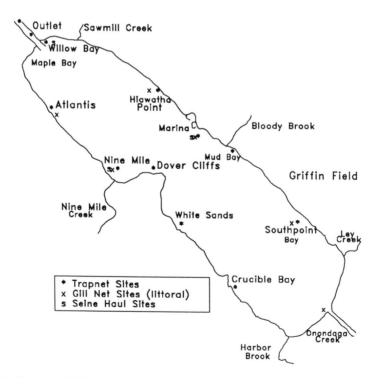

FIGURE 6.47. Distribution of fish sampling sites in Onondaga Lake, 1989–1991.

locations perpendicular to shore in water from 1 to 4 m (Figure 6.47). Gill nets were set either during the day or overnight, depending on the purpose of a particular set (2 hr set to gather fish for stomach samples, overnight sets for species assemblages), and checked daily. Gill netting conducted in the limnetic zone in 1991 is described subsequently. More than 50,000 fish were captured as part of the 1989–1991 sampling program.

A total of 28 species were collected in 1989, 31 in 1990 and 45 in 1991 (Table 6.27). The 1989–1991 data suggest an increasing trend in Onondaga Lake diversity from earlier surveys (Greeley 1928; Stone et al. 1946), though the enhanced sampling effort during recent years, coupled with a lack of quantification in historical collections, makes a comparative analysis difficult. The increased number of species recorded from the lake in the most recent studies (1989–1991; Table 6.27) probably reflects improvements in water quality associated with reductions in material loading from METRO and closure of the soda ash/chlor-alkali facility (Canale and Effler 1989; Effler 1987; Effler et al. 1990). Comparison of Shannon–Weiner diversity indices of Onondaga Lake with those calculated for other New York lakes (Wolf, Deer, Cranberry) shows that Onondaga Lake is currently quite similar in diversity to these other systems (Figure 6.48). However, it should be noted (as discussed subsequently) that in contrast to these comparison lakes, many of the species

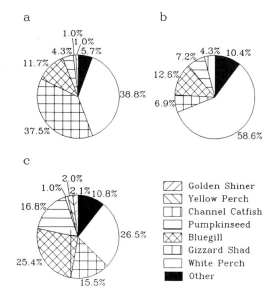

FIGURE 6.49. Fish community structure based on trap net catches in Onondaga Lake: (**a**) 1989, (**b**) 1990, and (**c**) 1991.

found in Onondaga Lake do not reproduce there.

The community structure apparently underwent little change from 1989 to 1990. These collections document a fish community dominated by planktivorous species including white perch and gizzard shad (Figure 6.49). The most conspicuous difference between the two years was the contribution of gizzard shad; this species represented 37% of the total catch in 1989 and only 7% of the total catch in 1990 (Figure 6.49). Insectivorous species such as pumpkinseed and bluegill (*Lepomis macrochirus*) were also abundant in these trap net catches. Bluegill and pumpkinseed numbers increased in trapnet catches to comprise nearly half of the total fish captured in 1991 (Figure 6.49).

Seasonality in community structure was a prominent feature of the Onondaga Lake system during the 1989–1991 study period (e.g., Figure 6.50). Large numbers of white perch dominated catches in early June 1990. By late August and early September of that year bluegill and gizzard shad comprised the majority of the trap net catch. Gizzard shad and white perch were dominant in the trap

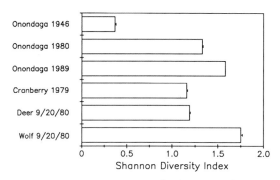

FIGURE 6.48. Comparison of Shannon diversity indices for Onondaga Lake and other selected New York (Adirondack Mountains region) lakes.

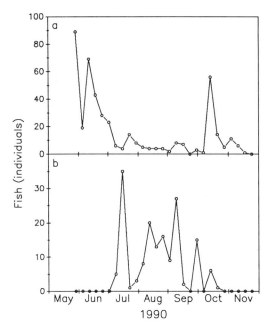

FIGURE 6.50. Temporal distributions of fish in trap net catches, Onondaga Lake, 1990: (**a**) pumpkinseed sunfish, and (**b**) channel catfish.

net catches in late September and early October of 1990. Pumpkinseed were common in trap net catches throughout the spring, summer, and fall (Figure 6.50a); this species spawns within the confines of the lake. However, species such as channel catfish did not appear in trap net catches until later in the season (Figure 6.50b); they are apparently unable to spawn in the lake. In 1990 the Shannon–Weiner diversity index was initially low in June due to the predominance of white perch in trap nets. Diversity of the inshore community gradually increased, reaching a peak in late July and then declined in October as gizzard shad began to dominate the catches.

Changes in the community structure were also evident during fall turnover. Some species that persist throughout the spring and summer were not as abundant or were absent from the lake in the fall (such as channel catfish and redhorse sucker (*Moxostoma macrolepidotum*)). This sudden decline is probably related to the reduction in dissolved oxygen levels during fall turnover (Chapter 5). The effects of diminished oxygen levels during fall turnover

are discussed in greater detail subsequently in this section. Migration between the lake and the Seneca River is an important contributor to the major seasonal changes observed in community structure (e.g., Figure 6.50).

Limited fishing pressure over the last several decades has probably contributed to the high biomass of certain species in recent years. Upon reopening of a consumptive fishery, angling success could be relatively high until standing crop is reduced (Chiotti 1981). Species that may be important to the establishment of a catch and release (nonconsumptive) warm water sport fishery for Onondaga Lake include smallmouth bass (*Micropterus dolomieu*) and largemouth bass (*M. salmoides*), which were relatively common throughout the lake during much of the summers of the 1989–1991 interval. The modest populations of walleye (*Stizostedion vitreum vitreum*), tiger muskellunge (*Esox masquinongy* × *Esox lucius*) and northern pike (*E. lucius*) may also support some fishing.

Angling activity is currently concentrated at the north end of the lake, particularly near the lake outlet. A few boat anglers fish for bass around old pier foundations and other structures with moderate success. The SUNY. ESF Student Chapter of the American Fisheries Society sponsored an 8 hr bass fishing derby on September 21, 1991, in which 22 anglers participated. A total of 134 bass were caught, many weighing 2 to 2.5 pounds; the derby winner weighed 4 lbs 1 oz. Other gamefish, including northern pike, channel catfish, walleye, and tiger muskellunge, appear to be rarely caught by lake anglers, perhaps because of the large standing crop of forage fishes, and the relatively low abundance of these species in the lake.

6.5.3.2 Fish Communities of the Seneca River

An investigation was conducted in 1991 to examine the fish community structure of the Seneca River near the Onondaga Lake outlet. Fish were sampled biweekly from June–October 1991. Four sites were monitored; one on the east side of Klein Island, one on the west side, and two in the main channel of the Seneca River, upstream and downstream of the lake inflow (Figure 6.51). Site designations correspond to navigation bouy numbers. The river was 55–65 m wide and 4.5–6.6 m deep at these locations. Experimental gill nets (50 m × 2 m) with mesh sizes in series from 3.8 cm to 10 cm were used for capturing fish. Heavy boat traffic and channelization of the river prohibited the use of trap nets. Two gill nets were set at each site for approximately two hours. The nets were deployed as bottom sets; one net was set parallel to the shoreline (about 5 m away from the shore) and the other perpendicular to flow. The perpendicular gill net stretched almost from shore to shore, to a depth covering a major portion of the river cross section. This deployment feature was also used to identify the direction of fish movement, based on the direction which they were gilled in the net.

A highly diverse fish community was encountered within the bounds of the sampling program. Twenty-three species were collected (Table 6.29); 21 were captured by gill net sampling, and two in minnow traps. The relative abundances of the species based on the total catch, are presented in Figure 6.52. White perch were the most abundant species, comprising 35% of the catch, followed by gizzard shad at 31%. A breakdown of the trophic categories of the catch is presented in Figure 6.53. The planktivorous species accounted for about two-thirds of the total riverine catch; the benthic piscivore-insectivore and littoral planktivore-insectivore species were the next most abundant trophic types. The relative abundance of the major species fluctuated widely from June to October (Figure 6.54; total catch for all four sites). White perch dominated the catch in June, August, and October (Figure 6.54a, c, and e). Gizzard shad dominated in July (Figure 6.54b) and golden shiner (*Notemogonis chrysoleucas*) in September (Figure 6.54d). The catch of channel catfish (*Ictalurus punctatus*) was greatest in June, but their numbers dropped in July, presumably because many of them

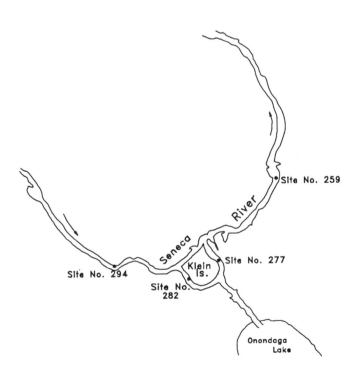

FIGURE 6.51. Seneca River/Lake Outlet fish sampling sites, 1991.

TABLE 6.29. Summary of fish surveys of Seneca River 1927–1991.

	1927	1978	1989	1991
1. Lake sturgeon	x			
2. Longnose gar	x			
3. Bowfin	x			
4. American eel	x			
5. Alewife	x			x
6. Gizzard shad			x	x
7. Channel catfish	x		x	x
8. Brown bullhead	x	x	x	x
9. Yellow bullhead	x			
10. Lake chubsucker	x			
11. Whitenosed redfin sucker	x			
12. White sucker	x			x
13. Shorthead redhorse	x			x
14. Carp		x	x	x
15. Goldfish		x		
16. Hornyhead chub	x			
17. Bridle shiner	x			
18. Blacknose shiner	x			
19. Mimic shiner	x			
20. Satinfin shiner	x			
21. Spottail shiner	x			
22. Spotfin shiner				x
23. Common shiner	x		x	
24. Golden shiner			x	x
25. Rosyface shiner	x			
26. Cutlips minnow	x			
27. Silvery minnow	x			
28. Slender minnow	x			
29. Chain pickerel	x	x		
30. Northern pike	x		x	x
31. Tiger muskellunge (hybrid)	x		x	
32. Banded killifish	x			
33. Brook silverside				x
34. Brook stickleback	x			
35. White perch		x	x	x
36. White bass	x			x
37. Smallmouth bass	x		x	x
38. Largemouth bass	x		x	x
39. White crappie			x	x
40. Black crappie	x		x	x
41. Rock bass	x		x	x
42. Bluegill	x		x	x
43. Pumpkinseed	x	x	x	x
44. Tesselated darter	x			
45. Logperch	x			
46. Yellow perch	x		x	x
47. Walleye	x			x
48. Freshwater drum			x	x

entered the lake (see the complementary increase in the lake population, Figure 6.50b).

Gizzard shad and white perch were most abundant at Sites 282 and 277 around Klein Island, proximate to the lake (Figure 6.51). Overall, the number of species of fish captured at these sites ranged from 6 to 10 per net during the study period. Although few small-

mouth bass and walleye were caught in the river during the summer, their contribution increased markedly in the September and October samples (Figure 6.54). The upstream site (No. 294) had a fairly diverse fish community, with 19 species captured during the study. No single species dominated the entire study period. Channel catfish, freshwater drum, walleye, golden shiners, and shorthead redhorse occurred in appreciable numbers. Gizzard shad was the dominant species in October, comprising 28% of the sample; walleye was the second most abundant species at 13%, followed by channel catfish and yellow perch, each contributing about 10%. Compared to the other three sites, the downstream site (No. 259) had fewer species on all sampling dates. Channel catfish, which represented only a small portion of the catch at that site during the early summer, increased significantly toward the fall. It was the second most abundant species in September, and the most abundant species in October, comprising 50% of the sample at the downstream site.

The fish community in the outlet and around Klein Island appears to represent an extension of the lake; gizzard shad and white perch dominated catches in this area. Some fish recaptured in the river were originally fin-clipped or tagged in the lake. Movements as

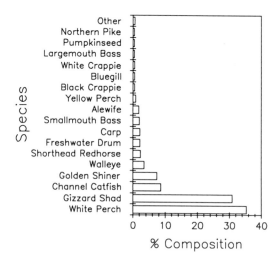

FIGURE 6.52. Fish community structure for Seneca River/Lake Outlet, based on all gill net catches over the June–October interval of 1991.

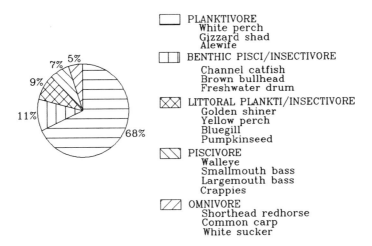

FIGURE 6.53. Fish community structure for Seneca River/Lake Outlet for the June–October interval of 1991, based on trophic categories.

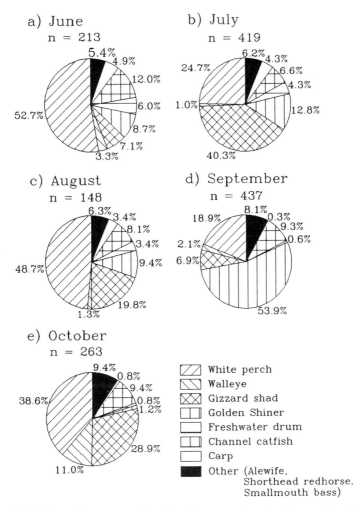

FIGURE 6.54. Monthly fish species composition in gill net catches for Seneca River/Lake Outlet, 1991.

great as 20 km were documented from anglers' tag returns from several areas along the river. The only trend that emerged from the analysis of the orientation of fish in the gill nets was marked downstream movement by both gizzard shad and white perch in September.

Only limited qualitative historic data on the fish community of the Seneca River exists, so that a detailed analysis of changes in the fauna is not possible. In the 1927 survey of the Oswego watershed Greeley (1928) reported 39 species of fishes in the Seneca River (Table 6.29). A limited survey conducted by the NYSDEC documented only six species, including white perch, brown bullhead (*Ameiurus nebulosus*), carp, goldfish (*Carassius auratus*), pumpkinseed sunfish, and chain pickerel (*Esox niger*). White perch were the most abundant, followed by carp and goldfish (NYSDEC data base, personal communication, Dan Bishop, NYSDEC). An electroshocking survey conducted by the NYSDEC in 1989 found 18 species (Table 6.29); bluegill were the most abundant, followed by carp, freshwater drum (*Aplodinotus grunniens*), and pumpkinseed sunfish (Figure 6.55).

6.5.4 Natural Reproduction

Abundance of juvenile fish among littoral habitat types is influenced by habitat characteristics and species adaptations (Meals 1991). Despite a highly modified littoral zone, many areas of Onondaga Lake support the nesting and rearing of juvenile fish. Recent sampling indicates that bluegill, pumpkinseed, and white perch reproduced successfully in 1989 and 1990, when thousands of young of the year fish were collected. However, sampling in 1992 revealed few young of the year fish for all species. This may be directly related to the unusual weather conditions that occurred in the spring that resulted in high water levels and unusually cold temperatures during spawning. Juvenile sunfish and largemouth bass have been collected where adults were observed to spawn. Densities of nests for *Lepomis* sp. as high as 0.56 nests · m^2 were observed by Sagalkin (unpublished) in 1991; a total of 1587 nests were counted. Other species reproduce within Onondaga Lake but without equivalent success. Chiotti (1981) reported that smallmouth bass were restricted to the northwest portion of the lake. Recent collections of smallmouth bass revealed significant populations of adults throughout the entire lake, but recruitment appears to be limited to the north end (Auer and Ringler, unpublished data).

A summary of levels of natural fish reproduction in Ononaga Lake based on observations for 1991 is presented in Table 6.30. Many species currently do not reproduce within Onondaga Lake. Only 16 of 48 species found in 1991 (Table 6.27) are known to reproduce in the lake. Walleye and northern pike likely immigrate into the lake from the Seneca River, and perhaps from other bodies of water such as Oneida and Cross Lakes. This hypothesis is based on the absence of juvenile pike and walleye in the lake, and the apparent lack of appropriate spawning habitat for these species within the lake. Fin clipped walleye believed to have originated in Oneida Lake have been collected by the NYDEC in Onondaga Lake (Dan Bishop, NYSDEC, personal communication).

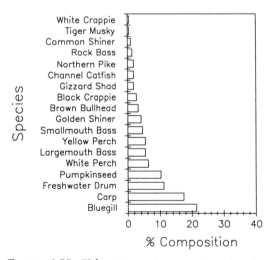

FIGURE 6.55. Fish community structure for the Seneca River (see Figure 6.51) based on an electroshocking survey (224 fish) by NYSDEC (NYSDEC data base, Cortland regional office) in 1989.

TABLE 6.30. Levels of natural fish reproduction in Onondaga Lake 1991, based on catches in shoreline seine hauls.

High success (>1000 juveniles)

White perch	Bluegill	
Gizzard shad	Brook silverside	Golden shiner?
Banded killifish	Pumpkinseed	

Moderate success (100–1000 Fish)

Largemouth bass	Yellow perch	Emerald shiner
Carp		

Low success (1–100 Fish)

Smallmouth bass	Northern pike	Brown bullhead
Black crappie	Spotfin shiner?	

No success or unknown (0 Fish)

Bowfin	White crappie	Redhorse sucker
Rudd	Rock bass	Bluntnose minnow
Brook stickleback	Redfin shiner	Central mudminnow
Fathead minnow	Creek chub	Common shiner
Spottail shiner	Logperch	Tesselated darter
Channel catfish	Walleye	White bass
Brown trout	Alewife	Longnose gar
Burbot	Rainbow trout	Rainbow smelt
Green sunfish	Chain pickerel	Freshwater drum

Anadromous/Catadromous spawners

White sucker	Sea lamprey	American eel

Hybrid—nonreproductive species

Tiger trout	Tiger muskellunge

TABLE 6.31. Summary of shoreline seine results, Onondaga Lake 1991.

Species	Number netted	% composition catch
Bluegill	4159	45.6
Gizzard shad	2436	26.7
Banded killifish	648	7.1
White perch	556	6.1
Creek chub	246	2.7
Brook silverside	227	2.5
Pumpkinseed	213	2.3
Largemouth bass	165	1.8
Yellow perch	109	1.2
Golden shiner	79	<1
Carp	78	<1
White sucker	43	<1
Tesselated darter	41	<1
Logperch	41	<1
Bluntnose minnow	32	<1
Spotfin shiner	23	<1
Emerald shiner	11	<1
Smallmouth bass	3	<0.1
Black crappie	2	<0.1
Brook stickleback	1	<0.1
Total	9113	100

A total of 9113 juveniles representing 20 species were collected by seine during the 1991 field season (Table 6.31). Substantial reproduction of white perch, bluegill, pumpkinseed, and gizzard shad was evident, as these species continually dominated the catches. Banded killifish (*Fundulus diaphanus*) and brook silversides (*Labidesthes sicculus*) were also present in significant numbers. Smaller numbers of juvenile carp, largemouth bass, smallmouth bass, yellow perch and white sucker (*Catostomus commersoni*) were collected (Table 6.31).

Catch rates varied among study sites (Figure 6.47) around the lake. Willow Bay, Maple Bay, and the Marina had the highest concentrations of juveniles, whereas few juveniles were collected at the Dover Cliffs and Griffin Field sites. Areas characterized by the presence of aquatic macrophytes and submerged structures (e.g., Maple Bay) supported the largest populations of juveniles. Few juveniles were taken at sites devoid of aquatic macrophytes

and other structures. Differences in fish abundance among sites may also be influenced by the amount of cover and turbidity during nesting. For example, for approximately a week in early June 1991, pumpkinseed sunfish nests were abandoned when heavy silt loads entered Maple Bay during a summer storm. Shoreline areas unprotected from the wind (e.g., Dover Cliffs, Wintersands, and Griffin Field) supported less macrophytes and juvenile fish. Krammer and Smith (1962) reported that year class strength of largemouth bass in Lake George, Minnesota, was primarily determined by wind. They found, wave action covered eggs with sand or removed them from optimal nesting areas.

6.5.5 Age, Growth, And Survivorship Patterns

Age and growth data are often used to (1) detect changes and diagnose environmental or ecological conditions influencing growth, (2) estimate age structure, year class strength, and mortality, and (3) calculate indices such as longevity and years required to reach sexual maturity (Hammers 1991). To describe age distribution, growth pattern, and mortality rates of dominant fish species in Onondaga Lake, at least 50 individuals of these species were examined. To determine growth rates, scales were removed from the left side of the body below the lateral line near the apex of the depressed pectoral fin. A minimum of 25 scales were removed from each fish and placed in scale envelopes with a record of length (mm), weight (g), date, and place of capture. Scale samples were mounted between glass slides and projected with a 40× Ken-A-Vision microprojector. Ages were determined by interpreting annual growth marks. Annuli were verified by comparison of recaptured fish tagged in 1988 and 1989, along with examination of size frequency histograms. Growth history was established by back calculation of length (Nielsen and Johnson 1983), which assumes a linear relationship between scale and body growth throughout the life of the fish. This assumption was tested by fitting the data using least squares regression. The following formula was employed for back calculation (Everhart and Youngs 1981)

$$L_n = C + [S_n/S(L - C)] \qquad (6.12)$$

in which L = total length of fish at capture, S = total scale radius, L_n = length of fish when annulus "n" formed, S_n = radius of annulus "n", and C = correction factor based on the species of fish (Carlander 1977). Length and weight are highly correlated, thus by knowing one the other may be predicted. Length and weight relationships of smallmouth bass, black crappie, and pumpkinseed observed in Onondaga Lake in 1992 appear in Figure 6.56a–c.

Growth rates of fish in Onondaga Lake were similar to those of other northeastern lakes, and rather uniform over the 1988–1990 period. Pumpkinseed sunfish in Onondaga Lake exhibited growth rates (Figure 6.57a) similar to those determined in 14 Pennsylvania lakes (Carlander 1977). Bluegill, white perch, and gizzard shad also had growth rates comparable to fish of other northeastern lakes.

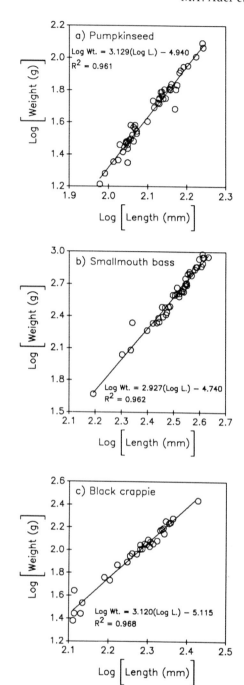

FIGURE 6.56. Relationship between fish weight and length in Onondaga Lake, 1992.

Pumpkinseed sunfish collected from 1988–1990 showed little change in average length at age by year class (Figure 6.57b). Yellow perch (Figure 6.57c) and white perch (Figure 6.57d)

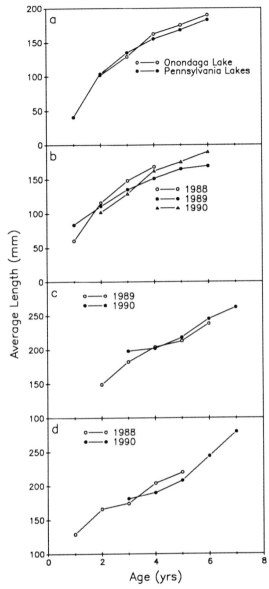

FIGURE 6.57. Comparison of fish length at age: (a) pumpkinseed sunfish in Onondaga Lake (1990) and selected Pennsylvania lakes (Carlander 1977), (b) pumpkinseed sunfish in Onondaga Lake in 1988, 1989, and 1990, (c) yellow perch, 1989, and 1990, and (d) white perch, 1988 and 1990.

ever these estimates assume a constant recruitment of all year classes, and that all ages used in the estimate are equally vulnerable to sampling gear. Since these assumptions are probably not met for most populations in Onondaga Lake, mortality rates presented here (Table 6.32) should be considered a first approximation. Instantaneous mortality rate (Z) can be determined from the natural logarithm of the slope of the descending limb of the catch curve (annual percentage mortality (S) is equal to e^{-Z}; see Figure 6.58). Mortality rates of two predatory fish, largemouth and smallmouth bass were estimated to be 15% and 18%, respectively (Table 6.32) These unusually low values may be an artifact of a small sample size. White perch apparently had a higher annual mortality (67%) than yellow perch (40%). The estimated annual

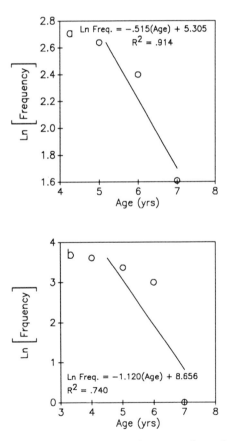

FIGURE 6.58. Catch curves for Onondaga Lake fish, 1990: (a) yellow perch, and (b) white perch.

also showed similar growth patterns from 1988–1990, although the maximum ages in the samples varied.

Mortality rates were determined from catch curves (e.g., Figure 6.58) based on the number of fish in each age class (Ricker 1975). How-

TABLE 6.32. Mortality and survivorship rates of selected populations in Onondaga Lake, 1992.

Species	Z*	S**	A***	Number of individuals
Largemouth bass	−0.16	0.85	0.15	33
Smallmouth bass	−0.19	0.82	0.18	26
White perch	−1.12	0.33	0.67	95
Yellow perch	−0.51	0.60	0.40	53
Bluegill	−2.01	0.13	0.87	86
Pumpkinseed	−1.31	0.27	0.73	48
Black crappie	−0.37	0.69	0.31	48
Gizzard shad	−0.22	0.80	0.20	58

* Z: Instantaneous rate of mortality
** S: Annual rate of survivorship
*** A: Annual rate of mortality (note A = 1 − S)

rates of mortality of bluegill and pumpkinseed were 87% and 73% respectively.

6.5.6 Movements of Lake Fishes

Fish movements were characterized primarily by marking and/or tagging and recovery of 29 species in 1990 and 1991; features of the program are described in Table 6.33. Fish were captured using trap nets as previously described, and marked according to site (Figure 6.47). Larger individuals of several species were tagged using numbered floy tags inserted below the dorsal fin, except gizzard shad, which rarely survived the tagging procedure. Small fish species such as brook silversides were not tagged. Commonly tagged species included white perch, bluegill, pumpkinseed, brown bullhead, and channel catfish. Fish not fitted with floy tags were marked by fin clipping according to a site-specific marking scheme. Signs informing anglers about the tagging program were posted around the lake and along the river to facilitate tag returns. A radio telemetry study was initiated in the early fall of 1991 to better understand fish movements.

Analysis of 1990 tag returns reflected only modest movement within the lake for several species. Most littoral species such as sunfish tended to remain within their original capture areas (Table 6.34). Some pumpkinseed sunfish were captured in areas devoid of macrophytes, but most were later recaptured in areas that

TABLE 6.33. Trapnet tagged fish versus recaptures, Onondaga Lake 1990 and 1991.

Species	Total captured	Number tagged	Tagged recaptures
a. 1990			
Bluegill	877	450	4
Pumpkinseed	449	250	1
White perch	3,521	328	6
Brown bullhead	84	72	1
Yellow perch	60	15	1
White sucker	128	28	1
Carp	101	87	0
Channel catfish	150	112	1
Black crappie	56	25	0
Smallmouth bass	22	11	0
Freshwater drum	23	6	0
Bowfin	20	13	0
Largemouth bass	41	40	1
Rock bass	1	1	0
Northern pike	6	3	0
Walleye	28	24	0
Longnose gar	5	2	0
Tiger musky	1	0	0
b. 1991			
Bluegill	11,985	937	32
Pumpkinseed	7,505	288	17
White perch	6,316	540	19
Brown bullhead	402	323	25
Yellow perch	395	14	16
White sucker	316	26	4
Carp	225	138	5
Channel catfish	199	143	11
Black crappie	130	49	2
Smallmouth bass	108	37	4
Freshwater drum	106	10	0
Bowfin	97	67	10
Largemouth bass	85	19	3
Yellow bullhead	26	14	2
Rock bass	18	10	2
White crappie	13	5	0
Northern pike	11	2	0
Walleye	6	3	0
Longnose gar	4	2	0
Tiger musky	2	1	0
Chain pickerel	1	1	0

did contain aquatic vegetation. Recaptures of pelagic fish (e.g., white perch) within the lake also appeared to be most frequent in areas containing macrophytes (Table 6.35). White perch are thought to rarely use macrophytes as cover, but often forage near plants nocturnally because of the presence of juvenile fish, a major food source (Smith 1985). Some

TABLE 6.34. Pumpkinseed sunfish movements within Onondaga Lake based on recapture of tagged individuals from various trapping sites 1990.

Capture	Tagging location			
	Willow Bay	Ninemile	Hiawath Point	Marina
Willow Bay	nm			
Willow Bay	nm			
Willow Bay	nm			
Willow Bay	nm			
Willow Bay	nm			
Willow Bay	nm			
Willow Bay	nm			
Ninemile		nm		
Griffin Field		x		
Hiawatha Point			nm	
Marina				nm
White Sands	x			

nm = denotes no movement

white perch were captured in areas without macrophytes, but were later captured in vegetated areas. A number of fish marked within Onondaga Lake were captured at various locations in the river. Gill nets set June–October 1991 in the Seneca River captured 4 white perch, 2 walleye, and a channel catfish that were marked previously in the lake.

Tag returns by anglers represent an important contribution to our knowledge of fish movements. Eighteen tags have been returned by anglers since 1990. Returns of fish have been reported in the lake outlet, as far up river as Baldwinsville (10 km), and as far down river as Fulton (25 km). Some returns have been reported even during winter months (February 21, 1991) by ice fishermen.

TABLE 6.35. White perch movements based on recapture of tagged individuals from Lake Trapping Sites, 1990. Abbreviations: Willow Bay (WB), Maple Bay (MB), Ninemile (NM), Dover Cliff (DC), Whit Sands (WS), Griffin Field (GF), Hiawatha Point (HP), and Marina (M).

Captured at	Recapture location							
	WB	MB	NM	DC	WS	GF	HP	M
Ninemile			nm					
Ninemile								
Griffin Field	x							
Ninemile			nm					
Willow Bay						x		
Maple Bay		nm						
Griffin Field	x							
Willow Bay	nm*							
Willow Bay	nm							
Marina								nm
White Sands			x					
Ninemile			nm					
Ninemile			nm					
White Sands							x	
Griffin Field							x	
White Sands					nm			
Marina								nm
Griffin Field							x	
Ninemile	x							
Willow Bay	nm							
Marina								nm
Willow Bay	nm							
Maple Bay			nm					
Maple Bay			nm					
H. Point						x		
White Sands	x							
H. Point	x							

*nm = denotes no movement

A tagged bluegill was recaptured in Phoenix, approximately 10 km from the tagging site. This fish was tagged on September 12, 1990 at the Marina (Figure 6.47) and recaptured on June 9, 1991. A yellow perch tagged October 2, 1990 at Maple Bay (Figure 6.47) was captured by an angler in Fulton May 30, 1991. This required the passage of the fish through the Phoenix lock on the Oswego River suggesting that movement of fish between Lake Ontario and Onondaga Lake is possible. Tagged channel catfish have been captured in Baldwinsville, and Phoenix. Angler tag returns also indicated movement of largemouth bass out of the lake, although the distance travelled was not as great as for other species.

Radio telemetry was used to follow fish movements during the fall turnover period in 1991. Radio signals are attenuated more rapidly in water than in air; the amount of attenuation is proportional to the conductivity of the water (Tyus 1982). The high conductivities of Onondaga Lake and the Seneca River (1600–2100 microS · cm^{-1}, and 700–900 microS · cm^{-1}, respectively), cause considerable loss of transmitter signal strength. Tests from boats and aircraft in 1991 indicated that fish could be radio tracked at depths <2 meters in Onondaga Lake and at 3 meters in the Seneca River. Seven individuals of four species were fitted with radio tags during October 1991. These tags were designed to last a minimum of nine months. Distances of fish

movement were measured as the shortest distance between the sites of capture and location by transmitter.

Features of the radio telemetry program, including preliminary results, are summarized in Table 6.36. The actual lengths traveled were undoubtedly greater than reported here. Six fish were monitored for 22 to 45 days between October 2 and November 22, 1991. Three fish (walleye, smallmouth bass, and a tiger muskellunge) captured in the lake moved out of the lake during the period of lowest dissolved oxygen in mid- to late October. These fish (w262, smb285, and tm337) were located in the Seneca River during late October. The tiger muskellunge (tm337) was found back in the lake on November 1, 1991, during a period of oxygen recovery in the lake. The walleye and smallmouth bass remained in the river. A second smallmouth bass (smb137) was captured in the Seneca River and released at the Onondaga Lake Marina, approximately 5 km away. This fish was located eight days later in the Seneca River near its initial capture site and remained in the river until the end of the study season. A smallmouth bass (smb238) captured in the river in mid-October remained in the river throughout the study. A second tiger muskellunge was radio tagged in the lake in early October and located 40 days later in the lake. Its location during the intervening period of low dissolved oxygen in the lake is unknown. The results of the tag return and preliminary radio telemetry programs support

TABLE 6.36. Radio tagged fish (n = 7) used in the preliminary telemetry project in Onondaga Lake, Autumn 1991.

Species/code	Length (mm)	Radio tagged date (1991)	Capture location	Method of capture	Distance from release site (km)
Smallmouth Bass/smb285	473	October 2	Maple Bay	Rod and reel	7.77
Smallmouth Bass/smb238	447	October 11	Seneca River	Gillnet	2.23*
Smallmouth Bass/smb137	451	October 11	Seneca River	Gillnet	7.06**
Tiger Muskellunge/tm337	765	October 10	Ninemile	Trap net	5.50
Tiger Muskellunge/tm810	820	October 13	Dover Cliffs	Trap net	2.89
Walleye/w262	610	October 8	Maple Bay	Gillnet	2.29
Channel Catfish/cc311	661	October 11	Dover Cliffs	Trap net	No data

* Caught and released in the Seneca River
** Caught in the Seneca River and released in Onondaga Lake

the position that the Seneca River is a corridor for fish movement into and out of Onondaga Lake.

6.5.7 Diets and Trophic Relationships

6.5.7.1 Background/Methods

Diet data represent an important tool in evaluating the effects of perturbations in aquatic communites (Ringler 1979, 1990). Growth and survival of juveniles are highly dependent on the availability of planktonic and benthic food, particularly in littoral habitats (Meals and Miranda 1991). Relatively few data are available on the food items for fishes in the Onondaga Lake system. The characterization of plankton and benthic macroinvertebrates of the lake presented earlier in this chapter provide some information, but stomach data still represent the best indication of trophic interactions. Knowledge of the kinds and quantities of prey eaten is essential to related modeling efforts. Forecasting and detecting shifts in trophic relationships in Onondaga Lake will be especially important in remediation and assessment. Fish diets in the Onondaga Lake system were not documented prior to the 1980s. The results of diet studies performed on the lake and the Seneca River in 1990 and 1991 are presented here.

Diet contents of white perch, yellow perch, pumpkinseed sunfish, largemouth bass, and alewife were obtained from seined and gill netted fish using a stomach pump for live specimens (Bannon and Ringler 1986; Seaburg 1957) or by removing the entire stomach from dead fish. Diets of trap netted fish were not analyzed because of the ability of many species to forage within the confines of the net. Stomach contents were preserved in 10% formalin in the field and transferred to 75% ethanol in the laboratory. Identifications were carried out to family or order level at 40 to 100X magnification. Dry weights of prey items were measured on subsamples of organisms of each taxon as described by Johnson and Ringler (1980).

A satisfactory approach to characterize the stomach contents of gizzard shad has not yet been developed, because of the unusual character of the stomach of this species. Their stomach is modified to form a gizzard comprised of cardiac and pyloric components (Bond 1979), which act as a physical mill, reducing small prey to a mass of particles in which taxa are difficult to distinguish. Small shad (>35 mm) in a variety of waters feed almost exclusively on microcrustaceans; larger specimens consume plankton, molluscs, insect larvae, and suspended or deposited detritus (Pierce et al. 1981). Because gizzard shad (>25 mm) are true filter feeders, their effects on zooplankton communities are not mediated by the size-selective mechanisms characteristic of most planktivorous fishes. The ways in which filter-feeding gizzard shad and particle feeders such as white perch interact to structure plankton communities are not well understood (Drenner et al. 1982). In some lakes, gizzard shad forage on bottom sediments throughout the daylight hours; the gizzard content of such fish is largely organic detritus with few identifiable organisms except Chlorophyta and Chrysophyta (Pierce et al. 1981).

6.5.7.2 Diet Descriptions: 1991 Findings

Diets for most species were described using percentage composition by weight, but for some cases only numeric data are available. Both methods are commonly used to describe fish diets since large but rare prey may contribute significantly to fish growth, and small abundant prey may require extended periods of foraging.

White perch adults (>160 mm) fed primarily on cladocerans in the lake and Seneca River during 1991. This was evident for both numeric and weight analyses (Figure 6.59a–d). Fish were relatively rare in the diets (Figure 6.59a) but their large size resulted in a contribution of 34% by weight in the Onondaga Lake sample (Figure 6.59b). Similarly, amphipods were captured infrequently but comprised 7.6% of the sample weight. In contrast to their significance in sunfish diets, chironomid

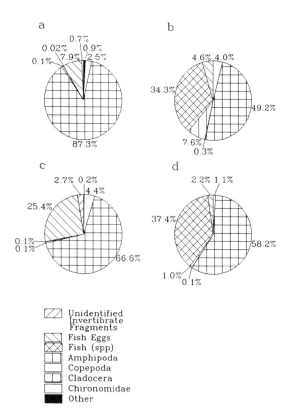

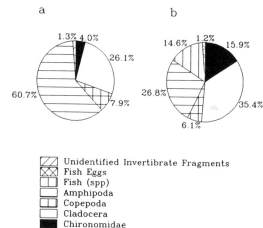

FIGURE 6.60. Diet composition of juvenile white perch in Onondaga Lake in 1991: (**a**) percent by abundance and (**b**) percent by dry weight.

FIGURE 6.59. Diet composition of adult white perch, 1991: (**a**) Onondaga Lake, percent by abundance, (**b**) Onondaga Lake, percent by dry weight, (**c**) Seneca River percent by abundance, and (**d**) Seneca River, percent by dry weight.

larvae and pupae contributed relatively little to the diets, e.g. 1% by dry weight in the river (Figure 6.59c), and 4% by dry weight in the lake. The similarities in diets of adult white perch in the river and lake (Figure 6.59a and c) suggest similarities in invertebrate fauna for the two water bodies. The mean and maximum number of prey items observed in the adult perch were not significantly different for the lake and river.

Diets of juvenile white perch were considered of particular importance because of the abundance of these fish in Onondaga Lake. Differences of prey composition based on numbers and weight were more pronounced for these smaller specimens (Figure 6.60a and b). Diet data also revealed some appreciable differences in feeding among sites (Figure

6.61). For example, amphipods (70%) dominated the sample from the Marina (Figure 6.61a) whereas cladocerans composed the largest component (48%) at Maple Bay (Figure 6.61b). Short-term changes in diets were also evident. During a two-week period in July, for example, juvenile white perch in the Marina area (Figure 6.62a and b) shifted from primarily chironomids (34%), amphipods (38%), and cladocerans (23%) to a diet consisting primarily of amphipods

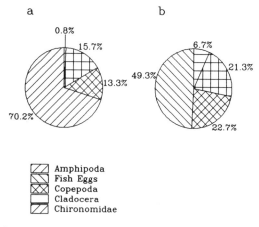

FIGURE 6.61. Diet composition (percent by dry weight) of juvenile white perch in Onondaga Lake in 1991: (**a**) Maple Bay and (**b**) Marina.

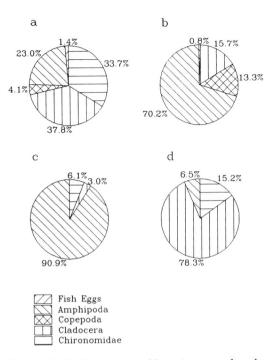

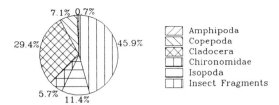

FIGURE 6.63. Diet composition (percent by abundance) of adult alewife in Onondaga Lake, 1990.

FIGURE 6.62. Diet composition (percent by dry weight) of juvenile white perch in Onondaga lake in 1991 for two dates at two sites: (**a**) Marina area, July 15–16, (**b**) Marina, July 31, (**c**) Willow Bay, July 15–16, and (**d**) Willow Bay, August 1.

influence food webs in several temperate lakes (e.g., Post and Cucin 1984). Stomach samples from juvenile perch seined biweekly at four sites (June–August) were compared in 1991 (Figure 6.64). The data, based on percentage composition by number, permit comparison among sites and demonstrate significant diet shifts as the fish grew. Chironomids were important food items for juvenile yellow perch, particularly at the Marina

(70%), cladocerans (16%), and copepods (13%). Amphipods appeared to play a more important role in diets in these 1991 samples than reported earlier. Somewhat less dramatic changes were recorded between mid-July and August 1 in Willow Bay (Figure 6.62c and d). These data suggest that future monitoring and analysis of trophic interactions in Onondaga Lake will require careful attention to spatial and temporal variability in foraging.

The planktivorous alewife currently represent a small component of the lake community. Although too few were collected for diet analysis in 1991, the 35 individuals examined in 1990 revealed a diet of copepods, cladocerans, and chironomids, in addition to a substantial fraction of (unidentified) invertebrate fragments (Figure 6.63).

Yellow perch are known to be efficient planktivores as well as effective insectivores and piscivores. Their foraging activity has been shown to shift prey size classes and

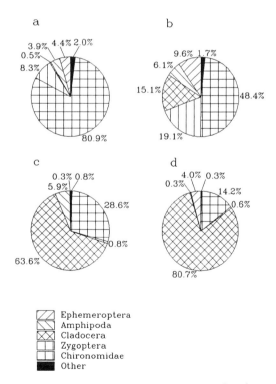

FIGURE 6.64. Diet composition (percent by abundance) of juvenile yellow perch at various locations in Onondaga Lake, June–August 1991: (**a**) Marina, (**b**) Willow Bay, (**c**) Ninemile, and (**d**) Maple Bay.

and Willow Bay sites (Figure 6.64a and b). Cladocerans were also important, particularly at the Ninemile and Maple Bay sites (Figure 6.64c and d). Baetid mayflies, which have not been found in the benthic samples taken in the lake (see earlier section of this chapter), comprised a significant portion of the diet at three of the four sites. Changes in prey type with size of yellow perch are presented in Table 6.37. The percentage of relatively large prey, such as zygopterans and amphipods, increases with fish size. Similarly, the percentage of fish stomachs in which such larger prey were found also increased as the fish grew.

Preliminary analysis of the diets of juvenile sunfishes were carried out from 1989 to 1991. Diets based on a collective sample of pumpkinseed sunfish seined at the Marina and Willow Bay are described on the basis of percent composition by number for 1991 in Figure 6.65a. The absence of molluscs from the diets of these fish is remarkable, as this is a significant food item for pumpkinseed sunfish in most northeastern lakes. These are some of the few observations of isopods in diets of Onondaga Lake fish. The significantly larger contribution of chironomids and small number of cladocerans in the Marina compared with the Willow Bay site (Figure 6.65b and c) closely parallels our findings for yellow perch described earlier (Figure 6.64a and b). Cladocerans, chironomids, and amphipods dominated the diet of juvenile largemouth bass, based on a collective sample of 72 specimens in 1991 from four sites (Figure 6.66). These specimens averaged 61 mm in length (31–94 mm range).

The food web of Figure 6.67 depicts the basic feeding relationships for the mid-summer

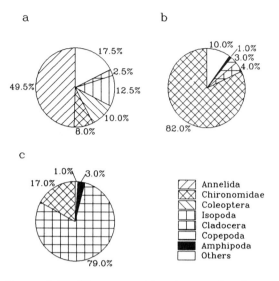

FIGURE 6.65. Diet composition (percent by abundance) of juvenile pumpkinseed sunfish in Onondaga Lake: (**a**) composite sample, (**b**) Marina, and (**c**) Willow Bay.

period in Onondaga Lake, based on the diet studies and current estimates of relative numbers of fishes. The hypereutrophic nature of the system is reflected in an apparently

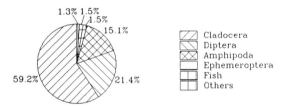

FIGURE 6.66. Diet composition (percent by abundance) of juvenile largemouth bass in Onondaga Lake in 1991.

TABLE 6.37. Diets of three size classes of juvenile yellow perch seined in Onondaga Lake in 1991 (4 sites, June–August).

Food organisms	<50 mm n = 39		50–70 mm n = 34		>70 mm n = 20	
	% Comp.	% of fish	% Comp.	% of fish	% Comp.	% of fish
Chironomid	33.6	72.0	59.6	70.6	28.45	35.0
Zygoptera	0.3	5.1	10.4	38.2	25.2	50.0
Cladocera	62.0	51.3	16.6	11.7	18.0	15.0
Amphipoda	0.4	7.7	6.3	35.3	15.5	35.0
Emphem.	1.2	20.5	5.6	23.5	12.2	25.0
Others	0.6	7.7	0.01	20.6	0.8	5.0

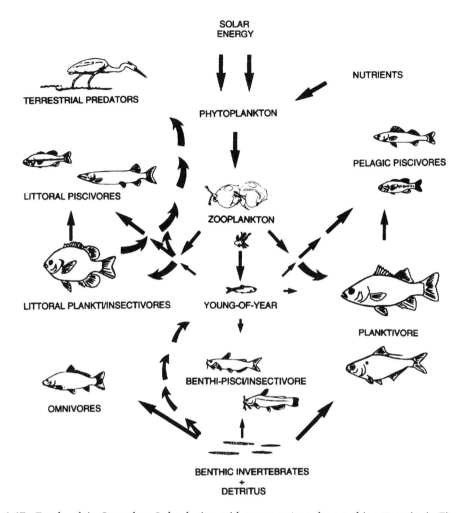

FIGURE **6.67.** Food web in Onondaga Lake during mid-summer (see also trophic categories in Figure 6.68).

disproportionally high numerical contribution of planktivorous fish (gizzard shad and white perch). The trophic categories of fishes in the lake are compared for three years in Figure 6.68a–c. A shift toward a greater contribution by littoral planktivores/insectivores since the early 1980s is suggested, perhaps associated with an increase in macrophyte growth.

6.5.8 Limiting Habitat Features, Lake Fishes

6.5.8.1 Dissolved oxygen Levels

Although recent studies have documented foraging by fishes in anoxic hypolimnetic waters (Rahel 1992), fishes in temperate

waters require aerobic conditions throughout their life history. In species such as salmonids any reduction in dissolved oxygen below saturation may result in reduced growth and survivorship, particularly for young individuals (Warren 1971).

Most work on responses of fish to hypoxic conditions has been done in the laboratory. Responses to hypoxic conditions include avoidance of low DO waters (Whitmore et al. 1960), increased rate of opercular movement, breathing at the surface, loss of feeding effort, loss of coloration, and loss of equilibrium or death (Gee et al. 1978; Petit 1973). Dissolved oxygen levels $<3.5\,\text{mg}\cdot\text{L}^{-1}$ at $15–20°\text{C}$ for 24 hr were fatal to most species tested (Moore 1942). The lethal concentration for adult

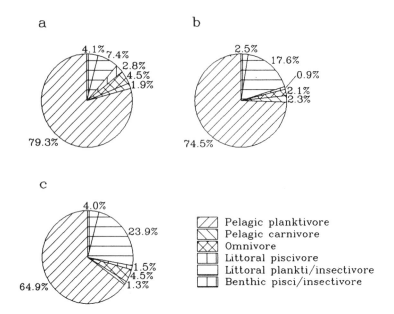

FIGURE **6.68.** Comparison of fish populations in Onondaga Lake for three years, according to trophic categories: (**a**) 1980, (**b**) 1989, and (**c**) 1990.

bluegill (5–20 g) was substantially lower, 0.75 mg · L⁻¹ at 25°C for 24 h (Moss and Scott 1961). Interactions between pollutants and effects of DO complicate the interpretations of such experiments (Carlson 1984, 1987).

Field analyses of the role of dissolved oxygen in lakes are relatively few, in contrast to the many studies in streams and streambeds (e.g., Coble 1961; Ringler and Hall 1975). Petit (1973) reported that many species of western Lake Erie were stressed at dissolved oxygen levels below 6.0 mg · L⁻¹. Price et al. (1985) predicted long term declines of striped bass (*Morone saxatilis*) in Chesapeake Bay in response to the loss of deep water habitat due to limiting dissolved oxygen concentrations. They described an oxygen-temperature "squeeze" in which the volume of hypoxic water expanded over a 10-yr period, forcing the bass into shallow areas or out of the upper Bay.

Despite the very low dissolved oxygen concentrations that develop in the upper waters of Onondaga Lake during fall turnover (e.g., below minimum concentration standard of New York State of 5.0 mg · L⁻¹ Effler et al. 1988; Chapter 5), extensive fish kills have not

been observed in the lake during that period. It has been hyothesized that fish leave the lake during this period, entering the Seneca River or the lower reaches of lake tributaries, to escape these stressful conditions. Limited recapture data for tagged fish (described earlier) support the hypothesis. To further test this hypothesis, and to characterize the impact of the low oxygen concentrations, the temporal and spatial patterns of net catches and dissolved oxygen concentrations were assessed in the upper waters of the lake during the low oxygen concentration period in the fall of 1991. Additionally, preliminary in situ incubation experiments were conducted to assess the ability of fish to survive low levels of dissolved oxygen for 1 to 2 days.

The dynamics of catches of fish using trap nets in 1990 reveal responses to the low oxygen concentrations during fall turnover. Oxygen concentrations in the upper waters of the lake remained rather close to saturation values during summer, but decreased in a generally progressive manner in early fall with the deepening of the epilimnion; a minimum of about 2 mg · L⁻¹ occurred with the onset of complete fall turnover in late

October (Figure 6.69a). Several fish species were absent in the catches during the period of minimum dissolved oxygen in October (Figure 6.69b–f). Shorthead redhorse sucker were found in net samples throughout the year except during fall turnover. Few redhorse suckers were trapped early in the year; catches of this species increased until July, but they disappeared from the nets as dissolved oxygen levels began to decrease during fall (Figure 6.69a and b). Other species not found in the nets during the period of minimum oxygen were channel catfish, gizzard shad, white perch, and smallmouth bass (Figure 6.69c–f). Numbers of fish caught varied throughout the season, but almost no captures were made for five weeks following the dissolved oxygen minimum (Figure 6.69b–f). Sampling by gill net in the Seneca River during the fall of 1991 indicated complimentary increases in these species during fall turnover. These observations support the position that the river, with its higher oxygen concentrations during fall turnover, serves as a refuge for Onondaga Lake fish.

The spatial distributions of fish and dissolved oxygen in the upper waters of the lake were characterized in mid-October (17–20) of 1991, a period of low oxygen concentration, and compared to observations made in mid-summer of the same year. Gill nets were deployed horizontally at depths of 1–3 m and 4–6 m at twelve locations representive of twelve compartments of the lake (Figure 6.70). Six nets were set each night; the total of 24 net nights per sampling period was divided equally between the two depth intervals. The schedule of depths and compartments for gill netting was randomized. Fifteen stations were monitored (depths of 0, 1, 3, and 4 m) for dissolved oxygen.

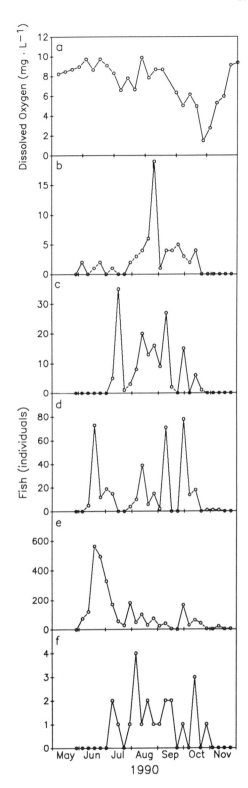

FIGURE 6.69. Temporal distributions in Onondaga Lake in 1990: (a) dissolved oxygen in the surface waters, and (b) trap net catches of redhorse sucker, (c) trap net catches of channel catfish, (d) trap net catches of gizzard shad, (e) trap net catches of white perch, and (f) trap net catches of smallmouth bass (modified from Tango and Ringler 1996).

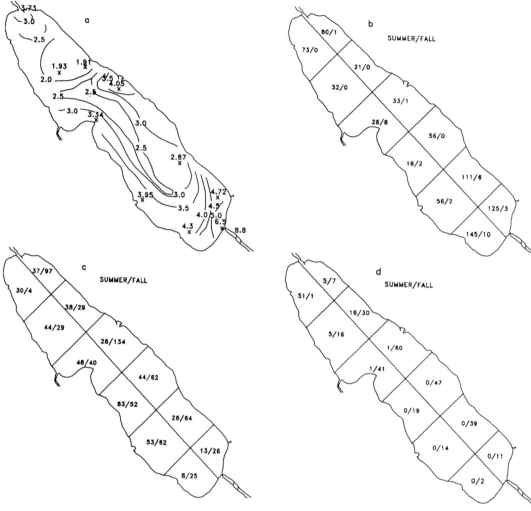

FIGURE 6.70. Spatial distributions in Onondaga Lake in 1991: (**a**) dissolved oxygen in the surface waters over the October 17–20 interval, and distributions of selected fish in the upper waters in the intervals July 16–20 and October 17–20 (format of results; July/October): (**b**) white perch, (**c**) gizzard shad, and (**d**) alewife.

The spatial distribution of dissolved oxygen in the surface waters of the lake during the mid-October study interval is depicted by the isopleths in Figure 6.70a. The isopleths were drawn based on the average values (n = 2) measured at the fifteen sites over the four-day period. This is the first documentation of longitudinal structure in oxygen concentrations in the lake during the critical fall turnover period. These results depict significant, but not major, spatial variations in the concentration of oxygen within the upper waters (Figure 6.70a). Measurements from a single deep water location (e.g., south deep for long-term

monitoring program, Chapter 5) appear to fairly represent the depressed oxygen concentrations and the contravention of the standard. The contribution of oxygen to the lake from the Onondaga Creek inflow is clearly depicted. Oxygen concentrations were higher in the south basin than in the north during this period. Higher concentrations were observed near the Marina and the outlet (Figure 6.70a).

Strong differences in the fish populations were observed between the mid-summer and fall catches (Table 6.38; Figure 6.70b–d). Predators such as smallmouth bass, channel catfish, and walleye were apparently strongly affected by the depressed concentrations of oxygen during fall turnover, as they were captured in mid-summer, but not in the fall sampling (Table 6.38). Of the planktivores, white perch apparently were impacted the most by the low oxygen concentrations of fall (Table 6.38; Figure 6.70b). In contrast, gizzard shad and alewife appeared to be more tolerant of these conditions, as they were caught in relatively higher numbers during the period of low dissolved oxygen (Table 6.38; Figure 6.70c and d). The modest longitudinal structure in oxygen concentrations during fall turnover (Figure 6.70a) apparently was important to the distribution of certain fish within the lake during that period. The higher catches of gizzard shad and alewife (Figure 6.70c and d) made near the Marina in the fall probably reflect the somewhat higher oxygen concentrations found in that area. However, catches of gizzard shad were somewhat higher in the northern end of the lake than in the south. Perhaps this reflects movements of shad to and from the Seneca River during that period.

An in situ experiment was conducted October 20−21, 1991 to examine the responses of fish held in small enclosures in the lake to the low levels of dissolved oxygen. Two littoral species (pumpkinseed and bluegill sunfish) and one pelagic species (white perch) were studied. Only adult white perch were tested, whereas adult and young-of-the-year sunfish were tested. Test individuals were collected from Maple Bay, a location where oxygen concentrations remained higher than in other regions of the lake during this period. Fish were incubated at Maple Bay and a site located between Maple Bay and Ninemile Creek (Maple Bay/Ninemile Creek, Figure 6.47; approximately 100 m offshore, depth of 3 m). The numbers of fish available for this experiment were small, because of the dramatic reductions in lake oxygen concentrations that were underway at the time; ≤4 individuals were placed in each enclosure. The enclosures were cylindrical wire minnow traps (25 cm diameter × 75 cm long, 6.3 mm mesh), deployed at depths of 1 to 3 m. Two to three replicates (traps) were used for each species. Observations of number of fish alive in the traps were made after 24 and 48 hr. Weather conditions remained stable throughout the experiment.

Results of the enclosures experiments are presented in Table 39. Oxygen concentrations in the upper waters of the lake during fall turnover of 1991 were low enough to cause mortality of a number of species common to the lake. Mortality of the test fish was greater at the Maple Bay/Ninemile site, particularly after 48 hr, apparently as a result of the lower oxygen concentration ($1.1 \, mg \cdot l^{-1}$). White perch appear to have been more sensitive to the low oxygen concentrations compared to

TABLE 6.38. Total gill net catches in Onondaga Lake, New York, in 1991, on twelve net nights, at each depth per sampling period.

Species	Surface nets (1−3 m)		Deep nets (4−6 m)	
	July 16−20	October 17−20	July 16−20	October 17−20
Predators				
Smallmouth Bass	50	0	19	0
Channel Catfish	11	0	28	0
Walleye	7	0	5	0
Planktivores				
White Perch	408	19	366	18
Gizzard Shad	214	519	234	98
Alewife	76	213	3	72
Others				
Common Carp	0	0	2	0
Freshwater Drum	1	0	0	1
Golden Shiner	0	0	0	1
Redhorse Sucker	0	0	1	0
Yellow Perch	0	0	0	5

TABLE 6.39. Mortality in enclosure experiments under natureal lake conditions at the time of fall turnover, 1991.

DO*/species		Mortality (%)			
		Ninemile/Maple Bay		Maple Bay	
		24 hr	48 hr	24 hr	48 hr
DO concentration (mg·L^{-1})		2.75	1.1	3.35	4.5
Species					
Pumpkinseed	No.				
YOY**	2	0		0	
adult	2	0		0	
YOY	2	50			0
adult	1	0			0
Bluegill					
YOY	2	50			0
adult	2	0			0
YOY	6		33		25
adult	6		0		0
Whith Perch					
adult	4	25		0	
adult	2		100		0

*DO: dissolved oxygen
**YOY: young of the year

the sunfish. Note, the greater tolerance of bluegill to low dissolved oxygen concentrations is consistent with the results of laboratory studies (Moore 1942; Moss and Scott 1961). Perhaps white perch have a higher oxygen requirement since they are pelagic; e.g., fast swimming species compared to the littoral dwelling sunfish. Alternatively, the sunfish population, with more restricted movement, may have better acclimated to this environmental stress through selection over several generations. The symptons of stress due to hypoxia for the species in this experiment were similar to those reported in the laboratory experiment of Petit (1973) for a variety of western Lake Erie fishes. Fish in the low dissolved oxygen waters of Onondaga Lake showed severe hemorrhaging in the enclosure experiments.

6.5.8.2 Mercury Contamination

6.5.8.2.1 Background

In 1970, the New York State Department of Environmental Conservation (NYSDEC), in concert with the Departments of Health (DOH) and Agriculture and Markets, closed Onondaga Lake to fishing due to excessive mercury (Hg) contamination of the fish. This action was based on the results of the initial analyses of fish flesh concentrations in the lake, conducted soon after the U.S. Department of Justice initiated legal action against Allied to reduce discharges of Hg. At the time of these initial collections this facility was discharging approximately 10 kg·d^{-1} to the lake (Chapter 1). Fish flesh in Onondaga Lake has been monitored for Hg contamination on a nearly annual basis since the initial effort in 1970, to stay apprised of prevailing conditions. The lake was reopened to recreational angling in 1986 on a "catch and release" basis. However, a health advisory, to eat no fish from the lake, prevails; signs are posted around the lake to remind anglers of the advisory.

The standard used for the 1970 closure was the U.S. Food and Drug Administration (FDA) interim action level of 0.5 ppm for Hg in fish flesh on a wet weight basis (FDA 1974). This criterion was used by FDA to prohibit interstate sales and shipment of fish products intended for human consumption which exceed the limit. In 1979, the FDA, on the basis of new evidence, set the action level at 1.0 ppm (FDA 1979). In 1984, the FDA modified the action level so that the primary consideration in evaluating Hg contamination would focus on methyl-Hg (FDA 1984), rather than total Hg, since this form is recognized as the most common and toxic of the Hg compounds. The majority of Hg in carnivorous fishes is in the methyl form (Eisler 1987) and, except for young fish, it comprises over 70% of the Hg burden (USEPA 1985). Grieb (1990) concluded that Hg in carnivorous fish was essentially methyl-Hg.

The emphasis of this section is on Hg contamination. However, less intensive analyses indicate contamination of the lake's fish flesh with other substances. Analysis of 37 fish samples (composite and individual) collected in 1980 from the lake, performed according to a USEPA interim method, indicated the presence of benzene, chlorobenzene, and dichlorobenzene contaminants in the flesh of a number of species (Hindenlong et al. 1981). Recall that these materials are known to enter the lake along the west shore as a result of disposal of related materials by the soda ash/ chlor-alkali facility, associated with the manufacture of chlorobenzene (Chapters 1 and 2). Contaminated species included channel catfish, walleye, northern pike, smallmouth bass, white perch, white sucker, and gizzard shad. Maximum concentrations (wet weight basis) observed for benzene, chlorobenzene, and dichlorobenzene were 0.2, 0.9, and 0.9 ppm, respectively (Hindenlong et al. 1981). The potential environmental and health risks of this contamination have not been established. Additionally, NYSDEC monitoring results indicate concentrations of PCB, chlordane and hexachlorobenzene are elevated, particularly in large, fatty specimens in the lake (Sloan et al. 1987). Dieldron was relatively high in smallmouth bass in 1983, but the levels were reduced in 1985 and 1986 samples (Sloan et al. 1987). Certain species have also been found to exceed FDA tolerance levels for PCB's (NYSDEC unpublished data).

Here we document the history of the Hg fish flesh contamination in Onondaga Lake for the 1970–1990 period, from the perspective of the related standards.

6.5.8.2.2 Methods for Mercury Monitoring

Fish captured in gill nets and seines were measured, weighed, tagged with identification numbers, and frozen in plastic bags at $-20°C$ until analysis. Scale samples were removed from some individuals for age determination. Fish sampling was more comprehensive following 1972, to support trend analysis. From 1972 to 1981, white perch and smallmouth bass were the principal species collected for the trend analysis. Walleye collections were begun in 1973. Smallmouth bass became the focus of Hg contamination in Onondaga Lake since movements of this species were believed to be localized in the lake.

Preparation of the fish for analysis varied in the early years of monitoring; generally the head and viscera were removed and the remainder ground in a food mill and homogenized by mixing in a commercial food mixer. Current preparation methods (post-1977) involve a standard filleting technique in which one whole side of the fish, scales removed but skin intact, is cut from behind the operculum to the tail. This fillet contains one pelvic fin and bones of half the ribs. Excluded from the fillet are the vertebral column and the dorsal, pectoral, anal and caudal fins. Fish less than 150 mm in length are analyzed with just the head and the viscera removed (Sloan 1985).

Mercury data have been obtained by the flameless atomic absorption spectrophotometric method, following the general procedure of Hatch and Ott (1968). Current instrumentation utilizes a MAS-50A Mercury Analyzer. The ground, homogenized tissue is digested using a mixture of sulfuric acid and hydrogen peroxide at 70°C for 2.5 hr (Armstrong and Sloan 1980). The procedure results in quantification of total elemental mercury. Results are expressed as micrograms total mercury per gram of wet fish flesh (ppm).

TABLE 6.40. summary of mercury analyses of unaged Onondage Lake fish collected in 1970.

Species	Number collected	Number analyzed	Average length (mm)	Length range (mm)	Average mercury (ppm)	Mercury range (ppm)
Bowfin	1	1	335	—	0.38	—
Alewife	3	2	180	160–208	3.28	2.30–3.77
Northern pike	1	1	523	—	3.74	—
Carp	9	5	508[b]	361–594	1.05	0.66–2.96
White sucker	21	6	206[c]	109–427	0.92	0.58–4.77
Brown bullhead	5 (3)[f]	5	269[b]	254–286	3.38	2.16–4.10
Channel catfish	3 (2)	3	534[a]	521–546	5.34	4.32–6.40
White perch	3	2	203	183–241	2.91	2.46–3.80
Bluegill	6	2	140	117–183	0.76	0.69–0.83
Smallmouth bass	8	8	243	213–274	1.96	1.04–2.72
Black crappie	9 (2)	6	253[d]	190–305	1.14	0.76–2.20
Yellow perch	24 (2)	10	255[e]	231–282	3.62	1.52–8.20
Freshwater drum	6 (3)	6	307	249–432	1.20	0.20–2.20

[a] 2 were measured
[b] 3 were measured
[c] 4 were measured
[d] 5 were measured
[e] 7 were measured
[f] Numbers in ( ) represent number of fish analyzed by DOH

6.5.8.2.3 Mercury Concentrations

Results of determinations for the concentration of Hg in the fish flesh found in Onondaga Lake are presented for 1970 and 1985 in Tables 6.40 and 6.41. The average concentration of Hg exceeded the 1.0 ppm level in all species in 1970 except for bowfin white sucker and bluegill. Substantial differences in the degree of contamination among fish species were indicated by these initial observations, though the sample sizes of fish analyzed were small. Interspecific differences in the level of contamination and long term trends are to be expected based on differences in metabolism, feeding habits, trophic level, and extent of

TABLE 6.41. Summary statistics of mercury concentrations on a wet weight basis on standard fillets of fish from Onondaga Lake collected in 1985.

Species	Number (N) analyzed	Length (mm)		Mercury concentration (ppm)			% of Fish exceeding (1 ppm Hg)	% Probability that a sample of fish will exceed (1 ppm Hg)
		Average	Range	Average	Median	Range		
Walleye	17	547	410–636	1.58	1.55	0.45–2.83	76	>99
Smallmouth bass (all)	46	341	253–440	1.20	1.17	0.56–2.18	67	>99
Smallmouth bass (≥12")	37	354	305–440	1.29	1.28	0.56–2.18	81	>99
Channel catfish	15	460	365–825	0.93	0.75	0.34–2.57	20	44
White perch	20	201	186–223	0.88	0.85	0.43–1.36	20	2
Northern pike	5	740	557–884	0.85	0.85	0.74–1.02	20	3
White sucker	20	406	352–467	0.76	0.75	0.31–1.49	10	<1
Carp	14	619	563–680	0.63	0.60	0.13–1.16	14	<1
Pumpkinseed	18	165	147–184	0.63	0.58	0.34–1.25	6	<1
Yellow perch	16	184	150–242	0.48	0.46	0.11–1.08	6	<1
Brown bullhead	12	264	230–299	0.27	0.22	0.11–0.72	0	<1

migration betweeen the Seneca River and the lake. The level of contamination was substantially reduced in 1985 (n = 336; Table 6.41). Most of the species averaged <1 ppm. However, two species of angling interest, walleye and smallmouth bass, had a high probability of exceeding the FDA action limit.

Mercury concentrations in the flesh of Onondaga Lake fish are positively correlated with fish length (Sloan et al. 1987), as would be expected based on age. This has been observed elsewhere (Scott 1974; Scott and Armstrong 1972). Related data for Onondaga lake are represented according to the linear regression format

$$[Hg] = m[L] + b \qquad (6.13)$$

where [Hg] = concentration of Hg in fish flesh (ppm), m = slope of the regression expression, [L] = length of fish (mm) and b = the y-intercept. Substantial variations in the strength of the relationships have occurred over the period of record. A summary of the linear regression analyses for smallmouth bass is presented in Table 6.42; summaries for white perch and walleye have appeared elsewhere (Sloan et al. 1987).

The data for smallmouth bass have received the greatest attention as this population is believed to be the best indicator of trends in Hg levels (Sloan et al. 1987). Regression lines for smallmouth bass from the period of record

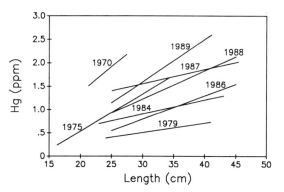

FIGURE 6.71. Relationships between mercury (Hg) fish flesh concentrations for smallmouth bass and specimen length in Onondaga Lake, for selected years in the 1970–1989 interval.

are presented in Figure 6.71; certain years do not appear on the plot to facilitate identification of major temporal patterns. Differences in the length of the specimens captured are apparent, particularly for 1970 and 1975. There have been significant changes (P < 0.05) in the apparent rate of accumulation (note the changes in the regression coefficients found in Table 6.42); the rate was high originally in 1970 and 1975, decreased in 1978 and 1979, and increased in the 1980s. Decreases in the accumulation rate were also observed for white perch and walleye in the 1970s (Sloan et al. 1987). The overall temporal pattern that emerges is an early general decline in Hg but increases in recent years to levels equal to or

TABLE 6.42. Linear regression relationships between length (X) and mercury concentrations (Y) in the flesh on a wet weight basis for smallmouth bass, white perch, and walleye collected from Onondaga Lake over several years (1970 to 1986).

Year	Number analyzed	Mean ± standard deviation		Correlation coefficient	Y-intercept	Slope
		Length (cm)	Mercury (ppm)			
1970	8	24.3 ± 2.1	1.96 ± 0.7	0.37	−0.83	0.11
1972	16	27.9 ± 3.1	1.26 ± 0.3	0.78	−1.98	0.12
1974	22	24.9 ± 4.0	0.81 ± 0.3	0.69	−0.67	0.06
1975	46	26.7 ± 3.7	1.09 ± 0.3	0.75	−1.06	0.08
1977	20	26.1 ± 2.6	0.87 ± 0.4	0.80	−2.56	0.13
1978	29	30.4 ± 5.9	0.63 ± 0.4	0.19	0.28	0.01
1979	50	31.3 ± 4.4	0.68 ± 0.1	0.63	−0.09	0.02
1981	50	33.0 ± 5.3	1.23 ± 0.3	0.70	−0.02	0.04
1983	50	32.5 ± 5.7	1.08 ± 0.4	0.70	−0.32	0.04
1984	48	31.0 ± 4.4	1.04 ± 0.3	0.47	0.01	0.03
1985	46	34.1 ± 4.3	1.20 ± 0.4	0.75	−1.09	0.07
1986	50	32.5 ± 5.0	1.05 ± 0.3	0.80	−0.70	0.05

greater than those observed in the mid-1970s (Figure 6.71; Table 6.42). Changes in the age structure of the smallmouth bass population may have contributed to some of the apparent changes in Hg accumulation rate (Sloan et al. 1987).

The database for Hg in smallmouth bass are presented from different perspectives in Figure 6.72. The mean concentration determined in each of the monitored years is plotted in Figure 6.72a; the error bars represent 95%

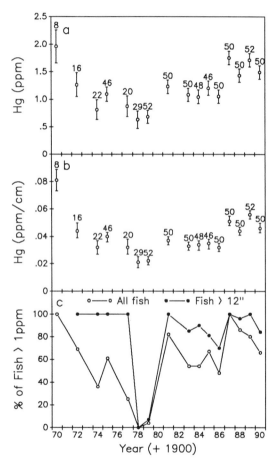

FIGURE 6.72. Long-term trends for mercury (Hg) concentrations in smallmouth bass in Onondaga Lake: (**a**) average concentration, with 95% confidence limits, sample sizes indicated, (**b**) average concentration per unit length, with 95% confidence limits and sample sizes indicated, and (**c**) % of smallmouth bass sampled with concentrations greater than 1 ppm, for two categories: all, and those greater than 12 in. long.

confidence intervals around the means, and the sample sizes are indicated. The sample sizes have been nearly uniform since 1979. These data have been normalized by fish length in Figure 6.72b. The status of the smallmouth bass population with respect to the 1.0 ppm FDA standard is represented for the period of record in Figure 6.72c. The same temporal pattern emerges throughout these various data presentations: initial decreases in concentrations through the late 1970s followed by increases in the 1980s. Two plateaus are apparent in the 1980s. The period 1981 through 1986, which appears similar to conditions in the mid-1970s, and the period 1987 through 1990, that reflects the highest level of contamination since the early 1970s.

The status of the fish flesh contamination with respect to the standard is represented as the percentage of collected fish that had concentrations >1.0 ppm (Figure 6.72c). Two sample categories were used: all collected smallmouth bass and those that were legal sized fish (>30.5 cm in length). In most years a smaller percentage of the legal sized fish were <1.0 ppm because of the higher concentrations usually observed in larger fish (Figure 6.72c; Table 6.42). All of the smallmouth bass collected in 1970 and 1987 exceeded the 1.0 ppm standard. In sharp contrast, in 1978, none of the smallmouth bass exceeded 1 ppm. Similar conditions were observed in 1979. With the exception of 1978 and 1979, a large percentage (>70%) of the legal sized smallmouth bass collected in Onondaga Lake over the period 1970–1990 have exceeded the FDA action level for Hg. More than 95% of the legal sized smallmouth bass exceeded the standard in three of the four years over the 1987–1990 interval. For the period 1981–1990, except in 1986, more than 50% of all smallmouth bass collected were above the FDA 1 ppm action level.

6.5.8.2.4 Interpretations

The flesh of Onondaga Lake fishes has been contaminated with Hg since at least 1970, and doubtless for a substantial period before monitoring was initiated. Levels continue to exceed the concentrations observed in other monitored systems in New York State (Sloan et al. 1987). Some fish from the Seneca River and linked systems are doubtless also influenced as several species actively migrate between the river and the lake, as discussed earlier. The continued contamination of fish flesh in the lake is not surprising in light of the elevated levels of Hg that persist in the lake's sediments (Chapter 8), and the high concentrations of Hg species that occur in the inflows to the lake and the lake's water column (Chapter 5; Bloom and Effler 1990).

Substantial changes in the level of contamination of smallmouth bass, the principal indicator species of the lake, are manifested in the database for the 1970–1990 period. In particular, after decreases in the concentrations of Hg in fish flesh in the 1970s, concentrations increased in the 1980s to levels observed in the early 1970s. Industrial loading of mercury decreased after 1970, first from improved housekeeping and alterations of processes, later from added wastewater treatment, and finally by discontinuation of chlor-alkali manufacture. The role that these events and changes in the lake conditions had in regulating the observed pattern in contamination of fish flesh remains unresolved. There is little hope of establishing cause and effect relationships and for developing effective remediation strategies for the Hg fish flesh contamination until the Hg cycle of the lake is understood and quantified. The central related management issue is the relative contribution of in-lake deposits versus continuing new inputs to the lake.

6.5.8.3 Epidemiology of Fishes

Few historical data exist on the general health of fishes from Onondaga Lake or its tributaries. Many environmental toxicants are known to cause adverse effects to fish at both the individual and population levels. Morphological abnormalities, such as cell and tissue changes, and certain tumors and diseases, may often serve as definitive evidence of adverse impact on the environment (Sindermann et al. 1980). Although mercury

levels in smallmouth bass have been monitored since 1970 (e.g., Sloan et al. 1987), information on occurrence of tumors and parasites is lacking. These baseline data can become valuable as long-term monitoring tools, especially when remediation efforts of a disturbed system are underway. Several studies have documented that elevated frequencies of tumors are associated with fish populations from industrially polluted waters (Bauman et al. 1987; Black 1983; Brown et al. 1977; Hirethota and Ringler 1989; Sindermann et al. 1980). A survey of more than 200 brown bullheads in 1989 from two Adirondack Lakes, Deer Lake and Arbutus Lake, which receive no industrial waste, revealed that none of the bullheads had gross external lesions (Hirethota et al. unpublished data).

A survey of external lesions and parasitic infestation of the fishes of the Seneca River revealed four types lesions and three types of external parasites. In comparison with other waters known to be contaminated in New York, e.g., Niagara River, and Irondequoit Bay, it appeared that the degree and occurrence of external abnormalities and the incidence of parasites was far lower in Seneca River fish.

The most common parasites in the Onondaga Lake system were the trematodes, followed by anchor worms and leeches. Both trematodes and anchor worms occurred mostly on golden shiners. Metacercarial stages of trematodes occur as black spots on the fins and body, while anchor worms (*Lernaea* sp.) were found primarily at the base of fins. The intensity of occurrence ranged from one to eight organisms per fish. Anchor worms occurred on approximately 8.5% of the golden shiners, and trematodes occurred on about 12%. About 5% of the channel catfish had leeches, generally attached to barbels or the lower jaw. The skin lesions (open wounds) on golden shiners (6%) were probably caused by previously attached anchor worms. Similar lesions were also found on fewer than 1% of the white perch. Bottom living fishes, such as ictalurids, are more prone to environmental contaminants due to their feeding habits and habitat requirements. Their barbels have sensory cells which help in location of food. Bottom living fishes from highly contaminated waters tend to have various types of external lesions and barbel deformities. Approximately 5% of channel catfish in the Seneca River had barbel deformities (shortened barbel to one or more completely missing barbels); 3% had skin lesions, and 1.5% had raised skin growth. This rate of deformities is negligible when compared with more than 30% bullheads with barbel deformities in the Buffalo River, NY (Hirethota and Ringler 1989).

6.5.9 Fish Communities and Habitats of the Onondaga Lake Tributaries

6.5.9.1 Stream Classification

The fish habitats provided by the tributaries to Onondaga Lake are of interest to lake remediation concerns because of the interest in reestablishing salmon runs and the influence these streams may have on the fish communities of the lake. A somewhat simplified schematic of the tributaries in the Onondaga Lake watershed is presented in Figure 6.73. The lower reaches of the tributaries are presently strongly influenced by anthropogenic impacts. These reaches have been straightened and channelized, and control structures impede the passage of fish in certain stream reaches. Discharges are received from CSOs during runoff events in Onondaga Creek, Ley Creek, and Harbor Brook (Chapters 1 and 3). Lower Ninemile Creek receives ionically enriched inputs from the area of the most recent waste beds (Chapter 3). These forms of pollution may influence longitudinal zonation in fish in these streams, thus limiting their distribution (Trial 1989).

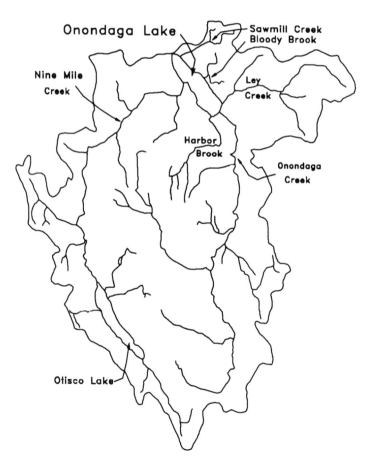

FIGURE 6.73. Tributaries in the Onondaga Lake watershed.

Onondaga and Ninemile Creek represent most of the stream fisheries habitat in the watershed (Figure 6.73). The main channel of Ninemile Creek, connecting Otisco to Onondaga Lake, has been extensively altered throughout most of its length. Stage height of this channel is largely controlled by a dam at the Otisco Lake outlet. Onondaga Creek, on the other hand, is a more typical stream with a network of tributaries with stage responding to meteorological conditions. Onondaga Creek has a flood control dam downstream of the confluence of the main stem and the West Branch, and within the City the channel has been altered to accommodate flood potentials.

Ninemile and Onondaga Creeks have substantial subwatersheds, 323 and 298 km^2 respectively, with long reaches of potential fish habitat above the influence of the urban area. Classification of stream habitats in Ninemile and Onondaga Creeks was conducted during the fall of 1990. The catchments of both streams have been engineered to a high degree. These types of channels are not included in the classification system used here (Rosgen 1985), although they have been, characterized.

The Rosgen system categorizes streams based on morphological characteristics including stream gradient, sinuosity, width/depth ratio, bed composition, channel entrenchment, channel confinement, and soil/landform features. Most streams fall into one of three categories (A, B, or C) based largely on stream gradient. Each category is subdivided using other factors so that streams are referred to, for example, as A3, B1, or C5. Stream type A has the highest gradient, B an intermediate gradient, and C the lowest of the three.

The variety and lengths of channel types, and therefore fish habitats, throughout the watersheds of the two creeks are presented in Table 6.43. The greatest variation occurs within the Onondaga Creek drainage, which includes headwaters as well as valley stream reaches. The morphological and hydrological differences between Onondaga and Ninemile Creeks are evident from the abundance of C channels and the lack of A and B channels within the Ninemile Creek watershed. Further differences are described subsequently.

TABLE 6.43. Length (km) of major stream class types designated for Onondaga and Ninemile Creeks[1].

	Channel type				Total
	A	B	C	M[2]	
Onondaga Creek					
Syracuse	0	0	0	9.9	9.9
Onondaga	4.0	16.8	33.9	1.6	56.3
LaFayette	5.9	8.9	22.9	0	37.7
Otisco	1.5	7.7	12.3	0	21.5
Tully	15.7	13.9	11.3	0	40.9
Total	27.1	47.3	80.4	11.5	166.3
Ninemile Creek					
Geddes	0	0	0.9	2.4	3.3
Camillus	0	0	12.6	1.6	14.2
Marcellus	4.1	6.7	24.8	0	35.6
Skaneatles	0	0.9	6.5	0	7.4
Total	4.1	7.6	44.8	4.0	60.5

[1] Not including tributaries to Otisco Lake
[2] Includes only major modifications of substantial lengths of stream channel and not structures such as culverts and bridges with associated rip rap

6.5.9.2 Stream Communities (1927–1991)

Stream fishes captured in seines by Greeley (1928) are compared to those collected as part of the 1990–1991 work in Table 6.44. The collection sites of the 1927 survey were listed as "Onondaga Lake Tributaries" in Greeley's (1928) table, but as "Onondaga Lake inlets" in his text. The 1927 survey evidently concentrated on the lower reaches of the streams, and was probably not representative of the headwaters. In this study a 400–600 V DC backpack electroshocker pulsed at 60–120 Hz was used to collect fish in 100 m sections blocked by 5 mm^2 mesh seines. An upstream and a downstream run was completed at each collection site.

Most sampling during 1990–1991 was conducted upstream of urban and industrial influences. The most downstream site sampled on Onondaga Creek was near the northern boundary of the City of Syracuse. This site is probably similar to those sampled in 1928. The similarity of fish communities between 1928 and today is remarkable, considering the urbanization of the region during the past 50 years. Possibly most of the sensitive species were lost by 1927, following extensive agri-

TABLE 6.44. Fish species collected at Onondaga Lake Inlets (1928) and in the tributaries of Ninemile Creek (1990) and Onondaga Creek (1991)[1].

Species	Onondaga Lake inlets 1928	Onondaga Creek 1991	Ninemile Creek 1990
1. Brown Bullhead		×	×
2. White sucker	×	×	×
3. Northern hog sucker	×	×	×
4. Carp	×		
5. Cutlips minnow	×	×	×
6. Eastern silvery minnow	×		
7. Common shiner	×	×	×
8. Golden shiner		×	×
9. Emerald shiner		×	×
10. Northern Redbelly dace	×		
11. Bluntnose minnow	×		×
12. Fathead minnow	×	×	×
13. Longnose dace	×	×	×
14. Blacknose dace	×	×	×
15. Fallfish	×	×	×
16. Creek chub	×	×	×
17. Brown Trout	2	×	×
18. Rainbow Trout			×
19. Brook Trout	×	×	×
20. Chain pickerel			×
21. Banded killifish			3
22. Brook stickleback	×	×	3
23. Slimy sculpin	×	×	
24. Smallmouth bass	×		×
25. Largemouth bass			×
26. Rock bass		×	×
27. Bluegill		×	×
28. Pumpkinseed		×	×
29. Iowa darter	×		
30. Fantail darter	×	×	×
31. Tesselated darter	×	×	×
32. Logperch	×		4
33. Yellow perch	×		×
Total Species	23	22	29

[1] Many common names have changed during the past 60 years. Species listed as the current common names
[2] According to stocking records, brown trout were extensively stocked in the upstream reaches of Onondaga Creek by 1928
[3] These species were collected in minnow traps during 1991 within 0.5 km of Onondaga Lake. White perch and gizzard shad were also collected, but not included in this table because they were most likely not residents of the creek
[4] In 1991, logperch were captured in the lower reach of Ninemile Creek

cultural development of the watershed during the 1800s. Three species found in 1927 but not in 1990–1991 were the northern redbelly dace (*Phoxinus eos*), Iowa darter (*Etheostoma exile*), and silvery minnow (*Hybognathus regius*). Northern redbelly dace and Iowa darter are characteristic of slow moving, warm streams and swamps, habitats not sampled in 1990–1991. Eight species in the recent collections were not found in 1927, although four were present in Onondaga Lake and one in the Seneca River during 1927. Rainbow trout (*Oncorhynchus mykiss*) in Ninemile Creek today are the result of stocking. Emerald shiners (*Notropis atherinoides*) and brown bullhead were found infrequently during the 1990–1991 surveys. Possibly those species were missed in earlier surveys, although seine collections by experienced ichthyologists often compare very favorably with modern techniques. Species diversity indices cannot be compared between years because the number of individuals were not reported by Greeley (1928). Overall, however, the species composition can be said to have changed remarkably little over the past 60 years.

6.5.9.3 Ninemile Creek Fish Communities and Habitats

The main stem (i.e., exclusive of tributaries) of Ninemile Creek (Figure 6.73) is 36.2 km in length. Marcellus Falls represents a natural demarcation in environment types. Upstream of the falls is predominantly a warmwater fish community, which reflects the release of epilimnetic water from Otisco Lake. The channel flows through farmland, with diverse structure and abundant natural riffles, pools and runs. The average gradient is 0.43 m · 100 m^{-1}. Smallmouth bass and Rock bass (*Ambloplites rupestris*) are abundant in this area, along with a small trout population (Preall 1985). The substantially higher (5–7°C) summer stream temperatures in these upstream areas (Murphy 1992) is the factor most likely limiting trout distribution. Below the falls, the gradient increases to 2.1 m · 100 m^{-1} (i.e., currents are faster) and the stream valley narrows. The stream is structurally less diverse with sections comprised of long, straight runs of high gradient and relatively few pools. This is primarily a coldwater fish community, reflecting the entry of cool groundwater below Marcellus Falls. This section is stocked annually with Brook trout (*Salvelinus fontinalis*) and Brook trout (*Salmo trotta*) and has been characterized as an excellent trout stream, with substantial natural reproduction (Preall and Ringler 1989).

Eleven sites, representing a variety of habitats, were sampled during June and July along Ninemile Creek in 1990 (Figure 6.74). These sites represent the upper 18 km of stream channel; the lower half of the stream was

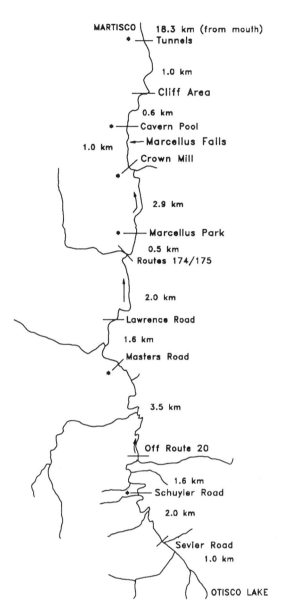

FIGURE **6.74.** Fish Sampling sites (11) on Ninemile Creek in 1990.

not sampled except for an electroshocking (by boat) sample in 1990 and minnow traps set near the mouth in 1991. The reach from Otisco Lake to Marcellus Falls is characterized by slow flowing water.

A total of 26 species was captured in Ninemile Creek during 1990 (Table 6.44). Brown trout dominated the sections below Marcellus Falls, with additional species including longnose dace (*Rhinichthys cataractae*), white sucker, and northern hog sucker (*Hypentelium nigricans*). Sites above Marcellus Falls were dominated numerically by longnose dace, hog sucker, creek chub (*Semotilus atromaculatus*), cutlips minnow (*Exoglossum maxillingua*), common shiner (*Notropis cornutus*), and tesselated darter (*Etheostoma olmstedi*). The site with the greatest species richness was Schuler Road (Figure 6.74), with 15 species. The least diverse sites (Cliff Area and Cavern Pool; Figure 6.74) were located 2.6 km downstream of Marcellus Falls; these sites had the greatest densities of brown trout.

Twelve individuals of five species were electroshocked by boat during October in the lower 1 km of Ninemile Creek (mouth to Rt. 690 overpass). These species were white sucker, carp, white perch, alewife, and gizzard shad. The degraded conditions of the stream in this reach (high siltation, erosion, and high turbidity) probably explain the low fish abundance and diversity compared to upstream sites.

6.5.9.4 Stream Habitat Analysis: Atlantic Salmon Restoration

Atlantic salmon were extirpated from the Onondaga Lake watershed by the late 1800s, probably as a result of the construction of mill dams and deforestation (Webster 1982). Changes in the water quality of the lake may have also contributed to the loss of this highly prized sport fish. Atlantic salmon require high quality water (e.g., dissolved oxygen concentrations of $\geq 6\,mg \cdot L^{-1}$) and temperatures of $16-17°C$ for optimum growth (Dwyer and Piper 1987; Neth and Barnhart 1983). The low oxygen concentrations that develop lakewide during fall turnover (described earlier, Chapter 5) may not be suitable for the survival and growth of juveniles of this species. Conditions in the lake are more favorable during the spring and early summer, when adults migrate upstream and most juveniles move downstream. Onondaga Creek and Ninemile Creek show potential for nursery and spawning habitat of Atlantic salmon.

The Habitat Suitability Index (HSI) that has been developed for Atlantic salmon (Trial

and Stanley 1984; Trial 1989) incorporates habitat features such as stream temperatures, substrate type, and water velocity in a weighted index that predicts suitability of a site for various life stages of the salmon. These variables have been shown to be significant by Chadwick (1985) and others. Each is rated on a scale from 0 to 1.0 (the best), and a geometric mean is used to express habitat suitability for fry, parr, and adults. A total HSI is developed from the geometric mean values for each of the three life stages. An example determination of HSI for a site on Ninemile Creek in 1990 is presented in Table 6.45.

6.5.9.5 Stream Description and Classification

The suitability of Ninemile Creek as habitat for Atlantic salmon was pursued in studies conducted in 1990 and 1991. Total HSI values ranged from 0 to 0.88 among six sites. Below Marcellus Falls showed the best potential habitat for all life stages of Atlantic salmon, mostly because of the lower temperature and higher velocities that prevail in that reach. This is consistent with the existing distribution of brown trout (which have similar habitat requirements) in the stream, as the highest densities of this species are found in the same area. Approximately 2.5 km ($30,000\,m^2$) of quality stream habitat is currently available to all Atlantic salmon life stages in Ninemile Creek (Murphy 1992). This high-quality habitat extends from Marcellus Falls downstream to the tunnels site (Figure 6.74). Based on findings of Gee et al. (1978) and Orciari et al. (1988), optimum fry densities would range from 1 to $2 \cdot m^{-2}$. Thus 30,000 to 60,000 Atlantic salmon fry would need to be stocked in Ninemile Creek to achieve optimum fry densities.

The utility of the HSI model was tested at five sites (Figure 6.74) covering a wide range of HSI ratings by stocking 100 marked Atlantic salmon fingerlings (Little Clear Lake Strain, Altmar Hatchery) at each site (i.e., total of 500 stocked) in October 1990. Electroshocking sampling at two sites (Cavern Pool and Marcellus Park, Figure 6.74) documented the

TABLE 6.45. Sample determination of Habitat Suitability Index (HSI) for Cavern Pool site, Ninemile Creek, 1990.

	Parameter	Value	Suitability Index
V1	Maximum temperature (°C)	20	1.0
V2	Average temperature (°C)	15	1.0
V3	Average turbidity (NTU)	est.	1.0
V4	Minimum oxygen (% saturation)	80	1.0
V5	Minimum pH	8.0	1.0
V6	Velocity (near bottom; cm · s^{-1})	29	0.8
V7	Predominant substrate (fry)	Boulder	0.7
V8	Depth (fry, cm)	34	1.0
V9	Velocity (parr, cm · s^{-1})	31.5	1.0
V10	Predominant substrate (parr)	Boulder	1.0
V11	Depth (parr, cm)	34	1.0
V12	Depth (reproduction, cm)	34	1.0
V13	Velocity (reproduction, cm · s^{-1})	31.5	0.3
V14	Spawning temperature (°C)	9	1.0
V15	Minimum pH in fall and winter	8.0	1.0
V16	Embryo incubation temp (winter, °C)	2	
V17	Stream order	3	1.0
V18	Predominant substrate (reproduction)	Boulder	0.8

Water Quality (WQ) HSI	1.0	Lowest of V1−V5
Fry HSI	0.82	(V6 × V7 × V8)$^{0.33}$
Parr HSI	1.0	(V9 × V10 × V11)$^{0.33}$
Reproduction (R) HSI	0.62	Lowest of: (V12 × V13 × V18)$^{0.33}$ V14, V15, V16, or V17
Total HSI	0.84	(WQ × Fry × Parr × R)$^{0.25}$

continued presence of the fingerlings in December 1990. Salmon were captured at all five of the stocked sites in March 1991. No movement between sites was evident based on a examination of marks (fin clips). The mean growth for fish captured at all five sites over the October–March interval was 20 mm. The growth was not statistically significant based on a student's t distribution (t = 1.55, p = 0.05), in part because sample sizes differed among sites. Salmon were captured at only four of the five sites electroshocked in June 1991 (n = 23); none were found at the most downstream site (Figure 6.74). All fish captured were beginning to smolt, but again no movement was observed among sites. The growth of salmon (mean increase of 50 mm) from stocking in October 1990 to June 1991 was significant, based on the student's t test (t = 2.72, p = 0.05). The recapture of fish shows promising habitat conditions in Ninemile Creek, as winter is the critical period for juvenile Atlantic salmon survival (Gibson and Myers 1988). No marked salmon were recaptured in gill nets in Onondaga Lake during 1991. However, this is not unexpected based on the small size of the stocking and natural mortality levels.

Growth and survival were documented above Marcellus Falls, despite being unsuitable according to the HSI analysis. Saunders (1987) reported the lower and upper temperature limits for Atlantic salmon at −0.7°C and 27.8°C, respectively, with optimum temperature for juvenile growth from 14°C–18°C. At all but one site above the falls, temperatures reached more than 27.8°C in 1991.

The existing fishery for brown trout in Ninemile Creek is one of the most intense in New York State. Angler interest in trout–salmon interactions, coupled with concerns about potential competition between reintroduced salmon and resident trout stimulated a diet study in Ninemile Creek (unpublished data, Ruby and Ringler). Significant overlap in the diets of brown trout and Atlantic Salmon was observed for Ninemile Creek. Diet similarity of salmon and brown trout was found to vary according to fish population density and habitat type. Although competition for food resources may play a role in restoration of salmon, it is likely that available physical habitat and water quality will be of primary importance. Atlantic salmon and brown trout live sympatrically in many European streams. The positive growth of salmon documented here in Ninemile Creek suggests they have the capacity to compete successfully for food. On the other hand, brown trout have successfully colonized the Onondaga Lake tributaries for a century, and are unlikely to be significantly displaced by Atlantic salmon.

6.5.9.6 Onondaga Creek Fish Communities and Habitat

The main stem and tributaries of Onondaga Creek were studied in 1991. Fish were col-

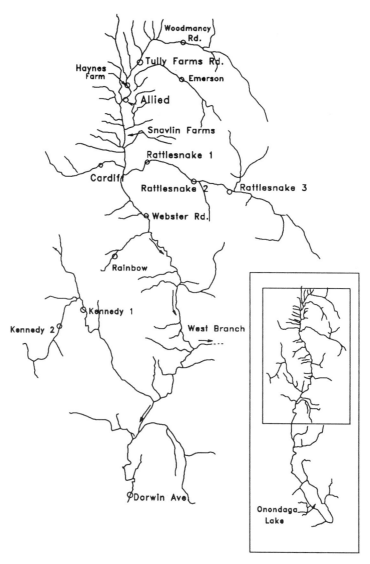

FIGURE 6.75. Fish sampling sites (4) on Onondaga Creek in 1991.

lected at 15 sites, including one site in a channelized section of the creek at Dorwin Avenue
(Figure 6.75). The west branch of Onondaga
Creek was not examined. Morphological characteristics and Rosgen classifications for the
sites are presented in Table 6.46. The Dorwin
Avenue site is an artificial channel that cannot
be classified using this system. A channel of
high gradient, such as Cardiff, is an A channel,
while Kennedy Creek, Rattlesnake Gulf, and
upper Onondaga Creek are B channels with
intermediate gradient. Low gradient streams
such as the lower reaches of Onondaga Creek
are C channels. The mean substrate particle
size (mm) was determined for three riffles at
each site (Wolman 1954); mean pebble size
increases with gradient (Table 6.46).

A length of stream equivalent to 30 bank
full widths was sampled at each site (Leopold
et al. 1964). A length of five bank full stream
widths usually encompassed a single pool-
riffle sequence in this system, so that in most
cases six pool-riffle sequences were sampled
at each site. Study sites were divided into
three groups: main stem, east side, and west

side of the valley. Sites along the main stem
demonstrate increased species richness
down stream (Table 6.47). Brown trout were
found throughout the creek, but significant
reproduction (documented by the capture
of more than five young of the year per 100 m
study section) was not evident below the
Snavlin Farm site (Figure 6.75). No reproduction was evident above Woodmancy Road
(Figure 6.75). Several impassible barriers
probably block migration to these upstream
sites.

The fish communities of the east and west
slopes of the valley are similar. With the exception of Kennedy Creek, the gradients are
high and fish passage likely difficult due to
waterfalls. This probably explains the absence
of brown trout at upstream sites of Rattlesnake
Gulf and Emerson Gulf. Brook trout are found
in Emerson Gulf. The use of chlorine at a
public swimming impoundment upstream of
the site at the unnamed creek in Cardiff may
explain the lack of fish there. Kennedy Creek,
particularly at Webb Road, has many features
typical of a high quality trout stream. It has a

TABLE 6.46. Morphological characteristics and Rosgen Stream classifications of 15 study sites along
Onondaga Creek.

Region/site	Mean width (m)	Maximum pool depth (m)	Mean substrate particle size (mm)	Rosgen Stream class
Main Stem				
Above Dorwin Avenue	13.2	0.65	12	—
Above Webster Road	6.0	>2	6	C3
Behind Snavlin Farm	4.7	>2	24	C1
Behind Haynes Farm	5.1	1.8	24	C3
On Allied Property	4.6	1.5	24	C4
Above Tully Farms Road	2.5	0.7	48	B3
Above Woodmancy Road	2.2	0.5	48	B3
East Slope of Valley				
Kennedy Creek at Webb Road	4.5	1.2	96	B2
Kennedy Creek at Sentinel Heights Road	2.4	0.7	48	B3
Unnamed Creek in Cardiff	1.3	0.5	96	A3
Rainbow Creek above Route 11A	2.4	0.6	48	A3
West Slope of Valley				
Rattlesnake Gulf above Tully Farms Road	4.1	1.1	48	B1-1
Rattlesnake Gulf at end of Hayes Road	2.5	1.5	48	B1
Rattlesnake Gulf below Barker Street	2.7	1.2	48	B3
Emerson Gulf above Woodmancy Road	1.5	0.7	48	A3

Both mean width and maximum pool depth were measured during a period of low stream stage

TABLE 6.47. Fish species richness and brown trout information for 15 study sites along Onondaga Creek, 1991.

Region/site	Species richness	Brown trout density (m^{-2})	Natural reproduction	Maximum size (mm)
Above Dorwin Avenue	15	0.0026	no	284
Above Webster Road	8	0.0073	no	207
Behind Snavlin Farm	7	0.0542	yes	495
Behind Haynes Farm	6	0.0676	yes	445
Allied Property	7	0.3358	yes	258
Above Tully Farms Road	5	0.4857	yes	259
Above Woodmancy Raod	6	0.0510	no	219
East Slope of Valley				
Kennedy Creek at Webb Road	7	0.1221	yes	348
Kennedy Creek at Sentinal Heights Road	4	0.2722	yes	232
Unnamed Creek in Cardiff	0	—	—	
Rainbow Creek above Route 11A	3	0.0347	yes	240
West Slope of Valley				
Rattlesnake Gulf at Tully Farms Road	4	0.0273	yes	320
Rattlesnake Gulf at end of Hayes Road	3	—	—	
Rattlesnake Guld below Barker Street	3	—	—	
Emerson Gulf above Woodmancy Road	4	—	—	

moderate gradient, with extensive instream cover, dense riparian vegetation, and consistent flows. The other tributaries experience very low flows by late summer.

Of the 23 species found during the first sampling effort, 15 were found at only one site, usually in low abundance (Table 6.47). The Dorwin Avenue site supported eight species not encountered in the rest of the watershed. Brown trout, white sucker, creek chub and blacknose dace were the major components of the fish communities at all sites. Slimy sculpin (*Cottus cognatus*) and longnose dace were the only other species found at more than three sites.

The distribution of fish within the watershed illustrates the effect of natural and artificial structures on fish communities. The flood control dam blocks fish movement between Dorwin Avenue and all upstream sites. The fish of Onondaga Lake could reach Dorwin Avenue but not the other sites, which may account for the richness and composition of the Dorwin Avenue fish community. Brown trout were found at all sites within the valley except for those above substantial natural waterfalls; Rattlesnake Gulf (above the Rattle-snake 1 Site) and Emerson Gulf have waterfalls of 10 m or higher.

The blacknose dace was the only species found at all sites that had fish, and was the most abundant fish in all tributaries except Kennedy Creek. Slimy sculpin were found in the main stem at all sites exception for the most upstream site.

Brown trout are broadly distributed in the watershed of Onondaga Lake. They were stocked at various times beginning in the late 1800s, but data collected as part of this effort show extensive natural reproduction by this resilient species. The main stem of the valley receives moderate fishing pressure in the early fishing season and consistent pressure from landowners and a few others throughout the summer. Pools from the Webster Road site to Tully Farms Road are probably fished several times each year. Trout larger than 250 mm were captured behind the farms in the deepest pool within each site. These large trout were found at seven sites (Table 6.47). The 4 km reach of Onondaga Creek above the "mud boils" input supports a substantial population of brown trout, with comparable densities (Table 6.47) to other high quality streams.

6.5.9.7 Comparison of Fish Communities of Ninemile and Onondaga Creeks

Two features are important in the comparison of the fish communities of Ninemile and Onondaga Creeks. The hydrology of the two systems is very different, and the fishing pressure on highly accessible Ninemile Creek is many times greater than on Onondaga Creek. Ninemile Creek receives more than 50 angler hours · 100 m², one of the highest rates in New York State (Earnest Lanteign, NYDEC, personal communication). As many as 10,000 brown trout are stocked annually in Ninemile Creek, creating a popular put and take fishery (Walter Zelie, Carpenter's Brook Fish Hatchery, personal communication). Onondaga Creek is lightly stocked and many stretches are difficult to reach; angling pressure is generally light.

A comparison of growth of 99 brown trout throughout Ninemile Creek (Murphy 1992) and 171 brown trout captured at two sites on Onondaga Creek (Ruby, unpublished) is presented in Table 6.48. The most obvious difference is the lack of fish older than three years in Ninemile Creek. This likely reflects the high levels of fishing pressure in this stream. Growth of brown trout in the main stem of Onondaga Creek is greater for fish ≥3 yr than for the east slope tributaries of the creek (Table 6.48). Brown trout in Ninemile Creek grow at about the same rate as fish in the Onondaga Creek main stem.

The analysis of fish communities reflects a difference in the channel morphology of the two streams. Except for the two sites with abundant brown trout, all sites in the upper half of Ninemile Creek had at least 10 species. In contrast, only two of the main stem sites of Onondaga Creek had as many as eight species. The greater species richness in Ninemile Creek reflects elements of both warm and cold water communities. Species diversity and evenness values are difficult to interpret. Diversity was consistently somewhat higher along Ninemile Creek (mean of 1.531 versus 1.225) due to the higher species richness. When only the Onondaga main stem sites were considered (mean of 1.410), the difference between streams was small. Evenness was the same when all sites were considered.

A technique developed by Tramer (1969) helps to compare fish communities between Ninemile and Onondaga Creeks. Diversity in rigorous and unpredictable ecosystems is expected to vary more with evenness, whereas in more stable systems diversity varies with richness. Much of the variability in diversity in both streams is attributable to richness (Figure 6.76), indicating both streams are stable systems. Data from the only Onondaga Creek site with rock bass, hog sucker, fall fish (*Semotilus corporalis*), cutlips minnow and common shiner (Dorwin Avenue) lie closest to a cluster representing the Ninemile Creek warm water community. Although the two streams are quite different in hydrologic char-

TABLE 6.48. Ages and mean back-calculated lengths (mm) of brown trout collected in Ninemile and Onondaga Creeks.

Stream/sites	Age				
	I	II	III	IV	V
Ninemile Creek (all sites) (n = 99)	92	154	216		
Onondaga Creek (main stem behind Haynes farm (n = 17)	87	160	230	324	402
Onondaga Creek (east slope, Kennedy 1) (n = 17)	87	145	189	227	

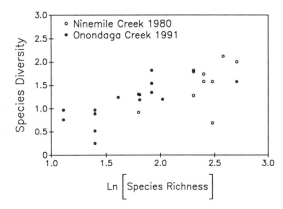

FIGURE 6.76. Relationships between species diversity and species richness in Ninemile Creek and Onondaga Creek.

acter, the relationship between species diversity and richness is remarkably similar in reaches with abundant brown trout (Figure 6.76).

6.6 Indicator Bacteria

6.6.1 Coliform Bacteria

6.6.1.1 Background

Indicator bacterial groups are commonly used by public health officials to reflect the potential presence of pathogenic organisms. The coliform group of bacteria is the principal indicator of suitability of water for various uses (APHA 1990). The fecal coliform (FC) group represents those coliforms common to the feces of warm blooded animals (including man). Tests are routinely conducted to assess both the concentration of the entire coliform group, herein described as total coliform (TC), and the concentration of FC. The TC test has been established longer, and is considered necessary to monitor drinking water supplies, as even the non-FCs (e.g., associated with soils) render the water potentially unsafe for consumption. The TC test is also routinely used to assess the adequacy of surface waters for recreation, although significant contributions by non-FCs can complicate effective management of these resources. The FC test has been used increasingly for this application because of its greater specificity. A general correlation exists between the concentration of FC and the occurrence of *Salmonella*. However, "relationships between TC and pathogens are generally not considered quantitative" (Thomann and Mueller 1987). Thomann and Mueller (1987) indicate that FC levels are about 20% of the TC concentration, although wide variablity is to be expected.

Bacteriological standards, based on both the TC and FC groups, have been developed for the regulation of public water supplies and the use of surface waters for recreational purposes. Standards applicable in New York State are presented in Table 6.49. State Sanitary Code (statutory authority: public health law No. 225 of the State of New York) standards for TC and FC must be met to open and operate a

bathing beach. The other standards are set by the New York State Department of Environmental Conservation for the B/C classifications that presently prevail for Onondaga Lake.

The urban setting of Onondaga Lake, and the entry of high loads of fecal coliforms associated with the inflow of dilute raw sewage during runoff events (Canale et al. 1993), have led to great concern for the levels of indicator bacteria that prevail in the lake. Here we review the coliform data base for the lake to document the temporal and spatial distributions that have prevailed, and the frequency and character of contravention of the public health standards. Additionally, the results of field and laboratory experiments conducted to determine the kinetics of important loss processes for FCs in the lake are presented. Subsequently, in Chapter 9, certain of the lake FC data, results of the kinetic studies, and loading data presented in Chapter 3, are used to support the development and validation of a mechanistic model for FCs in the lake.

6.6.1.2 Description of Monitoring Programs

The lake has been monitored for coliform bacteria by various parties since as early as the late 1960s. These programs have differed greatly with respect to tenure, sampling frequency, and sampling stations. Features of these programs are summarized in Table 6.50. The locations of the stations are presented in Figure 6.77. The first two (described hereafter as long-term) programs have had preset sampling days, in contrast to the other shorter-term efforts that were intended to document the temporal details associated with runoff events (Table 6.50). The same 10 sampling stations were utilized in the two short-term efforts. In both cases these data sets supported the development and testing of mechanistic FC models for the lake (Canale et al. 1993; Freedman et al. 1980). However, these station locations differed from most of those used in the longer-term/multiple station program (Program No. 2, Table 50; Figure 6.77). All samples were grab-type. Only near-surface samples were collected for the long-term/

TABLE 6.49. Bacteriological water quality standards of interest to Onondaga Lake.

Application	Indicator	Standard	Source
Bathing beaches	TC	"The total number of organisms of the total coliform group shall not exceed a logarithmic mean of 2400/100 ml for a series of five or more samples in any 30 day period, nor shall 20% of total samples during the period exceed 5000/100 ml. When the above prescribed standards are exceeded, the permit-issuing official shall cause an investigation to be made to determine and eliminate the source or sources of pollution, or	Chapter I. State Sanitary Code, Part 6, Subpart 6-2, bathing beaches (1988)
Bathing beaches	FC	The fecal coliform density from the five successive sets of samples collected daily on five different days shall not exceed a logarithmic mean of 200 per 100 ml. When fecal coliform density of any sample exceeds 1,000 per 100 ml, consideration shall be given to closing the beach and daily samples shall immediately be collected and analyzed for fecal coliform for at least two consecutive days"	Chapter I. State Sanitary Code, Part 6, Subpart 6-2, bathing beaches (1988)
Quality standards for class B and C waters	TC and FC	"The monthly median coliform value for 100 ml of sample shall not exceed 2400 from a minimum of five examinations, and provided that not more than 20% of the samples shall exceed a coliform value of 5000 for 100 ml of sample and the monthly geometric mean FC value for 100 ml of sample shall not exceed 200 from a minimum of five examinations. This standard shall be met during all periods when disinfection is practiced."	Section 701.19, Classifications and Standards of Quality and Purity (1986)

multiple station program; near-surface and 12 m samples were collected for the two storm impact studies. All the programs focused on the summer months.

Results from both the membrane filter technique and the multiple tube fermentation test are included in the overall database. All the data from the two storms impact studies

TABLE 6.50. Description of coliform monitoring programs for Onondaga Lake.

Program number	Tenure	Parameter	Responsible party	Stations	Frequency
1	Since 1977 Since 1977	TC FC	Drainage and Sanitation, Onondaga County	Mostly 1 (7 depths)	$2 \cdot mo^{-1}$
2	1975–1990	TC/FC	Department of Health, Onondage County/ NYSDEC	10, until 1987; 10 to 13 until 1990	Variable
3	1976	FC	Drainage and Sanitation, Onondaga County	10 (2 depths)	$\geqslant 1\, d^{-1}$ after 5 storm events
4	1987 June–August	FC	Drainage and Sanitation Onondaga County	10 (2 depths)	$\geqslant 1\, d^{-1}$ after 2 storms; $5\, d \cdot wk^{-1}$ in dry weather

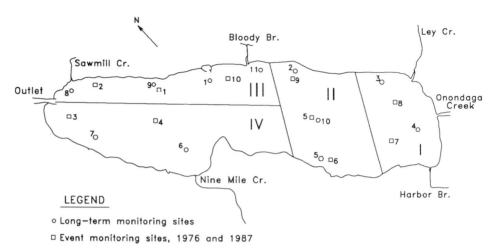

FIGURE 6.77. Sampling stations for coliform monitoring programs for Onondaga Lake; and segmentation of lake into four cells for analysis of data base from long-term programs.

are membrane filter results. Results from the other two programs are "mixed." Inclusion of both data types is not expected to significantly compromise historic analysis, as the two procedures are both standard methods, and described as equally effective and yielding comparable results (APHA 1985).

6.6.1.3 Detailed Monitoring Data: Summer 1987

6.6.1.3.1 Sampling

Intensive monitoring of the lake and tributaries was conducted in 1987 to support the development and validation of a fecal coliform model (Canale et al. 1993). The program was made up of dry weather and wet weather components. Dry weather samples were collected daily (weekdays) from the 10 lake stations (Figure 6.77); collections ranged from one to three times per day for two storms (Storm 1: June 23–26; Storm 2: July 15–16). Samples were collected at depths of 1 and 12 m (or 1 m above bottom at shallow stations). The statistical reliability of the FC results was represented by the 95% confidence limits developed by APHA (1990).

6.6.1.3.2 Dry Weather Conditions

Statistics describing selected features of the distribution of FC bacteria during the study

period of 1987 are presented in Table 6.51. Time plots of FC concentrations for each of the 10 stations are presented in Figure 6.78a–j. The abrupt increases observed lake-wide in late June and mid-July correspond to the only two storms that occurred over the monitoring period. Fecal coliform bacteria concentrations in the north end of the lake were generally low and invariant. The concentrations in the south end of the lake were higher and more variable, reflecting the impact of dry weather discharges of water contaminated with fecal material to that portion of the lake (Chapter 3). Dry weather discharges from Onondaga Creek and Harbor Brook result in a gradient in FC concentrations in the southern portion of the lake. For example, note the concentrations at stations 7 and 8, proximate to the stream mouths, are markedly higher than at stations 5, 6 and 9 (Table 6.51; Figure 6.78).

Open lake FC concentrations during dry weather periods (most of the summer) did not exceed (except immediately adjacent to the Onondaga Creek and Harbor Brook inflows) the current public health standards for contact recreation (Figure 6.78). It is interesting to note that in dry weather average fecal coliform concentrations at the 12 m depth were invariably higher than in the near-surface waters. Concentrations in excess of the public health standard were observed during dry weather at

TABLE **6.51a.** Average fecal coliform bacteria concentrations (cells $\cdot$ 100 ml^{-1}) for lake stations.

Station	Dry weather (1 m)	Dry weather (12 m)	Storm 1 (1 m)	Storm 1 (12 m)	Storm 2 (1 m)	Storm 2 (12 m)
1	7	10	222	73	102	120
2	8	13	>225	47	168	58
3	7	13	182	53	140	208
4	8	12	226	27	406	38
5	15	17	>280	233	160	88
6	8	16	>408	>314	646	66
7	26	59	>316	>316	502	194
8	20	138	>276	301	280	314
9	10	16	>379	71	30	156
10	11	18	>272	365	50	30

TABLE **6.51b.** Maximum fecal coliform bacteria concentrations (cells $\cdot$ ml^{-1}) for lake stations.

Station	Dry weather (1 m)	Dry weather (12 m)	Storm 1 (1 m)	Storm 1 (12 m)	Storm 2 (1 m)	Storm 2 (12 m)
1	25	25	620	135	340	440
2	20	25	>500	75	470	180
3	15	25	435	120	560	660
4	30	20	500	55	1900	100
5	75	35	>500	700	360	120
6	35	30	>500	700	1580	140
7	35	304	>500	>500	940	380
8	40	480	>500	500	820	520
9	20	25	>500	135	90	290
10	25	50	>500	545	110	60

Periods for Averaging and Determination of Maxima
Dry Weather: July 28–August 9, 1987
Storm 1: June 23–26, 1987
Storm 2: July 15–16, 1987

the 12 m depth of stations 7 and 8, sites influenced by the discharge of Onondaga Creek. Factors influencing this observation may include: incomplete mixing of the Onondaga Creek inflow with the lake (e.g., plunging of Creek inflow), bactericidal effects of light in the upper waters (e.g., Auer and Niehaus 1993), or the accumulation of persistent strains of fecal coliform bacteria in the bottom waters (cf. Thomann and Mueller 1987).

6.6.1.3.3 Wet Weather Conditions

The strong, relatively short-term, increases in FC concentrations that occurred throughout the lake in response to both storms are clearly manifested in the summerlong time plots of Figure 6.78a–j. The short-term temporal

structures for Storm 2, along with the 95% confidence bounds, are presented for the 1 m depth in Figure 6.79. Analysis of data from Storm 1 is complicated by uncertainty associated with the inclusion of "too- numerous-to-count" (TNTC) observations at peak concentrations early in the event at certain stations (e.g., Figure 6.78; also see columns 4 and 5 of Table 6.51a), attributable to inadequate dilution of samples.

A rainfall of 2.66 inches (as measured at Hancock Airport) occurred over a period of 8 hours for Storm 1; 0.95 inches fell within one hour. The storm had a return frequency of 7 years, based on an average precipitation of 0.38 inches per hour over a 7 hour period (Moffa and Associates, Personal Communication). Fecal coliform concentrations were 10

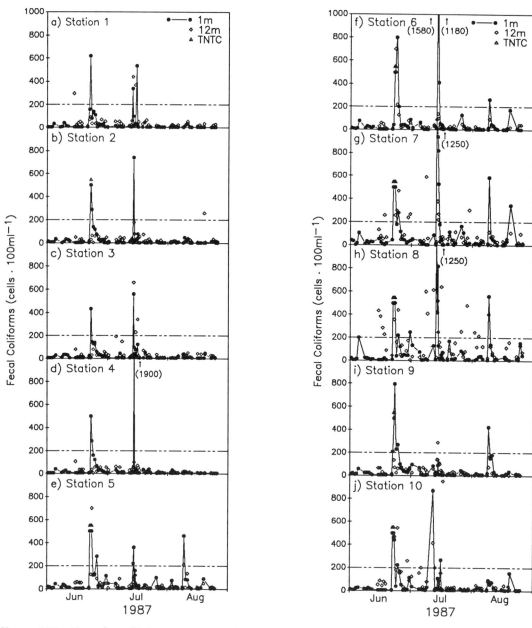

FIGURE 6.78. Time plots of FC concentrations for summer of 1987: (**a**) station 1, (**b**) station 2, (**c**) station 3, (**d**) station 4, (**e**) station 5, (**f**) station 6, (**g**) station 7, (**h**) station 8, (**i**) station 9, and (**j**) station 10.

to 30 times higher during the storm period than under dry weather conditions. Violations of the single observation public health standard (1000 cells · 100 ml^{-1}) may have occurred at as many as seven stations (Table 6.51b, column 4) on June 23. Concentrations fell below the 200 cells · 100 ml^{-1} level in 1 to

2 days in the northern portion of lake and in 2 to 3 days in the southern portion. Background concentrations were reached in about four days lake-wide. Concentrations were generally higher in the surface waters than at the 12 m depth. Increases in surface water concentrations in the southern portion of the lake

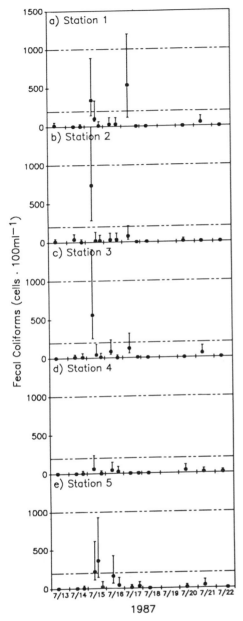

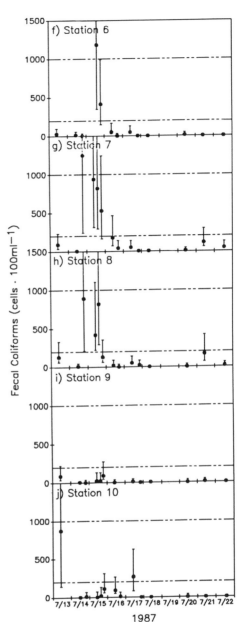

FIGURE 6.79. Time plots of FC concentrations, with 95% analytical confidence limits, for Storm 2 of 1987: (a) station 1, (b) station 2, (c) station 3, (d) station 4, (e) station 5, (f) station 6, (g) station 7, (h) station 8, (i) station 9, and (j) station 10.

were generally tracked by less dramatic peaks in the deeper water. However, concentrations at 12 m remained quite low in the north. This difference may reflect the magnitude of inflow at the south end which influences much of the water column, while loads to the northern part of the lake are smaller and probably more restricted to the surface waters.

A rainfall of 0.95 inches (as measured at Hancock Airport) occurred over a period of 5 hours for Storm 2; 0.72 inches fell in a one hour period. The storm had a return frequency of one year, based on an average precipitation of 0.72 inches per hour over a one hour period (Moffa and Associates, Personal Communication). The duration of the perturbation was shorter (Figure 6.78), consistent with the smaller magnitude of the storm. The expanded time scale of Figure 6.79 more completely resolves the abrupt increase and the subsequent progressive decreases that occur in response to a storm. Also clearly manifested in the time plots of Figure 6.79 is the substantial uncertainty that is inherent in the FC analysis (APHA 1990); as reflected by the rather broad 95% confidence bounds. Wet weather concentrations were 20 to 50 times higher than dry weather levels in the north end and 10 to 20 times higher at the south end. Violations of the single observation public health standard were observed at three stations in the south basin; logarithmic means of five consecutive samples were above $200 \cdot 100 \, ml^{-1}$ at two of these sites (stations 7 and 8, Figure 6.79). Two mid-lake stations (stations 9 and 10) showed little response to storm loads. Concentrations fell below $200 \, cells \cdot 100 \, ml^{-1}$ in less than 2 days at the south end. Background concentrations were reached in approximately two days lakewide. The timing of peaks at the north end for both storms was at the upper limit of that which could have been caused by northward transport of inputs at the south end (Owens 1989). This, coupled with the fact that two mid-lake stations (stations 9 and 10) showed little response in Storm 2, indicates that fecal coliform bacteria loads at the south end of the lake may not be entirely responsible for increases in concentration at north basin stations.

6.6.1.3.4 Wet Weather Observations in 1976

Fecal coliform concentrations were intensively monitored in the lake for five different storms during the summer of 1976 (Table 6.50) as part of a study of the impact(s) associated with these events (Freedman et al. 1980). It should be recalled that these efforts preceded the BMP program for the sewer system. Lake monitoring results are presented for the same

10 stations used in 1987 for a single 1976 event in Figure 6.80, as a basis for qualitative comparisons with the more recent event data (e.g., Figures 6.78 and 6.79). The event occurred on August 28, 1976, 2.08 inches of rainfall was received over a seven hour period (as measured at Hancock Airport); 0.74 inches fell in one hour. The severity of the storm, with respect to intensity and duration, was intermediate to that of the two 1987 events.

Violations of the single observation public health standard were observed at all stations except the site located in the central portion of the north basin (Figure 6.80). The same type of spatial picture emerges as demonstrated for the 1987 data sets (Figure 6.79), associated with the location of the largest FC loads at the south end of the lake. The impact on the south basin was more severe with respect to the degree and duration of contravention. Baseline concentrations were reached within three days after the event in the south basin, and within two days in the north (Figure 6.80).

The impact of this storm on FC levels in the lake appears to have been greater than observed for either sorm in 1987 (despite the limitations of the TNTC observations in storm 1 of 1987). Note that an expanded scale was utilized in Figure 6.80 to accommodate the higher FC concentrations encountered. A larger fraction of the lake failed to meet the single observation standard, and the margin of contravention was probably greater, in 1976. Five observations $\geq 8000 \, cells \cdot 100 \, ml^{-1}$ were made during the event. Further, background concentrations in the southern end of the lake were higher in 1976. The BMP program may have contributed to these apparent reductions in the degree of bacteriological contamination of the lake following runoff events.

6.6.1.4 Long-Term Monitoring Program

6.6.1.4.1 Spatial Segmentation of Lake, and Frequency of Sampling

Analysis of the long-term data sets is more tractable if the surface waters of the lake are segmented into several regions or cells, instead of considering each station (note these are the

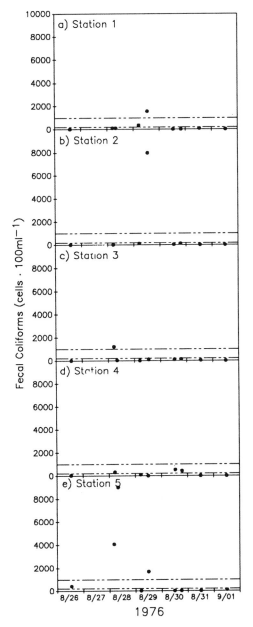

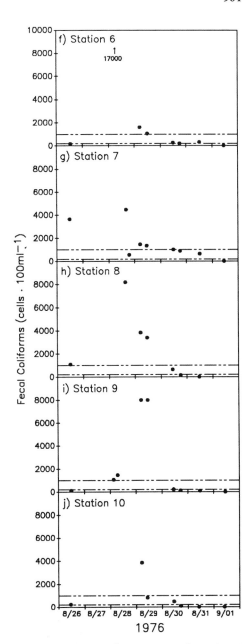

FIGURE **6.80.** Time plots of FC concentrations for storm of August 28, 1976: (**a**) station 1, (**b**) station 2, (**c**) station 3, (**d**) station 4, (**e**) station 5, (**f**) station 6, (**g**) station 7, (**h**) station 8, (**i**) station 9, and (**j**) station 10.

"L" stations of Figure 6.77) separately. In this analysis the upper waters of the lake have been segmented into four "cells" (Cells I–IV, Figure 6.77). While the details of the segmentation are necessarily somewhat arbitrary, its basic character is consistent with the localiza-tion of the principal sources of coliforms at the southeast end of the lake, the basic flow pattern and configuration of the lake (Chapter 4), and the emphasis of the monitoring pro-gram in the northeast portion of the lake (Cell III; also see subsequent description of sampling frequency). The segmentation in the southern basin of the lake (Cells I and II positioned along the main axis of the lake) is similar to the epilimnetic segmentation used for that portion of the lake in the recent FC model for the lake (Canale et al. 1993). Cell III includes the proposed bathing beach area(s), and for that reason, the sites that have been monitored most frequently. Cell IV is also remote from the principal sources of coliforms, but this area has received less intense monitoring.

This segmentation is supported by correlation analyses between stations from the different cells. Scatter plots of selected pairs of stations for paired TC and FC analyses are presented in Figure 6.81. The aggregating of the individual stations within each of the cells is supported, as concentrations were more similar within the segments than between cells. These analyses indicate that in general concentrations in Cell I > Cell II ⩾ Cell III ⩾ Cell IV. This is illustrated in greater detail subsequently.

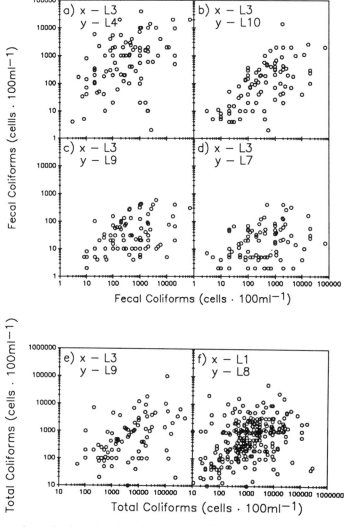

FIGURE 6.81. Scatter plots of selected pairs of stations for paired TC and FC analysis: (a) FC, L4 versus L3, (b) FC, L10 versus L3, (c) FC, L9 versus L3, (d) FC, L7 versus L3, (e) TC, L9 versus L3, and (f) TC, L8 versus L1.

Three different sampling frequencies are associated with the four cells (Figure 6.82). Cell III has been monitored the most frequently, to support potential bathing beach considerations. Generally this frequency was maintained at a higher level after 1980. Cells I and IV have been monitored at an equal frequency. The frequency of monitoring of these stations has been the lowest; temporal coverage of these areas decreased in the 1980s (Figure 6.82). The intermediate frequency of sampling for Cell II reflects the incorporation of the data base of Program No. 1 (Table 6.50; see Figure 6.77; surface samples only).

6.6.1.4.2 Data Analysis and Presentation

Long-term conditions (1975–1990) with respect to TC and FC levels are represented here in three different ways. First, annual log mean concentrations of TC and FC are presented in Figures 6.83 and 6.84. Observations from all stations within the cells are treated equally. Results from Programs 1 and 2 (Table 6.49) were combined for Cell II. Variability in time (within each summer) and space (between different stations within each cell) is represented by one standard deviation about the log mean. The standard deviation error bars appear only for the lower FC concentrations observed (e.g., Figure 6.84d and e), indicating the relative variability (according to this measure) was low. Concentrations incorporated in the standards for TC and FC (Table 6.49) are included on the plots, as is the period over

which the combined sewer overflow abatement "best management practices" (BMP) program was executed. Total rainfall over the May–September interval of each year is included for reference. This is an imperfect statistic to represent the interplay between indicator loading and in-lake concentrations, as the magnitude of loading is highly dependent on the intensity of the runoff event (e.g., Canale et al. 1993; Freedman et al. 1980). An alternate analysis is presented in Table 6.52; the percent of measurements for which indicator concentrations exceed values incorporated in the standards (Table 6.49) is presented for each cell for three time periods corresponding to before (1975–1982), during (1983–1985), and after (1986–1990) BMP. While neither of these modes of presentation of the data match the specific constraints of the standards (Table 6.49), each is valuable in identifying and describing trends in bacteriological water quality in the different areas of the lake.

The more intensive monitoring data for the stations of Cell III (usually Nos. L1, L8, and L9; Figure 6.77) were subjected to a third type of analysis that is consistent with the form of the standards for the state sanitary code (Table 6.49); the results are presented for the period of record in Figures 6.85 through 6.88. The distribution of the log mean TC concentration of five consecutive values is presented in Figure 6.85; the distribution of the percentage of the five consecutive values that exceeds 5000 coliforms · 100 ml^{-1} appears in Figure 6.86. Values are plotted on the date of the collection of the third sample of the "moving" series of five observations; an example of the period bounds is presented in Figure 6.85. The distribution of the log mean FC concentration of five consecutive samples is plotted in Figure 6.87, and a simple time plot of FC appears in Figure 6.88. The majority of the data sets from different years fail to meet the sampling frequency constraints of the standards (e.g., 5 samples in a month, Table 6.49); since 1981 samples have been collected about four times per month.

The limitations of sampling frequency need to be considered in review of the long-term data sets. The results of short-term monitoring

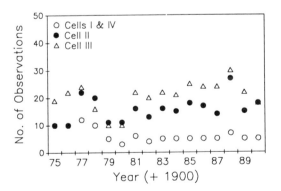

FIGURE 6.82. Summer sampling frequencies for coliforms for Cells I–IV for period of record.

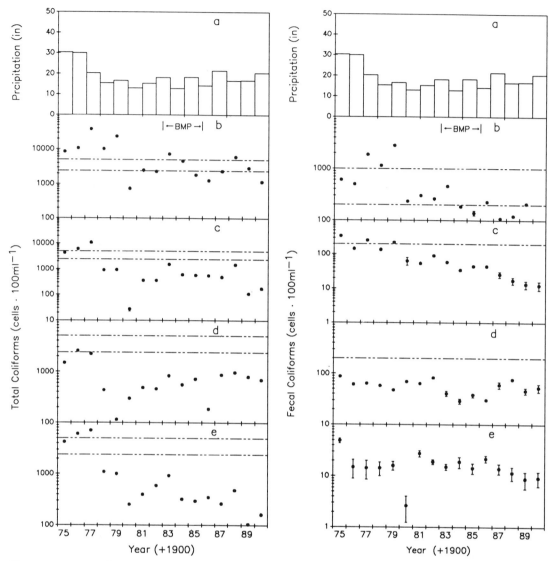

FIGURE 6.83. Trends in annual TC concentrations and attendant summer rainfall, 1975–1990: (a) rainfall, May–September, (b) log mean TC concentration in Cell I, (c) log mean TC concentration in Cell II, (d) log mean TC concentration in Cell III, and (e) log mean TC concentration in Cell IV.

FIGURE 6.84. Trends in annual FC concentrations and attendant summer rainfall, 1975–1990: (a) rainfall, May–September, (b) log mean FC concentration in Cell I, (c) log mean FC concentration in Cell II, (d) log mean FC concentration in Cell III, and (e) log mean FC concentration in Cell IV.

efforts (e.g., Figures 6.78, 6.79, and 6.80), focusing on the impact of storm-induced runoff events, indicate that concentrations of indicators return to background (dry weather) levels within three to four days. Routine monitoring conducted at greater intervals (particularly Cells I, II and IV in recent years, Figure 6.82) rarely will capture the maximum impact

and associated risk conditions, and may not detect any perturbations from some events. For example, since the 1980s it's likely the long-term monitoring results for Cells I, II, and IV reflect mostly background/dry weather conditions. The probability of incorporating wet weather impacts is of course greater in the years with more frequent rainfall events.

TABLE 6.52. Percent exceedance of selected TC and FC concentrations in Onondaga Lake for four lake segments and three time periods.

Indicator	Cell number	Statistics	Periods		
			1975–1982	1983–1985	1986–1990
FC	I				
		n	116	28	42
		% > 200	75	50	33
		% > 1000	45	29	9
	II				
		n	236	77	122
		% > 200	44	13	10
		% > 1000	14	1	0
	III				
		n	421	199	404
		% > 200	23	9	26
		% > 1000	3	1	8
	IV				
		n	113	26	43
		% > 200	7	4	0
		% > 1000	0	0	0
TC	I				
		n	115	30	47
		% > 2400	70	60	86
		% > 5000	65	22	49
	II				
		n	226	79	145
		% > 2400	45	10	10
		% > 5000	37	8	6
	III				
		n	383	203	461
		% > 2400	29	13	25
		% > 5000	17	5	13
	IV				
		n	102	27	54
		% > 2400	13	22	9
		% > 5000	5	4	0

Undoubtedly wet weather impacts are to some extent reflected in the more intensive monitoring program conducted for Cell III in recent years. The principal attributes of the long-term data are the large number of observations and the lengthy duration of measurements. Year-to-year differences are not emphasized, because of the sampling frequency constraints, and the large variability in summer rainfall. These data are used to reflect longer term changes in the lake, particularly with respect to background concentrations.

6.6.1.4.3 Observations and Interpretations

A gradient in coliform concentrations is apparent from Cell I to Cells III and IV, based on a log mean treatment of the summer data sets (Figures 6.83 and 6.84). The differences between Cells I and II have been the most striking. Concentrations in Cell II were distinctly higher than in Cell III for the period 1975–1979, however, this has not been recurring since that time. The higher concentrations observed in Cell III in 1987 and 1990 compared to Cell II (Figures 6.83 and 6.84) probably reflect the addition of station L11, located proximate to the mouth of Bloody Brook, to the monitoring program in those years. The log mean concentration of TCs was lower in Cell IV than Cell III in most years since 1982 (Figure 6.83d and e). In contrast, FC concentrations in Cell IV have been less than found

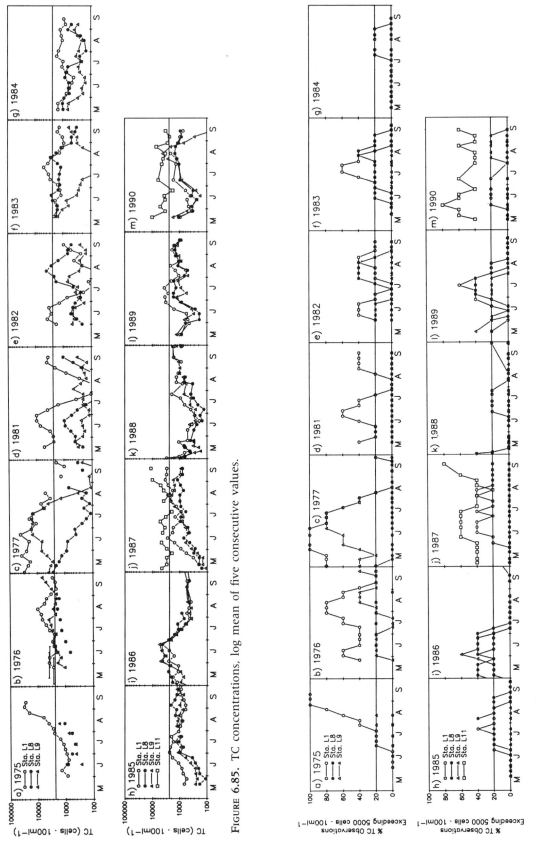

FIGURE 6.85. TC concentrations, log mean of five consecutive values.

FIGURE 6.86. Percent of five consecutive TC values >5000 · 100 ml⁻¹.

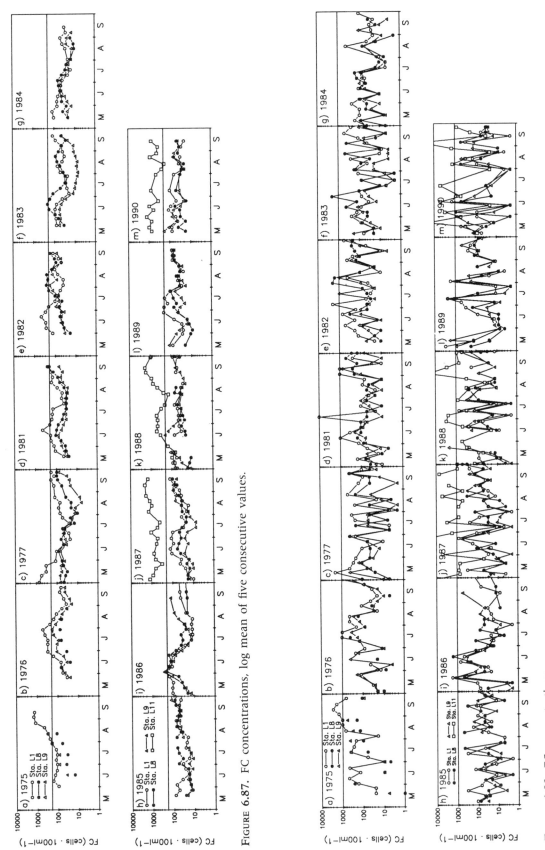

FIGURE 6.87. FC concentrations, log mean of five consecutive values.

FIGURE 6.88. FC concentrations.

in Cell III annually since the start of the program. With the exception of this feature of the distributions, the TC and FC data sets depict similar trends for the long-term monitoring program.

The highest TC concentrations observed in Cells II, III, and IV occurred in the first three years of the monitoring program (1975–1977); some of the highest concentrations observed in Cell I were also observed in that period. The relatively high levels of rainfall that occurred in those three years (Figures 6.83a and 6.84a) may have contributed to the elevated concentrations. However, the rainfall statistic does not perform well in explaining the observed variations in coliform concentrations for the period of record (Figures 6.83 and 6.84). Elevated concentrations of FCs were particularly high in Cell I the first 5 yrs (Figure 6.84b), consistent with the observations for TCs (Figure 6.83b). However, FC levels were not distinctly higher in the early years in the other cells. The summer concentrations of FCs have decreased in a nearly progressive manner in Cell II since the start of the monitoring program (Figure 6.84c). The picture for this Cell based on TC data is more erratic (Figure 6.83c). No abrupt and lasting reductions in TC or FC concentrations have been clearly manifested in the long-term data based in response to the BMP program. Improvements seem to have preceded the BMP program (e.g., late 1970s instead of 1983–1985).

The presentation of the data in the format of Table 6.52 depicts some of the same features of the spatial and temporal distributions of coliforms in the lake described in Figures 6.83 and 6.84. Rather distinct spatial gradients were observed for TCs and FCs in the 1975–1982 period with respect to percent exceedance of critical concentrations; according to Cell I > Cell II > Cell III > Cell IV. For example, 45% of the FC observations in that period from Cell I were greater than 1000 cells $\cdot$ 100 ml^{-1}, compared to 14, 3, and 0% for Cells II, III and IV, respectively. Conditions in Cell III were worse than in Cell II over the period 1986–1990. The addition of station L11 to Cell III over part of this period contributed to this

reversal. Substantial improvements in both TC and FC levels in Cell II were achieved, from the 1975–1982 interval to the 1986–1990 period, with respect to the limits of Table 6.52. A major improvement in FCs was also observed in Cell I over the same interval, however, this was not reflected in the TC data base. Improvements were apparent for some parts of the lake during the period of the execution of the BMP program (1983–1985), however, these conditions were not greatly different from those observed in the later interval. Recall that the more continuous treatment of this data base in Figures 6.83 and 6.84 does not support the position that the BMP program was solely responsible for the observed improvements.

The monitoring results for Cell III, presented in Figures 6.85 through 6.88, represent the most complete statement of the adequacy of bacteriological water quality for swimming for Onondaga Lake. Despite the distance that separates this area from the principal sources, clear contraventions of the TC standards are indicated for all years (Figures 6.85 and 6.86). Recall that the condition of 20% of TC samples exceeding 5000 $\cdot$ 100 ml^{-1} is a contravention (Table 6.49; boundary indicated on Figure 6.86). This second part of the TC standard is the more stringent for the Onondaga Lake data base in terms of occurrence and duration of contravention (compare Figures 6.85 and 6.86). On this basis the standard was not met in Cell III more than 50% of the summer in all years except 1984, 1986, and 1988. In contrast, the first part of the standard failed to be met more than 50% of the time in only four years (Figure 6.85). Some regionalization of bacteriological water quality is apparent within Cell III. Stations skewed towards the southern end (station 1, and station 11 in later years) often show higher concentrations of TCs and FCs.

Log mean FC concentrations within Cell III exceeded 200 $\cdot$ 100 ml^{-1} at times in the summers of 9 of the 13 years included in the analysis of Figure 6.87. The most severe contraventions observed recently have been associated with the inclusion of station 11 (Figure 6.87j, k, and m). Contraventions of the single observation public health standard of $1000 \cdot 100 \, \text{ml}^{-1}$ have been observed in 10 of 13 yrs (Figure 6.88). Based on the characteristics reflected in the intensive short-term program (Figure 6.78), these contraventions are probably largely associated with runoff events.

6.6.1.5 Determination of Fecal Coliform Loss Kinetics

6.6.1.5.1 Background

When the frequency of violation of fecal coliform standards is such that water use is impaired, as is the case in Onondaga Lake, water quality managers seek to develop remedial strategies. Mathematical models have been used in support of these efforts (e.g., Freedman et al. 1980; Palmer and Dewey 1984; Canale et al. 1993). These models are developed from mass balance principles, and external sources and the kinetics of fecal coliform loss processes are quantified.

The rate of loss of fecal coliform bacteria is assumed to be proportional to the bacterial concentration, e.g., follows first order kinetics (Thomann and Mueller 1987),

$$R = \frac{dC_{fc}}{dt} = -k \cdot C_{fc} \qquad (6.14)$$

in which C_{fc} = fecal coliform bacteria concentration (cfu $\cdot$ 100 ml^{-1}), k = overall first order loss coefficient (d^{-1}), R = rate of change in fecal coliform bacteria concentration (cfu $\cdot$ 100 ml$^{-1} \cdot$ d^{-1}), and t = time (d). The unit cfu represents "colony forming unit" in the membrane filtration method for fecal coliforms (APHA 1985); it's assumed each colony is formed from one cell, i.e., cfu $\cdot$ 100 ml^{-1} = cells $\cdot$ 100 ml^{-1}. A number of biotic and abiotic factors influence the fecal coliform bacteria death rate, including algal toxins, bacteriophages, nutrients, pH, predation, temperature, salinity, and irradiance (Bowie et al. 1985). Among these, irradiance (Gameson and Saxon 1967), temperature (Lantrip 1983), and the combined effects of irradiance and temperature (Mitchell and Chamberlin 1978) are generally considered most important.

The knowledge that bacterial cells may suffer injury or death on exposure to irradiance has been utilized in the design of disinfection devices. Bactericidal effects are strongest in the ultraviolet (UV) region ($\sim$260 nm; Lantrip 1983), however, visible irradiance has also been shown to be an agent in coliform mortality (Fujioka et al. 1981; Kapuscinski and Mitchell 1983; McCambridge and McMeekin 1981). Visible wavelengths take on particular significance in natural systems because (1) UV wavelengths represent only a small (<3%) fraction of total incident radiation (Jassby and Powell 1975; Kirk 1983; Lantrip 1983), and (2) UV radiation is rapidly attenuated in the water column, especially when dissolved organic matter is present (Jerlov 1968; Kirk 1983; Wetzel 1983).

Rates of biochemical reactions and thus, microbial growth rates, tend to increase as temperatures rise. High growth rates place added demands on nutrient reserves which may not be renewed in dilute, natural systems, leading to an increase in the death rate. Early observations that the rate of disappearance of coliform bacteria in rivers was greater in summer than in winter (cf. Kittrell and Furfari 1963) lead investigators to pursue a relationship between the death rate coefficient and temperature. Although several relationships have been published (Canale et al. 1973; Mancini 1978), attempts to develop a reliable function have been hampered by experimental artifacts inherent to laboratory measurements and by logistic difficulties in conducting controlled in situ experiments.

Sedimentation may be a major mechanism responsible for the disappearance of fecal coliform bacteria from surface waters (Mitchell and Chamberlin 1978; Gannon et al. 1983). Cells settle from the water column as discrete entities and as part of larger aggregates of fecal material, stormwater debris and other suspended solids (Schillinger and Gannon 1982). Gannon et al. (1983) observed that viable fecal coliform bacteria accumulated at the sediment surface in Ford Lake, Michigan, and concluded that sedimentation played an important role in the overall disappearance of fecal coliform bacteria from the water column.

The collective effects of sedimentation and irradiance- and temperature-mediated death on the overall loss coefficient are represented in equation (2) (cf. Thomann and Mueller 1987),

$$k = k_{de} + k_i + k_s \qquad (6.15)$$

in which k_{de} = rate coefficient for death in the dark; includes effects of temperature, salinity, predation, etc. (d^{-1}), k_i = rate coefficient for death as mediated by irradiance (d^{-1}), and k_s = rate coefficient for sedimentation loss (d^{-1}).

6.6.1.5.2 Methods

An abstracted version of the methods is presented here; experimental details appear elsewhere (Auer and Niehaus 1993). In situ experiments were conducted to evaluate the influence of irradiance; measurements on samples collected in situ were made to assess sedimentation losses. The effect of temperature on fecal coliform death was evaluated in laboratory experiments. Inocula for field and laboratory experiments were prepared using untreated domestic wastewater collected from nearby suburban treatment plants (to avoid potential toxicity effects from industrial discharges to METRO). Fecal coliform bacteria were enumerated using the membrane filter technique (APHA 1985; Geldreich 1981).

A two step approach was applied to assess settling losses. First, the association of fecal coliform bacteria with particles of various size classes was determined. Second, the sedimentation rate of particles in each size class was quantified. A serial screening technique (Schillinger and Gannon 1982; Gannon et al. 1983) was used for the first step. Water samples for this determination were collected from Onondaga Creek, the major source (Chapter 3), during storm overflow events. Sedimentation rates for particulate matter in the lake were determined using established sediment trap techniques (Bloesch and Burns 1980; Blomqvist and Hakanson 1981). Traps (cylindrical with 6 : 1 aspect ratios) were deployed at a depth 1 m below the thermocline, with the onset of storm overflow events, at two stations

aligned along the axis of the Onondaga Creek inflow (nearshore station, 300 m from creek mouth; offshore station, 3000 m from creek mouth). Sedimentation rates were determined for seven particle size classes, the boundaries of which corresponded to the size of the screens (0.45−1, 6−10, 10−20, 2−53, 53−102, and >102 μm). The sedimentation flux $(J, g \cdot m^{-2} \cdot d^{-1})$ was calculated according to

$$J = \frac{M}{A \cdot t} \qquad (6.16)$$

in which M = mass of sediment collected in the trap (g), A = area of the sediment trap opening (m^2), and t = deployment time (d). The sedimentation velocity of FC's $(v_{FC}, m \cdot d^{-1}$; associated with particles) was calculated according to

$$v_{FC} = \frac{J}{C_{ss}} \qquad (6.17)$$

in which C_{ss} = concentration of suspended solids $(g \cdot m^{-3})$ in the water column above the traps. The sedimentation loss rate coefficient $(k_s$ in equation (6.15)) is calculated according to

$$k_s = \frac{v_{FC}}{z_e} \qquad (6.18)$$

in which z_e = epilimnion depth (m), the distance across which the particles must settle. The values of J, v_{FL}, and k_s were determined for individual particle classes and for two aggregate classes: 0.45 − 10 μm (small) and >10μm (large).

In situ measurements of fecal coliform bacteria death rates in the lake were made at ambient levels of irradiance and temperature on three occasions. Inoculated samples were incubated in dialysis tubes, an approach which offers the advantage of exposing the bacteria to in-lake conditions of irradiance, nutrients, pH, predation, salinity, and temperature, while avoiding settling losses. Tubes were harvested daily from each of six depths for five days. The deepest depth (8 m) received less than 1% of the irradiance that penetrated the lake surface. The daily average irradiance at depth $(I_{z,avg}, cal \cdot cm^{-2} \cdot d^{-1})$ was calculated using Beer's Law, from estimates of daily average incident irradiance ($I_{o,avg}$, estimated from total incident solar irradiation according to Bannister (1974)) and the vertical attenuation coefficient (k_d, m) for photosynthetically active radiation (400–700 nm; see Chapter 7 for procedure to determine k_d).

Laboratory studies to examine the effect of temperature on fecal coliform bacteria death rates were conducted in the dark at six different temperatures, ranging from 10 to 35°C. The duration of the experiments ranged from 8 to 14 days. Sampling for fecal coliform bacteria analysis was conducted twice a day in the early phase of the experiments, and once per day in the later phase.

6.6.1.5.3 Irradiance and Death Rate

Bacterial cells were exposed to a wide range of irradiance conditions (~ 0–$230\ \text{cal} \cdot \text{cm}^{-2} \cdot \text{d}^{-1}$) through in situ incubation at various depths. Fecal coliform bacteria death rate coefficients ($k_{death} = k_{de} + k_i$) were calculated as the slope of a linear regression of the log fecal coliform bacteria concentration versus time. The relationship between the death rate coefficient and the average irradiance at the depth of incubation ($I_{z,avg}$) is illustrated in Figure 6.89. Death rate coefficients range from 0.5–4.57 d^{-1} and are well correlated with the average irradiance at depth (linear regression, $R^2 = 0.88$). Two data points in Figure 6.89, excluded from the regression, exhibited death rates significantly higher than those which would be expected based on measurements of visible irradiance. Both of these data points represent surface incubations for days of mostly clear, cloudless conditions. As ultraviolet radiation is strongly and preferentially scattered by haze (Kirk 1983), it is likely that these elevated death rates reflect a response to an increased receipt of ultraviolet energy.

The y-intercept of the regression line in Figure 6.89 is the dark death rate ($k_{de} = 0.73\ \text{d}^{-1}$), uncorrected for temperature effects, and the slope is a proportionality constant ($\alpha = 0.00824\ \text{cm}^2 \cdot \text{cal}^{-1}$) which relates the irradiance-mediated death rate coefficient to irradiance,

$$k_i = \alpha \cdot I \qquad (6.19)$$

A value for the irradiance-mediated death rate coefficient, representative of conditions over the entire epilimnion,

$$k_{i,e} = \alpha \cdot I_{e,avg} \qquad (6.20)$$

may be calculated using the depth-averaged epilimnetic irradiance,

$$I_{e,avg} = \frac{I_{o,avg}}{k_d \cdot z_e} \cdot [1 - e^{(-k_d \cdot z_e)}] \quad (6.21)$$

in which $I_{e,avg}$ = average irradiance in the epilimnion over the incubation period ($\text{cal} \cdot \text{cm}^{-2} \cdot \text{d}^{-1}$), and $I_{o,avg}$ = average irradiance immediately below the water surface over the incubation period ($\text{cal} \cdot \text{cm}^{-2} \cdot \text{d}^{-1}$). The value of $k_{i,e}$ is then determined by substitution,

$$k_{i,e} = \frac{\alpha \cdot I_{o,avg}}{k_d \cdot z_e} \cdot [1 - e^{(-k_d \cdot z_e)}] \quad (6.22)$$

The value of α is as specified above and values for $I_{o,avg}$, k_d, and z_e are measured directly for the system under study.

6.6.1.5.4 Temperature and Death Rate

The results from these experiments, and the three "dark" (from 8 m depth) values from the in situ incubations are presented in Figure 9.90. Death rates ranged from 0.43 d^{-1} to 0.81 d^{-1} with a mean of 0.61 $\pm$ 0.11 d^{-1}. There was no significant relationship between

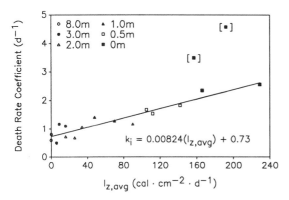

FIGURE 6.89. Relationship between the fecal coliform bacteria death coefficient and the average irradiance; bracketed observations not included in the analysis.

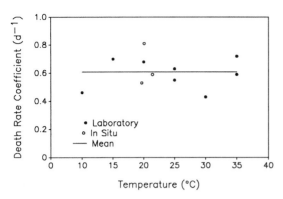

FIGURE 6.90. Relationship between the fecal coliform bacteria death coefficient and temperature.

death rate and temperature (analysis of variance on slope; null hypothesis: slope = 0).

A number of investigators have reported a similar lack of dependence of fecal and total coliform bacteria death rates on temperature. Mitchell and Chamberlin (1978) cite work in the Ohio River by Frost and Streeter (1924) which demonstrated that total coliform death rates were virtually identical at 5°C and 20°C. In a study of fecal and total coliforms in wastewater lagoons, Moeller and Calkins (1980) observed no temperature-related difference in the death rate coefficient among five temperature groups: <10, 10 to 14, 15 to 19, 20 to 24, and >24°C. Scarce et al. (1964) noted no clear trend in death rates measured at 5, 20, and 35°C in incubations of wastewater effluent in receiving stream water. Sieracki (1980), as cited by Lantrip (1983), showed that the death of *Escherichia coli*, an important component of the fecal coliform group, was unrelated to temperature over the range 9°C to 20°C. Lantrip (1983), however, observed a strong correlation between fecal coliform bacteria death rates and temperature in diffusion chamber studies in the Potomac River.

Mathematical models of coliform death in receiving waters generally apply a temperature adjustment to the dark death rate (cf. Bowie et al. 1985; Thomann and Mueller 1987). Canale et al. (1973) proposed a linear relationship between temperature and the death rate coefficient for modeling total coliform bacteria in Grand Traverse Bay, Lake Michigan. Laboratory studies of coliform death rates in Grand Traverse Bay water (Gannon and Meier, Unpublished Data) used in developing this relationship showed no difference in the death rate coefficient at 5, 10, and 15°C. The death rate coefficient increased in field measurements at ~19°C, but this may have been due to the bactericidal effects of light. Other rates used to build this relationship (e.g., Scarce et al. 1964) included the effects of intermittent irradiance and chlorination which make their suitability for this application suspect.

Functions based on the Arrhenius or van't Hoff equations, simplified as shown in Equation (6.26), are often utilized to specify the relationship between temperature and reaction rate coefficients,

$$k_{de,T} = k_{de,20} \cdot \Theta^{(T-20)} \qquad (6.23)$$

in which $k_{de,T}$ = the dark death rate coefficient at T = T°C (d^{-1}), $k_{de,20}$ = the dark death rate coefficient at T = 20°C (d^{-1}), Θ = a dimensionless constant which describes the relationship between the death rate coefficient and temperature, and T = water temperature (°C).

The relationship between the death rate coefficient and temperature and the fit of Equation (6.23) to those data are illustrated for five independent studies in Figure 6.91; the last panel of the figure pools the data from all studies. Although there is a tendency toward low values of k_{de} at low temperatures and high values at high temperatures, trends are difficult to discern, especially over the temperature range of interest for contact recreation, e.g., 15°C to 30°C. Figure 6.92 compares k_{de} values measured in the current study with the pooled data of other investigators. Death rates for fecal coliform bacteria measured here are comparable in magnitude to those observed in other systems, but it is difficult to justify the application of a temperature adjustment function.

Some investigators have suggested that a relationship between temperature and nutrients may mask the temperature effect on death rates in laboratory experiments. For example, Lessard and Sieburth (1983) noted a death

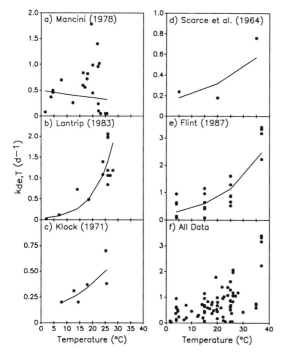

FIGURE 6.91. Temperature–death rate coefficient relationships for five independent studies.

mortality where enteric organisms persist until endogenous nutrient reserves are depleted. Temperature effects on the rate of nutrient utilization and variability in nutrient availability in natural systems may explain the observed scatter in temperature/death relationships. In addition, batch culture conditions may present a more favorable environment than that encountered in natural systems.

6.6.1.5.5 Sedimentation

The association of fecal coliform bacteria with particles of seven size classes (0.45–1, 1–6, 6–10, 10–20, 20–53, 53–102, and >102 μm) is presented for two occasions (Figure 6.93a). Although there was some variability between experiments, the majority of the fecal coliform bacteria were found to be associated with two particle classes: 0.45 to 1 μm and 6 to 10 μm. To simplify the analysis, all particles were assigned to one of two classes: small (0.45–10 μm) and large (>10 μm). On average, 90.5% of the fecal coliform bacteria were found to be associated with small particles and 9.5% were associated with large particles (Figure 6.93b).

Particle sedimentation velocities, calculated for each of the seven size classes using sediment trap samples collected at nearshore and offshore stations following a storm event, are presented in Figure 6.94a. In general, higher sedimentation velocities were noted at the

rate dependence on temperature in studies of *Escherichia coli* in seawater using diffusion chambers; no such response was observed in collateral batch culture experiments. Klock (1971) and Sjogren and Gibson (1981) have proposed a starvation mechanism for coliform

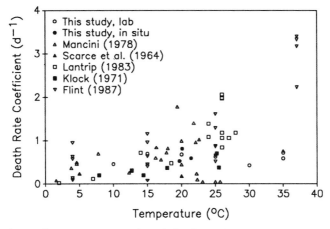

FIGURE 6.92. Comparison of temperature-mediated death coefficients from this study with those from other studies.

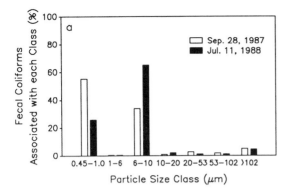

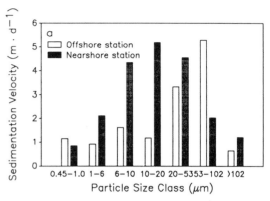

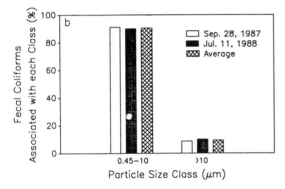

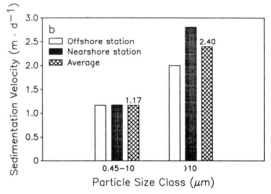

FIGURE 6.93. Association of fecal coliform bacteria with particles of (**a**) seven size classes, and (**b**) two size classes.

FIGURE 6.94. Particle sedimentation velocities based on (**a**) seven size classes, and (**b**) two size classes.

nearshore station. This may be due to the fact that this station, proximate to the mouth of Onondaga Creek, receives high density, terrigenous solids which settle rapidly. The particle assemblage at the offshore station, distant from terrigenous inputs, was composed of organic debris and algal cells which settle slowly. When the sedimentation velocities are examined according to the two-class system described above, less interstation difference is noted (Figure 6.94b). Size class specific sedimentation velocities, averaged for the two stations, are $1.17 \, m \cdot d^{-1}$ and $2.40 \, m \cdot d^{-1}$ for the small and large particle classes, respectively.

6.6.2 Pelagic Heterotrophic Bacteria

6.6.2.1 Background

Particulate and dissolved detrital material is received by lakes from both allochthonous and autochthonous sources. A substantial fraction of this material (particularly the dissolved components) is decomposed by pelagic heterotrophic bacteria and fungi. While the biochemical transformations associated with the activity of these microbes has been described as important in nutrient cycling and energy flux, relatively little is known about pelagic microflora and their metabolism (Wetzel 1983).

The distribution of the pelagic microflora usually reflects substrate limitation, e.g., increases in the biomass of this assemblage are observed in response to increases in detrital carbon. The seasonal average concentration of pelagic bacteria is higher in systems of elevated trophic state (see compilation of Wetzel (1983), Table 17.1, p. 491). Strong seasonal variations in the biomass of pelagic bacteria have been observed, which often appear to be correlated with the dynamics of phytoplankton activity; often bacterial increases lag behind pulses of phytoplankton by 5 to 10 days (Wetzel 1983).

Enumeration of bacteria to assess biomass is problematic. Direct microscopic enumeration is limited by the difficulty of differentiating live and dead cells, though advances in epifluorescence microscopy have improved this situation. Culturing bacteria in media is very selective; Wetzel (1983) indicates such methods underestimate natural populations by factors of from 1000 to 100,000. New methods, including new media, have been developed to improve recovery of the heterotrophic bacterial population (e.g., APHA 1985).

The topic of pelagic bacteria has received very little attention for Onondaga Lake. We present here the results of monitoring the upper waters of the lake for the spring to fall period of 1978, utilizing a culture approach. Heterotrophic bacteria in the water column of the lake were enumerated by the "pour plate" (standard plate count) method (APHA 1975). Pour plates were incubated at 35°C for 48 hr. Despite the limited scope of the pelagic bacteria data base presented here, it serves to demonstrate the strong dynamics of this population and the potential role it plays.

6.6.2.2 Distribution: 1978

Paired temporal distributions of the concentrations of heterotrophic bacteria and chlorophyll a (by method of Lorenzen (1976)), for the upper waters (average of 0.25 and 5.0 m results) of Onondaga Lake for the spring to fall interval of 1978 are presented in Figure 6.95a and b, respectively. The concentration of bacteria, according to this measurement protocol, was extremely dynamic, consistent with the literature reviewed by Wetzel (1983). Concentrations of bacteria ranged over two orders of magnitude. Most observations were between 50,000 and 200,000 cfu · 100 ml^{-1}. Abrupt increases (more than order of magnitude) were observed in mid-to-late April, mid-June, early August, and mid-August. The dynamics in the bacteria population (Figure 6.95a) did not seem to be strongly coupled to those of phytoplankton biomass (Figure

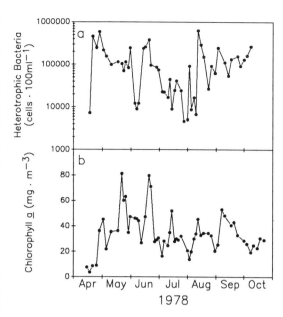

FIGURE 6.95. Temporal distributions in the upper waters during 1978: (a) concentration of heterotrophic bacteria and (b) chlorophyll a.

6.95b). For example, three of the four abrupt increases in bacteria *preceded* increases in chlorophyll a. The frequency of sampling was not high enough (51 observations in 6 months) to adequately evaluate the factors regulating bacterial dynamics in the lake, as the sampling interval was substantially longer than the life span of most bacteria (Wetzel 1983).

6.7 Summary

6.7.1 Phytoplankton

The phytoplankton community of Onondaga Lake has been monitored through a program of identification and enumeration since 1968. Bloom conditions were common through much of the spring–fall growing season in the early years of monitoring. A seasonality of forms common to many eutrophic/hypereutrophic lakes was observed: diatoms dominated in spring, green algae in early summer and early fall. Crytomonads were also common. Algae belonging to marine genera were found in small quantities, apparently as a result of the lake's unusually high salinity.

Cyanobacteria were lost from the assemblage by 1972, apparently in response to the increase in the nitrogen/phosphorus ratio that accompanied the phosphorus loading reductions that resulted from a ban on high phosphorus detergents. The cyanobacteria were largely replaced by chlorococcalean green algae. Species dominance during blooms varied year to year over the early 1970s to late 1980s interval; important forms included centric diatoms, flagellated greens, cryptomonads, and chlorococcalean green algae.

The recurrence of cyanobacteria was noted in the late 1980s; *Aphanizomenon flos-aquae* (a filamentous cyanobacterium) became a dominant in the summer assemblage in 1990. The reemergence of this nuisance form does not appear to be in response to changes in nutrient loading; it may reflect recent shifts in the zooplankton population to more efficient grazers. Concentrations of cryptomonads have remained relatively uniform since the late 1970s. The spring abundance of diatoms and the frequency of their occurrence in summer and fall have decreased over the same period. A decrease in the degree of eutrophy of the lake over the period of monitoring is indicated by the decline in intensity of phytoplankton blooms.

Chlorophyll a is the most widely used aggregate measure of phytoplankton biomass, despite limitations such as difference in cellular content among species, and dependence on environmental conditions. The summed concentration of chlorophyll *a*, *b*, and *c* (C_T), as measured by spectrophotometric analysis of acetone extracts of the pigments (Parsons et al. 1984), has been adopted as the primary measure of phytoplankton pigments and biomass in Onondaga Lake, following comparison to paired measurements made with two other methods. Phaeopigments, chlorophyll degradation products associated primarily with zooplankton grazing, were essentially absent from the water column of the lake in 1990, and presumably in other years. Phaeopigments undoubtedly contribute significantly to overall pigment deposition, associated with the settling of zooplankton fecal pellets. Another pigment method (Lorenzen 1967), that

has been used for Onondaga Lake since the 1970s, yields false low chlorophyll a (C) and false high phaeopigment (P) values. However, these data continue to have historic value for the lake as the summation of the pigment concentration (C + P) is highly correlated (R = 0.89) to C_T, according to the linear regression expression.

$$C_T = 0.87(C + P) + 2.5 \qquad (6.24)$$

Extremely high C_T concentrations have been observed in the upper waters of Onondaga Lake, particularly during the early spring phytoplankton bloom; values greater than $100\,\mu g \cdot L^{-1}$ are common, and levels greater than $200\,\mu g \cdot L^{-1}$ have been observed, during this period. Strong seasonal variations are observed, however, minima were distinctly lower ($<3\,\mu g \cdot L^{-1}$) and of longer duration over the 1987–1990 interval compared to earlier years. These periods have been described as "clearing" events because of the attendant dramatic increases in clarity (Chapter 7); such events, associated with precipitous decreases in phytoplankton biomass, have been documented in a number of other highly eutrophic lakes. Strong vertical heterogeneity in C_T (with the higher concentrations in the near-surface waters) is often observed during calm periods of phytoplankton blooms in Onondaga Lake. Substantial longitudinal differences in biomass have been observed on occasion in recent years, but no major recurring trend was indicated along the main axis of the lake. In earlier years (e.g., 1978) concentrations in the south basin were frequently greater than in the north basin. Decreases in the summer median and mean values of C + P since the late 1970s and mid-1980s reflect either, reduced primary production, or increased loss rates (e.g., grazing). However, Onondaga Lake remains highly eutrophic, e.g., summer average concentrations continue to exceed the lower boundary value(s) of eutrophy by a wide margin.

The strong seasonal changes in phytoplankton biomass observed in Onondaga Lake represent imbalances between growth and loss processes. A review of the physiological models describing growth and loss processes has been presented. The value of evaluating these models and determining the related kinetic coefficients on a system-specific basis to enhance the credibility of simulations of the phytoplankton dynamics was identified. The determination of kinetic coefficients describing the mediation of phytoplankton production by light and temperature, and phytoplankton respiration by temperature, in Onondaga Lake by laboratory experiments with natural phytoplankton populations was reviewed. The kinetic coefficients compare favorably with literature compilations of these values.

The settling velocity of phytoplankton pigments (C_T), as determined by a program of analysis of sediment trap collections (below the epilimnion) and the overlying water column, represents a basis to estimate the summation of the loss processes of phytoplankton settling and zooplankton grazing. The average value of this velocity was similar for each of the years in the 1987–1989 interval (range of 0.69 to 0.78 m · d^{-1}), but varied greatly within each year. Much of this seasonality is probably due to dynamics in zooplankton grazing.

The high concentrations of nitrogen species (ammonia and nitrate) maintained in the productive waters of the lake are non-limiting to phytoplankton growth. The high nitrogen/phosphorus ratio's that prevail in Onondaga Lake (≥ 29) indicate phosphorus is the limiting nutrient. Soluble reactive phosphorus concentrations in the lake are often sufficient to support nutrient-saturated algal growth. During mid-summer periods when SRP is depleted, concentrations of soluble organic phosphorus are sufficient to meet algal needs. Internal P reserves remain sufficient to sustain near maximum rates of algal growth over the entire summer, even when ambient supplies of phosphorus are depleted. Thus the Monod kinetic framework is presently inappropriate to represent P limitation of algae growth in Onondaga Lake.

6.7.2 Zooplankton

Changes in the zooplankton community of Onondaga Lake over the period 1969–1989 have been documented and evaluated. This represents an update of surveys conducted in 1969 (Waterman 1971) and 1978 (Meyer and Effler 1980). Previously undocumented monitoring results for summers in the intervals 1979–1981 and 1985–1989 are presented.

Species richness has increased in the lake, particularly since the closure of the soda ash/chlor-alkali facility in 1986. However, despite these increases, the zooplankton species assemblage of the lake remains "depauperate" compared to other lakes in the region, presumably as a result of the system's continued polluted condition. Zooplankton biomass was dominated by *Cyclops vernalis*, a cyclopoid copepod, until 1987. *Ceriodaphnia quadrangula* was the only important cladoceran during that period. Dominance has shifted to large-bodied cladocera and the calanoid copepod *Diaptomus siciloides* since 1987, coincident with the closure of the facility. Two smaller cladocerans, *Bosmina longirostris* and *Diaphanosoma leuchtenbergianum*, only became abundant over the 1986–1989 interval. Substantial year-to-year variations in summertime zooplankton biomass have occurred, however, no long-term trend has emerged over the 1978–1989 period.

The reemergence of the alewife, an effective planktivore, in the late 1970s apparently was largely responsible for the loss of large daphnids from the lake, observed in 1978. Reductions in the concentrations of CaCO$_3$ particles and salinity, that resulted from the closure of the soda ash/chlor-alkali facility (in the absence of effective size selective planktivores), were probably the "triggers" for the shift in the zooplankton community observed since 1987. The effects of zooplankton grazing on the seasonal dynamics of phytoplankton biomass were apparent in a number of years, however, the effect became much more pronounced over the 1987–1989 interval because of the greater efficiency of grazing by the large daphnids and the calanoid copepod. The "clearing events" observed annually over the 1987–1989 interval, associated with abrupt reductions in phytoplankton, are largely attributable to increased grazing pressure. The shift to large daphnids may have also contributed to the reemergence of *Aphanizomenon flow-aquae*, a large inedible filamentous cyanobacterium, as a late summer dominant of the phytoplankton community of the lake. The seasonality and longer term changes in the zooplankton community, as well as the relationship between the shift to large daphnids and the calanoid copepod and the coupled reductions in phytoplankton biomass and improved clarity, were clearly depicted by principal component analysis.

6.7.3 Macrophytes

Unusually low species richness and standing crop(s) of aquatic macrophytes were found in the 1991 aquatic macrophyte survey of Onondaga Lake. Only five species of submersed macrophytes were found. Emergent and floating-leaved macrophytes were absent, indicating a significant gap in the aquatic community. Only scattered locally abundant submersed aquatic macrophyte communities of low diversity, dominated by Sago Pondweed, were documented. Substrate conditions appear to greatly influence the distribution of plants. Populations were substantially reduced on the extensive oncolite deposits of the lake, probably because of the unstable character of these low density "stones" in wave action. The maximum depth of colonization observed was 3 to 4 m. This is consistent with prevailing light penetration conditions, as quantified by various empirical models. Other factors presently limiting species richness and standing crop include the elevated salinity and level of $CaCO_3$ precipitation in the lake.

Submersed plants have probably existed in the lake for some time in protected areas. The populations have probably expanded recently in response to increased light penetration, and reductions in salinity and calcium carbonate precipitation. Further expansion of the submersed macrophyte community is anticipated under the prevailing conditions. Yet greater increases can be expected if further increases in light penetration, reductions in salinity, and improvements in substrate conditions are achieved.

The growth of *Potamogeton pectinatus* in a greenhouse was directly correlated to in situ abundance of plants, but was not related to a visual classification of sediment types. Greenhouse growth experiments indicate Onondaga Lake littoral sediments are significantly less productive than a fertile reference sediment. Of eleven species surveyed in growth studies, initial selections for revegetation include *P. pectinatus, P. nodosus,* and *Sagittaria latifolia.*

6.7.4 Benthic Macroinvertebrates

The results of a single survey of four nearshore sites in June 1989 indicate the benthic macroinvertebrate community of Onondaga Lake has been severely impacted by pollution. The community has an unusually low species (taxon) diversity. The chironomid community had a particularly low number of species. Diversity along Onondaga Lake Park on the east shore (2 sites) was greater than found along the waste beds on the west shore (2 sites), but much less than reported for an unpolluted site on Cayuga Lake. The benthic macroinvertebrate community of the lake, and the chironomid assemblage in particular, is composed of forms known to be tolerant of pollution. The community is dominated by chironomids; other significant components include oligochaetes, and a single amphipod (*Gammarus fasciatus*). The overall population densities are not unusual. The degraded condition of the benthic invertebrate community of the lake probably reflects the combined effects of the continuing polluted conditions of the water column, as well as the alterations to the near-shore sediments caused by earlier industrial discharges. The occurrence of adult zebra mussels (*Dreissena polymorpha*) in Onondaga Lake was first observed in the summer of 1992.

The results of a single survey of two sites that bound the "mud boils" inputs to Onondaga Creek indicate the benthic macroinvertebrate community at the downstream (2.5 km) location has been greatly modified as a result of the sediment discharged from the mud boils. Diversity, richness, and density were significantly lower at the downstream site; e.g., the density at the upstream location was 2752 individuals $\cdot$ m^{-2}, compared to only 139 individuals $\cdot$ m^{-2} at the downstream site. The community was dominated by Ephemeroptera upstream and Oligochaeta downstream. Filter feeders were common upstream but rare downstream. Predators (visual) were absent at the downstream site, probably due to the low clarity. These results imply a biological impact of the "mud boils" greater than previously reported.

6.7.5 Fish

Prior to the 1870s, Onondaga Lake was characterized by large and consistent catches of Atlantic salmon as well as ciscoes, known locally as Onondaga Lake whitefish. The existence of ciscoes, which are coldwater, non-migratory fish, indicates that at least portions of the hypolimnion contained substantial levels of dissolved oxygen, conditions that presently do not exist in the lake. The fishery in the lake was markedly degraded by the time of the first scientific survey in 1927; Atlantic salmon and ciscoes were absent. The 12 species recorded represented a depauperate community compared to local lakes and the adjoining Seneca River. Fourteen species were found in the survey of 1946, but more than 90% of the catch was carp, a species known to be tolerant of perturbed conditions. White perch had colonized to the lake via the Mohawk–Hudson system by 1969. Twenty-two species were found in 1980 and 1981, and from 28 to 45 species were captured annually from 1989 to 1991. The increase, particularly in the most recent years, is at least in part a result of improved water quality associated with reductions in pollutant loadings to the lake since the 1970s.

The most intensive investigation of the fishery of Onondaga Lake was conducted from 1989 to 1991. Studies were directed at community structure of the lake, its tributaries and adjoining portions of the Seneca River, spatial and temporal features of the community, reproduction, movements, diet, epidemiology, and responses to low dissolved oxygen concentrations. More than 50,000 fish were captured in the field program. The lake supports a warm water fish community, dominated by the planktivores white perch and gizzard shad, a manifestation of the prevailing hypereutrophic conditions (i.e., a "bottom-up" system). Insectivorous sunfishes are abundant in the littoral zone. Sport fish of significance include smallmouth and large-mouth bass, channel catfish, and walleye. The diversity of the community, based on trap net catches, is currently comparable to other New York lakes. Fishing pressure is light, and recent fishing derbies have reported relatively high catch rates. Growth rates, age distribution, and mortality rates of several species are similar to those observed in other northeastern lakes.

The large number of species currently in Onondaga Lake is related, in part, to its connection to the Seneca River. Species that migrate from the river into the lake seasonally include channel catfish, walleye, and northern pike. Evidence of reproduction within the lake was found for only 16 species in 1991. Changing seasonal patterns of community structure provide strong evidence for the extent of migration between the lake and river. Tag returns by anglers document movement as far upstream in the Seneca River as Baldwinsville (10 km) and as far downstream as Fulton (25 km). Without the availability of the river and other regions that serve as refuges during the period of low dissolved oxygen in the fall, species diversity and population density in the lake would probably be substantially lower.

The absence of dissolved oxygen in the hypolimnion during the summer, and the unusually low dissolved oxygen levels experienced at fall turnover, represent significant constraints on fish life. Catches of piscivorous species, such as walleye and channel catfish, dropped precipitously during fall turnover, both in trap nets inshore and in offshore gill nets. Most species disappeared from the trap nets after fall turnover in 1990 and 1991. Gizzard shad and alewife appeared to be more tolerant of the degraded dissolved oxygen conditions of the fall mixing period. An in-situ experiment showed that low levels of dissolved oxygen (2–3 mg $\cdot$ L^{-1}) caused moderate mortality of sunfish and white perch over 48 hr. Several radio-tagged fish left the lake and moved into the Seneca River at fall turnover in October 1991.

Sunfish and bass have been observed spawning in the littoral zone of the northwest portion of the lake, where seining later revealed moderate numbers of juveniles. Other species with high levels of recruitment within the lake in recent years include white perch, gizzard shad, brook silversides and golden shiners. Age class analysis suggests that recruitment of several important species in 1992 was far below 1990–1991 levels. The importance of macrophytes in supporting the fish community in the lake is demonstrated by the localization of juveniles, and adults of certain species, in areas with abundant plants. Few juveniles were captured at sites devoid of macrophytes or other submerged structures. The shift in the fish community to increased representation by insectivores from 1980 to 1991 may reflect the hypothesized increase in macrophytes in the lake over the same period.

Foods consumed by fishes in Onondaga Lake include fish, fish eggs, zooplankton (primarily copepods and cladocerans), and benthic invertebrates, especially midge larvae (Chironomidae). Diets were generally comprised of relatively few taxa. The gizzard shad population is an important component in the lake's food web. However, quantification of its role is confounded by the unusual character of its stomach. Diets were examined with regard to variation among sites and seasons. The minor differences in diets of the same species in the lake and Seneca River suggested close similarities in the invertebrate fauna of these waters. Analysis of fish communities grouped by trophic category suggested a shift toward increased numbers of littoral planktivores, as well as insectivores, between 1980 and 1990.

Soon after the U.S. Department of Justice initiated legal action against the soda ash/chlro-alkali facility to reduce discharges of Hg, Onondaga Lake was closed to fishing (1970) due to Hg contamination of fish tissue. Nearly all fish tested in 1970 (13 species) had concentrations that exceeded the present FDA standard of 1.0 ppm. Fish tissue has been monitored for Hg almost annually since that time; smallmouth bass have been adopted as the primary indicator because of their residence in the lake. The concentration of Hg in fish from the lake is positively correlated with length for several important species that have been tested. The overall temporal pattern in Hg contamination for smallmouth bass is a decline in Hg concentration in the 1970s followed by increases in the late 1980s. More than 95% of the legal-sized smallmouth bass collected from the lake in 3 of 4 years during 1987–1990 exceeded the FDA standard. More than 50% of all smallmouth bass analyzed during 1981–1990 interval exceeded the standard. The lake was reopened to angling in 1986, but fish from the lake are not to be eaten. Limited monitoring indicates that some fish are also contaminated with benzene, various chloro-benzenes, chlordane, and PCBs. Contamination extends beyond the limits of the lake (and health advisories), as a result of documented migration that occurs between the lake and the Seneca River system.

Morphological abnormalities such as cell and tissue changes and tumors sometimes provide evidence of environmental impacts. Approximately 5% of the channel catfish in the Seneca River had barbel deformities, 3% had skin lesions, and 1.5% had raised-skin growth. The rate of deformities was low relative to well-known contaminated sites, such as the Buffalo River, where more than 30% of the bullheads had barbel deformities.

The feasibility of reestablishing Atlantic salmon in the Onondaga Lake system depends on the biological capacities of the major lake tributaries, Ninemile Creek and Onondaga Creek. These two streams were classified in 1990 and 1991; they represent most of the stream fish habitat for the lake's watershed. Sampling in 1990–1991 in Onondaga Creek above the urban area, and in Ninemile Creek above the wastebeds, documented 29 species of fish. The communities were remarkably similar to collections reported in 1927. Brown trout populations, first introduced in the 1870s, are well established in both streams. Brown trout in Ninemile Creek and the main stem of Onondaga Creek grow at about the same rate.

Ninemile Creek provides exceptional habitat for brown trout in stretches downstream of Marcellus Falls, an area subject to intensive fishing pressure. Sampling results following the trial stocking of juvenile landlocked salmon in Ninemile Creek, and application of a "Habitat Suitability" model, indicate that landlocked salmon will survive and grow in the stream. Ninemile Creek may be capable of producing 30,000 to 60,000 salmon fry. Competition for food between salmon and trout probably would not be a significant factor in the restoration of salmon to the Onondaga system. Restoration of salmon would require an on-going hatchery program coupled with remediation of the impacted lower reaches of the tributaries. Lake water quality conditions would also need to support passage of juvenile salmon in the spring, and adults in the early summer.

6.7.6 Indicator Bacteria

Long-term monitoring programs depict a continuing south to north gradient in coliform bacteria in the lake that apparently is largely a result of inputs that enter in the southern end of the lake. However, inputs from Bloody Brook may be responsible for elevated concentrations that have recently been documented in the vicinity of this inflow on the eastern shore. With few exceptions, very similar spatial and temporal trends emerge from the reveiw of the TC and FC data sets. Concentrations of indicator bacteria have decreased in the lake since the late 1970s, however, no dramatic and lasting improvements can be assigned specifically to the BMP program. More intensive monitoring conducted along the northeast shore indicates that the bathing beach standard continues to be contravened annually (1975–1990) in this area. The most stringent interpretations of the data base for this region indicates contravention for more than 50% of the summer in most years.

Higher frequency monitoring (nearly daily) conducted in 1987 indicates that the only region of the lake that may be unsuitable for contact recreation with respect to bacteriological standards during dry weather is the southern end proximate to the inflow of Onondaga Creek (the major source, Chapter 3), Ley Creek, and Harbor Brook. This same program clearly documented the contravention of standards for most of the south basin following storm induced runoff events in 1987. Concentrations of FC's return to baseline conditions within 2 to 4 days. Most of the north basin becomes unsuitable only following the most severe storms. There are indications that the degree of contamination of the lake following runoff events has been reduced since the mid-1970s as a result of related management efforts. However, the status of the lake with respect to public health standards remains essentially unchanged.

The rate of loss of fecal coliform bacteria can be quantified by first order kinetics, according to

$$R = \frac{dC_{fc}}{dt} = -k \cdot C_{fc} \qquad (6.24)$$

in which C_{fc} = fecal coliform bacteria concentration (cfu $\cdot$ 100 ml^{-1}), k = overall first order loss coefficient (d^{-1}), R = rate of change in fecal coliform bacteria concentration (cfu $\cdot$ 100 ml$^{-1} \cdot$ d^{-1}), and t = time (d). The value of k for Onondaga Lake is calculated by a summation expression that accommodates losses associated with death, mediated by temperature and light, and settling

$$k = k_{de,20} \cdot \Theta^{(T-20)} + \frac{\alpha \cdot I_{o,avg}}{k_d \cdot z_e}\left[i - e^{(-k_d \cdot z_e)}\right]$$

$$+ \frac{v_{FC}}{z_e} \qquad (6.25)$$

The parameters and coefficients of this expression are presented in Table 6.53, along with the value of kinetic coefficients determined through field and laboratory experiments.

The value of the dark death rate ($K_{d,20}$), determined in laboratory experiments, is lower (Table 6.53) than that reported in earlier work on Onondaga Lake, but quite similar to values reported for other systems for coliforms. No temperature effect was observed (i.e., Θ = 1.00), whereas Θ has been reported to be >1.00 elsewhere. The irradiance—mediated death rate was shown to be proportional to

TABLE 6.53. Parameters and kinetic coefficients, loss of fecal coliform bacteria in Onondaga Lake.

Parameter/coefficient	Description	Value
$K_{de,20}$	Dark death rate at 20°C	$0.73\,d^{-1}$
Θ	Temperature adjustment coefficient	1.00
T	Water temperature	—, °C
α	Irradiance proportionality constant	$0.00824\,cm^2 \cdot cal^{-1}$
$I_{o,avg}$	Average irradiance below water surface	(measured) $cal \cdot cm^{-2} \cdot d^{-1}$
k_d	Attenuation coefficient	(measured) m^{-1}
Z_e	Depth of epilimnion	(measured) m
v_{FC}	Weighted average sedimentation velocity for FC's	$1.38\,m \cdot d^{-1}$

irradiance in water column incubations of raw sewage in dialysis tubes. The proportionality constant ($\alpha = 0.00824\,cm^2 \cdot cal^{-1}$; Table 6.53) is within the range of the few observations reported elsewhere. Settling losses were quantified by determining the association of fecal coliform bacteria with particles of various size classes and measuring the sedimentation rates for those particle classes (using sediment traps). Ninety to 96% of the fecal coliforms in Onondaga Lake were associated with particles $0.45\,\mu m$ to $5\,\mu m$ in diameter. The overall sedimentation rate developed here ($1.38\,m \cdot d^{-1}$; Table 6.53) is a weighted average for particles of different size classes; it is about 2 times greater than that estimated earlier for Onondaga Lake through model calibration.

References

Ahn, T.S., Choi, S.I., and John, K.S. 1993. Phosphatase activities in Lake Soyang, Korea. *Verh Internat Limnol.* 25: 183–186.

Aladin, N.V. 1991. Salinity tolerance and morphology of the osmoregulation organs in cladocera with special reference to Cladocera of the Aral Sea. *Hydrobiologia* 225: 291–299.

Anderson, B.G. 1950. The apparent thresholds of toxicity to *Daphnia magna* for chlorides of various metals when added to Lake Erie water. *Trans Am Fish Soc.* 78: 96–113.

Andersson, G., and Cronberg, G. 1984. *Aphanizomenon flos-aquae* and large *Daphnia*—an interesting plankton association in hypereutrophic waters. In: S. Bosheim and M. Nicholls (eds), *Norsk Limnologforening*, Oslo, pp. 63–76.

Andrews, R.W., Biesinger, K.E., and Glass, G.E. 1977. Effects of inorganic complexing on the toxicity of copper to *Daphnia magna Water Res.* 11: 309–315.

APHA. 1990. *Standard Methods for the Examination of Water and Wastewater*, 16th ed. American Public Health Association, Washington, DC.

Armstrong, R., and Sloan. R. 1980. "Trends in Levels of Several Known Chemical Contaminants in Fish from New York State Waters". Technical Report 80-2, New York Department of Environmental Conservation, Bureau of Environmental Protection, Albany.

Arts, M.T., Evans, M.S., and Roberts, R.D. 1992. Zooplankton energy reserves in relation to algal species composition. Program and Abstract, Limnol. Oceanogr. Aquat. Sci. Meeting, Santa Fe, NM (abstract).

Auer, M.T., and Canale, R.P. 1982. Ecological studies and mathematical modeling and *Cladophora* in Lake Huron: the dependence of growth rates on internal phosphorus pool size. *J Great Lakes Res.* 8: 93–99.

Auer, M.T., and Canale, R.P. 1982a. Ecological studies and mathematical modeling of *Cladophora* in Lake Huron. 2. Phosphorus uptake kinetics. *J Great Lakes Res.* 8: 84–92.

Auer, M.T., and Canale, R.P. 1982b. Ecological studies and mathematical modeling of *Cladophora* in Lake Huron. 3. The dependence of growth rates on internal phosphorus pool size. *J Great Lakes Res.* 8: 93–99.

Auer, M.T., and Canale, R.P. 1986. Mathematical modelling of primary production in Green Bay (Lake Michigan, USA), a phosphorus- and light-limited system. *Hydrobiolog Bull.* 20: 195–211.

Auer, M.T., and Effler, S.W. 1989. Variability in photosynthesis: Impact on DO models. *J Environ Eng ASCE* 115: 944–963.

Auer, M.T., Kieser, M.S., Canale, R.P. 1986. Identification of critical nutrient levels through field verification of models for phosphorus and phytoplankton growth. *Can J Fish Aquat Sci.* 43: 379–388.

Auer, M.T., and Niehaus, S.L. 1991. Modeling fecal coliform bacteria: I. Field and laboratory determination of loss kinetics. *Water Res.* 27: 693–701.

Auer, M.T., Storey, M.L., Effler, S.W., Auer, N.A., and Sze, P. 1990. Zooplankton impacts on chlorophyll and transparency in Onondaga Lake, New York, USA. *Hydrobiologia* 200/201: 603–617.

Bannister, T.T. 1974. A general theory of steady state phytoplankton growth in a nutrient saturated mixed layer. *Limnol Oceanogr.* 19: 1–12.

Bannon, E., and Ringler, N.H. 1986. Optimal prey size for stream resident brown trout (*Salmo trutta*): Tests of predictive models. *Can J Zool.* 64: 704–713.

Baumann, P.C., Smith, W.D., and Parland, W.K. 1987. Tumor frequencies and contaminant concentrations in brown bullhead from an industrialized river and a recreational lake. *Trans Am Fish Soc.* 116: 79–86.

Beauchamp, W.W. 1908. *Past and Present of Syracuse and Onondaga County, New York.* S.J. Clarke, New York.

Bentzen, E., Taylor, W.D., and Millard, E.S. 1992. The importance of dissolved organic phosphorus to phosphorus uptake by limnetic plankton. *Limnol Oceanogr.* 37: 217–231.

Berman, T. 1970. Alkaline phosphates and phosphorus availability in Lake Kinnert. *Limnol Oceanogr.* 15: 663–674.

Biesinger, K.E., and Christensen, G.M. 1972. Effects of various metals on survival, growth, reproduction, and metabolism of *Daphnia magna J Fish Res Bd Canada* 29: 1691–1700.

Black, J.J. 1983. Field and laboratory studies of environmental carcinogenesis in Niagara River fish. *J Great Lakes Res.* 9: 326–334.

Blasland & Bouck Engineers. 1989. "Hydrogeologic Assessment of the Allied Waste Beds in the Syracuse Area: Volume 1." Submitted to Allied-Signal Inc., Solvay, NY.

Bloesch, J., and Burns, N.M. 1980. Critical review of sediment trap technique. *Schweiz Z Hydrol.* 42: 15–55.

Blomqvist, S., and Hakanson, L. 1981. A review of sediment traps in aquatic environments. *Arch Hydrobiol.* 91: 101–132.

Bloom, N.S., and Effler, S.W. 1990. Seasonal variability in the mercury speciation of Onondaga Lake (New York). *Water Air Soil Pollut.* 53: 251–265.

Bode, R.W., Novate, M.A., and Abele, L.E. 1989. Macroinvertebrate water quality survey of streams tributary to Onondaga Lake, Onondaga County, New York. Stream Biomonitoring Unit, New York State Department of Environmental Conservation, Albany, NY.

Bodola, A. 1966. Life history of the gizzard shad, *Dorosoma cepidianum* (LeSuerur), in western Lake Erie. *US Fish Wildlife Serv Fish Bull.* 65: 391–425.

Bond, C.E. 1979. *Biology of Fishes.* W.B. Saunders, Philadelphia.

Bousfield, E.L. 1973. *Shallow-Water Gammaridean Amphipoda of New England.* Cornell Univ Press, Ithica, NY.

Bowie, G.L., Mills, W.B., Porcella, D.B., Campbell, C.L., Pagenkopf, J.R., Rupp, G.L., Johnson, K.M., Chan, P.W.H., Gherini, S.A., and Chamberlain, C. 1985. "Rates, constants, and kinetics formulations in surface water quality modeling," 2d ed. EPA/600/3-85/040. U.S. Environmental Protection Agency, Athens, GA.

Brandlova, J., Bandl, Z., and Fernando, C.H. 1972. The cladocera of Ontario with remarks on some species and distributions. *Can J Zool.* 50: 1373–1403.

Brooks, J.L. 1957. The systematics of North American *Daphnia. Mem Conn Acad Arts Sci.* 13: 1–180.

Brooks, J.L., and Dodson, S.I. 1965. Predation, body size, and composition of the plankton. *Science* 150: 28–35.

Brown, E.R., Sinclair, T., Keith, L., Beamer, P., Kazdra, J.J., Nair, V., and Callaghan, D. 1977. Chemical pollutants in relation to disease in fish. *Ann NY Acad. Sci* 295: 535–546.

Canale, R.P., and Effler S.W. 1989. Stochastic phosphorus model for Onondaga Lake. *Water Res.* 23: 1009–1016.

Canale, R.P., Auer, M.T., Owens, E.M., Heidtke, T.M., and Effler, S.W. 1993. Modeling fecal coliform bacteria—II. Model development and application. *Water Res.* 27: 703–714.

Canale, R.P., Patterson, R.L., Gannon, J.J., and Powers, W.F. 1973. Water quality models for total coliform. *J Water Pollut Control Fed.* 45: 325–336.

Canfield, D.E., Jr., Langeland, K.A., Linda, S.B., and Haller, W.T. 1985. Relations between water transparency and maximum depth of macrophyte colonization in lakes. *J Aquat Plant Manag.* 23: 25–28.

Canton, S.P., and Chadwick, J.W. 1988. Variability in benthic invertebrate density estimates from stream samples. *J Freshwater Ecol.* 4: 291–297.

Carlander, K.D. 1977. *Handbook of Freshwater Fishery Biology*, Vol. 2. Iowa State University Press, Ames, IA.

Carlson, A.R. 1987. Effects of lowered dissolved oxygen concentration on the toxicity of 1,2,4-trichlorobenzene to fathead minnows. *Bull Environ Contam Toxicol.* 38: 667–673.

Carlson, R.W. 1984. The influence of pH, dissolved oxygen, suspended solids or dissolved solids upon the ventilatory and cough frequencies in the bluegill (*Lepomis macrochirus*) and brook trout (*Salvelinus fontinalis*). *Environ Pollut. (Series A)* 34: 149–169.

Chadwick, E.P.M. 1985. Fundamental research problems in the management of Atlantic salmon, *Salmo salar L.*, in Canada. *J Fish Biol.* 27 (Suppl. A): 9–25.

Chambers, P.A., and Kalff, J. 1985. Depth distribution and biomass of submersed aquatic macrophyte communities in relation to Secchi depth. *Can J Fish Aquat Sci.* 42: 701–709.

Chambers, P.A., Kalff, J., and Prepas, E.E. 1988. Underwater spectral attenuation and its effect on the maximum depth of angiosperm colonization. *Can J Fish Aquat Sci.* 45: 1010–1017.

Chapra, S.C., and Rechkow, K.H. 1983. *Engineering Approaches for Lake Management*. Volume 2. *Mechanistic Modeling*. Butterworth, Boston.

Chiotti, T.L. 1981. "Onondaga Lake Survey Report, 1980 and 1981." New York State Department of Environmental Conservation, Albany.

Coble, D.W. 1961. Influence of water exchange and dissolved oxygen in redds on survival of steelhead trout embryos. *Trans Am Fish Soc.* 90: 469–474.

Compton, B., Lazaroff, B., Nair, J.H. and Zweig. G. 1966. Onondaga Lake survey, 1964–65, Final report. Prepared for Onondaga County Dept. Public Works under Contract No. 153. Syracuse Univ Res Corp, Suracuse, NY.

Connors, S.M., Auer, M.T., and Effler, S.W. 1996. Phosphorus pools, alkaline phosphatase activity, and phosphorus limitation in hypereutrophic Onondaga Lake, NY. *Lake Reservoir Manag.* (in press).

Cotner, J.B., and Wetzel, R.G. 1992. Uptake of dissolved inorganic and organic phosphorus compounds by phytoplankton and bacterioplankton. *Limnol Oceanogr.* 37: 232–243.

Dean, W.E., and Eggleston, J.R. 1984. Freshwater oncolites created by industrial pollution, Onondaga Lake, New York. *Sediment Geol.* 40: 217–232.

DeBernardi, R., and Giussani, G. 1990. Are blue-green algae a suitable food for zooplankton? An overview. *Hydrobiologia* 200/201: 29–41.

DeMott, W.R., Zhang, Q., and Carmichael, W.W. 1991. Effects of toxic cyanobacteria and purified toxins on the survival and feeding of a copepod and three species of *Daphnia*. *Limnol Oceanogr.* 36: 1346–1357.

Dence, W.A. 1956. Concretions of the alewife, *Pomolobus psuedoharengus* (Wilson), at Onondaga Lake, New York. *Copeia* 3: 155–158.

DeNoyelles, F. 1968. A stained-organism filter technique for concentrating phytoplantkon. *Limnol Oceanogr.* 13: 562–565.

Devan, S.P., and Effler, S.W. 1984. History of phosphorus loading to Onondaga Lake. *J Environ Engr ASCE* 110: 93–109.

DiToro, D.M. 1980. Applicability of cellular equilibrium and Monod theory to phytoplankton growth kinetics. *Ecol Model.* 8: 201–218

DiToro, D.M., and Connolly, J.P. 1980. Mathematical Models of Water Quality in Large Lakes, Part 2: Lake Erie. EPA/600/3080–065. Environmental Research Laboratory, Office of Research and Development, U.S. Environmental Protection Agency, Duluth, MN.

Dobson, H.F.H., Gilbertson, M., and Sly, P.G. 1974. A summary and comparison of nutrients and related water quality in Lakes Erie, Ontario, and Superior. *J Fish Res Bd Can.* 31: 731–738.

Drenner, R.W., de Noyelles, F., Jr., and Kettle, D. 1982. Selective impact of filter-feeding gizzard shad on zooplankton community structure. *Limnol Oceanogr.* 27: 965–968.

Droop, M.R. 1968. Vitamin B_{12} and marine ecology. IV. The kinetics of growth, uptake, and inhibition in *Monochrysis lutheri*. *J Mar Biol Assoc UK* 48: 629–636.

Droop, M.R. 1973. Some thoughts on nutrient limitation. *J Phycol.* 9: 624–638.

Dumont, H.J., Van de Velde, I., and Dumont, S. 1975. The dry weight estimate of biomass in a selection of cladocera, copepoda, and rotifera from the plankton, periphyton, and benthos of continental waters. *Oceologia* (Berl.): 75–97.

Dwyer, W.P., and Piper, R.G. 1987. Atlantic salmon growth efficiency as affected by temperature. *Prog Fish Cult.* 49: 57–59.

Eadie, B.J. 1979. The cycle of $CaCO_3$ in the Great Lakes. Program and Abstracts, Amer Soc Limnol Oceangr (abstract).

Edmondson, W.T. (ed.) 1959. *Freshwater Biol.* John Wiley, New York 1248pp.

Effler, S.W. 1987. The impact of a chlor-alkali plant on Onondaga Lake and adjoining systems. *Water Air Soil Pollut.* 33: 85–115.

Effler, S.W., Auer, M.T., and Johnson, N. 1989. Modeling Cl concentration in Cayuga Lake, USA. *Water Air Soil Pollut* 44: 347–362.

Effler, S.W., Brooks, C.M., Auer, M.T, and Doerr, S.M. 1990. Free ammonia and toxicity criteria in a polluted urban lake. *Res J Water Pollut Contr Fed.* 62: 771–779.

Effler, S.W., and Driscoll, C.T. 1985. Calcium chemistry and deposition in ionically polluted Onondaga Lake, NY. *Environ Sci Technol.* 19: 716–720.

Effler S.W., Field, S.D., Meyer, M.A., and Sze, P. 1981. Response of Onondaga Lake to restoration efforts. *J Environ Eng ASCE* 107: 191–210.

Effler S.W., Wodka, M.C., Driscoll, C.T., Brooks, C., Perkins, M.G., and Owens, E.M. 1986. Entrainment-based flux of phosphorus in Onondaga Lake. *J Environ Engr.* 112: 617–622.

Eichler, L.W. 1990. "Assessment of Brant Lake". Rensselaer Fresh Water Institute Report #90-17, December 1990. Rensselaer Polytechnic Institute, Troy, NY.

Eichler, L.W., and Madsen, J.D. 1990a. "Assessment of Lake Luzerne." Rensselaer Fresh Water Institute Report #90-2, February 1990. Rensselaer Polytechnic Institute, Troy, NY.

Eichler, L.W., and Madsen, J.D. 1990b. "Assessment of Galway Lake". Rensselaer Fresh Water Institute Report #90-5, March 1990. Rensselaer Polytechnic Institute, Troy, NY.

Eichler, L.W., and Madsen, JD. 1990c. "Assessment of Eagle Lake". Rensselaer Fresh Water Institute Report #90-6, April 1990. Rensselaer Polytechnic Institute, Troy, NY.

Eisler, R. 1987. Mercury hazards to fish, wildlife, and invertebrates: a synoptic review. *US Fish Wildlife Serv Biol Rep.* 85(1.10).

Elser, J.J., Elser, M.M., MacKay, N.A., and Carpenter, S.R. 1988. Zooplankton-mediated transitions between N- and P- limited algal growth. *Limnol Oceanogr.* 33: 1–14.

Evans, M.S., Roberts, R.D., and Arts, M.A. 1992. Zooplankton control of algal biomass in a hypereutrophic lake. Program and Abstracts, Amer Soc Limnol Oceanogr. Aquat Sci Meeting, Sata Fe, NM. (abstract).

Everhart, W.H., and Youngs, W.D. 1981. *Principles of Fishery Science*, 2d ed. Cornell University Press, Ithaca, NY.

Falkowski, P.G., Dubinsky, Z., and Santostefano, G. 1985. Light-enhanced dark respiration in phytoplankton. *Verh Int Ver Limnol.* 22: 2830–2833.

Fassett, N.C. 1957. A Manual of Aquatic Plants, 2d ed. (Revised by E.C. Ogden). Univ. of Wisconsin Press, Madison.

FDA (U.S. Food and Drug Administration). 1974. Action level for mercury in fish and shellfish. *Fed Reg.* 39: 42738–42740.

FDA (U.S. Food and Drug Administration). 1979. Action level for mercury in fish, shellfish, crustaceans, and other aquatic animals. *Fed Reg.* 44: 3990–3994.

FDA (U.S. Food and Drug Administration). 1984. "Fish, Shellfish, Crustaceans, and Other Aquatic Animals—Fresh, Frozen or Processed—Methyl Mercury". Compliance Policy Guide 7108.7, USFDA, Washington, DC.

Field, S.D. 1980. Nutrient-saturated algal growth in hypereutrophic Onondaga Lake, Syracuse, N.Y. Ph.D Dissertation, Department of Civil Engineering, Syracuse University, Syracuse, NY.

Field, S.D., and Effler, S.W. 1978. Light attenuation in Onondaga Lake, NY, U.S.A., 1978. *Arch Hydrobiol.* 98: 409–421.

Field, S.D., and Effler, S.W. 1982. Photosynthesis-light mathematical formulations. *J Environ Engr ASCE* 108: 199–203.

Field, S.D., and Effler, S.W. 1983. Light-productivity model for Onondaga Lake, NY. *J Environ Engr ASCE.* 109: 830–844.

Fitzgerald, G.P., and Nelson, T.C. 1966. Extractive and enzymatic analyses for limiting or surplus phosphorus in algae. *J Phycol.* 2: 32–37.

Fox, W.S. 1930. The literature of *Salmo salar* in Lake Ontario and tributary streams. *Trans Royal Soci Can.*, Section II, Series III, 45–55.

Franco, D.A. 1983. Size-fractionation of alkaline phosphatase activity in lake water by membrane filtration. *J Freshwater Ecol.* 2: 305–309.

Freedman, P.L., Canale, R.P., and Pendergast, J.F. 1980. Modeling storm runoff impacts on a eutrophic urban lake. *J Environ Engr ASCE* 106: 335–349.

Frost, B.W. 1972. Effects of size and concentration of food particles on the feeding behavior of the marine planktonic copepod *Calanus pacificus*. *Limnol Oceanogr.* 17: 805–815.

Frost, W.H., and Streeter, H.W. 1924. Section 6. Bacteriological Studies. In: *Study of Pollution and Natural Purification of the Ohio River: 2. Report on Surveys and Laboratory Studies.* Public Health Bulletin No. 143, U.S. Public Health Service, Washington DC.

Fujioka, R.S., Hashimoto, H.H., Siwak, E.B., and Young, R.H.F. 1981. Effect of sunlight on survival of indicator bacteria in seawater. *Appl Environ Microbiol.* 41: 690–696.

Fulton, R.S. 1988. Resistance to blue-green toxins by *Bosmina longirostris*. *J Plankton Res.* 10: 771–778.

Fulton, R.S., and Paerl, H.W. 1987. Toxic and inhibitory effects of the blue-green alga *Microcystis aeruginosa* on herbivorous zooplankton. *J Plankton Res.* 9: 837–855.

Gage, M.A., and Gorham, E. 1985. Alkaline phosphatase activity and cellular phoshorus as a index of the phosphorus status of phytoplankton in Minnesota lakes. *Freshwater Biol.* 15: 227–233.

Gameson, A.L.H., and Saxon, J.R. 1967. Field studies on effect of daylight on mortality of coliform bacteria. *Water Res.* 1: 279–295.

Gannon, J.J., Busse, M.K., and Schillinger, J.E. 1983. Fecal coliform disappearance in a river impoundment. *Water Res.* 17: 1595–1601.

Garofalo, J.E., and Effler, S.W. 1987. Coating of zooplankton with calcium in Onondaga Lake , New York. *Environ Sci Technol.* 21: 604–606.

Gee, A.S., Milner, M.J., and Hemsworth, R.J. 1978. The effect of density on mortality in juvenile Atlantic salmon (*Salmo salar*). *J Anim Ecol.* 47: 497–505.

Geldreich, E.E. 1981. Membrane filter techniques for total coliform and fecal coliform populations in water. In: B.J. Dutka (ed.), *Membrane Filtration: Applications, Techniques, and Problems.* Marcel Dekker, New York, pp. 41–75.

Gibson, R.J. and Myers, R.A. 1988. Influence of seasonal river discharge on survival of juvenile Atlantic salmon, *Salmo salar. Can J Fish Aquat Sci.* 45: 344–348.

Gieskes, W.W., and Kraay, G.W. 1983. Unknown chlorophyll a derivatives in the North Sea and the tropical Atlantic Ocean revealed by HPLC analysis. *Limnol Oceanogr.* 28: 757–765.

Gilman, B. 1985. "An Inventory of the Aquatic Weedbeds of Honeoye Lake with Suggestions for their Management." Community College of the Finger Lakes, Canandaigua, NY.

Gliwicz, Z.M., and Siedlar, E. 1980. Food size limitation and algae interfering with food collection in *Daphnia. Arch Hydrobiolog.* 88: 155–177.

Goldman, J.C., and Carpenter, E.J. 1974. A kinetic approach to the effect of temperature on algal growth. *Limnol Oceanogr.* 19: 756–766.

Gorham, P.R., and Carmichael, W.W. 1988. Hazards of freshwater blue-green algae (cyanobacteria). In: *Algae and Human Affairs*, C.A. Lembi and Waaland, J.R. (ed.), Cambridge University Press, New York, pp. 403–431.

Graham, J.M., Auer, M.T., Canale, R.P., and Hoffmann, J.P. 1982. Ecological studies and mathematical modeling of *Cladophora* in Lake Huron: 4. Photosynthesis and respiration as functions of light and temperature. *J Great Lakes Res.* 8: 100–111.

Great Lakes Group, International Joint Commission. 1976. Waters of Lake Huron and Lake Superior, Vol. I. Summary and Recommendations, Windsor, Ontario, Canada.

Greeley, J.R. 1928. Fishes of the Oswego Watershed. pp 84–248 In: "A Biological Survey of the Oswego River System." Supplemental to Seventeenth Annual Report, State of New York Conservation Department. Albany, NY.

Grieb, T.M., Driscoll, C.T., Gloss, S.P., Schofield, C.L., Bowie, G.L., and Porcella, D.L., 1990. Factors affecting mercury accumulation in fish in the upper Michigan peninsula. *Environ Toxicol Chem.* 9: 919–930.

Hall, D.J., Threlkeld, S.T., Burns, C.W., and Crowley, P.H. 1976. The size efficiency hypthesis and the size structure of zooplankton communities. *Ann Rev Ecol Syst.* 7: 177–208.

Haller, W.T., Sutton, D.L., and Barlowe, W.C. 1974. Effects of salinity on growth of several aquatic macrophytes. *Ecology* 55: 891–894.

Hammer, U.T., and Heseltine, J.M. 1988. Aquatic macrophytes in saline lakes of the Canadian prairies. *Hydrobiologia* 158: 101–116.

Hammers, B.E. 1991. Comparison of methods for estimating age, growth and related population characteristics of white crappie. *North Am J Fish Manag.* 11: 492–498.

Haney, J.F. 1971. An in situ method for the measurement of zooplankton grazing rates. *Limnol Oceanogr.* 16: 970–977.

Hantke, B., and Melzer, A. 1993. Kinetic changes in surface phosphatase activity of *Syneda-Acus* (Bacillariophyceae) in relation to pH variation. *Freshwater Biol.* 29: 31–36.

Harris, G.P. 1986. *Phytoplankton Ecology: Structure, Function, and Flucuation.* Prentice Hall, New York.

Hart, R.C. 1988. Zooplankton feeding rates in relation to suspended sediment content: potential influences on community structure in a turbid reservoir. *Freshwater Biol.* 19: 123–139.

Hatch, W.R., and Ott, W.L. 1968. The determination of submicrogram quantities of mercury by atomic absorption spectrophotometry. *J Anal Chem.* 40: 2085.

Healey, F.P. 1973. Characteristics of phosphorus deficiency in Anabaena. *J Phycol.* 9: 383–394.

Healey, F.P., and Hendzel, L.L. 1979. Flourometric measurement of alkaline phosphatase activity in algae. *Freshwater Biol.* 9: 429–439.

Heath, R.T., and Cooke, G.D. 1975. The significance of alkaline phosphatase in a eutrophic lake. *Verh Internat Verein Limnol.* 19: 959–965.

Heichel, G.H. 1970. Prior illumination and the respiration of maize leaves in the dark. *Plant Physiol.* 46: 359–362.

Hernandez, I., Niell, F.X., and Fernandez, J.A. 1992. Alkaline phosphatase activity in *Porphyra umbilicalis* (L) Kutzing. *J Exp Mar Biol Ecol.* 159: 1–13.

Hindenlong, D.M., P. Rudewicz, and D.E. Smith. 1981. "Analysis of Onondaga Lake Fish Sampls for Benzene and Chlorobenzene." Allied Chemical Corporation, Memorandum to R.C. Price.

Hirethota, P. and N.H. Ringler. 1989. Fish populations as bioindicators of long-term contaminant related stress. In: *Advances in Ecology,* Academic Press, New York.

Hohn, M.H. 1951. A study of the distribution of diatoms (Bacillarieae) in western New York State. *Cornell Univ Agri Exp Sta Mem.* 303: 1–39.

Hurley, D.A. 1986. Fish populations of the Bay of Quinte, Lake Ontario before and after phosphorous control. *Can Spec Publ Fish Aquat Sci.* 86: 201–214.

Hutchinson, B.P. 1971. The effect of fish predation on the zooplankton of ten Adirondack lakes, with particular reference to the alewife, *Alosa pseudoharengus* on determinations of selective feeding. *Trans Am Fish Soc.* 105: 89–95.

Hutchinson, G.E. 1967. *A Treatise on Limnology.* Vol. II *Introduction to Lake Biology and the Limnoplankton.* Wiley, New York.

Hutchinson, G.E. 1975. *A Treatise on Limnology.* Vol. III: *Limnological Botany.* Wiley, New York.

Hynes, H.B.N. 1970. *The Ecology of Running Waters.* Univ. of Toronto Press, Toronto.

Jackson, D.F. 1968. Onondaga Lake, New York—An Unusual Algal Environment. In: *Algae, Man. and the Environment.,* D.F. Jackson (ed.). Syracuse University Press, Syracuse, NY, pp. 515–524.

Jackson, D.F. 1969. Primary productivity studies on Onondaga Lake, N.Y. *Verh Int Verein Theor Angew Limnol.* 17: 86–94.

Jacobsen, T.R. 1982. Comparison of chlorophyll *a* measurements by fluorometric, spectrophotometric and high pressure liquid chromatographic methods in aquatic environments. *Arch Hydrobiol Beih Ergebn Limnol.* 16: 35–45.

Jansson, M., Olsson, H., and Pettersson, K. 1988. Phosphatases: origin, characteristics and function in lakes. *Hydrobiologia* 170: 157–175.

Jassby, A.D., and Platt, T. 1976. Mathematical formulation of the relationship between photosynthesis and light for phytoplankton. *Limnol Oceanogr.* 21: 540–547.

Jassby, A.D., and Powell, T. 1975. Vertical patterns of eddy diffusion during stratification in Castle Lake, California. *Limnol Oceanogr.* 20: 530–543.

Jerlov, N.G. 1968. *Optical Oceanography.* Elsevier, Amsterdam, The Netherlands.

Johnson, D.L., Jiao, J., Des Sartos, S.G., and Effler, S.W. 1991. Individual particle analysis of suspended materials in Onondage Lake, New York. *Environ Sci Technol.* 25: 736–744.

Johnson, J.H., and N.H. Ringler. 1980. Diets of juvenile coho salmon (*Oncorhynchus kisutch*) and steelhead (*Salmo gairdneri*) relative to prey availability. *Can J Zool.* 58: 553–558.

Jones, J.G. 1971. Studies on freshwater bacteria: association with algae and alkaline phosphatase activity. *J Ecol.* 60: 59–75.

Jorgensen, E.G. 1968. The adaption of plankton algae: II. Aspects of the temperature adaption of *Skeletonema costatum. Physiologia Plantarum.* 21: 423–427.

Kadono, Y. 1982. Occurrence of aquatic macrophytes in relation to pH, alkalinity, Ca^{++}, Cl^-, and conductivity. *Jap J Ecol.* 32: 39–44.

Kapuscinski, R.B., and Mitchell, R. 1983. Sunlight-induced mortality of viruses and *Escherichia Coli* in coastal seawater. *Environ Sci Technol.* 17: 1–6.

Kilham, P., Kilham S.S., and Hecky, R.E. 1986. Hypothesized resource relationships among African planktonic diatoms. *Limnol Oceanogr.* 31: 1169–1181.

Kirk, J.T.O. 1983. *Light and Photosynthesis in Aquatic Ecosystems.* Cambridge University Press, New York.

Kirk, J.T.O. 1985. Effects of suspensoids (turbidity) on penetration of solar radiation in aquatic ecosystems. *Hydrobioly.* 125: 195–209.

Kittrell, F.W., and Furfari, S.A. 1963. Observations on coliform bacteria in streams. *J Water Pollut Control Fed* 35: 1361–1385.

Klock, J.W. 1971. Survival of coliform bacteria in wastewater treatment lagoons. *J Water Pollut Control Fed.* 43: 2071–2083.

Koenings, J.P., Burkett, R.D., and Edmundson, J.D. 1990. The exclusion of limnetic cladocera from turbid glacier-meltwater lakes. *Ecology* 71: 57–67.

Kramer, R.H., and L.L. Smith, Jr. 1962. Formation of year classes in largemouth bass. *Trans Am Fish Soc.* 91: 29–41.

Kutkuhn, J.H. 1957. Utilization of plankton by juvenile gizzard shad in a shallow praire lake. *Trans Am Fish Soc.* 87: 80–103.

Lampert, W. 1978. Climatic conditions and planktonic interactions as factors controlling the regular succession of spring algal bloom and extremely clear water in Lake Constance. *Verh Int Ver Limnol.* 20: 969–974.

Lampert, W. 1987. Laboratory studies on zooplankton-cyanobacteria interactions. *New Zealand J Mar Freshwater Res.* 21: 483–490.

Lampert, W., Fleckner, W., Rai, H., and Taylor, B.E. 1986. Phytoplankton control by grazing zooplankton: A study on the spring clear-water phase. *Limnol Oceangr.* 31: 478–490.

Lantrip, B.M. 1983. The decay of enteric bacteria in an estuary. Ph.D. Dissertation. The John Hopkins University, Baltimore, MD.

Laws, E.A., Bienfang, P.K., Ziemann, D.A., and Conquest, L.D. 1988. Phytoplankton population dynamics and the fate of production during the spring bloom in Auke Bay, Alaska. *Limnol Oceanogr.* 33: 57–65.

Lehman, J.T. 1980. Release and cycling of nutrients between planktonic algae and herbivores. *Limnol Oceanogr.* 25: 620–632.

Lehman, J.T. 1984. Grazing, nutrient relaease, and their impacts on the structure of phytoplankton communites. In: *Trophic Interactions within Aquatic Ecosystems*, D.G. Meyers and J.R. Strickler (ed.). Westview Press, Boulder, CO.

Leopold, L.B., M.G. Wolman, and Miller, J.P. 1964. *Fluvial Processes in Geomorphology*. Freeman, San Francisco, CA.

Lessard, E.J., and Sieburth, J.M. 1983. Survival of natural sewage populations of enteric bacteria in diffusion and batch chambers in the marine environment. *Appl Environ Microbiol.* 45: 950–959.

Likens, G.E., and Gilbert, J.J. 1970. Notes on quantitative sampling of natural populations of planktonic rotifers. *Limnol Oceanogr.* 15: 816–820.

Lorenzen, C.T. 1967. Determination of chlorophyll and phaepigments: Spectrophotometric equations. *Limnol Oceanogr.* 12: 343–346.

Lund, J.W.G. 1954. The seasonal cycle of the plankton diatom, *Melosira italica* (Ehr.) Kütz, subsp. *subarctica*. O. Müll. *J Ecol.* 42: 151–179.

Lund, J.W.G. 1971. An artificial alteration of the seasonal cycle of the plankton diatom *Melosira italica* subsp. *subarctica* in an English lake. *J Ecol.* 59: 521–533.

Lynch, M. 1980. *Aphanizomenon* blooms: Alternate control and cultivation by *Daphnia pulex*. In: *Evolution and Ecology of Zooplankton Communities*, W.C. Kerfoot (ed.), University Press of New England, Hanover, NH.

MacArthur, R., and MacArthur, J. 1961. On bird species diversity. *Ecology.* 42:594–598.

Madsen, J.D., Eichler, L.W., Sutherland, J.W., Bloomfield, J.A. Smart, R.M., and Boylen, C.W. 1992. "Submersed Littorial Vegetation Distribution: Field Quantification and Experimental Analysis of Sediment Types from Onondaga Lake, New York." Report submitted to Onondaga Lake Management Conference. U.S. Army Engineer Waterways Experienced Station, Lewisville Aquatic Ecosystem Research Facility, Lewisville, TX.

Madsen, J.D., Sutherland, J.W., Bloomfield, J.A., Roy, K.M., Eichler L.W., and Boylen, C.W. 1989. "Lake George Aquatic Plant Survey Final Report." New York State Department of Environmental Conservation, Albany, NY.

Magurran, A.E. 1988. *Ecological Diversity and Its Measurement*. Princeton University Press, Princeton, NJ.

Makarewicz, J.C., and Likens, G.E. 1979. Structure and functions of the zooplankton community of Mirror Lake, New Hampshire. *Ecol Monogr.* 49: 109–127.

Mancini, J.L. 1978. Numerical estimates of coliform mortality rates under various conditions. *J Water Pollut Control Fed.* 50: 2477–2484.

Marcotte, B.M., and Browman, H.I. 1986. Foraging behavior in fishes: perspectives on variance. In: *Contemporary Studies on Fish Feeding*, C.A. Simenstad and G.M. Caillet (ed.). W. Junk, pp. 25–31.

Marker, A.F.H., and Jinks, S. 1982. The spectrophotometric analysis of chlorophyll *a* and phaeopigments in acetone, ethanol and methanol. *Arch Hydrobiol Beih Ergebn Limnol.* 16: 3–17.

Mattox, K.R., and Stewart, K.D. 1984. Classification of the green algae: a concept based on comparative cytology. In *Systematics of the Green Algae*, D.E.G. Irvine and D.M. John (ed.), Academic Press, Orlando, FL, pp. 29–72.

McCambridge, J., and McMeekin, T.A. 1981. Effect of solar radiation and predacious microorganisms on survival of fecal and other bacteria. *Appl Environ Microbiol.* 41: 1083–1087.

McCauley, E., and Briand, F. 1979. Zooplankton grazing and phytoplankton species richness: field tests of the predation hypothesis. *Limnol Oceanogr.* 24: 243–252.

McIntosh, A.W., and Kevern K.P. 1974. Toxicity of copper to zooplankton. *J Environ Qual.* 3: 166–170.

McMullen, J.M. 1992. "Aquatic Vascular Plants in Onondaga Lake—A Comparison of Recent Finds to Historic Records." Report to the Onondaga Lake Management Conference, Syracuse, NY.

McQueen, D.J., Post, J.R., and Mills, E.L. 1986. Trophic relationships in freshwater pelagic ecosystems. *Can J Fish Aquat Sci.* 43: 1571–1581.

Meals, K.O., and Miranda, L.E. 1991. Variability in abundance of age-0 Centrarchids among littoral habitats of flood control resevoirs in Mississippi. *North Am J Fish Manag.* 11: 298–304.

Megard, R.O.1, Combs, W.S., Smith, P.D., and Knoll, A.S. 1979. Attenuation of light and daily integral rates of photosynthesis attained by planktonic algae. *Limnol Oceanogr.* 24: 1038–1050.

Merritt, R.W., and Cummins, K.W. (eds.) 1984. *An Introduction to the Aquatic Insects of North America.* 2d ed. Kendall/Hunt Publishing Company, Dubuque, IA.

Meyer, M.A. 1979. The temporal and spatial distribution of the limnetic zooplankton in Onondaga Lake, Syracuse, N.Y. Masters Thesis, Department of Civil Engineering Syracuse University, Syracuse, NY.

Meyer, M.A., and Effler, S.W. 1980. Changes in the zooplankton of Onondaga Lake 1969–1978. *Environ Pollut (Series A)* 23: 131–152.

Miller, G.L. 1978. "An Ecological Inventory of Aquatic Vegetation in the Major Lakes of Cayuga County New York and Recommendations for Their Management." Cayuga County Environmental Management Council, Auburn, NY.

Mills, E.L., and Forney, J.L. 1988. Trophic dynamics and development of freshwater pelagic food webs. In: *Complex Interactions in Lake Communities*, S.R. Carpenter (ed.), Springer-Verlag, New York.

Mills, E.L., Forney, J.L., Clady, M.D., and Schaffner, W.R. 1978. Oneida Lake. In: *Lakes of New York State*, Vol. 2, J.A. Bloomfield (ed.), Academic Press, New York.

Mitchell, R., and Chamberlin, C. 1978. Survival of indicator organisms. pp. 15–35, In: *Indicators of Viruses in Water and Food*, Berg, 6. (ed.). Ann Arbor Science Publishers, Ann Arbor, MI.

Moegenburg, S.M., and Vanni, M.J. 1991. Nutrient regeneration by zooplankton: effects on nutrient limitations of phytoplankton in a eutrophic lake. *J Plankton Res.* 15: 573–588.

Moeller, J.R., and Calkins, J. 1980. Bactericidal agent in wastewater lagoons and lagoon design. *J Water Pollut Control Fed.* 52: 2442–2451.

Monod, J. 1942. Recherches sur la croissance des cultures bacteriennes. Hermann, Paris.

Moore, R.B. 1967. "Microbiological Indicators of Pesticide Pollution in Aquatic Environments." Final report submitted in Division of Fishery Research, Bureau of Sports Fisheries and Wildlife, Department of the Interior, Contract No.14-16-0008-889, Washington, DC.

Moore, W.G. 1942. Field studies on the oxygen requirements of certain fresh-water fishes. *Ecology.* 23: 319–329.

Morgan, C.A., and Ringler, N.H. 1990. Preliminary analysis of the fishes of Onondaga Lake. Unpublished tables and figures presented at the Onondaga Lake Remediation Conference, Bolton Landing, NY, February 5–9,1990.

Moss, D.D., and Scott, D.C. 1961. Dissolved oxygen requirements of three species of fish. *Trans Am Fish Soc.* 90: 377–393.

Murphy, C.B. 1973. Effect of restricted use of phosphate-based detergents on Onondaga Lake. *Science* 181: 379–381.

Murphy, C.B. 1978. Onondaga Lake. pp 223–365 In: *Lakes of New York State*, Vol. II J.A. Bloomfield (ed.). Academic Press, NY.

Murphy, M.H. 1992. Suitability of Nine Mile Creek, New York for resoration of Atlantic salmon (*Salmo salar L.*). Master's Thesis, S.U.N.Y. College of Environmental Science and Forestry, Syracuse, NY.

Murray, K.R., and Ringler, N.H. 1988. Patterns of habitat utilization by trout and salmon in Lake Ontario tributaries. Annual Meeting, N.Y. American Fisheries Society, Binghamton, NY. January 29–30, 1988.

National Academy of Science and National Academy of Engineering. 1972. *Water Quality Criteria.* A Report of the Committee on Water Quality Criteria. Washington, DC.

Neill, W.E. 1984. Regulation of rotifer densities by crustacean zooplankton in an oligotrophic montane lake in British Columbia. *Oecologia (Berl.)* 61: 164–177.

Neth, P., and Barnhart, G. 1983. The landlocked salmon: its life cycle, its history, its management. *Conservationist*, NYDEC September–October, 1983.

Nielson, L.A., and Johnson, D.L. (eds.). 1983. *Fisheries Techniques.* American Fisheries Society, Bethesda, MD.

Noble, R.L., and Forney, J.L. 1969. Fishery survey of Onondaga Lake—Summer, 1969. Department of Conservation and Cornell University, Ithaca, NY.

O'Brien, J.W. 1979. The predator-prey interaction of planktivorous fish and zooplankton. *Am Sci.* 67: 572–581.

O'Gorman, R., Mills, E.L., and Degisi, J. 1991. Use of zooplankton to assess the movement and distribution of alewife (*Alosa pseudoharengus*) in south-central Lake Ontario in spring. *Can J Fish Aquat Sci.* 48: 2250–2257.

Onondaga County. 1971. Onondaga Lake Study. Water Quality Office, United States Environmental Protection Agency, Water Pollution

Control Research Series, Report #11060/ FAE 4/71.

Onondaga County. 1971–1991. Onondaga Lake Monitoring Program. Annual publication, Onondaga County, NY.

Orciari, R.D., Mysling, D.J. and Leonard, G.H. 1988. "Evaluation of stocking Atlantic salmon (*Salmo salar*) fry in Sandy Brook, Colebrook, CT." Connecticut Department of Environmental Protection, Final Report.

Palmer, M.D., and Dewey, R.J. 1984. Beach fecal coliforms. *Can J Environ Engr.* 11: 217–224.

Parsons, T.R., Maita, Y., and Lalli, C.M. 1984. *A Manual of Chemical and Biological Methods for Seawater Analysis.* Pergamon Press, New York.

Peckarsky, B.L. 1991. Freshwater Macroinvertebrates of Northeastern North America. Cornell University, Ithaca, NY.

Peckarsky, B.L., Fraissinet, P.R., Penton, M.A., and Conklin, D.J. 1990. *Freshwater Macroinvertebrates of Northeastern North America.* Cornell University Press, Ithaca, NY.

Pennak, R.W. 1978. *Freshwater Invertebrates of the United States.* Wiley, New York.

Pettersson, K. 1980. Alkaline phosphatase activity and algal surplus phosphorus as phosphorus-deficiency indicators in Lake Erken. *Arch Hydrobiol.* 89: 54–87.

Petit, G.D. 1973. Dissolved oxygen on survival and behavior of selected fishes of Western Lake Erie. *Bull Ohio Biol Surv.* 4: 1–79.

Pick, F.R. 1987. Interpretations of alkaline phosphatase activity in Lake Ontario. *Can J Fish Aquat Sci.* 44: 2087–2094.

Pielou, E.C. 1966. Species diversity and pattern diversity in the study of ecological succession. *J Theoret Biol.* 10: 370–383.

Pierce, E.C. 1969. *An Introduction to Mathematical Ecology.* Wiley, New York.

Pierce, R.J., Wissing, T.E. and Megrey, B.A. 1981. Aspects of the feeding ecology of gizzard shad in Action Lake, Ohio. *Trans Am Fish Soc.* 110: 391–395.

Porter, K.W. 1977. The plant-animal interface in freshwater ecosystems. *Am Sci.* 65: 159–170.

Post, J.R., and Cucin, D. 1984. Changes in the benthic community of a small precambrian lake following the introduction of yellow perch, *Perca flavescens. Can J Fish Aquat Sci.* 41: 1496–1501.

Poulet, S.A. 1973. Grazing of *Pseudocalanus minutus* on naturally occurring particulate matter. *Limnol Oceanogr.* 18: 564–573.

Preall, R.J. 1985. Comparison of actual and maximum growth rates for brown trout (*Salmo trutta L*) in three central New York streams. Master's Thesis, S.U.N.Y. College of Environmental Science and Forestry. Syracuse, NY.

Preall, R.J. and Ringler, N.H. 1989. Comparison of actual and potential growth rates of brown trout (*Salmo trutta*) in natural streams based on bioenergetic models. *Can J Fish Aquat Sci.* 46: 1067–1076.

Price, K.S., Flemer, D.A., Taft, J.L., Mackiernan, G.B., Nehlsen, W., Biggs, R.B., Burger, N.H., and Blaylock, D.A. 1985. Nutrient enrichment of Chesapeake Bay and its impact on the habitat of striped bass: a speculative hypothesis. *Trans Am Fish Soc.* 114: 97–106.

Rahel, F.J. 1992. Foraging in a Lethal Environment: Predation by Central Muklminnows (*Umbra Limi*) on *Chaoborus* in the anoxic hypolimnion of Wisconsin Lakes, Abstract. Paper presented at Annual Meeting American Fisheries Society, Rapid City, SD September 14–17, 1992.

Rai, H. 1980. "The Measurement of Photosynthetic Pigments in Freshwater and Standardization of Methods." Proceedings of the Workshop held at Plön, W. Germany, July 28/29, 1978, organized by Max-Planck Institut für Limnologic, Abt. Tropenökologie. *Arch Hydrobiol Beih Erebn Limnol,* Beiheft 14.

Rai, H., and Marker, A.G.H. 1982. "The Measurement of Photosynthetic Pigments in Freshwaters and Standardization of Methods." Proceedings of the 2nd Workshop held at Plön, W. Germany, July 17/18, 1980, organized by Max-Planck Institut für Limnologie. *Arch Hydrobiol Beih Erebn Limnol.* Beiheft 16.

Rawson, D.S. 1956. The net plankton of Great Slave Lake. *J Fish Res Bd Can* 13: 53–157.

Reckhow, K.H., and Chapra, S.C. 1983. *Engineering Approaches for Lake Management,* Volume 1: *Data Analysis and Empirical Modeling.* Butterworth, Boston.

Redfield, A.C., Ketchum, B.H., and Richards, F.A. 1963. The influence of organisms on the composition of seawater. In: M.N. Hill (Ed.), *The Sea.* Wiley Interscience, New York.

Reichardt, W. 1971. Catalytic mobilization of phosphat in lake water and by Cyanophyta. *Hydrobiologia* 38: 377–394.

Remane, A., and Schlieper, C. 1971. *Biology of Brackish Water,* Die Binnengewasser 25.

Resh, V.H., and Unzicker, J.D. 1975. Water quality monitoring and aquatic organisms: the importance of species identification. *J Water Pollut Cont Fed.* 47: 9–19.

Reynolds, C.S. 1984. *The Ecology of Freshwater Phytoplankton.* Cambridge University Press, New York.

Rhee, G-Y. 1978. Effects on N/P atomic ratios and nitrate limitations on algal growth, cell composition, and nitrate uptake. *Limnol Oceanogr.* 23: 10–25.

Ringler, N.H. 1979. Prey selection by benthic feeders. pp. 219–229. In: Predator-prey systems in fishery management. R. Stround and H. Clepper (eds.). Sport Fishing Institute, Washington DC, 504p.

Ringler, N.H. 1990. Fish foraging: Adaptations and patterns. pp. 1–30. In: *Advances in Fish Biology,* Academic Press (in press).

Ringler, N.H., and Hall, J.D. 1975. Effects of logging in water temperature and dissolved oxygen in spawning beds. *Trans Am Fish Soc.* 104: 111–121.

Ringler, N.H., and Wagner, B. 1988. Community Structure of Benthic Invertebrates in Cayuga Lake, New York: Comparison of Cargill Site 001 to Reference Site. Submitted to Cargill Salt Co., Ithaca, NY.

Rosgen, D.L. 1985. A stream classification system. In: "Riparian Ecosystems and Their Management; Reconciling ConflictingUses". Proceedings of the First North American Riparian Conference, April 16–18, Tucson, Arizona. GTR-RM120, pp. 91–95.

Russell-Hunter, W.D. 1970. *Aquatic Productivity: An Introduction to Some Basic Aspects of Biological Oceanography and Limnology.* Macmillan, New York.

Ruttner-Kolisko, A 1974. *Plankton Rotifers: Biology and Taxonomy.* Die Binnengewasser Suppl. 26.

Sandgren, C.D. [ed.] 1988. *Growth and Reproductive Strategies of Freshwater Phytoplankton.* Cambridge University Press (New York).

Saroff, S.T. 1990. "Proceedings of the Onondaga Lake Remediation Conference, Sagamore Conference Center, Bolton Landing, NY." February 5–8 1990. NYS Department of Law and NYS Department of Environmental Conservation, Albany, NY.

Saunders, D. 1987. Stock selection and thermal relation of Atlantic salmon. In: Atlantic salmon Workshop Report, University of Guelph, Guelph, Ontario, January 27–28, 1987: 6–8.

Scarce, L.E., Rubenstein, S.H., and Megregian, S. 1964. Survival of indicator bacteria in receiving waters under various conditions. In: *Proceedings of the 7th Conference on Great Lakes Research,* Publication No. 11, Great Lakes Research Division, The University of Michigan, Ann Arbor, MI, pp. 130–139.

Schanz, F., and Rai, H. 1988. Extract preparation and comparison of fluorometric, chromatographic (HPLC) and spectrophotometric determination of chlorophyll *a*. *Arch Hydrobiol.* 112: 533–539.

Schillinger, J.E., and Gannon, J.J. 1982. Coliform attachment to suspended particles in stormwater. The University of Michigan, Ann Arbor, MI.

Schindler, D.W. 1977. Evolution of phosphorus limitation in lakes. *Science* 191: 260–262.

Schleske, C.L., and Stoermer, E.F. 1971. Eutrophication, silica depletion, and predicted changes in algal quality in Lake Michigan. *Science* 173: 423–424.

Schneider, D.W. 1990. Direct assessment of the independent effects of exploitative and interference competition between *Daphnia* and rotifers. *Limnol Oceanogr.* 35: 916–922.

Scott, D.P. 1974. Mercury concentrations of white muscle in relation to age, growth, and condition in four species of fishes from Clay Lake, Ontario. *J Fish Res Bd Can.* 31: 1723–1729.

Scott, D.P., and Armstrong, F.A.J. 1972. Mercury concentration in relation to size in several species of freshwater fishes from Manitoba and north-western Ontario. *J Fish Res Bd Can.* 29: 1685–1690.

Seaburg, K.G. 1957. A stomach sampler for live fish. *Progr Fish Cult.* 19: 137–139.

Seger, E. 1979. The Fate of Heavy Metals in Onondaga Lake. Master's Thesis, Department of Civil Environmental Engineering, Clarkson College, Potdam, NY.

Shuman, M.S., and Woodward, G.P. 1977. Stability constant of carbon-copper-organic chelates in aquatic samples. *Environ Sci Technol.* 11: 809–813.

Siegfried, C.A. 1987. Large-bodied crustacea and rainbow smelt in Lake George, New York: Trophic interactions and phytoplankton composition. *J Plankton Res.* 9: 27–39.

Siegfried, G.A. 1991. The pelagic rotifer community of an acidic clearwater lake in the Adirondack Mountains of New York State. *Archiv für Hydrobiolog.* 122: 441–462.

Siegfried, C., Auer, N.A., and Effler, S.W. 1996. Changes in the zooplankton of Onondaga Lake: causes and implications. *Lake Reservoir Manag.* (in press).

Siegfried, C.A., Bloomfied, J.A., and Sutherland, J.W. 1989. Planktonic rotifer community structure in Adirondack lakes: Effects of acidity, trophic status, and related water quality characteristics. *Hydrobiologia.* 175: 33–48.

Siegfried, C.A., and Kopache, M.E. 1984. Zooplankton dynamics in a high mountain reservoir of southern California. *Calif Fish Game.* 70: 18–38.

Siegfried, C.A., and Quinn, S.O. 1987. "Lake George Clean Lakes 314 Study (Phase I—Diagnostic/Feasibility Study)". Final Report to

USEPA, New York State Deptartment of Environmental Conservation, Albany, New York.

Siegfried, C,A., and Sutherland, J.W. 1992. Zooplankton communities of acid lakes: changes in community structure associated with acidification. *J Freshwater Ecol.* 7: 97–112.

Siegfried, C.A., Sutherland J.W., Quinn, S.O., and Bloomfield, J.A. 1984. Lake acidification and the biology of Adirondack lakes I. Rotifer communities. *Verhangen Internat Verein Limnol.* 22: 549–558.

Sieracki, M. 1980. Effects of short exposures of natural sunlight on the decay rates of enteric bacteria and coliphage in a simulated sewage outfall microcosm. Master's Thesis, University of Rhode Island, Kingston, RI.

Simpson, K.W. 1982. Biological Survey of Onondaga Creek. New York State Department of Health, Albany.

Simpson, K.W., and Bode, R.W. 1980. "Common Larvae of Chironomidae (Diptera) from New York State Streams and Rivers: With Particular Reference to the Fauna of Artificial Substrates." Bulletin No. 439, New York State Museum, The State Education Department, Albany, NY.

Sindermann, C.J., Bang, F.B., Christensen, N.O., Dethlefsen, V., Harshbarger, J.C., Mitchell, J.R., and Mulcahy, M.F. 1980. The role and value of pathobiology in pollution effects monitoring programs. In: *Biological Effects of Marine Pollution and the Problems of Monitoring,* A.D. McIntyre and J.B. Pearce (Eds.). *Rapp P-v Reun Cons Int Explor Mer.* 179: 135–151.

Sjogren, R.E., and Gibson, M.J. 1981. Bacterial survival in a dilute environment. *Appl Envir Microbiol.* 41: 1331–1336.

Sloan, R.J. 1985. "Onondaga Lake Mercury Trend Analysis: Sampling Protocol". Bureau of Environmental Protection, Division of Fish and Wildlife, Albany, NY.

Sloan, R.J. 1987. "Toxic Substances in Fish and Wildlife Analyses Since May, 1982." Volume 6. Technical Report 87-4 (BEP). Division of Fish and Wildlife, Albany, NY.

Sloan, R.J., Skinner, L.C., Horn, E.G., and Karcher, R. 1987. An overview of mercury contamination in the fish of Onondaga Lake. Technical report 87-1 (BEP) Division of Fish and Wildlife, Albany, NY.

Smart, R.M. 1990. "Effects of Water Chemistry on Submersed Aquatic Plants: a Synthesis." Miscellaneous Paper A-90-4, U.S. Army Engineer Waterways Experiment Station, Vicksburg, MS.

Smith, C. Lavett. 1985. "The Inland Fishes of New York State." New York State Department of Environmental Conservation, Albany, NY.

Smith, H.M. 1892. Report on the fisheries of Lake Ontario. *Bull U.S. Fish Comm.* 10: 195–202.

Smith, V.H. 1983. Low nitrogen to phosphorus ratios favor dominance by blue–green algae in lake phytoplankton. *Science* 221: 669–671.

Sokol, R.R., and Rohlf, F.J. 1981. *Biometry.* W.H. Freeman, San Francisco, CA.

Sommer, U., Gliwicz, M., Lampert, W., and Duncan, A. 1986. The EG-model of seasonal succession of planktonic events in fresh waters. *Arch Hydrobiol.* 106: 433–471.

Sprules, W.G. 1975. Factors affecting the structure of limnetic crustacean zooplankton communities in central Ontario lakes. *Int Ver Theor Angew Limnol Verh.* 19: 635–643.

Starkweater, P.L., and Kellar P.E. 1983. Utilization of cyanobacteria by *Brachionus calyciflorus*: *Anabaena flos-aquae* (NRC-44-1) as a sole or complementary food source. *Hydrobiologia* 104: 373–377.

State of New York Conservation Department. 1928. "A Biological Survey of the Oswego River System." Supplement to the seventeenth annual report, 1927. J.B. Lyon, Albany, NY.

Stearns and Wheler Engineers. 1979. Onondaga Lake Storms Impact Study. Stearns and Wheler, Civil and Sanitary Engineers, Cazenovia, New York.

Stearns and Wheler Engineers. 1990. "Onondaga Lake Monitoring Program, 1988." Prepared for the Onondaga County Department of Drainage and Sanitation, Syracuse, NY.

Steele, J.H. 1962. Environmental control of phytosynthesis in the sea. *Limnol Oceanogr.* 7: 137–150.

Stone, U.B., and Pasko, D. 1946. "Onondaga Lake Investigation." Western District, New York State Conservation Department, Albany, NY.

Storey, M.L. 1989. Modeling primary production in Onondaga Lake, New York. Master's Thesis, Department of Biological Sciences, Michigan Technological University, Houghton, MI.

Storey, M.L., Auer, M.T., Barth, A.K., and Graham, J.M. 1993. Site-specific determination of kinetic coefficients for modeling algal growth. *Ecol Model.* 66: 181–196.

Svensson, J., and Stenson, J.A.E. 1991. Herbivoran impact on phytoplankton community structure. *Hydrobiologia* 226: 71–80.

Sylva, R.N. 1976. The environmental chemistry of copper (II) in aquatic systems. *Water Res.* 10: 789–792.

Sze, P. 1972. The phytoplankton of Onondaga Lake. Ph.D. Thesis, Cornell University, Ithaca, NY.

Sze, P. 1975. Possible effect of lower phosphorus coneartration on the phytoplankton in Onondaga Lake, New York. *Phycologia* 14: 197–203.

Sze. P. 1980. Seasonal succession of phytoplankton in Onondaga Lake, NY (USA). *Phycologia* 19: 54–59.

Sze. P. 1986. *A Biology for the Algae.* William C. Brown, Dubuque, IA.

Sze, P. 1992. *A Biology for the Algae,* 2d ed. William C. Brown, Dubugue, IA.

Sze. P., and Kingsbury, J.M. 1972. Distribution of phytoplankton ˙ in a polluted saline lake, Onondaga Lake, NY. *J Phycol.* 8: 25–37.

Taft, J.L., Loftus, M.E., and Taylor, W.R. 1977. Phospate uptake from phosphomonoesters by phytoplankton in Chesapeake Bay. *Limnol Oceanogr.* 22: 1012–1021.

Taggett, L.J. 1989. "Aquatic Plant Inventory of the Lake George Region". Rensselaer Fresh Water Institute Report #89-2, February 1989. Rensselaer Polytechnic Institute, Troy, NY.

Taggett, L.J., Madsen, J.D., and Boylen, C.W. 1990. "Annotated bibliography for species richness for submersed aquatic plants in worldwide waterways." Rensselaer Fresh Water Institute Report #90-9, Rensselaer Polytechnic Institute, April, 1990. Troy, NY.

Thomann, R.V., and Mueller, J.A. 1987. *Principles of Surface Water Quality Modeling and Control.* Harper and Row Publisher, New York.

Tilzer, M.M. 1983. The importance of fractional light absorption by photosynthetic pigments for phytoplankton productivity in Lake Constance. *Limnol Oceanogr.* 28: 833–846.

Tisa, M.S., and Ney, J.J. 1991. Compatibility of alewives and gizzard shad as reservoir forage fish. *Trans Am Fish Soc.* 120: 157–165.

Trainor, F.R., and Egan P.F. 1990. The implications of polymorphism for the systematics of *Scenedesmus. Br Phycol J.* 25: 275–279.

Tramer, E.J. 1969. Bird species diversity: components of Shannon's formula. *Ecology* 50: 927–929.

Trial, J.G. 1989. Testing Habitat Models for Blacknose Dace and Atlantic Salmon. Ph.D. Thesis, University of Maine, Orono, ME.

Trial, J.G., and Stanley, S.G. 1984. "Calibrating Effects of Acidity on Atlantic Salmon for Use in Habitat Suitability Models." Land and Water Resources Center, University of Maine, Completion Report, Proj. A054-ME.

Tyus, H.M. 1982. "Fish Radiotelemetry: Theory and Application for High Conductivity Rivers." U.S. Fish and Wildlife Service, Department of the Interior, Report No. FWS/OBS-82/83.

USEPA. 1985. "Ambient Water Quality Criteria for Mercury—1984." U.S. Environmental Protection Agency, Washington, DC. EPA 440/5-84-026.

Varderploeg, H.A., Eadie, B.J. Lieby, J.R., and Tarapchate, S.J. 1987. Contribution of calcite to the particle size spectrum of Lake Michyan sestom and its interaction with plankton. *Can J Fish Aquat Sci* 44: 1898–1914.

Vant, W.N., Davies-Colley, R.J., Clayton, J.S., and Coffey, B.T. 1986. Macrophyte depth limits in North Island (New Zealand) lakes of differing clarity. *Hydrobiologia* 137: 55–60.

Vollenweider, R.A. (ed.) 1974. *A Manual on Methods for Measuring Primary Production in Aquatic Environments.* Blackwell Scientific, Oxford, England.

Wallwork, J.A. 1970. *Ecology of Soil Animals.* McGraw-Hall, New York.

Warrer, C.E. 1971. Biology and Water Pollution Control. W.B. Saunders Co. Philadelphia, PA.

Waterman, G. 1971. Onondaga Lake Zooplankton. In: *Onondaga Lake Study.* Project No. 11060, FAE 4/71. Water Quality Office, Environmental Protection Agency, Onondaga County, Syracuse, NY, pp. 361–384.

Webster, D.A. 1982. Early History of the Atlantic Salmon in New York. *NY Fish Game J.* 29: 26–44.

Weidemann, A.D., and Barmrter, T.T., 1986. Absorption and scatlering coefficients in Irondepqusit Bay. *Limnol Ocearogr.* 31: 567–583.

Welschmeyer, N.A., and Lorenzen, C.J. 1985. Chlorophyll bungets: zooplankton quazing and phytoplankton growth in a temperate fjord and the Central Pacific Gyres. *Limnol Oceanogr.* 30: 1–21.

Wetzel, R.G. 1960. Marl ercrustation or hydrophytes in several Michyan lakes. *Oikos* 11: 223–236.

Wetzel, R.G. 1983. Limnology. Saunders, Philadelphia.

Whitmore, C.M., Warren, C.E., and Dourdoroff, P. 1960. Avoidance of reactions of salmonids and centrarchid fishes to oxygen concentrations. *Trans Am Fish Soc.* 89: 17–26

Wiederholm, T. 1984. Responses of aquatic insects to environmental pollution. In: *The Ecology of Aquatic Insects* V.H. Resh and D.M. Rosenberg, (Eds.). Praeger, New York, pp. 508–557.

Wiggins, G.B. 1977. *Larvae of the North American Caddisfly Genera (Trichoptera).* University of Toronto Press, Toronto.

Winner, R.W., Boesel, M.W., and Farrell, M.P. 1980. Insect community structure as an index of heavy metal pollution in lotic ecosystems. *Can J Fish Aquat Sci.* 37: 647–655.

Wolman, M.G. 1954. A method of sampling course river-bed material. *Trans Am Geophys.* 35: 951–956.

Wynne, D. 1977. Alterations in activity of phosphatases during the *Peridinium* bloom in Lake Kinneret. *Physiol Plant.* 40: 219–224.

Wynne. D. 1981. The role of phosphatases in the metabolism of *Peridinium cinctum* from Lake Kinneret. *Hydrobiologia* 83: 93–99.

7
Optics

Steven W. Effler and Mary Gail Perkins

7.1 Introduction

The behavior of light in water has important ecological and water quality implications. Incident light is the major source of heat in most northern temperate lakes. Light is essential for primary production; primary production is limited by light availability in all but the shallowest systems (Kirk 1983). High clarity (depth of visibility of submerged objects), low turbidity, absence of muddy appearance, and blue color are nearly ubiquitous desires of lake users. Restoration efforts focusing on the reduction of phytoplankton growth in culturally eutrophic lakes (e.g., reduction in nutrient loading) may be more appropriately described as directed at the assumed attendant improvements in the visual aesthetic qualities of water.

As light passes down into a water body, irradiance (the total radiant flux incident upon a unit area of surface; e.g., $\mu E \cdot m^{-2} \cdot s^{-1}$) diminishes and its spectral quality changes. Here we focus on the photosynthetically active radiation (PAR) wavelength interval (400–700 nm); which coincidently is the human vision waveband. Light is attenuated by two processes—absorption and scattering. The intensity of these processes is regulated by the composition and concentration of various attenuating materials. Light is absorbed in water by four components: water itself, dissolved yellow substances (gelbstoff), phytoplankton,

and tripton (inanimate particulate material). The spectral character of absorption by these materials is generally widely different (Kirk 1983). Scattering is caused almost entirely by particles. The presence of these particles increases the attenuation of light by lengthening the path that photons must traverse, thereby increasing the likelihood that the photons will be absorbed. Particle scattering varies little with wavelength in the PAR interval compared to absorption (Kirk 1983; Phillips and Kirk 1984). Light scattering versus absorption characteristics can vary greatly for different types of particles. For example, the relative absorption/scattering characteristics of phytoplankton are rather species specific (e.g., Bricaud et al. 1983; Kirk 1983). Tripton of terrigenous origin absorbs (e.g., Kirk 1985) and scatters light. However, newly precipitated calcium carbonate scatters light with little or no effect on absorption (Weidemann et al. 1985).

Wide differences in composition and concentration of light attenuating substances occur between lakes and within individual lakes with time causing major differences in the intensity of light absorption and light scattering, as specified by the magnitude of the absorption (a, m^{-1}) and scattering (b, m^{-1}) coefficients. These differences are manifested in a number of measurable optical properties that describe light penetration, the angular distribution of irradiance, and the spectral quality of irradiance (Kirk 1983). The values of commonly determined optical parameters, such as Secchi disc transparency (SD, m), the vertical attenuation coefficient for downward irradiance (K_d, m^{-1}), irradiance reflectance (R), and average cosine ($\bar{\mu}$) are largely a function of the composition of the water in which measurements are made. However, these properties also depend on the character of the light field. For example, they vary with solar altitude and depth (Kirk 1984a). These optical properties are described as *apparent* properties (Priesendorfer 1961). These should be distinguished from *inherent* properties, which are intrinsic properties of the medium itself (i.e., not affected by irradiance distribution), such as a, b, the beam attenuation coefficient (c; $c = a + b$), and the angular scattering function. It is widely assumed (e.g., Kirk 1981a; 1985) that the angular scattering function is similar for most lakes.

A number of optical measurements are made in situ to characterize the behavior of light in water. These measurements are often augmented by laboratory analyses of watercolumn samples to support empirical evaluations of materials regulating optical characteristics. The values of a and b are not amenable to direct measurement. Recent techniques estimate a and b from selected measured apparent optical properties (Kirk 1981b; 1989). In addition, laboratory techniques, in combination with field measurements of spectral irradiance, can be used to assess the contributions of different components to a (Kirk 1980; Weidemann and Bannister 1986). Some progress has also been made in partitioning b (Effler et al. 1987a, 1991a; Weidemann et al. 1985). Estimation and partitioning of a and b opens the way to understanding and mechanistically quantifying the roles various materials and related phenomena play in regulating apparent optical properties. Resolution of these roles is invaluable for effective management of degraded systems (e.g., low SD), as it establishes the feasibility of achieving substantial improvement and the appropriate focus for remedial activities.

The optical characteristics of Onondaga Lake are of major concern because of the degraded conditions of the visual aesthetics of the lake (e.g., Field and Effler 1983). Secchi disc measurements have been made routinely on the lake since the late 1960s. In recent years, the optics program for the lake has become more sophisticated. As related management concerns became more intense, and more powerful optical techniques became available (Kirk 1983), the complexity of the lake's optical regime emerged (Effler et al. 1984).

The goals of this chapter are to:

1. review the fundamentals of optics in the aquatic environment,

2. review the databases of apparent optical properties and selected attenuating components for Onondaga Lake, and document significant changes over the period of record (also see Perkins and Effler 1995),

3. identify and evaluate phenomena and components important in regulating the optical characteristics of Onondaga Lake, and

4. develop and test empirical and mechanistic frameworks (e.g., Effler and Perkins 1995) that relate light penetration in Onondaga Lake to routinely measured parameters and the concentrations of attenuating components.

7.2 Optical Measurements

A listing of underwater optical measurements made for Onondaga Lake since the late 1960s is presented in Table 7.1. A black-and-white quadrant disc was used for most of the Secchi disc measurements, but an entirely white disc was used for the earliest routine measurements (e.g., Onondaga County 1971). The use of two different discs should not significantly affect the comparability of SD measurements at the low values that prevailed during that period. The earliest in situ instrumentation measurements were made with a photometer, which measures illuminance (I_d, I_u). The two sensors were hemispheres. Paired measurements of downwelling illuminance and irradiance were made over the period 1980–1981.

TABLE 7.1. Optical measurements on Onondaga Lake since 1968; instrumentation, units, and period of measurements.

Parameter	Symbol	Units	Instrumentation	Period of measurement
A. Field				
1. Secchi disc transparency	SD	m	20 cm diameter white, or black and white, quadrant disc	Since 1968
2. Downwelling illuminance	I_d	ft-candle ⎤	Protomatic underwater	1978–1981
3. Upwelling illuminance	I_u	ft-candle ⎦	photometer	
4. Downwelling cosine irradiance	E_d	$\mu E \cdot m^{-2} \cdot s^{-1}$ ⎤	LiCor 192 SB sensors	E_d since
5. Upwelling cosine irradiance	E_u	$\mu E \cdot m^{-2} \cdot s^{-1}$ ⎦		1980, E_u Since 1985
6. Scalar irradiance	E_o	$\mu E \cdot m^{-2} \cdot s^{-1}$	LiCor 193 B senser	Since 1987
7. Spectral irradiance	$E_{d(\lambda)}$	$W \cdot m^{-2} \cdot nm^{-1}$	LiCor 1800 UW underwater spectroradiometer	
Downwelling				Since 1987
Upwelling				1991
B. Laboratory				
1. Total turbidity/acidified turbidity	T/T_{nc}	NTU	turbidimeter	Since 1985
2a. Chlorophyll a/phaeophytin	C_L/P	$mg \cdot m^{-3}$	spectrophotometer (Lorenzen 1967)	Irregularly since 1978
2b. Total chlorophyll	C_T	$mg \cdot m^{-3}$	spectrophotometer (Parsons et al. 1984)	Since 1987
3. Gelbstoff (dissolved humic substances) scans	$a_{Y(440)}$ and $a_{Y(\lambda)}$	m^{-1}	spectrophotometer (Davies-Colley and Vant 1987)	Since 1985
4. Particle scans	$a_{P(\lambda)}$	m^{-1}	spectrophotometer (Weidemann and Bannister 1986)	Since 1987
5. Individual particle analysis	—	—	SAX (Johnson et al. 1991)	1987

Flat irradiance sensors are described as cosine sensors. It is valuable to make irradiance measurements with one cosine sensor facing upwards (E_d) and another pointing downwards (E_u). Spherical irradiance sensors are described as scalar sensors, which measures irradiance (E_o) from essentially all directions. Irradiance measured with a scalar sensor is the best representation of light available to support phytoplankton productivity. The irradiance sensors are quantum sensors; they have nearly ideal quantum response (Figure 7.1). Note the associated units of measurements made with quantum irradiance sensors (Table 7.1). These sensors respond essentially equally to all photons in the PAR interval (i.e., regardless of the wavelength). This is desirable in studies concerned with plant production, because once photons are absorbed by a plant they are of equal energetic value to the plant,

regardless of their wavelength within the PAR range (Kirk 1983). Figure 7.1 compares the relative spectral response of the Protomatic photometer (the first instument used (1978)), to that of a quantum sensor (used since 1980). Note that the quantum sensor approaches ideal quantum response much more closely than the photometer. A spectroradiometer measures the spectral character of penetrating irradiance (i.e., as a function of wavelength), and thereby the influence of selective spectral attenuation by the various absorbing components.

Turbidity (T, NTU), as measured with a nephelometric turbidimeter, is a good estimator of b (Effler 1988; Kirk 1981b). Turbidity in a lake or reservoir is caused by a heterogeneous population of suspended particles, which may include clay, silt, finely divided organic and inorganic matter, phytoplankton, and

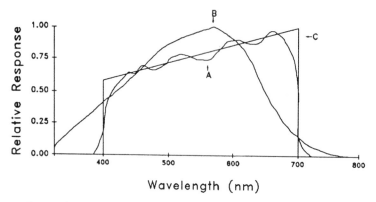

FIGURE 7.1. Comparison of relative spectral responses. A: response of LI-COR quantum sensor; B: response of Protomatic underwater photometer; C: ideal PAR quantum response.

other microscopic organisms. In general, this population of particles is a composite of sediments received as tributary inputs, resuspended lake deposits, and particles produced within the watercolumn of the lake (e.g., phytoplankton and $CaCO_3$ particles). The relationship between b and T has been described by Effler (1988) and Weidemann and Bannister (1986), as

$$T = \alpha \cdot b \qquad (7.1)$$

in which α = a constant (NTU · m). The value of α (usually represented as the mean determined from at least several paired values of T and b) has been found to fall in a rather narrow range of 0.80–1.27 (DiToro 1978; Effler 1988; Kirk 1981b; Weidemann and Bannister 1986). Furthermore, turbidity measurements of samples before (T) and after acidification (T_{nc}, NTU; pH = 4.3) have been used to partition b in hardwater lakes according to contributions of $CaCO_3$ particles versus all other particles (Effler et al. 1987a, 1991a; Weidemann et al. 1985). The acidification treatment has been found to selectively dissolve $CaCO_3$ particles in hard water lakes (Effler and Johnson 1987; Vanderploeg et al. 1987). The difference between the turbidities of the untreated (T) and treated (T_{nc}) sample is designated the $CaCO_3$ turbidity (T_c, NTU).

Chlorophyll a is the major absorbing pigment of phytoplankton. Phaeophytin, a chlorophyll degradation product, has similar light absorption characteristics (Tilzer 1983). Total chlorophyll (sum of chlorophylls a, b, and c; C_T); measured according to the method of Parsons et al. (1984), which includes (but does not differentiate) phaeopigments, has been adopted here (see Chapter 6) as the measure of the concentration of light absorbing phytoplankton pigments. Phaeopigments have not been found to contribute significantly to the pigment content of the watercolumn of Onondaga Lake (see Chapter 6).

Gelbstoff, often referred to as "yellow stuff" or humic substances, is composed of diverse organic materials that include humic and fulvic acids. These materials absorb strongly in the blue wavelength range (Kirk 1983). Gelbstoff has autochthonous (associated with phytoplankton production (see Bricaud et al. 1981)) and allochthonous (see Gjessing 1976) origins. In highly colored waters, most of the gelbstoff is received from the watershed, for example by extraction of water-soluble humic substances by rainfall, with subsequent delivery by tributary inputs (Kirk 1983). The light absorption properties of gelbstoff are determined rather easily by measuring the absorption spectrum of a filtered (e.g., 0.45 µm) water sample in long (e.g., 5 or 10 cm) pathlength cells (Table 7.1).

Phytoplankton and tripton are responsible for particle absorption. From a management perspective, it is important to partition the relative contributions of these two groups of particles, as they generally have different remedial strategies (e.g., nutrient loading versus erosion control). The light absorption properties of particles can be assessed by measuring the absorption spectrum of a suspension of concentrated particles (see Kirk 1985; Weidemann and Bannister 1986). This approach has been pursued in only a qualitative fashion here, by scanning suspensions contained in an opal glass cuvette (Table 7.1). Light absorption properties of particles were estimated based on determinations of spectral attenuation and optical theory.

Individual particle analysis (IPA) (see Table 7.1 and Chapters 3 and 5), by scanning electron microscopy equipped with automated x-ray analysis (SAX), provides detailed morphometric and elemental characterization of individual particles. IPA offers the opportunity to approximately partition scattering according to chemical groups and related origins (see Effler et al. 1991a; Weidemann et al. 1985). Particles are classified by computer according to chemical sorting criteria established by the user (see Johnson et al. 1991).

7.3 Optical Properties of Water

7.3.1 Basic Relationships

The rate of diminution of downward irradiance with respect to depth at z meters ($E_{d(z)}$, $\mu E \cdot m^{-2} \cdot s^{-1}$) is expressed in terms of the vertical attenuation coefficient ($K_{d(z)}$), which is defined by

$$K_{d(z)} = -\frac{1}{E_{d(z)}}\frac{dE_{d(z)}}{dz} = -\frac{d \ln E_{d(z)}}{dz}. \quad (7.2)$$

Even for monochromatic light, $K_{d(z)}$ changes slightly with depth because of changes in the angular distribution of the light field. For the PAR interval, vertical variations are also introduced by spectral shifts in irradiance with depth (Kirk 1983). These changes with depth are usually relatively minor for systems with high K_d values. It is common practice to ascribe an average value, K_d, to $K_{d(z)}$ for a portion of the, or the entire, illuminated interval of the water column. The diminution of downward irradiance with depth can then be described

by the following relationship, sometimes referred to as the Beer–Lambert law

$$E_{d(z)} = E_{d(o)}e^{-K_d \cdot z} \quad (7.3)$$

in which $E_{d(o)}$ = downward irradiance just below the surface ($\mu E \cdot m^{-2} \cdot s^{-1}$). The relationship may be applied over any illuminated depth interval. The symbol K_d is utilized here for the PAR interval; for monochromatic light, $K_{d(\lambda)}$ is used. Operationally, the value of K_d is determined as the slope of the regression of the natural logarithm of E_d on z, usually in the depth region of the 10% light level ($z_{0.1}$; i.e., $K_{d(z_m)}$ (Kirk 1981a)).

Kirk (1981a) developed the following expression based on a Monte Carlo analysis of the underwater light field to partition K_d at $z_{0.1}$ in turbid waters according to a and b, for vertically incident light.

$$K_d = (a^2 + 0.256ab)^{0.5} \quad (7.4a)$$

However, the value of K_d, an apparent optical parameter, depends on the angle of incidence of photons at the surface, and therefore time of day, time of year, and location. Equation (7.4a) was modified to accommodate the angle of incidence (Kirk 1984a), according to the following expression

$$K_d = \frac{1}{\mu_o}[a^2 + (0.473\mu_o - 0.218)ab]^{0.5} \quad (7.4b)$$

in which μ_o = cosine of incident photons just below the surface. The adjustment for nonsolar noon conditions becomes progressively more important as the relative contribution of a to K_d becomes greater (i.e., ratio b/a decreases). Equation (7.4b) can be used to normalize K_d measurements to solar noon (μ_o = 1.0) conditions.

Despite the limitations of the SD measurement (e.g., dependence on wave action, incident light conditions, observer, features of the disc (Priesendorfer 1986)), it continues to be the most widely used measure of the extent of light penetration. A simple inverse relationship has often been assumed (e.g., Poole and Atkins 1929; Beeton 1957) to exist between K_d and SD (i.e., $K_d \cdot SD$ = constant). However, it has been demonstrated that rather strong differences in the $K_d \cdot SD$ product occur between different lakes and with time in individual systems (Effler 1985). Review of theoretical treatments of the optics of the Secchi disc (e.g., Priesendorfer 1986; Tyler 1968) indicates there is no reason to expect constancy in the $K_d \cdot SD$ product. Based on contrast transmittance theory, it has been demonstrated that SD is inversely proportional to the sum of the beam attenuation (c, m^{-1}) and the vertical diffuse attenuation coefficients (Priesendorfer 1986; Tyler 1968)

$$SD = \frac{N}{K_d + c} \qquad (7.5)$$

in which N = a quasiconstant (dimensionless). Different values of N have been published (e.g., Holmes 1970; Tyler 1968). The value of N is now understood to vary somewhat, as it depends on certain uncontrollable conditions (e.g., wave action, cloud cover) that prevail during measurement. For example, Priesendorfer (1986) gives the range of N to be 8.0 to 9.6.

Two different parameters are used here to describe the angular distribution of the underwater light field: irradiance reflectance ($R_{(z)}$) and average cosine ($\bar{\mu}_{(z)}$). The value of $R_{(z)}$ is calculated as the ratio of the upward to the downward irradiance

$$R_{(z)} = \frac{E_{u(z)}}{E_{d(z)}} \qquad (7.6)$$

in which $E_{u(z)}$ = upward irradiance ($\mu E \cdot m^{-2} \cdot s^{-1}$) at depth z. Paired determinations of K_d and $R_{(z)}$ can be made from paired profiles of $E_{d(z)}$ and $E_{u(z)}$. Average cosine is the ratio of net downward irradiance and scalar irradiance

$$\bar{\mu}_{(z)} = \frac{E_{d(z)} - E_{u(z)}}{E_{o(z)}} \qquad (7.7)$$

in which $E_{o(z)}$ = scalar irradiance ($\mu E \cdot m^{-2} \cdot s^{-1}$) at depth z. Paired determinations of $\bar{\mu}_{(z)}$, $R_{(z)}$ and K_d can be made from simultaneous profiles of $E_{d(z)}$, $E_{u(z)}$, and $E_{o(z)}$, that is, with two cosine sensors and one scalar sensor. Depth profiles of $R_{(z)}$ and $\bar{\mu}_{(z)}$ demonstrate curvilinear distributions because the solar beam becomes more diffuse as it penetrates

further into water. With depth, the value of $R_{(z)}$ increases and the value of $\bar{\mu}_{(z)}$ decreases. The values level off (asymptotes are approached) once the asymptotic radiance (final angular) distribution is reached; usually in the vicinity of $z_{0.1}$. The asymptotic values of $R_{(z)}$ and $\bar{\mu}_{(z)}$ are designated R and $\bar{\mu}$, respectively. Both parameters are measures of how diffuse the light field is within the water. Higher R and lower $\bar{\mu}$ values correspond to more diffuse light conditions that result from greater relative contributions by scattering to overall attenuation.

Kirk (1981b) developed functions, based on a Monte Carlo analysis of the underwater light field, to obtain the ratio b/a and $\bar{\mu}$ from measurements of R at $z_{0.1}$ (or $z_{0.01}$). The value of a can then be calculated according to the Gershun–Jerlov equation (Jerlov 1976)

$$a = \bar{\mu} \cdot K_E \qquad (7.8)$$

in which K_E = attenuation coefficient for net downwelling irradiance ($E_d - E_u$). The value of b is calculated by

$$b = a \cdot (b/a) \qquad (7.9)$$

The values of a and b are spectral average values for the PAR interval, if quantum sensors are used to determine K_d and R. Subsequently, Kirk (1989) developed an alternate method to estimate b from paired K_d and R determinations; it involves three steps specified by the following relationships

$$b_{bd} \simeq 3.5R \cdot K_d, \qquad (7.10a)$$

$$b_b = b_{bd} \cdot \frac{1}{f(R)}, \text{ and} \qquad (7.10b)$$

$$b \simeq 53b_b \qquad (7.10c)$$

in which: b_{bd} = diffuse backscattering coefficient for downwelling light at $z_{0.1}$ (m^{-1}), b_b = backscattering (normal) coefficient (m^{-1}), and $f(R)$ = function describing relationship between the ratio b_{bd}/b_b and R (presented in Figure 7.8 of Kirk (1989); = $244.3R^3 - 146.1R^2 + 36.0R + 1.11$, according to our fit of those data). The relationship presented in Equation 7.10c assumes the same volume scattering function utilized by Kirk (1981a, 1981b) previously and found to be generally representative for most systems. Equations (7.9) and (7.10) represent two different ways to estimate b from paired determinations of K_d and R that Kirk (1989) has demonstrated yield similar results. The estimate(s) of b can also be checked against measurements of T (Weidemann and Bannister 1986; Effler et al. 1991a, 1991b). The value $\bar{\mu}$, and thereby a (equation (8)), can be checked against direct estimates of $\bar{\mu}$ using three quantum (two cosine, one scalar; equation (7)) sensors (e.g., Weidemann and Bannister 1986; Effler et al. 1991a, 1991b).

7.3.2 Partitioning a and b

The components of a and b are partitioned according to simple summations. A partitioning for a that can be supported by available analytical techniques (Kirk 1983; Weidemann and Bannister 1986) is

$$a = a_w + a_y + a_p \qquad (7.11)$$

in which a_w, a_y, and a_p = spectral average absorption coefficients (m^{-1}) for water, gelbstoff, and particles, respectively. Absorption by water, $a_{w(\lambda)}$, is known (Smith and Baker 1981). Absorption by gelbstoff $a_{y(\lambda)}$ is rather easily measured (Kirk 1976; Davies-Colley and Vant 1987). Absorption by particles $(a_{p(\lambda)})$ is more tenuous, mainly because of problems in efficient concentration of the particles. However, a number of investigators have recently reported such data (e.g., Bricaud and Stramski 1990; Kiefer and SooHoo 1982; Weidemann and Bannister 1986). The spectral average value, a_p, has been determined by difference (i.e., $a_p = a - (a_w + a_y)$; e.g., Effler et al. 1991a, 1991b). Absorption coefficients for PAR for the various components can be calculated from the respective absorption spectral and spectral irradiance $(E_{d(\lambda)})$ data from $z_{0.1}$, according to (Weidemann and Bannister 1986)

$$a_x = \frac{\int_{400}^{700} E_{d(\lambda)} a_{x(\lambda)} d\lambda}{\int_{400}^{700} E_{d(\lambda)} d\lambda} \qquad (7.12)$$

This equation weights the spectral absorption coefficients according to the irradiance spectrum at $z_{0.1}$ to obtain values of a_x that can be directly compared to a for PAR determined at $z_{0.1}$. The a_p component can be further partitioned between phytoplankton and tripton, according to

$$a_p = a_p' + K_a \cdot C \qquad (7.13)$$

in which a_p' = spectral average absorption coefficient for tripton (m^{-1}), K_a = chlorophyll-specific absorption coefficient $(m^2 \cdot mg^{-1}$ chlorophyll), and C = concentration of chlorophyll a $(mg \cdot m^{-3})$. A typical value for K_a is $0.010 \, m^2 \cdot mg^{-1}$ chlorophyll (Bannister and Weidemann 1984). Only recently have these particulate components been subject to separation by laboratory procedures (Bricaud and Stramski 1990).

In hard water lakes b has been partitioned according to (Effler et al. 1987a, 1991b; Weidemann et al. 1985)

$$b = b_{nc} + b_c \qquad (7.14)$$

in which b_{nc} = spectral average scattering coefficient for noncalcite particles (m^{-1}), and b_c = spectral average scattering coefficient for $CaCO_3$ particles. The partitioning can be supported by turbidity measurements (Effler et al. 1987a, 1991b; Weidemann et al. 1985). Partitioning of b_{nc} is more tenuous; separation of the phytoplankton and nonphytoplankton components is described by

$$b_{nc} = b' + K_b \cdot C \qquad (7.15)$$

in which b' = spectral average scattering coefficient for nonphytoplankton particles other than $CaCO_3$, and K_b = chlorophyll-specific scattering coefficient $(m^2 \cdot mg^{-1}$ chlorophyll). Weidemann and Bannister (1986) found the value of K_b to be usually in the range 0.05 to $0.12 \, m^2 \cdot mg^{-1}$ chlorophyll. Alternatively, a more complete resolution of the components of scattering can be supported through individual particle analysis (IPA) techniques. Both approaches are applied and evaluated here in an effort to partition attenuating components.

7.4 Historic Changes in Apparent Optical Properties

7.4.1 Secchi Disc Transparency

7.4.1.1 South Basin

Summary statistics for the 23-year SD record for the May–September interval for the south deep station of Onondaga Lake are presented in Figure 7.2. The May–September interval represents the principal water recreation period for this region. The population of Figure 7.2 includes more than 550 observations; however, the observations are not evenly distributed over the period of record (e.g., one observation in 1971, 71 observations in 1988). There were 10, or fewer, observations per year in the 1970–1977 interval, and again in 1983 and 1984. The occurrence of distinct maxima, described as "clearing events" (sudden and dramatic increases in SD (Auer et al. 1990)), manifested in the observations for a number of years (e.g., 1970, 1977, 1987–1990), can greatly skew such summary statistics if the number of observations is too small. For example, the distinctly higher mean values of SD in 1970 (n = 6) and 1977 (n = 10) are largely a result of the overrepresentation of "clearing events" in the populations of both years. Similarly, the means of 1974 and 1975 presented in Figure 7.2 may understate average condi-

tions, as the measurements may not have been frequent enough to capture such events in the respective data sets. In contrast, we have a high degree of confidence in the uniformly low SD conditions depicted for 1980, 1981, 1985, and 1986 (Figure 7.2) because of the large number of observations made in each year (n ≥ 30). Similarly we have great confidence in the highly dynamic conditions implied for the 1987–1990 interval because of the frequent measurements made in each year (n ≥ 29).

Important changes in the clarity of the lake emerge from the review of the data set of Figure 7.2, despite the uneven temporal coverage among the years. "Clearing events" occurred in most years from the late 1960s through the late 1970s. These events appear to have been absent in the 1980–1986 interval; SD was more uniformly low in this period. Since 1987 "clearing events" have been observed annually and with greater intensity; the maximum in each of these years exceeds all earlier observations (Figure 7.2). Zooplankton grazing plays an important role in promoting the occurrence of clearing events (Chapter 6). Average SD values were the lowest in the 1980–1986 interval; the highest mean values have occurred most recently.

The distribution of SD values for the 23-year period of record for the south deep station is presented for two time segments, 1968–1986 and 1987–1990, in Figure 7.3 (a and b). Particular attention should be paid to the position of the two distributions with respect to the SD value of 1.2 m, as that is the regulated (New York State Department of Health) minimum for opening a bathing beach in the State of New York. Clearly a substantial improvement in SD has occurred since 1987. The criterion was only met for about 20% of the summer observations in the south basin in the 1968–1986 interval (Figure 7.3a), compared to 50% of the observations since 1987 (Figure 7.3b). The distribution of SD became much broader over the 1987–1990 interval. This distribution has a bimodal character compared to the earlier distribution. The primary mode is centered around an SD value of between 0.9 and 1.2 m. The secondary

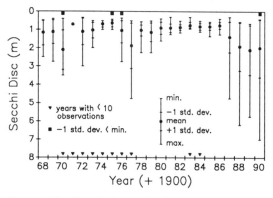

FIGURE 7.2. Trends in Secchi disc transparency in Onondaga Lake, for the May–September interval (1968–1990).

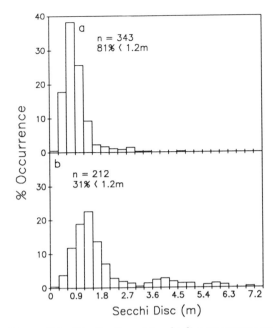

and early July. The SD maxima were 5.75 m
for both events. Relatively uniform SD values
persisted after these events in both years (Figure 7.4c and d). Note, however, that the late
summer (postevent) values in both 1987 (Figure 7.4c) and 1989 (Figure 7.4d) exceeded the
SD values observed for most of the monitoring
periods of 1978 (Figure 7.4a) and 1985 (Figure
7.4b). The dynamics in light attenuating materials responsible for the strong temporal vari-

FIGURE 7.3. Distribution of Secchi disc transparency
values for the south deep station of Onondaga Lake
for two different periods: (a) 1968–1986, and (b)
1987–1990.

mode is very broad; it corresponds to "clearing
event" conditions. Note that the recent "clearing events" represent a much greater fraction
of the population than the earlier events.

Detailed time plots of SD are presented
for the south basin for four selected years in
Figure 7.4a–d. Clarity was rather uniformly
low in 1978 and 1985 (Figure 7.4a and b).
Values were comparatively elevated in early
May in 1978. The peak for the year (SD =
2.25) was observed in this period; a secondary
maximum was observed in mid-August (Figure 7.4a). Clarity values were often even lower in 1985, the range was narrower, and the
maximum was only 1.1 m (Figure 7.4b). Clarity was low in 1987 through June (Figure
7.4c). A "clearing event" (note the abrupt
increase) developed in early July and lasted
most of the remainder of the month. The peak
SD value was 4.75 m. In contrast, two "clearing events" occurred in 1989 (Figure 7.4d).
The first event started in late May and lasted
until mid-June. The second event was much
shorter (~10 days); it occurred in late June

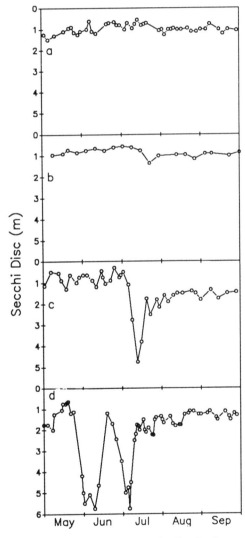

FIGURE 7.4. Detailed temporal distributions of
Secchi disc transparency for the south deep station
of Onondaga Lake for four selected years: (a) 1978,
(b) 1985, (c) 1987, and (d) 1989.

ations in SD in recent years are identified subsequently in this chapter.

7.4.1.2 Spatial Variability in Clarity

We evaluate spatial variability in SD here by comparing paired measurements made at the south and north deep stations. This variability is represented as the distribution of the ratio of the SD value at the north station (SD_N) to that of the south station (SD_S). Three different intervals for the period of record are considered, 1968–1970, 1978, and 1987–1989 (Figure 7.5a–c). More than 50% of the paired measurements in the 1968–1970 interval

were within 15% of each other (a reasonable upper bound for the precision of SD measurements). However, the population is somewhat skewed to ratios >1 (Figure 7.5a); the ratio was greater than 1.45 for almost 25% of the paired observations. The distribution in 1978 was narrower and more centrally positioned around a ratio value of 1 (Figure 7.5b), indicating that there was less overall difference between the sites than in the earlier (Figure 7.5a) measurements. However, the clarity at the north station still tended to be somewhat higher than at the south station; SD was lower at the north station for 11% of the observations, they were equal for 16% of the observations, and SD was lower at the south station for 73% of the observations. However, SD_N/SD_S was greater than 1.45 for only 2 of 45 paired observations. The distribution of the SD_N/SD_S ratio was the narrowest and most nearly equivalent for the two sites for the most recent period, 1987–1989 (Figure 7.5c). Seventy-five percent of the paired observations were within 15% of each other. The population was equally distributed between the two cases, $SD_N > SD_S$ and $SD_N < SD_S$.

Substantial spatial variation in SD has occurred in Onondaga Lake, though there are indications that substantial differences have become more infrequent in recent years (Figure 7.5a–c). In the late 1960s and in 1978, there was a clear tendency for the north station to have greater clarity than the south station. However, by the late 1980s, the clarity at these two sites had become nearly equivalent (Figure 7.5c). An analysis of clarity data (cooperative program, NYSDEC, and Onondaga County Health Department) for 10 stations on Onondaga Lake by Walker (1987) indicates clarity in the most probable future swimming area (northeast corner of the lake) is not significantly different from deep water zones of the lake.

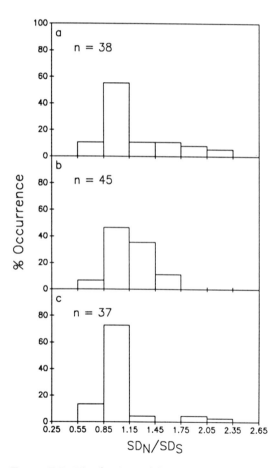

FIGURE 7.5. Distributions of the ratio of Secchi disc transparency values from the north and south deep stations of Onondaga Lake for three different periods: (a) 1968–1970, (b) 1978, and (c) 1987–1989.

7.4.2 Attenuation Coefficient (K_d)

The vertical distribution of E_d is presented for three different profiles as semilog plots in Figure 7.6a–c. They represent three distinct

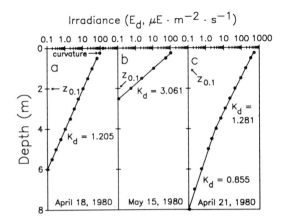

Irradiance (E_d, $\mu E \cdot m^{-2} \cdot s^{-1}$)

FIGURE 7.6. Profiles of downward irradiance (E_d) from the south deep station Onondaga Lake for three selected dates in 1980: (**a**) April 18, (**b**) May 15, and (**c**) April 21.

cases observed in Onondaga Lake that are common in the vast majority of systems. The value of E_d in all cases decreases progressively with depth. In the first case (Figure 7.6a), minor curvature is observed in the uppermost layers but linearity is established substantially above $Z_{0.1}$. The modest curvature and higher slope is largely associated with the spectral narrowing that occurs in the uppermost layers. This change in the rate of attenuation of PAR with depth is observed in most marine waters and in lakes with a relatively high degree of light penetration (Kirk 1983), and thus in Onondaga Lake only infrequently. The Beer–Lambert law (Equation (7.2)) still represents a good approximation for quantification of the diminuation of irradiance for this case (Figure 7.6a). The second case (Figure 7.6b) is the most commonly observed in Onondaga Lake. Here the ln E_d versus z relationship is more nearly linear throughout the depth range, though anomalous measurements associated with precision limitations in the uppermost waters are commonly encountered. The divergence of this case from the first case is due to the counteracting effects of the downward flux becoming more diffuse (Kirk 1983) and the much shallower depth of light penetration. The third case is observed only infrequently during calm periods (e.g., low turbulence), when the concentrations of light attenuating

components, particularly phytoplankton, are high. Distinctly different values of K_d are obtained in two, and sometimes more, layers (Figure 7.6c), corresponding to stratified concentrations of attenuating components. Semilog plots of E_u, E_d, and E_o are parallel for the lake, and K_d essentially equals K_E, as observed elsewhere (Kirk 1983; Weidemann and Bannister 1986).

Measurements to support the estimation of K_d are available for the south deep station for 10 years of the 1978 through 1990 interval. Photometer measurements were the only basis of K_d estimates (Table 7.1) for two years, 1978 and 1979. A strictly empirical approach was taken in evaluating the extent to which photometer-based estimates tracked those obtained with a PAR sensor irradiometer configuration. Paired estimates of K_d obtained with the two different instruments in 1980 and 1981 are compared in Figure 7.7. Despite the fundamental differences in these two instruments (e.g., spectral response (Figure 7.1)), results obtained were quite similar. Thirty-two of the 88 paired estimates of K_d were within 10%; 74 of the 88 paired estimates of K_d were within 20%. The best fit linear regression expression between the two

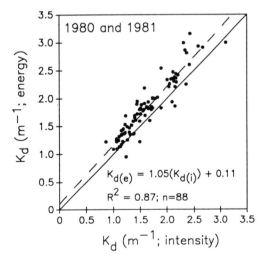

FIGURE 7.7. Comparison of K_d values determined from paired measurements with a photometer (K_d; intensity) and an irradiometer (K_d; energy) in Onondaga Lake in 1980 and 1981.

estimates of K_d is quite strong ($R^2 = 0.87$) and reflects a near equivalence (e.g., slope = 1.05). Because of this near equivalence we have included the photometer estimates of 1978 and 1979 in the overall historic K_d data set.

Figure 7.8 presents the distribution of K_d values for the 10-year period of record for two different time segments, 1978–1986 (1978–1981, 1985, and 1986; 6 years) and 1987–1990. The 1978 and 1979 determinations were converted to irradiometer estimates according to the relationship of Figure 7.7. Clearly a major shift to lower K_d values has occurred since 1987. Values of K_d most often (nearly 65% of the observations) fell in the interval 1.25 to 2.0 m^{-1} in the earlier period (Figure 7.8a). In the more recent four-year period, K_d values most often (>50% of the observations) were in the interval 0.75 to 1.25 m^{-1} (Figure 7.8b). Nearly 20% of the estimates of K_d since 1987 have been less than 0.75 m^{-1}, compared to approximately 1% for the 1978–1986 interval (Figure 7.8a and b).

Detailed time plots of K_d are presented for the April–October interval of each of the 10

years in Figure 7.9a–j. Estimates of K_d based on photometer readings are presented (Figure 7.9a and b). Normalization of K_d data to solar noon conditions reduced the measured values on average by only 4.5%, with little variation, because of the uniformity of the time of measurement in the data base (usually 1000 to 1200 hr). For these reasons, measured values are presented throughout this chapter. Selected summary statistics for each year are presented in Table 7.2. The most uniform conditions were observed in 1978 (Figure 7.9a). Rather distinct peaks in K_d, which were observed in the May–early June interval of all the other years, except 1990, were associated with the most intense phtoplankton bloom experienced annually. Abrupt changes in K_d have been common (Figure 7.9). Particularly abrupt changes in K_d occurred in 1987, 1988, and 1989 through mid-summer, associated with the clearing events (Figure 7.9 g–j). The late summer K_d values in these three years were distinctly lower than in earlier years. The late summer-early fall peak in K_d in 1990 (Figure 7.9j) was associated with a blue–green algae bloom. Maxima above 3.0 m^{-1} were documented in four different years (Table 7.2). Minima have been distinctly lower since 1987. The mean and median values since 1987 have been generally lower than in the earlier six years (Table 7.2). The highest mean and median values of K_d were observed in 1980, 1981, and 1986 (Table 7.2). Materials responsible for the dynamics observed in K_d (Figure 7.9) are evaluated subsequently in this chapter.

7.4.3 Relationship between K_d and SD

The relationship between K_d and SD in Onondaga Lake, based on 273 paired observations made in 10 years, is evaluated in Figure 7.10a. The slope of the linear least squares regression relationship (forced through zero) is equal to the product $K_d \cdot SD$. A commonly invoked value for the product, in the absence of paired measurements, has been 1.9, obtained by Beeton (1957) for Lake Huron. Poole and Atkins (1929) proposed a value of 1.7. A

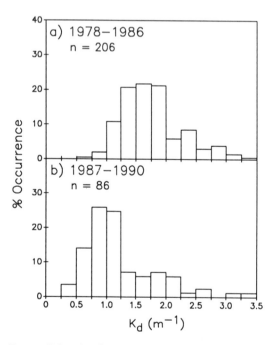

FIGURE 7.8. Distribution of K_d values for the south deep station of Onondaga Lake for two different periods: (a) 1978–1986 and (b) 1987–1990.

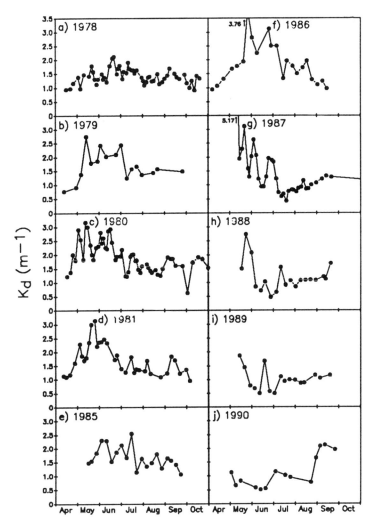

FIGURE 7.9. Temporal distributions of K_d for the south deep station of Onondaga Lake for ten years: (a) 1978, (b) 1979, (c) 1980, (d) 1981, (e) 1985, (f) 1986, (g) 1987, (h) 1988, (i) 1989, and (j) 1990.

TABLE 7.2. Selected summary statistics for K_d in Onondaga Lake for ten different years.

Statistic	Year									
	1978*	1979*	1980	1981	1985	1986	1987	1988	1989	1990
n	55	17	58	34	20	23	36	19	17	14
$\bar{x}$	1.40	1.69	1.87	1.72	1.67	1.80	1.41	1.09	1.01	1.14
$SD/\bar{x}$	0.19	0.31	0.28	0.32	0.23	0.40	0.62	0.47	0.41	0.49
MIN	0.91	0.77	0.63	0.95	1.06	0.93	0.42	0.48	0.50	0.51
MAX	2.12	2.73	3.16	3.13	2.55	3.76	5.17	2.74	1.87	2.12
MEDIAN	1.37	1.57	1.83	1.69	1.56	1.72	1.18	1.09	0.98	0.95

*Based on photometer measurements

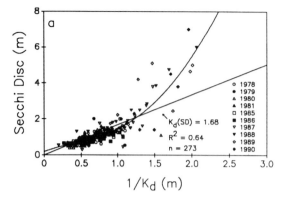

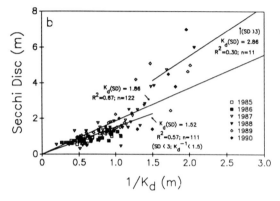

FIGURE 7.10. Evaluation of the relationship between K_d and SD in Onondaga Lake: (**a**) 10 years of data over the interval 1978–1990, with linear least squares and polynomial best fit expressions, and (**b**) 1985–1990 interval, with linear least squares regression relationships.

number of lakes have been found to have $K_d \cdot SD$ products that differ substantially from 1.9, and that vary strongly with time (Effler 1985; Effler et al. 1988). Variations in the product reflect changes in the relative contributions of a and b to attenuation (Effler 1985). For example, an increase in the value of the product indicates a decrease in the relative importance of the scattering process in attenuation (Effler 1985; Effler et al. 1988). Indeed, tracking temporal variations in $K_d \cdot SD$ can have diagnostic value in identifying shifts in attenuating processes (Effler 1985; Effler et al. 1988).

Clearly, invoking uniformity in this product over the entire range of light penetration conditions that have been documented for

Onondaga Lake in 10 years, or in recent individual years, is inappropriate (Figure 7.10a; also see Perkins and Effler 1995). The best fit linear least squares regression fit is not strong (R = 0.64), but more importantly, it systematically fails throughout much of the range of observations. It overestimates the $K_d \cdot SD$ product in the most commonly observed range of conditions (e.g., $1\,m^{-1} < K_d < 2\,m^{-1}$), and seriously underestimates the product during the clearing events encountered in the 1987–1990 interval (Figure 7.10a). This implies a major decrease in the relative contribution of scattering during the clearing events. A polynomial expression (SD = $-0.170\,(K_d^{-1})^4$ + $1.301\,(K_d^{-1})^3$ − $1.359\,(K_d^{-1})^2$ + $1.506\,(K_d^{-1}) + 0.193$; Figure 7.10a) provides a substantially improved fit ($R^2 = 0.89$) of the data base.

The last six years of the database are analyzed again in Figure 7.10b within the framework of the $K_d \cdot SD$ product, as this subset is supported by paired estimates of a and b and can thus be tested alternatively within the mechanistic framework of contrast transmittance theory (Equation (7.5); Priesendorfer 1986; Tyler 1968; subsequent treatment in this chapter). Note that a linear relationship for the entire range of observations remains inappropriate, and that the associated $K_d \cdot SD$ product value (1.86) is higher than that obtained for the 10-year data set (Figure 7.10a), probably as a result of the greater contribution clearing events have made to the more recent observations. A two-part linear least squares regression fit is also presented that better represents the disparate conditions that prevail at low and high levels of light penetration (Figure 7.10b). Note the substantially higher $K_d \cdot SD$ value (2.86) associated with the clearing events. The efficient grazing of abiotic particles, as well as phytoplankton, by nonselective zooplankton (see Chapter 6) leading up to and during these events probably contributes to these conditions (e.g., Vanderploeg et al. 1987).

Statistics describing the distributions of the values of the $K_d \cdot SD$ product in each of the 10 years of record are presented in Table 7.3. The mean and median values were higher over the

TABLE 7.3. The $K_d \cdot SD$ product for Onondaga Lake for individual years.

Year	$K_d \cdot SD$			n
	Mean	Median	Range	
1978	1.38	1.38	0.80–1.94	54
1979	1.39	1.24	0.50–2.96	14
1980	1.50	1.47	0.69–2.21	52
1981	1.51	1.45	1.08–2.29	31
1985	1.40	1.40	1.00–1.91	22
1986	1.42	1.34	0.83–2.45	22
1987	1.62	1.59	0.42–2.98	34
1988	1.70	1.51	1.20–2.90	18
1989	1.93	1.72	1.27–3.47	17
1990	1.81	1.88	0.83–3.57	14

1987–1990 period, since the regular occurrence of clearing events. The lowest mean values (1.38 and 1.39) were observed in 1978 and 1979; the highest (1.93) in 1989.

7.5 Components of Attenuation/Evaluation of Empirical Relationships

7.5.1 Turbidity

7.5.1.1 Distributions of T and T_c

The value of T is often assumed to be correlated to the concentration of suspended solids (SS, mg $\cdot$ L^{-1}). A limited number of paired observations (n = 13) made in 1990 supports this assumption for the upper waters of Onondaga Lake. The best fit linear least squares regression fit between these parameters was T = $(0.63 \cdot SS) - 0.02$; according to this expression, variations in SS explained 87% of the variability observed in T.

The temporal distributions of T and T_c at a depth of 1 m at the south deep station are presented for the April through October interval for the six-year period of record in Figure 7.11. Selected statistics describing these distributions are presented in Table 7.4. The distributions and magnitude of T_c indicate CaCO$_3$ is sometimes a significant component of T. The conditions reported for 1 m are generally representative of those found at $z_{0.1}$. Strong temporal variability in T and T_c was observed in

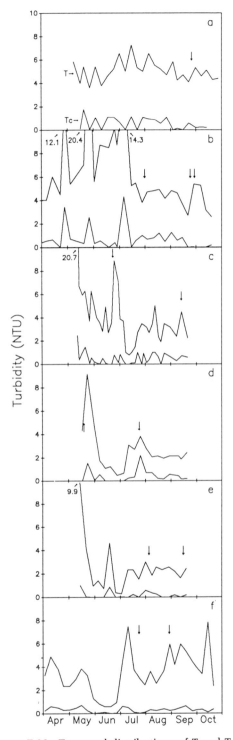

FIGURE 7.11. Temporal distributions of T and T_c at a depth of 1 m at the south deep station in Onondaga Lake, for the April–October interval of six years: (a) 1985, (b) 1986, (c) 1987, (d) 1988, (e) 1989, and (f) 1990. Occurrences of major (>1.5 in/24 hrs) rainfall events are indicated.

TABLE 7.4. Summary statistics for T, T_{nc}, and T_c for the south deep station of Onondaga Lake, 1985–1990.

Year		Turbidity (NTU)				n
		$\bar{x}$	σ	min	max	
1985	T	5.0	0.9	3.6	7.2	26
	T_{nc}	4.5	0.8	3.5	6.1	24
	T_c	0.5	0.5	0.0	1.7	24
	%Tc	10.1	9.3	0.0	32.1	24
1986	T	6.4	3.8	2.6	20.4	29
	T_{nc}	5.7	3.2	2.4	17.8	29
	T_c	0.8	1.0	0.0	4.3	28
	%Tc	11.2	4.4	0.0	30.2	28
1987	T	4.2	3.3	0.8	20.6	35
	T_{nc}	3.8	3.0	0.8	18.2	35
	T_c	0.5	0.5	0.0	2.5	35
	%Tc	11.9	10.5	0.0	37.4	35
1988	T	2.7	1.9	0.5	9.2	19
	T_{nc}	2.4	1.7	0.5	8.0	19
	T_c	0.2	0.5	0.0	1.8	19
	%Tc	5.7	11.1	0.0	47.6	19
1989	T	2.5	2.0	0.3	9.9	19
	T_{nc}	2.4	1.8	0.2	8.9	19
	T_c	0.2	0.3	0.0	1.0	19
	%Tc	8.0	7.0	0.0	19.1	19
1990	T	3.5	1.8	0.6	7.8	30
	T_{nc}	3.2	1.7	0.5	7.7	30
	T_c	0.3	0.2	0.0	0.7	30
	%Tc	8.6	5.2	0.0	18.6	30

The effect of the rainfall/runoff events can be direct, associated with the influx of large quantities of terrigenous turbidity, or indirect, through stimulation of phytoplankton growth and autochthonous precipitation of minerals (Johnson et al. 1991). Note that in several instances T_c also increased substantially after these events. It has been hypothesized that this may be in response to increases in appropriate nucleation surfaces for $CaCO_3$ precipitation provided from tributaries during these events (Johnson et al. 1991; Chapter 5). Peaks in T_c and T have coincided in a number of instances (e.g., three maxima in 1986 and two maxima in 1988). However, only during late July of 1988 (Figure 7.11d) could T_c be described as the principal component of T.

Vertical profiles of T and T_c within the epilimnion are presented for two selected days in Figure 7.12a and b, which exemplify the two most common cases for the lake. In the first case (Figure 7.12a) T and T_c are approximately uniformly distributed throughout the epilimnion. In the second, less common, case there is substantial stratification of turbidity within the epilimnion (Figure 7.12b). However, based on the position of $z_{0.1}$ in the stratified case and utilizing the homogeneity criterion of Weidemann and Bannister (1986) (T changes <20% from surface to $z_{0.1}$), this

each year. The most uniform conditions were observed in 1985 (Figure 7.11a). The highest maxima were observed in 1986 (Figure 7.11b) and 1987 (Figure 7.11c). The average T and T_c of Onondaga Lake has decreased since 1986 (Table 4). The lowest average T was observed in 1989 (2.5 NTU). The average value of T_c was also the lowest in that year (0.2 NTU, Table 4). The average contribution of T_c to T over the six-year period was about 10%. The seasonal minima in T have decreased greatly since 1986, consistent with the dramatic clearing events that were observed (e.g., Figure 7.11c and d).

The occurrences of rainfall events of >1.5 in. within 24 hr (as measured at Hancock Airport) are included as indicators of the potential contribution of allocthonous inputs to the distributions. Abrupt increases in T were observed immediately following most of these events.

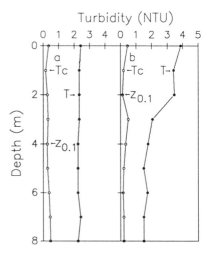

FIGURE 7.12. Turbidity profiles for the epilimnion at south deep station of Onondaga Lake: (a) August 22, 1988, and (b) July 11, 1988.

vertical distribution would not be expected to cause substantial heterogeneity in optical properties within the limits of $z_{0.1}$. Stratification of particulate material within the epilimnion is observed only during calm periods when turbulence is low.

7.5.1.2 Empirical Relationships

The relationship between T and SD has been described by (Effler 1988)

$$SD = \frac{N''}{T} \qquad (7.16)$$

in which N'' = coefficient that is a function of the value of b/a, N (see Equation (7.5)), and α (see Equation (7.1)). The value of N'' is therefore subject to temporal variation, particularly as the value of b/a changes. Effler (1988) has indicated that a reasonable range for N'' is 4.6 to 10.1, which corresponds to ranges in b/a, N, and α of 5 to 15, 8.0 to 9.6, and 0.8 to 1.27, respectively. The relationship between T (at $z_{0.1}$) and SD for Onondaga Lake is evaluated in Figure 7.13. The population is far from normally distributed; 80% of the observations are within the limits T $\geq$2 NTU and SD $\leq$2 m. The relationship is reasonably strong (R^2 = 0.82) at the scaling of Figure 7.13, that emphasizes conditions that have occurred during recent clearing events. However, the values of SD and T are not highly correlated within the range of conditions that usually prevail in the lake; e.g., T $\geq$2 NTU and SD $\leq$2 m. This is attributable to inaccuracies in measurements

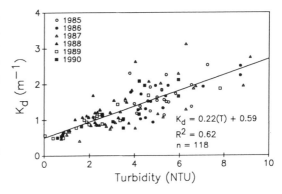

FIGURE 7.14. Evaluation of the relationship between K_d and T in Onondaga Lake.

of SD and T, as well as temporal variations in b/a. There is no evidence that the relationship between T and SD changed significantly in the lake over the period of record. The best estimate of N'' for the entire data set is 3.8, somewhat below the range indicated by Effler (1988). This is largely because the ratio b/a is frequently <5 (see subsequent sections of this chapter), below bounds used by Effler (1988) to develop the "typical" range of N''. The extremely high R^2 value is an artifact of forcing an intercept of zero.

The relationship between T (at $z_{0.1}$) and K_d for Onondaga Lake over the 1985–1990 interval is evaluated in Figure 7.14 for T $\leq$10 NTU; this includes all but four of the paired observations. Ranges of K_d and T of 0.5 to 2.2 m^{-1} and 0.5 to 7 NTU are well represented in this population. Despite the rather strong deviation of this empirical relationship from the form of the expression developed from theory (recall it is often reported that T $\simeq$ b; Equation (7.4)), a rather strong relationship is evident; changes in T explained 62% of the observed variability in K_d. A similar approach explained only 32% of the variability observed in Cross Lake for a smaller data set (Effler et al. 1989a).

7.5.2 Chlorophyll and Empirical Relationships with Optical Measurements

A detailed treatment of chlorophyll a and related pigment measures as a measure of

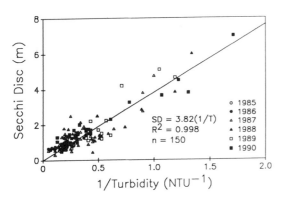

FIGURE 7.13. Evaluation of the relationship between SD and T in Onondaga Lake.

phytoplankton biomass in Onondaga Lake is presented in Chapter 6. An evaluation of the different analytical methods, an analysis of seasonal dynamics, and documentation of historic changes in pigment concentrations is also presented there. The goal here is to identify the important interplay between phytoplankton biomass and measures of light penetration in the lake. Empirical analyses of the relationship between the concentration of total chlorophyll (C_T (mg $\cdot$ m^{-3}), sum of chlorophylls *a*, *b*, and *c*; according to method of Parsons et al. 1984) and K_d and SD are presented here. The relationship between C_T and turbidity is also explored as a preliminary evaluation of the chlorophyll-specific scattering coefficient (K_b).

Temporal distributions of C_T (at 1 m), SD, and K_d during the May–September interval of 1988 are presented in Figure 7.15. Secchi disc transparency received the most complete temporal coverage (n = 44), followed by C_T (n = 25), and K_d (n = 19). Clearly the strong variations observed in SD (Figure 7.15c) were coupled to the dynamics in C_T (Figure 7.15a). Peaks in K_d and SD minima coincided with C_T maxima, and minima in K_d and SD maxima corresponded to minima in C_T. Secchi disc transparency was rather insensitive to the major changes in C_T experienced within the high biomass period of late May. The first clearing event in early June (Figure 7.15b) was not fully resolved in the K_d data (Figure 7.15c) because of the lower frequency of underwater irradiance measurements.

It has frequently been assumed that K_d is regulated by the quantity of phytoplankton biomass. This relationship has most often been represented by the empirical linear expression (e.g., Field and Effler 1983; Jewson 1977; Megard et al. 1979)

$$K_d = K_c \cdot C + K_w \qquad (7.17)$$

in which C = concentration of chlorophyll *a* (mg $\cdot$ m^{-3}), and K_c and K_w = constants determined from the linear regression of K_d on C (units of m^2 $\cdot$ mg^{-1} chlorophyll *a* and m^{-1}, respectively). The coefficient K_w has been described as representing the spectral average attenuation coefficient for water and

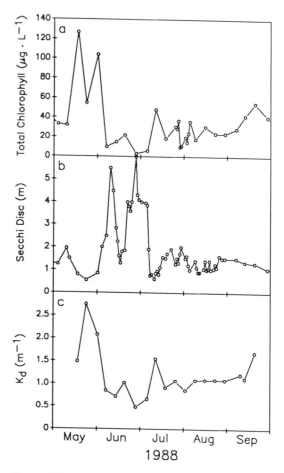

FIGURE 7.15. Temporal distributions of chlorophyll and measures of light penetration in Onondaga Lake in 1988: (**a**) chlorophyll (C_T), (**b**) SD, and (**c**) K_d.

its nonphytoplankton components. It has been recognized that temporal variations in nonphytoplankton components can compromise the application of such an empirical approach (e.g., Field and Effler 1983).

Field and Effler (1983) determined the coefficient values of Equation (7.17) to be K_c = 0.011 m^2 $\cdot$ mg^{-1} chlorophyll *a*, and K_w = 0.98 m^{-1} in Onondaga Lake, based on measurements (C = C_L; according to the method of Lorenzen (1967)) made in 1978; the expression explained 46% of the variability observed in K_d. This K_c value fell within the range of values reported at that time for both the PAR wavelength interval and for selected individual wavelengths within the PAR range (Field and Effler 1983). There were indications that nonphytoplankton attenuating components varied substantially with time. The relationship between K_d and C (= C_T) has been re-examined for the 1987–1990 database in Figure 7.16. The upper bound of C_T was set as $120 \, \text{mg} \cdot \text{m}^{-3}$, excluding only several very high observations in 1987. By linear least squares regression, the recent K_c is essentially the same as reported earlier by Field and Effler (1983), and Kw is $0.65 \, \text{m}^{-1}$. This estimate of K_w, based on recent data, is about one-third less than the earlier estimate. The similarity of the earlier findings (Field and Effler 1983) and this analysis (Figure 7.16) is somewhat misleading, in light of the systematic differences in pigment concentrations obtained with the two analytical methods (C_L versus C_T; see Chapter 6). The rather low R^2 (0.51) for the analysis of the most recent data (Figure 7.16) indicates continued substantial variations in the nonphytoplankton components of attenuation in the lake.

The simple linear regression is imperfect; in particular, the value of K_d is overestimated at lower concentrations of C_T. This may be due to the covariation of background attenuation (e.g., gelbstoff, tripton) with C_T. Note that the annual minima for K_d documented for the 1987–1990 interval (range of $0.42–0.51 \, \text{m}^{-1}$; Table 7.2) were substantially less than the K_w value of $0.65 \, \text{m}^{-1}$, indicating that nonphytoplankton attenuation is at least occasionally overestimated by this value. An alternate empirical relationship between C_T and K_d is presented in Figure 7.16, based on a two-part linear least squares regression fit of the data set and a demarcation C_T of $30 \, \text{mg} \cdot \text{m}^{-3}$. This regression is preferred, as it predicts K_d values for low C_T values that more closely approach recent observations (e.g., at $C_T = 0 \, \text{mg} \cdot \text{m}^{-3}$, $K_d = 0.58 \, \text{m}^{-1}$ (Table 7.2)).

An appropriate empirical expression to describe the relationship between SD and C, based on the approximate inverse relationship between K_d and SD, takes the form

$$1/SD = K_c' \cdot C + K_w' \qquad (7.18)$$

in which K_c' and K_w' = constants determined from linear regression of $1/SD$ on C (= C_T) (units of $\text{m}^2 \cdot \text{mg}^{-1}$ chlorophyll and m^{-1}, respectively). Basically Equation (7.18) is a SD-based corollary to the K_d expression of equation (7.17). The relationship is evaluated for the 1987–1990 interval, for $C_T \leqslant 150 \, \text{mg} \cdot \text{m}^{-3}$, in Figure 7.17. The population does not have a normal distribution; the relatively few very high C_T observations bias the regression analysis. Undoubtedly the relationship is limited by the inherent imprecision of the SD measurement, particularly at very high chlorophyll levels. The linear least squares regression y-intercept for the entire population (0.36), as in the K_d-C analysis case (Figure 7.16), is unrealistically high, implying an upper limit for SD of 2.8 that was annually greatly exceeded during the "clearing"

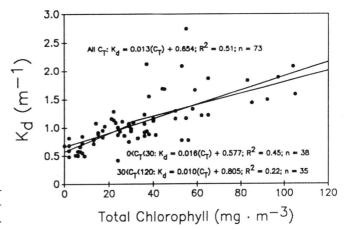

FIGURE 7.16. Evaluation of the relationship between K_d and C_T in Onondaga Lake over the period 1987–1990.

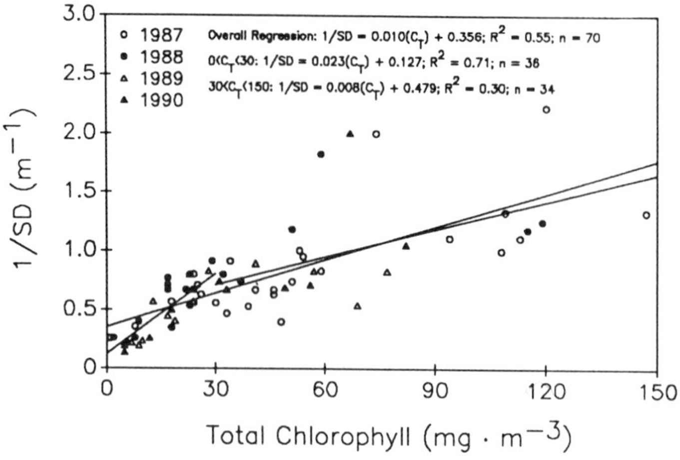

FIGURE 7.17. Evaluation of the relationship between SD and C_T in Onondaga Lake over the period 1987–1990.

event(s) in the 1987–1990 period (e.g., Figure 7.4c and d). A preferred empirical relationship is presented in Figure 7.17, in the form of a two-part linear least squares regression fit of the data set, with a demarcation C_T of $30\,mg \cdot m^{-3}$. This expression predicts SD values at low C_T concentrations consistent with recent observations (e.g., at $C_T = 0\,mg \cdot m^{-3}$, SD = 7.8 m) that are very similar to the maximum observed to date (7.5 m). It should be noted that changes in the empirical relationships developed in Figures 7.16 and 7.17 could occur in the future, associated with various limnological transformations, such as major shifts in the phytoplankton assemblage or changes in the concentrations of nonphytoplankton components of attenuation.

Paired turbidity and C data can be used to estimate the chlorophyll-specific scattering coefficient (K_b; Equation (7.16)). This has taken the form of T versus C plots, where K_b is the slope (e.g., Weidemann and Bannister 1986), under the assumption of uniform levels of nonalgal turbidity. Clearly a better turbidity statistic for Onondaga Lake is T_{nc}, as this eliminates the effect of the sometimes significant contribution of $CaCO_3$ particles to T. The plot of T_{nc} versus C_T for the upper waters (1 m) of Onondaga Lake for the 1987–1990 period is presented in Figure 7.18. The analysis should be considered a first approximation, as the relationship between T and b is imprecise for the lake (see subsequent section of this chapter), and the nonalgal component of T_{nc} undoubtedly varies. Assuming a background turbidity exclusive of phytoplankton and $CaCO_3$ of 0.3 NTU (e.g., b minimum of $0.3\,m^{-1}$), 68% of the points fall between slopes of 0.05 and $0.15\,NTU \cdot m^3 \cdot mg^{-1}$ chlorophyll (or $\sim m^2 \cdot mg^{-1}$ chlorophyll, assuming an equivalence between T and b). The average K_b value is estimated to be $0.098\,NTU \cdot m^3 \cdot mg^{-1}$ chlorophyll; the standard deviation is quite large, $0.074\,NTU \cdot m^3 \cdot mg^{-1}$ chlorophyll. These findings appear to be generally consistent with those of Weidemann and Bannister (1986) for Irondequoit Bay, though the ranges in T and C_T included in the Onondaga Lake database are somewhat larger than those observed in Irondequoit Bay. Forty-five of 55

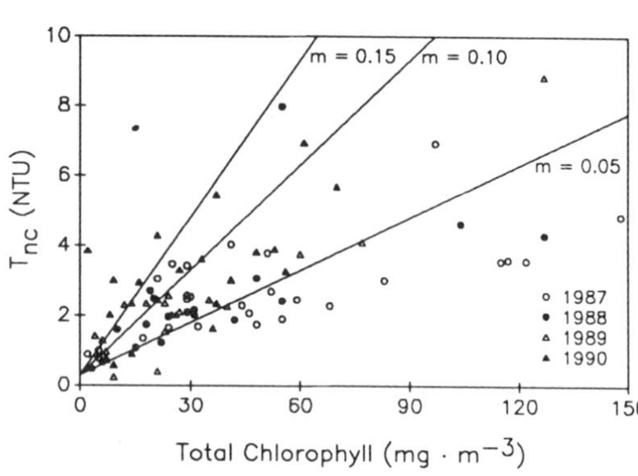

FIGURE 7.18. Noncalcium carbonate turbidity (T_{nc}) versus C_T in Onondaga Lake at 1 m, for the period 1987–1990. Lines indicative of different values for K_b.

observations (82%) fell in the range of slopes of 0.05 to 0.15 NTU · m³ · mg⁻¹ chlorophyll in Irondequoit Bay. Most of the C_T values > 100 mg · m⁻³ in Onondaga Lake fell below this envelope. Sources of variability in the relationship include variations in nonalgal turbidity, algal speciation, chlorophyll content of phytoplankton, and procedural errors. Note that a number of observations in 1990 were at or above the upper bound of the envelope (Figure 7.18); most of these corresponded to a period of dominance of gas vacuolate cyanobacteria similar to the observations of Weidemann and Bannister (1986). This analysis indicates that K_b was usually between 0.05 and 0.15 NTU · m³ · mg⁻¹ chlorophyll. This issue will be revisited subsequently by analyzing paired b and C_T data.

7.5.3 Gelbstoff

All gelbstoff absorption spectra for Onondaga Lake demonstrated exponential decreases with increasing wavelength, as documented in other lakes (Davies-Colley and Vant 1987; Effler et al. 1991a,b; Kirk 1976) and marine systems (Bricaud et al. 1981). A representative spectrum for the lake is presented in Figure 7.19. Absorption values for wavelengths >560 nm are omitted in Figure 7.19 because of the low precision of the spectrophotometer measurements at those low absorbance levels. Gelbstoff spectra have been described by the following function (Davies-Colley and Vant 1987)

$$a_{y(\lambda)} = a_{y(440)} \exp[S(440 - \lambda)] \quad (7.19)$$

in which $a_{y(\lambda)}$ = absorption due to gelbstoff at wavelength λ (400–700 nm) (m⁻¹), $a_{y(440)}$ = absorption due to gelbstoff at 440 nm (m⁻¹), and S = slope parameter (nm⁻¹). The wavelength of 440 nm has been selected as a reference wavelength (Davies-Colley and Vant 1987). The value of $a_{y(440)}$ essentially reflects the magnitude of absorption by gelbstoff; thus it can be used to simply describe dynamics and compare gelbstoff levels for different systems (e.g., Davies-Colley and Vant 1987).

The temporal distributions of gelbstoff ($a_{y(440)}$) at $z_{0.1}$ in Onondaga Lake over the spring–early fall interval of 1987–1990 are presented in Figure 7.20a–d; summary statistics for $a_{y(440)}$ appear in Table 7.5. Year-to-year differences were rather modest. On average, the highest levels were observed in 1989; the lowest in 1987. These were the highest and lowest runoff summers, respectively. The $a_{y(440)}$ levels were well within the values reported and reviewed by Davies-Colley and Vant (1987), and appear to be typical of productive waters.

Rather substantial variations in $a_{y(440)}$ were observed within the study period of each year, particularly in late spring and early summer. The least variable conditions occurred in 1990 (lowest coefficient of variation (C.V.) value; Table 7.5). The fact that several of the tributaries (particularly Ley Creek) have an $a_{y(440)}$ value as high, or greater than, the lake (unpublished data) indicates that terrigenous sources contribute significantly to gelbstoff levels in the lake. Paired distributions of $a_{y(440)}$ and chlorophyll have been analyzed elsewhere as an indication of the degree of autochthonous contributions. For example, correlated distributions indicate autochthonous sources are important (Davies-Colley and Vant 1987). However, this has been described as perhaps flawed, because gelbstoff removal reactions are probably slow compared to reductions in phytoplankton production and biomass following a bloom. The temporal distributions of C_T (e.g., Figure 7.15a and Chapter 6) and

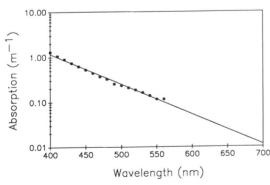

FIGURE 7.19. Representative absorption spectrum for gelbstoff in the upper waters of Onondaga Lake.

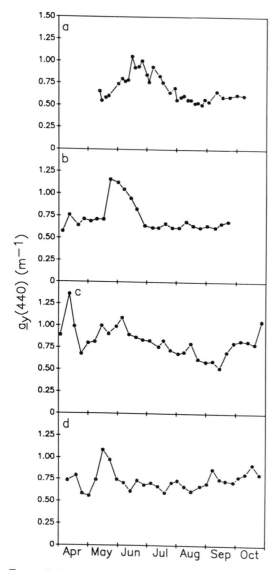

FIGURE 7.20. Temporal distributions of gelbstoff (as $a_{y(440)}$) at $z_{0.1}$ in Onondaga Lake: (a) 1987, (b) 1988, (c) 1989, and (d) 1990.

$a_{y(440)}$ were not significantly correlated (R = 0.28 for the four years of data in Figure 7.20). For example, no change in $a_{y(440)}$ occurred in 1988 (Figure 7.20b) in response to the first, and largest, peak in C_T (Figure 7.15a). The abrupt increase in $a(440)$ in late May 1988 coincided with the second peak in C_T, but also immediately followed a rainfall of >1.7 inches. Following the progressive decreases through June 1988, $a_{y(440)}$ remained relatively uniform, despite substantial variations in C_T. Gelbstoff is known to be absorbed to calcite (Reynolds 1978); the importance of this sink process in regulating the observed distributions in Onondga lake is unknown.

The mean S value for 88 observations (from 1987, 1988, and 1990; no scans performed in 1989) was $0.0166 \, nm^{-1}$; the standard deviation was $0.0016 \, nm^{-1}$. The mean falls within the lower portion of the range reported by Davies-Colley and Vant (1987) for 12 New Zealand lakes that includes a wide range of gelbstoff levels. The relative standard deviation reported here is high by comparison to their results.

The dependence of light penetration, as measured by SD and K_d, on gelbstoff (parameterized as $a_{y(440)}$) was evaluated empirically in a manner consistent with the analyses for T and C_T (e.g., Figures 7.13, 7.14, 7.16 and 7.17). Gelbstoff was found not to be an important regulator of light penetration in Onondaga Lake, as demonstrated by the lack of correlation between $a_{y(440)}$ and K_d, or SD^{-1}. However, gelbstoff is subsequently demonstrated to be important in establishing the upper bound (e.g., under conditions of minimal phytoplankton concentration) of spectral quality for the lake.

7.6 Regional Comparison

Comparison of the extent of light penetration and concentrations of attenuating components in Onondaga Lake to conditions in proximate lakes provides a valuable perspective for managers concerned with improving the clarity of Onondaga Lake. Most of the comparisons focus on the summer of 1988, when a number

TABLE 7.5. Summary statistics for gelbstoff (as $a_{y(440)}$) in Onondaga Lake, for four years.

Year	$\bar{a}_{y(440)}$ (m^{-1})	C.V.	Range, $a_{y(440)}$ (m^{-1})	n
1987	0.690	0.21	0.507–1053	35
1988	0.732	0.22	0.576–1.162	25
1989	0.826	0.20	0.522–1.359	30
1990	0.740	0.15	0.558–1.088	20

of optical features of several central New York lakes were characterized. Data from the other lakes were collected from deep water, and usually centrally located, sites. The comparisons are somewhat compromised by differences in the frequencies and periods of monitoring.

7.6.1 Components of Attenuation

The temporal distributions of T and T_c in Onondaga Lake in 1988 are compared to the distributions documented for five other central New York lakes over the same period in Figure 7.21a–f; related statistics are presented in Table 7.6. The modest "whiting" event of late July in Onondaga Lake (Figure 7.21a) coincided with much more pronounced events in Cross Lake (Figure 7.21b), Otisco Lake (Figure 7.21c), and Owasco Lake (Figure 7.21d). Calcium carbonate turbidity was present in Little Sodus Bay at the same time, though at somewhat lower levels (Figure 7.21e). The timing of the event in Little Sodus Bay was not resolved completely. Turbidity levels were very low in Skaneateles Lake by comparison (Figure 7.21f). The relative uncertainty in T and T_c in Skaneateles Lake is considered high; it remains uncertain whether T_c was actually present during the monitoring period.

The contribution of $CaCO_3$ to T is clearly not unique to Onondaga Lake in the central New York region (Figure 7.21). The peak T_c values in 1988 in Cross, Otisco, and Owasco Lakes exceeded the maximum value in Onondaga Lake by a factor of 2 to 3. Thus the whiting phenomenon has greater relative optical impact in these lakes (see Effler and Johnson 1987; Effler et al. 1987a). For example, T_c represented more than 30% of T in Otisco Lake and more than 60% of T in Owasco Lake. Perhaps most striking is the coincidence of the "whiting" events for the four different lakes. This was noted earlier for Otisco and Owasco Lakes (Effler et al. 1987b). On-going work by D.L. Johnson indicates the availability of various heteronuclei (surfaces) for $CaCO_3$ precipitation may be critical to the

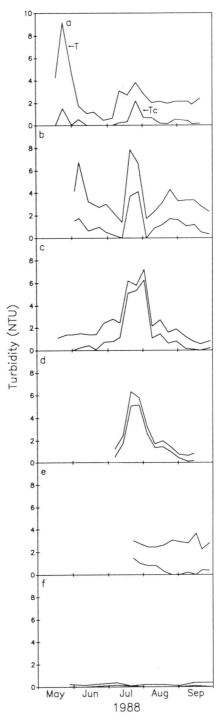

FIGURE 7.21. Comparison of the distributions of T and T_c in the upper waters of central New York lakes in 1988: (**a**) Onondaga Lake, (**b**) Cross Lake, (**c**) Otisco Lake, (**d**) Owasco Lake, (**e**) Little Sodus Bay, and (**f**) Skaneateles Lake.

TABLE 7.6. Comparison of turbidity conditions in Onondaga Lake to other Central New York systems, 1988.

Lake	T (NTU)		Tc (NTU)		Tc/T (100)		n	Source
	$\bar{x}$	Range	$\bar{x}$	Range	$\bar{x}$	Range		
Onondaga	2.66	0.5−9.2	0.41	0.0−2.2	13.6	0−57	19	
Cross (north)	3.66	1.4−7.9	1.27	0.0−4.1	29.6	0−62	17	Effler et al. 1989a
Otisco	2.34	0.6−7.2	1.23	0.0−6.3	32.7	0−92	20	Effler et al. 1989b
Owasco	2.40	0.7−6.4	1.78	0.1−5.2	63.2	0−89	11	Effler et al. 1990b
Little Sodus Bay	2.78	0.2−3.6	0.47	0.0−1.4	17.9	0−48	11	Effler et al. 1990a
Skaneateles	0.23	0.10−0.38	0.04	0.0−0.15	16.6	0−79	9	Effler et al. 1989c

magnitude of the "whiting" phenomenon in Onondaga Lake (see Johnson et al. 1991).

The temporal distribution of C_T in Onondaga Lake in 1988 is compared to the distributions documented for Skaneateles, Otisco, Owasco, and Cross Lakes over the same period in Figure 7.22; related statistics are presented for these systems and Little Sodus Bay in Table 7.7. The three distinct maxima in C_T in Onondaga Lake in April and May greatly exceeded any concentrations observed in the other lakes. The average concentration of C_T in Onondaga Lake also exceeded those documented for the other systems (Table 7.7), though the comparisons are somewhat flawed by the different study periods (Figure 7.22). Only during the clearing events of early and late June did C_T concentrations in Onondaga Lake approach values observed in Skaneateles, Owasco, and Otisco Lakes. The C_T values of Onondaga Lake and Cross Lake (another eutrophic system) were similar over the late May−October interval, though temporal variability was greater in Onondaga Lake. The order of decreasing seasonal average concentrations

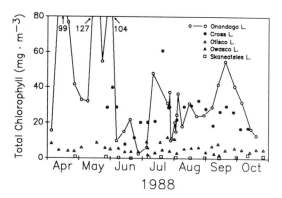

FIGURE 7.22. Comparisons of the distribution of total chlorophyll (C_T) in the upper waters of Onondaga Lake to concentrations measured in four other central New York Lakes, 1988.

was (Table 7.7) Onondaga Lake > Cross Lake > Litte Sodus Bay > Otisco Lake > Owasco Lake > Skaneateles Lake. The average C_T concentration in Onondaga Lake was nearly twice that of eutrophic Little Sodus Bay, nearly 6 times greater than in mesotrophic Otisco Lake, nearly 9 times greater than in oligo/

TABLE 7.7. Chlorophyll (C_T) levels in Onondaga Lake and other selected central New York lakes, 1988.

Lake	C_T		n	Source
	$\bar{x}$ (mg·m^{-3})	range (mg·m^{-3})		
Onondaga	35	3−127	25	
Cross	26	8−61	21	Effler et al. 1989a
Little Sodus Bay	19	9−39	10	Effler et al. 1990a
Otisco	6	4−12	25	Effler et al. 1989b
Owasco	4	2−6	11	Effler et al. 1990b
Skaneateles	0.9	0.1−1.9	14	Effler et al. 1989c

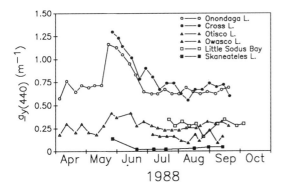

FIGURE 7.23. Comparisons of the distribution of gelbstoff ($a_{y(440)}$) in the upper waters ($z_{0.1}$) of central New York Lakes in 1988.

mesotrophic Owasco Lake, and more than 30 times greater than in oligotrophic Skaneateles Lake.

Gelbstoff levels and dynamics documented for the upper waters of Onondaga Lake in 1988 (see Figure 7.20b) are compared to conditions found in other central New York lakes during the same period in Figure 7.23. Summary statistics are presented in Table 7.8. The level of gelbstoff in Onondaga Lake was higher than observed in the other systems except in Cross Lake, where gelbstoff levels generally track those that prevail in the inflowing Seneca River (Effler et al. 1989a). The order of decreasing levels of gelbstoff for the other systems was, Sodus Bay ≈ Otisco Lake > Owasco Lake > Skaneateles Lake.

7.6.2 Clarity and Attenuation

The distributions of SD for Otisco Lake (Effler et al. 1987c) and Onondaga Lake are com-

pared for 1986 in Figure 7.24. The SD of Onondaga Lake was substantially less than that of Otisco Lake throughout the May–September period of 1986. Onondaga Lake met the swimming safety standard of 4 ft (~1.2 m) for only two (late September) of 29 observations. Otisco Lake always met this criterion. The average of the 17 observations in Otisco Lake was 2.6 m; the range was 1.65 to 3.8 m. The average SD in Onondaga Lake was 0.75 m; the range was 0.4 to 1.25 m. Note that "clearing events" were not observed in either lake in 1986.

Comparisons are expanded to include an additional three proximate lakes, Cross Lake, Owasco Lake, and Skaneateles Lakes, for 1988 in Figure 7.25. This is a substantially different case than the 1986 comparison because of the occurrence of two distinct clearing events in Onondaga Lake in 1988.

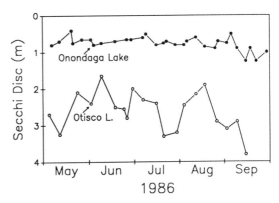

FIGURE 7.24. Comparison of Secchi disc transparency in Onondaga and Otisco (Effler et al. 1987c) Lakes for the May–September interval of 1986.

TABLE 7.8. Gelbstoff levels in Onondaga Lake and other selected central New York lakes, 1988.

Lakes	$a_{y(440)}$ (m^{-1})		n	Source
	$\bar{x}$	range		
Onondaga	0.732	0.576–1.162	25	—
Cross (north)	0.810	0.553–1.296	18	Effler et al. 1989a
Otisco	0.278	0.184–0.415	24	Effler et al. 1989b
Owasco	0.151	0.092–0.230	11	Effler et al. 1990b
Little Sodus Bay	0.290	0.161–0.345	10	Effler et al. 1990a
Skaneateles	0.047	0.023–0.138	7	Effler et al. 1989c

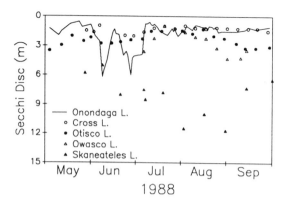

FIGURE 7.25. Comparison of Secchi disc transparency in Onondaga Lake to that observed in four other central New York lakes, for the May–September interval of 1988.

The Onondaga Lake distribution is presented as a line to fascilitate the comparisons; this continuous character is however supported by the large number of observations (n = 71). There are fewer observations for the other lakes. Summary statistics for the five lakes and Little Sodus Bay are presented in Table 7.9.

It should be recalled that the average SD value for Onondaga Lake is strongly influenced by the very high values observed during the clearing events. The maximum SD observed in Onondaga Lake was exceeded only by values measured in Skaneateles Lake (Table 7.9, Figure 7.25). On average, SD was higher in Onondaga Lake than in Cross Lake and Little Sodus Bay, but less than in the other three lakes. Onondaga Lake failed to meet the 1.2 m criterion for 35% of the

observations in 1988; Cross Lake failed for 79% of the observations; Owasco Lake failed for 1 of the 11 observations. The mid–to late July minima in SD in Otisco and Owasco Lakes (Figure 7.25) were associated with coincident whiting events (Figure 7.21c and d; Effler et al. 1989b, 1990b). Skaneateles Lake has the highest clarity of the Finger Lakes (Schaffner and Olgesby 1978) and the central New York region. Note that for a brief period during the first clearing event the SD of Onondaga Lake exceeded that of Skaneateles Lake. However, the SD of Skaneateles Lake was much greater throughout the remainder of the study period. In sharp contrast to 1986 (Figure 7.24), the SD of Onondaga Lake equaled or exceeded that of Otisco Lake from early June through early July. Clarity in Onondaga Lake was equal to, or greater than, values observed in Otisco and Owasco Lake during the whiting events of mid-July to early August (Figure 7.25). However, SD was much higher in Otisco Lake in May. The clarity of Onondaga lake was greater than, or equal to, that of Cross Lake for about 85% of the June through September interval.

The temporal distribution of K_d values in Onondaga Lake in 1988 is compared to distributions for four other proximate lakes in Figure 7.26. These distributions approximately mirror those presented for SD for the same lakes in Figure 7.25. Note that the temporal distribution of K_d (Figure 7.26) was not as well resolved as for SD (Figure 7.25) because of the lower number of observations (n = 19). Measurements commenced in May for four of the five lakes; they started in July in Owasco

TABLE 7.9. Comparison of summary statisics for Secchi disc transparency in 1988 for Onondaga Lake and five other central New York lakes.

Lake	n	Secchi disc (SD, m)			Monitoring interval	Source
		$\bar{x}$	min	max		
Onondaga	71	1.92	0.55	6.2	May–September	—
Cross	19	1.41	0.95	2.75	late May–September	Effler et al. 1989a
Otisco	21	2.38	1.3	3.5	May–September	Effler et al. 1989b
Owasco	11	2.66	1.0	4.3	July–September	Effler et al. 1990b
Skaneateles	11	8.15	5.0	11.7	late May–September	Effler et al. 1989c
Little Sodus Bay	12	1.75	1.3	2.35	July–October	Effler et al. 1990a

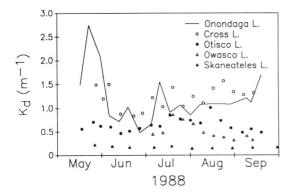

FIGURE 7.26. Comparison of K_d in Onondaga Lake to values determined for four other central New York lakes, for the May–September interval of 1988.

Lake. Summary statistics for the lakes and Little Sodus Bay are presented in Table 7.10.

The highest average values of K_d were observed in Onondaga and Cross Lakes (Table 7.10); the average for Little Sodus Bay was only slightly lower. The average value for Onondaga Lake was strongly influenced by the three particularly elevated values observed in May (Figure 7.26). The values of K_d in Cross Lake exceeded or equaled those in Onondaga Lake for most of the remainder of the summer period. Conditions were much more variable in Onondaga Lake (Figure 7.26). Only during the late June–early July clearing event in Onondaga Lake, and the late July–August "whiting" event in Otisco Lake, did K_d in Onondaga Lake approach values that prevailed in Otisco Lake (Figure 7.26). The average K_d value in Otisco Lake was slightly more than half of that in Onondaga Lake (Table

7.10). Light penetration was somewhat greater in Owasco Lake than in Otisco Lake (Figure 7.26). Values of K_d in Owasco Lake approached those in Onondaga Lake only during the late July whiting event (Figure 7.26). The greatest light penetration was observed in Skaneateles Lake (Figure 7.26; Table 7.10). The average value of K_d in Skaneateles Lake was $0.17 \, \text{m}^{-1}$; approximately seven times less than observed for Onondaga and Cross Lakes, nearly four times less than in Otisco Lake, and approximately three times less than found in Owasco Lake (Table 7.10).

It is important to note the apparent differences in relative light penetration indicated by SD (Figure 7.25) versus K_d (Figure 7.26). For example, recall that SD was greater in Onondaga Lake than in Otisco Lake for much of June in 1988 (Figure 7.25). Also the differences in SD between Onondaga Lake and Skaneateles Lake during the clearing events of June and early July were small compared to the relative differences in K_d that prevailed over the same period (Figure 7.26). The differences in the apparent response of these two measures of light penetration is a manifestation of their different dependencies on the relative contributions of absorption and scattering (Equations (7.4) and (7.5)).

The extent of light penetration in Onondaga Lake is poor by comparison to other local lakes considered desirable for contact recreation, such as Otisco, Owasco, and Skaneateles Lakes. This is largely a result of elevated concentrations of phytoplankton biomass that prevail in the lake for much of the year (Table 7.7; Figure 7.22) associated with its culturally eutrophic condition. This is consistent with

TABLE 7.10. Comparison of summary statistics for K_d in 1988 for Onondaga Lake and five other central New York lakes.

Like	n	$K_d \ (\text{m}^{-1})$			Monitoring interval	Source
		$\bar{x}$	min	max		
Onondaga	19	1.20	0.48	2.74	mid-May–September	—
Cross (north)	17	1.22	0.84	1.57	late May–September	Effler et al. 1989a
Otisco	19	0.64	0.47	1.01	mid-May–September	Effler et al. 1989b
Owasco	11	0.50	0.32	0.87	July–September	Effler et al. 1990b
Skaneateles	10	0.17	0.15	0.22	late May–September	Effler et al. 1989c
Little Sodus Bay	12	1.07	0.64	1.38	July–October	Effler et al. 1990a

the observations that the dynamics of K_d and SD in Onondaga Lake are largely regulated by the dynamics in C_T (Figures 7.16 and 7.17). The disparity in the comparative responses of SD and K_d during clearing events reflects a particularly strong reduction in turbidity (i.e., scattering) during these events. The occurrence of the whiting phenomenon in Onondaga Lake contributes irregularly in controlling the extent of light penetration. However, the phenomenon is decidedly more prominent in the other central New York lakes considered here, with the exception of Skaneateles Lake. Without the disparity in this phenomenon, the differences in light penetration (particularly SD) between Onondaga Lake and Otisco and Owasco Lakes would be substantially greater. The elevated gelbstoff levels in Onondaga Lake do not contribute greatly to the differences in light penetration that prevail for several of the lakes considered here.

7.7 Angular Distribution of Underwater Irradiance

Example depth profiles for $R_{(z)}$ and $\bar{\mu}_{(z)}$ in Onondaga Lake are presented in Figure 7.27. Note they show the general curvilinear distributions predicted by theory (Kirk 1981a); $R_{(z)}$ increases (Figure 7.27a) and $\bar{\mu}_{(z)}$ decreases (Figure 7.27b) with depth. Asymptotes were generally established at $z_{0.1}$ (Figure 7.27a and b), indicating the asymptotic radiance distribution was essentially established at this depth (Kirk 1983).

The temporal distributions of R and $\bar{\mu}_{KIRK}$ (from R and Kirk's (1981b) functions) for $z_{0.1}$, for the period 1985–1990, are presented in Figures 7.28 and 7.29. Estimates of $\bar{\mu}$ based on direct measurements (profiles of $\bar{\mu}_{(z)}$ according to Equation (7.7)) are also included for the period of collection of simultaneous profiles of $E_{d(z)}$, $E_{u(z)}$, and $E_{o(z)}$ (most of the 1987–1990 period) in Figure 7.29. The mean and range of the estimates R and $\bar{\mu}_{KIRK}$ for each of the years are presented in Table 7.11. Rather strong seasonal variability was observed, particularly for R, indicating abrupt changes in the relative

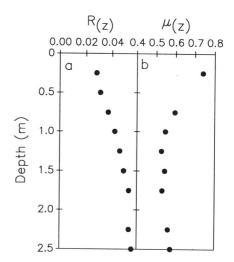

FIGURE 7.27. Example depth profiles of apparent optical properties in Onondaga Lake: (**a**) R_z and (**b**) $\bar{\mu}_z$.

magnitudes of a and b. Average cosine demonstrates progressively less response as the ratio b/a increases (Kirk 1981b). The value of R remained below 0.075 throughout the 1985–1990 period. Generally progressive increases in R were observed in 1986. Values of R were low and $\bar{\mu}_{KIRK}$ was high during most of the clearing events, reflecting a disproportionate reduction in b. Recall that SD is more sensitive to changes in b than K_d (see Effler 1985). The relatively high R values sustained through late summer of 1990 (Figure 7.28f) were probably associated with the dominance of filamentous blue–green algae during that period. For example, Weidemann and Bannister (1986) noted particularly high values of b (on a chlorophyll a-specific basis) when gas-vacuolate cyanobacteria were important in Irondequoit Bay.

Year-to-year differences in the angular distribution of underwater irradiance (Table 7.11) have been relatively minor; further, these may have been influenced by differences in the temporal coverage of the monitoring program. On average, the highest R values occurred in 1986, the lowest in 1988. The average values of $\bar{\mu}_{KIRK}$ were the highest in 1988 and 1989. The minimum values of R were substantially higher in 1985 and 1986,

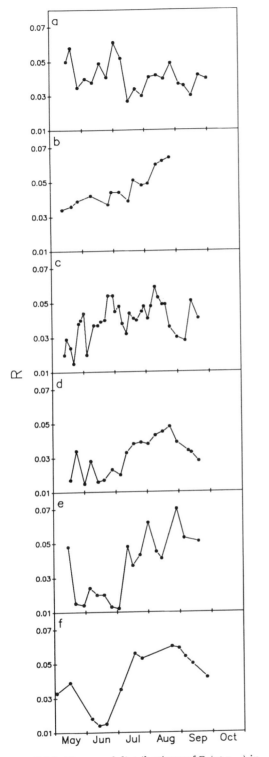

FIGURE 7.28. Temporal distributions of R (at $z_{0.1}$) in Onondaga Lake: (**a**) 1985, (**b**) 1986, (**c**) 1987, (**d**) 1988, (**e**) 1989, and (**f**) 1990.

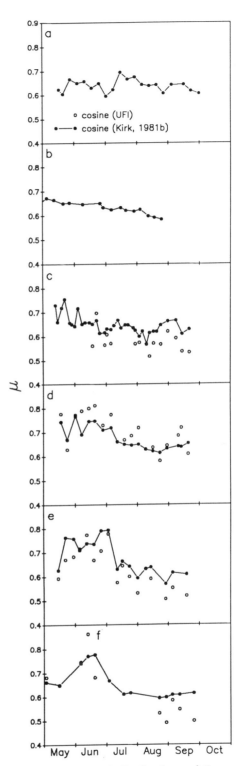

FIGURE 7.29. Temporal distributions of $\bar{\mu}_{KIRK}$ (at $z_{0.1}$) in Onondaga Lake: (**a**) 1985, (**b**) 1986, (**c**) 1987, (**d**) 1988, (**e**) 1989, and (**f**) 1990.

TABLE 7.11. Summary statistics of R and $\bar{\mu}_{KIRK}$ for Onondaga Lake, 1985–1990.

Year	R			$\bar{\mu}_{KIRK}$		
	$\bar{x}$	range	n	$\bar{x}$	range	n
1985	0.042	0.027–0.061	21	0.638	0.596–0.695	21
1986	0.047	0.028–0.071	19	0.626	0.566–0.673	19
1987	0.040	0.015–0.059	35	0.647	0.568–0.755	35
1988	0.031	0.015–0.048	19	0.676	0.615–0.773	19
1989	0.036	0.012–0.070	17	0.677	0.569–0.794	17
1990	0.041	0.014–0.060	13	0.655	0.594–0.776	13

apparently because of the absence of "clearing events" in these two years. The maxima in $\bar{\mu}_{KIRK}$ were substantially higher for the 1988–1990 interval. The average values of R and $\bar{\mu}_{KIRK}$ in Onondaga Lake for the six years of record were 0.04 and 0.65, respectively.

7.8 Checks on Optical Measurements and Estimates

7.8.1 Cosine ($\bar{\mu}$)

The percent difference between $\bar{\mu}_{KIRK}$ and the measured $\bar{\mu}$ was less than 20% for 60 of the 61 paired estimates, and less than 10% for 39 of the estimates. However, the differences between these two estimates did not appear to be random. The measured value of $\bar{\mu}$ was lower than $\bar{\mu}_{KIRK}$ through most of the study periods of 1987 and 1989, and for the early fall period of 1990; measured values of $\bar{\mu}$ were generally higher than $\bar{\mu}_{KIRK}$ in 1988. Most of the differences can probably be attributed to limitations in the measurements, including sensor calibration and imperfect configuration. The potential contribution of shortcomings in the original model analysis that supported the development of Kirk's (1981a, 1981b) functions is considered minor by comparison. Overall the match between the two estimates of $\bar{\mu}$ is considered good, and is supportive of the consistency of the optical measurements and the theory on which the estimates of a (Kirk 1981a, 1981b) will be based. Recall that $\bar{\mu}$ is linearly coupled

to the estimate of a, according to the Gershun–Jerlov equation (Equation (7.8)).

7.8.2 Turbidity versus b

Paired values of the T measurements and the b estimates at $z_{0.1}$ are presented in the plot of Figure 7.30. The average value of α (= T/b; Equation (7.1)) for the 118 paired values was 0.84. The slope of the best fit linear least squares regression fit of the data (an alternate estimate of α) is 0.91. Both of these representations of α are within the commonly observed range of 0.8–1.27 (DiToro 1978; Effler 1988; Kirk 1981b; Weidemann and Bannister 1986). However, only 52% of the pairs were within these bounds (see limits

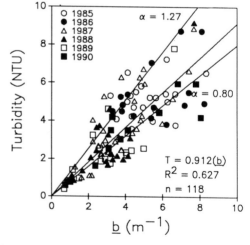

FIGURE 7.30. Evaluation of the relationship between T (at $z_{0.1}$) and b in Onondaga Lake for the period 1985–1990.

added to Figure 7.30). A significant number of the observations fell below the lower bound (e.g., expansion of the lower limit to 0.6 resulted in 77% of the pairs being bracketed). We suspect most of the variability in the T/b relationship to be associated with the T measurement, as concluded by Weidemann and Bannister (1986). The degree of match between T and b is considered good overall and supportive of the subsequently presented estimates of b (Kirk 1981b).

7.9 Estimates of a and b

7.9.1 Coupling of Inherent and Apparent Optical Properties, 1988

An example of the interplay between the distributions of the inherent optical properties, a and b (estimated according to Kirk (1981b)), and apparent optical properties is presented in Figure 7.31a–f. The major inflections in a (Figure 7.31a) and b (Figure 7.31b) tracked each other for the most part of 1988. This indicates that these transformations were probably regulated by attenuating particles that both scatter and absorb light (e.g., phytoplankton and/or tripton). Notable disparities in the distributions of a and b included a decrease in b in late May that was not matched by a, and increases in b in August while the value of a remained uniform. The changes in the relative contributions of b and a to attenuation are manifested in the changes in the angular distribution of underwater irradiance, as measured by R (Figure 7.31c) and $\bar{\mu}$ (Figure 7.31d). Abrupt changes in R and $\bar{\mu}$ were experienced in late May and early June, as a result of changes in b/a experienced during and after a phytoplankton bloom (Figure 7.15). Subsequently R increased in a largely progressive manner until late August, and decreased through September. The distribution of $\bar{\mu}$ approximately mirrored that of R (according to Kirk's functions (1981b)).

The distributions of a and K_d (Figure 7.31e) were highly correlated ($a = 0.685 \cdot K_d$, R = 0.98). This is consistent with theory, in light

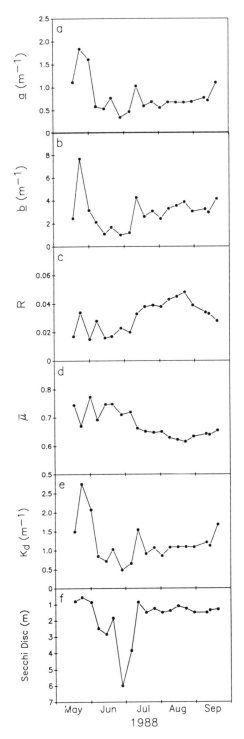

FIGURE 7.31. Temporal distributions of estimated inherent optical properties and measured apparent optical properties for Onondaga Lake in 1988: (**a**) a, (**b**) b, (**c**) R, (**d**) $\bar{\mu}$, (**e**) K_d, and (**f**) SD (modified from Effler and Perkins 1995 and Perkins and Effler 1995).

of the relative uniformity of $\bar{\mu}$ observed (recall, $a = \bar{\mu} \cdot K_d$). Note that peak light penetration (minimum K_d and maximum SD (Figure 7.31f)) occurred during the minima in a and b in late June and that the minimum penetration occurred in late May during maxima in a and b (associated mostly with a phytoplankton bloom). Intermediate degrees of light penetration were observed from late July through mid-September when intermediate levels of a and b prevailed. Secchi disc transparency is rather insensitive to substantial increases in a and b at low SD levels (Effler 1985, 1988; Effler et al. 1988). For example, note the relative insensitivity of SD to the increases in a and b at the beginning and end of the study period. Greater sensitivity of SD to a occurs at higher SD levels. For example, the rather substantial difference in the magnitude of the SD maximum ($\sim 6\,\text{m}$) and the secondary maximum ($\sim 3\,\text{m}$) apparently was associated mostly with a rather modest difference in a. In contrast to SD, K_d is less sensitive to changes in b. For example, K_d decreased from 2.6 to $2.0\,\text{m}^{-1}$ in late May, despite a decrease in b from 8 to $3\,\text{m}^{-1}$.

7.9.2 Distributions of a and b, 1985–1990

Temporal distributions of a and b in Onondaga Lake are presented for the May–October interval of 1985–1990 in Figures 7.32 and 7.33, respectively. Substantial seasonal variability was observed in all six years. The most uniform conditions for both inherent optical properties occurred in 1985 (Figures 7.32a and 7.33a). The most dynamic conditions were observed in the May to early July interval of 1987 (Figures 7.32c and 7.33c), associated with abrupt changes in phytoplankton biomass. Transformations in a (particularly, Figure 7.32f) and b (Figure 7.33f) were less abrupt in 1990 in this period compared to the preceding four years. The levels of a were distinctly lower and more uniform after mid-summer in 1987, 1988, and 1989 (Figure 7.32c, d, and e) compared to the earlier two years. The major late summer increases in a and b in 1990 coincided with a bloom of gas-vacuolate

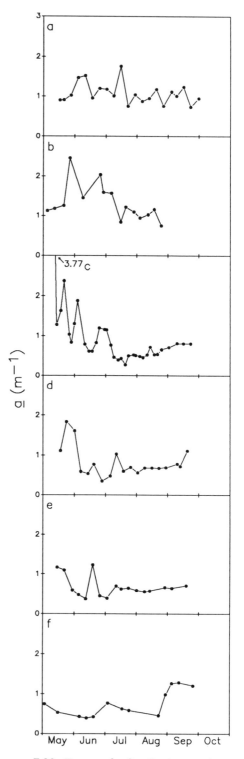

FIGURE 7.32. Temporal distributions of a in Onondaga Lake: (**a**) 1985, (**b**) 1986, (**c**) 1987, (**d**) 1988, (**e**) 1989, and (**f**) 1990.

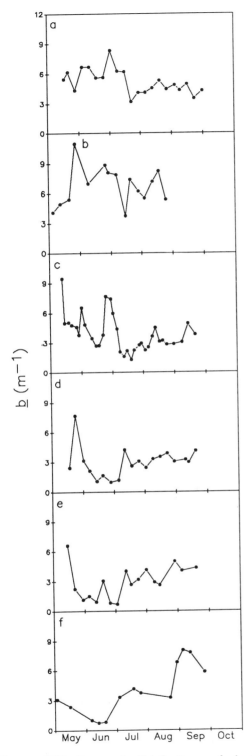

FIGURE 7.33. Temporal distributions of *b* in Onondaga Lake: (**a**) 1985, (**b**) 1986, (**c**) 1987, (**d**) 1988, (**e**) 1989 and (**f**) 1990.

cyanobacteria, a phenomenon that has not been observed in Onondaga Lake since the earlier 1970s (see Chapter 6). The disproportionate effect these organisms have on scattering has been noted earlier (Weidemann and Bannister 1986). Most of the major inflections in *a* and *b* coincide, though the relative changes in these two inherent properties often differ significantly, as manifested in measurements of R and $\bar{\mu}$ (Figures 7.28 and 7.29).

Annual summary statistics for *a* and *b* for the six years included in the 1985–1990 interval are presented in Table 7.12. The major shift to increased light penetration (Figures 7.3 and 7.8) documented since 1987 has clearly been a result of decreases in both *a* and *b*. The average decreases for the 1987–1990 interval compared to 1985 and 1986 have been 30 and 40% for *a* and *b*, respectively. The annual minima for *a* and *b*, observed during the clearing events (Figure 7.4c and d; Figure 7.9g–j; Figure 7.31) in 1987–1990, are much lower than minima observed in 1985 and 1986. The widest ranges in *a* and *b* values occurred in 1987; the lowest annual averages in 1989.

The values of *b* estimated according to the alternate method of Kirk (1989) were very similar to those presented here, determined by a protocol (see Equations (7.8) and (7.9)) developed earlier by the same researcher (Kirk 1981b). Ninety-eight percent of the paired (n = 123) estimates of *b* were within 15% of each other; 50% were within 5%. The value obtained with the more recently developed method (b_{1989}) tended to be only slightly higher than the estimate from the earlier approach (b_{1981}; $b_{1989} = 1.0056b_{1981} + 0.125$, $R^2 = 0.98$). Kirk (1989) earlier reported a high degree of agreement between the two methods for several test systems in Australia.

7.9.3 Relationship between SD and K_d and *c*

The relationship between clarity, K_d, and the beam attenuation coefficient (*c*) in Onondaga Lake is evaluated within the framework of contrast transmittance theory (Priesendorfer

TABLE 7.12. Summary statistics of a and b for Onondaga Lake, 1985–1990.

Year	a			b		
	$\bar{x}$	range	n	$\bar{x}$	range	n
1985	1.073	0.734–1.762	21	5.188	3.188–8.329	21
1986	1.174	0.552–2.458	19	6.120	3.303–10.97	19
1987	0.918	0.272–3.774	35	3.898	1.261–9.436	35
1988	0.812	0.346–1.837	19	2.994	1.003–7.676	19
1989	0.675	0.373–1.234	17	2.932	0.684–6.614	17
1990	0.748	0.393–1.288	13	3.912	0.735–8.044	13

1986; Tyler 1968; Equation (7.5)), as a plot of SD versus $(K_d + c)^{-1}$ (Figure 7.34). Linear least square regression fits are presented for the total population and for the commonly observed range of conditions (SD $\leq$ 3.0 m and $(K_d + c)^{-1} \leq$ 0.4 m). The slopes are estimates of N (see Equation (7.5)); the values determined (7.89 and 7.09 for the total population and the subset of more common values, respectively) fall slightly below the range of values described by Priesendorfer (1986) as typical.

Substantial scatter is observed, particularly for the clearing events, for which a tendency to underpredict clarity at the lowest levels of K_d and c (i.e., a and b) is indicated. Modest differences in the conditions under which measurements are made at these high levels of light penetration may explain much of this variability. However, the linearity observed

across the rather broad range of inherent and apparent optical properties (Figure 7.34) supports the contrast transmittance theory (Equation (7.5)). Note that the performance of the related framework (Figure 7.34) was distinctly better than the often invoked inverse relationship between K_d and SD ($K_d \cdot$ SD as a constant; Figure 7.10), over the entire range of conditions, as well as the most frequently observed (SD < 3 m) conditions. For example, the observed best fit $K_d \cdot$ SD relationship for SD < 3.0 m explained only 57% of the observed variability in these two parameters, whereas the contrast transmittance theory framework explained 66% of the variability over the same range of SD values. The relationships presented in Figure 7.34 represent a mechanistically sound, and reasonably reliable, basis to project the implications of changes in a and b, and thereby changes in the concentrations of related attenuating substances, on clarity in Onondaga Lake.

7.9.4 Relationship between b and Chlorophyll (C_T)

The relationship between b and C_T for the 1987–1990 data base is evaluated in the form of a plot of b versus C_T in Figure 7.35a, and in the form of a plot of b_{nc} versus C_T in Figure 7.35b. The latter analysis eliminates the effect of $CaCO_3$ particles. However, both analyses are included for comparison purposes (Figure 7.35a and b). Clearly the relationship between b and C_T is highly variable, as indicated earlier in the analysis of turbidity and C_T data. Contributing factors include uncoupled variations

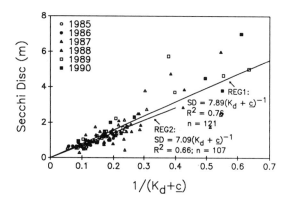

FIGURE 7.34. Application of contrast transmittance theory for SD in Onondaga Lake, based on the 1985–1990 data set. Slope is equal to N (equation (7.5), modified from Effler and Perkins 1995).

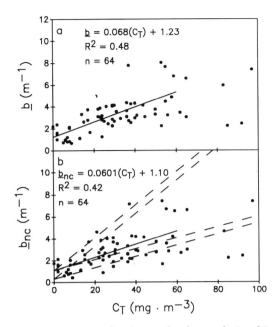

FIGURE 7.35. Evaluation of the relationship between scattering coefficients and chlorophyll (C_T) in Onondaga Lake, for summers, over the 1987–1990 period: (**a**) b and (**b**) b_{nc} (modified from Effler and Perkins 1995).

in nonphytoplankton components of b (b_c and b'; see Equations (7.14) and (7.15)), species-specific scattering properties of phytoplankton (Bricaud et al. 1983), and variations in the cellular content of C_T. Procedural errors probably contribute little to the observed scatter.

Linear least square regression fits for the C_T range of 0 to 60 mg · m^{-3}, the best represented portion of the C_T population, are presented. The slope of the b_{nc} versus C_T plot (Figure 7.35b) represents an estimate of K_b (= 0.060 m^2 · mg^{-1} chlorophyll); the intercept represents an estimate of b' (= 1.10 m^{-1}) (Equation (7.15)). Alternatively the value of K_b can be estimated as the median of ratios of the individual observations of b_{nc}/C_T over the 0–60 mg · m^{-3} range of C_T; the value determined in this manner was 0.105 m^2 · mg^{-1} chlorophyll. Note that the slope of the b versus C_T plot (m = 0.068; Figure 7.35a) differed by only 12% from K_b, indicating the dominant role phytoplankton played (particularly compared to CaCO$_3$ particles) in regulating b in the lake. The intercept of this plot is an estimate

of the nonphytoplankton b (1.22 m^{-1}). To support the development of a mechanistic optical submodel (subsequent section of this chapter) that couples C_T to K_d and SD, the relationship between b and C_T is specified according to $K_b = 0.060$ m^2 · mg^{-1} chlorophyll and a nonphytoplankton value of $b = 1.22$ m^{-1}.

Alternately, the relationship between scattering and C_T is evaluated from the perspective of the number of observations that fall within the K_b bound of 0.05 to 0.15 m^2 · mg^{-1} chlorophyll (Figure 7.35b; Weidemann and Bannister 1986), as pursued previously based on T data (Figure 7.18). Approximately 64% of observations fall within these bounds, if a b' is assumed to equal 0.3 m^{-1} (the minimum oserved). The envelope encompasses approximately the same percentage of observations when b' is fixed at the value determined from linear regression, however, more of the values fall below the lower bound.

7.10 Spectroradiometer Measurements

7.10.1 PAR Attenuation Coefficient

The attenuation coefficient for downwelling irradiance for PAR was also determined from the spectroradiometer scans ($K_{d(PAR)}$) (see Jerome et al. 1983). Paired measurements (n = 49) of K_d (from broad-band sensor) and $K_{d(PAR)}$, made over the 1987–1990 period, are compared in Figure 7.36. Estimates of K_d and $K_{d(PAR)}$ were quite similar considering the less than ideal conditions (e.g., variable cloud cover, wave action) that prevailed during some of measurements. The two measures of the vertical irradiance attenuation coefficient for PAR appear to approach equivalency; e.g., note the rather even distribution of observations about the equivalency line. Sixty percent of the paired observations were within 10% of each other. The slope of the best fit linear regression line forcing a y-intercept of zero, was 0.95. This near equivalent performance supports the response of the quantum sensors

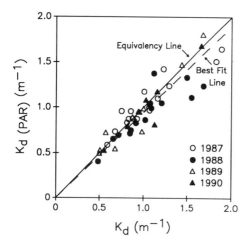

FIGURE 7.36. Comparison of values of the attenuation coefficient for downwelling irradiance for the PAR interval in Onondaga Lake based on measurements with a broad band sensor (K_d) and a spectroradiometer ($K_{d(PAR)}$).

depth distribution, in accordance with the Lambert–Beer Law. The relative rates of attenuation for different wavelengths are determined largely by the absorption spectrum of the lake waters (Kirk 1983).

"In non-productive oceanic waters generally, where water itself is the main absorber, blue and green light both penetrate deeply and to about the same extent, while red light, which water absorbs quite strongly, is attenuated much more rapidly" (Kirk 1983, p. 105). In more productive and/or gelbstoff-enriched waters (e.g., most lakes), blue light is attenuated the most strongly because of absorption by phytoplankton pigments and/or gelbstoff. Green light is usually the most penetrating, followed by red, except in highly colored

(for which a larger database exists) and their use to determine the attenuation coefficient for downwelling irradiation for the PAR interval.

7.10.2 Monochromatic Attenuation

A common way of representing the spectral character of the attenuation of downwelling irradiance is to present depth profiles of selected wavelengths; e.g., blue, green, and red light (Kirk 1983). Profiles of blue (440 nm), green (552 nm), and red (673) light, and PAR, measured with a spectroradiometer in Onondaga Lake, are presented for three selected days in Figure 7.37a–c. The July 1988 and May 1989 profiles (Figure 7.37a and c) were collected during phytoplankton blooms ($C_T = 38$ and $85\,\text{mg chlorophyll} \cdot \text{m}^{-3}$, respectively), dominated by green and cryptomonad algae, respectively. The profile for mid-July of 1987 was taken during a clearing event (SD = 4.75 m; $C_T = 2\,\text{mg} \cdot \text{m}^{-3}$). Measurements for these three days are used subsequently to illustrate other features of spectral quality of penetrating light in the lake. The monochromatic and PAR light profiles have a log-linear

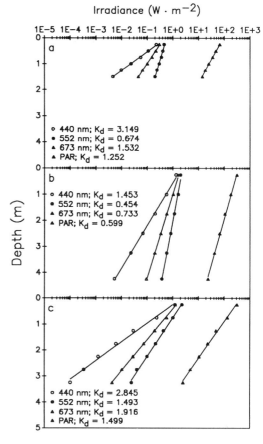

FIGURE 7.37. Depth profiles for downwelling irradiance for three wavelengths and the PAR interval for Onondaga Lake: (**a**) July 11, 1988, (**b**) July 13, 1987, and (**c**) May 22, 1989.

waters, where red light penetrates the best (Kirk 1983).

In all cases monitored to date in Onondaga Lake, the order of attenuation among these three wavelengths has been $K_{d(blue)} > K_{d(red)} > K_{d(green)}$ (e.g., Figure 7.37a–c). The gelbstoff levels that prevail in the lake are apparently adequate to maintain this ordering, even at the rather low phytoplankton biomass concentrations that prevail during the clearing events (see Figure 7.37b). The relative divergence of blue and red light attenuation from green increases during phytoplankton blooms (e.g., Figure 7.37a and c versus Figure 7.37b). Note that the value of $K_{d(PAR)}$ is intermediate to the values presented for the three wavelengths indicating its composite character.

Spectra for $K_{d(\lambda)}$ are presented for the same three days in Figure 7.38, along with a spectrum for Skaneateles Lake (August 5, 1991), the system of highest optical quality (e.g., light penetration, blue color) in central New York. This form of presentation fully resolves the wavelength dependence of $K_{d(\lambda)}$ within the PAR interval. The conditions for the mid-July 1987 and late May 1989 dates bound most of the conditions encountered during the 1987–1990 period. Clearly rather substantial shifts in the spectral features of attenuation occur as a result of the dynamics in phytoplankton biomass and assemblage. Strong preferential increases in absorption of the shorter (blue)

wavelengths occur during phytoplankton blooms. Peaks in $K_{d(\lambda)}$ in the vicinity of the secondary absorption peak of phytoplankton pigments (about 675 nm: Effler et al. 1991a; Kirk 1983; Weidemann and Bannister 1986) were clearly manifested during the blooms but absent during the clearing event (Figure 7.38). The striking disparity in the green interval for the two bloom days probably reflects differences in the absorption spectra of cryptomonads and green algae, and perhaps greater contributions from tripton on the May 1989 date. The spectrum for Skaneateles Lake sharply contrasts the Onondaga Lake spectrum during the clearing event (Figure 7.38), despite very similar chlorophyll concentrations. The highest values of $K_{d(\lambda)}$ in Skaneateles Lake are observed at the red end of the spectrum. The differences are largely a result of the higher gelbstoff levels that prevail in Onondaga Lake (Table 7.8). This comparison serves to demonstrate the important role gelbstoff plays in spectral quality. Even at the lowest phytoplankton concentrations, Onondaga Lake has a more green appearance compared to the blue color of Skaneateles Lake. This characteristic should be considered natural for Onondaga Lake, as the gelbstoff is apparently derived from terrigenous sources.

The temporal features of $K_{d(\lambda)}$ for the mid-June through August interval of 1987 are presented in the form of a surface in Figure 7.39. Strong seasonal dynamics are manifested in this mode of presentation. The most prominent features are the major peak in blue wavelengths and secondary peak at around 675 nm associated with a phytoplankton bloom in late June (~julian day 173), and the minima observed throughout all wavelengths during the clearing event of mid-July (~julian day 195). Oscillations in $K_{d(\lambda)}$ throughout all wavelengths, associated mostly with oscillations in the concentration of phytoplankton pigments, impart the "wavy" appearance to the surface when viewed along the wavelength axis. The "edge" of the surface at the end of the period gives the clearest representation of a $K_{d(\lambda)}$ spectrum (compare to Figure 7.38). Athough this mode of presentation has the advantage of completeness with respect to the

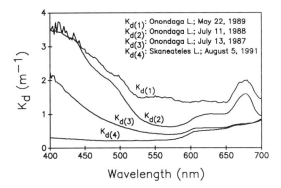

FIGURE 7.38. Spectra for $K_{d(\lambda)}$ in Onondaga Lake for three dates and a single typical spectrum for nearby Skaneateles Lake (modified from Perkins and Effler 1995).

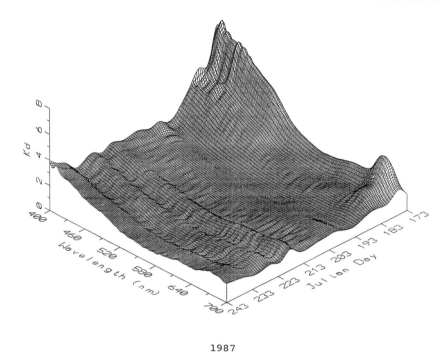

1987

FIGURE 7.39. Surface of $K_{d(\lambda)}$ for Onondaga Lake over the period mid-June–August of 1987.

$K_{d(\lambda)}$ spectra, the quantity of information tends to cloud resolution of details. The time plots of $K_{d(\lambda)}$ for blue (440 nm), green (552), and red (673 nm) light for the same period (Figure 7.40) provide a coarser spectral representation, but this mode of presentation facilitates clearer resolution of temporal structure. The rather broad peak in K_d for all three wavelengths, following by the minima during

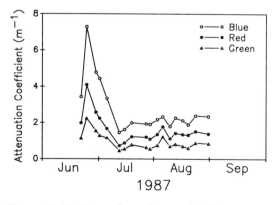

FIGURE 7.40. Time plots of $K_{d(\lambda)}$ for blue (440), green (552), and red (673) light in Onondaga Lake over the period mid-June–August of 1987.

the mid-July clearing event, and subsequent oscillations, are clearly resolved. Perhaps more importantly, the relative spectral attenuation is seen to vary with time, probably largely as a result of dynamics in phytoplankton and perhaps tripton.

Temporal changes in relative spectral attenuation among the same (3) primary wavelengths, and thereby the spectral quality of penetrating irradiance, are represented here by time plots of two ratios of $K_{d(\lambda)}$ values ($b/g = K_{d(440)}/K_{d(552)}$; $r/g = K_{d(673)}/K_{d(552)}$). Distributions of these ratios for all four years are presented in Figure 7.41. Variations in the two ratios track each other to a large extent because the selected blue and red wavelengths are both absorbed by phytoplankton pigments (Kirk 1983; Weidemann and Bannister 1986; Figure 7.38). Some differences in the response of the two ratios are to be expected based on phytoplankton composition and variations in gelbstoff and tripton. Strong seasonal and year-to-year variations in relative spectral attenuation, and therefore the spectral quality of penetrating irradiance, are implied by this analysis. The abrupt peaks in early August

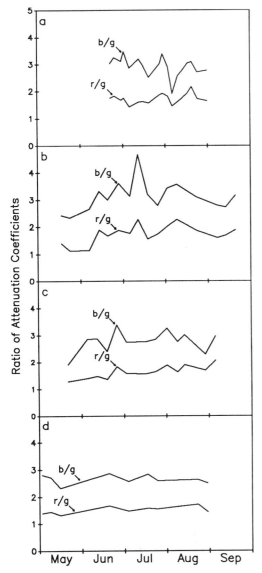

FIGURE 7.41. Time plots of attenuation ratios, b/g ($= K_{d(440)}/K_{d(552)}$) and r/g ($= K_{d(673)}/K_{d(552)}$), for the summers of: (**a**) 1987, (**b**) 1988, (**c**) 1989, and (**d**) 1990.

1987, mid-July 1988, and late June 1989 were in response to rapid increases in phytoplankton biomass.

7.10.3 Spectral Composition of Downwelling Irradiances

The spectral composition of downwelling irradiance narrows progressively with incre-

asing depth as a result of the preferential attenuation of both lower and higher wavelengths. Spectral distributions of downwelling irradiance in Onondaga Lake are presented for the same three days in Figure 7.42a–c. These examples indicate that certain differences in the composition of penetrating irradiance occur. Minima were apparent at about 675 nm on the two days for which phytoplankton blooms prevailed (Figure 7.42a and c), as a result of absorption by phytoplankton pigments. Distinct peaks in the distribution of penetrating irradiances were observed during the green algae bloom of July 1988 and the clearing event of July 1987, whereas the distribution was rather flat over the 550 to 650 nm range during the crypto-

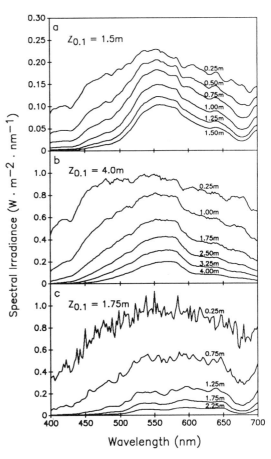

FIGURE 7.42. Spectral distribution of downwelling irradiance with depth in Onondaga Lake: (**a**) July 11, 1988, (**b**) July 13, 1987, and (**c**) May 22, 1989.

monad bloom of May 1989 (Figure 7.42). The peak irradiances at $z_{0.1}$ for these days were at wavelengths of about 560, 580, and 595 nm respectively.

The spectral composition of downwelling irradiance at $z_{0.1}$ for the three days is represented in Figure 7.43 as the wavelength distribution of the fraction of total irradiance contributed, in 10 nm increments, within the PAR interval. This mode of representation is qualitatively consistent with Figure 7.42; such distributions are utilized in the calculation of absorption coefficients for the various components from the respective absorption spectra (see Equation (7.12)).

7.10.4 Spectral Distribution of Emergent Flux

Approximately half of the upwelling irradiance that reaches the surface passes through into the atmosphere; this is referred to as emergent flux. The intensity and spectral distribution of the emergent flux greatly influences the appearance of a water body to an observer (Kirk 1985). The greater the emergent flux, the more muddy, turbid, or milky the water appears (Kirk 1985). The spectral distribution of the emergent flux determines the color of the water to the observer, an important aesthetic feature of water quality. The spectral composition of upwelling irradiance, just as

for downwelling irradiance, is regulated by the absorption spectrum of the water. Oceanic waters and lakes with low concentrations of gelbstoff and phytoplankton, such as Skaneateles Lake, appear blue since this waveband is least absorbed. At higher gelbstoff or phytoplankton pigment concentrations water appears green, because of the more intense removal of blue light. Water can appear muddy brown when high concentrations of gelbstoff-coated tripton prevail, because it is the long wavelength light that is least absorbed.

Upwelling irradiance spectra (normalized for total PAR) collected just below the surface and just above the surface (i.e., emergent flux plus reflected solar radiation) on two different occasions for Onondaga Lake and a single occasion for Skaneateles Lake, are presented in Figure 7.44a–c. The two Onondaga Lake cases correspond to a blue–green algae bloom (Figure 7.44a) and a clearing event (Figure 7.44b). Note that for all three cases the spectrum measured above the water is distinctly flatter than the upwelling spectrum measured just below the water surface, as a result of the inclusion of reflected solar radiation. The peak wavelength band above the water during the algae bloom was in the green interval (Figure 7.44a), consistent with the apparent color to an observer. During the clearing event the water color was blue–green as a result of the shift of emergent flux towards the blue wavelengths (Figure 7.44b). Skaneateles Lake appears decidedly more blue than the Onondaga Lake clearing event case because of the greater contributions of blue light to the emergent flux (Figure 7.44c).

7.11 Partitioning a

The estimated partitioning of a in Onondaga Lake into a_w, a_y, and a_p is presented for the summer of 1987, 1988, 1989, and 1990 in Figure 7.45a–d. A breakdown of the contributions of the three components of a for the four summers is presented in Table 7.13. The estimates for these components are subject to different levels of uncertainty. The principal

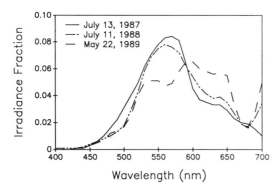

FIGURE 7.43. Wavelength distributions of the fraction of total irradiance contributed in 10 nm increments within the PAR interval in Onondaga Lake, at $z_{0.1}$, for three dates.

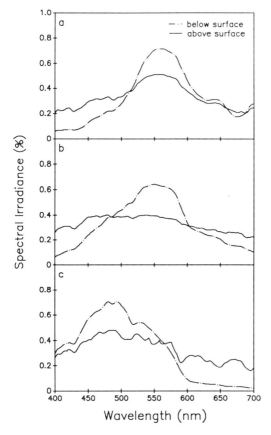

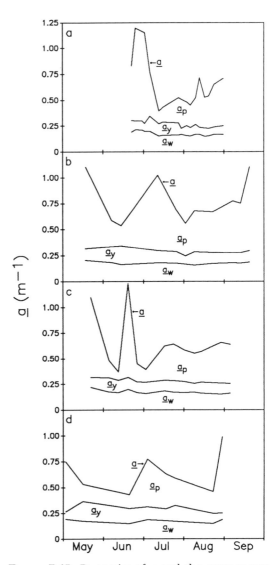

FIGURE 7.44. Spectral distributions of upwelling irradiance (just below water surface) and the sum of emergent flux and reflected solar radiation (just above water surface) for three cases: (**a**) Onondaga Lake, with a blue-green algae bloom, (**b**) Onondaga Lake during a clearing event, and (**c**) Skaneateles Lake.

FIGURE 7.45. Dynamics of a and the components a_w, a_y, and a_p in Onondaga Lake for the summer of four years: (**a**) 1987, (**b**) 1988, (**c**) 1989, and (**d**) 1990.

source of error for a_w is the measurement of $E_{d(\lambda)}$. Additional sources of error for a_y include the spectrophotometric measurement of absorbance of filtered water and the estimate of $a_{y(\lambda)}$ from this measurement. The uncertainty in a_p is much greater because it is determined as a residual here (according to Equation (7.11)). Because of the dependence on the estimated value of a (from broad band quantum sensors, according to Kirk (1981b)), the estimate of a_p is also influenced by potential errors associated with the measurement of E_d and E_u and the estimates of K_d and R. For these reasons, we have the least confidence in a_p. However, the approach for determining a_p

TABLE 7.13. Contributions of a_w, a_y, and a_p to a in Onondaga Lake for the summers of four years.

Year	% Contribution		
	a_w	a_y	a_p
1987	27.5	15.5	57.0
1988	24.2	16.5	59.3
1989	29.7	19.5	50.8
1990	28.4	21.0	50.5
overall 87–90	27.5	17.7	54.8

as a residual is supported by the findings of
Weidemann and Bannister (1986). They
observed good agreement between Kirk's
(1981b) a and the summation of independently
determined values of a_w, a_y, and a_p in
eutrophic Irondequoit Bay. The summations
in their work matched a to within 15% for
sixteen of seventeen comparisons, and to
within 10% for nine of the comparisons.

The dominant absorbing component in all
four years was a_p (Table 7.13). Only during
the clearing events (e.g., mid-July 1987) did
the values of a_p approach the lower magnitudes of the other components. Ninety-nine
percent of the variability observed in a was
explained by variations in a_p, according to
the linear expression, $a = 1.020 \cdot a_p + 0.273$.
This strong relationship is in part a manifestation of the relative uniformity in a_w and a_y
observed. The y-intercept reflects the sum of
a_w and a_y. Similarly, variations in a_p explained
98% of the temporal variations in a observed
in eutrophic Little Sodus Bay (Effler et al.
1991a). The strong variations in a_p in
Onondaga Lake largely tracked the dynamics
of phytoplankton biomass. The relationship
between a_p and C_T is evaluated in Figure 7.46.
Variations in C_T explained 65% of the variability in a_p estimates over the four summers.
The slope (K_a) (see Equation (7.13)),
$0.0084\,m^2 \cdot mg^{-1}$ chlorophyll, is typical of
chlorophyll-specific absorption coefficients

TABLE 7.14. Comparison of a_w values for Onondaga
Lake, 1987–1990.

Year	Mean	Std*	Min	Max	N
1987	0.172	0.020	0.148	0.212	17
1988	0.177	0.013	0.156	0.207	11
1989	0.173	0.018	0.151	0.220	13
1990	0.174	0.016	0.149	0.195	8
Overall	0.174	0.018	0.148	0.220	49

* Std. = standard deviation

reported for phytoplankton (Bannister and
Weidemann 1984). Variations in absorption
characteristics among the phytoplankton (see
Bricaud et al. 1983) and irregular contributions
of tripton probably contributed to the observed
scatter (Figure 7.46). The distinctly nonzero
intercept is representative of the average
absorption by tripton ($a_p' = 0.132\,m^{-1}$; see
Equation (7.13)) over the four-year database.
Reductions in this component may occur
during clearing events, associated with zooplankton grazing of nonphytoplankton particles. The absorption component, a_p, made
the smallest contributions to a in 1989 and
1990 (Table 7.13), the years in which the
lowest summer average chlorophyll concentrations were observed.

The average value of a_w for all the observations (n = 49) was $0.174\,m^{-1}$; the average
values for the four years were very similar
(Table 7.14). However substantial variations
in a_w were observed within each year (Table
7.14) as a result of shifts in the spectral quality
of downwelling irradiance (e.g., a_w range of
0.148 to $0.220\,m^{-1}$), associated mainly with
variations in concentrations of absorbing particles. The character of this feedback from
phytoplankton pigments is illustrated in Figure
7.47; the value of a_w is seen to increase as C_T
increases. This relationship is qualitatively
expected, as absorption by phytoplankton
pigments tends to shift the distribution of
penetrating irradiance to the higher wavelengths which are preferentially absorbed by
water itself. The average a_w value reported for
Onondaga Lake is compared to values reported
for other lakes in Table 7.15. These differences
reflect differences in the spectral quality of
penetrating downwelling irradiance among

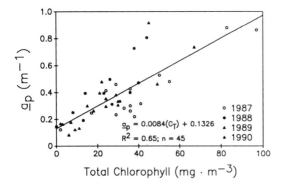

FIGURE 7.46. Evaluation of the relationship between a_p and chlorophyll (C_T) in Onondaga Lake
for summers over the 1987–1990 period (modified
from Effler and Perkins 1995).

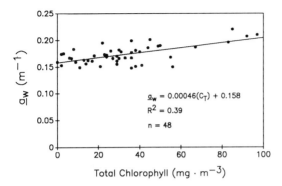

FIGURE 7.47. Evaluation of the relationship between a_w and chlorophyll (C_T) in Onondaga Lake for summers over the 1987–1990 period (modified from Effler and Perkins 1995).

TABLE 7.16. Comparison of a_y values for Onondaga Lake, 1987–1990.

Year	Mean	Std	Min	Max	N
1987	0.093	0.023	0.058	0.145	17
1988	0.117	0.026	0.079	0.180	11
1989	0.110	0.013	0.085	0.140	13
1990	0.121	0.040	0.066	0.192	8
Overall	0.107	0.028	0.058	0.192	49

the lakes. The value for Onondaga Lake falls about in the center of the limited observations presented.

The average value of a_y for all observations (n = 49) in Onondaga Lake was $0.107\,\mathrm{m}^{-1}$ (Table 7.16). The average value for the four years range from $0.093\,\mathrm{m}^{-1}$, in 1987, to $0.121\,\mathrm{m}^{-1}$ in 1990 (Table 7.16). Note that the sum of the average a_w (Table 7.14) and a_y values ($0.281\,\mathrm{m}^{-1}$) very nearly equals the y-intercept value ($0.273\,\mathrm{m}^{-1}$) determined from the regression of a on a_p. The substantial differences in a_y are due to variations in gelbstoff (Figure 7.24a–d) as well as spectral quality associated with the dynamics in phytoplankton biomass. Each of these two interactions are tested separately in Figure 7.48a and b. Neither relationship is particularly strong. The two influences are apparently uncoupled (recall $a_{y(440)}$ and C_T are not correlated in the lake). The value of a_y tends to

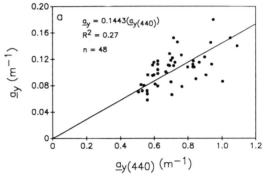

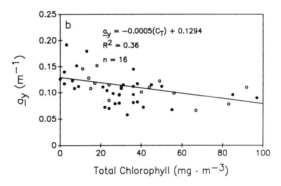

FIGURE 7.48. Testing of factors influencing the dynamics of a_y in Onondaga Lake for summers over the 1987–1990 period: (a) gelbstoff, as $a_{y(440)}$, and (b) chlorophyll (C_T) (modified from Effler and Perkins 1995).

TABLE 7.15. Values of a_w reported for different lakes.

Lake	a_w	Source
Lake Champlain	0.12–0.22*	Effler et al. 1991b
Irondequoit Bay	0.23	Weidemann and Bannister 1986
Oneida Lake	0.23	Weidemann and Bannister 1986
Onondaga Lake	0.174	This work
Lake Ontario	0.077	Effler et al. 1991a
Otisco Lake	0.16	Weidemann et al. 1985
Little Sodus Bay	0.168	Effler et al. 1991a

* Range of longitudinal values

increase as $a_{y(440)}$ increases (Figure 7.48a). However, a_y tends to decrease at higher C_T concentrations (Figure 7.48b), as the preferential absorption of the lower wavelengths by phytoplankton pigments leaves less light available for absorption by gelbstoff.

In an effort to isolate the effect of biomass, the relationship between a_y and C_T was reevaluated for the subset of observations corresponding to a narrow range of $a_{y(440)}$ ($0.65 \leqslant a_{y(440)} \leqslant 0.80$, n = 16; Figure 7.48b). Note that this subset contains one-third of the total observations and essentially covers the entire range of C_T. The relationship for the subset is quite similar to that obtained for the total population, but statistically stronger; it is utilized subsequently in simulating the impact of C_T on a_y. Other subsets were inappropriate to analyze the relationship because of the limited population size(s).

The mean value of a_y was somewhat lower than reported for Irondequoit Bay ($0.13\,m^{-1}$) and most locations assessed in Lake Champlain (Effler et al. 1991b), but more than found in Little Sodus Bay and adjoining Lake Ontario waters (Effler et al. 1991a), and in Otisco Lake (Weidemann et al. 1985). The component, a_y, on average represented from 15.5% (1987) to 21.0% (1990) of a for the four years (Table 7.11).

Examples of absorption spectra for particles obtained for Onondaga Lake on two different days during phytoplankton blooms are presented in Figure 7.49. These spectra are presented as qualitative data to identify certain recurring features associated with absorption by phytoplankton (note that no scaling appears on the y-axis). These were obtained from absorbance scans measured at 10 nm intervals on concentrated samples in an opal glass cuvette (see Weidemann and Bannister 1986). Selective absorption of blue light (peak at about 440 nm) and the secondary absorption maximum at 675 nm are characteristic of phytoplankton pigments. The character of these spectra is generally consistent with those obtained in other systems where phytoplankton pigments are the dominant component in absorption by particles (Kirk 1983, 1985; Weidemann and Bannister 1986). The

disparities in the two spectra may reflect differences in the composition of the phytoplankton assemblage, as well as differences in contributions from detritus and colored particles (characteristically these have spectra quite similar to gelbstoff). Spectra obtained from pure cultures (Kirk 1983) and natural assemblages (Weidemann and Bannister 1986) demonstrate substantial differences among phytoplankton groups (e.g., diatoms, blue−greens, etc.). Green algae were dominant in Onondaga Lake on July 11, 1988, and cryptomonads dominated on May 22, 1989 (Figure 7.49).

The $a_{(\lambda)}$ spectrum can be estimated from the $K_{d(\lambda)}$ spectrum and the value of b estimated from broad band sensor measurements (Kirk 1981b), as b can be assumed invariant with wavelength when scattering is predominantly by particles larger than the wavelength of light (Jerlov 1976). Phillips and Kirk (1984) observed approximate spectral uniformity for b based on simultaneous measurements of spectral beam transmittance, $c_{(\lambda)}$, and $K_{d(\lambda)}$. Thus Kirk's (1984) expression (Equation (7.4b)) can be applied to monochromatic light to estimate $a_{(\lambda)}$. The particle absorption spectrum can then be estimated as a residual, according to

$$a_{p(\lambda)} = a_{(\lambda)} - a_{w(\lambda)} - a_{y(\lambda)} \quad (7.20)$$

Spectra of $a_{(\lambda)}$ and $a_{p(\lambda)}$ determined from spectral irradiance and b, along with $a_{w(\lambda)}$

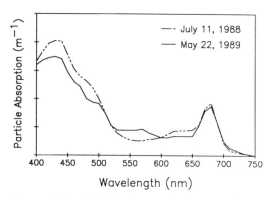

FIGURE 7.49. Partical absorption spectra characteristics in Onondaga Lake during phytoplankton blooms (two examples).

and $a_{y(\lambda)}$, are presented for July 11, 1988, July 13, 1987, and May 22, 1989 in Figure 7.50a–c. The $a_{p(\lambda)}$ spectra of July 11, 1988 (Figure 7.50a) and May 22, 1989 (Figure 7.50c) are characteristic of phytoplankton blooms; that is, a broad peak in the blue light region, and a sharper secondary peak in the red region at about 675 nm (Kirk 1983, 1985, Weidemann and Bannister 1986). The relatively high particle absorption in the green region on May 22, 1989 (Figure 7.50c) appears to be atypical of spectra reported in the literature; this may be characteristic of cryptomonads that were dominant on that day. These spectra imply a dominant role of phytoplankton in regulating a_p, though some

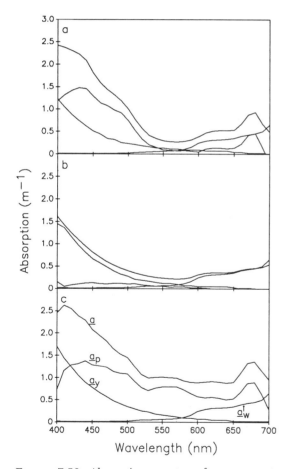

FIGURE 7.50. Absorption spectra of components (a_w, a_y, a_p) and a for the upper waters of Onondaga Lake: (**a**) July 11, 1988, (**b**) July 13, 1987, and (**c**) May 22, 1989.

contribution from tripton cannot be discounted. Note that the $a_{p(\lambda)}$ spectra during algae blooms (Figure 7.50a and c) essentially regulate the character of the $a_{(\lambda)}$ spectra, except at the far blue end of the PAR interval, where gelbstoff absorption contributes significantly. During clearing events (e.g., Figure 7.50b) absorption by particles is minor and overall absorption is regulated by gelbstoff ($a_{y(\lambda)}$) in the blue wavelengths, and water ($a_{w(\lambda)}$) in the red region.

7.12 Partitioning b

7.12.1 Background

Recall that the partitioning of b in Onondaga Lake has been supported by two rather disparate analytical efforts, turbidity measurements before and after acidification, and IPA techniques. The first method has been applied to other hardwater lakes in which $CaCO_3$ particles contribute significantly to the overall particle assemblage of the water column (Effler et al. 1987a, 1991a). This partitioning is limited to two components, T_c and T_{nc} (Equation (7.14)). The second method, IPA, more fully resolves the particle assemblage, however, the resulting data (elemental chemistry and morphometry) is farther removed from the implications of light attenuation than the simple turbidity measurements.

Johnson and Jiao (1992) developed an empirical turbidity apportionment model for Onondaga Lake, based on IPA data for samples (n = 36) collected at $z_{0.1}$ over the May–October 1987 interval. The fraction of measured turbidity (T) was apportioned according to six different particle types (or classes) as specified in Table 7.17 (also see Chapter 5 and Johnson et al. 1991). The "ORG" particle type represents particles with low x-ray emission; these are organic particles, mostly phytoplankton (other than diatoms) and detritus. The "SI" (silica) particle type represents mostly diatom frustules in Onondaga Lake, though quartz is also included. The "CA" category represents $CaCO_3$ particles and "nucleus" particles that are coated with

TABLE 7.17. Classification criteria for the six generic particle types.

Class	Elements	Net X-ray		Note
		Lower limit	Upper limit	
ORG		0.0	100.0	net X-ray
COMP		100.0	1000.0	count
		X-ray Fraction		
SI	Si	0.95	1.0	
	other	0.0	0.05	
CA	Ca	0.95	1.0	
	other	0.0	0.05	
CL	Si	0.15	0.88	total net
	Al	0.05	0.45	X-ray count
	Ca	0.0	0.15	>1000
	Fe	0.0	0.30	
	K	0.0	0.25	
	Mg	0.0	0.20	
	Ti	0.0	0.20	
	other	0.0	0.08	
MS	all other	0.0	1.0	

calcium carbonate precipitate. The "CL" features are alimino-silicate materials, that may include x-ray contributions from several elements (Table 7.17). These particles are clays that are derived from the watershed (Johnson et al. 1991). The "COMP" particles were considered to be features comprised of mixtures of organic and inorganic constituents. In a previous summary of the same data base (Chapter 5; Johnson et al. 1991), these feaures were combined with the ORG particles and designated as NL particles (see Chapter 5). They are considered separately here because they apparently have light-scattering properties intermediate between organic and inorganic particles. The "MS" (miscellaneous) class represents the sum of all other particles not included in the other five classes. A large fraction of these particles are aggregates of $CaCO_3$ and silica, the AGG particles discussed in Chapter 5.

The greatest morphometric uncertainty is associated with the ORG particles because of imaging problems with the scanning electron microscope (SEM), though the COMP class suffers from some of the same problems because of the inclusion of organic material. Unfortunately, because of the lake's hypereutrophic state, the ORG class is a major component of its particle assemblage (Johnson et al. 1991). The fraction of total particle cross-sectional area per unit volume of water (PAV) contributed by each of these six groups was measured for each water sample in the study. These fractional contributions to PAV by each particle type were represented as the percentage contribution made by each of six size classes. The resulting data matrix for each sample was used as input for the turbidity apportionment model.

The model that couples T and measured PAV is based on the concept of optically effective area of particles per unit volume of water (PAV_{eff}), as discussed by Owen (1974). An empirical relationship between T and PAV_{eff} was developed, incorporating some aspects of Mie Theory, to. account for three key attributes of the measurement data matrix:

1. the measured PAV of the organic particles underestimates the true area by a factor of about 4×,
2. different inorganic particle size distributions show different relationships between turbidity and cross sectional area, and
3. some particle types absorb light as well as scatter it.

The underestimation of "true" organic particle cross sectional area was quantified by

Jiao (1991) for samples collected in Onondaga Lake. Average adjustments to reflect "true" organic PAV were made, but it was not possible to further correct for the size distributions of the organic features. The particle size effect was examined by measuring PAV and T in serial dilutions of Onondaga Creek mud boil sediment suspensions which had been subjected to size fractionation using a series of membrane filters (Jiao 1991). Within a specified size distribution, there is a strong linear relationship between T and PAV (Figure 7.51a). However, across the size ranges, there is a dramatic difference in the slopes of the plots. The strikingly greater scattering efficiency of the smallest particle fraction ($<3\mu m$) is consistent with optical theory (Kirk 1983; Treweek and Morgan 1980). When the mud boil data (n = 26) and the Onondaga

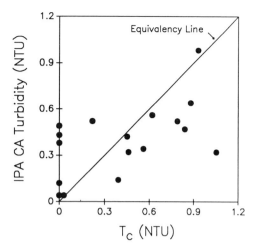

FIGURE 7.52 Evaluation of relationship between T_{CA} and T_C in the upper waters of Onondaga Lake in 1987.

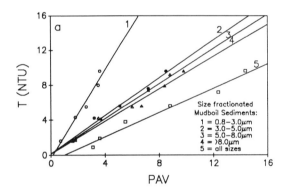

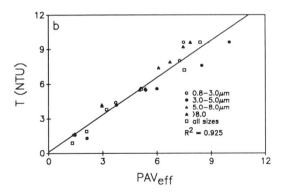

FIGURE 7.51 Relationships between T and: (a) PAV for "mud boil" particles for four nominal size classes, and an unfractionated (i.e., all size classes) sample, (b) PAV$_{eff}$ for "mud boil" particles.

Lake samples (n = 27) were combined, a correction to consider the "composite" (COMP) particles as absorbing a small fraction of light gave a better linear relationship between T and PAV$_{eff}$.

Thus, the empirical model corrects for estimates of "true" organic particle PAV (but does not include a size correction for them), adjusts for a greater scattering efficiency for small particles and a lesser scattering efficiency of large inorganic particles, and reduces the effective PAV of COMP particles to correct for their (hypothesized) partial absorption of light. When the model is applied to the mud boil particle results, the relationship between T and PAV$_{eff}$ is as shown in Figure 7.51b, for which the R^2 value is 0.925. The model coefficients were then employed to calculate the fraction of PAV$_{eff}$ contributed by each of the six particle types in a given Onondaga Lake sample for which a measured T was available.

The consistency of the estimates of turbidity associated with $CaCO_3$ determined by the two partitioning techniques is tested in Figure 7.52, as a plot of the turbidity associated with the CA group (multiplied by 1.25; T_{CaCO_3}) versus T_c. The factor, 1.25, accommodates the contribution of $CaCO_3$ from the MS particle group, developed from review of the IPA data. Despite the inherent limitations in T_c estimates, associated with it being determined as a residual (recall, $T_c = T - T_{nc}$), it is considered the primary measure in this case, as turbidity measurements are more directly related to b than the morphometric information provided on a particle-type basis by the IPA technique. The estimates were far from equivalent over the 1987 test period. One particularly strong outlier ($T_{CaCO_3} = 2.85$ NTU, $T_c = 0.79$ NTU), measured in late June following a major runoff event, was eliminated in the comparison. Coating of particles with $CaCO_3$ had been particularly pervasive at the time of that collection, apparently as a result of the introduction of a large quantity of particle surfaces with the runoff that served as nuclei for $CaCO_3$ precipitation (Johnson et al. 1991). The IPA technique could not differentiate the coatings, thereby overestimating the associated turbidity, whereas acidification left behind the nuclei that continued to impart substantial turbidity. On several occasions T_c was observed to be zero despite T_{CaCO_3} estimates being as high as 0.5 NTU (Figure 7.52). The T_c measurement becomes more imprecise at low values. The occurrence of thin $CaCO_3$ coatings on other particles may also have contributed to these observations. The value of T_c exceeded T_{CaCO_3} for most of the other observations. Though the performance of the IPA partitioning for this component was not particularly strong (Figure 7.52), its considered adequate to support an approximate resolution of the contributions of the six particle types to T in the upper waters of the lake in 1987. Factors contributing to the uncertainty in turbidity partitioning by IPA include the lack of consideration of potential particle-type differences in reflection and refraction (see Kirk 1983) and the imperfect accommodation of the influence of morphometric characteristics on scattering.

7.12.2 IPA Partitioning of T

The measured T in the upper waters of Onondaga Lake in 1987 (Figure 7.53a; originally presented in Figure 7.11c) is partitioned according to the specified six particle types in

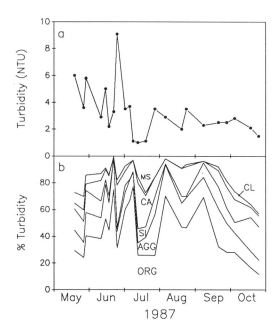

FIGURE 7.53. Temporal distributions in the upper waters (1m) of Onondaga Lake in 1987: (**a**) T, and (**b**) percent contribution of 6 particle classes (Table 7.17).

Figure 7.53b. The strong dynamics in particle concentrations and assemblage over this period have been described in a different format previously (Chapter 5; Johnson et al. 1991). These dynamics are a result of irregular runoff events (note particularly the abrupt changes that occurred in late June in response to a major storm), phytoplankton growth, the precipitation of $CaCO_3$, and other processes. The clearing event of mid-July was accompanied by a strong shift in the relative contribution of the various particle types (Figure 7.53b). The ORG particle type also became less important in the fall as T levels decreased. This particle class was estimated to represent the single largest contribution to T for most of the study period (Figure 7.53b).

The breakdown of estimated contributions to T, and therefore b, by the six particle types for the study period is presented in Table 7.18. The ORG particles were estimated to be the dominant fraction of T. Calcium carbonate was a significant component; recall it represents not only the CA group, but also contri-

TABLE 7.18. Estimated percent contributions of IPA particle types (6) to turbidity (T) in the upper waters of Onondaga Lake, May–October 1987.

Particle-Type*	Percent (%)
ORG	42
COMP	17
SI	9
CA	13
CL	3
MS	16

* See Table 7.17 for descriptions

butes importantly to the MS class. The very small contribution to T by the CL type particles is noteworthy, as it indicates terrigenous (including "mud boil") particles are not an important component of the light-scattering assemblage of particles in the deep water zones of the lake. This is in sharp contrast to conditions that prevail in the mouth of Onondaga Creek, where more than 65% (weight basis) of the sediment load is clays (Yin and Johnson 1984; Chapter 3).

The temporal distributions of the six specific turbidity components estimated from IPA are presented in Figure 7.54a–f. The estimated turbidity associated with these components, except CL, varied strongly during the study period. The minima in ORG (Figure 7.54a) and COMP (Figure 7.54b) turbidity coincided with the lake-wide clearing event in mid-July (maximum SD = 4.8 m), and are consistent with the abrupt decrease in the phytoplankton population that occurred in that period (Chapter 6; Auer et al. 1990). Note the SI group was also low during that interval (Figure 7.54c). High SI turbidity occurred during the major diatom growth periods of late spring and early fall (Chapter 6). The particularly high CA (Figure 7.54d) and MS (Figure 7.54e) turbidities estimated for the period immediately following the major storm of late June have been attributed to the coating of other particles with $CaCO_3$ (see Johnson et al. 1991). The distributions of CA and MS turbidity are correlated over the mid-June–mid-October interval. Note that the abrupt influx of terrigenous particles during the storm event is clearly manifested in a

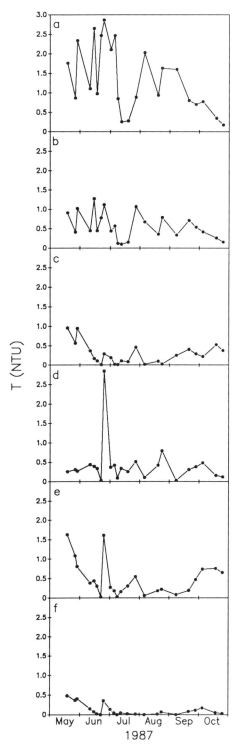

FIGURE 7.54 Temporal distributions of estimated turbidity associated with six paticle types: (a) ORG, (b) AGG, (c) SI, (d) CA, (e) MS, and (f) CL.

modest peak in CL turbidity (Figure 7.54f). The CL turbidity was also estimated to be higher during the relatively high runoff period of May.

The scattering associated with the various particle groups is calculated from the estimates of T partitioning by invoking the widely observed (DiToro 1978; Effler 1988; Kirk 1981b; Weidemann and Bannister 1986) linearity between b and T. Certain of the components of b are determined according to the ratio of the estimated turbidity of a particle type to T; the others are calculated as a residual. For example, the value of b_c is estimated from T_c and T measurements according to

$$b_c = (T_c/T) \cdot b \qquad (7.21)$$

The b_{nc} component is calculated by difference. An analogous approach is used to partition b based on the IPA partitioning.

7.13 Components of Attenuation, Models, and Analysis of Scenarios

7.13.1 Comparison to Previous Partitioning

The results of the suite of in-lake and laboratory optical analyses performed on Onondaga Lake, related analyses of these data, and the optical theory frameworks already presented, represent a basis for evaluating the contribution of the various attenuating constituents to overall attenuation and clarity. Earlier, Effler et al. (1984) developed essentially the same optical framework to elucidate the relative contributions of various constituents to overall attenuation in the lake. A largely speculative partitioning of "average" conditions in 1978 was presented (Effler et al. 1984). Limitations in the underlying optics program for that effort (Effler et al. 1984) included: (1) use of an illuminance meter, equipped with hemispheres directed upwards and downwards (resulted in some systematic errors in application of Kirk's (1981b) pro-

tocol to estimate a and b), (2) no spectral irradiance data, (3) limited gelbstoff data, and (4) shortcomings in the method (Lorenzen 1967) used to determine chlorophyll concentrations (see Chapter 6).

Despite the limitations, the earlier effort provides a valuable perspective for the more recent and fully developed findings presented here. Average values and coefficients developed here for the 1987–1990 interval are compared to the earlier estimates in Table 7.19. The values reported for a_w were similar. The value of a_w reported by Effler et al. (1984) was based on measurements made by Weidemann and Bannister on a single occasion in early August 1983 (Weidemann and Bannister 1986). This earlier value approaches the upper bound reported for the lake over the 1987–1990 interval, implying high C_T concentrations prevailed on the day of measurements in 1983 (Figure 7.47). Differences in responses of the two spectroradiometers used in these efforts could also cause differences in a_w. The value of a_y has been found to be much lower (~60%) than previously estimated (Table 7.19), undoubtedly associated with systematic error(s) (e.g., laboratory measurements, related calculations, and the lack of spectroradiometer data). The single value reported by Weidemann and Bannister (1986) for the date in 1983 is within the ranges reported annually over the 1987–1990 interval (Table 7.16).

Similar values for K_a were obtained (Table 7.19), though they were derived in different ways. The 1978 value was determined as the slope of the regression of a on C_L. More recent analyses of chlorophyll results using different methodologies indicate a better estimate of absorbing phytoplankton pigments is C_L + P (nearly equivalent to C_T; see Chapter 6). The average values of C_L and P for the population used to support partitioning in 1978 were 35.1 and 16.4 mg·m^{-3}, respectively. The use of C_L + P instead of C_L did not influence the value reported for K_a for 1978 greatly, however it yielded a false low y-intercept (i.e., false low nonphytoplankton component of a). The average absorption by phytoplankton pigments ($K_a \cdot C$) in 1978 was

TABLE 7.19. Comparison of partitioning of attenuation: Effler et al. (1984) versus "this chapter."

Parameter	Units	Value	
		Effler et al. (1984)	This chapter
K_d	m^{-1}	1.534	1.238
b	m^{-1}	6.31	3.500
a	m^{-1}	0.950	0.819
a_w	m^{-1}	0.203	0.174
a_y	m^{-1}	0.273	0.107
$\bar{C}^*$	$mg \cdot m^{-3}$	51.5 $(C_L + P)$**	42
K_a	$m^2 \cdot mg^{-1}$ chlorophyll	0.0094	0.0084
$\overline{K_a \cdot C}$	m^{-1}	0.330 (C_L) 0.484 $(C_L + P)$	0.353 (C_T)
a_P'	m^{-1}	—	0.133
a_p	m^{-1}	0.484	0.403
b_c	m^{-1}	1.210	0.354
b_{nc}	m^{-1}	5.100	3.082

$\bar{C}^*$ = Average chlorophyll concentration
$(C_L + P)$** = Average of the sum of chlorophyll a and phaeophytin concentrations determined according to the method of Lorenzen (1976)

underestimated (0.330 versus 0.484 m^{-1}; Table 7.19) as a result of the use of C_L instead of $C_L + P$. The magnitude of the underestimate ($K_a \cdot P$, 0.154 m^{-1} (Effler et al. 1984)) was incorrectly attributed to the nonphytoplankton absorbing particles. Thus absorption by nonphytoplankton organic and inorganic particles (tripton, a_p') was not accommodated in the earlier partitioning effort. In the 1987–1990 interval, the estimated value of a_p' (0.133 m^{-1}) was similar to the values reported for a_w and a_y. Note that this omission in the original partitioning analysis was largely compensated for by the false high value determined for a_y (Effler et al. 1984).

Rather substantial differences in the partitioning of b between b_c and b_{nc} have been reported for the two efforts; b_c was estimated to represent about 19% of b in 1978, while it was determined to be a substantially smaller fraction (11%) of b over the 1987–1990 interval. A decrease in b_c has probably occurred since closure of the soda ash/chlor-alkali, as indicated by reductions in average T_c since 1985 (Table 7.4). However, we have much greater confidence in the latter estimate(s), as it is based on turbidity measurements (see Effler et al. 1987a, 1991b;

Weidemann et al. 1985). The 1978 estimate was substantially more speculative. It was calculated as a percent of a residual, based on the observation by Yin (1984) that approximately 80% of the (inorganic) particles in the upper waters of the lake in 1981 were CaCO$_3$ (i.e., the critical morphometric features of the particles in regulating scattering efficiency were not considered) (Kirk 1983; Figure 7.51a). The average contribution of b_c to b determined subsequently for a portion of the summer of 1985 (10%; Table 7.4) may be more representative of the period before closure of the soda ash/chlor-alkali facility.

Much of the framework for partitioning attenuating components presented by Effler et al. (1984) has been used to evaluate a comprehensive suite (particularly over the 1987–1990 interval) of optical measurements and related water quality parameters, as described in this chapter. However, the originally presented speculative partitioning for 1978 conditions has been shown to have several important flaws, including (1) major overestimation of a_y, (2) failure to identify tripton (a_p') as a absorbing component, (3) underestimation of $K_a \cdot C$ (absorption by phytoplankton pigments), and (4) major uncertainty in, and probably overestimation of, b_c. The partitioning developed in this chapter, based on the 1987–1990 data base, supplants the partitioning presented for 1978 (Effler et al. 1984). The shortcomings in the earlier partitioning analyses, specifically the false high a_y, the false low $K_a \cdot C$, and probably false high b_c value (Table 7.19) lead Effler et al. (1984) to suggest that transparency goals for Onondaga Lake (e.g., SD routinely $>1.2\,\mathrm{m}$) may be attainable through other than nutrient-based restoration efforts. The improved partitioning developed here (Table 7.19) and subsequent evaluations using optical models *do not support this position*. Remediation of the continuing problem of poor clarity in Onondaga Lake must focus on reduction of phytoplankton growth.

7.13.2 Models for Attenuation and Attenuating Components

The critical importance of clarity, attenuation, and spectral quality to optical aesthetics and primary production make the development of reliable quantitative frameworks that specify the relationships between these features and attenuating components essential for effective lake management. These frameworks range from strictly empirical expressions between chlorophyll concentrations and K_d and SD (see Figures 7.16 and 7.17), to fully mechanistic frameworks that completely describe the underwater light field with stochastic techniques such as Monte Carlo calculations (Kirk 1981a). Here we identify three different optical submodels that can be used to estimate optical characteristics of Onondaga Lake from concentrations of attenuating components. The advantages and disadvantages of each are discussed, and the submodels are used to demonstrate the role the various constituents play and to evaluate selected scenarios.

Model I is strictly empirical; it includes the K_d-C_T (Figure 7.16) and SD^{-1}-C_T (Figure 7.17) linear regression expressions developed from the respective paired observations made over the 1987–1990 period. The routine incorporation of clearing events in this database has

greatly increased the ranges of the observations compared to earlier years, and thereby enhanced the credibility of these empirical relationships to support simulations of management interest (e.g., future lower C_T concentrations). Observed deviations from these relationships reflect variability in the pigment content of phytoplankton and coupled and uncoupled variations in other attenuating components (e.g., CaCO$_3$ particles, other tripton, gelbstoff). This model type cannot be used to evaluate the contributions of components other than phytoplankton to K_d and SD, and the spectral implications of the various attenuating components.

Model II utilizes a hybrid of empirical and mechanistic approaches (also see Effler and Perkins 1995). Attenuation (K_d) is partitioned according to spectral (PAR) average values of a and b (Equation (7.4)), which in turn are partitioned according to the summation of their spectral average components (Equations (7.11) and (7.14)). The dependencies of the three absorbing components, a_p, a_y, and a_w on the concentration of phytoplankton pigments (C_T) are specified by empirical linear regression expressions developed from paired measurements and estimates (Figures 7.46, 7.47, and 7.48b) for the lake. An approximate (e.g., study average) partitioning of a_p into contributions from tripton ($a'_p = 0.132\,\mathrm{m}^{-1}$, as intercept of a_p versus C_T linear regression analysis) and phytoplankton ($0.0084 \cdot C_T$) is provided (Figure 7.46). The dependencies of a_w and a_y on C_T reflect the influence of shifts in spectral quality of penetrating irradiance that accompany changes in phytoplankton biomass. The dependence of a_y on gelbstoff ($a_{y(440)}$) is also represented by a linear regression expression (Figure 7.48a). The value of the scattering coefficient is assumed to be invariant throughout the PAR interval (e.g., Phillips and Kirk 1984), thus b is not influenced by shifts in spectral quality. Note that Bricaud et al. (1981) report evidence that scattering in natural waters varies weakly with wavelength. The value of b can be partitioned according to b_c and b_{nc} (Equation (7.14)), based on turbidity measurements, or more speculatively by the partitioning of turbidity based on IPA analyses (e.g., Figure 7.53). The relationship between b and C_T is based on the linear regression relationship developed over the C_T range of 0–60 $mg \cdot m^{-3}$ (Figure 7.35a). Model II simulates the impact of changes in C_T on K_d in a more mechanistically sound manner than the strictly empirical approach of Model I. Further, unlike Model I, Model II can predict the impact of variations in gelbstoff and strictly scattering components (e.g., $CaCO_3$) on K_d. Related implications for clarity (SD) are simulated according to contrast transmittance theory utilizing the N value (7.89) determined specifically for Onondaga Lake (Figure 7.34). Because Model II only accommodates spectral average conditions, it cannot simulate the effects changes in absorbing components have on the spectral quality of penetrating irradiance.

Model III is the most mechanistic of the three. A key component is the software SOL5, a BASIC program designed by J.T.O. Kirk (1984b) to calculate downwelling irradiance ($E_{d(\lambda)}$) and $K_{d(\lambda)}$. The program has spectral capabilities—fifteen 20 nm wavebands (SOL5). In each waveband, Equation (7.4b)

$$K_d = \frac{1}{\mu_o}[a^2 + (0.425\mu_o - 0.19)ab]^{0.5} \quad (7.4b)$$

is used to calculate K_d for the euphotic zone (just below surface to $z_{0.01}$)) from specified values of b and the components of a. A set of spectral distribution data for incident solar flux for clear, sunny weather, taken from the literature (Bird et al. 1982) is incorporated in the program. The water surface is assumed to be flat. The value of μ_o is calculated using Snell's law from the solar altitude (i.e., time of day, time of year, and position). The spectral average value of b is entered into the program; this value is applied to the center of the PAR range and the program calculates b as a function of wavelength ($bf(\lambda)^{-1}$), as supported by Bricaud et al. (1981). This assumption with respect to the wavelenth dependency of b produces only minor differences in spectral average values (e.g., a, a_p and K_d) compared to the assumption of spectral uniformity in b. The values of $a_{w(\lambda)}$ are fixed in the program

(Smith and Baker 1981). The summed values of $a_{p(\lambda)}$ and $a_{y(\lambda)}$ are entered at increments of 20 nm (starting at 410 m). Spectral downwelling irradiance ($E_{d(\lambda,z)}$) is calculated (and printed out) for 10 depths (z), according to (Kirk 1984b)

$$E_{d(\lambda,z)} = E_{d(\lambda,o)}e^{-K_{d(\lambda)} \cdot z} \quad (22)$$

in which $E_{d(\lambda,o)}$, $E_{d(\lambda,z)}$ = downwelling irradiances just below the surface and at depth z in a 20 nm waveband.

Kirk (1984b) found excellent agreement between predictions of K_d with this model and a much more complex Monte Carlo model (e.g., Kirk 1981a, 1989) based on the inputs for inherent optical parameters. However, SOL5 is limited to downwelling irradiance (Kirk 1984b). More complex approaches are required (e.g., Kirk 1981a, 1984a, 1989) if information on the underwater radiance distribution or other detailed aspects of the light field are required.

The wavelength dependencies of a_p on C_T and a_y on $a_{y(440)}$ were developed from the Onondaga Lake data base to support development of relationships between the apparent properties of K_d and SD and the attenuating components of phytoplankton and gelbstoff. The relationship between a_p and C_T was evaluated empirically at 20 nm intervals (410, 430, etc.), based on 17 paired sets of measurements obtained in the summer of 1987 (n = 4), 1988 (n = 9), and 1989 (n = 4). The spectra of a_p were determined from K_d, a_y, and a_w spectra, and b as described previously. Empirical relationships determined between $a_{p(\lambda)}$ and C_T for Onondaga Lake for eight selected wavelengths throughout the PAR interval are developed in Figure 7.55a–h. The increases in $a_{p(\lambda)}$ with increases in C_T were best represented by polynomial expressions for blue and red light and by linear relationships for green light. All the relationships were strong; e.g., $R^2 \geq 0.77$. The value of $a_{y(\lambda)}$ was specified according to the relationship (Equation (7.19))

$$a_{y(\lambda)} = a_{y(440)} \exp[S(440 - \lambda)] \quad (7.19)$$

The mean S value for 88 observations of 0.0166 nm^{-1} was maintained for all simulations. For simulations of the impact of variation

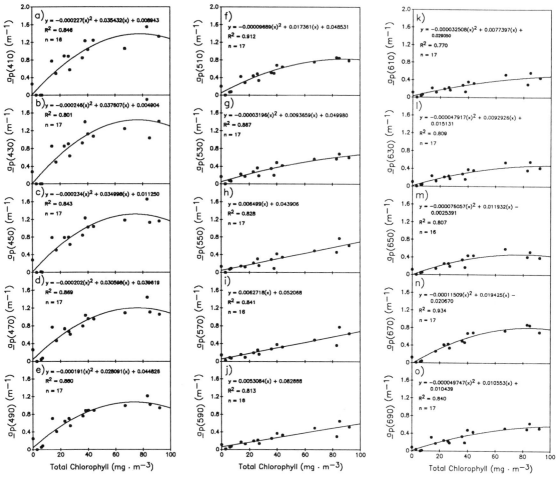

FIGURE 7.55. Evaluation of relationships between $a_{p(\lambda)}$ and C_T at (**a**) 410 nm, (**b**) 430 nm, (**c**) 450 nm, (**d**) 470 nm, (**e**) 490 nm, (**f**) 510 nm, (**g**) 530 nm, (**h**) 550 nm, (**i**) 570 nm, (**j**) 590 nm, (**k**) 610 nm, (**l**) 630 nm, (**m**) 650 nm, (**n**) 670 nm, and (**o**) 690 nm.

in C_T, the value of $a_{y(440)}$ was kept constant at $0.7\,\text{m}^{-1}$, the average of the 110 observations over the 1987–1990 interval. To evaluate the implications of changes in gelbstoff, C_T was maintained constant, and $a_{y(440)}$ was adjusted by multipliers of more than, and less than, one. Phytoplankton scattering was specified according to the value of K_b (Equation (7.15)) determined in Figure 7.35a. To evaluate implications of the variation in C_T, for both Models II and III, nonphytoplankton scattering was specified as the sum of b_c and b'.

7.13.3 Model Performance, Sensitivity, Scenarios, and Management Implications

The predicted K_d-C_T and SD-C_T relationships obtained by application of Model II are compared to the observations for the 1987–1990 database in Figure 7.56a and b. The model performed well in matching the observations and can be described as calibrated.

Sensitivity analyses were conducted to evaluate the relative importance of sources of variability to the overall variability in the K_d-

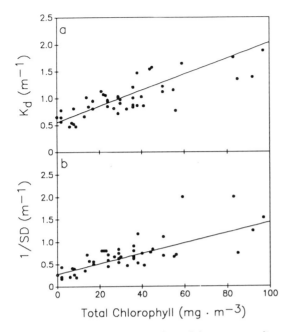

FIGURE 7.56 Performance of Model II: (a) prediction of K_d as a function C_T, and (b) prediction of SD as a function of C_T. Empirical relationships included for comparison (modified from Effler and Perkins 1995).

C_T and SD-C_T relationships (Figures 7.56). Uncertainty limits were set according to statistical measures of performance of the contributing component relationships to maintain a degree of objectivity in selection of the bounds for the sensitivity analyses. The uncertainty limits for sensitivity analyses on the a_p-C_T and b_{nc}-C_T relationships were selected as ± 1 standard error of estimate on the y-intercept (a_p' and b', respectively). The uncertainty limits for N were set as ± 1 standard error of estimate of the best estimate of N. The K_d-C_T part of the model is much more sensitive to the uncertainty in the a_p-C_T relationship than the SD-C_T predictions (Figure 7.57a and b). Conversely, the SD-C_T part of the model is much more sensitive to the uncertainty in the b_{nc}-C_T relationship (Figure 7.57c and d). The influences of uncertainties in other estimates (e.g., N; see Figure 7.57e) and relationships (e.g., a_w-C_T and a_y-C_T) incorporated in the model are minor by comparison. These analyses indicate

that the major sources of variability in the K_d-C_T and SD-C_T relationships in Onondaga Lake are variations in the a_p-C_T and b_{nc}-C_T relationships, respectively. Much of this is probably due to the dynamics of tripton in the lake.

The mechanistic framework of the model also supports quantitative analyses of implications of changes in nonphytoplankton attenuating components on measures of light penetration. As examples, model predictions of the relationships between K_d and gelbstoff concentration ($a_{y(440)}$) and SD and the tripton component of scattering ($b_c + b'$) for Onondaga Lake are presented in Figure 7.58a and b for two specified C_T concentrations. The specified C_T concentrations correspond approximately to the annual minimum and averages over the 1988 and 1990 interval. The SD $- a_{y(440)}$ and $K_d - (b_c + b')$ relationships are not shown, as these measures of light penetration are inherently less sensitive to these components (e.g., Figure 7.57). Both of the modeled relationships were predicted to be linear. The average and range of $a_{y(440)}$ observations (Figure 7.58a), and the best estimate and bound of $+1$ standard error of the estimate of b' (Figure 7.58b) are included for reference. The range of gelbstoff encountered corresponds to a modest change in K_d ($<0.1\,\mathrm{m}^{-1}$; Figure 7.58a). An approximate three-fold increase in gelbstoff from the average concentration would increase K_d by about $0.2\,\mathrm{m}^{-1}$, or about 50% of the increase associated with an increase (a rather commonly observed fluctuation) of C_T from 5 to $30\,\mathrm{mg} \cdot \mathrm{m}^{-3}$ (Figure 7.58a). Such an increase in gelbstoff could only occur as a result of anthropogenic influences. A three-fold decrease in gelbstoff concentration, which is unrealistic based on its terrigenous origins for this system, would reduce K_d by only about $0.1\,\mathrm{m}^{-1}$, though it would result in the water having a more blue color during clearing events (see Figure 7.38).

The predicted relationship between SD and tripton scattering (Figure 7.58b) reaffirms the sensitivity of clarity to scattering. Modest increases in $b_c + b'$, associated with terrigenous particle loading or $CaCO_3$ precipitation events, decrease SD markedly. At $C_T = 5\,\mathrm{mg} \cdot \mathrm{m}^{-3}$, an

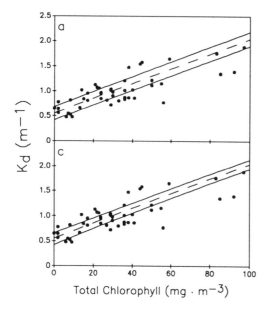

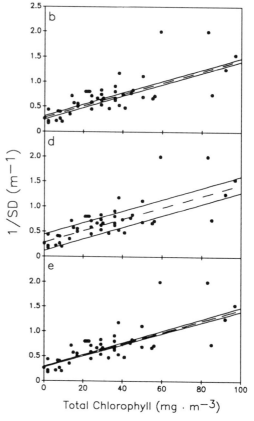

FIGURE 7.57 Sensitivity of Model II predictions of K_d-C_T and SD-C_T relationships to component uncertanties, as ± 1 standard error of estimate of coefficients: (a) a'_p, (b) a'_p, (c) b', (d) b', and (e) N (modified from Effler and Perkins 1995).

increase in tripton scattering of about $1.6\,\mathrm{m}^{-1}$ would reduce SD by the same amount as an increase of C_T from 5 to $30\,\mathrm{mg}\cdot\mathrm{m}^{-3}$. Such increases have been observed to occur irregularly for both b' and b_c in Onondaga Lake.

The relative contributions of the attenuating components to the distributions of K_d and SD observed in Onondaga Lake in 1988 are depicted in Figure 7.59a and b. The general approach is similar to that used often in mechanistic water quality modeling (Canale et al. 1994; DiToro and Connolly 1980; Martin et al. 1985), in which contributing processes (e.g., individual oxygen sink processes contributing to a depletion in dissolved oxygen) are sequentially suppressed to identify the relative importance of the various components to temporal distributions of measures of water quality. The SD analysis is presented for SD^{-1}

(Figure 7.59b) to make it more directly comparable to the K_d analysis and to avoid the potentially misleading character of results for scenarios of sequential decreases in absorption and/or scattering at low levels of attenuation (causing very large incremental increases in SD, e.g., see Equation (7.5)). The distributions of "observed" K_d and SD presented in Figure 7.59 differ somewhat from the actual measurements (Figure 7.31e and f), because they are generated with the model from the documented distributions of the attenuating components. The time distributions of the tripton components of absorption and scattering (non-CaCO$_3$) were first determined by difference (e.g., $a_p' = a - (a_w + a_y + K_a \cdot C_T)$). The sequence of suppression of attenuating components was $K_a \cdot C_T$ and $K_b \cdot C_T$ ($C_T = 0$, phytoplankton; i.e., simultaneously), a_p' and

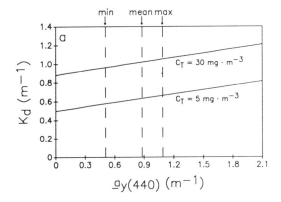

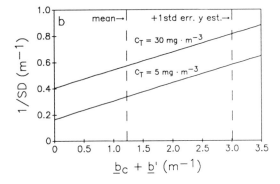

FIGURE 7.58 Predicted relationships; according to Model II; for two specified concentrations of C_T: (**a**) K_d as a function of gelbstoff ($a_{y(440)}$), and (**b**) 1/SD as a function of non-phytoplankton scattering ($b_c + b'$) (modified from Effler and Perkins 1995).

b' (tripton), b_c (CaCO$_3$, scatters without absorbing (Weidemann et al. 1985)), and a_y (gelbstoff). This order moves from the most manageable of the components, phytoplankton, to the least manageable, gelbstoff.

The partitioning between phytoplankton and noncalcite tripton is undoubtedly imperfect, particularly because of temporal variations in the scattering and absorption characteristics of phytoplankton in the lake. For example, we suspect that the tripton contribution in August and September may be overestimated (Figure 7.59). Despite these limitations, the analysis clearly demonstrates that the components primarily responsible for the limited light penetration in Onondaga Lake were phytoplankton and tripton other than CaCO$_3$. Tripton (other than CaCO$_3$) can be dominant, particularly after a major runoff event. Gelbs-

toff and CaCO$_3$ were unimportant by comparison (Figure 7.59). Note the greater relative contributions of a_y to K_d (Figure 7.59a) and b_c to SD (Figure 7.59b), consistent with theory. The partitioning analysis presented here for 1988 is generally representative of conditions for the entire 1987–1990 interval. Management efforts to improve the clarity of the lake should focus on reduction of phytoplankton growth (e.g., reduction in phosphorus loading (Auer et al. 1990)), not only because it is an important component responsible for the limited light penetration (Figure 7.59), but it is also the most subject to remediation. Improvements in clarity achieved through reductions in phytoplankton biomass (e.g., reductions in external phosphorus loading) would be compromised irregularly by particle inputs from major runoff events (e.g., Figure 7.59b).

Remediation of the degraded clarity conditions of Onondaga Lake should focus on the

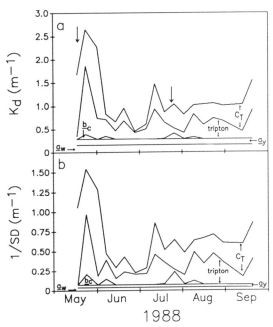

FIGURE 7.59 Model analysis of contributions of components to attenuation and clarity in Onondaga Lake, for the conditions of 1988: (**a**) K_d, and (**b**) 1/SD. "Tripton" refers only to the non-CaCO$_3$ components (modified from Effler and Perkins 1995).

attribute of duration (e.g., percent occurrence) of acceptable transparency, consistent with swimming safety concerns. Average SD transparency is not a valuable statistic with respect to the swimming goals for Onondaga Lake, particularly because of the extremely high SD values that are observed annually in the summer during "clearing events." For example, the summer average SD in Onondaga Lake over the 1987–1990 interval was 1.96 m, but the 1.2 m standard was only met for 50% of the observations.

Prevalence of the very high SD values observed during "clearing events" is not a realistic goal for Onondaga Lake. These values are an artifact of the very efficient grazing of all particles (phytoplankton and non-phytoplankton) by zooplankton. "Clearing events" are observed in a number of productive lakes (Lampert et al. 1986). Recall that Otisco and Owasco Lakes, two of the less productive lakes that support extensive contact recreation, never reach the SD maxima reported for Onondaga Lake during "clearing events."

The optics models can be used to establish the limits of attenuating components that will meet SD standards/goals for the lake. Model II and the data collected over the 1987–1990 interval are used here to identify C_T values that meet SD goals of 1.2 m (New York State Department of Health/bathing beach) and 2.0 m (Figure 7.56b). A SD of 1.2 m corresponds to a $C_T \approx 48\,\mu g \cdot L^{-1}$. Thus to maintain SD ≥ 1.2 m, the value C_T must remain < $48\,\mu g \cdot L^{-1}$. Values of SD were more than 1.2 m for 95% of the 1987–1990 observations when $C_T < 48\,\mu g \cdot L^{-1}$ (Figure 7.56b). Recall that SD may fail to meet the 1.2 m goal at $C_T < 48\,\mu g \cdot L^{-1}$ when nonphytoplankton scattering is high (e.g., terrigenous particles after a runoff event, or a "whiting" event). For example, SD = 1.2 m would not be met at $C_T = 30\,\mu g \cdot L^{-1}$ if $(b_c + b') = 3.5$ (Figure 7.58b). If the value of C_T was kept below $17\,\mu g \cdot L^{-1}$, a SD goal of 2.0 m could be met in Onondaga Lake most of the time (Figure 7.56b). Values of SD were more than 2.0 m for 70% of the 1987–1990 observations when $C_T < 17\,\mu g \cdot L^{-1}$ (Figure 7.56b).

7.14 Summary

7.14.1 Background

The visual aesthetic features of Onondaga Lake have been degraded for quite some time. Frequently the lake has had poor clarity and a turbid appearance. In this chapter the optical features of the lake have been documented, based on the analysis of a comprehensive suite of measurements of optical properties collected over a number of years. These measurements have been utilized, in combination with certain laboratory measurements and established optical theory, to determine regulating processes and materials. This has provided essential management information by establishing the appropriate focus for remedial activities and the feasibility of attaining certain optical attributes. Relationships have been developed that relate K_d and SD to routinely measured water quality parameters. These relationships have been used to predict optical characteristics for the lake for a range of concentrations of light attenuating materials (e.g., management scenarios).

7.14.2 Light Penetration, Attenuating Constituents, and Empirical Relationships

Review of the SD (23 yr) and K_d (10 yr) data bases indicates that the extent of light penetration in the lake has improved in recent years. The SD measurement is more closely coupled to the public's perception of water quality than is K_d. Abrupt increases in SD, termed "clearing events," occurred in most summers from the late 1960s through the late 1970s. These events apparently did not occur over the 1980–1986 interval. However, clearing events have occurred annually, and with greater intensity and duration, since 1987. The maximum SD value observed to date is 7.5 m. The 4 ft (~1.2 m) SD criterion for opening a public bathing beach in New York State was met for only 20% of the summer observations over the 1968–1986 interval. This increased to about 50% for the 1987–1990 interval. Clarity at the north station was frequently greater than at the south station in the late 1960s and in 1978. However, more spatially uniform clarity conditions are presently maintained.

A major shift to lower K_d values has occurred since 1987, consistent with the reported improvement in clarity. Sixty-five percent of the K_d observations made in the period 1978–1986 fell in the interval 1.25 to $2.0\,m^{-1}$. In the period 1987–1990 approximately 70% of the observations were less than $1.25\,m^{-1}$. The median value for the observations made in the 1978–1986 interval was $1.53\,m^{-1}$; the value for the 1987–1990 interval ($1.07\,m^{-1}$) was 30% lower.

The relationship between K_d and SD has widely been assumed to remain uniform, and has been represented by the $K_d \cdot SD$ product (= constant). The $K_d \cdot SD$ product has not remained uniform over the range of light penetration conditions encountered seasonally and year to year in Onondaga Lake, as a result

of substantial variations in the relative contributions of a and b to attenuation. The $K_d \cdot SD$ product increases during clearing events indicate decreases in the relative contribution of scattering, which is also supported by measurements of R and estimates of a and b. The shift in the attenuating processes is attributed to the efficient grazing of abiotic particles, as well as phytoplankton, by nonselective zooplankton prior to and during the events. Summer average values of the $K_d \cdot SD$ product increased in the late 1980s associated with the routine occurrence of clearing events, and perhaps reductions in the concentration of $CaCO_3$ particles. A polynomial expression that describes the relationship between SD and K_d for the entire data base is presented in Table 7.20 (Equation (7.23)).

Turbidity, an approximate surrogate measure of b, was found to be highly correlated to the concentration of suspended solids in the upper waters of the lake (Table 7.20; Equation

TABLE 7.20. Empirical relationships between attenuating materials and measures of light penetration for Onondaga Lake.

Equation number	Relationship	Symbols/comments
(7.23)	$SD = -0.170\,(K_d^{-1})^4 + 1.301\,(K_d^{-1})^3$ $-1.359\,(K_d^{-1})^2 + 1.506\,K_d^{-1} + 0.193$	SD = Secchi disc transparency (m) K_d = Vertical attenuation coefficient for for downward irradiance (m^{-1}) $R^2 = 0.89$; based on 10 yr paired data
(7.24)	$T = 0.63 \cdot SS - 0.02$	T = Turbidity (NTU), SS = conc. of suspended solids $(mg \cdot L^{-1})$; $R^2 = 0.87$; based on 13 paired observations in 1990
(7.25)	$SD = 3.82 \cdot T^{-1}$	$R^2 = 0.82$, n = 150; 1985–1990; poor correlation at $T \geqslant 2$ NTU and $SD \leqslant 2$ m
(7.26)	$K_d = 0.22 \cdot T + 0.59$	$R^2 = 0.62$, n = 118; 1985–1990
(7.27)	$SD^{-1} = 0.023 \cdot C_T + 0.127$; $C_T \leqslant 30$	$R^2 = 0.71$, n = 36; C_T = conc. of total chlorophyll $(mg \cdot m^{-3})$
	$SD^{-1} = 0.008 \cdot C_T + 0.479$; $C_T > 30$	$R^2 = 0.30$, n = 34
(7.28)	$K_d = 0.016 \cdot C_T + 0.58$, $C_T \leqslant 30$	$R^2 = 0.45$, n = 38
	$K_d = 0.010 \cdot C_T + 0.81$, $C_T > 30$	$R^2 = 0.22$, n = 35
(7.29)	$SD = 7.89\,(K_d + c)^{-1}$; $(K_d + c)^{-1} < 0.65$	$R^2 = 0.76$, n = 121
	$SD = 7.09\,(K_d + c)^{-1}$; $(K_d + c)^{-1} < 0.4$	$R^2 = 0.66$, n = 107
(7.30)	$a = 1.020 \cdot a_p + 0.273$	$R^2 = 0.99$, n = 45
(7.31)	$a_p = 0.0084 \cdot C_T + 0.133$	$R^2 = 0.65$, n = 45
(7.32)	$a_w = 0.00046 \cdot C_T + 0.158$	$R^2 = 0.39$, n = 48
(7.33)	$a_y = 0.00050 \cdot C_T + 0.129$	$R^2 = 0.36$, n = 16

(7.24)). Calcium carbonate (CaCO$_3$) particles represent a common, but usually minor, component of the turbidity (T) of the upper waters of the lake. The average magnitude of CaCO$_3$ turbidity (T$_c$) was lower in recent years (1988 –1990) compared to the 1985–1987 interval. Pronounced increases in T in the upper waters of the lake are observed after most major runoff events (e.g., >1.5 in. of rainfall), apparently associated with increases in both internally produced (e.g., phytoplankton and CaCO$_3$) and externally derived particles. Minima in T, as well as phytoplankton biomass (C$_T$), have become decidedly lower with the advent of the summer clearing event(s) in 1987.

The strong variations in SD and K$_d$ observed in the lake have been coupled to dynamics in T and C$_T$. For example peaks in K$_d$ and SD minima coincided with C$_T$ and T maxima, and minima in K$_d$ and SD maxima observed during clearing events corresponded to minima in C$_T$ and T. The important role zooplankton grazing has played in regulating the dynamics of C$_T$ are described in Chapter 6. Improvements in average light penetration conditions in the lake in recent years (1987–1990) reflect reductions in average C$_T$ concentrations and T over the same period. Empirical relationships developed between T and C$_T$ and measures of light penetration, specific to Onondaga Lake, are presented in Table 7.20. The SD-T relationship is of limited value in predicting changes in SD for modest changes in T at high levels of T. The strong performance of the K$_d$-T expression is surprising in light of its divergence from optical theory. The expressions for K$_d$-C$_T$ and SD-C$_T$ are two-part linear least squares regression fits for data collected over the 1987–1990 interval. The demarcation C$_T$ value for these relationships was 30 mg · m^{-3}. Increased dependency on C$_T$ was manifested at low C$_T$ concentrations for both measures of light penetration. These expressions represent an appropriate empirical basis to predict light penetration from values (or predictions) of C$_T$.

Gelbstoff, as measured by the absorption of filtered water at 440 nm ($a_{y(440)}$, m^{-1}), has varied seasonally and year to year in the upper waters of the lake in response to dynamics in

external loading and probably internal loss processes (e.g., coating of CaCO$_3$). There is no evidence that phytoplankton activity has been a significant source of gelbstoff to the lake. Gelbstoff absorption spectra for the lake demonstrated the widely observed exponential decrease with increasing wavelength. Gelbstoff is not presently an important regulator of light penetration in Onondaga Lake. However, it has an important influence on spectral quality during periods of very low phytoplankton biomass.

7.14.3 Regional Comparisons

The extent of light penetration and the concentrations of attenuating components in Onondaga Lake were compared to conditions documented for several other hard water systems in central New York, to place the degraded optical features of the lake in perspective for water quality managers (Table 7.21). The comparison systems represent a wide range of trophic states (Skaneateles Lake is oligotrophic and Cross Lake is nearly as eutrophic as Onondaga Lake). Distinct, essentially coincident, peaks in T$_c$ were observed in late July 1988 in the upper waters of Onondaga, Cross, Otisco, and Owasco Lakes, indicating the rather widespread occurrence of the "whiting" phenomenon in the hard water systems of this region. However, the magnitude of T$_c$ was substantially greater in Cross, Otisco, and Owasco Lakes than in Onondaga Lake.

Skaneateles Lake has the highest clarity and lowest K$_d$ of the Finger Lakes and the central New York region as a result of the low concentrations of attenuating components that are maintained (Table 7.21). The extent of light penetration in Onondaga Lake is poor by comparison to other local lakes considered desirable for contract recreation, such as Otisco, Owasco, and Skaneateles Lakes. This is largely a result of elevated concentrations of phytoplankton biomass that prevail in Onondaga Lake for much of the year, associated with its culturally eutrophic state. Comparisons of light penetration differ for SD and K$_d$ because of the differences in the relative contributions of absorption and scattering to attenuation

TABLE 7.21. Comparison of concentrations of attenuating components and measures of light penetration in Onondaga Lake and five other central New York systems, 1988, on a seasonal average basis.

Lake	Attenuating components				Light penetration	
	C_T $(mg \cdot m^{-3})$	T (NTU)	T_c (NYU)	$a_{y(440)}$ (m^{-1})	SD (m)	K_d (m^{-1})
Onondaga	35	2.7	0.4	0.73	1.92	1.20
Cross	26	3.7	1.3	0.81	1.41	1.22
L. Sodus Bay	19	2.8	0.5	0.29	1.75	1.07
Otisco	6	2.3	1.2	0.28	2.38	0.64
Owasco	4	2.4	1.8	0.15	2.66	0.50
Skaneateles	0.9	0.23	0.04	0.05	8.15	0.17

among the various lakes. On average, SD was higher in Onondaga Lake than in Cross Lake and Little Sodus Bay, somewhat lower than in Otisco and Owasco Lakes, and much less than in Skaneateles Lake (Table 7.21). The maximum reported for Onondaga Lake in 1988 (6.2 m) was exceeded only by observations from Skaneateles Lake. If the magnitude of the whiting phenomenon (as measured by T_c) in Otisco and Owasco Lakes had not been substantially greater than in Onondaga Lake, the differences in average SD between these systems would have been much greater. The relative differences in K_d between these systems was much greater (Table 7.21) because of the selective influence of the whiting phenomenon on scattering.

7.14.4 Angular Distribution of Irradiance, QA/QC, and Estimates of a and b

Depth profiles of R and $\bar{\mu}$ indicate the "asymptotic radiance distribution" (angular distribution of light intensity takes on a fixed form) has routinely been established at the 10% light level in Onondaga Lake. Abrupt temporal changes in R and $\bar{\mu}$ were common in the 1985–1990 period, reflecting abrupt changes in the relative contributions of absorption and scattering to attenuating. The value of R remained below 0.075 over the 1985–1990 period. The disproportionate decrease in

b during clearing events was manifested by particularly low values of R and high values of $\bar{\mu}$.

The veracity of the estimates of a and b presented, and the underlying program of optical measurements, were supported in two ways. First, direct determinations of $\bar{\mu}$ from measurements compared favorably with $\bar{\mu}_{KIRK}$ (the primary) estimates. This supports the veracity of the estimates of a, as $\bar{\mu}$ is linearly coupled to a ($a = \bar{\mu} \cdot K_d$). Secondly, the estimates of b were similar to the measured values of T. The value of α, according to the relationship $T = \alpha \cdot b$, has been reported elsewhere to fall in a rather narrow range of 0.80 to 1.27. The average value for Onondaga Lake, based on six years (1985–1990) of measurements and estimates, was 0.84.

Inflections in a and b generally tracked each other in the lake indicating the major role of particles in regulating the processes and absorption and scattering. The distributions of a and K_d have been highly correlated. The minima in K_d and maxima in SD observed during clearing events reflect minima in a and b. The average values of a and b decreased by 30 and 40%, respectively, from 1985 and 1986 to the 1987–1990 interval. The value of SD in Onondaga Lake can be reliably estimated from a and b, and therefore from concentrations of attenuating substances, within the framework of contrast transmittance theory. The specific relationships developed for Onondaga Lake are presented in Table 7.20

(Equation (7.29)). These expressions performed better than the $K_d \cdot SD$ product in relating attenuation to clarity.

7.14.5 Spectral Attenuation and Partitioning a and b

The order of attenuation of blue (440 nm), green (552 nm), and red (673 nm) light maintained in Onondaga Lake, over the 1987–1990 period was $K_{d(blue)} > K_{d(red)} > K_{d(green)}$. Thus penetrating irradiance is dominated by green light. However, substantial shifts in the details of the distributions of spectral attenuation, and therefore the spectral composition of penetrating and reflected irradiance, were common, apparently mostly as a result of variations in phytoplankton biomass and composition. A distinct peak in $K_{d(\lambda)}$ was manifested at about 675 nm (characteristic of absorption by photosynthetic pigments) during phytoplankton blooms. The lake usually has a turbid green appearance during phytoplankton blooms, associated with increased emergent irradiance dominated by green (the most penetrating) light. The appearance of the lake becomes more blue (e.g., green–blue) during the very low biomass conditions that prevail during clearing events, however, it remains distinctly more green than Skaneateles Lake because of the higher gelbstoff concentrations in Onondaga Lake. The blue color of Skaneateles Lake is an unrealistic goal for Onondaga Lake, as there is no evidence that the gelbstoff loadings to the lake are caused by cultural activity.

Absorption by particles, a_p, was the dominant absorbing component in the upper waters of the lake in each of the four summers over the 1987–1990 interval; it represented between 50 and 59% of a on a summer average basis. Only during clearing events did the magnitude of a_p approach the lower magnitudes of a_y and a_w. Ninety-nine percent of the variability of a was explained by variations in a_p, according to the linear expression presented in Table 7.20. The magnitude of a_p was largely regulated by the concentration of chlorophyll. The chlorophyll specific absorption coefficient determined by linear regression

analysis, was $0.0084 \, m^2 \cdot mg^{-1}$ chlorophyll (Equation (7.31); Table 7.20), well within the range of values reported for phytoplankton in the literature. The value of a_w increased as C_T increased (Equation (7.32); Table 7.20) as a result of the associated shift in penetrating irradiance to higher wavelengths. Variations in a_y occurred in response to changes in gelbstoff as well as spectral quality associated with the dynamics of phytoplankton biomass. The value of a_y tended to decrease as C_T increased, as described by Equation (7.33) of Table 7.20.

The value of b was partitioned in two different ways, based on turbidity measurements before and after acidification and on IPA techniques. The first approach was used to partition b according to b_c and b_{nc}. Over the 1985–1990 period, b_c was approximately 10% of b. However, the contribution of b_c was subject to strong temporal variability. Most of b_{nc} in the upper waters of the lake is associated with phytoplankton. The value of the chlorophyll-specific scattering coefficient (K_b) determined from linear regression analysis from paired b_{nc} and C_T data was $0.060 \, m^2 \cdot mg^{-1}$ chlorophyll; the best estimate of the average noncalcite and nonphytoplankton component (b') was about $1 \, m^{-1}$.

The second approach is more speculative, but it partitions b into a greater number of particle types. Six particle types were estimated to make the following contributions to b in the upper waters of the lake in 1987: organic (e.g., phytoplankton)—42%, composite (aggregate of organic and inorganic components)—17%, silica (e.g., diatoms)—9%, calcium carbonate—13%, clays—3%, and miscellaneous—16%.

7.14.6 Partitioning Components of Attenuation

An earlier attenuation partitioning analysis (Effler et al. 1984), based on the same framework utilized here, suggested that transparency goals for Onondaga Lake might be attainable through other than nutrient-based restoration efforts. The analysis presented here establishes that this earlier finding is incorrect. This is a result of shortcomings in the supporting optical measurements program for the earlier analysis. The findings of the more recent analysis establish that remediation of the continuing problem of poor clarity in Onondaga Lake must focus on reduction of phytoplankton growth.

7.14.7 Optics Modeling

A model that describes the regulation of the optical properties of clarity and light attenuation in Onondaga Lake by the various attenuating components has been developed and calibrated. The model is mechanistic, in that it partitions attenuation according to the processess of absorption and scattering, and partitions the components of a and b as summations. The dependencies of the component values on attenuating constituent concentrations incorporated in the model are those developed specifically for Onondaga Lake in this chapter. Clarity is predicted from K_d predictions based on contrast transmittance theory. The model performed well, as predictions of K_d and SD as a function of C_T matched observations well.

Sensitivity analyses conducted with the model indicated the observed variability in the K_d-C_T relationship is primarily a result of variability in the a_p-C_T relationship and secondarily to variations in the b-C_T relationship (e.g., variations in nonphytoplankton scattering and phytoplankton scattering per unit chlorophyll). The variability in the SD-C_T relationship is primarily associated with the variability in the b-C_T relationship.

The model was also used to generate relationships between K_d, SD, and other attenuating constituents. The K_d-$a_{y(440)}$ relationship indicates the full range of gelbstoff observations for the lake causes a change in K_d of less than $0.1\,m^{-1}$. The value of SD is predicted to change greatly for rather small changes in nonphytoplankton turbidity (e.g., terrigenous particle loading or "whiting" events).

Remediation of the degraded clarity conditions of Onondaga Lake should focus on the attribute of duration (e.g., percent occurrence) of acceptable transparency, consistent with swimming safety concerns. Prevelence of the very high SD values observed for brief periods during the "clearing events" over the 1987–1990 is not a realistic goal for Onondaga Lake. Management efforts to improve clarity should focus on the lake's cultural entrophication problems. Based on Model II and observations for the lake, the concentration of C_T must remain below $48\,\mu g \cdot L^{-1}$ to maintain SD $\geqslant$ $1.2\,m$. The model predicts that a goal of SD $>$ $2.0\,m$ could be met most of the time if C_T remained below $17\,\mu g \cdot L^{-1}$. Terrigenous particle inputs received during runoff events would continue to substantially reduce lake clarity on an irregular basis if they are not abated.

References

Auer, M.T., Storey, M.L., Effler, S.W., Auer, N.A., and Sze, P. 1990. Zooplankton impacts on chlorophyll and transparency in Onondaga Lake, New York, U.S.A. *Hydrobiology* 200/201: 603–617.

Bannister, T.T., and Weidemann, A.D. 1984. The maximum quantum yield of phytoplankton photosynthesis in situ. *J Plankton Res* 6: 275–292.

Beeton, A.M. 1957. Relationship between Secchi disc readings and light penetration in Lake Huron. *Trans Am Fish Soc* 87: 73–79.

Bird, R.E., Hulstrom, R.L., Kliman, A.W., and Eldering H.G. 1982. Solar spectral measurements in the terrestrial environment. *Appl Optics* 21: 1430–1434.

Bricaud, A., Morel, A., and Prieur, L. 1981. Absorption by dissolved organic matter of the sea (yellow substance) in the UV and visible domains. *Limnol Oceanog* 26: 43–53.

Bricaud, A., Morel, A., and Prieur, L. 1983. Optical efficiency factors of some phytoplankters. *Limnol Oceanogr* 28: 816–832.

Bricaud, A., and Stramski, D. 1990. Spectral absorption coefficients of living phytoplankton and nonalgal biogenous matter: a comparison between the Peru upwelling area and the Sargasso Sea. *Limnol Oceanogr* 35: 562–582.

Davies-Colley, R.J., and Vant, W.N. 1987. Absorption of light by yellow substances in freshwater lakes. *Limnol Oceanogr* 323: 416–425.

DiToro, D.M. 1978. Optics of turbid estaurine waters: Approximations and applications. *Water Res* 12: 1059–1068.

Effler, S.W. 1985. Attenuation versus transparency. *J Environ Engr Div ASCE* 111: 448–459.

Effler, S.W. 1988. Secchi disc transparency and turbidity. *J Environ Engr* 144: 1436–1447.

Effler, S.W., and Johnson, D.L. 1987. Calcium carbonate precipitation and turbidity measurements in Otisco Lake, NY. *Water Res Bull* 23: 73–79.

Effler, S.W., and Perkins, M.G. 1995. An optics model for Onondaga Lake. *Lake Reservoir Manag.* (in press).

Effler, S.W., Perkins, M.G., Carter, C., Wagner, B., Brooks, C., and Kent, D. 1989a. *Limnology and Water Quality of Cross Lake, 1988.* Cayuga County Health Department, Auburn, NY.

Effler, S.W., Perkins, M.G., Garofalo, J.E., and Roop, R. 1987c. Limnological Analysis of Otisco Lake for 1986. Water Quality Management Agency, Onondaga County, Syracuse, NY.

Effler, S.W., Perkins, M.G., Greer, H., and Johnson, D.L. 1987a. Effect of whiting on turbidity and optical properties in Owasco Lake, New York. *Water Res Bull* 23: 189–196.

Effler, S.W., Perkins, M.G., and Johnson, D.L. 1991b. Optical heterogeneity in Lake Champlain. *J Great Lakes Res* 17: 322–332.

Effler, S.W., Perkins, M.G., Kent, D., Brooks, C.M., Wagner, B., Storey, M.L., and Greer, H. 1990a. *Limnology of Lake Como, Duck Lake, Otter Lake, Parker Pond, and Little Sodus Bay, 1988.* Cayuga County Health Department, Auburn, NY.

Effler, S.W., Perkins, M.G., and Wagner, B.A. 1991a. Optics of Little Sodus Bay. *J Great Lakes Res* 17: 109–119.

Effler, S.W., Perkins, M.G., Wagner, B.A., and Greer, H. 1989b. *Limnological Analysis of Otisco Lake, 1988.* Water Quality Management Agency, Onondaga County, Syracuse, NY.

Effler, S.W., Perkins, M.G., and Wagner, B.A. 1989c. *Limnological Analysis of Skaneateles Lake, 1988.* Water Quality Management Agency, Onondaga County, Syracuse, NY.

Effler, S.W., Perkins, M.G., and Wagner, B.A. 1990b. *Optics of Owasco Lake; and the Influence of Sediment Input on Clarity.* Cayuga County Health Department, Auburn, NY.

Effler, S.W., Roop, R., and Perkins, M.G. 1987b. Whiting, an interference in interpreting transparency data. In: *Proceedings, Symposium on Monitoring Modeling and Mediating,* S. Nix and P. Black (ed.). American Water Resources Association, Syracuse, NY.

Effler, S.W., Roop, R., and Perkins, M.G. 1988. A simple technique for estimating absorption and

scattering coefficients. *Water Res Bull* 24: 397–404.

Effler, S.W., Wodka, M.C., and Field, S.D. 1984. Scattering and absorption of light in Onondaga Lake. *J Environ Engr ASCE* 110: 1134–1145.

Field, S.D., and Effler, S.W. 1983. Light attentuation in Onondaga Lake, NY, USA, 1978. *Arch Hydrobiol* 98: 409–421.

Gjessing, E.T. 1976. *Physical and Chemical Characteristics of Aquatic Humus.* Ann Arbor Science, Ann Arbor, MI.

Holmes, R.W. 1970. The Secchi disc in turbid coastal waters. *Limnol Oceanogr* 15: 688–694.

Jerlov, N.G. 1976. *Marine Optics* Elsevier, Amsterdam, The Netherlands.

Jerome, J.H., Bukata, R.P., and Burton, J.E. 1983. Spectral attenuation and irradiance in the Laurentian Great Lakes. *J Great Lake Res* 9: 60–68.

Jewson, D.H. 1977. Light penetration in relation to phytoplankton content of the euphotic zone of Lough Neagh, N. Ireland. *Oikos* 28: 74–83.

Jiao, J.F. 1991. Suspended Particles and Turbidity. Apportionment in Onondaga Lake and Owasco Lake, New York. Thesis. College of Environmental Science and Forestry, State University of New York, Syracuse, NY.

Johnson, D.L., and Jiao, J.F. 1992. An Empirical Turbidity Apportionment Model for Onondaga Lake. Report Submitted to the Central New York Regional Planning and Development Board, Syracuse, NY.

Johnson, D.L., Jiao, J.F. DosSontos, S.G., and Effler, S.W. 1991. Individual particle analysis of suspended materials in Onondaga Lake. *Environ Sci Technol* 25: 736–744.

Kiefer, D.A., and SooHoo, J.B. 1982. Spectral absorption by marine particles of coastal waters of Baja California. *Limnol Oceanogr* 27: 492–499.

Kirk, J.T.O. 1976. Yellow substance (gelbstoff) and its contribution to the attenuation of photsynthetically active radiation in some inland and coastal southeastern Australian waters. *Aust J Mar Freshwater Res* 27: 61–71.

Kirk, J.T.O. 1980. Spectral absorption properties of natural waters: contribution of the soluble and particulate fractions to light absorption in some inland waters of southeastern Australia. *Aust J Mar Freshwater Res* 31: 287–296.

Kirk, J.T.O. 1981a. A Monte-Carlo study of the nature of the underwater light field in, and the relationship between optical properties of, turbid yellow waters. *Aust J Mar Freshwater Res* 32: 517–532.

Kirk, J.T.O. 1981b. Estimation of the scattering coefficients of natural waters using underwater irradiance measurements. *Aust J Mar Freshwater Res* 32: 533–539.

Kirk, J.T.O. 1983. *Light and Photosynthesis in Aquatic Ecosystems.* Cambridge University Press, Cambridge.

Kirk, J.T.O. 1984a. Dependence of relationship between inherent and apparent optical properties of water on solar altitude. *Limnol Oceanogr* 29: 350–356.

Kirk, J.T.O. 1984b. Attenuation of solar radiation in scattering-absorbing waters: a simplified procedure for its calculation. *Appl Optics* 23: 3737–3789.

Kirk, J.T.O. 1985. Effects of suspensoids (turbidity) on penetration of solar radiation in aquatic ecosystems. *Hydrobiolog* 125: 195–208.

Kirk, J.T.O. 1989. The upwelling light stream in natural waters. *Limnol Oceanogr* 34: 1410–1425.

Lampert, W., Fleckner, W., Rai, H., and Taylor, B.E. 1986. Phytoplankton control by grazing zooplankton: a study on the spring clear-water phase. *Limnol Oceanogr* 12: 487–490.

Lorenzen, C.J. 1967. Determination of chlorophyll and phaeopigments: spectrophotometric equations. *Limnol Oceanogr* 12: 343–346.

Megard, R.O., Combs, W.S., Smith, P.D., and Knoll, A.S. 1979. Attentuation of light and daily integral rates of photosynthesis attained by planktonic algae. *Limnol Oceanogr* 24: 1038–1050.

Onondaga County. 1971. *Onondaga Lake Study.* Water Quality Office Environmental Protection Agency, Project No. 11060 FAE 4/71, 0-439-910. U.S. Government Printing Office, Washington, DC.

Owen, R.W. 1974. Optically effective area of particle ensembles in the sea. *Limnol Oceanogr* 19: 584–590.

Parsons, T.R., Marta, J., and Lalli, C.M. 1984. *Chemical and Biological Methods for Seawater Analysis.* Pergamon Press, New York.

Perkins, M.G., and Effler, S.W. 1995. Optical characteristics of Onondaga Lake: 1968–1990. *Lake Reservoir Manag* (in press).

Phillips, D.M., and Kirk, J.T.O. 1984. Study of the spectral variation of absorption and scattering in some Australian coastal waters. *Aust J Mar Freshwater Res* 35: 635–644.

Poole, H.H., and Atkins, W.R.G. 1929. Photoelectric measurements of submarine illumination throughout the year. *J Mar Biol Assoc UK* 16: 297–324.

Preisendorfer, R.W. 1961. Application of radiative transfer theory to light measurements in the sea. *Union Geod Geophys Inst Monogr* 10: 11–30.

Preisendorfer, R.W. 1986. Secchi disc science: visual optics of natural waters. *Limnol Oceanogr* 31: 909–926.

Reynolds, R.C., Jr. 1978. Polyphenol inhibition of calcite precipitation in Lake Powell. *Limnol Oceaogr* 23: 585–597.

Schaffner, W.C., and Oglesby, R.T. 1978. Limnology of Eight Finger Lakes: Hemlock, Canadice, Honeoye, Keuka, Seneca, Owasco, Skaneateles, and Otisco. In: *Lakes of New York State, Volume I.* J.A. Bloomfield (ed.). Academic Press, New York, NY, pp. 313–470.

Smith, R.C., and Baker, K.S. 1981. Optical properties of the clearest natural waters (200–800 nm). *Appl Opt* 20: 177–184.

Tilzer, M.M. 1983. The importance of fractional light absorption by photosynthetic pigments for phytoplankton productivity in Lake Constance. *Limnol Oceanogr* 28: 833–846.

Treweek, G.P., and Morgan, J.J. 1980. Prediction of suspension turbidities from aggregate size distribution. In: Advances in Chemistry Series No. 189 *Particulates in Water*, M.C. Kavanaugh and J.O. Leckie (eds). ACS, Washington DC.

Tyler, J.E. 1968. The Secchi disc. *Limnol Oceanogr* 13: 1–6.

Vanderploeg, H.A., Eadie, B.J., Liebig, J.R., Tarapchak, S.J., and Gloger, R.M. 1987. Contribution of calcite to the particle-size spectrum of Lake Michigan seston and its interactions with the plankton. *Can J Fish Aquat Sci* 44: 1898–1914.

Walker, W.W. 1987. *Analysis of Onondaga Lake Transparency Data.* Report submitted to the Upstate Freshwater Institute. Syracuse, NY.

Weidemann, A.D., and Bannister, T.T. 1986. Absorption and scattering coefficients in Irondequoit Bay. *Limnol Oceanogr* 31: 567–583.

Weidemann, A.D., Bannister, T.T., Effler, S.W., and Johnson, D.L. 1985. Particle and optical properties during $CaCO_3$ precipitation in Otisco Lake. *Limnol Oceanogr* 30: 1078–1083.

Yin, C.Q. 1984. Research of Sources and Budget of Mineral Sediments in Onondaga Lake, Central, New York. Thesis. College of Environmental Science and Forestry, State University of New York, Syracuse, NY.

Yin, C.Q., and Johnson, D.L. 1984. Sedimentation and particle class balances in Onondaga Lake, NY *Limnol Oceanogr* 29: 1193–1201.

8
Sediments

Steven W. Effler, Martin T. Auer, Ned Johnson, Michael Penn, and H. Chandler Rowell

8.1 Deposition

8.1.1 Background

Many materials of ecological and water quality significance become associated with, or incorporated in, particles that settle out of the watercolumn. This is often a major pathway in the cycling of these constituents. Collection of depositing particles, with properly designed and deployed sediment traps, (Bloesch and Burns 1980; Blomqvist and Hakanson 1981; Gardner 1980; Hargrave and Burns 1979; Rosa et al. 1991) is an accepted basis for quantifying the downward fluxes of depositing materials. Deposition rates are valuable indicators of metabolism, processes, and interactions in the overlying waters, and they may serve as key inputs to mass balance water quality models.

Approaches for quantifying the deposition rate of various materials in lakes have included material budgets (Kirchner and Dillon 1975; Larsen and Mercier 1976; Edmondson and Lehman 1981), sedimentary analyses coupled with dating techniques (Bortleson and Lee 1972; Bruland et al. 1974) and analysis of settling particles collected with sediment traps (reviewed by Bloesch and Burns 1980). The last of these is the most direct, has the greatest temporal resolution, and shows the greatest promise for identifying specific transport

mechanisms. The first two reflect net input of material to lake sediments. In small lakes, a sediment trap positioned below the epilimnion measures the net downward flux out of the epilimnion; this value represents a gross downward flux into the hypolimnion that should then be reduced, for all but totally conservative materials, into a net input to sediments. A comparison of fluxes determined from sediment trap studies with those determined by other techniques may provide insight into the extent of cycling in a hypolimnion. In large turbulent lakes, resuspension of sediments causes sediment trap collections to exceed the net downward flux out of the epilimnion (e.g., Charlton and Lean 1987).

Here we quantify the downward flux of several constituents from the upper waters of Onondaga Lake based on analyses of sediment trap collections. The downward flux of total phosphorus (TP) is documented for the spring–fall intervals of 1980 and 1987. The deposition rates of dry weight, particulate organic carbon (POC), and total Kjedahl nitrogen (TKN), are documented for the spring–fall intervals of 1985 and 1988. Deposition rates of particulate inorganic carbon (PIC) for 1985, 1988, and 1989 are presented. The flux of phytoplankton pigments (and related degradation products) is reported for the same interval over the 1987–1990 period. Phosphorus is of particular interest because of its critical role in regulating phytoplankton growth and the continuing management efforts to remediate certain of the problems of Onondaga Lake through reduction of external phosphorus loads (e.g., Canale and Effler 1989). The downward flux of dry weight is widely used to represent gross loading to the sediments. The downward flux of POC has been used as an indicator of the productivity of the overlying waters. The flux of phytoplankton pigments includes deposition of phytoplankton as well as related detritus incorporated in zooplankton fecal pellets, and thus is also expected to be a valuable indicator of primary productivity. Particulate inorganic carbon is assumed to be in the form of $CaCO_3$. Deposition of $CaCO_3$ is a particularly important issue for Onondaga Lake because of the potential impact industrial discharges of Ca^{2+} (Effler 1987) may have had on the deposition of this material and other constituents that may be associated with it. The deposition of TKN is an important feature of the T-NH_3 cycle (Chapter 9), and therefore, the NH_3 problems (Effler et al. 1990a) of the lake. Special attention is given to changes in deposition rates in recent years, as a reliable way of evaluating the impact of recent reductions in external loading of phosphorus and ionic waste.

8.1.2 Sediment Trap Methodology

Reviews and theoretical papers (Bloesch and Burns 1980; Blomqvist and Hakanson 1981; Gardner 1980; Hargrave and Burns 1979; Rosa et al. 1991) have agreed on some of the features of sediment trap design and mooring techniques that are critical to collection efficiency. The use of simple cylinders with aspect ratios (length: diameter) >5 has been recommended. Traps should be suspended vertically so as not to reduce the effective size of the orifice exposed to depositing material. The sediment traps used for Onondaga Lake collections were cylindrical, made of PVC, and had an aspect ratio of 6. The traps used before 1988 were 24 cm along and 4 cm in diameter; later, they were 45.7 cm long and 7.6 cm in diameter. The traps were deployed in clusters of three at depths of 10 m and 17.5 m in 20 m of water at the south deep station. Only 10 m samples have been analyzed to date. Air-filled polyethylene containers 5 m below the water surface (at a depth where vibration along the suspension line was probably minimal) kept the traps vertical. The traps have been deployed essentially continuously for the spring–fall interval of 1980, 1981, and 1985–1991. Collections have been made approximately weekly since 1985; more frequent collections were made in 1980. The 10 m depth usually remains below the epilimnion until late September or early October (Chapter 3). No chemicals were used to inhibit or prevent mineralization processes, for the reasons given by Bloesch and Burns (1980). Related limitations were minimized by the high frequency of collection (e.g., Rosa et al. 1991). Two different sets of traps were used in 1980 to evaluate the influence of collection frequency on phosphorus deposition for short (2–7 days) and long periods (12–21 days). After collection the contents of the traps have been frozen until analysis.

Calculations based on the experimental findings of Lau (1979) make it doubtful that material deposited in the traps was resuspended and lost during collection. A horizonal velocity of $24\,cm \cdot s^{-1}$ (at 15°C) near the collection orifice would have been necessary to induce resuspension of material from the traps. There was no visual evidence of loss of material during retrieval of the traps; upon collection the sediment traps always contained relatively transparent water overlying the particulate material.

8.1.3 Phosphorus

The temporal distributions of the rate of deposition of P ($mgP \cdot m^{-2} \cdot d^{-1}$) for the May–September interval of 1980 and 1987 are presented in Figure 8.1a and c. The precision of measurements was high. The average coefficient of variation for a limited number of replicate samples in 1980 was 11.1% (Wodka et al. 1985); the average for 1987, based on analyses of replicate samples for all sediment trap deployments, was 9.3%. Most of the variability was associated with collection of the depositing material. The average period of sediment trap deployment in 1980 was 4 d; the average in 1987 was 1 wk. Substantial variations in P deposition were observed in both years. The highest rates in 1980 occurred in mid- to late June, the lowest in mid-September to mid-October (Figure 8.1a). The average rate in 1980 was $48.4 \, mg \cdot m^{-2} \cdot d^{-1}$. Most of the P deposited in 1980 was organic (Figure 8.1b). About 30% of the P deposited was in an inorganic form. Wodka et al. (1985) speculated that a substantial fraction of the inorganic P may be associated with $CaCO_3$ (e.g., coprecipitation). Some temporal variability in the composition of depositing P occurred; the deposited P was more organic in June and July than in August. The maximum rates of deposition in 1987 occurred in early May and mid-June (Figure 8.1c). Minima occurred in late May, early July, and early September. The average rate of deposition in 1987 was $39.4 \, mg \cdot m^{-2} \cdot 1^{-1}$; 18% lower than observed in 1980.

Bacterial mineralization and zooplankton grazing and subsequent excretion in the traps

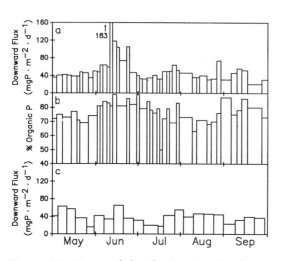

FIGURE 8.1. Temporal distributions in phosphorus deposition (sediment traps at 10 m depth) characteristics in Onondaga Lake: (**a**) downward flux in 1980, (**b**) percent of deposited phosphorus in organic form (1980), and (**c**) downward flux in 1987.

converts particulate organic P to soluble forms (Wodka et al. 1985). The extent of these conversions depend on the duration of trap deployments and the intensity of the cycling processes. By comparing the reduced deposition rates determined for longer (14 d) deployment periods in 1980 to the shorter-term deployment results, Wodka et al. (1985) estimated a 5% per day decay rate for particulate P (PP) collected by the traps. This apparent decay rate should be considered an approximate estimate; it is higher than the values reported for other systems (1–2% per day in the review of Bloesch and Burns (1980)) and the particulate organic mineralization rate (3% per day) reported by Johnson and Brinkhurst (1971). If the deposition rates in 1980 and 1987 are adjusted for the differences in deployment time, according to the Wodka et al. mineralization rate, nearly equal deposition rates are obtained in these two years.

Intuitively, it seems reasonable that the deposition rate would vary according to variations in overlying concentrations of particulate P (PP), increasing with increasing concentrations of PP above the sediment traps. The slope of a linear relationship between these parameters would be valuable in comparing this important feature of P behavior in different lakes. It has been assumed in the development of several seasonal mechanistic P models (Imboden 1974; Lung et al. 1976; O'Connor et al. 1975; Snodgras and O'Meila 1975) that P deposition from the epilimnion is a linear function of the PP concentration. While this may be true for long time-scales, it is not supported on short time-scales for Onondaga Lake. The relationship between PP and the P deposition rate for the study period of 1987 is evaluated in Figure 8.2. No significant relationship existed between these parameters during the study period of 1987. The same analysis conducted on the 1980 data found that a statistically significant relationship existed (99.5% confidence level); however, according to linear least squares regression, variations in PP explained only 15.6% of the dynamics observed in the rate of P deposition (Wodka et al. 1985). Invoking

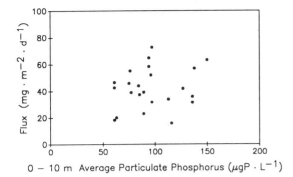

FIGURE 8.2. Evaluation of relationship between downward flux of phosphorus and particulate phosphorus concentration in the overlying water, Onondaga Lake, 1987.

for the different deployment intervals (i.e., 5% decay loss per day), the $\overline{PDR}$ in Onondaga Lake is still more than three times the average flux reported for any of the other systems. The level of temporal variability in Onondaga Lake is similar to that observed in other lakes (Table 8.1). Values of the apparent (net) phosphorus settling velocity (determined from an annual material budget approach), an alternate representation of P deposition, for Onondaga Lake also indicate an elevated rate of deposition has prevailed. The average value of the apparent settling velocity in Onondaga Lake over the 1975 to 1981 period was $51\,m \cdot yr^{-1}$ (Devan and Effler 1984), higher than any value included in the compilation (51 lakes) of DePalma et al. (1979). Wodka et al. (1985) speculated that the high P deposition rate in Onondaga Lake was in part a result of coprecipitation of P with rapidly settling $CaCO_3$ particles.

linearity between deposition rates of P and epilimnetic PP concentrations in seasonal mechanistic P models for the lake should be expected to produce uncertainty in model simulations (Chapter 9).

The average P deposition rates determined for Onondaga Lake in 1980 and 1987 are compared to sediment trap results for other lakes in Table 8.1. The temporal variability in P deposition rate (PDR) is described by $\sigma/\overline{PDR}$ (in which σ = standard deviation of PDR observations, and $\overline{PDR}$ = mean of PDR observations), based on data reported in each work. The $\overline{PDR}$ reported for Onondaga Lake is more than four times the average flux reported for any of the other systems. Adjusting

8.1.4 Phytoplankton Pigments

The temporal distributions of the downward flux of total chlorophyll (C_T, Chapter 6) for 1987 to 1990 are presented in Figure 8.3a–d. Summary statistics of this flux are presented for each of the four years in Table 8.2. The relationship between C_T and the sum of chlorophyll a plus phaeophytin, determined according to the Lorenzen (1967) method (i.e., $C_L + P$; see Chapter 6), was reevaluated for

TABLE 8.1. Comparison of P deposition rates (PDR, $mg \cdot m^{-2} \cdot d^{-1}$), determined with sediment traps, for different systems[a].

Lake	n	PDR			Deployment period (d)	Trophic state	Reference
		$\overline{PDR}$	$\sigma/\overline{PDR}$	Range			
Onondaga (1980)	44	48.3	0.58	21–183	4	Hypereutrophic	Wodka et al. 1985
Onondaga (1987)	20	39.4	0.36	15.8–64.9	7	Hypereutrophic	Doerr 1991
Lucerne	25	2.8	0.79	0.4–7.1	14	Mesotrophic	Bloesch et al. 1977
Rotsee	19	6.4	0.60	2.5–14.6	14	Eutrophic	Bloesch et al. 1977
Canadarago	12	8.5	1.19	1.0–35.2	14	Eutrophic	Fuhs 1973
Erie (37 m)	10	7.7	0.14	6.0–9.7	9	Eutrophic	Bloesch and Burns 1980
Sempach	39	5.5	—	1.0–30.0	Variable	Eutrophic	Gächter and Meyer 1990
Washington	—	—	—	0.3–10.0	—	—	Edmondson and Lehman 1981
Lawrence	21	1.3	0.66	0.1–3.2	14	Mesotrophic	White and Wetzel 1975
Zug	19	6.1	0.54	2–12	21	Eutrophic	Bloesch and Sturm 1986

[a] Modified from Wodka et al. 1985

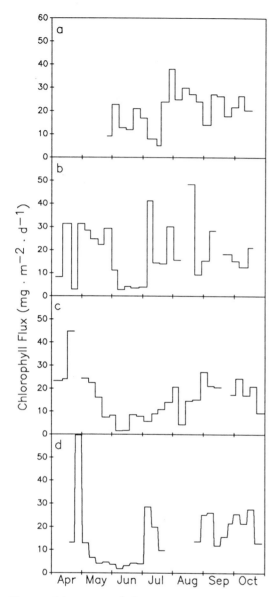

evaluating the optical impacts of phytoplankton biomass in the lake (Chapter 7)).

Extremely strong dynamics in downward flux are observed in the lake (Figure 8.3); note the broad range of fluxes during each year (Table 8.2). It is valuable to evaluate these dynamics in downward flux in concert with those of watercolumn chlorophyll (Chapter 6). In 1987, the two peaks in chlorophyll flux during the period of late May through June (Figure 8.3a) coincided with peaks in C_T in the overlying watercolumn. A drop in C_T to the study period minimum in July corresponded to the lowest chlorophyll flux of the season ($5.1 \, mg \cdot m^{-2} \cdot d^{-1}$). The relatively uniform fluxes observed in the August to October period of 1987 (Figure 8.3a) corresponded to a period of stable concentrations of C_T in the watercolumn. Coincident peaks in downward flux and C_T occurred during periods of April and May in 1988 (Figure 8.3b) and 1989 (Figure 8.3c). A trough in both downward flux and C_T occurred during August 1989 (Figure 8.3c). The highly dynamic character of downward flux in 1988 (Figure 8.3b) from July through September was not paired with dramatic changes in C_T. Note that the downward flux was lower for most of the April–May period of 1990 (Figure 8.3d) when C_T levels were lower than observed in previous years.

The relationship between downward flux and the average C_T above the traps over the deployment period is tested in the same way as presented for phosphorus (Figure 8.2) in Figure 8.4. The substantial scatter indicates highly variable deposition characteristics for phytoplankton during the spring–fall period.

sediment trap collections based on paired measurements. The values of C_T and $C_L + P$ were found to be nearly equivalent. Therefore, C_T has been retained as the measure of downward flux of pigments in the lake (recall that C_T has been adopted as the primary measure of phytoplankton biomass (Chapter 6) and it serves as the empirical basis for

TABLE 8.2. Downward flux of phytoplankton pigments (C_T) in Onondaga Lake, 1987–1990.

Year	n	Flux ($mg \cdot m^{-2} \cdot d^{-1}$)		
		$\bar{x}$	c.v.*	Range
1987	21	20.4	0.4	5.1–38.0
1988	26	18.7	0.7	2.8–48.3
1989	28	15.8	0.6	1.5–44.7
1990	24	15.7	0.8	1.7–59.6

* c.v. (coefficient of variations) = std. dev./$\bar{x}$

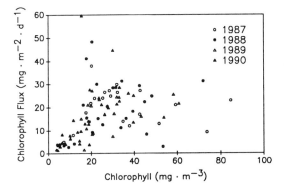

FIGURE 8.4. Evaluation of relationship between downward flux of chlorophyll and chlorophyll concentration in the overlying water, Onondaga Lake, 1987–1990.

8.1.5 Dry Weight and Inorganic Carbon

Temporal distributions of the downward flux of suspended solids in Onondaga Lake are presented for 1985 and 1988 in Figure 8.5a and b, respectively; the distributions for PIC for 1985, 1988, and 1989 appear in Figure 8.6a, b, and c, respectively. The flux of $CaCO_3$ is estimated according to the stoichiometry, inorganic carbon : $CaCO_3$ equal to 12 : 100. The limits of the bars in the plot represent ± 1 standard deviation, based on analysis of replicate samples. The average coefficient of variation for dry weight in 1985 was 6.3%; in 1988 it was 9.7%. Similarly for PIC it was 10.0% in 1985, 12.4% in 1988, and 13.1% in 1989. These levels of precision compare favorably with values reported in the literature (e.g., Bloesch and Burns 1980; Rosa et al 1991). Average fluxes and ranges observed for the mid-May to mid-October interval for dry weight and PIC are presented in Table 8.3.

The continuous deposition of $CaCO_3$ observed during the deployment periods of the

This variability has been represented in terms of settling velocity (V_s, m · d^{-1}) in Chapter 6. The variable relationship between C_T and downward flux (i.e., variable V_s) can be attributed to the influences of size, shape, density, and physiological state on phytoplankton settling (Bowie et al. 1985). Further, dynamics in macrozooplankton activity may contribute greatly to the scatter observed in the relationship between deposition and standing crop, as their fecal pellets may contribute significantly to deposition, though the pellets are generally not found in samples from the water column because of their rapid deposition (Welshmeyer and Lorenzen 1985).

The average downward flux of phytoplankton pigments and related detritus decreased in a progressive manner over the 1987–1989 period; downward fluxes were essentially equivalent in 1989 and 1990. A decrease of 23% has been observed from 1987 to 1989. These results could be interpreted as indicating a reduction in primary productivity over the same interval. However, the changes are not so great that they cannot be explained by a combination of natural (e.g., meteorologically based) variability in primary productivity, modest shortcomings in monitoring, and increased respiratory losses of net primary production from increased zooplankton grazing over the same period.

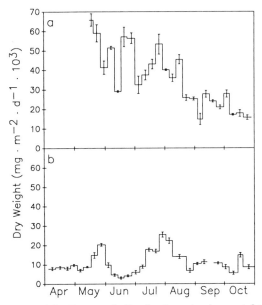

FIGURE 8.5. Temporal distributions in dry weight deposition in Onondaga Lake: (**a**) 1985, and (**b**) 1988.

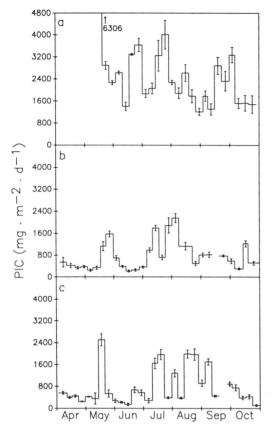

FIGURE 8.6. Temporal distributions in downward flux (sediment traps at 10 m depth) of PIC in Onondaga Lake: (**a**) 1985, and (**b**) 1988, and (**c**) 1989.

three years (Figure 8.6) is consistent with the continuously oversaturated conditions that have prevailed in the water column of the lake with respect to the solubility of calcite (Effler and Driscoll 1985). The degree of oversaturation did not change substantially from 1985 to 1988 and 1989 (Chapter 5), a

period that bracketed the closure of the soda ash/chlor-alkali facility.

Dry weight and PIC fluxes are reported together here because of the dominant role $CaCO_3$ played in overall deposition in these years. Utilizing the inorganic carbon:$CaCO_3$ stoichiometry presented above, $CaCO_3$ represented approximately 60% of the total dry weight deposition reported for the study periods of each year. The importance of $CaCO_3$ is further evidenced in the highly correlated temporal distributions observed for dry weight and PIC fluxes in 1985 and 1988 (r = 0.75 in 1985; r = 0.92 in 1988).

Major seasonal variations were observed for both dry weight and PIC deposition, e.g., the variations exceeded the uncertainty in flux associated with sampling and analytical precision for a number of the deployments. Further, the distributions were dissimilar for the different years. Peaks in the fluxes of both parameters were observed in mid-May, late June, and late July in 1985 (Figures 8.5a and 8.6a). Late May peaks were separated from the primary mid-July to mid-August peaks of 1988 by the study period minimum (Figure 8.5b and 8.6b). The seasonality in PIC deposition was similar in 1988 and 1989 (Figure 8.6b and c). Temporal variations in $CaCO_3$ deposition have not been observed to be highly correlated to the attendant degree of oversaturation in Onondaga Lake (Effler and Driscoll 1985, Chapter 5), or in Lake Constance (Stabel 1986).

The most conspicuous difference between 1985 and the later years is the much higher deposition rates in 1985. The average downward flux of dry weight for the mid-May to mid-October interval of 1985 was 3.3 times

TABLE 8.3. Downward fluxes, Onondaga Lake, mid-May to mid-October.

Year	Dry weight (mg·m^{-2}·d^{-1})		Total inorganic carbon (mg·m^{-2}·d^{-1})	
	$\bar{x}$	Range	$\bar{x}$	Range
1985	37,900	14,900–65,800	2,560 (21,300)*	1200–6300
1988	11,500	3,000–25,400	870 (7,250)*	220–2200
1989	—	—	950 (7,920)*	140–2500

( )* Estimated $CaCO_3$ flux

higher than measured for the same period in 1988 (Table 8.3). The relative decrease for PIC (or CaCO$_3$) was similar (factor of 3; decrease from 2560 to 870 mg $\cdot$ m^{-2} $\cdot$ d^{-1}). The differences in PIC flux between 1988 and 1989 were minor (within 10%) by comparison, supporting the position that a major change in downward flux has occurred since 1985. The minimum dry weight flux observed in early September 1985 was only exceeded in five of the deployment intervals in 1988 (Figure 8.5). The minimum PIC flux observed in late August–early September 1985 was only exceeded in four of the deployment intervals in 1988 and six in 1989 (Figure 8.6).

The average CaCO$_3$ deposition calculated here for 1985 (Table 8.3) is substantially greater than the value determined from the approximate average downward flux of Ca (0.12 mols Ca $\cdot$ m^{-2}d^{-1}; equal to 12,000 mg $\cdot$ CaCO$_3$ $\cdot$ m^{-2} $\cdot$ d^{-1}) reported by Effler and Driscoll (1985) for 1980. The extent to which this apparent dissimilarity for two years (1980 and 1985), before closure of the soda ash/chloralkali facility, is real, is presently unknown. Some year-to-year variability is to be expected in hard water systems, because of natural variability in such influencing factors as the external loading of Ca^{2+} and alkalinity, availability of nuclei for precipitation, and primary productivity. For example, White and Wetzel (1975) reported an approximately fivefold difference in CaCO$_3$ precipitation and a threefold difference in deposition for two consecutive years in hard water Lawrence Lake. Stratigraphic variations of up to twofold in sediment concentrations of CaCO$_3$ in Onondaga Lake also suggest year-to-year variations in deposition (see later section of this chapter). An analysis of the complete suite of archived sediment trap samples is expected to fully resolve the details of changes in deposition over the 1980–1990 period, and the apparent disparity in CaCO$_3$ fluxes before closure of the facility.

The major decrease in downward flux reported from 1985 to 1988, is largely attributable to a reduction in the deposition of CaCO$_3$ (Table 8.3). Recall that decreased downward fluxes of CaCO$_3$ may also cause

reductions for other components that become associated with CaCO$_3$ precipitate in the watercolumn (e.g., materials that serve as nuclei or are coprecipitated). The decrease in deposition is consistent with the reduced depletion of DIC from the upper waters of the lake over the same period (Chapter 5). Recall that the rates of deposition determined by mass balance analysis for 1985 and 1989 matched quite closely the fluxes reported here (Figure 8.6) for those years (Chapter 5). The primary cause of the conspicuous decrease in CaCO$_3$, and thereby sediment dry weight, since 1985 is the major decrease in water column concentrations of Ca^{2+} (16 to 4.3 mmoles) that resulted from closure of the soda ash/chlor-alkai facility (Chapter 5). However, the potential role the apparent coincident reductions in primary productivity (see subsequently presented fluxes for POC) may have played needs to be considered. Several authors have argued that CaCO$_3$ precipitation is stimulated by increasing levels of primary productivity (Effler 1984; Effler and Driscoll 1985; Otsuki and Wetzel 1972; Stabel 1986). Analysis of the complete suite of archived sediment trap samples is also expected to resolve the influence of primary productivity on CaCO$_3$ deposition in the lake.

The temporal characteristics of CaCO$_3$ deposition apparently differ for the systems where the phenomenon is observed. The dynamics documented for Lake Constance by Stabel (1986) have a much stronger "event-like" character than observed in Onondaga Lake. Two or three "events" occurred annually in the summer in Lake Constance over the period 1980–1982; increases from near zero to more than 5 gCa $\cdot$ m^{-2} $\cdot$ d^{-1} occurred, but the period of deposition was brief (e.g., <2 weeks) for each event. In contrast, CaCO$_3$ deposition is much more invariant in hard water Lawrence Lake; biweekly collections remain within about a factor of three of each other over the entire summer period (White and Wetzel 1975). The character of the dynamics observed in Onondaga Lake (Figure 8.6) is more similar to that of Lawrence Lake.

It is valuable to place the magnitude of the deposition rates presented here in a comparative perspective with values reported for other systems; selective listings of downward fluxes for dry weight and $CaCO_3$ are presented in Table 8.4. The list for $CaCO_3$ fluxes is presently very short. The dry weight deposition reported for Onondaga Lake in 1985 exceeds all the values listed for other systems in Table 8.4a, including the wide range given by Rosa et al. (1991). The lower flux reported for Onondaga Lake in 1988 exceeds all the values except the upper bounds of ranges presented for Bay of Quinte, limnocorals in the Bay of Quinte, and the lakes range presented by Rosa et al. (1991). The very high rates of deposition reported here for Onondaga Lake are consistent with the elevated rates of net sedimentation documented subsequently in this chapter for the lake. The dominant role $CaCO_3$ deposition plays in dry weight deposition in the lake has been demonstrated. Calcite deposition in Green Lake, Fayetteville, was found to be more than 80% of the total dry weight deposition. Calcuim carbonate deposition rates in Onondaga Lake in all three years greatly exceeded the limited observations encountered in the literature (Table 8.4b). The Green Lake flux is expected to be false low as a result of the use of funnels for sediment traps (Rosa et al. 1991), though the effect is probably small due to the quiescent

TABLE 8.4. Comparison of literature deposition rates (sediment traps) to values reported here for Onondaga Lake[a].

Location	Tropic state	Deposition rate ($mg \cdot m^{-2} \cdot d^{-1}$)	Source
a. Dry Weight/Summer			
Blelham Tarn	—	9,000	Jones 1976
Canadarago Lake	Eutrophic	1,000–5,000	Fuhs 1973
Lawrence Lake	Oligotrophic	500	White and Wetzel 1975
Windermere	Mesotrophic	500	Pennington 1974
Ennerdale Water	Oligotrophic	600	Pennington 1974
Wastwater	Oligotrophic	500	Pennington 1974
Char Lake	Oligotrophic	1,000	Charlton 1975
Königsee[†]	Oligotrophic	650	Stabel 1985
Überlinger See[†]	Meso-/eutrophic	1,250–3,100*	Stabel 1985
Lake Lucerne[†]	Mesotrophic	3,150	Stabel 1985
Rotsee[†]	Eutrophic	2,700	Stabel 1985
Bay of Quinte	Eutrophic	5,000–18,000	Johnson and Brinkhurst 1971
Limnocorals in Bay of Quinte	—	2,600–20,400	Charlton 1975
Lake Constance	Eutrophic	450–1,140	Stabel 1986
Wintergreen	Hypereutrophic	3,700–6,100*	Molongoski and Klug 1980
lakes (range)	—	100–30,000	Rosa et al. 1991
Onondaga Lake	Hypereutrophic		
1985		37,900	This study
1988		11,500	This study
b. $CaCO_3$			
Lake Constance	Mesotrophic	200–560	Annual	Stabel 1986
Green Lake, Fayetteville	Oligotrophic	658	Annual	Brunskill 1969
Lawrence Lake	Oligotrophic	170–560	Summer	White and Wetzel 1975
Onondaga Lake	Hypereutrophic			
1985		21,300	Late spring to	This study
1988		7,250	early fall	This study
1989		7,920		This study

[†] Average for year
* Range for different years
[a] Modified from Charlton 1975

character of this meromictic system. Onondaga Lake is more productive than most of the lakes included in Table 8.4, and almost certainly has the highest Ca^{2+} concentrations. However, it is interesting to note that the degree of oversaturation of the epilimnion of the lake with respect to calcite is not unusual compared to other productive hard water lakes (Chapter 5).

8.1.6 Organic Carbon and Nitrogen

Temporal distributions of the downward flux of POC in Onondaga Lake are presented for 1985 and 1988 in Figure 8.7 (a and b, respectively); the distributions for TKN for these two years appear in Figure 8.8a and b, respectively. The limits of the bars in the plot represent ±1 standard deviation, based on analysis of replicate samples. The average coefficient of variation for POC in 1985 was 13.4%; in 1988 it was 14.8%. Measurements for the first three deployments in 1985 were particularly uncertain (Figure 8.7a). The average coefficient of variation for TKN in

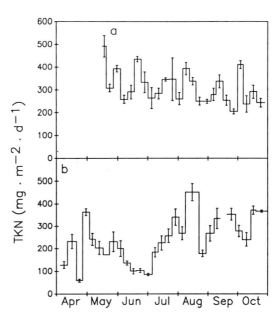

FIGURE 8.8. Temporal distributions in downward flux (sediment traps at 10 m depth) of TKN in Onondaga Lake: (**a**) 1985, and (**b**) 1988.

1985 was 8.4%; in 1988 it was 10.9%. As with the dry weight and PIC fluxes, these levels of precision compare favorably with values reported in the literature (e.g., Bloesch and Burns 1980; Rosa et al. 1991). The downward flux of POC was particularly high from mid-May to early June in 1985 (Figure 8.7a). The minima in POC and TKN deposition in 1988 were substantially lower than observed in 1985 (Figures 8.7 and 8.8). Average fluxes and ranges observed for the mid-May to mid-October interval of the two years for POC and TKN are presented in Table 8.5. Particulate organic carbon represented only about 3% of the dry weight deposition in 1985, and 6.5% in 1988.

TKN fluxes are reported here with POC results, as these constituents are generally both associated primarily with phytoplankton cell mass and related detritus in highly productive lakes such as Onondaga Lake. The temporal distributions of POC and TKN deposition were correlated in each year (R = 0.74 in 1985, R = 0.67 in 1988). Note also that similar reductions in deposition rate occurred for these constituents from 1985 to 1988 (36% for

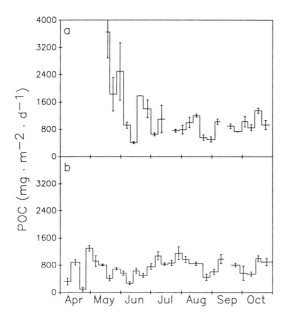

FIGURE 8.7. Temporal distributions in downward flux (sediment traps at 10 m depth) of POC in Onondaga Lake: (**a**) 1985, and (**b**) 1988.

TABLE 8.5. Downward fluxes, Onondaga Lake, mid-May to mid-October.

Year	POC ($mg \cdot m^{-2} \cdot d^{-1}$)		TKN ($mg \cdot m^{-2} \cdot d^{-1}$)		C/N* (molar)
	$\bar{x}$	Range	$\bar{x}$	Range	
1985	1180	410–3650	317	210–490	4.3
1988	750	100–1300	230	60–450	3.8

* Average value for study period

POC, 27% for TKN), and that the C/N ratio remained largely unchanged (Table 8.5). The average C/N ratio of the deposited material was substantially less than the Redfield ratio (~6.6). The C/N ratio of the depositing material in Onondaga Lake is low compared to those reported for Blelham Tarn (~9; Jones 1976) and Wintergreen Lake (average ~8.5; Molongoski and Klug 1980), though it was similar to the ratio of the overlying seston in Blelham Tarn. The ratio varied greatly in Onondaga Lake in both 1985 and 1988 (Figure 8.9a and b). Temporal variations of similar magnitude apparently are common (e.g.,

Jones 1976; Stabel 1985), but not well understood.

The deposition rate of organic carbon has been commonly used to reflect the level of primary productivity of the overlying water column, based on the simple reasoning that the downward flux increases with enhanced primary production. However, quantitative experimental data to test such a relationship appears to be scarce (Stabel 1985). The relationship was only qualitatively supported for four test systems in central Europe (Stabel 1985). The relationship between POC and C_T deposition in 1988 (Figure 8.10) supports POC deposition as an indicator of the downward flux of phytoplankton and related detritus in Onondaga Lake. Variations in the cellular content of pigments probably explains much of the scatter in the relationship, though allochthonous inputs of POC may contribute irregularly to downward flux.

The study period average POC fluxes for Onondaga Lake for 1985 and 1988 are compared to deposition rates reported for other lakes in Table 8.6. The fluxes for hypereutrophic Onondaga Lake appear to be consistent with the limited deposition/trophic state information. The only higher flux than the Onondaga Lake 1985 value found was reported for another hypereutrophic system, Wintergreen Lake (Table 8.6). Note that this flux is 6 to 12 times greater than reported for oligotrophic and mesotrophic systems. The implications of these high POC deposition

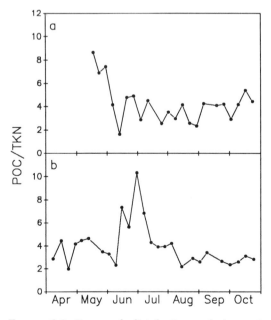

FIGURE 8.9. Temporal distributions of the ratio POC/TKN (C/N) in deposited sediments (sediment trap at 10 m depth) in Onondaga Lake: (a) 1985, and (b) 1988.

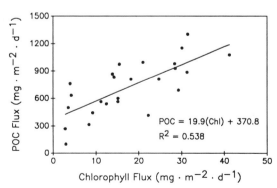

$$POC = 19.9(Chl) + 370.8$$
$$R^2 = 0.538$$

FIGURE 8.10. Evaluation of the relationship between the downward fluxes of POC and chlorophyll in Onondaga Lake in 1988.

TABLE 8.6. Comparison of literature deposition rates (sediment traps) for POC to values reported here for Onondaga Lake.

Location	Trophic state	Deposition rate $(\text{mgC} \cdot \text{m}^{-2} \cdot \text{d}^{-2})$	Source
Blelham Tarn	—	220	Jones 1976
Königsee[†]	Oligotrophic	100	Stabel 1985
Werlinger[†]	Meso-/eutrophic	150–160*	Stabel 1985
Lake Lucerne[†]	Mesotrophic	160	Stabel 1985
Rotsee[+]	Eutrophic	420	Stabel 1985
Third Sister	—	110–180*	Kelly and Chynoweth 1981
227 (Ontario)	—	230	Rudd and Hamilton 1978
Frains	—	300–400	Kelly and Chynoweth 1981
Mendota	Eutrophic	810	Kelly and Chynoweth 1981
Wintergreen	Hypereutrophic	1500	Fallon et al. 1980
Onondaga Lake	Hypereutrophic		Molongoski and Klug 1980
1985		1180	This study
1988		750	

[†] Yearly average
* Range based on different years

rates for the oxygen resources and hypolimnetic accumulations of reduced substances in the lake are detailed in Chapter 5. The lower flux observed in 1988 is similar to the value reported for eutrophic Lake Mendota. The reduction in downward flux from 1985 may simply reflect increasing diversion of organic C to support the increased zooplankton grazing (Chapter 6) and reduced coprecipitation of organic particles with $CaCO_3$.

8.2 Surficial Sediments

8.2.1 Background

The surficial sediment layer in lakes, only a few centimeters in thickness, may contain more organic matter, nutrients, and anthropogenic pollutants than the entire watercolumn (Avnimelech et al. 1984). Surficial sediments play an important role in mediating water quality. Dissolved oxygen depletion (Lucas and Thomas 1972), and the cycling of ammonia nitrogen (Effler et al. 1990a) and phosphorus (Nurnberg 1987), as well as toxic

materials such as mercury (Nakanishi et al. 1989) and PCBs (Baker et al. 1985), are phenomena largely governed by sediment processes. Further, distributions of surficial sediment characteristics can reflect the operation of physical sedimentary processes and localized impacts associated with external sediment loading. The distributions of the major, and selected minor, chemical constituents of the surficial sediments of Onondaga Lake are documented here and evaluated from the perspective of sediment sources and depositional environment (also see Auer et al. 1995). The analysis is intended to identify impacts of past activities and to facilitate an understanding of the potential for these sediments to influence future water quality.

Surficial sediment samples were collected from Onondaga Lake during July and August of 1987 using a PONAR dredge. Seventy sites (Figure 8.11a), covering a spectrum of regions and depths, were sampled to insure an accurate representation of the character and distribution of sediment types. The gravimetric and chemical analyses conducted are described elsewhere (Johnson 1989).

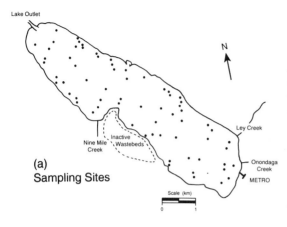

(a)
Sampling Sites

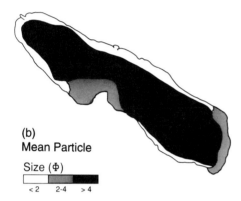

(b)
Mean Particle

Size (Φ)

< 2   2-4   > 4

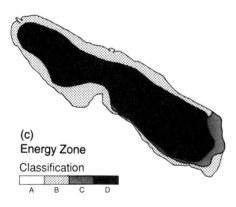

(c)
Energy Zone

Classification

A   B   C   D

FIGURE 8.11. (a) Surficial sediment sampling locations in Onondaga Lake, (b) distribution of mean particle size, and (c) distribution of energy zone classification (modified from Auer et al. 1995).

8.2.2 Lake Morphometry and Sediment Types

Morphometry plays an important role in determining the physical distribution, and therefore the composition, of sediments. Although the random rain of sedimentary material should assure sediment homogeneity in open, flat areas of uniform water depth (Downing and Rath 1988), sediments are not uniformly distributed over entire lake bottoms. Sediment accumulation generally increases with water depth, associated in part with sediment focusing (movement to low energy environments) which results from resuspension by wave action or other mixing conditions (Downing and Rath 1988). High sediment accumulation areas may also be observed proximate to mouths of tributaries that carry high sediment loads.

Morphometric influences are evidenced in Onondaga Lake through differences in sediment character among the littoral, littoriprofundal, and profundal zones. The littoral (shallow; $<\sim4.5\,m$) and littoriprofundal (slope; $4.5-12\,m$) zones are limited to a narrow strip along the lakeshore (Figure 1.2, Chapter 1). The greatest portion ($>50\%$) of the lake bottom is occupied by the profundal (deep; $>\sim12\,m$) zone, a broad, flat, plain (Figure 1.2, Chapter 1). Littoral sediments contain shell fragments, sand, and (along the eastern and north half of the western shores) oncolites. Oncolites are ovoid concretions (several millimeters to 15 cm in diameter) made up mostly of $CaCO_3$ (Dean and Eggleston 1984). The character, distribution, and origins of these structures are described subsequently in this section. The littoral zone offshore of the METRO discharge and that adjacent to the Allied-Signal wastebeds are dominated by fine muds. Profundal sediments are a malodorous (H_2S; organics) mud ranging in color from gray to black. The dark color of the sediments is attributed to the presence of amorphous iron sulfides (Yin and Johnson 1984), common in productive lakes with anoxic hypolimnia (Wetzel 1983). Littoriprofundal sediments have characteristics of both the littoral and profundal zones: coarse, gritty muds in the shallower waters and fine, mal-odorous mud at greater depths.

8.2.3 Energy Zones and Sediment Particle Size Distribution

Sediment particle size distributions are me-diated by the availability of particle sizes and the local hydraulic energy regime, for example, the energy available at the sediment–water interface for transport or resuspension (Sly et al. 1982). Coarse sediments (e.g., sands) are deposited in high energy regions characteristic of the nearshore environment. Fine grained sediments (organic particulates, silts and clays, and inorganic precipitates such as calcite) settle out in deep, quiescent regions called depositional basins. "Depositional basin" is utilized here in reference to sediment grain size. Much of the fine sediment initially deposited in high energy zones, can be resu-spended and transported to the depositional basin(s). The location and size of depositional basins are significant features of a lake because many pollutants associated with organic particles (e.g., nitrogen, phosphorus, trace metals) tend to accumulate in these regions.

The boundaries of the depositional basin (as utilized herein) are established through an analysis of the spatial distribution of particle sizes (e.g., Sly et al. 1982; Thomas et al. 1972, 1976), described as an energy zone analysis. A summary of particle size descriptors is pre-sented in Table 8.7. In general, the sediment particle size distribution follows a typical size–depth relationship (Figure 8.11b). Large par-ticles (small Φ values; Table 8.7, equation (1)) are associated with high energy, nearshore environments, whereas the deeper areas of the lake, unaffected by tributaries or wave action, are dominated by small particles (high Φ values). Sediments at the extreme southern end of the lake have a smaller mean particle size than other shallow water sediments (Figure 8.11b), presumably due to the de-position of small, high density, terrigenous solids near the mouths of Ley and Onondaga Creeks. A similar delta-like structure is ob-served near Ninemile Creek, associated not only with terrigenous solids, but an abundance of $CaCO_3$ particles (Yin and Johnson 1984) that were precipitated in the stream, as a result of the discharge of ionic waste from the soda ash manufacturing process, and delivered to the lake. Portions of this $CaCO_3$ sludge delta were dredged in the late 1960s.

Statistical representation of the particle size distribution facilitates characterization of the surficial sediments (e.g., Sly et al. 1982). For example, samples with low standard devia-tions about the mean particle size are well sorted, that is, a relatively pure population; high standard deviations indicate a mixed population (Perloff and Baron 1976). Near-shore areas (high energy) and deep basins typically have well-sorted sediments (Thomas et al. 1972). The skewness (Table 8.7) of a particle size distribution describes the tendency of the mean toward large or small sizes (Sly et al. 1982). Populations with positive skew-

TABLE 8.7. Specification of particle size descriptors.

A. Calculation of Φ

$$\Phi = -\log_2 \text{(mean particle size, mm)} \quad (8.1)$$

B. Calculation of Skewness and Kurtosis (Sly et al. 1982)

1). Mean Particles Size

$$\bar{X} = \Sigma(X)/n \quad (8.2)$$

2). Standard Deviation about the Mean

$$S = [X_2/(N-1)]^{0.5} \quad (8.3)$$

3). Skewness

$$M_3 = \frac{X_3}{2 \cdot (N-2) \cdot S^3} \quad (8.4)$$

4). Kurtosis

$$M_4 = \frac{X_4}{(N-4)S^4} - 3 \quad (8.5)$$

Where:

X = individual particle size
$\bar{X}$ = mean particle size
n = number of samples
$$X_n = \sum_{i=1}^{k} (X_i - \bar{X})^n \cdot \int (X_i)$$
X_i = class midpoint
N = total number of frequencies = $\sum \int (X_i)$
k = number of class intervals

ness are dominated by large particles, those with negative skewness by small particles, and those with skewness near zero represent a mixture of sizes (Thomas et al. 1972). Kurtosis (Table 8.7) describes the "peakedness" or degree of sorting of the particle size distribution. Sediments with positive kurtosis are well sorted (strongly peaked) and those with negative kurtosis are more normally distributed (Sly et al. 1982). Particle populations composed of sands or clays tend toward strongly peaked distributions; the addition of silt shifts the profile more to a normal distribution (Thomas et al. 1972).

A plot of skewness as a function of kurtosis may be used to illustrate major sediment types and to define associated energy zones (Figure 8.12a) and the boundaries of depositional basins (Sly et al. 1982; Thomas et al. 1972, 1976). Zone A corresponds to well-sorted sands of nearshore environments, Zone B represents a mix of sands, silts, and clays, Zone C is a mix as in Zone B, but with more small particles, and Zone D corresponds to well-sorted clays of deep water environments (Figure 8.12a). The transition in sediment texture from Zone A through Zone D reflects the gradient in energy from the turbulent, low deposition regions of nearshore areas to the more quiescent, depositional basins of the profundal zone (Thomas et al. 1972). The effective depth of wave influence on sediment distributions may be marked by a well-defined change in the slope of the mean particle size–water depth relationship (Sly et al. 1982). In Onondaga Lake, this boundary occurs at a depth of approximately 6 m (Figure 8.12b). Based on this, it is concluded that lake regions with depths in excess of 6 to 8 m (65–71% of the lake area) represent the depositional basin of the lake. Thus pollutants associated with the fine sediments characteristic of the depositional basin are likely to be widely distributed over the bottom of Onondaga Lake and not confined to a few isolated deepwater locations.

The classification of Onondaga Lake sediments by energy zone and the location of these zones in the lake are illustrated in Figures 8.12a and 8.11c, respectively. Zone D sedi-

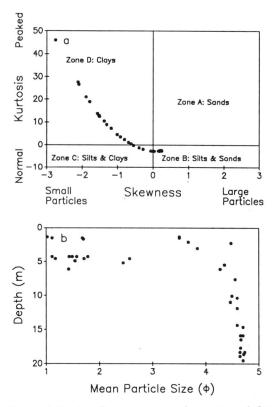

FIGURE 8.12. (a) Skewness versus kurtosis, and (b) mean particle size versus depth (from Auer et al. 1995).

ments, well-sorted populations with a small, mean particle size, are dominant within the depositional basin of the lake. Zone B sediments, poorly sorted sands and silts, are next most abundant and represent the comparatively high energy environment of nearshore waters (maximum depth 7 m; mean depth 4 m). Zone C sediments, poorly sorted silts and clays, were found only in the southern end of the lake and adjacent to the Allied Corporation wastebeds in water between 2 m and 6 m deep. No Zone A sediments (well-sorted sands) were observed. The abundance of poorly sorted materials (Zone B sediments) in the nearshore environment (silts and clays among the larger sand particles, oncolites, and shell fragments) and the absence of well-sorted sands (Zone A sediments) suggests that energy levels are generally low in Onondaga Lake compared to much larger lakes (e.g., Sly et al. 1982; Thomas et al. 1972). It should be

TABLE **8.8A.** Correlation coefficient (R) matrix for surficial sediment characteristics.

	D	MC	VS	CaCO$_3$	CL	TC
Depth (D)	1	0.805	0.722	−0.427	0.303	0.366
Moisture Content (MC)		1	0.853	−0.549	0.412	0.411
Volatile solids (VS)			1	−0.761	0.624	0.335
Calcium Carbonate (CaCO$_3$)				1	−0.982	0.253
Clastics (CL)					1	−0.594
Total Carbon (TC)						1
Inorganic Carbon (IC)						
Organic Carbon (OC)						
Chemical Oxygen Demand (COD)						
Nitrogen (N)						
Phosphorus (P)						

TABLE **8.8B.** Correlation coefficient (R) matrix for surficial sediment characteristics.

	IC	DC	COD	N	P
Depth (D)	−0.288	0.522	0.693	0.871	0.462
Moisture Content (MC)	−0.329	0.653	0.802	0.888	0.603
Volatile solids (VS)	−0.526	0.773	0.847	0.863	0.740
Calcium Carbonate (CaCO$_3$)	0.873	−0.598	−0.699	−0.594	−0.604
Clastics (CL)	−0.897	0.495	0.594	0.465	0.512
Total Carbon (TC)	0.369	0.503	0.363	0.462	0.389
Inorganic Carbon (IC)	1	−0.617	−0.550	−0.428	−0.523
Organic Carbon (OC)		1	0.819	0.788	0.815
Chemical Oxygen Demand (COD)			1	0.903	0.783
Nitrogen (N)				1	0.698
Phosphorus (P)					1

noted that sand is not found in large quantities in the watershed of Onondaga Lake (Onondaga County 1971). Further, the addition of fill and dredge waste and the lowering of the lake have altered the nearshore sediments.

8.2.4 Chemical Characteristics

The chemical characteristics of surficial sediments can be used to: (1) identify sources, (2) characterize transport and deposition of sedimented material, (3) compare sediment character among lakes, and (4) guide sampling programs for studies of exchange at the sediment–water interface. The sediment composition for a suite of chemical parameters is examined here: those that comprise the bulk of the sediment mass (moisture, calcium carbonate, clastics, and organic matter) and others that are minor components by weight, but which may be important in mediating conditions in the water column (nitrogen and

phosphorus). Correlations among the various parameters (including water depth) are computed to aid in the identification of sediments with common characteristics; these results are presented as a correlation matrix in Table 8.8. The chemical characteristics of the surficial sediments of Onondaga Lake are compared with those reported for a large number of lakes in North America. Data sources for this comparison are listed subsequently.

8.2.4.1 Moisture Content

Water constitutes the single largest component of the surficial sediments of Onondaga Lake, accounting for more than half of the wet weight of most sediments. Moisture content ranged from 36 to 74% of the wet weight (WW) with a mean of 62% WW. In Onondaga Lake, high moisture content (~70% WW) is typical of profundal sediments in the low energy, depositional basin, and low moisture content (~50% WW) is associated with coarse, shallow water sediments (Figure

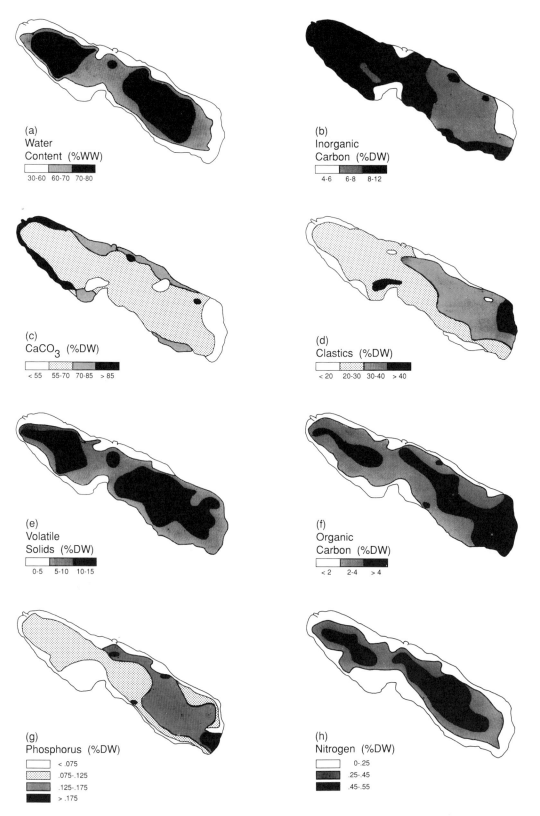

FIGURE 8.13. Distribution of (a) moisture content, (b) inorganic carbon, (c) CaCO₃, (d) clastics, (e) volatile solids, (f) organic carbon, (g) phosphorus, and (h) nitrogen (from Auer et al. 1995).

8.13a). The moisture content of the transitional (littoriprofundal zone) sediments ranged from 52 to 73% WW. Moisture content showed a strong positive correlation with volatile solids, chemical oxygen demand, and the nitrogen and phosphorus content of the surficial sediments (Table 8.8; also compare Figure 8.13a with Figure 8.13f, 8.13g, and 8.13h).

8.2.4.2 Inorganic Carbon and Calcium Carbonate

8.2.4.2.1 Basin Wide Distributions

Perhaps the most unusual feature of the chemistry of the surficial sediments of Onondaga Lake is the relative contributions of the organic and inorganic carbon fractions to total carbon content. Inorganic carbon accounts for 4 to 11% of the dry weight (DW) (mean 8%DW) of the surficial sediments, and represents approximately 74% of the total carbon, despite the lake's hypereutrophic state. The average fraction of total carbon as inorganic C was 25% for 27 calcareous Minnesota lakes included in the survey (46 lakes) of Dean and Gorham (1976), three times less than the mean for Onondaga Lake. Concentrations of inorganic carbon are highest in the north end of the lake, but differences within the depositional basin are not great (Figure 8.13b).

The inorganic component of total carbon in Onondaga Lake sediments is primarily $CaCO_3$. Inorganic carbon and $CaCO_3$ are strongly correlated and the slope of their regression line (m = 8.28) corresponds almost exactly to the stoichiometric ratio of $CaCO_3$ to C (8.33). Despite the substantial allochthonous load of $CaCO_3$ particles received from Ninemile Creek during the operation of the soda ash/chlor-alkali facility, autochthonous production was found to be substantially greater (Yin and Johnson 1984). The spatial distribution of $CaCO_3$ in Onondaga Lake sediments is illustrated in Figure 8.13c. Calcium carbonate content was negatively correlated with clastics, volatile solids and nitrogen, and phosphorus (Table 8.8). The highest concentrations of $CaCO_3$ were found in nearshore areas (especially in the north end) where shell fragments and oncolites are abundant. The depositional basin had lower concentrations. Low concentrations of inorganic carbon (Figure 8.13b) and $CaCO_3$ near Ley, Onondaga, and Ninemile Creeks reflect dilution due to the input of clastics. On a dry weight basis, $CaCO_3$ is the dominant component of the surficial sediments at most sites in Onondaga Lake, ranging from 30 to 92%DW with a mean of 66%DW.

Calcium carbonate concentrations in the surficial sediments of Onondaga Lake are high by comparison to values reported in the literature for other lakes. The mean content of Onondaga Lake exceeds the maximum measured in 219 systems in Florida (Brenner and Binford 1988), Minnesota (Dean and Gorham 1976), the Laurentian Great Lakes (Thomas et al. 1972), as well as five calcareous lakes in Wisconsin (Williams et al. 1971), calcareous Lake Kinneret in Israel (Serruya 1971), and Lake Balaton in Hungary (Pettersson and Bostrom 1986). Stratigraphic (below surficial layer) analyses indicate the sediments of the lake historically were naturally rich in $CaCO_3$ (Rowell 1992; see subsequent sections of this chapter). However, the higher $CaCO_3$ concentrations of the upper sediments (including the surficial sediments) are almost certainly mostly a result of the anthropogenic effects of the industrial discharge of Ca^{2+} from the soda ash/chlor-alkali plant.

Recall that the entire water column remained oversaturated with respect to calcite before the closure of the facility (Effler and Driscoll 1985) and these conditions have continued since its closure (Chapter 5). The higher deposition rate of $CaCO_3$ observed during the operation of the facility (Figure 8.6) supports the position that the unusually high concentrations of $CaCO_3$ in the surficial sediments are a result of the industrial discharge. The extremely high $CaCO_3$ concentrations that prevail have important implications for interpreting the concentrations of other constituents, as the $CaCO_3$ serves to dilute the sediment content of other components (see subsequent sections of this chapter).

8.2.4.2.2 Oncolites

8.2.4.2.2.1 Background. Oncolites, or lacustrine algal pisoliths, are ovoid lobate or flattened cryptalgal structures that are not attached to substrate and are characterized by nonplanar (concentric) lamination. Jones and Wilkinson (1978) report that these concretions are growing presently in many calcareous lakes throughout the midwestern United States, though only two lakes are specifically identified (Ore Lake and Littlefield Lake). These researchers report that the oncolites are found on early pure, fine-grained marl platforms along the shore. They are composed mostly of $CaCO_3$ (described as low-Mg calcite) and a small percentage (<5%) of organic matter (e.g., acid-insoluble residual). The concentric laminae are alternating porous and dense layers, which surround a nucleus. It has been speculated that the porous layer is formed as a result of photosynthetically induced precipitation, associated with the activity of filamentous cyanobacteria, while the dense layer reflects nonbiologically mediated precipitation (described as accretion; Jones and Wilkinson 1978).

The special role cyanobacteria play in the precipitation of $CaCO_3$ in certain calcium-enriched lacustrine environments has been identified by several researchers (e.g., Jones and Wilkinson 1978; Pentecost and Bauld 1988; Thompson and Ferris 1990). Filamentous cyanobacteria have been associated with the formation of the oncolite concretions (Dean and Eggleston 1984, Jones and Wilkinson 1978). Nuclei identified from Ore Lake and Littlefield Lake oncolites consisted primarily of carbonate debris, mostly cortical fragments from older nodules (Jones and Wilkinson 1978). Gastropods, *Chara* stems, and woody twigs were also observed to serve as nuclei (Jones and Wilkinson 1978). The diameter of these concretions was observed to range up to several centimeters (Jones and Wilkinson 1978).

8.2.4.2.2.2 Oncolites in Onondaga Lake. Dean and Eggleston (1984) described the occurrence of oncolites in Onondaga Lake and speculated on how and when the structures formed. They found much of the nearshore area to be covered with oncolites. The Onondaga Lake oncolites have a relatively uniform composition of 92% $CaCO_3$ (low-Mg calcite) and 3% organic matter. These concretions apparently

range to larger sizes (e.g., up to 15 cm in diameter) in Onondaga Lake (Dean and Eggleston 1984) than observed in the Michigan lakes (Jones annd Wilkinson 1978). The oncolites in Onondaga Lake were described as mostly light gray in color, though some have a green color, apparently due to active growth of cyanobacteria. Upon dissolution of the $CaCO_3$ in the oncolites with dilute acetic acid, an insoluble gelatinous-like residue remained, with similar dimensions to those of the original concretions, composed of cyanobacteria filaments (Dean and Eggleston 1984). The Onondaga Lake concretions are characterized by concentric laminae, with features (Dean and Eggleston 1984) that appear to be similar to those described for oncolites in Michigan lakes (Jones and Wilkinson 1978). Some of the oncolites in Onondaga Lake have nuclei of whole snail shells or fragments of shells; however, the vast majority have stems of charophytes as nuclei (Dean and Eggleston 1984).

Dean and Eggleston (1984) contend that the formation of oncolites in Onondaga Lake is a result of the Ca^{2+} enriched discharge of ionic waste by the adjoining soda ash manufacturer. Although charophytes are common in other hard water lakes in central New York (e.g., Effler and Rand 1976; Effler et al. 1985) they are not presently found in Onondaga Lake (Chapter 6), nor were they observed during the field studies of Dean and Eggleston (1984). They report, through personal communication, that charophytes were not present in the lake at least as far back as 1925. Dean and Eggleston (1984) reasoned that prior to the introduction of soda ash manufacturing Onondaga Lake probably supported a healthy standing crop of charophytes. "With the introduction of soda ash manufacturing and the discharge of $CaCO_3$ wastes into the lake, the increased salinity and rate of $CaCO_3$ precipitation eliminated the charophytes" (p. 229, Dean and Eggleston 1984). Coatings of $CaCO_3$ were built around dead plants; the encrusted stems eventually broke into fragments that behaved as sediment particles and substrates for growth of cyanobacteria (Dean and Eggleston 1984). The cyanobacteria promoted the precipitation and continued growth of the oncolites, perhaps through a combination of physical and metabolically mediated processes (e.g., producing a high pH microenvironment; Dean and Eggleston 1984).

The coverage of much of the littoral zone of Onondaga Lake with oncolites may significantly limit macrophyte and macrobenthos communities in affected areas. Some of the atypical features of macrophyte growth and distribution in the lake described in Chapter 6 may be a manifestation of these unusual concretions. Further, the reemergence of filamentous cyanobacteria as a significant component of primary production in the lake (Chapter 6) could promote additional oncolite growth.

8.2.4.3 Clastics

Clastics are terrigenous solids, operationally defined as that fraction of the sediment remaining after ignition to drive off water, volatile solids, and carbonates (Dean and Gorham 1976). In Onondaga Lake this fraction is composed mainly of clays, though other silica-enriched particles may contribute (Yin and Johnson 1984). The clastic content of surficial sediments in Onondaga Lake ranged from 5 to 59%DW with a mean of 26%DW. Sediments from 97 Florida and 46 Minnesota lakes have much higher clastic concentrations (average 59%DW for both surveys) than Onondaga Lake sediments. The low Onondaga Lake concentrations are in part a manifestation of the diluting effect of the high deposition rate of $CaCO_3$.

The spatial distribution of clastic materials (Figure 8.13d) reflects their terrigenous origin (Chapter 3), and the dominant circulation pattern in the lake. Regions high in clastics were located near tributary mouths and low clastic regions were found in the shallow, nearshore regions dominated by shell fragments and oncolites. The large plume of high clastic sediments at the south end of the lake reflects the influence of Onondaga Creek, the major source of suspended solids and clay to Onondaga Lake (Yin and Johnson 1984). The principal source of those solids is mud boil inputs in the upper reaches of the Onondaga Creek watershed (Effler et al. 1991). The clas-

tics content of the profundal plain (Zone D in Figure 8.11c) averages ~30%DW; these are allochthonous materials. The balance of the sediment, approximately 70%DW, has a largely autochthonous origin, that is, generated internally through the precipitation of $CaCO_3$ and sedimentation of organic matter produced in the watercolumn.

8.2.4.4 Organic and Nutrient Content

Most sedimentary organic and nutrient material in lakes originates from autochthonous primary production (Auer and Canale 1986). It has been estimated (Freedman et al. 1980) that allochthonous inputs account for only ~5 to 10% of the total organic loading to Onondaga Lake. Three measures of organic content (volatile solids, chemical oxygen demand (COD), and organic carbon) and the two key macronutrients (phosphorus and nitrogen) are examined here.

The distributions of volatile solids, organic carbon, phosphorus and nitrogen in the surficial sediments of Onondaga Lake are illustrated in Figure 8.13e–h. High concentrations of each analyte were found in the depositional basin. However, elevated levels of volatile solids, organic carbon and phosphorus were also observed in near-shore areas at the south end of the lake ("lobes" in Figure 8.13e–g). The position of these lobes reflects the localized entry of these constituents from METRO. The "lobes" of phosphorus (Figure 8.13g) and of organic carbon (Figure 8.13f) along the eastern shore of the lake reflect northward transport of materials introduced at the south end of the lake. A similar distribution pattern is noted for clastic materials (Figure 8.13d) that originate largely from Onondaga Creek.

The mean and range values for these parameters are presented in Table 8.9. The interrelationships between the distributions of these materials are depicted by the correlation matrix of Table 8.8; the correlation coefficients range from 0.70 to 0.90. These high correlations reflect largely the same origins for these materials, algae and algal detritus.

Although commonly measured in sediment studies, volatile solids (or loss on ignition) do

TABLE 8.9. Mean and range values for measures of organic content and nutrients in Onondaga Lake.

Parameter	Mean	Range
Volatile solids[1]	8.0	3.0–13.0
Organic carbon[1]	3.1	0–5.0
Chemical oxygen demand[2]	70	16–125
Phosphorus[3]	943	160–2250
Nitrogen[4]	3.2	0.4–5.4

[1] % DW
[2] $mgO_2 \cdot gDW^{-1}$
[3] $mgP \cdot kgDW^{-1}$
[4] $gN \cdot kgDW^{-1}$

not precisely separate organic and inorganic matter because some weight loss is associated with decomposition or volatilization of minerals (APHA 1985). Chemical oxygen demand and organic carbon provide better estimates of organic content; the former has been correlated with sediment oxygen demand (Gardiner et al. 1984), and the latter is widely used as an indicator of primary production in lakes.

All the organic and nutrient parameters are negatively correlated with $CaCO_3$, reflecting the low organic and nutrient content of the nearly pure $CaCO_3$ sediments found in the nearshore. The concentration of organics in shallow water sediments is typically low for lakes due to the resuspension and transport of this low density material from these environments. The volatile solids concentrations found in Onondaga Lake are low compared to values reported in the literature for productive lakes. This apparent anomaly is a manifestation of dilution, associated with the high rate of deposition of $CaCO_3$. This dilution is also apparent for the other measures of organic content and nutrients evaluated.

The concentration of phosphorus in the surficial sediments ranged from 160 to $2250 \, mgP \cdot kgDW^{-1}$ with a mean of 943 mgP $\cdot kgDW^{-1}$. Again, because of dilution, surficial sediment phosphorus concentrations are not unusually high (similar to values reported for 5 calcareous lakes in Wisconsin by Williams et al. 1971), despite the fact that: (1) phosphorus loading rates to Onondaga Lake are excessive (Heidtke and Auer 1992), (2) phosphorus release rates are similar to those of other highly

productive systems (Auer et al. 1993), and (3) phosphorus settling rates (sedimentation traps; Wodka et al. 1985) are very high.

Notable differences in the distributions of nitrogen and phosphorus are apparent proximate to the METRO discharge. In part, this may reflect the more conservative behavior of nitrogen (Effler et al. 1990a) in the lake compared to phosphorus (Wodka et al. 1985). The spatial distribution of nitrogen in the surficial sediments of Onondaga Lake (Figure 8.13h) corresponds very well to that of the depositional basin.

The chemical oxygen demand of the surficial sediments ranged from 16 to $125 \, mgO_2 \cdot g \, DW^{-1}$ with a mean of $70 \, mgO_2 \cdot gDW^{-1}$. These values are in the range of those reported for Green Bay (Gardiner et al. 1984) and Milwaukee Harbor (SWRPC 1987).

8.2.5 Generalized Sediment Distribution

Sediment collection sites were assigned to groups based on relationships among their physical and chemical characteristics (cluster analysis; Massart and Kaufman 1983). This analysis was designed to yield an integrated and comprehensive picture of the surficial sediments of Onondaga Lake. Four sediment types (Groups 1–4) were identified; their distribution in Onondaga Lake and their average physical and chemical characteristics are presented in Figure 8.14.

Group 1 sediments occur in shallow water and are distributed in a band around the perimeter of the lake (but are absent in the extreme southern end; Figure 8.14). These sediments have characteristically large particles dominated by $CaCO_3$, (due to the contribution of oncolites), and are low in moisture, clastics, organic matter, nitrogen, and phosphorus; they cover approximately 23% of the lakes sediment area. Group 2 and Group 3 sediments are found in deep water and are characterized by small particle size and high moisture, phosphorus, nitrogen, and organic content. These sediments are intermediate to Groups 1 and 4 in $CaCO_3$ and clastics content. Group 3

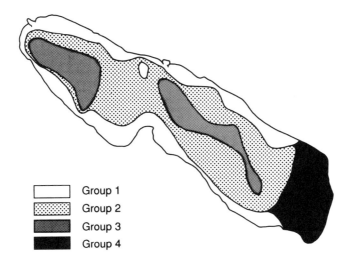

SEDIMENT GROUP

PARAMETER	1	2	3	4
Depth (m)	3.0	13.8	14.9	5.2
Mean Particle Size (Φ)	1.7	4.6	4.7	4.1
Water Content (%WW)	48.7	70.3	71.5	61.7
Volatile Solids (%DW)	4.2	10.2	11.1	6.7
CaCO$_3$ (%DW)	83.3	66.8	68.4	57.5
Clastics (%DW)	12.5	23.0	20.6	32.9
Organic Carbon (%DW)	1.3	3.9	3.9	4.5
Inorganic Carbon (%DW)	9.4	7.8	8.3	6.6
Total Carbon (%DW)	10.7	11.7	12.2	11.0
Nitrogen (%DW)	0.12	0.42	0.47	0.33
Phosphorus (%DW)	0.04	0.11	0.13	0.13
COD (mgO$_2$·gDW^{-1})	31.4	83.5	107.0	86.6

FIGURE 8.14. Distribution of generalized sediment types and listing of mean values for parameters of each type (from Auer et al. 1995).

sediments are differentiated from Group 2 sediments by having higher phosphorus and COD concentrations. These sediment types are distributed throughout the deepwater depositional basin of the lake, with Group 3 sediments found in slightly deeper regions (Figure 8.14). Group 3 sediments are more completely sorted and are probably found in areas of preferential settling. Groups 2 and 3, together, cover approximately 63% of the lake bottom. Group 4 sediments are found in the southern end of the lake and reflect tributary (mostly Onondaga Creek) and METRO influences. Average depth, particle size, moisture content, volatile solids, total carbon, nitrogen, and COD are all intermediate to those of Group 1 and Groups 2 and 3. Group 4 sediments are higher in phosphorus and organic carbon than the sediments of other groups, probably due to the influence of METRO and the three tributaries that discharge at the south end of the lake. The high clastic, low CaCO$_3$ composition of these sediments can also be attributed to tributary influences. These sediments cover approximately 14% of the lake's sediment area.

8.3 Sediment Stratigraphy

8.3.1 Introduction

The interpretation of sedimentary sequences and diagenetic processes in lakes is the subject of paleolimnology (Wetzel 1983). The completeness of the paleolimnological record, or the extent to which sediment stratigraphy is representative of deposition, is influenced by sediment supply, mode of sediment transport, the extent of reworking of deposited sediment (e.g., subaqueous erosion, bioturbation, and other mixing processes), and diagenesis. Mixing processes are influenced by climatic variables, basin morphometry, and environmental factors (e.g., presence of oxygen) that limit the occurrence of bottom dwelling organisms (Ludlam 1969; Dearing and Foster 1986; Wetzel 1983). The extent of diagenesis is constituent-specific and influenced by attendant environmental conditions.

A variety of methods can be used to reconstruct past changes in environmental conditions based upon the sedimentary record. A multidisciplinary approach was adopted by Rowell (1992, 1995) in a paleolimnological investigation of Onondaga Lake. The study included sedimentology and radiometric dating techniques, and analyses of sediment chemistry, fossil diatoms, and pollen remains. Emphasis was placed upon reconstructing the water quality and pollutant load history of the lake, particularly with respect to changes

in the rate of sediment accumulation, trophic status, salinity, and level of industrial pollution (Rowell 1992, 1995). The precultural conditions of the lake were examined in less detail.

Earlier efforts that have documented aspects of the sediment stratigraphy in Onondaga Lake are summarized in Table 8.10. These studies were generally more limited in scope and vertical resolution than the more recent efforts of Rowell (1992, 1995), which are summarized here. Selected features of these earlier studies (Table 8.10) will be addressed in the following sections.

The results of Rowell (1992) are based on core samples collected from the profundal sediments of the lake (Figure 8.15) during the summer of 1988. Two types of core samples were collected; gravity cores that retrieved a sediment section approximately 2 m in length, and piston cores approximately 5 m in length. An intact sediment–water interface was recovered with the gravity corer. However, the interface was not recovered with the piston corer, a problem encountered earlier by Effler (1975; Effler et al. 1979) with the same equipment, related to the low degree of compaction of the upper sediments. A single piston core from the south deep station (Figure 8.15) was analyzed by Rowell (1992), mostly to support sedimentological, pollen, and ^{14}C analyses of precultural (deep) sediments. The piston corer sampled the sediment at a less than ninety degree angle for the analyzed core (Rowell 1992), which compromises its quantitative value. Gravity cores were collected and analyzed from the south deep and north basin (Figure 8.15) stations. Most of the analyses were conducted on a single core from each site. Loss on ignition analyses were done on replicate cores. Analytical procedures are documented by Rowell (1992).

A time-line of selected historical events that either impacted or reflected an impact upon Onondaga Lake is presented in Table 8.11. A number of the historic events depict a worsening of the lake's water quality during the past century, for example, the disappearance of the commercial fishery by 1898, pollution of lake ice by 1901, and the ban on swimming in the lake in 1920. Generally increasing dis-

TABLE 8.10. Summary of previous sediment stratigraphy analyses of Onondaga Lake.

Source	Sample sites	Vertical resolution	Measured parameters
Onondaga County (1971) and	8 cores lakewide		Visual description
Murphy (1978)	20 cores lakewide	≥0.3 meters	Chloride in pore water
USEPA (1973)	43 cores lakewide	≥6 inches	Visual description, Hg, volatile solids, percent solids
Effler* (1975) and Effler et al. (1979)	3 cores	≥5 cm	Ca, K, Mg, Fe, Mn, P, Cd, Cr, Cu, Ni, Pb, Zn, volatile solids, particle size, x-radiography, percent solids, specific gravity, Cs-137, Pb-210
NYSDEC (1990)	43 cores lakewide	≥3 inches	Hg, percent solids
	3 cores	≥3 inches	Organic compounds, Al, Sb, As, Ba, Cd, Ca, Cr, Co, Cu, Pb, Fe, Mg, Mn, Ni, K, Ag, Na, V, Zn, cyanide
Effler et al. (1990b)	2 cores	1 cm	Chloride in pore water

* Core sites located in Figure 8.16

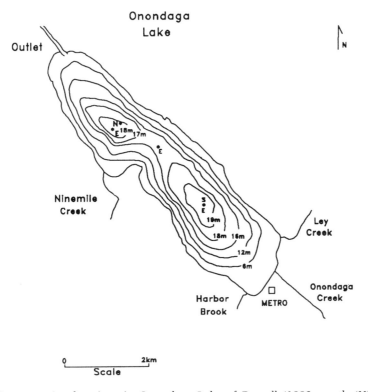

FIGURE 8.15. Sediment coring locations in Onondaga Lake of Rowell (1992; north (N) and south (S) basins) and Effler (1975; E).

TABLE 8.11. Selected historical events for Onondaga Lake.

Data	Event
1988	End of permitted Hg loading to Onondaga Lake from LCP
1986	Closure of soda ash/chlor-alkali facility
1981–1986	Rise in Hg levels in fish flesh
1981	METRO tertiary treatment starts, reduction in TP loading
1979	METRO upgrade to secondary treatment
1978	Further reductions of Hg loading from chlor-alkali process
1976	85%–90% reduction in metal loadings to the lake from Crucible Steel
1972 and 1975	Hurricanes strike Syracuse
1971	Onondaga County ban on high phosphate detergents
1970–1979	Drop in Hg levels in fish flesh
1970	95% reduction in Hg loading to lake from chlor-alkali process
1953	Doubling of Hg loading to lake ($\sim 10 \, kg \cdot d^{-1}$) from chlor-alkali process
1946	Start of Hg loading to lake ($\sim 5 \, kg \cdot d^{-1}$) from chlor-alkali process
1940s	Solvay waste deposited along Ninemile Creek
1928	METRO primary treatment started
1920s	Oil City established; Solvay waste deposited at Saddle Point
1920	Lake closed to public swimming
1901	Ice cutting banned due to water impurities
1900–1910	Commercial salt production ends; industrial development (e.g., Crucible Steel) begins
1898	Whitefish disappear from Onondaga Lake
1890s	Solvay waste deposited at south end of Onondaga Lake
1884	Start of Solvay Process Company soda ash production
1822	Lake lowered to drain Syracuse swamps; lake surface reduced by 20%
1794	First commercial production of salt near Onondaga Lake
1786	Ephraim Webster, first white settler, settles near lake on Onondaga Creek
1779	Indian crops and villages within the Onondaga Lake watershed burned by Sullivan Military Expedition

charges of domestic and industrial (e.g., soda ash/chlor-alkali and steel manufacturing) wastes were received by the lake until the 1970s. Industrial discharges of Hg and heavy metals were greatly reduced thereafter (Table 8.11). The closure of the soda ash/chlor-alkali plant in 1986 resulted in the abrupt reduction in ionic waste (Ca^{2+}, Na^+, Cl^-) loading to the lake, associated with soda ash production (Chapter 3). Major reductions in the loading of TP associated with domestic wastes have been achieved from improved treatment at METRO and a ban on high phosphorus detergents (Chapter 3). Certain of these events are coupled to stratigraphic "horizons" (levels marked by significant changes in constituent composition) in the lake's sediments. The sequence of horizons serves not only to establish cause and effect, but to support (1) reconstruction of environmental quality with respect to historic events, and (2) estimation of interval average sediment accumulation

rates (a quantitative assessment of environmental impact on the lake).

8.3.2 Sedimentology

8.3.2.1 Sediment Structure and Appearance

8.3.2.1.1 Sedimentary Units

A variety of terms have been used to describe the types of sediments found in the deepwater basins of lakes. "Gyttja" is applied to organic sediments in which the organic carbon content is less than 50% and the humic material content is low (Wetzel 1983). Gyttja is observed over the complete range of trophic states. "Sapropel" is a form of gyttja produced when anaerobic reducing conditions create a bluish-black, sulfide and methane-rich sediment (Wetzel 1983; Cole 1983; Reineck and Singh 1980). Sapropel is commonly observed in culturally eutrophic lakes (Cole 1983). A third term, applicable to either gyttja or sapropel sediment, is "marl." Marl has been widely and loosely used to refer to various types of calcareous (e.g., $CaCO_3$) clays (American Geological Institute 1962). Marl sediments typically form in temperate climate lakes where calcareous rocks are common within the watershed (Collinson 1978; Kelts and Hsu 1978).

Two sedimentary units were recognized within the sediment column from deepwater locations in Onondaga Lake. The top unit is a black clay exhibiting the appearance and sulfide smell of a sapropel (Figure 8.16). Numerous gas bubbles are present. These bubbles are probably responsible for the failure of seismic profiling methods to penetrate the lake sediments (Onondaga County 1971, Mullins personal communication 1988). The substantial ebullitive flux (Chapter 5) that prevails in the lake is consistent with the presence of bubbles in the sediments. This upper unit was previously reported in cores from Onondaga Lake (Onondaga County 1971) and

included in the description of surface material from the lake's profundal sediment (Johnson 1989). Alternating black and dark brown laminations become visible within this unit after exposure of the sediment to the air. The dark brown layers are probably attributable to the oxidation of reduced iron sulfides within the sediment (Kemp et al. 1976; Onondaga County 1971).

A dark gray clay comprises the bottom unit (Figure 8.16). Laminations are visible, although they become less distinct over the upper portion of the unit. Occasional gastropod shell and wood fragments occur throughout the unit. The sediment is progressively darker as the upper unit is approached. The boundary between the two units at the south deep station is defined by a number of distinct white laminations (Figure 8.16).

Both of the units described above are calcareous clays and may be called marls. Onondaga County (1971) used the term marl to refer to a grayish white clay found in the near-shore areas and at the north end of the lake. It has not previously been used in reference to the deepwater sediments described above. The terms gyttja and sapropel have been adopted here to describe those sedimentary units (Figure 8.16) in order to distinguish them from the "marl" sediments identified by Onondaga County (1971) in the near-shore areas.

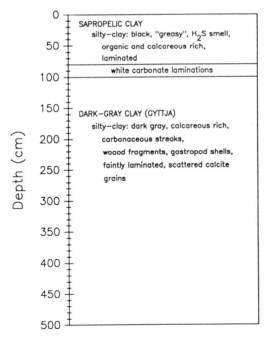

FIGURE 8.16. Sedimentary units in piston core from south site on Onondaga Lake, with qualitative descriptions.

8.3.2.1.2 Laminations

The preservation of distinct laminations within the sediment of Onondaga Lake indicates very little or no mixing of the sediments following deposition (Ludlam 1969), and specifically, a lack of bottom dwelling fauna and associated bioturbation (Saarnisto 1986). The absence of bottom dwelling fauna is consistent with the anoxic conditions that prevail in the bottom waters of the lake for much of the year (Effler et al. 1988; Chapter 5). X-radiographs showing the laminations are presented by Rowell (1992). Numerous thin microlaminations, ranging up to 1 mm in thickness, occur in groups separated by, and in some cases overlying, layers of less dense material up to

1 cm thick (macrolaminations). A similar pattern has been reported for the sediments of other productive lakes (Kelts and Hsu 1978; Bodbacka 1986). The macrolamination couplets in these systems have been attributed to yearly depositional cycles of clays, calcite, and silica (diatoms). The microlaminations have been attributed to the occurrence of individual productivity events. In Lake Zurich, diatoms and organic matter comprise the less dense layers; the dense layers are made up mostly of autochthonously produced calcite (Kelts and Hsu 1978). Clay laminae may form in response to runoff related loadings of eroded watershed material (Bodbacka 1986). Such laminae could form a vertically irregular pattern associated with the stochastic character of metereological events. These irregularities, caused by variations in watershed erosion, can be exacerbated by cultural influences (Saarnisto 1986). Underflows laden with sediments and downslope transport of sediment (Ludlam 1969; Saarnisto 1986) may also be responsible for the formation of sediment laminae.

Several recent lake characteristics and phenomena have probably contributed to the formation of laminations in Onondaga Lake, including: (1) strong temporal variations in the autochthonous production and deposition of phytoplankton (e.g., Figure 8.3) and $CaCO_3$ (e.g., Figure 8.6), (2) high loading events of clastics (clays) from Onondaga Creek during runoff events (Effler et al. 1992; Chapter 3), and (3) the strong seasonality of the depth of entry of the Ninemile Creek underflow (Owens and Effler 1989; Chapter 4) that was enriched in $CaCO_3$ particles (Yin and Johnson 1984) before closure of the soda ash/chlor-alkali plant. The exact composition and phenomena responsible for the formation of the laminations in the sediments of Onondaga Lake have yet to be fully resolved.

Visual inspection of cores and analysis of x-radiographs indicates the laminations of the lower unit are decidedly less distinct and more evenly spaced than those found in the upper unit. These differences are almost certainly anthropogenically based, as the dominant lamination forming processes have become more intense since colonial times.

The distinct white laminations within the black clay that were visible at the boundary between the two units in the south deep core are composed of calcite. In x-ray they have the same appearance as the other laminations found within the top unit. These (bounding) laminations were not observed between the two units in the north basin. Subsequently presented paleolimnological data (e.g., the *Ambrosia* horizon) indicates that the transition from the underlying dark gray gyttja to the overlying sapropel in Onondaga Lake represents an early (probably from the late 1700s through the early 1800s) sediment response to anthropogenic activities within the watershed.

8.3.2.2 Major Minerals

The dominant mineral constituents within the profundal Onondaga Lake sediments were identified by x-ray diffraction analysis (Rowell 1992). A listing of the detected minerals is presented in Table 8.12. These minerals were found at all sedimentary depths. The dominant x-ray peak was associated with calcite. Other carbonate minerals detected include dolomite, aragonite, and possibly rhodochrosite. Dolomite may be formed diagenetically (Jones and Bowser 1978). Onondaga Lake also receives this mineral with tributary inflow (Yin and Johnson 1984; Chapter 3), as it is common in the watershed (Winkley 1989). Aragonite exists in the gastropod shells and shell frag-

TABLE 8.12. Minerals present in the deep water sediments of Onondaga Lake.

Mineral	Element composition
Carbonate Minerals	
calcite	$(CaCO_3)$
dolomite	$(CaMg(CO_3)_2)$
aragonite	$(CaCO_3)$
rhodochrosite (?)	$(MnCO_3)$
Detrital Minerals	
quartz	(SiO_2)
clays and aluminosilicates	(K, Mg, Fe, Al, Si)
gypsum	$(CaSO_4 \cdot 2H_2O)$
Diagenetic Minerals	
pyrite	(FeS_2)

ments occasionally found in the bottom unit. Rhodochrosite is a diagenetic mineral occasionally found in carbonate rich sediments (Jones and Bowser 1978).

The dominant role of calcite within the Onondaga Lake sediments is supported by the stoichiometry of Ca and inorganic carbon reported herein for surficial, as well as deep, sediments. Based on the Mg/Ca ratio (<2) of the lake waters in recent years (Chapter 5), the primary carbonate precipitate of the lake is expected to be calcite (Miller et al. 1972). Yin and Johnson (1984) reported the following composition for the mineral component of the surficial sediments of Onondaga Lake (south deep station): calcite, 38.5 ± 14.7%; illite, 29.7 ± 10%; quartz, 10.4 ± 3.6%; kaolinite, 7.3 ± 3.2%; dolomite, 4.2 ± 1.5%; and other minerals, 10%.

8.3.2.3 Gross Sediment Composition

Stratigraphic profiles of total solids (Penn 1992, unpublished data) for the uppermost sediment in the southern basin are shown in Figure 8.17. The solids concentration increases from 14% to 33% over the 0 to 15 cm interval. Effler (1975) reported comparable maximum solids percentages (35%–40%) from deeper sediments at the Saddle Point area of the lake. The decrease in solids percentage below 15 cm depth infers differential sediment compaction,

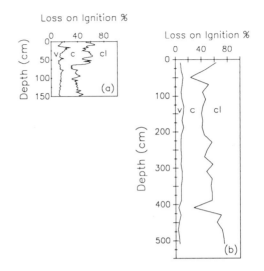

FIGURE 8.18. Stratigraphic profiles of volatile solids (v), carbonate (c), and noncarbonate clastic (cl) material for (a) a gravity core, and (b) a piston core from the south site on Onondaga Lake, based on ignition losses.

possibly due to variations in $CaCO_3$ content (Figure 8.18) as supported by the correlation (% solids = 0.32 × % dry weight $CaCO_3$ + 14.51; r = 0.76) from 15 to 60 cm depth (Penn, personal communication 1993).

The downward increase in solids content demonstrates the degree of undercompaction within the upper Onondaga Lake sediments. From Figure 8.17, the surface sediment is compacted less than half (0.14/0.33 = 0.42; the "compaction factor") as much as the sediment at 15 cm depth. This undercompaction must be considered when comparing differences between sediment accumulation rates determined over different stratigraphic intervals (see Table 8.19).

The percentages of volatile solids, total carbonate, and noncarbonate clastic material in the sediment were determined by the loss on ignition method (Dean 1974). Profiles for the south site for both the gravity and piston cores are shown in Figure 8.18. The total carbonate and noncarbonate clastic components were generally dominant (Figure 8.18). The average concentration of volatile solids (dry weight basis) within the upper 1 m of sediment is about 15%, ranging from 10% up to a maximum of 30% (Figure 8.18). Below a depth

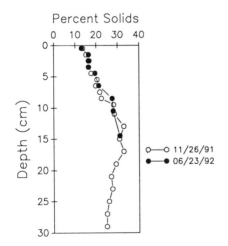

FIGURE 8.17. Stratigraphic profiles of total solids from the south site on Onondaga Lake.

of 1 m, the values average 12%, with a range from 5 to 15%. These ranges encompass the ignition loss values reported by Effler (1975). The volatile solids data for the tops of the cores are similar to the values reported for surficial sediment samples from the lake (Figure 8.13e). Recall that the volatile solids test overestimates (2.5–5×) the concentration of organic carbon (e.g., Bengtsson and Enell 1986) in Onondaga Lake sediments (e.g., Table 8.4; Figure 8.13f).

Paired carbonate (as $CaCO_3$; calculated according to stoichiometry of $CO_3:CaCO_3 = 60:100$) and calcium profiles determined in the south and north deep gravity cores are presented in Figure 8.19a–d, respectively. The highly correlated character of the calcium carbonate and calcium profiles ($r \geq 0.7$) and the stoichiometry maintained between total inorganic carbon and calcium throughout the south deep piston core (Figure 8.20a–d) reflects the historical dominance of $CaCO_3$ in the carbonates of the sediments. Sharp rises in $CaCO_3$ content are observed in the upper unit, starting at a depth of 60 cm at the south site (Figure 8.19b), and at a depth of 45 cm at the north site (Figure 8.19a). These increases are attributed to increases in $CaCO_3$ deposition caused primarily by the Ca^{2+} discharge from soda ash production (1884; recall the 3x reduction in $CaCO_3$ deposition following closure of the facility (Table 8.3)). The average $CaCO_3$ concentrations above the rise at the south station, based on calcium carbonate measurements, are 41%; based on calcium data, they are 48%. Calcium carbonate concentrations are somewhat greater at the north station, averaging 49% based on carbonate, and 62% based on calcium. The higher $CaCO_3$ concentrations obtained from calcium analyses more closely match $CaCO_3$ results reported earlier for the surficial sediments (Figure 8.13c). The spatial differences may reflect greater dilution with clastics in the south basin (e.g., Figure 8.13d), emanating mostly from Onondaga Creek (Chapter 3), and greater tributary contributions of $CaCO_3$ to the north basin from Ninemile Creek. These near surface $CaCO_3$ concentrations are somewhat lower than reported earlier in this chapter for surficial sediments.

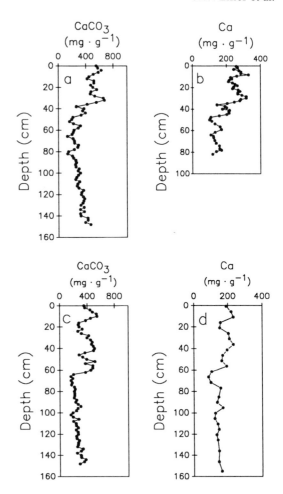

FIGURE 8.19. Stratigraphic profiles from gravity core in Onondaga Lake: (**a**) $CaCO_3$ (from carbonate/ignition) from north site, (**b**) $CaCO_3$ (from carbonate/ignition) from south site, (**c**) Ca from north deep, and (**d**) Ca from south site (modified from Rowell 1995).

The lower concentrations observed at these two sites below the rises (i.e., lower portions of upper unit) are comparable to values reported for the surficial profundal sediments of a number of Minnesota hard water lakes (Dean and Gorham 1976; Jones and Bowser 1978). A generally increasing trend in $CaCO_3$ content with increasing depth through the bottom sedimentary unit is manifested in the carbonate, calcium, and total inorganic carbon profiles of Figure 8.20a–c. The high concentrations of $CaCO_3$ in the deep sediments in-

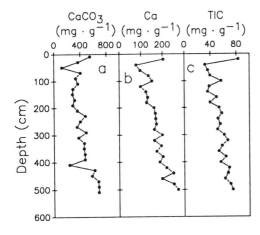

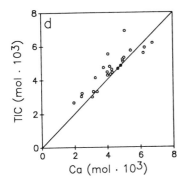

FIGURE 8.20. Sediment stratigraphy of calcium carbonate in a piston core from the south site: (**a**) $CaCO_3$ (from carbonate/ignition), (**b**) Ca, (**c**) TIC, and (**d**) evaluation of relationship between TIC and Ca, indicating primarily $CaCO_3$ composition.

dicates $CaCO_3$ deposition occurred naturally in the lake before the influence of man. Undoubtedly both allochthonous (e.g., Yin and Johnson 1984) and autochthonous (e.g., Effler and Driscoll 1985) sources contributed to the $CaCO_3$ deposits. Autochthonous production is the principal origin for most calcareous systems (Jones and Bowser 1978). The increasing $CaCO_3$ trend with depth may reflect greater deposition, but lower levels of dilution from terrigenous inputs (e.g., clastics), perhaps associated with shifts in vegetative cover (e.g., Kelts et al. 1989), is a more likely explanation.

The stratigraphy of the clastics fraction is generally negatively correlated with the carbonate fraction in the Onondaga Lake sediments (e.g., r = −0.95; Figure 8.18c). The

clastics fraction is dominated by clay minerals (e.g., illite, kaolinite) that are received in runoff from the watershed (e.g., Yin and Johnson 1984). The clastics fraction in the sampled interval of the deepwater sediments of Onondaga Lake ranged from about 20% (at a depth of 500 cm) to 80% (at a depth of 66 cm; Figure 8.18c).

8.3.3 Chemical Stratigraphy

8.3.3.1 Presentation and Constituent Groups

The stratigraphy for a number of chemical constituents is presented and evaluated in this section for both the south and north sites. The profiles for the north site cover a smaller depth interval (0–82 cm for the north versus 0–158 cm for the south), but the vertical structure for this site is more completely resolved (analysis of sections at a 2 cm interval vs. 5 cm for the south). All concentrations are in units of $mg \cdot g^{-1}$ (parts per thousand) dry weight.

The constituents presented can be placed in five groups based on the shapes of their stratigraphic profiles. The relationships between these groups is in part depicted by the correlation matrix of Table 8.13. These groups, with the exception of Fe, correspond to the carbonate, "conservative," mobile, enriched, and nutrient element groups of Kemp et al. (1976). Unlike considerations within the water column, "conservative" here refers to insoluble materials that do not appear to have been subjected to diagenetic transformations. In this case they are mostly soil constituents from the watershed. The mobile group identified in this analysis of Onondaga Lake sediments includes Mn. The enriched group includes Hg and other heavy metals.

Calcium carbonate, as a dominant sedimentary constituent, has a profile shape that is largely independent of the other elements (Table 8.13). The $CaCO_3$ profile is shown again here (as Ca; Figure 8.21a and b) for reference in evaluating other constituents profiles. A similar Ca profile was reported for the saddle location by Effler (1975). Recall that the horizons of Ca increase in the sediments (45 cm at

TABLE 8.13. Correlation matrix of selected constituent profiles in Onondaga Lake sediments at the North Basin Site.

Constituent	Ca	Al	Mg	K	Fe	Mn	Be	V	Hg	Cr	Cd	Ag	Mo	Cu	Ni	Zn	Pb	TOC
Ca																		
Al	-0.88																	
Mg	-0.91	0.94																
K	-0.85	0.99	0.90															
Fe	-0.90	0.94	0.91	0.92														
Mn	-0.79	0.69	0.83	0.69	0.71													
Be	-0.76	0.83	0.79	0.81	0.79	0.55												
V	-0.86	0.97	0.89	0.94	0.95	0.62	0.83											
Hg	0.27	-0.35	-0.44	-0.35	-0.30	-0.42	-0.25	-0.25										
Cr	0.39	-0.46	-0.56	-0.46	-0.42	-0.57	-0.20	-0.36	0.87									
Cd	0.41	-0.48	-0.57	-0.47	-0.45	-0.56	-0.27	-0.39	0.85	0.95								
Ag	0.25	-0.31	-0.37	-0.32	0.27	-0.39	-0.14	-0.22	0.85	0.87	0.88							
Mo	0.17	-0.22	-0.30	-0.20	-0.22	-0.35	-0.07	-0.17	0.35	0.48	0.48	0.49						
Cu	0.45	-0.55	-0.62	-0.57	-0.47	-0.67	-0.27	-0.45	0.77	0.92	0.86	0.79	0.45					
Ni	0.40	-0.45	-0.57	-0.45	-0.37	-0.58	-0.20	-0.33	0.84	0.95	0.91	0.82	0.46	0.89				
Zn	0.47	-0.46	-0.58	-0.45	-0.36	-0.74	-0.25	-0.32	0.66	0.77	0.73	0.66	0.50	0.83	0.79			
Pb	0.41	-0.41	-0.55	-0.40	-0.33	-0.71	-0.17	0.28	0.73	0.83	0.81	0.71	0.58	0.85	0.83	0.94		
TOC	0.07	-0.28	-0.27	-0.22	-0.22	0.10	-0.08	-0.28	0.58	0.64	0.64	0.66	0.37	0.55	0.58	0.36	0.40	

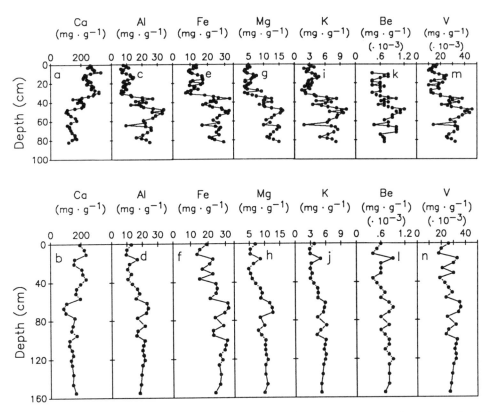

FIGURE 8.21. Sediment profiles from gravity cores in Onondaga Lake at the north (n) and south (s) sites: (a) Ca(n), (b) Ca(s), (c) Al(n), (d) Al(s), (e) Fe(n), (f) Fe(s), (g) Mg(n), (h) Mg(s), (i) K(n), (j) K(s), (k) Be(n), (l) Be(s), (m) V(n), and (n) V(s).

the north site (Figure 8.19c) and 60 cm at south deep (Figure 8.19d)) can be attributed to the opening of the soda ash facility in 1884 (Table 8.11). The increases in CaCO₃ deposition had a "diluting" effect on the concentrations of other constituents over the same sediment depth interval. Thus the manifestations of domestic and industrial waste contamination evident in the upper sediments in subsequently presented profiles would be substantially more severe in the absence of the diluting effect provided by the industrially enhanced deposition of $CaCO_3$.

The second group includes Al, Mg, K, Fe, Be, and V. The high correlations among these constituents indicate they share common sources and pathways to the lake's sediment (Cowgill and Hutchinson 1966). These materials are primarily associated with soils in the watershed, including various detrital

minerals (i.e., clay minerals, aluminosilicates and dolomite); they are delivered to the lake largely as a result of weathering and erosion (Cowgill and Hutchinson 1966; Engstrom and Wright 1984; Kemp et al. 1976). Industrial loadings represent another potential source of some of these materials (e.g., Onondaga County 1975). Manganese may be considered an outlier of this group. The dissimilarity of its profile (Table 8.13) indicates that other lake processes have also affected its stratigraphic distribution.

The distributions of mercury and other heavy metals (Cr, Cd, Ag, Mo, Cu, Ni, Zn, and Pb) in the sediments of Onondaga Lake are similar to each other (Table 8.13; Mo distribution the most dissimilar), indicating similar histories of industrial inputs and, in some cases, the same origin(s). The stratigraphy of heavy metals is expected to be highly coupled to the history of loading to the lake, as these materials are relatively insoluble in water compared to other constituents. The heavy metals may be particularly immobile in Onondaga Lake because of the high sulfide concentrations that develop annually in the hypolimnion (Effler et al. 1988), as heavy metal sulfides have particularly low solubilites (Stumm and Baccini 1978).

Total organic carbon and P (south station only) make up the nutrient element group. The deposition of these materials is primarily linked to primary productivity (as driven by external phosphorus loading) in the overlying water column (Figure 8.10) and secondarily to external loading of organics. These materials are subject to substantial diagenetic transformations following deposition. However, increases in TP concentrations within lake sediments have been correlated elsewhere to increases in anthropogenic loadings of P (Engstrom et al. 1985; Kemp et al. 1976; Schelske et al. 1986; Shapiro et al. 1971). The positive correlation between TOC and the heavy metals in the lake's sediments (Table 8.13) probably reflects the coincident histories of industrialization and cultural eutrophication, although deposition of metals can be substantially influenced by the presence of organic matter in the water column (Sly 1977).

8.3.3.2 Stratigraphy of Individual Constituents

8.3.3.2.1 Calcium and Soil Constituents

Sediment profiles of Ca and the soil constituents are presented for the north site and south station in Figure 8.21a–n. Opposite trends (r ≤ −0.76; Table 8.13) are clearly manifested for these two groups. Note in particular the lower concentrations of the soil constituents in the upper 30 cm of the north site, over which the highest Ca concentrations are observed. The similarities in certain portions of the profiles for the various soil constituents are striking; for example, for the north basin site, minima at 78, 64, 42, and 30 cm, and maxima at about 48 and 14 cm. The highly structured profiles of the soil con-

stituents (particularly manifested in the more detailed resolution for the north basin station) probably reflect the influences of natural variations in runoff, anthropogenic effects on erosion, and variations in the rates of $CaCO_3$ deposition. The assignment of a horizon(s) associated with the "mud boil" input (Chapter 3) is confounded by the lack of detailed history of related loadings from this source and the "masking" influence of the $CaCO_3$ deposition.

Iron and Mn are usually considered to be subject to diagenetic transformations, associated primarily with redox reactions (Bengtsson and Enell 1986; Engstrom and Wright 1984; Engstrom et al. 1985; Kemp et al. 1976). There is evidence that Fe redox behavior is important in regulating P release from Onondaga Lake's sediments and that FeS is formed (Chapter 5). However, the high degree of correlation of Fe with the other soil constituents, even in the upper sediments, indicates diagenetic processes do not greatly influence the overall Fe distribution in Onondaga Lake's profundal sediments. The Mn stratigraphy is the most disparate of this group of constituents (Table 8.13). Declines in Mn sediment concentration over time have been associated with increases in lake bottom reducing conditions (Engstrom et al. 1985; Mackereth 1966). Perhaps for Onondaga Lake the fraction of Mn involved in diagenetic processes is substantially greater than for Fe.

Changes in the contributions of autochthonous $CaCO_3$ and allochthonous inputs of clastics can be illustrated by the stratigraphy of the molar Mg/Ca ratio. Ratios of the other erosion-based constituents to Ca could also be used. Magnesium is assumed to be supplied principally from the watershed (e.g., by erosion). Calcium is derived largely from autochthonous precipitation and deposition of $CaCO_3$ (Yin and Johnson 1984). Thus increases in erosion would be manifested as increases in the Mg/Ca ratio. Conversely, increases in precipitation and deposition of $CaCO_3$ would be reflected in decreases in the ratio. The profiles of the molar Mg/Ca ratio (Figure 8.22a and b) indicate that major changes in the relative sediment contributions of external loading and internal production have oc-

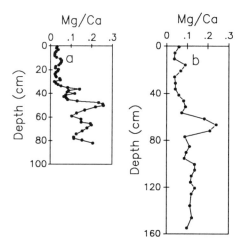

FIGURE 8.22. Stratigraphic profiles of the molar Mg/Ca ratio in gravity cores from Onondaga Lake: (a) north site, and (b) south site.

curred. Peaks in the Mg/Ca ratio, and therefore erosion inputs, are clearly manifested at both sites. The ratio averages 0.11 below the south basin peak (~65 cm depth) and 0.15 below

the principal north basin peak (~50 cm depth). Above the peaks the north basin ratio is instead lower than the south basin. Impacts that can be inferred for the entire pelagic zone based on these profiles include: (1) increases in erosion inputs (the peaks) associated with the early clearing and settlement of the adjoining area (early 1800s) and lowering of the lake's level in 1822, and (2) subsequent increases in the autochthonous production of CaCO$_3$ (reductions in ratio) related to the opening of the soda ash facility. Other horizons discussed subsequently are consistent with this interpretation.

8.3.3.2.2 Mercury and Other Heavy Metals

Sedimentary profiles of Hg and eight other heavy metals are presented for the north and south basin sediments in Figure 8.23a–r. Dramatic increases in the concentrations of these materials, which reflect the chronology of industrial loading, are apparent in the

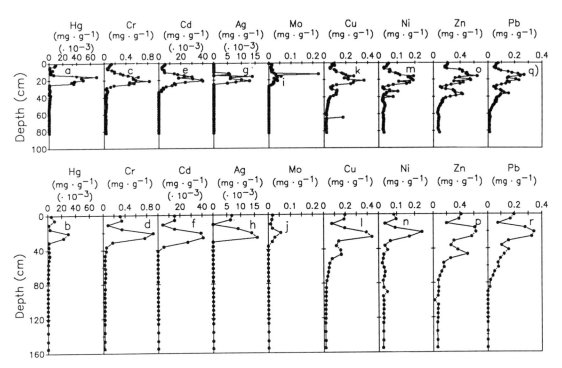

FIGURE 8.23. Stratigraphic profiles for gravity cores from Onondaga Lake at the north (n) and south (s) sites: (a) Hg(n), (b) Hg(s), (c) Cr(n), (d) Cr(s), (e) Cd(n), (f) Cd(s), (g) Ag(n), (h) Ag(s), (i) Mo(n), (j) Mo(s), (k) Cu(n), (l) Cn(s), (m) Ni(n), (n) Ni(s), (o) Zn(n), (p) Zn(s), (q) Pb(n), and (r) Pb(s) (modified from Rowell 1992).

upper sedimentary unit at both sites. The major reductions in concentrations in the uppermost sediments, resulting in background levels being approached for several heavy metals, reflect loading reductions from the adjoining steel producer (heavy metals; Table 8.11) and the soda ash/chlor-alkali facility (Hg; Table 8.11) and subsequent burial from continued sediment deposition.

Two different historic patterns emerge. Progressive increases in Cu, Zn, and Pb start deeper in the sediments than observed for the other heavy metals (Figure 8.23), indicating earlier sources, perhaps associated with other industries and anthropogenic origins. The decreases observed for these materials before development of the principal peak probably reflect dilution from increased $CaCO_3$ deposition. In contrast single peaks for Hg, Cr, Cd, Ag, and Mo, occur (Figure 8.23). The stratigraphy of Ni is most like the first group, though the deeper peaks are more irregular (Figure 8.23m and n).

The elevated Hg concentrations in the Onondaga Lake sediment are a manifestation of the history of Hg input to the lake. The abrupt increase at a sediment depth of 24 cm in the north basin is attributable to the opening of the chlor-alkali process at Allied Chemical in 1946 (Table 8.11). The sharp decrease at 14 cm is most likely associated with the abrupt (~95%) reduction in Hg discharge from the facility (~10 kg·d^{-1} to <0.5 kg·d^{-1}) achieved in 1970. Based on

these age assignments, the average sediment accumulation rate calculated for the 1946–1970 interval in the north basin is about

$$0.42 \, \text{cm} \cdot \text{yr}^{-1} \left(= \frac{24 - 14 \, \text{cm}}{1970 - 1946} \right).$$

Mercury concentrations in the sediments of Onondaga Lake have been reported in three other studies (EPRI 1987; NYSDEC 1990; USEPA 1973). Selected features of the sampling programs and findings are presented in Table 8.14. Multiple core samples were collected and analyzed in two of the studies (NYSDEC 1990; USEPA 1973) in efforts to describe the distribution of Hg in the sediments throughout the lake basin. The USEPA (1973) concluded that more than 90% of the surface sediments of the lake were contaminated with Hg, assuming 0.1 mg·kg^{-1} was the upper limit for background concentrations. However, the NYSDEC (1990) subsequently determined 0.5 mg·kg^{-1} to be a more representative background concentration for the system. Mercury was found not to be uniformly distributed in the sediments of the lake (NYSDEC 1990; USEPA 1973); for example, concentrations were higher in the near-surface profundal sediments than in littoral sediments, and subsurface sediment concentrations were higher off the mouth of Onondaga Creek and in the southwest corner of the lake.

Areal isopleths of total Hg concentration from the USEPA (1973) and NYSDEC (1990) studies are presented for an upper sediment

TABLE 8.14. Description/findings of earlier mercury studies of Onondaga Lake sediments.

Features	Study		
	USEPA (1973)	EPRI (1987)	NYSDEC (1989)
1. Date(s) of sampling	1972	1984/1985	1986/1987
2. Number: profundal/littoral	37/6	2/2	40/0
3. Core lengths (cm)	7.6–154.9	—	up to 137
4. Surface samples (no.)	6	4	—
5. Sectioning interval (cm)	7.6–30.5; typically 15.2	—	Mostly 7.6 and 15.2
6. Maximum concentration (mg·kg^{-1} dry wt.)	122.9*	1.74	85.37
7. Volume contaminated sediment ($\times 10^6 \, \text{m}^3$)**	3.7	—	5.4

*Reported from other samples collected in 1970
**Concentrations of Hg > 1 mg·kg^{-1}

depth interval in Figure 8.24a and b. The USEPA (1973) concluded from such distributions (e.g., Figure 8.24a; at various depths) that the prime Hg sources have been the discharges of the chlor-alkali plant via Ninemile Creek and the East Flume at the plant site at the southwest corner of the lake. The USEPA (1973) estimated the total mass of Hg in the sediments of the lake to be about 13,150 kg, or approximately 18% of the estimated load from the facility up to that time (Chapter 1). The differences in the distributions of Hg determined by the two studies (e.g., Figure 8.24a vs. b) have been described as pronounced (PTI 1991). Note also the much lower maximum concentration reported by EPRI (1987) for their surficial sampling program (Table 8.14). The NYSDEC (1990) estimated the volume of sediment that

exceeded $1 \, mg \cdot kg^{-1}$ was $5.4 \times 10^6 \, m^3$; this exceeded the estimate of the USEPA by a factor of 1.44. Factors contributing to the differences reported by USEPA (1973) and NYSDEC (1990) may include: differences in sampling and analytical techniques, changes in the lake over the 14 intervening years and uncertainties associated with limitations in spatial and vertical resolution of samples. The last of these is perhaps most important; for example, contours by the USEPA (1973) were developed from concentrations of homogenized vertical sections, applied to the center of the sections, and interpolation between sites on the horizontal. The vertical sectioning of both studies was very broad compared to the scale of vertical structure established for this constituent for the profundal sediments of the lake (Figure 8.23a and b). Thus the

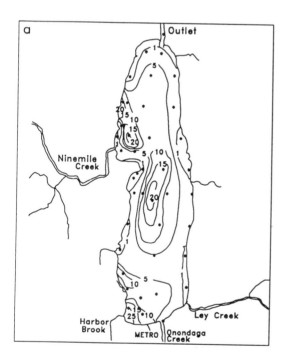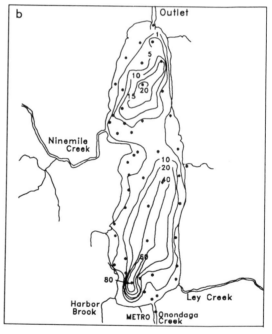

* Sampling Stations

Scale (m): 0 — 1000 — 2000

FIGURE 8.24. Areal contours of total Hg concentrations ($mg \cdot kg^{-1}$ dry weight) in sediments of Onondaga Lake: (a) 3 inches below surface in 1972, and (b) 0–3 inches depth interval in 1986 (modified from PTI 1991).

sectioning represents a major source of uncertainty in the developed distributions and
related calculations of sediment content.

With the exception of Ag, the heavy metals
with single sedimentary peaks depict more
gradual increases and decreases in the loading
of these materials. These constituents are
associated with the production of specialty
steel at the adjoining steel mill. A reduction in
metal loadings from the facility of 85 to 90%
was reported in 1976, associated with the
construction of a treatment plant. However it
appears this reduction was achieved over
several years (decreases in sediment concentrations less abrupt than for Hg; Figure 8.24).
The vertical positions of the decreases of Hg
and the other heavy metals in the sediments
of Onondaga Lake support the chronology of
reductions in loading from the soda ash/chlor-
alkali and steel production facilities (Table
8.11).

Effler et al. (1979) found a similar stratigraphy for heavy metals in the saddle area,
based on a sectioning interval of 5 cm. The
stratigraphic utility of these profiles was
questioned (e.g., upward metal migration
was hypothesized (Effler et al. 1979)). This
interpretation has subsequently been shown
to be incorrect because of the overestimation
of the sediment accumulation rate (see subsequent section).

The maximum concentrations of Hg and
other heavy metals encountered in the
Onondaga Lake sediments represent a severe
degree of contamination. The maximum Hg
concentration measured in the north basin
core was $70 \mu g \cdot g^{-1}$, the maximum for the
south basin was $30 \mu g \cdot g^{-1}$ (Rowell 1992;
Figure 8.24a and b). The NYSDEC (1990) and
USEPA (1973) reported maxima of 85.37 and
$122.9 \mu g \cdot g^{-1}$, based on analysis of many
more cores (Table 8.14). The maxima for
Onondaga Lake are high compared to values
reported for other sediments contaminated
from chlor-alkali plants (Gotterman et al.
1983; Saroff 1990). Effler et al. (1979) reported that higher Cu concentrations were
found only in lakes routinely treated with
CuSO$_4$ for algae control. Higher Zn concentrations were reported for a bay of Great Slave

Lake that received mining wastes (Mudroch et
al. 1989). Foerstner (1977) reported higher Zn
and Pb values from the Detroit River. The
maximum concentrations of Cd, Cr, and Ni
found in Onondaga Lake exceeded those
reported in the literature reviewed by Effler
et al. (1979). Comparison with Lake Erie
maxima for Hg, Pb, Cu, Cd, and Zn (Table
8.15) serves to demonstrate the much greater
level of contamination in Onondaga Lake.
Recall that the level of contamination would
be even greater (i.e., higher concentrations) if
it were not for the diluting effect of the high
amounts of CaCO$_3$ deposition. Long and
Morgan (1990) have considered the potential
impacts of sediment contamination for a
number of pollutants.

8.3.3.2.3 Phosphorus and Organic Carbon

Statigraphic profiles of TOC are presented for
the north and south sites in Figure 8.25a and
b. A profile for TP for the south site appears as
Figure 8.25c. The deeper peaks in TOC (75 to
60 cm at the north site; 100 to 85 cm at the
south site) appear to fall in the transition zone
between the gyttja and sapropel. These peaks
probably reflect increased productivity associated with the early development of the area.
The onset or lengthening of anaerobiosis
during summer stratification probably also
favored preservation of an increased fraction
of deposited organic material (e.g., Bortleson
and Lee 1972). The subsequent decrease in
TOC concentrations (Figure 8.25a and b) most
likely reflects dilution, first by allochthonous
inputs (Figure 8.21c–j), and later by CaCO$_3$

TABLE 8.15. Comparison of maximum concentrations of selected elements in the sediments of
Onondaga Lake and Lake Erie[a].

Element	Maximum concentration ($\mu g \cdot g^{-1}$)	
	Onondaga Lake	Lake Erie
Hg	70	1.4
Pb	350	150
Cu	450	85
Cd	40	5
Zn	660	350

[a] From Kemp et al. 1976

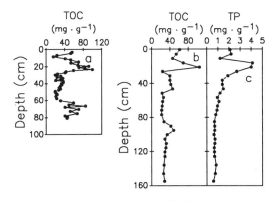

FIGURE **8.25.** Stratigraphic profiles for gravity cores from Onondaga Lake: (**a**) TOC at the north site, (**b**) TOC at the south site, and (**c**) TP at the south site (modified from Rowell 1992).

(Figure 8.21a and b). Strong peaks in TOC are manifested at about 20 cm at both sites. These almost certainly are manifestations of increasing domestic waste discharge. Settleable organic components (allochthonous) probably contributed directly. Nutrients contributed indirectly by accelerating cultural eutrophication (autochthonous). The subsequent reductions in sedimentary TOC concentrations (minima at about 10 cm depth) largely reflect improvement in domestic waste treatment and dilution by $CaCO_3$ deposition (Figure 8.20). The nearsurface increases in TOC and TP may be artifacts of incomplete diagenesis of the most recently deposited organic material, though higher concentrations of these materials may persist after burial and more complete diagenesis because of recent reductions in $CaCO_3$ dilution associated with closure of the soda ash facility (Table 8.11).

A similarly shaped TP profile was reported for the saddle site by Effler (1975). The TP and TOC distributions are correlated (r = 0.79) for the south station; there are, however, notable differences. The deeper TOC peak is not observed for TP. Perhaps diagenesis was more complete for P than TOC, though this is not expected (e.g., Shapiro et al. 1971). Concentrations of TP remained uniform ($\sim$700 µg $\cdot$ g^{-1}) up to 80 cm. Kemp et al. (1976) reported a similar background TP concentration ($\sim$800 µg $\cdot$ g^{-1}) for Lake Erie. The TP concentration

increases progressively, starting at 80 cm (Figure 8.25c), which coincides approximately with the *Ambrosia* horizon. The increasing trend becomes more pronounced at 35 cm depth, reflecting accelerating cultural eutrophication. The peak TP concentration is observed at about 20 cm, as observed for TOC. The major reduction (minimum at $\sim$10 cm) in TP concentration reflects in part the decrease in loading, particularly in the early 1970s (Chapter 3).

8.3.4 Diatoms

8.3.4.1 Background

Diatoms frustules and crysophyte cysts accumulate in lake sediments. Stratigraphic analysis of these remains is a widely used paleolimnological technique (e.g., Agbeti and Dickman 1989; Dixit et al. 1992; Fritz and Battarbee 1986; Klee and Schmidt 1987; Stoermer et al. 1987; Wolin et al. 1988), as the diatom assemblage is sensitive to environmental conditions (Battarbee 1986; Patrick 1977; Sicko-Goad and Stoermer 1988). However, transport during settling, postdepositional resuspension and species-specific differences in the robustness of the skeletal remains can bias the paleolimnological record (Anderson 1990; Battarbee 1986; Haberyan 1990).

The diatom stratigraphy presented here represents the detailed analysis of a gravity core collected at the south deep station (Rowell 1992). A piston core from the same station and gravity core from the north basin site have been analyzed in less detail (Rowell 1992).

8.3.4.2 Total Diatom Concentrations

The stratigraphy of total diatom frustules per gram of sediment is presented for the south deep gravity core in Figure 8.26. The total concentration of diatom frustules increases from 10 million per gram to 60 million per gram above the 100 cm depth within the sediment. This range of frustule concentrations is comparable to frustule concentrations reported from sediments in Lake Erie (Stoermer

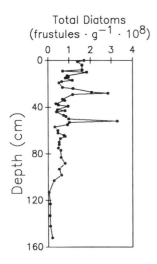

Total Diatoms
(frustules · g^{-1} · 10^8)

FIGURE 8.26. Stratigraphic profile of total diatom frustules per gram of dry sediment in a gravity core from the south site on Onondaga Lake.

et al. 1987), Lake Ontario (Stoermer et al. 1985a and b), and Lake Superior (Stoermer et al. 1985c). In these Great Lakes, the increases in concentrations of diatom frustules are associated with periods of increased nutrient loadings to the lakes resulting from human activities within the watersheds (Stoermer et al. 1985a). Wider sample to sample variation in the frustule concentrations occurs in the upper sediments of Lake Erie (Stoermer et al. 1987) than is found in sediments from less intensely impacted regions of the Great Lakes (Stoermer et al. 1987; Stoermer et al. 1985a, b, and c). The range in variation in Lake Erie, from 30 million to 150 million frustules per gram, is less than observed over the top 60 cm of sediment in Onondaga Lake (30 million to 350 million frustules per gram), but greater than the 30 million to 100 million frustules per gram range in variation for Onondaga Lake sediment below 60 cm depth (Figure 8.26). The increases in both concentrations and variability in concentrations of diatoms in the sediments of Onondaga Lake is consistent with the cultural eutrophication of the lake (e.g., Hickman et al. 1990).

8.3.4.3 Most Abundant Taxa

Two hundred and sixteen diatom taxa were identified. The total number of chrysophyte cysts per sample was also noted. Most of the taxa form only a minor portion of the sediment assemblage. Fifty-four species, grouped as the thirty-seven most abundant taxa, characterize the assemblage (Table 8.16). Each of these taxa constitutes at least two percent of the total assemblage at some point in its vertical distribution.

8.3.4.3.1 Planktonic Taxa

Depth distributions of planktonic taxa within the Onondaga Lake sediment are presented in Figures 8.27 to 8.30. The data are presented as both abundances per gram of sediment (a

TABLE 8.16. Most abundant diatom taxa in sediments of Onondaga Lake.

Taxa	Planktonic/benthic
Achnanthes spp.	B
Actinocyclus normanii fo. *subsalsa* (#1)	P
A. normanii fo. *subsalsa* (#2)	P
Amphora spp.	B
Asterionella formosa	P
chrysophyte cysts	P
Cocconeis spp.	B
Cyclotella atomus	P
C. comta	P
C. meneghiniana	P
C. spp.	P
Cymbella spp.	B
Denticula spp.	B
Diatoma tenue	P
Epithemia spp.	B
Eunotia spp.	B
Fragilaria crotonensis	P
F. spp.	P
Gomphonema olivaceum	B
G. spp.	B
Mastogloia smithii	B
Melosira granulata	P
M. spp.	P
Navicula integra	B
N. viridula v. *avenacea*	B
Nitzschia palea	B
Rhoicosphenia curvata	B
Rhopalodia spp.	B
Stephanodiscus hantzschii	P
S. cf. *neoastraea*	P
S. niagarae	P
S. spp.	P
Surirella brebissonni	B
Synedra cyclopum v. *robustum*	(Epizooic) P
S. pulchella	B
S. spp.	P
Tabellaria fenestrata	P

representation of productivity and flux) and as relative contributions (%) to the total assemblage (an indication of ecological significance). The differences in the profiles for these two modes of data presentation are not great above a sediment depth of 100 cm.

Chrysophyte cysts dominate the assemblage below 80 cm depth. Their concentration rises in conjunction with the general increase in total diatom frustules around 100 cm depth (Figure 8.26). The chrysophyte cysts drop sharply, both in concentration (Figure 8.27a) and as a relative component of the entire

diatom and chrysophyte cyst assemblage (Figure 8.27b), at 80 cm depth. Chrysophyte cysts occur most abundantly under oligotrophic to mesotrophic conditions (Cronberg 1986). Stoermer et al. (1987, 1985a and b) report a drop in the relative numbers of chrysophyte cysts in Lake Erie and Lake Ontario sediments following anthropogenic influence.

The depth distributions of the dominant *Cyclotella* species are shown in Figure 8.27c–l. The occurrence of *C. comta* is restricted to below 70 cm depth (Figure 8.27c and d). *Cyclotella*

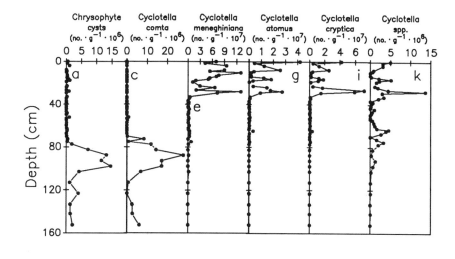

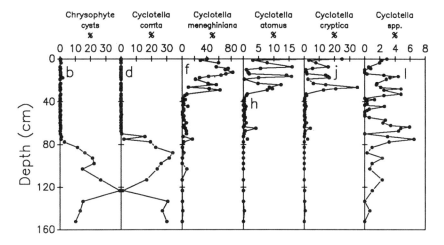

FIGURE 8.27. Diatom and chrysophyte cyst stratigraphic profiles in a gravity core from the south site on Onondaga Lake in number per gram of dry sediment (c), and contribution to total assemblage (%): (**a**) chrysophyte cysts (c), (**b**) chrysophyte cysts (%), (**c**) *Cyclotella compta* (c), (**d**) *C. comta* (%) (**e**) *C. meneghiniana* (c), (**f**) *C. meneghiniana* (%), (**g**) *C. atomus* (c), (**h**) *C. atomus* (%), (**i**) *C. cryptica* (c), (**j**) *C. cryptica* (%), (**k**) *C. spp.* (c), and (**l**) *C. spp.* (%).

meneghiniana, C. atomus, and *C. cryptica* dominate the diatom assemblage above 30 cm depth (Figure 8.27e−j). Less abundant species of *Cyclotella* occur throughout the section above 100 cm depth (Figure 8.27k and l). This latter group includes *C. glomerata,* which was reported in abundance among the phytoplankton of Onondaga Lake (Sze 1975, 1980). *Cyclotella comta* is known to occur abundantly under relatively unperturbed, oligotrophic and mesotrophic conditions (Stoermer 1984; Agbeti and Dickman 1989). *Cyclotella meneghiniana, C. atomus,* and *C. cryptica* are dominant under extremely perturbed conditions (Stoermer 1984). *Cyclotella cryptica* is described as a brackish water species (Reimann et al. 1963) found in harbors and nearshore areas of the Great Lakes subject to chloride contamination (Stoermer and Yang 1969). *Cyclotella meneghiniana* also grows well under saline conditions (Tuchman et al. 1984b).

The distributions of the abundant *Stephanodiscus* species are presented in Figure 8.28. *Stephanodiscus hantzschii* (including *S. hantzschii* fo. *tenuis*) and *Stephanodiscus* spp. (including *S. minutulus* and the related species *Cyclostephanos dubius*) occur as several peaks above 80 cm depth (Figure 8.28a−d). *Stephanodiscus* cf. *neoastraea* occurs most abundantly above 15 cm depth (Figure 8.28e and f). *Stephanodiscus niagarae* is restricted in occurrence to below 55 cm depth (Figure 8.28g and h). The assemblage of *S. hantzschii, S. hantzschii* fo. *tenuis, S. minutulus,* and *C. dubius* is associated with culturally enriched lakes, including lakes subject to inputs of sewage effluent (Battarbee 1986). *Stephanodiscus hantzschii* has been used as a marker of cultural eutrophication and salinification in lake sediments both in Europe and North America (Hakansson and Stoermer 1984). *Stephanodiscus niagarae* is tolerant of a wide range of productivity conditions, but decreases in abundance under conditions of industrial pollution and high chlorides (Theriot and Stoermer 1981). *Stephanodiscus neoastraea* has been associated with mesotrophic conditions in sediments reported from the Mondsee, Austria (Klee and Schmidt 1987).

Many of the remaining abundant species from Onondaga Lake (Table 8.16) have also been reported from the Great Lakes (Stoermer et al. 1987). *Actinocyclus normanii* fo. *subsalsa* is abundant above 40 cm depth (Figure 8.29a and b). It is found in shallow regions in the Great Lakes which receive large nutrient and conservative ion loadings. In Europe, it is widely distributed in eutrophic estuaries (Stoermer et al. 1987). *Asterionella formosa* is abundant between 80 cm and 45 cm depth (Figure 8.29c and d). It is tolerant of a wide range of trophic conditions. *Diatoma tenue* (consisting mostly of *D. tenue* var. *elongatum*) is abundant above 80 cm depth, with peak abundance between 55 cm and 30 cm depth (Figure 8.29e and f). It is associated with eutrophic conditions in the Great Lakes (Stoermer et al. 1987) and may reach maximum abundance under saline conditions (Tuchman et al. 1984a). *Fragilaria crotonensis* is abundant between 80 cm and 30 cm depth (Figure 8.29g and h). It is widely distributed in the Great Lakes with no particular relation to trophic conditions (Stoermer et al. 1987). *Melosira granulata* occurs most abundantly above 100 cm depth (Figure 8.29i and j). It is considered an indicator of eutrophication, but its distribution in lakes may also be influenced by physical factors (Stoermer et al. 1987). *Tabellaria fenestrata* occurs abundantly between 80 and 60 cm depth (Figure 8.29k and l). It is widely distributed in the Great Lakes under a wide variety of ecological conditions (Stoermer et al. 1985c).

8.3.4.3.2 Benthic and Epiphytic Taxa

Benthic and epiphytic (attached) diatoms comprise a significant portion of the total diatom assemblage of the sediments of Onondaga Lake (Table 8.16). They, or the plants to which they were attached, may have been washed into open waters by waves and currents (Battarbee 1986). The dominant species are representatives of the genera *Navicula, Nitzschia, Synedra, Entomoneis, Rhoicosphenia, Mastagloia* (*M. smithii*), *Denticula, Epithemia, Eunotia, Rhopalodia, Surirella, Achnanthes* (including *A. minutissima*), *Cym-*

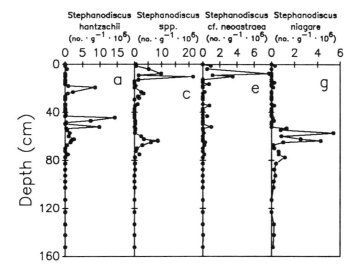

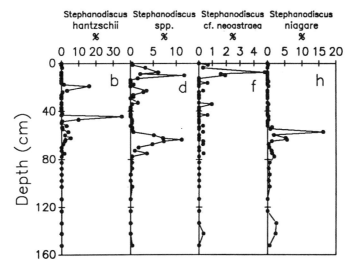

FIGURE 8.28. Diatom stratigraphic profiles in a gravity core from the south site on Onondaga Lake in numbers per gram of dry sediment (c), and contribution to total assemblage (%): (a) *Stephanodiscus hantzschii* (c), (b) *S.* hantzschii (%), (c) *S.* spp. (c), (d) *S.* spp. (%), (e) *S.* cf. *neoastraea* (c), (f) *S.* cf. *neoastraea* (%), (g) *S. niagarae* (c), and (h) *S. niagarae* (%).

bella, Amphora, Cocconeis, Gomphonema (including *G. olivaceum*), and *Fragilaria*.

Navicula integra and *Nitzschia palea* are present above 80 cm depth, but are most abundant between 20 and 40 cm depth (Figure 8.30a–d). *Navicula integra* is often found in polluted water (Patrick and Reimer 1966). *Nitzschia palea* is found under a wide range of conditions and is considered an indicator of freshwater pollution (Palmer 1977). *Synedra* species are abundant between 100 cm and 30 cm depth (Figure 8.30e and f). Members of this genus occur under a wide range of ecological conditions (Patrick and Reimer 1966). *Synedra pulchella* (including *S. pulchella* var. *lanceolata*) increases in abundance from 80 to 50 cm depth, then decreases in abundance above 30 cm depth (Figure 8.30g and h). It has

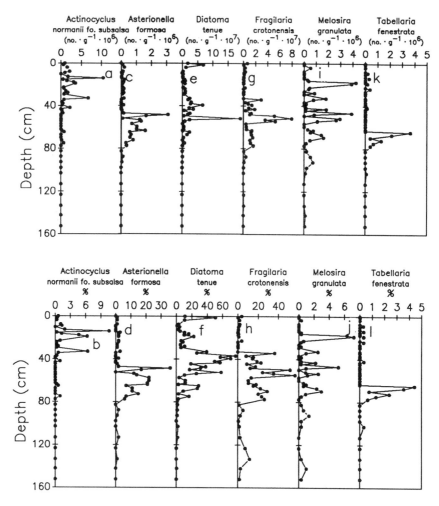

FIGURE 8.29. Diatom stratigraphic profiles in a gravity core from the south site on Onondaga Lake in numbers per gram of dry sediment (c), and contribution to total assemblage (%): (a) *Actinocyclus normanii fo. subsalsa* (c), (b) *A. normanii fo. subsalsa* (%), (c) *Asterionella formosa* (c), (d) *A. formosa* (%), (e) *Diatoma tenue* (c), (f) *D. tenue* (%), (g) *Fragilaria crotonensis* (c), (h) *F. crotonensis* (%), (i) *Melosira granulata* (c), (j) *M. granulata* (%), (k) *Tabellaria fenestrata* (c), and (l) *T. fenestrata* (%).

been reported in waters of high mineral content, slightly brackish to brackish (Patrick and Reimer 1966). Another species of this genus, *S. cyclopum* var. *robustum*, occurs only between 80 cm and 55 cm depth (Figure 8.30i and j). *Entomoneis alata* peaks in abundance between 20 and 40 cm depth (Figure 8.30k and l). It is rarely found in freshwater and is regarded as an indicator of mesohaline to marine conditions (Patrick and Reimer 1966).

The relative abundance of benthic and epiphytic diatoms versus planktonic forms is presented as a ratio in Figure 8.31. From 80 cm

to 50 cm depth the ratio declines fourfold. Declines in the ratio are evident at 80, 58, and 17 cm depths. An increase occurs between 38 and 17 cm depths. Several factors may have contributed to this pattern in the relative abundance of benthic and epiphytic taxa. The two most probable reasons for declines in the ratio include: (1) increases in planktonic diatom productivity, and (2) lowering of the level of the lake, which resulted in a significant decrease in the area of the littoral zone. The elimination of submerged macrophyte surfaces in response to increased $CaCO_3$

deposition (Dean and Eggleston 1984; Chapter 6), may also have contributed to the decline at 60 cm. The rise in the ratios between 38 and 17 cm depths is attributable to the increase in abundances of the pollution tolerant benthic diatoms *Navicula integra* and *Nitzschia palea*.

8.3.4.3.3 Successional Sequence of Dominant Taxa

The sequence of the stratigraphic changes in diatom species dominance in Onondaga Lake

sediments is summarized in Figure 8.32. Many of the diatom taxa occur initially between 100 cm and 80 cm depth. A number are present deeper in the sediment column (Table 8.17).

Chrysophyte cysts and *Cyclotella compta*, identified with mesotrophic conditions, cease to compose a significant part of the assemblage above 80 to 70 cm depth. The relative abundance of benthic and epiphytic taxa also declines above this depth. Recall that the sediment concentration of total phosphorus

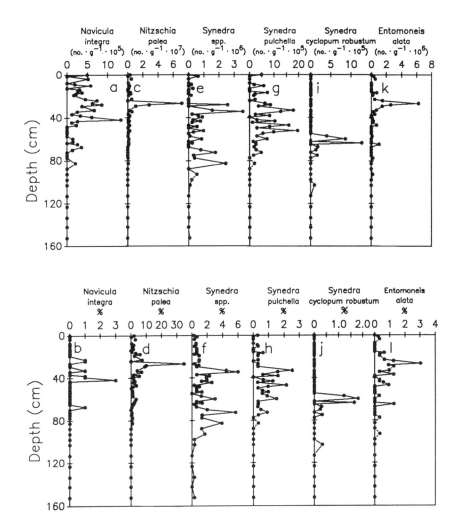

FIGURE 8.30. Diatom stratigraphic profiles in a gravity core from the south site on Onondaga Lake in numbers per gram of dry sediment (c), and contribution to total assemblage (%): (**a**) *Navicula integra* (c), (**b**) *N. integra* (%), (**c**) *Nitzschia palea* (c), (**d**) *N. palea* (%), (**e**) *Synedra* spp. (c), (**f**) *S.* spp. (%), (**g**) *S. pulchella* (c), (**h**) *S. pulchella* (%), (**i**) *S. cyclopum robustum* (c), (**j**) *S. cyclopum robustum* (%), (**k**) *Entomoneis alata* (c), and (**l**) *E. alata* (%).

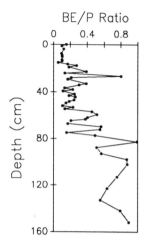

BE/P Ratio

FIGURE 8.31. Profile of the ratio of benthic and epiphytic diatoms to planktonic diatoms (BE/P) in a gravity core from the south site on Onondaga Lake (modified from Rowell 1995).

TABLE 8.17. Taxa occurring between 160 cm and 520 cm depth within the sediment of Onondaga Lake.

Taxa	Planktonic/benthic
chrysophyte cysts	P
Mastogloia smithii	B
Cyclotella comta	P
Cyclotella meneghiniana	P
Stephanodiscus niagarae	P
Fragilaria crotonensis	P
Asterionella formosa	P
Gomphonema spp.	B
Cymbella spp.	B
Epithemia spp.	B
Amphora spp.	B
Eunotia spp.	B

starts to increase at this depth (Figure 8.25c). *Tabellaria fenestrata* rises sharply in abundance and the saline indicator, *Synedra pulchella*, increases in abundance. Above 55 to 60 cm depth, the level of a sharp increase in sediment carbonate content (Figure 8.19), *T.*

fenestrata, Synedra cyclopum var. *robustum*, and the salt intolerant species, *Stephanodiscus niagarae*, cease to be present and the anthropogenic impact indicator species, *Stephanodiscus hantzschii*, first exceeds 2% of the assemblage.

At 45 to 50 cm depth, the start of initial peaks in Pb, Zn, and Cu (Figure 8.23), both *S. hantzschii* and *Asterionella formosa* decline in abundance while the brackish water species,

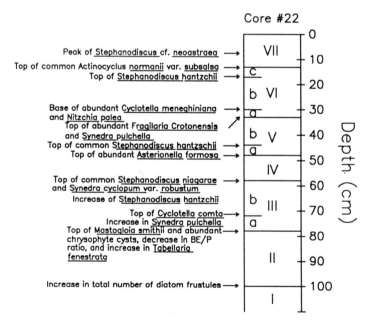

FIGURE 8.32. Schematic of sequence of the stratigraphic changes in diatom species dominance in a gravity core from the south site on Onondaga Lake.

Synedra pulchella, and the pollution associated species, *Navicula integra*, reach their highest concentrations. At 30 cm depth, a spike in the pollution diatom, *Nitzschia palea*, comprises 35% of the assemblage. Above 30 cm depth, the interval of highest concentrations of phosphorus, organic carbon, and metals, *S. pulchella* and *Fragilaria crotonensis* decline while *Cyclotella meneghiniana*, *C. cryptica*, *C. atomus*, and *Actinocyclus normanii* fo. *subsalsa* increase dramatically. All of these latter species are associated with severe nutrient and salt pollution conditions. Above 15 cm depth, *C. cryptica*, and *A. normanii* fo. *subsalsa* cease to dominate the assemblage, *Stephanodiscus hantzschii* does not occur, and the (questionably) mesotrophic

form, *Stephanodiscus* cf. *neoastraea*, reaches a peak in abundance.

Apart from the paleoenvironmental significance of this stratigraphic sequence, the diatom stratigraphy is valuable as a tool for correlation between sediment cores taken from different locations in the lake. To facilitate this application, the sequence shown in Figure 8.32 is divided into biostratigraphic zones (Schoch 1989). These zones are defined in Table 8.18. The stratigraphic sequence was the same in both the gravity cores from the south and north basin sites. Zones VI and VII were missing from the south site piston core, indicating the upper 35 to 40 cm of sediment had not been recovered.

TABLE **8.18.** Definition of diatom stratigraphic zones.

Zone	Definition
VII	Base at the top of Zone VI; Top undefined (at top of core).
	An abundance peak of *Stephanodiscus* cf. *neoastraea* occurs within this zone.
VI	Base at the top of Zone V; top of this zone is defined by a dramatic drop in occurrence of *Actinocyclus normanii* var. *subsalsa*.
	Subzone C: top of this subzone corresponds with top of Zone VI.
	Subzone B: top of this subzone is defined by a drop in occurrence of *Stephanodiscus hantzschii*.
	Subzone A: top of this subzone is defined by the base of the abundant occurrence of *Cyclotella meneghiniana* and *Nitzschia palea*.
V	Base at the top of Zone IV; top of this zone is defined by the top occurrence of abundant *Fragilaria crotonensis* and by the top occurrence of abundant *Synedra pulchella*.
	Subzone B: the top of this subzone corresponds to the top of Zone V.
	Subzone A: the top of this subzone is defined by the top of the common occurrence of *Stephanodiscus hantzschii*.
IV	Base at the top of Zone III; top of this zone is defined by a dramatic drop in the occurrence of *Asterionella formosa*.
III	Base at the top of Zone II; top of this zone is defined by the top occurrence of common *Stephanodiscus niagarae* and the top occurrence of *Synedra cyclopum* var. *robustum*.
	Subzone B: top of this subzone corresponds with the top of Zone III. *Stephanodiscus hantzschii* increases in abundance and *Tabellaria fenestrata* abruptly declines in abundance within this subzone.
	Subzone A: top of this subzone is defined by the top of *Cyclotella comta*. *Synedra pulchella* increases in abundance within this subzone.
II	Base at the top of Zone I; top of this zone is defined by the top occurrence of abundant chrysophyte cysts, the top occurrence of *Mastogloia smithii*, a drop in the ratio of benthic-epiphytic/planktonic diatoms, and a rise in occurrence of *Tabellaria fenestrata*.
I	Base undefined; top at rise (more than doubling) in the concentration of total diatom frustules per gram of sediment.

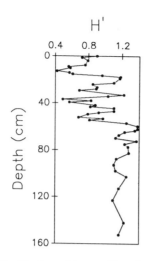

FIGURE 8.33. Stratigraphic profile of Shannon–Weaver diversity index (H') for the diatom assemblage in a gravity core from the south site on Onondaga Lake (modified from Rowell 1995).

8.3.4.4 Total Assemblage Analysis

8.3.4.4.1 Assemblage Diversity

The stratigraphic distribution of Shannon–Weiner diversity indices (H'; Wetzel 1983) for the south deep site in Onondaga Lake is shown in Figure 8.33. Low index values reflect low species diversity. Below 60 cm depth, the indices average 1.2, with a range between 1.0 and 1.4. Above 60 cm, the indices average 0.9, with a range between 0.4 and 1.2. These values are low compared to the Shannon–Weaver index values for sediment diatom assemblages reported by Tuchman et al. (1984a) for salt impacted Fonda Lake in Michigan (values decreasing from around 2.3 to 1.7) and by Cooper and Brush (1991) for colonially impacted Chesapeake Bay (values descending from 3.8 to 2.3).

The lower indices, which occur above 60 cm depth, indicate that greater environmental stress was placed upon the Onondaga Lake diatom assemblage over this interval (Cooper and Brush 1991; Patrick 1977). This is consistent with other paleolimnological evidence presented in this chapter of increasingly polluted conditions over time. Documentation

of the water column diatom assemblage in Onondaga Lake over recent years (Onondaga County 1969–1991; Sze 1975, 1980) supports the paleolimnological evidence of limited diversity. While more than thirty-five species of diatoms have been documented from the water column of Onondaga Lake (Hohn 1951; Onondaga County 1971; Onondaga County 1969–1991), the assemblage has been dominated annually by only a few species (Sze 1975, 1980; Chapter 6).

8.3.4.4.2 Ecological Analysis

The taxa within the entire sediment diatom assemblage may be grouped according to environmental associations (Duthie and Sreenivasa 1971; Haworth 1969). One hundred and forty of the 216 taxa identified (Rowell 1992) have been reported as preferentially occurring under particular trophic and salinity conditions (e.g., Beaver 1981; Patrick and Reimer 1966). All of these taxa are reported to be alkalinity tolerant. This, along with the deep sediment content of $CaCO_3$, indicates that the lake is a naturally alkaline system. The relative numbers of diatoms associated with the three trophic categories are shown in Figure 8.34. Oligotrophic and mesotrophic taxa prevail below 80 cm depth, although eutrophic taxa are also well represented. Mesotrophic and eutrophic taxa dominate in the 80 to 30 cm depth interval. Eutrophic taxa dominate above 30 cm depth. A slight resurgence of oligotrophic taxa occurs at 10 cm and above.

The relative numbers of diatoms associated with different levels of salinity are shown in Figure 8.35. The percentage of salt tolerant diatoms rises to over 90% at around 55 cm depth (Figure 8.35a). Between 80 and 55 cm, it ranges from 50 to 70 percent. Below 80 cm about 40% of the diatoms are salt tolerant. Although minor in overall abundance, the distributions of taxa designated as halophilous (preferring saline conditions) and halophobous (intolerant to saline conditions) show an almost mutually exclusive relationship between levels of sediment containing these two catagories; most halophobous forms occur

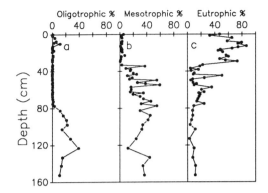

FIGURE 8.34. Stratigraphic profiles of relative numbers of diatoms associated with the three trophic categories in a gravity core from the south site on Onondaga Lake, as percentages: (a) oligotrophic, (b) mesotrophic, and (c) eutrophic (modified from Rowell 1995).

below 55 cm depth (Figure 8.35b), most halophilous forms occur above 55 cm (Figure 8.35c). Tuchman et al. (1984a) report an increase in halophilous diatoms with increases in salt loading to Fonda Lake, Michigan, and a decrease in the Shannon–Weaver diversity index for the sediment diatom assemblage. The increase in halophilous and salt tolerant diatoms within the sediments of Onondaga Lake is also accompanied by a drop in the diversity of the diatom assemblage (Figure

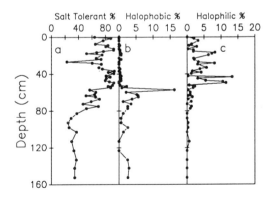

FIGURE 8.35. Stratigraphic profiles of relative numbers of diatoms associated with different levels of salinity in a gravity core from the south site on Onondaga Lake, as percentages: (a) salt tolerant, (b) halophobic, and (c) halophilic (modified from Rowell 1995).

8.33). It appears that the increase in salinity of Onondaga Lake contributed significantly to the stress placed upon the lake's diatom assemblage.

The vertical sequences manifested in this analysis indicate that both the productivity and salinity of Onondaga Lake increased over time. Below 80 cm depth, the ecological associations suggest that mesotrophic lake conditions prevailed. Above 80 cm depth, productivity in the lake increased, and above 60 cm depth, salinity rose markedly and diversity declined. There are indications salinity increased above 80 cm, probably as a result of salt pollution from early colonial salt manufacture. Further increases in productivity are implied above 30 cm depth. In the upper 10 cm, however, there are indications of reduced productivity.

The concentration of diatom frustules is much lower below 100 cm depth than above, which also implies that lake conditions were much less productive. Despite this increase in frustule abundance, however, the trophic state preference of the individual taxa found below 100 cm are similar to the preference of those found between 100 cm and 80 cm depth. Species of *Cyclotella* and *Stephanodiscus* associated with unperturbed oligotrophic and mesotrophic conditions (Stoermer 1984; Battarbee 1986) were not found in any abundance in the Onondaga Lake sediment, even to a depth of more than 5 m.

8.3.5 Isotopic Dating

8.3.5.1 Background

Three radioactive isotopes commonly used in dating sediments, carbon-14 (^{14}C), cesium-137 (^{137}Cs), and lead-210 (^{210}Pb), have been utilized for Onondaga Lake. Carbon-14 is produced in the atmosphere and is absorbed by living organisms. It has a half-life of approximately 5570 years (Olsson 1986; Schoch 1989) and can be used to date samples up to 50,000 years old (Schoch 1989). Due to the increased likelihood of contamination and its long half-life, it is not reliable for dating recent sediments; that is, deposited since the advent of anthropogenic alterations of lake watersheds (Olsson 1986). Cesium-137 is included in the fallout from nuclear explosions. It has a half-life of 30 years and can be used in the Northern Hemisphere to distinguish sediment horizons for 1954 and 1964 (Ritchie and McHenry 1990; Olsson 1986). The horizon of maximum ^{137}Cs activity corresponds to 1963–1964, when the maximum radioactive fallout from atmospheric atomic bomb testing occurred (Ritchie and McHenry 1990). Cesium-137 adsorbs readily to organic particles and has a tendency to diffuse within the sediment (Ritchie and McHenry 1990; Olsson 1986; Krishnaswami et al. 1971). Lead-210 comes from radium in the earth, which decays to radon gas and escapes into the atmosphere before decaying to [210]Pb. Lead-210 adsorbs to particles in the atmosphere, which settle upon the lake surface and into the sediment. It has a half-life of 22.26 years and is useful for dating sediments up to 150 years in age (Olsson 1986).

8.3.5.2 Carbon-14

Two ^{14}C dates have been obtained from the deeper Onondaga Lake sediments. Ages of 2960 ± 50 years before present (YBP) at 175 cm depth and 5490 ± 70 YBP at 465 cm depth were determined from the south site piston core. The uncertainty range of the results corresponds to the 68.3% confidence limit (one standard deviation).

Since the piston core did not recover the upper approximately forty centimeters of sediment immediately beneath the sediment–water interface, the depths should be adjusted to 215 cm and 505 cm before estimating the long-term sediment accumulation rates. Further, since the angle of entry of the piston core apparently deviated from the vertical, the calculated rates should be regarded as upper bounds of the average true accumulation rate(s) for these depth intervals (Rowell 1992). The calculated rates are 0.12 cm · yr^{-1} for the interval from 505 cm to 215 cm depth and 0.07 cm · yr^{-1} for the interval from 215 cm to 0 cm depth. These values are similar to those reported for a number of other lakes in nor-

theastern United States and Canada (e.g., Cotter and Crowl 1981; McAndrews 1981; Miller 1973; Spear and Miller 1976; Procter 1978).

The ^{14}C ages reported here are in uncalibrated radiocarbon years, that is, the results have not been adjusted to correspond with standard dendrochronological (tree ring) calibration curves (Olsson 1986). Correlation of the ^{14}C ages with an independently derived age control is desirable because of contaminating factors that may bias the ^{14}C dating results. The "hard water effect," which occurs in lakes with carbonate-rich (marl) sediments, may result in ^{14}C values older than the true sediment age (Turner et al. 1983). "Old" carbon, washed in from the watershed, can also lead to anomolously old ^{14}C ages (Olsson 1986), especially if the ^{14}C analysis is conducted on sediments with very low concentrations of organic carbon (Figure 8.25). The effects of these biases can only be evaluated through corroboration of the ^{14}C dating results by another stratigraphic dating method. Pollen stratigraphy is often used for this purpose (Olsson 1986). The ^{14}C ages presented here are subsequently evaluated in relation to the lake's pollen stratigraphy.

8.3.5.3 Cesium-137

Effler (1975) did not detect any peaks in ^{137}Cs activity (dpm · g^{-1}) in his analysis of Onondaga Lake sediments because the uppermost sediment layers were not recovered in his cores. Distinct horizons of ^{137}Cs were found in gravity core samples from both basins collected in 1988 (Figure 8.36), as the uppermost sediments were recovered. The ^{137}Cs activity peak occurs at 21 cm depth in the south basin core and at 19.7 cm depth in the north basin core. Based upon these horizons, the average sediment accumulation rates may be calculated for the interval 1964 to 1988 as 0.88 cm · yr^{-1} and 0.82 cm · yr^{-1} for the south deep and north basin sites, respectively.

The rise in ^{137}Cs activity at the top of each profile (Figure 8.36) may be a manifestation of diffusion. Cesium-137 activity generally decreases steadily following the 1964 peak

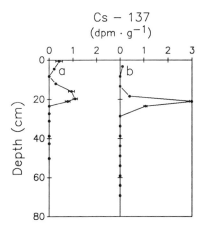

FIGURE 8.36. Stratigraphic profiles of ^{137}Cs activity for gravity core from Onondaga Lake: (**a**) north site, and (**b**) south site.

and approaches the limits of detection at depths corresponding to the mid-1980s (Ritchie and McHenry 1990). The sharpness of the ^{137}Cs peaks, however, especially at the south site, suggests that diffusion has not greatly altered the ^{137}Cs profiles.

8.3.5.4 Lead-210

Lead-210 activity (dpm · g^{-1}) profiles obtained from gravity cores collected from the north and south basins in 1988 are presented in Figure 8.37a and b. The profiles are nonlinear.

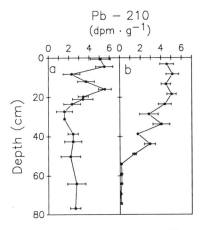

FIGURE 8.37. Stratigraphic profiles of ^{210}Pb activity for gravity core from Onondaga Lake: (**a**) north site, and (**b**) south site.

Nonlinear ^{210}Pb profiles have been interpreted through the constant rate of supply (CRS) model (Appleby and Oldfield 1978; Oldfield and Appleby 1984; Olsson 1986). The model assumes a constant flux of ^{210}Pb to the sediments. Nonlinearity in profiles is generally attributable to variations in sedimentation rates, differences in compaction, and bioturbation. The primary cause in Onondaga Lake is probably variations in accumulation rates, though the lower degree of compaction of the uppermost sediments may also have contributed. Unsupported (derived directly from the atmosphere) ^{210}Pb activities are often determined by subtracting a uniform deep sediment activity, assuming it is representative of background (supported) activity. The use of the CRS model for the profiles of Figure 8.37 is confounded by the apparent horizontal, as well as vertical, variations in background (deep) levels of ^{210}Pb; that is, no time horizons or accumulation rates have been determined from these data.

Earlier Effler (1975; Effler et al. 1979) reported ^{210}Pb sedimentary profiles for the south and north basins, and an intermediate deepwater location referred to as the "saddle" site (Figure 8.16). The saddle site profile was distinctly nonlinear and similar to the profile reported here for the north basin. Profiles measured for the south and north basins were approximately log-linear (Effler 1975). The constant initial concentration (CIC) model has commonly been used to determine sediment accumulation rates when log-linear ^{210}Pb profiles are observed (Oldfield and Appleby 1984; Olsson 1986). The model assumes both a constant flux of ^{210}Pb to the sampled sediments and a constant rate of sediment accumulation. Early applications of the ^{210}Pb technique (e.g., Goldberg 1963; Krishnaswami et al. 1971), through the period of Effler's (1975) work, invoked this model. This approach, along with the failure to collect the upper sediments for both the north and south sites, led to the overestimates of net accumulation rates (5–9 cm · yr^{-1}; Effler 1975; Effler et al. 1979). These older south and north basin profiles of [210]Pb (Effler 1975, 1979) are more properly interpreted as evidence of vertical variations in background ^{210}Pb. The large vertical intervals between sediment samples (10 cm; Effler 1975) contributed to the misinterpretation of the ^{210}Pb stratigraphy of this earlier work.

8.3.6 Pollen Analysis

8.3.6.1 Background

Pollen grains are abundant, widely dispersed, and usually well preserved in lake sediments. These characteristics make pollen analysis one of the most useful tools in sediment chronology and paleoenvironmental interpretation (Moore and Webb 1978; Berglund and Ralska-Jasiewiczowa 1986). There are two major categories of pollen, those forms produced by trees (arboreal pollen), and those produced by herbaceous plants (nonarboreal pollen). In eastern North America, major changes in vegetation (and therefore pollen composition) accompanied the northward retreat of glacial climate and the clearing of forest land by colonial farmers (Delcourt and Delcourt 1987; Faegri and Iversen 1989). The ages of the former have been independently established through radiometric carbon dating and the age of the latter by historical documentation. The pollen record may be biased by the different modes of pollen dispersal, and variations in the degree of preservation of the various forms (Faegri and Iversen 1989).

As the glaciers retreated from New York State, a cleared landscape was left upon which vegetation began to encroach (Muller 1977). A spruce-pine-fir forest became established before 7500 B.C., followed by a pine-dominated forest after 7500 B.C. and deciduous-dominated forests after 6500 B.C. (Cox 1965). Based upon pollen diagrams from sites in New England, New York, and Pennsylvania, a regional pollen stratigraphic zonation has been developed (Deevy 1939; Miller 1973; Cotter and Crowl 1981). It has been tied to radiometric carbon dates at a number of sites (Muller 1977; Spear and Miller 1976; Miller 1973). Features of this zonation are manifested in the pollen sedimentary profiles from Onondaga Lake presented here. A dif-

ferent but correlatable set of pollen zones has been developed in southeastern Canada (McAndrews 1981).

A decline in the percentages of forest tree pollen and an increase in *Ambrosia* (ragweed) pollen mark the onset of colonial watershed clearing (Faegri and Iversen 1989). The start of this clearing proximate to Onondaga Lake is imprecisely documented. The first settlers entered the area during the 1780s and by 1797 the State had divided the adjoining land into "marsh lots" (Smardon et al. 1989), but the extent of clearing of vegetation is unclear. It seems plausible that the operation of the salt industry on the "marsh lots" would have had an effect on pollen. Therefore, the first increase in *Ambrosia* pollen in the Onondaga Lake sediments has been attributed here to approximately the year 1800.

The pollen diagrams (a customary form of presention for pollen stratigraphic data (Berglund and Ralska-Jasiewiczowa 1986)) reported from sites proximate to Onondaga Lake all exhibit the pollen zones of Deevy (1939) and Miller (1973), and include a recognizable forest clearing (*Ambrosia*) horizon. These sites include Cicero Swamp north of the City of Syracuse (Cox 1959), Sandy Ridge Bog twelve miles north of Syracuse (McCulloch 1939), Mud Pond south of Oswego (Shipman 1970), and Crusoe Lake west of Syracuse (Cox and Lewis 1965). The pollen analysis of Onondaga Lake sediments reported by Onondaga County (1971) did not result in a pollen diagram.

8.3.6.2 Palynology of Onondaga Lake

8.3.6.2.1 Pollen Zonation

The pollen diagram for Onondaga Lake, based on analysis of the piston core sample (sectioning interval of 10 cm) from south deep, is presented in Figure 8.38a and b. The pollen generic names and their common name equivalents are listed in Table 8.19.

Among the total pollen grains counted, unknowns made up a significant percentage (5%−45%; Figure 8.38b). These consist of grains too badly corroded (degraded) to be classified and grains unrecognizable at the

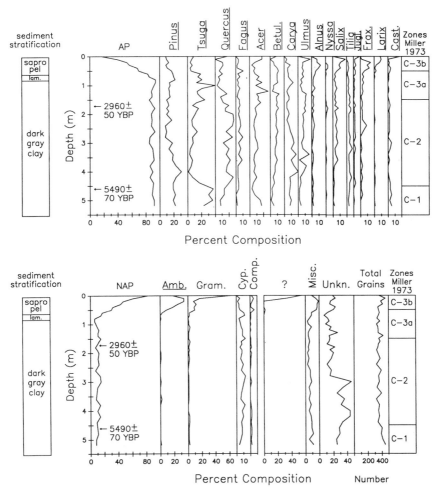

FIGURE 8.38. Stratigraphic profiles of pollen (pollen diagram) for piston core from the south site on Onondaga Lake (see Table 8.18 for full names). (a) Arboreal pollen. Pollen zones after Miller (1973) and Deevy (1939). (b) Nonarboreal pollen, plus special categories. Pollen zones after Miller (1973) and Deevy (1939).

angle of view. The reason for the high percentage of corroded grains in Onondaga Lake is uncertain. Perhaps they reflect the input of pollen grains originally deposited on soils on the watershed (e.g., Faegri and Iversen 1989). Following the recommendations of Faegri and Iversen (1989) for samples with high percentages of corroded pollen grains, the number of unknowns (as well as the catagories of miscellaneous and "?") has not been included in the calculation of the pollen percentages shown in Figure 8.38. Unknowns are included in the plot of total grains counted per sample (Figure 8.38b).

The upper pollen zone and subzones of Deevy (1939), as applied by Miller (1973) in north central New York State, are easily distinguishable on the Onondaga Lake pollen diagram (Figure 8.38). At the base of the diagram, hardwood tree pollen is dominant over coniferous pollen. The spruce, fir, and pine dominated A and B Zones of Deevy (1939) were not encountered. The high hemlock pollen percentages at the bottom of the core (Figure 8.38) indicate the deepest zone sampled with the piston corer to be Zone C-1. The top of this zone is marked by a sharp decline in hemlock pollen (Miller

TABLE 8.19. Pollen generic names and their common name equivalents.

Common name	Generic name	Abbreviation*
Arboreal Pollen		
Pine	*Pinus*	
Fir	*Abies*	
Spruce	*Picea*	
Hemlock	*Tsuga*	
Oak	*Quercus*	
Beech	*Fagus*	
Maple	*Acer*	
Hickory	*Carya*	
Elm	*Ulmus*	
Alder	*Alnus*	
Black Gum	*Nyssa*	
Willow	*Salix*	
Basswood	*Tilia*	
Black Walnut	*Juglens*	Jugl.
Ash	*Fraxinus*	Frax.
Larch	*Larix*	
Chestnut	*Castanea*	Cast.
Birch family	*Betulaceae*	Betul.
Nonarboreal Pollent		
Ragweed	*Ambrosia*	Amb.
Grass family	Gramineae	Gram.
Sedge family	Cyperaceae	Cyp.
other NAP	Compositae	Comp.

*Used in Figure 8.38

1973). Based upon radiometric carbon dates, Davis (1981) reported the age of this decrease to be 4800 YBP across northeastern North America. Turner et al. (1983) suggested that the age is slightly younger (4200 corrected carbon YBP) in southern Ontario. The uncorrected carbon-14 date of 5490 YBP obtained in this study below the top of Zone C-1 is compatible with these ages for the C-1/C-2 pollen zone boundary (Figure 8.38).

Zone C-2 is marked by lower percentages of hemlock pollen and higher percentages of the hardwood trees, particularly oak (Deevy 1949; Figure 8.38). Miller (1973) placed the top of this zone after the decline in oak pollen, which occurred in the center of the increase in his hemlock pollen profiles. The top of Zone C-2 is placed at around 150 cm depth in Figure 8.38a, also after the initial decrease in oak pollen and within the increase in hemlock pollen. The age of this transition has been given by Cox (1965) as around 2000

YBP and by Miller (1973) as slightly younger. The increase of hemlock pollen below this boundary starts at around 3000 YBP (Davis 1981; Delcourt and Delcourt 1987), an age supported by an unadjusted C-14 date in Ontario (Turner et al. 1983) and by the C-14 date at 175 cm depth reported here of 2960 YBP. Cox and Lewis (1965) identified subzones of Zone C-2 based upon the hemlock profile, but these are not recognizable on the hemlock profile from Onondaga Lake. The 10 cm sample interval used may be too broad to clearly resolve these subzones.

8.3.6.2.2 Anthropogenic Effects

The base of Zone C-3b (Deevy 1939; Miller 1973) is recognizable by the decrease in arboreal pollen (Figure 8.38a) and the increase in grass and *Ambrosia* pollen (Figure 8.38b). It coincides with a change in sediment character as well (Figure 8.17). This boundary represents the start of colonial clearing (~1800) in the Onondaga Lake watershed. Earlier declines in some of the forest tree pollen (elm, oak, maple) might be an indication of Native American activity (Burden et al. 1986). The rise in nonarboreal pollen is accompanied by an increase in alder pollen, an early succession tree that increases in numbers following forest clearing (e.g., Burden et al. 1986). This trend is followed by increases in maple, elm, oak, ash, and willow pollen, indications of secondary forest tree growth (Faegri and Iversen 1989). Hickory and beech pollen continued to decline, perhaps in response to continued selective cutting.

The direct interpretation of cultural events from the pollen profiles within Zone C-3b is speculative. The immediate Onondaga Lake watershed has been highly developed into an urban area. It is quite possible that the pollen relationships may be due to interacting effects upon vegetation that are not readily apparent (Faegri and Iversen 1989). For instance, a sharp increase in grass pollen in the sample at the very top of the pollen diagram (Figure 8.38b) suggests an increase in the amount of watershed clearing. However, much of the tree pollen is diluted in percentage by this increase; that is, the actual tree vegetation may not have fallen as precipitously as suggested by the pollen diagram. Oak, hemlock, black walnut, and chestnut pollen percentages actually increase (Figure 8.38). Furthermore, the top of the stratigraphic section of the pollen diagram is known to be incomplete and a very high percentage of an unidentified (questionably pollen) taxa (see "?" in Figure 8.38b) may be biasing the upper two samples (0–20 cm). Increased depth resolution in these uppermost sediments of the lake (i.e., including layers not successfully recovered for this analysis) is desirable to delineate pollen chronology within the period of severe anthropogenic impact.

8.3.6.2.3 Sediment Accumulation Rate

Pepperage or sour gum pollen (*Nyssa sylvatica*; also known as black gum (Murphy 1978)) has been reported present at sediment depths below 180 cm at the saddle of Onondaga Lake (Onondaga County 1971). Newspapers from 1875 to 1880 showed advertisements seeking pepperage wood for use as hollowed out pipes to transport salt brine. This has been taken as evidence that the local supply of pepperage was exhausted (Onondaga County 1971). Therefore the decrease in this pollen in the lake sediments would represent approximately the years 1875 to 1880. Based on this information it was "surmised that the upper two meters of black sediments in the lake have been deposited since 1875 to 1880" (Onondaga County 1971, p. 86). This corresponds to an average sediment accummulation rate over that interval of about $2 \, cm \cdot yr^{-1}$.

Small percentages of sour gum pollen were found in the sediment collected from the south basin in 1988 (Figure 8.38a). This pollen was not observed in the upper 30 cm of sediment, but was found in the next 10 cm interval. Based on the same logic utilized in the Onondaga County (1971) study, this level may be dated at around 1880. The depth of occurrence of this horizon, corrected for the lost upper 40 cm (~70 to 80 cm), is considerably shallower than found at the saddle of the lake by Onondaga County (1971), indicating a lower average accumulation rate.

8.3.7 Synthesis of Stratigraphy and Sediment Accumulation Rates

A summary of stratigraphic horizons identified in the sediments of Onondaga Lake for the south and north basins is presented in Figures 8.39 and 8.40, respectively. Paleolimnological analysis has resulted in the identification of six major stratigraphic intervals:

1. a bottom interval representing lake conditions prior to cultural eutrophication; no age date has been assigned to the top of this interval;
2. an interval containing the first indications of lake eutrophication (i.e., increased numbers of diatoms, increased TOC concentration); a date of 1822 is assigned to the top of this interval based on the decline in the benthic/planktonic diatom ratio, which probably resulted from the lowering of the lake's water surface; the *Ambrosia* pollen horizon (about 1800) also falls within this interval;
3. an interval containing indications of watershed clearing; a date of 1884 is assigned to the top of this interval based upon the increase in $CaCO_3$ concentrations as a reflection of the onset of industrial waste loadings associated with the production of soda ash and the correlation to the decline in abundance of *Nyssa* pollen;
4. an interval containing indications of increased industrial activity; a date of 1946 is suggested for the top of this interval based upon the increase in Hg concentrations, which resulted from increased Hg loadings to the lake;
5. an interval containing indications of intense anthropenic pollution of Onondaga Lake; a date of 1971 is suggested for the top of this interval based on the decline in Hg and TP concentrations related to decreased loadings to the lake; and
6. a top interval reflecting reductions in lake pollution.

Sediment accumulation rates calculated from the assigned horizon dates and the 1964 ^{137}Cs horizons shown in Figures 8.39 and 8.40 are

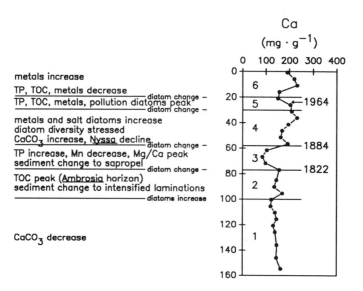

FIGURE 8.39. Stratigraphic horizons in gravity core from the south site on Onondaga Lake.

presented in Table 8.20. Those that include the upper sediments require adjustment for undercompaction to estimate a true sediment accumulation rate, that is, that can be compared to rates obtained for deeper compacted sediment intervals (Table 8.20). The average sediment "compaction factor" of the uppermost interval of Table 8.20 is about 0.7 (Figure 8.21). The adjusted sediment accumulation rate for the uppermost interval of Table 8.19 (for the time interval 1964 to 1988) is approximately $0.6 \, cm \cdot yr^{-1}$. The rates presented in Table 8.20 represent average rates of sediment accumulation over the indicated time intervals, and are based on the assumption that no

sediment compaction occurred due to coring. Annual rates probably varied within these intervals on a stochastic basis (meteorological variations), and perhaps systematically in some cases (e.g., with accelerated productivity and changes in production of soda ash).

The average accumulation rates increase upward for each succeeding time interval. The rates in the south basin are somewhat higher than in the north basin. The pattern of upward increase in sediment accumulation rates is commonly observed in anthropogenically impacted lakes (e.g., Oldfield and Appleby 1984; Appleby and Oldfield 1978; Oldfield et al. 1978). The range of sediment accumulation

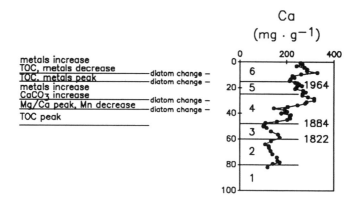

FIGURE 8.40. Stratigraphic horizons in gravity core from the north site on Onondaga Lake.

TABLE 8.20. Rates of sediment accumulation for Onondaga Lake.

Time period	Sediment accumulation rate ($cm \cdot yr^{-1}$)	
	South	North
1964 to 1988	0.88*	0.83*
1884 to 1964	0.46	0.35
1822 to 1884	0.32	0.21
5490 YBP to 1800	≤0.08	

*Approximately $0.6\, cm \cdot yr^{-1}$ when adjusted for compaction

rates for Onondaga Lake sediments ($0.08-0.60\, cm \cdot yr^{-1}$) approaches the range reported by Krishnaswami and Lal (1978; $0.01-0.90\, cm \cdot yr^{-1}$) for twenty lakes representing a wide range of trophic state. For instance, the low value of $0.01\, cm \cdot yr^{-1}$ is for Lake Superior (Krishnaswami and Lal 1978), the largest and least productive of the Great Lakes of North America (Stoermer et al. 1985c). Proctor (1978) reported rates as low as $0.08\, cm \cdot yr^{-1}$ from the western New York Finger Lakes. The estimated recent sediment accumulation rate for Onondaga Lake of approximately $0.6\, cm \cdot yr^{-1}$, adjusted for compaction (Table 8.20), equals some of the highest rates reported for other hypereutrophic lakes (Wetzel 1983). Lake Mendota, a culturally eutrophic lake, was also estimated to have a sedimentation rate of $0.6\, cm \cdot yr^{-1}$ (Krishnaswami and Lal 1978). Higher sediment accumulation rates do occur in impoundments that receive high allochthonous sediment loads (e.g., Ritchie et al. 1973). Based on the recent decreases in sediment deposition associated with the closure of the soda ash/chlor-alkali facility (Table 8.3), the sediment accumulation rate is expected to decrease, probably to between 0.2 and $0.25\, cm \cdot yr^{-1}$.

8.4 Summary

8.4.1 Deposition

Deposition is a major pathway in the cycling of many constituents in lakes. Sediment trap collections have been made in 1980, 1981, 1985–1991, to support the determination of deposition rates of various substances of water quality and sedimentary concern. Traps were simple cylinders, with an aspect (length/diameter) ratio of 6. Traps were deployed at 10 m (i.e., below the upper mixed layer until fall), usually for intervals of one week.

A summary of deposition rates ($mg \cdot m^{-2} \cdot d^{-1}$) determined for dry weight, particulate inorganic carbon (PIC), P, chlorophyll (C_T), particulate organic carbon (POC), and TKN, for the subset of analyzed trap samples is presented in Table 8.21. Strong temporal variations were observed annually for each of the constituents. The high rates of POC, TKN and C_T deposition (Table 8.21) reflect the hypereutrophic status of the lake; the fluxes of POC and TKN are generally consistent with those reported for other highly productive systems in the literature. Reductions in POC, TKN, and C_T deposition rates may indicate a decrease in primary productivity, but more probably reflect the effects of increased grazing and reduced co-precipitation of organic particles with $CaCO_3$.

The deposition rates of dry weight, $CaCO_3$ (determined from PIC), and P in Onondaga Lake (Table 8.21) are extremely high compared to values reported in the literature for other lakes. The downward flux of dry weight is driven largely by $CaCO_3$ deposition. Calcium carbonate represented approximately 60% of dry weight deposition in 1985 and 1988 (Table 8.21). Further, the temporal distributions of the dry weight and $CaCO_3$ fluxes were highly correlated within each of these two years. The decrease in the deposition rate(s) of $CaCO_3$, and therefore dry weight, is largely attributable to the reduction in Ca^{2+} loading associated with the closure of the soda ash/chlor-alkali facility. Further, the reduction in the deposition rate of PIC ($CaCO_3$) from 1985 to 1988 is consistent with the observed decrease in alkalinity (DIC) depletion in the epilimnion over the same period (Chapter 5). The average deposition rates of P in 1980 and 1987, after normalization for different trap deployment periods, were approximately equal (Table 8.21). The very high fluxes of P probably reflect not only the enrichment of the overly-

TABLE 8.21. Deposition rates in Onondaga Lake from sediment traps.

Parameter	Year	$(mg \cdot m^{-2} \cdot d^{-1})$		Period
		$\bar{x}$	Range	
Dry weight	1985	37,900	14,900–65,800	mid-May to mid-October
	1988	11,500	3,000–25,400	mid-May to mid-October
PIC	1985	2,560	1,200–6,300	mid-May to mid-October
	1988	870	220–2,200	mid-May to mid-October
	1989	950	140–2,500	mid-May to mid-October
CaCO$_3$*	1985	21,300	10,000–52,500	mid-May to mid-October
	1988	7,250	1,830–18,200	mid-May to mid-October
	1989	7,920	1,170–20,800	mid-May to mid-October
P	1980	48.3**	21–183	May to September
	1987	39.4	16–65	May to September
C$_T$	1987	20.4	5–38	June to October
	1988	18.7	3–48	April to October
	1989	15.8	2–45	April to October
	1990	15.7	2–60	April to October
POC	1985	1,180	410–3,650	mid-May to mid-October
	1988	750	100–1,300	mid-May to mid-October
TKN	1985	317	210–490	mid-May to mid-October
	1988	230	60–450	mid-May to mid-October

* CaCO$_3$ determined from inorganic C according to stoichiometry CaCO$_3$: inorganic C = 100:12
** Average deployment period of 4d

ing waters, but the association of P with inorganic particles (approximately 30% of deposited P in 1980 was in an inorganic form), in particular CaCO$_3$.

8.4.2 Surficial Sediments

The dimensions of the low energy environment ("depositional basin") of Onondaga Lake were delineated (based on an analysis of the spatial distribution of particle sizes), distribution maps for selected constituents were developed, and four sediment categories were established, based on physical and chemical characterization of 70 surficial sediment grab samples collected from throughout the lake. The lake's low energy environment corresponds to depths >8 m; it occupies about 63% of the bottom area. High energy environments, where sediment resuspension is expected to occur, correspond to depths <6 m, which occupy a relatively narrow zone along the lakeshore.

Oncolites, algal pisoliths (cryptalgal structures) composed mostly of CaCO$_3$ (92%) and ranging up to 15 cm in diameter, dominate the nearshore deposits for most of the eastern, northern, and northwest shores of the lake. Most of these concretions have fragments of charophyte stems as nuclei. It has been contended that the oncolites formed in Onondaga Lake as a result of the discharge of Ca^{2+} enriched ionic waste by the adjoining soda ash/chlor-alkali facility, and that charophytes were eliminated as a result. The coverage of the littoral zone with these concretions may limit macropyte (Chapter 6) and macrobenthos communites in affected areas.

The points of entry of certain constituents are manifested clearly in the distribution maps. A sludge delta of CaCO$_3$ adjoins the mouth of Ninemile Creek as a result of the high loading (caused by the upstream discharge of Ca^{2+} enriched ionic waste) received from this tributary during the operation of the soda ash/chlor-alkali facility. The localized entry of clastics from Onondaga Creek, and organic carbon and P from METRO are also apparent.

The concentrations of organic carbon and nutrients in surficial sediments of the low energy environment are unusually low by comparison to other productive systems. This is an artifact of the diluting effect of the very high deposition rate of $CaCO_3$. The $CaCO_3$ content of the surficial sediments of the central lake basins (55–70%) matches quite closely the concentrations measured in depositing material collected with sediment traps. Organic carbon concentrations are also consistent with the composition of sediment trap collections. The depositing $CaCO_3$ and organic carbon have mostly autochthonous origins. The remaining fraction of surficial sediments of the depositional basin is mostly clastics that have allochthonous origins. On a lakewide basis (including high energy near shore zones and depositional basins), the concentrations of organic carbon, N, and P are negatively correlated with $CaCO_3$ concentrations.

Cluster analysis was used to gather the surficial sediments of Onondaga Lake into four groups. Shallow water sediments have characteristically large particles (e.g., oncolites) and are as much as 90% $CaCO_3$. The shallow water group forms a narrow band, enriched in $CaCO_3$, that extends around most of the lake except the southern shore. Deepwater sediments have higher water content and organic concentrations, and lower $CaCO_3$ content. Two deepwater sediment groups were identified: one is slightly more enriched in phosphorus and COD and occupies the deepest waters. The fourth group is the sediments in the southern end of the lake that reflect the influence of the Onondaga Creek inflow and the METRO discharge.

8.4.3 Sediment Stratigraphy

The sediment stratigraphy of the profundal sediments of Onondaga Lake has been documented through a multidisciplinary paleolimnological analysis of core samples collected in 1988 from the south and north basins of the lake. Two stratigraphic units are clearly manifested; a bottom dark gray clay (gyttja), characterized by fine laminations and relatively low organic content, and a top gaseous black clay (sapropel), containing numerous macrolaminations and microlaminations and higher concentrations of organic material. These laminations, and the sharpness of various constituent horizons found in analysis of the upper unit of the core samples, indicate minimal sediment mixing (e.g., bioturbation) has occurred in the centers of the lake basins. A transitional unit between the lower and upper unit reflects early sediment response to cultural impacts on the watershed. The transition occurs between about 95 to 75 cm below the sediment–water interface in the south basin. It is estimated the transition corresponds to the late 1700s to the mid-1800s.

Characterization of the lower unit establishes the sediments of the lake were naturally enriched in $CaCO_3$, in the form of calcite, indicating $CaCO_3$ precipitation was probably common in the upper waters of the lake. Calcium carbonate represents approximately 60% of the dry weight sediment deposited about 5000 years ago (~ 470 cm depth). The relative contribution of terrigenous inputs to the deposition apparently became greater, as noncarbonate clastics represent about 60% of sediment deposited just before colonial development of the region. The remainder is mostly $CaCO_3$. The distributions of $CaCO_3$ and clastics are negatively correlated within the sampled sedimentary record of the lake. The sedimentary assemblage of diatom frustules and chrysophyte cysts indicates the lake was mesotrophic before colonial development, with much lower salinity than the concentrations observed in recent years during the operation of the soda ash/chlor-alkali facility.

Interpretation of the degree of impact of anthropogenic inputs from constituent sediment stratigraphy is made more difficult by the diluting effect of increases in $CaCO_3$ deposition that has occurred. The location of stratigraphic horizons, the assigned dates, and the interval average sediment accumulation rates for four intervals are presented in Table 8.22. Horizons for other constituents, such as *Ambrosia* pollen, and Hg, support the approximate chronology presented. Variations in accumulation rates occurred within each of these intervals. Substantial changes in these

TABLE 8.22. Sedimentary stratigraphic horizons, associated dates, and interval average sediment accumulation rates at the south site on Onondaga Lake*.

Sedimentary depth (cm)	Horizon/event	Interval	Accumulation rate $(cm \cdot yr^{-1})$
0	Core sample collected (1988)	1964–1988	0.6*
21	C_S-137/peak fallout from bomb testing	1884–1964	0.46
60	Abrupt increase in $CaCO_3$ concentration soda ash production starts in 1884	1822–1884	0.32
80	Abrupt decrease in ratio of benthic-epiphytic to planktonic diatoms/ lowering of lake level in 1822	5490YBP–1822	0.08
5000	5490 year before present (YBP), by C-14 dating		

* Adjusted to account for sediment undercompaction

rates over the last 140 to 150 years, and variation in background activity, are largely responsible for the lack of success in the application of the Pb-210 dating technique. The accumulation rate value for the uppermost sediment interval has been adjusted to account for undercompaction (Table 8.22). Accumulation rates have increased progressively since colonial times, primarily as a result of increased deposition of $CaCO_3$, associated mostly with the industrial discharge of Ca^{2+} waste, and secondarily due to cultural eutrophication and increased watershed erosion. Modest differences were apparent between the two deep-water stations, possibly as a result of localized allochthonous inputs. The most recent accumulation rate, estimated at $0.6 \, cm \cdot yr^{-1}$ (Table 8.22), is high compared to most other lakes because of the combined effects of domestic and industrial waste discharges. Based on recent reductions in deposition documented with sediment traps (Table 8.21), a reduction of more than 50% in the sediment accumulation rate is expected.

The recent histories of Hg and heavy metal loadings to the lake, emanating mostly from the adjoining soda ash/chlor-alkali and steel production facilities, respectively, are clearly manifested in sedimentary profiles as well-developed peaks that have maxima at a sediment depth of about 20 cm (early to mid-1960s). The maxima observed for these constituents (Table 8.23) represents an extreme level of contamination, despite the diluting effect of the high rate of $CaCO_3$ deposition.

The subsequent sharp decreases in the concentrations of these contaminants are attributable to reductions in industrial loadings from these facilities in the early 1970s. Increases in the concentrations of certain of these contaminants may be manifested in the uppermost sediments in the future as a result of decreases in $CaCO_3$ deposition.

The diatom frustule and chrysophyte cyst sedimentary record clearly depicts the accelerated eutrophication and increased salinity of the lake that followed colonial settlement. Progressive increases in primary productivity were indicated (e.g., loss of chrysophyte cysts, shift from mesotrophic to eutrophic and pollution tolerant diatom assemblages) until the uppermost (~10 cm) layer. Apparent improvements reflected in the most recent sediments are consistent with improvements documented over the same period from watercolumn monitoring (Chapter 6), in

TABLE 8.23. Maximum concentrations of mercury and other heavy metals in a core sample from the south site on Onondaga Lake.

Constituent	Maximum concentration $(\mu g \cdot g^{-1})$
Hg	70
Cr	950
Cd	40
Cu	450
Ni	250
Zn	660
Pb	350

response to reductions in pollutant loading (Chapter 3). The stratigraphy of elemental indicators of primary productivity (TOC and TP) support this basic temporal pattern. Increases in the salinity of the lake are indicated soon after colonial settlement by a shift to greater representation by salt tolerant diatoms (from about 40%−50 to 70%). This initial shift may reflect early pollution inputs from the adjoining salt production operations. The shift to salt tolerant forms became much more pronounced (at a sedimentary depth of ~55 cm), indicating yet greater lake salinity after soda ash production began. The distributions of halophilous and halophobus diatoms further support the position that a major inflection in salinity occurred at that time; most halophobous forms occur below 55 cm depth, and most halophilous forms are found above 55 cm. The diatom evidence presented here does not support the position that the lake was naturally highly saline (e.g., Cl concentration $\geqslant 500\,mg \cdot L^{-1}$) before development of the region.

References

Agbeti, M., and Dickman, M. 1989. Use of lake fossil diatom assemblages to determine historical changes in trophic status. *Can J Fish Aquat Sci.* 46: 1013−1021.

American Geological Institute. 1962. *Dictionary of Geological Terms.* Dolphin Books, New York.

American Public Health Association (APHA). 1985. *Standard Methods for the Examination of Water and Wastewater*, 16th ed. American Public Health Association, Washington, DC.

Anderson, N.J. 1990. Spatial pattern of recent sediment and diatom accumulation in a small, monomictic, eutrophic lake. *J Paleolimnol.* 3: 143−160.

Appleby, P.G., and Oldfield, F. 1978. The calculation of lead-210 dates assuming a constant rate of supply of unsupported lead-210 to the sediment. *Catena* 5: 1−8.

Auer, M.T., and Canale, R.P. 1986. Mathematical modelling of primary production in Green Bay (Lake Michigan, USA), a phosphorus—and light-limited system. *Hydrobiolog Bull.* 20: 195−211.

Auer, M.T., Johnson, N.A., Penn, M.R., and Effler, S.W. 1993. Measurement and verification of rates of sediment phosphorus release for a hypereutrophic urban lake. *Hydrobiologia* 253: 301−309.

Auer, M.T., Johnson, N.A., Penn, M.R., and Effler S.W. 1995. Pollutant sources depositional environment and the surficial sediments of Onondaga Lake, New York. *Environ Quality* (in press).

Avnimelech, Y., Mcttenry J.R., and Ross J.D. 1984. Decomposition of organic matter in lake sediments. *Environ Sci Technol.* 18: 5−11.

Baker, J.E., Eisenreich, S.J., Johnson, T.C., and Halfman, B.M. 1985. Chlorinated hydrocarbon cycling in the benthic nepheloid layer of Lake Superior. *Environ Sci Technol.* 19: 854−861.

Battarbee, R.W. 1986. Diatom analysis. In: B.E. Berglund (Ed.), *Handbook of Holocene Palaeoecology and Palaeohydrology*, John Wiley and Sons, New York, pp. 527−570.

Beaver, J. 1981. *Apparent Ecological Characteristics of Some Common Freshwater Diatoms.* Ministry of the Environment, Ontario.

Bengtsson, L., and Enell, M. 1986. Chemical analysis. *In*: B.E. Berglund (Ed.), *Handbook of Holocene Palaeoecology and Palaeohydrology*, John Wiley and Sons, New York, pp. 423−454.

Berglund, B.E., and Ralska-Jasiewiczowa, M. 1986. Pollen analysis and pollen diagrams. In: B.E. Berglund (Ed.), *Handbook of Holocene Palaeoecology and Palaeohydrology*, John Wiley and Sons, New York, pp. 455−484.

Bloesch, J., and Burns, N.M. 1980. A critical review of sediment trap technique. *Schweiz Z Hydrol.* 42: 15−55.

Bloesch, J., Stadelmann, P., and Bufirer, H. 1977. Primary production, mineralization, and sedimentation in the euphotic zone of two Swiss lakes. *Limnol Oceanogr.* 22: 511−526.

Bloesch, J., and Sturm, M. 1986. Settling flux and sinking velocities of particulate phosphorus (PP) and *particulate* organic carbon (POC) in Lake Zug, Switzerland. In: S.G. Sly (Eds.), *Sediments and Water Interactions*, Springer-Verlag, New York.

Blomqvist, S., and Hakanson, L. 1981. A review of sediment traps in aquatic environments. *Arch Hydrobiol.* 91: 101−132.

Bodbacka, L. 1986. Sediment accumulation in Lakes Lilla Ullfjarden and Stora Ullfjarden, Sweden. *Hydrobiologia* 143: 337−342.

Bortleson, G.C., and Lee, G.E. 1972. Recent sedimentary history of Lake Mendota, Wisconsin. *Environ Sci Technol.* 6: 799−808.

Bowie, G.L., Mills, W.B., Porcella, D.B., Campbell, C.L., Pagenkopf, J.R., Rupp, G.L., Johnson, K.M.,

Chan, P.W.H., Gherini S.A., and Chamberlain, C. 1985. Rates, Constants, and Kinetic Formulations in surface Water Quality Modeling. EPA-6-00/3-85-640, U.S. Environmental Protection Agency, Athers. GA.

Brenner, M., and Binford, M.W. 1988. Relationships, between, concentrations of sedimentary varables and trophic state in Florida lakes. *Can J Fish Aquat Sci.* 45: 294–300.

Bruland, K.W., Bertine, K., Koide, M., and Goldberg, E.D. 1974. History of metal pollution in southern California coastal zone. *Environ Sci Technol.* 8: 425–432.

Brunskill, G.J. 1969. Fayetteville Green Lake, New York. II. Precipitation and sedimentation of calcite in a meromictic lake with laminated sediments. *Limnol Oceanogr.* 14: 830–847.

Burden, E.T., McAndrews, J.H., and Norris, G. 1986. Palynology of Indian and European forest clearance and farming in lake sediment cores from Awanda Provincial Park, Ontario. *Can J Earth Sci.* 23: 43–55.

Canale, R.P., and Effler, S.W. 1989. Stochastic phosphorus model for Onondaga Lake. *Water Res.* 23: 1009–1016.

Charlton, M.N. 1975. Sedimentation: measurements in experimental enclosures. *Verh. Internat Verein Limnol.* 11: 267–272.

Charlton M.N, and Lean, D.R.S. 1987. Sedimentation, resuspension, and oxygen depletion in Lake Erie (1979). *J Great Lakes Res.* 13: 709–723.

Cole, G.A. 1983. *Textbook of Limnology.* C.V. Mosby, St. Louis.

Collinson, J.D. 1978. Lakes. In: H.G. Reading (Ed.), *Sedimentary Environments and Facies,* Elsevier, New York, pp. 61–79.

Cooper, S.R., and Brush, G.S. 1991. Long-term history of Chesapeake Bay anoxia. *Science* 254: 992–996.

Cotter, J.F.P., and Crowl, G.H. 1981. The Paleolimnology of Rose Lake, Potter Co., Pennsylvania: A comparison of palynologic and paleo-pigment studies. In: R.C. Romans (Ed.), *Geobotany II,* Plenum Press, New York, pp. 91–116.

Cowgill, U.M., and Hutchinson, G.E. 1966. La Aguada de Santa Ana Vieja: The history of a pond in Guatemala. *Arch Hydrobiol.* 62: 335–372.

Cox, D.D. 1959. Some postglacial forests in central and eastern New York State as determined by the method of pollen analysis. *NY State Museum and Sci Serv Bull.* No. 377.

Cox, D.D. 1965. Postglacial Vegetation in New York State. In: *Palynology in New York.* Inter. Assoc Quat Res Guidebook, Neb Acad Sci, Lincoln, NB.

Cox, D.D., and Lewis, D.M. 1965. Pollen studies in the Crusoe Lake Area of prehistoric indian occupation. *NY State Museum and Sci Serv Bull* No. 387.

Cronberg, G. 1986. Blue-green algae, green algae and chrysophyceae in sediments. In: B.E. Berglund (Ed.), *Handbook of Holocene Palaeoecology and Palaeohydrology,* John Wiley and Sons, New York, pp. 507–526.

Davis, M.B. 1981. Mid-Holocene hemlock decline: Evidence for a pathogen or insect outbreak. In: R.C. Romans (Ed.), *Geobotany II,* Plenum Press, New York, p. 253.

Dean, W.E. 1974. Determination of carbonate and organic matter in calcareous sediments and sedimentary rocks by loss on ignition: Comparison with other methods. *J Sediment Petrol.* 44: 242–248.

Dean, W.E., and Eggleston, J.R. 1984. Freshwater oncolites created by industrial pollution, Onondaga Lake, New York. *Sediment Geol.* 40: 217–232.

Dean, W.E., and Gorham, E. 1976. Major chemical and mineralogical components of profundal surface sediments in Minnesota lakes. *Limnol Oceanogr.* 21: 259–284.

Dearing, J.A., and Foster, I.D.L. 1986. Lake sediments and palaeohydrological studies. In: B.E. Berglund (Ed.), *Handbook of Holocene Palaeoecology and Palaeohydrology,* John Wiley and Sons, New York, pp. 67–90.

Deevy, E.S. 1939. Studies on Connecticut lake sediments. I. A postglacial climatic chronology for Southern New England. *Am J Sci.* 237: 691–724.

Deevy, E.S. 1949. Biogeography of the Pleistocene. *Bull Geol Soc Amer.* 60: 1315–1416.

Delcourt, P.A., and Delcourt, H.R. 1987. *Long-Term Forest Dynamics of the Temperate Zone,* Springer-Verlag, New York, 439p.

DePalma, L.M., Canale, R.P., and Powers, W.F. 1979. A minimum cost surveillance plan for water quality trend detection in Lake Michigan. In: D. Scavia and A. Robertson (Eds.), *Perspectives on Lake Ecosystem Modeling,* Ann Arbor, MI, pp. 223–246.

Devan, S.P., and Effler, S.W. 1984. The recent history of phosphorus loading to Onondaga Lake. *J Environ Eng Div.* ASCE 110: 93–109.

Dixit, S.S., Smol, J.P., Kingston, J.C., and Charles, D.F. 1992. Diatoms: powerful indictors of environmental change. *Environ Sci Technol.* 26: 23–32.

Doerr, S.M. 1991. Development and Application of a Mass Balance Model for Total Phosphorus in a

Eutrophic Lake. Master's Thesis. Syracuse University, Syracuse, NY.

Downing, J.A., and Rath, L.C. 1988. Spatial patchiness in the lacustrine sedimentary environment. *Limnol Oceanogr.* 33: 447–458.

Duthie, H.C., and Sreenivasa, M.R. 1971. Evidence for the eutrophication of Lake Ontario from the sedimentary diatom succession. Proc 14th Conf. Great Lakes Res, 1–13.

Edmondson, W.T., and Lehman, J.T. 1981. The effects of changes in the nutrient income on the condition of Lake Washington. *Limnol Oceanogr.* 26: 1–30.

Effler, S.W. 1975. A Study of the Recent Paleolimnology of Onondaga Lake. Doctoral Dissertation, Department of Civil Engineering, Syracuse University, Syracuse, NY, 155p.

Effler, S.W. 1984. Carbonate equilibria and the distribution of inorganic carbon in Saginaw Bay. *J Great Lakes Res.* 10: 3–14.

Effler, S.W. 1987. The impact of a chlor-alkali plant on Onondaga Lake and adjoining systems. *Water Air Soil Pollut.* 33: 85–115.

Effler, S.W., Brooks, C.M., Addess, J.M., Doerr, S.M., Storey, M.L., and Wagner, B.A. 1991. Pollutant loadings from Solvay waste beds to lower Ninemile Creek, New York. *Water Air Soil Pollut.* 55: 427–444.

Effler, S.W., Brooks, C.M., Auer, M.T., and Doerr, S.M. 1990a. Free ammonia and toxicity criteria in a polluted urban lake. *Res J Water Pollut Cont Fed.* 62(6): 771–779.

Effler, S.W., Doerr, S.M., Brooks, C.M., and Rowell, H.C. 1990b. Chloride in the pore water and water column of Onondaga Lake, NY, U.S.A. *Water Air Soil Pollut.* 51: 315–326.

Effler, S.W., and Driscoll, C.T. 1985. Calcium chemistry and deposition in ionically enriched Onondaga Lake, New York. *Environ Sci Technol.* 19: 716–720.

Effler, S.W., Hassett, J.P., Auer, M.T., and Johnson, N. 1988. Depletion of epilmnetic oxygen and accumulation of hydrogen sulphide in the hypolimnion of Onondaga Lake, NY, U.S.A. *Water Air Soil Pollut.* 39: 59–74.

Effler, S.W., Johnson, D.L., Jiao, J.F., and Perkins, M.G. 1992. Optical impacts and sources of suspended solids in Onondaga Creek, U.S.A. *Water Resour Bull.* 20: 251–262.

Effler, S.W., Perkins, M.G., Brooks, C.M., and Owens, E.M. 1985. A Review of the Limnology and Water Quality of Owasco Lake. Upstate Freshwater Institute, Submitted to Cayuga County, Auburn, NY.

Effler, S.W., and Rand, M.C. 1976. The Cazenovia Lake Study. I. Initiation. Department of Civil Engineering, Syracuse University, Syracuse, NY.

Effler, S.W., Rand, M.C., and Tamayo, T.A. 1979. The effect of heavy metals and other pollutants on the sediments of Onondaga Lake. *Water Air Soil Pollut.* 12: 117–134.

Engstrom, D.R., Swain, E.B., and Kingston, J.C. 1985. A palaeolimnological record of human disturbance from Harvey's Lake, Vermont: Geochemistry, pigments and diatoms. *Freshwater Biol.* 15: 261–288.

Engstrom, D.R., and Wright, H.E. 1984. Chemical Stratigraphy of Lake Sediments as a Record of Environmental Change. In: E.Y. Haworth and J.W.G. Lund (Eds.), *Lake Sediments and Environmental History,* University of Minnesota Press, Minneapolis, pp. 11–67.

EPRI. 1987. Measurement of bioavailable mercury species in fresh water and sediments. Prepared by Battelle Pacific Northwest Laboratories, Richland, WA. Electric Power Research Institute, Palo Alto, CA.

Faegri, K. and Iversen, J. 1989. *Textbook of Pollen Analysis.* John Wiley and Sons, New York.

Fallon, R.C., Harrets, S., Hanson, R.S., and Brock, T.D. 1980. The role of methane in internal carbon cycling in Lake Mendota during summer stratification. *Limnol Oceanogr.* 25: 357–360.

Foerstner, U. 1977. Metal concentrations in freshwater sediments-natural background and cultural effects. In: H.L. Golterman (Ed.), *Interactions Between Sediments and Fresh Water,* Dr. W. Junk B.V. Publishers, The Hague, 94–103.

Freedman, T.L., Canale, R.P., and Pendergast, G.F. 1980. Modelling storm runoff impacts on a eutrophic urban lake. *J Environ Engin Div.* ASCE, 106: 335–349.

Fritz, S.C., and Battarbee, R.W. 1986. Sedimentary diatom assemblages in freshwater and saline lakes of the Northern Great Plains, North America: Preliminary results. In: 9th Diatom Symposium, 1986, pp. 265–271.

Fuhs, G.W. 1973. Improved device for the collection of sedimenting matter. *Limnol Oceanogr.* 18: 989–993.

Gächter, R., and Meyer, J.S. 1990. Mechanisms controlling fluxes of nutrients across the sediment/water interface in a eutrophic lake. In: R. Baudo, J.P. Giesy and H. Muntau (Eds.), *Sediments: Chemistry and Toxicity of In-Place Pollutants,* Lewis Pub, Ann Arbor, MI.

Gardner, W.D. 1980. Sediment trap dynamics and calibration: a laboratory calibration. *J Mar Res.* 38: 17–39.

Gardiner R.D., Auer, M.T., and Canale, R.P. 1984. Sediment oxygen demand in Green Bay (Lake Michigan). In: M. Pirbazari and J.S. Devinny (Eds.), *Environmental Engineering: Proceedings of the 1984 Specialty Conference, ASCE*, New York, pp. 515–519.

Goldberg, E.D. 1963. Geochronology with ^{210}Pb. In: *Radioactive Dating*. International Atom Energy Agency, Vienna, pp. 121–131.

Golterman, H.L., Sly, P.G., and Thomas, R.L. 1983. Study of the Relationship between Water Quality and Sediment Transport. UNESCO, Paris.

Haberyan, K.A. 1990. The misrepresentation of the planktonic diatom assemblage in traps and sediments: Southern Lake Malawi, Africa. *J Paleolimnol.* 3: 35–44.

Hakansson, H., and Stoermer, E.F. 1984. Observations on the type material of *Stephanodiscus hantzschii Grunow* in Cleve and Grunow. *Nova Hedwigia* 39: 477–495.

Hargrave, B.T., and Burns, N.M. 1979. Assessment of sediment trap collection efficiency. *Limnol Oceanogr.* 24: 1124–1136.

Haworth, E.Y. 1969. The diatoms of a sediment core from Blea Tarn, Langdale. *J Ecol.* 57: 429–439.

Heidtke, T.M., and Auer, M.T. 1992. Partitioning phosphorus loads: Implications for the restoration of a hypereutrophic lake. *J Water Resour Plan Manag ASCE.* 118: 562–579.

Hickman, M., Schweger, C.E., and Klarer, D.M. 1990. Baptiste Lake, Alberta—a late Holecene history of changes in a lake and its catchment in the southern boreal forest. *J Paleolimnol.* 4: 253–267.

Hohn, M.H. 1951. A Study of the Distribution of Diatoms (Bacillarieae) in Western New York State. Memoir 308, Cornell University Agricultural Experiment Station, Ithaca, NY.

Imboden, D.M. 1974. Phosphorus models of lake eutrophication. *Limnol Oceanogr.* 19: 297–304.

Johnson, D.L., Jiao, J., DosSontos, S.G., and Effler, S.W. 1991. Individual particle analysis of suspended materials in Onondaga Lake, New York. *Environ Sci Technol.* 25: 736–744.

Johnson, M.G, and Brinkhurst, R.O. 1971. Benthic community metabolism in Bay of Quinte and Lake Ontario. *J Fish Res Bd Can.* 28: 1715–1725.

Johnson, N.A. 1989. *Surficial Sediment Characteristics and Sediment Phosphorus Release Rates in Onondaga Lake, NY*. Master's Thesis, Department of Civil Engineering, Michigan Technological University, Houghton, MI.

Jones, B.F., and Bowser, C.J. 1978. The mineralogy and related chemistry of lake sediments. In: A.

Lerman (Ed.), *Lakes: Chemistry, Geology, Physics,* Springer-Verlag, New York, pp. 179–236.

Jones, J.G. 1976. The microbiology and decomposition of seston in open waters and experimental enclosures in a productive lake. *J Ecology.* 64: 241–278.

Jones, F.G., and Wilkinson, B.H. 1978. Structure and growth of lacustrine pisoliths from recent Michigan marl lakes. *J Sediment Petrol.* 48: 1103–1110.

Kelly, C.A., and Chynoweth, D.P. 1981. The contributions of temperature and the input of organic matter in controlling rates of sediment methanogenesis. *Limnol Oceanogr.* 26: 891–897.

Kelts, K., and Hsu, K.J. 1978. Freshwater carbonate sedimentation. In: A. Lerman (Ed.), *Lakes: Chemistry, Geology, Physics,* Springer-Verlag, New York, pp. 295–324.

Kelts, K., Zao, C.K., Lister, G., Qing, Y.J., Hong, G.Z., Niessen, F., and Bonani, G. 1989. Geological fingerprints of climate history: a cooperative study of Qinghai Lake, China. *Eclogae Geol Helv.* 82: 167–182.

Kemp, A.L.W., Thomas, R.L., Dell, C.I., and Jaquet, J.M. 1976. Cultural impact on the geochemistry of sediments in Lake Erie. *J Fish Res Bd. Can.* 33: 440–462.

Kirchner, W.B., and Dillon, P.J. 1975. An empirical method of estimating the retention of phosphorus in lakes. *Water Resour Res.* 11: 182–183.

Klee, R., and Schmidt, R. 1987. Eutrophication of Mondsee (Upper Austria) as indicated by the diatom stratigraphy of a sediment core. *Diatom Res.* 2: 55–76.

Koide, M., Bruland, K.W., and Goldsberg, E.D. 1973. Th-228/Th-232 and Pb-210 geochronologies in marine and lake sediments. *Geochimica Cosmochimica Acta* 37: 1171–1187.

Krishnaswami, S., and Lal, D. 1978. Radionuclide limnochronology. In: A. Lerman (Ed.), *Lakes: Chemistry, Geology, Physics,* Springer-Verlag, New York, pp. 153–177.

Krishnaswami, S., Martin, J.M., and Meybeck, M. 1971. Geochronology of lake sediments. *Earth Planetary Sci Lett.* 11: 407–414.

Larsen, D.P., and Mercier, H.J. 1976. Phosphorus retention capacity of lakes. *J Fish Res Bd Can.* 32: 1742–1750.

Lau, J.L. 1979. Laboratory study of cylindrical sediment traps. *J Fish Res Bd Can.* 36: 1288–1291.

Long, E.R., and Morgan, L.G. 1990. The Potential for Biological Effects of Sediment-Sorbed Contaminants Tested in the National Status and Trends Program, NOAA Technical Memorandum NOS OMA 52, 175p.

Lorenzer, C.J. 1967. Determination of chlorophyll and phaeopigments: spectrophotometric equations. *Limnol Oceanogr.* 12: 343–346.

Lucas, A.M., and Thomas, N.A. 1972. Sediment oxygen demand in Lake Erie's central basin, 1970. In: *Project Hypo, An Intensive Study of the Lake Erie Central Basin Hypolimnion and Related Surface Water Phenomena*, Canada Centre for Inland Waters, Paper No. 6, Burlington, Ontario, Canada, pp. 45–51.

Ludlam, S.D. 1969. Fayetteville Green Lake, New York. III. The laminated sediments. *Limnol Oceanogr.* 14: 848–857.

Lung, W.S., Canale, R.P., and Freedman, P.L. 1976. Phosphorus models for eutrophic lakes. *Water Res.* 10: 1101–1114.

Mackereth, F.J.H. 1966. Some chemical observations on postglacial lake sediments. *Phil Trans Roy Soc B.* 250: 165–213.

Massart, D.L., and Kaufman, L. 1983. *The Interpretation of Analytical Chemical Data by Use of Cluster Analysis*, John Wiley & Sons, New York.

McAndrews, J.H. 1981. Late quaternary climate of Ontario: temperature trends from the fossil pollen record. In: W.C. Mahaney (Ed.), *Quaternary Paleoclimate*, Geo Abstracts, Norwich, England, pp. 319–333.

McCulloch, W.F. 1939. A postglacial forest in Central New York. *Ecology* 20: 264–271.

Miller, G., Irion, G., and Foerstner, U. 1972. Formation and diagenesis of inorganic Ca-Mg carbonates in the lacustrine environment. *Naturwissenschäften* 59: 158–164.

Miller, N.G. 1973. Late-glacial and postglacial vegetation change in Southwestern New York State. *NY State Museum Sci Serv Bull.* No. 420.

Molongoski, J.J., and Klug, M.J. 1980. Quantification and characterization of sedimenting particulate organic matter in a shallow hypereutrophic lake. *Freshwater Biol.* 10: 497–506.

Moore, P.D., and Webb, J.A. 1978. *An Illustrated Guide to Pollen Analysis*, John Wiley and Sons, New York.

Mortimer, C.H. 1942. The exchange of dissolved substances between mud and water in lakes (Part III and IV). *J Ecol* 30: 147–201.

Mudroch, A., Joshi, S.R., Sutherland, D., Mudroch, P., and Dickson, K.M. 1989. Geochemistry of sediments in the Back Bay and Yellowknife Bay of the Great Slave Lake. *Environ Geol Water Sci.* 14: 35–42.

Muller, E.H. 1977. Late glacial and early postglacial environments in Western New York. *Ann New York Acad Sci.* 288: 223–233.

Murphy, C.B. 1978. Onondaga Lake. In: J.A. Bloomfield (Ed.), *Lakes of New York State, Volume II, Ecology of the Lakes of Western New York*, Academic Press, New York, pp. 224–366.

Nakanishi, H., Ukita, M., Sekine, M., and Murakami, S. 1989. Mercury pollution in Tokuyama Bay. *Hydrobiologia* 176/177: 197–211.

NYSDEC 1990. Mercury Sediments—Onondaga Lake. Engineering Investigations at Inactive Hazardous Waste Sites, Phase II Investigation, New York State Department of Environmental Conservation, Volume 1.

Nurnberg, G.K. 1987. A comparison of internal phosphorus loads in lakes with anoxic hypolimnia: Laboratory incubation versus in situ hypolimnetic phosphorus accumulation. *Limnol Oceanogr.* 32: 1160–1164.

O'Connor, D.J., DiToro, D.M., and Thomann, R.V. 1975. Phytoplankton models and eutrophication problems, In: C.S. Russell (Ed.), *Ecological Modeling in a Resource Management Framework*. Resour Future, pp. 149–210.

Oldfield, F., and Appleby, P.G. 1984. Empirical testing of ^{210}Pb-dating models for lake sediments. In: E.Y. Haworth and J.W.G. Lund (Eds.), *Lake Sediments and Environmental History*, Univesity of Minnesota Press, Minneapolis, pp. 93–124.

Oldfield, F., Appleby, P.G., and Battarbee, R.W. 1978. Alternative ^{210}Pb dating: Results from the New Guinea Highlands and Lough Erne. *Nature* 271: 339–342.

Olsson, I.U. 1986. Radiometric dating. In: B.E. Berglund (Ed.), *Handbook of Holocene Palaeoecology and Palaeohydrology*, John Wiley and Sons, New York, pp. 273–312.

Onondaga County. 1969–1991. Onondaga Lake Monitoring Report. Annual report. Onondaga County, Syracuse, NY.

Onondaga County. 1971. Onondaga Lake Study. United States Environmental Protection Agency Research Series Report #11060 SAE 4/71.

Onondaga County. 1975. Industrial Waste Monitoring Program 1974—Appendices, County of Onondaga Department of Drainage and Sanitation, O'Brien and Gere Engineers.

Otsuki, A., and Wetzel, R.G. 1972. Coprecipitation of phosphate with carbonates in a marl lake. *Limnol Oceangr.* 17: 763–767.

Owens, E.M., and Effler, S.W. 1989. Changes in stratification in Onondaga Lake, New York. *Water Resour Bull.* 25: 587–597.

Palmer, C.M. 1977. Algae and Water Pollution, Environmental Protection Agency, EPA-600/9-77-036.

Patrick, R. 1977. Ecology of freshwater diatoms and diatom communities. In: D. Werner (Ed.), *The Biology of Diatoms* University of California Press, Berkeley, pp. 284–332.

Patrick, R., and Reimer, C.W. 1966. *The Diatoms of the United States*, Volume 1, Monographs of the Academy of Natural Sciences of Philadelphia, No. 13.

Perloff, W.H., and Baron, W. 1976. *Soil Mechanics: Principles and Applications*, John Wiley & Sons, New York.

Pennington, W. 1974. Seston and sediment formation in five Lake District lakes. *J Ecol.* 62: 215–251.

Pentecost, A., and Bauld, J. 1988. Nucleation of calcite on the sheaths of cyanobacteria using a simple diffusion cell. *Geomicrobiol J.* 6: 129–135.

Pettersson, K., and Bostrom, B. 1986. Phosphorus exchange between sediment and water in Lake Balaton, pp. 427–435. In: P.G. Sly (Ed.), *Sediments and Water Interactions*, Springer-Verlag, New York.

Proctor, B.L. 1978. Chemical Investigation of Sediment Cores from Four Minor Finger Lakes of New York. Ph.D. Thesis, State University of New York at Buffalo, Buffalo.

PTI Environmental Services. 1991. Onondaga Lake RI/FS Work Plan.

Reimann, B.E.F., Lewin, J.M.C., and Guillard, R.R.L. 1963. *Cyclotella cryptica*, a new brackish-water diatom species. *Phycologia* 3: 77–81.

Reineck, H.E., and Singh, I.B. 1980. *Depositional Sedimentary Environments*. Springer-Verlag, Berlin.

Ritchie, J.C., and McHenry, J.R. 1990. Application of radioactive fallout cesium-137 for measuring soil erosion and sediment accumulation rates and patterns: A review. *J Environ Qual.* 19: 215–233.

Ritchie, J.C., McHenry, J.R., and Gill, A.C. 1973. Dating recent reservoir sediments. *Limnol Oceanogr.* 18: 254–263.

Rosa, F., Bloesch, J., and Rathke, D.E. 1991. Sampling the settling and suspended particulate matter (SPM). In: A. Mudroch and S.C. Macknight (Eds.), *Handbook of Techniques for Aquatic Sediments Sampling*, CRC Press, Ann Arbor, MI.

Rowell, H.C. 1992. Paleolimnology, Sediment Stratigraphy, and Water Quality History of Onondaga Lake, New York. Ph.D. Thesis, State University of New York, College of Environmental Science and Forestry, Syracuse, NY.

Rowell, H.C. 1995. Paleolimnology of Onondaga Lake: the history of anthropogenic impacts on water quality. *Lake Reservoir Manag.* (in press).

Rudd, J.W.M., and Hamilton, R.D. 1978. Methane cycling in a eutrophic shield lake and its effects on whole lake metabolism. *Limnol Oceanogr.* 23: 337–348.

Saarnisto, M. 1986. Annually laminated lake sediments. In: B.E. Berglund (Ed.), *Handbook of Holocene Palaeoecology and Palaeohydrology*, John Wiley and Sons, New York, pp. 343–370.

Saroff, S.R. (Ed.). 1990. Proceedings of the Onondaga Lake Remediation Conference, New York State Department of Law and Department of Environmental Conservation.

Schelske, C.L., Conley, D.J., Stoermer, E.F., Newberry, T.L., and Campbell, C.D. 1986. Biogenic silica and phosphorus accumulation in sediments as indices of eutrophication in the Laurentian Great Lakes. *Hydrobiologia* 143: 79–86.

Schoch, R.M. 1989. *Stratigraphy Principles and Methods*. Van Nostrand Reinhold, New York.

Serruya, C. 1971. Lake Kinneret: the nutrient chemistry of the sediments. *Limnol Oceanogr.* 16: 510–521.

Shapiro, J., Edmondson, W.T., and Allison, D.E. 1971. Changes in the chemical composition of Lake Washington, 1958–1970. *Limnol Oceanogr.* 16: 437–452.

Shipman, L. 1970. Two Pollen Profiles from North-Central New York State, South of Lake Ontario. Masters Thesis, State University of New York, Oswego.

Sicko-Goad, L., and Stoermer, E.F. 1988. Effects of toxicants on phytoplankton with special reference to the Laurentian Great Lakes. In: M.S. Evans (Ed.), *Toxic Contaminants and Ecosystem Health: A Great Lake Focus*, John Wiley and Sons, Inc., New York, pp. 19–51.

Sly, P.G. 1977. Sedimentary environments in the Great Lakes. In: H.L. Golterman (Ed.), *Interactions Between Sediments and Fresh Water*, Dr W. Junk B.V. Publishers, The Hague, pp. 76–82.

Sly, P.G., Thomas, R.L., and Pelletier, B.R. 1982. Comparison of sediment energy-texture relationships in marine and lacustrine environments. *Hydrobiologia* 91: 71–84.

Smardon, R., Drake, B.J., and Kalinoski, R. 1989. Report to the City of Syracuse on the Oil City Remediation Workshop, Vol. 1., IEPP Pub. No. 89–001.

Snodgras, W.J., and O'Meila, C.R. 1975. Predictive model for phosphorus in lakes. *Environ Sci Technol.* 9: 937–944.

Southeastern Wisconsin Regional Planning Commission (SWRPC). 1987. *A Water Resources Management Plan for the Milwaukee Harbor Estuary*.

Southeastern Wisconsin Regional Planning Commission Planning Report No. 37, Waukesha, WI.

Spear, R.W., and Miller, N.G. 1976. A radiocarbon dated pollen diagram from the Allegheny Plateau of New York State. *J Arnold Arboretum.* 57: 369–403.

Stabel, H.H. 1985. Mechanisms controlling the sedimentation sequence of various elements in prealpine lakes. In: W. Stumm (Ed.), *Chemical Process in Lakes*, John Wiley & Sons, New York.

Stabel, H.H. 1986. Calcite precipitation in Lake Constance. Chemical equilibrium, sedimentation, and nucleation by algae. *Limnol Oceangr.* 31: 1081–1093.

Stoermer, E.F. 1984. Qualitative characteristics of phytoplankton assemblages. In: *Algae as Ecological Indicators*, Academic Press Inc. London, pp. 49–67.

Stoermer, E.F., Kociolek, J.P., Schelske, C.L., and Conley, D.J. 1987. Quantitative analysis of siliceous microfossils in the sediments of Lake Erie's central basin. *Diatom Res.* 2: 113–134.

Stoermer, E.F., Rociolek, J.P., Schelske, C.L., and Conley, D.J. 1985c. Siliceous microfossil succession in the recent history of Lake Superior. *Proceed Acad Natural Sci Philadelphia* 137: 106–118.

Stoermer, E.F., Wolin, J.A., Schelske, C.L., and Conley, D.J. 1985a. An assessment of ecological changes during the recent history of Lake Ontario based on siliceous algal microfossils preserved in the sediments. *J Phycol.* 21: 257–276.

Stoermer, E.F., Wolin J.A., Schelske, C.L., and Conley, D.J. 1985b. Postsettlement diatom succession in the Bay of Quinte, Lake Ontario. *Can J Fish Aquat Sci.* 42: 754–767.

Stoermer, E.F., and Yang, J.J. 1969. Plankton Diatom Assemblages in Lake Michigan, Special Report No. 47, Great Lakes Research Division, University of Michigan, Ann Arbor.

Stumm, W., and Baccini, P. 1978. Man-made chemical perturbation of lakes. In: A Lerman (Ed.), *Lakes: Chemistry, Geology, Physics*, Springer-Verlag, New York.

Sze, P. 1975. Possible effect of lower phosphorus concentrations on the phytoplankton in Onondaga Lake, New York, U.S.A. *Phycologia* 14: 197–204.

Sze, P. 1980. Seasonal succession of phytoplankton in Onondaga Lake, New York (U.S.A.). *Phycologia* 19: 54–59.

Theriot, E., and Stoermer, E.F. 1981. Some aspects of morphological variation in *Stephanodiscus niagarae* (Bacillariophyceae). *J Phycol.* 17: 64–72.

Thomas, R.L., Jaquet, J.-M., and Kemp, A.L.W. 1976. Surficial sediments of Lake Erie. *J Fish Res Bd Can.* 33: 385–403.

Thomas, R.L., Kemp, A.L.W., and Lewis, C.F.M. 1972. Distribution, composition and characteristics of the surficial sediments of Lake Ontario. *J Sediment Petrol.* 42: 66–84.

Thompson, J.B., and Ferris, F.G. 1990. Cyanobacterial precipitation of gypsum, calcite, and magnesite from natural alkaline lake water. *Geology* 18: 995–998.

Thompson, R.L., Kemp, A.L.W., and Lewis, C.F.M. 1972. Distribution, composition and characteristics of the surficial sediments of Lake Ontario. *J Sediment Petrol.* 42: 66–84.

Tuchman, M.L., Stoermer, E.F., and Carney, H.J. 1984a. Effects of increased salinity on the diatom assemblage in Fonda Lake, Michigan. *Hydrobiologia.* 109: 179–188.

Tuchman, M.L., Theriot, E., and Stoermer, E.F. 1984b. Effects of low level salinity concentrations on the growth of *Cyclotella meneghiniana Kutz.* (Bacillariophyta). *Arch Protistenk.* 128: 319–326.

Turner, J.V., Fritz, P., Darrow, P.F., and Warner, B.G. 1983. Isotopic and geochemical composition of marl lake waters and implications for radiocarbon dating of marl lake sediments. *Can J Earth Sci.* 20: 599–615.

USEPA 1973. Report of Mercury Source Investigation, Onondaga Lake, New York and Allied Chemical Corporation, Solvay, New York, United States Environmental Protection Agency.

Welshmeyer, N.A., and Lorenzen, C.J. 1985. Chlorophyll budgets: Zooplankton grazing and phytoplanktkon growth in a temperate Fjord and the Central Pacific Gyres. *Limnol Oceanogr* 30: 1–21.

Wetzel, R.G. 1983. *Limnology*, 2nd ed., Saunders College Publishing, New York.

White, W.S., and Wetzel, R.G. 1975. Nitrogen, phosphorus, particulate and colloidal carbon content of sedimenting seston of a hard-water lake. *Verh Int Ver Limnol.* 19: 330–339.

Williams, J.D.H., Syers, J.K., Shulka S.S., Harris, R.F., and Armstrong D.E. 1971. Levels of inorganic and total phosphorus in lake sediments as related to other sediment paramters. *Environ Sci Technol.* 5: 1113–1120.

Winkley, S.J. 1989. The Hydrogeology of Onondaga County, New York. Masters Thesis. Syracuse University, Syracuse, NY.

Wodka, M.C., Effler, S.W., and Driscoll, C.T. 1985. Phosphorus deposition from the epilimnion of Onondaga Lake. *Limnol Oceanogr.* 30: 833–843.

Wolin, J.A., Stoermer, E.F., Schelske, C.L., and
Conley, D.J. 1988. Siliceous microfossil succession
in recent Lake Huron sediments. *Arch. Hydrobiol.*
114: 175–198.

Yin, C., and Johnson, D.L. 1984. An individual
particle analysis and budget study of Onondaga
Lake sediments. *Limnol Oceanogr.* 29: 1193–1201.

9
Mechanistic Modeling of Water Quality in Onondaga Lake

Steven W. Effler, Susan M. Doerr, Martin T. Auer, Raymond P. Canale, Rakesh K. Gelda, Emmet M. Owens, and Thomas M. Heidtke

9.1 Background and Evolution of Model Frameworks

9.1.1 Purposes and the Mass Balance Approach

Water quality modeling serves two important, and at times disparate, purposes: (1) to support basic research, by providing a quantitative framework for the synthesis of scientific data, and (2) to support effective management of water resources by providing reliable predictive frameworks. These research and management purposes are mutually consistent on a long-term basis, as research advancements have led to improved capabilities and credibility of mechanistic management models (Chapra and Reckhow 1983; Thoman and Mueller 1987). Empirical modeling (e.g., Reckhow and Chapra 1983) can be valuable in certain instances, but is often inadequate for more complex water quality problems. Further it provides only very limited theoretical insight. Models developed for Onondaga Lake and the adjoining Seneca River, as documented here, are mechanistic, as they explicitly accommodate key mechanisms underlying the dynamics of various aspects of the water quality of these systems. Model testing, to establish the credibility of each model, and

application for selected management scenarios are also documented in this chapter. The models presented here are intended to meet both the research and management purposes stated above. The models will serve to identify and test hypotheses for ongoing research and guide related programs. They will undoubtedly evolve as research on the lake continues. In the shorter term, the models are expected to support management decisions related to the reclamation of this highly polluted lake and the protection of the adjoining river system.

The models developed here are "mass balance" models; i.e.,"within a finite volume of water, mass is neither created nor destroyed" (p. 4, Chapra and Reckhow 1983). Within the models this principle is manifested through mass balance equations that accommodate transfers of the constituents of interest across the system's boundaries and important related transformations (or reactions) occurring within these boundaries (Chapra and Reckhow 1983). The generalized mass balance expression is (Chapra and Reckhow 1983):

$$[\text{accumulation}] = [\text{loading}] \pm [\text{transport}]$$
$$\pm [\text{reactions}] \tag{9.1}$$

Accordingly, if losses (or sinks) are greater than sources of a constituent, over the boundaries and time period of interest, the mass of that constituent decreases. "If sources are in balance with sinks, the mass remains at a constant level and the system is said to be at a steady-state or dynamic equilibrium" (p. 5, Chapra and Reckhow 1983). The mathematical expression of the conservation of mass (equation (9.1)), provides the framework for calculating (e.g., predicting) the response of a water body to external stimuli, in particular, the impact of man's actions on water quality.

The preceding chapters have documented the components necessary to support mass balance model analyses (equation (9.1)) of various constituents of water quality interest in Onondaga Lake. Thus, to a large extent, this chapter builds on the preceding chapters. The time course of the concentrations of important water quality constituents have been documented over various time frames of interest in Chapters 5 and 6. The deposition and incorporation of some of these constituents in the lake's sediments were described in Chapter 7. Detailed external loading analyses were presented in Chapter 3. Advective and diffusive transport processes within Onondaga Lake were characterized, and the development of three hydrodynamic models was documented, in Chapter 4. Reactions, transformations, and cycling of various constituents within the lake have been documented in Chapters 5 and 6; additional transformations are quantified in this chapter as part of model development. Indeed, the models developed here serve to synthesize much of the available scientific information on Onondaga Lake presented in earlier chapters.

9.1.2 Model Structure

9.1.2.1 Complexity/Credibility Issues

Very important model design decisions with respect to the structure of each water quality model must be made. Certain design features are established by the constituent to be modeled. For example, key processes affecting the cycling of nonconservative substances must be accommodated. However, the important operative descriptor here is "key." Often the best performing model only accommodates the principal regulating mechanisms, not

every mechanism/process known to influence the material. Other important factors influencing model design are system-specific characteristics, particularly morphometry, and the designated management/research issue(s) and application(s). Related decisions concerning model resolution, or the degree to which space and time are segmented, must therefore be made. The temporal and spatial scales of a model are often interrelated according to the constituent type being modeled (Chapra and Reckhow 1983). For example, indicator bacteria models generally have relatively fine temporal and spatial scales (e.g., Canale et al. 1993a), while the scales are more coarse for conservative substances (e.g., Chapra and Reckhow 1983; O'Connor and Mueller 1970).

There is an interplay between the complexity of model structure and model performance characteristics (credibility) that has been addressed by a number of investigators (e.g., Auer and Canale 1986; Auer and Effler 1989; Reckhow and Chapra 1983; Whitehead et al. 1981). The relationship between model complexity and model credibility has been illustrated by Auer and Canale (1986) according to Figure 9.1. The presented figure is a qualitative idealization of the relationship; the details of the relationship are material and

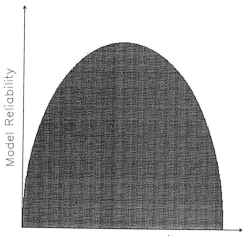

FIGURE 9.1. Idealized relationship between model complexity and model reliability (modified from Auer and Canale 1986).

problem-specific. Regardless, Figure 9.1 makes a valuable point. Mechanistic models can be too simple and they can be overly complex; in both cases the credibility of model calculations is compromised. Overly simple models do not provide an adequate description of mechanisms and/or the system. Models that are too complex compromise the accuracy of predictions by the introduction of uncertainties associated with the estimation/specification of a large number of coefficients. The goals of the modeling effort are particularly important to the issue of model complexity. Model credibility is important to both research and management goals, but it is most critical when the model is to support very costly management decisions, such as construction of treatment facilities. Various investigators (e.g., Reckhow and Chapra 1983; Whitehead et al. 1981) argue a model intended for management purposes should only be as complex as is necessary to credibly address the designated management issue. The potential benefit of more complex models can in some cases be masked in futuristic management applications due to random variations of important environmental (particularly meteorological) forcing conditions (e.g., Bierman and Dolan 1986; Lam et al. 1987).

Increased model complexity is often implicit in research modeling, as more mechanisms and processes are introduced in an effort to more closely mimic reality. The goal is to improve, or at least maintain, the accuracy of model calculations with the increased complexity (e.g., a systematic shift in the relationship of Figure 9.1). The mechanistic water quality models presented here are of intermediate complexity. They have evolved from an integrated program of intensive field monitoring and process studies (e.g., see Chapters 3 through 8). These models accommodate the key regulatory processes, which in most cases have been independently quantified by experimentation. Detailed treatments of the development of the external loads for each of the models were presented in Chapter 3. The models are intended to support the important threshold management decisions for this polluted lake, and to serve as frameworks for future research. These models will evolve into more complex research tools of perhaps even greater management and engineering utility, as the understanding of the system and material cycling advances.

9.1.2.2 Scales of Time and Space/ Comparative Descriptions of Onondaga Lake Models

Selected features of the time and space segmentation of the water quality models for Onondaga Lake and the Seneca River presented in this chapter are listed in Table 9.1. Reaction terms are introduced as part of the development of each model (see subsequent sections). Taken together, these models represent a diverse array of model structures and complexities that match the associated management issues as well as system characteristics and constituent behavior. The first four models have year-round capabilities. The structure of the chloride (Cl^-) model is mechanistically the most simple; no reactions need to be accommodated for this conservative (unreactive) substance, and the lake is treated as a completely mixed system. The principal water quality interest is the long-term trend in Cl^- concentration, but performance at the seasonal time scale is important in supporting

TABLE 9.1. Time and space scales for Onondaga Lake water quality models.

Model	Time	Space	Reference
Chloride	Seasonal	1 segment	Doerr et al. 1994
Phosphorus	Seasonal	2 vertical layers, no longitudinal segmentation	Doerr et al. 1995a
Nitrogen	Seasonal	2 vertical layers, no longitudinal segmentation	Canale et al. 1995a
Oxygen	Seasonal	2 vertical layers, no longitudinal segmentation	Gelda and Auer 1995
Fecal coliform	Days	2 vertical layers, 8 surface segments, 3 bottom segments	Canale et al. 1993a
Seneca River (oxygen)	Steady state	2 vertical layers, 25 longitudinal segments in each layer	Canale et al. 1995b

the hydrologic features of the other water quality models as well as in evaluating the origins of the strong seasonality observed in Cl⁻ concentration. The completely mixed assumption is supported for Cl⁻ in the longitudinal plane of the lake, but is imperfect in the vertical (e.g., plunging inflows, salinity stratification; see Chapter 4). However, despite this limitation, the simplified structure of the model serves both the long-term and seasonal goals of the model well, and in apportioning external Cl⁻ loads to evaluate sources of the enrichment (Doerr et al. 1994).

9.1.2.2.1 Physical Framework for Total Phosphorus, Nitrogen, and Dissolved Oxygen Lake Models

The total phosphorus (TP), nitrogen (N), and dissolved oxygen (DO) models for the lake all utilize the same physical framework. The lake is modeled as two vertical completely mixed layers. The dimensions of the layers are fixed (constant volume assumption for lake supported in Chapter 3); the demarcation depth bounding the layers is 8.5 m (Figure 9.2a; for all but deepwater discharge management scenarios for METRO). This depth approximately delineates the observed epilimnion/hypolimnion boundary (thermocline depth) for much of the summer period (Figure 9.2b). Water quality management concerns for Onondaga Lake focus primarily on conditions in the upper waters; e.g., degraded esthetics, public health concerns associated with the entry of raw sewage, and the violation of water quality standards intended to protect aquatic life. The lower waters are largely of concern as a source of, or conduit for, feedback of pollutants to the upper waters. The completely mixed assumption for the upper mixed layer (UML) is supported for these constituents by longitudinal and vertical measurements. Paired measurements of TP, T-NH₃, and DO made at the centers of the south basin (routine monitoring site) and north basin (e.g., Figure 9.3) indicate no recurring significant gradients in these constituents prevail in the lake, thereby supporting the completely mixed assumption. Concentrations of these

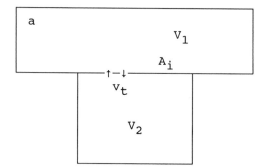

A_i = interface area
V_1 = volume of UML
V_2 = volume of LML

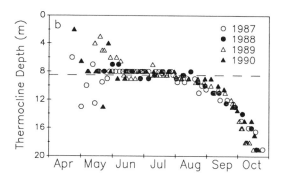

FIGURE 9.2. (a) Physical framework for TP, N, and DO models for Onondaga Lake, for segment boundary of 8.5 m. (b) Observations of thermocline depth in Onondaga Lake for 4 years.

constituents (particularly TP and T-NH₃) remain approximately vertically uniform within the epilimnion of the lake except during periods of "secondary stratification" (Chapter 5); these periods are short-lived and they have a stochastic distribution (Chapter 4). The completely mixed simplification eliminates the need to model longitudinal transport processes (advection and dispersion) for these constituents.

The completely mixed assumption is more imperfect for the lower mixed layer (LML) of Onondaga Lake for these constituents; major vertical gradients have been documented, associated with sink-source processes operating at the sediment–water interface (Chapter 5). However, the vertical distribution of concentrations of these constituents have recur-

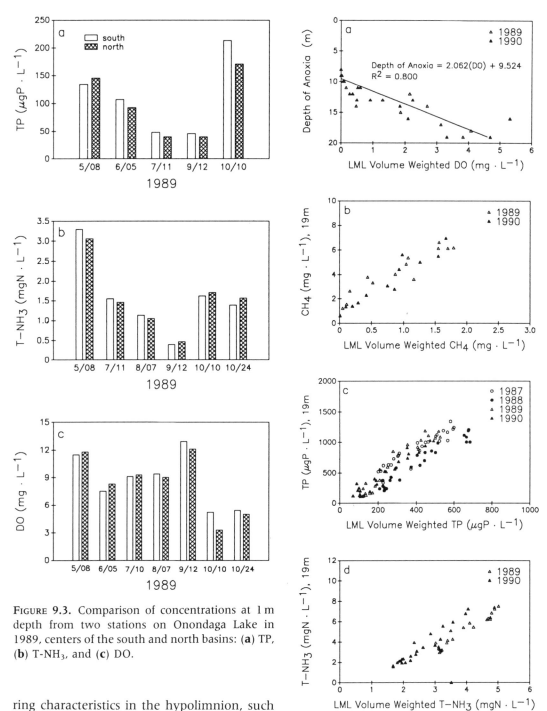

FIGURE 9.3. Comparison of concentrations at 1 m depth from two stations on Onondaga Lake in 1989, centers of the south and north basins: (**a**) TP, (**b**) T-NH$_3$, and (**c**) DO.

ring characteristics in the hypolimnion, such that the volume-weighted concentration of the lower layer is a good (empirical) predictor of various features of the vertical structures. This is illustrated in Figure 9.4. Thus important features of the vertical structure of these constituents within the hypolimnion can be

FIGURE 9.4. Relationship between volume-weighted concentrations and specific depth characteristics in the LML of Onondaga Lake: (**a**) depth of anoxia, (**b**) CH$_4$ concentration at a depth of 19 m, (**c**) TP concentration at a depth of 19 m, and (**d**) T-NH$_3$ concentration at a depth of 19 m.

delineated from volume-weighted concentrations (see Figure 9.4). Further, volume-weighted concentrations can be used to model vertical transport across the two layers.

The fixed two-layer framework (Figure 9.2a) is attractive because of its simplicity and ease of application. The seasonality in the occurrence of stratification and the magnitude of mixing-based exchange between the layers during stratified periods is accommodated through the seasonality in the net vertical exchange coefficient (e.g., Chapra and Reckhow 1983; Thoman and Mueller 1987). This coefficient incorporates the effect of all vertical exchange processes; e.g., primarily diffusion and entrainment (Chapter 4). The procedure for estimating the coefficient from temperature profiles is described subsequently in the TP model section.

9.1.2.2.2 Other Features

The management concerns for TP (trophic state), N (NH_3 toxicity), and DO (oxygen resources) in Onondaga Lake focus on seasonal transformations (Table 9.1) over the spring to fall interval. This is in sharp contrast to the shorter term (days) management concerns for public health indicator bacteria following runoff events during the contact recreation months of late spring and summer. Abrupt increases in fecal coliform bacteria concentrations occur in the lake and persist for several days, with a distinct south to north gradient, in response to major runoff events (Canale et al. 1993a; also see Chapter 6). Thus finer temporal and spatial resolution is necessary in the fecal coliform model to address the management issues of duration and longitudinal extent of public health risks (e.g., Canale et al. 1993a). The specifics of spatial segmentation and calculations of transport processes for the fecal coliform model were described in Chapter 4 and reviewed by Canale et al. (1993a).

Mass balance models, particularly for DO, are widely used to establish the waste assimilative capacity of streams and rivers, and for waste load allocation (Bowie et al. 1985; Krenkel and Novotny 1979; Thoman and

Mueller 1987). The Seneca River is greatly impacted by the inflow from polluted Onondaga Lake (Canale et al. 1995b; Effler 1987; Effler et al. 1984a). Leading management alternatives for reclamation of the lake include increased direct discharge of municipal waste effluent to the river (Onondaga Lake Management Conference 1993). Thus a reliable water quality management model for the river system is required to provide a quantitative basis to protect its resources. Unlike the lake models that are dynamic (or time variable), the water quality model for the Seneca River presented here (also see Canale et al. 1995b) is a steady-state model (Table 9.1). An unusual two-layer vertical segmentation has been adopted in the river model (described subsequently) to accommodate the strong stratification that occurs proximate to the lake inflow (Canale et al. 1995b; Effler et al. 1984a) that is set up by the unusual bidirectional flow regime in the lake's outlet (Chapter 4).

9.2 Chloride Model

9.2.1 Introduction

The large salinity loads received by Onondaga Lake and the elevated salinity of the lake during the operation of the soda ash facility, as well as the reductions in loading and lake concentrations following closure, have been documented (Doerr et al. 1994; Effler et al. 1990b, Chapters 3 and 5). Impacts of the high salinity have been manifested with respect to the hydrodynamics (Chapter 4), chemistry (Chapter 5), and biology (Chapter 6) of the lake, and have in part extended out into the adjoining river system (e.g., Effler et al. 1984a,b; Canale et al. 1995b; also see later sections of this chapter). Recall that NaCl was mined from brine wells adjoining the lake from colonial times to the early 1900s (see Chapter 1 for historic review) and that it has been the contention of certain investigators (Rooney 1973; O'Brien and Gere Engineers 1973) that as much as 50% of the salinity of the lake before closure of the facility was natural. Chloride (Cl^-) has been demonstrated to be the single largest component, and a good surrogate measure, of salinity in Onondaga Lake (Effler et al. 1986c, Chapter 5). Chloride is often used as a hydrodynamic tracer and to test hydrologic budgets (Chapra and Reckhow 1983; Effler et al. 1984b; Richardson 1976; Thomann and Mueller 1987). Validation of a Cl^- model for the lake thus would support the evaluation of past and continuing issues related to the origins of salinity, as well as support certain features of other water quality models for the lake.

The detailed distribution (e.g., vertical and temporal) of Cl^- in the lake has been simulated for 1980 and 1981, within the framework of a hydrodynamic model of the complex stratification regime of the lake that prevailed before closure of the facility (Owens and Effler 1989, Chapter 4). Here we document the validation of a seasonal 1 layer mass balance Cl^- model for Onondaga Lake for the period 1973–1991, as a test of both hydrologic and Cl^- budgets over that period. Subsequently, another mass balance conservative substance model is used to characterize the complex hydrodynamic regime of the lake outlet (see Chapter 4 also) and adjoining portions of the Seneca River. The one-layer Cl^- model and the supporting loading analysis presented here are used to: (1) explain the major long-term and seasonal changes in Cl^- concentration observed in the lake, (2) apportion external loads of Cl^- to the lake before and after closure of the soda ash/chlor-alkali facility, and (3) evaluate the extent to which anthropogenic sources have contributed to the Cl^- content of the lake.

9.2.2 Model Development

9.2.2.1 Model Framework and Solution Technique

The model is a simple mass balance of Cl^- for a completely mixed system. No reaction terms are necessary, due to the conservative behavior of Cl^-. The model calculates changes in Cl^- concentration as a function of time, based on time series of Cl^- loads and hydrologic export, according to the relationship

$$V \cdot \frac{d[Cl^-]}{dt} = W_{Cl} - (Q \cdot [Cl^-]) \quad (9.2)$$

where V = lake volume (m^3), $[Cl^-]$ = concentration of Cl^- $(mg \cdot L^{-1})$, W_{Cl} = sum of Cl^- loads $(kg \cdot d^{-1})$, Q = export flow $(m^3 \cdot d^{-1})$, and t is time (d). The constant volume assumption is supported for Onondaga Lake (Chapter 3). The export flow is calculated as the sum of the inflows. The export of Cl^- is determined as the product of export flow and the simulated lake concentration. Precipitation and evaporation are not accommodated, as the imbalance between these processes is less than 1% of the annual inflow (Chapter 3).

The completely mixed assumption does not accommodate the stratification regime of the lake, particularly when salinity stratification prevailed. During periods of plunging of the dense ionic inflow (see Chapter 4), the Cl^- concentration in the outflow of the lake is less than the average Cl^- concentration of the entire lake. Accurate simulation of the vertical distribution of Cl^- in the lake during these periods requires a hydrodynamic model that mechanistically accommodates the plunging process and mediating influences of meteorological conditions (e.g., Owens and Effler 1989, Chapter 4). The simplicity of the approach adopted here has advantages in testing mass and hydrologic budgets for the lake over the rather lengthy period of monitoring.

The mass balance expression of equation (9.2) was solved numerically using the second order Runge–Kutta numerical method on a daily time step. Continuous model simulations were made for the 19-year period, 1973–1991.

9.2.2.2 Lake Concentrations and Loads

Different data bases were drawn upon to develop a time series of volume-weighted Cl^- concentrations for the lake for the 1973–1991 period (Table 9.2). Sampling was conducted over the spring–fall interval of each year at the routine south basin monitoring station. Volume-weighting was conducted according to the hypsographic data of Owens (1988). Four hundred sixty-nine volume-weighted concentrations are included in the 1973–1991 data base, based on approximately 6800 depth measurements of Cl^-.

TABLE 9.2. Databases for Onondaga Lake Cl⁻ concentrations.

Years	Depth interval (m)	Frequency	Source
1973–1979	3	$2\,mo^{-1}$	D&S*
1980–1981	1	$2–3\,wk^{-1}$	UFI†
1982–1984	3	$2\,mo^{-1}$	D&S
1985–1988	1	$1\,wk^{-1}$	UFI
1988–1991	2	$1\,wk^{-1}$	UFI

*Department of Drainage and Sanitation, Onondaga County, Syracuse, NY
†Upstate Freshwater Institute, Syracuse, NY

The temporal distribution of the volume-weighted concentration of Cl⁻ in Onondaga Lake over the 1973–1991 period is presented in Figure 9.5. The major reduction in Cl⁻ concentration that occurred following the closure of the soda ash/chlor-alkali facility in early 1986 is clearly depicted (Figure 9.5; also see Effler et al. 1990b, and Chapter 5). The average concentration over the 1973–1985 period was $1585\,mg \cdot L^{-1}$; the average for 1990 and 1991 was $430\,mg \cdot L^{-1}$. Thus the reduction from the past 13 years of operation of the facility was 73%. The lower concentrations that have prevailed since closure are still extremely high compared to levels found in other, even highly impacted, lakes (e.g., Wetzel 1983). Rather strong seasonality in Cl⁻ concentration has prevailed (Figure 9.5,

Chapter 4), with progressive increases observed during the summer period, when tributary flows to the lake are lower (Chapter 3). The vertical stratification of Cl⁻ and other ionic waste constituents that contributed significantly to overall density stratification (Effler et al. 1986a) before closure (e.g., see inset of Figure 9.5), has been greatly reduced (Effler et al. 1990b; Chapters 4 and 5).

The development of the daily time series of Cl⁻ loads for the 1973–1991 interval was documented in Chapter 3 (also see Doerr et al. 1994). Hydrologic loads through 1990 were also documented in Chapter 3; these loads were expanded through 1991 by Doerr et al. (1994). Recall that the relative contributions of sources have changed substantially over this period, reflecting the discharge of ionic waste from the soda ash/chlor-alkali facility via Ninemile Creek (until mid-1981), discharge of a large fraction of this waste via the METRO outfall (mid-1981 through early 1986), and conditions following closure of the facility. Loads associated with upward diffusion from the enriched deepwater sediments and minor tributaries are included. However, loads from the recently found seeps in lower Onondaga Creek (Chapter 3) and from the Seneca River were not accommodated in the long-term simulations, because estimates are available for only short intervals. The model

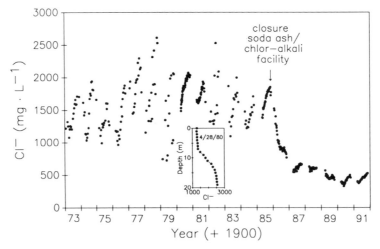

FIGURE 9.5. Temporal distribution of volume-weighted concentration of Cl⁻ in Onondaga Lake, 1973–1991. Depth profile of Cl⁻ for a single day as an inset (modified from Doerr et al. 1994).

was used to investigate the significance of these two additional loads. Inputs of groundwater within lake basins, in cases where they are significant, are usually received along the margins of the lake (e.g., Winter 1978). Such inputs were not identified over the period of model testing.

9.2.3 Model Performance

The model fit to the documented distribution of Cl^- in Onondaga Lake over the 1973–1991 period is presented in Figure 9.6. The model performed very well in simulating the important features of this distribution, including: (1) the very high concentrations observed before closure of the soda ash/chlor-alkali facility, (2) year-to-year variations in concentration before closure, (3) the strong seasonality in concentration before closure, and (4) the major and rapid reduction in concentration following closure of the facility. The model is described as validated, as model predictions generally closely matched observations. The model performed particularly well in tracking the strong reduction in concentration that occurred over the last year of operation of the facility and the following year (1985–1986), an interval for which a particularly detailed lake data set exists (Figure 9.5). The high performance of

the model supports the flow data base, lake bathymetry, features of the tributary and lake monitoring programs, estimates of Cl^- loading, and the simple model framework.

The most severe deviations between observations and predictions occurred for 1974, 1976, 1979, and 1982 (Figure 9.6). Lake data are particularly suspect in 1979 and 1982; for example unrealistic abrupt changes in lake concentration were reported for both years. Further, data from an alternate source differ greatly in several instances in these years (Walker 1991). The insets of Figure 9.6 depict other systematic limitations in model performance. The expanded plot for 1980 illustrates a shortcoming of not accommodating salinity stratification in the simple model framework. Note the model underpredicts concentrations and overpredicts the rate of increase in concentration during the May–July interval. The underprediction of concentrations is due to not accommodating the plunging inflow phenomenon (e.g., Owens and Effler 1989). The overprediction of the rate of increase in Cl^- is because summertime inputs of ionic waste entered the upper waters (Effler et al. 1986a; Owens and Effler 1989) and were discharged to the river without mixing with the lower more dense layers. The expanded scaling provided for the 1988–1991 interval

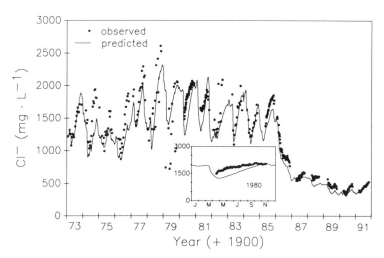

FIGURE **9.6.** Comparison of model simulations to observed Cl^- concentrations in Onondaga Lake, 1973–1991. Seasonal performance presented for 1980 as an inset. Performance for 1988–1991 interval presented on an expanded scale as an inset (modified from Doerr et al. 1994).

as an inset in Figure 9.6 depicts a recurring tendency to slightly underpredict Cl^- concentration since closure of the facility, indicating the existence of a load presently not accounted for.

Model performance has been quantified (e.g., Thomann 1982) through regression analysis of model predictions versus observations (Figure 9.7), and as percent relative error, according to the expression

$$r_e = \frac{|X - [Cl^-]_p|}{X} \cdot 100 \qquad (9.3)$$

where r_e = percent relative error, X = observed Cl^- concentration $(mg \cdot L^{-1})$, and $[Cl^-]_p$ = predicted Cl^- concentration $(mg \cdot L^{-1})$. The best fit linear regression line describing the relationship between model predictions and observations deviates just slightly from equivalency; overall, predictions explained nearly 90% of the observed variability in lake Cl^- concentration (Figure 9.7). The average r_e (calculated daily) over the 1973–1991 period was 14.1%. A breakdown of yearly average r_e values appears in Table 9.3. The year-to-year differences in model performance depicted in Table 9.3 support qualitative observations that can be made from Figure 9.6 for the period

TABLE 9.3. Yearly average r_e values for Onondaga Lake Cl^- model, 1973–1991.

Year	Number of observations	Average percent relative error
1937–1991	19	14.1
1973	17	3.8
1974	17	21.5
1975	15	11.9
1976	16	20.4
1977	16	10.3
1978	15	15.8
1079	14	45.5
1980	57	16.7
1981	37	10.9
1982	13	16.2
1983	14	6.8
1984	14	4.7
1985	30	10.3
1986	38	13.6
1987	43	6.2
1988	25	18.9
1989	30	18.5
1990	31	11.6
1991	27	4.2

before closure of the soda ash/chlor-alkali facility. The r_e statistic is particularly valuable in identifying the significant relative errors in two years following closure (underpredictions in 1988 and 1989) that were not clearly manifested in the graphical presentation of model performance (Figure 9.6). Five of the 19 modeled years had an average r_e less than 10%; 11 years had an average r_e less than 15% (Table 9.3).

9.2.4 Sensitivity Analyses

Sensitivity analyses were performed on the validated model to investigate the implications of uncertainty in various of the model inputs. An analysis was conducted to determine the impact of the possible omission of a saline load of the magnitude indicated by the most recent (1992) observations just upstream of the mouth of Onondaga Creek (Chapter 3). The existing Onondaga Creek loads from 1973–1991 were multiplied by a factor of 1.66 to account for the increase in concentration observed downstream in Onondaga Creek in the summer of 1992. No hydrologic load was

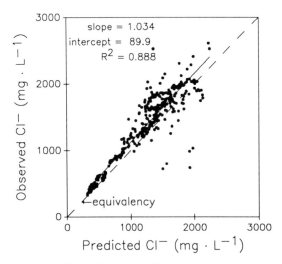

FIGURE 9.7. Evaluation of Cl^- model performance for Onondaga Lake; comparison of observed versus predicted Cl^- concentrations (modified from Doerr et al. 1994).

added because of the relatively high concentrations of Cl^-, but low flow, of these seeps (Chapter 3). This additional load had a very minor effect on predictions before closure of the soda ash/chlor-alkali facility (Figure 9.8), as the relative increase in load is minor. However, after closure the effect is significant (Figure 9.8), because of the greater relative importance of the load. These higher predictions and the validated model predictions essentially bound the observations over the 1987–1991 period (Figure 9.8). The r_e over the 1988–1991 period was reduced by 2.0% by the inclusion of this additional load.

The implications of not considering loading from the Seneca River in the model, and its interplay with the additional downstream loading from Onondaga Creek, was investigated with the validated model for a portion of 1991 (Figure 9.9). The performance of the validated model appears in Figure 9.9a. The distributions of hydrologic and Cl^- loading from the Seneca River to the lake were determined by E.M. Owens (personal communication) for the summer of 1991, based on a Cl^- mass balance analysis on the river and lake outlet (discussed further subsequently). Accommodation of these river loads decreases model calculations of lake Cl^- concentration (Figure 9.9b), because of the lower concentrations that prevail in the Seneca River (e.g.,

Effler et al. 1984b). The effect of including the downstream Onondaga Creek load (as in Figure 9.8) in the model over the same interval of 1991, without accommodating river loading, results in overprediction of lake Cl^- concentration (Figure 9.9c). Inclusion of both the Onondaga Creek and Seneca River loads results in a model simulation (Figure 9.9d) nearly equivalent to the model validation run (Figure 9.9a). Apparently the adequate performance of the Cl^- model for the validation run since closure of the soda ash/chlor-alkali facility (Figure 9.6), which does not consider the downstream Onondaga Creek and Seneca River loads, is a manifestation of an approximate balancing between the enriching and diluting influences of these inputs, respectively. Year-to-year differences in the performance of the model for this interval are to be expected (e.g., Figure 9.6), as loading from the river apparently varies with hydrologic loading to the lake from its tributaries (Chapter 4). Continued monitoring and analyses of these loads will be necessary to document temporal details of their distributions, and to fully delineate the compensating effects of these inputs on the Cl^- concentration of the lake, as implied by Figure 9.9.

Other possible sources of model uncertainty investigated include: (1) hydrologic, and coupled Cl^-, loading from Ninemile Creek,

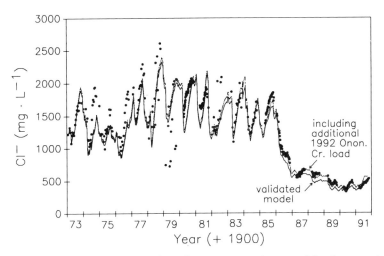

FIGURE 9.8 Sensitivity analysis of Cl^- model predictions to an increased load approximately equal to recently identified additional Onondaga Creek input (modified from Doerr et al. 1994).

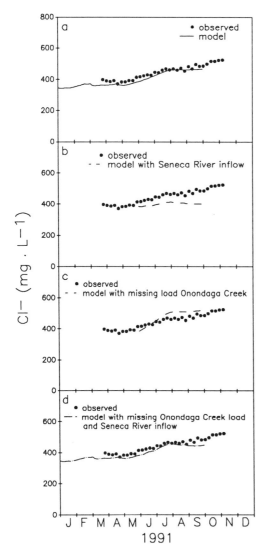

FIGURE 9.9. Sensitivity analyses demonstrating compensating effects of Seneca River inflow and downstream Onondaga Creek Cl⁻ loading on Onondaga Lake Cl⁻ concentration in 1991: (**a**) validated model, (**b**) inclusion of Seneca River hydrologic and Cl⁻ loads, (**c**) inclusion of downstream Onondaga Creek Cl⁻ load, and (**d**) inclusion of both the river and downstream Onondaga Creek loads (modified from Doerr et al. 1994).

for both before and after closure of the soda ash/chlor-alkali facility, (2) Cl⁻ loading from METRO following closure of the industrial facility, and (3) omission of evaporation/precipitation in the hydrologic budget. Recall that the poorest rated gauge in the lake inflow

monitoring program is on Ninemile Creek at Lakeland, the major source of Cl⁻ for most of the period of record (Chapter 3). The selected limits of ±20% represent reasonable worst case bounds of errors in daily flow measurement at this location. Errors in flow and Cl⁻ loading are coupled for this input. The model was run at these limits for 1981 and 1989. The limits of ±15% in Cl⁻ loading from METRO following closure represent reasonable worst case bounds associated with the irregular inflow of enriched water from the waste-bed lagoon (Chapter 3). This sensitivity analysis was conducted for 1990. The effect of including evaporation/precipitation was tested for 1987 and 1989, the second driest year and a near average runoff year, respectively, for the 1973–1990 interval (Chapter 3). The sensitivity of model projections to these sources of uncertainty is represented by the change in r_e from the model validation (Table 9.4). The relative importance of these sources of uncertainty can be compared to the earlier analysis for the additional Onondaga Creek load (Figure 9.8). The performance of the model had only a modest level of sensitivity to the specified levels of uncertainty in input loads. Model performance was essentially insensitive to the inclusion of evaporation/precipitation in the hydrologic budget.

TABLE 9.4. Influence of uncertainty in model inputs on Cl⁻ model performance, in terms of percent relative error (r_e).

Scenario for sensitivity analysis	Change in r_e*
1. Ninemile Creek, flow and loading	
a. 1981	
(1) −20%	−2.8
(2) +20%	+2.0
b. 1989	
(1) −20%	−2.4
(2) +20%	+2.3
2. METRO, 1990 load	
a. −15%	−2.4
b. +15%	+1.8
3. Evaporation/precipitation, inclusion	
a. 1987	+0.4
b. 1989	−0.1

* Compared to validated model (Table 9.3)

9.3 Total Phosphorus Model

9.3.1 Introduction

Phosphorus has been established to be the most critical nutrient for plant growth in Onondaga Lake (Chapter 6). The history of external P loading to the lake since the early 1970s and apportionment among the sources have been documented (Chapter 3). Despite major reductions in external loading, concentrations of TP in the lake continue to greatly exceed levels considered indicative of eutrophy (Chapter 5), and phytoplankton growth is not substantially limited by prevailing concentrations of available P (Chapter 6). Water quality implications of the cultural eutrophication of Onondaga Lake include: (1) periods of excessive concentrations of phytoplankton (Chapter 6), (2) periods of poor water clarity (Chapter 7), and (3) severe hypolimnetic oxygen depletion and subsequent lakewide depletions during the fall mixing period (Chapter 5). Numerous lake P models have been developed (see review of Bowie et al. 1985) because of the central role this nutrient plays in the eutrophication process (Hutchinson 1973).

Often values for the coefficients used in models for P, and other nonconservative substances, are developed through model calibration. The credibility of coefficient values is usually supported by comparison to literature compilations (e.g., Bowie et al. 1985) and through model verification; i.e., demonstrating the model, with the same coefficient values developed for calibration, continues to match observations for a significantly different set of environmental conditions. This mode of coefficient determination introduces inherent, and largely unquantifiable, uncertainty into model simulations. Further, the opportunity for verification does not always exist (e.g., small interannual variability in phosphorus concentrations), thereby eliminating the practical testing of the calibrated model and the various coefficients determined through the calibration process. Regardless of the application of the verification process, the use of

coefficients determined through calibration continues to draw criticism from the scientific community. In some cases, models have incorporated coefficients that are not amenable to direct experimental determination, thus eliminating any possibility of testing the conceptual and computational treatment of related processes. Even when appropriate experimental and analytical techniques have been available to independently establish key model coefficients, they have often not been utilized.

The development and testing of a comprehensive mechanistic seasonal TP model for Onondaga Lake is documented here (also see Doerr et al. 1995a). The model accommodates the principal features of the phosphorus cycle for stratified, eutrophic lakes, and all model inputs and kinetic coefficients can be independently determined. This approach is said to produce a "zero degree of freedom" model because "tuning" of model coefficients is eliminated from the model testing process—all inputs and kinetic coefficients are independently measured. Under these conditions, the model can be described as *validated* if an acceptable fit of model output to field data is achieved. The "zero degree of freedom" modeling approach represents a rigorous test of the integrity of the conceptual framework of the model, as well as the quality of input data and kinetic coefficients. This approach requires an integrated program of model development and laboratory and field studies, and places an unusual emphasis on supporting studies. An effort has been made in all the subsequently presented models to independently determine as many model coefficients as possible, thereby minimizing the use of literature values.

9.3.2 Model Structure

The conceptual framework of the original TP model (Canale et al. 1993b) is presented in Figure 9.10a. The atypical transport processes that occur presently, associated with the plunging underflow phenomenon (Chapter 4), and would occur with the adoption of a management option for a deep water discharge of the METRO effluent, are illustrated

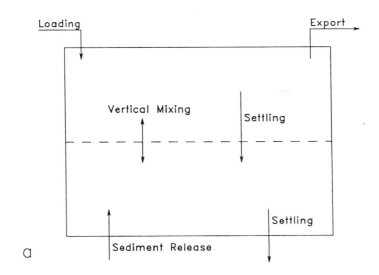

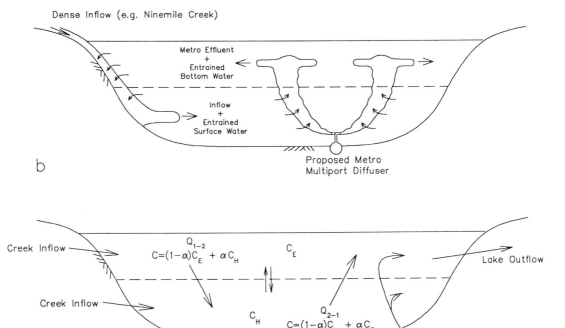

FIGURE 9.10. (**a**) TP model framework, with simple transport framework (modified from Doerr et al. 1995a), (**b**) transport processes associated with plunging underflow and deepwater discharge of buoyant inflow (modified from Doerr et al. 1995b), and (**c**) modified transport framework for two layer water quality models to accommodate processes of (b: modified from Doerr et al. 1995b).

in Figure 9.10b. Related modifications, adopted for the transport portion of the TP model (Doerr et al. 1995a), and the N (Canale et al. 1995a) and DO (Gelda and Auer 1995) models, to accommodate these complexities are presented in Figure 9.10c. The kinetic frameworks of these models (described subsequently) remain unchanged.

The simple framework of Figure 9.10a is expected to be appropriate for most stratifying lakes. According to this framework, the upper mixed layer (UML, essentially the epilimnion) has two sources of phosphorus, the sum of external TP loads received from METRO (a point source) and tributaries (nonpoint sources), and internal loading from the enriched lower mixed layer (LML, essentially the hypolimnion) mediated by vertical mixing. The upper layer has two sink processes, export of TP through the lake outlet to the Seneca River, and settling losses of the particulate fraction of TP to the LML. The LML has two sources of TP, TP released from lake sediment, and TP deposited from the UML. Two sink processes operate in the LML, vertical-mixing-based upward transport to the UML and incorporation of depositing P into the underlying sediments.

Recall that some of the shoreline inflows to Onondaga Lake, particularly Ninemile Creek, are negatively buoyant because of ionic enrichment, and plunge in the lake (Chapter 4; Figure 9.10b). The proposed deepwater discharge of METRO (see Owens and Effler (1995), and subsequent portions of this chapter for specific features of this proposal) would be positively buoyant and tend to rise in the water column of the lake as a plume. These buoyancy effects induce a vertical cycling of water (advective transport) and associated heat and mass by entraining ambient lake water (Figure 9.10b). Plunging shoreline inflows entrain ambient lake water into the plunging flow, effectively transporting lake water from shallow to deep layers (Figure 9.10b). Conversely, momentum and buoyancy-induced mixing of a submerged buoyant inflow would transport lake water from deep layers upward (Figure 9.10b). The modified transport framework for the water quality models (Figure 9.10c) retains the desired simplicity of the fixed boundary two-layer representation, yet accommodates the inflow and coupled advective transport processes associated with the atypical buoyancy effects. Inflows directly enter either of the two layers or are split between the layers. The advective transport from the upper layer to

the lower layer associated with the entrainment of water from the upper layer into a plunging inflow is designated Q_{1-2} (Figure 9.10c). Upward advective transport from the bottom layer to the top layer, Q_{2-1} (Figure 9.10c), results from the entrainment of ambient lake water from the lower layer into the rising plume from the diffuser. These advective transport components are also adjusted to maintain the fixed layer water balance constraints of the framework (calculated by difference). The modifications in the transport framework (Figure 9.10c) represent an increase in complexity that make the water quality models more mechanistically realistic, by accommodating the prevailing plunging inflow phenomenon (Chapter 4). Further, this framework provides for the alterations in vertical transport that would accompany the proposed deepwater discharge (Owens and Effler 1995). Time series of volume-weighted temperatures of the two layers, inflows to the two layers, and exchange flow between the layers, that drive the modified transport framework are provided as output files from a hydrothermal model (e.g., Owens and Effler 1989, Chapter 4; also see subsequent sections of this chapter). Estimates of the exchange flow Q_{1-2} for 1989 from the hydrothermal model are presented in Figure 9.11a. The magnitude(s) of the exchange flow associated with plunging were substantial in 1989 from the perspective of the summed total inflow (as the ratio Q_{1-2} to total inflow), particularly during the low watercolumn stability periods of the year (Figure 9.11b).

The TP mass balance equations for the UML and LML for the simpler framework of Figure 9.10a are:

$$
V_1 \cdot \frac{d[TP]_1}{dt}
$$

$$
= \underset{\text{(loading)}}{W_p} \; - \; \underset{\text{(export)}}{Q \cdot [TP]_1}
$$

$$
- \underset{\text{(settling out of UML)}}{vel_{PP} \cdot f_1 \cdot [TP]_1 \cdot A_i}
$$

$$
+ \underset{\text{(vertical transport)}}{v_t \cdot A_i \cdot ([TP]_2 - [TP]_1)}
$$

$$
(9.4a)
$$

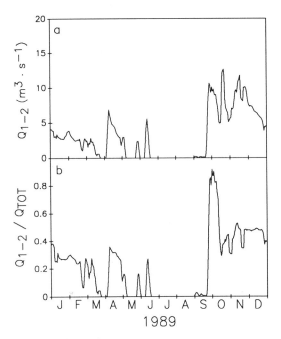

FIGURE 9.11. Time distributions of exchange flow from UML to LML in Onondaga Lake in 1989, associated with plunging entraining tributary inflow (as simulated by hydrothermal/stratification model): (a) Q_{1-2}, and (b) as the fraction of the total inflow to the lake, Q_{1-2}/Q_{TOT}.

$$V_2 \cdot \frac{d[TP]_2}{dt}$$

$$= \underbrace{R_{sed} \cdot A_i}_{\substack{[\text{sediment} \\ \text{release}]}} + \underbrace{vel_{PP} \cdot f_1 \cdot [TP]_1 \cdot A_i}_{\substack{[\text{settling out of} \\ \text{UML}]}}$$

$$- \underbrace{vel_{PP} \cdot f_2 \cdot [TP]_2 \cdot A_i}_{\text{(settling out of LML)}}$$

$$- \underbrace{v_t \cdot A_i \cdot ([TP]_2 - [TP]_1)}_{\text{(vertical transport)}} \qquad (9.4b)$$

in which: W_P = sum of the external TP loads $(mgP \cdot d^{-1})$, Q = export flow from the lake $(m^3 \cdot d^{-1})$, $[TP]_1$ = TP concentration in the UML $(\mu g \cdot L^{-1})$, $[TP]_2$ = TP concentration in the LML $(\mu gP \cdot L^{-1})$, V_1 = volume of the UML $(7.98 \times 10^7 \, m^3)$, V_2 = volume of the LML $(5.10 \times 10^7 \, m^3)$, vel_{PP} = particulate phosphorus (PP) settling velocity $(m \cdot d^{-1})$, f_1 = ratio of PP to TP concentrations in the UML, f_2 = ratio of PP to TP concentrations in the LML, A_i = area of interface between the UML and LML $(7.57 \times 10^6 \, m^2)$, R_{sed} = sediment release rate $(mgP \cdot m^{-2} \cdot d^{-1})$, and t = time (d). The

solution of these two simultaneous equations is approximated numerically using the Euler integration approach. Export is calculated as the product of the summed tributary flow and $[TP]_1$. The mass balance equations for the two layers for the more complex transport framework of Figure 9.10c are:

$$V_1 \cdot \frac{d[TP]_1}{dt}$$

$$= W_{1P} - (Q_1 + Q_{2-1} - Q_{1-2}) \cdot [TP]_1$$

$$- vel_{PP} \cdot f_1 \cdot [TP]_1 \cdot A_i$$

$$+ v_t \cdot A_i \cdot ([TP]_2 - [TP]_1)$$

$$- Q_{1-2} \cdot \alpha([TP]_1 + [TP]_2)$$

$$+ Q_{2-1} \cdot \alpha([TP]_1 + [TP]_2) \qquad (9.5a)$$

Q_1 = sum of inflows to UML $(m^3 \cdot d^{-1})$, Q_{1-2} = flow from UML to LML $(m^3 \cdot d^{-1})$, Q_{2-1} = flow from LML to UML $(m^3 \cdot d^{-1})$ $(Q = Q_1 + Q_{2-1} - Q_{1-2})$, W_{1P} = sum of external P loads to UML $(mgP \cdot d^{-1})$, and α = dimensionless weighting factor in the range 0–1 $(\alpha = 0.5$ established subsequently).

$$V_2 \cdot \frac{d[TP]_2}{dt}$$

$$= W_{2P} + R_{sed} \cdot A_i + vel_{PP} \cdot f_1 \cdot [TP]_1 \cdot A_i$$

$$- vel_{PP} \cdot f_2 \cdot [TP]_2 \cdot A_i$$

$$- v_t \cdot A_i \cdot ([TP]_2 - [TP]_1)$$

$$- Q_{2-1} \cdot \alpha([TP]_1 + [TP]_2)$$

$$+ Q_{1-2} \cdot \alpha([TP]_1 + [TP]_2) \qquad (9.5b)$$

W_{2P} = phosphorus load to LML $(mgP \cdot d^{-1})$

9.3.3 In-Lake Processes

9.3.3.1 Settling

Only the PP fraction of TP settles. The settling term in the mass balance expressions (equation (9.4)) has two parts, the PP settling velocity, vel_{PP}, and the PP:TP ratio for the UML and LML, f_1 and f_2. Particulate P is lost from the UML via settling, according to vel_{PP}, applied only to that fraction of TP which is particulate, f_1. This settling occurs across the interface between the layers, A_i (Figure 9.2a).

The short-term PP settling velocities $(vel_{PP'})$ were calculated from watercolumn P data (TP and TDP) and downward P flux data determined from sediment trap collections made at a depth of 10 m in 1987 (Chapter 8). The measured PP flux is the product of $vel_{PP'}$ and the PP concentration in the watercolumn above the trap. Thus $vel_{PP'}$ was determined from measurements according to

$$vel_{PP'} = \frac{J_{PP}}{PP} \qquad (9.6)$$

in which J_{PP} = measured downward flux of PP $(mg \cdot m^{-2} \cdot d^{-1})$, and PP = volume weighted PP concentration overlying the sediment traps $(mgP \cdot m^{-3})$.

The time series of J_{PP} in 1987 was presented in Chapter 8, and appears here again as Figure 9.12a. The time series of PP for the UML (i.e., volume-weighted concentration for the 0–8 m depth interval) in 1987, determined from paired TP and TDP profiles of 1 m depth resolution, is presented in Figure 9.12b. The estimated time series of $vel_{PP'}$ appears in Figure 9.12c. The individual values of $vel_{PP'}$ for each trap deployment period were calculated using the average PP value of the days(2) of deployment and collection. Rather strong dynamics in the velocity of deposition of P-containing particles from the lake's upper waters are indicated (Figure 9.12c). These variations were probably mostly a result of the dynamics of plankton populations in the lake (e.g., Auer et al. 1990). The value of vel_{PP} was estimated in two ways from the information of Figure 9.12, as the quotient of the average values of J_{PP} (47.2 mgP · m^{-2}, for the period corresponding to the PP data base of Figure 9.12b) and PP (74 μgP · L^{-1}), and as the average value of the time series of $vel_{PP'}$ (Figure 9.12c). The respective estimates of the model coefficient were 200 and 228 m · yr^{-1} (0.55 and 0.62 m · d^{-1}). A value of 214 m · yr^{-1}, the average of these estimates, was adopted for the TP model.

The temporal distributions of the ratios f_1 and f_2 (PP:TP) are presented in Figure 9.13. Comprehensive paired measurements of TP and TDP directly support the distributions of 1987. The distributions of 1989 and 1990 were based on paired profiles of TP and SRP, and the strong relationship observed between SRP

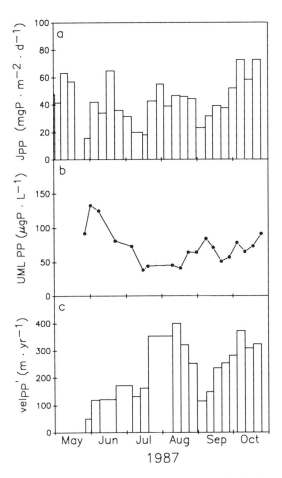

FIGURE 9.12. Time series in Onondaga Lake for the spring–fall interval of 1987: (**a**) downward flux of PP(J_{PP}), (**b**) PP concentration in the upper mixed layer, and (**c**) settling velocity of PP ($vel_{PP'}$).

and TDP at the limited depths of paired measurements (TDP = 1.04 (SRP) + 16.16, R^2 = 0.992, P concentrations in μg · L^{-1} (Canale et al. 1993b)). The ratios in 1988 were based solely on paired TP and TDP measurements at depths of 1 m (UML) and 16 m (LML). In general, the seasonal trend in the UML for the four years was a low ratio in early spring increasing to values between 0.6 and 0.8 in summer, followed by lower ratio values in fall. In the LML the ratio was relatively high in spring (0.4–0.8) and decreased during summer as soluble P accumulated, and increased in late fall to approximately 0.2.

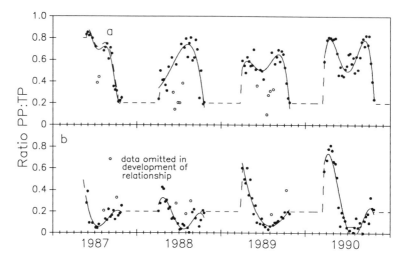

FIGURE 9.13. Time series of the ratio of PP:TP in Onondaga Lake for the two model layers: (**a**) UML, f_1, and (**b**) LML, f_2. Polynomial fits for time series included (modified from Doerr et al. 1995a).

Polynomials were fit to the temporal distributions of the ratios (Figure 9.13) to support the model. Certain data were not considered in development of the polynomials (see Figure 9.13). These data are not under question, but instead correspond to clearing events (Auer et al. 1990, Chapter 6). The algal-zooplankton dynamics in Onondaga Lake responsible for the abrupt changes in the ratio are not modeled in the TP model framework. A more complex model would be required. To support year-round model simulations, the polynomials for portions of each year were made continuous by extending f_1 and f_2 values of 0.2 from late fall to spring of the following year (Figure 9.13). Note a "clip" was added to the 1990 fit of the distribution of f_2 to avoid negative values, by using the average of the values observed over that period.

9.3.3.2 Vertical Transport

The transport of phosphorus from the enriched hypolimnion to the epilimnion is determined by the magnitude of vertical mixing (v_t) and the concentration gradient between the two layers (equation (9.4)); also see Wodka et al. 1983). In the fixed two-layer framework utilized here, the value of v_t accommodates the vertical exchange processes of diffusion and entrainment. The temporal

distribution of v_t was determined for the spring to fall intervals of four different years, 1987, 1988, 1989, and 1990, using a temperature balance (or model) on the LML, according to

$$V_2 \cdot \Delta T_2 = [v_t \cdot A_i \cdot (T_1 - T_2)] \cdot \Delta t \quad (9.7)$$

in which ΔT_2 = the change in hypolimnion (LML) temperature (°C), T_1 = volume-weighted temperature of the epilimnion (UML; °C), T_2 = volume-weighted temperature of the hypolimnion (LML; °C), and Δt is the interval over which v_t is determined (d). Polynomials were fit to the temporal distributions of T_1 for each year, as illustrated for 1989 in Figure 9.14a. These polynomials were used as input (T_1) to the temperature model for the LML, and v_t was adjusted to match the temporal distribution of T_2 (Figure 9.14b). This iterative calculation procedure is an improvement on related noniterative estimative techniques presented by other investigators (e.g., Chapra and Reckhow 1983; Effler and Field 1983; Wodka et al. 1983). Values of T_1 and T_2 were based on measured profiles for the simple model framework (Figure 9.10a), and on temperature profile predictions made with the stratification model (Owens and Effler 1995) for the more complex transport framework version (Figure 9.10c).

The distributions calibrated from the four years of T observations are presented in

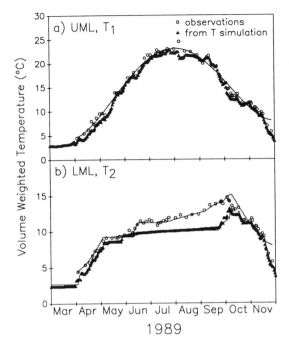

FIGURE 9.14. Methodology for estimation of v_t; based on observed T_2 for simple transport framework and simulated T_2 for complex framework: (**a**) Polynomial fit to time series of T_1 in Onondaga Lake in 1989, from observations and hydrothermal model simulations, and (**b**) match of observed and predicted (hydrothermal model) T_2 distributions with those determined from calculated distribution of v_t (modified from Doerr et al. 1995a).

Figure 9.15. Alternate estimates for 1989, based on simulations with the calibrated/verified stratification model, and consistent with the framework of Figure 9.10c, are also included. The distributions have recurring

temporal features widely reported elsewhere for north temperate lakes (e.g., Chapra and Reckhow 1983; Effler and Field 1983; Thomann and Mueller 1987; Wodka et al. 1983), and described earlier for Onondaga Lake, outside of the fixed-layer constraints invoked here, in Chapter 4. Minimal vertical mixing is indicated in midsummer (Figure 9.15), when stratification is strongest; vertical mixing is higher in spring and fall when stratification is weaker (Chapter 4). The details of the distributions differ from year to year as a result of meteorological variability (Effler et al. 1986d; Owens and Effler 1989). Upward transport is of particular interest during summer stratification because of the strong vertical gradient in P concentrations (see Chapter 5). A value of $v_t = 5\,m \cdot d^{-1}$ was invoked for the nonmonitored months (late fall to early spring) for all years (Figure 9.15). This assumption is generally representative for turnover periods, but overestimates vertical mixing during periods of ice cover (Owens and Effler 1995).

Estimates of v_t for the fall mixing period of 1990 were subject to refinement by varying the magnitude of v_t to achieve the observed vertical uniformity in oxygen concentration. Somewhat higher values of v_t were necessary in the fall of 1990 to maintain vertical uniformity in oxygen than for temperature (see "adjust" of Figure 9.15), based on the highly reactive character of DO in the lake (Chapter 5). Any associated changes in the temporal distributions of v_t were checked to assure maintenance of the temperature model per-

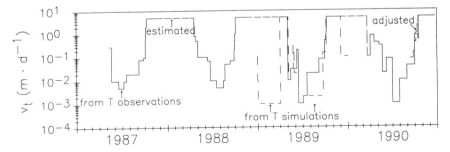

FIGURE 9.15. Distributions of v_t in Onondaga Lake for portions of four years, based on observations. Alternate 1989 distribution corresponds to the more complex transport framework, based on hydrothermal model simulations (modified from Doerr et al. 1995a).

formance. The distributions of v_t presented in Figure 9.15 were utilized in the simple transport framework versions of the TP, N, and DO models.

9.3.3.3 Sediment Release

The average anoxic P release rate reported for Onondaga Lake sediments by Auer et al. (1993) of $13.3 \, mgP \cdot m^{-2} \cdot d^{-1}$, based on laboratory core experiments (at 8°C) conducted with intact sediment–water interfaces, has been adopted in the TP model (R_{sed}) as the rate for the entire LML (see equation (9.4)). Uniform application of R_{sed} over area A_i is consistent with modeling convention (e.g., Imboden 1974; Lung et al. 1976), results of chemical characterization of the surficial sediments of the lake (Chapter 8), and the observed linearity of the rate of hypolimnetic accumulation of P in the lake (Chapter 5). Features of the release experiments of Auer et al. (1993) are reviewed in Chapter 5, as are the findings of more recent experiments (Penn 1994) and system analyses (Driscoll et al. 1993), which indicate that the model's representation of release is a simplification of temporally and spatially more complex phenomena.

The value of R_{sed} is adjusted for hypolimnion temperatures that deviate from 8°C according to the Arrhenius relationship:

$$R_{sed,T} = 13.3 \cdot \Theta^{(T-8)} \qquad (9.8)$$

$R_{sed,T} = R_{sed}$ at T ($mgP \cdot m^{-2} \cdot d^{-1}$), and $\Theta = $ a dimensionless coefficient (= 1.06). The form of

this temperature adjustment is widely used in modeling (e.g., Bowie et al. 1985), and is used in describing other sediment–water interactions accommodated in other Onondaga lake water quality models (see subsequent sections of this chapter). The temperature dependence of sediment P release has not received a great deal of research attention. A modeling analysis of the cycling of P in Shagawa Lake (Chapra and Canale 1991) indicated a temperature dependence of phosphorus release in that lake. The value of Θ adopted here is consistent with values used for other sedimentary processes by other investigators (Bowie et al. 1985).

The release rate $R_{sed,T}$ is applied from mid-May through October in the TP model, a period that exceeds somewhat the duration of hypolimnion anoxia (see Chapter 5). Outside of this interval (e.g., during oxic periods) the TP model has no sedimentary P release. The dynamics of the process during ice-cover have yet to be determined. Recall that the most recent experiments indicate some release occurs even under oxic conditions, though the flux is substantially less than during summer anoxia (Penn 1994).

9.3.4 Loads and Lake Concentrations

The model is to be tested for the May 1987–October 1990 period. A detailed description of external TP loads, and an apportionment of the contributions between METRO and tributaries, was presented for this period in Chapter 3. A comparative summary of average

TABLE 9.5. Comparative summary of daily average TP loading and flow rates for TP model testing years, 1987, 1988, 1989, and 1990.

Year	METRO		Tributaries		Total	
	TP load $kg \cdot d^{-1}$	flow $10^3 \cdot m^3 \cdot d^{-1}$	TP load $kg \cdot d^{-1}$	flow $10^3 \cdot m^3 \cdot d^{-1}$	TP load $kg \cdot d^{-1}$	flow $10^3 \cdot m^3 \cdot d^{-1}$
1987*	284	262	67	479	351	741
1988	209	268	113	674	322	942
1989	164	278	122	1001	286	1279
1990	173	315	143	1464	316 (5)**	1770 (54)

* From mid-May to end of year
** Additional loads associated with estimated net inflow to lake from Seneca River, according to Owens (1993)

external TP loading and flow rates for the four years is presented in Table 9.5. A rather broad range of tributary flows, associated with natural variations in runoff (Chapter 3), and METRO TP loading, associated with changes in P treatment (Chapter 3), are included in the four years. Note that the estimated additional loads to the lake associated with inflow from the Seneca River (Chapter 4) are included for 1990 in Table 9.5. The inflow was estimated based on a conservative substance mass balance analysis conducted around the lake and river (Owens 1993). Material loads were estimated from these flows and water quality monitoring data from the river.

Concentrations of TP for the two model layers were determined from profiles collected from the routine monitoring south basin station over the spring–fall intervals of 1987 through 1990. The profiles were based on a 1 m sampling interval in 1987 and on a 2 m depth interval in the other three years. Concentrations of TP were measured according to the ascorbic acid technique (APHA 1985). Time distributions of the volume-weighted concentrations are presented for the four years in Figure 9.16a–d. The bars represent the range of observations within each layer. Recall the progressively increasing concentrations observed with depth in the hypolimnion during summer stratification (Chapter 5). The wide range in concentration in the hypolimnion is reflected in Figure 9.16. The average concentrations of the two model layers for the four years are presented in Table 9.6. Together, the broad ranges in TP loading, flow (Table 9.5), and observed lake concentrations, represent a good test for the performance of the TP model.

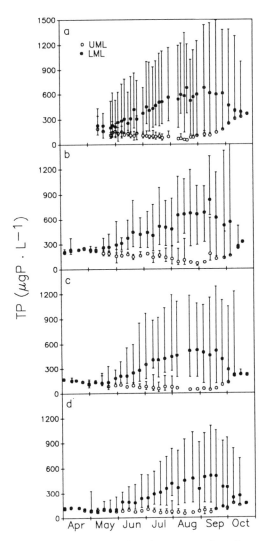

FIGURE 9.16. Time distributions of volume-weighted concentrations of TP in two layers of Onondaga Lake: (**a**) 1987, (**b**) 1988, (**c**) 1989, and (**d**) 1990 (modified from Doerr et al. 1995a). Dimensions of vertical bars equal to range of observations in layers.

9.3.5 Model Performance

Simulations of the TP model for the entire four-year period are compared to observations in Figure 9.17a; alternate predictions for 1989 utilizing the more complex transport framework of Figure 9.10c are included for comparison. Expanded plots of the simulations and observations for 1990 are presented for the two layers in Figure 9.17b and c, that more

fully resolve performance. Detailed comparisons have been provided for each of the years elsewhere (Canale et al. 1993b). The model performs very well in simulating the strong seasonality observed annually in the lake in both the UML and LML. Specifically, the simulations track the progressive increases in TP observed in the LML, and the largely progressive decreases in the UML through mid-August and the subsequent increase with the

TABLE 9.6. Average TP concentrations measured in two Onondaga Lake layers over the mid-May to mid-September interval for four years, and comparison to TP model predictions.

Year	UML* (µg·L⁻¹)		LML** (µg·L⁻¹)	
	Observed	Predicted	Observed	Predicted
1987	105	112	431	496
1988	140	128	494	492
1989	74	90	347	334
1990	76	76	296	301

*UML: upper mixed layer
**LML: lower mixed layer

approach to fall turnover. Further, the model performs well in simulating initial TP concentration measurements made in subsequent springs, extending over more than 5 months (the late fall–spring interval) without lake monitoring data. The model also performs well over the substantial differences in concentrations that prevailed for the four years, as reflected in Table 9.6. Note the performance of the modified, more hydrodynamically realistic, version of the model was essentially equivalent (Canale et al. 1993b) to that of the simpler model for 1989 (Figure 9.17a). Differences in performance of the two frameworks are largely a manifestation of the different distributions of v_t used; recall the simpler framework is based on observed temperature

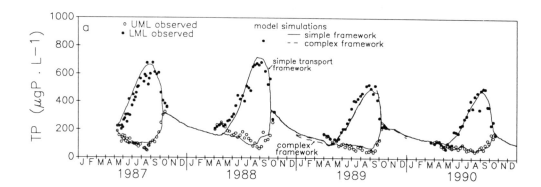

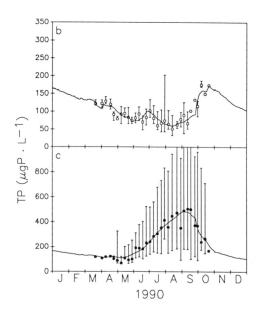

FIGURE 9.17. Simple transport framework TP model performance for Onondaga Lake: (a) TP, 1987–1990 period (modified from Doerr et al. 1995a), (b) UML, 1990, and (c) LML, 1990. Complex transport framework version performance presented for 1989 in (a) for comparison.

profiles, while the more complex transport framework relies on temperature predictions from a separate hydrothermal (stratification) model. Both TP models can be fairly described as validated, and are supported for application for evaluation of management scenarios. Either framework can be used for scenarios of reduced loading from METRO. The more complex hydrodynamic framework (see Figure 9.10b) is appropriate for a deepwater discharge alternative for METRO.

Modest differences in year-to-year performance are apparent (e.g., Table 9.6; Figure 9.17). The model failed to capture the UML minimum in mid-August of 1987 and slightly overpredicted LML concentrations for much of the early summer. The worst model performance was for the UML in 1988; the model underpredicted TP concentrations in the UML until late summer and overpredicted concentrations thereafter. Much of the early summer underprediction is attributed to underprediction of the initial spring concentration (Figure 9.17). Recall that the load from Onondaga Creek was unusually high in 1988, associated with sewer breaks (Chapter 3). Loads from this tributary may not have been accurately assessed by the biweekly monitoring program the preceding winter. The model underpredicted epilimnion concentrations in late spring and early summer of 1989, and overpredicted late summer concentrations. Concentrations in the LML were well tracked in 1989. The best model performance was observed for 1990 (Figure 9.17), a year in which the initial TP concentration was simulated with good accuracy.

9.3.6 Sensitivity Analyses

Analyses were conducted with the simpler TP model to establish the sensitivity of the predictions to reasonable levels of uncertainty in model inputs and coefficients. The base year for the analyses was 1990. The primary emphasis is placed on the UML here because this is the focus of water quality concerns for the lake.

Model predictions for the UML are moderately sensitive to the timing of loads, as indicated by the comparison of model output for the detailed daily loads versus simulations for the simplification of temporally uniform hydrologic and TP loads at the annual average values (Figure 9.18a). This simplified loading scenario was adopted as the base case for the other sensitivity analyses.

The predictions of the TP model for the UML are relatively insensitive to reasonable levels (e.g., $\pm 10\%$) of uncertainty in the coefficients that quantify the cycling of phosphorus in the lake, as demonstrated for vel_{PP} and R_{sed} in Figure 9.18 (b and c, respectively). Similar insensitivity to v_t and f_1 at the $\pm 10\%$ level was demonstrated by Canale et al. (1993b). Predictions are also relatively insensitive to reasonable levels of uncertainty ($\pm 10\%$ for summed tributary flow) in hydrologic loading (USGS flow measurements, and hydrologic budget; Chapter 3), and therefore, coupled estimates of TP loading (Canale et al. 1993b). Simulations were somewhat more sensitive to uncertainties in phosphorus loading (Canale et al. 1993b).

Substantial uncertainty exists concerning the magnitude of R_{sed} for certain management scenarios, including major reductions in water-column TP concentrations (and therefore deposition) resulting from the diversion of METRO, and injection of oxygen into the lake's hypolimnion. Substantial reductions in R_{sed} would be accompanied by increases in f_2. The influence of these uncertainties is bracketed here by assuming $R_{sed} = 0\,mg \cdot m^{-2} \cdot d^{-1}$, in combination with two cases of the PP:TP ratio, existing (1990) values of f_1 and f_2, and $f_2 = f_1$ (at 1990 values). This range of conditions has a dramatic effect on predictions for the LML (Figure 9.18d), but has only a modest effect on the UML during summer (Figure 9.18e) because of the limited vertical mixing between the layers. Inclusion of an R_{sed} value of $3\,mg \cdot m^{-2} \cdot d^{-1}$ for presently observed oxic periods, as indicated in the findings of Penn (1994), has only a minor effect on simulations for the UML. The relative impact of these uncertainties on model predictions increases somewhat for scenarios of major reductions in external loading (Canale et al. 1993b).

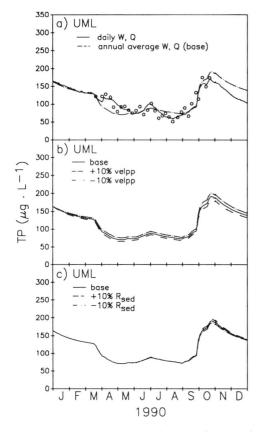

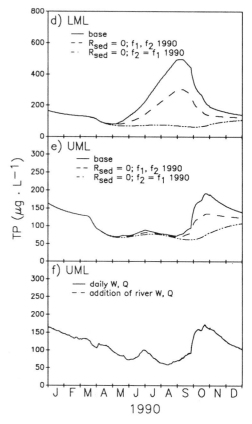

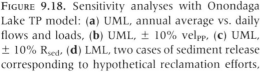

FIGURE 9.18. Sensitivity analyses with Onondaga Lake TP model: (**a**) UML, annual average vs. daily flows and loads, (**b**) UML, ± 10% vel$_{PP}$, (**c**) UML, ± 10% R$_{sed}$, (**d**) LML, two cases of sediment release corresponding to hypothetical reclamation efforts, (**e**) UML, two cases of sediment release corresponding to hypothetical reclamation efforts, and (**f**) UML, without Seneca River inputs vs. with river inputs (modified from Doerr et al. 1995a).

The accommodation of the inputs from the Seneca River (Owens 1993) had no perceptible effect on the performance of the TP model in 1990 (Figure 9.18f). This is because the TP concentration of the river (e.g., Effler et al. 1995) was similar to that found in the epilimnion of the lake in 1990. Continuation of such a river load following a major reduction or elimination of TP input from METRO would tend to enrich the lake in TP, because under that scenario the river concentration would exceed that of the lake. Reclamation plans for the lake should include elimination of the river inflow phenomenon (also see subsequent related material on the river and Canale et al. 1995b).

9.4 Nitrogen Model

9.4.1 Background

Recall that various forms of N have water quality implications with respect to plant nutrition, oxygen resources, and toxicity to aquatic life (Chapter 5). Further, the N cycle is relatively complex (e.g., compared to the P cycle) because of the large number of chemical species of N (Harris 1986) and biochemical processes (Sprent 1987) involved in the cycle, and the great sensitivity of these processes to ambient environmental conditions (Hutchinson 1957; Sprent 1987; Wetzel 1983).

The extremely high external loading rates of N species to Onondaga Lake were documented in Chapter 3. METRO is the major source of N to Onondaga Lake. This discharge represents 75% of the total N load, and 90% of the T-NH$_3$ load, received by the lake (Chapter 3). The second largest source of T-NH$_3$ ($\sim$3%) is the waste-bed area of Ninemile Creek.

The concentrations of T-NH$_3$, NH$_3$ (free ammonia), and NO$_3^-$ that are maintained, and the concentrations of NO$_2^-$ that develop, in the epilimnion of Onondaga Lake are unusually high compared to concentrations reported for other stratifying lakes in the literature (Brooks and Effler 1990; Effler et al. 1990a, Chapter 5). Very high concentrations of T-NH$_3$ develop in the hypolimnion, and NO$_x$ (NO$_3^-$ plus NO$_2^-$) is lost from this layer (Chapter 5). The concentrations of T-NH$_3$ and NO$_3^-$ in the upper waters continuously exceed levels associated with N limitation of phytoplankton growth (Chapter 6). The principal water quality concern presently for N in Onondaga Lake is the recurring contravention of the national chronic toxicity criteria, and violation of related New York State standards, for NH$_3$ for salmonid and nonsalmonid fish, at all depths over substantial portions of the spring–summer interval (Effler et al. 1990a, Chapter 5). The NH$_3$ problems of the lake are a manifestation of the very high concentrations of T-NH$_3$ (Effler et al. 1990a).

A N model is developed and tested here (also see Canale et al. 1995a) that is intended to support evaluation of various remediation strategies to eliminate violations of the NH$_3$ standard(s) for Onondaga Lake. Thus the principal focus is the prediction of seasonal and annual variations of T-NH$_3$ in the upper waters of the lake as a function of ambient environmental forcing conditions and waste discharge. It is necessary to consider the important features of the overall N cycle, including other key constituents, to support credible simulations of T-NH$_3$ (e.g., Bowie et al. 1985). The focus of this lake N model appears to be unusual, if not unique. Other efforts have primarily been concerned with the interplay with phytoplankton growth (see review of Bowie et al. 1985).

9.4.2 Conceptual Framework

A conceptual representation of the nitrogen cycle accommodated in the N model for Onondaga Lake appears as Figure 9.19. Individual processes and parameters involved in the N cycle, and quantitatively coupled according to mass balance in the model, are represented by linkages and compartments (Figure 9.19). Sources and sinks for each of the components, particulate organic N (PON), dissolved organic N (DON), T-NH$_3$, TKN (= organic N + T-NH$_3$), NO$_x$, and TN (= TKN + NO$_x$), are presented. PON is partitioned into two components, the phytoplankton fraction (p-PON) and the detrital fraction (d-PON; i.e., PON = p-PON + d-PON). The p-PON component is formed strictly from internal production (i.e., assumed no significant ex-

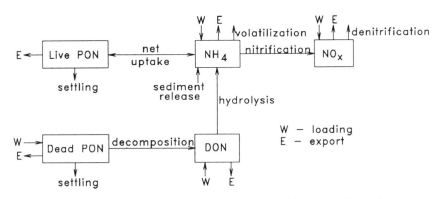

FIGURE 9.19. Conceptual N model for Onondaga Lake (modified from Canale et al. 1995a).

ternal sources), and the other four components have external sources (Figure 9.19). Internal sources of d-PON are assumed to be insignificant. All components are subject to discharge from the lake to the river.

Two pathways of loss to the atmosphere exist (Figure 9.19). Free ammonia is subject to loss through volatilization (e.g., Effler et al. 1991b). Also a portion of the N_2 gas formed through denitrification (loss of NO_x) in the anoxic hypolimnion can be lost via gas ebullition (Chapter 5). T-NH_3 is taken up by phytoplankton to support growth, that is, converted to p-PON. DON is produced by decomposition of d-PON. T-NH_3 is produced from p-PON when phytoplankton net growth is negative (e.g., zooplankton grazing losses of phytoplankton exceed gains). Both forms of PON are subject to settling. T-NH_3 is produced from hydrolysis of DON and is released from the lake's sediments. T-NH_3 is converted to NO_x through the nitrification process. It is assumed that NO_x is not presently assimilated by phytoplankton in Onondaga Lake (Figure 9.19), because of the availability of T-NH_3, the N species preferred for energetic reasons (Wetzel 1983). The transformations incorporated in the Onondaga Lake N model (Figure 9.19) include features that have been incorporated in a number of models (e.g., Baca and Arnett 1976 as cited by Bowie et al. 1985; Lean and Knowles 1987; Brezonik 1972; DiToro and Connolly 1980). However, there is no effort here to be mechanistically replete. Certainly other known, but poorly quantified, processes could be accommodated. Rather, the emphasis here is to accurately quantify the important source and sink processes for T-NH_3.

9.4.3 p-PON and Phytoplankton Net Growth

The net effect of phytoplankton production and several loss processes on T-NH_3 uptake/release is accommodated in the N model (Figure 9.19). Phytoplankton models consider growth, and several sink processes including respiration, excretion, settling, grazing, and nonpredatory mortality (or decomposition) losses. A general equation, which includes all of these processes and forms the basis for most phytoplankton models (Bowie et al. 1985), can be expressed as:

$$\frac{dP}{dt} = (G - D - e_x - s - m - z_g) \cdot P \quad (9.9)$$

in which P = phytoplankton biomass (e.g., mg chlorophyll $\cdot$ m^{-3}; here represented by CHL), G = gross growth rate (d^{-1}), D = respiration rate (d^{-1}), e_x = excretion (d^{-1}), s = settling rate (d^{-1}), m = nonpredatory mortality rate (d^{-1}), and z_g = loss rate due to grazing (d^{-1}).

Several environmental factors, such as temperature, light, and nutrients, affect the kinetics of the processes of growth, death, and grazing. Many phytoplankton models combine several of the processes of equation (9.9) in a single term, thereby simplifying the expression (Bowie et al. 1985). The simplified version of equation (9.9) used here is:

$$\frac{dP}{dt} = (G_{NET} - s) \cdot P \quad (9.10)$$

in which G_{NET} = net growth, = $G - D - e_x - m - z_g$ (d^{-1}). The magnitude of G_{NET} varies seasonally in the lake, because of temporal changes in mitigating environmental factors, as manifested in the seasonality in P. The seasonality in G_{NET} is not mechanistically modeled, but instead is determined by calibration. Further research, particularly in quantifying zooplankton grazing and related forcing conditions, and the dynamics of nutrient availability, are needed to develop a predictive model for Onondaga Lake.

The time distributions of G_{NET} determined for the UML and LML for 1989, by calibration, are presented in Figure 9.20a. These distributions have been adopted in the N model. The performance of the calibration in matching the observed distributions of CHL is shown in Figure 9.20b. The temporal distributions of CHL observed in 1989 are generally typical of those observed since 1987, though substantial interannual variability in the magnitude of the fall bloom is observed (Chapter 6). This is turn causes substantial interannual differ-

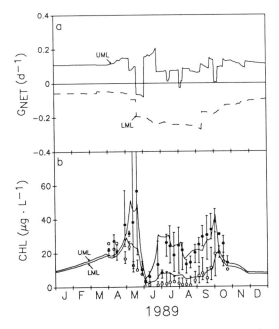

FIGURE 9.20. Phytoplankton, Onondaga Lake, 1989: (a) temporal distributions of G_{NET} for UML and LML, by calibration, and (b) performance of G_{NET} distribution in matching observed distributions of CHL for the two model layers (modified from Canale et al. 1995a).

ences in the depletion of T-NH$_3$ in the upper waters during fall. Recall that most of the temporal structure in CHL, and therefore G_{NET} (Figure 9.20a; e.g., minima in early to mid-summer), is regulated by zooplankton grazing (Chapter 6), as only a modest degree of phosphorus limitation develops in the productive layers (Chapter 6, Connors et al. 1995). Note the major decreases in G_{NET} that emerge during the clearing event(s) (Figure 9.20a). The negative values of G_{NET} for the LML depict a T-NH$_3$ recycle pathway.

The stoichiometry of p-PON and CHL was determined directly from paired measurements of PON and CHL made in the upper waters (1 m) of the lake over the 1989–1991 interval. The analysis is presented in Figure 9.21. The p-PON/CHL ratio is estimated as the slope (=5.2) of the linear least squares regression of PON on CHL. The intercept (= 0.15 mgN · L^{-1}) represents an estimate of the average d-PON for the conditions of the

measurements. The observations are nearly evenly distributed over the range of measurements. The relationship is rather strong; variations in CHL explained 60% of the variability in PON. The relationship is used to generate estimates of CHL (e.g., Figure 9.19b) from predictions of p-PON. Much of the variability in the relationship is probably attributable to variations in the CHL content of the lake's phytoplankton and in the concentrations (e.g., loading) of d-PON.

The mass balance expressions for p-PON in the UML and LML are

$$V_1 \frac{d[\text{p-PON}]_1}{dt} = G_{NET,1} \cdot [\text{p-PON}]_1 \cdot V_1$$
$$- Q \cdot [\text{p-PON}]_1 - \text{vel}_N \cdot A_i \cdot [\text{p-PON}]_1$$
$$+ v_t \cdot A_i \cdot ([\text{p-PON}]_2 - [\text{p-PON}]_1) \quad (9.11a)$$

$$V_2 \frac{d[\text{p-PON}]_2}{dt} = G_{NET,2} \cdot [\text{p-PON}]_2 \cdot V_2$$
$$+ \text{vel}_N \cdot A_i \cdot [\text{p-PON}]_1 - \text{vel}_N \cdot A_i \cdot$$
$$[\text{p-PON}]_2 + v_t \cdot A_i \cdot ([\text{p-PON}]_1 - [\text{p-PON}]_2) \quad (9.11b)$$

in which $[\text{p-PON}]_1$, $[\text{p-PON}]_2$ = p-PON concentrations of the UML and LML, respectively (mg · L^{-1}); $G_{NET,1}$, $G_{NET,2}$ = net growth of phytoplankton in UML and LML, respectively; and vel_N = average settling velocity of N (m · d^{-1}). The value of vel_N used in the N model (for both p-PON and d-PON) is 214 m · yr^{-1} (0.586 m · d^{-1}); equal to vel_{PP} of

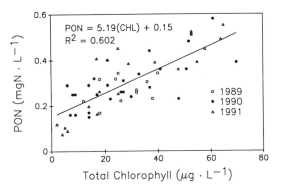

FIGURE 9.21. Determination of the p-PON/CHL ratio for Onondaga Lake (modified from Canale et al. 1995a).

the TP model. The equivalence for the settling process is a unifying feature of the TP and N models for the lake. This value of vel_N has independent support, because it is quite close to the estimate based on limited paired measurements of PON and the downward flux of N ($225 \, m \cdot yr^{-1}$, Canale et al. 1993c). Export for each of the modeled N species is calculated as the product of the summed tributary flows and the constituent concentration of the UML.

9.4.4 Other Compartments of N Model

9.4.4.1 d-PON

According to the adopted model framework (Figure 9.19), the d-PON fraction of PON is received from external loads, settles, and is subject to conversion to DON through decomposition processes. The major source of d-PON is METRO. Total organic N (TON) in the METRO effluent during 1989 (calculated as the difference of unfiltered TKN and T-NH3) was determined routinely. Onondaga County reported a TON loading rate of $1900 \, kg \, N \cdot d^{-1}$ from METRO for 1989. Thirty-eight percent of the TON from all sources was reported to be PON (Onondaga County 1990; i.e., d-PON in the framework of Figure 9.19). This partitioning (d-PON/DON) was adopted for all external loads in the model. The mass balance equations for d-PON in the two layers are:

$$V_1 \frac{d[\text{d-PON}]_1}{dt} = W_{\text{d-PON}} - Q \cdot [\text{d-PON}]$$
$$+ v_t \cdot A_i \cdot ([\text{d-PON}]_2 - [\text{d-PON}]_1)$$
$$- vel_N \cdot A_i \cdot [\text{d-PON}]_1$$
$$- K_{\text{decomp},T} \cdot V_1 \cdot [\text{d-PON}]_1 \qquad (9.12a)$$

$$V_2 \frac{d[\text{d-PON}]_2}{dt} = vel_N \cdot A_i [\text{d-PON}]_1$$
$$+ v_t A_i \cdot ([\text{d-PON}]_1 - [\text{d-PON}]_2)$$
$$- vel_N \cdot A_i \cdot [\text{d-PON}]_2$$
$$- K_{\text{decomp},T} \cdot V_2 \cdot [\text{d-PON}]_2 \qquad (9.12b)$$

in which $[\text{d-PON}]_1$, $[\text{d-PON}]_2$ = d-PON concentrations of the UML and LML, respec-

tively (mg $\cdot$ L^{-1}); $W_{\text{d-PON}}$ = loading of d-PON ($g \cdot d^{-1}$), and $K_{\text{decomp},T}$ = decomposition rate at temperature T (d^{-1}).

The influence of temperature on $K_{\text{decomp},T}$ is expressed in an Arrhenius format, according to:

$$K_{\text{decomp},T} = K_{\text{decomp},20} \cdot \Theta^{T-20} \qquad (9.13)$$

The N model adopts values of $0.1 \, d^{-1}$ for $K_{\text{decomp},20}$, and 1.08 for Θ.

9.4.4.2 DON

The sources of DON in the model are external loading and decomposition of d-PON. Sinks of DON accommodated include export and ammonification (hydrolysis). The mass balance equations for the two layers are:

$$V_1 \frac{d[\text{DON}]_1}{dt} = W_{\text{DON}} - Q \cdot [\text{DON}]_1$$
$$+ K_{\text{decomp},T} \cdot V_i \cdot [\text{d-PON}]_1$$
$$- k_{\text{hyd},T} \cdot [\text{DON}]_1 \cdot V_1 + v_t \cdot A_i \cdot$$
$$([\text{DON}]_2 - [\text{DON}]_1) \qquad (9.14a)$$

$$V_2 \frac{d[\text{DON}]_2}{dt} = K_{\text{decomp},T} \cdot V_2 \cdot [\text{d-PON}]_2$$
$$- k_{\text{hyd},T} \cdot [\text{DON}]_2 \cdot V_2 + v_t \cdot A_i \cdot$$
$$([\text{DON}]_1 - [\text{DON}]_2) \qquad (9.14b)$$

in which $[\text{DON}]_1$ = DON concentration of UML (mg $\cdot$ L^{-1}), $[\text{DON}]_2$ = DON concentration of LML (mg $\cdot$ L^{-1}), W_{DON} = loading of DON ($g \cdot d^{-1}$), and $k_{\text{hyd},T}$ = hydrolysis or ammonification rate constant at temperature T (d^{-1}).

The influence of temperature on $k_{\text{hyd},T}$ is expressed in an Arrhenius format, according to:

$$k_{\text{hyd},T} = k_{\text{hyd},20} \cdot \Theta^{T-20} \qquad (9.15)$$

The N model adopts values of $0.005 \, d^{-1}$ for $k_{\text{hyd},20}$, and 1.08 for Θ.

9.4.4.3 Ammonia

The sources of ammonia include loadings from METRO and the four major tributaries, ammonification, zooplankton (PON) excretion, and sediment release. The sinks of ammonia in the model are algal uptake, nitrification, export to the river and volatilization (Figure 9.19).

The mass balance equations for T-NH$_3$ in the two layers are:

$$V_1 \frac{d[\text{T-NH}_3]_1}{dt} = W_{\text{T-NH}_3} - Q \cdot [\text{T-NH}_3]_1$$
$$+ k_{\text{hyd,T}} \cdot [\text{DON}]_1 \cdot V_1 - G_{\text{NET,1}} \cdot$$
$$[\text{p-PON}]_1 \cdot V_1 - k_{\text{fn,T}} \cdot (A_s - A_i) \cdot$$
$$[\text{T-NH}_3]_1 - K_{\text{L,NH}_3} \cdot A_s \cdot ff \cdot [\text{T-NH}_3]_1$$
$$+ v_t \cdot A_i \cdot ([\text{T-NH}_3]_2 - [\text{T-NH}_3]_1) \quad (9.16)$$

$$V_2 \frac{d[\text{T-NH}_3]_2}{dt} = k_{\text{hyd,T}} \cdot [\text{DON}]_2 \cdot V_2$$
$$- G_{\text{NET,2}} \cdot [\text{p-PON}]_2 \cdot V_2 - k_{\text{fn,T}} \cdot A_i \cdot$$
$$[\text{T-NH}_3]_2 + v_t \cdot A_i \cdot ([\text{T-NH}_3]_1$$
$$- [\text{T-NH}_3]_2) + S_{\text{r,T}} \cdot A_i \quad (9.16b)$$

in which $[\text{T-NH}_3]_1$ = T-NH$_3$ concentration of the UML (mg $\cdot$ L^{-1}), $[\text{T-NH}_3]_2$ = T-NH$_3$ concentration of the LML (mg $\cdot$ L^{-1}), $W_{\text{T-NH}_3}$ = external loading of T-NH$_3$ (g $\cdot$ d^{-1}), $k_{\text{fn,T}}$ = nitrification rate constant at T°C (m $\cdot$ d^{-1}), A_s = surface area of the lake (12 × 10^6 m^2), $K_{\text{L,NH}_3}$ = surface transfer coefficient for NH$_3$ (m $\cdot$ d^{-1}), ff = fraction of T-NH$_3$ as NH$_3$, and $S_{\text{r,T}}$ = sediment release rate of T-NH$_3$ (g $\cdot$ m^{-2} $\cdot$ d^{-1}). The development of model inputs for nitrification, and sediment release and volatilization of T-NH$_3$, is treated below.

9.4.4.3.1 Nitrification

A number of researchers have reported nitrification to be localized at the sediment–water interface (Cavari 1977; Curtis et al. 1975; Hall 1986). Results of laboratory microcosm experiments indicate no significant nitrification occurs within the watercolumn of Onondaga Lake. The kinetics of nitrification in the lake were quantified based on sediment flux of T-NH$_3$ (depletion) determined in laboratory experiments with intact sediment cores. A film transfer approach (analogous to reaeration, Bowie et al. 1985) was used to describe the kinetics of the process, to reflect the localization of the process at the sediment–water interface. Accordingly, the rate of nitrification is quantified by

nitrification rate (g T-NH$_3$ N $\cdot$ d^{-1})
$$= k_{\text{fn}} \cdot A \cdot [\text{T-NH}_3] \quad (9.17)$$

The film transfer nitrification coefficient (k_{fn}) is determined by dividing the measured areal flux (g $\cdot$ m^{-2} $\cdot$ d^{-1}) by the overlying bulk concentration (mg $\cdot$ L^{-1}).

The film transfer treatment assumes that diffusion-based transport of T-NH$_3$ from the overlying bulk liquid across the stagnant fluid layer (film) immediately overlying the sediments is the rate limiting step for nitrification. The laboratory measurements of T-NH$_3$ flux suggest a value of K_{fn} = within a range of 0.08 to 0.3 m $\cdot$ d^{-1}. The value of K_{fn} determined from calibration in 1989 was 0.135 m $\cdot$ d^{-1}, similar to that determined for the Seneca River (Canale et al. 1995b). Temperature corrections were made according to the relationship

$$K_{\text{fn,T}} = K_{\text{fn,20}} \cdot \Theta^{(T-20)} \quad (9.18)$$

in which Θ = 1.06. Zison et al. (1978) reported a range of Θ values for nitrification of 1.0548 to 1.0997.

The model sets nitrification to zero for temperatures below 10°C, because nitrifying bacteria do not multiply below 10°C (Thomann and Mueller 1987). The model uses the observed oxygen concentration at 19 m to determine the oxic and anoxic period. For 1989 the anoxic period was May 24 to October 10. The upper mixed layer is always oxic. The effective oxic area contributing to nitrification in the upper layer is approximated by ($A_s - A_i$); and for the lower layer during the oxygenated periods the area is A_i.

9.4.4.3.2 Sediment Release and Volatilization

The sediment release rate of T-NH$_3$ ($S_{\text{r,T}}$) is applied across the entire sediment area of the LML. The value determined by model calibration was 70 mgN $\cdot$ m^{-2} $\cdot$ d^{-1} ($S_{\text{r,8}}$). The flux determined directly from laboratory experiments on cores (with intact sediment–water interfaces) collected from a single deepwater location in the south basin of the lake was 92 mgN $\cdot$ m^{-2} $\cdot$ d^{-1} (Penn et al.

1993). Results of one of the experiments, depicting the progressive enrichment of the water overlying the core with T-NH$_3$, are presented in Figure 9.22. Part of the difference between these laboratory results and the calibration value may reflect spatial variations in the release rate across the sediment area. Temperature corrections were made according to the Arrhenius relationship

$$S_{r,T} = S_{r,8} \cdot \Theta^{(T-8)} \qquad (9.19)$$

in which $\Theta = 1.085$.

Free ammonia is subject to loss to the atmosphere via the process of volatilization (Effler et al. 1991b, Thomann and Mueller 1987). The fraction of T-NH$_3$ that exists as NH$_3$ (ff) is dependent primarily on pH and temperature, and, to a lesser extent, ionic strength (Effler et al. 1990a). The protocol for calculation of ff was described in Chapter 5 and by Effler et al. (1990a). The temporal distribution of ff used in the UML in the model corresponds to the conditions documented for the upper waters of the lake in 1989 (Chapter 5). The average (e.g., spring to fall) value of K_{L,NH_3} reported for the lake (Effler et al. 1991b) of $0.17\,\text{m}\cdot\text{d}^{-1}$ was adopted in the model. These simplifying assumptions for ff and K_{L,NH_3} in the model are supported by the fact that the volatilization sink for T-NH$_3$ (equation (9.16a)) in Onondaga Lake is minor (Effler et al. 1991b).

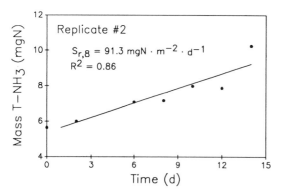

FIGURE 9.22. Accumulation of T-NH$_3$ in water overlying sediment core sample from Onondaga Lake, to support laboratory estimate of T-NH$_3$ release rate ($S_{r,8}$; mgN $\cdot$ m^{-2} $\cdot$ d^{-1}).

9.4.4.4 Nitrogen Oxides

The N model simulates the sum of nitrate and nitrite, referred to as NO$_x$. The sources of NO$_x$ are loadings from METRO and tributaries, and nitrification. The sinks of NO$_x$ are denitrification and export to the river (Figure 9.19). Algal uptake of NO$_x$ is considered to be presently negligible because ammonia is favored by phytoplankton for energetic reasons (Wetzel 1983), and the T-NH$_3$ concentration never falls to limiting levels. The management version of the N model accommodates the switch to utilization of NO$_x$ for low T-NH$_3$ concentrations (described subsequently).

The mass balance equations for NO$_x$ in the two layers are:

$$V_1 \frac{d[NO_x]_1}{dt} = W_{NO_x} - Q \cdot [NO_x]_1 + k_{fn,T} \cdot$$

$$(A_s - A_i) \cdot [T\text{-}NH_3]_1 + v_t \cdot A_i \cdot ([NO_x]_2 - [NO_x]_1) \qquad (9.20a)$$

$$V_2 \frac{d[NO_x]_2}{dt} = k_{fn,T} \cdot A_i \cdot [T\text{-}NH_3]_2$$

$$- k_{fd,T} \cdot A_i \cdot [NO_x]_2 + v_t \cdot A_i \cdot ([NO_x]_1 - [NO_x]_2) \qquad (9.20b)$$

in which $[NO_x]_1 = NO_x$ concentration of the UML (mg $\cdot$ L^{-1}), $[NO_x]_2 = NO_x$ concentration of the LML (mg $\cdot$ L^{-1}), W_{NO_x} = loading of NO$_x$ (g $\cdot$ d^{-1}), and $k_{fd,T}$ = denitrification rate constant at temperature T (m $\cdot$ d^{-1}).

Denitrification, as with the nitrification process, is localized in the sediments (Seitzinger 1988). Thus this process has also been modeled according to film transfer theory; the film transfer coefficient is designated $k_{fd,T}$. The denitrification process is the sole sink for NO$_x$ in the hypolimnion. The value of $k_{fd,T}$ was determined by model calibration. Temperature corrections were made according to the Arrhenius relationship

$$k_{fd,T} = k_{fd,20} \cdot \Theta^{(T-20)} \qquad (9.21)$$

The model uses $k_{fd,20} = 0.4\,\text{m}\cdot\text{d}^{-1}$ (from calibration). This corresponds to a first order rate coefficient value (= $k_{fd,20} \div$ mean depth; consistent with a watercolumn process) of about $0.06\,\text{d}^{-1}$, that falls in the range of values reported by Bowie et al. (1985). A value of Θ = 1.06 was adopted, which falls in the range of values reported by Bowie et al. (1985).

9.4.4.5 Total Kjeldahl Nitrogen and Total Nitrogen

The concentration of total Kjeldahl N [TKN], defined as the sum of the concentrations of organic N and T-NH$_3$, is simply calculated in the N model as the sum of predicted concentrations of PON (= [p-PON] + [d-PON]), DON, and T-NH$_3$.

$$[TKN] = [PON] + [DON] + [T\text{-}NH_3] \quad (9.22)$$

The concentration of total N [TN] in the system is the sum of the concentrations of organic and inorganic forms of N, i.e.,

$$[TN] = [PON] + [DON] + [T\text{-}NH_3] + [NO_3^-] + [NO_2^-] \quad (9.23)$$

Concentrations of TKN have been determined analytically, thus offering another level of testing predictions of the individual species.

9.4.5 Model Inputs

The simple transport framework version of the N model is tested for the conditions documented in 1989 and 1990. The N model was retested for 1989 within the more complex hydrodynamic framework (Figure 9.10c). The development of daily loads of T-NH$_3$, organic N, and NO$_x$ to the lake for 1989 and 1990 was documented in Chapter 3. T-NH$_3$ loads were

greater and NO$_x$ loads less in the summer of 1990 compared to 1989, as a result of the greater degree of nitrification achieved at METRO in 1989 (Chapter 3). However, greater dilution of the METRO discharge was afforded by the higher tributary flow in 1990 compared to 1989 (Table 9.5, Chapter 3).

The monitoring program for N species and downward flux of PON in Onondaga Lake supporting testing of the N model for the two years is described in Table 9.7. The data base (Table 9.7) is from a weekly monitoring program conducted at the routine south basin site over the spring–fall interval. The characteristics of the detailed depth profiles of T-NH$_3$ and NO$_x$ have been described in Chapter 3, and by Brooks and Effler (1990). The distributions of TKN, PON, and DON specified for the lake are based on a more vertically limited program (Table 9.7). The depths of 1 and 16 m are assumed to be representative of the UML and the LML. Partitioning of the PON and DON species was based on analytical partitioning of dissolved and particulate TKN. Originally this was done by making paired measurements of TKN and filtered TKN. The corresponding estimate of particulate TKN, calculated as the residual of the two measurements, was subject to substantial uncertainty because of the relatively small difference of these two concentrations. The partitioning of PON and DON in the lake probably became more reliable (starting in August 1989) when particulate TKN was measured directly. The downward fluxes of PON were measured in 1989

TABLE 9.7. Features of Onondaga Lake monitoring program supporting testing of the N model.

N species/flux	Depths/depth interval (m)	Remarks/references
T-NH$_3$	2(i)*	Chapter 5, Brooks & Effler (1990)
NO$_x$	2(i)	Chapter 5, Brooks & Effler (1990)
TKN	1, 16	Chapter 5
PON	1, 16	= [Particulate TKN] − [T-NH$_3$][†]
DON	1, 16	= ([TKN] − [T-NH$_3$]) − [PON][†]
TN	1, 16	= [TKN] + [NO$_x$]
Downward flux, PON (1989)	10	Sediment trap collections, Chapter 8

2(i)* Vertical profiles of 2 m depth intervals
[†] since August 1989

698   S.W. Effler et al.

(unpublished data, UFI) through analysis
of weekly sediment trap (see Chapter 8)
collections.

The distributions of T-NH₃ and NOₓ in the
UML were distinctly different for the two
model testing years (Figure 9.23), offering a
good test to the simulation capabilities of the
model. Concentrations of T-NH₃ were much
higher in the spring and early summer, and
again in October 1989 compared to 1990
(Figure 9.23a). Concentrations of NOₓ were
substantially higher in late summer of 1989
(Figure 9.23b).

9.4.6 Model Performance

Model simulations of N species and deposition
in Onondaga Lake, obtained with the simpler
transport framework, are compared to observations for 1989 in Figure 9.24 to demonstrate
calibration. Model simulations track the
dynamics of PON well (Figure 9.24a). The
predictions of the components (p-PON and d-
PON), for which there are no direct measurements, are included. The dynamics of PON

largely reflect temporal patterns in phytoplankton biomass (compare to Figure 9.20a),
according to the established p-PON/CHL
stoichiometry (Figure 9.21). Further, the
predicted patterns of PON and CHL deposition
(Figure 9.24b) approximately match observations. Predictions of DON approximately
match the observations for the UML, but
overestimate the concentrations of the LML
somewhat in late summer (Figure 9.24c).

The distributions of T-NH₃ are particularly
well matched in both layers (Figure 9.24d). As
was the case for the TP model, the dimensions
of the vertical bars on the observations of T-
NH₃ (and NOₓ) reflect the range of concentrations observed in the two layers. The fit in the
LML layer is largely the result of the calibration
procedure. However, the simulations for the
UML, the principal management focus of the
N model, are much more independent of
calibration (e.g., "tuning") procedures. The
progressive depletion of T-NH₃ in the UML
until the fall mixing period is well simulated,
though the clear trough and secondary peak
manifested in the predictions for early summer
(associated with the clearing event) were not
as clearly resolved in the observations (Figure
9.24d). The seasonality of NOₓ is also well
tracked by the model simulations (Figure
9.24e); specifically the uniform concentrations
in both layers through early summer, followed
by the progressive increase in the UML and
decrease in the LML. As in the case of T-NH₃,
the simulations of NOₓ for the UML are more
independent of the calibration process than
those for the LML. The model simulated well
the dynamics of TKN in both layers (Figure
9.24f) and the relative uniformity in TN
concentrations in both layers (Figure 9.24g).

The more complex transport framework
version of the N model performed essentially
equally well to the simpler version. This is
illustrated for T-NH₃ in the UML (Figure
9.24d), the principal management focus of the
model. As described earlier in a similar comparison of the effect of transport frameworks
on the TP model predictions, the origins of the
modest differences in the spring–fall interval
are the different temperature profiles (observed
versus similations) on which vertical mixing

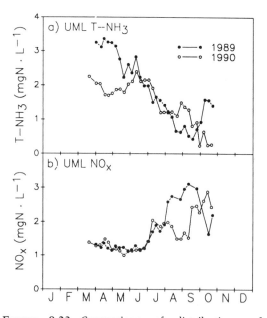

FIGURE 9.23. Comparison of distributions of
volume-weighted concentrations in the UML in
1989 and 1990: (a) T-NH₃, and (b) NOₓ (modified
from Canale et al. 1995a).

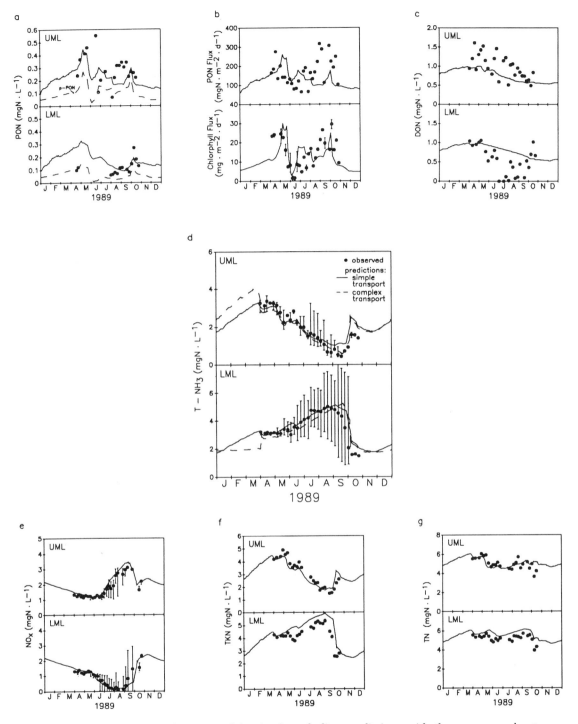

FIGURE 9.24. Calibration performance of the simple transport framework version of the N model for Onondaga Lake, 1989: (**a**) PON and p-PON concentrations, (**b**) deposition, PON and CHL, as a flux, (**c**) DON concentration, (**d**) T-NH₃ concentration, in-cluding predictions with the more complex transport framework version, (**e**) NOₓ concentration, (**f**) TKN concentration, and (**g**) TN concentration (modified from Canale et al. 1995a).

estimates are based. The larger differences during ice cover (Figure 9.24d), for which no data are available, are attributable to the much more limited vertical mixing predicted for that period by the hydrothermal model (Owens and Effler 1995) compared to the high levels assumed in the simpler framework. The differences during ice cover are not significant to subsequent predictions during the critical water quality period of spring to fall (e.g., Figure 9.24d). However, the predictions of the complex transport version appear to be more realistic for the ice-cover period (unpublished data), and thus are considered a model enhancement.

The simple transport framework version of the N model was additionally tested for the environmental and loading conditions of 1990 (Figure 9.25). The same G_{NET} distribution established for 1989 was invoked for 1990, consistent with model verification testing. The magnitude of G_{NET} was slightly reduced to accommodate the reduction in summer TP concentration from 1989 to 1990 (96 to 76 $\mu g \cdot L^{-1}$) according to a relationship developed by Auer et al. (1986). The basis and protocol for this adjustment is described later in the "model application" section. Significant differences in the distributions of CHL were observed in the lake in 1989 and 1990, particularly in early fall (Chapter 6). Thus net exchange of T-NH$_3$ with the phytoplankton is less accurately represented in 1990, because G_{NET} is based on 1989 conditions. The related shortcomings manifest themselves in the imperfect fit of CHL deposition (Figure 9.25b) and PON (Figure 9.25a), and the larger deviations in T-NH$_3$ for the UML from mid- to late summer of 1990 (Figure 9.25d) compared to 1989 (Figure 9.24d). Despite these modest shortcomings, the N model generally performed well in predicting the distributions of the various N species (Figure 9.25).

Most important, the predictions of T-NH$_3$ in the UML continue to perform well through early summer (Figure 9.25d). This is particularly important to the free ammonia issue for the lake, because this is the period of the greatest margins of violations of standards (Chapter 5, Effler et al. 1990a). Epilimnetic T-NH$_3$ is somewhat over-predicted in late summer, and more so during the fall mixing period, consistent with the underprediction of the phytoplankton sink of T-NH$_3$ for this period of 1990 associated with invoking the G_{NET} distribution of 1989. This shortcoming is not considered major, as managers will instead focus on the critical early summer period when model performance is still high. In Chapter 5 it was demonstrated that the major year-to-year differences in T-NH$_3$ concentrations observed during the fall in recent years (1988–1993) have been a result of interannual differences in the intensity of the fall phytoplankton bloom; e.g., greater T-NH$_3$ depletion in the UML in fall 1990 (Figure 9.23a) as a result of the more intense phytoplankton bloom (Chapter 6). Based on the generally good performance of the N model for 1990 conditions, particularly for T-NH$_3$ in the UML during the critical period, the N model can fairly be described as verified, for the prevailing nutrient rich (e.g., saturated) conditions of the lake. Further, it supports the use of a generalized G_{NET} distribution (e.g., from 1989) to accommodate the role of phytoplankton activity as a sink for T-NH$_3$ in the lake. The protocol for application of the N model to substantially reduced trophic state conditions for the lake is described subsequently.

9.4.7 Sensitivity Analyses

Sensitivity analyses are conducted here with the simpler transport framework version of the N model to establish the impact of reasonable levels of uncertainty in model coefficients and natural variability in model inputs on simulations of T-NH$_3$ for the UML of the lake. The influence of variations in tributary flow is reflected in the input of a reasonable range of flows. For this analysis documented daily distributions for the lake's tributaries (see Chapter 3) for a dry year (1987; second lowest annual runoff over a 20-yr period) and a wet year (1976; second highest annual runoff over a 20-yr period) were input into the N model. Initial conditions for this analysis are those calculated for the

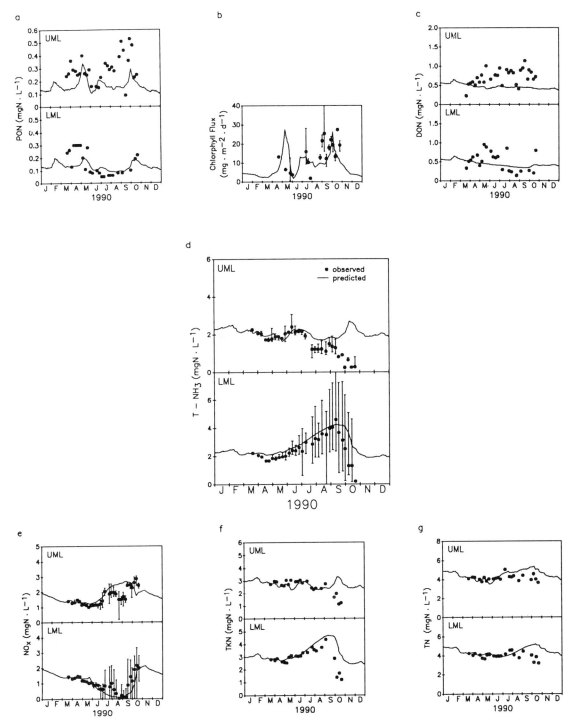

FIGURE 9.25. Verification performance of N model for Onondaga Lake, 1990: (**a**) PON concentration, (**b**) deposition CHL, as a flux, (**c**) DON concentration, (**d**) T-NH$_3$ concentration, (**e**) NO$_x$ concentration, (**f**) TKN concentration, and (**g**) TN concentration (modified from Canale et al. 1995a).

end of 1989. The base year for comparison was 1990. Other sensitivity analyses were conducted using 1989 as the base year. The results of the tributary flow sensitivity analysis (Figure 9.26a) indicate substantial year-to-year differences in T-NH₃ concentrations are to be expected for the upper waters of the lake, under prevailing METRO N loading conditions, as a result of natural variations in runoff.

Model simulations are moderately sensitive to variations in loading (±10%) from METRO (or errors in loading estimates from the facility of the same magnitude; Figure 9.26b). The distribution of T-NH₃ in the UML is rather insensitive to realistic levels of uncertainty in the sediment release rate (e.g., calibrated value vs. laboratory measurement; Figure 9.26c). A reasonable level of uncertainty in $k_{fn,T}$ (0.5X, 2X) only has a significant impact on the T-NH₃ concentration of the UML in late summer (Figure 9.26d), after the most critical period of high T-NH₃ concentrations (Effler et al. 1990a, Chapter 5). Nitrogen

model predictions of T-NH₃ in the UML were essentially insensitive to the uncertainties of the other coefficients. Similar to the case for TP (Figure 9.17f), the accommodation of the inputs from the Seneca River (Owens 1993) had only a minor effect on the performance of the T-NH₃ model in 1990 (Figure 9.26e).

9.5 Dissolved Oxygen Model

9.5.1 Introduction

An adequate supply of dissolved oxygen (DO) is necessary for protection and propagation of aquatic life. Regulators set standards for minimum DO concentrations in streams, rivers, and lakes in order to protect aquatic life and related uses of these systems. Mechanistic mass balance DO models have long been recognized as valuable tools for the analysis of river water quality (Streeter and Phelps 1925;

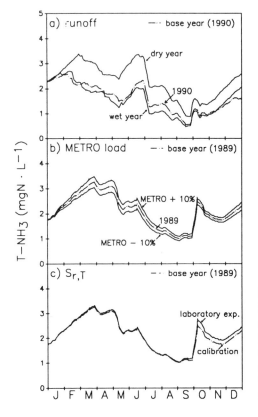
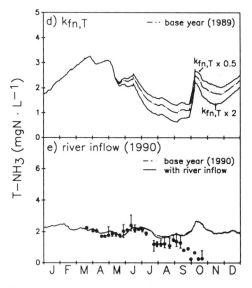

FIGURE 9.26. Sensitivity analyses with N model for Onondaga Lake for T-NH₃ concentrations in the UML: (a) tributary runoff, dry year (1987) and wet year (1976) daily flows, (b) ± 10% METRO T-NH₃ loading, (c) $S_{r,T}$, laboratory (92 mgN · m⁻² · d⁻¹) vs. calibration (70 mgN · m⁻² · d⁻¹), (d) 0.5 × $k_{fn,T}$, 2 × $k_{fn,T}$, and (e) without Seneca River inputs (1990) vs with inputs (modified from Canale et al. 1995a).

Biswas 1981; Gromiec et al. 1983; also see subsequent sections of this chapter); they are routinely utilized as a quantitative basis for issuing permits for the discharge of treated wastewater to streams and rivers (Krenkel and Novotny 1979; Thomann and Mueller 1987). Comparatively few attempts have been made to model DO in lakes (Bella 1970; DiToro and Connolly 1980; Lam et al. 1987; Snodgrass and Dalrymple 1985; Symons et al. 1967). This disparity in the frequency of DO modeling efforts between lotic and lentic systems reflects the much more frequent occurrence of acute DO problems in streams and rivers.

The extremely degraded conditions of the oxygen resources of Onondaga Lake were documented in Chapter 5. Dissolved oxygen is lost rapidly from the lake's hypolimnion (Effler et al. 1986b), and severe lakewide depletions occur during the fall mixing period (Effler et al. 1988; Gelda et al. 1993), associated with the oxidation of metabolic by-products of anaerobic metabolism that accumulate in the lake's hypolimnion (Chapter 5). Standards for DO are violated lakewide in most years during the fall mixing period. These conditions are manifestations of hyper-eutrophy (Effler et al. 1988, Chapter 5), and thus discharges of treated (METRO) and untreated (CSOs) domestic waste to the lake (Canale and Effler 1989; Devan and Effler

1984, Chapter 3). Presently most of the warm water fish that inhabit the lake exit to the Seneca River during the fall mixing period (Chapter 6). These conditions represent a severe degradation for a lake that once supported a cold-water fishery (Chapter 6).

A mechanistic mass balance DO model is developed and tested here that is intended to support evaluation of remediation strategies to improve the oxygen resources of the lake (also see Gelda and Auer 1995). The model is capable of year-round simulations and is tested for conditions documented in the lake in 1989 and 1990. Features of the capabilities of this lake DO model that contrast river DO models include accommodation of: (1) stratification/vertical mixing, (2) wind-induced atmospheric reaeration, and (3) sedimentary release and subsequent oxidation of reduced substances.

9.5.2 Conceptual Framework

A conceptual representation of the DO budget simulated in the DO model for Onondaga Lake is presented in Figure 9.27. Oxygen sources included in the model are phytoplankton photosynthesis, atmospheric reaeration, and external loading. External loading is an insignificant source on a lakewide basis, but it may provide localized refuge for fish during the fall mixing period (Chapter 6).

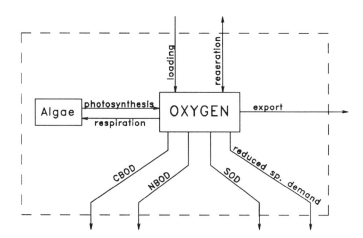

FIGURE 9.27. Conceptual DO model for Onondaga Lake (modified from Gelda and Auer 1995).

Photosynthesis is limited mostly to the UML because of limited light penetration (Chapter 7). Sinks of DO accommodated are export, phytoplankton respiration, exertion of carbonaceous biological oxygen demand (CBOD) and nitrogenous biological oxygen demand (NBOD; nitrification), sediment oxygen demand (SOD), and oxidation of reduced chemical species. SOD is assumed to be exerted only in the hypolimnion. Dissolved species are exchanged across the boundary between the two layers, driven by concentration gradients, and mediated by vertical mixing (Figure 9.15). As in the cases of the TP and N models, both the simplified (Figure 9.10a) and more complex (Figure 9.10c) transport regimes were accommodated in this DO modeling effort.

9.5.3 Sink and Source Processes

9.5.3.1 Surface Transfer/Reaeration

9.5.3.1.1 Background

The liquid side of the air–water interface controls exchange for gaseous substances of low solubility including oxygen (O'Connor 1983). The flux of oxygen at the interface is described by

$$J = K_{L,DO} \cdot ([DO]_s - [DO]) \quad (9.24)$$

in which J = DO flux ($g \cdot m^{-2} \cdot d^{-1}$), $K_{L,DO}$ = liquid film (surface) transfer coefficient for DO ($m \cdot d^{-1}$), $[DO]_s$ = saturation concentration of DO in water ($g \cdot m^{-3}$), and $[DO]$ = DO concentration of water ($g \cdot m^{-3}$). This process represents a source of DO when $[DO]_s >$ $[DO]$, and a sink when $[DO] > [DO]_s$. The depth-averaged liquid film transfer coefficient for DO is termed the reaeration coefficient, K_a (d^{-1}):

$$K_a = \frac{K_{L,DO}}{H} \quad (9.25)$$

in which H = mean depth (m). The magnitude of $K_{L,DO}$ (and thus K_a) is importantly influenced by internal turbulence, which acts to reduce the thickness of the diffusion layer. Surface turbulence in lakes is regulated primarily by wind (O'Connor 1983), in con-

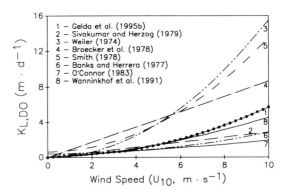

FIGURE 9.28. Comparison of published empirical relationships between the liquid film transfer coefficient ($K_{L,DO}$), and wind speed (U_{10}) (modified from Gelda et al. 1995b).

trast to the flow-induced turbulence that prevails in most streams and rivers.

A number of empirical and theoretical expressions have been developed to simulate reaeration in lakes (Bowie et al. 1985; O'Connor 1983; Daniil and Gulliver 1991). These functions generally relate $K_{L,DO}$ and wind speed and yield a wide range of values of $K_{L,DO}$ (e.g., $K_{L,DO} \approx 0.2-4.3 \, m \cdot d^{-1}$ for wind speed = $5 \, m \cdot s^{-1}$; Figure 9.28). There is a paucity of determinations of $K_{L,DO}$ that can be described as representative of whole lake conditions. Heretofore, only the relationship of Wanninkhof et al. (1991) was based on whole lake determinations of $K_{L,DO}$. The relationship of Wanninkhof et al. (1991) was based on the results of whole-lake tracer gas studies (Wanninkhof et al. 1985, 1987, 1991) obtained from five lakes ranging in surface area from 0.13 to 450 km².

A whole-lake approach has been used to develop a credible relationship between wind speed and $K_{L,DO}$ for Onondaga Lake (Gelda et al. 1995b), that takes advantage of the lakewide oxygen depletion and subsequent recovery during the fall mixing period (Chapter 5). This analysis is summarized here.

9.5.3.1.2 Reaeration in Onondaga Lake and Wind Speed

The reaeration coefficient can be determined from a mass balance on DO during the post-turnover recovery period. The mass balance, assuming that the gains of DO through inflow and losses to outflow are negligible, is given by

$$V\frac{d[DO]}{dt} = (V \cdot K_a \cdot ([DO]_s - [DO])) - V \cdot S$$

(9.26)

where S represents the net sum of all DO sources and sinks:

$$S = P - R - k_d \cdot CBOD - k_n \cdot NBOD - SOD/H$$

(9.27)

in which CBOD = carbonaceous biochemical oxygen demand $(gO_2 \cdot m^{-3})$, k_d = rate constant for CBOD oxidation (d^{-1}), NBOD = nitrogenous biochemical oxygen demand $(gO_2 \cdot m^{-3})$, k_n = rate constant for NBOD oxidation (nitrification; d^{-1}), P = rate of phytoplankton gross photosynthesis $(gO_2 \cdot m^{-3} \cdot d^{-1})$, R = rate of phytoplankton respiration $(gO_2 \cdot d^{-3} \cdot m^{-1})$, SOD = rate of sediment oxygen demand $(gO_2 \cdot m^{-2} \cdot d^{-1})$, and V = lake volume (m^3). The value of S is assumed to be constant during the recovery period; this has been supported by observations (Gelda et al. 1995b).

The relationship between wind and $K_{L,DO}$ is generally described empirically, often taking the form (Bowie et al. 1985):

$$K_{L,DO} = \alpha \cdot U_{10}^{\beta}$$

(9.28)

in which α, β = empirical coefficients, and U_{10} = wind speed $(m \cdot s^{-1})$ at the standard height (10 m) above the water. The reaeration expression therefore is

$$K_a = \frac{\alpha \cdot U_{10}^{\beta}}{H}$$

(9.29)

Substituting equation (9.29) into equation (9.26) yields:

$$\frac{d[DO]}{dt} = \frac{\alpha \cdot U_{10}^{\beta}}{H} \cdot ([DO]_s - [DO]) + S$$

(9.30)

Determination of the value of S has been described by Gelda et al. (1995b). The value of β has commonly been set between 0.5 and 2.0, depending on the range of wind speeds considered. Here, the coefficient has been set,

$\beta = 1.0$ for $U_{10} \leq 3.5 \, m \cdot s^{-1}$ and $\beta = 2.0$ for $U_{10} > 3.5 \, m \cdot s^{-1}$. Equation (9.30) was then solved numerically to obtain values for α, which minimize the root mean square error (RMSE) for [DO] observations over the post-turnover period.

The optimization analysis was first applied to the 1989 data base; Gelda (1995b) estimated the value of S to be $0.289 \, g \cdot m^{-3} \cdot d^{-1}$ during the fall mixing period of 1989. Daily average wind speeds, measured at the National Weather Service at Hancock Airport, Syracuse, and corrected to a reference elevation of 10 m, were utilized as model input. Gelda et al. (1995b) found that the 24 hour averaging time for wind was appropriate for the overall DO model goals, and that the wind fetch (i.e., wind direction) could be ignored. The value of α was determined to be 0.2 for $U_{10} \leq 3.5 \, m \cdot s^{-1}$ and 0.057 for $U_{10} > 3.5 \, m \cdot s^{-1}$. The results of the calibration are presented in Figure 9.29a; model performance was excellent (RMSE = 0.77). The empirical relationship determined for Onondaga Lake,

$$K_{L,DO} = 0.2 \, U_{10}^{1.0}, \quad U_{10} \leq 3.5 \, m \cdot s^{-1} \quad (9.31a)$$

$$K_{L,DO} = 0.057 \, U_{10}^{2.0}, \quad U_{10} > 3.5 \, m \cdot s^{-1}$$

(9.31b)

was verified using the lake data base for the recovery period of 1990 ($S = 0.164 \, g \cdot m^{-3} \cdot d^{-1}$ (Gelda et al. 1995b)). The fit of the relationship to the 1990 field data was excellent (Figure 9.29b; RMSE = 0.52). The relationship developed for Onondaga Lake by this whole-lake approach (equation (9.31)) has been added to Figure 9.28 for comparison to the other relationships available. The most similar relationship is that of Wanninkhof et al. (1991), also developed from a whole-lake approach. The Wanninkhof et al. (1991) relationship performed essentially equally well in matching the observed recovery in DO in both years (Figure 9.29). All the other relationships of Figure 9.28 performed poorly in matching the Onondaga Lake data.

The credibility of the $K_{L,DO} - U_{10}$ relationship (equation (9.31)) presented here, and developed in detail by Gelda et al. (1995b), is established by: (1) its performance in matching

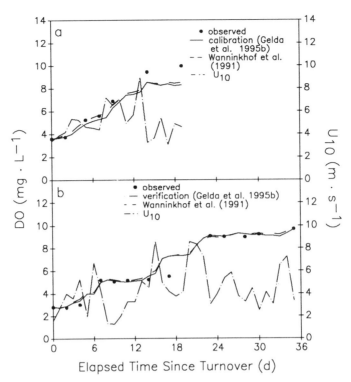

FIGURE 9.29. Performance of film transfer co-efficient-wind relationships of Gelda et al. (1995b) and Wanninkhof et al. (1990) in matching observed fall oxygen recovery in Onondaga Lake: (a) 1989, calibration for Gelda et al. (1995b), and (b) 1990, verification for Gelda et al. (1995b). Time distributions of daily average values of U_{10} also included.

the observed recovery in DO in the fall mixing period of two different years, and (2) its near-equivalence with the relationship of Wanninkhof et al. (1991), which is the only other expression based on whole-lake measurements. The analysis implies that whole-lake experiments should be conducted to determine $K_{L,DO}$, or the relations of Wanninkhof et al. (1991) or Gelda et al. (1995b) should be considered, in related studies of other inland lakes.

The dependence of $K_{L,DO}$ on temperature is described by an Arrhenius expression

$$K_{L,DO,T} = K_{L,DO,20} \cdot \Theta^{(T-20)} \quad (9.32)$$

in which $\Theta = 1.024$ (Bowie et al. 1985; Thomann and Mueller 1987).

9.5.3.2 Loads and Export

External input is not presently an important source of oxygen to Onondaga Lake; however, it is accommodated for the sake of completeness and to support evaluation of management scenarios that include highly oxygenated inputs. The DO model quantifies oxygen loading for METRO and four major tributaries (Onondaga Creek, Ninemile Creek, Ley Creek, and Harbor Brook) by using annual average concentrations of DO of $9.4\,mg \cdot L^{-1}$ for the METRO effluent and $9.8\,mg \cdot L^{-1}$ for the tributaries (Onondaga County 1990). The loading of oxygen is calculated according to:

$$W_{DO} = Q_{METRO} \cdot [DO]_{METRO}$$
$$+ \; Q_{Trib} \cdot [DO]_{Trib} \quad (9.33)$$

in which W_{DO} = loading of DO $(gO_2 \cdot d^{-1})$, Q_{METRO} = discharge from METRO $(m^3 \cdot d^{-1})$, $[DO]_{METRO}$ = DO concentration in METRO discharge $(mg \cdot L^{-1})$, Q_{Trib} = summed flow into lake from four tributaries $(m^3 \cdot d^{-1})$, and $[DO]_{Trib}$ = DO concentration in four tributaries $(mg \cdot L^{-1})$. Daily flow data from the USGS and

METRO, already incorporated in loading estimtes for TP and N species (Chapter 3), were also utilized to support these calculations.

As in the TP and N models, export is calculated as the product of the DO concentration of the UML ($[DO]_1$) and the summed inflows of the tributaries and METRO

$$export = Q \cdot [DO]_1 \qquad (9.34)$$

9.5.3.3 Photosynthesis and Respiration/Phytoplankton Submodel

The phytoplankton submodel developed as part of the N model, described previously in this chapter, is also utilized in the DO model. The output of the N model provides time series of G_{NET}, CHL, and $[T-NH_3]$ for the two layers. This information is utilized in the DO model to estimate net photosynthetic production of DO, and to estimate DO consumed in the processes of respiration and nitrification.

Net photosynthetic oxygen production per unit CHL is calculated from G_{NET} and phytoplankton/photosynthesis stoichiometry according to:

$$a_{OP} = G_{NET} \cdot a_{CP} \cdot a_{OC} \qquad (9.35)$$

in which a_{OP} = oxygen produced per unit weight CHL produced, a_{CP} = phytoplankton carbon to CHL ratio, and a_{OC} = stoichiometric oxygen equivalent of phytoplankton carbon. The value of a_{CP}, calculated as the ratio of the downward fluxes of POC and CHL obtained from analysis of sediment trap collections (Chapter 8), is 42.7. Stumm and Morgan (1981) report a value of 2.7 for a_{OC}. The daily average net photosynthetic production of DO (P_{DO}; $g \cdot m^{-3} \cdot d^{-1}$) is calculated according to:

$$P_{DO} = \frac{a_{op} \cdot [CHL]}{1000} \qquad (9.36)$$

in which [CHL] = CHL concentration ($\mu g \cdot L^{-1}$; $mg \cdot m^{-3}$). Including the above stoichiometric information, the expression for the epilimnion ($P_{DO,1}$) becomes

$$P_{DO,1} = 0.115\, G_{NET,1} \cdot [CHL]_1 \qquad (9.37)$$

The daily average net photosynthetic production of DO in the hypolimnion ($P_{DO,2}$) is calculated according to

$$P_{DO,2} = 0.115\, G_{NET,2} \cdot [CHL]_2 \cdot f_{OL} \qquad (9.38)$$

in which f_{OL} = rate multiplier (≤ 1.0) for respiration processes. The multiplier f_{OL} accommodates the widely observed reduction in biochemical oxygen consuming processes as DO concentrations decrease (Bowie et al. 1985). A Monod-type expression has been adopted for f_{OL} in the DO model (e.g., Bowie et al. 1985)

$$f_{OL} = \frac{[DO]}{[DO] + K_{s,DO}} \qquad (9.39)$$

in which $K_{s,DO}$ = half-saturation constant ($mg \cdot L^{-1}$). The model uses a value of $K_{s,DO}$ = 3.5 $mg \cdot L^{-1}$, that was determined from model calibration.

9.5.3.4 Carbonaceous Biochemical Oxygen Demand

The decomposition of organic matter in the watercolumn through aerobic, heterotrophic respiration is quantified by the concentration of carbonaceous biochemical oxygen demand. This is presently not an important sink for DO in Onondaga Lake; however, it is accommodated for the sake of completeness. The CBOD submodel incorporates loadings from METRO and the tributaries and sinks that include export to the river and deoxygenation associated with the aerobic decomposition. The mass balance equations for CBOD are presented below.

$$V_1 \frac{d[CBOD]_1}{dt} = W_{CBOD} - Q \cdot [CBOD]_1$$
$$- k_{CBOD,T} \cdot [CBOD]_1 \cdot V_1$$
$$+ v_t \cdot A_i \cdot ([CBOD]_2 - [CBOD]_1) \qquad (9.40a)$$

$$V_2 \frac{d[CBOD]_2}{dt} = -k_{CBOD,T} \cdot f_{OL} \cdot [CBOD]_2 \cdot V_2$$
$$+ v_t \cdot A_i \cdot ([CBOD]_1 - [CBOD]_2) \qquad (9.40b)$$

in which [CBOD] = concentration of ultimate carbonaceous biochemical oxygen demand ($mg \cdot L^{-1}$), W_{CBOD} = loading of ultimate CBOD ($g \cdot d^{-1}$), and $k_{CBOD,T}$ = deoxygenation coefficient for CBOD at temperature T (d^{-1}). The dependence of k_{CBOD} on T is described by an Arrhenius expression

$$k_{CBOD,T} = k_{CBOD,20} \cdot \Theta^{(T-20)} \quad (9.41)$$

in which the temperature correction coefficient, Θ, is equal to 1.047 (Bowie et al. 1985). The ultimate CBOD was assumed to be 1.33 × the 5-day CBOD value. A value of $k_{CBOD,20} = 0.28\,d^{-1}$ was determined through laboratory measurements, based on application of the Thomas Slope Method (Metcalf & Eddy 1979). The nitrogenous component of biochemical oxygen demand (NBOD) in the lake was quantified previously in this chapter, as the nitrification process ($k_{fn,T}$) of the N model.

The loading of ultimate CBOD from METRO and four major tributaries (Onondaga Creek, Ninemile Creek, Ley Creek, and Harbor Brook) was approximated with annual average concentrations of 5-day BOD (Onondaga County 1990) and the assumed relationship between ultimate and 5-day values. Thus, W_{CBOD} is calculated according to

$$W_{CBOD} = Q_{METRO} \cdot [CBOD_{METRO}]$$
$$+ Q_{Trib} \cdot [CBOD_{Trib}] \quad (9.42)$$

in which $[CBOD]_{METRO}$ = average concentration of ultimate CBOD in METRO effluent $(mg \cdot L^{-1})$, and $[CBOD]_{Trib}$ = average concentration of ultimte CBOD in the four tributaries $(mg \cdot L^{-1})$. The model uses annual average ultimate CBOD concentrations for METRO as $16.0\,mg \cdot L^{-1}$, and $2.5\,mg \cdot L^{-1}$ for the four tributaries (Onondaga County 1990).

9.5.3.5 Sediment Oxygen Demand

Organic matter delivered to the lake bottom by sedimentation is stabilized through aerobic and anaerobic processes, which directly or indirectly exert a demand on the oxygen resources of a lake (see Hatcher (1986) for a review). The aggregate oxygen demand exerted at the sediment–water interface, which is a manifestation of a number of biological and chemical processes (Chiaro and Burke 1980; DiToro et al. 1990; Walker and Snodgrass 1986), is described as the sediment oxygen demand (SOD, $g \cdot m^{-2} \cdot d^{-1}$). The oxygen consumption by SOD is given by:

SOD oxygen consumption $(gO_2 \cdot d^{-1})$

$$= f_{OL} \cdot SOD_T \cdot A_{SOD} \quad (9.43)$$

in which SOD_T = SOD rate at T $(gO_2 \cdot m^{-2} \cdot d^{-1})$, and A_{SOD} = sediment area across which SOD is exerted (m^2). SOD is included as a sink term only for the LML. It is assumed that the SOD exerted in the UML is negligible. This is consistent with the much lower organic content reported for the sediments of this layer (Chapter 8). The value of A_{SOD} corresponds to that portion of the sediments in the LML that is overlaid with oxic water; that is, the value varies with time in Onondaga Lake (see Chapter 5). An expression describing the dynamics of A_{SOD} in the lake is described subsequently. Adjustments of SOD for temperature are made according to the following Arrhenius expression

$$SOD_T = SOD_8 \cdot \Theta^{(T-8)} \quad (9.44)$$

in which SOD_8 = measured SOD rate at 8°C $(g \cdot m^{-2} \cdot d^{-1})$, and $\Theta = 1.065$. This value of Θ falls about in the center of the range reported by Bowie et al. (1985).

Reliable techniques for the direct field or laboratory measurement of SOD have been developed (e.g., Bowman and Delfino 1980; Gardiner et al. 1984; Hatcher 1986). Values for SOD in Onondaga Lake were obtained from laboratory measurements on 15 intact sediment cores collected from three deepwater stations located along the main axis of the lake in 1990 and 1991 (Penn et al. 1993). Two of the stations correspond approximately to the routinely monitored sites in the centers of the north and south basins. The sampling and laboratory protocol was similar to that developed by Gardiner et al. (1984). Cores were incubated at 8°C, a representative LML temperature. Anoxic water overlying the cores at the time of collection was replaced by filtered, well-oxygenated, surface lake water prior to commencement of the SOD experiments. Stirring was provided to eliminate oxygen gradients in the overlying water, but was controlled to avoid disturbing the sediment–water interface. Following acclimation, the depletion of DO from the overlying water was continuously monitored. The depletion rate was used to calculate the SOD (Penn et al. 1993). The results of one of the experiments are presented in Figure 9.30; the slope is converted to SOD utilizing the volume of the

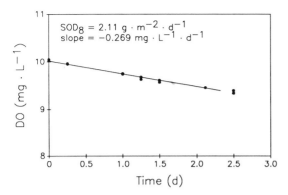

FIGURE 9.30. Example experimental results for SOD determination in Onondaga lake.

overlying water and the area of the core. The average measured SOD from these experiments ($1.68 \pm 0.56 \, g \cdot m^{-2} \cdot d^{-1}$) was adopted in the DO model. There were no significant differences in the rate for the three sampling sites (Penn et al. 1993). The measured values fall in the range of values reported for other systems of similar trophic state (e.g., Bowie et al. 1985; DiToro et al. 1990).

9.5.3.6 Reduced Chemical Species Submodel

High concentrations of oxygen-demanding reduced chemical species, produced as by-products of anaerobic metabolism, develop in the hypolimnion of Onondaga Lake (Addess and Effler 1995; Effler et al. 1988; Effler et al. 1990a, Chapter 5), as a result of the lake's advanced state of cultural eutrophication (Effler et al. 1988). Oxidation of these accumulations of reduced species has been identified as the principal cause of the severe lakewide depletion of DO observed annually during the fall mixing period (Effler et al. 1988). The accumulations are quantitatively coupled to the SOD process (Gelda et al. 1995a, Chapter 5). The reduced species (primarily CH_4, H_2S, NH_4^+, and Fe^{2+}) are produced in the anaerobic sediments. These materials exert their oxygen demand in the form of SOD until anoxia is established in the overlying water; thereafter the substances are released to the overlying water and progressively accumulate through the remainder of the stratification period.

A simple reduced species submodel is included in the Onondaga Lake DO model to accommodate the oxygen demand generated from the anaerobic processes, based on the measured SOD. The NH_4^+ component of SOD is handled separately in the DO model through the process of nitrification. The upward flux of the other reduced species (CH_4, H_2S, and Fe^{2+}) from the sediments is simulated according to the expression

upward flux of reduced species

$$= f \cdot SOD_T \cdot A_{RS} \cdot f_{OL} \qquad (9.45)$$

in which f = fraction of SOD_T associated with the upward fluxes of CH_4, H_2S, and Fe^{2+}, and A_{RS} = sediment area within hypolimnion across which reduced species are released to the watercolumn (m^2). A value of 0.72 has been adopted for f in the DO model. This is approximately the value obtained in a partitioning analysis of SOD that was based on the observed rates of hypolimnetic accumulation of the contributing species (Gelda et al. 1995a).

The oxygen demand "equivalent" associated with the reduced species (exclusive of NH_4^+) is computed according to the oxidation pathways of the individual species

$$[RS_{obs}] = a_1 \cdot [CH_4] + a_2 \cdot [H_2S] + a_3 \cdot [Fe^{2+}] \qquad (9.46)$$

in which $[RS_{obs}]$ = computed concentration of oxygen "equivalent" reduced species ($mgO_2^* \cdot L^{-1}$), $[CH_4]$ = observed CH_4 concentration ($mg \cdot L^{-1}$), $[H_2S]$ = observed H_2S concentration ($mgS \cdot L^{-1}$), $[Fe^{2+}]$ = observed Fe^{2+} concentration ($mg \cdot L^{-1}$), and a_1, a_2, and a_3 = stoichiometric coefficients to convert concentrations to oxygen demand equivalents. The values of the stoichiometic coefficients are derived from the assumed pathways of oxidation reviewed in Chapter 5, and presented below

$$CH_4 + 2O_2 \rightarrow CO_2 + 2H_2O$$
(DiToro et al. 1990;
Sweerts et al. 1991) $\qquad (9.47)$

$$HS^- + 2O_2 \rightarrow SO_4^{2-} + H^+$$
(Sweerts et al. 1991) $\qquad (9.48)$

$$4Fe^{2+} + O_2 + 6H_2O \rightarrow 4FeOOH + 8H^+ \text{ (Sweerts et al. 1991)} \qquad (9.49)$$

then, from equation (9.47), $a_1 = 4.0$; from equation (9.48), $a_2 = 2.0$; and from equation (9.49), $a_3 = 0.143$. The time distributions of [RS_{obs}] have been computed from the distributions of the individual reduced species concentrations (Chapter 5). Further, the partitioning of SOD from hypolimnetic accumulation rates of the individual species has been estimated (Gelda et al. 1995a, Chapter 5), according to these coefficient values. The contribution of Fe^{2+} sedimentary releases to SOD is insignificant (Gelda et al. 1995a; Chapter 5).

The value of A_{RS} corresponds to the sediment surface overlain with anoxic water. Thus,

$$A_i = A_{SOD} + A_{RS} \qquad (9.50)$$

The time course of this partitioning of A_i is specified in the DO model according to conditions observed in 1989 and 1990. The observed time course of the depth of anoxia (D_A) is described by the following polynomial

$$D_A = c_0 + c_1 t + c_2 t^2 \qquad (9.51)$$

The values of c_0, c_1, and c_2 were 110.8, -0.9298 and 0.002113 for 1989; for 1990 the values were 127.4, -1.0405 and 0.002270; t = julian day. Utilizing the available hypsographic data for the lake (Owens 1988), the time course of A_{RS} (and thereby A_{SOD}) can then be calculated by

$$A_{RS} = b_0 + b_1 D_A + b_2 D_A^2 + b_3 D_A^3 \quad (9.52)$$

The values of b_0, b_1, b_2, and b_3 are 1.2×10^7, -892765.2, 65538.9, and 2631.50.

Inputs of RS into both layers, associated with the dissolution of a small fraction of rising methane-enriched gas bubbles (ebullition; Addess and Effler 1995, Chapter 5), have been included for completeness, though this process does not have a significant effect on the lake's oxygen budget.

The mass balance equations for reduced species can be written as:

$$V_1 \frac{d[RS]_1}{dt} = W_{1,eb} - Q \cdot [RS]_1$$

$$- k_{RS} \cdot [RS]_1 \cdot V_1$$

$$+ v_t \cdot A_i \cdot ([RS]_2 - [RS]_1) \qquad (9.53a)$$

$$V_2 \frac{d[RS]_2}{dt} = W_{2,eb} + f \cdot SOD_T \cdot A_{RS}$$

$$- k_{RS} \cdot f_{OL} \cdot [RS]_2 \cdot V_2$$

$$+ v_t \cdot A_i \cdot ([RS]_1 - [RS]_2) \qquad (9.53b)$$

in which [RS] = simulated reduced species concentration (in oxygen demand equivalents; $mgO_2^* \cdot L^{-1}$), k_{RS} = oxidation rate of reduced species (d^{-1}), $W_{1,eb}$ = redissolution rate of CH_4 gas in UML ($3.76 \times 10^5 gO_2^* \cdot d^{-1}$), and $W_{2,eb}$ = redissolution rate of CH_4 gas in LML ($2.38 \times 10^5 gO_2^* \cdot d^{-1}$). The coefficient k_{RS} is a "lumped" first order oxidation rate for all the reduced species (exclusive of NH_4^+); the value determined from calibration is $0.25 d^{-1}$.

9.5.3.7 Overall DO Model Mass Balance

The complete mass balance equations for DO for the two layers of the DO model are:

$$V_1 \frac{d[DO]_1}{dt} = W_{DO} - Q \cdot [DO]_1$$

$$+ K_{L,DO,T} \cdot A_s \cdot ([DO]_s - [DO]_1)$$

$$+ P_{DO,1} \cdot V_1 - k_{CBOD,T} \cdot [CBOD]_1 \cdot V_1$$

$$+ a_{ON} \cdot k_{fn,T} \cdot (A_s - A_i) \cdot [T\text{-}NH_3]_1$$

$$- k_{RS} \cdot [RS]_1 \cdot V_1 + v_t \cdot A_i \cdot$$

$$([DO]_2 - [DO]_1) \qquad (9.54a)$$

$$V_2 \frac{d[DO]_2}{dt} = P_{DO,2} \cdot V_2 \cdot f_{OL}$$

$$- k_{CBOD,T} \cdot f_{OL}[CBOD]_2 \cdot V_2$$

$$- a_{ON} \cdot k_{fn,T} \cdot A_i \cdot [T\text{-}NH_3]_2$$

$$- f_{OL} \cdot SOD_T \cdot A_{SOD} - k_{RS} \cdot f_{OL} \cdot [RS]_2 \cdot V_2$$

$$+ v_t \cdot A_i \cdot ([DO]_1 - [DO]_2) \qquad (9.54b)$$

in which a_{ON} = stoichiometric oxygen equivalent of $T\text{-}NH_3$ for the nitrification process (= 4.57 gO_2/gNH_3-N). Values of $P_{DO,1}$ and $P_{DO,2}$ are based on output from the phytoplankton submodel of the N model.

9.5.4 DO Model Performance

9.5.4.1 Submodels

The predicted average CBOD (5-day) value for the UML of the lake for both 1989 and 1990 was about $0.5 \, mg \cdot L^{-1}$; the predicted value for the LML was lower. These concentrations fall below the commonly acknowledged ($2.0\,\text{mg} \cdot \text{L}^{-1}$ (APHA 1992)) detection limit of the BOD test. The most appropriate laboratory measurement for this model component is the BOD test conducted on filtered (to eliminate contributions from phytoplankton respiration), nitrification-inhibited (to eliminate contributions from NBOD), lake water. Such lake measurements are not available for 1989 and 1990. However, the average of 8 observations made in the summer of 1991, $0.85\,\text{mg} \cdot \text{L}^{-1}$, is generally supportive of the very low predicted concentrations for this insignificant oxygen sink.

Simulations of the volume-weighted concentrations of reduced species for the LML are compared to observations in 1989 and 1990 in Figure 9.31a and b. The increasing wide range of concentrations in reduced substances that develop vertically in the hypolimnion (Chapter 5) are clearly depicted by the increasing dimensions of the vertical bars in the plots. The simulations of Figure 9.31, based on an independent estimate of f, generally track well

the temporal distributions of [RS] in both years; specifically the progressive accumulation during summer stratification, and the subsequent decline during the fall mixing period with the approach to the onset of complete fall turnover. The overprediction of RS in both years is not considered to be major in light of the magnitude of the uncertainty that accompanies the measurement of SOD (Penn et al. 1993). The predicted near absence of RS in the UML (not shown) is consistent with observations (Chapter 5). Only very low concentrations ($\leq 1.5\,\text{mgO}_2^* \cdot \text{L}^{-1}$) of RS were predicted for the UML at the onset of complete fall turnover (when mixing-based upward transport of RS was maximal).

9.5.4.2 DO

The model was validated for two years, 1989 and 1990, for the simpler transport framework, using the G_{NET} distribution determined for 1989 for both years. The fit of predictions to observations for the UML and LML for 1989 and 1990 appears in Figures 9.32 and 9.33, respectively. The performance of the more complex transport framework in 1989 is included for comparison in Figure 9.32. The DO model performs well in simulating the primary features of the distribution of DO in Onondaga Lake. In particular, the model simulates well the rapid depletion of DO from the hypolimnion following the onset of stratification, and the lakewide depletion of DO during the fall mixing period. The performance of the simple and more complex transport frameworks was essentially equivalent over the period of measurements in 1989 (Figure 9.32). The simulations with the more complex framework version of the model are more consistent with the limited number of observations for the period of ice cover (Figure 9.32). The depletion and recovery during the fall of 1989 was simulated better than the 1990 conditions, in particular, the predicted minimum remained significantly above the observed minimum in 1990 (Figure 9.33a). This shortcoming in 1990 is coupled to the overprediction of DO in the UML in September. Hypolimnetic depletion of DO was

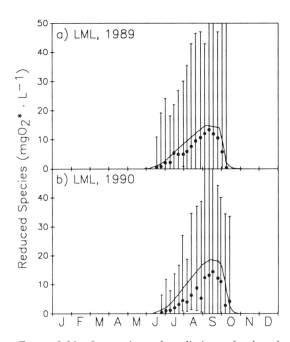

FIGURE 9.31. Comparison of predictions of reduced species (RS; in dissolved oxygen equivalents) to observed in the LML of Onondaga Lake: (**a**) 1989, and (**b**) 1990 (modified from Gelda and Auer 1995).

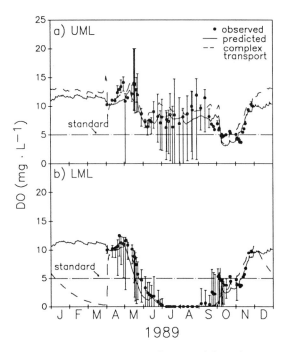

FIGURE 9.32. Comparison of DO model predictions of dissolved oxygen concentrations to observations in Onondaga Lake in 1989: (**a**) UML, and (**b**) LML. Simulations from both simple (modified from Gelda and Auer 1995) and complex transport framework versions presented.

also better tracked in 1989 (Figure 9.32b). Seasonal dynamics within the UML from late spring through summer were simulated substantially better in 1989 (compare Figures 9.32a and 9.33a).

An empirical relationship for the vertical limits of anoxia (from the LML volume-weighted DO concentration statistic; e.g., Figure 9.4) was incorporated in the DO model framework to support simulations of the dynamics of the upper boundary of anoxia in the lake. The predictions (simple framework) matched the observed distributions of the anoxic boundary well for both years (Figure 9.34a and b), though midsummer extremes in the vertical extent of anoxia were somewhat underpredicted in both years.

9.5.5 Sensitivity Analyses

Analyses were conducted with the validated simple transport framework version of the

DO model to establish the sensitivity of model predictions to uncertainty and variations in inputs. The analyses reviewed here focus on DO predictions in the UML during the critical water quality period of fall (e.g., lakewide violations of DO standards). The base year is 1989. The limits evaluated for k_{fn} ($\times 0.5$, $\times 2$) and k_{RS} ($\pm 50\%$) were quite broad, yet the oxygen resources during the fall mixing period are only modestly sensitive to this level of uncertainty (Figure 9.35a and b). The lower reaction rates resulted in reduced oxygen depletion; higher rates caused greater depletion. The limits of the sensitivity analysis for SOD represent ± 1 standard deviation of the experimental observations (Penn et al. 1993). The fall oxygen "sag" is sensitive to this rather broad range of SOD values (Figure 9.35c). The lower SOD results in a reduced "sag" during fall mixing, the predicted minimum for this case is about $5\,mg \cdot L^{-1}$. The higher SOD bound causes the fall minimum to decrease to less than $3.0\,mg \cdot L^{-1}$. These same sensitivity

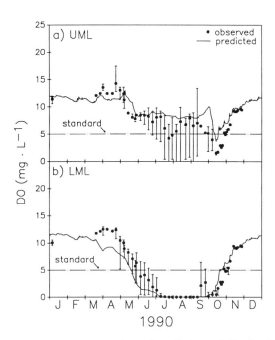

FIGURE 9.33. Comparison of predictions of dissolved oxygen concentrations (from simple transport model version) to observations in Onondaga Lake in 1990: (**a**) UML, and (**b**) LML (modified from Gelda and Auer 1995).

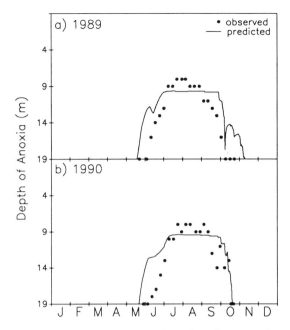

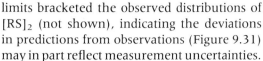

FIGURE 9.34. Comparison of predicted versus observed depths (upper bound) of anoxia in Onondaga Lake: (**a**) 1989, and (**b**) 1990.

limits bracketed the observed distributions of $[RS]_2$ (not shown), indicating the deviations in predictions from observations (Figure 9.31) may in part reflect measurement uncertainties.

The influence of wind speed on the rate of recovery from the fall DO "sag" is demonstrated in Figure 9.35d. For this analysis the wind speed has only been adjusted for the period following the onset of complete fall turnover, to isolate the interplay of reaeration with this recovery period. It is inappropriate to vary wind speed in well-stratified portions of the fall mixing period because this would affect vertical mixing in the DO model, which is specified (not predicted) from the observed dynamics in the stratification regime. Wind speed is clearly important to the recovery from the fall oxygen depletion. Recovery is much more rapid at an average wind speed of $4.5\,\mathrm{m \cdot s^{-1}}$ than it is for $0.45\,\mathrm{m \cdot s^{-1}}$ (Figure 9.35d). The time course of the approach to fall turnover (and the meteorological conditions that drive it) is almost certainly important to the character of the observed sag. For example, a particularly severe case would be the rapid entrainment of the lower (reduced species-enriched) waters followed by calm conditions.

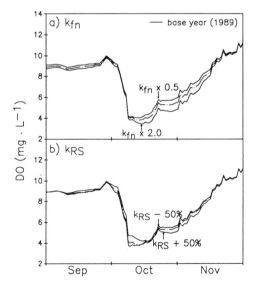

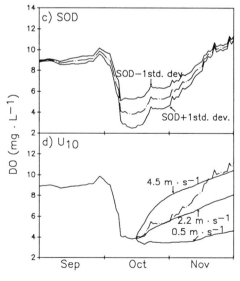

FIGURE 9.35. Sensitivity analyses with Onondaga Lake DO model (simple transport framework version) for UML for September to November period: (**a**) $0.5 \times k_{fn,T}$, $2.0 \times k_{fn,T}$, (**b**) $K_{RS} \pm 50\%$, (**c**) SOD ± 1 standard deviation, and (**d**) U_{10}, 4.45, 2.23, and $0.45\,\mathrm{m \cdot s^{-1}}$ (modified from Gelda and Auer 1995).

9.6 Fecal Coliform Bacteria Model

9.6.1 Introduction

Contamination of lakes and rivers by fecal material increases the risk that people using those waters for contact recreation will contract waterborne communicable diseases. Water quality managers seek to regulate fecal coliform loads so that water quality standards are not violated. An empirical approach (e.g., comparison of measured loads and observed in-lake concentrations) is of little value because the receiving water response varies dramatically with the character of the discharge event and attendant environmental conditions. The range in storm intensities and the spectrum of environmental conditions that may be encountered preclude such an approach. Instead, credible, mechanistic models are appropriately applied, simulating the time–course of disappearance of fecal coliform bacteria following a contamination incident, and establishing cause-effect relationships between sources of bacteria and watercolumn concentrations.

Although this topic has been treated in documents supporting the field of water quality modeling (cf. Bowie et al. 1985; Thomann and Mueller 1987), detailed descriptions of the development and application of such models for bacterial pollution in lakes is rare (cf. Freedman et al. 1980; Palmer and Dewey 1984). Further, the transient nature of pollution events, the sensitivity of the fecal coliform group to environmental conditions, and inherent problems in obtaining accurate and precise quantification of organism abundance, make this a unique and particularly challenging problem.

The extremely high loads of fecal coliform bacteria that are received by Onondaga Lake during runoff events from the tributaries along its southern shore, particularly Onondaga Creek, were described and quantified in Chapter 3 (also see Canale et al. 1993a). The contraventions of public health standards that occur in the lake, especially in the south basin, following the loading events, were docu-

mented in Chapter 6. Transport processes that mediate the nonuniform distributions of the fecal coliform bacteria were described and quantified within the framework of a hydrodynamic transport model in Chapter 4. Kinetic formulations and the development of coefficient values that quantify the losses of fecal coliform from the water column of the lake were presented by Auer and Niehaus (1993), and reviewed in Chapter 6. Here we review the development and testing of a mechanistic fecal coliform bacteria model that synthesizes this information into a predictive framework, which is intended to support related management efforts to remediate the continuing violation of public health standards (also see Canale et al. 1993a). Key model inputs and components, in the form of submodels, are reviewed, and sensitivity analyses are conducted to identify key environmental conditions regulating the response of the lake to loading events.

9.6.2 Field Monitoring

The data base supporting model development was collected in 1987; sampling locations are shown in Figure 9.36 (Chapters 3 and 6). Intensive sampling was conducted for two summer storm events (Storm 1: 22–29 June and Storm 2: 14–21 July). Dry weather sampling was conducted throughout the summer; a subset (12–19 August) was selected for use in model validation. Discrete (grab) samples were collected in all cases. Lake samples were collected at depths of 1 and 12 m; surface grab samples were collected near the mouth of the tributaries. Fecal coliform bacteria concentrations (colony forming units, cfu · 100 ml^{-1}) were determined by the membrane filter technique (APHA, 1985); statistical reliability of the analysis (95% confidence limits) was determined according to APHA (1992).

9.6.3 Model Development

Spatial and temporal dynamics in bacteria concentrations are modeled by conducting a mass balance over a series of interconnected,

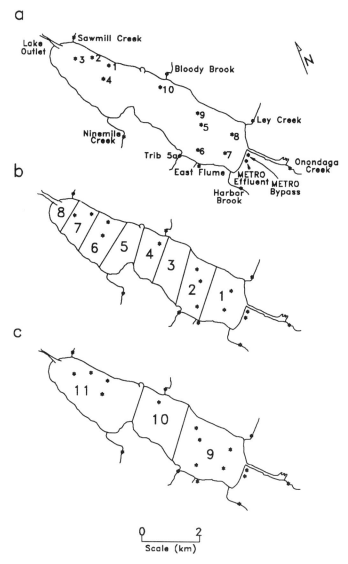

FIGURE 9.36. Onondaga Lake sampling locations and model cell configuration, supporting fecal coliform bacteria (FC) model: (a) lake and tributary sampling stations, (b) model cell configuration, surface layer, and (c) model cell configuration, bottom layer (modified from Canale et al. 1993a).

completely mixed volumes (model cells), which represent segments of the surface and bottom layers of the lake. An ideal network of model cells for Onondaga Lake would capture the gradient in bacteria concentrations along the major (longitudinal) axis of the lake, would produce negligible numerical dispersion, and would be consistent with the distribution of field sampling stations. Three configurations were tested; an 11-cell network of 8 surface layer cells and 3 bottom layer cells was selected (Figure 9.36b and c; also see Chapter 4).

In-lake fecal coliform concentrations are mediated by loading from tributary and point source discharges and by mass transport (advection and dispersion) and kinetic losses within the lake. A mass balance equation which accommodates these processes was written for each model cell:

Change in = Loading + Advective
Mass   Exchange
   + Dispersive − Kinetic
     Exchange   Losses   (9.55)

$$V_i \frac{dC_i}{dt} = W_i + \sum_j [Q_{ij} \cdot (C_j - C_i)]$$

$$+ \sum_j \left[\frac{E_{ij} \cdot A_{ij}}{L_{ij}} \right] \cdot C_i - V_i \cdot K \cdot C_i$$

in which A_{ij} = interfacial area between cells i and j (m^2), C_i = fecal coliform bacteria concentration in cell i (cfu · m^{-3}), C_j = fecal coliform bacteria concentration in adjacent cell j (cfu · m^{-3}), E_{ij} = dispersion coefficient (m^2 · d^{-1}), k = overall first order loss coefficient (d^{-1}), L_{ij} = distance between centers of cells i and j (m), Q_{ij} = net advective flow between cells i and j (m^3 · d^{-1}), t = time (d), V_i = volume of cell i (m^3), W_i = fecal coliform loading to cell i (cfu · d^{-1}); includes tributary inputs and settling to bottom layer cells from the surface layer. The expression is solved numerically (4th Order Runge−Kutta) to yield fecal coliform bacteria concentrations as a function of time. A conversion factor is included so that model output is expressed in the widely used concentration units (cfu · 100 ml^{-1}).

9.6.4 Submodels

Each of the major factors influencing fecal coliform dynamics in Onondaga Lake (loads, mass transport, and kinetics) is, in turn, impacted by variability in environmental conditions. Loads vary with tributary flow and bacteria concentration, and may be modified by in-stream processes such as settling and decay. Mass transport results from energy inputs due to wind action and tributary inflow and is influenced by lake geometry and the physical properties of water itself. Kinetic losses include settling and death, the latter mediated by conditions of light and temperature. Submodels describing the influence of environmental conditions on each of these major model components were developed and independently validated. The overall model framework is illustrated in Figure 9.37.

9.6.4.1 Loading Submodel

Fecal coliform bacteria loads were estimated for the ten tributary and point source discharges identified in Figure 9.36a (Chapter 3). A time series of hourly loads was calculated by interpolation of paired measurements of flow and concentration made for each tributary.

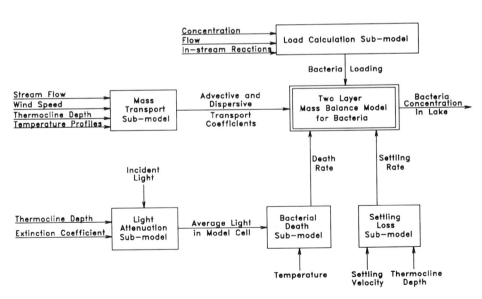

FIGURE 9.37 Overall framework and inputs for the Onondaga Lake FC model (modified from Canale et al. 1993a).

Adjustments were made to the raw data base to minimize inaccuracies resulting from insufficient sampling immediately prior to or following a wet weather event. Changes in concentration due to settling and death in the portion of Onondaga Creek between the monitoring station and the lake (Figure 9.36) were determined using a one-cell model; resulting concentrations were used in loading calculations.

An example of output from the loading submodel is provided in Figure 9.38, which illustrates the time series of loads calculated for Onondaga Creek for Storm 2; additional fecal coliform loading data are presented in Chapter 3. The rapid rise of the pollutograph is a reflection of extensive urban development in the lower reaches of the watershed. The impervious land surface conditions, short time of concentration, and numerous CSO inputs combine to affect an abrupt rise to a relatively large peak loading. The tail of the time series essentially mirrors that of the recession side of the hydrograph, where flow and bacteria loads exhibit a more gradual decrease over an extended period as a result of residual inputs from the upland, rural portion of the basin. Similar output was generated for each of the ten tributaries for both storms and the dry weather period. Recall that Onondaga Creek, with the greatest number of CSOs, is the largest source of bacteria, contributing approximately 80% of the load (Chapter 3; Canale et al. 1993a).

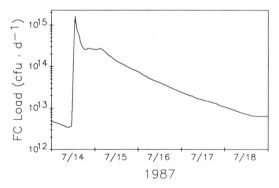

FIGURE 9.38. Time-course of FC loading to Onondaga Lake from Onondaga Creek during Storm 2 (modified from Canale et al. 1993a).

9.6.4.2 Mass Transport Submodel

Mass transport considerations are important in modeling transient constituents such as fecal coliform bacteria that may be rapidly lost from the water column due to settling and death. High rates of mass transport facilitate the movement of contaminated water to sites (e.g., swimming beaches) far distant from the point of discharge. Under conditions of weak mass transport, fecal coliform bacteria may be lost from the water column before they are transported any significant distance. Features of the transport submodel described here are developed in more detail in Chapter 4.

Dispersive exchange (E_{ij}) occurs horizontally between surface cells and between bottom cells, and diffusive exchange (E_{ij}) occurs between surface and bottom cells. Horizontal dispersion varies with the thickness of the surface layer (epilimnion depth) and shear velocity (Csanady 1973; Fischer et al. 1979), as described by

$$E_{ij} = C_1 \cdot u^* \cdot z_e \qquad (9.56)$$

in which C_1 = empirical dispersion parameter (dimensionless), u^* = shear velocity ($m \cdot d^{-1}$), and z_e = epilimnion depth (m). Shear velocity is a function of wind speed, according to

$$u^* = U_{10} \cdot \left[C_d \cdot \frac{density_{air}}{density_{water}} \right]^{0.5} \qquad (9.57)$$

in which C_d = empirical drag coefficient (dimensionless). A value of C_d (0.004) was determined by calibration using a dynamic water circulation model and field observations of drogue movement (Chapter 4). Calculated values for the horizontal dispersion coefficient, based on computed three-dimensional circulation, were related empirically to shear velocity and epilimnion depth, leading to values of C_1 = 5000 in the turbulent surface waters and C_1 = 500 in the quiescent bottom layer (Chapter 4).

Vertical mass transport is governed by the magnitude of concentration gradients between surface and bottom layers and by the vertical diffusion coefficient, which is related to vertical density gradients (Chapter 4). An average value for the vertical diffusion coef-

TABLE 9.8a. Mass transport and kinetic coefficients.

Coefficient	Value	Units	References
Vertical diffusion (offshore)	0.408	$m^2 \cdot d^{-1}$	Owens 1989
Vertical diffusion (Cell 1)	40.8	$m^2 \cdot d^{-1}$	Owens 1989
Horizontal dispersion (surface)	5000	dimensionless	Owens 1989, Chapter 4
Horizontal dispersion (bottom)	500	dimensionless	Owens 1989, Chapter 4
Dark death rate	0.73	d^{-1}	Auer and Niehaus 1993, Chapter 6
Irradiance proportionality constant	0.00824	$cm^2 \cdot cal^{-1}$	Auer and Niehaus 1993, Chapter 6
Sedimentation velocity	1.38	$m \cdot d^{-1}$	Auer and Niehaus 1993, Chapter 6

TABLE 9.8b. Period average environmental conditions (1987).

Case	Storm 1	Storm 2	Dry
Date	22–29 June	14–21 July	12–19 Aug.
Rainfiall (in.)	2.66	0.95	0.00
Return frequency (yr)	5	1	—
Incident irradiance ($cal \cdot cm^{-2} \cdot d^{-1}$)	166	215	240
Attention coefficient (m^{-1})	1.65	0.60	0.96
Thermocline depth (m)	6	6	6
Surface water temperature (°C)	22.6	23.8	23.5
Wind speed ($m \cdot s^{-1}$)	2.8	2.7	2.6

ficient ($0.408 \, m^2 \cdot {}^{-1}$) was calculated for the stratified period (Owens 1989) and applied to all model cells except cell 1 immediately adjacent to Onondaga Creek (Figure 9.36). Increased vertical exchange, perhaps associated with energy inputs from storm flows, or plunging of the ionically enriched inflow (Chapter 3), appears to occur in this region; a value of $40.8 \, m^2 \cdot d^{-1}$ was used for cell 1. Physical data and coefficients relating to mass transport calculations in Onondaga Lake are summarized in Table 9.8.

9.6.4.3 Kinetics Submodel

A number of biotic and abiotic factors have been reported to influence the fate of fecal coliform bacteria in lakes. Among these, light (Gameson and Saxon 1967), temperature (Lantrip 1983), and the combined effects of light and temperature (Mitchell and Chamberlin 1978) are considered most important. In addition, bacteria settle from the water column as discrete particles and as part of larger aggregates of fecal material, stormwater debris, and other suspended solids (Gannon et al. 1983; Schillinger and Gannon

1982). Losses due to death and settling are addressed through separate submodels (Figure 9.37; Chapter 6).

Kinetic mechanisms are represented collectively by the overall loss coefficient, k,

$$k = k_{de} + k_i + k_s \qquad (9.58)$$

in which k_{de} = dark death rate coefficient (d^{-1}), k_i = light-mediated death rate coefficient (d^{-1}), and k_s = sedimentation loss coefficient (d^{-1}). The light-mediated death rate coefficient is given by

$$k_i = \alpha \cdot I_{avg} \qquad (9.59)$$

in which α = light proportionality constant ($cm^2 \cdot cal^{-1}$), and I_{avg} = irradiance ($cal \cdot cm^{-2} \cdot d^{-1}$). Note that irradiance in this case is measured in energy units, and may include wavelengths (particularly near the surface) outside of the PAR (400–700 nm) waveband (Chaper 7). This differs from irradiance considerations that focus on phytoplankton growth. Light is averaged over the layer depth according to:

$$I_{avg} = \frac{I_{o,avg}}{k_d \cdot z}[1 - e^{(-kd \cdot z)}] \qquad (9.60)$$

in which I_{avg} = average irradiance in the layer $(cal \cdot cm^{-2} \cdot d^{-1})$, $I_{o,avg}$ = average incident irradiance $(cal \cdot cm^{-2} \cdot d^{-1})$, and z = layer depth, e.g., model cell thickness (m). The sedimentation loss coefficient is calculated from:

$$k_s = \frac{v_{FC}}{z} \qquad (9.61)$$

in which v_{FC} = sedimentation velocity of fecal coliform bacteria $(m \cdot d^{-1})$. Substituting equations (9.59), (9.60), and (9.61) into equation (9.58) yields the kinetics sub-model (Auer and Niehaus 1993, Chapter 6)

$$k = k_{de} + \alpha \cdot \frac{I_{o,avg}}{k_d \cdot z} \cdot [1 - e^{(-k_d \cdot z)}] \qquad (9.62)$$

$$+ \frac{v_{FC}}{z}$$

Note that irradiance approaches zero in the bottom layer of Onondaga Lake (and other productive lakes), thus the k_i (center) term drops out of equation (9.62) when modeling bottom cells.

Kinetic coefficients and associated environmental data are summarized in Table 9.8. The dark death rate (k_{de}) was held constant, that is, it does not vary with temperature. This is consistent with the experimental findings for Onondaga Lake, but differs from assumptions and limited findings reported by other researchers (Auer and Niehaus 1993). A weighted-average settling velocity (v_{FC}), representative of the two dominant particle size classes in Onondaga Lake (Auer and Niehaus 1993, Chapter 6), was used here (Table 9.8) to avoid the need to track and simulate two separate fecal coliform populations.

9.6.5 Model Performance

The model was applied for the two wet weather events and the dry weather period (Canale et al. 1993a). These cases represent a spectrum of loads, ranging from the background inputs of dry weather conditions, through those of a common summer storm (Storm 2, return frequency 1 year), to a severe wet-weather event (Storm 1, return frequency 5 years). Simulations were conducted without further adjustment of the kinetic and mass

transport coefficients described above. Data for comparison to model output were derived from the 10-station lake monitoring program (Chapter 6). Average concentrations were computed for model cells containing more than one sampling station to define the gradient in fecal coliform bacteria levels along the major axis of the lake. In the accompanying figures, field measurements are displayed as filled circles bracketed by bars that represent the 95% confidence intervals for the analytical technique. Dashed horizontal lines indicate the single occurrence (maximum) and average concentrations (5-day mean), which represent violations of public health standards.

A comparison of field data and model output for Storm 1 is presented in Figure 9.39. The upper row of illustrations represents epilimnetic cells and the bottom row hypolimnetic cells. Model simulations and data both show high bacteria concentrations $(1300-1900 \text{ cfu} \cdot 100 \text{ ml}^{-1})$ in the south end of the lake proximate to CSO discharges and lower concentrations in the north end $(400-600 \text{ cfu} \cdot 100 \text{ ml}^{-1})$. The decline in concentration with distance northward occurs as dispersion, settling, and decay act upon bacteria discharged at the south end of the lake. Good agreement is achieved between predicted and observed peak concentrations and the time course of attenuation in contamination following the storm event. The general character of north–south and top–bottom differences is also well described.

Model output and field data for Storm 2 are compared in Figure 9.40. Spatial and temporal trends in fecal coliform bacteria concentrations are similar to those for Storm 1. The highest concentrations are again noted in the south end of the lake, but maximum levels, 800 to $1300 \text{ cfu} \cdot 100 \text{ ml}^{-1}$ in the south and $200-300 \text{ cfu} \cdot 100 \text{ ml}^{-1}$ in the north, are somewhat lower for this smaller storm (Chapter 6). The attenuation period, that is, time for return to background levels (approximately 3 d), is similar to that for Storm 1. Again, the model successfully predicts peak values and die-away patterns as well as north–south and top–bottom differences.

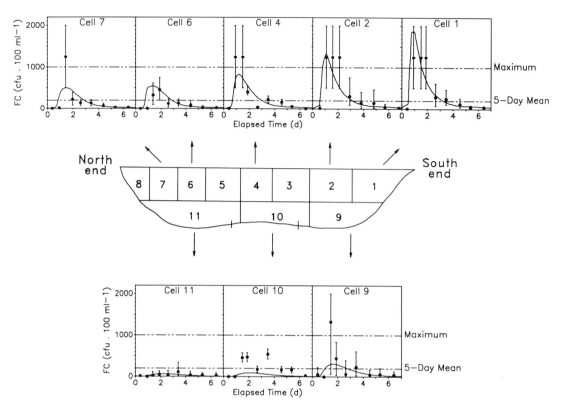

FIGURE 9.39. Measured and model-predicted FC concentrations for Storm 1. Points are measured values with bars indicating 95% confidence inter-vals; solid lines are model output; dashed lines are water quality standards as described in text (modified from Canale et al. 1993a).

Model performance for the dry-weather period is illustrated in Figure 9.41. Bacteria loads during this interval were low and quite steady compared with those observed during wet weather events (Chapter 3). Fecal coliform bacteria concentrations were low in the lake as well ($<50 \, \text{cfu} \cdot 100 \, \text{ml}^{-1}$). A small north–south gradient was evident, reflecting the influence of the input of fecal pollution at the south end of the lake during nonevent periods (Chapter 3). The model performed well in simulating the modest south to north gradient, but generally underpredicted concentrations in the hypolimnion. The high concentrations in bottom waters may be a result of direct hypolimnetic inputs from Onondaga Creek (recall the recently discovered salt springs (Chapter 3) may cause the stream to plunge (Chapter 4) at certain times) or may reflect reduced death rates in the low light environment of the bottom waters.

There is no widely accepted statistical technique for evaluating the quality of model output of this type. However, examination of Figures 9.39 to 9.41 indicates that model output is generally consistent with the observed lake response to storm-induced fecal coliform bacteria loads. These comparisons demonstrate the ability of the model to predict spatial and temporal dynamics in fecal coliform bacteria levels in Onondaga Lake under dry weather conditions and following wet weather events. Model credibility has been enhanced because the kinetic and mass transport coefficients utilized here were derived independently from the lake response data base used for model validation. It is concluded that the model is validated and can be used with confidence to test the effectiveness of related management and abatement schemes over a lakewide spatial scale.

9.6.6 Sensitivity Analyses

The importance of environmental conditions in mediating the disappearance of fecal coliform bacteria following a discharge event suggests that variability in those conditions may impart some uncertainty to predictions of future bacteria levels in the lake. Sensitivity analyses were conducted to illustrate the effect of changes in environmental factors (light levels and mass transport) on fecal coliform bacteria concentrations (Canale et al. 1993a). Baseline model runs made using the loading and environmental conditions of Storm 2 are shown in the accompanying figures as dashed lines. Simulations using environmental conditions bracketing those of Storm 2 are shown as solid lines. The system response at the south and north ends of the lake is illustrated using model output for Cells 1 and 7 (see Figure 9.36b), respectively.

The system response to variation in wind mixing conditions is shown in Figure 9.42a. Storm 2 wind speeds averaged $2.7\,\mathrm{m\cdot s^{-1}}$; strong winds in this simulation are three times greater than that speed, while mild winds are 20% of the measured value. These values bracket most daily average wind speeds experienced at Onondaga Lake (Owens and Effler 1989) during the recreational season. Strong winds serve to increase horizontal dispersion, promoting mass transport along the major axis of the lake. This reduces the north–south concentration gradient, resulting in lower south end concentrations and higher levels in the north. Mild winds reduce dispersion and horizontal mass transport, holding bacteria in the south end near their point of origin. This results in lower concentrations in the north where recreation areas are proposed. North end concentrations can be expected to vary by as much as an order of

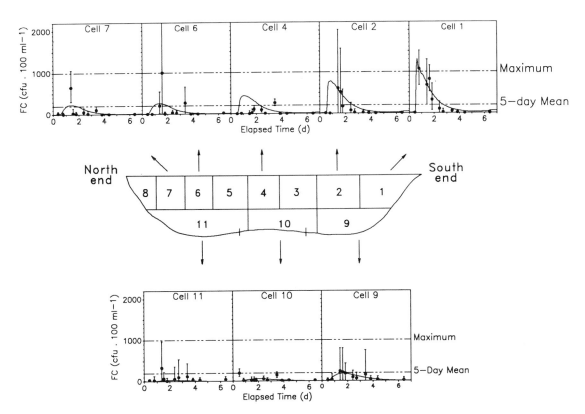

FIGURE 9.40. Measured and model-predicted FC concentrations for Storm 2 (modified from Canale et al. 1993a).

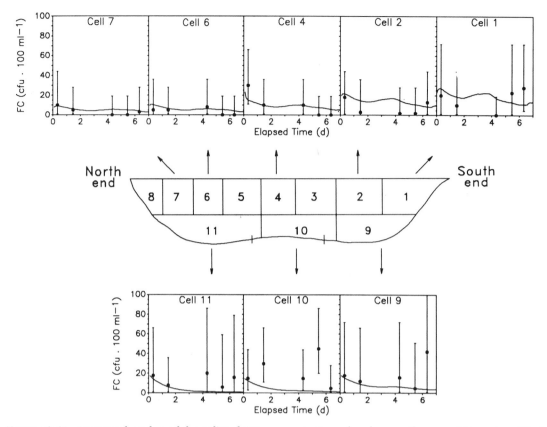

FIGURE 9.41. Measured and model-predicted FC concentrations for dry-weather conditions (modified from Canale et al. 1993a).

magnitude for a given loading, depending on wind speeds in the poststorm period.

The lake response to changes in light conditions is illustrated in Figure 9.42b. Incident light and light extinction interact to determine the amount of light in the watercolumn and thus determine the light-mediated death coefficient. Low light conditions simulated

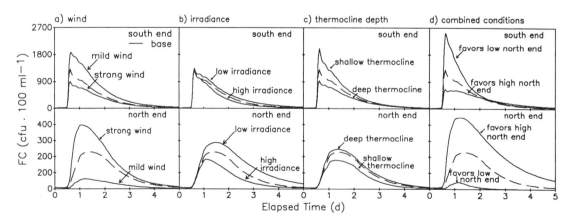

FIGURE 9.42. Sensitivity analyses: (a) wind speed, (b) irradiance, (c) thermocline depth, and (d) combined conditions (modified from Canale et al. 1993a).

here are typical of a cloudy day ($I_{o,avg}$ = 12 cal · cm^{-2} · d^{-1}) with a high degree of watercolumn light attenuation (k_d = 2 m^{-1}). High light conditions correspond to clear skies ($I_{o,avg}$ = 300 cal · cm^{-2} · d^{-1}) and a low degree of watercolumn light extinction (k_d = 0.5 m^{-1}). These conditions represent reasonable upper and lower bounds (see Chapter 7). During Storm 2, incident light averaged 217 cal · cm^{-2} · d^{-1} and k_d averaged 0.6 m^{-1}. Low light conditions result in higher concentrations due to a reduction in the light-mediated death coefficient. Note that light conditions have their greatest impact, not in determining the magnitude of peak concentrations (loads do this), but in modifying the rate of poststorm dieoff. Light conditions can have a significant impact on the time course of recovery at the north end of the lake.

The lake response to variability in thermocline depth (thickness of the surface layer) is illustrated in Figure 9.42c. Thermocline depth during Storm 2 was 6 m. The sensitivity analysis is run with shallower (4 m) and deeper (8 m) thermocline depths, a typical range for Onondaga Lake over the summer period (Chapter 4). The response to changes in thermocline depth is difficult to predict intuitively. For example, a shallow thermocline reduces the volume of the surface layer and thus the amount of water available to dilute tributary loads, leading to higher concentrations. However, a shallow thermocline also has greater settling losses (smaller values of z in equation (9.61)) and higher death rates (increased values of average light). A deep thermocline offers more dilution, but also smaller kinetic losses. It can be seen from Figure 9.42c that the relative importance of dilution and kinetic losses varies with location in the lake. At the south end, where the largest loads are received, dilution is the dominant process and a shallow thermocline serves to increase concentrations while a deep thermocline decreases concentrations. At the north end, kinetics dominate and a shallow thermocline results in reduced concentrations while a deep thermocline increases concentrations.

Analysis of model sensitivity to environmental conditions indicates that substantial variability in peak concentrations and the time course of recovery may be expected over the range of wind speeds, light conditions, and thermocline depths observed for Onondaga Lake. It is important from a management perspective to consider the collective impact of environmental conditions on the likelihood of violating water quality standards at locations far distant from the major sources of contamination. The combined effects of these factors were examined in two additional simulations: those that would favor high concentrations in the north end (high wind, low light, and a deep thermocline) and those that would favor low concentrations in the north end (low wind, high light, and a shallow thermocline). The results of those simulations, presented in Figure 9.42d, suggest that environmental conditions can dictate the likelihood of water quality violations. For example, conditions favoring high north end concentrations lead to a violation of the standard for 5-day average concentrations (200 cfu · 100 ml^{-1}), while those favoring low north end concentrations result in bacteria levels only slightly above background concentrations. At the south end, the single occurrence standard (1000 cfu · 100 ml^{-1}) is violated for conditions favoring low concentrations in the north, but not for conditions favoring high concentrations in the north. Thus, it can be seen that the impact of environmental conditions must be considered in developing a management plan for fecal coliform bacteria in Onondaga Lake.

9.7 Water Quality Model for the Seneca and Oswego Rivers

9.7.1 Background and System Description

The Seneca River/Three Rivers system is an important part of the Onondaga Lake water quality story. The river system not only receives the lake outflow but is also a source of water and material loading to the lake (Nauman and Owens 1993; Owens 1993, Chapter 4). The Seneca River is severely impacted by the inputs from Onondaga Lake (Canale et al. 1995b; Effler 1987; Effler et al. 1984a). Further, leading remediation alternatives for the domestic waste problems of the lake include diversion of a portion, or all, of the wastewater effluent from METRO for discharge to the river (see subsequent sections). A reliable mechanistic water quality model for the river system is necessary to support quantitative evaluations, and to support regional planning of domestic waste treatment and disposal, particularly with respect to the METRO diversion alternatives. This section reviews the characteristics of the Seneca River/Three Rivers System, including features of the degradation caused by the inflow from Onondaga Lake, and documents the development and testing of a water quality model for portions of the river system.

The Seneca River is a large river that drains about 8960 km^2 of central New York to the Oswego River, and subsequently Lake Ontario (Figure 9.43a). The average annual flow at Baldwinsville, NY (8 km upstream of the point of entry of the Onondaga Lake outflow (Figure 9.43)), over the period 1951 to 1991 was 96.3 m$^3 \cdot$ s^{-1}. Much of the river is part of the Barge Canal. The Seneca River combines with the Oneida River, at the Three Rivers junction, to form the Oswego River, which enters Lake Ontario at Oswego. The annual average flow at Oswego over the 1933 to 1991 period was 188.6 m$^3 \cdot$ s^{-1}. The area of the Oswego River drainage basin is about 13,200 km^2 (includes Seneca River). The overall system is sometimes referred to as the Three Rivers system. The Three Rivers system is a multiple-use river system; uses include navigation, hydroelectric power generation, fishing, contact recreation, and waste disposal. The natural flow and mass transport characteristics of the river system have been greatly altered (e.g.,

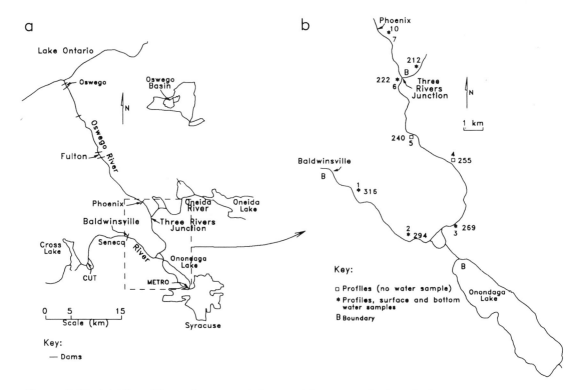

FIGURE 9.43. (a) Three Rivers System and Oswego River watershed, and (b) study section, roughly bounded by the box of (a), with sampling and experiment locations (modified from Canale et al. 1995b).

dams and locks) to support navigation and hydroelectric power generation.

This region of New York has been attractive to industries because of the availability of surface waters, particularly the Three Rivers System, for economic waste disposal (Calocerinos & Spina 1984). Upstream point and nonpoint loads make the Seneca River (and Cross Lake, Figure 9.43) hypereutrophic above the point of entry of Onondaga Lake. Summertime water quality conditions that prevailed in the Seneca River, upstream of the influence of Onondaga Lake, before the invasion of the zebra mussels in 1992 and 1993 (Effler and Seigfried 1994; Effler et al. 1994), are represented in the distribution plots for 1990/1991 in Figure 9.44. The major shifts in water quality and related implications since the zebra mussel invasion are discussed subsequently. Before the zebra mussels were present in high numbers (e.g., 1990/1991) the river supported very high concentrations of TP (Figure 9.44f) and phytoplankton biomass (CHL, Figure 9.44d), had poor clarity (Figure 9.44e, mostly associated with nonalgal turbidity (Canale et al. 1995b)), distinctly alkaline pH values (Figure 9.44i), relatively low concentrations of available phytoplankton nutrients (Figure 9.44g and h), and DO concentrations that represented 75 to 85% saturation conditions (Figure 9.44b and c).

9.7.2 Salinity Stratification and Water Quality in the Seneca River

Recall that an unusual bidirectional stratified flow regime is set up in the Onondaga Lake outlet to the Seneca River during periods of low flow (Chapter 4). Often relatively dense lake surface water exits along the bottom of the outlet and river water flows into the lake in the top of the channel (Chapter 4). Two conditions promote this phenomenon: the lake's elevated salinity, in part reflecting the continued loading of ionic waste from soda ash production (Doerr et al. 1994), and the absence of a natural hydraulic gradient between the lake and river, brought about by

channelization of the river sytem and lowering of the lake to support navigation. The salinity-based stratification set up in the outlet extends out into the river (Canale et al. 1995b; Effler et al. 1984a).

Features of the salinity stratification that existed in the river during summer before closure of the soda ash/chlor-alkali facility are compared to those that have prevailed since closure of the chemical plant for a position about 8 km downstream of the lake inflow in Figure 9.45a. Substantial variability within these periods (before and after closure) occurred because of the important interplay with river flow (e.g., source of dilution (Canale et al. 1995b; Effler et al. 1984a)). Salinity stratification was much stronger in the river before closure of the soda ash/chlor-alkali plant (Figure 9.45a), associated with the higher salinity of Onondaga Lake (Doerr et al. 1994; Effler et al. 1990b, Chapter 5), which was caused by the greater loading from the operating facility (Doerr et al. 1994; Effler and Driscoll 1985, Chapter 3). The extent of the difference in salinity stratification implied by this comparison (Figure 9.45a) is made more impressive by the greater flow (i.e., greater turbulence within the upper river layer) that prevailed on the 1982 date. The specific conductances of the upper layer, largely reflective of the upstream river conditions, are similar for the two profiles (Figure 9.45a). The specific conductance in the lower layer in 1982 was nearly $3\times$ that measured in 1991. Note that the specific conductances of the upper waters of the lake nearly match the lower layer values observed in the river, clearly depicting the lake origins of the salinity enriched lower river layer.

The negative implications of the stratification phenomenon, over the range of conditions of Figure 9.45a, for the river's oxygen resources are apparent from the paired oxygen profiles; oxygen was greatly depleted in the lower stratified layer in both cases. These represent violations of the minimum DO standard $(4.0\,\text{mg}\cdot\text{L}^{-1})$ for class B (as well as other classes) waters in New York State. The oxygen depletion in the stratified layer has been attributed to the isolation of the lower layer

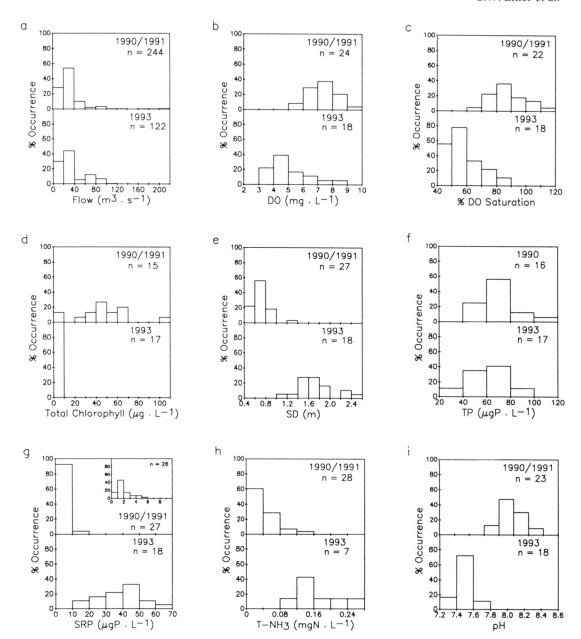

FIGURE 9.44. Distributions of summertime water quality in the Seneca River, just below Baldwinsville, in 1990/1991 and 1993: (**a**) flow, (**b**) [DO], (**c**) % DO saturation, (**d**) [CHL], (**e**) Secchi disc trans-parency, (**f**) [TP], (**g**) concentration of SRP [SRP], (**h**) [T-NH$_3$], and (**i**) pH (modified from Effler et al. 1995).

(containing the Onondaga Lake outflow) from the oxygen sources of reaeration and photosynthesis, combined with the continued exertion of oxygen consuming processes (Effler et al. 1984a). The similarity of the oxygen profiles for the disparate magnitudes

of salinity stratification of Figure 9.45a implies the level of stratification that persists since closure of the soda ash/chlor-alkali facility continues to exceed a threshold that results in limited vertical exchange and coupled degradation of the river's oxygen resources.

Features of the longitudinal range of chemical stratification and related impacts on water quality in the river are illustrated (July 30, 1991) for the ameliorated conditions of 1991 in Figure 9.45b. Here vertical differences in Cl⁻ concentration reflect salinity strati-

fication. Note that the stratification extends upstream as well as downstream from the point of entry of Onondaga Lake. The upstream movement, or "salt wedge" effect, is commonly observed in stratified estuaries (Thomann and Mueller 1987), but is unique

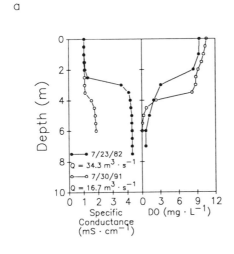

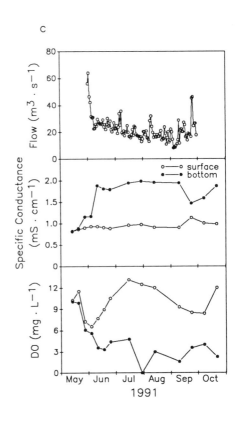

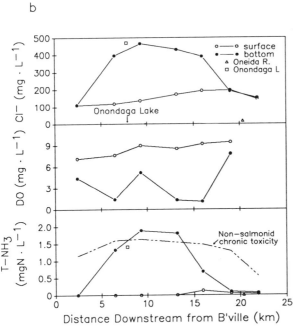

FIGURE 9.45. Stratification and water quality in the Seneca River: (a) stratification for specific conductance and dissolved oxygen, July 30, 1991 and July 23, 1982, 8 km downstream of Onondaga Lake inflow, (b) longitudinal extent of stratification and violations of DO and NH₃ standards, July 30, 1991, and (c) interplay of stratification and DO violations with river flow, 1991, at site No. 4 (Figure 9.43b; modified from Canale et al. 1995b).

for inland rivers. The concentration of Cl$^-$, a conservative substance, increases in the upper waters and decreases in the lower waters progressively downstream of the lake inflow as a result of vertical mixing. On July 30, 1991, the stratification extended more than 8 km downstream of the lake. The salinity stratification promotes the extension of Onondaga Lake's problems into the Seneca River. Strong DO depletion occurs in the lower layer of the river upstream, as well as downstream, of the point of entry of the lake, causing violations of the DO standard over substantial distances in both sections of the river (Figure 9.45b). The high T-NH$_3$ concentrations retained in the lower layers adjoining the inflow of the lake, and the attendant pH and temperature conditions (not shown), cause violations of the state's NH$_3$ standard (Figure 9.45b; see Chapter 5 for protocol for calculations). The extent of stratification and related impact on oxygen resources of the river was greater before closure of the soda ash facility (Effler et al. 1984a). In low flow periods during the operation of the facility the stratification persisted 14 km downstream to the dam in Phoenix (Figure 9.43); the turbulence provided by passage over the dam broke up the stratification. Presently the stratification is broken up before the confluence with the Oneida River (Figure 9.43b).

The occurrence of stratification, and the coupled depletion of DO in the lower layer, is limited to periods of low (but common in summer) river flow, as illustrated for a downstream site (No. 4, Figure 9.43b) for 1991 in Figure 9.45c. A critical flow for this site appears to be about 80 m$^3 \cdot$ s^{-1} (Figure 9.45c); above this flow the attendant turbulence is great enough to break up the stratification. The critical flow is somewhat higher closer to the entry point of Onondaga Lake. Stratification prevailed at this site for at least 5 months in 1991, and violations (e.g., <4 mg $\cdot$ L^{-1}) of the DO standard occurred in the lower layer on about 60% of the days over the June to October interval. Substantial year-to-year variations in the duration of stratification and the occurrences of coupled water quality violations doubtless occur in the river as a result of the large annual variations in river

flow that are common to this region. The conditions presented for 1991 in Figure 9.45c probably approach worst case (with respect to duration) for the present water quality of Onondaga Lake, as the flow at Baldwinsville was less than the 30Q10 (minimum 30 day average flow with recurrence interval of 10 years) value (17.6 m$^3 \cdot$ s^{-1}) for a substantial portion of the summer. Undoubtedly the duration of the stratification phenomenon in the river, and the coupled degradation of oxygen resources, was longer before closure of the soda ash/chlor-alkali facility, because the river threshold flow was higher.

9.7.3 Water Quality Model

9.7.3.1 Model Framework

A water quality model has been developed for the river system that extends from Baldwinsville on the Seneca River to Oswego on the Oswego River, covering 19 km of the Seneca River and the entire length of the Oswego River (Figure 9.43a; a total length of river of 55 km). Supporting process and field studies, and model testing, documented here, have focused on the Baldwinsville to Phoenix reach, because this is presently the most degraded section of the river system (e.g., Effler et al. 1984a), and it is expected to be the most impacted by some of the leading management options under consideration for remediation of Onondaga Lake (see subsequent treatment).

The water quality model uses a multiple box or multiple segment approach (e.g., Shanahan and Harleman 1984). The river is divided into a number of segments; the concentrations within each segment are assumed to be uniform. The generalized mass balance expression for DO in each segment in the river is

$$V_i \frac{d[DO]_i}{dt} = \text{reaeration} + (\text{photosynthesis}$$

$$- \text{respiration}) - \text{oxidation of CBOD}$$

$$- \text{oxidation of NBOD}$$

$$- \text{sediment oxygen demand}$$

$$+ \text{oxygen inputs}$$

$$\pm \text{oxygen transport} \qquad (9.63)$$

in which V_i = volume of model segment i (m^3), $[DO]_i$ = concentration of DO in segment i ($mg \cdot L^{-1}$); CBOD = carbonaceous biochemical oxygen demand; and NBOD = nitrogenous biochemical oxygen demand ($mg \cdot L^{-1}$). Sources of DO include reaeration, photosynthesis (net; i.e., gross photosynthesis minus respiration), inputs from tributaries or effluents, and oxygen transported into segment i from adjoining segments. Oxygen sinks include oxidation of CBOD, oxidation of NBOD (nitrification), oxygen demand exerted by sediments at the interface (SOD), and oxygen transported out of segment i to adjoining segments. The kinetic expressions are presented below, in a format consistent with equation (9.63).

$$V_i \frac{d[DO]_i}{dt} = [K_a \cdot ([DO] - [DO]_i)$$

$$+ (P_g - D) \cdot [CHL]_i - k_{CBOD} \cdot [CBOD_u]$$

$$- k_n \cdot [NBOD_u] - SOD/H] \cdot V_i$$

$$+ DO \text{ inputs}$$

$$\pm \text{ (advective and dispersive transport)}$$
$$(9.64)$$

in which P_g = gross photosynthesis ($mgO_2 \cdot \mu g$ $CHL^{-1} \cdot d^{-1}$). Temperature adjustments for the kinetic processes were made according to the Arrhenius format

$$k_{x,T} = k_{x,20} \cdot \Theta^{T-20} \qquad (9.65)$$

where $k_{x,T}$ and $k_{x,20}$ = values of kinetic coefficient x at temperatures T and 20°C, and Θ = dimensionless temperature coefficient. The values of Θ for the various coefficients appear in Table 9.9.

A similar mass balance approach was utilized to simulate the concentration of $CBOD_u$ in a

TABLE 9.9. Values of theta (Θ) for Seneca River oxygen model kinetic coefficients.

Coefficient	Θ
K_a	1.024
P_g	1.03
D	1.05
k_{CBOD}	1.04
k_n	1.08
SOD	1.06

submodel of the oxygen model. The mass transport processes in this complex system are determined separately with a transport submodel (described subsequently); outputs from this submodel serve as inputs for the biochemical models. The range in diurnal variation in DO, driven by plant metabolism, and modulated by reaeration, was estimated from predictions of a phytoplankton production submodel (described subsequently) according to a modified formulation of the "delta" method (Chapra and DiToro 1991) that accounts for stratification.

9.7.3.2 Monitoring Program

An intensive program of field measurements, sampling, and laboratory analyses was conducted in 1990 and 1991 for the Baldwinsville to Phoenix reach to support the development, testing, and application of the water quality model. The goals were to: (1) characterize the prevailing water quality in the river system, (2) develop an understanding of the processes that regulate these conditions, and (3) document environmental forcing and system boundary conditions. The design of the monitoring program (Table 9.10) was guided by the findings of earlier studies of the system (e.g., Calocerinos & Spina 1984; Effler 1982; Effler et al. 1984a) and the needs of the water quality model (equation (9.64)).

Sampling stations for the monitoring program are shown in Figure 9.43b, along with buoy numbers. Seven sites were positioned along the study reach. These stations, and a site near the mouth of the Oneida River that establishes boundary conditions for this inflow, were monitored routinely over the May to October interval of 1990 and 1991. The boundary conditions of the Onondaga Lake inflow are established through an ongoing comprehensive monitoring program (e.g., Effler et al. 1988, 1990a; Onondaga County 1971–1990). The river stations were monitored weekly over the late spring to early fall interval in 1990, and in May and June of 1991; monitoring was conducted biweekly over the July–October interval of 1991. In situ profile measurements of DO, temperature, specific conductance, and pH were made

Table 9.10. Monitoring program for Seneca river (1990 and 1991).

Parameters	Depths	Comments/justification
Lab	0.5 m below surface/ 0.5 m above bottom	
$CBOD_5$*		Oxygen sink, $CBOD_u^1$ in model
$T-NH_3$, $NO_3 + NO_2$		Toxicity status, nitrification, nutrients
Cl^-		Conservative, hydrodynamic tracer, modeled parameter
Turbidity		Light attenuation
TP		Trophic state
Chlorophyll		Trophic state, P/R submodel
SRP		Nutrient, P/R submodel
SF_6**		Inert insoluble tracer gas, support determination of K_a
Field		
DO	0.5 m intervals	Modeled parameter, water quality status
Temperature	0.5 m intervals	DO% saturation, stratification, ammonia toxicity
Specific conductance	0.5 m intervals	Stratification, measure of salinity (conservative)
pH	0.5 m intervals	Ammonia toxicity
Underwater irradiance	0.25 m intervals	Calculation of light attenuation coefficient (k_d, m^{-1}, P/R submodel
Secchi disc	—	Light attenuation, water quality
Incident irradiance	—	Hourly integrated, continuously measure, P/R submodel
Flow		Transport submodel

* Carbonaceous biochemical oxygen demand on filtered (0.45 μm) samples, inhibited for nitrification; assumed ultimate CBOD ($CBOD_u$) = $1.3 \cdot CBOD_5$
** Sulfur hexafluoride

(Table 9.10) at all stations to assess the occurrence and character of stratification. Profiles of underwater irradiance, to support determination of the attenuation coefficient for downwelling irradiance (k_d, m^{-1}), were collected routinely at site No. 2 (Figure 9.43b) in 1991 and irregularly at all the stations in 1990. Samples for laboratory analyses were collected routinely from two depths, which were within each of the two layers during periods of stratification, at each station.

Additionally, in situ diurnal measurements of DO, temperature, pH, and specific conductance were made at two depths on five dates over the July–September interval of 1991. Each station was usually visited 8 times (e.g., 3 hr return frequency) within a 24 hr period. These measurements supported calculations of daily average conditions, as well as established the range of DO concentrations within a day to support testing of the water quality model. Concentrations of sulfur hexafluoride (SF_6) were measured in samples collected at the routine monitoring sites of Figure 9.43b and below the Phoenix dam over the interval of the last three diurnal surveys of

1991 to support direct estimates of K_a (see subsequent treatment).

9.7.3.3 Transport Submodel

A mass transport submodel was developed to simulate the complex flow and transport patterns that exist in the Baldwinsville to Phoenix reach of the river system. This transport submodel established the array of linked segments to which a mass balance equation of the form of equations (9.63) and (9.64) was applied. The transport submodel defined the interaction of adjacent segments through the processes of horizontal and vertical advection, longitudinal dispersion, and vertical diffusion. Details of the development of the mass transport submodel, and its application in the estimation of the reaeration coefficient, are described by Naumann and Owens (1995). Consistent with the approach described earlier, the magnitude of the individual processes in the mass transport submodel was determined independently of the water quality model.

The most important feature of the mass transport submodel with regard to the unusual water quality conditions in the Seneca River is its use of two layers to describe the stratified conditions. As indicated in Figure 9.45a, the vertical profiles of specific conductance, dissolved oxygen, and other physical and chemical parameters may be approximated by two completely mixed layers. The use of a layered approach is very unusual in a water quality model of a nonestuarine river, but it is critical for the simulation of the stratified conditions that prevail here during low flow. The surface and bottom layers are then further divided into an array of longitudinal segments.

Features of the transport submodel and its application to the Seneca River are presented in Figure 9.46, including: (a) morphometric characteristics (Figure 9.46a), (b) description of the flow pattern in the river proximate to the inflow of Onondaga Lake (Figure 9.46b), (c) longitudinal and vertical segmentation of the transport (and water quality) model, with advective and entrainment flow patterns identified (Figure 9.46c), (d) temporal distributions of system inflows for the study period (Figure 9.46d), (e) calibration of the transport model for the third diurnal survey (July 29–30, 1991; Figure 9.46e), (f) application to support estimate of the SF_6 mass transfer coefficient (Figure 9.46f), and (g) longitudinal distribution of the vertical turbulent diffusion coefficient (E_v; $m^2 \cdot d^{-1}$), specified according to the diffusivity ratio ($E_v \div$ molecular diffusion coefficient; Figure 9.46g).

The geometry of the river channel (Figure 9.46a) was derived from earlier measurements (O'Brien & Gere Engineers 1977) and measurements made as part of this study. A weak motion in the upstream direction occurs in the stratified "salt wedge" upstream of Onondaga Lake, as described by Arita and Jirka (1988). The upstream flow is balanced by entrainment losses from the "salt wedge" to the overlying layer. The upstream extent of the stratified flow is limited, at least in part, by a shallow region ~3.5 km upstream from the lake inflow. Stratified (two layer) flow occurs downstream of the point of entry of the dense lake inflow. Vertical turbulent diffusion tends to break up the stratified flow regime downstream (Figure 9.46b).

The Baldwinsville to Phoenix study reach was partitioned into 25 longitudinal segments of about 0.9 km length (Figure 9.46c). The depth of each segment in the upper layer was approximately 3 m throughout the study reach, and for the five diurnal surveys, consistent with the results of field profiles in the salinity stratified portion of the reach. The geometries of the transport model segments (950) were determined from river channel morphometric data (Figure 9.46a). For low flow conditions, such as those experienced in 1991 (Figure 9.46d), the volumes of the model segments are constant, due to the very small slope of the water surface. Equal velocities of flow are assumed for the two layers upstream of the "salt wedge." The magnitude of the "salt wedge" flow was determined as part of the calibration of the transport model. The flow in the bottom layer immediately downstream of the lake is water that has flowed from Onondaga Lake. Five inflows that are significant in terms of either discharge or pollutant loading were included in the Baldwinsville to Phoenix reach, the Onondaga Lake outflow, flow from the Oneida River, and inputs from three wastewater treatment plants (WWTP). The Oneida River was assumed to enter the surface layer, while the entry of the three WWTP discharges was based on actual outfall elevations. The extension of the water quality model downstream of Phoenix to Oswego was accommodated by 25 more longitudinal segments, of two layers each. Significant density stratification has not prevailed in this reach, even before closure of the soda ash/chlor-alkali facility.

River and Onondaga Lake tributary flows were low during the summer of 1991 (Figure 9.46d). Though flows in the Seneca River were less than the 30Q10 ($17.6\,m^3 \cdot s^{-1}$) for portions of the study period, they remained above the 7Q10 of $12.1\,m^3 \cdot s^{-1}$. Note that the inflow from METRO to the lake represented nearly 50% of the total at times in 1991 (Figure 9.46d). The value of the longitudinal dispersion coefficient determined for the Seneca River from an instantaneous dye release during low flow conditions in September 1991 was $1.5 \times 10^6\,m^2 \cdot d^{-1}$.

a

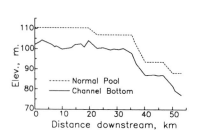 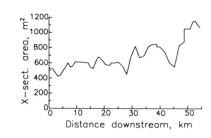

b

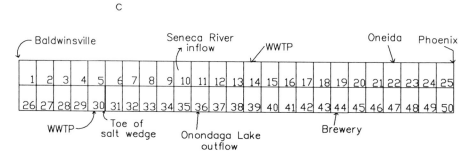

c

FIGURE 9.46. Seneca River transport submodel: (a) river morphometry, distance is from Baldwinsville, (b) flow pattern proximate to Onondaga Lake inflow, and (c) model framework (modified from Canale et al. 1995b).

The transport submodel was applied to salinity for the purpose of determining the magnitude of vertical turbulent diffusion between the two layers over the reach from Baldwinsville to Phoenix. An example of the performance of the transport model in matching the longitudinal and vertical distribution of salinity according to the two-layer framework, is presented for the July 29–30, 1991, survey in Figure 9.46e. The lowest

d

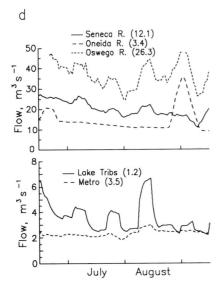

e

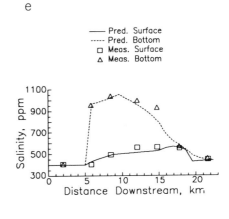

f

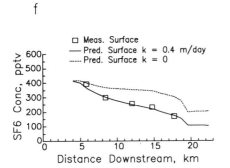

g

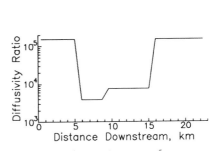

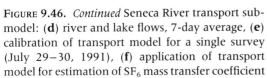

FIGURE 9.46. *Continued* Seneca River transport sub-model: (**d**) river and lake flows, 7-day average, (**e**) calibration of transport model for a single survey (July 29–30, 1991), (**f**) application of transport model for estimation of SF_6 mass transfer coefficient (July 29–30, 1991), and (**g**) longitudinal distribution of vertical dispersion along study reach. Distances on × axis extend from Baldwinsville (modified from Canale et al. 1995b).

vertical diffusion occurs in the "salt wedge" region, while the highest occurs outside the region of salinity stratification. The application of the model to salinity is described in detail by Naumann and Owens (1995).

Following this calibration procedure, the transport submodel was applied to the dissolved gas SF_6 (Figure 9.46f), which was deliberately injected into the river at Baldwinsville for a portion of the study period to assess gas exchange at the air–water interface. The application of the submodel to SF_6 is described in detail by Naumann and Owens (1995), and is summarized in the next section. The same transport submodel was also applied to simulate each of the mass constituents in the water quality model, using the magnitude of the transport processes determined in the salinity simulations.

9.7.3.4 Development of Kinetic Coefficients

A summary of the development of the kinetic coefficients for the river water quality model is presented as Table 9.11. Additional descriptive information is provided in this section.

9.7.3.4.1 SOD

The distribution of SOD along the study reach was established (Figure 9.47) through a combination of field and laboratory studies. Portions of the reach have little or no sediment deposits and thus significant SOD is not exerted in these sections. The longitudinal profile of SOD was based on COD analyses of sediment samples collected at 8 locations over the longitudinal extent of sediment occurrence, based on the empirical relationship developed by Gardiner et al. (1984; Table 9.11). This relationship was supported for the river system by direct determination of SOD on two intact core samples (Figure 9.47), using the Gardiner et al. (1984) methodology. A longitudinal SOD profile consistent with the model segmentation was developed by interpolation.

9.7.3.4.2 k_n

Recall that a number of researchers have reported nitrification to be localized at the sediment–water interface (Cavari 1977; Curtis et al. 1975; Hall 1986). Results of laboratory microcosm experiments are consistent with these observations and indicate no significant nitrification occurs within the water column of the Seneca River. Thus the kinetics of nitrification in the river have been developed in the same manner outlined previously in this chapter for Onondaga Lake; that is, based on laboratory determinations of T-NH$_3$ depletion from water overlying intact river sediment cores, and application of the film transfer approach to estimate k_{fn} (Table 9.11). The value of k_n is then estimated as the

quotient of k_{fn} and the depth of the lower river layer (Table 9.11). The film transfer coefficient (k_{fn}; equation (9.67), Table 9.11) for nitrification was found to be in the range of 0.08 to 0.33 m · d^{-1}. A model value of 0.135 m · d^{-1} was selected from this range by comparison with observed T-NH$_3$ profiles in the river. The corresponding value for k_n was estimated to be 0.021 d^{-1} (for an average river depth of 6.4 m). This value is lower than many reported for other streams and rivers, but generally consistent with the observation that lower values are associated with deeper systems. According to equation (9.68) (Table 9.11), lower values of k_n are expected as H increases.

9.7.3.4.3 k_{CBOD}

It is important to differentiate among the oxygen sinks of decay of CBOD, phytoplankton respiration, and the process of nitrification. Thus, the estimation of k_{CBOD} was based on laboratory BOD analyses of filtered (0.45 µm) nitrification-inhibited samples. Samples represented a realistic mixture of METRO (1 part) and the Seneca River (4 parts) under critical low flow conditions, corresponding to the Onondaga Lake remediation option of full diversion of METRO to the river. A value of $k_{CBOD} = 0.11$ d^{-1} was determined, using the Thomas slope method (Metcalf & Eddy 1979). A nearly equivalent value (0.1 d^{-1}) was estimated for present conditions, based on calibration of the CBOD submodel against CBOD profiles observed in the river in 1991. For the same mixture of METRO and the river, but for unfiltered samples, a value for $k_{CBOD} = 0.243$ d^{-1} was determined.

9.7.3.4.4 K_a

Direct experimental determination of K_a was deemed necessary because of the substantial uncertainty of estimates based on empirical expressions for rivers of this great depth and low velocity of flow (e.g., Bowie et al. 1985), and the absence of a clearly defined DO "sag" in the surface waters that could support estimates through model calibration. The inert, relatively insoluble, gas SF$_6$ was con-

TABLE 9.11. Summary of development of kinetic coefficients.

Coefficient	Components/description	References
SOD	1. Field survey to establish the distribution of river deposits	
	2. SOD determinations (2) on intact core samples (Figure 9.46)	Gardiner et al. 1984
	3. COD river sediment profile (n = 10, Figure 9.46)	
	4. SOD river profile from COD profile, according to:	Gardiner et al. 1984
	$$SOD = (7.66 \cdot COD)/(157 + COD) \qquad (9.66)$$	
k_n	1. Determination of T-NH_3 flux (J; $mg \cdot m^{-2} \cdot d^{-1}$) on intact core samples (3)	
	2. Determination of film transfer coefficient (K_{fn}; $m \cdot d^{-1}$)	
	$$K_{fn} = \frac{J}{[T\text{-}NH_3]} \qquad (9.67)$$	
	3. Determination of first order nitrification rate constant (k_n; d^{-1})	
	$$k_n = \frac{K_{fn}}{H_l} \qquad (9.68)$$	
	where H_l = river depth of lower layer (m)	
k_{CBOD}	1. Laboratory BOD analyses; filtered (0.45 µm) and nitrification inhibited; DO monitored daily	APHA 1985
	2. k_{CBOD} determined from results according to Thomas slope method	Metcalf & Eddy Inc. 1979
	3. Calibration of field measurements	
K_a	1. Continuous injection of SF_6 downstream of Baldwinsville (Figure 9.42) for 5 weeks	
	2. Downstream monitoring of SF_6	Wanninkhof et al. 1987
	3. Oxygen transfer coefficient ($K_{L,DO}$; $m \cdot d^{-1}$) determined from SF_6 exchange coefficient ($K_{L,SF}$; $m \cdot d^{-1}$)	
	$$K_{L,DO} = 1.38 \cdot K_{L,SF} \qquad (9.69)$$	
	4. Reaeration coefficient (K_a; d^{-1}) for upper cells of river model	
	$$K_a = \frac{K_{L,DO}}{H_u} \qquad (9.70)$$	
	where H_u = depth of upper layer (m)	
P/R	1. Laboratory measurements of gross photosynthesis and respiration with Seneca River phytoplankton to develop productivity irradiance (P-I) curve	Storey et al. 1993, 1995
	2. Field verification of P-I curve by in situ light—dark bottle technique	Storey et al. 1995, Vollenweider 1974, Auer et al. 1986
	3. Chlorophyll-light attenuation relationship for Seneca River	
	$$k_d = 0.0056CHL + 1.53 \qquad (9.71)$$	
	where k_d = light attenuation coefficient and CHL = chlorophyll concentration ($\mu g \cdot L^{-1}$)	
	4. Photosynthesis submodel; accommodates influences of light, temperature, and phosphorus concentration	Storey et al. 1993, 1995
	5. Calculation of diurnal ranges in DO ("delta") using "delta" method	Chapra and DiToro 1991

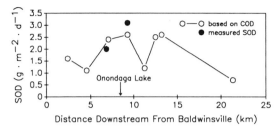

FIGURE 9.47. Longitudinal distribution of SOD along study reach (modified from Canale et al. 1995b).

tinuously injected at Baldwinsville for several weeks during the study period, and its concentration in the water was measured periodically at points downstream. The gas SF_6 was selected over the more widely used propane for this large river because of its much lower analytical detection limits. Wanninkhof et al. (1987, 1990) have documented the successful use of SF_6 to assess gas exchange for a range of surface water systems.

The value of the SF_6 surface mass transfer coefficient ($K_{L,SF}$; $m \cdot d^{-1}$) was determined with the mass transport model (e.g., Figure 9.46f) that had been validated for salinity (Figure 9.46e; Naumann 1993). The mass transfer coefficient for oxygen ($K_{L,DO}$) is determined directly from $K_{L,SF}$ (Table 9.11) based on the differences of the molecular diffusion coefficients for the two gases. Substantial longitudinal differences in $K_{L,SF}$ were not identified in the experimental results. However, temporal differences were observed during the study period; for example, the values of $K_{L,DO}$ determined for the last 3 model validation surveys were 0.55, 0.90, and $0.48\,m \cdot d^{-1}$, respectively. The values for the first 2 surveys were estimated, from an empirical $K_{L,SF}$-flow relationship, to be 0.83 and $0.76\,m \cdot d^{-1}$. The value adopted for critical low flow conditions was $0.55\,m \cdot d^{-1}$. Variations in wind conditions probably also contribute to dynamics in $K_{L,DO}$ for this deep slow moving river (e.g., like a lake). The reaeration coefficients (K_a) for the upper model cells are calculated directly from the gas exchange coefficients (Table 9.11; equation (9.70)).

Reaeration at the dams is described by the gas transfer efficiency, E, defined by

$$E = \frac{[DO]_d - [DO]_u}{[DO]_s - [DO]_u} \qquad (9.72)$$

in which $[DO]_d$ = DO concentration downstream of the dam ($mg \cdot L^{-1}$), and $[DO]_u$ = DO concentration upstream of the dam ($mg \cdot L^{-1}$). For a structure that provides no gas transfer (i.e., $[DO]_d = [DO]_u$), the value of E = 0. Owens (1993) tracked SF_6 in 1991 upstream and below the Phoenix dam. The value of E was computed for 6 different days, and the average, adopted in the water quality model, was 0.48. The value appeared to be independent of river flow over the range of 1991 measurements (23 to $90\,m^3 \cdot s^{-1}$). Samide's (1993) estimate of the value of E for the lower Fulton dam (Granby) of 0.75 has been adopted. No reaeration (E = 0) has been assumed for other dams (Figure 9.43) over the modeled portion of the system (upper Fulton and Minetto), because there is no requirement to pass any of the river flow over these structures, instead of through adjoining power generating facilities.

9.7.3.4.5 Photosynthesis/Respiration

The source-sink character of the algal component of the DO mass balance of productive rivers and streams varies among days and changes within days due to natural variation in incident light (Auer and Effler 1989). These influences, as well as the effects of temperature and nutrient availability, were accommodated with the following phytoplankton production submodel

$$P_n = P_{g,max,20} \cdot \frac{I_a}{K_I + I} \cdot \frac{[SRP]}{K_P + [SRP]} \cdot \Theta_P^{(T-20)} - R_{20}\Theta_R^{(T-20)} \qquad (9.73)$$

where P_n = CHL − specific rate of net photosynthesis ($mgO_2 \cdot \mu g\ CHL^{-1} \cdot d^{-1}$), $P_{g,max,20}$ = maximum CHL-specific rate of gross photosynthesis (= $0.6\,mgO_2 \cdot \mu g\ CHL^{-1} \cdot d^{-1}$) at 20°C, K_I = half-saturation coefficient for irradiance (= $180\,\mu E \cdot m^{-2} \cdot s^{-1}$), R_{20} = CHL-specific respiration rate (= $0.04\,mgO_2 \cdot \mu g\ CHL^{-1} \cdot d^{-1}$) at 20°C, [SRP] = concentration of soluble reactive phosphorus ($\mu g \cdot L^{-1}$), K_p = half-saturation coefficient for SRP (assumed $0.7 \mu g \cdot L^{-1}$), Θ_P = dimensionless temperature coefficient for photosynthesis (Table 9.9), and Θ_R = dimensionless temperature coefficient for respiration (Table 9.9).

Recall that the Monod-type nutrient limitation kinetics, adopted here, were not found to be an appropriate representation of nutrient availability to phytoplankton in Onondaga Lake (Chapter 6). Kinetic coefficients describing the photosynthesis-light relationship (P-I curve; e.g., Auer and Effler 1989) were determined through laboratory experiments with the natural phytoplankton assemblage of the river (Table 9.11) and verified by field incubations at several depths in the river (Storey et al. 1995). Site-specific determination of these coefficients enhances model credibility (Auer and Canale 1986; Storey et al. 1993). Light and photosynthesis were integrated over both time (hourly) and depth (0.25 m depth intervals) to support accurate calculations (e.g., Auer and Effler 1990). Values of k_d for sites without direct measurements were estimated from concentrations of CHL, according to a system-specific empirical relationship (Table 9.11), based on the subset of paired observations. The large background attenuation ($1.53 \, m^{-1}$, equation (9.71), Table 9.11) reflects the high nonalgal turbidity of the river, a situation that is common to many large rivers.

9.7.3.5 Model Performance

Simulations of the water quality model are compared to observations in Figure 9.48a–d. Model predictions in general closely match measurements. Comparisons are shown here for one of the diurnal surveys of 1991 for CBOD, representative of conditions and model performance for all five surveys. Note there was little structure in the distribution of $CBOD_u$ (e.g., Figure 9.48a). Precision of the $CBOD_u$ measurements is poor at the low concentrations that presently prevail in the river. The $CBOD_u$ submodel was calibrated to the observed distribution of $CBOD_u$ for the first diurnal survey ($k_{CBOD} = 0.10 \, d^{-1}$). Simu-

lations of the calibrated $CBOD_u$ submodel matched the observed distributions of the other four diurnal surveys reasonably well (e.g., Figure 9.48a).

Model performance for DO is presented for all five diurnal surveys for the upper and lower layers in Figure 9.48b and c, respectively. Simulations of daily average and diurnal values appear. The observed diurnals are equal to the dimensions of the bars that reflect the range about the daily average. Daily average DO concentrations remained near saturation in the upper layer (Figure 9.48b), offering little in the way of a test of the model. However, the unique depletions in the lower layer both upstream and downstream of the lake inflow (Figure 9.48c), and the observed diurnal variations in both layers (Figure 9.48b) offer good tests of model performance. The model performed well in simulating the upstream and downstream DO sags of the lower layer (Figure 9.48c). Further, the predictions of diurnal variations in the upper layer tracked the observations well for most of the surveys (Figure 9.48b). The diurnal variation of survey No. 3 was overpredicted, probably as a result of a nonuniform vertical distribution of phytoplankton in the upper waters (e.g., the volume-weighted concentration of CHL may have been less than the near-surface concentration measured). The success of the upper layer diurnal predictions supports the validity of the delta method that uses K_a and P_g (Chapra and DiToro 1991). The significant diurnal variations observed in the lower layer, beyond the depth of light penetration, reflects propagation from the upper layer, mediated by vertical mixing; as well as diurnal variation in the upper waters of Onondaga Lake discharged through the outlet to the river. The model cannot presently accommodate diurnal variations in the Onondaga Lake boundary condition. Therefore the model uses the average of measured diurnal variations in the lower layer to calculate the range in concentration.

Calibration procedures were used in the transport submodel to determine components of the flow budget and mixing processes, and in the $CBOD_u$ submodel. However, the

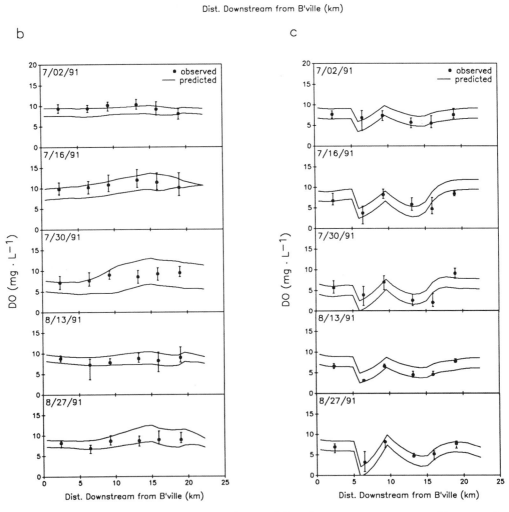

FIGURE 9.48. Seneca/Oswego Rivers water quality model performance: (**a**) CBOD$_u$ calibration, (**b**) top layer DO validation, and (**c**) bottom layer DO validation (modified from Canale et al. 1995b).

framework and coefficients of these submodels remained fixed in the modeling of DO. All other model inputs were established by measurements or the outcome of experiments. Based on the high performance of the model for DO (Figure 9.48b and c) for all five surveys, it is considered validated.

9.7.3.6 Sensitivity Analyses

Analyses were conducted with the river water quality model to establish the sensitivity of model simulations to reasonable levels of uncertainty in model coefficients. Probably the most uncertain model coefficient value is for k_{CBOD}, in part, because of the low concentrations of CBOD that presently prevail in the

river system. The calibration fit for CBOD for the first survey of 1991 is presented in Figure 9.49a, along with simulations corresponding to the value of $0.243\,d^{-1}$ determined on unfiltered samples. The higher value degrades the performance of the CBOD submodel considerably.

The absence of significant longitudinal structure in DO concentrations for the upper layer of the river under the monitored conditions of 1990 and 1991 confounds representative depiction of the potential impact of uncertainties in coefficients for potential future scenarios of reduced oxygen resources. This was remedied in this analysis by adopting a full METRO diversion option as the base case. Permit loadings were applied for all three

a

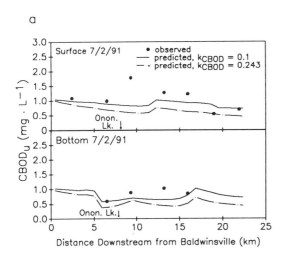

b

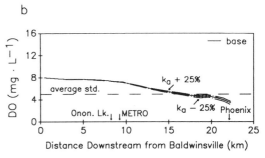

c

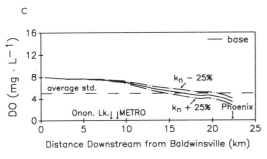

d

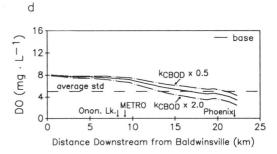

FIGURE 9.49. Sensitivity analyses for river water quality model for Baldwinsville to Phoenix reach: (a) for CBOD submodel, $k_{CBOD} = 0.243$, (b) DO model, $K_a \pm 25\%$, (c) DO model, $k_n \pm 25\%$, and (d) DO model, $k_{CBOD} \times 0.5$, and $k_{CBOD} \times 2.0$ (modified from Canale et al. 1995b). Conditions specified for (b), (c), and (d) include 7Q10 river flow, full METRO diversion, and destratified river.

domestic waste treatment facilities (Baldwinsville, METRO, and Wetzel Road), the 7Q10 river flow was used, METRO was input into cell 11, just downstream of the lake inflow (Figure 9.46c), and de-stratified (see subsequent treatment) river flow conditions were adopted.

The results of sensitivity analyses conducted with this base case for K_a, k_n, and k_{CBOD} appear in Figure 9.49b–d. The sensitivity limits do not have detailed statistical bases, but they are generally representative of the level of uncertainty believed to be associated with the determination of the various coefficients. Model predictions of DO, for the highly stressed loading conditions invoked, were relatively insensitive to the specified uncertainty limits for K_a (Figure 9.49b), somewhat more sensitive to k_n (Figure 9.49c), and highly sensitive to the uncertainty in k_{CBOD} (Figure 9.49d).

9.7.4 Invasion of the Zebra Mussel: Impact and Implications

9.7.4.1 Description of the Zebra Mussel

The zebra mussel (*Dreissena polymorpha*) is a bivalve mollusk that was native to southern Russia (Ludyanskiy et al. 1993). It took about 200 years for this invader to spread throughout western Russia and all of Europe (Ludyanskiy et al. 1993). It was introduced to the Great Lakes of North America in the mid-1980s, probably via water ballast from a foreign ship (Ludyanskiy et al. 1993; Mackie 1991). The zebra mussel is spreading much more rapidly in North America (Ludyanskiy et al. 1993), in part, because of a lack of natural ecological constraints (Ludyanskiy et al. 1993) and its high reproductive rate (Ramcharan et al. 1992b). In 1993 Ludyanskiy et al. reported the invader had spread to all the Great Lakes and eight major adjoining river systems, including the Three Rivers System. Zebra mussels prefer solid substrates for colonization; rock substrates support particularly dense populations in infested systems (Effler and Siegfried 1994; Mackie 1991), where high concentrations of food particles (e.g., phytoplankton) prevail (Schneider 1992). The environmental impact of the zebra mussel invasion of North America has been the subject of a great deal of speculation, extrapolation, and in some cases, documentation.

Selected metabolic functions and related features of material cycling of the zebra mussel are illustrated in Figure 9.50. Zebra mussels

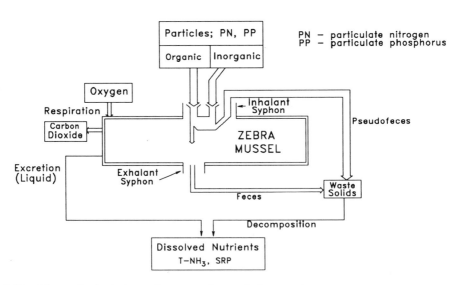

FIGURE 9.50. Schematic depicting zebra mussel metabolism and related features of material cycling (modified from Effler et al. 1995).

are efficient filter feeders; a particle removal efficiency of 100% has been reported for particles greater than 1 μm. It is this characteristic of the zebra mussel that is responsible for the improved clarity benefit attributed to its invasion of particle-rich systems (Leach 1993; Ludyanskiy et al. 1993; Reeders and Bij de Vaate 1992). Particles not selected for consumption, inorganic and certain organic particles, are embedded in mucus in the mussels and discharged (via the inhalant siphon, Figure 9.50) as pseudofeces (waste solids). Portions of the consumed material, not incorporated into tissue, are discharged as feces (waste solids, via the exhalant syphon, Figure 9.50), or excreted (dissolved waste). Reeders and Bij de Vaate (1992) observed that more than 90% of the solids discharged by zebra mussels in turbid eutrophic Dutch systems were psuedofeces. Mussel excretions are enriched in T-NH$_3$ (Quigley et al. 1993) and undoubtedly P. An important net effect of zebra mussel activity is the conversion of particulate (i.e., unavailable to phytoplankton) forms of nutrients to dissolved (i.e., available) forms (Figure 9.50). The zebra mussel uses oxygen (respiration), to support its basic maintenance and active feeding (Schneider 1992), and releases carbon dioxide as a by-product of this activity.

9.7.4.2 Invasion of Seneca River: Impact and Implications

Biological surveys (SCUBA) conducted in the summer and fall of 1993 (e.g., Effler and Seigfried 1994; unpublished data from Siegfried), found that zebra mussels occupied all available (rock) substrate along the bottom of the Seneca River from just below Cross Lake to Phoenix (Figure 9.43); undoubtedly the range of the invasion extends throughout the Barge Canal/Three Rivers System. Particularly high density populations were found in a 1.4 km reach just downstream of Cross Lake, known as the CUT (Figure 9.43a); population densities ranged from about 30,000 to 60,000 individuals · m^{-2} (Effler and Seigfried 1994). The extremely high densities in this reach apparently reflect the ideal sub-

strate (bedrock, exposed from channelization of this section as part of the construction of the Barge Canal in 1915) and high concentrations of food emanating from Cross Lake (Effler and Seigfried 1994). The size structure of the population indicated the vast majority of the zebra mussel infestation developed in 1992 and 1993 (Effler and Seigfried 1994), consistent with the first sightings for the system (fall 1990) listed in the review of Ludyanskiy et al. (1993).

Major depletions of DO occurred upstream of Baldwinsville in the summer of 1993, as a result of the respiration demands of the severe zebra mussel infestation (Effler and Seigfried 1994). An example DO profile is presented in Figure 9.51; note the greatest depletion occurred over the CUT, where the most dense zebra musssel populations existed (Effler and Seigfried 1994). The pattern downstream of the CUT (e.g., failure of DO to recover, Figure 9.51) reflects decreasing population densities of the mussels, which tracked the reductions in rock substrate, and the very limited reaeration that occurs in this channelized system (e.g., Canale et al. 1995b).

The metabolic activity of the invading zebra mussel (Figure 9.50) has caused major shifts in several important features of the water quality of the Seneca River upstream of the Onondaga Lake inflow (Figure 9.44). Note that the flow conditions (Figure 9.44a) for the two populations of water quality

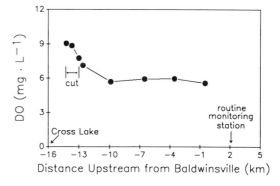

FIGURE 9.51. Typical DO profile for a portion of the Cross Lake to Baldwinsville reach of the Seneca River, late summer of 1993 (modified from Effler and Seigfried 1994).

measurements being compared in Figure 9.44, 1990/1991 and 1993, were similar, thus the changes since the major infestation cannot be attributed to diffrences in flow (dilution). Oxygen concentrations decreased greatly from 1990/1991 to 1993 (Figure 9.44b) as a result of the respiration demands of the severe zebra mussel infestation. The median DO concentration at Baldwinsville in 1993 was only $4.7\,mg \cdot L^{-1}$; a number of the observations from the river monitoring program (Figure 9.44b) depicted violations of the New York State standard of a minimum DO within a day of $4.0\,mg \cdot L^{-1}$. Additional, more intensive, monitoring over a 3 week period in August 1993 documented numerous violations of the standard (Effler et al. 1995). Chlorophyll concentrations decreased (Figure 9.44d) and clarity (Secchi disc transparency) increased (Figure 9.44e), as a result of the efficient filter feeding of the zebra mussel, and concentrations of available nutrients (Figure 9.44g and h) increased. The potential negative implication of increases in T-NH_3 on NH_3 toxicity concerns (e.g., associated with METRO diversion alternatives to the river) is mitigated by the attendant shift to lower pH values (see Chapter 5; Effler et al. 1994).

The changes in water quality brought about in this large river by the invading zebra mussel are impressive (Figure 9.44). This organism has converted the Seneca River (at Baldwinsville, NY) from a turbid, phytoplankton-rich, nutrient-depleted system, with slightly undersaturated oxygen concentrations, to a system with high clarity, low phytoplankton concentrations, enriched in available nutrients, with greatly undersaturated oxygen concentrations (Figure 9.44). Taken together, these changes represent a "signature" of the activity of the zebra mussel (Figure 9.50). This signature, and coupled upstream quantitative biological and water quality analyses (Effler and Seigfried 1994), represent compelling evidence of the central role this invasion has played in bringing about the changes in basic water quality features documented in Figure 9.44.

The changes in the water quality of the Seneca River caused by the zebra mussel invasion (Figure 9.44) present challenges for the effective management of the river system as well as Onondaga Lake. Of particular concern is the degradation of oxygen resources. This new DO sink has eliminated the assimilative capacity of this portion of the Seneca River for oxygen-demanding wastes, thereby confounding remediation efforts for Onondaga Lake focusing on METRO. Persistance of the infestation at the 1993 levels would require remediation to avoid continued violations of state water quality standards in the river. Under the conditions of 1993, existing wastewater discharge permits issued by the New York State Department of Environmental Conservation no longer protect the oxygen resources of the river.

There is little objective basis to assess the stability of zebra mussel infestations, and coupled water quality impacts, in rivers. The most dense lake infestations in Europe have been observed to be the least stable; e.g., peak densities for 1 to 2 years followed by a population crash (Ramcharan et al. 1992a). This may not apply for river systems, such as the Seneca River, where the upstream food supply remains stable. Ultimately the severity of the Seneca River infestation can be attributed to the cultural eutrophication of upstream portions of the river and Cross Lake (Effler and Carter 1987), as this is the primary food source that supports the invaders. Numerous effluent discharges, as well as nonpoint sources, contribute to the phosphorus load received by the river system upstream of Cross Lake.

The impact of zebra mussel infestation on oxygen resources documented here for the Seneca River (Figure 9.44b and c) may approach a worst case because of man's alterations of this river system, which have provided abundant food (cultural eutrophication), ideal substrate (e.g., bedrock surfaces of the CUT), and eliminated natural reaeration capacity (channelized flow). Undoubtedly zebra mussel invasion will result in some degradation of the oxygen resources of many other streams and rivers. Many streams and rivers in well-developed areas have little or no waste assimilative capacity to lose if oxygen resources are to continue to be protected. The costs of increased waste treatment or direct remedial action (e.g., oxygen injection) to counteract this impact of the zebra mussel invasion could be substantial. These potential costs have not been included in the analyses of economic impact of the zebra mussel invasion of North America reviewed by Ludyanskiy et al. (1993).

The zebra mussel invasion also offers a challenge to water quality scientists and engineers to develop an integrated quantitative understanding of these organisms and their impact on water quality that ultimately can be incorporated into the framework(s) of mechanistic management models. Recently Effler et al. (1994) have presented an approach to accommodate this new oxygen sink that can be handled within existing model frameworks. The existing Seneca River model, as documented herein (and by Canale et al. 1995b), should be modified (at least model inputs) and revalidated, to establish its continuing capabilities as a management tool for the recent major shifts in river conditions.

9.8 Application of Models

9.8.1 Introduction

In this section the predictive capabilities of the models are utilized to address selected management issues and options for Onondaga Lake and the adjoining Three Rivers system. In some cases they are used to assess the impact of existing pollutant loads and to partition the contributions of selected processes to prevailing conditions. However, the greater emphasis is placed on application of the models to evaluate selected management options for METRO from the perspective of meeting water quality standards and goals for the lake and protecting the water quality of the adjoining river system. Credible application of models for futuristic scenarios (i.e., a priori predictions) requires: (1) appropriate loading information, (2) appropriate assumptions for ambient environmental conditions, and

(3) model frameworks that accommodate regulating processes realistically over the full range of applications.

For model application to support management decisions, it is necessary to have established water quality goals and standards to compare model output to, and the critical environmental conditions under which the goals and standards must be met. The present status with respect to the various water quality constituents for the river and lake on the issues of standards and critical conditions is perhaps best described as uneven; well established in certain cases (e.g., DO standard(s), and critical temperature and flow conditions in rivers), and still evolving in others. Despite imperfections with respect to certain of these requirements that are identified herein, these models, and the information set they are based on, are more than adequate to support important management decisions regarding the reclamation of important features of this polluted lake.

Items addressed in this model application section include:

1. evaluation of anthropogenic contributions to historic and present Cl^- concentrations in Onondaga Lake,

2. development and linkage of management TP, N, and DO models for the lake (see Effler and Doerr 1995),

3. application of the lake management models to evaluate the adequacy of selected options for METRO to meet water quality standards and goals for the lake (see Effler and Doerr 1995),

4. evaluation of the implications of control of CSO loads with respect to the lake's trophic state and concentrations of public health indicator bacteria after runoff events,

5. demonstration of the need to destratify the Seneca River proximate to the lake inflow to meet water quality standards in the river, and

6. application of the Seneca/Oswego River water quality model to evaluate the adequacy of treatment alternatives for diverted wastewater to protect the oxygen resources of the river system.

9.8.2 Anthropogenic Contributions to Lake Cl⁻ Concentrations

The Cl⁻ model and available information on sources of continuing Cl⁻ loading (Chapter 3) are used here to predict what the Cl⁻ concentration of Onondaga Lake would have been over the 1986 to 1991 interval without the continuing inputs of saline waste from soda ash production. This model analysis incorporates the recently quantified loads from downstream Onondaga Creek and the Seneca River used in the sensitivity analysis of Figure 9.9. The effect of the waste input from the soda ash/chlor-alkali facility is eliminated here by using available Cl⁻ concentration data from just above the waste beds (average of $54 \, mg \cdot L^{-1}$; Effler et al. 1991a, Chapter 3) to estimate the Ninemile Creek load, and by eliminating the contribution of the lagoon inputs to the METRO load (Chapter 3). Model projections for this scenario are presented in Figure 9.52 for the 1986–1991 interval, invoking these reduced loads since the closure of the facility. The average lake concentration for the 1987–1991 interval would have been about $245 \, mg \cdot L^{-1}$; 43% lower than the observed level. The projected average concentration for 1990 and 1991, a period that arguably represents a new steady-state for the lake as the lake had flushed about 13 times

since the closure of the facility, is about $230 \, mg \cdot L^{-1}$. Thus the continuing inputs from the soda ash/chloralkali facility represent about 50% of the prevailing lake Cl⁻ concentration. The projected average concentration without this source ($\approx 230 \, mg \cdot L^{-1}$) represents about 15% of the preclosure concentration (average of $1585 \, mg \cdot L^{-1}$ for the period 1973–1985). Therefore, the soda ash/chloralkali facility's contribution before its closure was about 85%, much more than certain earlier estimates (O'Brien & Gere Engineers 1973; Rooney 1973), but consistent with the estimate of Effler and Driscoll (1985). The possibility that Cl⁻ inputs presently received from other tributaries in the southern basin originate in part from some of the older waste beds (Chapter 1), or that Cl⁻ inputs may be leached directly into the lake from these deposits along the bordering shoreline, cannot be eliminated. Thus the contribution of the soda ash/chlor-alkali facility presented above is probably conservative.

Evaluation of other anthropogenic contributions is less straightforward. Little additional change in lake concentration (e.g., $<10 \, mg \cdot L^{-1}$) would be achieved by elimination of the METRO discharge (presently under consideration for other water quality issues), as the background concentration of METRO is nearly equal to the above projected value. Elimination of the METRO discharge would probably be accompanied by increased backflow from the Seneca River into the lake (Chapter 4), thereby diluting lake Cl⁻ levels. The river inflow is itself an anthropogenic effect associated in part with the lowering of the lake and the regulation of the river elevation for navigation (Canale et al. 1995b; Chapter 4; also subsequent sections of this chapter). The extent to which the present Onondaga Creek load is anthropogenic is open to debate. Issues that bear on this include: (1) the origins of the "mud boils" (Chapter 2), (2) the extent to which anthropogenic inputs from the Tully Valley brine mine area occur, (3) possible contributions from the adjoining waste beds (Chapter 1), (4) the effects of the use of NaCl for street deicing in Syracuse, and (5) the effects of well drilling, changes in

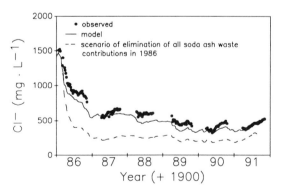

FIGURE 9.52. Model projection of time-course of Onondaga lake Cl⁻ concentrations for the scenario of the elimination of Cl⁻ load associated with the soda ash/chloralklai facility with the closure of the facility (modified from Doerr et al. 1994).

channel configuration, and lowering of the lake, on groundwater flow (e.g., seeps) adjoining the creek.

The concentrations that would prevail in the lake without anthropogenic loads are probably high (e.g., as high as $200 \, mg \cdot L^{-1}$; Figure 9.52), compared to levels observed in most lakes (Wetzel 1983). This is a result of Onondaga Lake's hydrogeologic setting (Chapter 2). The reductions in Cl^- concentrations that would result from elimination of continuing loading of Cl^- associated with soda ash production (mostly from the waste-bed area of Ninemile Creek) would greatly reduce the duration and extent of the bidirectional flow phenomenon in the outlet by reducing the density difference between the lake and the river (Chapter 4). This in turn would reduce the occurrence and extent of density stratification in the Seneca River proximate to the lake inflow (Canale et al. 1995b), thereby ameliorating the associated negative impact on the oxygen regime (Figure 9.45). Subsequently, the Seneca River water quality model is applied to demonstrate the dominant role density stratification presently plays in the occurrence of water quality violations in the lower layer of the river (also see Canale et al. 1995b).

9.8.3 Development and Linkage: Lake TP, N, and DO Management Models

9.8.3.1 General, Built-In Kinetics, and Physical Conditions

Personal computer-based management models to support a priori predictions of lake water quality have been developed from the tested TP, N, and DO models. The programs retain the basic model structures (simpler hydrodynamic framework adopted), and values for kinetic coefficients and certain forcing conditions. However, other forcing conditions and state variables of management interest become user-defined. Additionally, user-friendly interfaces have been added to facilitate selection of program options and use of the models by managers. Graphical output of

seasonal distributions of material concentrations in both model layers, and related summary statistics, are provided. Plot files depicting validation or calibration/verification performance are provided for comparison purposes. Conditions specified in the three lake management models are described in Table 9.12. Conditions incorporated in the management models are of two types, *hardwired* or *built-in*, which are not subject to change by the user, and *default*, which the user can override. Both types incorporated in the three models are specified in Table 9.12. Kinetic coefficient values developed for the models and used in testing are retained in the management models; these are also listed in Table 9.12.

Physical characteristics not subject to management action are built-in to the management models (Table 9.12). Several of these have been handled identically for the three models because of the unifying manner in which each treats the stratification/vertical mixing regime. Average water temperatures of the two layers are daily averages for 1989, calculated by interpolation between measurements. The temporal distribution of v_t is that determined for 1989 (Figure 9.15). Daily average wind speeds determined as the average of observations made at Hancock Airport in 1989 and 1990, are specified in the DO model. Wind speed is subject to substantial natural year-to-year variations in this region (Effler et al. 1986d; Owens and Effler 1989), which in turn can have significant impacts on measures of water quality (Lam et al. 1987; Owens and Effler 1989). Thus the unavoidable specification of environmental forcing conditions necessary to support management model (a priori) predictions inherently introduces a degree of uncertainty. Earlier sensitivity analyses provided some insight into the relative importance of this form of variability on model predictions. For example, realistic variations in v_t have a relatively minor effect on constituent concentrations for the UML layer (Canale et al. 1995a; Doerr et al. 1995a) compared to the effect of variations of wind on DO concentrations during the fall mixing period (Figure 9.35d). Subsequently, the

TABLE 9.12. Conditions specified in lake management models; TP, N species, and DO.

	Description	TP	N species	DO
a. Built-in parameters				
1. Physical				
(a) layer temperature	interpolated temperatures	1990 (LML)	1989	1989
(b) vertical mixing, v_t		1990	1989	1989
(c) settling, velocity, PP:TP		$vel_{PP} = 214\,m \cdot yr^{-1}$ polynomial 1990	$vel_N = 214\,m \cdot yr^{-1}$	—
(d) wind speed	Hancock Airport, daily average	—	—	average of 1989 and 1990; U_{10}
2. Loadings, Tributary				
(a) flows and loads	daily; protocol in text of this chapter and Chapter 3	1987 and 1990	1976, 1986, 1987	1986
3. Stoichiometry				
(a) fraction T-NH$_3$, as NH$_3$		—	1989	—
(b) p-PON/CHL		—	5.2	—
(c) DO model		—	—	$a_{CP} = 42.7,\ a_{OC} = 2.7$
4. Standards/Limits		—	"probability of contravention" distribution	—
5. Kinetics				
(a) sediment release		$R_{scd,8} = 13.3\,mg \cdot m^{-2} \cdot d^{-1}$	$S_{r,8} = 70\,mg \cdot m^{-2} \cdot d^{-1}$	—
(b) other		—	$k_{dccomp,20} = 0.1\,d^{-1}$	$K_{L,DO} = f(U_{10})$, eq. 9.30
		—	$k_{hyd,20} = 0.005\,m \cdot d^{-1}$	$k_{CBOD} = 0.28\,d^{-1}$
		—	$k_{fn,20} = 0.135\,m \cdot d^{-1}$	$K_{S,DO} = 3.5\,mg \cdot L^{-1}$
		—	$k_{fd,20} = 0.4\,m \cdot d^{-1}$	f = 0.72
		—	$K_{L,NH3} = 0.17\,m \cdot d^{-1}$	—
(c) net phyto. growth	base "shape" 1989	—	Figure 9.20a	—
b. Default				
1. METRO				
(a) flow	monthly avg. 1988–1990	X	X	—
(b) load		1990 annual average	polynomial 1988–1990	X
2. tributary flow and load		1990	1986	1990 monthly average
3. fraction reduction on trib load		0–1, 0 = default		—
4. trophic state		—	75 μgP·L^{-1}	—
5. DON/TON METRO		—	0.62	TP75.tsf file from N model
6. SOD		—	—	1.68 g·m^{-2}·d^{-1}

inherent limitation of the N management model in predicting year-specific conditions, because of natural variations in tributary flow, is demonstrated.

Initial concentrations of modeled constituents are determined by running the models through several consecutive annual cycles internally until steady-state time distributions of the constituents are attained.

9.8.3.2 Trophic State/Phytoplankton Submodel Linkages

Changes in the phytoplankton sink for T-NH$_3$ and source for DO that would result from changes in trophic state need to be accommodated to support futuristic projections for scenarios that include major reductions in P loading. The Monod-type relationship between the summer average epilimnetic TP concentration and gross photosynthesis developed by Auer et al. (1986) (Figure 9.53), for the trophic state gradient that persists across Green Bay, Lake Michigan, has been adopted. The Monod character of the relationship is quite important. Summer average TP concentrations $>40\,\mu g \cdot L^{-1}$ are essentially saturating (Auer et al. 1986). In other words, little reduction in growth is achieved until summer average TP concentrations decrease below this saturating threshold. The 1989 (G$_{NET}$) distribution has

been adopted in the N management model as generally representative of the recurring "shape" (seasonality) for the lake. The magnitude of G$_{NET}$ is "scaled down" for reduced TP concentrations according to the response of the Auer et al. (1986) relationship (see example in Figure 9.53).

The Onondaga Lake TP, N and DO management models are linked to the extent that output of one serves as input to the other, in the sequence, TP $\rightarrow$ N $\rightarrow$ DO. This is the appropriate sequence of operation of the three models for a selected management scenario. The TP model predicts the summer average TP concentration of the UML. This becomes one of the inputs to the N model, which generates the distribution of G$_{NET}$, and predicts distributions of CHL and T-NH$_3$; these become inputs to the DO model. Two pathways are provided to accommodate trophic state (TP concentrations) conditions in the DO management model. The user may select one of five permanently stored files that correspond to summer average TP concentrations of 10, 30, 40, 60, and 75 $\mu g \cdot L^{-1}$, or generate files for specific management options (with different TP concentrations) by serial operation of the management TP and N models.

The kinetics of phytoplankton uptake of N have been modified for the UML in the N management model to accommodate scenarios in which the concentration of T-NH$_3$ would decrease to levels which favor NO$_3^-$ uptake as an alternate source of N. The switch from T-NH$_3$ to NO$_3^-$ uptake is fixed at a T-NH$_3$ concentration of 50 $\mu g \cdot L^{-1}$. The respective mass balance equations for T-NH$_3$ and NO$_x$ are modified according to the following expressions (note the inclusion of a preference factor (pref) for T-NH$_3$).

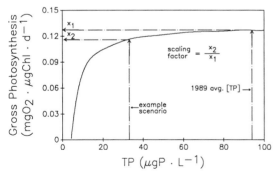

FIGURE 9.53. Gross photosynthesis per unit chlorophyll *a* as a function of summer average TP concentration, developed across trophic state gradient of Green Bay Lake Michigan by Auer et al (1986). Example of use to develop "scaling factor" to adjust G$_{NET}$ in the N model and SOD in DO model (modified from Effler and Doerr 1995).

Ammonia

$$V_1 \frac{d[\text{T-NH}_3]_1}{dt} = W_{\text{T-NH}_3} - Q \cdot [\text{T-NH}_3]_1$$
$$+ K_{\text{hyd,T}} \cdot [\text{DON}]_1 \cdot V_1 - G_{\text{NET,1}} \cdot$$
$$[\text{p-PON}]_1 \cdot V_1 \cdot (\text{pref}) - K_{\text{fn,T}} \cdot (A_s - A_i) \cdot$$
$$[\text{T-NH}_3]_1 - K_{\text{L,NH}_3} \cdot A_s \cdot (\text{ff} \cdot [\text{T-NH}_3]_1)$$
$$+ v_t \cdot A_i \cdot ([\text{T-NH}_3]_2 - [\text{T-NH}_3]_1) \quad (9.74a)$$

$$V_2 \frac{d[T\text{-}NH_3]_2}{dt} = K_{hyd,T} \cdot [DON]_2 \cdot V_2$$

$$- G_{NET,2} \cdot [p\text{-}PON]_2 \cdot V_2 - K_{fn,T} \cdot A_i \cdot$$

$$[T\text{-}NH_3]_2 + v_t \cdot A_i \cdot ([T\text{-}NH_3]_1 - [T\text{-}NH_3]_2)$$

$$+ S_{r,T} \cdot A_i \qquad (9.74b)$$

Nitrogen Oxides

$$V_1 \frac{d[NO_x]_1}{dt} = W_{NO_x} - Q \cdot [NO_x]_1 +$$

$$k_{fn,T} \cdot (A_s - A_i) \cdot [T\text{-}NH_3]_1 - G_{NET,1} \cdot$$

$$[p\text{-}PON]_1 \cdot V_1 \cdot (1 - pref) +$$

$$v_t \cdot A_i \cdot ([NO_x]_2 - [NO_x]_1) \qquad (9.75a)$$

$$V_2 \frac{d[NO_x]_2}{dt} = k_{fn,T} \cdot A_i \cdot [T\text{-}NH_3]_2$$

$$- k_{fd,T} \cdot A_i \cdot [NO_x]_2 - v_t \cdot A_i \cdot ([NO_x]_1$$

$$- [NO_x]_2) \qquad (9.75b)$$

in which pref = preference factor for T-NH$_3$ by phytoplankton = 1.0 for $[T\text{-}NH_3] >$ $50\,\mu g \cdot L^{-1}$, = 0.0 for $[T\text{-}NH_3] \leqslant 50\,\mu g \cdot L^{-1}$. Similar algal preference formulations are described by Bowie et al. (1985).

9.8.3.3 Standards and Water Quality Goals

9.8.3.3.1 Background and TP Goal(s)

Water quality managers need to consider model output associated with various management options from the perspective of the acceptability of predicted concentrations with respect to established standards and goals. The water quality standards that are in place in the State of New York for the parameters of the Onondaga Lake models are of two basic types, *numerical* and *narrative*. Well-established numerical standards for DO (e.g., daily average $\geqslant 5\,mg \cdot L^{-1}$, Chapter 5) and fecal coliform bacteria (Chapter 6) exist. The USEPA (1985) established criteria for NH$_3$ to protect against its toxic effects (Chapter 5), which have been adopted by the State of New York as its standards. The NH$_3$ standards issue is complex by comparison to standards for DO and indicator bacteria, in part because the concentration of NH$_3$ is not measured directly but is calculated as a fraction of the analyte T-NH$_3$

(Effler et al. 1990a, Chapter 5). Recall the fraction of T-NH$_3$ that exists as NH$_3$, and the toxicity of NH$_3$ to fish, are both highly dependent on pH and temperature (Chapter 5). A reasonable approach in the development of such a standard for a particular lake is to quantify the temporal distribution of T-NH$_3$, which incorporates the seasonality in pH and temperature, that corresponds to the NH$_3$ standard. This approach is utilized in this section to develop a "probability of contravention" distribution for T-NH$_3$ in the UML of Onondaga Lake. The USEPA (1985) provided guidelines for modifying the national criteria on a site-specific basis where it was warranted, based on: (1) unique sensitivities of species at specific sites, (2) unique water quality characteristics, or (3) both species sensitivity and water quality characteristics. Subsequent testing of the water quality effect for 11 sites indicated only modest deviation from the national criteria (Hansen 1989).

In contrast to the other parameters, there is no numerical standard for TP in the State of New York. Instead there is a narrative standard "...none in amounts that will result in growths of algae, weeds and slimes that will impair the best usages," and a recently established guidance value (open to some regulatory discretion) of $20\,\mu g \cdot L^{-1}$, as the mean summer epilimnetic TP concentration (New York State Department of Environmental Conservation (NYSDEC) 1993). The guidance value is based on empirical relationships with aesthetic effects for primary and secondary recreation (e.g., Kishbaugh 1993). Additional guidance is provided by published trophic state limits (e.g., Auer et al. 1986; Chapra and Dobson 1981; Vollenweider 1975, 1982; see Chapter 5) and the TP response curve developed by Auer et al. (1986), presented here as Figure 9.53. The lower bound of eutrophy in these works range in TP concentrations from 20 to $37.5\,\mu g \cdot L^{-1}$. Thus management scenarios that fail to reduce the summer average concentration of TP in the UML to less than $40\,\mu g \cdot L^{-1}$ are not expected to substantially change the lake's trophic state, the prevailing nearly saturated state of P availability to phytoplankton (Chapter 6), and related degraded features of water quality in the lake (Chapters 5–7). In contrast, a management scenario that resulted in summer average TP concentrations $\leqslant 30\,\mu g \cdot L^{-1}$ would be accompanied by significant reduction in phytoplankton growth. Increasing responsiveness with respect to reductions in phytoplankton growth are expected for further reductions in TP concentrations (Figure 9.53), consistent with distinctly mesotrophic characteristics (Auer et al. 1986; Vollenweider 1982).

9.8.3.3.2 T-NH₃ Limits: Probability of Contravention of NH₃ Criteria

Probability of contravention distributions are presented with the T-NH₃ output of the Onondaga Lake N management model, to form a basis of reference in evaluating the acceptability of predicted concentrations in meeting regulatory standards. The probability of contravention distributions developed herein for the UML faithfully represent the form of the seasonality that emerges from the observed seasonality in pH and temperature (Effler et al. 1990a, Chapter 5). Here we review the development of the probability of contravention distribution incorporated in the N model and factors that affect the details of the distribution.

Establishment of a seasonal probability of contravention distribution, or a seasonal standard, requires specification of daily or period average values for temperature and pH. Substantial interannual variability in the distributions of these parameters is observed in the lake (Chapters 4 and 5), particularly for pH in the upper waters, associated in some instances with natural meteorological variability and coupled variations in phytoplankton growth. Thus a probabilistic approach, in the spirit of the critical conditions concept applied to DO in rivers, was used to define a seasonally variable probability of contravention distribution for T-NH₃. First, considerations involved in the selection of an appropriate data base are reviewed.

Application of the EPA criteria to the Onondaga Lake problem requires important regulatory decisions concerning the selection of the appropriate pH data base for the probabilistic analysis. At least two important issues emerge, the depth range from which pH data are considered, and the use of laboratory versus field pH. Recall (Chapter 5) that significant vertical structure in pH within the epilimnion (highest pH near the surface) is not unusual in the lake, particularly during calm periods of high phytoplankton production. Further, laboratory measurement of pH, as opposed to field measurements, apparently often introduces some bias (false high values), associated with degasing of CO_2 from the samples, except at the highest pH (i.e., most critical) values (Chapter 5).

The potential implications of these issues are illustrated by the comparison of the temporal distributions of T-NH₃ concentrations in 1991 (Figure 9.54) that represent the USEPA chronic toxicity criterion for non-salmonid fish (final chronic 73 value, FCV_n) based on: (1) laboratory pH and field temperature measurements made at a depth of 1 m, (2) field pH and temperature measurements made at a depth of 1 m, and (3) average field pH (i.e., from average H^+ concentrations) and temperature over the 0 to 8 m (e.g., 9 individual depth measurements) depth interval. The most stringent limit for T-NH₃

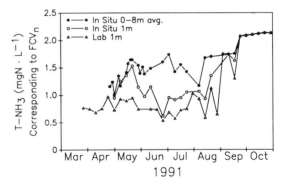

FIGURE 9.54. Temporal distributions of T-NH₃ concentrations in 1991 that represent the USEPA chronic toxicity criterion for free ammonia for nonsalmonids, based on: laboratory pH and field temperature measurements made at a depth of 1 m, field pH and temperature measurements made at a depth of 1 m, and average field pH and temperature over the 0 to 8 depth interval.

(highest pH values and temperatures; see [NH₃] and related criteria calculation protocol in Chapter 5), and the most protective for the lake, would emerge from the distribution derived from laboratory pH data from near surface (1 m) waters (Figure 9.54). The distribution based on field measurements of pH over the same depth is significantly less stringent for much of the spring–fall interval (Figure 9.54). However, the effect on facility requirements to meet specified limits may not differ significantly, as this distribution converges with the one based on laboratory measurements at critical (the lowest) T-NH₃ concentration limits (i.e., the highest pH values). The most lenient, and least protective, limit of the three alternatives emerges from depth averaging (Figure 9.54), largely because depth averaged H⁺ concentrations tend to be higher and temperatures lower than values measured near the surface. Selection of the more stringent limit associated with conditions in the upper waters (e.g., field measurements at 1 m) seems more consistent with the protective spirit of standards set for other parameters, because fish would thereby be protected throughout the depths of the epilimnion, instead of only at the "average" conditions within that depth interval.

The data base selected for the probabilistic analysis, and the basis for the development of the probability of contravention distributions incorporated in the N management model, are laboratory pH and field temperature measurements made at a depth of 1 m over the spring–fall intervals of 1988–1992 (143 paired measurements). The corresponding determinations of the values of FCVₛ (salmonid final chronic value) and FCVₙ for the data base appear in Figure 9.55 (as points), along with the probability of contravention distributions fit to the values. On any date, the value of the FCV is described by a probability distribution that reflects interannual variability in pH and temperature. Graphical analysis indicates that a normal distribution (bell-shaped curve) adequately portrays this variability. The location and shape of the distribution are defined by its mean and standard deviation. Values for the mean for each date

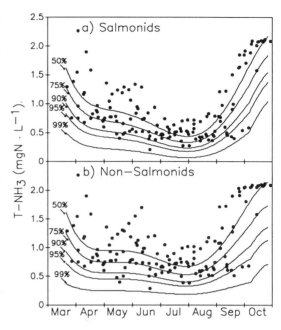

FIGURE 9.55. Distributions, as family of curves of "probability of contravention" of free ammonia criteria for 1 m depth in Onondaga Lake: (a) salmonids, and (b) nonsalmonids. Criteria values, as points, correspond to observed conditions over 1988–1992 interval.

were estimated by fitting a trend line using polynomial regression. The spread of the points about that line was used to estimate the standard deviation. A distribution of FCV values, reflecting the interannual variability in environmental conditions, is thus generated for each date in the monitored interval. The distributions may be used to identify daily values for the FCV, which will insure protection with a specified level of confidence, e.g., 95% of the time. Seasonal distributions of five different levels of confidence (ranging from 50 to 99%) of avoiding contravention of the FCVₛ and FCVₙ have been included in Figure 9.55(a and b). Note the more stringent limits for midsummer, for salmonids (Figure 9.55a) versus nonsalmonids (Figure 9.55b), and for a greater level of confidence of avoiding contravention of the criteria. The 50% confidence distributions would allow for contraventions about 50% of the time. The 95% confidence distributions, which allow for contraventions about 5% of the time, were incorporated in the N management model output. In 1994 the state regulatory agency (NYSDEC) established the in-lake standard for T-NH$_3$ as 0.77 mgN $\cdot$ L^{-1}, based on critical ambient conditions they specified as pH = 8.0 and temperature = 25°C. This in-lake standard corresponds approximate to a confidence level of 50%, according to the probabilistic analysis and data base of Figure 9.55.

9.8.3.4 Loads and Flows

Users of the management models are provided with an input screen for entering monthly average discharge flow and effluent concentrations from METRO; TP concentrations for the TP model, T-NH$_3$, NO$_x$, and TKN concentrations for the N model, and BOD (5-day) for the DO model. Organic N (TON) in the effluent is partitioned into PON (d-PON) and DON concentrations in the N model either according to the ratio DON : TON = 0.62, the average reported for 1989 and 1990, or by a user specified ratio. Predictions of T-NH$_3$ are rather insensitive to realistic variations in this partitioning. Recall that predictions of DO for the lake, under prevailing conditions, are not very sensitive to BOD loads. Default conditions for METRO inputs are identified in Table 9.12.

Two tributary flow and TP loading regimes are provided in the TP model to reflect the influence of natural variability in runoff on TP concentrations in the lake. The regimes correspond to conditions documented in 1987 and 1990, which represent a wide range in tributary flow and for which particularly detailed TP loading information is available (Chapter 3). The third highest tributary flow over the 1971–1990 (20 years) interval occurred in 1990; the second lowest over that interval was observed in 1987 (Chapter 3). Nonpoint loading reduction scenarios are accommodated in the TP model with a fractional reduction multiplier (e.g., 0.1× corresponds to a 10% reduction in nonpoint, or tributary, loading) that is applied uniformly throughout 1990 or 1987 loads.

Three flow regimes are provided in the N management model to reflect the important influence (Figure 9.26a) of natural variability

in runoff on T-NH$_3$ concentrations. Selection of certain of these runoff cases was made to be generally consistent with the accepted approach for identifying critical stream and river flow conditions (e.g., return frequency of one in ten years). The three runoff options in the model include (see Chapter 3):

1. 1986, a year that approaches the median flow for the 1971–1990 period,
2. 1976, the second highest flow year over the 1971–1990 period, and
3. 1987, the second lowest flow year over the 1971–1990 period (e.g., one in ten year return frequency).

The corresponding tributary loads of the N species for these flow regimes are calculated, with the exception of the Ninemile Creek T-NH$_3$ and NO$_x$ loads, using temporally uniform concentrations equal to the observed averages. This simplification is supported by observations and the fact that METRO is presently the dominant source of N (particularly T-NH$_3$) species (Chapter 3). The Ninemile Creek T-NH$_3$ and NO$_x$ loads are calculated according to documented (Effler et al. 1991a, Chapter 3) flow-concentration relationships.

9.8.3.5 Selection of Management Model Options/Inputs to Support Testing of Management Scenarios

9.8.3.5.1 Flow: Implications of A Priori Predictions

Predictions of the TP management model for the UML of Onondaga Lake for default loading conditions (Table 9.12) are presented for both the high and low tributary flow options in Figure 9.56. The predictions for the wet and dry years are quite similar for the critical summer period; average TP concentrations over the mid-May to mid-September interval are 72 and 76 µg $\cdot$ L^{-1}, respectively. This reflects uniformity in the METRO input, and the compensating effect of slightly decreasing TP concentration with increasing flow observed for the tributaries (Chapter 3). The dry year (1987) option is utilized in subsequent projections for selected management scenarios to maintain consistency with the distinctly critical

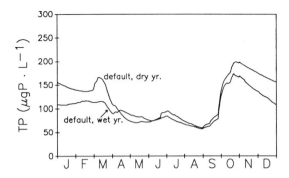

FIGURE 9.56. Predictions of the TP management model for the UMLL of Onondaga lake for default loading conditions; high and low tributary flow cases included.

conditions for T-NH₃ concentrations in the lake associated with low tributary flows.

Simulations of T-NH₃ concentrations in the UML generated by the N management model, using default inputs, for the three tributary flow options appear in Figure 9.57. The critical role tributary flow plays in diluting the METRO loading, demonstrated earlier with sensitivity analyses, is clearly manifested in this output. The low flow ("dry year") case in the management model has the greatest value for evaluating management scenarios, as it can be described as a *reasonable* worst case. Recall that the 1987 daily flows incorporated were the second lowest over a 20 year period (Chapter 3). Further, the G_{NET} conditions

adopted (1989) reflect representative (but not the highest; Chapters 5 and 6) phytoplankton growth during late summer and early fall, which avoids an elevated sink effect on T-NH₃ concentrations in the following spring (Chapter 5). The low flow (1987) option of the N management model has been adopted in the subsequent evaluation of management scenarios. Regulators may specify different tributary flow conditions for final design conditions. Predictions for the "average" flow year fall below the "wet" simulations in December and January because tributary flow was substantially higher in December 1986 than in 1976.

The temporal distribution of the concentration of T-NH₃ and performance of the N model for the calibration year of 1989 (previously as Figure 9.24d) have been added to Figure 9.57 to illustrate the inherent limitations of a priori predictions compared to a posteriori simulations. The bounds of the a priori management model predictions bracket the 1989 observations for most, but not all of the spring to fall interval; measurements fall below the envelope for much of August and September. Most important for planning purposes, the critical high T-NH₃ concentrations of late spring to early summer of 1989 are accommodated by the dry year case of the management model. In spring the observations are close to the upper bound of the a priori predictions, established by the low tributary runoff conditions of 1987, but the observations of midsummer correspond more closely to the high runoff conditions of 1976 specified in the management model. The variable fit of the a priori predictions is consistent with the documented dynamics of tributary flow in 1989 (relatively low flow in spring and high flow in late summer) and, to a lesser extent, the particularly effective seasonal nitrification achieved at METRO in the summer of 1989 (Chapter 3).

The model calibration for the spring to fall interval of 1989 (Figure 9.57, shown earlier in Figure 9.24d) is an example of a posteriori simulation; that is, based on measurements of environmental forcing conditions that actually prevailed during the period of mea-

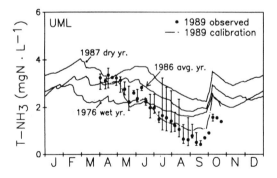

FIGURE 9.57. Predictions of T-NH₃ for the UML of Onondaga Lake from N management model, for three tributary flow scenarios. Observations of T-NH₃ (bars represent range within UML) and N model calibration for 1989 included for comparison (modified from Effler and Doerr 1995).

surements of T-NH$_3$ in 1989. In contrast, the a priori predictions of the N management model (e.g., futuristic, such as projecting in 1988 what to expect, before it happens, in 1989) necessarily invoke assumptions for certain environmental forcing conditions that can only be known in detail "after the fact." Natural variations in tributary flow have important implications for T-NH$_3$ concentrations in the UML of Onondaga Lake (Figure 9.57, also see Figure 9.26a). As demonstrated here, a priori predictions from management models have inherent limitations for year-specific conditions (e.g., Bierman and Dolan 1986) compared to a posteriori simulations (e.g., validation, calibration/verification efforts; Figures 9.6, 9.17, 9.24, 9.25, 9.32, and 9.33). Thus benefits of certain features of model frameworks clearly manifested in comparative evaluations of a posteriori simulations may be masked for management (a priori) applications.

9.8.3.6.2 Sediment Feedback

Prevailing sediment feedback in Onondaga Lake has been quantified either through laboratory experiments conducted on intact core samples (P release and SOD; Penn et al. 1993) or by model calibration (T-NH$_3$). These fluxes implicitly accommodate prevailing conditions in the overlying water column (e.g., phytoplankton growth, and organic deposition). However, the possibility cannot be discounted that these fluxes are not presently in equilibrium with water column conditions because of recent changes in deposition (e.g., Chapter 8). The existing Onondaga Lake water quality models do not simulate the changes in sediment feedback (i.e., there is no mechanistic sediment submodel) that would occur in response to major improvements in overlying water quality. Modeling of sediment feedback is receiving increasing attention (e.g., Chapra and Canale 1991; DiToro et al. 1990; Snodgrass 1987). However, documentation of successful performance of mechanistic sediment models through model postaudit evaluation does not yet appear to have reached the peer-reviewed

literature. Onondaga Lake represents an ideal test system to pursue research necessary to support the development of a credible mechanistic sediment submodel. A unique opportunity for model postaudit may present itself as part of the remediation of the cultural eutrophication problems of this lake.

The added capability of a mechanistic sediment submodel would contribute little to the important management decisions that need to be made related to trophic state and the exceeding high T-NH$_3$ concentrations that prevail in the lake, because upward transport of P and T-NH$_3$ from the enriched LML is small compared to continuing external loads (see Figures 9.18e and 9.26d). The large quantities of TP and T-NH$_3$ transported to the upper waters of the lake during fall mixing are largely flushed out of the system before the productive months of the subsequent year (Chapter 3).

In contrast, SOD plays a central role in the oxygen problems of the lake of rapid loss of hypolimnetic DO (Effler et al. 1986c) and lakewide depletions during the fall mixing period (Effler et al. 1988; Gelda et al. 1995a). This is demonstrated in Figure 9.58, as the relationship between SOD and the fall minimum DO, generated by application of the DO management model, for the scenario of full bypass of the METRO discharge (see subsequent sections). The rather broad band is empirically based, it reflects the substantial variability in this feature of the oxygen resources documented for the lake over the 1987 to 1992 period (Chapter 5), that appears to be driven by natural meteorological variability. According to the relationship of Figure 9.58, the SOD would need to decrease to about $1.3 \, \text{g} \cdot \text{m}^{-2} \cdot \text{d}^{-1}$ to avoid contravention of the $5 \, \text{mg} \cdot \text{L}^{-1}$ standard during the fall mixing period for an average year. Even at this reduced SOD, violations would occur in some years, associated with natural variations in meteorological forcing conditions. Yet greater reductions in SOD (e.g., to about $0.6 \, \text{g} \cdot \text{m}^{-2} \cdot \text{d}^{-1}$) would be required to avoid violations in all years (Figure 9.58). Assuming equilibrium conditions between sediment–water processes and the productivity of the

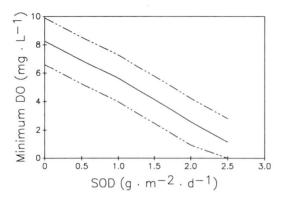

FIGURE 9.58. Predicted relationship between SOD and fall minimum DO in Onondaga Lake, for the scenario of full bypass of the METRO discharge around the lake, by application of the DO management model. Solid line for average conditions. Dimensions of dashed line represent approximate magnitude of natural variability estimated from recent observations.

lake's watercolumn prevail, reductions in SOD are expected only following reductions in POC deposition (e.g., DiToro et al. 1990), such as would accompany reductions in watercolumn phytoplankton productivity. According to the relationship of Figure 9.53, significant reductions would not be expected unless the summer average TP concentration of the epilimnion decreased to $<40\,\mu g \cdot L^{-1}$.

The existing mechanistic simulation models of SOD generally consist of three submodels (DiToro et al. 1990; DiToro and Connolly 1980; Klapwijk and Snodgrass 1986; Snodgrass 1987). First, the deposition of organic matter to the bottom muds must be predicted. Second, the diagenesis of two or three functional groups of organic matter, based on their resistance to microbial breakdown (multi-G models; Berner 1980) in the muds, is simulated. Last, the diffusive flux of inorganic by-products to the overlying watercolumn is predicted. The weak link with respect to simulation of the time course of the response of the sediments to changes in overlying water quality is the time response of the diagenesis component. The existing sediment models have not been demonstrated to accurately project the time for the sediment–water exchange to

attain a new equilibrium following significant reductions in phosphorus loadings.

Adjustments of SOD were made for the subsequently described management scenarios in which the summer average TP concentration $<40\,\mu g \cdot L^{-1}$ according to the TP concentration-gross photosynthesis functionality of Auer et al. (1986), adopted to support adjustments in the magnitude of the G_{NET} distribution (Figure 9.53). This should be considered a first (probably conservative) approximation for changes in the *steady-state* value of SOD to be expected for reductions in trophic state. Recall, it is not known whether recent SOD measurements reflect steady-state conditions nor has a reliable basis to estimate the time to reach a steady-state SOD value following a reduction in watercolumn primary production been established for the lake. The approach adopted here to estimate new steady-state SOD values assumes the most recent value of $1.68\,g \cdot m^{-2} \cdot d^{-1}$ is representative of steady-state conditions, and it invokes a proportionality between SOD and gross photosynthesis (e.g., coupled by POC deposition).

9.8.4 Evaluation of Management Scenarios for METRO and CSOs: Onondaga Lake

9.8.4.1 Surface Discharges: Lake and Diversion

Some of the leading management alternatives under consideration for METRO and the CSOs to remediate the impacts of these inputs on Onondaga Lake have been selected to illustrate the utility of the TP, N, and DO management models in supporting related deliberations. There are three salient features of the alternatives: one is the level of treatment at METRO, the second is treatment of the CSO P load, and the third is the location of the discharge(s). A rather wide range of P and T-NH$_3$ treatment at METRO is evaluated here. METRO discharge options include the Seneca River (e.g., "diversion," both upstream and downstream of the lake outlet), Lake Ontario, the surface waters of the lake, and a deepwater position (e.g., hypolimnion). Alternatives other deepwater discharge are addressed first; these are identified in the schematic of Figure 9.59. The alternatives are lumped into four main "cases" numbered 1 through 4 (Figure 9.59). Subcases within each of the cases are identified alphabetically, e.g., subcases 3a, 3b, and 3c. Loading data developed to support these evaluations are from a range of sources, including historic performance data for METRO (Stearns & Wheler 1993b) and from other facilities presently operating processes incorporated here as components of the selected scenarios. The fine features of these alternatives may change with continued input from design engineers and regulators. However, they are generally representative and more than adequate to demonstrate the utility of these management tools and identify the most promising alternative(s).

Seasonal distributions, as monthly averages, of the METRO flow, and TP and N species concentrations in the effluent of the facility are specified for the various cases in Figure 9.60. A consistent set of BOD concentrations (e.g., Stearns & Wheler 1993a, b) were used to support application of the DO management model. These distributions are not shown because predictions are relatively insensitive to this range of concentrations of this relatively minor DO sink. The projected annual average

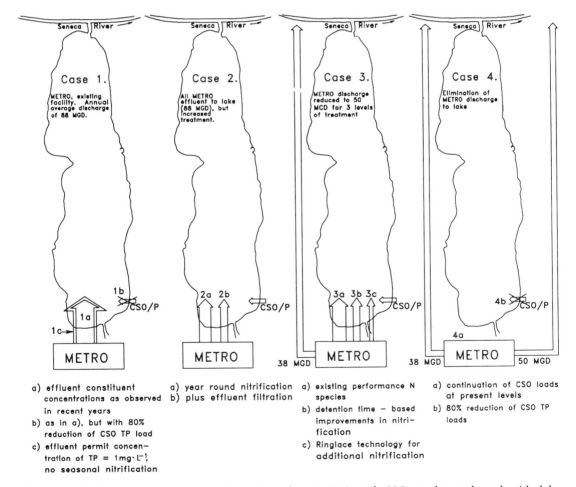

FIGURE 9.59. Selected management alternatives for METRO and CSOs to be evaluated with lake management models; as Cases (1–4) and subcases (modified from Effler and Doerr 1995).

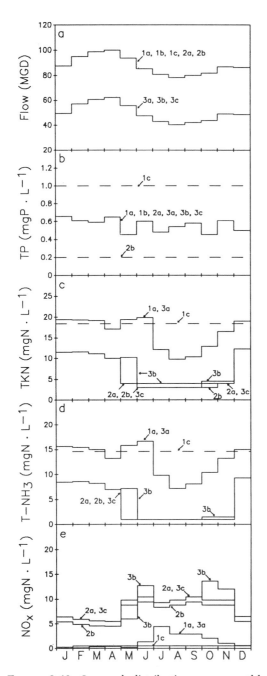

FIGURE 9.60. Seasonal distributions, as monthly averages, of METRO flow and effluent concentrations for cases of Figure 9.59: (a) flow, (b) TP concentration, (c) TKN concentration, (d) T-NH₃ concentration, and (e) NO$_x$ concentration (modified from Effler and Doerr 1995).

flow developed for METRO (Stearns & Wheler 1992) was 88 MGD (3.85 m³·s⁻¹) (Figure 9.60); this is about 20% higher than the average flow reported for the facility in 1989. Case 1 addresses the continued discharge of the entire effluent flow from the existing facility (Figure 9.59). Case 1a assumes performance characteristics for TP, N species, and BOD removal consistent with recent loading conditions (Stearns & Wheler 1993b, very similar to the default inputs of the management models), which reflect seasonal nitrification (Figure 9.60) and TP removal beyond that required by the existing permit (1 mg·L⁻¹). Case 1b represents an 80% reduction in the CSO TP load that has been estimated to correspond to a reduction in the tributary TP load of about 16%. This level of CSO loading reduction is believed to be achievable by storage of CSO storm discharges (e.g., 1 year storm) followed by treatment at METRO. Case 1c assumes a TP concentration in the effluent of 1 mg·L⁻¹, consistent with the permit, and no seasonal nitrification (Figure 9.60), consistent with original design goals and earlier performance (Chapter 3). Comparing the model simulations for Cases 1a and 1c demonstrates the benefits achieved through efforts made to optimize the performance of the existing facility.

Case 2 represents continued discharge of all of METRO's effluent to the lake, but with substantial (and costly) additional treatment (Figure 9.59). Case 2a reflects the addition of year-round nitrification. This greatly reduces the effluent concentrations of T-NH₃ and TKN and increases NO$_x$ concentrations during the non-summer months (Figure 9.60). Nitrogen species concentration data used in Case 2a (Figure 9.60) are consistent with those reported for a 170 MGD operating facility in Baltimore (City of Baltimore 1992). The TP concentration of the effluent is assumed not to be affected (i.e., same distribution as Case 1a) for this scenario. Case 2b corresponds to the addition of both year-round nitrification and effluent filtration. The 0.2 mg·L⁻¹ TP concentration of the effluent assumed for this case (Figure 9.60) is consistent with the NPDES permit and operating results at the 170 MGD Baltimore wastewater treatment plant (City of Baltimore 1992). The temporal distributions of T-NH$_3$ and NO$_x$ for Case 2b are nearly identical to those of Case 2a. This level of treatment probably approaches the present practical technological limit of treatment for a facility the size of METRO.

Case 3 represents an alternative that includes diversion of 38 MGD of wastewater presently received by METRO from the western part of its service area to a treatment plant on the Seneca River (Figure 9.59) located at the existing Baldwinsville facility (Figure 9.43). Accordingly, METRO would discharge 50 MGD to the lake, instead of 88 MGD (Stearns & Wheler 1993a). The three subcases correspond to varying levels of nitrification. The present loading conditions for TP specified by Stearns & Wheler (1993b), used previously in Cases 1a and 2a, were applied for Case 3 (Figure 9.60). Case 3a represents a worst case, and corresponds to the existing seasonality in N species concentrations (Figure 9.60, used also for Case 1a). Case 3b reflects an increased level of nitrification (Figure 9.60) that is consistent with the increased detention time within METRO associated with this partial bypass scenario (Stearns & Wheler 1993a). Case 3c corresponds to the additional nitrification that could be achieved by use of the innovative Ringlace technology (Figure 9.60).

Case 4 represents full bypass of the METRO discharge around Onondaga Lake (Figure 9.59). This has included a single pipe and two-pipe alternatives to the Seneca River (Onondaga Lake Management Conference 1993). These differences have no effect on the lake, but they do have implications for the river (treated subsequently). Case 4a corresponds to continued inputs of TP from the CSOs at existing levels, accommodated in the default tributary loading conditions. Case 4b represents the same reduction in the CSO TP load specified in Case 1b.

9.8.4.1.1 TP Predictions

The predictions of TP for the UML of Onondaga Lake from the TP management model for the evaluated scenarios (Figures 9.59 and 9.60)

are presented as a family of temporal distributions in Figure 9.61, and as summer average concentrations in Table 9.13.

Redundant results, for example, the equivalency of Cases 1a and 2a and Cases 3a, b, and c, are not included in the table. The concentrations of TP in the lake in summer would be nearly 60% greater if METRO operated at its permit limit (Case 1c), instead of the higher performance achieved in recent years (Case 1a, Table 9.13). However, little if any benefit with respect to phytoplankton growth in the lake is achieved by this reduction because the lower concentrations remain saturating (Figure 9.53). The reduced loading may benefit downstream systems such as Lake Ontario. The predictions for Cases 1a and 1c are consistent with lake observations made under similar METRO loading conditions in recent years (Figure 9.16a; also see Chapter 5). Note that elimination of most of the CSO TP load (Case 1b), which could only be achieved at great cost, would result in a relatively minor re-

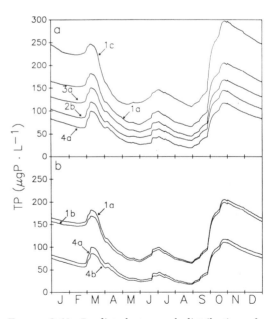

FIGURE 9.61. Predicted seasonal distributions for TP in the UML of Onondaga Lake: (**a**) Cases 1a, 1c, 2b, 3a, and 4a, and (**b**) Cases 1a, 1b, 4a, and 4b, to depict relative importance of CSO loading control (modified from Effler and Doerr 1995).

TABLE 9.13. Predicted summer average (mid-May to mid-September) TP concentrations for the UML of Onondaga Lake for selected METRO/CSO scenarios.

METRO/CSO scenarios	Case*	Summer average** [TP] ($\mu g \cdot L^{-1}$)
1. Existing METRO [TP] in effluent at permit concentration ($1 \, mg \cdot L^{-1}$), 88 MGD	1c	126
2. Existing METRO, at prevailing [TP] in effluent, 88 MGD	1a	79
3. Existing METRO, with 80% reduction in CSO load	1b	76
4. Upgrade METRO to effluent filtration, all effluent to lake surface, 88 MGD	2b	46
5. Reduced METRO discharge to lake surface (50 MGD), at prevailing [TP]	3a	59
6. Complete METRO bypass of lake, CSO load remains	4a	30
7. Complete METRO bypass of lake, with 80% reduction in CSO load	4b	26

*See Figure 9.59
**Mid-May through mid-September

duction in lake concentration (Figure 9.61b; Table 9.13), at the prevailing METRO loading rate, and would not be expected to significantly influence phytoplankton growth (Figure 9.53).

Substantial numerical reductions in lake TP concentrations would result from either partial bypass (reduction in METRO discharge flow to 50 MGD, Case 3) or upgrading METRO to effluent filtration (Figure 9.61a; Table 9.13). However, neither option, even in combination with the 80% reduction in CSO loading scenario (not shown), could reduce TP concentrations adequately to significantly reduce phytoplankton growth (Figure 9.53) and achieve mesotrophic conditions.

The only option(s) evaluated that would achieve substantial reductions in phytoplankton growth in Onondaga Lake are those that include full bypass of METRO's effluent around the lake. The predicted summer average TP concentration for complete bypass, without reduction of the CSO load (Case 4a), is $30 \, \mu g \cdot L^{-1}$. For the wet year option of the management model (not shown), the predicted average concentration was $33 \, \mu g \cdot L^{-1}$. These concentrations are within the mesotrophic range according to certain researchers (Auer et al. 1986; Vollenweider 1982).

Results for the scenario of control of the CSO load combined with full METRO bypass of the lake (Case 4b) are particularly instructive and significant with respect to management of the lake's cultural eutrophication problems. The CSO load has increased in relative importance with the elimination of the METRO load. Numerically the CSO load represents a larger fraction of the total load, and, more importantly, within this lower range of lake concentrations (Figure 9.61b; Table 9.13), phytoplankton growth is responsive to reductions in TP concentrations (Figure 9.53). At these lower concentrations the lake would also be responsive to other forms of reductions in tributary loading, such as might result from nonpoint land-use control measures. The upper bound of TP loading reduction that could be achieved through an aggressive land-use control program (see Chapter 3) is probably somewhat less than that invoked for CSO control in these scenarios. Further reductions in TP in the UML may be achievable, if sedimentary P can be immobilized, thereby reducing internal loading to the upper waters. The TP management model for Onondaga Lake predicts the guidance value of $20 \, \mu g \cdot L^{-1}$ can only be approached through elimination of the METRO input. The TP model indicates reductions in tributary loads (CSOs and perhaps land-use management) are probably also needed, though this would be best evaluated as part of model post-audit that would follow diversion of METRO.

The first quantitative insights into the sedimentary response to major changes in overlying water quality to be expected for Onondaga Lake have emerged from the recent research of Penn (1994). Penn (1994) developed a long-term, linked sediment and watercolumn model for TP for the lake. The model has a single lake layer (i.e., completely mixed system) and one homogeneous sediment layer. The influx of labile P (determined by extraction procedures, Penn 1994) to the sediments is modeled as a function of the watercolumn TP and verified using sediment trap measurements (Penn 1994). Diagenesis of labile P in the sediment is modeled using first order kinetics with experimentally determined rate coefficients (Penn 1994). Sediment P release is calculated as the depth-integrated rate of diagenesis, and matches well laboratory measurements made on intact sediment cores (Penn 1994). The model predicts the time to a new steady-state for TP in the lake is about 30 y. This delay in lake response is attributable to a "lag" in sediment response that has been observed elsewhere (Ahlgren 1977). The annual average rate of sediment release for a METRO diversion scenario, after the "lag," was predicted to decrease to $\sim$5 mg $\cdot$ m^{-2} $\cdot$ d^{-1} (Penn 1994). The model predicted an initial rapid ($\sim$1 y) decrease in the epilimnetic summer average TP concentration to $\sim$37 μg $\cdot$ L^{-1} with the diversion of METRO, with further decreases to $\sim$30 μg $\cdot$ L^{-1} over the following 30 year period. This modeling effort is highly significant, as it provides independent support for the TP model developed herein, and it quantifies the further long-term improvements that can be expected following diversion (e.g., $\sim$7 μg $\cdot$ L^{-1}).

9.8.4.1.2 T-NH₃ Predictions

The predictions of T-NH$_3$ for the UML of Onondaga Lake for the evaluated scenarios (Figures 9.59 and 9.60) are presented as a family of temporal distributions in Figure 9.62. The newly established standard for T-NH$_3$ in the upper waters of the lake of 0.77 mgN $\cdot$ L^{-1} is included for references. Selected statistics reflecting the predicted status of the UML with respect to the standard are presented in Table 9.14; the margin of violation is represented as the ratio [T-NH$_3$]/standard.

Progressive reductions in predicted lake concentrations of T-NH$_3$ (Figure 9.62), and associated margins of violation of the standard (Table 9.14), track the reductions in METRO

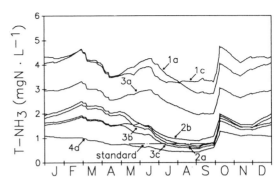

FIGURE 9.62. Predicted seasonal distributions for T-NH$_3$ in the UML of Onondaga Lake for indicated management alternatives, with non-salmonid standard for reference (modified from Effler and Doerr 1995).

effluent concentrations (Figure 9.60) and loading for the various scenarios. The present seasonal nitrification achieved at METRO results in a modest improvement in the status of the upper waters with respect to T-NH$_3$, but the effect is not substantial compared to the interannual variations that can arise from natural variations in tributary runoff (e.g., Figures 9.26a and 9.57). Substantial reductions in the margin(s) of violation are indicated for the addition of year-round nitrification at METRO (Cases 2a and 2b) and for reductions in METRO's discharge flow combined with increased nitrification (Cases 3b and 3c). However, violations remain nearly continuous over the spring to late summer period, for all the options except for the total bypass option (Case 4, Figure 9.62). The maximum margins of violation are above 2.0 for all the options except total bypass (Table 9.14). Maintenance of a year-round T-NH$_3$ concentration in the METRO effluent of 1.0 mgN $\cdot$ L^{-1} for a surface lake discharge would also meet the standard. However, this level of treatment has not been demonstrated for a facility the size of METRO in this climate.

9.8.4.1.3 DO Predictions

Predictions of DO corresponding to selected management scenarios are presented for the UML, the LML, and the depth interval of

TABLE 9.14. Predicted status of UML of Onondaga Lake with respect to non-salmonid T-NH₃ standard for selected METRO scenarios[†].

METRO/CSO scenarios	Case	$\left[\dfrac{[\text{T-NH}_3]}{\text{stand.}}\right]_{\max}$[*]	$\left[\dfrac{[\text{T-NH}_3]}{\text{stand.}}\right]_{\text{avg}}$[**]
1. Existing METRO, N-species in effluent consistent with no seasonal nitrification, 88 MGD	1c	5.1	4.5
2. Existing METRO, at prevailing N species in effluent, 88 MGD	1a	5.5	4.6
3. Upgrade METRO, year-round nitrification, 88 MGD	2a	2.2	1.3
4. Reduce METRO discharge to lake, prevailing N species in effluent, 50 MGD	3a	3.9	3.2
5. Reduce METRO discharge to lake, detention-time based improvements in nitrification, 50 MGD	3b	2.1	1.4
6. Reduce METRO discharge to lake, process innovation for improved nitrification, 50 mGD	3c	1.8	1.1
7. Complete METRO bypass of lake	4a	0.9	0.7

[†] Standard = 0.77 mgN·L⁻¹
[*] Maximum margin of violation over the mid-May to mid-September interval
[**] Average margin of violation over the mid-May to mid-September interval

anoxia in Figure 9.63a−c. These scenarios were selected because they bracket the responses for all the scenarios evaluated (Figure 9.59). The predictions should be considered substantially more speculative than those for TP and T-NH₃ because of the lack of predictive capabilities for SOD, the major oxygen sink in the lake. Further, the DO model simulations assume the SOD values are representative of steady-state conditions (e.g., SOD is in equilibrium with the existing downward fluxes of POC). The only change in SOD invoked was for Case 4; according to the predicted summer average TP concentration of 30 µg · L⁻¹ (Table 9.13). The steady-state value of SOD for this case was estimated (Figure 9.53) to be 1.52 g · m⁻² · d⁻¹.

The DO simulations are best reviewed from a comparative perspective, as details of the distributions are subject to significant interannual variability, driven by natural meteorological variability (e.g., Owens and Effler 1989). Modest improvements in the oxygen resources of the lake are predicted for reductions in pollutant loading from METRO. Oxygen concentrations are predicted to remain higher in the upper waters during summer and fall turnover for the full bypass scenario(s) (Figure 9.63a). The rate of hypolimnetic depletion is predicted to be slightly lower (Figure

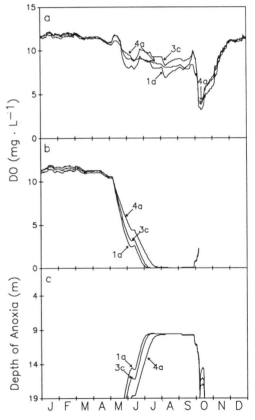

FIGURE 9.63. Predicted seasonal distributions of DO in Onondaga Lake for selected scenarios: (**a**) UML, (**b**) LML, and (**c**) upper bound of anoxia (modified from Effler and Doerr 1995).

9.63b). The onset of anoxia is predicted to be delayed by about one half month for the full bypass option, though the late summer depth interval of anoxia is not expected to change (Figure 9.63c). The lower DO concentrations predicted for the UML in summer for Case 1a (compared to both Cases 3c and 4a) are associated with the higher BOD and T-NH₃ loads and reduced tributary dilution for this scenario. The modest improvements for the full bypass (Case 4a) option are largely due to the slight (and speculative) reduction in SOD. According to the first approximation estimate of SOD for this case, lakewide violations are still expected during the fall mixing period (Figure 9.63a), though they may occur in fewer years (e.g., see Figure 9.58).

9.8.4.2 Deep-Water Discharge

9.8.4.2.1 Description of Proposal

General specifications of the proposed discharge were contained in a planning document (Onondaga County 1994). The proposal called for discharge of the METRO effluent through a 1 km long, 275 cm diameter outfall pipe to a diffuser located at a depth of 14 m. The peak flow to be accommodated is $7.0 \, m^3 \cdot s^{-1}$ (160 MGD). The outfall is to be operated by gravity. The minimum head available (e.g., corresponding to 25-yr-flood conditions) to drive the discharge by gravity has been estimated to be 1.85 m (Owens and Effler 1995). The proposal also called for supersaturation of the effluent with $9100 \, kg \cdot d^{-1}$ of DO, resulting in an effluent DO concentration of $27 \, mg \cdot L^{-1}$ at an average flow of $3.9 \, m^3 \cdot s^{-1}$ (88 MGD). The stated purpose of the supersaturation of the effluent is to maintain oxic conditions in the hypolimnion, thereby preventing the recycle of P from the sediments (Onondaga County 1994). The basis of expectations for water quality remediation is the effective isolation or "trapping" of the enriched effluent in the lower layers of the lake, thereby reducing nutrient and toxic species concentrations in the upper layer.

The seasonality of the temperature of the METRO effluent and Onondaga Lake at a depth of 14 m is depicted for the 1987 to 1993

interval in Figure 9.64a. The variability within the two distributions is largely meteorologically driven (e.g., Effler and Owens 1986; Effler et al. 1986d; Owens and Effler 1989; Chapter 4). The discharge is warmer than water temperatures at this lake depth year-round, the difference is particularly great in midsummer ($\geq 10°C$; Figure 9.64a) during the period of intense lake stratification. Primarily because of these temperatures (though the METRO discharge would also be somewhat less saline than the lake), the discharge would be buoyant (Figure 9.64b) and tend to rise in the water column of the lake as a plume that entrains ambient lake water (Figure 9.10b). The entraining plume rises in the water column to a depth, or somewhat above (associated with the momentum of the rising plume), where the plume density equals the ambient lake density.

Discharge of treated wastewater to a receiving water through a submerged diffuser is a common practice along the ocean coasts and the Great Lakes (e.g., Fischer et al. 1979;

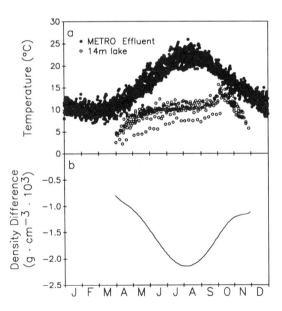

FIGURE 9.64. Time distributions: (a) temperatures of the METRO effluent and Onondaga Lake at a depth of 14 m, 1987–1993, and (b) density difference between METRO effluent and Onondaga Lake at a depth of 14 m (modified from Owens and Effler 1995).

Wright et al. 1982). As the hydrologic contribution of the METRO discharge to the total inflow to Onondaga Lake (Chapter 3) is unprecedented for lakes in the United States, the proposed deepwater discharge of this effluent in the lake is unprecedented. At an average discharge flow of $3.9 \, m^3 \cdot s^{-1}$, the existing hypolimnion (assumed to be the volume below a depth of 8.5 m, or $47 \times 10^6 \, m^3$ (Owens 1987)) would be flushed in about 5 months, or roughly the duration of summer stratification (Chapter 4). Thus, lakewide stratification conditions can be expected to be modified by this deepwater discharge.

9.8.4.2.2 Impact on Lake Stratification and Mixing

Owens and Effler (1989, also see Chapter 4) calibrated a one-dimensional mixed-layer stratification model for Onondaga Lake to simulate conditions that prevailed before the soda ash/chlor-alkali facility closed in 1986. The effort focused on the impact of the plunging inflows, made dense from the facility's saline waste discharge, on the lake's stratification regime. Recall that the model performed well in simulating the complex stratification regime of the lake that prevailed before closure of the facility, and in supporting identification of related impacts (Chapter 4). The earlier model (Owens and Effler 1989) was enhanced in this study in several ways (Owens and Effler 1995). First, the method of solution of the mixed-layer conservation equations was modified to yield a more accurate numerical solution of related equations for this layer, and a modified expression for the turbulent diffusion coefficient was employed (Owens and Effler 1995). Other enhancements include improvements in the plunging inflow submodel, year-round simulations (i.e., inclusion of an ice cover submodel), and addition of a submodel (e.g., Wright et al. 1982) to accommodate the effects of the proposed buoyant diffuser discharge. An analysis based on the densimetric Froude number supported the appropriateness of the one-dimensional assumption in simulating the impact of the proposed deepwater discharge.

The enhanced model was successfully calibrated for six consecutive (1987–1992) years of stratification conditions, since closure of the soda ash/chlor-alkali facility (Owens and Effler 1995). The model performed well in simulating the dimensions and temperatures of layers, and the timing/duration of stratification, but was less successful in simulating the more subtle effects of dense saline inflows that linger from the chemical facility (Owens and Effler 1995).

A number of diffuser configurations were investigated for depths of discharge at 14 m (Onondaga County 1994) and 18 m. A preliminary "selected" diffuser design, which resulted in (near-) maximum trapping of the METRO discharge in the lower layer (i.e., achieves a high degree of near-field mixing), is a diffuser length of 75 m, with 35 ports, each with a diameter of 25 cm (Owens and Effler 1995). All subsequently presented model simulations for METRO deepwater discharge scenarios are based on this "selected" design. Reasonable variations in the details of this design do not substantially affect these predictions.

Simulations of thermal structure that would result from a METRO discharge via the "selected" diffuser design at depths of 14 and 18 m are compared to observations and calibration ("hindcast") predictions for a late summer date in 1991 in Figure 9.65. The originally specified 14 m (Onondaga County 1994) discharge would yield a 3-layer stratification structure in late summer (Figure 9.65). This would compromise the stated water quality goals of the deepwater discharge, as the lowermost layer would become largely isolated from the effects of the diffuser. Anoxia would develop in this bottom layer (e.g., Chapter 5), and oxygen-demanding reduced substances and phosphorus would be mobilized within this layer. An 18 m discharge largely avoids this shortcoming (Figure 9.65). However, at this, or a deeper, depth the potential mobilization of the sediments and associated contaminants could become a regulatory issue. An 18 m discharge would require a pipe length of about 1.5 km, instead of the proposed (Onondaga County 1994)

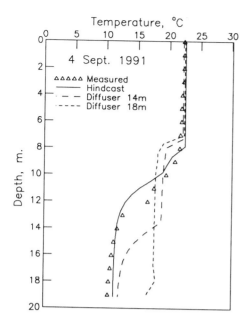

FIGURE 9.65. Temperature profiles for Onondaga Lake for September 4, 1991, including measurements, calibrated model simulation (hindcast), and model simulations for scenarios of METRO deepwater discharge (selected diffuser design) at depths of 14 and 18 m.

1 km (see bathymetric map of Chapter 1). All subsequently presented model simulations for METRO deepwater discharge scenarios are based on a discharge depth of 18 m.

Several of the impacts of a deepwater discharge on the lake's stratification regime emerge from review of Figure 9.65. This discharge of the warm (e.g., Figure 9.64a) METRO effluent would: (1) warm the lake's hypolimnion (e.g., bottom waters about 5°C warmer in late summer), (2) alter the dimensions of the stratified layers (e.g., decrease depth of upper mixed layer), (3) reduce the magnitude of overall density stratification, and (4) increase mixing and homogeneity within the hypolimnion (Figure 9.65). Further, the duration of stratification is predicted to be ~4 week shorter than prevailing conditions for the deepwater discharge scenario (Owens and Effler 1995). The increased temperatures of the hypolimnion would increase reaction rates of sedimentary processes, and may be of concern to habitat considerations. Based on the

substantial alterations to a number of basic features of the stratification regime predicted with the stratification model (Owens and Effler 1995), the impact of the proposed deepwater discharge on these fundamental physical characteristics of the lake can be fairly described as "profound."

The predicted temporal variation of the "intrusion depth interval" for the period of April 1989 through March 1990, shown in Figure 9.66 (Owens and Effler 1995), is indicative of the seasonality that would prevail for the position of vertical entry of the METRO discharge in the lake's water column. It can be seen that during spring (e.g., April 1989) and fall (e.g., October and November 1989) turnover, the METRO effluent would often rise to the water surface, due to the general lack of stratification during these periods. Thus, during these periods there is no "trapping" benefit for the deepwater discharge. At the beginning of summer stratification, the intrusion depth interval would extend from 9 to 12 m; progressively the interval would become more shallow (Figure 9.66) thereafter, due to warming and mixing of the hypolimnion. By August the interval would extend to within about 5 m of the surface. During ice cover a deepwater METRO discharge would become nonbuoyant after only a small vertical rise, resulting in the trapping of the effluent slightly above the diffuser depth.

9.8.4.2.3 Impact on Lake Water Quality

The linkages between the multilayer hydrothermal model (Owens and Effler 1995) described in the preceding section, the modified (Figure 9.10c) two-layer water quality models, and the vertical diffusion submodel for the two-layer framework, are illustrated in Figure 9.67. Simulations of the stratification/mixing regime with the hydrothermal model are driven by meteorology, inflow hydrology and density (e.g., temperature and salinity (Chapter 4)) and, for METRO deepwater discharge scenarios, diffuser configuration (Owens and Effler 1995). Time series of volume-weighted temperatures of the two layers, inflows to the two layers, and exchange flows between the

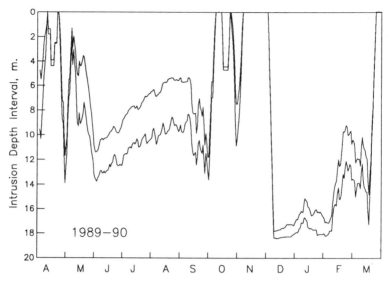

FIGURE 9.66. Predicted seasonal distribution of intrusion depth interval in Onondaga Lake for METRO effluent discharged at a depth of 18 m (selected diffuser design) for April 1989 through March 1990 (modified from Owens and Effler 1995).

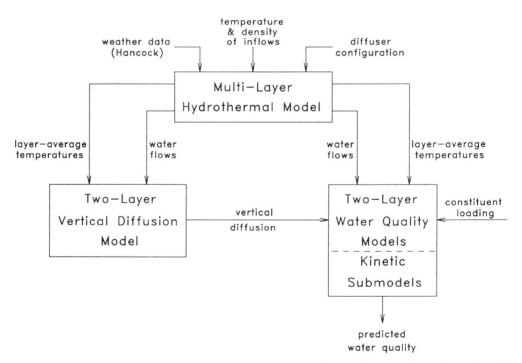

FIGURE 9.67. Interfacing of hydrothermal/stratification model with the more complex transport framework version of the water quality models, and the vertical diffusion submodel (modified from Doerr et al. 1995b).

layers (according to a user-specified layer boundary) are generated as output files from the hydrothermal model, which drive the two-layer water quality models and vertical diffusion submodel (Figure 9.67).

A comparative tracer substance analysis was conducted with the multilayer hydrothermal model (the basis of evaluation) and the modified transport framework of the two-layer models to determine α (see Figure 9.10c), and to establish the veracity of the modified framework in simulating the water quality impact(s) of the proposed deepwater discharge. Loading conditions for the tracer analysis were set to mimic those for phosphorus (P). The flow rates of the tributaries and METRO were set to their 1989 average values. The concentration of the tributary inflows was set to 110 ppm, while that for METRO was specified as 550 ppm, approximately proportional to average TP concentrations for these inputs. Forcing meteorological and tributary hydrology and density conditions were those measured in 1989 (e.g., Owens and Effler 1995). The simulation(s) were for the deepwater discharge. Simulations were initialized

during spring turnover at a tracer concentration of 195 ppm, the steady-state concentration for a completely mixed system.

The predictions of the hydrothermal model were compressed into two layers (volume weighting of multilayer output) according to the boundary depth of 8.5 m used in the original water quality model simulations. Predictions with the two-layer model were made according to the linkages of Figure 9.67 for various specified values of α, with the goal of obtaining the optimal match with the multilayer simulations. The best closure of the two models was achieved with α ≥ 0.5 (e.g., Figure 9.68). Performance of the two-layer frame-work was insensitive to increases in α above 0.5, as this influence was compensated for by changes in the computed distribution of the diffusion coefficient (see Figure 9.67, diffusion submodel). A value of α = 0.5 was adopted for all the modified water quality models (see Figure 9.10c). The veracity of the modified two-layer framework (Figure 9.10c) is supported by its near equal performance to that of the multilayer model (Figure 9.68). Performance for the upper layer was par-

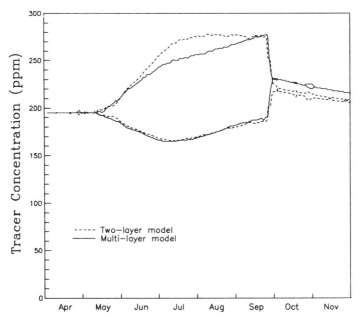

FIGURE 9.68. Tracer analysis with the hydrothermal/stratification model and modified two-layer transport framework (α = 0.5) of the water quality models (modified from Doerr et al. 1995b).

ticularly good during summer stratification. Simulations for the lower layer during the summer and the entire water column during fall turnover were within 10 and 5%, respectively (Figure 9.68).

Comparisons of the performances of the original and modified lake TP, N, and DO models have been presented previously in this chapter (see Doerr et al. 1995b); they were nearly equivalent over the spring to fall interval (Figures 9.17a, 9.24d, and 9.32). This supports the modified (more complex) transport (Figure 9.10c) version of the water quality models, and the interfacing of the hydro-thermal (Owens and Effler 1995) and water quality models (Figure 9.67). Further, it indicates the omission of the plunging phenomenon in the simpler transport version framework (Figure 9.10a) did not have a significant impact on model predictions.

The water quality projections presented here (also see Doerr et al. 1995b) represent a screening level effort, intended to evaluate the promise of the proposal, particularly from the perspective of the feasibility of realizing certain water quality benefits. No effort is made to arrive at the set of critical discharge and ambient environmental conditions, that may be necessary to support regulatory decision(s) and final engineering design. The deepwater discharge is operated continuously, and METRO discharge flows and constituent concentrations are those measured in 1989, for those simulations except as specified otherwise. All presented predictions correspond to the fourth consecutive year of simulations, to represent steady-state conditions. The base cases, presented for comparison purposes, are the retest simulations made with the modified transport versions of the water quality models (Figures 9.17a and 9.24d). A demarcation depth of 4.5 m for the two model layers was adopted to fairly reflect the "trapping" benefit of the deepwater discharge for the entire summer period (see Figure 9.66). This change from the 8.5 m boundary used in the testing and application of the simpler transport framework versions of the models is a manifestation of one of predicted impacts of this alternative on the lake's stratification regime (Owens and

Effler 1995). The G_{NET} distribution of the N model was recalibrated for the 4.5 m boundary to maintain the same observed distribution of CHL in the UML.

The "trapping" and upper bound kinetic benefits of the deepwater discharge alternative for TP concentrations in the UML are depicted in Figure 9.69. The epilimnetic summer average TP concentrations are presented in Table 9.15. Note the predicted earlier abrupt enrichment of the upper waters in September for the deepwater discharge scenarios (Figure 9.69) is coupled to the shortening of the period of stratification predicted by the hydrothermal/stratification model (Owens and Effler 1995). The "trapping" benefit of the deepwater discharge yields about a 25% reduction in the epilimnetic summer average TP concentration (Table 9.15). The kinetic benefits are attributed to major reductions in sediment release of P that may attend the maintenance of oxic conditions in the hypolimnion (see Chapter 5). The values invoked for R_{sed} ($3\,mg \cdot m^{-2} \cdot d^{-1}$) and f_2 ($=f_1$), and thus the simulation for the kinetic benefit case (Figure 9.69 and Table 9.15), are considered highly optimistic. Approximately a 50% reduction in the epilimnetic TP concentration is predicted. However, feasibility concerns in maintaining oxic

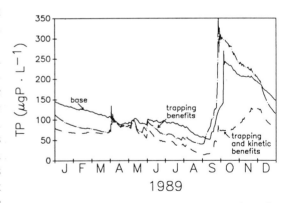

FIGURE 9.69. Simulations of TP concentrations in the upper mixed layer of Onondaga Lake with the more complex transport framework version TP model, for the conditions of 1989. Deepwater discharge scenarios compared to base (existing) conditions. Predicted trapping and potential kinetic benefits indicated.

TABLE 9.15. TP model predictions of epilimnetic summer average concentrations for selected cases.

Case number	Description	TP Conc. ($\mu g \cdot L^{-1}$)
1	Base case, existing conditions	83
2	Deepwater discharge ("trapping benefit")	63
3	Deepwater discharge, kinetics changed consistent with maintenance of oxia ("trapping" and kinetic benefits)	42
4	Case 2, with METRO effluent TP concentration = 0.2 mg/L*	50

* Associated with addition of effluent filtration

conditions (see subsequent treatment) make this projection even more speculative. An additional alternative of increased P treatment to maintain an effluent concentration of 0.2 mg · L^{-1} (effluent filtration, Figure 9.60b), in combination with the "trapping" benefit, yielded a predicted summer average epilimnetic TP concentration of 50 $\mu g \cdot L^{-1}$ (Table 9.16). None of the deepwater scenarios was adequate to significantly reduce phytoplankton growth in the upper waters of the lake (e.g., Auer et al. 1986).

Simulations with the modified N model presented here (Figure 9.70) address three issues: (1) the desirability of operating the deepwater discharge only during summer stratification instead of year-round (Figure 9.70a), (2) the benefit of trapping the METRO discharge during summer in the hypolimnion and of maintaining oxia in the hypolimnion to reduce T-NH$_3$ in the epilimnion (Figure 9.70b), and (3) the feasibility of meeting the T-NH$_3$ standard established for the lake (Figure 9.70c). The T-NH$_3$ standard for the upper waters of the lake established by the state are included for reference. The deleterious effect of operating the deepwater discharge year-round on epilimnetic T-NH$_3$ concentrations is a result of the trapping of much of the T-NH$_3$ enriched METRO effluent in the lower layer during winter (Figure 9.66; Owens and Effler 1995). Distribution of the accumulated T-NH$_3$ throughout the water column with the onset of spring turnover causes the abrupt increase

in concentrations in the upper layer (Figure 9.70a). A seasonal (e.g., summertime) operation scheme for the deepwater discharge (i.e., use of existing shoreline discharge for winter months) would reduce the spring turnover concentration (Figure 9.70a), by allowing a portion of the wintertime METRO load to exit the lake before spring.

The "trapping" benefit for the summertime-only operation scenario is illustrated in Figure 9.70b. Further reductions in T-NH$_3$ in the upper waters, particularly in late summer, are predicted with the maintenance of oxia in the hypolimnion (Figure 9.70b; with summertime operation of deep discharge). This is a manifestation of reduced upward transport of T-

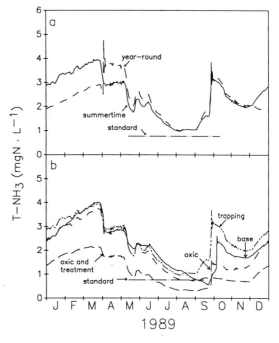

FIGURE 9.70. Simulations of T-NH$_3$ concentrations in the upper mixed layer of Onondaga Lake with the more complex transport framework version N model, for the conditions of 1989: (**a**) evaluation of effect of year-round versus summertime only operation of a deep-water discharge (modified from Doerr et al. 1995b), and (**b**) depiction of the benefits of "trapping" the METRO effluent and sequential additions of maintaining oxia in the hypolimnion, and addition of year-round nitrification at METRO. State T-NH$_3$ standard for lake included for reference.

NH$_3$ in summer, caused by a reduction in the hypolimnetic pool of T-NH$_3$ associated with the operation of the nitrification process. However, year-round nitrification treatment at METRO (effluent concentrations according to 3c of Figure 9.60), in combination with summertime operation of a deepwater discharge and maintenance of oxia in the hypolimnion, would be inadequate to meet the established T-NH$_3$ standard (Figure 9.70b).

The more complex transport framework version of the DO model is utilized here to determine the DO concentration that would be required in the METRO effluent to maintain oxic conditions in the hypolimnion throughout the summer, and the higher effluent concentrations that would be needed to maintain a hypolimnetic concentration of $5\,mg \cdot L^{-1}$ (e.g., support fish survival). Model simulations that maintain these specified conditions for 1989 METRO effluent characteristics are presented in Figure 9.71. The METRO effluent DO concentrations necessary to maintain oxia and a $5\,mg \cdot L^{-1}$ minimum in the hypolimnion are 70 and $130\,mg \cdot L^{-1}$, respectively. The required effluent DO concentration to maintain oxia is about 2.5× the value specified by Onondaga County (1994). The needed effluent concentrations of DO decrease as the oxygen demand of the effluent (primarily nitrogenous

and carbonaceous components) decreases. For the case of year-round nitrification at METRO (3c of Figure 9.60), the effluent concentrations that are predicted to meet the hypolimnion targets of oxia and $5\,mg \cdot L^{-1}$ are 40 and $100\,mg \cdot L^{-1}$. The feasibility of routinely maintaining these highly supersaturated concentrations in the METRO effluent needs to be established.

9.8.5 Fecal Coliform Bacteria/ CSOs

The FC model is in the form of a personal computer management tool. Three test periods are incorporated, the dry weather period, and Storms 1 and 2. Recall that Storm 1 was a very severe, rather low return frequency (1 in 5 year) event, while Storm 2 corresponded to a more common event, observed approximately annually. The important environmental conditions of mass transport (wind), light (incident irradiance and k_d), and depth of the upper mixed layer (Canale et al. 1993a) that prevailed during each model test period are included, along with reasonable bounding conditions. User-friendly menu-driven interfaces are provided to facilitate selection of program options and use of the model for management applications. Tributary FC loads documented for each tributary for the three test periods can be adjusted on the screen by multipliers (5×, 1×, 0.5×, 0.1×, 0.05×, 0×). Other menu options include selection among the three periods, and the measured versus bounding ambient environmental conditions. Model validation performance can be viewed for each test period, for each model cell, as graphs. The same graphical support is provided for predictions, corresponding to different loading conditions or the specified variations in ambient environmental conditions.

The FC model is applied here to demonstrate its utility as a planning tool and to identify the approximate magnitude of loading reduction that would be required to avoid violating standards. Recall that the related water quality (bathing beach) standards for FC bacteria concentrations are (Chapter 6): (1) logarithmic mean of $\leqslant 200\,cfu \cdot 100\,mL^{-1}$

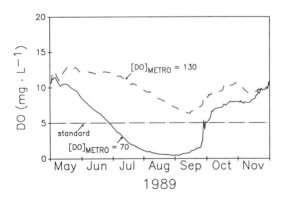

FIGURE 9.71. Simulation of DO concentrations in the lower mixed layer of Onondaga Lake with the more complex transport framework version DO model, for a deep-water discharge and the conditions of 1989. Predictions reflect indicated DO concentrations in the METRO effluent.

over 5 days, and (2) maximum <1000 cfu · 100 mL^{-1}. Ultimately, regulators will need to make fundamental decisions concerning the severity of the storm and range of environmental conditions that need to be accommodated. Earlier the substantial variability in the features of the response to storm events to be expected due to variations in ambient environmental conditions was demonstrated through sensitivity analyses (Figure 9.42). Statistical analyses of limnological, meteorological, and runoff data could be used to support the development of an objective probabilistic basis for establishing an adequately protective goal (e.g., Monte Carlo techniques). A more arbitrary approach has been adopted here.

Model predictions are presented for the southernmost cell (No. 1; Figure 9.36) and a northern cell (No. 7; e.g., area where a bathing beach might be located) for Storm 2 conditions in Figure 9.72a and b. Recall that the most severe impact is manifested in the southern end of the lake. Families of curves are presented that correspond to various levels of tributary loading. The model validation is included for reference, indicating clear contravention of both standards in the southern end of the lake (Figure 9.72a), and the absence of violations in the north cell (Figure 9.72b). The scenarios simulated correspond to percent loading reductions on each of the southern tributaries with CSOs (Onondaga Creek, OC, Harbor Brook, HB, Ley Creek, LC) of 50, 90, and 95. Recall that these tributaries (particularly Onondaga Creek) are the appropriate foci for remediation, as they contribute >90% of the FC load (Canale et al. 1993a, Chapter 3). The 50% reduction is predicted to be adequate for the ambient environmental conditions that prevailed during Storm 2. Substantially lower concentrations of FCs are predicted for the northern portion of the lake in response to loading reductions for the southern tributaries.

Cell 1 is the appropriate focus for remediation of lakewide problems. Remedial actions taken to meet standards in this area should be protective for the upper waters of the entire lake, except perhaps immediately

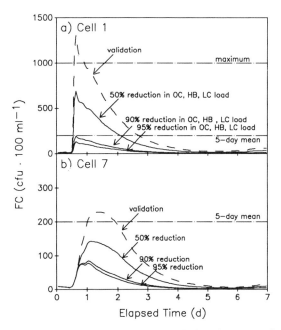

FIGURE 9.72. Predicted temporal distributions of the concentration of FCs in the epilimnion of Onondaga Lake for the conditions of Storm 2, for model validation and various management scenarios of % reductions in external loading from Onondaga Creek (OC), Harbor Brook (HB), and Ley Creek (LC): (a) cell 1 (south end of the lake), and (b) cell 7 (north basin).

adjacent to other tributaries. The implications of variability in the mediating environmental conditions, as they may affect the level of loading reduction required to avoid violation of standards, is illustrated in Figure 9.73. Here predictions are presented for cell No. 1 for Storm 2 for both prevailing and "critical" (Figure 9.42; low wind, low irradiance, shallow thermocline) environmental conditions. This combination represents highly conservative, and therefore protective, conditions (e.g., low return frequency). Note that the 50% loading reduction is no longer adequate to meet the standards, but that the 90% reduction is still protective. Margins of safety in the standard desired, feasibility, and cost, will undoubtedly enter into the final decision regarding the loading reductions required. Preliminary application of the model indicates at least a 50% loading reduction for a one year storm would be needed to meet standards lakewide.

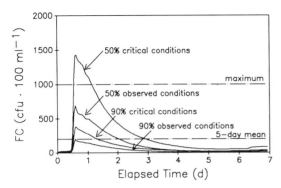

FIGURE 9.73. Predicted temporal distribtuions of the concentration of FCs in the epilimnion of the south end of Onondaga Lake for Storm 2, for 50 and 90% reductions in external loads from Onondaga Creek, Harbor Brook, and Ley Creek; for observed and critical ambient environmental conditions.

Increasing levels of loading reduction, approaching 90%, will be needed to assure avoidance of violations for particularly critical (but low return frequency) environmental conditions.

9.8.6 Applications of Seneca River Water Quality Model

9.8.6.1 Lower Layer DO Problem: Evaluation of Contributing Processes

The Seneca River water quality model, validated for the conditions of 1991, is used here to investigate the processes that contribute to the observed major sags in DO (water quality violations) in the lower layers of the river proximate to the point of entry of Onondaga Lake. The relative contributions of the biochemical processes are depicted in Figure 9.74. The reference conditions selected here are the prevailing stratification and biochemical processes, average point source loading, boundary and primary productivity conditions (e.g., incident light, light extinction, phytoplankton biomass) observed in the summer of 1991, and a critical low flow (7Q10). Note that under these critical flow conditions DO depletions are even more severe than observed in 1991 (compare to Figure 9.48c).

Each of the contributing biochemical sinks for DO has been sequentially suppressed in the model analysis (Figure 9.74). The processes of nitrification, sediment oxygen demand, and phytoplankton respiration all contribute importantly to the major depletions of DO observed in the lower layer of the river during low flow periods. Oxidation of CBOD is the least important oxygen sink. The nitrification sink of this analysis corresponds to a TKN concentration of $3.18 \, mgN \cdot L^{-1}$ in the Onondaga Lake outflow, a typical mid-summer value. Progressive decreases during the summer period, and substantial year-to-year variations (Figure 9.57), are observed for $T-NH_3$ concentrations in the lake outflow (Chapter 5). Earlier, Effler et al. (1984a) had attributed the DO depletions in the lower layer of the river largely to respiration and decay of phytoplankton released from Onondaga Lake.

The only sink of oxygen amenable to reduction, related to the Onondaga Lake inflow, without diverting a portion or all of the METRO discharge from the lake, is nitrification. Year-round nitrification at METRO would reduce the concentration of TKN in the upper waters

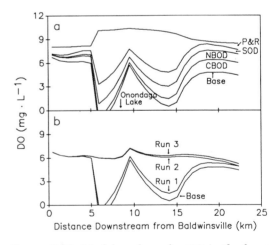

FIGURE 9.74. Model analyses for DO in the lower layer of Seneca River: (a) components of DO depletion, (b) projected improvements for scenarios; Run 1, year-round nitrification of METRO with stratification, Run 2, ''de-stratification,'' Run 3, ''de-stratification'' with year-round nitrification at METRO (modified from Canale et al. 1995b).

of the lake to perhaps $1.0\,mg \cdot L^{-1}$, thereby reducing the nitrogenous oxygen demand in the lower river layer. This reduction, that would only be achieved at great cost, would fall far short of eliminating the violations of the DO standard in the river (Run 1, Figure 9.74b), though it would eliminate violations of the NH_3 standard in the river. The phytoplankton related respiration component of the lower layer depletion could be reduced by diverting the METRO discharge to the Seneca River downstream of the outlet. However, even at a much reduced concentration of CHL in the lake outflow (e.g., CHL = 10–$20\,\mu g \cdot L^{-1}$, consistent with the diversion), violations of the DO standard would probably continue at critical low flow.

The dominant factor responsible for the contraventions of water quality standards in the Seneca River adjoining the Onondaga Lake inflow is the reduced vertical mixing associated with the occurrence of density (mostly salinity-based) stratification. Elimination of this stratification (i.e., E_v values observed upstream and downstream of the bounds of stratification, Figure 9.46) would eliminate the violations in DO and T-NH_3 standards observed in the lower river layer under the river conditions of 1990 and 1991 (Figure 9.74). Year-round nitrification would result in only a minor increase in oxygen concentration beyond that achieved by de-stratification alone (Run 3, Figure 9.74).

Recall that the occurrence of stratification in the river would be greatly reduced, if not eliminated, by eliminating the continuing anthropogenic loading of ionic waste to Onondaga Lake (see Chapter 4, and Figure 9.51 of this chapter). However, it may not be practical to remediate this loading. Artificial means of inducing turbulence to prevent the development of stratification could be used. Regardless of the means used to accomplish the task, it is clear that de-stratification will be necessary to eliminate prevailing water quality violations in the river. Further, it is desirable because it would eliminate material loading from the Seneca River to the lake. Thus de-stratification of the river will be an integral component of any management plan

aimed at remediating the water quality problems of Onondaga Lake through diversion of the METRO load to the Seneca River. This conclusion is manifested in the manner in which the river water quality model is subsequently applied to evaluate impacts associated with selected remediation alternatives.

9.8.6.2 Management Model: River Water Quality

The river water quality model is in the form of a user-friendly personal computer management tool. Main menu options direct the user to either a review of the system, a summary of model performance for various parameters and survey dates (1991), or to a session of management model projections. Four river flow (stratified/unstratified)/bypass pipe (1 or 2) combinations are supported by the model. However, the analysis summarized in Figure 9.74b indicates only the unstratified scenarios (2) deserve the attention of water quality managers for evaluating METRO bypass alternatives. All projections presented in this section invoke the absence of density stratification. The two pipe alternatives fix one discharge point at the existing Baldwinsville wastewater treatment plant (Figure 9.46c, permitted presently for 9 MGD), the location of the other pipe is user-specified. The METRO discharge can be apportioned between the lake, the Baldwinsville site, and a downstream location by the user. Leading alternatives include bypass of the lake with 38 MGD of untreated wastewater to an expanded facility at the Baldwinsville site (i.e., 47 MGD following expansion). This condition has been invoked for all two-pipe scenarios investigated here. The projected future 88 MGD flow for the existing METRO facility (Stearns and Wheler 1992) constrains the flow for the second pipe (from METRO to the river system, e.g., Case 4, Figure 9.59) to 50 MGD.

The user has the option, according to separate menus, to specify model inputs, including: (1) boundary conditions for upstream (Baldwinsville) Seneca River, the Onondaga Lake outflow, and Oneida River (Figure 9.43), (2) treatment plant loads along the river system, (3) river and wastewater treatment plant flows, (4) certain kinetic coefficients, and (5) certain environmental conditions. Default critical conditions are provided that are based on statistical analyses of monitoring data for boundary conditions, treatment plant permits, 7Q10 river flow, a water temperature of 25°C, and the independent estimates of model coefficients. Model predictions are provided as graphical output on the computer screen for both model layers for Cl$^-$, CBOD$_u$, and DO (average, minimum, and maximum).

9.8.6.3 Evaluations of METRO Options: DO

The analysis of management options presented here focuses on the DO resources of the river that prevailed before (e.g., 1990 and 1991) the recent major shift in water quality brought about by the zebra mussel invasion

(Figure 9.44). Critical default environmental conditions for flow (7Q10) and temperature (25°) have been invoked in all the simulations. Onondaga Lake boundary concentrations of TKN and DO are determined from lake model simulations as the average values in the month of August.

A number of options, with respect to level of treatment and position of discharge along the river system, are evaluated here for the full bypass of the METRO loads around Onondaga Lake (Case 4, Figure 9.59). Recall that application of the lake management models established the need for full bypass to meet water quality standards and goals for the lake (e.g., Table 9.14, Figure 9.62). Features of the management alternatives that have been evaluated herein, with respect to impact on the river system's oxygen resources, with the river water quality model are specified in Table 9.16. Seven alternatives are evaluated,

TABLE 9.16. Configurations, effluent concentrations, and descriptors for river DO management model runs, METRO bypass scenarios.

Scenario number	Number of discharges to Seneca River	Baldwinsville discharge			METRO discharge			Description
		Q (MGD)	CBOD$_u$ (mg·L^{-1})	TKN (mg·L^{-1})	Q (MGD)	CBOD$_u$ (mg·L^{-1})	TKN (mg·L^{-1})	
A*	1	47	38	16	—	to Lake	—	Partial bypass, at existing permit limits
B*	1	9	38	16	88	23	32	Full bypass, at existing permit limits, METRO discharge to cell 11
C*	1	9	38	16	88	23	32	Full bypass, at existing permit limits, METRO discharge to cell 22
D*	2	47	38	16	50	23	32	Full bypass, at existing permit limits, METRO discharge to cell 11
E*	2	47	13	3	50	13	3	Full bypass, year-round nitrification, METRO discharge to cell 11
F**	2	47	8	3	50	8	3	Full bypass, year-round nitrification, further reductions in CBOD$_u$, METRO to cell 11
G*	0	9	38	16	0	—	—	Full bypass to Lake Ontario, existing facilities at existing permit limits

* Wetzel Road facility at existing permit conditions (3.5 MGD, CBOD$_u$ = 68 mg·L^{-1}. TKN = 32 mg·L^{-1})
** Wetzel Road facility with effluent concentrations same as Baldwinsville and METRO

and each is identified alphabetically (Table 9.16). It should be noted that the two existing domestic wastewater treatment plants (WWTP, Figure 9.43) on the adjoining portions of the Seneca River, the Baldwinsville and Wetzel Road facilities, are also operated by Onondaga County. Assumptions concerning the Wetzel Road discharge are included in Table 9.16 as footnotes. Loading data (Table 9.16) correspond either to existing permit limits, performances documented for existing facilities, or in one case, a hypothetical level of treatment for $CBOD_u$. The details of these alternatives may change with continued input from design engineers and regulators. A limited number of permutations are addressed here. However, the array of alternatives addressed is adequate to demonstrate the utility of the model for management applications and to identify promising alternatives to protect the oxygen resources of the river, while allowing full bypass of the METRO discharge around the lake.

A single partial bypass scenario (A, Table 9.16) has been included for comparison; it is not considered a viable option because it would fail to meet standards and goals for the lake. Two other single discharge alternatives are addressed, but each is for full bypass of METRO. In one case the discharge (88 MGD) is proximate to the inflow of Onondaga Lake into the Seneca River (scenario B of Table 9.16; enters cell 11, Figure 9.46c), the other discharges 16 km downstream at the Three Rivers Junction (scenario C of Table 9.16; enters cell 22, Figure 9.46c). Effluent concentrations for all the single discharge scenarios are specified at existing permit levels. The first two-discharge alternative evaluated (D of Table 9.16) is also for effluent concentrations at permit values. The first discharge is from the existing Baldwinsville WWTP site. The second discharge is to cell 11. The two other scenarios for two discharges to the Seneca River assume the same discharge locations, but invoked lower effluent concentrations for $CBOD_u$ and TKN, corresponding to increased levels of waste treatment. The effluent concentrations specified for scenario E for both the Baldwinsville and METRO

facilities (Table 9.16) are consistent with year-round nitrification. Scenario F is more speculative, as the feasibility of achieving the specified lower $CBOD_u$ value is not well established for large facilities. The last scenario (G) corresponds to full diversion of the METRO discharge around Onondaga Lake and the river system (e.g., to Lake Ontario).

The adequacy of existing levels of treatment, as specified by the individual facility discharge permit levels, and the potential benefit of two (split) discharges versus one, and variations in the position of the discharge, are demonstrated in Figure 9.75 (scenarios A, B, C, and D; Table 9.16). Predictions of daily average DO concen-

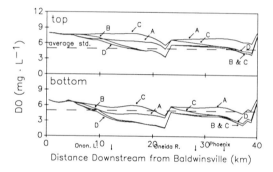

a) daily average

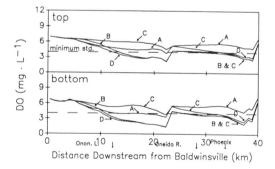

b) daily minimum

FIGURE 9.75. Predicted longitudinal profiles of DO for two layers of the Seneca and Oswego (to Fulton) Rivers for critical environmental conditions for selected scenarios of waste treatment and discharge (see Table 9.17) related to reclamation alternatives for Onondaga Lake: (a) daily average DO, and (b) daily minimum DO. Upstream Seneca River boundary conditions before zebra mussel invasion.

trations are presented for four scenarios in Figure 9.75; the standard of $5 \, mg \cdot L^{-1}$ is included for reference. The corresponding simulations for the daily minimum DO concentrations appear in Figure 9.75; the standard of $4 \, mg \cdot L^{-1}$ is included for reference. Predictions are presented for a 40 km river length extending from below the Baldwinsville dam on the Seneca River to downstream of the second dam in Fulton, on the Oswego River (see Figure 9.43).

Oxygen concentrations are predicted to be lower in the bottom layer than in the upper layer in all the scenarios (e.g., Figure 9.75) evaluated here, despite invoking (density) destratified conditions in these model analyses. This is consistent with the limited vertical mixing that is presently observed in unstratified portions of the system during low flow conditions (Figure 9.46), and the extent of DO stratification documented in these river sections in 1990 and 1991. Recall this magnitude of DO stratification is small compared to that observed in the river in recent years in density-stratified (mostly salinity-based) sections. Differences in the daily minimum and average values are predicted to be greater for the upper layer than the lower layer.

Existing levels of treatment, as specified by discharge permits, are inadequate to meet water quality standards for DO in the river system (under critical environmental conditions) for METRO bypass options (Figure 9.75). Even the partial diversion option is predicted to cause violations upstream of the Phoenix dam of both the average (Figure 9.75a) and minimum (Figure 9.75b) standards, though the margins of violation are reduced compared to the full bypass options. The most severe stress to the oxygen resources of the system is predicted upstream of the Phoenix dam for full bypass alternatives B and D. Violations of the daily average and minimum standards are predicted for both layers, upstream and downstream of Phoenix (Figure 9.75). The longitudinal minimum in DO ($\sim 2 \cdot mg \cdot L^{-1}$) is predicted to occur just upstream of the Phoenix dam. More than 10 km of the river are predicted to be in

violation of DO standards for these two scenarios under critical environmental conditions (Figure 9.75). Some recovery is provided by the dams at Phoenix and Fulton.

Note that the predicted DO profiles for the two discharge/one discharge total bypass options of scenarios B and D diverge at the point of entry of the Baldwinsville discharge (~ 4 km), but converge downstream about 18 km below Baldwinsville, as a result of their equivalent overall demands for oxygen. The profiles for alternatives B and C track each other until the entry of the METRO bypass at cell 11 (scenario B), as do alternatives A and D. The profiles of scenarios A and C cross below the Phoenix dam. Alternative C protects the Seneca River, but causes violations in the Oswego River downstream of Phoenix. This more costly (added pipe length of ≥ 12 km) option significantly reduces the margin and duration of violations (Figure 9.75) by taking advantage of the locations of dams on the river system (Figure 9.42).

Simulations for scenarios of increased treatment (E and F) are presented in Figure 9.76; the results of scenario D are included for reference. A single discharge (88 MGD) scheme, as illustrated in Figure 9.75, would not significantly affect the outcome for the two-discharge alternatives considered here. The reductions in loading invoked in scenario E, consistent with year-round nitrification, are predicted to be adequate to avoid violation of standards under critical conditions. However, there is no margin of safety with respect to the daily average standard for the bottom layer (Figure 9.76a). Only a relatively small benefit would be realized if the $CBOD_u$ concentration of the domestic waste discharges could be reduced further to $8 \, mg \cdot L^{-1}$ (scenario F, Figure 9.76). Interestingly, the predicted oxygen resources for this critical river reach for scenario G (discharge to Lake Ontario) match closely predictions obtained for scenarios (E and F) of discharge of an additional 88 MGD of highly treated domestic waste effluent.

The analyses presented herein indicate that it is feasible to completely bypass METRO to the Seneca River of 1990/1991 and protect

a) daily average

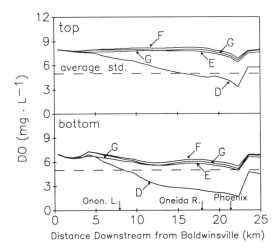

b) daily minimum

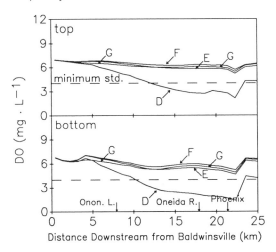

FIGURE 9.76. Predicted longitudinal profiles of DO for two layers of the Seneca and Oswego (past Phoenix) Rivers for critical environmental conditions for selected scenarios of waste treatment and discharge (see Table 9.17) related to reclamation alternatives for Onondaga Lake: (**a**) daily average DO, and (**b**) daily minimum DO. Upstream Seneca River boundary conditions before zebra mussel invasion.

the river system against violations of DO standards, by providing advanced treatment (e.g., year-round nitrification) for the diverted wastewater. Refinements of this analysis undoubtedly will emerge as related engineering design work evolves. However, as has been

reviewed previously, the oxygen resources of the Seneca River had been severely degraded by 1993 from the zebra mussel infestation. A single DO profile collected over the Baldwinsville to Three Rivers Junction reach on August 10, 1993, from 0800 (upstream) to 1030 (downstream) hours, that is generally characteristic of late summer monitoring data of 1993, is presented in Figure 9.77. The poor upstream oxygen conditions (also see Figure 9.44) persisted throughout the reach. Under these conditions the river has no assimilative capacity for oxygen-demanding wastes. Discharges would be required to have no net oxygen demand, an extraordinarily stringent requirement for a large facility. Violations of the daily minimum standard are apparent in this monitoring data (Figure 9.77); violations of the minimum and average standard probably occurred through much of the reach on this day. Continued modeling and support studies will need to establish the credibility of the river water quality management model for the entire range of environmental conditions depicted in Figure 9.44.

The equally important water quality issue of the potential for free ammonia toxicity in the river associated with bypass of the METRO discharge around the lake has not been

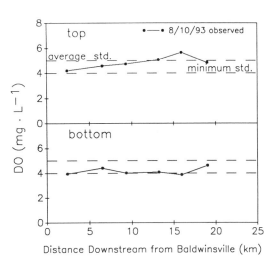

FIGURE 9.77. Longitudinal DO profiles for two layers of the Seneca River, for August 10, 1993 (after zebra mussel invasion).

addressed here. Important features of such an analysis would include: (1) specification of critical environmental conditions, (2) establishment of in-river T-NH$_3$ standards, and (3) a credible T-NH$_3$ (sub)model for the river. New York State policy for critical flow conditions for the free ammonia issue is the use of the 30Q10 flow (usually ~1.2 to 1.3 × 7Q10). The in-river summertime T-NH$_3$ standard established by the NYSDEC for conditions before the zebra mussel invasion (critical pH of 8.35 and temperature of 25°C) is 0.37 mgN · L^{-1}. The wintertime value is 0.75 mgN · L^{-1}.

9.9 Summary

9.9.1 Description of Models

This chapter reviews the development, testing, and preliminary application of five mass balance water quality models for Onondaga Lake, and a water quality model for adjoining portions of the Seneca River and Oswego River. These models are intended to support ongoing research of these impacted systems, by providing quantitative frameworks for the synthesis of related scientific information, and to support basic management decisions for these resources by providing reliable predictive frameworks. These models, and this chapter, build on the material documented in preceding chapters, including constituent loading (Chapter 3), in-lake constituent concentrations (Chapters 5 and 6), lake hydrodynamics (Chapter 4), and reactions, transformations and material cycling (Chapters 5, 6, and 8).

The models represent a diverse array of model structures and complexities, that match the associated management issues, as well as system characteristics and constituent(s) behavior. The models presented here are generally of intermediate complexity. They have evolved from an integrated program of intensive field monitoring and process studies (see Chapters 3–8). These models accommodate the key regulating processes, which in most cases have been independently quantified by experimentation. An effort has been made

to avoid overly simple or complex frameworks that can produce unreliable predictions.

Design features of the Onondaga Lake and river water quality models are summarized in Table 9.17. The structure of the chloride (Cl$^-$) model is the most simple, because no reactions need to be considered for this conservative substance, and the lake is treated as a completely mixed system. The total phosphorus (TP), nitrogen species (particulate organic N (PON), dissolved organic N (DON), ammonia (T-NH$_3$), nitrogen oxides (NO$_x$), TKN, and total N (TN)), and dissolved oxygen (DO) models for the lake all utilize the same physical framework; two vertical completely mixed layers, coupled by vertical mixing. Two different physical/transport frameworks have been developed. The simple version is appropriate for most stratifying lakes. The more complex version includes exchange flow between the layers to accommodate entrainment-based transport associated with plunging inflows and the management scenario of a rising buoyant METRO effluent discharged to the deep waters of the lake. Kinetically, the simplest of the these two-layer models is the TP model. The most complex is the N model, because of the number of N species and processes that need to be accommodated (Table 9.17) to adequately describe the behavior of T-NH$_3$. The DO model has carbonaceous BOD (CBOD) and reduced species (RS, sum of oxygen demands associated with CH$_4$, H$_2$S, and Fe^{2+}) submodels.

Finer temporal and spatial resolution is incorporated in the fecal coliform bacteria (FC) model for the lake to address the management issues of duration and longitudinal extent of public health risks lakewide following runoff events. The FC model is the most hydrodynamically complex of the lake models, to accommodate longitudinal transport between the multiple top and bottom layer segments of the model. The hydrodynamic features of the model were specified from an independently validated transport submodel (Chapter 4).

The Three Rivers system has multiple uses including navigation, power generation, waste assimilation, fishing, and contact recreation. Related reductions in natural turbulence

TABLE 9.17. Design features of Onondaga Lake and river water quality models.

Model	Time	Space		Processes	
		Vertical layers	Longitudinal segments	Hydrodynamics	Others
Cl/Lake	Seasonal	1	1	None	None
TP/Lake	Seasonal	2	1	Vertical mixing	Settling, sediment release
N/Lake	Seasonal	2	1	Vertical mixing	PON: phyto growth/respiration, settling, net exchange with T-NH$_3$ through uptake, predation, and other loss processes
					DON: decay of PON, ammonification
					T-NH$_3$: ammonification, PON excretion, sediment release, PON (phyto) uptake, nitrification, volatilization
					NO$_x$: nitrification, denitrification, phyto uptake at low T-NH$_3$ <50 μgN$\cdot$L^{-1}
					TKN = PON + DON + T-NH$_3$
					TN = TKN + NO$_x$
DO/lake	Seasonal	2	1	Vertical mixing	DO: phyto production/respiration, reaeration, demands from CBOD, NBOD, SOD, and RS
					RS: SOD as a source, oxidation as a sink
FC/lake	Days	2	8—top layer 3—bottom layer	Vertical mixing, Advective and dispersive transport	Settling, dark death rate, light mediated death
river	Steady-state	2	25—top layer 25—bottom layer Baldwinsville to Phoenix; 25—top layer 25—bottom layer Phoenix to Oswego	Vertical advection and diffusion, longitudinal advection and dispersion	DO: reaeration, phyto production/respiration, demands from CBOD, NBOD, SOD, T-NH$_3$—nitrification CBOD: decay

have made the system more susceptible to anthropogenic impacts. The natural reaeration potential of the system is low. The Seneca River adjoining Onondaga Lake has been severely impacted by pollution inputs from the lake. Chemical stratification occurs in the river proximate to the lake during the low flow periods common to summer, associated with the elevated salinity of the lake. This stratification promotes oxygen depletion in the lower layer of the river to concentrations that often represent violations of New York State water quality standards. In contrast to the dynamic lake models, the river water quality model is a steady-state tool. Two-layer vertical segmentation is necessary to accommodate the pronounced density stratification, and coupled water quality problems. Twenty-five segments in each of the two layers are utilized to resolve longitudinal structure in water quality between Baldwinsville and Phoenix, New York. A mass balance transport submodel, based on conservative constituents (Cl$^-$ and salinity), has been developed to simulate the complex flow and transport patterns that prevail in the Seneca River during summer low flow periods. The transport submodel defines the hydrodynamic interaction of adjacent segments through the processes of horizontal and vertical advection, longitudinal dispersion, and vertical diffusion. The DO mass balance of the river model accommodates the source/sink processes included in most river models (Table 9.17).

9.9.2 Model Testing

The water quality models have undergone comprehensive performance testing. A summary of selected features of this program is presented in Table 9.18. The lake N model and the CBOD submodel of the river model were tested according to calibration/verification procedures. Calibration procedures were used to develop the time distribution of net phytoplankton growth incorporated in the N model and the river transport submodel. All other models were validated (Table 9.18). Model validation refers to the successful simulation of observations, without the "tuning" of model coefficient values that is common to the calibration process; instead coefficients are independently determined through measurements or experimentation.

The Cl⁻ model performed very well in simulating features of the distribution of Cl⁻ concentration in the lake documented over the 1973 to 1990 period, including the: (1) high concentrations before closure of the soda ash/chloralkali plant, (2) year-to-year variations in concentration before closure, (3) strong seasonality before closure, and (4) major and rapid reduction in concentration following closure.

The TP model performed well in simulating the distribution of TP concentrations documented in the lake over the 1987 to 1990 period, specifically the strong seasonality in the upper and lower layers observed annually,

the initial TP concentrations of spring, and the substantial differences in concentrations observed among the four years.

The N model performed well in tracking the dynamics of the various N species and the downward flux of chlorophyll and PON over the spring to fall interval of the model calibration year of 1989. The N model continued to perform well in simulating T-NH₃, the principal management focus of the model, and the other species through the critical mid-summer period of 1990, the verification year. The performance of the N model in 1990 did not match that of the calibration year thereafter, largely because of adoption of the phytoplankton net growth distribution established for 1989.

The lake DO model (and reduced species submodel) performed well in simulating important features of the oxygen regime of the lake in both 1989 and 1990, including the rapid depletion of DO from the lower layer and subsequent accumulation of oxygen-demanding reduced substances, and the lakewide depletion of DO during the fall mixing period. Predictions of the vertical limits of anoxia, which incorporated an empirical relationship, also performed well. Model performance was better in 1989 than 1990.

The FC model successfully predicted peak concentrations and die-away patterns, as well as north–south and top–bottom differences, for both a commonly observed (e.g., one-year return frequency) storm and a more severe

TABLE 9.18. Summary of Onondaga Lake and Seneca River water quality model testing.

Model	Submodel	Type of testing	Period tested
Cl/lake	—	Validation	1973–1991
TP/lake	—	Validation	1987–1990
N/lake		Calibration/verification	1989–1990
	Phytoplankton	Calibration	1989–1990
DO/lake		Validation	1989–1990
	Reduced species	Validation	1989–1990
	CBOD	—	—
FC/lake		Validation	3–7 day periods; 2 storm, 1 dry, summer 1987
	Transport/circulation	Validation	8 droque release experiments (Chapter 4)
river			
	DO	Validation	5–24 hr surveys, summer 1991
	CBOD	Calibration/verification	5–24 hr surveys, summer 1991
	transport	Calibration	5–24 hr surveys, summer 1991

wet weather event. Background distributions for dry weather conditions were also well simulated.

The river water quality model was tested against measurements made during five 24 hour surveys conducted in the summer of 1991 (Table 9.18). The CBOD submodel was calibrated to the distribution observed in the first survey; simulations of the calibrated model matched the other surveys reasonably well. The river DO model performed well in simulating the recurring upstream and downstream DO sags in the lower layer adjoining the lake inflow, and in tracking the general magnitude of the diurnal variations in the upper layer of the river.

The invasion of the river system by zebra mussels, and in particular, the development of dense populations downstream of Cross Lake, has caused major shifts in water quality in the Seneca River, as measured at Baldwinsville. Concentrations of DO were reduced to levels in the summer of 1993 that often represented water quality violations. Advantage should be taken of this perturbation, to continue testing of the water quality model, and to establish its credibility over a broader range of conditions.

9.9.3 Sensitivity Analyses

Two types of sensitivity analyses have been performed. The first is to identify the uncertainty in model predictions that is to be expected from the level of uncertainty in the determination of model inputs. The second is to identify the level of variability in predictions and observations that is to be expected as a result of natural variability in environmental forcing conditions.

Sensitivity analyses with the Cl^- model demonstrated that the adequate performance of the model since closure of the soda ash/chloralkali facility, which does not consider the downstream Onondaga Creek and Seneca River loads, is a manifestation of an approximate balancing between the enriching and diluting influences of these inputs, respectively.

The emphasis of sensitivity analyses for the TP model has been placed on the UML, because this is the focus of related water quality concerns for the lake. Simulations of TP in the UML in summer are moderately sensitive to the timing and uncertainty in tributary loads, but largely insensitive to the levels of uncertainty in the determinations of model coefficients and a wide range of sediment release scenarios.

Sensitivity analyses with the N model have focused on $T\text{-}NH_3$ in the UML of the lake, because of concerns for free ammonia toxicity. These analyses demonstrate that substantial year-to-year differences in $T\text{-}NH_3$ concentrations are to be expected for the upper waters of the lake, under prevailing METRO loading conditions, as a result of natural variations in runoff (dilution provided by tributary flow). Predictions of $T\text{-}NH_3$ in the UML are relatively insensitive to realistic levels of uncertainty in the rates of hydrolysis of DON and nitrification, and other model coefficients. Model performance in late summer and early fall is sensitive to interannual variations observed in phytoplankton growth.

Sensitivity analyses with the DO model focused on DO in the UML during the critical fall mixing period. Simulations were rather insensitive to realistic levels of uncertainty in the rate coefficients describing nitrification and oxidation of reduced species, but were moderately sensitive to reasonable levels of uncertainty in phytoplankton respiration and SOD. Simulations are highly sensitive to specified wind speeds. Thus substantial year-to-year differences in the distribution of DO in this interval, as has been observed, are to be expected as a result of natural variations in wind speed.

Predictions of the FC model for the upper waters are sensitive to prevailing environmental conditions, particularly light levels, mass transport, and thermocline depth. Factors favoring high concentrations of FC in the north end of the lake and relatively low concentration in the south, following a specified storm event, are high wind, low light, and a deep thermocline.

Sensitivity analyses were conducted for DO predictions with the river quality model, under a hypothetic increased load case, to evaluate the implications of reasonable levels of uncertainty in model coefficients. Model predictions were relatively insensitive to the limits for K_a, somewhat more sensitive to k_n, and highly sensitive to uncertainty in k_{CBOD}.

9.9.4 Model Application

9.9.4.1 Onondaga Lake

The predictive capabilities of the tested models have been utilized to address selected water quality management issues concerning Onondaga Lake and the adjoining river system. Application of the lake Cl^- model and available loading information has demonstrated that about 50% of the prevailing lake Cl^- concentration is the result of continuing saline waste inputs from the soda ash/chlor-alkali facility. The projected average Cl^- concentration without this source is approximately 230 mg $\cdot$ L^{-1}, or about 15% of the average concentration over the 1973 to 1985 period before closure, indicating the facility's contribution before its closure was 85%.

The development of user-friendly management model versions of the lake TP, N, and DO models, intended to support futuristic (a priori) predictions, was reviewed. Specified model inputs were identified. A Monod-type relationship developed between summer average TP concentration and gross photosynthesis across the trophic state gradient of Green Bay, Lake Michigan, was adopted as the basis to modify "net growth" in the N model, for scenarios that would result in reduced summer average TP concentrations in the lake. This functionality was also adopted to support a first approximation of future reductions in the steady-state value of SOD (DO model) associated with a significant change in trophic state.

Lake water quality standards and goals, important in the analysis of the results of model predictions for management scenarios, were reviewed. New York State has recently established a guidance value for the summer average epilimnetic TP concentration in lakes of 20 μg $\cdot$ L^{-1}. An approach is reviewed for the development of seasonal limits of T-NH_3 concentration in the UML of Onondaga Lake to protect against exceeding free ammonia (NH_3) standards. Utilizing a 4 year data base of pH and temperature measurements and a probabilistic approach, "probability of contravention" distributions for T-NH_3 concentration were developed for a range of confidence levels. The state standard established in 1994 for the summertime in-lake T-NH_3 concentration is 0.77 mgN $\cdot$ L^{-1}, which corresponds to a confidence level of about 50% for the occurrence of violations.

Two tributary flow and TP loading regimes are provided for the TP management model that represent a rather wide range of flow conditions. Tributary loading reduction scenarios are accommodated by multipliers applied uniformly through a year. Three tributary flow regimes, high, near-median, and low, are accommodated in the N model. Preliminary application of the N management model for default inputs identified the low tributary flow case (1987 daily flows), the second lowest annual tributary flow over a 20-yr period, as a reasonable worst case for evaluation of management scenarios. The inherent limitations of the a priori predictions of the N management model, associated mostly with natural variations in tributary flow, were illustrated by comparing management model performance to calibration performance for T-NH_3 in the UML for 1989. Application of the DO management model identified the substantial reductions in SOD required for the reclamation of the oxygen resources of the lake. A mechanistic sediment submodel would be required to predict both the time course and magnitude of the changes in sediment releases that would occur in response to reductions in lake primary productivity. Quantifying changes in sediment feedback would contribute little to the important management issues related to TP (trophic state) and T-NH_3 (NH_3 toxicity) in the upper waters of Onondaga Lake.

Leading management alternatives under consideration to remediate the impacts of METRO and the CSOs on Onondaga Lake were evaluated with the TP, N, and DO management models. Cases included higher levels of treatment (e.g., year-round nitrification and effluent filtration) with continued effluent discharge to the lake, partial bypass of wastewater around the lake with increased treatment at METRO, full bypass of effluent around the lake, discharge of the METRO effluent to the hypolimnnion, and 80% reduction of the TP load from CSOs. Effluent loadings for these scenarios were in most cases developed from performance data of existing facilities.

Several options, including partial bypass, hypolimnetic discharge of METRO, and the addition of effluent filtration at the existing METRO facility, would substantially reduce the summer average TP concentration in the UML of Onondaga Lake. However, the only alternative(s) evaluated that would achieve the decreases in TP concentrations necessary to significantly reduce phytoplankton growth in the lake are those that include full bypass of METRO's effluent around the lake.

Several management options, including year-round nitrification at METRO, partial bypass, and hypolimnetic discharge of METRO would substantially reduce the margin of violation of the T-NH₃ standard in the UML of Onondaga Lake. However, substantial violations would continue for all the options evaluated except total bypass.

The reductions in TP and T-NH₃ concentrations in the upper waters that would accompany the deepwater discharge of the METRO effluent would be achieved at the price of profoundly altering the stratification/mixing regime of the lake from its natural state. Combination of this alternative with increased levels of treatment presently available for large-scale facilities would also fail to meet the T-NH₃ standard and significantly reduce phytoplankton growth in the lake. Thus the deepwater discharge option has little to offer as an interim or long-term measure for these lake problems.

The application of the TP and N models described herein indicates full bypass of the METRO discharge around the lake will almost certainly be necessary to approach or meet the established TP quidance value ($20\,\mu g \cdot L^{-1}$) and the T-NH₃ standard. Reductions in tributary loading (CSOs and perhaps land-use management) will probably also be necessary to reach the TP goal. Expensive control of CSO TP loading should not be considered without elimination of the METRO input, because of the relatively small contribution CSOs make to the existing TP loading. Remediation of the T-NH₃ load from the waste bed area of Ninemile Creek may be necessary to meet a salmonid standard.

Management model predictions presented for DO are substantially more speculative than those for TP and T-NH₃ because of the lack of predictive capabilities for SOD and uncertainty in the assumption that steady-state conditions with respect to SOD presently prevail. Only modest improvements in the oxygen resources of the lake are predicted for the full bypass scenario. Hypolimnetic anoxia is predicted to continue, though the onset will be somewhat later, and lakewide violations of the DO standard during the fall mixing period are predicted, though it can be expected to occur in fewer years. Supersaturation of a deepwater discharge of the METRO effluent could augment the limited oxygen resources of the lake's hypolimnion (e.g., maintain oxia), and reduce internal loading of TP and T-NH₃. However, the feasibility of maintaining the high effluent DO concentration(s) required needs to be established, and the trade-offs with the impact on the lake's stratification regime evaluated.

Application of the fecal coliform (FC) management model indicates at least a 50% loading reduction from tributaries with CSOs (Onondaga Creek, Harbor Brook, and Ley Creek) would be necessary to meet standards lakewide for a one year (return frequency) storm. Increasing levels of loading reduction for those sources, approaching 90%, may be needed to assure avoidance of violations for particularly critical environmental conditions.

9.9.4.2 Seneca/Oswego River

The river water quality model, validated for the Baldwinsville to Phoenix reach for the conditions of 1991, has been used to delineate the processes that contribute to the observed major sag in DO in the lower layers of the Seneca River proximate to the lake inflow.

The processes of nitrification, SOD, and phytoplankton respiration all contribute importantly to the DO depletion. The dominant factor responsible for the DO sag, and the associated violations of standards, is the reduced vertical mixing caused by the occurrence of (mostly chemical) stratification. De-stratification will be necessary to eliminate the prevailing water quality violations in the lower layer of the river, and prevent loading from the river to the lake. Thus it will be a integral component of any management plan aimed at remediating the water quality problems of Onondaga Lake through bypass of the METRO load around the lake.

The river water quality model was utilized to evaluate the impact of selected (7) management alternatives, for treatment and discharge of bypassed METRO loads, on the DO resources of the river system of 1990 and 1991. Analyses were conducted for de-stratified (density) and critical flow and temperature conditions. Alternatives included one and two discharge options, at various levels of treatment. Two discharge options positioned one at the point of entry of the existing Baldwinsville WWTP (4 km upstream of the lake inflow), and the other at varying distances downstream. Lower DO concentrations are predicted for the bottom river layer compared to the top layer for all scenarios, because of the limited vertical mixing that prevails in the system, even under de-stratified conditions.

The maximum impact on DO was predicted to occur just upstream of the Phoenix dam for all alternatives that discharged to the river either upstream of, or proximate to, the inflow of the lake to the river. The existing level of treatment at METRO (for a diverted discharge), as specified by the discharge permit, is inadequate to meet water quality standards for DO in the river system. Increasing treatment to effluent levels typical of year-round nitrification, for discharges (2) from an expanded (47 MGD total) Baldwinsville WWTP and at METRO (50 MGD), to be discharged proximate to the point of entry of the lake inflow, was adequate to avoid violations of DO standards. Thus it appears feasible to completely bypass METRO to the Seneca River of 1990/1991 and

protect the river system against violations of DO standards, by providing advanced treatment for the diverted wastewater.

The degradation in the oxygen resources of the Seneca River manifested in the summer of 1993, brought about by the severe upstream infestation of zebra mussels, has eliminated the assimilative capacity of the river for oxygen-demanding wastes. The duration of the infestation at its present level, the feasibility of remediation of the impacts, and the potential for recurrence, thus are important issues to consider in evaluating management plans that include diversion of the METRO discharge to the river system.

References

Addess, J.M., and Effler, S.W. 1995. Methane production and cycling in Onondaga Lake. *Lake Reservoir Manag* (in press).

Ahlgren, J. 1977. Role of sediments in the progress of recovery of a eutrophicated lake. In: H.L. Golterman (Ed.), *Interactions Between Sediments and Fresh Water*. Dr. W. Junk B.V. Publisheres, The Hague, pp. 372–377.

APHA (American Public Health Association), American Water Works Association and Water Pollution Control Federation. 1985, 1987, 1992. *Standard Methods for the Examination of Water and Wastewater*, 16th, 17th, and 18th eds., APHA, Washington, DC.

Arita, M., and Jirka, G.H. 1987. Two-layer model of saline wedge. *J Hydraulic Engin ASCE* 113: 1229–1263.

Auer, M.T., and Canale, R.P. 1986. Mathematical modeling of primary production in Green Bay (Lake Michigan, USA), a phosphorus and light-limited system. *Hydrobiolog Bull* 20: 195–211.

Auer, M.T., and Effler, S.W. 1989. Variability in photosynthesis: Impact on DO models. *J Environ. Engin* 115: 944–963.

Auer, M.T., and Effler, S.W. 1990. Calculation of daily average photosynthesis. *J Environ Engin* 116: 412–418.

Auer, M.T., Effler, S.W., Heidtke, T.M., and Doerr, S.M. 1992. Hydrologic Budget Considerations and Mass Balance Modeling of Onondaga Lake. Submitted to the Onondaga Lake Management Conference, Syracuse, NY.

Auer, M.T., Johnson, N.A., Penn, M.R., and Effler, S.W. 1993. Measurements and verification of rates of sediment phsophorus release for a hypereutrophic urban lake. *Hydrobiologia* 253: 301–309.

Auer, M.T., Kieser, M.S., and Canale, R.P. 1986. Identification of critical nutrient levels through field verification of models for phosphorus and phytoplankton growth. *Can J Fish Aquat Sci* 43: 379–388.

Auer, M.T., and Niehauss, S.L. 1993. Modeling fecal coliform bacteria. I. Field and laboratory determination of loss kinetics. *Water Res* 27: 693–701.

Auer, M.T., Storey, M.L., Effler, S.W., Auer, N.A., and Sze, P. 1990. Zooplankton impacts on chlorophyll and transparency in Onondaga Lake, New York, USA. *Hydrobiologia* 200/201: 603–617.

Banks, R.B., and Herrera, F.F. 1977. Effect of wind and rain on surface reaeration. *J Environ Engin ASCE* 103: 489–504.

Bella, D.A. 1970. Dissolved oxygen variations in stratified lakes. *J San Engin ASCE Div* 96: 1129–1146.

Berner, R.A. 1980. *Early Diagenesis* Princeton University Press, Princeton, NJ.

Bierman, V.J., and Dolan, D.M. 1981. Modeling of phytoplankton-nutrient dynamics in Saginaw Bay, Lake Huron. *J Great Lakes Res* 7: 409–439.

Bierman, V.J., and Dolan, D.M. 1986. Modeling phytoplankton in Saginaw Bay: post-audit phase. *J Environ Engin ASCE* 112: 415–429.

Biswas, A.K. (Ed.). 1981. *Models for Water Quality Management* McGraw-Hill, New York.

Bowie, G.L., Mills, W.B., Porcella, D.B., Campbell, C.L., Pagenkopf, J.R., Rupp, G.L., Johnson, K.M., Chan, P.W.H., Gherini, S.A., and Chamberlain, C. 1985. Rates, Constants, and Kinetic Formulations in Surface Water Quality Modeling. EPA-600/3-85-640, U.S. Environmental Protection Agency, Athens, GA.

Bowman, G.T., and Delfino, J.J. 1980. Sediment oxygen demand: a review and comparison of laboratory and in-situ systems. *Water Res* 14: 491–499.

Brezonik, P.L. 1972. Nitrogen: sources and transformations in natural waters. In: H.E. Allen and T.R. Kramer (Eds.), *Nutrients in Natural Waters* John Wiley and Sons, New York, pp. 1–50.

Broecker, H.C., Petermann, J., and Siems, W. 1978. The influence of wind on CO_2 exchange in a wind-wave tunnel. *J Marine Res* 36: 595–610.

Brooks, C.M., and Effler, S.W. 1990. The distribution of nitrogen species in polluted Onondaga Lake, NY. USA. *Water, Air, Soil Pollut* 52: 247–262.

Calocerinos & Spina. 1984. Three Rivers Study. Water Quality Assessment. Phase I. Submitted to

Onondaga County, New York, Department of Drainage and Sanitation. Syracuse, NY.

Canale, R.P., Auer, M.T., Owens, E.M., Heidtke, T.M., and Effler, S.W. 1993a. Modeling fecal coliform bacteria. II. Model development and application. *Water Res* 27: 703–714.

Canale, R.P., Doerr, S.M., and Effler, S.W. 1993b. Total Phosphorus Model Report—Onondaga Lake, 1987–1990. Submitted to Onondaga Lake Management Conference, Syrcuse, NY.

Canale, R.P., and Effler, S.W. 1989. Stochastic phosphorus model for Onondaga Lake. *Water Res* 23: 1009–1016.

Canale, R.P., Gelda, R.K., and Effler, S.W. 1993c. A Nitrogen Model for Onondga Lake. Submitted to Onondaga Lake Management Conference, Syracuse, NY.

Canale, R.P., Gelda, R.K., and Effler, S.W. 1995a. Development and testing of a nitrogen model for Onondaga Lake, NY. *Lake Reservoir Manag* (in press).

Canale, R.P., Owens, E.M., Auer, M.T., and Effler, S.W. 1995b. Validation of water quality models for Seneca River, NY. *J Water Resour Plan Manag* 121: 241–250.

Cavari, B.Z. 1977. Nitrification potential and factors governing the rate of nitrification in Lake Kinnert. *Oikos* 28: 285–290.

Chapra, S.C., and Canale, R.P. 1991. Long-term phenomenological model of phosphorus and oxygen for stratified lakes. *Water Res* 25: 707–715.

Chapra, S.C., and DiToro, D.M. 1991. Delta method for estimating primary production, respiration and reaeration in streams. *J Environ Engin ASCE* 117: 640–655.

Chapra, S.C., and Dobson, H.F.H. 1981. Quantification of the lake typologies of Naumanm (surface growth) and Thienemanm (oxygen) with special reterence to the Great Lakes. *J Great Lakes Res* 7: 182–193.

Chapra, S.C., and Reckhow, K.H. 1983. *Engineering Approaches for Lake Management, Volume 2: Mechanistic Modeling.* Butterworth Publishers, Boston, MA.

Chiaro, P.S., and Burke, D.A. 1980. Sediment oxygen demand and nutrient release. *J Environ Engin* 106: 177–195.

City of Baltimore. 1992. NPDES Monthly Reports of Effluent Concentrations. Submitted to the State of Maryland.

Connors, S.D, Auer, M.T., and S.W. Effler. 1995. Phosphorus pools, alkaline phosphatase activity, and phosphorus limitation in hypereutrophic

Onondaga Lake, NY. *Lake Reservoir Manag* (in press).

Csanady, G.T. 1973. *Turbulent Diffusion in the Environment.* D. Reidel, Dordrecht, The Netherlands.

Curtis, E.J.C., Durrant, K., and Harman, M.M.I. 1975. Nitrification in rivers in the Trent Basin. *Water Res* 9: 255–268.

Daniil, E.I., and Gulliver, J.S. 1991. Influence of waves on air–water gas transfer. *J Environ Engin ASCE* 117: 522–540.

Devan, S.P., and Effler, S.W. 1984. History of phosphorus loading to Onondaga Lake. *J Environ Engin Div ASCE* 110: 93–109.

DiToro, D.M., and Connolly, J.P. 1980. Mathematical Models of Water Quality in Large Lakes. Part 2: Lake Erie. EPA-600/3-80-065, U.S. Environmental Protection Agency, Duluth, MN.

DiToro, D.M., Paquin, R.P., Subburamu, K., and Gruber, D.A. 1990. Sediment oxygen demand model: methane and ammonia oxidation. *J Environ Engin* 116: 945–986.

Doerr, S.M., Canale, R.P., Auer, M.T., Effler, S.W. 1995a. Development and testing of a total phosphorus model for Onondaga Lake, NY. *Lake Reservoir Manag* (in press).

Doerr, S.M., Effler, S.W., and Owens, E.M. 1995b. Forecasting impacts of a hypolimnetic wastewater discharge on lake water quality. *Lake Reservoir Manag* (in press).

Doerr, S.M., Effler, S.W., Whitehead, K.A., Auer, M.T., Perkins, M.G., and Heidtke, T.M. 1994. Chloride model for polluted Onondaga Lake. *Water Res* 28: 849–861.

Driscoll, C.T., Effler, S.W., Auer, M.T., Doerr, S.M., and Penn, M.R. 1993. Supply of phosphorus to the water column of a productive hardwater lake: controlling mechanisms and management considerations. *Hydrobiologia* 253: 61–72.

Effler, S.W., 1982. A Preliminary Water Quality Analyses of the Three Rivers System, Upstate Freshwater Institute, Inc, Syracuse, NY.

Effler. S.W. 1987. Impact of a chlor-alkali plant on Onondaga Lake and adjoining systems. *Water Air Soil Pollut* 33: 85–115.

Effler, S.W., Brooks, C.M., Addess, J.M., Doerr, S.M., Storey, M.L., and Wagner, B.A. 1991a. Pollutant loading from Solvay waste beds to lower Ninemile Creek, New York. *Water Air Soil Pollut* 55: 427–444.

Effler, S.W., Brooks, C.M., Auer, M.T., and Doerr, S.M. 1990a. Free ammonia toxicity criteria in a polluted urban lake. *Res J WPCF* 62: 771–781.

Effler, S.W., Brooks, C.M., Auer, M.T., and Doerr, S.M. 1991b. Closure to discussion of: free ammonia toxicity criteria in a polluted urban lake, Effler et al. 62: 771 (1990); Discussion by R. Ott, N.G. Hatala, E.C. Moran 63: 278 (1991). *Res J WPCF* 63: 280–283.

Effler, S.W., Brooks, C.M., Whitehead, K., Wagner, B., Doerr, S.M., Perkins, M.G., Siegfried, C.A., and Canale, R.P. 1995. Impact of zebra mussel invasion on river water quality. *Res J Water Pollut Contr Fed* (in press).

Effler, S.W., and Carter, C.F. 1987. Spatial Variability and Selected Physical Characteristics and Processes in Cross Lake, N.Y. *Water Resour Bull* 23: 243–249.

Effler, S.W., and Doerr, S.M. 1995. Water quality model evaluations for scenarios of loading reductions and diversion of domestic waste effluent around Onondaga Lake. *Lake Reservoir Manag* (in press).

Effler, S.W., Doerr, S.M., Brooks, C.M., and Rowell, H.C. 1990b. Chloride in the pore water and water column of Onondaga Lake, NY, USA. *Water Air Soil Pollut* 51: 315–326.

Effler, S.W., and Driscoll, C.T. 1985. A chloride budget for Onondaga Lake, New York, USA. *Water Air Soil Pollut* 27: 29–44.

Effler, S.W., and Field, S.D. 1983. Vertical diffusivity in the stratified layers of the mixolimnion of Green Lake, Jamesville, NY. *J Freshwater Ecol* 2: 273–286.

Effler S.W., Field S.D., and Perkins, M.G. 1984b. Onondaga Lake and the dynamics of chloride in the Oswego River. *Water Air, Soil Pollut* 23: 69–80.

Effler, S.W., Hassett, J.P., Auer, M.T., and Johnson, N. 1988. Depletion of epilimnetic oxygen and accumulation of hydrogen sulfide in the hypolimnion of Onondaga Lake, NY, USA. *Water Air, Soil Pollut* 39: 59–74.

Effler, S.W., McCarthy, J.M., Simpson, K.W., Unangst, F.J., Schafran, G.C., Schecher, W.D., Jaran, P., Shu, H.A., and Khalil, M.T. 1984a. Chemical stratification in the Seneca/Oswego Rivers (NY). *Water Air Soil Pollut* 21: 335–350.

Effler, S.W., and Owens, E.M. 1986. The density of inflows to Onondaga Lake, U.S.A., 1980 and 1981. *Water Air Soil Pollut.* 28: 105–115.

Effler, S.W., and Owens, E.M. and Schimel, K.A. 1986a. Density stratification in ionically enriched Onondaga Lake, USA. *Water Air Soil Pollut* 27: 169–180.

Effler, S.W., Owens, E.M., Schimel, K.A., and Dobi, J. 1986d. Weather-based variations in thermal stratification. *J Hydraul Engin Div ASCE* 112: 159–165.

Effler, S.W., Perkins, M.G., and Brooks, C. 1986b. The oxygen resources of the hypolimnion of ionically enriched Onondaga Lake, NY, USA. *Water Air Soil Pollut* 29: 93–108.

Effler, S.W., Schimel, K., and Millero, F.J. 1986c. Salinity, chloride, and density relationships in ion enriched Onondaga Lake, NY. *Water Air Soil Pollut* 27: 169–180.

Effler, S.W., and Siegfried, C.A. 1994. Zebra mussel (*Dreissena polymorpha*) populations in the Seneca River, New York: Impact on oxygen resources. *Environ Sci Technol* 28: 2216–2221.

Fischer, H.B., List, E.J., Imberger, J., Koh, R.C.Y., and Brooks, N.H. 1979. *Mixing in Inland and Costal Waters*. Academic Press, Orlando, FL.

Freedman, P.L., Canale, R.P., and Pendergast, J.F. 1980. Modeling storm overflow impacts on a eutrophic lake. *J Environ Engin Div ASCE* 106: 335–349.

Gameson, A.L.H., and Saxon, J.R. 1967. Field studies of effect of daylight on mortality of coliform bacteria. *Water Res* 17: 1595–1601.

Gannon, J.J., Busse, M.K., and Schillinger, J.E. 1983. Fecal coliform disappearance in a river impoundment. *Water Res* 17: 1595–1606.

Gardiner, R.D., Auer, M.T., and Canale, R.P. 1984. Sediment oxygen demand in Green Bay (Lake Michigan). In: M. Pirbazari and J.S. Devinny (Eds.), *Environmental Engineering: Proceedings of the 1984 Specialty Conference*, ASCE, New York, NY, pp. 514–519.

Gelda, R.K., and Auer, M.T. 1995. Development and testing of a dissolved oxygen model for a hypereutrophic lake. *Lake Reservoir Manag* (in press).

Gelda, R.K., Auer, M.T., and Effler, S.W. 1995a. Determination of sediment oxygen demand by direct measurement and by inference from reduced species accumulation. *Marine Freshwater Res* 46: 81–88.

Gelda, R.K., Canale, R.P., Auer, M.T., and Effler, S.W. 1993. A Dissolved Oxygen Model for Onondaga Lake. Submitted to Onondaga Lake Mangement Conference, Syracuse, NY.

Gelda, R.K., Auer, M.T., Effler, S.W., and Chapra, S.C. 1995b. Determination of reaeration coefficients: a whole lake approach. *J Environ Engin* (in press).

Gromiec, M.J., Loucks, D.P., and Orlob, G.T. 1983. Stream quality modeling. In: G.T. Orlob (Ed.), *Mathematical Modeling of Water Quality: Streams, Lakes, and Reservoirs*. John Wiley & Sons, New York, pp. 227–273.

Hall, G.H. 1986. Nitrification in lakes. In: J.I. Prosser (Ed.), *Nitrification* IRL Press, Washington, DC, pp. 127–156.

Hansen, D.J. 1989. Status of the development of water quality criteria and advisories. In: *Water Quality Standards for the 21st Century* Proceedings of the National Conference. USEPA. March 1–3, Dallas, TX, pp. 163–169.

Harris, G.P. 1986. *Phytoplankton Ecology, Structure, Function and Fluctuation*, Chapman and Hall, New York.

Hatcher, K.J. 1986. Introduction to Part 1: Sediment oxygen demand processes. In: K.J. Hatcher (Ed.), *Sediment Oxygen Demand Processes, Modeling and Measurement*. Institute of Natural Resources, University of Georgia, Athens, pp. 113–138.

Hutchinson, G.E. 1957. *A Treatise of Limnology. Vol. I. Geography, Physics and Chemistry*. John Wiley and Sons, New York.

Hutchinson, G.E. 1973. Eutrophication: the scientific background of a contemporary practical problem. *Amer Sci* 61: 269–279.

Imboden. D.M. 1974. Phosphorus model of lake eutrophication. *Limnol Oceanogr* 19: 297–304.

Kishbaugh, S.A. 1993. Applications and limitations of qualitative lake assessment data. Paper presented at the North American Lake Management Society Conference, on December 3, Seattle, WA.

Klapwijk, A., and Snodgrass, W.J. 1986. Lake oxygen model I: Modeling sediment–water transfer of ammonia, nitrate and oxygen. In: P.G. Sly (Ed.), *Sediments and Water Interactions*, Springer-Verlag, New York, pp. 243–250.

Krenkel, P.A., and Novotny, V. 1979. River water quality model construction. In: H.V. Shen (Ed.), *Modeling of Rivers*, John Wiley and Sons, New York.

Lam, D.C.L., Schertzer, W.M., and Fraser, A.S. 1987. Oxygen depletion in Lake Erie: modeling the physical, chemical, and biological interactions, 1972 and 1979. *J Great Lakes Res* 13: 770–781.

Lantrip, B.M. 1983. The decay of enteric bacteria in an estuary. Ph.D. Thesis, School of Hygiene and Public Health, The Johns Hopkins University, Baltimore, MD.

Leach, J.H. 1993. Impacts of the zebra mussel (*Dreissena polymorpha*) on water quality and fish spawning reefs in Western Lake Erie. In: T.F. Nalepa and D.W. Schloesser (Eds.), *Zebra Mussels: Biology, Impacts, and Control* Lewis, pp. 381–398.

Lean, D.R.S., and Knowles, R. 1987. Nitrogen transformation in Lake Ontario. *Can J Aquat Sci* 44: 2133–2143.

Ludyanskiy, M.L., McDonald, D., and MacNeill, D. 1993. Impact of the zebra mussel, a bivalve invader. *Bioscience* 43: 533–544.

Lung, W.S., Canale, R.P., and Freedman, P.L. 1976. Phosphorus models for eutrophic lakes. *Water Res* 10: 1101–1114.

Mackie, G.L. 1991. Biology of exotic zebra mussel, *Dreissena polymorpha*, in relation to native bivalves and its potential impact on Lake St Clair. *Hyrdobiologia* 219: 251–268.

Metcalf & Eddy, Inc. 1979. *Wastewater Engineering: Treatment Disposal and Reuse*, 2nd ed. McGraw-Hill, New York.

Mitchell, R., and Chamberlin, C. 1978. Survival of indicator organisms. In: G. Berg (Ed.), *Indicators of Viruses in Water and Food*. Ann Arbor Science Publishers, Ann Arbor, MI.

Naumann, W.R. 1993. Determination of Surface Gas Transfer in a Stratified River. Master's Thesis, Syracuse University, Syracuse, NY.

Naumann, W.R., and Owens, E.M. 1995. Mass Transport and Surface Gas Transfer in a Stratified Inland River. *Environ Engin ASCE* (in preparation).

New York State Department of Environmental Conservation. 1993. New York State Fact Sheet for Phosphorus: Ambient Water Quality Value for Protection of Recreational Uses. Bureau of Technical Services and Research, Albany, NY.

O'Brien & Gere Engineers Inc. 1973. Environmental Assessment Statement, Syracuse Metropolitan Sewage Treatment Plant and the West Side Pump Station and Force Main. Submitted to Onondaga County, Syracuse, NY.

O'Brien & Gere Engineers Inc. 1977. Seneca-Oneida-Oswego Rivers Water Quality Survey and Preliminary Modeling. Central New York Regional Planning Development Agency, Syracuse, NY.

O'Connor, D.J. 1983. Wind effects on gas–liquid transfer coefficients. *J. Environ Engin ASCE* 109: 731–752.

O'Connor, D.J., and Mueller, J.A. 1970. A water quality model of chlorides in Great Lakes. *J San. Engin ASCE* 96: 955–975.

Onondaga County. 1980–1991. Onondaga Lake Monitoring Report. Annual Report. Onondaga County, Syracuse, NY.

Onondaga County. 1994. Draft Municipal Compliance Plan/Draft Environmental Impact Statement. Onondaga County Wastewater Collection and Treatment System, Onondaga County, Syracuse, NY.

Onondaga Lake Management Conference. 1993. Ononaga Lake: A Plan for Action. Onondaga Lake Management Conference, Syracuse, NY.

Owens, E.M. 1988. Bathymetric Survey and Mapping of Onondaga Lake, New York. Submitted to Central New York Regional Planning and Development Board, Syracuse, NY.

Owens, E.M. 1989. A model of transport of fecal bacteria in Onondaga Lake, New York. Technical Report of No. 38. Upstate Freshwater Institute. Submitted to the Central New York Regional Planning and Development Board, Syracuse, NY.

Owens, E.M. 1993. Estimation of Inflow from Seneca River, Onondaga Lake, 1990–1991. Submitted to Upstate Freshwater Institute, Syracuse, NY.

Owens, E.M. 1993. Reaeration at the Phoenix Dam. Submitted to the Upstate Freshwater Institute, Syracuse, NY.

Owens, E.M., and Effler, S.W. 1989. Changes in stratification in Onondaga Lake, New York. *Water Resour Bull* 25: 587–597.

Owens, E.M, and Effler, S.W. 1995. Forecasting impacts of a hypolimnetic wastewater discharge on lake stratification and mixing. *Lake Reservoir Manag* (in press).

Palmer, M.D., and Dewey, R.J. 1984. Beach fecal coliforms. *Can J Civ Engin* 11: 217–224.

Penn, M.R. 1994. The deposition, diagenesis and recycle of sedimentary phosphorus in a hypereutrophic lake. Ph.D Thesis, Michigan Technological University, Houghton, MI.

Penn, M.R., Pauer, J.J., Gelda, R.K., and Auer, M.T. 1993. Laboratory Measurements of Chemical Exchange at the Sediment–Water Interface of Onondaga Lake. Submitted to the Onondaga Lake Management Conference, Syracuse, NY.

Quigley, M.A., Gardiner, W.S., and Gordon, W.M. 1993. Metabolism of the zebra mussel (*Dreissena polymorpha*) in Lake St. Clair of the Great Lakes. In: T.F. Nalepa and D.W. Schloesser (Eds.), *Zebra Mussels: Biology, Impacts, and Controls* Lewis, Ann Arbor, MI, pp. 295–306.

Ramcharan, C.W., Padilla, D.K., and Dodson, S.I. 1992a. A multivariate model for predicting population fluctuations of *Dreissena polymorpha* in North America Lakes. *Can J Fish Aquat Sci* 49: 150–158.

Ramcharan, C.W., Padilla, D.K., and Dodson, S.I. 1992b. Models to predict potential occurrence and density of the zebra mussel, *Dreissena polymorpha*. *Can J Fish Aquat Sci* 49: 2611–2620.

Reckhow, K.H., and Chapra, S.C. 1983. *Engineering Approaches for Lake Management*. Volume 1. *Data*

Analysis and Empirical Modeling. Butterworth Publishers, Boston, MA.

Reeders, H.H., and Bij de Vaate, A. 1992. Bioprocessing of polluted suspended matter from the water column by zebra mussel (*Dreissena polymorpha Pallas*). *Hydrobiologia* 239: 53–63.

Richardson, W.L. 1976. An evaluation of the transport characteristics of Saginaw Bay using a mathematical model for chloride. In: R.P. Canale (Ed.), *Modeling Biochemical Processed in Aquatic Ecosystem.* Ann Arbor Science Publishers, Ann Arbor, MI, pp. 113–140.

Rooney, J. 1973. Memorandum (cited by U.S. Environmental Protection Agency, 1974, Environmental Impact Statement on Wastewater Treatment Facilities Construction Grants for the Onondaga Lake Drainage Basin).

Samide, H. 1993. Letter to S.W. Effler (Upstate Freshwater Institute), Regarding Three Rivers System Model Inputs (January 28, 1993).

Shanahan. P., and Harleman, D.R.F. 1984. Transport in lake water quality modeling. *J Environ Engin ASCE* 110: 42–57.

Schillinger, J.E., and Gannon, J.J. 1982. Coliform Attachment to Suspend Particles in Stormwater. School of Public Health, the University of Michigan, Ann Arbor, MI.

Schneider, D.W. 1992. Bioenergetic model of zebra mussel, *Dreissena polymorpha*, growth in the Great Lakes. *Can J Fish Aquat Sci* 49: 1406–1416.

Seitzinger, S.P. 1988. Denitrification in freshwater and coastal marine ecosystems: Ecological and geochemical significance. *Limnog Oceanogr* 33: 702–724.

Sivakumar, M., and Herzog, A. 1979. A model for the prediction of reaeration coefficient in lakes from wind velocity. In: W.H. Graf and C.H. Mortimer (Eds.), *Hydrodynamics of Lakes.* Elsevier, New York.

Smith, D.I. 1978. Water Quality for River-Reservoir Systems. U.S. Army Corps of Engineers, Hydrologic Engineering Center, Davis, CA.

Snodgrass, W.J. 1987. Analysis of models and measurements for sediment oxygen demand in Lake Erie. *J Great Lakes Res* 13: 738–756.

Snodgrass, W.J., and Dalrymple, R.J. 1985. Lake Ontario oxygen model. 1. Model development and application. *Environ Sci Technol* 19: 173–179.

Sprent, J.I. 1987. The Ecology of the Nitrogen Cycle, University Press, Cambridge.

Stearns and Wheler. 1992. Addendum No. 1 to Final Report, Projected Flows and Loadings, METRO Engineering Alternatives. Prepared for

Department of Drainage and Sanitation, Onondaga County, Syracuse, NY.

Stearns and Wheler. 1993a. METRO Performance Pharacteristics Representing Seasonal Nitrification for Alternatives 3, 4, 5, 6, 7, and 8 (Work Plan Amendment). Prepared for Department of Drainage and Sanitation, Onondaga County, Syracuse, NY.

Stearns and Wheler. 1993b. Letter to J.A. Bloomfield (including monthly average METRO wastewater flows and effluent concentrations). Prepared for Department of Drainage and Sanitation, Onondaga County, Syracuse, NY.

Storey, M.L., Auer, M.T., Barth, A.K., and Graham, J.M. 1993. Site-specific determination of kinetic coefficients for modeling algal growth. *Ecol Model* 66: 181–196.

Storey, M.L, Auer, M.T., and Rijkenboer, M. 1995. Laboratory determination and field verification of a photosynthesis–light response curve for the Seneca River, New York. *Freshwater. Ecol* (in review).

Streeter, H.W., and Phelps, E.B. 1925. A study of the pollution and natural purification of the Ohio River. III. Factors concerned in the phenomena of oxidation and reaeration. *Pub. Health Bull* No. 146. U.S. Public Health Service.

Stumm, W., and Morgan, J.J. 1981. *Aquatic Chemistry,* Wiley-Interscience, New York.

Sweerts, J.P.A., Bar-Gilissen, M.J., Cornelese, A.A., and Cappenburg, T.E. 1991. Oxygen consuming processes at the profundal and littoral sediment–water interface of a small meso-eutrophic lake (Lake Vechtan, The Netherlands). *Limno Oceanogr* 36: 1124–1133.

Symons, J.M., Irwin, W.H., Clark, R.M., and Robeck, G.G. 1967. Mangement and measurement of DO in impoundments. *J San Engin Div ASCE* 93: 181–209.

Thomann, R.V. 1982 . Verification of water quality models. *J Environ Engin Div ASCE* 108: 923–940.

Thomann, R.V., and Mueller, J.A. 1987. *Principles of Surface Water Quality Modeling and Control.* Harper Row, New York.

USEPA. 1985. Ambient Water Quality Criteria to Ammonia—1984. EPA 440/5–85–001, Office of Research and Development, Cincinnati, OH.

Vollenweider, R.A. 1974. *A Manual on Methods for Measuring Primary Production in Aquatic Environments,* 2d ed. IBP Handbook No. 12. Blackwell Scientific, Oxford.

Vollenweider, R.A. 1975. Input-output models with special reference to the phosphorus loading concept in limnology. *Schweiz Z Hydrol* 33: 58–83.

Vollenweider, R.A. 1982. *Entrophication of Waters: Monitoring, Assessment and Control.* Organization of Economic Cooperation and Development, Paris, France.

Walker, R.R., and Snodgrass, W.J. 1986. Models for sediment oxygen demand in lakes. *J Envir Engin* 112: 25–43.

Walker, W.W., Jr. 1991. Compilation and Review of Onondaga Lake Water Quality Data. Submitted to Onondaga County Department of Drainage and Sanitation, Syracuse, NY.

Wanninkhof, R., Leduidl, J.R., and Broecher, W.S. 1985. Gas exchange–wind speed relation measured with sulfur hexafluoride on a lake. *Science* 227: 1224–1226.

Wanninkhof, R., Leduidl, J.R., and Broecher, W.S., and Hamilton. M. 1987. Gas exchange on Mono Lake and Crowley Lake, California. *J Geophys Res* 92: 14567–14580.

Wanninkhof, R., Leduidl, J.R., and Crusius, J. 1991. Gas transfer velocities on lakes measured with sulfur hexafluoride. In: S.C. Wilhelms and J.S. Gulliver (Eds.), *Air Water Mass Transfer*, Symposium volume of the second international conference on gas transfer at water surfaces, Minneapolis, MN.

Weiler, R.R. 1974. Carbon dioxide exchange between water and atmosphere. *J Fish Res Bd Can* 31: 392–402.

Wetzel, R.G. 1983. *Limnology*, 2d ed. Saunders College Publishing, New York.

Whitehead, P., Beck, B., and O'Connell, E. 1981. A systems model of streamflow and water quality in the Bedford Ouse River System-II. Water quality modeling. *Water Res* 15: 1157–1171.

Winter, T.C. 1978. Ground-water component of lake water and nutrient budget. *Verh Internat Verein Limnol* 20: 438–444.

Wodka, M.C., Effler, S.W., Driscoll, C.T., Field, S.D., and Devan, S.D. 1983. Diffusivity-based flux of phosphorus in Onondaga Lake. *J Environ Engin Div ASCE* 109: 1403–1415.

Wright, S.J., Wong, D.R., Zimmerman, K.E., and Wallace, R.B. 1982. Outfall diffuser behavior in stratified ambient fluid. *J Hydrol Div ASCE* 108: 483–501.

Zison, S.W., Mills, W.B., Diemer, D., Chen, C.W. 1978. Rates, Constants and Kinetic Formulations in Surface Water Quality Modeling, by Tetra Tech., Inc. for USEPA, ORD, Athens, GA, ERL, EPA 600-3-78-105.

10
Synthesis and Perspectives

Steven W. Effler

10.1 Impact of the Soda Ash/Chlor-alkali Facility on Onondaga Lake and Adjoining Systems: Update

10.1.1 Background

In late 1985 and early 1986 an analysis of the impact of the soda ash/chlor-alkali facility on Onondaga Lake and adjoining systems was conducted by Effler (1987) in anticipation of the closure of the plant. Effects of the ionic waste discharge of the facility identified (Effler 1987), included: (1) ionic enrichment, (2) altered stratification regime of the lake, (3) altered exchange between the lake and river, (4) altered hydrodynamics in the river downstream of the lake, (5) precipitation and deposition of large quantities of $CaCO_3$, and (6) altered chemistry of lake sediments regulating P availability. It was concluded (Effler 1987) that mercury and benzene and chlorobenzene wastes from the facility had contaminated the lake sediments and fish of the lake. Deleterious impacts of the facility identified by Effler (1987) included: the elimination of fish habitat, exacerbation of the problem of limited O_2 resources of the lake, contamination of fish flesh, exacerbation of the problem of low transparency of the lake, and severe O_2 depletion in the lower waters of the river system.

Documented changes in related characteristics of the lake and adjoining portions of the Seneca River brought about by reductions in waste loading that accompanied the closure of the facility (February 1986), and continued research on related lake processes, as described in earlier chapters, offer an excellent opportunity to update the impact analysis. Shortcomings in previous interpretations are also identified. It is emphasized that this should be considered an update, rather than a final postaudit analysis of the facility's impact, as related research continues.

10.1.2 Ionic and Ammonia Waste Loading

An abrupt decrease in ionic waste loading occurred as a result of the closure of the soda ash/chlor-alkali plant (Chapter 3). The average annual summed load of Cl^-, Na^+, and Ca^{2+}, the major ionic waste constituents, for the 1974–1985 interval (i.e., before closure) was 1.2×10^9 kg, corresponding to a daily average loading rate of 3.3×10^6 kg $\cdot$ d^{-1}. By 1989 the summed load was 0.14×10^9 kg, or about 12% of the preclosure load. By 1989 the annual loading of Cl^-, Na^+, and Ca^{2+} had decreased by 79, 67, and 70% from the average documented for the 1974–1985 period (Chapter 3), as a result of the closure. The plant was the single largest source of Cl^- to Lake Ontario before it closed (Effler et al. 1985a). During the 1970s this facility contributed 48, 37, and 10% of the U.S. tributary, total tributary, and grand total Cl^- loading, respectively, to Lake Ontario.

The substantial residual of ionic waste loading that has continued (Chapters 2 and 3) following closure was not anticipated by Effler (1987); thus, related impacts identified in that earlier analysis have been ameliorated, but not eliminated. Most of the ionic waste from soda ash manufacture that continues to be released enters Ninemile Creek in the area of the most recently active waste beds. This is manifested as dramatic increases in Cl^-, Na^-, and Ca^{2+} concentrations downstream of the waste beds (Lakeland station, e.g., Effler et al. 1991, Chapter 3) and the constant relationships maintained among these three constituents in the stream below the beds (Chapters 2 and 3). The average Cl^- concentration at Amboy (located on Ninemile Creek upstream of the waste beds) was less than 7% (54 mg · L^{-1}) of the average observed at Lakeland (821 mg · L^{-1}), over a 12-month interval in 1989 and 1990 (Effler et al. 1991). A strong relationship exists between Cl^- (and Na^+ and Ca^{2+}) concentration and flow ("dilution model") at the Lakeland station (Chapter 3), consistent with a load that emanates largely as leachate and contaminated ground water. The relationships between the three primary ionic constituents in the creek below the beds continue to be essentially equivalent to those of the Solvay waste bed overflow during the operation of the facility (Chapters 2 and 3), establishing the origin of these constituents as waste generated from soda ash production. The ionic waste from soda ash production also continues to enter the lake via METRO, associated with irregular reception of enriched water from a lagoon adjoining the waste beds that continues to fill. Approximately 55% (50% from Ninemile Creek and 5% from METRO) of the total external Cl^- load that continues to be received by Onondaga Lake has its origins as soda ash waste (Chapter 3).

The estimates for Na^+ and Ca^{2+} are somewhat less certain, about 42 and 30%, respectively (Chapter 3).

Effler et al. (1991) also found a significant load of $T-NH_3$ enters Ninemile Creek in the area of the most recent waste beds. The concentrations of $T-NH_3$ and Cl^- were highly correlated in the creek, implying the same origin(s) for these constituents (Effler et al. 1991). Ammonia, an important intermediary reactant in the Solvay process, was largely recycled. However, this recently discovered load indicates the recovery of $T-NH_3$ was incomplete. This source of $T-NH_3$ presently represents only about 3% of the total external load (Effler et al. 1991). However, it would probably represent the single largest source of $T-NH_3$ to the lake if METRO were diverted. This source is not critical to meeting the free ammonia standard for the lake for nonsalmonids, but it could be critical if a salmonid goal for the lake was established (see Chapter 9).

The "mud boils," located along Onondaga Creek (33 km upstream of the lake) in Tully Valley, had not been identified as a potential impact of the soda ash/chlor-alkali facility by Effler (1987). The occurrence of the "mud boils" (effusion of soft sediment brought to the surface by artesian discharge (Shilts 1978)) has been attributed (Getchell 1983) to solution mining activities at nearby evaporite (NaCl) deposits by the soda ash manufacturer. However, there is continuing contention concerning this issue (e.g., Kosinski 1985; Tully 1983). The "mud boils" are a significant source of sediment (Effler et al. 1992) to the lake, as well as Cl^- and Na^+ (Chapter 3). Approximately 8% of the present total external Cl^- load and 10% of the Na^+ load are supplied from the "mud boils". Thus the soda ash facility may be responsible for more than 60% of the continuing Cl^- load and more than 50% of the Na^+ load (Chapter 3). Breaks in the brine line (not previously identified by Effler (1987)), that extends from Tully Valley to the facility on the western shore of the lake, have caused short-term increases in Cl^- and Na^+ concentrations and related fish kills in Onondaga Creek.

10.1.3 Lake Concentrations

The concentrations of Cl^-, Na^+, and Ca^{2+} decreased rapidly following the closure of the soda ash/chlor-alkali facility (Chapter 5) in response to the abrupt decreases in loading and the rapid flushing rate of the lake (Chapters 3 and 9, Doerr et al. 1994). The contribution of these three constituents to the lake's salinity have decreased since closure of the facility from more than 85% to less than 70%. The volume-weighted average salinity of the lake for the 1968–1985 interval before closure was about 3 parts per thousand (‰). By 1990 the salinity had decreased 60% to 1.2‰. An error in the presentation of the system-specific equation of state for Onondaga Lake, developed by Effler et al. (1986a), was corrected in the treatment presented here in Chapter 5.

Effler (1987) had reviewed earlier analyses (O'Brien & Gere 1973; Rooney 1973), which had been based on limited data, that indicated the soda ash/chlor-alkali facility's contribution to the Cl^- concentration in the lake before closure was 50 to 60% (i.e., 40–50% originated from natural sources). Effler (1987) contended the contribution from the facility was much higher, about 80%, based on the results of a Cl^- mass balance conducted at a yearly time step over the 1970 to 1981 period by Effler and Driscoll (1986; they concluded ≥85%). The average Cl^- concentration in the lake over the 1973 to 1985 period was $1585 \, \mu g \cdot L^{-1}$; the average for 1990 and 1991 was $430 \, \mu g \cdot L^{-1}$ (Doerr et al. 1994, Chapter 9). Thus, the reduction from the last 13 years of operation of the facility was 73%, establishing that the contribution of the facility during its operation was substantially more than the earlier estimates of O'Brien & Gere (1973) and Rooney (1973).

Doerr et al. (1994) developed and validated a completely mixed dynamic Cl^- mass balance model for Onondaga Lake for the period 1973–1991 (also see Chapter 9), based on daily time series of surface inflows and estimates of Cl^- loads, carried by those inflows (Chapter 3). The model performed well in simulating: (a) the major decrease in lake concentration observed following closure of the facility, and (2) the year-to-year and seasonal variations that were clearly manifested before closure (Doerr et al. 1994, Chapter 9). The small natural diffusion-based contribution of Cl^- from the sediments, described by Effler et al. (1990), was included. The model was used to demonstrate the compensating effects of the Cl^- enriched seeps in lower Onondaga Creek (Chapter 3) and the dilute backflow from the Seneca River (Chapter 4) had on lake concentrations. The model predicts the average lake concentration would be about $230 \, mg \cdot L^{-1}$, or about 50% of the prevailing concentration, if the continuing inputs of soda ash waste were eliminated (Doerr et al. 1994, Chapter 9). This indicates that 85% of the preclosure Cl^- concentration ($1585 \, mg \cdot L^{-1}$) was attributable to the ionic waste discharge from the soda ash/chlor-alkali facility, consistent with the earlier analysis of Effler and Driscoll (1986). This estimate of the facility's contribution to the Cl^- concentration of the lake is considered conservative, as direct inputs from the older waste beds bordering the lake are not quantified, and loadings from the mud boils were not eliminated in the model simulation.

Effler (1987) indicated that the elevated salinity of the lake before the closure of the soda ash/chlor-alkali facility probably exerted selective pressure on various biological assemblages. Species of the lake known to be tolerant of salinity, particularly zooplankton (Meyer and Effler 1980), were identified. Species diversity has been reported to be strongly reduced at the salinity levels that prevailed in the lake before closure (Remane and Schleiper 1971). Recent paleolimnological analysis (Rowell 1992, Chapter 8) has demonstrated that the diatom community of the lake shifted to a distinctly more salinity tolerant assemblage during the operation of the facility. The dominant macrophyte in the lake presently, *Potamogeton pectinatus*, is known to be tolerant of high salinity levels (Chapter 6). Earlier, the USEPA (1974, p. 34) indicated "The chloride/salinity levels of the lake (before closure) approach concentrations at which one would expect to find the smallest species diversity; that is, the chloride/salinity level is near the upper limit for freshwater organisms and near the lower limit for marine species." The negative impact the facility's ionic waste discharge had on diversity/species richness is depicted by the clear increases in these characteristics of the lake's biota that have emerged since closure. Major increases in the diversity of the zooplankton and fish communities have been documented since closure of the facility (Chapter 6). The changes in the zooplankton assemblage following the closure are particularly well resolved. More efficient grazers, intolerant of the earlier salinity levels, emerged soon after closure. The aesthetic benefits of this shift in the community were discussed in Chapter 7.

10.1.4 Plunging Underflow(s), Spent Cooling Water, and Impacts on Lake Stratification and Oxygen

Effler and Owens (1986) established the occurrence of the plunging underflow phenomenon for inflows to Onondaga Lake (primarily Ninemile Creek) made more dense by the inclusion of large quantities of ionic waste from soda ash production, and the net cycling of water from lower stratified layers to the upper layers of the lake (as an overflow) when the deeper cooling water intake was used by the facility. The difference in the densities of Ninemile Creek and the lake has decreased greatly since closure of the facility because of the major reduction in the concentrations of ions, generated as waste from soda ash production, in the creek (Chapters 3 and 4). However, Ninemile Creek's density remains greater than the surface waters of the lake most of the time (Chapter 4) because its ionic content continues to exceed that of the lake (Chapters 3, 5, and 9). The underflow phenomenon continues to occur, but less frequently and to a much reduced extent (Chapter 4). The artificial internal P loading that accompanied the use of the deep cooling water intake by the facility was identified by Effler and Owens (1987). Major reductions in external P

loading, including diversion of METRO, would not have achieved significant limitation of phytoplankton growth in the lake as long as the use of the deepwater intake during the warmer months of the year continued. The summer heat income of the lake has decreased by about 9% since the discontinuation of the discharge of spent cooling water to the lake that accompanied closure of the facility (Chapter 4).

Several features of the stratification/mixing regime of the lake were found to be altered by the operation of the soda ash/chlor-alkali facility, particularly associated with the underflow phenomenon caused by the ionic waste discharge (Effler 1987). This portion of the analysis was supported by the detailed seasonal partitioning of the thermal and chemical components of stratification (also see Effler et al. 1986b) for two years and the application of a calibrated/verified mechanistic mathematical model of density stratification (Effler 1987). The analysis of the impact of the facility on the stratification/mixing regime has since been expanded, through analysis of stratification and related water quality data for additional years before and after closure (Effler and Perkins 1987, Chapter 4) and further application of the stratification model (Owens and Effler 1989, Chapter 4).

An analysis by Effler and Perkins (1987) found that the lake failed to turn over (or turnover was extremely brief) in spring of several years before closure of the facility because the substantial chemically based antecedent density stratification that had developed under the ice, caused by the ionic waste discharge, was not overcome by the turbulent kinetic energy inputs during the spring of those years. Several recurring impacts of the facility on the lake's stratification regime have emerged from the analysis of 14 years of monitoring data (Chapter 4). The plunging inflow caused the lake to be stratified with respect to salinity, which increased overall density stratification (salinity component superimposed on the natural thermal component). Salinity stratification was established first; it was the dominant component through the early portion of the stratification period.

The salinity component decreased progressively. Salinity stratification was reestablished (i.e., after onset of complete fall turnover) in most years, because of the increased tendency for plunging of the ionic waste discharge during periods of low watercolumn stability. This caused interrupted and abbreviated turnover periods, thereby extending the period of stratification. Substantial year-to-year differences in the stratification regime of the lake, including the contribution of the salinity component, occurred, probably largely driven by natural variations in meteorological conditions (Chapter 4). The salinity component represented from 13 to 50% of overall density stratification over the May to September interval for the 1968 to 1986 period (before closure of the facility; Chapter 4).

Subsequent to the earlier analysis reviewed by Effler (1987), the mechanistic stratification model was further applied by Owens and Effler (1989) to assess the impact of the soda ash/chlor-alkali facility on the lake's stratification/mixing regime and predict changes that would occur following closure. The analysis was placed in the perspective of natural variability in meteorological conditions, by utilizing a 30 year meteorological data base available for the region. The analysis indicated the facility had a profound impact on the regularity of turnover and the duration of stratification. These findings have been supported by the documented changes in these conditions since closure of the facility (Chapter 4). Model simulations indicated both spring and fall turnover were not regular occurrences in the lake during the period of ionic waste discharge. The average duration of stratification was predicted to be approximately 7 months before closure, consistent with observations for the lake (Chapter 4). The average duration of stratification following closure was predicted to be 4 months; this appears to be somewhat of an underestimate, as the average observed since closure has been about 5 months (Chapter 4). Other more subtle impacts of the facility resolved with the model were reductions in the depth of the upper mixed layer, increases in the maximum density gradient in early summer, and depression of hypolimnion

temperature (Owens and Effler 1989, Chapter 4).

The stratification regime of Onondaga Lake is presently usually similar to those of other dimictic central New York lakes of similar dimensions (Chapter 4). However, the failure of the lake to completely turn over in the spring of 1993 may reflect the effect of the continued occurrence, albeit ameliorated, of the plunging underflow phenomenon associated with the continued input of ionic waste.

Continued monitoring since closure (Chapter 5) and further analysis of earlier data bases (Effler and Perkins 1987) support the position (Effler 1987) that the soda ash/chlor-alkali facility exacerbated the lake's problem of limited oxygen resources. The extension of the period of stratification, that resulted from ionic waste discharge, increased the period the lower layers were isolated from oxygen sources and thereby the duration of anoxia. As a result of the closure of the soda ash/chlor-alkali facility the lower layers remain oxygenated longer in spring and early summer, and are replenished earlier in fall (Chapter 5). The impact on oxygen resources was even worse in years the lake failed to turn over, as this eliminated the replenishment of the stratified layers with oxygen (Effler and Perkins 1987).

10.1.5 Seneca River

The unusual bidirectional flow conditions that often prevail in the lake outlet (Seneca River flowing into the lake on the outlet surface; Onondaga Lake exiting out the bottom of the channel; Effler 1987; Seger 1980; Stewart 1978) are a result of two anthropogenic effects: (1) the elevated density of the lake caused by ionic waste inputs, and (2) the nearly equal surface elevations of the lake and river associated with the use of the system for navigation and hydropower (Chapter 4). Density stratification set up in the outlet continues out into the river, both upstream and downstream of the point of entry of the outlet (Canale et al. 1994; Effler et al. 1984a). Before closure of the soda ash/chlor-alkali facility, during low flow periods, salinity-based stratification extended 14 km downstream to the Phoenix dam, where it was broken up by the river flow passing over the dam (Effler et al. 1984a). The upstream boundary of the "salt wedge," unique for inland rivers (Canale et al. 1994), at times extended several km upstream toward Baldwinsville. Salinity-based density stratification, and the attendant limited vertical mixing, has a severe negative impact on the river by causing major depletions of oxygen in the lower stratified layer. Violations of the NYS minimum DO standard of $4.0 \, mg \cdot L^{-1}$ are observed in the lower layer, and conditions approaching anoxia have been documented (e.g., Effler et al. 1984a). The oxygen depletion in the lower layer has been attributed to the isolation of this layer (containing the Onondaga Lake outflow) from oxygen sources, combined with the operation of various oxygen demanding processes (Effler et al. 1984a). Further, chemical stratification promotes the extension of Onondaga Lake's problems into the river; for example, high ammonia concentrations retained in the lower river layer adjoining the lake inflow represent violations of the State's free ammonia standard (Chapter 9).

Bidirectional flow conditions in the outlet and salinity-based density stratification in the river, with its negative implications for dissolved oxygen, have continued to occur since closure of the soda ash/chlor-alkali facility (Canale et al. 1994, Chapter 9). However, these impacts have been ameliorated because of the reduction in ionic waste loading that accompanied the closure. Under low flow conditions in 1991, salinity stratification was observed to extend about 8 km downstream of the lake and more than 1 km upstream (Canale et al. 1994, Chapter 9). Violations of dissolved oxygen standards continued to occur in the lower layer (Canale et al. 1994, Chapter 9) before the zebra mussel invasion (e.g., 1990 and 1991). The reductions in the lake's salinity, brought about by the closure of the facility, have reduced the longitudinal extent of salinity stratification and the threshold river flow at which the phenomenon occurs, and thereby have lessened the extent of coupled

impacts on the river's oxygen resources (Chapter 9). The elevated salinity of the lake compared to the river has remained the principal driving force for the persistence of the stratification phenomenon in the river (Chapter 4). The salinity difference between these systems would be further reduced, if not essentially eliminated, and the continuing negative impact on the river's oxygen resources would be further ameliorated, by the elimination of ionic waste inputs from soda ash production (see Cl^- model of Chapter 9, also Doerr et al. 1994). However, a more limited, thermally based density stratification and coupled oxygen depletion in the river would probably continue to occur irregularly, associated with natural temperature differences between the river and the upper waters of the lake.

The water quality model developed and validated for the Seneca River, and the supporting calibrated transport model (Canale et al. 1994, Chapter 9), provides a quantitative basis to evaluate the biochemical and hydrodynamic processes responsible for the severely impacted oxygen resources of the river adjoining the Onondaga Lake inflow. Application of the water quality model has demonstrated that the dominant factor responsible for the contravention of water quality standards in this reach, before the zebra mussel invasion, was the reduced vertical mixing caused by the occurrence of salinity-based density stratification (Canale et al. 1994, Chapter 9). Major improvements in the water quality of the Onondaga Lake outflow would continue to result in dissolved oxygen violations. However, elimination of density stratification would eliminate the violations in water quality standards in the lower river layer, under the river conditions that prevailed before the major zebra mussel impact of 1993 (Canale et al. 1994, Chapter 9). This recent modeling effort for the river (Canale et al. 1994) supersedes the earlier modeling efforts of Effler (1982) and Effler et al. (1982), that incorrectly concluded that destratification of this portion of the river would cause dissolved oxygen violations throughout the watercolumn of the river.

10.1.6 Calcite Oversaturation, Precipitation, and Deposition, and Interactions with Phosphorus Cycling

The entire watercolumn of Onondaga Lake was found to be oversaturated with respect to the solubility of calcite ($CaCO_3$) over the spring to fall interval before the closure of the facility (e.g., Effler and Driscoll 1985; Effler 1987), indicating a continuous tendency for the precipitation of calcite, and, once formed, no tendency for dissolution. These conditions have been described as extraordinary (Effler 1987; particularly for the hypolimnion), and have been attributed to the very high concentrations of Ca^{2+} maintained in the lake from the ionic waste discharge (Effler and Driscoll 1985). Despite the major (60–70%) reduction in lake concentrations of Ca^{2+} brought about by the decreased loading that accompanied closure of the facility, the lake remains oversaturated with respect to calcite (Driscoll et al. 1994, Chapter 5). Contrary to the speculation of Yin and Johnson (1984), there is no evidence of dissolution of a significant fraction of deposited $CaCO_3$ in Onondaga Lake before or after closure. Significant shifts in the distributions of inorganic carbon and pH have occurred in response to the decrease in Ca^{2+} concentrations that are manifestations of reduced $CaCO_3$ precipitation. The depletion of DIC and alkalinity from the epilimnion in summer has decreased, a clear indication that $CaCO_3$ precipitation/deposition was greater during the operation of the soda ash/chloralkali facility (Driscoll et al. 1994). Further, the distribution of pH values in the upper waters is narrower and values are shifted higher (Driscoll et al. 1994). These changes reflect the maintenance of higher buffering capacity and reduced H^+ production associated with $CaCO_3$ precipitation, respectively, since closure (Driscoll et al. 1994).

Earlier work by Yin and Johnson (1984), referred to by Effler (1987), established the important contribution $CaCO_3$ particles made to the particle population of the lake before closure, and the dominant role autochthonous production (i.e., in-lake precipitation) played in maintaining this $CaCO_3$ particle population. Research conducted since closure indicates $CaCO_3$ particles remain a significant component of the particle population of the upper waters of the lake (Johnson et al. 1991, Chapter 5). Various other particle types provide solid surfaces that serve as nucleation sites (i.e., heterogeneous nucleation) for $CaCO_3$ precipitation in Onondaga Lake, including clays, organic particles, and diatom frustules (Johnson et al. 1991). This enhances the rate of deposition of most of the coated materials. Runoff events apparently enhance $CaCO_3$ precipitation in Onondaga Lake, as a result of the influx of terrigenous particles that serve as nucleation sites (Johnson et al. 1991), thereby coupling the "mud boil" and $CaCO_3$ precipitation phenomena. Wodka et al. (1985) hypothesized that calcite formed on the surfaces of phytoplankton in the upper waters of the lake, thereby enhancing the deposition of these organic particles (Effler 1987). However, continued application of individual particle analysis techniques (e.g., Johnson et al. 1991) does not support this as a major particle association of settling phytoplankton in the lake. This does not eliminate the possibility that the high $CaCO_3$ precipitation rate induced by the Ca^{2+} discharge from the facility may have exacerbated the lake's problem of limited oxygen resources by enhancing the deposition of degradable organic particles other than phytoplankton.

The enhanced precipitation of $CaCO_3$ caused by the Ca^{2+} discharge of the soda ash/chloralkali facility probably has had a deleterious impact on the clarity of Onondaga Lake (e.g., Effler et al., 1984b), associated with increases in light scattering (e.g., Effler et al. 1987, 1991; Weidemann et al. 1985). However, the potential role of $CaCO_3$ in light attenuation in the lake was overestimated in the earlier partitioning analysis of Effler et al. (1984b). Further, the turbidity associated with $CaCO_3$ in Onondaga Lake is modest compared to that observed in other hardwater lakes in the region during natural $CaCO_3$ precipitation ("whiting") events (e.g., Effler et al. 1987; 1991; Weidemann et al. 1985). Thus the direct impact of the facility on the lake's clarity associated with increased light scattering from $CaCO_3$ precipitation is not considered substantial. The much more profound negative impact the facility had on the lake's clarity during its operation emerged following its closure (Chapters 6 and 7). Dramatic and recurring improvements in clarity have been observed since closure (Chapter 7) due to reductions in phytoplankton biomass caused by increased zooplankton grazing (Chapter 6). The increased grazing efficiency reflects the major shift in the zooplankton assemblage since closure of the facility, made possible by the attendant reductions in salinity (Chapter 6). The selective pressure placed on the zooplankton assemblage by the ionic pollution of the lake apparently minimized "top-down" sinks on the phytoplankton, thereby exacerbating the lake's clarity problem.

The reduction in $CaCO_3$ precipitation in the upper waters of the lake since closure, indicated by the decrease in the extent of depletion of alkalinity and DIC, is corroborated by the threefold reduction in the downward flux of particulate inorganic carbon (PIC) over the same period, determined from analysis of sediment trap collections (Chapter 8, Driscoll et al. 1994). The earlier $CaCO_3$ downward fluxes reported from Ca measurements of 1980 sediment trap collections by Effler and Driscoll (1985) are under review, and may be revised as part of a more complete analysis of archived trap samples from a number of years. The ionic waste discharge from the soda ash/ chlor-alkali facility clearly increased precipitation and deposition of $CaCO_3$ in the lake (Driscoll et al. 1994), and thereby accelerated the "filling-in" of the lake (Effler 1987, Chapter 8). Calcium carbonate is the single largest component of the lake's sediments (Effler et al. 1979; Rowell 1992, Chapter 8). However, the misinterpretation of sedimentary Pb-210 profiles by Effler et al. (1979) (see Chapter 8) led to a major overestimation of the lake's sedimentation rate and thereby overstatement of the impact of the facility in "filling-in" the lake (Effler 1987).

The Ca^{2+} discharge from the facility has altered the near-shore sediments of Onondaga Lake through the formation of $CaCO_3$ enriched deposits. A large delta of soft sediment, composed mostly of $CaCO_3$, has formed at the mouth of Ninemile Creek. Most of this $CaCO_3$ was probably formed in the creek, between the point(s) of entry of the Ca^{2+} waste from the facility's waste beds, and the creek mouth. Portions of the delta were dredged in the late 1960s. Much of the near-shore zone of the lake (particularly for depths <2 m) is covered with oncolites (Chapters 6 and 8), ovoid crytalgal structures or concretions, with concentric laminations. These concretions are composed mostly of $CaCO_3$ (92%) and range in size from a few millimeters to as much as 15 cm in diameter (Dean and Eggleston 1984). Dean and Eggleston (1984) contend that the formation of oncolites in Onondaga Lake is a result of the Ca^{2+} enriched discharge of ionic waste from the soda ash/chlor-alkali facility, and that these concretions started to form in Onondaga Lake when the ionic waste was introduced (see Chapter 8). Oncolites do occur naturally in certain hardwater lakes (Chapter 8). Effler (1987) hypothesized that the oncolite deposits significantly limited the macrophyte and macrobethos communities in affected areas of Onondaga Lake. The unusually low species diversity of macrobenthos and macrophytes, and the low standing crop of macrophytes, reported for the oncolite deposit areas (Chapter 6), supports this hypothesis. Results reported in Chapter 6 (also see Madsen et al. 1992) imply that macrophytes have difficulty in establishing and remaining on oncolite sediments. This appears to be related more to the physical character of these deposits (e.g., highly susceptible to movement associated with normal wave action) than their chemical makeup (Chapter 6, Madsen et al. 1992).

Continuing research supports earlier evidence (Wodka et al. 1985) that coprecipitation of P with $CaCO_3$ is a significant pathway in the cycling of P in Onondaga Lake. Recent studies (Driscoll et al. 1993, Chapter 5) continue to indicate that Ca^{2+}, instead of Fe^{2+}, concentrations regulate P activity in the pore water of the upper sediments of Onondaga Lake, probably through the solubility of a metastable Ca−P mineral. However, contrary to earlier speculation (Effler 1987; Effler et al. 1985b), this atypical regulatory mechanism has not resulted in an unusually low magnitude of feedback of P from the lake sediments; the prevailing sediment release rate of P is typical of values reported for other eutrophic systems (Auer et al. 1993).

10.1.7 Mercury Contamination

The contamination of the sediments of Onondaga Lake with Hg, which presumably resulted primarily from the large load received from the soda ash/chlor-alkali facility (e.g., ~75,000 kg from 1946 to 1970, Chapter 1), was originally documented by the USEPA (1973), and reviewed by Effler (1987). Additional sedimentary Hg analyses have subsequently been reported (EPRI 1987; NYSDEC 1990; Rowell 1992). A high degree of contamination was found in each case. Core samples were analyzed from ≥40 sites for two of these studies (NYSDEC 1990; USEPA 1973) in efforts to describe the lake-wide distribution of Hg in the sediments. Mercury is not uniformly distributed in the lake's sediments (NYSDEC 1990; USEPA 1973). The differences in the distributions of Hg obtained by the two most comprehensive studies (NYSDEC 1990; USEPA 1973; that is, volume of contaminated sediment) have been described as pronounced (PTI 1991). The more important finding from the management perspective has been the clear paleolimnological evidence that the highest concentrations of Hg in the profundal sediments have been buried (10 to 20 cm below the surface; Rowell 1992, Chapter 8), perhaps reducing recycle to the water column from this large reservoir.

Trends in Hg fish flesh contamination in Onondaga Lake over the 1970−1984 interval, based on nearly annual monitoring conducted by NYSDEC, were reviewed by Effler (1987). Most of the adult fish in the lake continued to exceed the FDA standard of 1.0 ppm in the early 1980s. Analysis of the fish tissue monitoring data was extended through 1990 in Chapter 6. The overall temporal pattern that

emerges for smallmouth bass, the primary indicator, is a decline in Hg concentration in the 1970s followed by increases in the 1980s, particularly in the late 1980s. More than 95% of the legal-size (>12 inches in length) smallmouth bass collected from the lake in 3 of 4 years during 1987−1990 exceeded the FDA standard. Despite the lack of progressive improvements in Hg contamination (Chapter 6), the lake was reopened to angling in 1986. However, according to FDA guidelines enforced by the New York State Health Department, a consumption advisory was issued indicating fish from the lake are not to be eaten.

Watercolumn measurements of Hg, an important component in the evaluation of the lake's fish tissue contamination problem, were not available at the time of preparation of the earlier impact analysis of the facility (Effler 1987). Since that time the findings of three different studies of the distribution of Hg species in Onondaga Lake have been reported (Bloom and Effler 1990; Robertson et al. 1987; Wang 1993). Measurements of total Hg (Hg_T) were conducted in all three studies. Additional species were monitored in the work of Bloom and Effler (1990) and Robertson et al. (1987), including methyl Hg, the primary form that accumulates in fish. Concentrations reported in these different studies matched quite well; generally the concentrations are an order of magnitude greater than observed in relatively pristine systems (Bloom and Effler 1990, Chapter 5), indicating a high degree of contamination.

A Hg_T budget for the lake, supported by tributary monitoring, showed that drainage flows were the major source of Hg_T to the lake and that the major sink was outflow from the lake (Chapter 5). Only a small component of the watercolumn inputs of Hg_T appear to be derived from sediment release (Chapter 5). Rather, the major sources of Hg_T to the lake are Ninemile Creek (~50%) and METRO (~25%). Most of the Ninemile Creek load enters the stream between Amboy and Lakeland (Chapter 5). Contrary to earlier speculation (Effler 1987), the results of the Hg_T budget analysis (Chapter 5) suggest that the continued contamination of fish tissue in

Onondaga Lake is largely a result of continued external loading, instead of mobilization of sedimentary Hg. This is qualitatively consistent with the burial of sedimentary Hg documented in Chapter 8. Ongoing studies by the soda ash/chlor-alkali facility's consultants, which address a larger suite of Hg species and in-lake processes, may yield a more complete understanding of the Hg cycle, including the contamination of fish flesh, in the lake.

10.2 The Polluted State of Onondaga Lake: How Bad Is It?

The foregoing chapters and section of this chapter have documented the severe degradation of Onondaga Lake, and 'he adverse impacts on adjoining portions of the Seneca River, brought about by the historic and ongoing use of the lake and bordering environs for the disposal of municipal and industrial waste. The lake's cold-water fishery was eliminated before the turn of the century and fish from the lake cannot be eaten. Resorts that once flourished are no more; the lake is often not fit for contact recreation.

An impressive list of numerical standards, intended to protect the fishing and contact recreation resources of surface waters, continues to be violated in Onondaga Lake; these violations are summarized in Table 10.1. Oxygen is lost extremely rapidly from the lake's hypolimnion in spring. At times in late summer, anoxia (the lack of oxygen) extends to within 6 to 7 m of the lake surface. Perhaps the most telling manifestation of the severity of the degraded state of the oxygen resources of the lake is the lakewide (i.e., including the upper layers) depletion observed annually during the fall mixing period. Violations of state standards of minimum dissolved oxygen (DO) concentrations are observed nearly lakewide in most years, and for as long as 2 to 3 weeks in some years. This severity of impact on DO appears to be unique for lakes in the United States. The exodus of most fish from the lake during the period of severe lakewide depletion in fall, documented in Chapter 6, is clear evidence that this is a resource limiting condition. The severely degraded oxygen resources of the lake are a manifestation of cultural eutrophication caused by excessive phosphorus (P) loading (Figure 10.1), emanating principally from METRO.

TABLE 10.1. Violations/exceedance of numerical standards/guidance value for State of New York, in Onondaga Lake.

Constituent/attribute	Resource/use	Standard/guideline	Primary responsibility	Support documentation
Free ammonia (NH_3)	fishing	toxicity; standard function of pH and temperature; differ for salmonid and nonsalmonid fisheries	METRO	Chapters 3, 5, and 9
Nitrite (NO_2^-)	fishing	toxicity; $<100\,\mu g\,NO_2^-\cdot L^{-1}$ for nonsalmonid, $<20\,\mu g NO_2^-\cdot L^{-1}$ for salmonid	METRO	Chapter 5
Dissolved oxygen (DO)	fishing	$\geqslant 5\,mg\cdot L^{-1}$, daily average; $\geqslant 4\,mg\cdot L^{-1}$, minimum within a day	METRO	Chapter 3, 5, 6, and 9
Mercury (Hg) in fish flesh	fishing	FDA standard of <1 ppm	?	Chapters 5, and 6
Clarity (Secchi disc transparency, SD)	swimming	standard for opening public bathing beach; $\geqslant 4$ ft (or 1.2 m)	METRO	Chapters 3, 5, 6, and 7
Fecal coliform (FC) bacteria	swimming	log mean $\geqslant 200\,FC\cdot 100\,ml^{-1}$ over 5 days, single observations <1000 $FC\cdot 100\,ml^{-1}$	combined sewer overflows (CSOs)	Chapters 3, 6, and 9
Total phosphorus (TP)	swimming	guidance value; epilimnetic summer average $\leqslant 20\,\mu g\cdot L^{-1}$	METRO	Chapter 3, 5, 6, 8, and 9

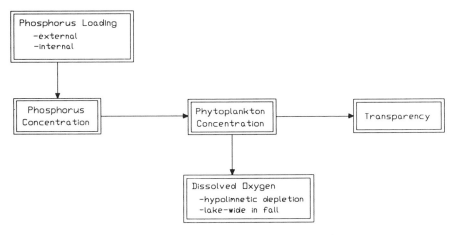

FIGURE 10.1. Phosphorus loading and manifestations of cultural eutrophication.

The lake's poor clarity is also largely a manifestation of cultural eutrophication (Figure 10.1). Thus remediation of the clarity problem must also focus on reduction of phytoplankton growth through reduction in P loading. The clarity standard (Table 10.1) for swimming safety has not been met about 20% of the summer period since 1987, though substantial interannual differences have been observed. The average total phosphorus (TP) concentration in the epilimnion of Onondaga Lake in the summer is indicative of a highly eutrophic lake, and it exceeds the New York State guidance value of $20\,\mu g \cdot L^{-1}$ and the upper boundary associated with mesotrophy by a factor $\geqslant 3$ (Table 10.1).

Standards for free ammonia (NH_3) and nitrite (NO_2^-), intended to protect fish and other aquatic animals from the toxic effects of these species, are violated in the upper waters of the lake for more than three months in the spring to early fall interval, and often by an extremely wide margin. Violations of the NH_3 standard also occur in the hypolimnion. The NH_3 violations are attributed almost entirely to the METRO discharge of total ammonia (T-NH_3), that represents about 90% of the external load. There is less detailed evidence linking the lake's NO_2^- problem to METRO, but there appears to be little question concerning METRO's responsibility. This issue draws less attention than NH_3 because the high chloride (Cl^-) concentration of the lake probably mitigates against the toxic effects of this specie.

The fecal coliform bacteria standard(s) for swimming usage, intended to protect against the transmission of disease organisms, is violated in the lake's south basin following significant runoff events, and lakewide following major storms. Unlike the other water quality violations associated with domestic waste, attributable to the continuous discharge of treated wastewater from METRO (Table 10.1), the indicator bacteria violations are a result of the irregular discharge of dilute untreated sewage from the combined sewer overflow system to lake tributaries (particularly Onondaga Creek) that enter the south basin.

The recent finding (Wang 1993, Chapter 5), that continuing external inputs of Hg may be responsible for the persisting fish flesh contamination, has important management implications. This bodes well for the overall lake reclamation initiative as external loads are much more amenable to an affordable remediation solution than sediment-based solutions (e.g., dredging).

The extent of the degradation of Onondaga Lake is not fully depicted by its status with respect to numerical standards (Table 10.1). Certain of the impacts are not amenable to simple quantification. In particular, discharges from the soda ash/chlor-alkali facility have degraded habitats within the lake as described in Chapter 6 and earlier in this chapter.

Responsibility, in a legal sense, for these impacts is presently being pursued through litigation. Several pollution issues for the lake (e.g., organics, heavy metals), some coupled to the soda ash/chlor-alkali facility, have not been considered in this volume because of limited data or the incompleteness of related ongoing studies.

Testimony to the U.S. Senate has described Onondaga Lake as one of the most polluted lakes in the United States; perhaps the most polluted (U.S. Senate Committee on the Environment and Public Works, Subcommittee on Water Resources, Transportation and Infrastructure 1989). Hennigan (1990) has described the lake as "America's dirtiest lake." In reality there is no widely accepted basis to quantitatively rate and compare the degree of pollution of different lakes. The best answer to the question, "Is Onondaga Lake the most polluted lake in the United States?", is probably with another question, "Do you know of a more polluted lake?" The pursuit of appropriate superlatives to describe the polluted state of the lake is not a meaningful endeavor. It can be said that the impact of the discharge of municipal and industrial wastes to Onondaga Lake has been profound, that the ecology of the lake has been severely impacted, and that the fishing and swimming uses of the lake have been lost. The severely degraded state of the lake is manifested through the routine violation, often by a wide margin, of a number of water quality standards (Table 10.1), as well as through deterioration of various habitat features that are not represented by numerical standards. Clearly the prevailing conditions in Onondaga Lake are inconsistent with the national goal to make all surface waters fishable and swimmable by July 1988 (Section 101, F.W.P.C.A. 1972).

10.3 But Has Not the Quality of Onondaga Lake Improved?

Major reductions in pollutant loadings to Onondaga Lake have been achieved over the late 1960s to early 1990s interval, through implementation of management actions and by closure of the adjoining soda ash/chlor-alkali plant (Chapter 3). The lake has responded positively to these reductions. The average concentration of TP in the METRO effluent was reduced about twentyfold, and the overall TP load to the lake was reduced about twelvefold. The average TP concentration in the facility's effluent in the 1990s has been about $0.6\,mg \cdot L^{-1}$, substantially less than the present permit requirement of $1.0\,mg \cdot L^{-1}$. METRO presently achieves some seasonal nitrification, though the facility was not designed to do so. Execution of a Best Management Practices program on the sewer system by Onondaga County reduced dry and wet weather loads of indicator bacteria from the urban tributaries to the lake. The discharge of Hg to the lake was reduced greatly (95%) soon after the U.S. Department of Justice took action against the facility in 1970, and was reduced further with the closure of the chlor-alkali operation (Linden Chemical) in 1988. The major reductions in ionic waste (principally Cl^-, Na^+, and Ca^{2+}) loading from soda ash production that accompanied the closure of the soda ash/chlor-alkali facility in 1986 have been described. The closure also marked the discontinuation of the use of the lake as a source of cooling water.

Some of the more noteworthy changes in the physical, chemical and biological characteristics of Onondaga Lake documented in this volume, and the identified causes, are summarized in Table 10.2. Note that none of the improvements/changes listed have changed the status of the lake with regard to the occurrence of violations of standards. However, the margins and durations of violations have improved in some cases. For example, the degree of violation of the mercury (Hg) fish flesh standard has decreased since the earliest measurements. The clarity standard (Table 10.1) is violated less frequently, and the concentrations of TP during summer in the upper predictive layers are lower (Table 10.2).

It is especially important to correctly couple the documented lake responses to the appropriate forcing condition (i.e., specific reduction in external material loading). Some of the

TABLE 10.2. Changes in Onondaga Lake characteristics in response to reductions in external loading.

Aspect impacted	Characteristic/constituent/component	Description of change	Cause(s)
Hydrodynamics (Chapter 4)	Plunging underflow	Occurrence and extent of phenomenon greatly reduced	Reduced salinity of inflows
	Stratification/mixing regime	Chemical stratification mostly eliminated, more normal mixing regime	Reduced salinity of inflows
	Heat content	Average summer heat income reduced by ~9%	Discontinued use of lake as source of cooling water
	Density and oxygen stratification in the Seneca River	Longitudinal extent and duration of phenomenon reduced	Reduced salinity loading
Chemistry (Chapter 5)	Salinity (mostly Cl^-, Na^+, Ca^{2+})	Reduced from 3 to 1.2‰ since closure of soda ash/chlor-alkali facility	Reduced salinity loading
	Oxygen	Reduced period of hypolimnetic anoxia	Shorter duration of stratification with reduced salinity of inflows
	Alkalinity and DIC	Reduced depletion of alkalinity and DIC from epilimnion	Reduced precipitation of $CaCO_3$ with reductions in Ca^{2+} loading
	pH	Slight increase in epilimnion and narrower distribution	Reduced precipitation of $CaCO_3$
	TP	Reduction in TP, 1987–1990; undoubtedly also before but poorly defined	Reductions in external TP loading
Biology (Chapters 6 and 8)	Phytoplankton	1. Loss of filamentous blue-greens by late 1970s 2. Reemergence of filamentous blue-greens by late 1980s 3. Chlorophyll minima lower and of longer duration after 1987 4. Shifts in diatom assemblage	1. P detergent ban 2. Probably secondary effect of reduction in salinity 3. Secondary effect of reduction in salinity, increased zooplankton grazing 4. Changes in anthropogenic salinity and nutrient loads
	Zooplankton	Species richness/diversity increased since close of soda ash/chlor-alkali facility, now more efficient grazers	Reductions in salinity loading
	Fish	1. species richness/diversity increased since early 1980s 2. reduction in the margin of violation of FDA Hg standard in 1970s	Reductions in pollutant loading, particularly salinity Probably from reduction in loading from chlor-alkali process
Optics (Chapter 7)	Light penetration (e.g., clarity)	Improved since 1987	Increased zooplankton grazing of algae, because of reduced salinity loading
Sediments (Chapter 8)	Deposition rate	Reduced downward flux of $CaCO_3$, and total solids since closure of soda ash/chlor-alkali facility	Mostly because of reduction in salinity (particularly Ca^{2+}) loading

responses have obvious direct couplings to loading reductions of certain constituents, while other changes are secondary or indirect effects. Particularly striking changes in the lake have been brought about by the major reduction in ionic waste loading from soda ash production that resulted from the closure of the chemical plant in 1986. The character and magnitudes of these changes give insight into the profound impacts these discharges had on the lake and adjoining portions of the Seneca River during the operation of the facility (see preceeding sections of this chapter). It is especially important to separate the effects of this industrial discharge from those of the municipal waste effluent being discharged from METRO. Superficial review of the character of some of the changes could lead to the false conclusion that various of the reported improvements reflect remediation of the impacts of domestic waste discharges. Plankton (phytoplankton and zooplankton) biomass and composition, clarity, and dissolved oxygen are coupled to external P loading in most lakes, according to the conceptual linkages presented in Figure 10.1. However, the evidence presented in this volume, and synthesized in earlier sections of this chapter, instead indicates that the rather substantial changes documented for these characteristics in Onondaga Lake have been largely a result of the reduction in the massive saline waste load received from the soda ash/chlor-alkali facility. For example, the reduced standing crop of phytoplankton, with more pronounced minima (described as "clearing events"), and the coupled increases in water clarity, largely reflect increased grazing by zooplankton associated with the reduction in salinity, not increased nutrient limitation.

The major reductions in P loading have been manifested as reductions in TP concentrations in the upper waters of the lake. However, rather detailed analysis of recent monitoring data (Chapter 6) indicates the lake continues to be nearly saturated with respect to the availability of P to support phytoplankton growth. Yet further reductions in P loading would be necessary (Chapter 9) to achieve coupled improvements in primary

productivity-related aspects of water quality, according to the linkages of Figure 10.1. This relationship between phosphorus concentration and the responsiveness of measures of primary productivity is qualitatively illustrated in Figure 10.2. A Monod-type relationship (see Chapter 6) is depicted; salient features include a saturated interval for high TP concentrations, below which increased responsiveness (decreases in measures and manifestations of primary productivity) occurs. Historically (e.g., late 1960s), Onondaga Lake was far out on the saturated phytoplankton growth plateau of Figure 10.2. The major reductions in loading from METRO since the late 1960s have moved conditions along the plateau (to the left), but have not been adequate to achieve substantial nutrient limitation (to get around the "elbow" of Figure 10.2) and related improvements in water quality.

Early reductions in P loading associated with the detergent ban apparently did have the beneficial effect of eliminating nuisance filamentous blue-green forms from the phytoplankton assemblage in the early 1970s (Table 10.2). However, the recent reemergence of this group of algae may be a secondary effect of reduced ionic waste loading, in response to increased grazing pressure (Table 10.2). The increase in species richness/diversity of the fish community since the early 1980s reflects reductions in pollutant loadings; most probably the primary effect has been the lake's reduced salinity since closure of the soda ash/chlor-

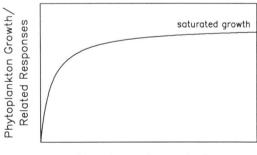

FIGURE 10.2. Relationship between phosphorus concentrations and primary productivity.

alkali facility. Reductions in mercury (Hg) fish flesh concentrations in the 1970s probably reflect the earlier major loading reductions from the chlor-alkali process. Other changes brought about by closure of the facility (Table 10.2) have been described previously in this chapter.

In summary, Onondaga Lake has responded positively to reductions in pollutant loading (Table 10.2). Unusual hydrodynamic characteristics have been ameliorated, modest improvements in the oxygen resources have been observed, increased biological diversity prevails because of decreases in salinity, coupled reductions in phytoplankton biomass and increases in clarity have occurred, and precipitation of $CaCO_3$ and deposition of solids have decreased. These improvements fall far short of meeting related standards. However, they serve to demonstrate the responsiveness of the system to reductions in pollutant loading. Major reductions in domestic waste loading to the lake will be necessary (see Chapter 9) to eliminate the violations identified in Table 10.1.

10.4 The METRO Discharge Is Too Much for a Small Lake

The signatures of the impacts of domestic waste discharge on a lake have rarely been so clear (Table 10.1). At the heart of the issue is the uniquely large contribution the METRO discharge makes to the overall hydrologic budget of Onondaga Lake. Over a 19-year period (1971–1989) METRO's discharge represented nearly 20% of the inflow received from all sources (Figure 10.3a). The METRO inflow represented an even greater (30%) fraction of the total inflow during August, a critical water quality interval, for the same 19-year interval (Figure 3b). During low flow summer intervals METRO often represents the single largest source of water to the lake. These contributions are unique for lakes in the United States. The next largest discharge to a lake in New York State is the Ithaca

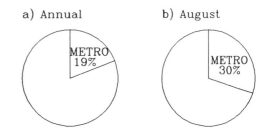

a) Annual   b) August

FIGURE 10.3. METRO's contribution to hydrologic budget of Onondaga Lake.

WWTP (10 MGD), which represents <2% of the total inflow received by Cayuga Lake. The METRO discharge is simply much too great for existing related water quality standards and the TP guidance value to be met in Onondaga Lake (Chapter 9).

The existing METRO facility was never intended to reclaim the resources of Onondaga Lake, but instead to *abate* the extremely degraded conditions that prevailed (e.g., Onondaga County 1971) before its construction. Resignation to this is clearly depicted in the United States Environmental Protection Agency's (USEPA 1974) *environmental impact statement*, which endorsed the construction of METRO, by essentially acknowledging the lake would remain polluted with respect to domestic waste pollutants (e.g., "The proposed project will not completely eliminate the pollution of Onondaga Lake waters" (p. 99, USEPA 1974)). The continuing discharges of large quantities of industrial waste from the adjoining soda ash/chlor-alkali plant undoubtedly contributed to this resignation.

It is clear the USEPA (1974) believed METRO would do little for the lake's cultural eutrophication problems—"this cannot be considered a final solution to the problem because other major sources of phosphorus will continue to feed the lake. These sources can provide enough phosphorus to support algal growth indefinitely. The best reason for instituting phosphorus removal at the MSSTP (METRO) is protection of Lake Ontario" (p. 194, USEPA 1974). While there apparently was a lack of understanding that the new facility would continue to be the dominant source of P, there was recognition that the major reduction in loading that would be achieved with tertiary treatment at the new METRO facility would be inadequate to significantly reduce primary production in the lake (e.g., see Figure 10.2).

The dominant role the METRO input plays in TP loading to Onondaga Lake is illustrated in Figure 10.4; other fractions of the existing loading that may be subject to elimination or reduction are also identified. About 30% of the existing total TP load, emanating from tributaries, is probably not subject to further reduction (Figure 10.4). Estimates of the manageable fractions of the tributary load associated with CSOs and land use practices are presented as ranges, reflecting reasonable ranges of success of related programs and uncertainties in loading estimates. The indicated potential loading reduction for land use management is particularly speculative and optimistic, incorporating as an upper bound a 20% reduction in rural nonpoint loading. Pursuit of these tributary loading reductions only, cannot be supported as a management strategy, as the modest reduction in lake P concentrations would not greatly limit phytoplankton growth. It becomes appropriate to address these sources only in the context of eliminating, or greatly reducing, the METRO load, which represents about 60% of the existing total load (Figure 10.4).

Recall that the present TP load from METRO (Figure 10.4) corresponds to an average effluent concentration of $0.6\,mg \cdot L^{-1}$ TP. The present technical limit of treatment for P for large facilities is an effluent TP concentration of about $0.2\,mg \cdot L^{-1}$ ("effluent filtration"). The lake TP model (Chapter 9) predicts that the application of this technology at METRO (with continued discharge to the lake) would result in a summer average TP concentration in the upper waters of the lake of about $45\,\mu g \cdot L^{-1}$. This is a major reduction from prevailing concentrations, but substantially above the State guidance value of $20\,\mu g \cdot L^{-1}$ and the level needed to greatly limit phytoplankton growth. Even in combination with tributary reduction programs, the TP concentrations would remain excessively high. The METRO effluent needs to be diverted in order to achieve (further, see Table 10.2) distinct improvements in all the features of water quality related to the trophic state of the lake.

The previous, less stringent ammonia toxicity standard for New York State Class B and C waters (applicable to Onondaga Lake) was T-$NH_3 \le 2.0\,mg$ (as NH_3) $\cdot L^{-1}$ at a pH $\ge$ 8.0. This standard applied in New York State until 1991 when the USEPA's (1985) new, and more stringent, national criteria were adopted by the State. The USEPA (1974) apparently had little confidence that the present METRO facility would even meet the older, less stringent, standard; "Conditions in the lake should be carefully monitored to determine whether or not contravention continues after the improved MSTP is put into operation" (p. 100, USEPA 1974). Appli-

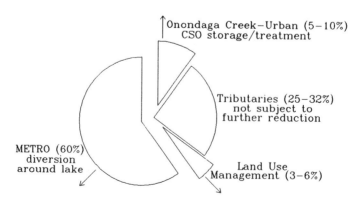

FIGURE 10.4. Partitioning external phosphorus loading to Onondaga Lake; identification of manageable fractions (modified from Effler et al. 1995a).

TABLE 10.3. Description of diversion projects in the United States and Europe.

Lake	Location	Physical description		Reduction in external TP load (%)	Year completed	Reference
		Surface area (km²)	Flushing rate (yr⁻¹)			
	USA					
Irondequoit B.	NY	7.0				
Kegonsa	WS	12.7	2.5	—	1958	Sonzongi et al. 1975
Minnetonka	MN			—	1972	Smith and Shapiro 1981
Sammamish	WA	19.8	0.6	33	1968	Welch 1980
Saratoga	NY					
Washington	WA	88	0.4	66	1968	Edmondson and Lehman 1981
Waubesa	WS	8.3	0.9	—	1958	Sonzongi et al. 1975
Onondaga?	NY	12.0	4.5	60		
	Europe					
Bergundasjön	Sweden	4.2	1.4	95	1976	Ryding 1981
Fuschl	Austria	2.7	0.33	77	1978	Haslauer et al. 1984
Glaningen	Sweden	0.8	5	60	1973	Forsberg and Ryding 1980
Gjersjoen	Norway	2.7	0.33	75	1971	Fargeng and Nilsson 1981
Malnsjön	Sweden	1.1	0.7	85	1974	Ryding 1981
Norrviken	Sweden	2.7	1.3	88	1969	Ahlgren 1978
Vesijarri	Finland	110	0.2	—	1976	Keto 1982

cation of the N model for the lake (Chapter 9) indicates the addition of year-round nitrification at METRO (assuming a performance achieved at other large-scale facilities, Chapter 9) would have easily met the old standard, but would be inadequate to meet the new standard(s), again, because the METRO discharge is too great for Onondaga Lake.

10.5 Diversion of METRO

10.5.1 Other Diversion Projects

A number of projects have been executed in the United States and Europe in the past 25 years to divert municipal wastewater effluent from lakes to adjoining systems with more assimilative capacity. The goal of these efforts has largely been to reclaim resources lost to the impacts of cultural eutrophication. A listing of some of these projects, and certain key descriptors, is presented in Table 10.3. The diversion of METRO would be larger than any of those identified for the listed United States projects (Table 10.3) with respect to magnitude of flow diverted.

10.5.2 The Seneca River Option

The diversion of municipal waste effluent around Onondaga Lake is not a new concept. The original plans for METRO (primary treatment plant completed at the existing site in 1960), called for the effluent to be pumped around the lake, combined with the Ley Creek plant effluent, and discharged to the Seneca River. The lake bypass was deferred, and the effluent was discharged to the lake, awaiting lake studies that would evaluate the need to divert to the river. The lake discharge was also viewed as a cost saving measure (Syracuse *Post Standard*, November 12, 1959). A subsequent study apparently became the scientific basis for the rejection of the diversion alternative. The study (SURC 1966) concluded, without use of water quality models, that "little beneficial effect on the quality and general composition of the south end of Onondaga Lake would be achieved by the diversion," and that "the concentration of inorganic salt . . . would undoubtedly rise if the MSTP flow were diverted" (cover letter, SURC 1966). The second conclusion implied the METRO discharge to the lake should be continued in order to mitigate (by dilution) the impact of the ionic waste discharge from soda ash production. This reasoning (SURC 1966), and the resignation of regulators to the limits of water quality improvement to be expected for the upgraded METRO (USEPA 1974), prompts the question, to what extent did the continuing discharges of the soda ash/chlor-alkali plant constrain lake reclamation efforts related to domestic waste discharge(s)?

The assimilative capacity of the Seneca River for oxygen demanding wastes was compromised well before the earliest considerations of diversion of METRO, by its incorporation into the barge canal navigation system (constructed in the early 1900s), and the subsequent development of hydroelectric power generation facilities. The resulting channelization of the river eliminated much of the system's natural turbulence. This reduced the river's capacity for reaeration and made it susceptible to the development of salinity stratification, and coupled severe oxygen depletion (violations of State standards) in the lower river depths, from the ionically enriched Onondaga Lake outflow (Chapter 9).

The Seneca River upstream of the Onondaga Lake inflow has been, until recently (e.g., 1993), enriched in phytoplankton and extremely turbid due to upstream point and nonpoint loading; oxygen concentrations tended to be only slightly undersaturated. By 1993 major changes in the water quality of the river were manifested, including reduced phytoplankton concentrations and improved clarity, as a result of the severe infestation of the system by zebra mussels (Chapter 9). However, the most notable shift in water quality from the perspective of potential diversion of METRO to the river has been the deterioration of the river's oxygen resources brought about by zebra mussel respiration. Violations of the State DO standards were common in the Seneca River upstream of the lake inflow in the late summer of 1993 (Effler and Siegfried 1994; Effler et al. 1995b) and 1994. Under these conditions the Seneca River has no assimilative capacity for additional oxygen-demanding waste. Presently there is little insight concerning the expected duration of the infestation at the level, and coupled

water quality impact, documented for 1993 and 1994.

The severity of the zebra mussel invasion in the river and coupled impact on oxygen resources is a manifestation of at least two anthropogenic alternations of the river. First, the cultural eutrophication of upstream portions of the river system, including Cross Lake, which provides copious quantities of food particles to support mussel growth. Second, the previously described channelization not only eliminated the natural reaeration potential of the river to recover from this new sink for oxygen, but it provided surfaces that have promoted particularly dense populations of the mussels. The most dense populations of zebra mussels ($30,000-60,000$ individuals $\cdot$ m^{-2}), and the greatest oxygen depletion, occurs in a $1.4\,km$ long artificial channel (dug as part of the canal system construction in about 1915), because the exposed bedrock is ideal attachment substrate for these invaders.

Application of a validated river water quality model has demonstrated that the Seneca River before the zebra mussel invasion could assimilate an upgraded (with nitrification) METRO discharge and protect the river's oxygen resources, if the river were destratified (Chapter 9). However, at the present level of zebra mussel infestation, an even higher level of treatment, perhaps combined with supersaturation of the effluent with O_2 and/or diversion of the discharge farther downstream (e.g., below Phoenix), would probably be required to protect the river system. An effective zebra mussel management program could reduce the level of treatment required for the management option of diverting METRO to the Seneca River. The issue of the acceptability of the diversion with respect to free ammonia toxicity standards in the river also deserves careful consideration.

10.5.3 The Lake Ontario Option

Diversion much farther downstream, directly into Lake Ontario (Figure 10.5), was one of the discharge alternatives considered for the existing METRO facility (USEPA 1974). This diversion would probably be about $50\,km$

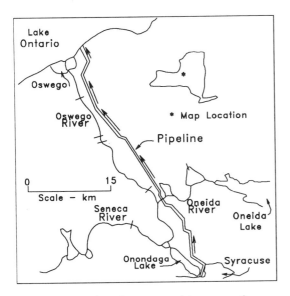

FIGURE 10.5. Diversion geographic perspective.

long, depending on the details of outfall siting and availability of land; it would be the longest diversion around a lake in the United States. This diversion alternative was considered to have less environmental impact than the Onondaga Lake and river discharge options by the USEPA in 1974; "no probable adverse effects on Lake Ontario water quality" (p. 130, USEPA 1974), but was eliminated for cost reasons. A clear precedent for a Lake Ontario discharge of this magnitude is presented by the input of the Van Lare WWTP effluent, serving the City of Rochester, about 90 km to the west of the City of Oswego. This facility discharges an average of 110 MGD to the hypolimnion (31 m depth) through a diffuser, 4.8 km off shore. The Lake Ontario option for the METRO diversion is superficially attractive for at least two reasons: (1) it diverts the waste discharge around two systems that both presently demonstrate clear manifestations of severe environmental stress, and (2) loading received by the Seneca River proximate to Onondaga Lake largely makes its way to Lake Ontario anyway.

An important consideration at the time of the earlier (USEPA 1974) impact analysis was P loading because of the concern for increasing cultural eutrophication of Lake Ontario. Recall

that it was the EPA's (1974) view that "The best reason for instituting phosphorus removal at the MSTP (METRO) is protection of Lake Ontario" (p. 194, USEPA 1974). To this end, an Onondaga Lake discharge had an additional benefit, a significant fraction of the P, and other nonconservative pollutants, discharged to the lake would be retained within the lake (immobilized in the sediments), thereby further reducing the load to Lake Ontario associated with the METRO discharge. Thus the placement of the METRO discharge into the southern end of Onondaga Lake relegated the lake to serve essentially as a "polishing pond" for the METRO effluent, to further protect Lake Ontario.

The distributions of monthly P loads from the Van Lare and METRO facilities in 1992 are compared in Figure 10.6. The METRO load was substantially less (45%) than the Van Lare load in that year; because both the discharge flow and TP concentration of the effluent were lower for METRO. Diversion of the METRO discharge to the river system or directly to Lake Ontario will increase the TP loading to Lake Ontario from the Onondaga Lake watershed from the present level because of reduced losses to the sediments of Onondaga Lake. However because of the achievements in TP treatment accomplished at METRO in the 1990s (substantially below the $1.0\,mg \cdot L^{-1}$ permit value), diversion of METRO would not significantly increase the TP loading to the river and Lake Ontario compared to the level

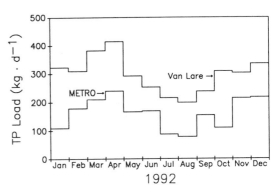

FIGURE 10.6. Comparison of monthly total phosphorus loads for 1992: METRO and Van Lare.

associated with its earlier operation at the permit value.

Emphasis in the environmental management of Lake Ontario has shifted away from control of external P loading in recent years for at least two reasons: (1) the external TP loading reduction goal for Lake Ontario has been achieved (USEPA and NYSDEC 1994), and (2) there is increasing concern by fishery biologists that reductions in primary productivity in Lake Ontario associated with control of external P input may be contributing to reductions in sport fish production. More timely issues for Lake Ontario presently include sources and fate of toxic substances (e.g., DePinto 1994), fisheries, and exotic species. Serious consideration of the Lake Ontario diversion option for METRO would dictate consideration of these potential interactions, and undoubtedly other additional issues.

10.6 Where Do We Go from Here?

This question needs to be addressed on at least two levels: first from the perspective of research and education, and second from the view of effective lake management. To this researcher, the appropriate pathway in the first case is clear, but the second is decidedly less certain. This volume has been prepared as a contribution to research and education, with an eye to providing the necessary technical support for lake managers. The myriad of environmental problems that prevail and the strong signatures of degradation manifested offer a unique opportunity for education and research. The lake's size and location make it tractable for both of these activities. In many ways it is the ultimate classroom to teach limnological and related engineering principles, to retest paradigms, characterize basic phenomena and processes, and test research hypotheses. The opportunity to carefully document the time course of response of the system to management action(s) should not be missed. Regardless of the path taken by lake managers with respect to remediation of

the lake's problems, they must assume their responsibilities with regards to the "lessons" offered by its degraded state. These "lessons" need to be shared with students and researchers through the encouragement and support of education and research programs that utilize the lake. Hopefully, the "lessons" learned here will serve to prevent similar pollution problems elsewhere.

The role of research in supporting effective management decisions for this extremely polluted system is straightforward. The system and its problems must be understood well enough to establish the underlying causes and to predict with confidence the adequacy of management proposals for remediation. The information sets and understanding of the lake's various problems, as documented in this volume, arguably can be described as uneven. For example, the understanding of the impacts related to the discharge of METRO, a classic case of cultural eutrophication, is superior to that related to certain aspects of the soda ash/chlor-alkali industry. Unfortunately, there is no widely accepted objective basis to decide when the knowledge base is complete and certain enough to support effective management action. In the end, in this country, this is decided by environmental regulators, or, if protested by the polluters, the courts. Major environmental management decisions have been made for other systems with much less data and simpler analytical frameworks than documented for Onondaga Lake in this book. At the time of the submission of this manuscript, debate between the polluters and the regulators continues within the framework of litigation. The research community needs to be ready to continue to answer the remaining legitimate technical questions that emerge from this process.

It is of course technically feasible to remediate all of Onondaga Lake's problems described in this volume. Perhaps the most important remaining questions concerning the reclamation of Onondaga Lake are not in the area of scientific and engineering research, but rather in the realms of economics, politics, and the law. Tough questions that need to be addressed include: (1) how much will it cost?,

(2) who will pay? and (3) ultimately, is it worth it? It appears that the "lessons" that will emerge from the continuing Onondaga Lake reclamation effort will not be limited to the disciplines of limnology and environmental engineering.

References

Ahlgren, I. 1978. Responses of Lake Norrviken to reduced nutrient loading. *Verh Int Ver Limnol* 20: 846–850.

Auer, M.T., Johnson, N.A., Penn, M.R., and Effler, S.W. 1993. Measurement and verification of rates of sediment phosphorus release for a hyper-eutrophic urban lake. *Hydrobiologia* 253: 301–309.

Bloom, N.S., and Effler, S.W. 1990. Seasonal variability in the mercury speciation of Onondaga Lake (New York). *Water Air Soil Pollut* 53: 251–265.

Canale, R.P., Owens, E.M., Auer, M.T., and Effler, S.W. 1994. Validation of a water quality model for Seneca River, NY. *J Water Resour Plan Manag ASCE* 121: 241–250.

Dean, W.E., and Eggleston, J.R. 1984. Freshwater oncolites created by industrial pollution, Onondaga Lake, New York. *Sediment Geol* 40: 217–232.

DePinto, J.V. 1994. Role of mass balance modeling in research and management of toxic chemicals in the Great Lakes: the Green Bay mass balance study. *Great Lakes Res Review* 1: 1–8.

Doerr, S.M., Effler, S.W., Whitehead, K.A., Auer, M.T., Perkins, M.G., and Heidtke, T.M. 1994. Chloride model for polluted Onondaga Lake. *Water Res* 28: 849–861.

Driscoll, C.T., Effler, S.W., Auer, M.T., Doerr, S.M., and Penn, M.R. 1993. Supply of phosphorus to the water column of a productive hardwater lake: controlling mechanisms and management considerations. *Hydrobiologia* 253: 61–72.

Driscoll, C.T., Effler, S.W., and Doerr, S.M. 1994. Changes in inorganic carbon chemistry and $CaCO_3$ deposition in Onondaga Lake, New York. *Environ Sci Technol* 28: 1211–1218.

Edmondson, W.T., and Lehman, J.T. 1981. The effect of changes in nutrient income on the conditions of Lake Washington. *Limnol Oceanogr* 26: 1–29.

Effler, S.W. (Ed.). 1982. *A Preliminary Water Quality Analysis of the Three Rivers System*, Upstate Freshwater Institute, Syracuse, NY.

Effler, S.W. 1987. The impact of a chlor-alkali plant on Onondaga Lake and adjoining systems. *Water Air Soil Pollut* 33: 85–115.

Effler, S.W., Baker, J., Carter, C., Hsu, H.S., Juran, P., Khalid, M.T., McCarthy, J.M., Shafran, G.C., Schechler, W.D., Unangst, F.J., and Yin, C.Q. 1984a. Chemical stratification in the Seneca/Oswego Rivers (NY). *Water Air Soil Pollut* 21: 335–350.

Effler, S.W., Brooks, C.M., Addess, J.M., Doerr, S.M., Storey, M.L., and Wagner, B.A. 1991. Pollutant loadings from Solvay waste beds to lower Ninemile Creek, New York. *Water Air Soil Pollut* 55: 427–444.

Effler, S.W., Brooks, C.M., and Whitehead, K.A. 1995a. Domestic waste inputs of nitrogen and phosphorus to Onondaga Lake, and water quality implications. *Lake Reservoir Manag* (in review).

Effler, S.W., Brooks, C.M., Whitehead, K., Wagner, B.A., Doerr, S.M., Perkins, M.G., Siegfried, C.A., and Carale, R.P. 1995b. Impact of zebra musrel invasion on river water quality. *Res J Water Pollut Control Fed* (in press).

Effler, S.W., Devan, S.P., and Rodgers, P. 1985a. Chloride loading to Lake Ontario from Onondaga Lake, NY. *J Great Lakes Res* 11: 53–58.

Effler, S.W., Doerr, S.M., Brooks, C.M., and Rowell, H.C. 1990. Chloride in the pore water and water column of Onondaga Lake, NY. U.S.A. *Water Air Soil Pollut* 51: 315–326.

Effler, S.W., and Driscoll, C.T. 1985. Calcium chemistry and deposition in ionically enriched Onondaga Lake, New York. *Environ Sci Technol* 19: 716–720.

Effler, S.W., and Driscoll, C.T. 1986. A chloride budget for Onondaga Lake New York, U.S.A. *Water Air Soil Pollut* 27: 29–44.

Effler, S.W., Driscoll, C.T., Wodka, M.C., Honstein, R., Devan, S.P., Juran, P., and Edwards, T. 1985a. Phosphorus cycling in ionically polluted Onondaga Lake, New York. *Water Air Soil Pollut* 24: 121–130.

Effler, S.W., Johnson, D.L., Jiao, J.F., and Perkins, M.G. 1992. Optical impacts and sources of suspended solids in Onondaga Creek, U.S.A. *Water Resour Bull* 28: 251–262.

Effler, S.W., Juran, P., and Carter, C.F. 1982. Onondaga Lake and dissolved oxygen in Seneca River. *J Environ Engin ASCE* 109: 945–951.

Effler, S.W., and Owens, E.M. 1986. The density of inflows to Onondaga Lake, U.S.A., 1980 and 1981. *Water Air Soil Pollut* 28: 105–115.

Effler, S.W., and Owens, E.M. 1987. Modifications in phosphorus loading to Onondaga Lake, U.S.A., associated with alkali manufacturing. *Water Air Soil Pollut* 32: 177–182.

Effler, S.W., Owens, E.M., and Schimel, K.A. 1986. Density stratification in ionically enriched Onondaga Lake, U.S.A. *Water Air Soil Pollut* 27: 247–258.

Effler, S.W., and Perkins, M.G. 1987. Failure of spring turnover, in Onondaga Lake, NY. U.S.A. *Water Air Soil Pollut* 34: 285–291.

Effler, S.W., Perkins, M.G., Greer, H., and Johnson, D.L. 1987. Effects of whiting on turbidity and optical properties in Owasco Lake, NY. *Water Resour Bull* 23: 189–196.

Effler, S.W., Perkins, M.G., and Wagner, B.A. 1991. Optics of Little Sodus Bay. *J Great Lakes Res* 17: 109–119.

Effler, S.W., Rand, M.C., and Tamayo, T.A. 1979. The effect of heavy metals and other pollutants on the sediment of Onondaga Lake. *Water Air Soil Pollut* 12: 117–134.

Effler, S.W., Schimel, K., and Millero, F.J. 1986b. Salinity, chloride, and density relationships in ion enriched Onondaga Lake, NY. *Water Air Soil Pollut* 27: 169–180.

Effler, S.W., and Siegfried, C.A. 1994. Zebra musrel (*Dreissena polymorpha*) populations in the Seneca River, New York: impact on oxygen resources. *Environ Sci Technol* 28: 2216–2221.

Effler, S.W., Wodka, M.C., and Field, S.D. 1984b. Scattering and absorption of light in Onondaga Lake. *J Environ Engin* 110: 1134–1145.

EPRI. 1987. Measurement of bioavailable mercury species in fresh water and sediments. Prepared by Battelle Pacific Northwest Laboratories, Richland, WA. Electric Power Research Institute, Palo Alto, CA.

Fargeng, B.A., and Nilssen, J.P. 1981. A twenty year study of eutrophication in a deep soft-water lake. *Verh Int Ver Limnol* 21: 412–424.

Forsberg, C., and Ryding, S.-O. 1980. Eutrophication parameters and trophic state indices in 30 Swedish waste-receiving lakes. *Arch Hydrobiol* 89: 189–207.

Getchell, F.A. 1983. Subsidence in the Tully Valley, New York. Masters Thesis, Syracuse University, Syracuse, NY.

Haslauer, J., Moog, O., and Plum, M. 1984. The effect of sewage removal on lake water quality (Fuschlsee, Austria). *Arch Hydrobiol* 101: 113–134.

Hennigan, R.D. 1990. America's dirtiest lake. *Clearwaters* 19: 8–13.

Johnson, D.J., Jiao, J., DeSantos, G., and Effler, S.W. 1991. Individual particle analysis of sus-

pended materials in Onondaga Lake, NY. *Environ Sci Technol* 25: 736–744.

Keto, J. 1982. The recovery of Lake Vesijarvi following sewage diversion. *Hydrobiologia* 86: 195–199.

Kosinski, A.J. 1985. "Analytical Critique" of Summary Consultants Report: Relationship of Brining Operations in the Tully Valley to the Behavior of Groundwater and Geological Resources. Albany, NY.

Madsen, J.D., Eichler, L.W., Sutherland, J.W., Bloomfield, J.A., Smart, R.M., and Boylen, C.W. 1992. "Submersed Littoral Vegetation Distribution: Field Quantification and Experimental Analysis of Sediment types from Onondaga Lake, New York." Report submitted to Onondaga Lake Management Conference. U.S. Army Engineer Waterways Experienced Station, Lewisville Aquatic Ecosystem Research Facility, Lewisville, TX, May 1992.

Meyer, M.A., and Effler, S.W. 1980. Changes in the zooplantiton of Ornondaga Lake, 1969–1978. *Environ Pollut* (Series A) 23: 131–152.

NYSDEC. 1990. Mercury Sediments—Onondaga Lake. Engineering Investigation at Inactive Hazardous Waste Sites, Phase II Investigation, New York State Department of Environmental Conservation, Volume 1, Albany, NY.

O'Brien & Gere Engineers Inc. 1973. Environmental Assessment Statement, Syracuse Metropolitan Sewage Treatment Plant and West Side Pump Station and Force Main. Submitted to Onondaga County, Syracuse, NY.

Onondaga County. 1971. Onondaga Lake Study. Project No. 11060, FAE 4/71. Water Quality Office, United States Environmental Protection Agency, Water Pollution Control Research Series.

Owens, E.M., and Effler, S.W. 1989. Changes in stratification in Onondaga Lake, New York. *Water Resour Bull* 25: 587–597.

PTI Environmental Services. 1991. Onondaga Lake RI/FS Work Plan. Bellevue, WA.

Remane, A., and Schlieper, C. 1971. *Biology of Brackish Water*, John Wiley and Sons, New York.

Robertson, D.E., Sklarew, D.S., Olsen, K.B., Bloom, N.S., Crecelino, E.A., and Apt, C.W. 1987. Measurement of Bioavailable Mercury Species in Fresh Water and Sediments. Final Report to Electric Power Research Institute, Palo Alto, CA, project EA-5197 by Battelle Pacific Northwest Laboratory.

Rooney, J. 1973. Memorandum (cited by U.S. Environmental Protection Agency, 1974, Environmental Impact Statement on Wastewater

Treatment Facilities Construction Grants for the Onondaga Lake Drainage Basin).

Rowell, H.C. 1992. Paleolimnology, Sediment Stratigraphy, and Water Quality History of Onondaga Lake, New York. Dissertation, State University of New York, College of Environmental Science and Forestry. Syracuse, NY.

Ryding, S-O. 1981. Reversibility of man-induced eutrophication. Experiences of a lake recovery study in Sweden. *Int Revue Ges Hydrobiol* 66: 449–503.

Seger, E.S. 1980. The Fate of Heavy Metals in Onondaga Lake. Master's Thesis. Clarkson College, Potsdam, NY.

Shilts, W.W. 1978. Genesis of mud boils. *Can J Earth Sci* 15: 1953–1968.

Smith, V.H., and Shapiro, J. 1981. Chlorophyll-phosphorus relationships in individual lakes, their importance to lake restoration strategies. *Environ Sci Technol* 15: 444–451.

Sonzogni, W.C., Fitzgerald, G.P., and Lee, G.F. 1975. Effects of wastewater diversion on the Lower Madison Lakes. *J Water Pollut Cont Fed* 47: 535–542.

Stewart, K.M. 1978. Infrared imagery and thermal transects of Onondaga Lake, NY. *Verh Internat Verin Limnol* 20: 496–501.

SURC (Syracuse University Research Corporation). 1966. Onondaga Lake Survey, 1964–1965. Final Report. Prepared for Onondaga County Department of Public Works. Syracuse, NY.

Tully, W.P. 1983. Summary Consultants Report. Relationship of Brining Operations in the Tully Valley to the Behavior of Groundwater and Geological Resources. Submitted to Allied Chemical Corporation.

USEPA. 1973. Report of Mercury Source Investigation, Onondaga Lake, New York, and Allied Chemical Corporation, Solvay, New York. United States Environmental Protection Agency, Cincinnati, OH.

USEPA. 1974. Environmental Impact Statement on Wastewater Treatment Facilities Construction Grants for the Onondaga Lake Drainage Basin. United States Environmental Protection Agency. Region II. New York, NY.

USEPA. 1985. Ambient Water Quality Criteria for Ammonia—1984. Office of Water Regulations and Standards Criteria and Standards Division, Washington, DC.

USEPA, and NYSDEC. 1994. Lakewide Impacts of Critical Pollutants on United State Boundary Waters of Lake Ontario. New York State Department of Environmental Conservation, Albany, NY.

U.S. Senate, Committee on Environmental and Public Works, Subcommittee on Water Resources. Transportation and Infrastructure. 1989. Onondaga Lake Restoration Act of 1989. Hearing 101–80, Government Printing Office, Washington, DC.

Wang, W. 1993. Patterns of total mercury concentration in Onondaga Lake. Master's Thesis. Syracuse University, Syracuse, NY.

Weidemann, A.D., Bannister, T.T., Effler, S.W., and Johnson, D.L. 1985. Particle and optical properties during $CaCO_3$ precipitation in Otisco Lake. *Limnol Oceanogr* 30: 1078–1083.

Welch, E.B., Rock, C.A., Howe, R.C., and Perkins, M.A. 1980. Lake Sammamish response to wastewater diversion and increasing urban runoff. *Water Res* 14: 821–828.

Wodka, M.C., Effler, S.W., and Driscoll, C.T. 1985. Phosphorus deposition from the epilimnion of Onondaga Lake. *Limnol Oceanogr* 30: 833–843.

Yin, C., and Johnson, D.L. 1984. An individual particle analysis and budget study of Onondaga Lake sediments. *Limnol Oceanogr* 29: 1193–1201.

Index

A

Absorption of light, 535
  due to phytoplankton, 576
Absorption spectrum
  of gelbstoff, 555–556
  particle, estimation of, 578–579
  of a suspension, 539
Accumulation rate, for sediment, 653
Advection
  induced by inflow, 681
  transport by
    of fecal coliform, 254
    of heat and mass, 240
    of inflow, 201
    vertical transport downward by, 259
Aesthetics, optical, mouth of Onondaga Creek, 120–121
Age of fish of Onondaga Lake, 463–466
  and mercury contamination, 480–482
Air-water interface, oxygen exchange across, 272
Alewife concretions, 293–294, 370
Alkaline phosphatase activity (APA), 410, 416, 419–420
Alkalinity
  METRO effluent measurements, 179

Ninemile Creek measurements, 171–172, 194
Onondaga Creek measurements, 177
of Onondaga Lake, 265
  upper waters, 285
Allied Chemical Corporation
  chlor-alkali process used by, 14
  history of salt mining by, Tully Valley, 61–64
  National Pollution Discharge Elimination System permit, 10
  waste bed overflow, as a source of ionic components in Onondaga Lake, 69
  waste from production processes, 10–14
Ammonia, 299–302
  effect on nitrification, 298
  equilibrium with ammonium ion, equation, 295
  formation of, 331
  interactions with zooplankton, 432–433
  levels of
    correlation with chloride ion levels, Ninemile Creek, 168
    effluent discharge by METRO, 16
  modeling, 694–696

Ninemile Creek, 130–133, 168
  and oxygen depletion, 127–128
  loading of, by the soda ash/chloralkali facility, 789–790
  oxidation of, 350
  sources of, 137, 195
  total
    phytoplankton sink for, 747
    predicting from models, 759
    probability of contravening standards for, 749–751
    Seneca River, 728
    setting standards for, 748
    simulation of, with METRO deepwater discharge, 767–768
  toxicity of, 294–295, 302–305
    associated with bypass of METRO discharge, 775–776
    and standards for, 799
Ammonium ion
  nitrification of, to form nitrate, 326
  plant use of, 294
Amphipods, white perch diet of, 470–471
Anabaena, 387
Anaerobic conditions, effects in the hypolimnion, 272, 709
Analysis methods, for total

phosphorus and total
inorganic phosphorus,
141
Anions, in fresh waters, 263
Annual loading, estimates,
180–183
Annual variability, in oxidation
of organic matter, 344
Anoxia
ferrous ion and hydrogen
sulfide release in response
to, 317–318
loss of oxidized microzone,
316
See also Oxygen depletion
Anthropogenic effects, on pollen,
stratigraphic analysis, 652
Anthropogenic loadings
of chloride ion, applications of
models, 743–744
of phosphorus, 632
Anthropogenic sources, ionic
constituents of, 183–185
Aphaizomenon flos-aquae, 387
Appalachian Upland, 32
Apparent optical properties, 536
coupling with inherent optical
properties, 565–566
Aquifers
defined, 84
Onondaga Creek source water,
42, 44
Areal diversity, patterns in
macrophyte distribution,
440–441
Areal hypolimnetic oxygen
deficit (AHOD), 273,
280–283, 351–352, 369
Areal productivity, temporal
distributions of, 408
Area per unit volume, projected,
for particles, 365–367
Arrhenius equations, for
temperature
and film transfer coefficients,
696–697
and kinetic coefficients, 729
and reaction rate coefficients,
512
and sedimentation rate
coefficients, 686
and SOD, 708
Artesian flow, defined, 85
Artesian pressure head, and mud
boil formation, 67

Atlantic salmon
historic record of catches in
Onondaga Lake, 519
reestablishing, constraints on,
520–521
Attenuation coefficient (K_d)
changes in, historical record,
544–546
for photosynthetically active
radiation, 569–570,
584–592
Attenuation of light
components of, 549–556,
586–588, 592–594
regional comparison, 557–
559
evaluation of models for,
588–592
monochromatic, 570–573
partitioning of, results, 585
spectral, and partitioning of
absorption and of
scattering, 596
See also Vertical attenuation
Average cosine
as a measure of angular
distribution of the
underwater light field,
540, 595–596
temporal distributions,
Onondaga Lake, 565

B
Bacteria
indicator, 494–515, 521–522
pelagic heterotrophic, 514–
515
See also Cyanobacteria; Fecal
coliform (FC) bacteria
Bathymetric map, 3
Beam attenuation coefficient (c),
relationship with SD and
K_d, 567–568
Beer-Lambert law, 570
application to change in
downward irradiance
with depth, 539, 545–546
Beer's law, for calculating
average irradiance change
with depth, 510–511
Benthic macroinvertebrates,
446–453, 518–519
Benzene, contamination of fish
flesh by, 479

Best Management Practices
(BMP) program, 10
for combined sewer overflows,
187–188, 503–505
in rural areas, effect on
phosphorus loading, 156
Best Practical Technology (BPT)
program, Allied Chemical
Corporation participation,
10
Bias, minimizing, for loading
estimation, 113
Bicarbonate ion, in Onondaga
Lake, 265
Bidirectional flow regime, 212
Binuclearia, 386
Biological oxygen demand. *See*
Carbonaceous biological
oxygen demand;
Nitrogenous biological
oxygen demand;
Sediment oxygen demand
Biology, 384–522
Biomass
algal
carbon to phosphorus ratio
and, 318
effect of zooplankton
grazing on, 430
of pelagic bacteria, seasonal
variation in, 514–515
phytoplankton, 416
seasonal changes in, 516–
517
and productivity, 399–408
summer, and median pigment
concentration, 406
zooplankton, 424–426
See also Phytoplankton biomass
Bottom friction factor,
adjustment to match
experimental and
calculated conditions, 235
Bristol Laboratories, load on
County sewage treatment
facilities, 15
Brunt-Vaisala frequency, 246
Buoyant wedge, 258
location of tip, and rate of
outflow, 212
of river inflow into the outlet
channel, 209–210

C
Calanoids, 421

Calcite
    interaction with phosphorus
        cycling, 795–797
    role of, in sediments, 627
    saturation index, isopleths,
        1980, 282
Calcium
    phosphates, 319–320
    stratigraphy of, 632–633
Calcium carbonate
    contribution to turbidity,
        557–558
        two methods for estimating,
            581–582
    deposition of, 1980–1990,
        286–289, 369–370
    deposits of, 164, 283–284,
        443–444, 605–609
        effects of soda ash/
            chlor-alkali facility,
            795–797
    oversaturation, patterns of,
        289–290
    particles of, 363
        effects on zooplankton
            communities, 433–435
    stratigraphy of, 629
    summary, 370
    in surficial sediments, 617
Calcium chloride, titration with,
    MINEQL, 320–321
Calcium ion, concentration of
    in METRO effluent, 177–178
    in Ninemile Creek, 169, 171
    in Onondaga Creek, 176–177
    in Onondaga Lake, 265, 285
    before and after soda ash
        facility closing, 369–370
Calculation techniques
    for loading estimation, 111,
        113–114
    from loadographs, 151–152
Carbon. See Dissolved inorganic
    carbon; Organic carbon
Carbon-14
    for dating sediments, 648
Carbonaceous biological oxygen
    demand (CBOD)
    kinetic coefficient, k_CBOD,
        734
    mass balance approach to
        simulating, 729
    modeling, 704, 707–708
Carbon dioxide
    accumulation of, after

        stratification, 330
    partial pressure of, 284
    and measured pH, 292
Carbon to nitrogen ratio, 331
    of sedimenting material,
        609–610
Carbon to phosphorus ratio, 344
    and algal biomass, 318
Cardiff Well, 61
Cations, in fresh waters, 263
Cazenovia Lake, ice-out data
    from, 249
Cesium-137, for dating
    sediments, 648–649
Chemical characteristics
    equilibrium calculations for
        mercury, 353–354
    stratigraphy, 629–637
    of surficial sediments,
        615–620
Chemical oxygen demand, in
    surficial sediments, 620
Chemical species
    models for effects on water
        quality, 776
    reduced, submodel of, 709–
        710
Chemistry, 263–374
Chemocline, relationship with
    thermocline, 223
Chironomids
    sunfish diet of, 472
    white perch diet of, 470–471
    yellow perch diet of, 471
Chlor-alkali process, 14, 27
Chloride ion
    concentration of, METRO
        effluent, 178–179
    conservation of, equations
        describing, 238
    effect on toxicity of nitrite ion,
        306
    in Ninemile Creek, 165–168
    in Onondaga Creek, 172–176
    in Onondaga Lake, 264–265
    paired comparisons, verifying
        one-dimensional
        assumption for
        Onondaga, 215
    and relative density, 206–207
    as a surrogate for salinity,
        266–267
    vertical distributions of,
        216–218
Chloride mass balance, estimate

    of inflow from the Seneca
        River from, 99
Chloride model, for water
    quality, 672–679, 776
Chlorobenzene
    contamination of fish flesh by,
        479
    contamination of groundwater
        by, 58–59
Chlorophyll
    concentration of
        changes after zebra mussel
            infestation, 742
        correlation with principal
            components, 429
    optical measurements of, 551–
        555
    relationship with particulate
        phosphorus, 308–309
    total
        relationship with b, 568–
            569
        relationship with SD and K_d,
            551–555, 603
        temporal distributions of
            downward flux of, 603–
            604
Chlorophyll a
    as a measure of phytoplankton
        biomass, 516, 538
    optical measurement of, 551–
        555
    regional comparisons, 558–
        559
Chromium, stratigraphy of, 634
Chrysophyte cysts, 638
    stratigraphic profiles, 639
    succession of, 643–645
Chrysophytes, census of
    1968–1971, 389
    1972–1977, 391
    1978–1990, 393
Circulation, lake, 227–238
Circulation model
    application to Onondaga Lake,
        232–238, 259–261
    prediction of horizontal
        velocities, 256
Cladocerans, 421
    assemblage, 1969–1989, 423,
        424
    salinity tolerance of, 435–436
    temporal dynamics of, 427
    white perch diet of, 469, 470–
        471

yellow perch diet of, 471
Clarity
    changes after zebra mussel
        infestation, 742
    and cultural eutrophication,
        799
    remediation of degraded
        conditions, techniques
        identified by analysis,
        591–592
    spatial variability in, 544
Clastics, in surficial sediments,
    619
Clean Water Act (1972), 22–23
Clearing events, 516
    exploitation of, by
        macrophytes, 442
    omission from the phosphorus
        model, 684
    SD values during, 592
    after spring bloom, 403–404
        effect of zooplankton
            biomass on, 430–431
        soluble reactive phosphorus
            levels during, 416
Climate, 5
Coliform bacteria, 494–514,
    521–522. See also Fecal
        coliform (FC) bacteria
Colony forming unit (cfu), for
    fecal coliform bacteria,
    509
Combined sewer overflows
    (CSOs)
    as dispersed point sources, 110
    number, location and usage, 9
    prediction from management
        models, 768–769
    total phosphorus
        measurements, 156–157
Community dynamics
    of fish, in Onondaga Creek,
        490–492
    in tributaries of Onondaga
        Lake, 484–494
    of zooplankton, 424–429
Community structure
    of benthic macroinvertebrates,
        447–450
    of fish, Onondaga Lake, 456–
        462
    of fish, Seneca River, 458–462
Compaction factor, for
    sediments, 654

Complete-mixed flushing, 107–
    109
Complexity, of a model, versus
    credibility, 668–669
Conductance, specific
    Ninemile Creek
        measurements, 170
    Onondaga Creek
        measurements, 174
Confined aquifer, defined, 84
Conservative group of elements,
    defined, 629
Constant initial concentration
    (CIC) model, for lead-210
    profiles, 649–650
Constant rate of supply (CRS)
    model, 649
Cooling water
    density of intakes and
        discharge, 202–203
    discharge of, as buoyant
        overflow, 258
    use of
        and chloride loading, 174,
            192
        and phosphorus loading,
            158–159
Copepods, 421
    assemblage, 1969–1989, 423,
        424
    salinity tolerance of, 435–436
    seasonal dynamics of, 427
Copper
    cladoceran sensitivity to, 433
    stratigraphy of, 634
Core samples
    for examining sediment
        stratigraphy, 622
    phosphorus release
        experiments with, 323–
        324
Coriolis acceleration, 231
Correlation matrix
    constituent profiles for
        sediments, 630
    for surficial sediment
        characteristics, 615
Cortland County, land use in, 22
Cosine sensors, for irradiance
    measurements, 537
Credibility
    of coefficients used in models,
        for total phosphorus, 679
    of a model, versus complexity,

668–669
Critical condition, for rate of flow
    from the Seneca River to
    the lake, 212
Cryptomonads, 386
    census, 1968–1971, 389
        1972–1977, 391
        1978–1990, 393
    concentrations of, 516
    as particles, 362
Cultural development, 5–9
Cyanobacteria, 386–387
    bloom of, 1990, 566–567
    census, 1968–1971, 389
        1972–1977, 391
        1978–1990, 393
    loss of, after 1972, 516
    nitrogen fixation by, 431
    place in seasonal succession,
        388
    role of, in precipitation of
        $CaCO_3$, 618
    toxicity of, 430

D

Darcy's Equation, 56, 83
    Ninemile Creek application,
        90–91
Decomposition
    anoxic processes, 326–329
    processes of, Onondaga Lake,
        329–330
Deformities, in fishes from
    contaminated waters, 520
Degradation, historic patterns,
    9–10
Demography, 21–22
Denitrification, 326–327, 331
    modeling of, 696–697
Density, benthic
    macroinvertebrates,
    below the mud boils, 452
Density, water
    differences between inflows
        and surface waters, 201,
        202
    empirical equation relating to
        temperature and ionic
        strength, 267–268
Density gradient, after the end of
    soda ash manufacture,
    252
Density stratification, 219–224
    and chloride ion

concentration, 266
factors affecting, 225, 258–259
resolution of, selected dates, 220
Depositional basins, 613–615
Deposition of sediments, 600–611, 655–656
Desulfovibrio, sulfur reduction by, 327
Detergents, high-phosphorus
measured effects of regulation, 162–163
regulation of, 9–10, 144
Development, economic
manufacturing
eighteenth through twentieth centuries, 26–27
twentieth century, 9
resort era, 9
waterfront, current plans for, 23
Diatom frustules
concentration, per gram of sediment, 637–638
as particles, 362
Diatoms, 386
benthic taxa, 640–643
census, 1968–1971, 389
1972–1977, 390
1978–1990, 393
epiphytic taxa, 640–643
oligotrophic, stratigraphic profile, 647
role in seasonal succession, 388
spring abundance, since 1970, 516
stratigraphic analysis, 637–647
stratigraphic profiles, 639, 641–643
stratigraphic zones, defined, 645
Dichlorobenzene, contamination of fish flesh by, 479
Dieldrin, contamination of fish flesh by, 479
Diet, of fish, 469–473
Diffuser installation, to modify thermal effects of cooling water discharge, 202–203
Diffusivity ratio, 241

Dilution model, 192
for loading estimation, 114
chloride ion, 166, 167, 172–173
major ions other than chloride ion, 176–177
sulfate ion, 170
Dinoflagellates, census of
1968–1971, 389
1972–1977, 391
1978–1990, 393
Discharges, 97–196
measurements and estimates, 97–99
Disequilibrium, in surface water dissolved oxygen, 274–276
Dispersion
calculation of, 254–257
simulating, 254
Dispersion coefficient, width-averaged, longitudinal, 256–257
Dissolved inorganic carbon (DIC), 286–292, 347
relationship with electron transfer reactions, 346–347
relationship with phosphorus release, 318–319
summary, 369
Dissolved inorganic nitrogen (DIN), 349
Dissolved organic carbon (DOC), relationship with mercury levels, 356
Dissolved organic nitrogen, modeling, 694
Dissolved organic phosphorus (DOP), 410, 416
Dissolved oxygen (DO), 272–283, 368–369
correlation of
with fish populations, 473–478
with population components, 519–520
with principal components, 429
depletion of
in the fall, 350–351
by zebra mussels, 741–742
models of options, 772–776
phytoplankton source of, 747

predicting from models, 759–761
redox chemistry and, 372–373
and stratification, Seneca River, 728
summary, 368–369
validation of model, Seneca/Oswego Rivers model, 737–738
Dissolved oxygen model, 702–713, 776
predictions of, for deepwater METRO discharge, 768
validation of, 711–712
Distributions
of macrophytes, field survey, 437–441
of major phytoplankton, by period, 1978–1990, 392–396
of sediment content, generalized, 620–621
spatial
of fish and of dissolved oxygen, 476
of phytoplankton, 405–406
spectral, of emergent flux, 574
temporal, profiles for phosphorus, 308–309
vertical
of phytoplankton, 404–405
and primary productivity, 407–408
profiles for phosphorus, 308–309
volume-weighted, of nitrogen species in the UML, 698
See also Temporal distributions
Diurnal patterns, in pH and in ammonia concentrations, 300–301
Diversity, benthic macroinvertebrates, below the mud boils, 451–452
Dolomite particles, as tracers for mud boil sediments, 126
Domestic waste discharge, 14–18, 27–28
diversion of METRO, 805–808
d-PON, 694
Drogue observations, for circulation model calibration, 233–238

Droop kinetics, effect of nutrient availability on growth, 410, 418–419

E

Eddy viscosity, 229–230
  calculation constant adjustment, to match experimental and calculated conditions, 235
Education, in environmental problems, 808
Effective contributing period (ECP), for stream pollutographs, 189
Einsele/Mortimer model, 318
Ekman spiral in constant depth, 231
Ekman-type model, 228–232, 259
Electron budget
  comparison among lakes, 348
  hypolimnetic, 346–347, 373
Electron micrographs, CA particles (calcium-rich), 365
Emergent flux, spectral distribution of, 574
Employment, patterns of, Onondaga County, 21–22
Energy zones, and sediment particle size, 613–615
Enriched elements, defined, 629
Ensele/Mortimer model, for phosphorus release, 316–317
Environment, conditions limiting growth, 415, 473–484
  of macrophytes, 441–445
Environmental Protection Agency (EPA), standards for ammonia, 302
Epidemiology, fish, 483–484
Epilimnetic flushing, 109
Epilimnion
  major ion composition, 1980 measurement, 268
  mixing in the, 246
  outflow at the level of, 245
  perfectly mixed model for, 670
  phosphorus in, 309–310
Erosion, effect on Mg/Ca ratio, 632–633

Euglenoids, census of
  1968–1971, 389
  1972–1977, 391
  1978–1990, 393
Eutrophication
  cultural, 307
  causes·of, 798
  and reduced chemical species concentrations, 709–710
  zebra mussel infestation associated with, 742
  rate of, after colonial settlement, 658
Eutrophic diatoms, stratigraphic profile, 647
Eutrophy, demarcation from mesotrophy, using chlorophyll *a* measurements, 406–407
Evaporation, 99–100
Export flow (Q), model framework, 673
  for dissolved oxygen, 706–707

F

Fecal coliform (FC) bacteria
  horizontal gradient within the lake, 215–216
  monitoring of, 111, 186–189, 494, 521–522
    intensive, 508–509
    limited data for 1987, 190
    results for selected years, 191
    summer of 1976, 500, 501
    summer of 1987, 496–500
  standards for, 799
  transport of, 227
    simulation, 252–257, 260–261
Fecal coliform bacteria model, 714–723, 776
  as a management tool, 768–770
Ferric ion, precipitation of, 336–337
Ferrous ion
  linkage to hydrogen sulfide, 372
  release from sediment, 316
Ferrous sulfate, phosphorus precipitation using, METRO, 142

Film transfer denitrification coefficient, 696–697
Film transfer nitrification coefficient, 695
Final acute value (FAV), EPA calculation for ammonia, 302–305
Final chronic value (FCV), EPA calculation for ammonia, 302–305
Finite element grid, application to Onondaga Lake, 232–233, 259
First flush phenomenon, 157
First order kinetics, for loss of fecal coliform bacteria, 509, 521
Fishery, 453–494, 519–521
  disappearance of, 9, 27, 798
  interactions with zooplankton, 431–432
  mercury levels in fish from Onondaga Lake, 352
    in 1970, 480
    in 1985, 480
  historic perspective on, 478–479
  protection of, criteria for setting water quality standards, 749–751
  toxicity of ammonia to, 432–433
Flagellate greens, as particles, 362
Flow rates
  correlation with phosphorus load, 163
  Onondaga Creek, 148–149
  outflow channel, and stratified flow with the Seneca River, 208–209
  relationships with constituent concentrations of inflows, 191–192
Fluid flow, 83–85
Flushing, rates of, 108–109, 191, 200
  and rate of response to soda ash facility closure, 287
Flux-gradient method
  for estimating the diffusion coefficient, 238, 260
  for estimating the vertical transport, 257

elationship with mercury concentration, 357
drogeology, 32–93
basic concepts, 83–88
of Ley Creek, 49–50
of Ninemile Creek Valley, 54–55
Oil City area, 46–48
See also Hydrology of Onondaga Lake
Hydrographs, 104–105
and total phosphorus time series, 151
HYDROLAB, for pH measurements, 285
Hydrologic flow load, by major tributary, 183
Hydrologic loading
of chloride ion, 674
natural variation in, and S values, 271–272
Hydrology of Onondaga Lake, 97–110, 189–191
load estimates, 100–105
Hydrostatic pressure, and lake levels, 107
Hydrostratigraphic units
defined, 32
Onondaga County, 69
Hydrothermal model, predictions of, for deepwater discharge by METRO, 765–766
Hyper-Lake Iroquois, 36
Hyperturbidity, effect of, on benthic macroinvertebrates, 450–451
Hypolimnetic accumulation, of ammonia, 127–128, 331
Hypolimnetic oxygen demand (HOD), 351–352, 373
Hypolimnion
accumulation in
of methane, 339
of soluble reactive phosphorus, 321–323
anoxic, iron reduction in, 328–329
major ion composition, 1980 measurement, 268
net photosynthetic production of DO in, 707
organic budget for, 344–346

oxygen demand in, 351–352
oxygen depletion during the summer, 272, 473
soluble reactive phosphorus in, 312–313
as a source of cooling water, 158–159
variation in density stratification and in oxygen concentration, 277
volume-weighted model for, 671–672

I

Illuminance, measures of, 1980–1981, 536–537
Individual particle analysis (IPA), 359, 539
for 1981–1982, 367–368
Onondaga Lake, 1981–1982, 367–368
partitioning
of *b* using, 579–580
of turbidity using, 582–584
Industrial development, 5–9
Industrial discharge, separation from effects of METRO discharge, 802–803
Inflows, 201–205, 257–258
gauged and ungauged, 97–99
near-field mixing at, 245
particles, composition of, 118–120
from the Seneca River, 208
Inherent optical properties, 536
coupling with apparent optical properties, 565–566
In-lake processes, modeling, for phosphorus, 682–686
Inland Fisher Guide Facility (IFG), PCB contamination at, 50
Inner Harbor project, 23
Inorganic carbon, 283–294
1980–1990, 369–370
in sediments, 605–609
in surficial sediments, 617
Inputs, nitrogen model, 697–698
Integral models, distinguishing characteristic of, 244
Internal load, phosphorus, estimation of, 321

Ionic constituents
concentrations in lake inflows, 191–194
loading of
external, 164–185
from the soda ash/chloralkali facility, 789–790, 791–792
volume-weighted concentrations, Onondaga Lake, 264–265
Ionic strength, chloride ion as a predictor of, 266–267
Ion ratios
for chloride, sodium and calcium, Ninemile Creek, 168–169
Ninemile Creek, Onondaga Creek, and Onondaga Lake, 72–73
Iron
accumulation of, 336–337
oxidation of, 350
precipitation of phosphorus with, 311
reduction of, in anoxic sediment, 328–329
stratigraphy of, 632
See also Ferric ion; Ferrous ion
Irradiance
angular distribution of, 595–596
downward, rate of diminution of, 539
downwelling, spectral composition of, 573–574
effect on fecal coliform bacteria, 509, 511, 722–723
modeling, 718–719
as a function of depth, 535
Irradiance reflectance (R), 536
and average cosine, 540
depth profiles and temporal distributions, Onondaga Lake, 562–563
Isopleths
for ammonia fraction and ammonia distribution, 300
for calcium ion, pH, alkalinity distribution, 282
for chloride ion distribution, 216, 218

FLUX software
  interpolation to provide
    annual loading
    estimations, 180–181
  for loading estimation, 113–
    114, 180, 185
    phosphorus in minor
      tributaries, 159
    phosphorus in Ninemile
      Creek, 152–153
    phosphorus in Onondaga
      Creek, 154–155
    suspended solids, 115–116
    time series, 160–161
Food web, Onondaga Lake, mid-
  summer, 473
Froude number, 200, 238
  and variation of layer depth
    with outlet channel
    position, 211

G
Gas transfer efficiency (E), for
  reaeration at dams, 736
Gauged inflows, Onondaga Lake,
  97
  by tributary, 98
Gelbstoff
  absorption spectrum, 555–556
  changes in a_y associated with,
    577–578
  levels of, regional comparison,
    559
  measuring, 538
Geographic information system
  (GIS), 22
Geologic time scale, 88
Geology
  bedrock, 32–35
    summary, 89
  of Ley Creek, 48–49
  of Ninemile Creek, 51–55
  of Oil City, 45–46
  of Onondaga Lake, 36–41
  surficial, summary, 89
  Tar Bed area, 59–60
  See also Hydrogeology
Gershun-Jerlov equation, 540
  optical cosine, 564
Glaciation, effect on
  physiography of
  Onondaga County, 35–36
Grab samples, analysis using
  replicability of, total

phosphorus
  measurements, 149–150
for suspended solid
  measurement, 115–118
Gradient, vertical, in phosphorus
  concentration, 159
Great Lakes Critical Programs Act
  (1990), 23
Green algae, 385, 516
  census, 1968–1971, 389
  1972–1977, 390
  1978–1990, 392–393
  Chlorococcalean, 386
  flagellated, 385
  place in seasonal succession,
    388
Groundwater
  defined, 84
  effect on hydrologic loads,
    100–101
  industrial contamination of,
    summary, 47
  interactions with lakes, 79–81
    Onondaga Lake, 81–83,
      92–93
  from the Lockport Formation,
    69
  as a source of chloride ion
    Ninemile Creek, 169
    Onondaga Creek, 174
  as a source of nitrate, 326
  transportation of, 33–34, 40–
    41
  water source for brining
    operations, 64
Grouped model, for loading
  estimates, 137
Growth
  gross rate of (G), 409–410
  patterns of, fish of Onondaga
    Lake, 463
Gypsum
  as a sulfate ion source, 171,
    176–177, 192
  in the Vernon Shale, 35
Gypsum molds, 33–34
Gyttja, organic sediments, 624–
    625, 657

H
Habitat, analysis of, Atlantic
  salmon restoration, 488
Habitat Suitability Index (HSI)
  for Atlantic salmon, 488, 521

determination
  Ninemile C
Health advisory, o.
  from Onond.
    478–479
Heat
  estimating content, .
    vertical transpor
    239
  from incident light, no\
    lakes, 535
  transfer of, and water su\
    mass, 244–245
Heat conservation equation, .
Heavy metals
  interactions with zooplankto
    433
  pretreatment of industrial
    waste to remove, 10
  stratigraphy of, 631–632,
    633–636
HELP infiltration model, 56
Heterogeneous nucleation, 363–
  364
High-pressure liquid
  chromatography (HPLC),
    for measuring
    chlorophylls, 400–401
Holomictic lakes, hydrogen
  sulfide in, 334–335
Horizontal mass transport model,
  252–257, 260–261
Hydraulic conductivity, defined,
  83
Hydraulic gradient, defined, 83–
  84
Hydraulic head, defined, 83–84
Hydrodynamic model
  equations for, 228
  outlet flow predictions from,
    213
  and transport, 200–261
Hydrogen sulfide, 357
  accumulation of, 332–336
    rates, 344
    during summer
      stratification, 372–373
  early reports of, 9
  effect on cooling water use for
    industrial purposes, 13–
    14
  oxidation of, 349
  product of sulfate reduction,
    327

for dissolved oxygen distribution, 273–274, 276–277

for ion distribution, 270

for mercury distribution in sediments, 634–635

for pH, dissolved oxygen, and nitrogen species distribution, 296–297

for sulfur species distribution, 333

Isotherms, for Onondaga Lake, 1980, 218

Isotopic dating, of sediments, 647–650

J

Jackknife computation method, FLUX, 115–116

K

Kinetic coefficients, 718
  for loss of fecal coliform bacteria, 521–522
  Seneca River model, 734–737
  values of
    Green Bay, 415
    Seneca River, 729
    site-specific, 411–412, 414
    system-specific, 517
Kinetic energy, for turnover, 224
Kinetics
  of fecal coliform loss, 509–514, 718–719
  models of, 776
  of phytoplankton growth, 409–421
    nitrogen uptake, 747–748
  See also Reaction rate coefficients

L

Lake flushing, 107
Lakefront Development, Action Plan for, 23
Lake Iroquois, 89
  hydrogeologic history, 36
Lake level, factors affecting, 106–110, 191
Lake Mendota
  hydrogen sulfide in the hypolimnion of, 335
  methane budget of, 342–343
Lake Ontario, diversion of

METRO effluent to, 806–808

Lambert-Beer law. See Beer-Lambert law

Laminations, in sediments, 625–626

Landfills, near Ley Creek outlet, 50

Land use, Onondaga Lake watershed, 22

Land Use Master Plan, 23, 24–25

Laplace's Equation, for describing groundwater motion, 83–85

Lead, stratigraphy of, 634

Lead-210, for dating sediments, 648–649

Ley Creek
  geology and hydrogeology of, 48–50
  hydrogeology of, 90
Light
  penetration of
    and attenuating materials, empirical relationships, 593
    and growth of macrophytes, 441–442
    See also Absorption; Attenuation of light; Beer-Lambert law; Beer's law
Light attenuation coefficient (K_c), 441. See also Attenuation
Limestone quarrying, 8
Limnological parameters, and Pearson correlation coefficients, principal components scores, 429
Linden Chemical and Plastics, plant closure, 10
Linear regression, for mercury concentration, as a function of fish age/length, 481–482
Liquid film transfer coefficient (K_LDO), for dissolved oxygen, 704–706
Loading budget
  for fecal coliform bacteria, 716–717
  reductions in, 800–803
  for suspended solids, 114–118

for total phosphorus, model, 686–687

Loading estimation, log-log relationships, 114, 132–133
  alkalinity and daily average flow, Ninemile Creek, 172
  for chloride ion, 168
  from flow, 117–118
  for phosphorus, 145, 148
Loadographs, for total phosphorus, 151–152
Location, geographic, 1–3, 26
Lock exchange hydraulics problem, 210
Lockport Formation, contribution to groundwater, 69
Loss on ignition method, 627
Loss processes, 410–411
Loss rate coefficients, for settling and grazing, 411
Lower mixed layer (LML), phosphorus model, 681

M

Macrophytes, aquatic, 436–446, 518
  distribution of, and areal diversity, 440–441
Magnesium, stratigraphy of, 632
Management actions, effect on total phosphorus loading, 162–163
Management models, 780–782
  for total phosphorus, nitrogen and dissolved oxygen, 745–754
  use of, 808 -809
Management Plan, Lake Onondaga Lake Management Conference, 23
Manganese, stratigraphy of, 632
Manganese ion
  accumulation of, 336–337
  oxidation of, 350
  reduction of, in anoxic sediment, 328–329
  release of, from sediment, 330
Marl, 625
Mass balance model, 110
  for carbonaceous BOD, 707–708

for chloride in a completely mixed system, 673
for chloride ion, 180, 266, 791–792
data to support, 194
for dissolved oxygen, 704–705
for d-PON, 694
for fecal coliform bacteria, 716
generalized, Seneca River, 728–729
for nitrogen oxides, 696–697
for phosphorus, in a transport framework, 682
for phytoplankton growth, 409
for p-PON in the upper and lower mixed layers, 693–694
for reduced chemical species, 710
for total ammonia, 695
for water quality, 667–668, 672
Mass conservation, multiple-box model, 254
Mass content, estimating, for a vertical transport model, 239
Masson Equation, for estimating apparent equivalent volume of salts, 269
Mass transport
coefficients of, 718
and evaporative losses, 99
model for fecal coliform bacteria, 717–718
Seneca River model, 730–734
See also Transport models
Material loading, 110–189
Measurement
of age, from fish scale growth marks, 464
of ammonia and ammonium ion, 295, 296
of chlorophyll a, 400–403
of discharges, 97–99
of fecal coliform bacteria, 186
of hydrogeologic parameters, 85–87
of mercury, 354–357
for monitoring mercury levels, 479–480
optical, 536–538, 551–555
of particulate matter, 360–361
of pH, 292–293

of phytoplankton kinetic coefficients for Onondaga Lake, 412–415
of productivity, oxygen evolution technique for, 408
of sediment deposition, trap methodology, 601
of water quality, 87–88
Mechanistic model, of optical properties, 586–588
summary, 597
Median pigment concentration, as a measure of summer biomass conditions, 406
Membrane filter technique, 186
Mercuric ion, reduction to mercury, 352
Mercuric sulfide, 352
Mercury, 352–359
chlor-alkali process use of, 14, 27
contamination levels, 478–483
in fish from the lake, 9, 19, 520, 799
reduction in levels of, pretreatment of industrial waste, 10
stratigraphic profile for, 633–636
summary, 373–374
See also Methyl mercury
Mercury budget, 357–359
and contamination by soda ash/chloralkali facility, 797–798
Meromictic lakes
conditions creating, 224
hydrogen sulfide concentrations in, 334
Mesotrophic diatoms, stratigraphic profile, 647
Mesotrophy, demarcation from oligotrophy, using chlorophyll a measurements, 406
Metabolism, of zebra mussels, 741
Metalimnia, oxidation of methane in, 349
Meteorological data, for forcing conditions, 248
Methane

accumulation of, 337–339
budget of, comparison with other lakes, 341–343
ebullition of, 339–341
relationship with phosphorus release, 318–319
release from sediments, 373
Methanogenesis, 327–328, 330, 346–347
free energy of, 325
proportion of anoxic decomposition attributed to, 326
Methyl mercury, 352–353, 355, 374
standards in terms of, 479
Metropolitan Development Association (MDA), 23
Metropolitan Sewer District (METRO), 9
bypass of, phosphorus from, 157–158
design and plans, 15–18
diversion of manufacturing ionic waste to, 165
effects of operation on nitrogen load, 139–140
effluent from
amount of discharge to Onondaga Lake, 803–804
changes in composition, 194
contribution to Onondaga Lake inflow, 101
effect on lake flushing, 109–110
major ionic constituents, 177–185
mercury content, 358
phosphorus content, 141–144
loading of carbonaceous BOD from, 708
management scenarios, modeling of, 754–768, 780–781
nitrogen loading from, 128–130, 691
oxygen loading from, 706
phosphorus loading attributed to, 162–163
plant layout, 17–18
plunging inflow from, 204
proposed deep-water discharge from, 761–768

shift of ionic constituents to,
from industrial waste
beds, 182–183
Microcystis, 387
Microzone, oxidized, 316
Mie theory, 580–581
Milliequivalents, for measuring
water quality, 87–88
MINEQL model, 284
calculations on chemical
equilibrium for mercury,
353–354
iron sulfide oversaturation
during anoxic periods,
318
oversaturation with respect to
calcium-phosphate
minerals, 319–320
partial pressure of carbon
dioxide, and pH, 292–293
titration of lake using, 291–
292
Mineralization, 331–332
and phosphorus mobilization,
318–319
Mineral particles, interactions of
zooplankton with, 433–
435
Minerals in sediments, 626–627
profiles for, 628–629
Mixed layer depth, prediction of
models, 252, 260
Mixing analysis for identifying
ion sources, 77–79
in Ninemile Creek, 90–91
Mobile elements, defined, 629
Mobility rates, organic
pollutants, estimated, 48
Model design, 667–672, 776–
777
dissolved oxygen, 703–704
fecal coliform bacteria, 714–
716
mass-balance, 110
Seneca River transport
submodel, 732
Model performance
chloride ion, mass balance
model, 675–676
fecal coliform bacteria, mass
balance model, 719–720
nitrogen species, mass balance
model, 698–701
reduced chemical species, mass

balance model, 710–712
total phosphorus, mass
balance model, 687–689
verification, 778–779
water quality model, 737–739
Moisture content, of surficial
sediments, 615–617
Molybdenum, stratigraphy of,
634
Monitoring
bacteriological, 494–496, 714
long-term program, 500–
509
findings, studies from 1968 to
1992, 20, 28
Onondaga Creek
for major ionic constituents,
176–177
for total phosphorus, 146
programs for, 111
Seneca River, 729–730
of total phosphorus, 314
tributary programs, 1968
through 1990, 112
uses and limitations of, 18
Monod-type function for growth
rate, 415
effect of light on, 409–410,
412–413
effect of nutrients on, 409–
410, 418, 737
multiplier for respiration
processes, 707
relationships between
photosynthesis and total
phosphorus, 747
Monomictic conditions, and
ionic pollution, 224
Monte Carlo analysis
of nutrient-saturated areal net
production, Green Bay,
412
of the underwater light field,
539
Morphology
and fish communities,
Ninemile Creek versus
Onondaga Creek, 493
Onondaga Creek, 491–492
Morphometry, 1–3, 200–201
and sediment types, 612–613
Seneca River, 732
summary, 26
Mortality rates

fecal coliform bacteria
effect of irradiance on, 511
effect of temperature on,
511–513
factors affecting, 509
fish of Onondaga Lake, 465–
466, 477–478
Mud boils, 64–68, 790
contribution to suspended
solids, 126–127
effect of, on benthic
macroinvertebrates, 450–
451
sodium and chloride ion loads
attributed to, 184
as a source of suspended solids,
122–123
Tully Valley, 45, 65–67, 92,
192
Mud volcanoes, 64–68, 92
Multiple-box model for
transport, 253–254, 260–
261, 728–740

N

National Pollution Discharge
Elimination System
(NPDES), permit to Allied
Chemical Corporation, 10
National Weather Service, data
of, for predicting
evaporative losses, 100
Near-field
boundary of, 204
mixing in, 217–218, 245
motion and transport in, 201
variables affecting mixing in,
205
Near-shore zone, benthic
macroinvertebrates of,
446–450
Net phytoplankton growth, rate
of, 409
Nickel, stratigraphy of, 634
Ninemile Creek
contribution to Onondaga
Lake inflow, 101
fish communities
Atlantic salmon restoration,
521
and habitats, 487–488
geology of, 50–57
hydrogeology of, 50–57, 90–
91

ionic constituents of, 165–172
leachate from waste beds
      discharged into, 56
nitrogen loadings, 130–133
phosphorus loadings, 144–
      145, 152–153
suspended solids loading, 116
Nitrate ion
   depletion from the
      hypolimnion, 296
   in the epilimnion, 296–298
   loss of, in decomposition, 330–
      332
   plant use of, 294
   sources of, for denitrification,
      326
Nitrification, 349–350
   of ammonium ion, 326
   effect on, of ammonia, 298
   kinetic coefficient, k_n, 734
   at METRO, modeling effects of,
      770–771, 774–776
   seasonal variation in, 129–130
   at the sediment-water
      interface, 695
Nitrite ion
   depletion of, from the
      hypolimnion, 296, 331
   in the epilimnion, 296–298
   oxidation to nitrate ion, 349–
      350
   as a pollutant, 127–128, 799
   toxicity of, 305–306
Nitrobacter, toxicity of ammonia
      to, 298
Nitrogen, 294–307
   gaseous, from anoxic
      decomposition, 326
   measuring inputs of, 127–140
   modeling levels of, 690–702
   and phytoplankton growth,
      517
   predictions of models, for
      deepwater METRO
      discharge, 767–768
   summary, distributions and
      species, 370–371
   See also Dissolved organic
      nitrogen; Organic
      nitrogen; Total Kjeldahl
      nitrogen
Nitrogen cycle, 295
   modeling, 691–692
Nitrogen loading, 134–140

total and partitioned, from
      METRO, 195
Nitrogenous biological oxygen
      demand (NBOD),
      modeling, 704
Nitrogen oxides
   modeling levels of, 696–697,
      698
   sources of, 137
Nitrosomonas, oxidation of
      ammonia to nitrite by,
      349–350
Nonpoint sources, defined, 110
Nucleation, catalysis of, by
      foreign solids, 363
Nutrients
   availability of, and growth
      rate, for phytoplankton,
      409–410
   changes after zebra mussel
      infestation, 742
   element group, defined, 629
   limiting, defined, 415
   in surficial sediments, 619–
      620
   and temperature relationship
      with mortality rate, 512–
      513
Nutrient saturation, 415
Nutrient status, 415–421

O
Oil City, industrial and
      commercial complex, 45,
      90–91
Oligotrophic diatoms,
      stratigraphic profile, 647
Oligotrophy, demarcation from
      mesotrophy, using
      chlorophyll a
      measurements, 406–407
Oncolites, in surficial sediments,
      612, 617–619
Onondaga County Sanitary
      Sewer and Public Works
      Commission, creation and
      powers of, 15
Onondaga Creek
   ammonia and nitrogen oxides
      in, 133–134
   benthic macroinvertebrates of,
      450–451
   contribution to Onondaga
      Lake inflow, 101

fish communities and habitat,
      490–492
geology and hydrogeology of,
      41–48
ionic constituents, 172–177
phosphorus loadings, 153–
      157
as a source of fecal coliform
      bacterial contamination,
      189
suspended solids load, 116
total phosphorus
      measurements, 145–152
Onondaga Creek Flood Control
      Dam, effect on
      communication between
      Onondaga Creek and the
      water table, 44
Onondaga Lake Management
      Conference (OLMC), 28
   Citizens Advisory Committee
      of, 26
   intergovernmental
      cooperation through, 23
   Management Plan, 23
Onondaga Lake Monitoring
      Program, 19
Onondaga Lake Study, 19
Onondaga Limestone, 32
Onondaga Trough, 41–48
   water flow through, 82–83
Optics, 535–597
Ordination, for interpreting SAX
      data, 125–126
Organic carbon
   anoxic decomposition of, 324–
      352
   oxidation of, by denitrification,
      326
   See also Particulate organic
      carbon; Total organic
      carbon
Organic materials
   content of surficial sediments,
      619–620
   relationship of deposition to
      methane production, 328
   storage and disposal of at Oil
      City, 45–48
   studies identifying pollution
      from, 21
   from the Tar Beds, 59–60
Organic nitrogen, 298
   sources of, 137

See also Dissolved organic
nitrogen; Particulate
organic nitrogen
Oswego River watershed, 724–
743
Otisco Lake
areal hypolimnetic oxygen
deficit, 281
ionic constituents of water in,
263
phosphorus, oxygen, ferrous
ion, and pH
measurements, above
sediment, 316
stratification regime of,
comparison with
Onondaga Lake, 223
temperature after spring
turnover, 225
upstream load to Ninemile
Creek, phosphorus, 153
Outflow, 257–258
at the epilimnion, 245
near-field mixing at, 245
Outlet channel
chloride concentration in,
variation with depth, 206
density variation in, 206–207
steady-state model of flow
regime in, 258
Outlet flow conditions, 205–213
Outwash deposits, glacial, 39–41
Overflow, from low density
inflow, 201
Oversaturation, with respect to
calcite, 285–286
Oxidation
reactions of reduced species,
347–352
summary, 343–347
Oxygen
deficit of
attributed to phytoplankton
biomass, 399
after failure of spring
turnover, 225
related to ionic loading, 164
related to stratification, 220,
224
depletion of, in the stratified
saline layers, Seneca
River, 725–728
dissolved, 272–283, 368–369
as an electron acceptor, in

aquatic environments,
325
Oxygenation, artificial, predicted
effect on phosphorus
release from sediment,
316–317
Oxygen evolution technique, for
measuring gross and net
productivity, 408
Oxygen sag, 275
modeling of, Seneca/Oswego
Rivers, 737–738
processes leading to, 770
sensitivity of, to sediment
oxygen demand, 712–
713

P

Paleolimnological analysis, 653–
655
Palynology, 650–653
Parasitic infestation, in fishes
from contaminated
waters, 483–484
Park system, development of,
23–26, 29
Particle analysis, individual,
125–127
Particle area per unit volume
(PAV), 580
Particle per milliliter (PNV), 361
Particles
chemistry of, 359–368, 374
classification of, 360–361
and fecal coliform bacteria
association with, 513–
514
size of, in sediment, 613–615
Particulate organic carbon
(POC), 365–366
change in, 1980–1990, 289
temporal distributions, 609–
611
Particulate organic nitrogen
p-PON, and phytoplankton net
growth rate, 692–693
ratio to dissolved organic
nitrogen, 694
Particulate phosphorus (PP), 307
correlation with phosphorus
deposition, 602–603
settling velocities, 682–683
Partitioning
of absorption, 574–579, 596–

597
of constants comprising
turbidity, 541
of scattering, 579–584, 596–
597
of turbidity, 582–584
Pearson correlation coefficients,
between collection scores
and limnological
variables, 427–429
Pelagic heterotrophic bacteria,
514–515
pH
and ammonia standards, 749–
751
changes after soda ash facility
closure, 290–292
diurnal variation in, 300–301
effect on phosphate adsorption
on calcite, 317
and fraction of T-NH_3 as
ammonia, 300
and inorganic carbon, 283
measurement, implication for
equilibrium calculations,
292–293
Ninemile Creek, 171–172
seasonal variation in, 285,
369–370
summary, 369
Phaeophytin, light absorption
characteristics of, 538
Phaeopigments, measurement of
to determine phytoplankton
settling losses, 411
to monitor phytoplankton,
399–400, 516
Phosphorus
chemistry of, 307–324
deposition of, 601–603
effluent standard for, Great
Lakes watershed, 16
as a limiting nutrient, N:P
ratio, 415–421, 517
loading of, 140–164, 195
environmental impact
analysis, 807–808
Onondaga Creek, 194
removal of
and disappearance of
cyanobacteria, 392
by precipitation, 16
with waste bed overflow,
204

stratigraphy of, 632
    profile, 636–637
summary, 371–372
in surficial sediments, 620
total
    framework for modeling,
        670
    mechanistic model for, 679–
        690
    standards for, 748–749
    stratigraphy of, 659
See also Dissolved organic
    phosphorus; Particulate
    phosphorus (PP); Soluble
    reactive phosphorus
Phosphorus cycling, interaction
    with calcite, 795–797
Phosphorus deposition rate
    (PDR), 603
Photic zone, oxygen production
    in, 272
Photosynthesis, and dissolved
    oxygen, 704
    modeling, 707
    in the Seneca River, 736–737
Photosynthetically active
    radiation (PAR), 535
    attenuation coefficient for,
        $(K_d(PAR))$, 569–570
    vertical variation in, 539
Physical framework
    for management models, 745–
        747
    for models, total phosphorus,
        nitrogen, and dissolved
        oxygen, 670–672
See also Model design
Phytoplankton, 384–421
    biomass of, 416
        effect on light penetration,
            regional comparisons,
            561–562
        and saturation index, 285
    blooms, pH at the time of,
        369–370
    and fraction of $T-NH_3$ as
        ammonia, 300
    interactions with zooplankton,
        430–431
    net growth of, and the
        nitrogen model, 692–694
    phosphorus as a critical
        nutrient for, 307
    phyla, differences among, 385

pigments from, sedimentation
    of, 603–605
and saturation of dissolved
    oxygen, 275, 296
and seasonal variations in N
    species, 371
summary, 515–517
P-I curves (photosynthesis and
    light), for phytoplankton,
    409–410
'Piezometer, 85
Piper plots, 69, 70
    to demonstrate groundwater
        mixing, 88
    Ninemile Creek surface water,
        77
    Onondaga Creek surface
        water, 78
    Onondaga Lake
        at 6m, 78
        at 18 m, 80
    Tully Brines, 86
Plankton, taxa in Onondaga Lake
    sediments, 638–640
Plant metabolism, and pH,
    epilimnia, 290–292
Plunging inflow
    effect on spring turnover, 226
    during the fall, 222–223
    and inflow water density, 201,
        245
    from METRO, 203–205
    from Ninemile Creek, 203–
        205, 216, 218
        stratification from, 220
    phosphorus model, 681
    during soda ash manufacture,
        109, 204, 257–258, 270
    and flux gradient
        estimation, 240
Plunging underflow
    effects of the soda ash/
        chloralkali facility, 792–
        793
    modeling, 680–681
Point sources
    for chloride ion, identifying in
        Onondaga Creek, 173–
        175
    defined, 110
    mud boils as, 122
Pollen, in sediments, analysis of,
    650–653
Polls, opinions of recreational

use of Onondaga Lake, 26
Polychlorinated biphenyls
    contamination of fish flesh by,
        479
    in landfills near Ley Creek, 50
PONAR dredge, for collecting
    surficial sediments, 611
Population density, benthic
    macroinvertebrates, 448–
        449
Population diversity of fish
    Onondaga Lake, 1927–1991,
        454
    Seneca River, 1927–1991, 460
Population dynamics, for
    particles, 361–365
Potentiometric head, defined, 84
Pour plate method, for
    enumerating
    heterotrophic bacteria,
        515
Precipitation
    concentrations of mercury in,
        358
    hydrologic responses to,
        variables affecting, 104–
        105
Precision, of pigment analysis,
    chlorophyll and
    phaeophytin, 403
Prediction
    from a circulation model,
        actual wind/stratification
        conditions specified, 236–
        237
    of discharges, based on gauged
        inflows, 99
    of evaporative losses, from
        windspeed, 99
    of the finite element
        circulation model, 232
    of lake circulation, 228
    from tested models, 780–782
See also Model performance
Pressure, hydrostatic, and
    methane ebullition, 340–
        341
Primary production
    and deposition rate for organic
        carbon, 610
    indicators for, 658–659
    and light, 535
    nutrients originating in, 619–
        620

Principal components analysis, zooplankton community, 427–429
Productivity
  areal, temporal distributions of, 408
  and biomass, 399–408
  measurements of, 407–408
Prokaryotes, cyanobacteria, 386–387
Public health
  compliance with standards protecting, measurements of, 111
  concerns about
    fecal coliform bacterial contamination, 186–189
    with recreational use of the lake, 26, 29
  effects of mercury pollutant, 9
  issues in modeling water quality, 672
  and use of lake water for ice and for recreation, 9, 27

R
Radio telemetry, for following fish movements, 468–469
Reaction rate coefficients, relationship with temperature, 512. See also Kinetic coefficients; Kinetics
Reaeration coefficient, Seneca River, determination of, 734, 736
Recharge area, defined, 84
Recovery curve, for estimating hydraulic conductivity, 86
Recreational use of surface water, 494
Redfield composition, 325, 610
Redlich equation, Debye-Huckel limiting law slope, 268
Redox reactions
  regulation of phosphorus release from sediment, 315
  relation to anoxia and ferrous ion mobilization, 372
  summary, 372–373
Reduced chemical species oxidation of, 349–352
  simulated volume-weighted

concentrations, 711
Regional variation
  of heavy metals, in sediments, 636
  of optics, summary, 594–595
  of total phosphorus, Onondaga Creek, 147–148
  in values of a_W for different lakes, 576–577
Regression analysis, chloride ion, model versus observation, 676
Release ratio, for phosphorus (($DIC+CH_4$):SRP), 318–319
Representative elemental volume (REV), 86–87
Reproduction, of fish, Onondaga Lake, 462–463
Restratification, after turnover, 278
Reynold's number, of lake circulation flows, 229
Richness, of benthic macroinvertebrates, below the mud boils, 452
Rosgen Stream classifications, Onondaga Creek, 491
Rotiera, assemblage, 1969–1989, 423
Rotifers, 421
  assemblage change, 1969–1989, 424
  seasonal and annual dynamics of, 426–427
Runge-Kutta numerical method, for solving a mass balance equation, 673
Runoff
  and fecal coliform bacterial levels, 714
  and turbidity, 121
Rural area, phosphorus load from, 155–156

S
S (summation of seven ions), 266
Salina Group, 32–33, 89–91
  bedrock salt of, 44
Salinity (S)
  chloride model for, 672–679
  distributions of, 270–272
  interaction of zooplankton

communities with, 435–436
  loading reduction, after soda ash/chlor-alkali facility closure, 218
  responses of macrophytes to, 442–443
  before and after soda ash facility closure, 218, 368
  and stratification
    temperature inversions, 214
    thermal, 216–219, 263–272
    water quality of the Seneca River, 725–728
  summary, 368
  transport submodel application, Seneca River, 733–734
  See also Chloride ion
Salmon, Atlantic, restoration to Onondaga Lake, 488, 520–521
Salt content
  geologic sources of, 33
    Onondaga Creek, 44
  historic accounts, 164
  salt springs, 8
Salt hoppers, 33–34
Salt mining, Tully Valley, 61–64, 91–92
Salt production, establishment of, 6
Samples, flow-weighted, composite, 111
Sampling
  frequency of, and detection of exceedance of standards, 504–505
  time-intensive, 110–111
Sapropel, 624–625, 657
Saturation index (SI)
  carbonate, calcium, calcite system, 284
  mercury, 355–357
SAX
  analysis of inflow particles, 118–119
  for suspended solids analysis, 114, 125–127
Scalar sensors, for irradiance measurements, 537
Scales, time and space, for water quality models, 669–670

Scanning electron microscopy, for suspended solids analysis, 114, 125–127, 539

Scattering
estimating from T partitioning, 584
of light in water systems, 535

Scattering coefficient, chlorophyll-specific (K_b), 554

Schmidt number, turbulent, 256

SCUBA biological surveys, 741

Seasonal variation
in biomass of pelagic bacteria, 514–515
in community structure
fish, 457–458, 519–520
fish of the Seneca River, 461
in density stratification, 221–223
in dissolved oxygen, 273–274
in the effects of deepwater discharge from METRO, 762–763
in hydrologic loading, 103–104, 190–191
in lake levels, 106
in nitrification, 129–130
in phytoplankton abundance, 384
in S (seven-ion summation), 270–271
in the turbulent diffusion coefficient, 241
in zooplankton abundance, 424–425
See also Turnover

Secchi disc transparency (SD), 536, 539–540
changes in, historical observations, 542–544
correlation of
with population components, 450–451
with principal components, 429
and macrophyte distribution, 441
measuring
Onondaga Creek, 120–121
Onondaga Lake, 361–362
relationship of
with c, 567–568

with chlorophyll a concentrations, 554
with K_d, 546–549, 567–568
with turbidity, 551
summary, 592–594
temporal distribution of measurements in 1988, 552
regional comparison, 559–560

Sediment, 600–659
evaluation of plant growth on, 445–446
feedback in modeling, 753–754
phosphorus release from, 315–324
modeling, 686
profiles for metallic elements, 631
release rates
mercury, 358
phosphorus, 321–324
sources of, Onondaga Creek, 44–45, 67

Sedimentation
and fecal coliform bacteria disappearance, 509–510, 513–514
index of (SI), and calcium deposition, 290
loss rate coefficient (k_s), 510
velocity of, 510

Sedimentation flux, defined, 510

Sedimentology, 624–629

Sediment oxygen demand (SOD), 272, 351–352, 708–709, 734
modeling, 704, 753–754

Segmentation, transport model, 254–255

Seiche
horizontal heterogeneity caused by, 215
internal, period of, 241

Semet Residue Ponds, 58–59

Seneca River
density difference from Onondaga Lake, 205–206
diversion of METRO effluents to, 805–806
effect on, of the soda ash/ chloralkali facility, 793–794

EPA dismissal of effluent discharge to, 16
flow from Onondaga Lake, 258
planned treated domestic effluent discharge to, 15

Seneca River/Oswego River system model, 723–743
applications to water quality, 770–776, 781–782

Sensitivity analysis, 779–780
chloride ion model, 676–678
dissolved oxygen model, 712–713
fecal coliform bacteria model, 721–723
nitrogen model, 700–702
Seneca/Oswego Rivers model, 739–740
total phosphorus model, 689–690

Sequential discrete procedure, 152

Settling, of phytoplankton, measurement of, 414–415

Settling velocity (V_s)
for chlorophyll, 605
for phytoplankton pigments, 517

Sewage contamination, nitrite as an indicator of, 298

Sewage treatment
disposal of sludge, with Solvay waste, 13
modeling effects of, 754–757

Sewer overflows, total phosphorus loading from, 156–157. See also Combined sewer overflows (CSOs)

Shannon-Weiner diversity index, 449–450
for fish, 457, 458
stratigraphic profile, 646

Silicon, diatom requirement for, 386

Silver, stratigraphy of, 634, 636

Simulations
during and after closing of soda ash manufacture, 248–250
using the verified model for turnover and

stratification, 250–252
Single layer model, to describe
    the epilimnion of a
    stratified lake, 228
Slug test, 85–86
Soda ash production, 11–14
    effect on Onondaga Lake and
        surrounding systems,
        789–798
Sodium carbonate, 11–14
Sodium ion concentration
    Ninemile Creek, 169
    Onondaga Lake, 265
        1980, 285
SOL5 software, for calculating
    downwelling irradiance,
    587
Soluble reactive phosphorus
    (SRP), 145, 307, 410, 416
    in the hypolimnion, 321–323
    in Ninemile Creek, 153
Solvay process
    brining operations associated
        with, 91–92
    description, 11–14, 27
    ionic enrichment of
        wastewater from, 165
    See also Allied Chemical
        Corporation
Spatial distributions,
    homogeneous, and
    stratification, 214–216
Species diversity
    of benthic macroinvertebrates,
        449–450
    and depth, filed survey, 438–
        440
    of fish in tributaries, 1927 and
        1990–1991, 485–487
    increase in, after closure of the
        soda ash/chloralkali
        facility, 802–803
    and lake salinity, 265, 791–
        792
    versus richness, 493–494
Species richness
    versus diversity, 493–494
    in Ninemile Creek versus
        Onondaga Creek, 493
    Onondaga Creek, 492
    in selected New York lakes,
        438
    of zooplankton, 517
Spectral distributions, of

emergent flux, 574
Spectral response, of the
    Protomatic photometers,
    537
Spectrophotometry, for
    measuring chlorophylls,
    400–402
Spectroradiometer, 537
Spectroradiometry, 569–574
Spring
    salinity stratification
        established in, 221–222
    bloom during, 403
Standards
    for ammonia, elimination of
        violations, 691
    bacteriological, for public
        water supplies, 494
    for fecal coliform bacteria,
        768–769
    model output to examine,
        748–751
    violations of, in Onondaga
        Lake, 798
State of Onondaga Lake, The
    (1993), Onondaga Lake
    Management Conference,
    23
Steady-state model, 228–232,
    258
    use for simulating horizontal
        mass transport, 260–261
Steady states, succession of, to
    simulate time-variable
    wind conditions, 228
Steel manufacturing,
    management of iron and
    chromium in effluent
    from, 10
Storm Water Management
    Model (SWMM), 157
Stratification
    and ammonia concentrations,
        298
    density, 219–224
        factors affecting, 258–259
    duration of, predictions from
        vertical transport model,
        250–252
    effect on
        of proposed METRO deep-
            water discharge, 761–762
        of the soda ash/chloralkali
            facility, 792–793

elimination of, to improve
    water quality, 771
and inflow density, 203–204
and mixing within layers, 246
modeling of, 238–252, 762–
    763
    UFILS4, 242–252
and oxygen depletion in the
    hypolimnion, 330
salinity, and temperature
    inversions, 214
seasonality of, accommodation
    with vertical exchange
    coefficients, 672
Stratified flow models, boundary
    conditions, 229
Stratigraphic profiles
    for heavy metals, 633
    for pollen, 650–652
    ratio of (benthic and epiphytic)
        diatoms to planktonic
        diatoms, 644
    for total solids, 627–629
Stratigraphy, of sediments, 622–
    655, 657–659
Streamflow, effect on circulation,
    simulation of, 236, 237
Stream function, prediction of,
    232
Subbasins, 3–5
Succession
    of diatom species, in sediment,
        643–645
    seasonal, 388
Sulfate ion
    concentration of
        Onondaga Creek, 176–177
        Onondaga Lake, 35
    correlation of
        with chloride, sodium and
            calcium ions, 169
        with phosphorus release,
            317
    depletion of, 332–336
    gypsum as a source of, 171,
        176–177, 192
    reduction of, 330
        anoxic, 327
        as a proportion of anoxic
            decomposition, 326
        free energy associated with,
            325
    in the hypolimnion, 265
    relationship with oxygen

deficit, 265
relationship with oxygen
    deficit and with
    phosphorus loading, 164
Sulfides
    heavy metal, in sediments, 632
    iron, in surficial sediments,
        612
    mercury complex, 374
    metal, precipitation of, 330
    See also Hydrogen sulfide
Sulfur, iron, and hydrogen
    sulfide in sediment, 318
Summer bloom, late, dominant
    organisms in, 388
Summer heat income, 226–227,
    258
Surface flows, average, to
    Onondaga Lake, 102–103
Surface velocity, prediction of,
    232
Surficial sediments, 611–622,
    656–657
Survivorship, patterns of, fish of
    Onondaga Lake, 463
Suspended solids, 114–127
    dry weight of, 605–609
    in Onondaga Creek, 194–195
    relationship with flow, 192
    See also Particle entries
Syracuse Formation, 32–33
    intersection of Ninemile Creek
        valley with, 53–54
Syracuse Intercepting Board,
    history of sewage
    treatment by, 14
Syracuse Research Corporation,
    water quality studies, 18–
    19

T
Tagging, to monitor movements
    of lake fishes, 466–469
Tar beds, 58–61, 91
Taylor-Fischer theory, for
    dispersion, 260–261
Temperature
    cooling water for industrial
        processes, 13–14
    artificial cycling from spent
        water, 158–159
    correlation with principal
        components, 429
    and density stratification,
        213–227

and dissolved oxygen, at 18 m,
    226
effect on fecal coliform
    bacteria, 509
and growth rate
    for phytoplankton, 409
    specific, model of, 411, 413–
        414
and methane production, 328
optimum, for salmon, 489
water, and relative density,
    206–207
Temperature of maximum
    density, change with
    salinity, 270
Temporal distributions
    for a and b, 566–567
    for ammonia and ammonium
        ion, selected years, 300
    for areal productivity, 408
    for attenuation, by
        wavelength, 572–573
    for calcium ion and dissolved
        inorganic carbon, 286–287
    for chloride ion
        METRO effluent, 178–179
        Ninemile Creek, 167
        Onondaga Creek, 172–173,
            175
    for chlorophyll, 552–555
    for C:N in deposited
        sediments, 610
    for dissolved phosphorus
        species, 417
    for estimates of G(N), 418
    for exchange flow from the
        UML to the LML, 682
    for fecal coliform bacterial
        count, 187–188
    for fraction of ammonia,
        selected years, 301
    for gelbstoff, 555–556
    for heterotrophic bacteria, 515
    for inorganic carbon and dry
        weight in sediment, 605–
        609
    interannual, for nitrogen
        species, 298–299
    for mercury, 355–357
    for methane ebullition, 340–
        341
    for nitrate ion, 330–331
    for particulate organic carbon,
        609–611
    for phosphorus deposition,

601–602
    for phytoplankton, 403–405
        net growth of, 692–693
    for specific turbidity
        components, 583–584
    for vertical turbulent diffusion
        coefficient, 243
    for volume-weighted
        concentration of chloride
        ion, 674
Tertiary sewage treatment
    initiation of, 9–10
    phosphorus reduction due to,
        143
    upgrade of METRO to, 16
Thermal plumes, of cooling
    water and METRO
    discharge, 202–203
Thermal stratification, 200
    and salinity, 216–219
Thermocline
    base for the surface layer at,
        227
    depth of, and fecal coliform
        bacteria, 723
    observations, 1987–1990, 670
    relationship with chemocline,
        223
Thermodynamic considerations
    for decomposition of organic
        carbon, 325
    versus observed sequence,
        346–347
    stable form of mercury, 374
Thomas slope method, for
    determining K_CBOD,20,
    708
Three RIvers system, 724–743,
    776–777
Time scale, of changes in
    pollutant concentrations,
    193–194
Time series, for loading
    estimation, 114
    alkalinity, 180
    chloride ion, 180, 673–675
    fecal coliform bacterial, 188
    phosphorus, 150–152, 153–
        154, 160–162, 683–684
    SAX data, 126
Total coliform (TC) test, 494
    observations, with intensive
        monitoring, 508
Total dissolved hydrocarbon,
    contamination in

Onondaga Lake, 61
Total dissolved phosphorus
    (TDP), 307, 410
Total dissolved solids (TDS),
    Onondaga Trough, 44
Total inorganic phosphorus
    (TIP), 140–141
Total Kjeldahl nitrogen (TKN)
    in sediments, 609–611
    and total nitrogen, 697
Total organic carbon (TOC)
    stratigraphic profile, 636–637
    stratigraphy of, 632, 659
Total petroleum hydrocarbon
    (TPH), groundwater, 48
Total phosphorus (TP)
    as a measure of trophic state,
        307
    monitoring of, 111
Toxicity
    of ammonia, 294–295, 302–
        305, 775–776, 799
        to *Nitrobacter*, 298
        to zooplankton, 432–433
    of cyanobacteria, 430
    of nitrite ion, 305–306
    of nitrogen species, 302–306
    of substances in Onondaga
        Lake, 808
Transfer, surface, of dissolved
    oxygen, 704–706
Transparency, correlation with
    principal components,
    429. *See also* Clarity; Secci
    disk transparency
Transport, and hydrodynamics,
    200–261. *See also* Mass
    transport; Transport
    models; Vertical transport
Transportation
    canal system development, 6
    and development, 26–27
    railroads, 6
    State Thruway Authority
        Barge Canal facility, 23
Transport models
    for deepwater METRO
        discharge, performance
        of, 766
    for nitrogen, 698–701
    for Onondaga Lake, 236, 259–
        261, 776
    for the Seneca River,
        application of, 733
    for suspended solids, 126

for total phosphorus, 680–681
    performance of, 688–689
    See also Mass transport;
        Vertical transport
Treatment, of industrial waste,
    Onondaga County, 10
Trellis diagram, benthic
    macroinvertebrates, two
    sites, 452
Trends
    in major groups of
        phytoplankton, 1978–
        1990, 397–399
    in mercury concentrations,
        smallmouth bass, 482
Tributaries, 3–5, 97–196
    carbonaceous BOD loading
        from, 708
    classification of, 484–485
    density of inflows from, 205
    geology and hydrogeology of,
        41–57
    minor, phosphorus loading,
        159–160
    Onondaga Lake, hydrogeology
        of, 89–91
Trophic boundaries, phosphorus
    concentrations
    corresponding to, 310
Trophic interactions, of
    zooplankton, 430–432
Trophic state
    changes in, 406–407
    linkages with the
        phytoplankton model,
        747–748
Tully Brine Field Areas, 62–64
Tully brines, as a source of ionic
    components, Onondaga
    Lake, 69
Tully Valley
    salt mining in, 61–64
    water table of, 44
Tumors, in fishes from
    contaminated waters,
    483–484, 520
Turbidity (T, NTU)
    apportionment of, model,
        579–580
    correlation with principal
        components, 429
    historical observations, 549–
        551
    measuring, 537–538
    as a function of total

chlorophyll, 554–555
    Onondaga Creek, 120–121
    and particle area per unit
        volume, 580–581
    relationship with SD and K_d,
        603
Turbulent diffusion, vertical
    transport by, 259
Turbulent kinetic energy budget,
    246, 260
Turnover
    defined, 213
    effect of soda ash/chloralkali
        facility on, 792–793
    fall
        conditions for, 220
        dissolved oxygen levels
            preceding, 274, 276
        factors affecting, 218–219
        salinity stratification after,
            218
    spring
        effect of ash facility, 276–
            277
        effect of soda ash, 223
        failure of, 224–226
        after soda ash facility
            closure, 280

U
UFILS4 stratification model,
    242–252
Unconsolidated deposits,
    underlying Onondaga
    Lake, 36–39
Ungauged inflows, Onondaga
    Lake, 97–99
Unit area load (UAL) for
    phosphorus
    in Ninemile Creek, 153, 195
    in Onondaga Creek, 146, 155,
        195
Upper mixed layer (UML)
    nitrogen model, comparison of
        flow options, 752–753
    phosphorus model, 681
        comparison of flow options,
            751
        predictions of, after sewage
            treatment changes, 757–
            759
Upstream sources, of suspended
    solids, 122–127
Urban load, of phosphorus,
    Onondaga Creek, 154

U.S. Geologic Survey (USGS), gauging stations, Onondaga watershed, 97–99

V
Van't Hoff equations, for relationship between reaction rate coefficients and temperature, 512
Variables, for using management models, 751
Verification
  of model performance, 778–779
  simulated stratification regime, 246–248, 260
Vernon Shale, brines in, 34
Vertical attenuation coefficient (K_d), 539
  for downward irradiance, 536
  methods for estimating values of, 552–554
  relationship with SD and c, 567–568
  summary, 592–594
  temporal distribution of
    in 1988, 552
    regional comparisons, 560–562
  See also Attenuation of light
Vertical diffusion coefficient, variation in, due to changing meteorological conditions, 241
Vertical distribution
  of dissolved oxygen, 273–274
  of phosphorus, profiles for, 308–309
  of phytoplankton, 404–405
  and primary productivity, 407–408
Vertical transport
  diagnostic model of, 238–242, 259–260
  modeling of, 238–252
    for phosphorus, 684–686
    predictive model of, 242–252
    by turbulent diffusion, 259
Volatilization, and sediment release rate for total ammonia, 695–696
Volume, representative elemental, 86–87

Volume scattering function, 540–541

W
Waste
  nature of, Solvay process, 11–13
  sources of, 10–18, 27–28
Waste-bed overflow, 69
Waste beds
  ammonia load from, 140, 192
  chloride
    sodium, and calcium ion load, from, 192
  chloride loading, from lagoon adjacent to, 178–179
  erosion into Onondaga Lake, 83
  hydrogeology of, 55–57
  ionic load from, 218
    chloride, sodium and calcium, 181–185, 195–196
  Ninemile Creek flow through, 53
  phosphorus load from, interaction of calcium carbonate with phosphorus, 145
WATEQ model, 284
Water, optical properties of, 539–541
Water balance, data needed to establish, 210
Water column, phosphorus content of, 311–313
Water level, macrophyte responses to changes in, 444
Water quality, 87–88
  bacteriological standards for, 495
  dissolved oxygen as a determining factor, 272–283
  effect on, of deepwater METRO discharge, 763–768
  effects of stratification on, 214
  interactions with zooplankton communities, 432–436
  and salinity, Seneca River, 725–728
  and salmon habitat, 488–489

Water quality model, 254, 728–740
  application of, 771–772
  mechanistic, 667–782
  summary, 776–782
Watersheds, 3–5
  Onondaga Creek, stratigraphy of, 42–43
Water surface drag coefficient, adjustment to match experimental and calculated conditions, 235
Water surface mass, and heat transfer, 244–245
Water table, defined, 84
Well logs, as a source of geologic data, 36–38, 52–53
Whiting phenomenon, 557–558
Windpower, and spring turnover, 277–278
Wind speed
  effect of on the fall dissolved oxygen sag, 713
  and fecal coliform bacterial disappearance, 721–722
  and reaeration, 704–706
Wind stress
  calculation of, equations for, 229
  effects on heat and mass transport, 201
  time variation of, modeling, 228

X
X-ray maps, CA particles (calcium-rich), 365

Y
Yearly model, for nitrogen estimation, 138
Young's rule, 268

Z
Zebra mussel, 740–743, 775, 779, 806
Zero degree of freedom modeling, 679
Zinc, stratigraphy of, 634
Zooplankton, 421–436
  biomass of, 424–426
  changes in the community of, 1969–1989, 422–424, 517–518
  effect of grazing on algal biomass, 430